ENVIRONMENTAL GEOTECHNICS
VOLUME 2

PROCEEDINGS OF THE SECOND INTERNATIONAL CONGRESS ON ENVIRONMENTAL GEOTECHNICS / OSAKA / JAPAN / 5-8 NOVEMBER 1996

Environmental Geotechnics

Edited by
MASASHI KAMON
Kyoto University, Japan

VOLUME 2

A.A. BALKEMA / ROTTERDAM / BROOKFIELD / 1996

Organized by The International Society for Soil Mechanics and Foundation Engineering and The Japanese Geotechnical Society

The texts of the various papers in this volume were set individually by typists under the supervision of each of the authors concerned.

Published by
A.A. Balkema, P.O. Box 1675, 3000 BR Rotterdam, Netherlands (Fax: +31.10.4135947)
A.A. Balkema Publishers, Old Post Road, Brookfield, VT 05036-9704, USA (Fax: +1.802.2763837)

For the complete set of three volumes, ISBN 90 5410 848 7
For Volume 1, ISBN 90 5410 849 5
For Volume 2, ISBN 90 5410 850 9
For Volume 3, ISBN 90 5410 851 7

Environmental Geotechnics, Kamon (ed.) © *1996 Balkema, Rotterdam. ISBN 90 5410 848 7*

Table of contents

Geotechnics of waste landfill

Geotechnical recycling and reuse of waste materials

Remediation of contaminated ground

Geo-environmental risks and regulations

Geotechnics of waste landfill

Environmental Geotechnics, Kamon (ed.) © *1996 Balkema, Rotterdam. ISBN 90 5410 848 7*

Offshore waste disposal by Osaka Bay Phoenix Project

Shinsuke Aburatani, Yosuke Hayashi & Teruhiro Nishikawa
Osaka Bay Regional Offshore Environmental Improvement Center, Japan

ABSTRACT: In 1982, Osaka Bay Phoenix Project was launched to operate waste disposal project on reclaimed land in sea areas in conjunction with port and harbor development.

Standards are set up in compliance with relevant laws and regulations concerning the quality of waste accepted at disposal sites. Accordingly, only non-hazardous and safe waste is accepted, and combustible waste is incinerated while non-combustible waste is crushed prior to disposal.

This paper outlines the Osaka Bay Phoenix Project, as well as the properties of reclaimed land, based on the results of soil exploration tests conducted at the controlled disposal zone of Amagasaki Offshore Disposal Site, which is in an advanced stage of formation after about 6 years of landfilling.

In sum, an almost uniform soil quality has been achieved despite the inclusion of diverse waste materials; as a whole, the reclaimed land is equivalent to loose alluvial sandy soil.

1. Introduction

Waste generated in daily life and various industrial activities must be disposed of appropriately while recycling, re-use and reduction are promoted at the same time. In the inland areas of Kansai, it is becoming extremely difficult for individual local administrative bodies and waste disposal companies to secure final waste disposal sites, due to lack of land availability. Under such circumstances, Osaka Bay Extensive Regional Improvement Project, or the commonly called Osaka Bay Phoenix Project, was launched in 1982 to construct, by reclaiming part of Osaka Bay, stable, long-term sites for appropriate final waste disposal that serve the wide spread administrative districts. The Phoenix Project is Japan's only extensive regional waste disposal project.

The "phoenix," the popular name of the Project, carries two meanings: first, it refers to the mythical bird that rose from its own ashes in Arabic mythology; second, it refers to a subtropical date-palm tree. This name was chosen for the Project in the hope that sites reclaimed by waste will be born again into highly serviceable land rich in greenery.

Land reclamation by final waste disposal in sea areas has already been carried out on a large scale in Tokyo Bay and other locations. These projects, however, have met problems such as land subsidence and gas generation as a result of the decay and decomposition of combustible waste materials, thereby obstructing land utilization after landfill completion. In a survey report on a waste disposal site in Tokyo Bay, for example, the site is expected to become stable enough for land utilization 20 years after landfill completion (Matsumoto, *et al.*, 1987).

In the Osaka Bay Phoenix Project, only waste materials which conform to stringent standards, mainly treated burned ash, are accepted. For this reason, the disposal sites of the Phoenix Project are expected to have different properties than conventional offshore waste disposal sites.

This paper evaluates the soil properties of the controlled disposal zone of the Amagasaki Offshore Disposal Site, where landfill has reached almost 50% of its design capacity.

2. Outline of the Osaka Bay Phoenix Project

(1) Districts served by, and disposal sites created by, the Phoenix Project

At present, 171 municipalities in six prefectures in the Kansai region (Fig. 1) are designated as districts to be served by extensive regional waste disposal of the Phoenix Project. These service districts jointly occupy about 38% of the total area but account for about 92% of the total population in the six Kansai prefectures, indicating a high degree of population concentration. Waste materials are collected at eight loading facilities in Amagasaki, Kobe, Harima, Tsuna, Osaka, Sakai, Izumiotsu and Wakayama (under construction) and transported to two final

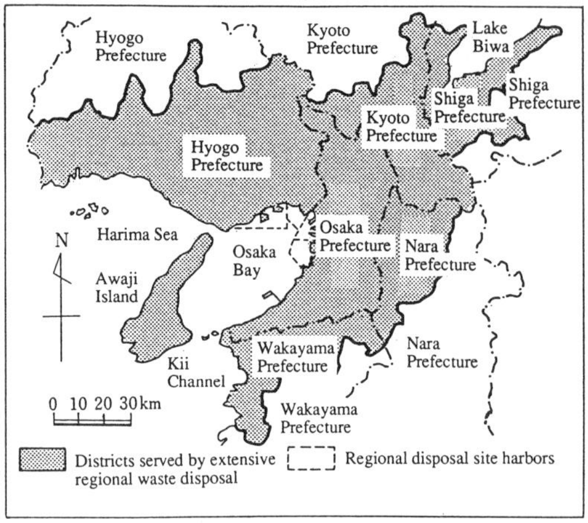

Fig 1. Districts Served by Extensive Regional Waste Disposal Sites and Regional Disposal Site Harbors.

Table 1. Disposal Site Scale.

Disposal Site	Area	Fill Volume
Amagasaki Offshore Disposal Site	113 ha	15 million m^3
Izumiotsu Offshore Disposal Site	203 ha	30 million m^3

disposal sites off Amagasaki in Hyogo Prefecture and off Izumiotsu in Osaka Prefecture (Table 1).

(2) Accepted waste and acceptance standards

Types of waste accepted into the offshore disposal sites of the Phoenix Project include domestic waste, waterworks and sewage sludge, industrial waste, surplus soil and dredged earth and sand. Acceptance standards have been set: for example, waste must be non-hazardous; combustible waste must be incinerated until ignition loss of up to 15% is achieved; sludge must be dehydrated by incineration until ignition loss of up to 15% or a water content in per cent of dry weight of up to 85% is achieved; and non-combustible waste must be crushed into pieces 30 cm or less. Waste is inspected upon signing a contract or when transporting waste, to check conformity to the standards.

(3) Measures for environmental preservation

In the Phoenix Project, great importance is attached to measures for environmental preservation. For example, waste water treatment facilities are installed at the disposal sites to treat surplus water; the immediate marine environment is periodically monitored in terms of water quality, bottom quality and aquatic life, to prevent the disposal sites from adversely affecting the environment.

At the loading facilities, large dust collectors, such as cyclone separators, are installed to prevent the scattering of waste during loading from transportation

Table 2. Status of Waste Acceptance at Two Offshore Disposal Sites.

(Unit: 1,000 m^3)

Disposal Site		Capacity	Accepted Volume	Filling Rate
Amagasaki	Controlled disposal zone	4,000	2,199	55%
	Stable disposal zone	11,000	5,948	54%
	Total	15,000	8,147	54%
Izumiotsu	Controlled disposal zone	10,000	2,836	28%
	Stable disposal zone	20,000	4,358	22%
	Total	30,000	7,194	24%
Grand total		45,000	15,341	34%

N.B. The accepted volume at the stable disposal zones include waste deriving from the Great Hanshin-Awaji Earthquake of January 17, 1995 (1.1 million m^3).

vehicles to barges. To minimize traffic congestion and related pollution, routes for transportation vehicles to the loading facilities are specified, and land environmental monitoring is carried out along roads near loading facilities located in large cities (Kobe, Amagasaki, Osaka, Sakai and Izumiotsu).

(4) Status of waste acceptance

Waste acceptance commenced in January, 1990 at the Amagasaki Offshore Disposal Site, and in January, 1992 at the Izumiotsu Offshore Disposal Site.

Table 2 summarizes the status of waste acceptance as of the end of September, 1995.

3. Offshore Waste Disposal Site Construction

(1) Facility planning

a) Planning

Since the Amagasaki Offshore Disposal Site is located adjacent to the first-grade Yodo River and the second-grade Nakajima River, normal lines on the side of the rivers were drawn 5° outside from the Nakajima River normal lines so as not to affect the river flow.

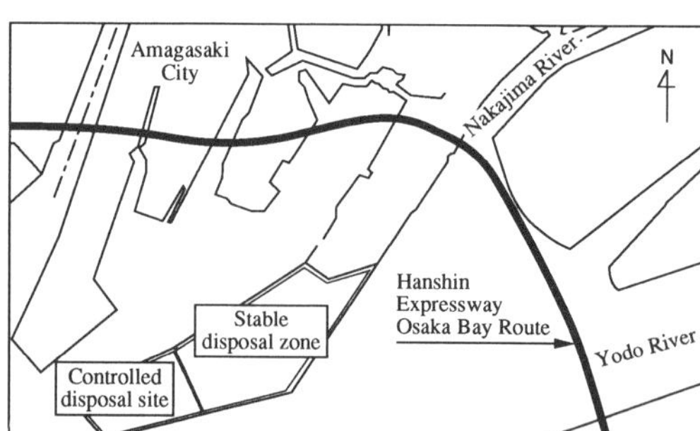

Fig 2. Zoning at Amagasaki Offshore Disposal Site.

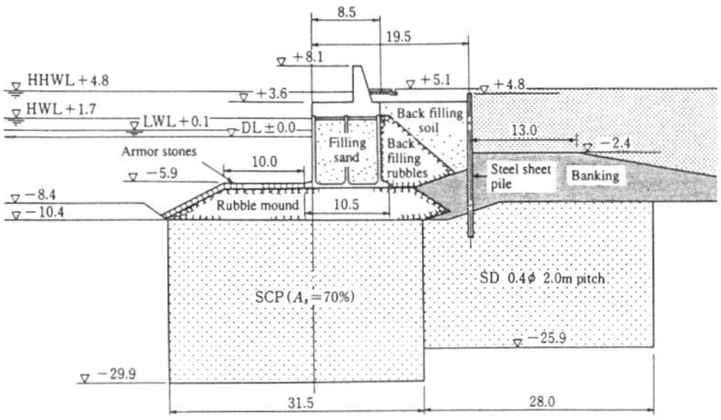

Fig 3. Cross Section of Controlled Disposal Zone Revetment (Amagasaki Offshore Disposal Site).

Photo 1. Landfill Using a Pontoon Bridge.

b) Revetment structure

The Amagasaki Offshore Disposal Site is situated in the Amagasaki-Nishinomiya-Ashiya Port in the inner part of Osaka Bay, on a 20 m-or-so layer of soft alluvial clay, beneath which are diluvial layers.

Fig. 3 illustrates a cross section of the revetment built around the Disposal Site. This particular revetment structure was selected in consideration of safety, ease of construction, influence on the surroundings, cost and duration of construction, maintainability and manageability.

In the controlled disposal zone (for waste from which effluent is discharged when coming into contact with water), steel sheet piles were installed as water barriers behind the revetment so that sea and rain water inside the Disposal Site would not flow into the open sea.

(2) Reclamation

a) Waste disposal method

Since the original ground of the Amagasaki Offshore Disposal Site was a soft alluvial clay layer, it was anticipated that direct dumping of voluminous waste would cause damage to the ground due to circular sliding. For this reason, as well as to achieve uniform soil composition and consolidation, the thin reclamation method has been adopted, using open-bottom barges (in the stable disposal zone only) and a pontoon bridge. The pontoon bridge consists of 11 linked pontoons, each one 8 m long and 22 m wide. Dump trucks drive directly onto the pontoon bridge and dump waste from one of the 10 available chutes either side; after about every two dumpings by each dump truck at each chute, the pontoon bridge is shifted. This process is repeated several times to form a thin layer of 2 m or so without damaging the ground (Photo 1).

When filling reaches the critical depth of the pontoon bridge (about –2 m), soil filling is performed by bulldozers to achieve the prescribed height.

In the controlled disposal zone, to reduce water treatment and discharge load, prevent offensive odor or environmental pollution in general, and improve soil conditions for future land utilization, layers of domestic and industrial waste alternates with a layer of surplus soil at a ratio of 7 to 3 (about 1 m soil lining is planned on the surface of reclaimed land).

Table 3. Actual Waste Acceptance in the Controlled Disposal Zone of Amagasaki Offshore Disposal Site (January, 1990 - September, 1995).

Waste Category	Type	Disposal Volume (1,000 m^3)	Percentage in Total (%)
Domestic waste	Burned ash	874	39.8
	Non-combustible/ large refuse	82	3.7
	Total	956	43.5
Industrial waste	Waterworks sludge	66	3.0
	Sewage sludge (mostly incinerated)	302	13.7
	Sludge	48	2.2
	Others	156	7.1
	Total	572	26.0
Surplus soil as lining	Surplus soil	671	30.5
Grand total		2,199	100.0

b) Accepted waste and waste composition

Table 3 indicates the composition of waste accepted up to September, 1995, in terms of type and nature.

4. Interim Evaluation of Reclaimed Land

Up to the present, there have been few cases of land reclamation using waste, in particular, burned ash and dehydrated sludge. To ensure appropriate future land utilization, it is necessary to thoroughly understand the geotechnical properties of offshore disposal sites of the Phoenix Project.

Therefore, soil exploration was carried out at the controlled disposal zone of the Amagasaki Offshore Disposal Site, which was still in the formation process (reclaimed to D.L. +2 m), to clarify its physical and dynamic properties.

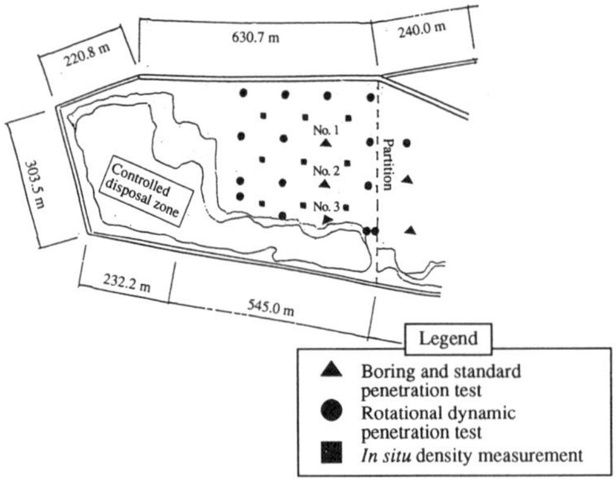

Fig 4. Soil Exploration Locations and Test Items.

(1) Soil exploration locations and test items
Soil exploration locations were designated mainly on the part of the zone where land formation has almost been completed (Fig. 4), and items of tests to be conducted of these locations were also determined.

(2) Rotational dynamic penetration test
a) Objectives
Since land reclaimed with waste is likely to contain hard and large materials, such as concrete masses and metallic objects, the standard penetration test may not be able to collect accurate data.

The rotational dynamic penetration test, involving three actions, namely hitting, pushing and rotating, has been developed to collect data while disregarding obstructive materials (Hashimoto *et al.*, 1993). Fig. 5 outlines a rotational dynamic penetration test system.

Rotational dynamic penetration tests were conducted to measure the strength (Nb value) of the Amagasaki Offshore Disposal Site, whose soil was estimated to be inhomogeneous. The Nb value corresponds to the N value obtained in the conventional standard penetration test.

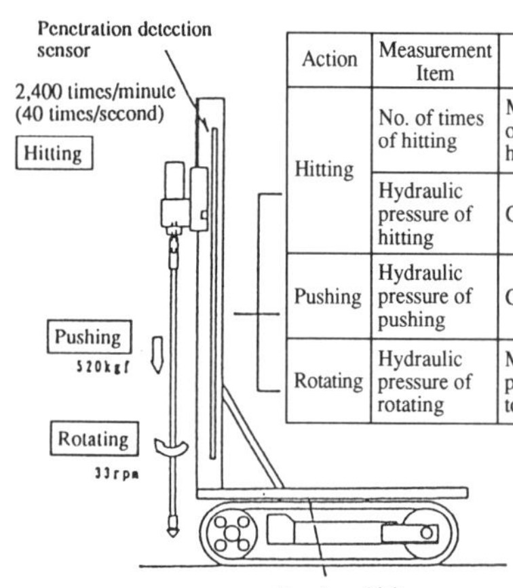

Action	Measurement Item	Measuring Method
Hitting	No. of times of hitting	Measure number of times of hitting.
	Hydraulic pressure of hitting	Check action.
Pushing	Hydraulic pressure of pushing	Check action.
Rotating	Hydraulic pressure of rotating	Measure hydraulic pressure required to keep 33 rpm.

Fig 5. Rotational Dynamic Penetration Test System.

b) Conversion of Nb value into N value
Among data obtained, Ro, Cn and Cd are applied to the basic formula of Nb value (1), and the formula (2) is established.

$$NB = \left\{Nh - (\frac{Rp - Ro}{Ro}) \times Cn\right\} / Cd \qquad (1)$$

$$= \left\{Nh - (\frac{Rp - 11}{11}) \times 10\right\} / 10 \qquad (2)$$

Nb: Corrected number of times of hitting (ground strength evaluation value per 10 cm)
Nh: Measured number of times of hitting (10 cm intervals)
Rp: Rotational pressure (representative value between 10 cm measurement sections)
Ro: Standard rotational pressure when skin friction is not applied (Set at 11.)
Cn: Rotational pressure conversion coefficient (Set at 10.)
Cd: Number of times of hitting conversion coefficient (Set at 10.)

The formula (3) expressing conversion of Nb value into N value is given below:

$$N \text{ value} = 3 \times Nb \text{ value} \qquad (3)$$

(3) Soil properties
a) Physical and chemical properties
① According to *in situ* density test results, the average water content in per cent of dry weight, wet density and dry bulk density of the reclaimed land are 27.7%, 1.8 g/cm^3 and 1.4 g/cm^3, with large fluctuations.
② Soil particle density (specific gravity; Fig. 6)
ρs = 2.68 - 2.79; this is slightly higher than in general soil (2.6 - 2.7).
③ Natural moisture content (Fig. 6)
Wn = 11.5 - 53.6%; the reclaimed land has a relatively high water content in per cent of dry weight, considering that the soil is classified as sandy or gravel based on grain composition.
④ Grading properties (Fig. 7)
The reclaimed land has a 5 - 20% fine grain content and a 30 - 70% gravel content, with uniformity coefficients concentrated between 20 and 70.
⑤ pH value (Fig. 6)
The reclaimed land has a considerably high pH level of 9.0 - 11.1. This is presumably attributable to the inclusion of burned ash, and waterworks and sewage sludge.
⑥ Ignition loss (Fig. 6)
Li = 3.72 - 15.19%. Organic substances contained in waste are believed to form a small percentage. The ignition loss approximately conforms to the acceptance standards.

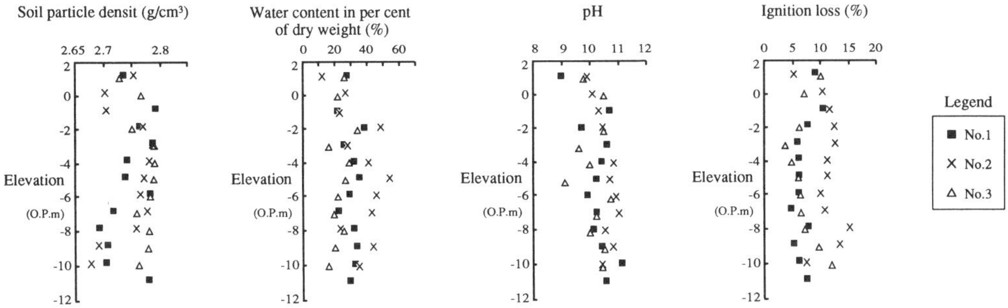

Fig 6. Soil Properties by Depth.

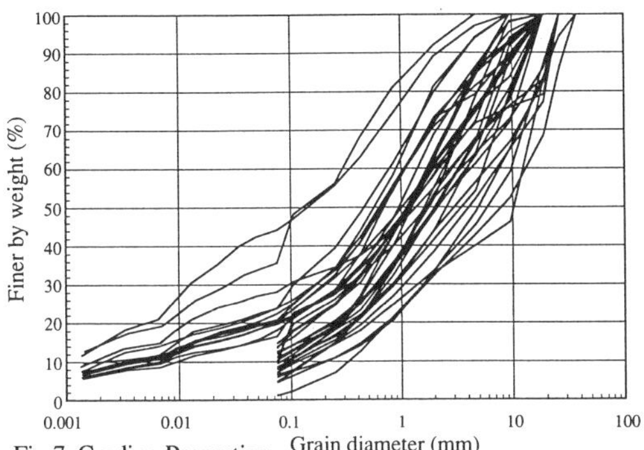

Fig 7. Grading Properties.

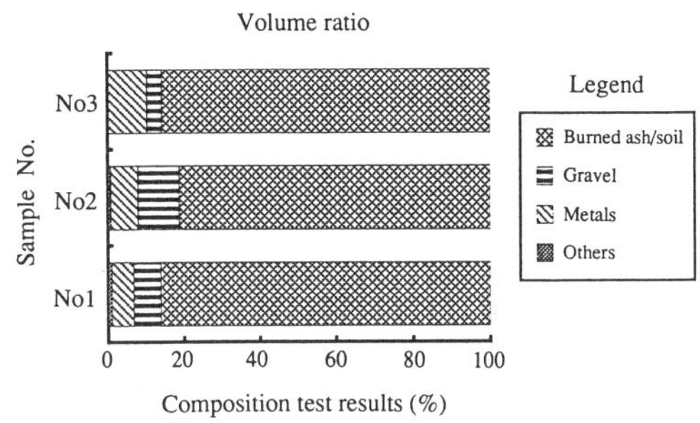

Fig 8. Ground Composition.

b) Composition (volume ratio; Fig. 8)
Approximately, burned ash and soil account for 85% and gravel and metals account for about 15% of the total waste. Other waste materials, such as concrete, ceramic, waste plastic, rubber and wood chips, together form less than 1%.

c) Ground strength
① The average N value obtained in standard penetration tests is 8.4 (No. of data: 40). Since these tests also include data obtained when the edge of the sampler touched metals and other mixed materials, a value based on frequency distribution is adopted as an average value.
N value = 4.5
② Nb values (per 10-cm penetration) obtained in rotational dynamic penetration tests are summarized in Table 4.

Table 4. Nb Value Frequency.

Range	No. of Values	Frequency (%)
0 - 3	2,421	82.2
4 - 6	373	12.7
7 - 9	57	1.9
10 - 59	80	2.7
60 or above	16	0.5
Total	2,947	100.0

Since values in the 0 - 3 range account for 82%, this range indicates representative ground strength. The average of these values is therefore 1.5, which is converted into an N value (N value = 3 × Nb value), 4.5.
③ Comparison with other formulas
For comparison, N value 4.5 is applied to other formulas.
- In Meyerhof's N – ø: Provided that sand condition is loose and N value is 4 - 10, ø = 30 - 35°
- In Osaki's formula: $\phi = \sqrt{20N} + 15 = 24.5°$
- In Dunham's formula: Provided that gradation is good and grains are round, $\phi = \sqrt{12N} + 20 = 27.3°$

d) Summary
① Geotechnical evaluation of the reclaimed land based on test results

Grading test data are plotted out on a triangular chart; most data are classified into gravel or sandy soil (Fig. 9).

In terms of ignition loss, the reclaimed land is compared to general soil as represented by decomposed granite (3.77%), sludge (7.97%) and alluvial soil (5.9%); the reclaimed land is similar to these types of soil.

As well, with rotational dynamic penetration test results taken into consideration, the reclaimed land is considered equivalent to loose alluvial sandy soil with an N value of 4.5 and ø = 24.5 - 35°.

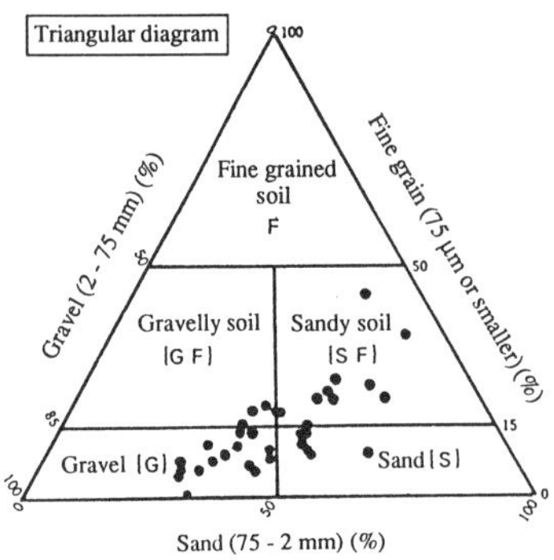

Fig 9. Soil Classification.

② Comparison with other offshore disposal sites
In comparison with data concerning other offshore disposal sites reclaimed with combustible waste (Shimizu, *et al.*, 1994), the Amagasaki Offshore Disposal Sites is found to be more stable than an offshore disposal site whose landfill was completed 20 years ago. The results of comparison in terms of gas generation are given below:

The concentrations of gases generated at the Amagasaki Offshore Disposal Site are: methane (CH_4) 0.045%; ammonia (NH_3) 0.004%; and carbon dioxide (CO_2) 0.038%. Gas generation is extremely low at Amagasaki.

Disposal Site	No. of Years since Landfill Completion	Methane Gas Concentration (%)
Site A	20 years	1.7
Site B	15 years	8.8
Site C	10 years	32.6
Site D	5 years	33.0
Amagasaki	—	0.045

5. Conclusion

For effective land utilization, it is essential to identify properties of the reclaimed land. The properties of land reclaimed by burned ash and other types of waste in the Phoenix Project and other interim findings related to the Project are summarized below.

① While burned ash and sludge have a measure of influence on the reclaimed land, it roughly corresponds to loose alluvial sandy soil with an N value of 4.5 and $\phi = 24.5° - 35°$.

② The reclaimed land shows smaller than expected inhomogeneity in soil content. This is presumably because of the method of transportation, which involves mixing during loading to barges via chutes at loading centers and the thin reclamation method on the site after unloading from barges.

③ The rotational dynamic penetration test method adopted in soil exploration is considered sufficiently effective in evaluating the strength of reclaimed ground which consists of grains of varying diameter.

The above results are based on data obtained through on-site observation, consisting mainly of soundings. In our further studies, an indoor N value test, a large-scale triaxial compression test and a liquefaction test will be conducted to determine the relation between N value (Nb value) and dynamic constants (C, ϕ etc.).

Subjects for future efforts include the implementation of follow-up surveys for clarifying chronological changes in the reclaimed land and the development of economical and speedy survey techniques suitable for the ground comprised of waste materials that is discontinuous in terms of composition and stratum formation.

Acknowledgements

The authors would like to express their deepest gratitude to the Waste Land Reclamation Research Committee (chaired by Professor Tamotsu Matsui of Osaka University), established within the Osaka Bay Regional Offshore Environmental Improvement Center, for its supervision of the studies and tests discussed above and its valuable advice on this paper.

References

Hashimoto, T., Harada, K., *et al.* (1993); Development of Rotational Dynamic Penetration Test Equipment, *The Proceedings of The 28th Japan National Conference on Soil Mechanics and Foundation Engineering*, pp. 303-304.

Matsumoto, Y., *et al.* (1987); Geotechnical Study of Refuse Landfill in Tokyo Port (3), *The Proceedings of The 22nd Japan National Conference on Soil Mechanics and Foundation Engineering*, pp. 39-42.

Shimizu, K. *et al.* (1994); Geotechnical Study of Refuse Landfill in Tokyo Port (15), *The Proceedings of The 29th Japan National Conference on Soil Mechanics and Foundation Engineering*, pp. 73-76.

Environmental Geotechnics, Kamon (ed.) © *1996 Balkema, Rotterdam. ISBN 90 5410 848 7*

Investigation, monitoring system and stability analysis of an old canyon landfill

A.M. Blengino & S.Veggi
Studio Geotecnico Italiano s.r.l., Milano, Italy

M. Manassero
Technical University of Torino, Italy

A. Rossello
Azienda Municipalizzata Igiene Urbana, AMIU, Genova, Italy

C. Spanna
IG Ingegneria Geotecnica s.r.l., Torino, Italy

ABSTRACT: The municipal landfill system of the city of Genova (Northern Italy) is located in a rather steep valley close to the urban area. The plant is divided in two different storing areas : at the moment the older part is almost filled whereas the second section located in the lower part of the valley is at the beginning of the storing activities. In order to check the stability conditions of the old part of the facility and to plan correctly the drainage systems that can be necessary in the future, investigations and monitoring programs have been implemented and are illustrated in the note.

1. BACKGROUND

The operation of a municipal waste landfill at Monte Scarpino began in 1968. At that time, compliance with the National Regulations did not require any kind of either sealing systems to be placed as a barrier between the waste and the natural soil or leachate and biogas extraction systems.

With the aim of remediating the site, following the introduction of revised National Regulations, the Genova Municipality decided to recover the existing part of stored waste and develop a new section of the Scarpino landfill facility by implementing the controlled disposal of the waste.

Waste disposal at Scarpino is in a steep sided valley formed by the erosive action of the stream "Rio Cassinelle": the valley rises 220 m to a maximum elevation of 570 meters above sea level (see Fig. 1).

More than eight million cubic meters of municipal waste are stored in the landfill cell now in activity, situated in the upper part of the valley (i.e. 450 to 570 meters above sea level). Waste is dumped in the valley from the bottom to the top in sub-horizontal waste layers, 10 to 20 meters high as can be seen in Fig. 2. Each layer is separated by earthen layers 0.50 to 1 meter thick.

In the lower part of the valley, preparation works for a new cell with a total capacity of two million cubic meters, able to deal with the city's needs up to about 1999, are now almost complete.

At present, the design of a final third section

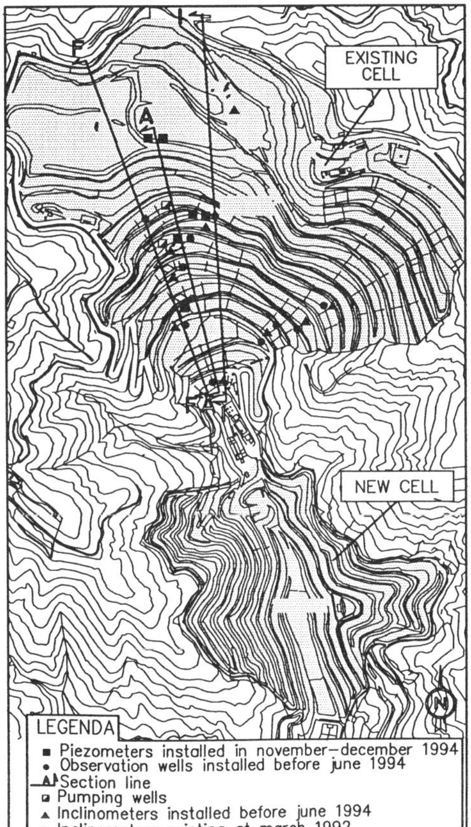

Fig. 1: Plan view of Scarpino landfill with monitoring instrumentation

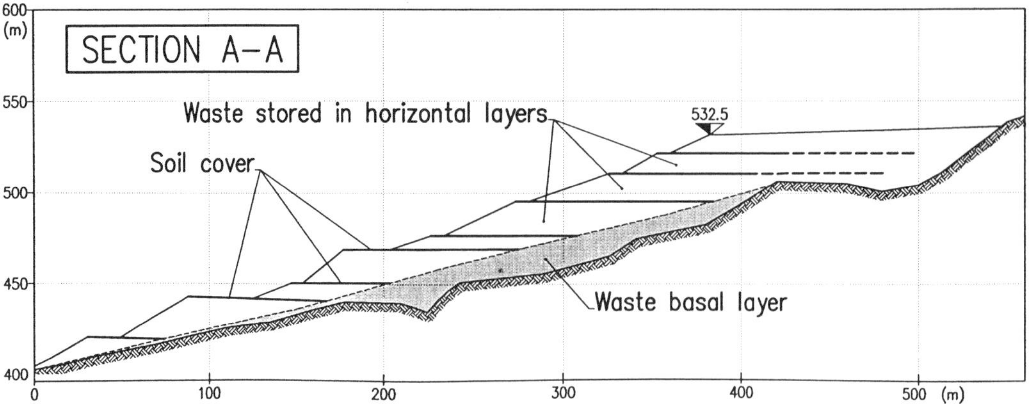

Fig. 2: Vertical section A-A of Scarpino landfill: sketch of the storing system

is being developed. The aim is to join the two existing sections each other in order to exploit the whole capacity of the canyon.

1.1 Geology

From a geological point of view the subsoil shows two different conditions: on the left orographic side of the river the subsoil is constituted by a serpentinite diabasi layer; the right side is characterized by the presence of slates.

The superficial part of both formations is weathered.

1.2 Sealing system

No sealing system has been placed in the old section of the landfill, except for an upper zone which has been developed more recently, where a layer of compacted clayey silt has been placed. The reasons of this, are, basically, the lack of regulations, controls and environmental awareness at the time of initial storing. Concern about the implications of the lack of sealing system has been low for many years because of the permeability of the subsoil considered low enough within the old conceptual approach to the landfill design.

1.3 Gas extraction

A plant for the extraction and combustion of biogas has been planned, designed and built only recently.

1.4 Drainage conditions

The formation of leachate is due to the natural water content of the waste, infiltration of rainfall and seepage of groundwater below the landfill from the shallower and weathered subsoil layer.

The leachate drainage system is constituted by a slotted steel pipe 1.5 meters in diameter, about 100 meters in length, placed along the lower part of the fall line of the valley, and by two pumping wells.

A first monitoring system was installed inside the waste body consisting of eleven observation wells and twelve inclinometers. Periodical topographical surveys were also carried out.

The observation wells were slotted for their full length.

The data collected from the observation wells, confirmed also by the data recorded from the pumping wells, showed a leachate surface level rather high and very sensitive to changes in weather conditions.

In order to check the current stability of the landfill and to carry out the design of the future landfill enlargement, a series of stability analyses was performed on representative sections using short-term strength parameters for the waste (effective intercept cohesion c' = 10 kPa and friction angle Φ' = 30°); it was found that the safety factors were very close to the limit values. Taking into account the possible decrease with time of waste strength (Jessberger & Kockel, 1993), the possible increase of leachate level and the remaining amount of waste to be stored before closure, the risk of instability would further on increase with time.

However, inclinometer and topographical data did

not give the perception of a so small margin of safety. The former did not indicate the presence of any significant movements along localized surfaces inside the waste body while the displacements of the surface indicated by the latter were in the range considered acceptable on the basis of literature indications (here not reported for lack of space) by Grisolia et al., 1995 and Jessberger & Kockel, 1995, since the creep behaviour of this kind of material.

In order to investigate further the actual safety margin, additional, different instrumentation was installed in the waste body.

2. RECENT INVESTIGATIONS AND MONITORING

Additional investigation and an extended monitoring system include 16 piezometers installed at different heights in 6 boreholes, as described below.

2.1 Boreholes

Six boreholes were drilled inside the waste with continuous sampling: four boreholes reached the rocky natural subsoil.

A detailed stratigraphy of the encountered materials was prepared during drilling operations.

From the examination of the stratigraphy, the presence of horizontal waste strata of variable thickness (about 10 meters) alternate with soil layers is evident (see Fig. 3).

This stratigraphy is confirmed by the storing procedures known to have been used at the site.

Inside the banks, levels of mixed waste and soil were observed corresponding to the periodical covers.

In the zone at the bottom of the landfill, a different waste layer parallel to the natural slope of the site has been observed, corresponding to the waste placed during the initial filling operations at the site.

During the boring operations, four infiltration tests were performed into the waste located near the bottom of the landfill. The results obtained from these tests showed a range of k values from $3*10^{-7}$ to $2.5*10^{-6}$ m/s. Some details of the tests are given in Table 1.

In more superficial layers the values of permeability should be higher than those measured in these tests, which compares with the lower limit of the typical range reported in the literature (Oweis & Khera, 1986, 1990; Oweis et al., 1985; Beaven & Powrie, 1995). This is due to the considerable thickness of the stored waste which results in significant densification of the lower waste and to the age of the deepest waste (biochemical degradation).

On the basis of these considerations, it is believed that at shallower depths permeability values may be in the range of $1*10^{-6}$ to $1*10^{-5}$ m/s.

Table 1. Permeability, k, from Lefranc tests.

Borehole No.	Test	Depth (m)	K (m/s)
1	falling head	38.80-39.30	$3.00*10^{-7}$
2	const. head	34.70-35.80	$7.02*10^{-7}$
3	falling head	34.70-35.80	$1.16*10^{-6}$
4	falling head	31.50-32.45	$2.56*10^{-6}$

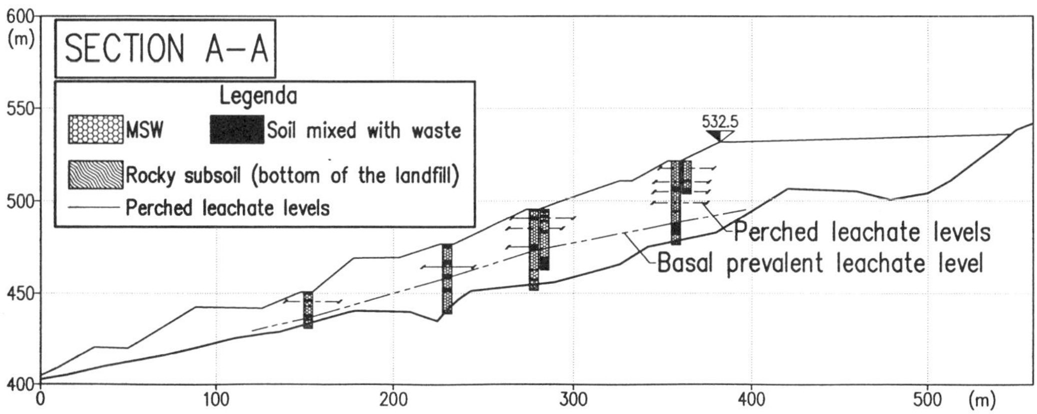

Fig. 3: Piezometric leachate levels in the waste body: section A-A

2.2 Piezometers

During boring operations some perched levels of leachate have been found. Their presence is confirmed by the water contents measured in the core samples.

Thereafter, the installation procedures and the type of the piezometers were designed with the aim of measuring each single leachate level taking into account the need for the installation of simple instruments of easy utilization, long life, resistance to high temperatures (40 to 60 °C) and to leachate and biogas contact. A sketch of the installed equipment is shown in Fig. 4. The measurements from the piezometers have indicated the presence of a prevalent basal leachate level. The upper minor perched levels are supported by the cover soils between the banks and in some cases by the denser layers of waste.

Recent measurements highlight the weather-dependency of the piezometric heads of the perched leachate levels, whereas the basal leachate level remains more or less constant with time.

The piezometric heads recorded in the old observation wells are higher than those measured in the piezometers recently installed in the basal leachate level because the old fully slotted wells have linked around them the different aquifers in the waste body.

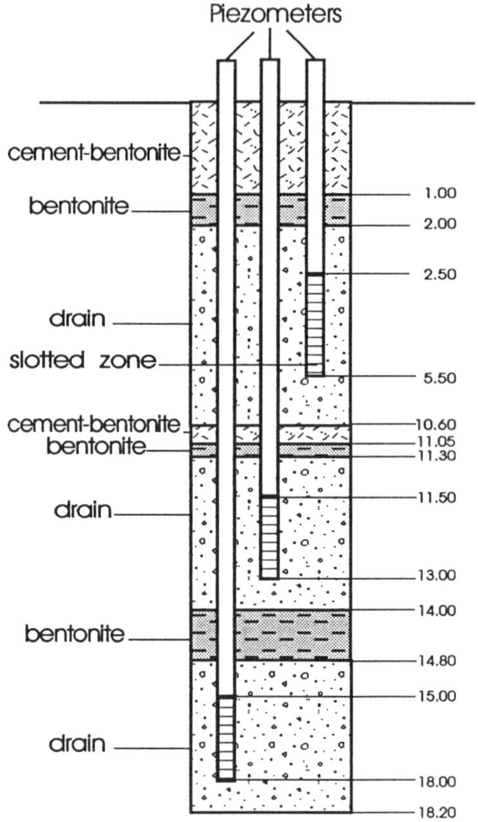

Fig. 4: Sketch of the installation of piezometers

2.3 Biogas monitoring

Sampling and laboratory tests were carried out for the verification of biogas production inside the waste body in the months of June and October 1994.

Samples were taken inside the piezometers and leachate pumping wells.

The following tests have been carried out:

- gascromatographic analyses of gas samples: oxigen (O_2), nitrogen (N_2), methane (CH_4), carbon dioxide (CO_2) and hydrogen sulfide (H_2S) were detected;
- leachate temperature;
- pH and Redox potential of leachate;
- atmospheric pressure.

The results of these investigations are reported in Table 2.

In the upper part of the cell recently filled, the volumetric concentrations of gas correspond to values (from 50 to 60%) typical of biogas production processes, in a stable phase of methane formations, as confirmed by pH and Redox potential measurements.

In the other zones the concentration decreases becoming equal to zero in the oldest areas.

Based on the measured methane concentrations, the morphological situation and the age of the waste, it is possible to suppose that the presence of biogas in the recently filled cells induces additional pore pressures in the range of 0.03 to 0.1 kPa (Kavazanjian et al., 1995).

3. STABILITY ANALYSES

A new series of stability analyses (Macksimovic, 1979) has been carried out assuming the input data obtained from the most recent monitoring program illustrated above.

Table 2. Results of tests on samples from piezometers. Comparison of the data recorded in the months of June and October 1994. Weather conditions: June: external temperature = 25°C, barometric pressure = 965 kPa
October: external temperature = 13°C, barometric pressure = 970 kPa

PIEZOMETERS	1		2		3		4		5		6	
BIOGAS	**Oct 94**	**June 94**	**Oct 94**	**June 94**	**Oct 94**	**June 94**	**Oct 94**	**June 94**	**Oct 94**	**June 94**	**Oct 94**	**June 94**
O_2 (%v/v)	21.68	21.2	11.22	4.67	21.70	1.19	15.27	4.50	1.62	1.92	0.19	8.17
N_2 (%v/v)	78.32	77.9	47.32	17.96	78.30	56.16	55.79	18.02	5.89	7.19	0.59	30.64
CH_4 (%v/v)	<0.1	0.3	26.21	57.98	<0.1	17.59	21.57	56.90	55.22	53.21	57.02	37.79
CO_2 (%v/v)	<0.1	0.6	15.25	19.39	<0.1	25.06	7.37	20.57	37.27	37.68	42.20	23.40
H_2S (%v/v)	<0.1	<0.1	<0.1	<0.1	<0.1	<0.1	<0.1	<0.1	<0.1	<0.1	<0.1	<0.1
LEACHATE												
pH	7.84	8.46	7.16	7.49	8.36	7.81	7.22	8.76	7.19	8.3	7.65	-
Temperature (C)	23	25	29	30	31	35	28	45	38	49	36	-
Redox potential(-V)	0.404	0.296	0.343	0.278	0.317	0.24	0.340	0.241	0.223	0.235	0.183	-
Depth from G.L. (m)	12.65	12.70	13.8	18.80	9.25	10.7	16.60	14.50	12.9	9.9	1.4	-

In particular, the waste have been characterized by the following long-term mechanical strength parameters obtained from back-analyses on landslides movements occurred in the past in a limited zone of the facility, accordingly to literature data and to results of cone penetration tests performed on the waste:

- effective intercept cohesion: c' = 0.5 kPa;
- friction angle: $\Phi' = 27°$.

The unit weight of the waste has been assumed equal to 10 kN/m^3.

Fig. 5 indicates the most representative sections on which the stability analyses have been carried out.

The most critical potential slip surfaces among those tested are also indicated in the same figure together with the assessed safety factors, S_F. The S_F obtained from two dimensional stability analyses have been weighted according to the indications of Lambe and Whitman, 1979 in order to take into account, even though in an approximate manner, the tridimensional geometry of the problem. The resulting critical S_F are ≥ 1.48, which is acceptable both with respect to the National ($S_F \geq 1.3$) and Regional ($S_F \geq 1.4$) Regulations.

However, as already mentioned, considering the possible further decrease with time of waste strength, the possible increase of leachate level, the remaining amount of waste to be stored and, mostly, the peculiar regional hydrogeological conditions and the climatic characteristics (i.e.high rainfall in short times) a series of drainage and reinforcement works have been planned in order to keep an acceptable safety level with time. The executive design of this improvement system is now in progress.

4. CONCLUSIONS

It is possible to make the following considerations on the basis of the results of the investigations and the stability analyses illustrated in this paper:

- stability analyses based on the basis of the most recent and reliable investigation data and on the first results from the monitoring systems showed that the global safety factors of the considered landfill satisfy the National and Regional Regulations, even if they are very close to these limit values;
- the safety factors are higher than those from computations carried out using the data collected from the previous monitoring system;
- the global stability results are consistent with the indications from inclinometers data and the topographical surveys;
- the safety factors are particularly dependent from the input data assumed for the leachate level: thereafter, the efficiency of monitoring system must be maintained in the long-term placing, if necessary, new devices in order to have a continuous control of the maximum leachate level admitted for satisfactory safety conditions;
- considering the predicted variation with time of the parameters that influence stability (par. 3), it is anticipated that drainage works will need to be implemented in the future. The long-term monitoring of the landfill must enable the extent and the timing of any additional reinforcement works that may be carried out or required to be optimized.

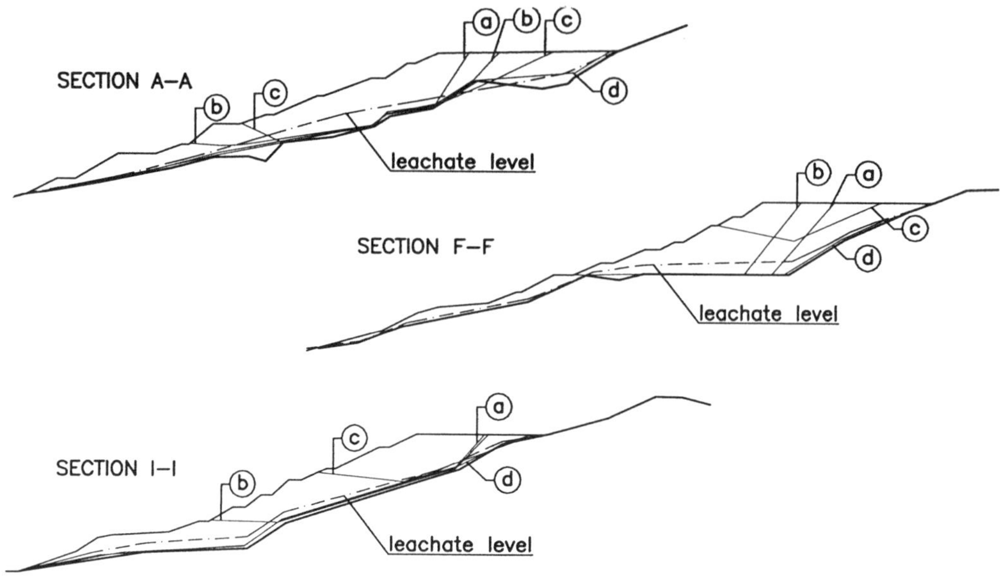

SAFETY FACTORS (Macksimovic Method)				
Slip surface	Section A–A	Section F–F	Section I–I	3D Analysis
a	1.38	1.87	1.34	1.48
b	1.37	1.74	1.42	1.48
c	1.61	3.16	2.15	2.08
d	1.80	1.96	1.41	1.71

Fig. 5: Stability analyses: most representative sections and slip surfaces

REFERENCES

Beaven R.P. & Powrie W. 1995. Hydrogeological and Geotechnical Properties of Refuse Using a Large Scale Compression Cell. *Sardinia '95*: 761-768. Cagliari: CISA.

Grisolia M., Napoleoni Q. & Tancredi G. 1995. The Use of Triaxial Tests for the Mechanical Characterization of MSW. *Sardinia '95*: 761-768. Cagliari: CISA.

Jessberger H.L. & Kockel R. (1993). Determination and Assessment of the Mechanical Properties of Waste Materials. *Sardinia '93*. Cagliari: CISA.

Kavazanjian E., Matasovic N., Bonaparte R., Schmertmann G. 1995. Evaluation of MSW Properties for Seismic Analysis. *The Geoenvironment 200*: 1126-1141. New Orleans: ASCE

Lambe T.W.,& Whitman R.W. 1979. Soil Mechanics. John Wiley & Sons.

Macksimovic M. 1979. Limit Equilibrium for Non-Linear Failure Envelope and Arbitrary Slip Surface. *Numerical Methods in Geomechanics*: 769-777. Aachen

Oweis I.A. & Khera R. 1986. Criteria for Geotechnical Construction on Sanitary Landfills. *Environmental Geotechnology*. Fang H.Y. ed.:205-222.

Oweis I.A. & Khera R.P. 1990. Geotechnology of Waste Management. London: Butterworths.

Oweis I.A., Mills W.T., Leung A. & Scarino J. 1985. Stability of Sanitary Landfills. *Geotechnical Aspects of Waste Management*: ASCE.

ACKNOWLEDGEMENTS

This paper is published by kind permission of AMIU of Genova.

Environmental Geotechnics, Kamon (ed.) © 1996 Balkema, Rotterdam. ISBN 90 5410 848 7

Quality control of dynamic compaction in municipal solid waste fills

A. Bouazza
Monash University, Department of Civil Engineering, Melbourne, Vic., Australia
W.F. Van Impe & W. Haegeman
Ghent University, Soil Mechanics Laboratory, Belgium

ABSTRACT: Compaction of municipal and industrial solid waste is important in landfill operations because increasing the density of the material increases both the amount of material that may be put into the landfill, thus increasing its life, and the strength of the waste, thus enhancing the stability of the waste pile. In addition densified waste undergoes less settlement, which is important for maintaining the integrity of the cover and the liner system. The control of the compaction is also of prime importance since it guarantees the optimum life time of the disposal construction. However, in the case of dynamic compaction of waste fills it is very difficult to obtain by means of traditional testing methods, a reasonable evaluation of the stiffness of the fill. This is mainly due to the wide heterogeneity of the waste.

In the present paper a review is made about the principal methods, such as the SPT, CPT, PLT and the WAK tests, used in the field to assess the compaction of a waste fill. The evaluation of dynamic compaction must ideally be fast to reduce compaction equipment downtime, to produce results in the field for immediate assessment, to be customisable for investigating any zone of interest and to directly evaluate the improvement of the properties as a function of depth without recourse to empirical correlations. In this respect, the Spectral Analysis of Surface Waves (SASW) which is a seismic technique for in situ evaluations for low strain stiffness moduli is presented. This method has been successfully applied to evaluate the dynamic compaction in a waste disposal site situated at Maldegem (Belgium).

1 INTRODUCTION

In recent years, the scarcity of land space necessary for new urban development has seen a renewed interest from local authorities on the extension of existing landfills life to allow storage of more waste. In this case, ground improvement techniques are generally used. The most commonly methods used are : preloading (Charles et al , 1981 ; Kissida & Beaton, 1991) and dynamic compaction (Gambin, 1981, Welsh, 1983, Lukas, 1985, 1992, Gifford et al., 1992, Van Impe, 1994, Van Impe & Bouazza, 1996).

Indeed, increasing the density of the waste increases both the amount of material that may be put into the landfill, so increasing its life and its strength. In addition densified waste undergoes less settlement, which is important for maintaining the integrity of the cover and the liner system. However, one has to bear in mind that much of the initial settlement of young waste fills is not due to consolidation as defined in soil mechanics. Biodegradation of the organic waste and other chemical processes contribute significantly to surface settlement.

In the case of dynamic compaction, the control of the compaction is also of paramount importance since it guaranties the optimum life time of the disposal construction. However the wide heterogeneity of the waste which can cover a large spectrum of materials (table 1) makes it very difficult to obtain a reasonable evaluation of the fill stiffness.

The focus of this paper is on in-situ tests for quality control of the densification of domestic solid waste fills, in order:

- to review briefly the various in-situ tests used to assess the compaction of a domestic solid waste fill.
- to describe a case history where SASW technique was used to control the quality of the densification process.

Table 1. Waste components as weight percentage for different cities (remodified from Van Impe & Bouazza, 1996).

components	cities																	
	1	**2**	**3**	**4**	**5**	**6**	**7**	**8**	**9**	**10**	**11**	**12**	**13**	**14**	**15**	**16**	**17**	**18**
metals	1	1	2	3	3	1.5	-	5	1	1	3	2	5	2.5	3	4	1	-
paper, board	25	5	8	12	33	8	14	22	9	20	22	10	24	31	9	19	2	12
plastics	-	1	4	5	-	6.5	-	-	8	11	8	3	4	9.5	4	7	3	-
leather,wood,rubber	7	1	1	-	7	1	-	3	-	4	2	6	-	4	-	4	1	-
textiles	3	-	1	-	10	1	-	-	4	4	-	3	4	5	5	-	-	-
organic materials	44	45	81	74	15	79	56	20	72	23	52	61	50	28	45	59	71	59
glass	1	1	2	4	10	2	-	6	1	7	4	1	5	9	1	2	1	-
others	19	46	1	2	22	1	30	46	5	30	9	14	8	11	33	5	21	-

1-Bangkok (Thailand), 2- Pekin (China), 3- Viçosa (Bresil), 4- Nairobi (Kenya), 5-Hong Kong, 6-Bandung city (Indonesia), 7- Madras (India), 8- New York (USA), 9- Algiers (Algeria), 10- Zagreb (Croatia), 11- Meruelo (Spain), 12- Istanbul (Turkey), 13- Beyrouth (Lebanon), 14- Geneva (Switzerland), 15- Dakar (Senegal), 16- Athens (Greece), 17-Cochabamba (Bolivia), 18- Moscow (Russia)

2 BACKGROUND

Some available in situ tests have wide applicability, while others are limited to special projects or soil conditions. The applicability of in situ tests is a function of mobilisation and cost/accuracy considerations. Mobilisation and access depend on the project location and must be addressed on a site by site basis. The relationship between test costs and accuracy dominates in many instances the use of in-situ testing. Fig. 1 indicates that rudimentary tests such as hand penetrometers, dynamic cone tests, SPT, are in case of "soils" relatively inexpensive but have low relevancy. On the other hand, tests with relatively higher accuracy, such as the CPTU, SASW, etc... are costly but much more relevant to perform.

The principal in-situ tests commonly used to assess the compaction of a domestic solid waste fill are the plate loading test (PLT), the standard penetration test (SPT), the cone penetration test (CPT) and the wave activated stiffness test. (WAK). In more exceptional cases the pressuremeter test (PMT) was used. Only recently, proposals were made (Van Impe, 1994, Haegeman & Van Impe, 1995) to extensively use SASW in order to evaluate waste stiffnesses.

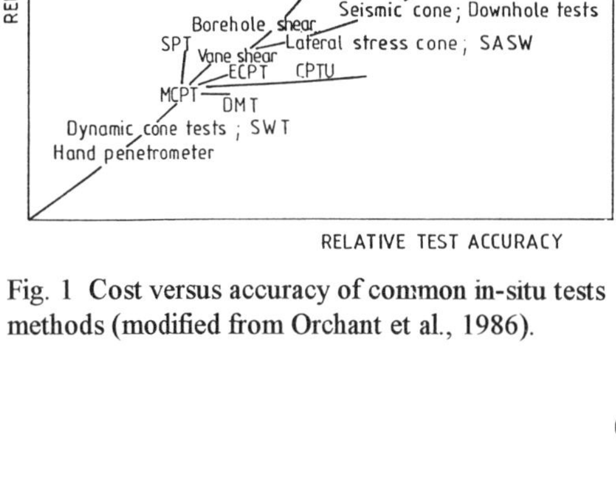

Fig. 1 Cost versus accuracy of common in-situ tests methods (modified from Orchant et al., 1986).

2.1 *Plate load test (PLT)*

Plate loading tests are usually performed on top of landfills after closure, once the final sealing cover has been installed. The usual aim is to evaluate the bearing capacity for potential use of the landfill area. In some cases, tests have been run during landfill operations to analyse the mechanical behaviour of the waste material. Typical results of PLT conducted at the Appels landfill, Dendermonde (Belgium) are shown in fig. 2. In the case of deep densification the PLT provide information only on the upper zone of the landfill. In this respect, it seems a useful tool to show the stiffness improvement at or near the ground surface. Another adaptation of the PLT is the "skip test", here a heavy duty waste disposal skip is used to simulate low levels of loading (≈ 45 kPa). To obtain meaningful results from a skip loading test several precautions have to be observed (Mc Entee, 1991).

2.2 *Penetration tests*

Penetration testing, using either the SPT or the CPT is the most frequently used means to evaluate deep compaction in soils. In the case of SPT there is considerable uncertainty in assigning material properties based on this type of test, since there are no published correlations between domestic solid waste properties and blow counts. Gifford et al. (1992) pointed out that the SPT can be at best used to show

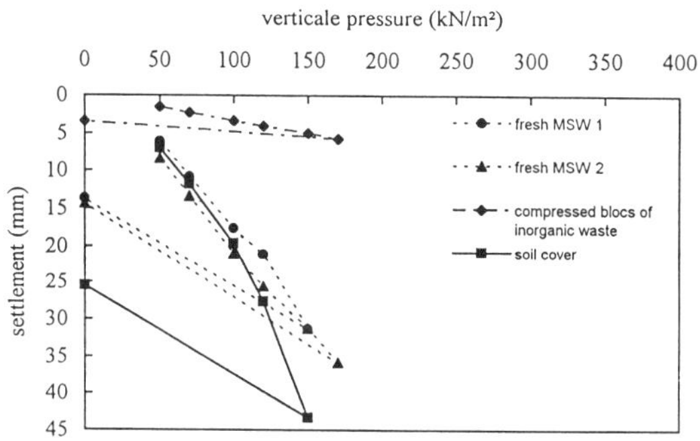

Fig. 2 Plate load tests at the Appels landfill, Dendermonde (Belgium).

trends of improvement at greater depths but should be coupled with other tests.

Referring to the CPT family (fig. 3), the most popular and widely used type is still the mechanical cone (CPT.M.). Following the new trends and the international standard, an increasing number of electric (CPT.E.) and piezo (CPTU) cones are available. However, for non text book materials such as domestic waste it is much more difficult for the practising engineers to to use such test results in a meaningful way.

Nevertheless, cone penetration tests may, in some cases, be useful, in combination with other test methods, for the investigations of the waste body (for ex : to localise qualitatively areas with waste materials reducing the stability of a landfill). Typical results of the CPT conducted on domestic and industrial solid waste at Maldegem (Belgium) are shown in fig. 4. In waste material relatively erratic CPT readings are obtained but a general trend can be observed. Average sleeve friction, cone resistance and friction ratio can be identified in the penetration logs for different landfills.

This type of test could show difficulties to penetrate the waste fills because of the relatively large sizes of inclusions in the fills.

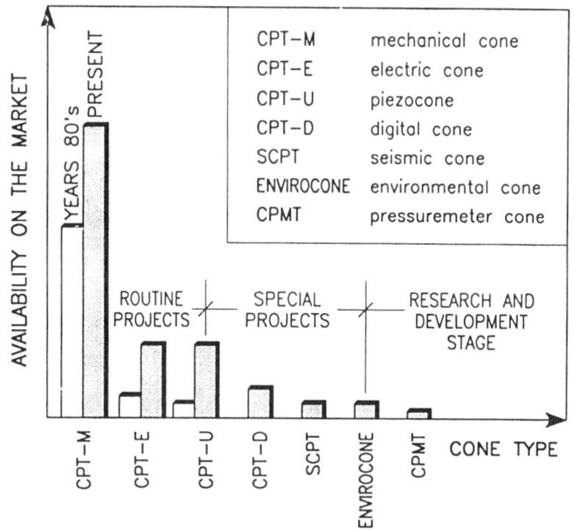

Fig. 3 The CPT family (from Pane et al., 1995).

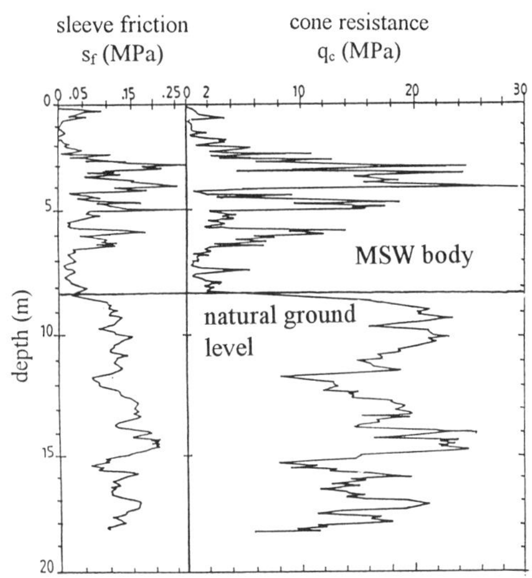

Fig. 4 CPT results in one location at the Maldegem landfill.

2.3 *Wak test*

The wave activated stiffness test (WAK) is used to evaluate stiffness of material under footings (Briaud et al, 1990 ; Maxwell & Briaud, 1994). In the case of dynamic compaction, the WAK test can be performed using an instrumented sledge hammer to impact the dropped weight while it is resting on the ground. This type of test measures improved stiffness only beneath the level of the dynamic compaction craters. This non intrusive test allows to evaluate a much larger volume as compared to the very locally obtained SPT and CPT data.

2.4 *Other methods*

Long term, expensive, but very worthwhile settlement monitoring of the whole waste compacted body can include effects caused by biodegradation. The settlements could be measured at different landfill levels so that the improvement distribution with depth could be known. Direct observation of ground improvement work in progress can also be a wealthy source of information.

3 SPECTRAL ANALYSIS OF SURFACE WAVES

The in-situ evaluation of any waste body improvement must ideally be fast to reduce equipment downtime, the produce results in the field for immediate assessment, to be customisable for investigating any zone of interest and to directly evaluate the improvement of the properties as a function of depth without recourse to empirical correlations. In this respect, the use of the Spectral Analysis of Surface Waves (SASW) is very attractive, (Van Impe, 1994, Haegeman & Van Impe 1995, Haegeman, 1995). The suggestion of determining the low strain stiffness by means of actificially generated surface waves of Rayleigh type has quite an old origin for soils. It was initially conceived to evaluate pavement stiffness and thickness (Van der Poel, 1951) and then this concept was proposed for geotechnical analysis by Ballard (1964). Since then the SASW technique has been applied in several fields (Addo & Robertson, 1992 ; Mancuso & Vinale, 1993, Kavazanjian et al, 1994, Haegeman & Van Impe, 1995, Haegeman, 1995).

Methods employing wave propagation principles in determining low strain dynamic properties are either intrusive or non-intrusive. Intrusive methods require penetration or borehole drilling and determine compression (V_p) and shear (V_s) wave velocities directly. Tests included in this category are the uphole, downhole, crosshole, bottom hole, and seismic cone penetration. Non-intrusive methods determine body wave velocity profiles from the ground surface. These include surface reflection, surface refraction, steady state vibration and the Spectral Analysis of Surface Waves. Surface reflection and refraction are generally more suited to geophysical exploration work and do not generally provide accurate profiles of seismic velocities for geotechnical engineering purposes. To varying extents, the last two non-intrusive methods use surface wave dispersion to indirectly determine compression and shear wave velocities. The advantages of these measurements are that they are made at known levels of stress and strain (< 10^{-4} %) and they involve essentially no soil disturbance. The SASW involves no boreholes since the equipment is placed on the ground surface, is extremely rapid and uses the latest technology in waveform analysis.

3.1 *Testing procedure*

Surface waves are generated by applying a dynamic vertical load to the ground surface. The propagation of these waves along the surface is monitored with two receivers placed at distances of d_1 and d_2 from the source. Additionally , distance d_2 is usually equal to two times d_1. In principle, it should be possible to use a single source/receiver spacing for the entire test. However, practical considerations such as attenuation and near-field effects dictate that several different source/receiver spacings must be used. Typically distances between receivers of 1, 2, 4, 8 and 16 m are used if the soil profile is to be evaluated to a depth of about 10 m. For pavement sites, these distances range from 0,15 m to 1 m. The spacing between the receivers is varied according to the wavelengths (L_R) being measured. At each spacing, one can typically measure wavelengths ranging from twice the receiver spacing to less than one-third of the receiver spacing.

Surface wave frequencies used in SASW testing range from several Hertz to about 500 Hz for most soil sites. In this case, velocity transducers with natural frequencies between 1 and 4.5 Hz work well as receivers. When rock or some other stiff material such as concrete is at the surface, frequencies on the order of 10 to 50 kHz must be generated, and accelero-meters are used as receivers. In both instances, the most common types of sources are either simple hammers (small, hand-held hammers or sledge hammers) or dropped weights weighing from 200 to 1500 N. Electromagnetic vibrators in conjunction with sinusoidal or random input motion can also be used as sources.

A dual channel Fast Fourier Transform (FFT) dynamic signal analyser is used to record and analyse the motions at any two velocity transducers. The ability to calculate transforms rapidly in the field, is an essential part of the SASW method, allowing operators to immediately assess the quality of the data being collected and, if necessary, modify the arrangement of source and receivers or other test parameters accordingly. This data can be easily transferred to a microcomputer for further analysis as desired.

4 CASE HISTORY AT MALDEGEM (BELGIUM)

4.1 *Site description*

The test site consisted of an about 20 years old municipal waste and industrial disposal. It is situated at Maldegem, 30 kms from Ghent (Belgium). Its area can be considered as around 30000 m^2. Borehole data were available at the site and showed organic components, plastic, paper, rubber... to even concrete and asphalt bolders. The thickness of the disposal varied from 4.5 to 8 m and was covered with natural soil of 0.5 to 2 m thickness which however was partly removed and partly mixed with the waste during recent profiling of the site. The reclamation consisted of profiling under controlled circumstances, transport and processing of some of the non municipal waste, isolating the former waste disposal from the surroundings and restoration of the environment. The choice of ground improvement by heavy tamping was made in order to decrease the volumetric amount of waste and above all to reduce the differential settlements of the new cover liner. SASW was selected for compaction control. A test site (60 m x 60 m) with an approximated waste disposal thickness of 8 m was chosen for a trial. In the present case of the Maldegem municipal waste fill, the craters, due to the various impacts, were filled with waste disposal material coming from the surrounding ground level, during relevelling of the treated test area. The ground levelling machines were immediately following the dynamic compaction crane. SASW tests were performed at 12 locations at the trial site before and after compaction.

For reasons of local practice the weight of the pounder was limited up to 12.9 ton and the drop height equal to 10 m. The spacing of the impact points and the number of blows per point were chosen as test variables.

5 RESULTS AND DISCUSSIONS

The two seismic receivers were consecutively placed at spacings of 0.5, 1, 2 and 4 m. Elastic stress waves were then generated by the impact of a sledge hammer on a distance of 0.5, 1, 2 and 4 m from the first receiver for each corresponding receiver spacing. For each receiver spacing, five or more signals were averaged in the frequency domain, where the phase of the cross-spectrum and coherence were also calculated. The phase delay and receiver spacing were then used to calculate the Rayleigh wave velocities as a function of frequency. The procedure was repeated for all the spacings . At the end of testing, dispersion points from the individual spacings are combined into a single dispersion curve for that well choosen location. To evaluate the effectiveness of the dynamic compaction (fig. 5), the low strain dynamic shear modulus and corresponding Young modulus (E_y) were calculated. These results were related to a conventional large strain plate modulus (E_{PLT}) taking into account a decay to 30 % of such modulus with respect to the small strain SASW modulus, because of the much larger strains which occur in a plate load test.

Some typical field results of the SASW tests are the in-situ Rayleigh-wave dispersion curves shown in Fig.5. In each case, the dispersion curves measured before and after dynamic compaction are presented. The general trend is that in depth the shear wave velocites after compaction significantly increased with respect to the initial values. The shear modulus increment ranged from about 50 to 100 % and was the highest for the zone with the impact points at 2 m distance. Obviously, the

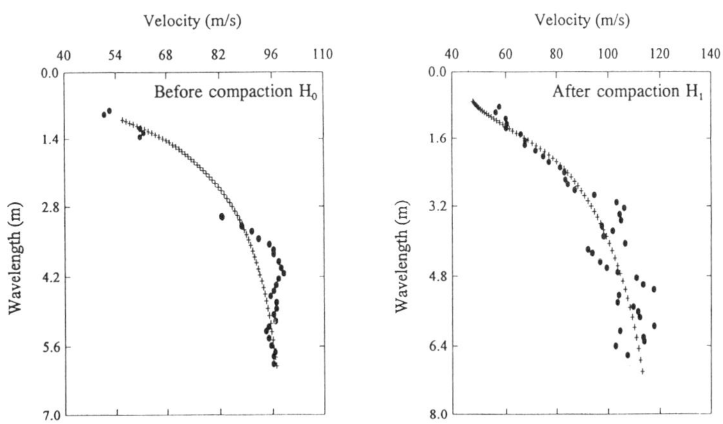

Layer	v_s (m/s)		E_y (MPa)		E_{PLT} (MPa)	
	Before	After	Before	After	Before	After
upper layer	40	50	1.92	3.00	0.71	1.11
deeper layer	115	140	15.87	23.52	5.88	8.71

Fig. 5 Typical dispersion curve obtained at the Maldegem landfill before and after compaction.

top layer, 0.3 to 0.5 m in thickness, which after a period of heavy rainfall was fully saturated during compaction, was hardly compacted. Here, an extra sand layer of 0.3 to 0.4 m with a surface compaction was required in order to improve later on the surface compaction. The large differences of the dispersion curves in some sections are indications of the non-homogeneity of the existing waste materials. Most SASW control tests were performed two or three weeks after backfilling the craters (± 2 m of depth).

6 CONCLUSIONS

The possibilities for succesfully applying the SASW method in evaluating the heavy tamping efficiency in waste fills are important. Significant and consistent improvements in Rayleigh and shear wave velocities (for the Maldegem waste site) were measured. These wave velocities may be directly converted into a low strain stiffness modulus without recourse to empiricism. The easy handling of the test and the simplicity of the required field equipment assure that the SASW test is quite cost-effective.

REFERENCES

Addo, K.O. & Robertson, P.K. 1992. Shear wave velocity measurement of soils using Rayleigh waves. *Can. Geotech. Jour.*, 29:558-568.

Ballard, R.F., Jr. 1964. *Determination of soil shear moduli at depths by in-situ vibratory techniques.* Paper n°4-691, U.S. Army waterways Experiment Station, Vicksburg, MS, USA.

Briaud, J.L., Liu, M.L & Lepert, P. 1990. The Wak test to check the increase in soil stifness due to dynamic compaction. *Geotechnics of Waste Fills Theory and Practice*, ASTM STP 1070: 107-122.

Charles, J.A., Burford, D. & Watts, K.S. 1981. Field studies of the effectiveness of Dynamic consolidation. *Proc. 10th Int. Conf. on Soil Mechs. Fndns Engng.*, Vol. 3, Stockholm:617-622.

Gambin, M.P. 1981. Consolidation dynamique et construction en zones de décharge. *Revue Française de Géotechnique*, 14bis:97-101.

Gifford, G.P., Landva, A.O. & Hoffman, V.C. 1992. Improvement of a demolition debris landfill by compaction. *Proc. Int. Sym. on Environmental Geotechnology*, Cesme: 391-399.

Haegeman, W. 1995. Non destructive evaluation of a road construction by the SASW method. *Proc. 10th Danube-Eur. Conf. on Soil Mechs and Fndns Engng.*, vol. 1, Mamaia-Constantza:105-112.

Haegeman, W. & Van Impe, W.F. 1995. Developments in SASW testing. *Proc. 11th Eur. Regional Conf. on Soil Mechs and Fndns Engng., Danish Geotechnical Society, Bulletin 11*, vol. 1, Copenhagen: 133-140.

Kavazanjian, E. Jr., Snow, M.S., Matasovic, N., Poran, C.J. & Satoh, T. 1994. Non-intrusive Rayleigh wave investigations at solid waste landfills. *Proc. 1st Int. Congress on Environmental Geotechnics*, Edmonton: 707-712.

Kissida, J. and Beaton, N.K. 1991. Landfill park: From eyesore to asset . *Civil Engineering.* 61 (6):49-51.

Lukas, R.G. 1985. Densification of a decomposed landfill deposit. *Proc. 10th Int Conf. on Soil Mechs and Fndns Engng.*, Vol. 3, San Fransisco:1725-1728.

Lukas, R.G. 1992. Dynamic compaction engineering considerations. *Grouting Soil Improvement and Geosynthetics, ASCE, Geotech. Spec. Publ. n° 30*, Vol. 2: 940-953.

Mancuso, C. & Vinale, F. 1993. Use of SASW in earth dam investigation. *Proc. Geotech. Engng. of Hard Soils-Soft Rocks*, vol.2, Athens:1291-1298.

Maxwell, J.C. & Briaud, J.L. 1994. 53 WAK tests on spread footings. *Vert. and Horiz. Def. of Fndns and Embankments, ASCE, Geotech. Spec. Publ. n° 40*, vol.1:143-152.

McEntee, J. 1991. Site investigation. In Fleming, G. (ed.), *Recycling Derelict Land*:64-87. London, ICE

Orchant, C.J., Trautmann, C.H. & Kulhawy, F.H. 1986. In-situ testing to characterize electric transmission line routes. *Proc. Use of In-Situ Tests in Geotech. Engng, ASCE, Geotech. Spec. Publ. n° 6*: 869-886.

Pane, V., Manassero, M., Brignoli, E. & Soccodato, C. 1995. Cone penetration testing in Italy. *Proc. Int. Symp. on Cone Penetration Testing*, vol.1, Linköping:101-114.

Van Der Poel, C. 1951. Dynamic testing of road constructions. *Jour. App. Chem.*, vol. 1:7.

Van Impe, W.F. 1994. Municipal and industrial waste improvement by heavy tamping. *Proc. Geotech. Aspects in the Design and Control of Waste Containment Systems*, Milazzo:90-125.

Van Impe, W.F. & Bouazza, A. 1996. Densification of waste fills by dynamic compaction. *Can. Geotech. Jour.* (submitted for publication).

Welsh, J.P. 1983. Dynamic deep compaction of sanitary landfill to support superhighway. *Proc. 8th Eur. Regional Conf. on Soil Mechs and Fndns Engng.*, vol. 1, Helsinki:319-321.

Environmental Geotechnics, Kamon (ed.) © *1996 Balkema, Rotterdam. ISBN 90 5410 848 7*

Geotechnical characterization of municipal solid waste: Choice of design parameters

M.Grisolia & Q.Napoleoni
University of Rome 'La Sapienza', Engineering Faculty, Italy

ABSTRACT: Results of stress and strain tests carried out on Municipal Solid Waste material, both in situ and in laboratory, and some direct observations made on actual sanitary landfills, have shown that the geotechnical behaviour of the material is strongly conditioned by the large deformations undergone by the material. Some criteria useful to make stability analysis and forecast of landfill settlements are presented and discussed.

1 INTRODUCTION

To verify the stability conditions of a controlled sanitary landfill is a basic design element. Possible failures in the waste heap are events of clear environmental danger and may interrupt and put out of order the control systems of the landfill itself.

Waste is, on the whole, a highly deformable medium, whose deformation and processes last a very long time and may cause dangerous settlement and distorsion (Jessberger and Kockel 1993; Grisolia and Napoleoni 1995; Fang 1993; Christensen et al. 1994).

Municipal Solid Waste (MSW) stability problem can be brought back to the principles of classical geotechnics applied to embankments. As a matter of fact, waste, although presenting some analogies with soils, has its own characters which hinder the direct application of geotechnical methods.

Anyway, alternative models are not available at the moment and the problem needs be studied and looked into.

2 GEOTECHNICAL BEHAVIOUR OF MSW

Waste, and in particular municipal solid waste, may be assumed as a multiphase medium consisting of solid, liquid, and gaseous elements, like a soil.

Yet, there are some essential differences between waste and soils. As to the solid phase in particular, materials can be distinguished, from the geotechnical point of view, into three classes (Grisolia et al. 1995a):

A) stable inert materials;
B) highly deformable materials;
C) readily degradable materials.

Class A group those materials such as glass, metal elements, masonry debris, etc. whose overall mechanical behaviour can be assumed as that of a natural coarse-grained and highly heterogeneous soil.

Class B group materials, such as fabric, paper and plastic, especially in sheets or fragments, may dispose themselves so as to undergo marked settlements even under modest loads and an overall mechanical anisotropy.

Class C group materials as organic matters vegetable, food refuse, etc., which cause deep and important physico-chemical transformations to occur in waste in a relatively short time; on the contrary, in soils any trasformation is the product of very slow natural processes. Organic matter decomposition undoubtedly affects the overall mechanical behaviour because it causes a volume reduction together with production of gases and other very low consistency materials.

Liquid and gaseous phases contribute to make the overall structure still more complex. We can generally say that material is unsaturated and the gaseous phase consists both of air –that was already present at the time of emplacement– and gases produced by the decomposition of biodegradable matters.

If the average composition of a waste heap is taken into account, one can assume that highly deformable and degradable elements are volumetrically prevalent over other components in the first stage of the emplacement. Only few points of the so-called "stable inerts" may be in contact with one another (Fig. 1a). Under load, the "highly deformable" elements such as paper, fabric, plastic sheets and organic matter, are, therefore, those which will cause the waste initial remarkable decrease in volume.

During this initial phase, displacement of stable inert elements is great and the elements tend to gradually place themselves into their final arrangement. Contacts between these elements will still be largely hindered by deformable materials and organic matter. Deformations in the heap will be still marked and long lasting because of constant load slow deformations of the "creep" type and of the proceding of organic matter transformations. At the end of the processes, direct contacts between stable inert elements will be re-established and a behaviour assimilable to that of the natural soils will be set up (Fig. 1c).

Moreover, owing to deformations, fibrous elements such as fabric, plastic sheets, etc. will arrange themseleves so as to create reinforced structures within waste, which thus will take a behaviour similar to that of a reinforced soil (Cowland et al. 1993; Kolsch 1993).

2.1 *Shear Strength of Waste*

If the mechanical behaviour of waste is assimilated to that of a "soil", the shear strength within the waste heap will depend mostly upon the "effective" stress at the contact between stable inert elements. In this case, strength is of the friction type and will be defined through a certain value of the friction angle φ. Easily degradable organic matter and highly deformable elements in the heap largely control number and type of contacts.

Thus, if reference is made to an average ideal sample, the waste shear strength may be correlated with the unit weight (Carucci et al. 1991; Cowland et al. 1993).

Natural and recomposed waste samples, of both great and small size, were tested by using CD triaxial tests. Such tests are included in a comprehensive research programme on the waste geotechnical characterization based on in situ and lab geotechnical measurements (Grisolia et al. 1992).

The stress-strain curves of some MSW samples, reconstituted in laboratory and consolidated with cell pressures σ_3 = 50 kPa; σ_3 = 100 kPa; σ_3 = 100 kPa are plotted in Fig. 2. From the obtained results, it is possible to observe that even if high deviatoric stresses had been applied and large deformations had occurred, failure connected with general "plasticization" conditions of the material was not reached in any of the samples tested.The mechanical behaviour of waste is governed by the deformations that the material undergoes, according to the processes described in the previous paragraph.

Thus, in problems connected with strength, it makes no sense to express waste shear strength by means of parameters deriving from the definition of rigid–plastic model commonly adopted in soil mechanics.

2.2 *Compressibility*

Deformations resulting from loading a multiphase medium depend on how its constitutive elements arrange themselves to contrast the stresses they undergo. In medium- to coarse-grained soils, deformations are essentially associated with slippage within grains, which rearrange themselves to new equilibrium conditions. Volume variations of single elements resulting from loading are virtually negligible.

In waste, many solid elements can vary in shape

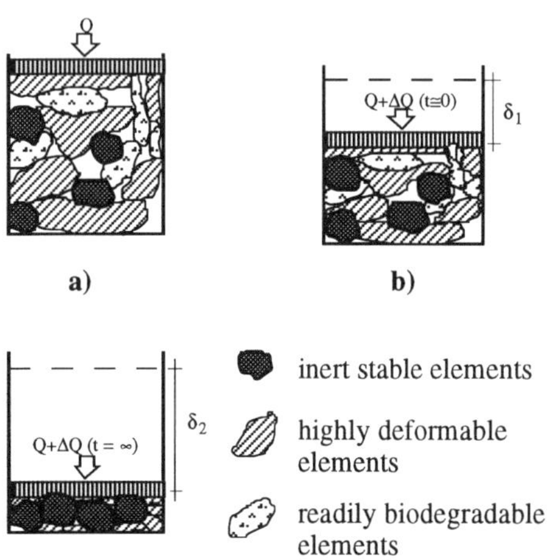

Figure 1. Schematic representation of the various phases of deformation of a MSW sample

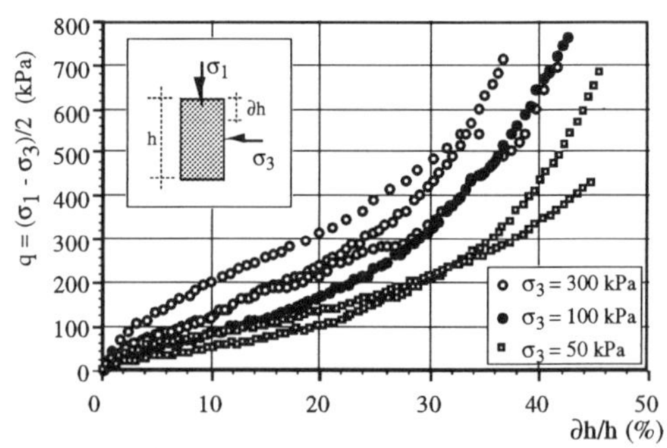

Figure 2. CD triaxial laboratory tests on reconstituted MSW samples.

and volume both because of their intrinsic deformability and of transformations resulting from biochemical decay processes. Therefore, we can reasonably assume the curve plotted in Fig. 3 as the theoretical MSW compressibility curve (Grisolia and Napoleoni 1995).

In order to make the proposed model of practical application, we should consider the following aspects of the process: i) the waste initial composition; ii) the role played by fluids; iii) the effects of the decomposition of organic matter.

The first aspect regards the influence of the different types of waste solid elements on immediate settlements.

We must consider that it is not easy to determine the composition of waste in a landfill and that is not a standard criterion for a technical classification of the material. (Grisolia et. al 1995b).

The presence of fluids modifies the properties of solid materials causing a different global rheological behaviour. Some waste components rich in micropores (paper, wood, fabric) can absorb large amounts of fluids which, under loads and/or in time, may cause changes in their physical state and mechanical properties. There is a limit condition corresponding to the maximum amount of fluids that a material can absorb before leachate begins to form (Cancelli and Cossu 1984). Under such limit condition the "absorbing" elements are generally saturated. Water content is not constant and varies with stress level, time, and decay rate. The substantial difference between initially saturated and unsaturated materials is that a larger amount of fluids accelerate degradation and decay processes in some substances and materials (e.g. paper, cloths etc.). (Grisolia et al. 1992).

Decomposition of organic matters necessarily implies a volume reduction in the mass, owing to the degradation of materials which, from an initial solid state, turn into liquid and gaseous fluids. Respective processes are difficult to schematize and forecast because they are governed by complex physical, chemical and biological factors. Amount and velocity rate of degradation phenomena depend on quantity and type of organic matter, and on new environmental conditions as they occur. Among these, the following are to be mentioned:

a) the way the initial aerobic conditions change to anaerobic conditions within the waste heap;

b) the evolution of transformations occurring in anaerobic conditions, associated with production of fluids and the passage to a so-called "metanigeous" phase;

c) the way of migration and extraction of gas

The process may be controlled, at least in theory, on the basis of the changes in temperature, chemical and biological characters of leachate, and quantity and quality of biogas occurring in landfill (Edger and Noble 1992; Herig 1991).

Some measurements of long duration settlements under constant load, carried out on a pilot landfill at lab scale, have provided interesting data (Boni et al.

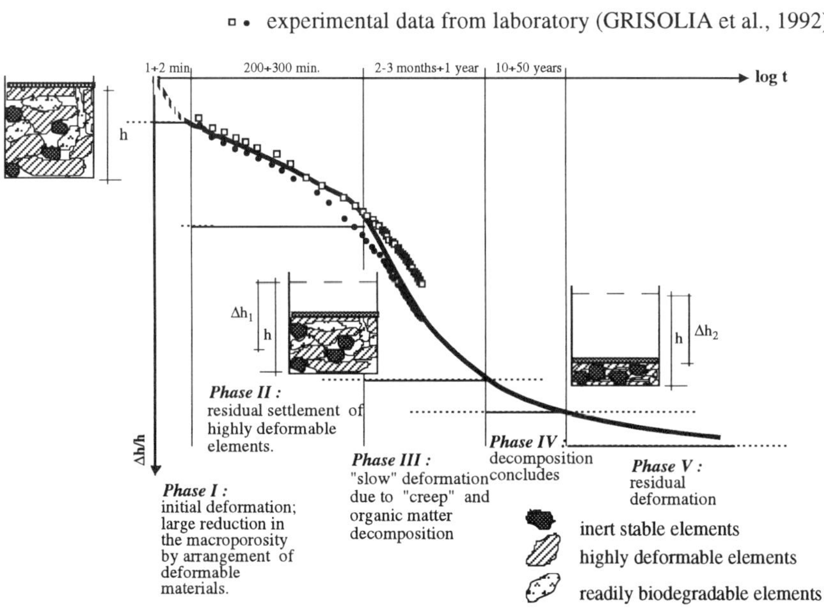

Figure 3. MSW compressibility curve.

1995). The diagram of Fig.4 shows the trend of settlements observed in time. The values of settlements have been correlated with other factors controlled during the tests, such as quantity and ways of water addition and of recirculating leachate, temperature trend, quantity and characters of the produced leachate.

Measurements carried out on a controlled landfill have provided interesting data (Gasparini et al. 1993). The elaboration of available data highlighted marked behaviour differences, particularly with regard to the velocity rate of deformations in the various parts of landfill. These differences have been explained with the age of waste in relation to the processes of decomposition for organic matters and on the amount and characteristics of extracted gases.

3 STABILITY ANALYSIS

In Geotechnics, "limit equilibrium methods" are commonly used in order to verify the stability of the external front of an embankment. The application of these methods to the study of the stability conditions of sanitary landfills raises problems which must be thoroughly evaluated.

As described in the previous paragraph, the waste mechnical behaviour under load is conditioned by the great deformations undergone by materials, so that it is not possible to recognize any plasticization domain in the material. This behaviour is confirmed by experimental evidence. Thus, it was observed that excavations with almost vertical walls carried out in landfills are stable even when they reach remarkable heights. Centrifuge tests carried out on models confirmed this behaviour (Jessberger and Kockel 1993).

Hence in practical cases, in examining the stability conditions of a landfill, attention must be paid to the possible deformations that may occur rather than to the possibility that failure may actually be reached. In order to apply this condition in practice, the design must verify that applied stress levels do not produce excessive deformations.

In Geotechnics, this complex aspect of the study can be dealt by adopting appropriate conservative safety coefficients, or by setting appropriate stress limit values restricting deformations within acceptable limits. For this purpose, triaxial tests were designed in such a way as to identify the stress values required to obtain prefixed values of axial deformations (Grisolia et al. 1995b). By setting these stress values as limit values, it was possible to obtain virtual shear stress parameters (cohesion and friction angles) versus measured deformations (Fig. 5). The diagram shows that waste mechanical behaviour may be represented by using "friction" and "cohesion" variables depending on deformation variability. Similar schematizations have been suggested also by other researchers (Jessberger and Kockel 1993; Cowland et al. 1993). It is worth noting that trends of c' and φ' parameters, even if the assumptions made are simple, are not in contrast with the complex models which can reasonably be hypotesized for waste. The fact that the shear strength angle increases as deformations increase and then tends to stabilize on an almost constant value, agrees with the hypothesis that -as deformations increase- the behaviour of soil is more and more controlled by the so-called "inert stable" elements.

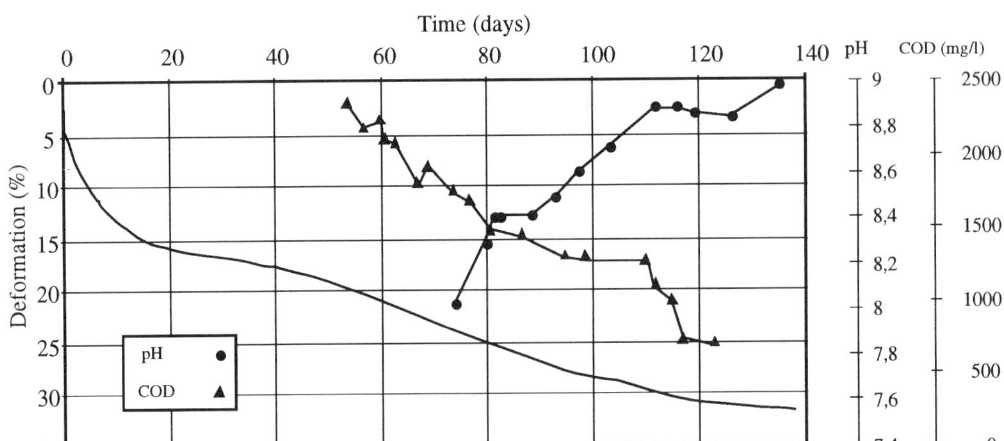

Figure 4. Typical settlement vs time curves in oedometric tests for MSW samples (adapted from Boni et al.1995)

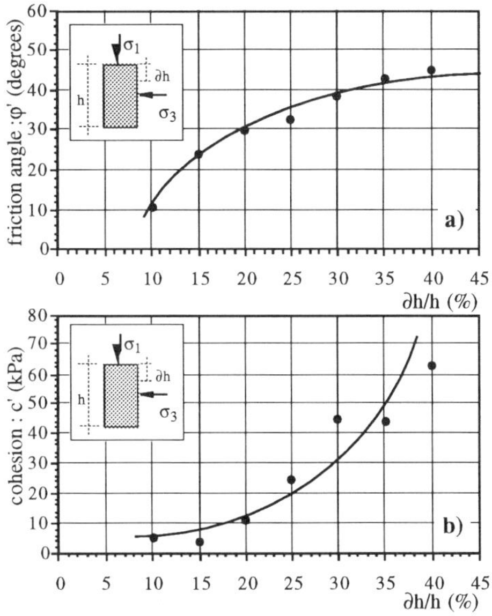

Figure 5. Triaxial test: friction angle (a) and cohesion (b) vs sample vertical strain

Moreover, the cohesion trend represents the strength contribution given by the "reinforcing" fibrous elements. Fibrous elements will begin to operate after great deformations - generally greater than those acceptable in actual cases - occur.

In the case of tested waste, to assume as tolerable deformations in the order of 10-15% means to attribute equivalent strength values to the material expressed by a practically null value of the cohesion c' and a shear strength angle of 22° (Grisolia et al. 1995b).

In practical applications, the actual correspondence between deformations measured on the volume element and those occurring in situ must be carefully evaluated. For this purpose and in order to establish practically acceptable values, it is necessary to collect data and measurements on the actual behaviour of operating landfills. Finally, it is necessary to consider that data cannot be easily transposed from a situation to another, as the waste mechanical behaviour depends on specific factors, such as initial composition, placement techniques, possible pre-treatments, time span, stress state, environmental conditions, etc.

4 SETTLEMENT PREDICTION

At present, there are not specific models for calculating MSW landfill settlements. Some elementary models deriving from Soil Mechanics methods (Sowers 1973; Zimmermann et al. 1977) do not seem totally reliable. To give practical utility to the model described in the previous paragraphs, it is necessary to investigate and to control some essential aspects of the waste deformation and transformation processes. The first step to make is to separate immediate settlements from long term ones.

Immediate settlements generally depend on the initial composition of waste, on the way it was placed in landfill and on the possible waste pre-treatments. As a matter of fact, it is not possible to evaluate how much these factors contribute to total settlements because in ultimate analysis, during landfill construction, compressibility characteristics of each waste layer vary with the stress level and in time. Measurements themselves become practically impossible to interpret, so that settlement forecast refers, at present, to the reaching of the designed final shape. In-time settlements depend on viscous deformations ("aging") and upon the transformation processes of organic matter. This phenomenon may be controlled by means of models deriving from the superposition of a consolidation effect directly correlated with loads and an "aging" effect due to quick creep phenomena and to other phenomena occurring in time (Van Meerten et al. 1995). These latter include, although conceptually separated, also phenomena linked to the transformation processes of organic matters.

The proposed model does not allow, however, to make reliable forecasts. Difficulties in applying the model rise from the impossibility to include in a few parameters all the factors playing a role in the process. Yet, the model may be useful to interpret and to extrapolate direct measurements of settlements. In interpreting these measurements, global factors such as waste initial composition, age, height, temperature and humidity conditions, etc. should be considered. The systematic collection of empirical correlations linking waste age and landfill maximum height with evolution of settlement rate in time can be seen as an example of global approach (Yen and Scanlon, 1975; Sohn and Lee, 1994). In order to apply this forecast method, one has to have at least disposal specific measurements of settlements, calibration, and control. Landfill designs should therefore be based on an "observational" approach, which implies the possibility of modifying initial hypothesis to adjust them to each case.

5 FINAL REMARKS AND PROSPECTS

Direct experiences on several controlled landfills confirmed the bearing of various geotechnical aspects in designing and cultivating landfills. The possibility of verifying the landfill stability conditions and of forecasting possible settlements is a primary exigency.

Waste cannot be assimilated directly to a soil, and it is necessary to define more advanced models on the waste behaviour.

The collection of data on landfill behaviour on the basis of ad-hoc conceived control measures and systems must be considered of primary importance.

The procedures usually applied in soil mechanics, although not completely appropriate, are a starting point to study and to interpret waste mechanical behaviour.

Acknowledgement

The paper includes a few initial findings derived from a research program, that is in progress. The program is jointly conducted by the Department of Hydraulics, Roads and Transport and the Faculty of Engineering, University of Rome "La Sapienza", at the Center for Studies, Geology Techniques for the C.N.R. - Rome. Research is 40% funded by the Ministry of Public Education.

REFERENCES

Boni, M.R.; Delle Site, A.; Misiti, A. & Rolle, E. 1995. Aerobic stabilization of MSW: pilot plant experiences. Proc. XI Int. Conf. on Solid Waste Management. Philadelphia (PA).

Cancelli, A. & Cossu, R. 1984. Problemi di stabilità negli scarichi controllati. *Ingegneria Ambientale*, 13: 11-12.

Carucci, A.; Gabrielli, B. & Grisolia, M. 1991. Stability of slopes in a sanitary landfill. Proc. of 3rd International Landfill Symposium - Sardinia '91, 2: 1161-1170. Cagliari (I): CISA.

Christensen, T.H.; Cossu, R. & Stegmann, R. 1994. *Landfilling of Waste - Barriers.*. London: E e FN SPON, (UK).

Cowland, J.W.; Tang, K.Y. & Stegmann, R. 1993. Density and strenght properties of hong kong refuse. Proc. of 4th International Landfill Symposium - Sardinia '93, 2: 1434-1446. Cagliari (I): CISA.

Edgers, L. & Noble, J. 1992. A biological model for long term settlement in landfill. *Environmental Geotechnology*, 177-184. M.A.Usmes and Y.B. Acar (eds) Rotterdam: Balkema.

Fang, H.Y. 1993. Engineering behaviour of urban refuse, compaction control and slope stability analysis of landfill.*Waste Disposal by Landfill - Green '93*. R.W. Sarsby (ed). 47-72. Rotterdam: Balkema.

Gasparini, P.A.; Grisolia, M. & Saetti, G.F. 1993. Survey of a controlled landfill deformability for a correct reclamation. Proc. of 4th International Landfill Symposium - Sardinia '93, 2: 1435-1456 Cagliari (I): CISA.

Grisolia, M. & Napoleoni Q. 1995. Deformability of Waste and settlements of sanitary landfills. Proc. World Congress on Waste Management ISWA '95. Wien.

Grisolia, M.; Napoleoni, Q. & Tancredi, G. 1995a. Contribution to a technical classification of municipal solid waste. Proc. of 5th International Landfill Symposium - Sardinia '95, 2: 703-710. Cagliari (I): CISA.

Grisolia, M.; Napoleoni, Q. & Tancredi, G. 1995b. The use of triaxial tests for the mechanical characterization of municipal solid waste. Proc. of 5th International Landfill Symposium - Sardinia '95, 2: 761-767. Cagliari (I): CISA.

Grisolia, M.; Napoleoni, Q.; Sirini, P. & Tancredi, G. 1992. Geotechnical behaviour of sanitary landfill based on laboratory and in situ test. *The Journal of Resource Management and Technology*,. 20 (n.4): 197-203 Dep. of Syst. University of Pennsylvania, Philadelphia (PA) USA.

Herig, H.J. 1991. Prediction of gas production from laboratory scale test. Proc. of 3rd International Landfill Symposium - Sardinia '91, 1: 87-114. Cagliari (I): CISA.

Jessberger, H.L. & Kockel, R. 1993. Determination and assessment of the mechanical properties of waste. In:*Waste Disposal by Landfill - Green '93*. R.W. Sarsby (ed), 313-322. Rotterdam: Balkema.

Kolsh, F. 1995. Material values for some mechanical properties of domestic waste. Proc. of 5th International Landfill Symposium - Sardinia '95, 2: 711-729. Cagliari (I): CISA.

Sohn, K.C. & Lee, S. 1994. A method for prediction of long term settlement of sanitary landfill. Proc. First Int. Congr. on Environmental Geotechnics, 807-811. Edmonton (Can): Bi Tech.

Sowers, G.F. 1973. Settlements of waste disposal fills. Proc. VIII Int. Conf. on Soil Mech. and Found. Eng.Moscow.

Van Meerten, J.J.; Sellmeijer, J.B. & Pereboom D. 1995. Prediction of landfill settlements. Proc. of 5th International Landfill Symposium - Sardinia '95, 2: 823-831. Cagliari (I): CISA.

Yen, B.C. & Scanlon, B. 1975. Sanitary landfill settlement rates. *Journ. of Geotech. Eng.* 105, n.GT5: 475-487. ASCE.

Zimmermann, R.E.; Chen, W.W.H. & Franklin, A.G. 1977. Matematical model for solid waste settlements. Proc. 1st Conf. on Geot. Practice for Disposal of Solid Waste Materials. ASCE, Ann. Arbor.

Environmental Geotechnics, Kamon (ed.) © *1996 Balkema, Rotterdam. ISBN 90 5410 848 7*

Coastal reclamation system of solid waste – North Port of Osaka City

Kiyoshi Hatano
Port and Harbor Bureau, City of Osaka, Japan

Junji Okada, Hiroyasu Hanafusa, Kiyoshi Taniguchi & Hideichi Hisaki
Nakabori Soil Corner Co. Ltd, Osaka, Japan

ABSTRACT: In the North Port waste disposal site of Osaka City, the coastal reclamation system which is taxed many men's ingennity has been applied to dispose as much waste as possible in limited area. General waste and industrial waste have been disposed to the first section of the site by sandwitch method. Dredged soil have been disposed to the second and third section by two– ponds method. Surplus land soil have been used as banking and filling.

Waste water treatment, using natural oxidation pond, high– rate oxidation pond, flocculator, stabilization pond and membrane, and gas treatment, using gas pipes and gas combustion apparatus, have been performed for the emvironmental preservation.

We have made effort to improve sampling and test method, and then the geotechnical engineering properties of waste ground have got clear. After reclamation, the applications of the waste disposal site is planned depending on its engineering properties.

1 INTRODUCTION

It is extremely difficult to construct the final waste disposal site inland, because the land in Osaka City is used its maximum, there are 2.6 million people on 220 km ^{2} area, so that the coastal waste disposal site is being constructed in the North Port of Osaka City.

There are two sections, the Northern Part of North Port waste disposal site and the Southern Part of the site in North Port.

In these site, the total reclamation system of solid waste is established, it treats not only general waste and industrial waste, but also dredged soil and surplus land soil which aren't recognized as waste under the law.

This paper presents reclamation systems, countermeasures for environmental preservation, engineering properties of waste ground and plan for land use in the Northern Part of the site in which waste acception has already finished.

2 AMOUNT OF GENERATED AND DISPOSED WASTE

For 25 years from 1974 to 1998, 3.3449 billion m ^{3} waste has been generated in Osaka City. 62% of the waste, 2.0726 billion m ^{3}, was reclaimed. In the Northern Part of the North Port waste disposal site, 28.55 million m ^{3} waste including 7.02 million m ^{3} (25 %) general waste, 2.09 million m ^{3} (7 %) industrial waste, 18.18 million m ^{3} (64 %) dredged soil and 1.26 million m ^{3} (4 %) surplus land soil from construction works, was disposed (the amount in parentheses is the percentage for the total amount of the waste in the Northern Part of the site).

3 COASTAL RECLAMATION SYSTEM OF SOLID WASTE

The waste disposal site is divided to three sections as shown in Fig.1.

The first section is accepted general waste and industrial waste, the second and third section are accepted dredged soil. Surplus land soil is used as banking and filling in these two sections.

The arrangement and the flow chart of the reclamation system at the Northern Part of the site are shown in Fig. 2 and Fig. 3.

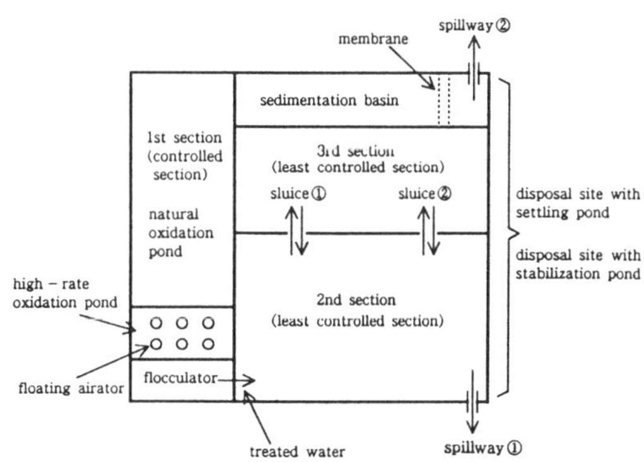

Fig.1 Coastal reclamation system of solid waste

The seabed is composed of the alluvial deposit (o.p. −7 m to −35 m) and the diluvial deposit which is consist of clay layers and gravel layers alternatively (Hatano 1995).

3.1 *First Section*

3.1.1 Outline

The first section is the 0.42km ^{2} controlled waste disposal site which has been accepted 9.11 million m ^{3} waste for 15 years from 1973 to 1987. After waste acception, 1 m thick cover was constructed by soil. The waste water was supplied to the Southern Part of the site through the undersea pipelines, and treated there.

3.1.2 Reclamation Method

The waste was reclaimed from one corner to another, while there was water area. After that, the waste landfill was constructed to 3 m thick, and then the 50 cm cover soil was placed it on. This reclamation method is called sandwitch method. This method is necessary for environmental preservation such as waste reduction, litter control, reduction of pollution by waste water and sanitation. From this view point, the law obliges to use this method.

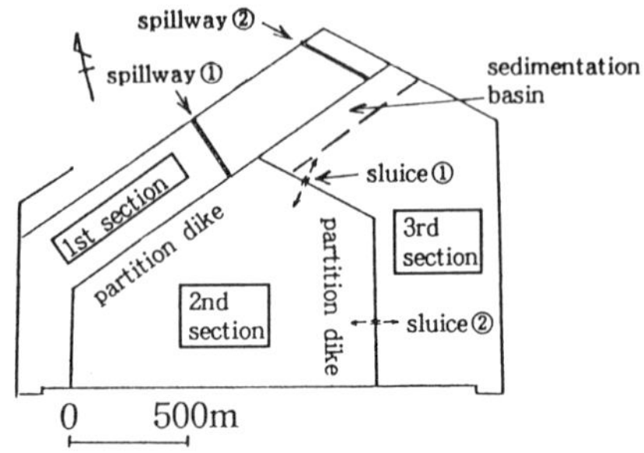

Fig.2 Nothern Part of North Port waste disposal site

3.1.3 Countermeasure for Environmental Preservation

The poisonous materials such as heavy metals must be treated to non−poisonous materials for disposing to the site, if its eluent is over the standard value. Therefore, from the view point of the environmental preservation, the follows must be considered.

1. organic waste water and its discharge
2. poisonous gas out of waste

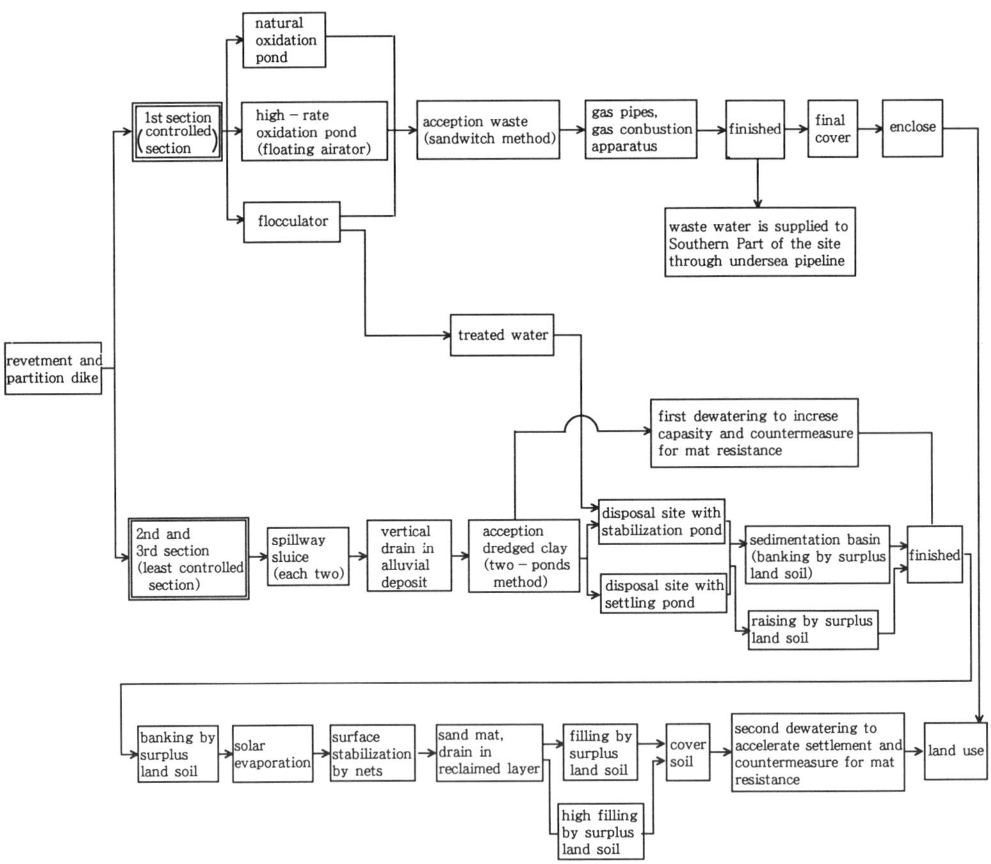

Fig.3 Flow chart of reclamation system at North port waste disposal site

Table.1 Effect of airator

	BOD (μg/cm³)	COD (μg/cm³)	T－N (μg/cm³)
Raw	150 － 400	150 － 300	160 － 170
After treatment	20 － 40	30 (summer) 60 － 80	80 － 90 (winter － spring) 20 － 40 (summer － autumn) under 10 (summer)

1) Waste Water Treatment

The water area in the site plays a roll of the natural oxidation pond. The waste water is oxidized biochemically in the high－rate oxidation pond, using floating airators. Further more, the waste is treated by flocculators. After that, the waste water is oxidized and flocculated in the second section. Finally it is drained to the surrounding sea through the spillway.

According to Mori's report (1995), the effect of the airator in the high－rate oxidation pond is confirmed as shown in Tab. 1.

2) Gas Treatment

The sectional plan of the first section is shown in Fig. 4. The gas pipes are latticed at intervals of 50m on each layers. The gas is burned by the combustion apparatus at the end of the pipes, and it is diffused in the air.

According to Fukuyama et al. (1995), the concentration of methane in the gas pipe where gas was well generated was 17 to 45 % after 6 years since disposal had been finished, on the other hand, in the inland waste disposal site where wet area has been buried by waste before the North Port waste disposal site, the concentration of methane was 48 to 72 % after 11 years since disposal had been finished.

They said, it showed that the former was semi aerobic sanitary landfill, and that the latter was anaerobic landfill.

3.2 *Second and third Section*

3.2.1 Outline

The second and third sections are the 1.36km ^{2} least controlled waste disposal site, which has been accepted 19.44 million m ^{3} of dredged soil and surplus land soil for 13 years from 1975 to 1987. The dewatering method was carried out two times, during the reclamation and after that. Then banking was built by surplus land soil. Passing through the period of solar evaporation(Hatano et al. 1995), surface soil stabilization was carried out using nets. The sand mat, plastic－board drains, filling and cover soil were constructed in order.

3.2.2 Reclamation Method

1) Dredged Soil

Reclamation was carried out by two－ponds method . Two sections were reclaimed alternatively. While dredged soil is accepted to the second section, the third section plays a roll of the settling pond, on the other hand, when the third section is the receiver, the second section is the settling pond. It is controlled by sluices to keep the residence time long. On the next step, the sedimentation basin was held in a part of the third section, other place was reclamed until water level, then banking was built. Advantages of the two－ponds method is follows(Hatano et al. 1994),

1. Flocculation is forced in the settling pond, so that it can be expected that the concentration of suspended solids (SS) in drainage is low. It has the drainage controll easy.
2. When one pond is completed, it can begin to accept dredged soil without waiting for completion of revetme－nt. The pond plays a roll of both receiver and settling pond.
3. Banking of sedimentation basin which is need for the final treatment, is constructed on the way of disposal. It is easier for banking than to be constructed before disposal , because dredge soi plays a roll of counter－weight fill.
4. The sedimentation basin makes the capasity of disposal greater, because dredged soil can be received up to the water level

2) Surplus Land Soil

Surplus land soil is used effectively in the dredged soil disposal site, as shown in Fig. 5.

3.2.3 Countermeasures of Environmental Preservation

The second section accepts the treated water from the

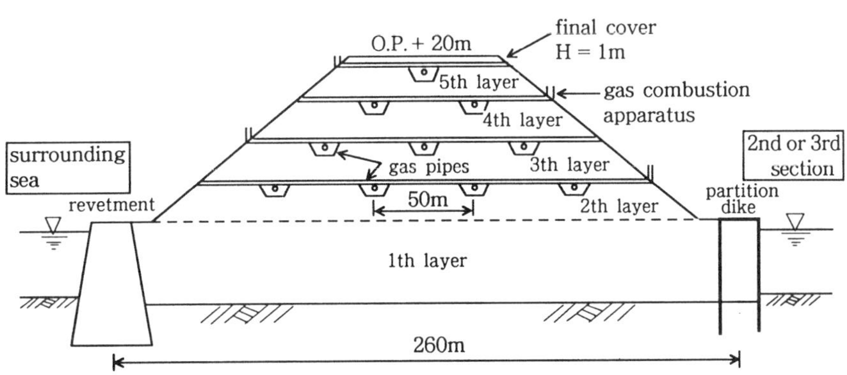

Fig4. Sectional plan of 1st section of Nothern Part

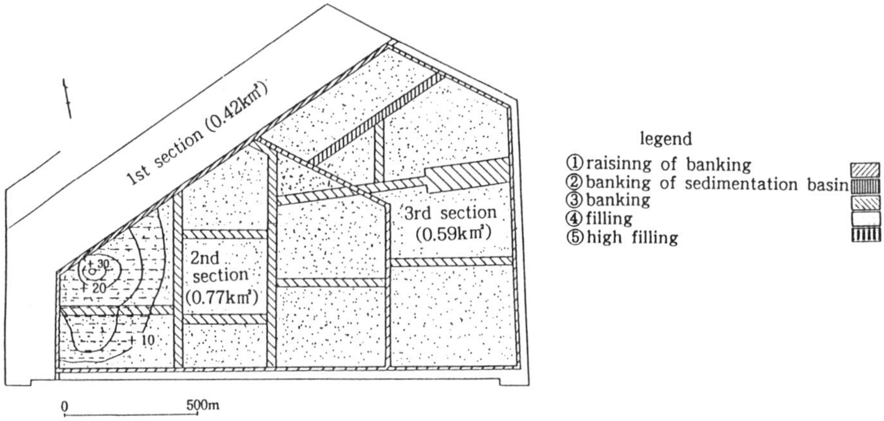

Fig.5 Area where applies surplus land soil in Nothern Part

first section and dredged soil from pump dredger. Follows must be considered for environmental preservation.

1. outflow of fine particles of dredged soil.
2. outflow of eutrophic water.

The countermeasure are follows.

1) Two-Ponds Method

The site is not only the dredged soil disposal site, but also the settling pond, so that the concentration of SS in outflow can be kept low.

2) Sedimentation basin

The sedimentation basin has worked effectively during disposal. For example, turbidity was over 150ppm , when clay suspension entered the sedimentation basin. It decreased to under 30ppm in drainage(Hatano et al. 1994).

3) Stabilization Pond

In the dredged soil disposal site, organic matters were decomposed biochemically.

Mori has investigated water quality of the site for 5 years from 1983 to 1987. The result is that COD was 4.5 to 44.5 μ g/cm ^{3} , T-N (total nitorogen) was 4.5 to 44.5 μ g/cm ^{3} , T-P(total phosphorus) was 0.1 to 1.3 μ g/cm ^{3} . T-N value was large in winter, small in summer. T-P was opposite against T-N. Concerned with COD, suspended COD(p-COD) was majority. When p-COD was high, a large number of phytoplanktons exist up to 1.5 m depth. The deeper water including little phytoplanktons was drained to surrounding sea through spillway. The phytoplanktons were shutted by membranes. Due to this operation , the COD of the drainage was seldom over 20 μ g/cm ^{3} .

4 ENGINEERING PROPERTIES OF WASTE GROUND

4.1 *First Section*

4.1.1 The Composition of General Waste at Receiving

The results of the segregation test of the dried waste showed that the noncombustible material, such as metals, glasses, stones, ceramics, and soil accounted for 62 to 85 % (75 % on average)and combustible material such as paper,woods, grass, plastics and rubbers dominated 25 % on the average.

4.1.2 Gas Generated from Waste Ground

The gas combustion apparatuses in 104 places have been installed. The gas generation has been decreasing gradually and the gas temperature has been getting lower since the September 1987 when acceptance of waste was finished. The concentration of methane gas becomes thinner and the composition of the gas generated from the waste ground can be comparable with atmospheric one.

4.1.3 Engineering Properties of Waste Ground

1) Method for Extracting the Undisturbed Sample

We devised the sampler shown in Fig. 6 because it is difficult to extract the undisturbed sample of solid waste through the sampler used conventionally(Hatano et al. 1990).

2) Results of the Extraction of the Sample

The extraction rate which can be calculated from dividing the length of sample by squeezed length showed 71 to 81 % when the squeezed length was 80 cm, on the other hand, it showed 94 to 98 % when it was 50 cm (Hatano et al.1991). This result means that we can extract undisturbed sample of solid waste sufficiently if the squeezed length is 50 cm.

3) Test Method

The mechanical test was carried out under the condition in which the diameter of every specimen was set to be equal to the inner diameter of sampling tube which was made by transparent acrylic plastic. We made the specimen from the sample in the sampling tube which was frozen and cut into the decided length. We set the frozen specimen into experimental apparatus and investigated it after defrosting.

The water permeability test and box shear test were carried out in the field.

Test method was based upon the Japanese Industrial Standard.

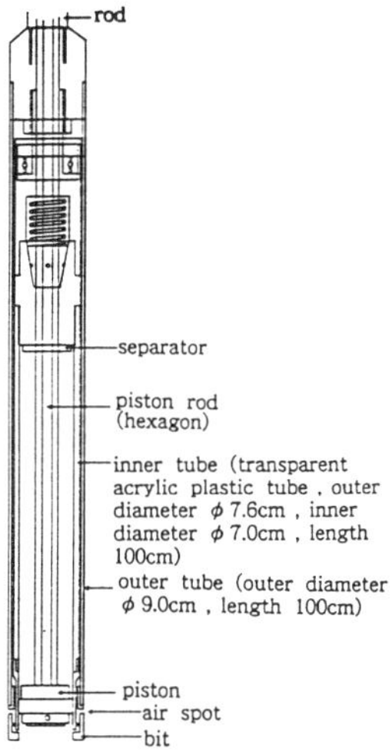

Fig.6 Fixed double tube sampler

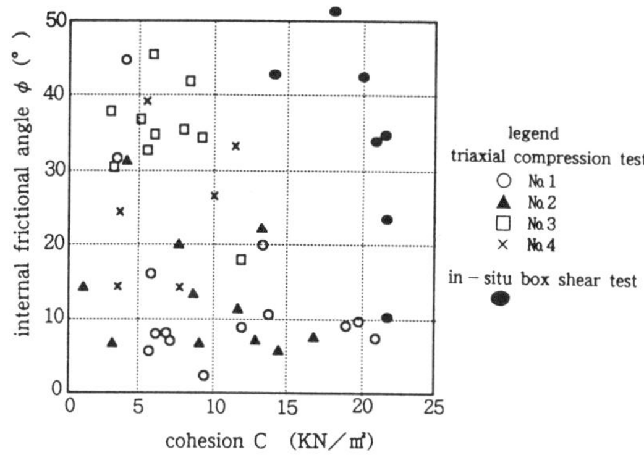

Fig.7 Cohesion~internal friction angle

4) Results

(1) Physical and Chemical Properties

The results of segregation test reveal that the composition of the waste ground is complicated because of the variety of waste which was put into the site. Therefore the unit volume weight and void ratio of waste ground varies widely and the ignition loss shows 10 to 30 %(Hatano et al. 1990). It shows that the ground contains organic matter heavily .

(2) Mechanical Properties

The compression index, C_c, showed small value near 0.2. The coefficient of consolidation C_v', showed large value in the range from 1.16×10^{-7} to 1.16×10^{-8} cm^2/s (Hatano et al. 1993).

The coefficient of permeability, k, showed 10^{-2} to 10^{-6} cm/s in the laboratory test and 10^{-2} to 10^{-3} cm/s in the field experiment. This implies that the former is dominated by vertical water permeability and the latter is dominated by the horizontal one. It reveals that the waste layer have an anisotropy.

Fig. 7 exhibits the result of triaxial compression test and box shear test in the field.

5) Settlement of Waste Layer

We assumed that compressin settlement occurred above the groundwater level, consolidation settlement occurred under this level and decomposition settlement occurred for the waste of paper,woods and cloth .

We studied the settlement at No. 1 and No. 2 site(Hatano et al. 1991).

(1) Calculation Method for Settlement

① Compression and Consolidation Settlement

The final amount of consolidation settlement was calculated from the equation described as follows,

$$S_f = \frac{e_o - e_f}{1 + e_o} H \qquad (4.1)$$

where, S_f ; final amount of settlement
e_o ; initial void ratio
e_f ; final void ratio
H ; thickness of waste layer

The consolidation equation used for settlement process is expressed as follows, and we differentiated it.

$$\frac{\partial u}{\partial t} = \frac{C_v'}{M_v} \frac{\partial M_v(\partial u/\partial z)}{\partial z} \qquad (4.2)$$

where, u ; excess pore water pressure
M_v ; coefficient of volume compressibility
t ; time
z ; depth
C_v'; coefficient of consolidation

We assumed that the final settlement calculated under the consolidation settlement appeared at once above the groundwater level. And we called it compression settlement.

② Decomposition Settlement

There were no appropriate datas for decomposition rate in our case, so we applied equation (4.3) written in another paper. It needs 70 years to finish decomposition completely in this equation (JSCE 1981).

$$C_{si} = 2.30 t_i^{0.56} \qquad (4.3)$$

Where, C_{si} ; decomposition rate (%)
t_i ; time (month)

We calculated the decomposition settlement by the equation (4.4) to (4.6) described as follows,

$$P_{oi} = P_i / (1 - \frac{C_{si}}{100}) \qquad (4.4)$$

$$S_f = \sum_{i=1}^{n} S_{fi} = L \sum_{i=1}^{n} P_{oi} \qquad (4.5)$$

$$S = \sum_{i=1}^{n} \frac{C_{si}}{100} \cdot S_{fi} \qquad (4.6)$$

Where, P_{0i} ; volume content of paper et al. at reclaimed time

S_f ; final amount of decomposition settlement from the reclaimed time

P_i ; volume content of paper et al. of each layer at any time

S ; amount of decomposition settlement of all layers at any time

S_{fi} ; final amount of decomposition settlement of each layer from reclaimed time

L ; thickness of each layer(=50cm)

(2) Result of Calculation for Settlement

The rate of calculated amount of settlement to measured value exhibits 66 % at No. 1 site and 92 % at No. 2 site.

4.2 *Second and third Section*

4.2.1 Engineering Properties of Reclaimed Ground

1) Physical and Mechanical Properties of Reclaimd Ground

The reclaimed ground is very soft that the thickness is about 17 m and the void ratio, e, near the surface is more than 4.

2) Effect of Alluvial Deposit Inprovement during Reclamation

The increase of the amount of settlement is 2.7 m in the case sand drains exist against the case sand drains not exist. This added the acceptance amount more 2 million m^3 (calculated from the effective depth, 1.5 m, for sand mat 1.2m thick, and the area, 1.36 km^2)(Hatano et al. 1993).

3) Settlement after Reclamation

We installed the plastic board drains in the position of triangle with 1.3 m side to improve the ground. The consolidation settlement of reclaimed layer and alluvial layer almost finished after two years and three months since cover soil had been operated (Hatano et al. 1995).

4) Liquefaction

The possibility of liquefaction is very small because the dredged soil is clay and the filling above that is surplus land soil which sand have a variety of grain size, and the cover soil is sand which have good grain size distribution. The liquefication phenomena didn't break out in the 1995 Hyogoken–Nanbu Earthquake.

5 PLAN FOR LAND USE

The Northern Part has been developed widely along the plan named "Techno Port Osaka". The institutions for sport,recreation, technical improvement, training, education and trading are under construction (Hatano et al. 1990).

6 PROPOSITION

For the future, waste should be controlled and utilized efficiently or recycled. Intermediate treatment, like incineration,should be done to make the organic matter less in the waste which is put into the waste disposal site. Then the site can be prolonged duration of disposal and be used in the early days.

REFERENCES

Fukuyama, J. , and Masuda, J. 1995. Gas generation and counterm easures at landfill site. The Japan society of waste management experts.vol6.No.4:24–32.

Hatano, K., 1995. Maishima (North Section of North Port) construction and reclam ation work. Submarine ground–Osaka Bay – Kansai Branch of the Japanese Society of SMFE: 336–342.

Hatano, K. ,Taniguchi, K., Okada,J. , and Hanafusa,H. 1995. Effect of natural dry shrinkage in reclaimed site used dreadged clay. Proc. of 30th Japan National Conference on SMFE:295–296.

Hatano, K., Okada, J. , Hanafusa,H., and Taniguchi,K. 1994: Reclamation method for dredged clay and surplus land soil. Proc. First National Symp. on Evironm ental Geot chnology: 149–156.

Hatano, K. ,Okada, J., Hanafusa,H. , and Taniguchi ,K.1994. Embankment of sedimentation basin at coastal reclaimed site used dredged clay. Proc. of 29th Japan National Conference on SMFE:223–224.

Hatano, K., Taniguchi, K., Okada,J. , and Yoshimura, . N. 1990. Undisturb sumpring method of wastes. Proc. of 25th Japan National Conference on SMFE:255–256.

Hatano, K.,Taniguchi, K., Okada, J., and Yoshimura,N.1991. Recovery ratio of undisturb sampling of wastes. Proc. of 26th Japan National Conference on SMFE:81–82.

Hatano, K., Hisaki,H.,Okada, J., and Hanafusa,H. 1990. Soil tests of wastes. Proc. of 25th Japan National Conference on SMFE:245–246.

Hatano,K. , Okada,J. , and Hisaki, H., 1993. Properties of wastes. Proc. of 28th Japan National Conference on SMFE :337–338.

Hatano, K.,Okada, J.,Hanafusa, H., and Taniguchi, K. 1991. Geotechical properties of waste disposal site by sea–filling. Proc.Symp. on Land Use of Waste Landfill:117–122.

Hatano, K.,Okada, J., Hanafusa, H., and Taniguchi, K. 1993. Effects of sand drain method in reclaimed site used dreged clay. Proc. of 28th Japan Natiomnal Conferenceon SMFE : 2423–2424.

Hatano, K., Hanafusa, H., and Okada, J. 1995. Method of recramation used dredged clay and surplus land soil for land use. Proc. of the 50th annual conference of JSCE.3–(B):1512–1513.

Hatano,K. 1990. The latest problems on environmental geotechnology. Text of lecture in Kansai Branch of the Japanese Society of SMFE:21–36.

JSCE 1981 . Planning and investigation report for waste disposal site and environmental preservation:293–295.

Mori, S. 1995. Appearance of plankton in the aerated lagoon and stabilization ponds. The Japan society of waste management experts.vol6,No.4:40–51

Environmental Geotechnics, Kamon (ed.) © *1996 Balkema, Rotterdam. ISBN 90 5410 848 7*

Ground improvement of municipal waste landfill by dynamic compaction

Y.M.Kim
Teajon National University of Technology, Korea

Abstract : This paper is case study of the ground improvememt project which was carried out for waste landfill. The project site is located near to Kapchun stream, in Taejon, center of Korea. The wastes at the landfill is composed of the municipal wastes, demolished building debris, coal ashes, and industrial wastes dumped between 1983 and 1989. The wastes was lacated in 0.3m ~ 4.5m below the ground surface and full thickness of wastes in about 3m ~ 7m. The dynamic compaction technique based on the results of the test compaction at two representative locations was imployed and field measurements were carried out for the ground improvement assessments and quality control for the dynamic compaction technique. From the results of field measurements and laboratory tests, it was found that the dynamic compaction was successful: the strength was increased satisfactorily (about 100% ~ 200% of original penetration resistance); and waste landfill was compressed considerably (about 15% of full thickness). Also, it is expected that the dynamic compaction is a good method for remedical action at man-made landfills, e.g. municipal waste landfills.

1. INTRODUCTION

Generally, waste landfill is supposed to be constructed with facilities, including water protection, gas collection, leachate and so on. Even though a long range plan for the construction of wasted landfill is needed which considers the classification of waste materials and soil cover and landfill uses after construction, in Korea, waste landfill has been constructed without those considerations and the locations of the landfills have been scattered near the highly populated metropolitan area.

Since various kinds of wastes were dumped and piled carelessly, problems of excessive settlement and bearing capacity are expected when the landfill sites are considered for the construction of roads and buildings.

This research is a case study of the ground improvement for the waste landfill area using dynamic compaction technique(DC), or dynamic consolidation.

The project site is located near Kapchun stream, Teajon and the wastes are composed of the municipal waste, demolished building debris, coal ashes, and industrial wastes dumped between 1983 and 1989. Since large settlement and low bearing capacity were expected, DC technique was first selected to compact the waste deposits as much as possible. After compaction of the waste deposits good quality soil layer over laid on the waste deposits. Soil cement method was used for the soil layers, which is located on the top of the waste deposits, in order to minimize the unbalanced settlement of the waste deposits.

The DC is a ground improvement method which uses impact energy by falling heavy hammer(pounder). This technique was introduced by Louis Menard in 1970.

Application of the DC for waste landfills has been reported in foreign countries. It is believed that this is the first project for the application of this technique in Korea. For the future development of this technique, the results of the tests are summarized here.

2. GROUND CONDITIONS

The waste landfill is divided by two sections as can be seen in Fig. 1. Site A(from STA. 4+000m to STA. 4+600m) was landfilled with 6.4m~6.8m of waste layer plus 4.0m~4.5m of top soil in July 1989. Site B(STA. 4+780m to STA. 5+240m) was landfilled with 3.7m~ 5.0m of waste layer plus 0.3m to 2.2m of soil layer on the top of the waste layer in 1983. The thickness of soil layer in site A is

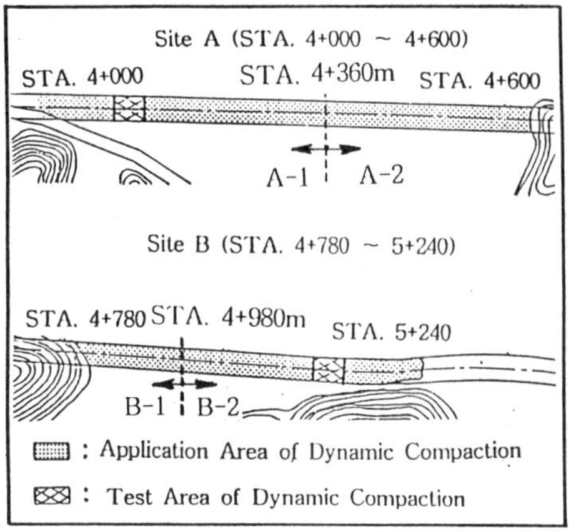

Fig. 1. Application of dynamic compaction area

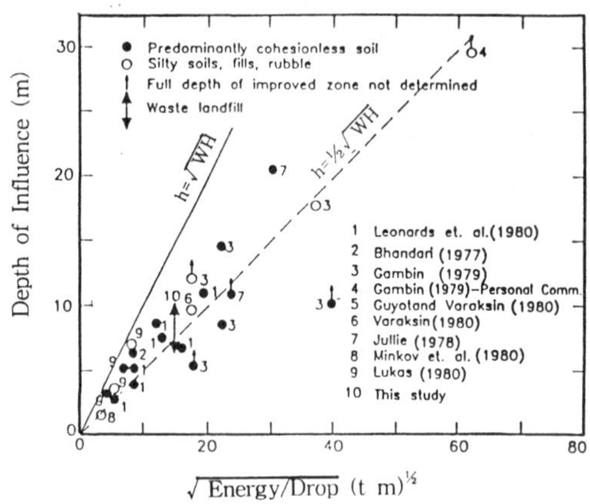

Fig. 2. Relationship between impact energy and depth of Influence (Mitchell, 1981).

Table 1. Ground conditions

Site	Top soil		Landfill		Alluvial Soil		Weathered Rock	
	Thick -ness (m)	Cont -ent	Thick -ness (m)	Cont -ent	Thick -ness (m)	Cont -ent	Thick -ness (m)	Cont- ent
A	4.0 ~4.5	Silty Sand, Gravel	6.4 ~6.8	*	2.8 ~3.5	Sandy Gravel	14.0 ~14.3	Silty Sand
B	0.3 ~2.2	Silty Sand, Gravel	3.7 ~5.0	*	more than 4.3	Sandy Gravel	14.0 ~14.3	Silty Sand

* : Municipal wastes, demolished building debrish, industrial wastes, coal ashes.

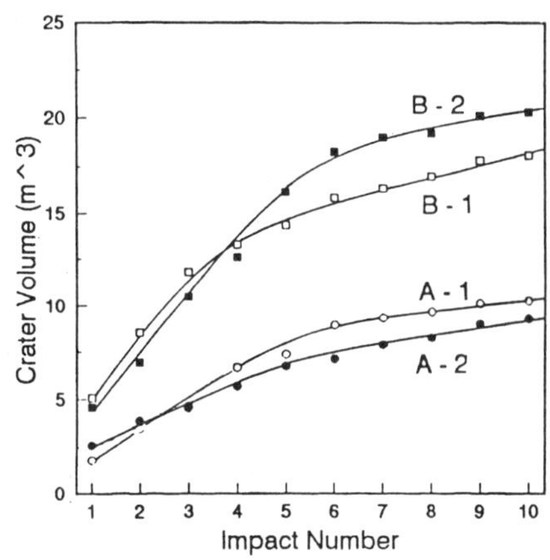

Fig. 3. The relationship between impact number and crater volume(1st-step)

comparatively unform and site B varies irregularly. The ground conditions of each site are shown in Table 1. Site A is divided by A-1 and A-2 at STA. 4+360m and Site B is divided by B-1 and B-2 at STA. 4+980m.

3. APPLICATION DYNAMIC COMPACTION

For construction plan and validity of the DC in the waste landfill, the DC test was carried out. From the dynamic compaction test, the depth of improvement, impact energy, impact number, impact space and so on were analyzed.

The pounder weight(W) and the fall distance(H) depend on the thickness(h) of the layer to compacted. h is determined by the relationship..

$$h = \alpha \sqrt{WH}$$

The condition of DC is as follows in this case.

- pounder weight (W) : 12 ton
- diameter of pounder (Φ) : 1.5 m
- fall distance (H) : 18 m

The coefficient of improvement α is assumed as 0.5. The results of DC test is showed in Fig. 2 (Mitchell, 1981). As can be seen in Fig. 2, the range of α is between 0.4 and 0.7 and that range is similar ordiinary soils.

The number of impact is 7 ~ 8 times in the 1st and 2nd step and 5 times in the 3rd step. Fig. 3 shows the relationship between impact number and crater volume. As can be seen in Fig. 3, the optimum number of impact is about 7 or 8. The crater volume of site B is grater than that of site A. It seems that Site B was landfilled ealier than site A and the thickness of soil layer on the top of the waste layer was rather thin.

The application of DC in this work is shown in table 2. The space of impact is 3m x 3m and the average impact energy per unit area is 480 t-m/m^2. This value of impact energy is similar to Janpan case(Yamazaki et al.).

Table 2. The application of dynamic compaction.

Site	Impact Energy (t-m)	Impact Space	Impact Number	Plan
A	216	3m x 3m grid	1st : 8 2nd : 8 3rd : 5	waste layer+ 1.5m upper layer soils
B-1	216	3m x 3m grid	1st : 7 2nd : 7 3rd : 3 4th : 3	waste layer +1.0m~1.5m upper layer soils
B-2	216	3m x 3m grid	1st : 7 2nd : 7 3rd : 3 4th : 3	waste layer+ 1.0m~2.5m upper layer soils

4. RESULTS OF DYNAMIC COMPACTION

The results of DC are confirmed by Standard Penetration Test(SPT), Dynamic Cone Penetration Test(DCPT), and elastic wave velocity(P-wave, S-wave) and so on. The SPT and DCPT were conducted for each about 50m distance from STA. 4+000m to STA..5+240m, and elasic wave velocity were mesured in the area of DC test and the vicinity of STA. 4+100m for each temping. The results of DC are as follow.

Table 3. Results of standard penetration tests (N - value)

Site	Before DC	After DC	Increment Ratio(%)
A-1	4 ~ 24 (14)	10~ 27 (18)	133
A-2	5 ~ 30 (13)	14~ 47 (24)	185
B-1	3 ~ 34 (10)	15~ 37 (22)	220
B-2	3 ~ 25 (14)	12~ 29 (17)	121

( ) : mean value

Table 4. Results of dynamic cone penetration tests (N_d - value)

Site	Before DC	After DC	Increment Ratio(%)
A-1	2 ~ 25 (11)	6 ~ 31 (17)	155
A-2	2 ~ 29 (7)	6 ~ 23 (11)	157
B-1	1 ~ 22 (5)	6 ~ 22 (9)	180
B-2	1 ~ 9 (3)	6 ~ 21 (10)	333

( ) : mean value

Table 3 is showed that N value is varied from 10 ~ 14 to 17 ~ 24 and increment ratio of N value is 121 ~ 220% due to DC work. But the N value was varied some large range at any points.

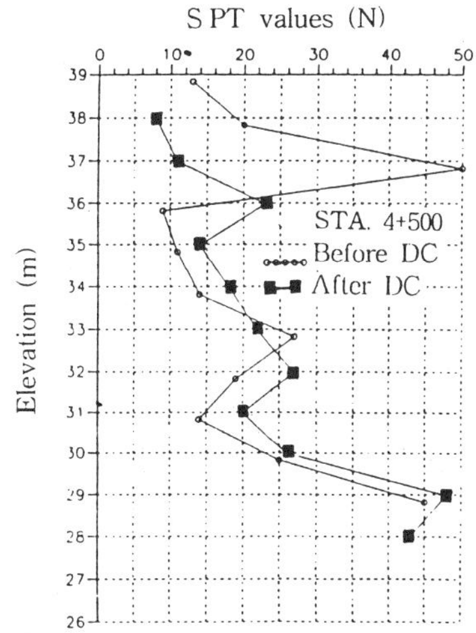

Fig. 4. The Value of SPT

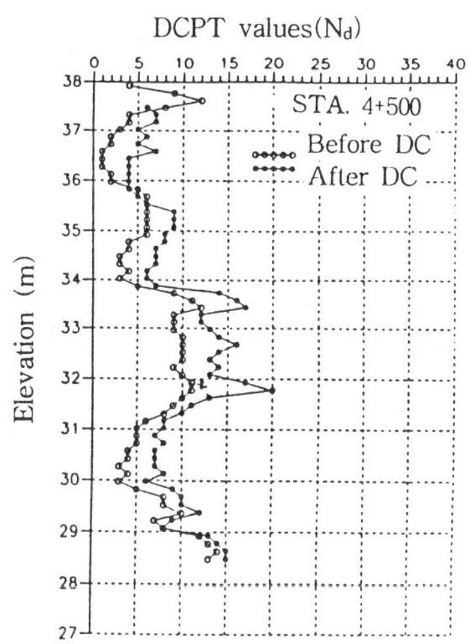

Fig. 5. The Value of DCPT

Table 5. Results of Elastic Wave Velocity (P-wave, S-wave)

Site	Before DC		After DC		Increment Ratio	
	P*	S**	P	S	P	S
A-1	210-240	129-140	271-307	162-176	28.5%	26.0%
A-2	243-273	158-187	309-340	196-226	25.8%	22.9%
B-1	170-174	103-118	208-217	133-141	23.5%	24.0%
B-2	166-205	100-125	218-258	130-166	28.3%	31.6%

* : P-wave, ** : S-wave

The field volume change is measured by borhole settlement measuring system and surface survey. Fig. 6 and Table 6 show the results of volume change before and after dynanic compaction in this filed.

Table 6. The results of volume change*

Site	Site Elevation (m)			Thickness of Waste Landfill (m)		
	Before DC	After DC	Variat -ion	Before DC	After DC	Variat -ion
A-1	38.452	37.999	0.453	5.5	4.9	0.6
A-2	38.218	37.452	0.758	5.1	4.6	0.5
B-1	37.578	36.965	1.149	3.2	3.2	0.5
B-2	36.772	35.253	1.519	2.7	2.7	1.3

* : Mean value

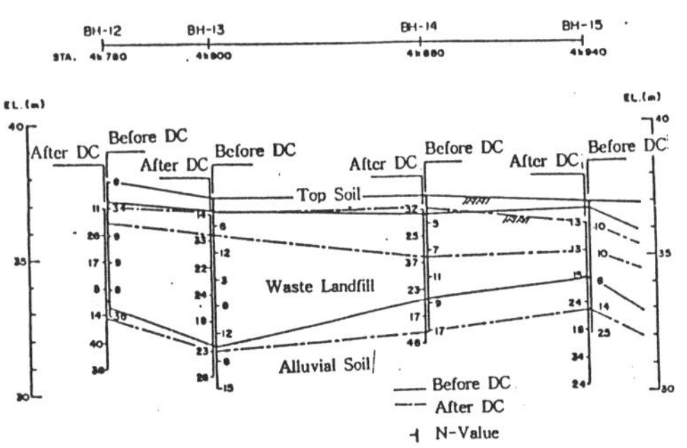

Fig. 6. Typical soil profiles in site B-1 (Before and After DC).

5. INFLUENCE OF IMPACT VIBRATION

The vibration due to impact energy is measured. The relationship of Peak Particle Velocity(PPV) and frequency are compared with standard of USBM. Fig. 7 shows the realtionship between scaled ditance and peak particle velocity. The PPV from the estimated formula is 25% and 24% of Mayne's formula(1985) at distance D = 10m and D = 20m, respectivly.

6. CONCLUSIONS

The magnitude of α is about 0.4 ~ 0.7 that is considered α = 0.5 is reasonable waste landfill as like this condition. From the results of field measurements, it was found that the dynamic compaction was successful : the strength was increased satisfactorily (about 100% ~ 200% of original penetration resistance); and waste landfill was compressed considerably (about 15% of full

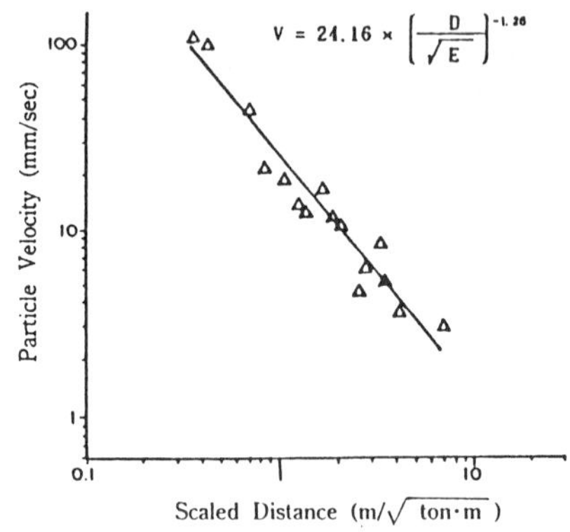

Fig. 7. Relationship between estimated PPV and scaled distance

depth). The PPV from the estimated formula is about 25% of general soils at the 10m ~ 20m distance. It is expected that the dynamic compaction is a good method for remdical action at man-made landfills, e.g. municipal waste landfills

REFERENCES

1. Kim, Y.M., et al., "Assessment of Dynamic Deep Compaction Applied to Waste Landfill", *Proceeding Korean Society of Civil Engineerings*, Vol.13, No.5, 1993, pp.209~222.

2. Mayne, P.W., "Ground Vibrations during Dynamic Compaction", *Vibration Problems in Geotechnical Engineering*, ASCE, 1985, pp.247~265.

3. Menard, L. and Broise, Y., "Theoretical and Practical Aspects of Dynamic Consolidation", *Geotechnique*, Vol.25, No.1, 1975, pp.3~18.

4. Michell, J.K., "Soil-Improvement - state of Art Report", *proceedings of the Tenth International Conference on Soil Mechanics and Foundation Engineering*, Vol.4, Stockholm, 1981, pp.509~520.

5. Yamazaki, Y., et al., "Improvement of Waste Ground Treated by Dynamic Consolidation", *The Foundation Engineering & Equipment*, Vol.19, No.6, 1991, pp.109~115.

Environmental Geotechnics, Kamon (ed.) © 1996 Balkema, Rotterdam. ISBN 90 5410 848 7

Case study on the disposal of industrial waste in Sakai No. 7-3 project

Toshiaki Ninomiya & Masao Nakagawa
Osaka Industrial Waste Management Public Corporation, Japan

Kojiro Nakaseko
Osaka University, Japan

Tadashi Hashimoto & Mitsuharu Fukuda
Geo-Research Institute, Osaka, Japan

ABSTRACT: This paper is concerned with the disposal of industrial solid wastes by landfilll on the soft ground in the Osaka area and the case history of the Sakai No.7-3 section project.

It is difficult to ensure large areas for disposing solid wastes in Japan. Therefore two geotechnical problems take place in determining disposal site. One of them is how to increase the volume of the vessel for disposal and the other problem is how to prevent contaminated ground water leaking out of the disposal field.

For these purposes, in proceeding with the Sakai No.7-3 Project, a specific treatment system brought into full play. This system was to displace the top layer of soft clay by weight of the filled materials and to replace soft ground by waste materials.

This method contains many difficult problems on controlling the construction of a solid vessel for disposal, however, the result of the ongoing investigation of this project has concluded that the thickness distribution of soft ground to be left changed smoothly and the remaining layer had important role in protection against polluted ground water flowing out from the disposal pond.

1. INTRODUCTION

It is difficult to maintain a large yard for filling industrial solid wastes. Therefore some disposal sites were located in offshore sites far from urban areas. However, since the thick soft clay generally lying under the sea bed in Japan may settle due to the weight of solid wastes and so on, it would be improved after completion of landfilling for facilitating as foundation structures.

Considering these ground conditions and ultimate land use after completion of disposal, it is imperative to improve the soft ground or force it to consolidate before filling. Although displacement treatment of soft clay by solid wastes is unique in Japan, it has taken important role at the Sakai No.7-3 disposal project. This method has effectiveness that the thin clay layer to be left underneath solid wastes decreases consolidation settlement and the heaved thick clay around the front of solid wastes is easy to be improved in lump. In addition, portion of clay layer surrounding solid wastes has a role of preventing polluted ground water from leaking out from the disposal ponds. However it is difficult to control the appropriate thickness of soft ground.

In this paper, thickness distribution properties of soft clay to be left after being pushed out by the solid wastes is shown and the role of the soft layer as a barrier against polluted ground water infiltrating outside the ponds is proven. As will be shown later, however, the numerical analyzing on how to control the constant thickness of the remaining soft clay is difficult procedure. This paper presents the result of simplified analysis that indicates to fill in narrow area to make displaced zone deeper and the remaining clay thinners, however, advance numerical study will be required if this method is generally used.

2. HISTORY OF THE DISPOSAL PROJECT

The Sakai No.7-3 industrial solid waste disposal site is located south-west of the Osaka area, and this project is being conducted as disposal sea. Fig.1 shows the location of the Sakai No.7-3 district where is about 280 ha in area. This project was established in 1971 by the Osaka Prefecture government. Landfilling began in 1974 using sand and construction waste materials. Since 1978, industrial waste materials were accepted for disposal. Figures 2 and 3 (Fujita, 1988) show the annual change of the properties related to the kinds of industrial waste materials and their quantity.

There was a rapid increase in the quantity of industrial waste materials from 1974 when the kinds of wastes to be accepted were expanded.

The sorts of waste materials are composed of sand and construction waste materials, sludge

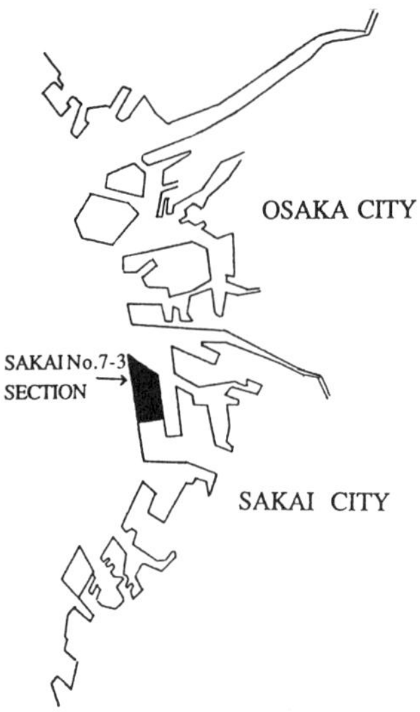

Fig.1 Location of the Sakai No.7-3 section

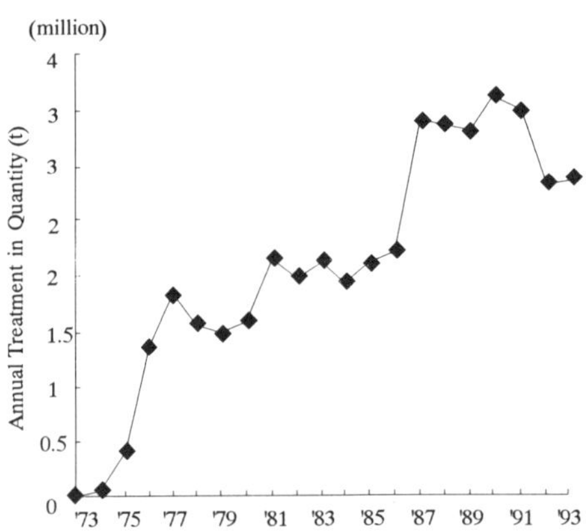

Fig.2 Trend of treatment in quantity

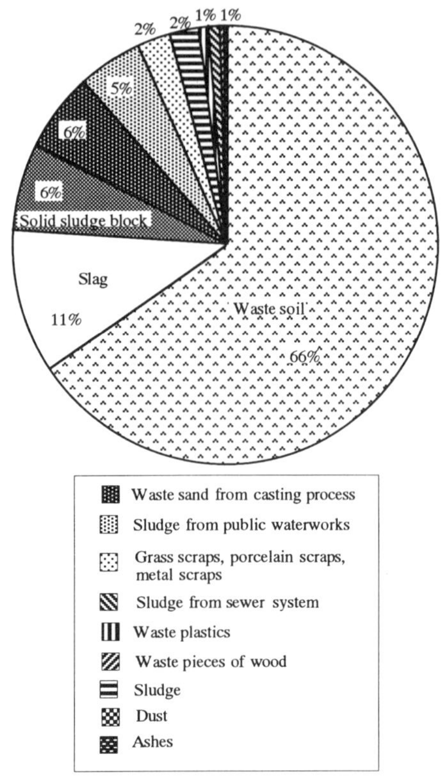

Fig.3 Kinds of accepted wastes in 1993

from filtration plants, slag, glass and metal scraps, waste sand from casting processes, ashes waste plastics and dust. Of these materials, sand and construction waste material constituted most of the quantity and its weight ratio reached to about 65% of the total weight according to the statistics recorded in 1993. The second largest material was slag, however, its weight ratio constituted only about 10%.

Fig.2 shows that peak of acceptance occurred among 1989 and 1991, but recently the annual accepted quantity returned to the value in the earlier stage.

The disposal site has 27 ponds, each of which is surrounded by the long partition banks. Landfiling has been executed using selected wastes one pond by one pond .

Fig.4 shows the whole shape of the Sakai No.7-3 district divided by 27 ponds with the ponds named A - Z. The length and width of each pond is about 350m by 300m. The A pond was the first completed site with industrial wastes in 1980 and two ponds (Z and Y) located at the tip of disposal site are now being filled.

Recently, the first completed reclamation subsections of the reclaimed ponds have been improved and utilized for several structures for promoting civil entertainment even before the completion of the whole project.

3. SPECIAL TREATMENT OF SOFT CLAY SOIL

There are three disposal waste sites in the Osaka Bay areas. Except Sakai No.7-3 project, two fields are conducted on the ordinary procedure to landfill directly on the top of the soft clay layers.

On the other hand, the specified displacement treatment that is called "partially thrusting and substitution of disposal materials for natural surface soft clay (Nakaseko, 1991)" was applied to construct a vessel for filling waste materials on this project. This system is that the dumped solid

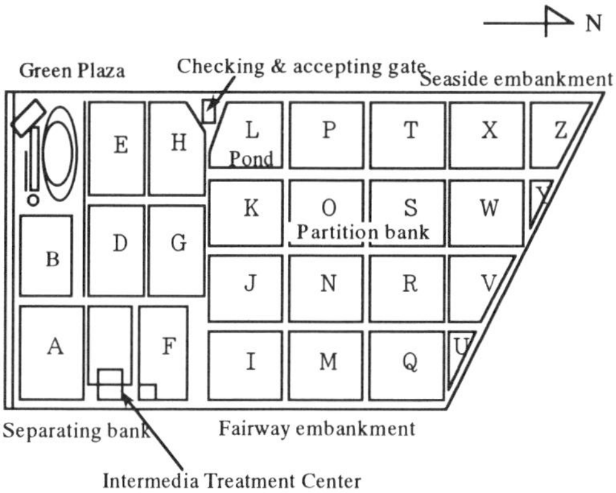

Fig.4 Schematic figure of the Sakai No.7-3 district

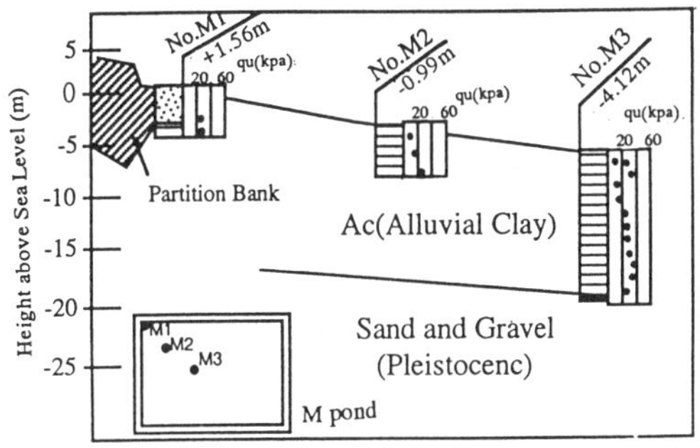

Fig.5Soil profile of the surface layer

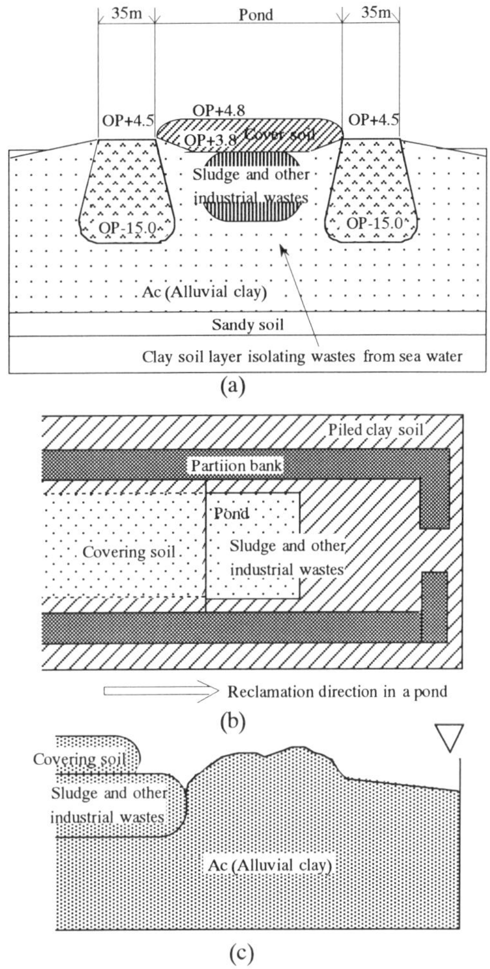

Fig.6 Disposing method

waste compulsively displaces the top zone of the soft clay and the well solid soil covers the surface of the landfill layer at the end. The remaining clay layer at the bottom of the pond and the heaved thick clay zone takes the role of the barrier wall for cutting off polluted ground water in the ponds. The polluted water is collected for purification in the disposal site.

However the thickness of the remaining clay layer is not constant due to the different kinds of filling materials and the different types of ways of spreading. Therefore it is difficult to estimate the distribution pattern of the clay thickness underneath the foundation after the completion of reclamation before filling.

Fig.5 shows the typical soil profile before this project started. The thickness of the top soft clay is about 15m, and increases slightly towards the offing. Pleistocene sand and gravel layer lie below this top layer.

Table.1 shows soil properties of the top soil layer obtained by soil tests and unconfined compressive strength ranges from 6 to 48 kpa. These soil properties have the same trend as in Osaka Bay area.

Fig.6 shows the operating process of the displacement method. As understanding on this figure, the most important problems of this operation are in relating how to ensure that the remaining thin clay layer acts as barrier against contaminated ground water leaking and how to make it thinner for decreasing consolidation settlement after completion of disposal.

Since bulldozer is major operating machine for spreading waste materials, the spreading width and the dumping height are the control factors.

Fig.7 shows how to ensure the thickness of the soft clay. Matsui (Matsui, 1978) derived an equation(1) governing extra soft clay flows according to the equilibrium condition.

Table 1 Soil properties of the top clay layer

	Item	Range
Physical properties	Specific Gravity ρ_s (gr/cm^3)	2.64～2.71
	Natural Water Content W_n (%)	23～97
	Liquid Limit W_L (%)	27～111
	Plastic Limit W_p (%)	18～43
	Wet Density ρ_t (gr/cm^3)	1.45～2.01
	Void Ratio e	0.61～2.61
	Gravel Fraction (%)	0 ～24
	Sand Fraction (%)	2 ～48
	Silt Fraction (%)	14～30
	Clay Fraction (%)	14～68
Mechanical properties	Unconfined Compressive Strength q_u (kpa)	6～48
	Internal Friction Angle τ (°)	0 ～10.5
	Cohesion C_u (kpa)	6～22
	Consolidation Yield Stress P_y (kpa)	21～39
	Compression Index Cc	0.72～1.09

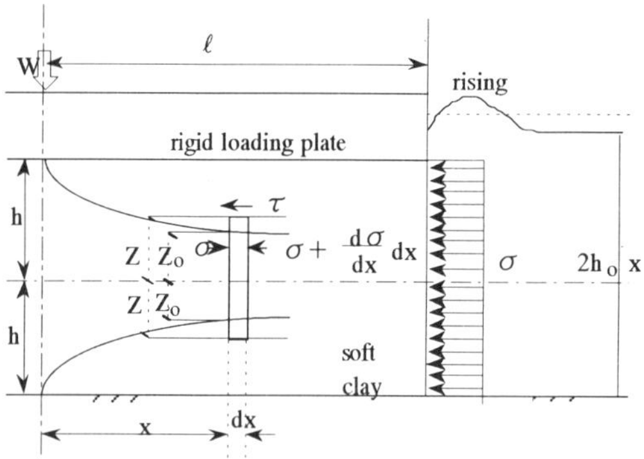

Fig.7 Mechanical and deformation condition

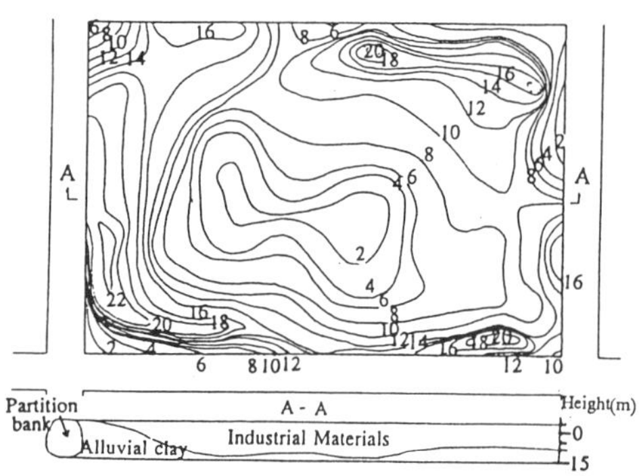

Fig.8 Distribution of the clay thickness left underneath disposal

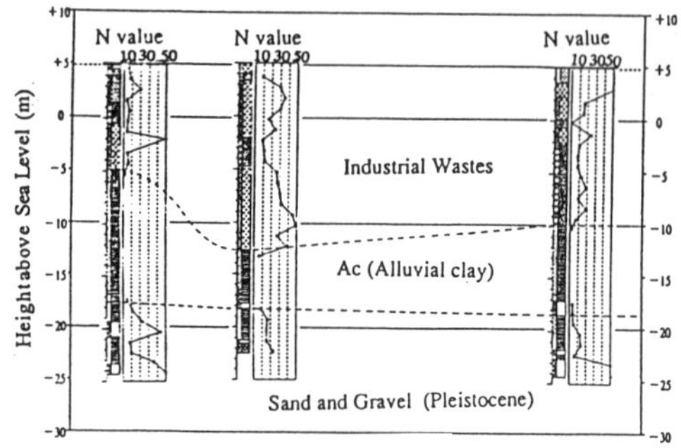

(a) Soil profile

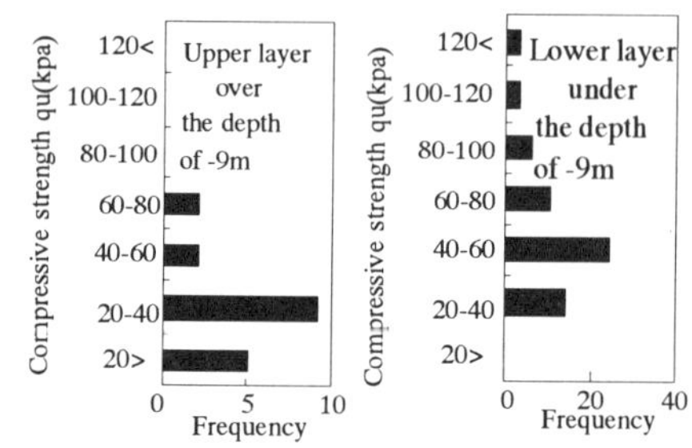

(b) Compressive strength of the alluvial clay

Fig.9 Soil profile and compressive strength(qu) after completion of filling at the M pond

$$S = \left\{ 4\ell(\gamma_o h_o - \tau_{yo}) + W - \sqrt{\{4\ell(2\gamma_o h_o + \tau_{yo}) - W\}^2 + 24\gamma_o \tau_{yi} \ell^3} \right\} / 6\gamma_o \ell \quad (1)$$

where S: settlement of ground surface, $2h_o$: initial thickness,

γ_o:unit weight, τ_{yo} :yield stress outer loading zone, τ_{yi} :yield stress under loading zone, W: load, 2ℓ : length of loading area.

This equation shows the relationship between settlement, loading length and load. However, the interrelation is very complex. Therefore equation (1) is changed as follows;

$$\frac{S}{h_o} = \left\{ 4\left(\gamma_o - \frac{\tau_{yo}}{h_o}\right) + \frac{2}{h_o}\left(\frac{W}{2\ell}\right) - \sqrt{\left\{4\left(2\gamma_o + \frac{\tau_{yo}}{h_o}\right) - \frac{2}{h_o}\left(\frac{W}{2\ell}\right)\right\}^2 + 24\frac{\gamma_o \cdot \tau_{yi}}{h_o}\ell} \right\} / 6\gamma_o \quad (2)$$

On considering the effect of operation in the case of using bulldozer, weight W and spreading length ℓ are bound as shown in equation (3).

$$\frac{W}{2\ell} = C \tag{3}$$

And further on this premise equation (4) is extended as follows;

$$\frac{S}{h_o} = \left\{ \alpha + \frac{2c}{h_o} - \sqrt{\left(\beta - \frac{2c}{h_o}\right)^2 + \gamma \cdot \ell} \right\} / 6\gamma_o \tag{4}$$

where α , β , γ , γ_0 and h_0 are constant.

Therefore the ratio of settlement of ground surface to the thickness of soft clay has distinct connection with the load intensity (c) at the surface of the soft clay and the length (ℓ) of spreading. The interrelationship among these factors contained in this equation shows that if the spreading length decreases, the settlement ratio increases. Therefore spreading should be carried out in so small area as possible that the load intensity increases.

4. PROPERTIES OF THE REMAINING SOFT LAYER

Fig.8 shows the distribution of the thickness of the remaining soft clay at the M pond searched by using boring investigation, and Fig.9 shows the typical soil profile after completion of filling the M pond. These figures indicate the some properties as follows;

1) Thickness of the remaining clay layer is around 2m at the center of the pond and increases more and more in all direction.
2) The thickness of the raised zone near along embankment is more than original thickness of 15 m and maximum thickness reaches to more than 20m.
3) The remaining clay layer under -9m above the sea level has the average compressive strength (q_u) of 76 kpa. On the other hand, the average compressive strength (q_u) of the shallower clay zone above -9m almost of which volume was removed is 32 kpa.

Other ponds have same distribution pattern of the remaining soft clay layer. However, the thickness of the thinnest zone in each pond is variable. As typical cases minimum thickness of the remaining clay layer are indicated as follows;

I pond	about 6 m
J pond	about 8 m
K pond	about 4 m
T pond	about 5 m

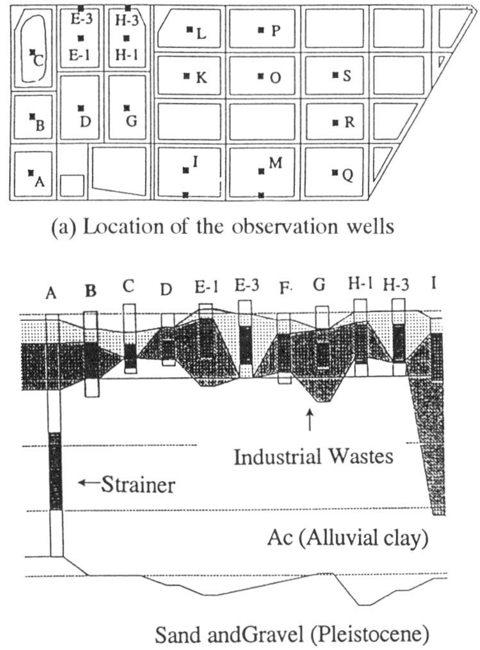

(a) Location of the observation wells

(b) Depth distribution of the observation wells

Fig.10 Location and depth distribution of the observation wells

5. GROUND WATER QUALITY

The remaining thin clay layer must have the important role of preventing the polluted ground water from infiltrating out of the ponds. Fig.10 and 11 show the location map of the observation wells and the results of the chemical analysis of the ground water obtained at the A pond and the B pond. Both are located in the ponds which come in contact. The observation well (A) has the strainer installed in the remaining clay and other wells have ones in the industrial waste zones. It is made clear that the quality of the ground water in the remaining clay layer under the vessel for disposal is clean on the view point of chemical components. As simply showing for examples, the pH obtained at the A pond is around 7. On the other hand, the pH at the B pond is near 10. The pH properties obtained at another wells installed in the disposal have the same value as at the observation well B.

Therefore the result of the chemical testing proves that the remaining thin clay layer is effect to take a role of the barrier wall.

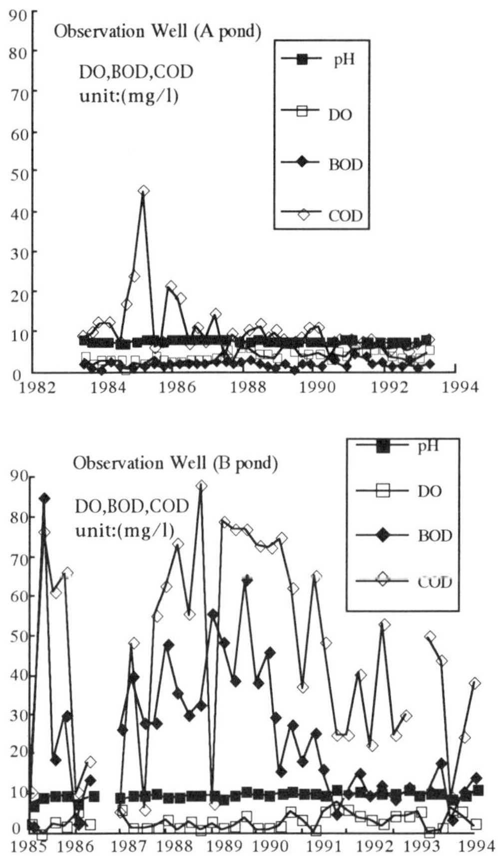

Fig.11 Comparison of the results obtained at the A pond and the B pond

6. IMPORTANT ROLE OF THE RAISED EXTRA SOFT CLAY

As shown in Fig.7, the disturbed extra soft clay rose along the embankment. These weak and thick clay layers appeared each pond. As these soft thick clay layers will cause large settlement and differential settlement compared with underneath embankment and at the center of the ponds, these weak zones are designed to be improved using vertical drain system. On this procedure, the tip of drain piles is necessary not to contact the aquifer lying under the raised thick clay, because if connecting the aquifer to the industrial wastes, the contaminated ground water will flow out through these drain piles.

Therefore the vertical drains were installed until the depth of 2 m over the bottom of the soft clay on this project.

7. CONCLUSION

1) Displacement method was effectively applied for the Sakai 7-3 project.
2) Thickness of the remaining thin soft layer range from 2m to 8m, therefore it is included that maximum thickness of about 7 to 13m were replaced by industrial waste materials.
3) Since improved areas by vertical drain was restricted around the embankment, therefore budget needed for improving could be saved.

8. ACKNOWLEDGMENT

The authors wish to express their gratitude to Mr. Mukuo and Mr. Kato who contributed their experimental skill and sustained effort to the accomplishment of the experimental program.

9. REFERENCES

Toyohide Fujita (1988) : Appropriate treatment of waste materials derived from construction site in Osaka prefecture, KISOKO , pp.23-29

Koji Nakaseko, Tadashi Hashimoto, Kuniaki Takami and Junichi Nagaya (1991) : Geotechnical engineering properties of the disposal area at the Sakai district in Osaka Bay and application of improvement methods for the difficult ground. The Japanese Geotechnical Society "Symposium on Utilization of disposed area after completion", pp.123-128

Tamotsu Matsui (1976) : Mechanism of Deformation of extra soft clay, Geotechnical Society "Symposium on extra soft clay", pp.25-34

Environmental Geotechnics, Kamon (ed.) © 1996 Balkema, Rotterdam. ISBN 90 5410 848 7

Industrial waste disposal and utilization of waste landfill in Japan

Katsutada Onitsuka – *Saga University, Japan*
Masashi Kamon & Takeshi Katsumi – *Kyoto University, Uji, Japan*
Junji Okada – *Nakabori Soil Corner Corporation, Suita, Osaka, Japan*
Mitsuhiro Nambu – *Oyo Corporation, Suita, Osaka, Japan*
Satoshi Ono – *Chuo-kaihatsu Corporation, Suita, Osaka, Japan*

ABSRACT: In metropolitan regions of Japan which are overcrowded with highly concentrated industries, enormous quantities of industrial wastes are generated and dumped thereby resulting in increased demand on ground. It is imperative to release pressure on land by creating large scale disposal sites, to accept these industrial wastes after intermediate treatment. This would be followed by ground investigations and tests soon after completion of landfill, so as to effectively utilize the landfill location by suitable ground improvement. This paper is based on the activities of the Sub-Committee on Land Use of Waste Reclaimed Site functioning under the Committee on Treatment and Potential Use of Industrial Waste (Japanese Geotechnical Society).

1. INTRODUCTION

In Japan, the annual rainfall is 1750mm. The topography is steep and geology is complicated. Of the total area of about 380,000km^2, 74% is of mountainous region and 11% is made up of plateau. Only 15% is habitable plane land. Most people of 120 million are living in these planes. Especially three metropolitan regions, namely Tokyo, Osaka and Nagoya regions, have nearly two thirds of population of Japan. As such most of the secondary and tertiary industries are concentrated in these regions. Concequently enormous waste would be generated.

Since in these areas it is very difficult to build satisfactorily waste disposal sites on the land alone, large scale coastal disposal sites are been planned and constructed. On the other hand, in mountainous and flat regions, relatively small land disposal sites are only possible. There are specific cases of disposal sites construted in the valley, from where ground water flows down to the flat areas. It is quite possible that ground water is contaminated affecting both the environment and the ecological system, when the geomembrane liner is punctured.

In order to meet heavy demands for acceptable land both for commercial and/or for industrial, in addition to building sites for dwelling the promotion of effective land use of finalized disposal sites is very urgent and inevitable.

This paper reports and discusses the following aspects : final disposal treatment, geotecnical properties of waste landfill and environment at disposal site, ground improvement, utilization of waste landfill in Japan. This paper is based on the activities of the Sub-Committee on Land Use of Waste Reclaimed Site in the Committee on Treatment and Potential Use of Industrial Waste (Japanese Geotechnical Society) as shown in Table 1.

2. WASTE DISPOSAL

Discharge of industrial waste has been increasing every year. Because of recent recycling and reduction of wastes the discharge volume since 1985 has practically remained the same. The annual discharge in 1992 was 403,000 Gg, 40% of which was recycled and 22% was disposed of. It is said that remaining space as the final disposal site, however, lasts only for a period of next 2.3 years space (Kankyo Sangyo Shinbunsya, 1995).

The final disposal sites are classified into following three types based on isolation methods adopted, 1)strictly controlled landfill site, 2)controlled landfill site, 3)least controlled landfill site. At the strictly controlled landfill site, hazardous wastes are handled and effectively isolated. At the controlled landfillsite, seeping water flow is prevented with geomembrane liner. Decomposable and nonhazardous wastes are disposed.

Table 1. Members of Research Committee on Treatment and Potential Use of Industrial Wastes (Japanese Geotechnical Society)

Chair: M. Kamon (Kyoto University)

Sub-committee on Utilization of Industrial Wastes		*Sub-committee on Utilization of Construction By-products*		*Sub-committee on Land Use of Waste Reclaimed Site*	
Chief:		*Chief:*		*Chief:*	
Y. Nagasaka	(Geogyro Co., Ltd.)	M.Yamada	(Osaka City University)	K. Onitsuka	(Saga University)
Members:		*Members:*		*Members:*	
K. Fukushima	(Toagosei Co., Ltd.)	T. Kawachi	(Obayashi Corp.)	K. Inoue	(Kansai Soil Research Center)
K. Higaki	(Taisei Corp.)	M. Kamiya	(Hokkaido Inst. of Tech.)	K. Kubota	(Kisojiban Consultants Co., Ltd.)
T. Hirata	(Kagoshima Nat. Col. of Tech.)	K. Kataoka	(Chuken Consultant Co.)	M. Nambu	(Oyo Corp.)
S. Horiuchi	(Shimizu Corp.)	T. Katsumi	(Kyoto University)	J. Okada	(Nakabori Soil Corner Co., Ltd.)
T. Kanda	(Fujita Corp.)	T. Kunito	(Seiko Kogyo Co., Ltd.)	M. Okada	(Toyo Constr. Co., Ltd.)
A. Kasai	(Onoda Chemico Co., Ltd.)	K. Mori	(Kumagai Gumi Co., Ltd.)	S. Ono	(Chuo-Kaihatsu Co., Ltd.)
Y. Nakano	(Konoike Constr. Co., Ltd.)	N. Ogawa	(PWRI, Min. of Constr.)	M. Otsubo	(Kyusyu University)
H. Omori	(Chichibu Onoda Cement Corp.)	P.O. Tran Duc	(Hazama Corp.)	K. Sonoi	(Nikken-Sekkei Ltd.)
H. Sakai	(Railway Technical Res. Inst.)	K. Saitou	(Ohmoto Gumi Co., Ltd.)	K. Terada	(Takenaka Corp.)
T. Yamaguchi	(Sumitomo Constr. Co., Ltd.)	K. Sasaki	(JDC Co.)	M. Tsuchihiro	(Kajima Corp.)
		M. Tokunaga	(Nippon Koei Co., Ltd.)		

At the least controlled landfill site, nonputrefactive and nonhazardous wastes are disposed. The ratio of least controlled and controlled landfill sites to the number of final disposal sites of 2,600, are 61% and 38% respectively. The ratio of strictly controlled type is hardly 1%, which by number wise totals to 37 sites. Further the final disposal sites are also classified as land disposal sites and coastal disposal sites based on location of the site. The former is 99% and latter is only 33 sites which consists of both least controlled and controlled types without strictly controlled type.

Since what distinguishes between three disposal type of sites, is prevention of seeping water, seepage control work is only primarily discussed here. At the land disposal site, both geomembrane sheet and clay liners are used as barriers. At the coastal disposal site, clay formations at the bottom are utilized. Steel sheet pile and steel pipe sheet are often used at the revetment for stopping water seeping through as shown in Fig.1 (Kamon, 1995). The effect of water control by geotextile sheet, 1.5-2.0mm in thickness at the land disposal site and clay layer are inferior to the double layer systems commonly used in U.S.A. which are relatively foolproof system. In utilizing clay layer as liner, both layer thickness and coefficient of permeability might not comply with the requirement. When a vertical drain is driven upto the sea bottom clay layer to enhance the possibility of early ultimate land use, the parametric evaluation is still vague. Even when the steel pile is driven through the sea bottom clay layer, it is equally necessary to make sure whether contaminated water seepages to supporting layer and not along interface between pile and soil. It is imperative to establish a building standard for final disposal site for environmental preservation.

3. PROPERTIES OF WASTE RECLAIMED GROUND

In utilizing the waste land, the following problems will be encountered such as ground settlement, inadquate bearing capacity, release of gas and leachate, liquefaction of ground, corrosion of steel, deterioration of concrete, apart from impairment of plant growth. Hence, it is needed to investigate and know in advance the constituents of waste land and its physical-chemical-mechanical-dynamic properties.

3.1 Investigations on waste landfill

Chemical investigations and chemical tests are more

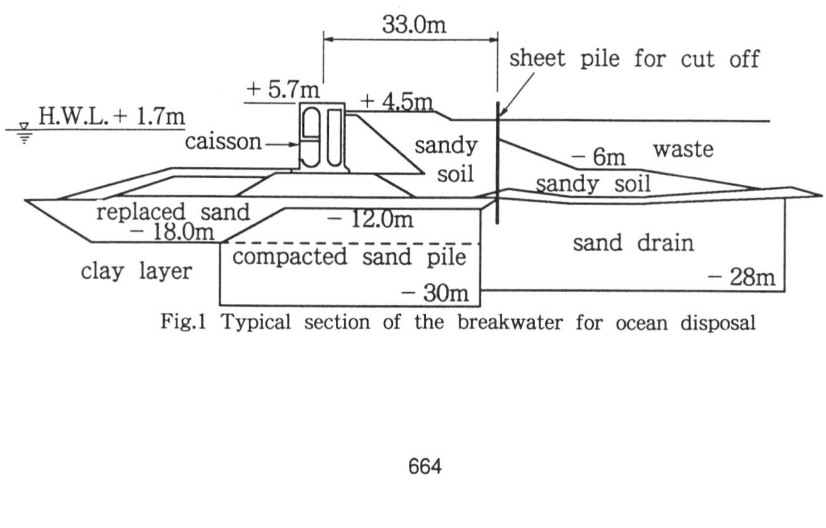

Fig.1 Typical section of the breakwater for ocean disposal

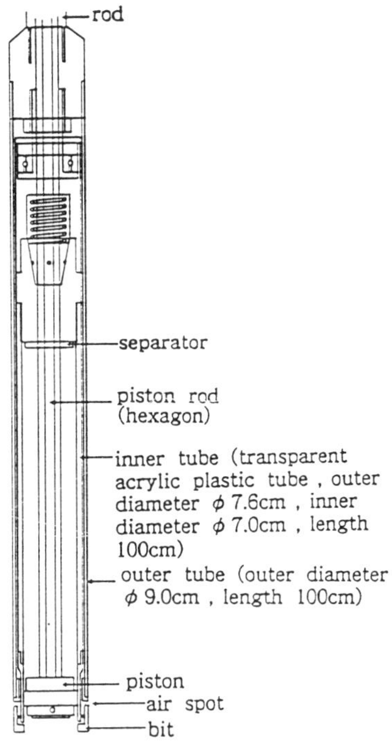

Fig.2 Fixed double tube sampler

important in waste landfill than conventional tests on soils. Fig.2 (Hatano et al. 1991) shows a double tube sampler which is capable for sampling of undisturbed waste materials. Air is used instead of drilling mud to prevent waste flowing away. The inner tube is made of a transparent acrylic plastic tube to see the sampled waste.

3.2 Settlement of waste landfill

At controlled disposal site in Tokyo bay which adopts incineration of domestic waste, the landfill layer depth is about 30m. Stabilization indices such as ground temperature, settlement of the ground, gas concentration etc. are shown against time in Fig.3 (Matsumoto et al. 1987). It can be seen that the waste land will be stable at 20 years after the completion of landfill. The coefficient of permeability of this reclaimed land and another waste land where Toyama airport was constructed, is shown in Fig.4. These values exhibit 100 to 1000 times of usual clays.

3.3 Environment of waste landfill

Due to exposure to generated gas, pH of pore medium and organic content, plants wither away at waste landfill site. Percentage of methane gas and

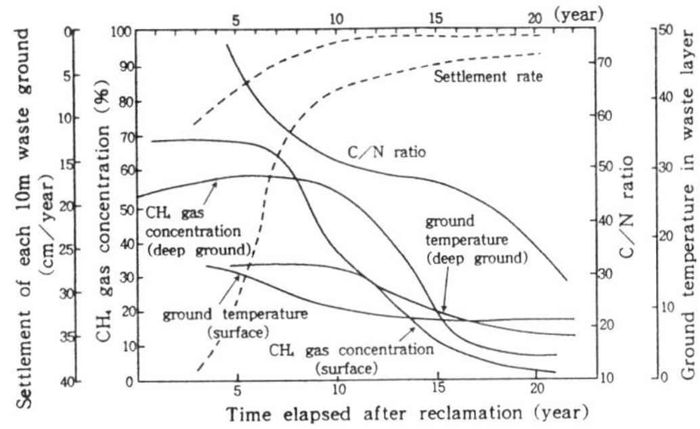

Fig.3 Stabilization levels of waste ground

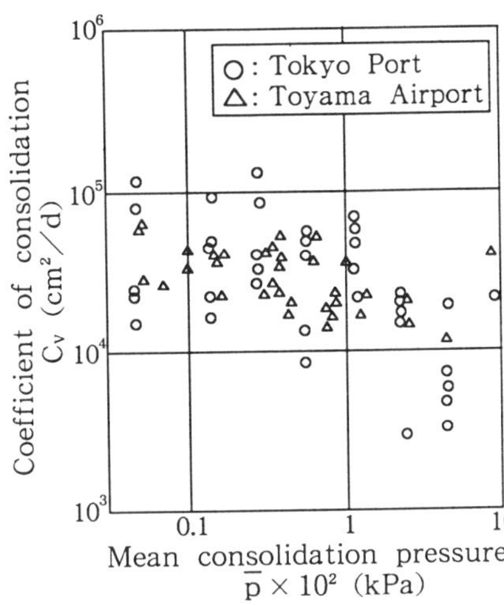

Fig.4 Settlement rates of refuse materials

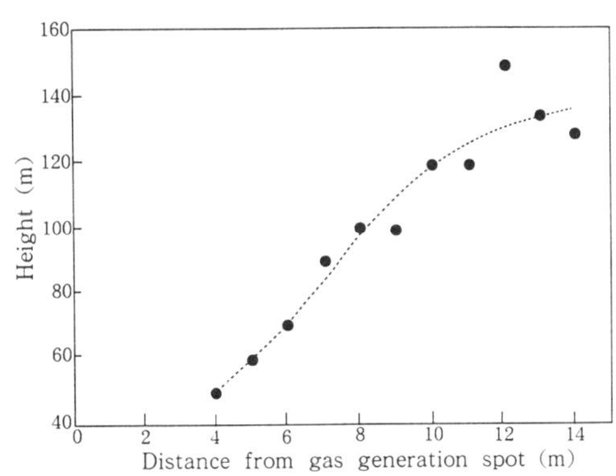

Fig.5 Gas generation and height of tall golden – rod

carbonic acid gas in the pores are more than air. There are some plants sensitive to this methane gas, such as tall golden-rod and arrowroot. The state of gas development can be known from the height of tall golden-rod by using Fig.5 (Nagano et al. 1989).

As countermeasure to gas development, 1)overlaying soil 2)gas cut off by geotextile sheet 3)setting of gas collection pipe 4)environmental preservation for soil organism 5)selection of durable plants for gas can be several alternative measures. It is suggested that overlaid soil is necessary at least to the extent of 1.5m depth for tall trees of deep root, 0.9m depth for tall trees of shallow root and 0.6m depth for low trees (FCTC,1994).

4. GROUND IMPROVEMENT OF WASTE LANDFILL

4.1 Ground improvement method

For utilizing of waste landfill, ground should be improved to have adequate bearing capacity and results in allowable settlement. There are many ground improvement methods such as replacement method, pre-loading method, heavy tamping (Dynamic Compaction) (DC) method, sand compaction pile (SCP) method and asphalt injection method. At the disposal site where neither vibration nor noise are tolerated, DC and SCP methods are suitable.

4.2 Evaluation of improvement effect

The improvements achieved by three techniques namely, heavy tamping method, sand compaction pile method and pre-loading method are examined here. At Toyama Airport extension construction noncombustible general waste land, of area of 0.11 km^2 was improved. Table 2 (Kamon,1991) shows the order of improvement in resistance to compression and compaction, namely by compression ratio, for above three methods. SCP method and DC method are very effective in improvement as shown in Fig.6 (Kamon, 1991). Higher the compression ratio, higher the improvement effect. Some properties after improvement got enhanced by 2 to 4 times of earlier value. N values increased from 8 to 20, deformation modulous from 294 to 392 N/cm^2, K_{30} from 20 to 39N/cm^2.

In this construction by SCP method of replacement ratio, 50% for the runway and DC method for landing strip to gain compression ratio, 30% were employed.

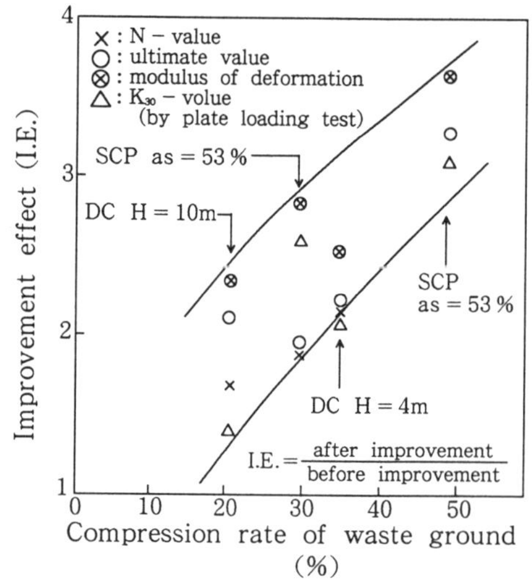

Fig.6 Improvement effect of waste ground

Table 2. Compressive rate of waste by compaction

	content	settlement (S)	thickness of waste layer (L)	compression rate (α)	remarks
SCP 1st section	a = 53 % V①= 217 m³	S = 9.9cm V②= − 12.5 m³	L = 2.92cm V③= 421 m³	0.487	$\alpha = \frac{V① + V②}{V③}$
SCP 2nd section	a = 34 % V①= 358 m³	S = 23.2cm V②= − 33.1 m³	L = 6.41m V③= 1083 m³	0.299	$\alpha = \frac{V① + V②}{V③}$
DC 1st section	Er = 4.9MNm／m²	S = 139.1cm	L = 4.00m	0.348	α = S／L
DC 2nd section	Er = 4.9MNm／m²	S = 207.5cm	L = 10.0m	0.208	α = S／L
PL 1st section	H = 7.0m	S = 55.0cm	L = 3.65m	0.151	α = S／L
PL 2nd section	H = 9.0m	S = 66.9cm	L = 4.25m	0.157	α = S／L

SCP : sand compaction pile method
DC : heavy tamping method
PL : pre − loading method
a : improvement rate of sand pile
V① : unit volume of used sand
V② : unit upheaval of ground surface
V③ : unit volume of waste
Er : tamping energy
H : height of filling by pre − loading

Fig.7 (Muranushi,1995) displays relationship between compaction energy and averaged surface settlement for 6 waste landfill sites including the Toyama Airport improved by DC method. It can be seen that the waste landfill will settle by 1.0 to 2.5m for compaction energy of 4.9 MNm/m^2.

5. LAND USE OF WASTE LANDFILL

5.1 Degree of land use

Degree of waste land use is classified into three types, namely, low degree such as farm and park etc., intermediate degree such as lowrise building etc. and highrise degree such as intermediate and high building. In order to know the state of land use, comparison between municipal waste disposal sites (Data A, Ministry of Public Health and Welfare, 1985) with industrial waste disposal sites (Data B) are shown in Table 3. According the table, at nearly 70% of sites of both types, land use is obtained as planned. The low degree of land use such as farm and park is 66% and intermediate-high degree land use such as house and school building is 34%. These are evident from the Data A. Especially the Data B were obtained from only local public bodies, hence, nearly half of the sites are coastal disposal sites and land use for such as harbor facilities are many.

5.2 Example of high degree of land use

As an example of high degree of land use, one school building (Matsufuji et al. 1992) is introduced here. The disposal site area in Kyusyu is 23,000m^2 and was filled up to less than 10m depth with non-combustible municipal waste. The facilities were a three storey school building and a gymnassium. The soil properties are that the ignision loss is 14 to 17 % at high value points. Water quality exhibits locally high sulfate concentration which

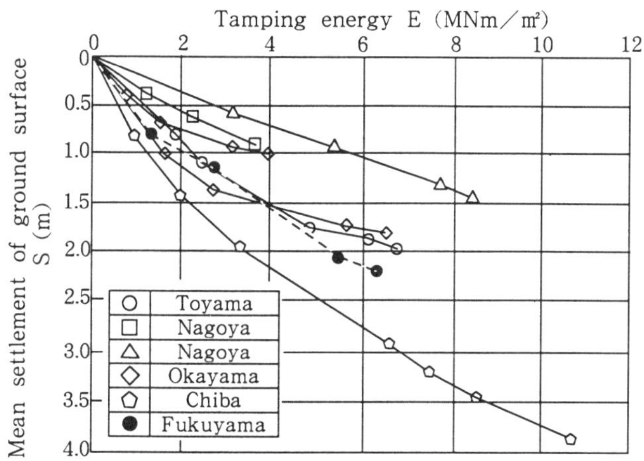

Fig.7 Tamping energy and settlement

Table 3. Ultimate land use of final disposal site

			A	B
rate of ultimate land use (%)			70 (446／657)	67 (18／27)
rate of facilities of ultimate land use (%)	low level use	agricultural land wood land	52	21
		park , open space , disaster prevention land	14	36
	medium or high level use	harbor facilities	0	29
		housing , school , factory	34	14
	total		100	100

A : general waste disposal site completed landfill, investigation by the Ministry of Public Welfare
B : industrial waste disposal site during landfill, investigation by the Japanese Geotechnical Society

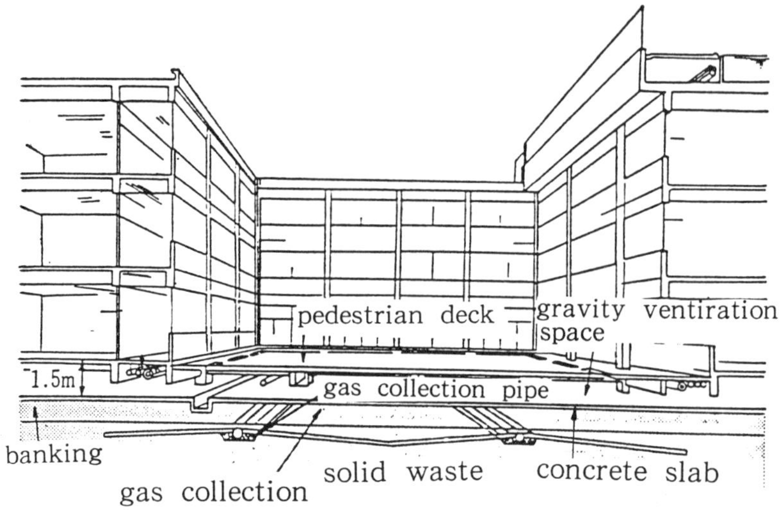

Fig.8 Gas countermeasure under school house

means high corrosion. There are sites where methane gas generation is more than 10% and maximum volume developted is 4,000 cm^3/day. The gas countermeasure as shown in Fig.8 was carried out in the school building. Eventhough it was a high degree land use, it took 15 years for the completetion of the landfill site before being used as planned.

5.3 Promotion of early effective utilization

Geotechnical properties of waste disposal land has been determined to evalluate the effectiveness of ground improvement method. Presently, early land use has been technically possible. According to the previous Data B, planning of land use was already done at the 60% disposal sites when the sites were planned (16% during landfilling, 24% after landfilling).

However, these results are restricted to public and large scale disposal sites. Hence the upper ratio will reduce for middle and small scale disposal sites. In order to promote the effective land use, the planning of land use before landfill should be made. The masterplanning including disposal site plan, environment impact assessment, construction, management, investigation, tests, ground improvement and land use plan should have been considered and finalized.

6. CONCLUSION

In Japan, finalization of structure design standard for industrial waste disposal site is urgent matter in the view of environmental preservation. The land use planning prior to landfill and the masterplanning of disposal site to land use should be established with both the promotion of early stabilization of industrial waste land and early effective utilization of the land.

REFERENCES

Frontier Construction Technology Center, 1994, Creation of excellent environment by technology (in Japanese).

Hatano,k., Okada,J., Hanafusa,H. and Taniguchi, K. 1991, Geotechnical properties of waste disposal site by sea-filling, Proc. Symp. on Land Use of Waste Landfill, pp.117-122 (in Japanese).

Kamon, M. 1991, Compaction of waste,Tsuchi-to-Kiso, JSSMFE, Vol.39, No.4, pp.61-68 (in Japanese).

Kamon,M. 1995, Waste management in Geotechnical Engineering, Proc. Regional Symposium on Infrastructure Development in Civil Engineering, pp. 154-166.

Kankyo Sangyo Shinbunsya, 1995, Annual Report on Waste, pp.18-64 (in Japanese).

Matsufuji, Y. and Hanashima M., et al. 1992, Construction of school buildings on an old landfill site, Tsuchi-to-Kiso, Vol.40, No.6, pp.17-22 (in Japanese).

Matusumoto, Y. et al. 1987, Geotechnical study of refuse landfill in Tokyo Port(3) - Stabilization indices for refuse randfill, The 22th Japan National Conference on Soil Mechanics and Foundation Engineering, pp.39-42 (in Japanese).

Ministry of Public Health and Welfare, Japan, 1985, Investigation report on regional final disposal sites and utilization of waste landfills (in Japanese).

Muranushi, S. 1995, Volume reduction in waste by heavy tamping method, The 8th Environmental Plan Center New Technology Conference, pp.26-30 (in Japanese).

Nagano, S. and Hanashima M., et al. 1989, Study on stabilization of waste landfill (No.2), Waste Treatment Countermeasure National Council, pp.55-58 (in Japanese).

Environmental Geotechnics, Kamon (ed.) © *1996 Balkema, Rotterdam. ISBN 90 5410 848 7*

Effects of soil improvement on the waste reclaimed land

K.Sasaki, S.Suguri & T.Fukushima
JDC Corporation, Kanagawa, Japan

K.Ueshita
Chubu University, Aichi, Japan

ABSTRACT: The effects of soil improvement was studied for a municipal landfill which was treated with the Heavy Tamping and Surcharge improvement methods about 15 years ago. The soil in the treated areas was found in all cases to be quite stable,and no problems, such as settlement or landfill gas, were evident.

1. INTRODUCTION

FUJIMAE municipal landfill site is located in the southern part of NAGOYA City. The land has been functioning as approximately 20.4 ha FUJIMAE Distribution Center since 1983, after the completion of the landfill, improvement effects including soil improvement. Municipal waste was dumped at this site from 1968 to 1973. Landfill improvement works were carried out from 1978 to 1983. The soil improvement methods of heavy tamping and surcharge were applied to parts of the site.

Some waste materials in the landfill tend to corrode and decompose over time. This may cause environmental problems due to uneven settlement of the ground and accumulation of gas underground. In FUJIMAE, no such problems have been reported in over 10 years, since Distribution Center started operation.

This report introduces the latest investigation results for properties, the condition of waste, landfill gas, underground water, etc.

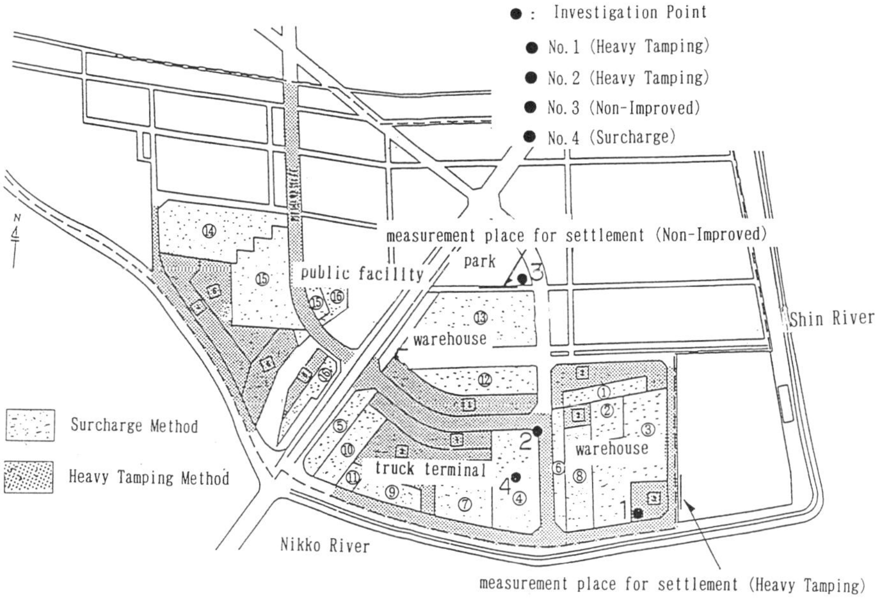

FIG. 1 Soil improved area and the latest investigation points

2. OUTLINE OF SOIL IMPROVEMENT AND INVESTIGATION

The landfill site is located in the reclaimed land between the Shin River and the Nikko River. The original ground level was the same as sea level and it was swampy. A thin silt layer covered the surface. Alluvium soils consisting of 4 meter thick silty-sand layer and 5 meter thick silt layer underlie surface site layer, followed by a diluvium stratum consisting of medium finesand.

Both layers of Alluvium were very soft (with N-value less than 3), and the properties of the soil were constant.
The original groundwater level was within 1.5 to 2 meters below sea level at Nagoya Port. The waste filled reclaimed land, reaching 4 to 5 meter higher than the original ground.
Two kinds of soil improvement methods are used, after completion of disposal and soil backfill. Heavy tamping was applied to the main road area and to parts of the facility area, surcharge was also applied to parts of the facility area. A description of each of these methods follows.

HEAVY TAMPING

A 12-ton heavy weight is dropped 20 m to consolidate the layer of waste up to 25 percent.
Due to dynamic pressure and the impact of the blows, landfill gas was released through temporary cracks created during tamping.

SURCHARGE(PRELOADING)

Surcharge embankments, twice the height of the waste layer, act to consolidate the waste layer in one month.
Landfill gas was released through gas pipes laid at 30 meter intervals.

The most recent study of this area was conducted at the end of 1993. The properties of the soil, landfill gas, and groundwater were examined at four bore-holes, two in the area improved by heavy tamping, one in an area improved by surcharge, and one in a non-improved area. The level of the main road was measured to check for uneven settlement. Interviews with users of the facilities were also conducted. FIG.1 shows the soil improvement area and the most recent investigation points.

3. RESULTS OF INVESTIGATION

3.1 *Properties of the Soil*

Standard-penetration tests and bore-hole load tests were conducted. FIG.2 shows the results of standard penetration tests at each point, as well as previous data for point No.1. The areas improved by heavy tamping are well-consolidated from the upper layer to the lower layer. The effects of surcharge are very evident from the ground to 2 meter below the surface, but much less evident a depths greater than 2 meters. By comparing with the previous data at point No.1, it was observed that the area improved by heavy tamping has steadily become more stable in the last 15 years.

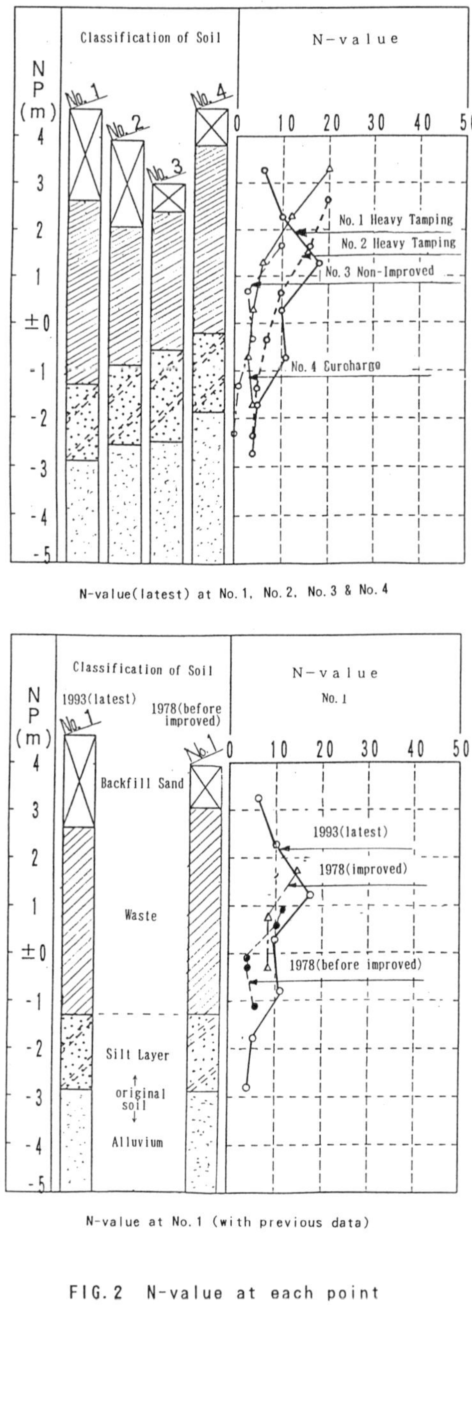

FIG. 2 N-value at each point

TABLE. 2 Detail of the waste and the latest data for landfill gas and underground water

Bore-hole Point			No. 1	No. 2	No. 3	No. 4
Soil inprovement method			Heavy Tamping	Heavy Tamping	Non-improved	Surcharge
Classification of the waste wt %						
Corrosive(wood, grass, paper etc.)			12.57 %	17.06 %	8.20 %	11.37 %
Non-Corrosive(stone, concrete, glass etc.)			87.43 %	82.94 %	91.80 %	88.63 %
Method to release landfill gas			Trough temporary cracks caused by the blows of Heavy Tamping		Non	Trough the gas pipes laid in
Situation and condition of the waste						
Thickness of the waste layer			3.65 m	2.95 m	2.95 m	4.00 m
Degree of Consolidation			high	high	low	high(up. only)
Depth of covered soil			1.95 m	1.85 m	0.60 m	0.80 m
Underground water table (①)			GL-2.25 m	GL-2.30 m	GL-2.30 m	GL-2.40 m
Thickness of the waste layer above ①			0.30 m	0.45 m	1.70 m	1.60 m
Thickness of the waste layer below ①			3.35 m	2.50 m	1.25 m	2.40 m
C/N			42	69	27	63
Volume of Landfill Gas		(ℓ/hr)	0.6	2.0	560	21.0
Composition of Landfill Gas						
Hydrogen sulfide	H_2S	(Vppm)	0.009	0.072	0.015	0.005
Oxygen	O_2	(%)	13.6	7.9	8.5	21.8
Nitrogen	N_2	(%)	49.2	25.3	41.9	77.0
Carbon dioxide	CO_2	(%)	3.8	13.7	3.1	0.38
Methane	CH_4	(%)	33.46	53.58	46.41	0.86
CO_2/CH_4			0.11	0.26	0.07	0.44
Analysis of Underground water within the waste						
Water Temperature		(℃)	20.0	19.5	18.5	20.5
Air Temperature		(℃)	12.3	13.5	15.0	14.5
PH		(mg/ℓ)	7.2(21℃)	7.0(16℃)	6.7(19℃)	7.0(16℃)
BOD		(mg/ℓ)	16	110	3.3	13
COD		(mg/ℓ)	94	370	17	43
TOC		(mg/ℓ)	88	36	21	37
CO_3^{2-}		(mg/ℓ)	less than 4	less than 4	less than 4	less than 4
HCO_3^-		(mg/ℓ)	2200	4400	1300	940
SO_4^{2-}		(mg/ℓ)	14	26	1.9	3.3
Cl^-		(mg/ℓ)	330	2500	78	160
Mg^{2+}		(mg/ℓ)	13	94	70	16
Ammonia nitrogen		(mg/ℓ)	220	790	29	120
Sulfite nitrogen		(mg/ℓ)	less than 0.02	less than 0.02	less than 0.02	less than 0.02
Nitrate nitrogen		(mg/ℓ)	less than 0.1	less than 0.1	less than 0.1	less than 0.1
Electric conductivity		(mS/m)	410	1300	180	320
Oxidation-reduction potential		(mv)	-115	-121	-73	-58

TABLE1 shows the results of bore-hole load tests using a pressure meter. This test was conducted to examine the soil strength horizontally. Even though the results varied very widely, the effect of soil improvement is clear, as the following demonstrates.
According to the initial soil investigation data in the waste layer, the pressure limit(P1) was less than 0.3 MPa and the modulus of deformation(Ep) was between 0.1 and 2.0 MPa. TABLE1 shows that the strength of soil in the waste layer is 1.5 to 2 times stronger than these initial values. Results also show that heavy tamping is slightly more effective than surcharge for the soil improvement.

Wooden chips from the sample of bore-hole No.1 showed growth rings that maintained the same shape. The wood from other bore-holes did not maintain their original ring shapes. The progress of corrosion is related to these differences.

TABLE. 1 Results of In-bore-hole Load Test

Bore-hole	Depth N.P. ±(m)	Classification of Soil	Pl (MPa)	Ep (MPa)
No. 1	+4.5~+2.7	Backfill Sand	0. 2	8. 4
	+2.7~-1.4	Waste	0. 5	7. 2
	-1.4~-2.8	Silt Layer	0. 2	2. 4
No. 2	+3.8~+2.2	Backfill Sand	1. 1	8. 4
	+2.2~-1.0	Waste	0. 4	2. 5
	-1.0~-2.6	Silt Layer	0. 3	1. 8
No. 3	+3.0~+2.4	Backfill Sand	——	——
	+2.4~-0.5	Waste	0. 3	3. 9
	-0.5~-2.5	Silt Layer	0. 1	0. 8
No. 4	+4.5~+3.6	Backfill Sand	——	——
	+3.6~-0.3	Waste	0. 3	3. 5
	-0.3~-1.7	Silt Layer	0. 2	1. 9

Pl : Pressure Limit Ep : Modulus of Deformation

N.P. = M.S.L. −1.4119 m

3.2 *Waste, Landfill gas, and, Ground water*

Waste is classified as either corrosive or non-corrosive. Eight to seventeen percent of waste consisting of wood and like material was corrosive, while remaining waste(stone, glass. etc.) was non-corrosive. The composition of the waste was somewhat different at each sampling point and at each sampling time.
The most recent investigation results for landfill gas and ground water in the waste layer are shown in TABLE2. The pressure of landfill gas was very low. The volume of landfill gas at the non-improved point is for higher than at improved points. In the analysis of ground water, the volume of ammonia, nitorogen, and COD at the area improved by heavy tamping is larger than at other areas. The progress of corrosion was relatively slow in the area consolidated by heavy tamping.

3.3 ***Measuring Settlement and Interviews.***

Settlement of the roads was measured at the curbstone along a straight pedestrian side walk, 100 meters in length, at two locations, one in the tamping area and one in the non-improved area. The results for the tamping areas were +1.4cm and −2.1cm, and the results for the non-improved area were +2.2cm and −7.0cm. The uneven settlement in the heavy tamping area was measured to be about one third of that in the non-improvement area.
The users of facilities in FUJIMAE Distribution Center responded that no damage due to ground settlement, and no fool odors, have occured so far. The government officers in charge of road maintenance also answered that the road had consistently been in good condition.

4. CONCLUSION

This most recent investigation was made 25 years after the beginning of waste disposal and 15 years after the completion of soil improvement. The effects of soil improvement have been studied by comparing the investigation results at each point in the areas improved by heavy tamping or surcharge with the non-improved area. In general, improved areas are more stable than the non-improved area, showing that improvements have satisfied functional demands, and were well designed initially.
We plan to continue this investigation over the long term, in order to use this data when selecting suitable improvement methods, time, and appropriate uses for landfill areas.

Environmental Geotechnics, Kamon (ed.) © 1996 Balkema, Rotterdam. ISBN 90 5410 848 7

Centrifuge model tests on leachate collection shafts in landfills

O. Syllwasschy, D. König & H. L. Jessberger
Ruhr-University Bochum, Germany

ABSTRACT: In centrifuge model tests leachate collection shafts have been investigated with special attention to the mechanical interaction between shafts in different designs and municipal solid waste. The results show that in combination with a stiff concrete lining the use of gravel between shaft lining and waste is not always advantageous under every load.

1 INTRODUCTION

The problems of waste-structure interaction are investigated in several research projects at the Institute for Soil Mechanics and Foundation Engineering at the Ruhr-University Bochum. In this paper the results of centrifuge model tests on leachate collection shafts inside landfills will are presented. In the past a lot of these shafts have been built within a landfill. Nowadays damages like inclinations of the shaft axis and deformation and destruction of the shaft lining have been observed (Prühs (1988), Dany & Ney (1992)). Due to these damages several shafts have to be renewed or repaired. Their stability and the worker health and safety place can not be guaranteed any longer. In contrast to shafts in rock or soil the leachate collection shafts suffer shear stress due to very large settlements of the waste by negative skin friction and symmetrical and asymmetrical stresses as well.

Surface settlements of waste landfills have been found to reach 25 % or more of the total height. Horizontal movements inside the landfill can be caused by uneven deposition or decomposition of the waste. So a municipal solid waste landfill has to be regarded as a system with a large deformation potential that can produce horizontal and vertical loads on buried structures.

A shaft inside a landfill can be designed by different principles. Stiff shafts are often constructed of lining materials such as concrete, high density polyethylene (HDPE) or glass fibre-reinforced plastics. In most cases gravel is placed between shaft and waste with the idea to reduce and equalize symmetrical and asymmetrical forces, to collect and remove leachate and to protect the lining surface from aggressive media and high temperatures.

The stability of an unlined shaft in the waste and, considering a stiff lining, the amount and distribution of shear stresses caused by negative skin friction, and the interaction between lining, gravel and waste are investigated with the help of centrifuge model technique.

2 CENTRIFUGE MODEL TECHNIQUE

The principle of centrifuge model technique is illustrated in figure 1. A model is produced simulating a prototype situation. This model (scale 1:n) will be accelerated in a centrifuge to a g-level with n-times the earth gravity (ng). Then the state of stress of the model at ng-acceleration will be comparable to that of the prototype.

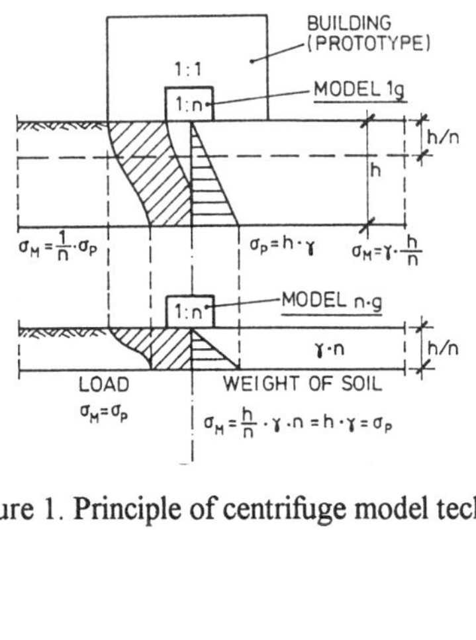

Figure 1. Principle of centrifuge model technique

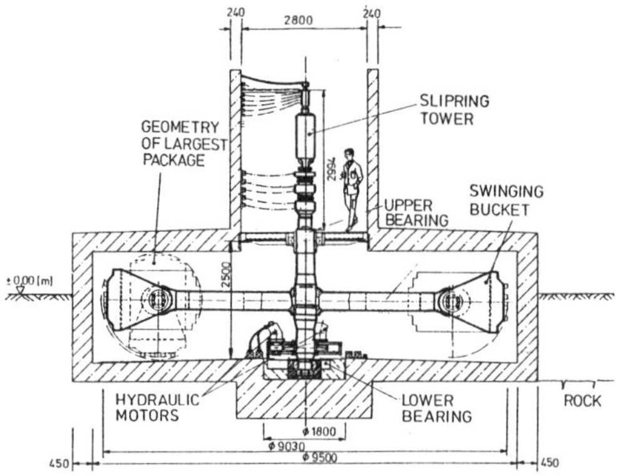

Figure 2. The Bochum Geotechnical Centrifuge Z1

The tests described in this paper were performed in the Bochum Geotechnical Centrifuge Z1. Figure 2 shows a cross section of this 500 g-tons centrifuge (Jessberger & Güttler, 1988).

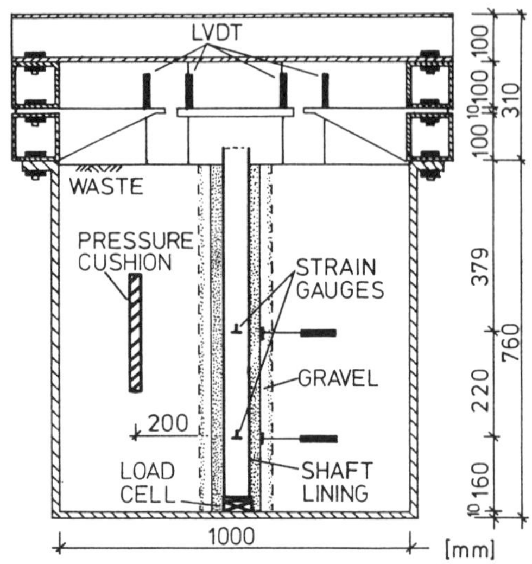

Figure 3. Test design

3 TEST DESIGN

In the centrifuge model tests a stiff concrete shaft is simulated by a brass tube with a diameter of 75 mm and a wall thickness of 0.5 mm. The vertical and tangential strains are measured in two sections by eight strain gauges each on the outer surface and on the inner surface. At the bottom of the shaft a load cell is installed to measure the sum of the forces acting on the shaft lining in vertical direction. At different depths several LVDTs are installed to measure the vertical settlements and horizontal deformations of waste and gravel ring.

The shaft model is placed in the middle of a strong box (Fig. 3) surrounded by the waste or, if necessary, by gravel and waste. The waste is built up in multiple layers and compacted by hand to a unit weight of about 10 kN/m³. In differnt tests the thickness of the gravel between lining and waste was varied from R/r = 0 and R/r = 1,47 to R/r = 2,13, where R is the outer radius from the shaft axis to the outer surface of the gravel mass and r is the outer radius of the shaft lining.

The complete model fixed in the centrifuge bucket is accelerated to the selected g-level of 40 g. In this way a shaft of about 25 m in depth and 3 m in diameter is simulated.

To simulate asymmetrical horizontal loads air pressure can be applied inside a cushion which is embedded inside the waste near the lining

The waste body consolidates due to the acceleration of the model. Shear stresses caused by negative skin friction are induced into the lining and measured when nearly no increase in settlement can be observed. Effects caused by decomposition have to be neglected.

In another test series a shaft without shaft lining is modelled to investigate the behaviour of waste under large radial deformations. These models are accelerated up to 100 g and an unsupported shaft of about 46 m in depth and 11 m in diameter is simulated.

4 MODEL WASTE

To model a waste body the original municipal solid waste from a landfill is used. Due to large waste components it has to be so prepared as to stand the use in centrifuge model tests.

The preparation may consist of several steps of processing. From the municipal solid waste very large particles like wooden branches, bricks or carpets are separated. Thereafter it is milled about four to five times in an electric milling machine (shredder) to crush the waste components. After this preparation the largest particles, e.g. wooden sticks and plastic foils, reach a size of about 32 mm, but the waste composition still remains similar to the original material.

With both materials, the original waste and model waste, triaxial compression tests have been conducted. The results show a very similar stress-strain behaviour of both materials (Figure 4).

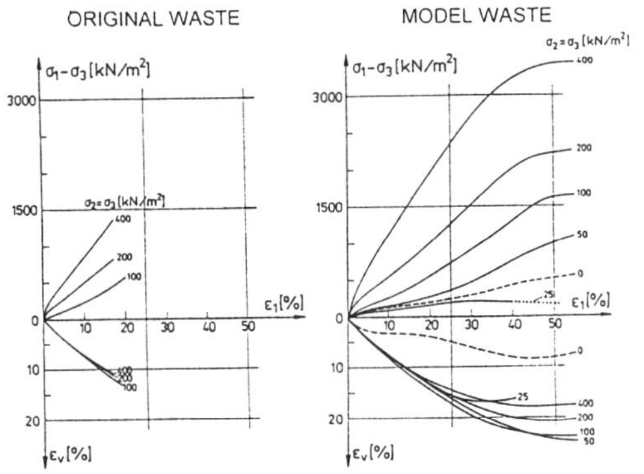

Figure 4. Comparison of original and model wastes

Municipal solid waste is very heterogenous in comparison to soils as it consists of a large variety of materials like plastics, soils, textiles, wood, paper, etc. The definition of a failure law (e.g. Mohr-Coulomb) is difficult as mixed waste can undergo large deformation without showing the tendency of reaching a critical state or failure (Jessberger & Kockel, 1993). A limit state can just be defined for very large deformations. It is not a shear failure but a critical state. So instead of a failure criteria a deformation criteria is used to describe the shear strength of waste.

The results of triaxial shear tests show a very large vertical deformability. A limit state can only be attained at strains ε_1, of about 40 to 50 %. Therefore the shear parameters of the model waste are illustrated in terms of vertical deformation ε_1 from large triaxial tests (∅ 300 mm) in figure 5.

The used model waste can acquire high shear strength but only under large vertical strain. The maximum angle of inner friction can be determined as φ= 54° at a vertical strain of about ε_1= 55 %. Cohesion shows its maximum at ε_1= 8 % with c= 17 kN/m² and decreases slowly to values of 10 kN/m² with larger deformation.

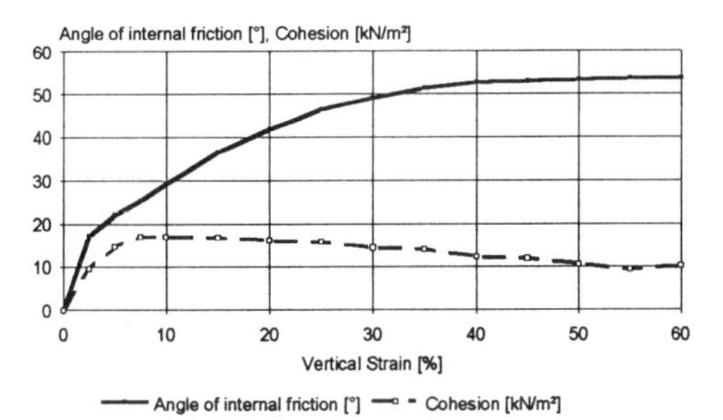

Figure 5. Deformation-dependent shear parameters of model waste

5 TEST RESULTS

5.1 UNSUPPORTED SHAFT SYSTEMS

Results from the shaft system without lining are shown in figure 6. The reduction of the diameter ΔD of the shaft, measured at two g-levels - 40 g and 100 g -, related to the initial diameter D is plotted versus the dimensionless depht z'/D. The dimensionless depth z'/D is calculated for the actual depth z' of the measuring section and the diameter D of the shaft model.

At 40 g the model simulates a prototype shaft of 24 m in height and 4.4 m in diameter, at 100 g it is 46 m in height and 11 m in diameter.

The shaft model without lining shows even at an acceleration of 100 g no tendencies to failure. Large deformations can be observed, from about 2 % of the diameter near the surface (z= 5.5 m at 100 g) up to 12 % near the bottom (z= 39.0 m at 100 g) of the shaft.

As Kockel (1995) and Reuter & Nolte (1995) indicate, it is possible to calculate landfill slope stability with deformation-dependent shear parameters if the strain or deformation behaviour of the material in the slope is known. To calculate the shaft stability of this centrifuge model test a design procedure from Walz & Hock (1987) based on a plasticity method is used. It is assumed that the deformation dependent shear parameters correspond to the vertical deformations of the waste. With the measured vertical deformation of 15 % the corresponding shear parameters can be estimated as

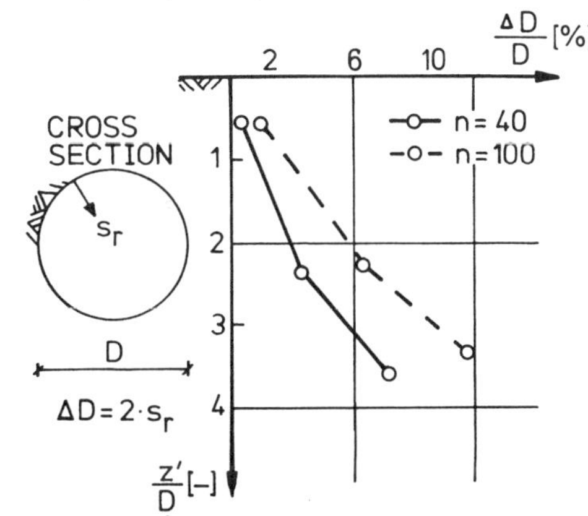

Figure 6. Radial deformation of an unsupported shaft

$\varphi = 37°$ and $c = 17\ kN/m^2$ from figure 5. With these parameters and the prototype dimensions at 100 g the minimum radial earth pressure e_{ahr}, necessary to stabilize the excavation, is calculated to be zero. This means the excavation is stable without an additional support like an unlined shaft.

5.2 STIFF LINING WITHOUT AND WITH GRAVEL UNDER SYMMETRICAL LOADS

The stress and strain behaviour of a stiff shaft with and without gravel between lining and waste under symmetrical loads is discussed. Of special interest is the behaviour of the gravel surrounding the shaft. In figure 7 the shear stresses of two different systems are shown. One system is a shaft without gravel (R/r= 0) and the other is a shaft surrounded by gravel with a thickness of about half the shaft's diameter (R/r= 2,13).

The distribution of the vertical stress in the lining is calculated from the vertical strains of the lining measured by strain gauges at different depths and from readings of the load cell at the bottom of the shaft. These stresses are caused by the self-weight of the shaft, which can be calculated easily, and negative skin friction of the gravel or waste. After separation of the stresses caused by the self-weight of the shaft the shear stresses acting on the lining due to negative skin friction are calculated by derivation of the vertical stress distribution.

Figure 7 shows the shear stresses versus the dimensionless depth determined from tests modelling a shaft without gravel and modelling a shaft with gravel between shaft and waste.

In both cases the shear stress increases with depth. The maximum shear stress is concentrated on the end of the shaft. A comparison between shear stress acting on the lining without and with gravel shows that the placement of the gravel leads to higher shear stresses.

This may be due to the following facts:

- The test results from deformation measurements indicate that the gravel body deforms vertically in the same way as the waste body does. The measured horizontal movement is about 15 % of the gravel thickness. With the assumption of a constant volume (constant unit weight) of the gravel the radial extension can be calculated to be about 11 % (constant distribution over depth of the shaft). This indicates an increase of the volume of the gravel.

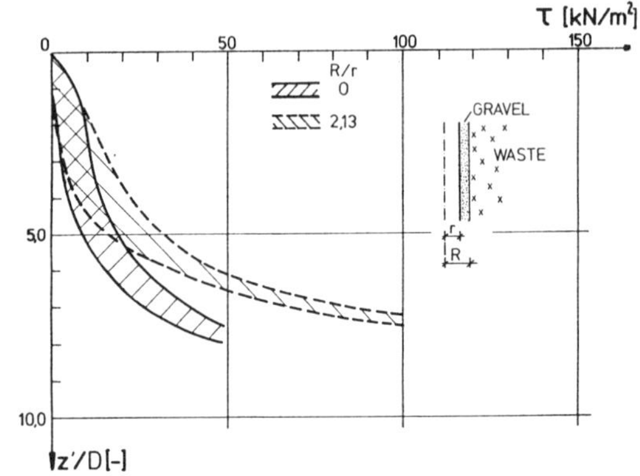

Figure 7. Shear stresses of different shaft systems

- A higher unit weight of the gravel in comparison to waste leads to higher vertical and therefore to higher radial stresses close to the shaft.

- The friction angle between gravel and shaft lining is larger than between waste and shaft lining.

- The radial deformation of the stiff lining is very small. Due to this the bearing capacity can not be activated significantly, so there is no reduction of the radial stress inside the gravel under the arching effect.

The measurements of the tangential strains of the lining indicate higher radial stress acting on the lining by the placement of gravel in comparison to direct contact between shaft and waste.

The radial stress σ_r acting on the lining is calculated from the tangential strains of the lining measured by strain gauges and the stress ratio K which is determined by the ratio of the radial stress to the theoretical vertical stress in the waste:

$$K = \frac{\sigma_r}{\sigma_v} = \frac{\sigma_r}{\gamma_{waste} \cdot h} \tag{5.1}$$

The stress ratio K can be derived from the tests without gravel between lining and waste. In this case K equals 0.3. This value is comparable to the stress ratio at rest conditions measured by laboratory tests on different samples of municipal solid waste. The placement of the gravel leads to an increase of the stress ratio K to 0.43.

5.3 STIFF LINING WITHOUT AND WITH GRAVEL UNDER ASYMMETRICAL LOADS

In the following, the additional stress and strain of the lining, caused by the asymmetrical load applied by the cushion at 3 bar air pressure besides the shaft, are discussed. The circumferential distribution of the tangential strains measured at the outer surface of the lining on two different sections is shown in figure 8 for the systems without and with gravel between lining and waste. The distribution of the tangential strains indicates a deformation of the cross section of the lining to an elliptical form. In the upper section MQ 1 this distribution is more distinctive than in the lower section MQ 2 which is more distant from the cushion. Strains of the lining without gravel are larger than the strains of the lining with gravel.

Figure 9 shows the relationship between the spread of the bending strains measured in two sections and the thickness of the gravel ring. The bending strains are calculated from the strains of the outer ε_{ta} and the inner ε_{ti} surface of the lining:

$$\varepsilon_b = \frac{\varepsilon_{ta} - \varepsilon_{ti}}{2} \tag{5.2}$$

For R/r = 2.13 the reduction of the bending strains by the placement of the gravel is obvious.

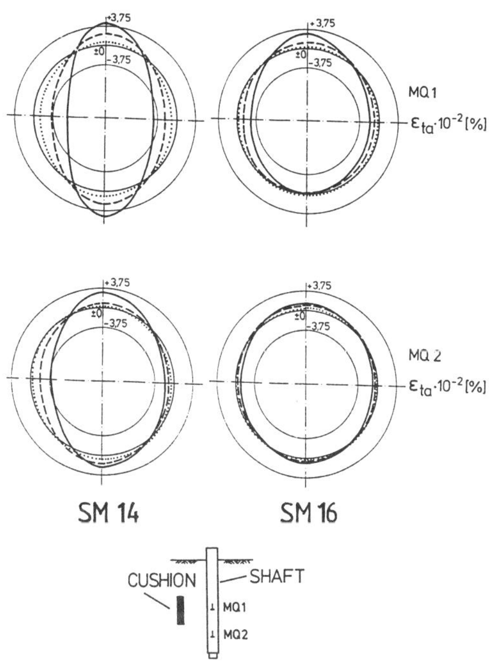

Figure 8. Distribution of tangential strains of test SM 14 (R/r=0) and SM 16 (R/r=2,13)

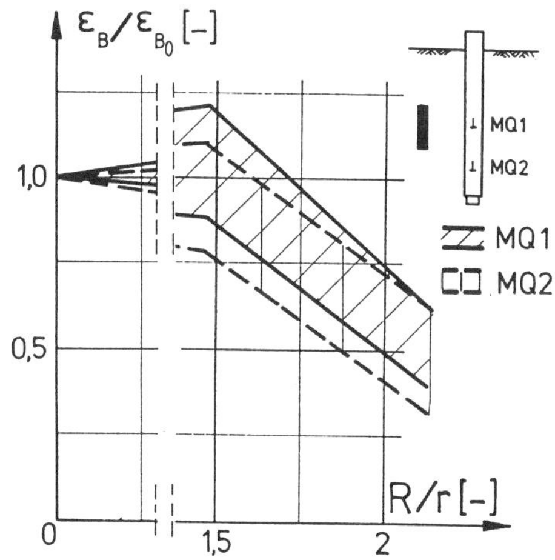

Figure 9. Strain reduction due to gravel thickness

Note that in both cases the distance of the cushion from the shaft system - i.e. the outer surface of the gravel ring or, if there is no gravel placed between shaft and waste, the lining of the shaft - is the same. From this fact it can be assumed that the loads on the shaft systems are comparable and the reduction of the bending strains is caused by the bearing capacity of the gravel.

6 CONCLUSIONS

The behaviours of different shaft systems in waste have been investigated by centrifuge model tests. The modelling of an unlined shaft at a depth of 46 m and 11 m in diameter demonstrates an important bearing capacity of the waste which is activated by large deformations. The stability of the unlined shaft can be calculated using a simple plasticity approach and deformation-dependent shear parameters to describe the shear strength of the waste.

On the other hand tests modelling a shaft with direct contact between stiff lining and waste show that the radial stress of the waste acting on the lining is characterized by K_0 - conditions. The ratio between radial stress determined by centrifuge model tests and the theoretical vertical stress is found to be 0.3, which is similar to stress ratios K at rest conditions in laboratory tests. It has to be the aim of further work to investigate the dependency of radial stresses on the stiffness of the lining.

The placement of gravel between lining and waste has a contradictory influence on the stress and strain behaviour of the lining under symmetrical and asymmetrical loads. Under symmetrical loads the placement of the gravel leads to an increase of the radial stress and to an increase of the shear stress acting on the lining due to negative skin friction. Under asymmetrical loads the nonuniform distribution of the radial stresses is reduced by the placement of the gravel between lining and waste about 50 %, if the outer radius of the area filled with gravel is proportional to the outer radius of the shaft larger than 2.

These results confirm the conclusions from field observations of existing shafts in landfills. Due to the variation of the boundary conditions in the field there is still some uncertainty concerning the design of such leachate collection shafts. The presented results show that centrifuge model tests are a powerful tool to investigate the interaction behaviour of shaft lining, gravel and waste, which is necessary to formulate design criteria.

REFERENCES

Prühs, H. 1988. Schächte in Deponien. Veröffentlichungen des Grundbauinstitutes der LGA Bayern 51: 221-235

Dany, H. & Ney, P. (1992). Sanierung eines Schachtes. Durchführung und erste Ergebnisse eines Meßprogramms am Beispiel der Deponie Mechernich. Mitteilungen des IGB der TU Braunschweig 37: 225-248

Jessberger, H. L., Güttler, U. 1988. Geotechnische Großzentrifuge - Modellversuche im erhöhten Schwerefeld. Geotechnik, Volume 2.

Jessberger, H. L. & Kockel, R. 1993. Determination and assessment of the mechanical properties of waste materials. 4th Intern. Landfill Symposium, Sardinia. Vol. 2. 1383-1392

Kockel, R. (1995). Scherfestigkeit von Mischabfall in Hinblick auf die Standsicherheit von Deponien. Schriftenreihe des Instituts für Grundbau. Heft 24. Ruhr-Universität Bochum.

Reuter, E & Nolte, H. C.(1995). Conception of stability analysis of great landfills in case of the central landfill of Hannover. 5th International Landfill Symposium, Sardinia. Vol. 2. 801-812.

Walz, B. & Hock, K. 1987. Berechnung des räumlich aktiven Erddrucks mit der modifizierten Elementscheibentheorie. Forschungs- und Arbeitsberichte aus den Bereichen Grundbau, Bodenmechanik und Unterirdisches Bauen. Bergische Universität - Gesamthochschule Wuppertal.

Environmental Geotechnics, Kamon (ed.) © 1996 Balkema, Rotterdam. ISBN 90 5410 848 7

Geotechnical properties of aged fly ash in reclaimed land of thermal power plant

Kunio Takahashi – *Port and Harbor Research Institute, Japan*
Koji Suzuki – *DRAM Engineering Inc., Tokyo, Japan*
Kazuo Ozasa – *Center for Coal Utilization, Japan*
Tatsuro Okumura – *Okayama University, Japan*
Masanori Shima – *Toa Corporation, Japan*

ABSTRACT: Standard penetration and piezocone tests were conducted on fly ash deposits located in two coal–fired power plants, as well as laboratory tests which include triaxial compression, cyclic triaxial and consolidation tests on undisturbed samples. Followings were obtained as the characteristics of the aged fly ash investigated:– 1) this material can be classified into intermediate and non–plastic soil, 2) comparing with common Japanese marine clay, it has higher permeability, lower compressibility and almost the same wet density, 3) high static strength is obtained from piezocone tests even in the layers where N–values obtained from standard penetration test range from 0 to 1, on the other hand liquefaction resistance is low.

1 INTRODUCTION

Coal ash left after the burning in combustion devices in Japan, mainly at coal–fired power plants, reaches 6.4 million ton per year in total(CCUJ, 1995). About 90% of it is constituted from fly ash. And a half of total amount of fly ash goes into recycle while the rest is disposed into closed areas as waste. Recent years, coal is reconsidered as a energy resource in place of petroleum because of a large amount of deposits, easy mining and development in burning technology that makes less influence to the environment. This circumstance leads to an estimation saying that coal ash from many industries in Japan will expand to more than 10 million ton in the year 2000. Therefore, to promote its utilization is recognized as a urgent issue to be solved.

In construction projects, using fly ash as a filling material is an effective answer for the issue. From this viewpoint, an investigation was carried out from Nov. of 1994 to Jan. of 1995 to clarify the characteristics of the aged fly ash in disposal areas in two coal–fired power plants, which are called plants *A* and *B* in this paper. The authors use the term "aged" for the fly ash elapsed over several months after disposal.

Standard penetration and piezocone tests were conducted in the investigation, as well as laboratory tests which include triaxial compression, cyclic triaxial and consolidation tests on undisturbed samples. This paper firstly reports the physical properties of the aged fly ash followed by the description of the static strength obtained from in–situ and laboratory tests. Consolidation characteristics and liquefaction resistance are also presented.

2 PHYSICAL PROPERTIES

The thickness of the fly ash deposits in the investigation sites in plants *A* and *B* are 10m and from 20 to 30m, respectively. Fly ash was spread by bulldozer in plant *A* from Mar. of 1992 to Oct. of 1993. In plant *B*, on the other hand, fly ash was pumped out in slurry state for the lower part and was spread by bulldozer for the upper part. Reclamation in plant *B* was commenced in 1979 and still going.

Particle density, ρ_s, wet density, ρ_t, water content, w_N and grain size distribution are summarized in Fig. 1. The values of ρ_s in both plants range from 2.1 to 2.3t/m^3, which is about 80% of that of ordinary soil particles, and the values of ρ_t are consequently as the same as those of marine clay usually found in Japan. Water content of plant *A* is approximately constant just around 50%, while that of plant *B* increases with depth from 50 to 100%. It is pointed out that silt particle content is predominant in both plants, from which the fly ash investigated can be classified into intermediate soil. Liquid and plastic limits tests were tried but it was found that

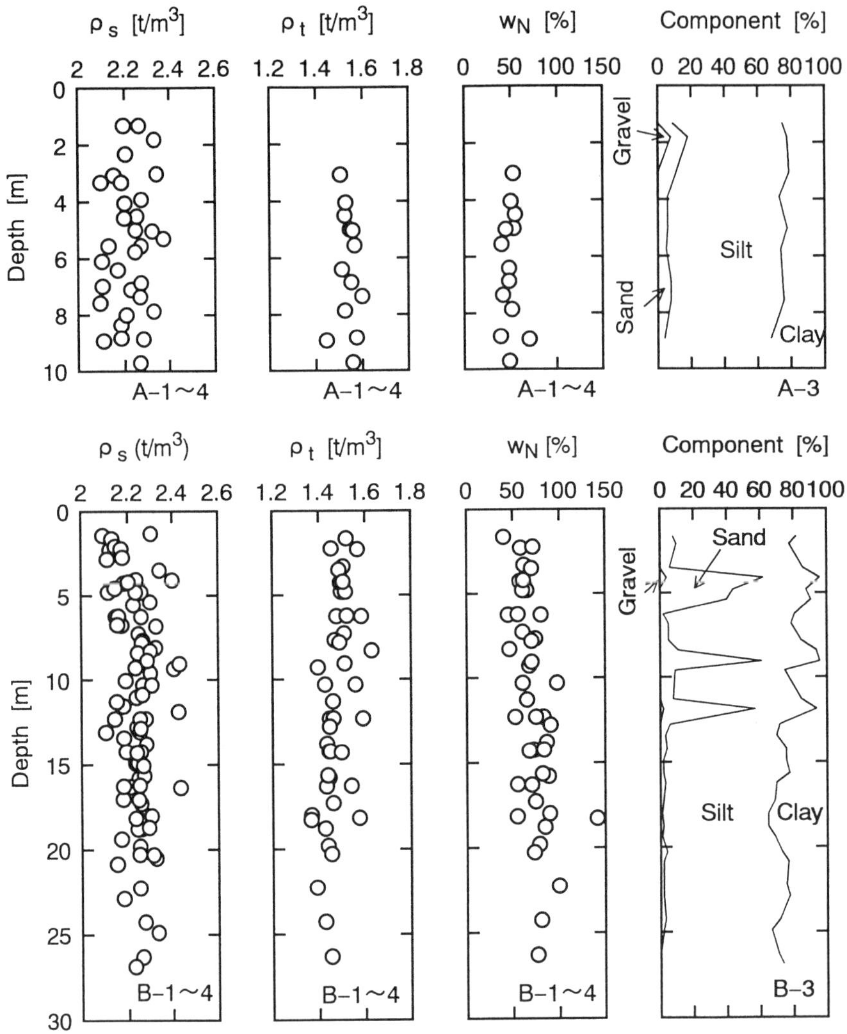

Fig. 1. Physical properties of the fly ash deposits

this material was non–plastic. Three parts are found in plant *B* where sand content is relatively high because of the reclamation with bottom ash, grain size distribution of which is similar to sand.

3 STATIC STRENGTH FROM IN-SITU AND LABORATORY TESTS

3.1 *N–value and cone resistance*

Fig. 2 presents N–values obtained from standard penetration test and cone resistance, q_T in both plants. Compacted by bulldozer, N–values above the ground water level in plant *A* are larger than those in submerged depth. N–values remarkably range from 0 to 1 in the lower parts of plant *B*. Since Reimond sampler used in the standard penetration test was filled with clean fly ash for each measurement, any evidence showing poor drilling technique such as remoulded fly ash was not found. This suggests these low N–values are reliable. The values of q_T in these parts reach 1 to 2 MPa, suggesting that the static strength of this material is expected to be high even in the layers where N–values are extremely low.

Values of q_T are plotted versus N–values in Fig. 3.

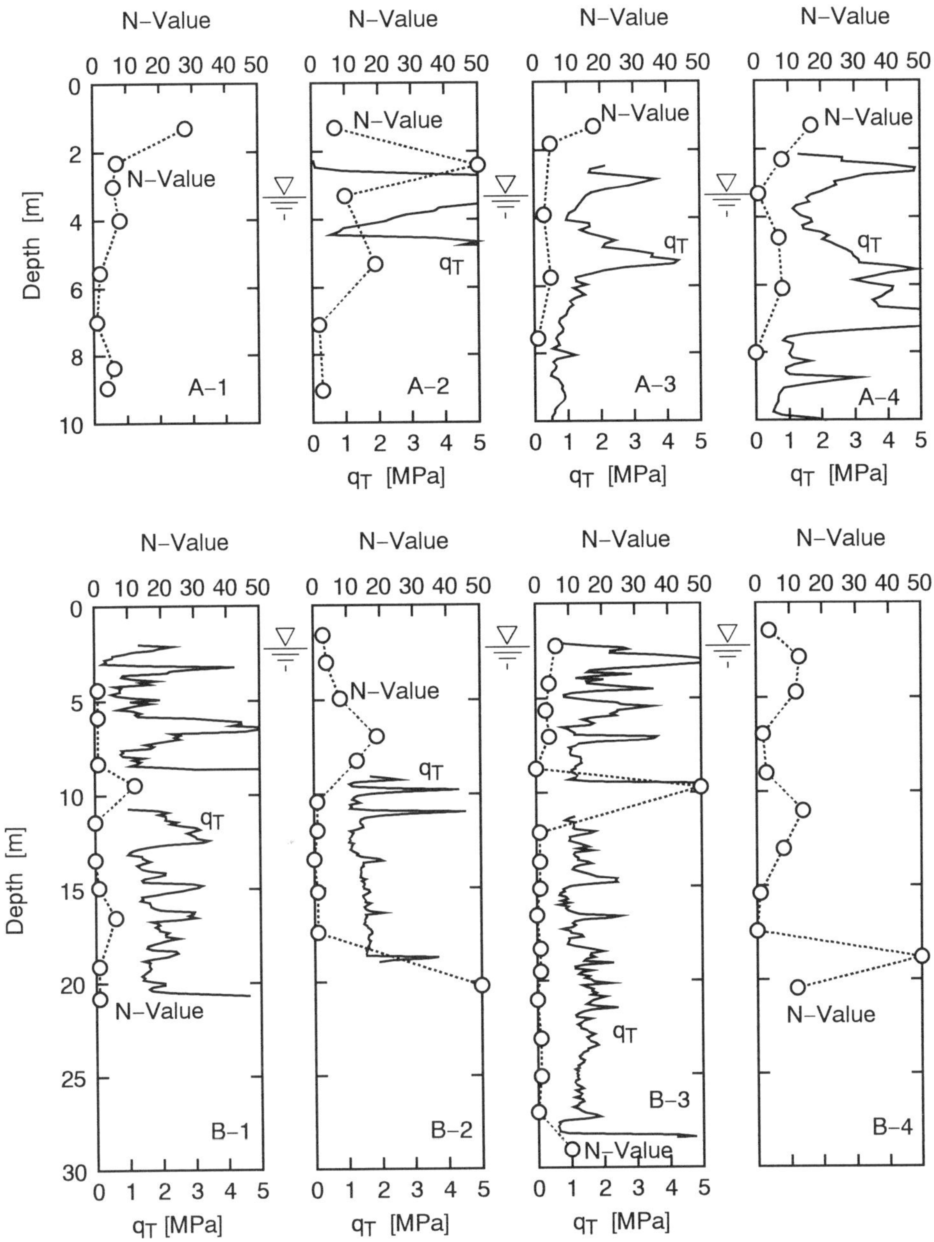

Fig. 2. N–values and values of q_T of fly ash deposits in both plants

Since N–values mostly indicated from 0 to 1, it was difficult to find out any meaning relation between both values. This fact yields following conclusions: 1)accurate static strength and proper design parameter of fly ash deposit can not be well estimated from standard penetration test, while this test is usually used to determine internal friction angle of sandy soil, 2)peizocone is a effective investigation tool for this kind of material to estimate static strength.

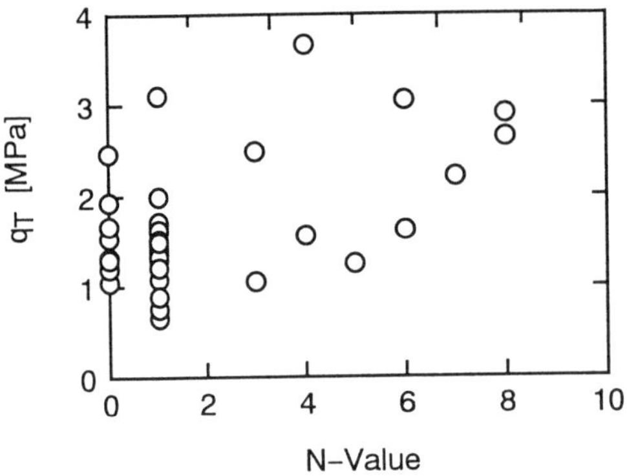

Fig. 3. Values of q_T plotted vs. N-values

3.2 *Triaxial compression test results*

Undisturbed samples for triaxial compression test were obtained by modified Bishop sampler (Hanzawa, et al., 1980) at plant *B*. The tube of this sampler has a 600mm length, a 53mm inner diameter and a 3.5mm thickness. The specimen extruded from the tube was directly mounted to test device with only cutting its bottom end and without trimming. On the other hand, stationary piston type thin wall sampler, which is widely used in Japan, was used for undisturbed sampling in plant *A*.

Triaxial compression test was carried out to investigate static strength characteristic of the aged fly ash. The specimens are at first isotropically consolidated and then sheared at a strain rate of 1.0%/min with applying a 200kPa back pressure through both stages. This test was conducted in undrained condition to investigate pore pressure behavior during shear. Examples of stress~strain curves given by the test are presented in Fig. 4. 38 specimens out of 42 did not show any clear peak point or gave gradually increasing curve until axial strain reached to 15%. Fig. 5 presents effective stress paths of each specimen from plant *B*. Significant pore pressure generation was observed in every specimen, which is consistent with the triaxial test results conducted by Yasuhara, et al(1991) for fly ash in loose state.

The stress paths in Fig.5 firstly go to left, which means significant pore pressure generation, and then change their directions. The behavior of the aged fly ash during shear in undrained condition is similar to that of cohesive soil when the early stage during shear is concerned, and is similar to that of sand in the later stage of the shear.

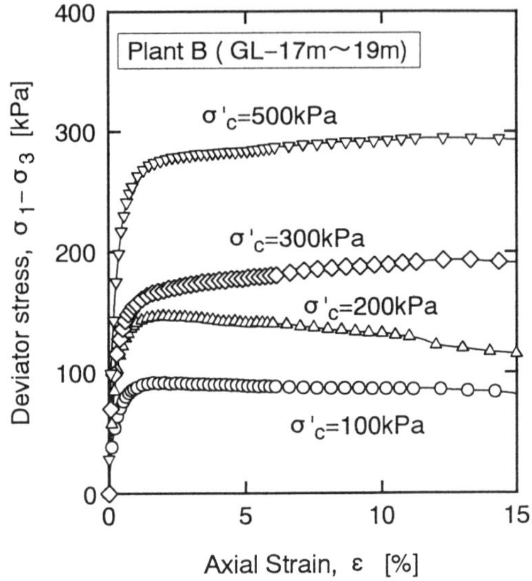

Fig. 4. Examples of stress~strain curves from triaxial compression tests

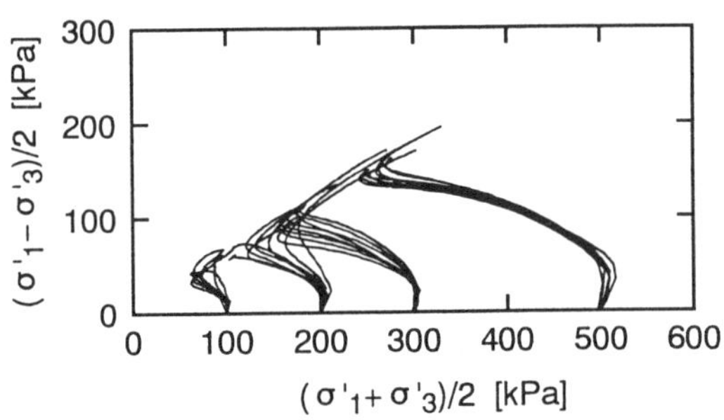

Fig. 5. Stress paths obtained from triaxial compression test for plant *B* samples

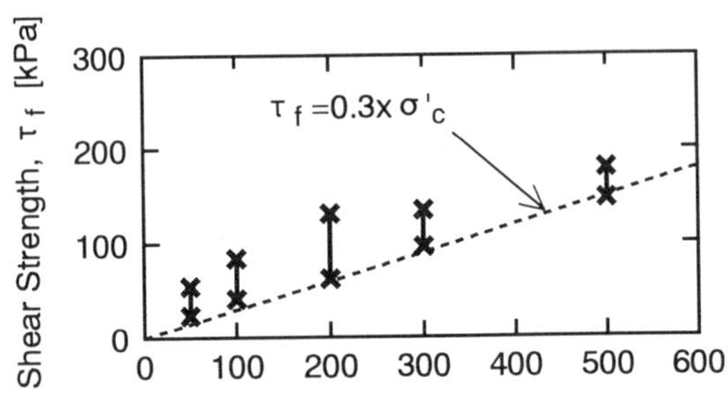

Fig. 6. Values of shear strength, τ_f plotted vs. consolidation stress, σ'_c

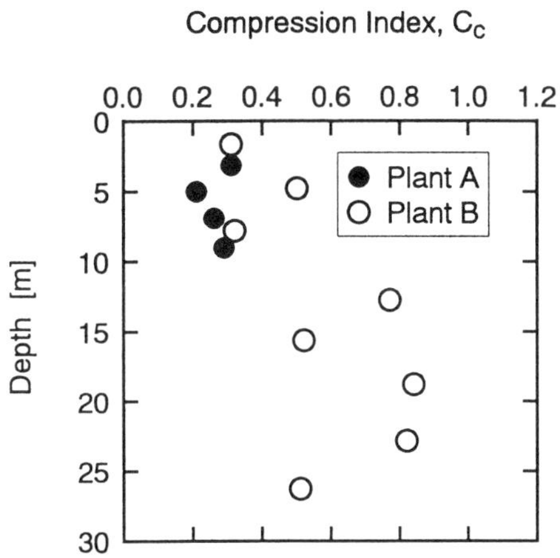

Fig. 7. Change in compression index, C_c through depth

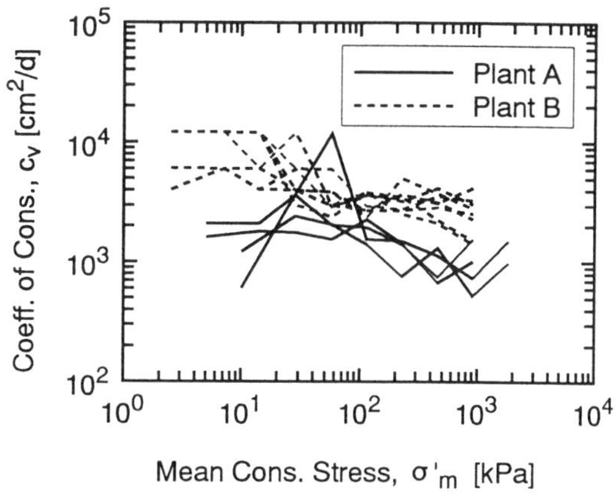

Fig. 8. Coefficient of consolidation, c_v obtained from standard oedometer test

The values of shear strength, τ_f are plotted versus consolidation stress, σ'_c in Fig. 6. Equation (1) was obtained between τ_f and σ'_c as lower limit as indicated in the figure. This equation is almost equal to the $\tau_f \sim \sigma'_c$ relation usually observed in Japanese marine clay.

$$\tau_f = 0.3 \times \sigma'_c \quad (1)$$

4 CONSOLIDATION CHARACTERISTICS

Fig. 7 shows change in the values of compression index, C_c from standard oedometer tests through depth. Diameter of the specimens are 60mm for plant *A* and 35mm for plant *B*. Values of compression index for plant *A* range from 0.2 to 0.3, and those for plant *B* range from 0.3 to 0.8 showing smaller values at the upper part than at the lower part. As previously described, plant *A* and the upper part of plant *B* were reclaimed by spreading with bulldozer, while the lower part of plant *B* was reclaimed by pumping. It is suggested that the difference in the values of C_c according to depth reflects the reclamation procedure.

Values of coefficient of consolidation, c_v obtained from the test are plotted versus mean consolidation stress, σ'_m in Fig. 8. These values range from 1,000 to 10,000 cm^2/d, which are 10 to 100 times larger than typical marine clay found in Japan. When this material is subjected to loading such as embankment, it is important that shear strength increment induced by consolidation should be taken into consideration because of higher permeability than usual cohesive soil. It is well known that underestimation in coefficient of consolidation easily occur when standard oedometer test is applied to the material with high permeability. Because of this reason, oedometer test with long drainage pass is needed to evaluate proper value of c_v for this type of material.

5 LIQUEFACTION RESISTANCE

Cyclic triaxial test was also conducted to investigate liquefaction resistance of the aged fly ash. In the case of plant *B*, the specimens for the test were prepared by the same manner as for the triaxial compression test. On the other hand, the specimen was artificially made in the case of plant *A* because of the difficulty in trimming. After mounting the specimen on the test apparatus, carbonic acid gas was firstly circulated through the specimen followed by the circulation of deaired water to accomplish full saturation. The specimen was then isotropically consolidated at σ'_c of 100kPa with application of a 200kPa back pressure($B \geqq 0.97$, where B is pore pressure coefficient, is achieved for each specimen).

It is assumed that liquefaction occurred when double amplitude of axial strain, ε_{da} reached 5%. The values of cyclic stress ratio, $\sigma_d/2\sigma'_c$ where σ_d is cyclic deviator stress, was plotted versus number of cycles, N_c in Fig. 9 together with those of Toyoura standard sand in loose (relative density, D_r=50%) and medium dense (D_r=75%) states (Yasuda, 1988).

Liquefaction resistance of the aged fly ash in plant

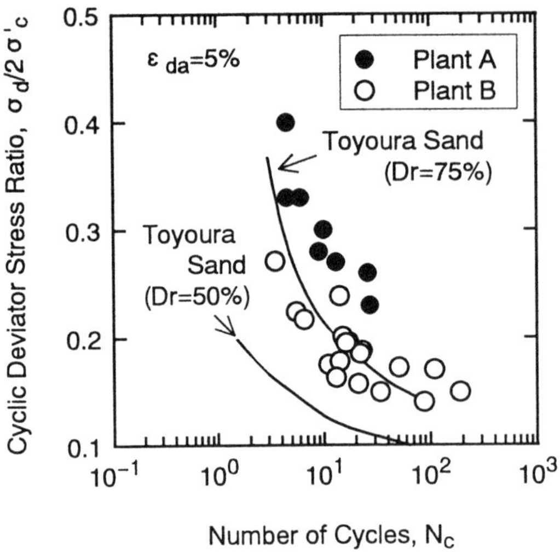

Fig. 9 Values of cyclic stress ratio, $\sigma_d/2\sigma'_c$ plotted vs. number of cycles, N_c

B is approximately the same as that of Toyoura standard sand in medium dense state, which shows its high liquefaction potential as reported by Miki, et al(1985). On the other hand, that of the aged fly ash in Plant *A* is greater than these values because the values of ρ_t of the reconstituted specimens (1.6 ~ 1.7t/m^3) were larger than that of the original values at the site (1.5 ~ 1.6t/m^3). These values of liquefaction resistance at Plant *A* are supposed to decrease when specimens which have the same ρ_t values as the original one are prepared.

It is commonly said that the higher fine (clay and silt) particle content produce the higher liquefaction resistance, however, liquefaction resistance of the aged fly ash, which has high silt particle content, is not very high as indicated in Fig. 9. This suggests that liquefaction potential should be related to whether or not the material is non–plastic.

6 CONCLUSIONS

Results of the series of field and laboratory tests are presented for the aged fly ash deposits in disposal areas in two power plants. From the study, the following conclusions can be made:

1)The aged fly ash found at the sites contains much silt particle and is classified into intermediate and non–plastic material.

2)For this type of material, piezocone test is much more suitable for in–situ investigation technique to obtain static strength than standard penetration test.

3)The relation between undrained static strength, τ_f and consolidation stress, σ'_c was conservatively given by $\tau_f = 0.3 \times \sigma'_c$.

4)Compression index, C_c of the aged fly ash depends on reclamation procedure. Coefficient of consolidation, c_v is significantly higher than that of typical marine clay in Japan, and strength increase gained by consolidation should be taken into account in practice.

5)Liquefaction resistance is almost the same as that of Toyoura standard sand in medium dense state, which indicate that the liquefaction resistance of the aged fly ash is low. It is suggested that plasticity is also an important factor to evaluate liquefaction potential of the material which has high fine particle content.

REFERENCES

Center of Coal Utilization, Japan. 1995. Investigation Report on Coal Ash in Japan(in Japanese).

Hanzawa, H., Matsuda, E. and Hirose., M. 1980. Evaluation of sample quality of sandy soil obtained by the modified Bishop sampler. Soils and Foundations, Vol. 20, No.3, pp. 17–31.

Miki, G., Imai, G., Akitani, H., Saito, T. and Doi, Y. 1985. Liquefaction Potential of Coarse Fly Ashes. Proc. of the 20th Japan National Conference on Soil Mechanics and Foundation Engineering. Vol. 1. pp. 613–614(in Japanese).

Yasuda, S. 1988. From investigation to countermeasure for liquefaction. Kajima Shuppankai. (in Japanese)

Yasuhara, K., Hyodo, M. and Hirao, K. 1991. Liquefaction Property of Coal Fly Ash as Reclamation Material. Tutchi to Kiso, Vol. 39, No. 2, pp. 5–10(in Japanese).

Geotechnical recycling and reuse of waste materials

Environmental Geotechnics, Kamon (ed.) © 1996 Balkema, Rotterdam. ISBN 90 5410 848 7

Slurry composition effects on soil-cement mixtures

H. Akagi & H. Nakazato
Waseda University, Tokyo, Japan

ABSTRACT : Soil-cement mixtures were prepared using three types of materials : sandy soil excavated from a shield tunnelling construction site in Tokyo, clay slurry used for stabilizing the shield tunnel cutting face, and cement. These soil-cement mixtures were planned to be used for the filling of the lower part (invert section) of the circular shield tunnel cross-section in place of cement mortar. A number of unconfined compression tests were carried out to investigate the time-dependent variation of compressive strength of soil-cement mixtures. The strength of soil-cement mixture was found to be satisfactory as the new invert material and to be influenced by the contents of the clay slurry. It could be seen that the density of soil-cement mixtures measured immediately after mixing and compaction was directly related to the compressive strength.

1 INTRODUCTION

When tunnels are constructed by the slurry shield tunnelling method, excavated soil is not used. Since the excavated clay soil is added to the initial volume of clay slurry used for the cutting face stability of shield machine in the case of clays, the volume of clay slurry becomes more than the amount needed. The evcavated soil and excessive clay slurry generally have high water and a fine soil particle contents. Consequently, these materials are removed as industrial waste.

Recently, attempts to create new materials for the construction of shield tunnels from the waste products described above are being carried out by the Teito Rapid Transit Authority in Japan (Sukegawa, 1994). The new materials were produced by mixing excavated soil, excessive clay slurry, and cement. The new materials have been used for the filling of the lower section, i.e., the invert section of a circular cross-sectional shield tunnel in place of cement mortar. These filling materials need to have high compressive strength, workability and durability. The new materials have had compressive strength high enough to satisfy these requirements. The workability, i.e., flow characteristics of that material have been shown to be dependent on the types of excavated soil. The long-term and chemical durability was proven to be about the same as conventional cement mortar. However, the effects of clay slurry types on the strength of the new materials have yet to be demonstrated.

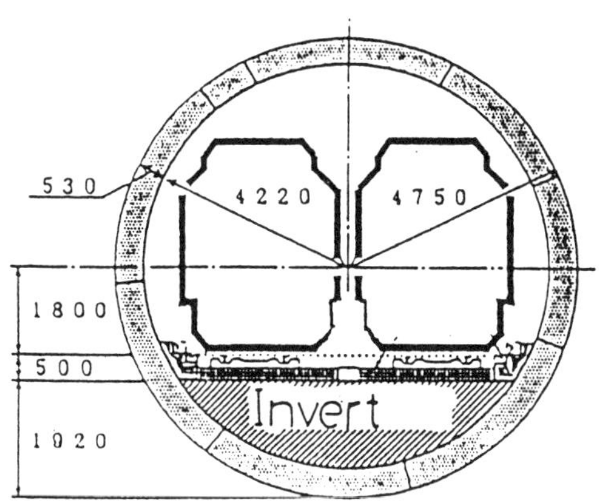

Fig. 1. A typical cross-section of a shield tunnel (unit: mm).

In this experimental study, various types of clay slurry were prepared and mixed with excavated sandy soil and cement. A number of unconfined compression tests were carried out using these soil-cement mixture samples. The effects of clay slurry composition on the strength of the soil-cement mixtures were investigated.

2 DEVELOPMENT OF NEW TUNNEL INVERT MATERIALS

According to Sukegawa(1994), the outline of newly developed tunnel invert materials are described as follows. Figure 1 shows a typical cross-section of a shield tunnel used for a double track subway train tunnel. The area of application of new tunnel invert material is about 11 % of the whole tunnel cross-section. The following requirements should be satisfied for the tunnel invert materials:

1) to support the load and bear the vibration when a train is actually running the tunnel,
2) to keep their mechanical properties, when immersed and/or subjected to temperature and humidity fluctuations, and
3) to be able to be fit on upper subgrade concrete.

The compressive strength of invert materials 28 days after mixed and compacted should be greater than about 6 MPa. The experimental results are shown in Fig.2. The unconfined compressive strength, q_{u28}, of four types of soil-cement mixtures are plotted against their water-cement ratio (W/C). The compressive strengths of the mixtures become greater than 6 MPa if water-cement ratio is small enough. These new materials have been successfully used in the practical construction of subway tunnels.

Figure 3 demonstrates the relationships between unconfined compressive strength q_{u28} and their fine soil particle contents F kg/m^3, which are smaller than 0.075 mm. In spite of the same fine particle content, different values for the unconfined compressive strength are obtained depending on the types of fine soil particles contained in the mixtures. This figure clearly indicates the dependency of the strength of soil-cement mixture on the consistency properties of fine soil particles.

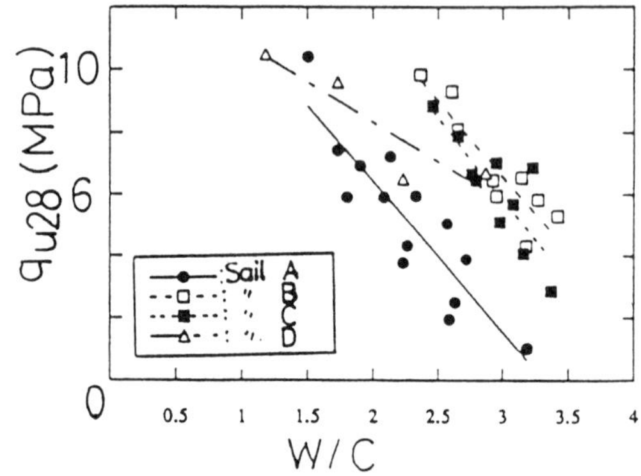

Fig. 2. Relationships between compressive strength q_{u28} and water cement ratio (W/C) (Sukegawa (1994)).

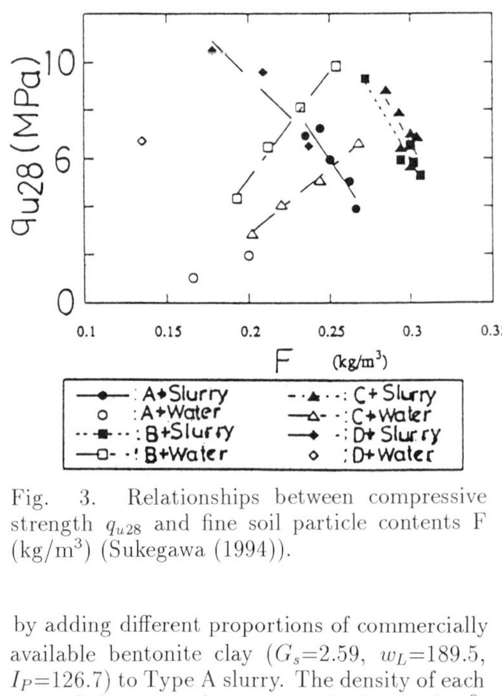

Fig. 3. Relationships between compressive strength q_{u28} and fine soil particle contents F (kg/m^3) (Sukegawa (1994)).

3 EXPERIMENTAL PROCEDURES

The following soil samples were used in this study. Sandy soil was taken from a shield tunnelling construction site in Tokyo. The physical properties of sandy soil are as G_s=2.66 and D_{50}=0.18 mm.

Since the consistency of fine soil particles affects the strength of the soil-cement mixtures, five types of clay slurry were used. Type A slurry was prepared by mixing the clay (G_s=2.64, w_L=107.2, I_P=57.5) obtained off-shore from Tokyo Bay with distilled water. Four types of bentonite clay slurry B, C, D, E were prepared by adding different proportions of commercially available bentonite clay (G_s=2.59, w_L=189.5, I_P=126.7) to Type A slurry. The density of each type of slurry was kept constant at 1.2 g/cm^3. Blast furnace cement was used and 200 kg of cement was added to 1 m^3 of other materials. The values of contents of clay slurry a=(weight of clay slurry)/((weight of clay slurry)+(weight of sandy soil)) were 0.3, 0.4 and 0.5. Table 1 shows the types of clay slurry and their bentonite mixing ratio b=(weight of bentonite slurry)/((weight of bentonite slurry)+(weight of Type A slurry)) used in this study.

The sequence of experimental procedures was as follows:
1)to measure the water content and bulk density of the sandy soil,
2)to calculate the weight of the three materials for mixing,
3)to mix the sandy soil and clay slurry for 2 min.,
4)to add cement materials to the mixtures prepared above and mix them for 10 min.,
5)to compact soil-cement mixtures into a mould for the specimen of unconfined compression tests (Diameter = 50 mm, Height = 100 mm),
6)to measure the bulk density and water content of the soil-cement mixtures,
7)to keep the samples under a high humidity condition for the required period for curing, and
8)to carry out the unconfined compression tests at 1 day, 3 days, 1 week, 3 weeks and 4 weeks after mixing and compaction.

4 STRENGTH OF SOIL-CEMENT MIXTURES

Figure 4 shows the time-dependent variations of unconfined compressive strength of soil-cement mixtures, whose clay slurry content a is 0.3. The unconfined compressive strength increases as the curing time increases. When bentonite mixing ratio b becomes greater, the compressive strength becomes lower. If the mixture contains a slight amount of bentonite clay, the compressive strength decreases remarkably.

Similar relationships for a = 0.4 and 0.5 are shown in Figs. 5 and 6. Figure 7 shows the relationships between compressive strength measured after 28 days of being mixed and compacted q_{u28} and clay slurry content a. The values of their compressive strength become smaller than those in the case of a = 0.3. The decrease in compressive strength due to the increase of b was not so evident as in a = 0.3.

The existence of bentonite clay in the soil-cement mixtures reduced their compressive strength compared with the other types of fine clay particle, i.e., Type A clay slurry. This reduction in compressive strength is assumed to be due to the weak skeleton of micro-structures of the soil-cement mixtures, which contains bentonite. However, all measured values of the compressive strength after 28 days after of being mixed and compacted are beyond 6 MPa, which is the required value for tunnel invert materials.

Figure 8 shows the relationship between compressive strength after 28 days q_{u28} and the bulk density ρ_t measured immediately after mixing and compaction. The bulk density can deter-

Table 1. Types of clay slurry

Type	A	B	C	D	E
Bentonite b	0.0	0.25	0.50	0.75	1.0

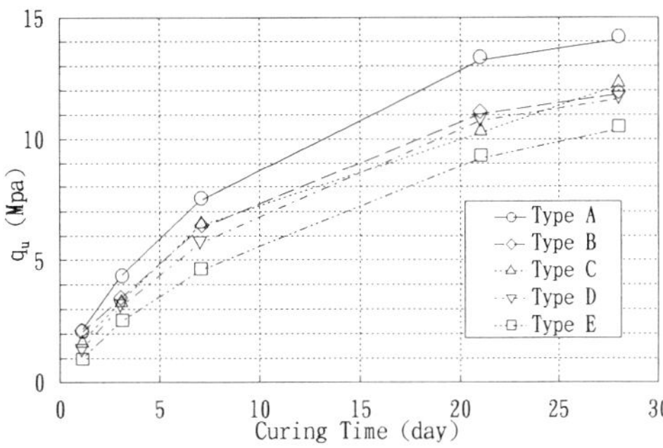

Fig .4. Time-dependent variations of compressive strength (a=0.3).

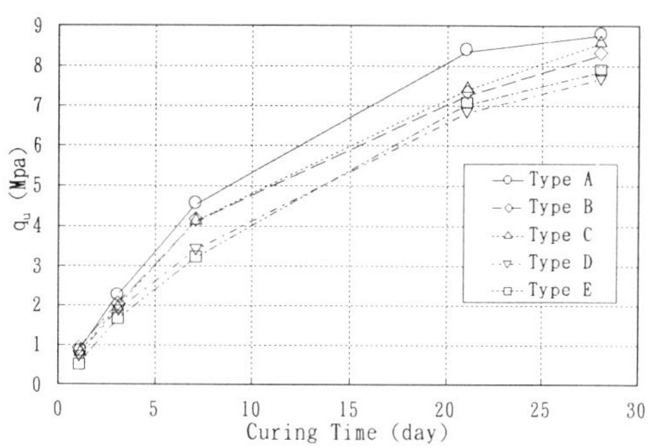

Fig. 5. Time-dependent variations of compressive strength (a=0.4).

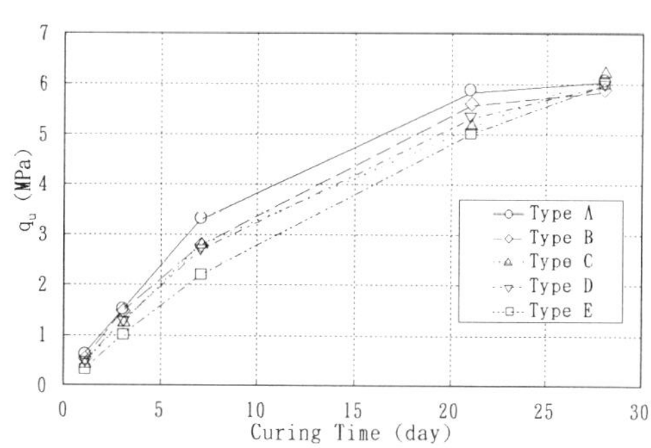

Fig. 6. Time-dependent variations of compressive strength (a=0.5).

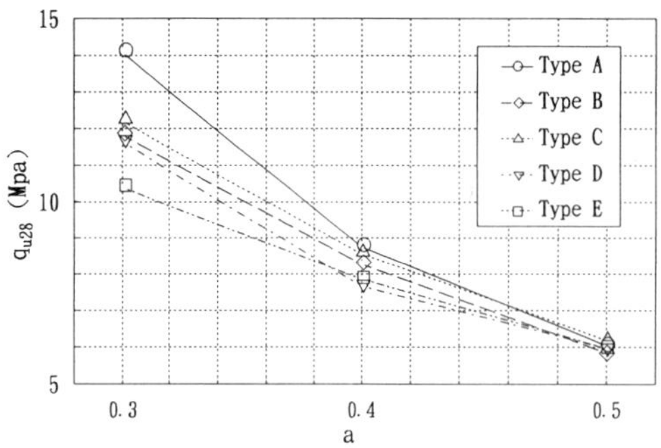

Fig .7. Compressive strength after 28 days, q_{u28}, versus clay slurry content a.

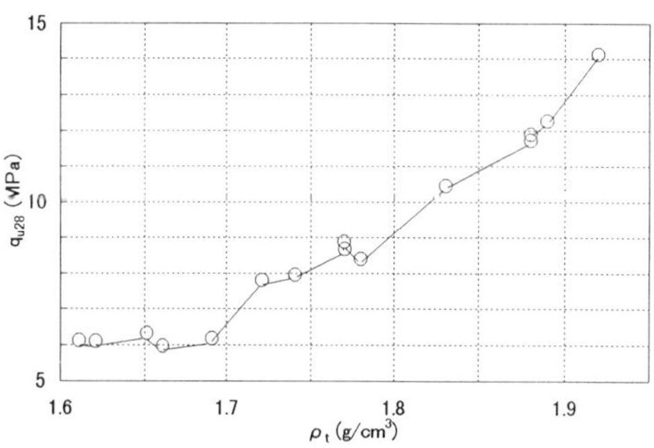

Fig. 8. Compressive strength after 28 days, q_{u28}, versus bulk density ρ_t.

mine the compressive strength after 28 days. Therefore, it is possible to control the compressive strength of tunnel invert materials after filling by measuring the bulk density immediately after mixing and compaction.

5 CONCLUSIONS

The following conclusions were derived from the experimental results:

(1) The unconfined compressive strength of new soil-cement mixtures are greater than the required value 6 MPa for tunnell invert materials.

(2) The existence of bentonite in the soil-cement mixtures reduces their compressive strength compared with the other types of fine clay particles. This reduction of compressive strength is assumed to be due to the weak skeleton of microstructures of the soil-cement mixtures, which contains bentonite.

(3) The bulk density can determine the compressive strength 28 days after mixing and compaction. Therefore, it is possible to control the compressive strength of tunnel invert materials after filling by measuring the bulk density immediately after mixing and compaction.

ACKNOWLEDGMENTS

The authors wish to express their gratitude to the Teito Rapid Transit Authority, who provided the test samples used in this study. Thanks are also due to Chemical Grout Company for their financial supports for this study.

REFERENCE

Sukegawa, T. 1994, Development of invert material using excavated soil from slurry shield, *Proc.JSCE*, No.504/VI-25, 107-116 (in Japanese).

Environmental Geotechnics, Kamon (ed.) © 1996 Balkema, Rotterdam. ISBN 90 5410 848 7

Mud ashes from a purification station being re-used

L.Curtil, C.Coulet & J.D.Rakotondramanitra
L2M2S, IUT Génie Civil, Villeurbanne, France

ABSTRACT : A test campaign has been carried out in order to know what might be the use of mud ashes from a purification station in the field of the building industry and public works. These ashes are indeed thrown away into dumping areas. We have measured the physical features of three different samples with a clearly defined composition, made Proctor CBR tests and oedometric tests in order to determine their capacity to be used in embankment work. These mud ashes from purification stations are mainly made of calcium and aluminium silicates. The study of the physical features of the material has shown that it could be compared to very fine sand particles. What is more these ashes are very wet so they would very unexpectedly change their consistency in case of tiny variations in the water content. Thus these soils are normally out of use in this state but a temporary deposit or previous drainage can be foreseen. The Proctor tests have shown that these was a 60% optimum. This figure is very high but fine soils with a high moisture content have already been used successfully after a detailed study of consolidation. Oedometric tests have been carried out. Consolidation tests have shown that in spite of rather long lengths of time, no stabilization appeared. On the contrary, the soil settling remains very weak which is therefore encouraging. With the lack of interest about this potential soil, these ashes have been tested, they have been mixed with the sand and cement. Much better results have been obtained.

1. INTRODUCTION

Our laboratory has been interested in the re-use of waste in the field of public works for over ten years. Indeed our consumer society produces more and more waste every year. These residues are creating a problem. Their dumping is more and more controlled and more and more expensive. In 2002, a European law will impose that "only ultimate waste might be put in dumping areas" (CEE 91/271).
At the request of a producer of mud ashes from a purification station, we have searched what might be the valorization of this residu in public works and more particulary in embankment work.

2. MUD ASHES FROM A PURIFICATION STATION.

The studied purification station in Lyon (France) produces about 10,000,000 kg of mud from liquid waste and rainwater treatment each year. These muds are burnt to ashes and today 5,000,000 kg of mud ashes are thrown away into dumping areas.
A test campaign has been carried out on three samples which have been taken at three different periods during the year : mud ash 1 during a dry period ; mud ash 2 during a wet period ; mud ash 3 during a very rainy period.
Chemical analysis of these three samples show that they are chemically different. But all are mainly made of calcium and aluminium silicates. They also have iron, magnesium, potassium, sodium, sulfur and heavy metals in small quantities. But they are environnemtally safe.
These ashes have arrived at the laboratory as they were very wet, the water contents ranging from 90 to 130 %. This very high water contents have a penalizing effect for a potential use in the field of the building industry and public works.
The total specific surface measured with Blaine air Permeability apparatus ranges from 4000 to 8000 cm^2/g. These very high surfaces mean that the particles are very fine and that consequently there is a great exchange surface. Comparatively cements have a Blaine surface from 2500 to 4500 and coal fly ashes from 2200 to 4000 cm^2/g.

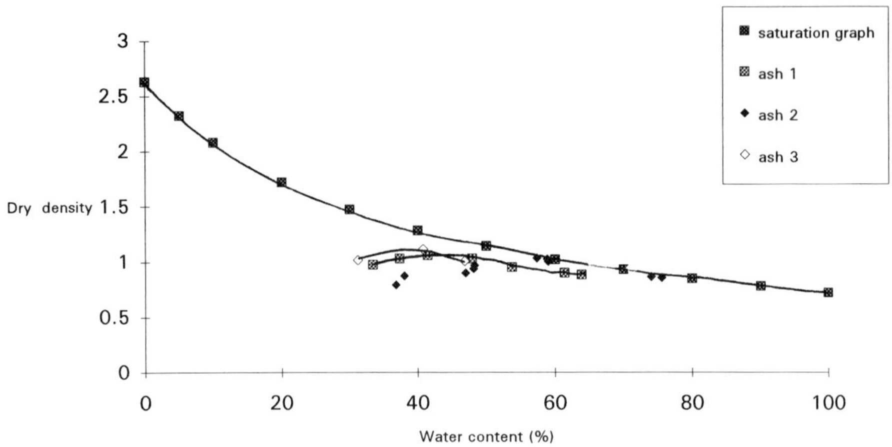

FIGURE 1 : PROCTOR TEST

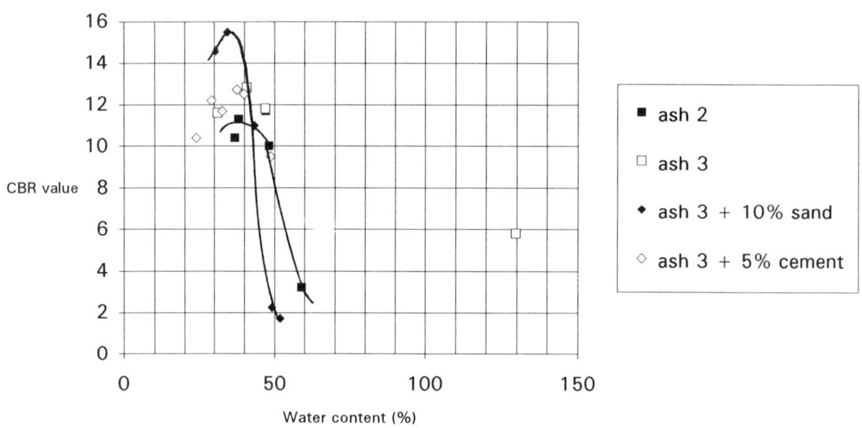

FIGURE 2 : CBR TEST ON ASH 2 AND ON ASH 3

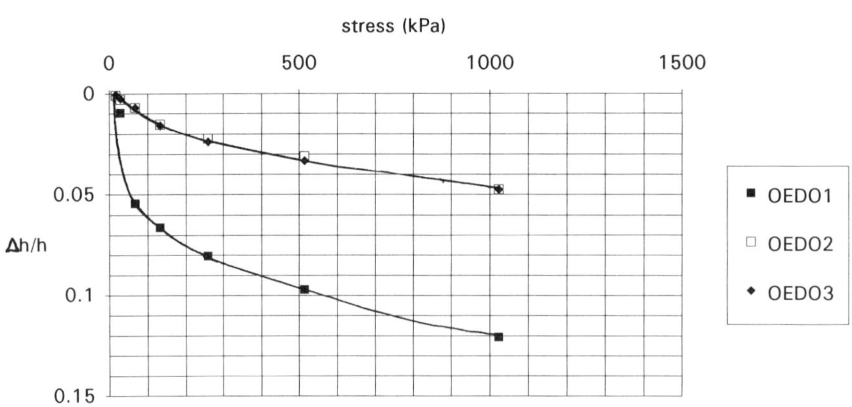

FIGURE 3 : oedometric test settlement-stress relationship

Particle density is about 2.65 that is to say a very usual density. Particle size distribution was given by sieving and by sedimentation next. Granulometry of these particles is about from 1 to 200 μm. So these ashes could be compared to coal fly ashes.

3. THE STUDY OF A USE OF MUD ASHES AS SOIL.

The first idea was to re-use these ashes only as soil in the field of public works and more particularly in embankment.
So we have been studying the physical features of the materials (sand equivalent value, Atterberg limits, Blue methylen value). These tests have enabled to determine that three ashes could be compared to very fine sand particles. Permeability in the order of 10^{-7} m/s confirms this result.
The Proctor tests have been carried out on the three samples (figure 1) in order to determine optimum moisture content to be compacted and optimum density obtained after compaction. Proctor graphs are rather flat and optimum water contents varies from 45 to 60% from ash to ash and dry density γ_d/γ_w is about 1. These results show that ash is an insensitive material to moisture contents (flat graphs). The ideal water contents for good compaction for road works is very high in comparison with an ordinary soil (about 12% regularly) and it fits about half the initial moisture contents. The dry density is close to 1.1 which is rather weak.
The CBR tests have been carried out after the Proctor tests on ash 2 and on ash 3 (figure 2). The CBR value has been measured. The bearing capacity of soil is about 12 which is weak. These tests have shown that optimum water contents is weaker than optimum Proctor's (around 10 at 20% less).
The oedometric tests have been implemented on ash 1 from values obtained at Proctor test. Two samples of ash 1 were prepared at Proctor optimum that is to say w=45% and γ_d/γ_w=1.06 (oedo1 and oedo2) and a sample of ash 1 was prepared at w=33% and same γ_d/γ_w(oedo3). These three samples have been then saturated with water. After that, they have been loaded between 17 and 1024 kPa and we have measured the settlement of these samples. So we have been able to plot a number of graphs :
- settlement - stress relationship (figure 3)
- void ratio as a function of logarithm of stress at the maximum settlement (figure 4)
- and the height of the sample according to log of time in minutes for each stress (figure 5).
The figure 3 allows to calculate oedometric modulus (E'). It has been calculated for a stress of 100 kPa and a delta of 100 :
E' = 12000 kPa which is quite acceptable.
The figure 4 shows oedometric compressibility of the three samples (oedo 1 to 3). The different stress exerted on samples (17 to 1024 kPa) are marked in abscissa with a log scale. The vacuum factor e is the ratio of vacuum volume to total volume of solid particles of soil. Three conclusions can be drawn from this graph :
* we can see that this graph is made up of two straight lines. The change of incline occurs at a stress of 45 kPa for oedo 2 and oedo 3 which indeed corresponds to the prestressed exerted during the compaction of the sample in the oedometer.
* the slope of the second straight line (highest stress) gives compression factor Cc :
Cc = 0.10 for oedo 2 and oedo 3 and Cc = 0.14 for oedo 1 which is rather weak.
* the sample oedo 1 reacts differently but the slope is roughly the same and only initially exerted prestress has been different.
The other graphs are similar to figure 5. At a given loading, we have the sample's height in terms of log of time in minutes. It is a consolidation test. A number of observations can be made :
→ the soil settlement is weak ;
→ the relation between the height of the sample and the log of time is linear and in spite of rather long lengths of time (two months) no stabilization appears.
The unloading of oedometers shows that the settlement decreases. The samples then slightly swell but they do not recover the initial value (figure 6). The oedometric compressibility graph of the three samples during loading as well as during unloading (figure 7) enables to point out that we have not the same vacuum factor (weaker slope).
Oedometric tests have also been carried out on ash 2. Three samples have been similarly loaded. The results are the same : soil settlement is weak and no stabilization appears.
The shearbox tests on ash 1 and ash 2 and ash 3 have been brought.
The results are as follows :
ash 1 : Shear strength = 56 kPa and $\varphi = 26°$
ash 2 : Shear strength = 16 kPa and $\varphi = 36°$
ash 3 : Shear strength = 25 kPa and $\varphi = 36°$
It means that the cohesion is good. Friction angle is quite reasonably good. Soil resistance is then right.

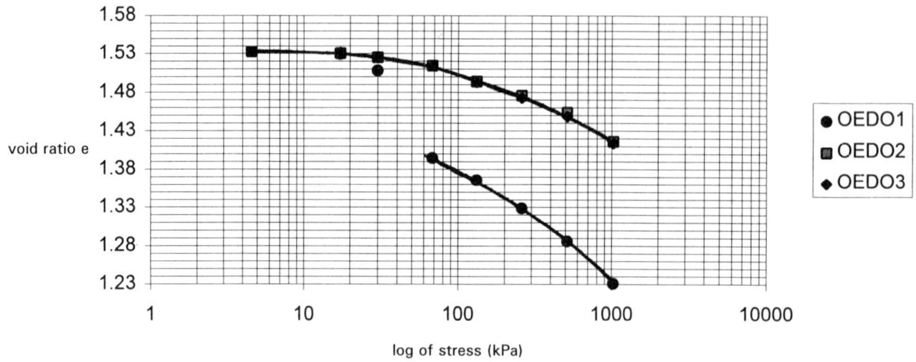

FIGURE 4 : oedometric test
void ratio - log of stress at the maximum settlement

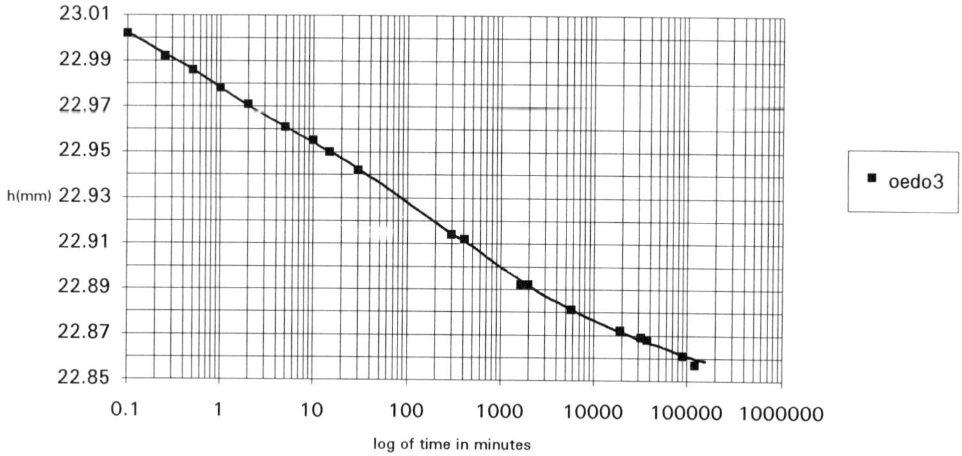

FIGURE 5 : consolidation test
1024 kPa loading

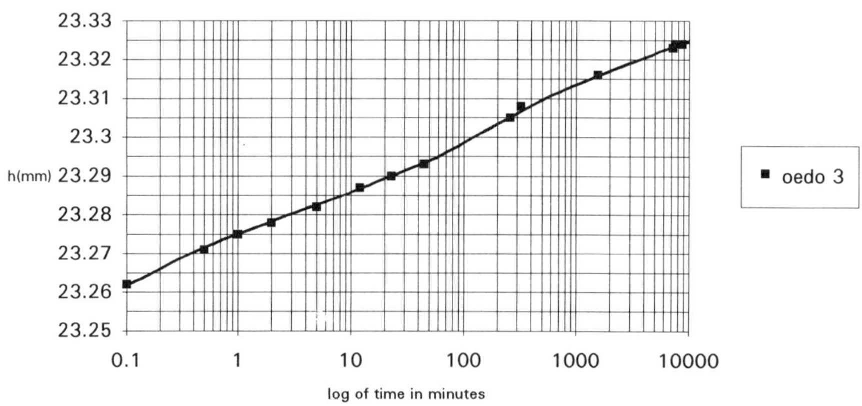

FIGURE 6 : consolidation test
unloading 68kPa-0.5kPa

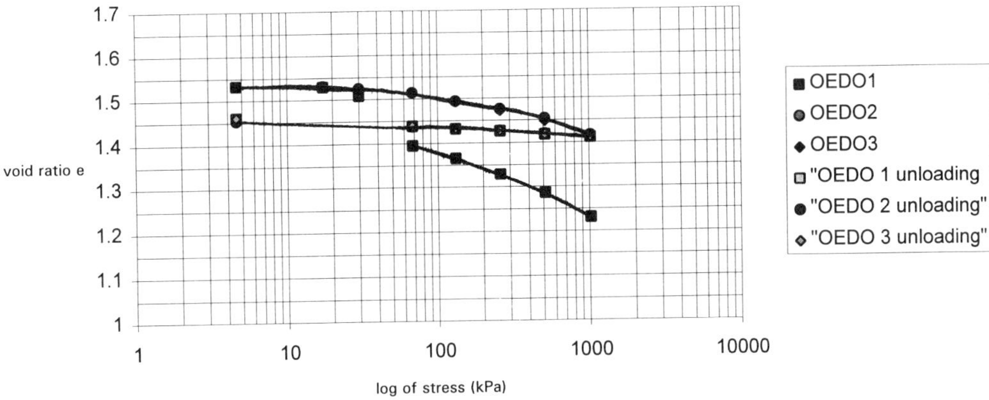

FIGURE 7 : oedometric test
void ratio - log of stress

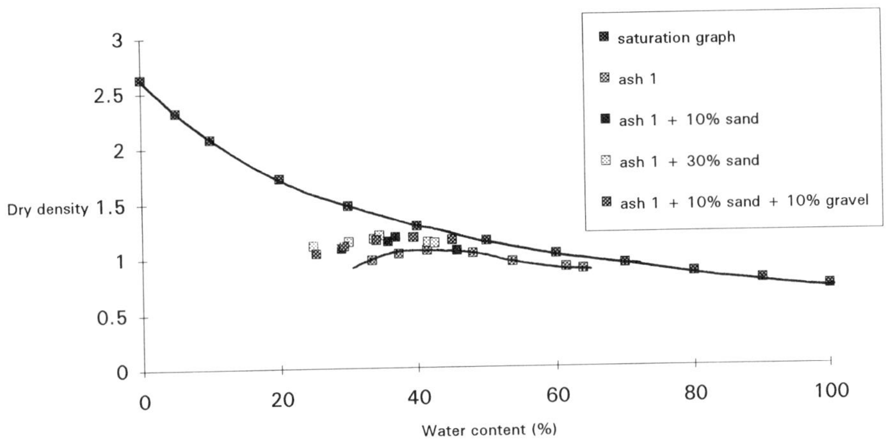

FIGURE 8 : PROCTOR TEST ON ASH 1

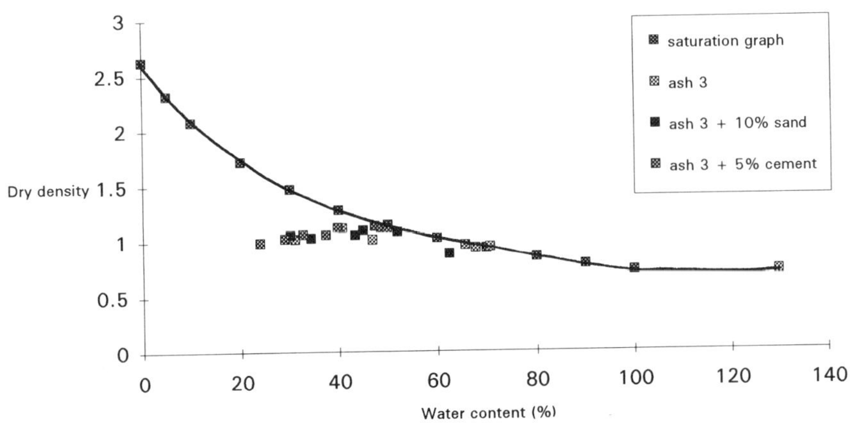

FIGURE 9 : PROCTOR TEST ON ASH 3

4. THE STUDY OF A POTENTIAL VALORIZATION OF MIXED ASHES AS SOIL.

Proctor tests have been carried out on different mixtures :

- 100% ash 1 ; 90% ash 1 + 10% sand ; 70% ash 1 + 30% sand and 80% ash 1 + 10% sand + 10% fine gravel (figure 8)
- 100% ash 3 ; 90% ash 3 + 10% sand and 95% ash 3 + 5% cement (figure 9)

We can see on these graphs that with ashes mixed with sand, fine gravel and cement, optimum water contents decreases up to 10% and dry density (γ_d/γ_w) increases slightly (up to 1.25). These results are therefore very encouraging.

CBR tests carried out after Proctor tests on ash 3 with sand and with cement have enabled to prove that the addition of sand or cement allows to improve the lift of ashes. Adding cement (figure 2) permits to obtain water insensitive soil and to decrease the moisture content.

Shearbox tests on mixtures have obtained the following results :

ash 3 + sand :

Shear strength = 56 kPa and $\varphi = 27°$

ash 3 + cement :

Shear strength = 0 kPa and $\varphi = 50°$

Ash 3 + cement test is disrupted by setting effect from which interpretation is made difficult. It seems as if cement addition increased cohesion.

5. CONCLUSION

To finish with the account of our work, we came to the conclusion that :

The use of mud ashes from a purification station is quite conceivable in Public works. However this use will lead to technical restraints. The study of the physical features of the material has shown that it could be compared to very fine sand particles. What is more these ashes are very wet so they would very unexpectedly change their consistency in case of tiny variations in the water content.Thus these soils are normally out of use in this state and a temporary deposit or previous drainage can be certainly foreseen before the use.

This soil should be carried out at CBR water content.

Soil settlement study has unfortunately shown no stabilization but settlement remains very weak.

Thus ashes could be used in embankment either alone carefully or mixed with sand or gravel or cement.

They could be used for the layer of a road surface. In fact, coal fly ash with pratically the same characteristics has already been used in foundation layer (Gravel + cement made with 91% fly ash, 4% quicklime and 5% gypsum) [Andrieux..., 1992] and in base layer (85% gravel 0/20, 12% fly ash and 3% slaked lime) [Andrieux, 1978] [Gaspar, 1975].

To understand mud ashes in the presence of water better, it would be judicious to determine the CBR factor after immersion and to carry out triaxial tests on undrained samples in order to know if the characteristics of mud ashes remain acceptable.

It is to be hoped to make a full-sized test with a number of unfavorable conditions to evaluate the real behavior of these ashes. Lixiviation tests would also be necessary on this experimental embankment to know things about the risks of pollution.

Another study has been carried out concurrently in order to know the capacity of mud ashes to be incorporated in mortars or concretes. This one gives good results. So these mud ashes could also be used in the field of the building industry.

REFERENCES

ANDRIEUX P. - *Utilisation des cendres volantes en technique routière* - Colloque international sur l'utisation des sous produits et déchets dans le Génie Civil , volume II, p333-338 - Paris 28 au 30 novembre 1978.

ANDRIEUX and COLOMBEL - *Utilisation des cendres volantes en technique routière : les graves cendres volantes* - Valorisation des déchets et sous produits dans les travaux de Génie Civil - Paris 15 au 17 décembre 1992.

GASPAR L. - *Les cendres volantes et le laitier granulé dans la construction routière en Hongrie* - Bulletin de liaison du Laboratoire des Ponts et Chaussées n°86, p120-121, 1975.

Environmental Geotechnics, Kamon (ed.) © 1996 Balkema, Rotterdam. ISBN 90 5410 848 7

Viscoelastic model applied on plastic light-weight material blocks

H. El Ghoche & C. Coulet
Claude Bernard University, Lyon, France

J. Perrin
INGEVAL Consulting Office, Veyssilieu, France

ABSTRACT : Plastbloc is a new process of light-weight embankments by using parallelepipedal blocks which are made up of plastic waste.
The experimental and theoretical studies of these blocks enabled us to elaborate a nonlinear viscoelastic behaviour law. This nonlinear model shows clearly the real behaviour, but it is difficult to use it because it needs the identification of a great number of parameters.
The purpose of this paper is to simplify the model by linearizing it into a particular domain of stresses.
This linearization was carried out in two ways :
- adoption of linear forms of the elastic and the delayed deformations according to stress.
- adoption of a linear form of the elastic deformation and a constant delayed one according to stress (linear viscoelasticity).The comparison between the experimental and theoretical results shows that the linear model describes the reality as well as the nonlinear one, in a field of stresses which can reach 200 kPa.

1 INTRODUCTION

Industrial countries produce increasing quantities of waste, among which plastics represent 20% by volume. This plastic waste creates serious environmental and economical problems.

10 millions tons are produced in the EEC countries every year (Ademe, 1994). The annual French production is 2 millions tons (Ademe, 1993).

Processes by either dumping or incineration are proven expensive.

On the other hand, the new installations and great public works oblige us to confront several technical problems, such as construction on poor caracteristics soils, that the Geotechnics is called to find efficient solutions.

The new patent process "PLASTBLOC" of light-weight embankments, made up of plastic waste, contributes seriously to the elimination of this waste and brings good geotechnical solutions.

2 PLASTBLOC PROCESS

Plastbloc is a new process which can be used to build light-weight embankments. It reuses waste plastic material of several origins which has been compacted in order to form blocks of $1 m^3$ in volume and a density of 0.4, 0.5 or 0.6 (Fig. 1).

Fig. 1 : Photo of a real block.

The raw material is constituted mainly by polyethylene films and bags and does not need any classification or treatment. But it has to be non pollutant.

Because the compression, during the fabrication, is unidirectional, this material is presented aproximatelly as a stack of identical layers. So, we can consider it an axial orthotropic material. The revolution axis is identical to the unidirectional compression during the fabrication.

The interesting lightness of blocks is created by the emprisoned air volume between the plastic sheets. Then, the material is alveolar with open or closed cells (EL GHOCHE, 1992).

3 EXPERIMENTAL STUDIES

The experimental studies of the behaviour of the material began by compression and creep tests on full size blocks and completed by several kinds of tests on reduced size samples.

The dimensions of real blocks or samples were carefully chosen in order to neutralize the microscopic details and be applied the Continum Mechanics principles.

The carried out tests are :
- compression and creep tests corresponding to the revolution axis direction of material.
- cyclical and monodirectional tests with constant deformation velocity.
- biaxial tests.

For all kind of test, blocks of the same or different densities were tested according to a path of applied stresses by steps.
This approach helps us to understand further the behaviour of this discontinuous and anisotropic material.

The examination and analysis of the experimental results enabled us to identify the material behaviour and to find all necessary caracteristics and parameters in order to begin the theoretical modelling (EL GHOCHE, 1994):

- The material behaviour is a viscoelastic and nonlinear one. The deformation variation according to compression and creep tests under constant stress can be considered linear according to time logarithm (Fig. 2) :

$$\varepsilon = \varepsilon_0 + A\ln(t/t_0)$$

ε : unidirectional deformation

ε_0 : instantaneous deformation
A : slope of the deformation straight line according to time logarithm
t : time

t_0 : time unit

In this scale, the slope and the ordinate at origin (instantaneous deformation) are variable according to the initial density and applied stress (Fig. 3a and 3b).
- The behaviour is very dependant on the initial density (Fig. 4).
- Many complex rheological phenomena were observed (hardening, aging, recovery,...).

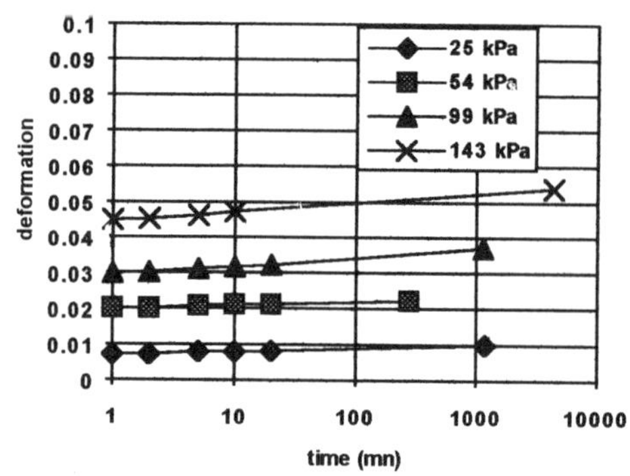

Fig. 2 : Deformation curves according to logarithm time. Initial density : 0.5.

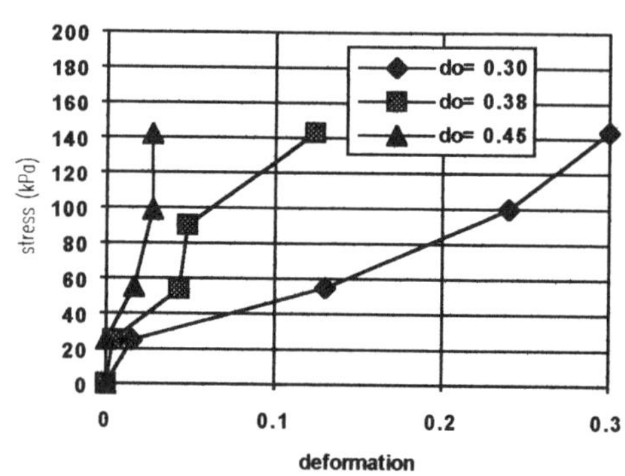

Fig. 3a : Stress - elastic deformation curves for different densities.

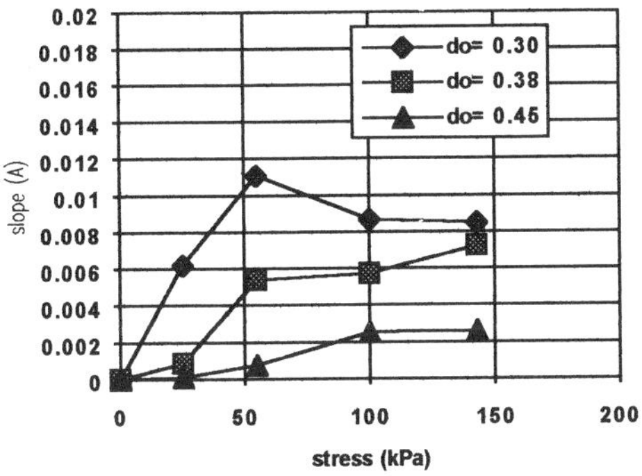

Fig. 3b : Slope(A) - stress curves for different initial densities (do).

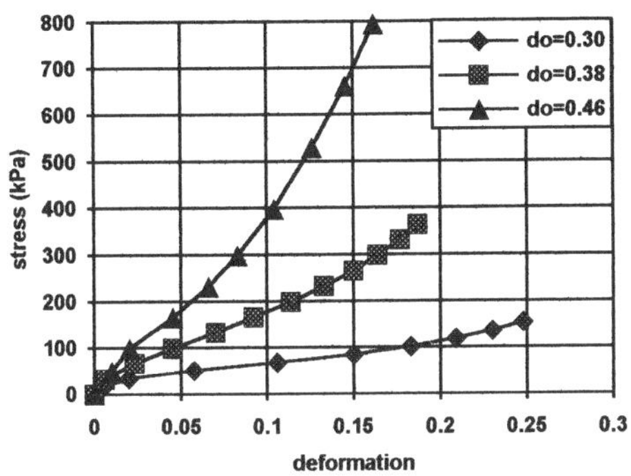

Fig. 4 : deformation-stress curves with constant deformation velocity ($\varepsilon^{\bullet} = 0.017 mn^{-1}$).

- The proper plastic deformation can be neglected in comparison with the total one. Then, the deformation is the direct result of air volume variation by compressibility or expulsion.

4 THEORETICAL STUDIES

The physics reality shows that the material is anisotropic and discontinuous. The only way that enables us to apply the Continum Mechanics principles is to assume that the material is homogeneous and continous. Furthermore, we admit that there is neither no slippery nor ungluing between the elementary plastic sheets. Then, the equivalent body is a continous, axial orthotropic and viscoelastic one (MANDEL, 1978).

In the general case, the behaviour can be presented by the following tensorial relation :

$$f(\sigma, \sigma^{\bullet}, \varepsilon, \varepsilon^{\bullet}, \mathbf{D}) = 0$$

σ : stress tensor

$\sigma^{\bullet}$: velocity stress tensor

ε : deformation tensor

$\varepsilon^{\bullet}$: velocity deformation tensor

D : fourth order tensor

But in this paper we study the uniaxial behaviour according to the revolution axis. In fact, the use of Plastbloc process up to now has been considered interesting only in this particular direction.

The uniaxial and nonaging behaviour can be described by the following analogical model :

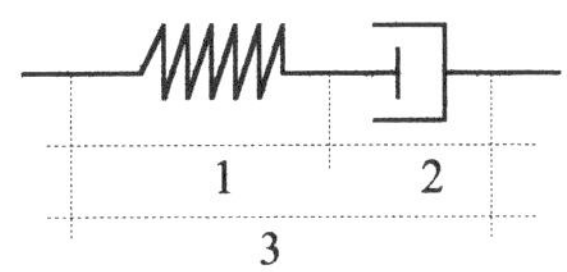

(1) : ε_0 ; (2) : ε_1 ; (3) : ε

$$\varepsilon = \varepsilon_0 + \varepsilon_1$$

$$\sigma = E\varepsilon_0 = \eta\varepsilon_1^{\bullet}$$

The spring presents the elastic behaviour and the dash-pot describes the viscous one.

The experimental results enabled us to elaborate the behaviour law. After necessary analysis and calculations, the model is presented in a nonlinear differential equation (EL GHOCHE, 1994):

$$\varepsilon^{\bullet} = \left(\frac{d_0 \times A}{\left(d_p(\sigma^* + B) - A\right)^2} - m\frac{(d_p - d_0)}{d_0}(\sigma^* + 1)^{(m-1)} \times K \right)\sigma^{*\bullet} + \frac{(d_p - d_0)}{d_0}\left[(\sigma^* + 1)^m - 1\right] Exp\left\{ - \left[\frac{\varepsilon - 1 + \frac{d_0(\sigma^* + B)}{d_p(\sigma^* + B) - A} + \frac{(d_p - d_0)}{d_0}\left[(\sigma^* + 1)^m - 1\right] \times K}{\frac{(d_p - d_0)}{d_0}\left[(\sigma^* + 1)^m - 1\right]} \right] \right\}$$

d_0 : initial density

d_n : plastic density

$\sigma^* = \dfrac{\sigma}{P_a}$, P_a is the atmospheric pressure

$m = a * d_0 + b$

A, B, K,a,b : parameters

This model, elaborated in small deformations, does not take into account the hardning due to the evolution of the density during loading. It describes very well the reality, but it's complicated because the differential equation is complex and needs the identification of a large number of parameters.

In fact, it's necessary to simplify this model and to present a simple and acceptable implement in a particular field of stress up to 200 kPa.

5 SIMPLIFIED MODEL

The simplification of the nonlinear model was carried out in two ways :

- Adoption of linear forms of the elastic and the delayed deformations according to stress.
- Adoption of a linear form of the elastic deformation and a constant delayed one according to stress. In other words, the linear viscoelasticity is applied.

In any case, the initial density is taken as a parameter and the model is presented as following :

$$\varepsilon^{\bullet} = (\frac{d_p - d_0}{(d_0)^n}) \times \frac{1}{P} \times \sigma^{\bullet} + A \times \exp(\frac{-(\varepsilon - \varepsilon_0)}{A})$$

$$\varepsilon_0 = \frac{d_p - d_0}{(d_p)^n} \times \frac{1}{P} \times \sigma$$

ε_0 : instantaneous deformation

$\varepsilon^{\bullet}$: unidirectional deformation velocity

A : slope of the deformation straight line according to time logarithm

$n = a*d_0 + b$

$m = c*d_0 + e$

d_0 : initial density

In the first case, the slope "A" of the deformation straight line in logarithmic scale time under constant stress is taken as following :

$$A = \frac{d_p - d_0}{(d_0)^m} \times \frac{1}{p} \times \sigma$$

In the second one, the value of the same slope "A" is introduced in the model as the average of all values calculated under several stresses, but for one initial density :

$$A = \frac{d_p - d_0}{(d_0)^{m'}} \times \frac{1}{p}$$

$m' = c'*d_0 + e'$

The experimental results enabled us to identify all necessary parameters :

$n = -17.2*d_0 + 4.3$

$m = -27.0*d_0 + 5.0$

if $d_0 \leq 0.5$

$m' = -0.04 *d_0 + 0.02$

if $d_0 > 0.5$

$m' = -0.001 *d_0 + 0.001$

5.1 Validation

The comparison of theoretical and experimental results was possible by numerical resolution of the differential equation. We used the fourth order Runge-Kutta method and introduced into the model a path of applied stresses by steps. The loading change was taken as a linear variation according to time during one minute (Fig. 5).

The concordance between theoritical and experimental results is good and the two simplified forms of model give us reasons of satisfaction in comparison with the reality and the nonlinear model (Fig.6a, 6b, 7a, 7b, 8a, 8b).

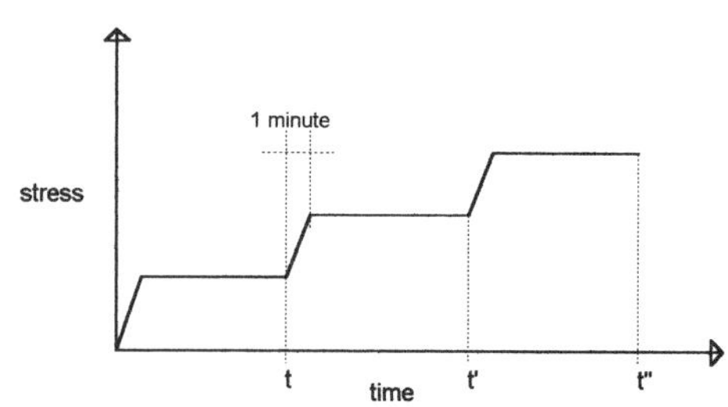

Fig 5 : path of applied stresses introduced in the model

Note : exp : experience.
nl.th. : non linear theory
lls.th. : linear elasticity and linear slope theory
lcs.th. : linear elasticity and constant slope theory

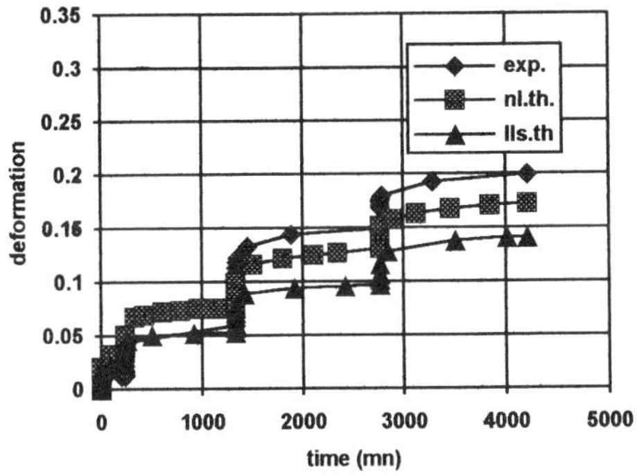

Fig. 6a :Deformation-time curves, do=0.384.Path of applied stresses is : 25-50-100-145 kPa.

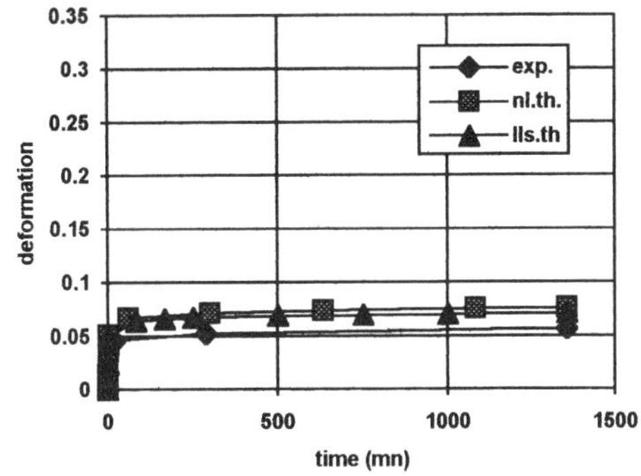

Fig.7b : Deformation-time curves, do=0.447. applied stresses is : 145 kPa.

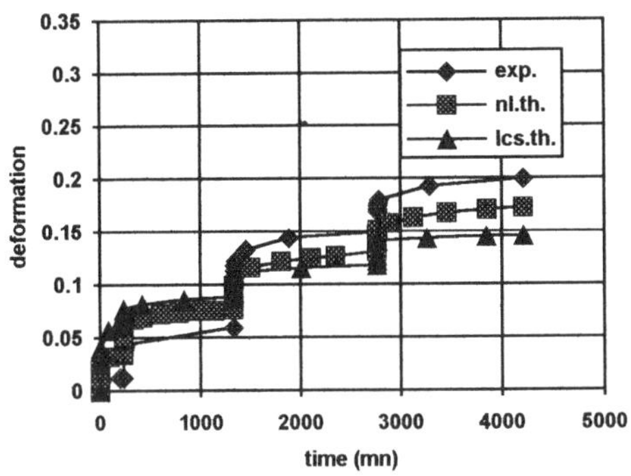

Fig.6b : Deformation-time curves, do=0.384.Path of applied stresses is : 25-50-100-145 kPa.

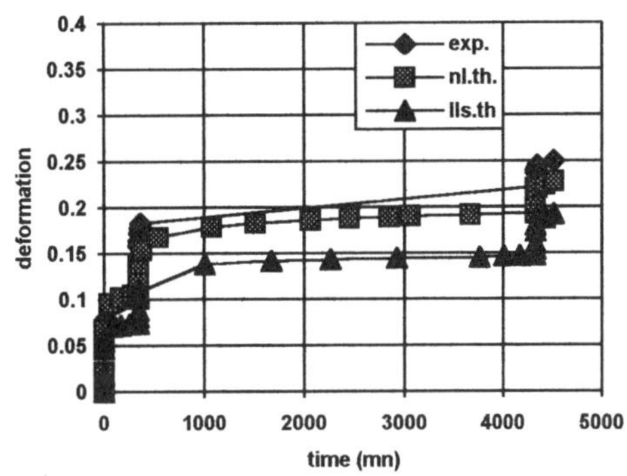

Fig.8a : Deformation-time curves, do=0.354.Path of applied stresses is : 50-100-145 kPa.

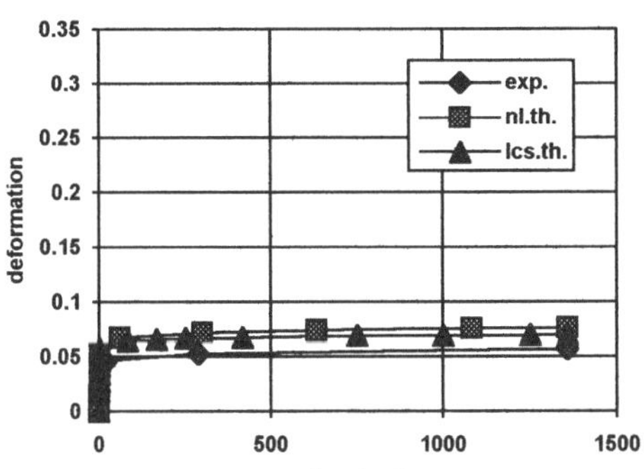

Fig.7a : Deformation-time curves, do=0.447. applied stresses is : 145 kPa.

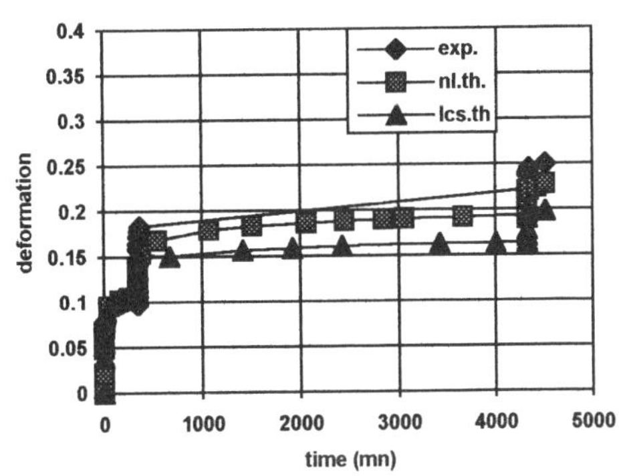

Fig.8b : Deformation-time curves, do=0.354.Path of applied stresses is : 50-100-145 kPa.

6 CONCLUSION

In a particular field of stress up to 200 kPa, the simplified model in its two forms gives good satisfaction. The results are acceptable and the difference between theoretical and experimental ones does not overstep the bounds of uncertainty both in physical measurments and caracteristics variability due to the material anisotropy.

However, and in several cases of application, this simplified model helps us to conceive and design structures by the use of this new process. Indeed, many projects were conceived and realized in France.

REFERENCES

1- Notes de Conjoncture des Activités aux Déchets. Données et Références. Ademe, Décembre 1993.

2- Les Déchets en Chiffres. Ademe, 1994.

3- EL GHOCHE H. - "Experimentation and Modelling of Compressible Plastic Materials for Light-Weight Embankments. Determination of a Nonlinear Viscoelastic and Axial Orthotropic Model". Ph.D. thesis report. Claude Bernard University.Lyon - France. December 1992. 274 pages.

4- EL GHOCHE H., COULET C., CAMBOU B. - "Experimentation and Modelling of Compressible Plastic Materials for Light-Weight Embankments. Elaboration of a Nonlinear Viscoelastic and Orthotropic Model. Journal " Revue Française de Géotechnique".N° 66, first semester 1994.

5- MANDEL S.;" Propriétés Mécaniques des Matériaux. Rhéologie - Plasticité". Editions Eyrolles, 1978.

Environmental Geotechnics, Kamon (ed.) © *1996 Balkema, Rotterdam. ISBN 90 5410 848 7*

Utilisation of electric arc furnace steel slag in road construction

Ann-Marie Fällman & Jan Hartlén
Swedish Geotechnical Institute, Linköping, Sweden

ABSTRACT: Pre-treated steel slag from scrap based electric arc furnace production of low alloyed steel was studied in order to make an environmental assessment for utilisation in road construction. Laboratory tests and field tests (test road) were carried out on the 0-300 mm and 11-300 mm fractions. The major concern was the leaching of chromium. The exclusion of fine particles in the slag increased leaching of chromium and vanadium in the test road due to changes in factors controlling leaching. Laboratory experiments had indicated an improvement by screening. A paved road using the steel slag fraction 0-300 mm in the subbase is also reported.

1. INTRODUCTION

Slag from scrap based low alloyed steel production in electric arc furnaces is utilised for construction purposes in many countries. It is used both as road base material and as aggregates in pavement (Ye et al. 1995). In Sweden, the material is landfilled due to environmental concerns and lack of regulations regarding its environmental properties for utilisation purposes. However, the utilisation of slags and ashes in construction works, when safe environmental properties have been established, is of high priority since this will save not only geological materials but also landfill space.

In 1992, a study was initiated with the purpose of investigating the environmental properties of processed steel slag at a scrap based electric arc furnace steel plant in Sweden. The annual amount of slag produced was 70,000 tonnes. The study was required in order to apply for a concession regarding the total environmental permit for the plant from the Swedish National Franchise Board for Environmental Protection. The environmental investigation consisted of four test levels: total composition, laboratory leaching tests, lysimeter tests and a test road. The physical properties of the material were also tested.

It was concluded from the laboratory investigation that the major environmental concern regarding the slag was the concentration of chromium in the leachate due to the toxicity of Cr(VI) (Fällman 1992).

In this paper, the results from the laboratory study and test road are presented and discussed, with the focus being on chromium, vanadium and lead. Special attention is given to the environmental effect due to the pretreatment of the slag by screening off the finest particles prior to utilisation. The environmental impact assessment and the full scale road design will also be presented.

2. MATERIALS AND METHODS

2.1 *Materials*

The processed slag suitable for use is part of the stream of residues from scrap based production of low alloyed steel in an electric arc furnace. The slag was emptied below the furnace, excavated, and transported while still hot to an intermediate storage area for cooling by water sprinkling. The slag was screened in combination with magnetic separation in two steps, resulting in two slag fractions: non-magnetic 0-300 mm from the first magnetic separation step and 0-70 mm from the second magnetic separation of the magnetic minus 70 mm fraction from the first separation step. The magnetic fractions were recycled to the process. Only the non-magnetic slag fraction of 0-300 mm was used in this study. Half the test road was built with the 0-300 mm material and the other half with the 11-300 mm fraction. The <11 mm fraction sieved off comprised one third of the 0-300 mm material.

The total composition was analysed with ICP-AES and MS (Inductively Coupled Plasma Emission Spectrometry and Mass Spectrometry). The digestion tech-

Table 1. Total composition of steel slag.

Substance	Unit	Laboratory tests	Test road 2)
Al	%	2.29	2.19
Ca	%	23.2	22.1
Fe	%	25.0	24.2
K	%		0.046
Mg	%		4.52
Mn	%		3.93
Na	%		0.037
P	%		0.46
S	%		0.126
Si	%	6.05	5.77
Ti	%		0.28
As[1]	ppm	3.8	5.3
Ba	ppm		728
Be	ppm		<1.2
Cd[1]	ppm	0.38	0.45
Co[1]	ppm		5.8
Cr	ppm	7340	7760
Cu[1]	ppm	142	166
Hg[1]	ppm		<0.41
Mo	ppm		20.6
Nb	ppm		198
Ni[1]	ppm	31	45
Pb[1]	ppm	7.0	21.5
Sn	ppm		17.5
Sr	ppm		178
V	ppm	1590	1210
W	ppm		410
Zn[1]	ppm	169	244
Zr	ppm		55.2

1) Digestion in Teflon bomb in nitric acid.
2) Fällman and Hartlén (1994)

niques were lithium metaborate melt (ASTM D 3682-91 but with $LiBO_2$ instead of $Li_2B_4O_7$) for the main components, and nitric acid in Teflon bomb for certain trace elements (SS 02 81 83). The total composition of the materials in the laboratory study and in the test road (Fällman and Hartlén 1994) is shown in Table 1.

2.2 *Sampling*

A representative sample for laboratory testing was selected by setting aside every sixth bucket of slag during five days of production. The total amount of 65 tonnes of slag was screened and samples representing each of the five days were selected.

The test road was built in December 1992, and the same material was sampled for laboratory study.

2.3 *Physical properties*

The grain size distribution of the 0-300 mm slag was determined by sieving, as shown in Table 2.

Table 2. Particle size distribution.

Particle size, mm	Fraction, %
< 0.1	2
0.1 - 1	2
1 -10	29
10 - 100	52
100 - 300	15

The specific gravity of the slag was 3.76 g/cm^3 for the 8-11.2 mm and the 11.2-16 mm fractions. The Swedish abrasion value was 1.78 (FAS Method 231-95) and the Nordic ball mill value was 14.0 (FAS Method 259-95). These parameters were determined by The Swedish National Road and Research Transport Institute (Eriksson and Höbeda, 1992) who commented that the ball mill value was highly related to the abrasion value due to porous particles being broken down into finer material. If the slag is to be used in a surface course, the porous particles should be eliminated by thorough crushing. These physical properties showed that the slag could be used both as base course and subbase material (Eriksson and Höbeda, 1992).

2.4 *Leaching tests*

The potential for leaching was tested based on NEN 7341 (availability test), where leaching is carried out in two steps. Both steps are carried out at a leachate to solid ratio (L/S) of 100 for 3 hours. In the first step the pH was held constant at a value of 7 and in the second step at a value of 4. The slag was ground to <125 μm. This test was also carried out on <4, 4-31.5 and >31.5 mm fractions of the material. The oxidised availability for leaching of the test road material was also determined. In this test, also based on NEN 7341, oxidised conditions are controlled with the addition of hydrogen peroxide, H_2O_2 (30 %), until a stable value of the redox potential is obtained. Leaching in relation to L/S was carried out in 0.9 m high by 0.1 m diameter columns. Standard leachant (de-mineralised water acidified to pH=4 by nitric acid) was added with inflow in the bottom of the columns. The flow rate was approximately L/S=0.1 per day maintained by a peristaltic pump. The leaching tests are further described in Fällman and Aurell (1996).

2.5 *Analysis of leachates*

The leachates from laboratory column tests were analysed for metals on AAS-HGA (Atomic Adsorption Spectrophotometer with graphite furnace), SO_4 gravimetrically, and Cr(VI) by calorimetric method after filtration on 0.45 μm. Leachates from availability tests were analysed for metals on ICP-AES and ICP-MS.

Leachates from the test road were collected every month. At the same time the samples for Cr(VI) and pH were taken directly from the tipping bucket. The flow proportional test samples of leachate were before

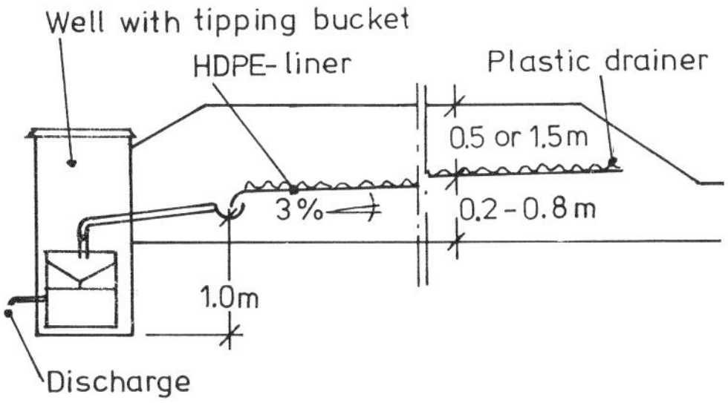

Figure 1. Leachate collection unit in the test road.

the monthly collection stored in bottles where contact with air could not be excluded. These leachates were analysed after digestion with HNO_3 on ICP-AES for metals and S-tot, except for Pb and Cd which were analysed on AAS-HGA. Cr(VI) was analysed separately without digestion.

2.6 *Test road*

A test road was built with a road base of steel slag in a 50 m long section of a forest road, 5 m in width. The average height of the fill was 2 m. The road was not paved. The test road was divided into two sections: one with the original slag of 0-300 mm and a second of grain size 11-300 mm where material <11 mm had been separated. Two leachate collection units were placed in each section at 0.5 and 1.5 m below the road surface. The principles for the leachate collection unit are shown in Figure 1. The leachates were collected on an HDPE membrane of 1x4 m^2 at 3 % slope covered with a draining geotextile. The membrane was connected to a plastic drain at the short side from which the leachate flowed into a tipping bucket flow-measuring device with flow proportional leachate sampling. Collection of leachate was carried out from February 25, 1993 to September 27, 1994.

3. RESULTS AND DISCUSSION

3.1 *Laboratory and test road leachates*

The potentially leachable fraction as measured in the availability test on the total sample and on the three fractions <4 mm, 4-31.5 mm and >31.5 mm showed that the leaching of chromium, aluminium, sulphur and magnesium decreased upon an increase in grain size, see Figure 2. Leaching from the smallest fraction was greater compared to the other two fractions for zinc, copper, lead and cadmium, which all showed similar results. The difference between the fractions was not as evident for the other substances. The apparent

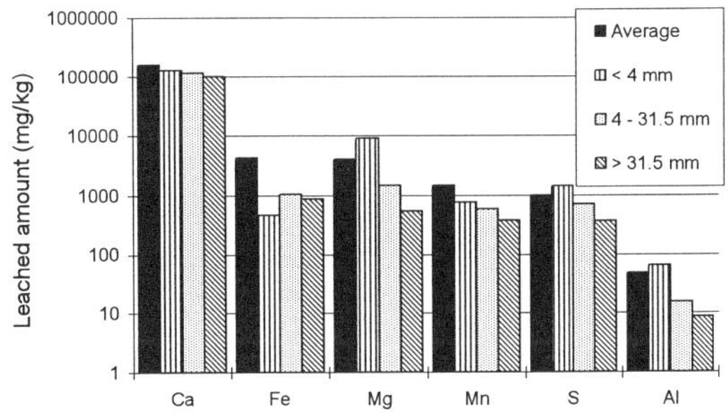

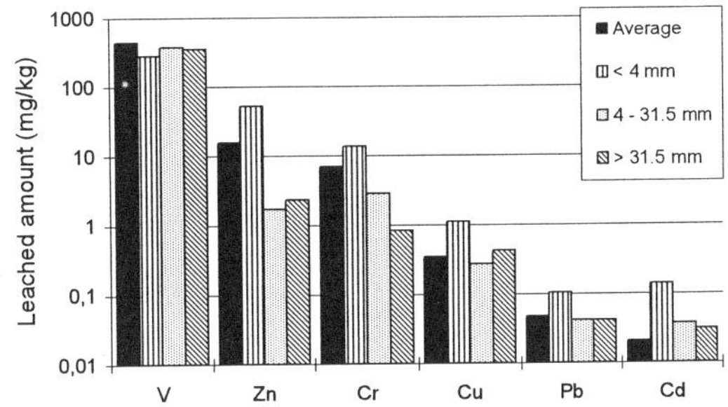

Figure 2 Potential for leaching (availability test) of steel slag screened to different grain sizes.

disagreement between the average and fraction values for vanadium, calcium and cadmium may be caused by the use of different laboratory samples. For iron and manganese, there may have been an oxidation of the fractionated samples.

The conclusion from these laboratory results was that screening of the material before utilisation to exclude particle sizes < 4 mm would improve the environmental properties. This was confirmed for copper in column tests with material of 4-300 mm and 0-300 mm respectively, where the concentrations in the leachates decreased more rapidly for the screened material than for the unscreened (Fällman 1992). However, no effect of the screening was seen for chromium in the same experiment (see Figure 5).

The pH values in the leachates from the test road only showed similarities with the column tests values of 11.4-12.4 (Fällman 1992, Fällman and Hartlén 1994) for the unscreened material. Lower and more varying pH levels were obtained from the screened part of the road (see Figure 3). A possible explanation was thought to be the greater exposure of the material to air in the screened portion where less fines were present, resulting in a lower water retention capacity of the material. Carbon dioxide in the air may react with calcium in the slag and form calcite, which has an equilibrium pH of approximately 8.4. This has been observed for other materials (see e.g. Comans and Meima, 1994).

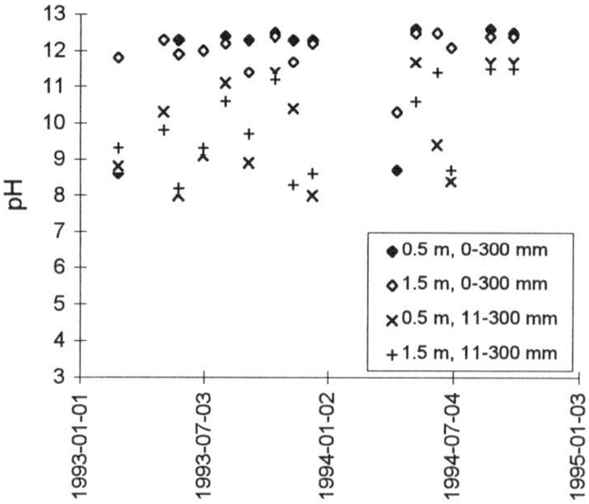

Figure 3 pH in the leachates from the four leachate collection units in the test road.

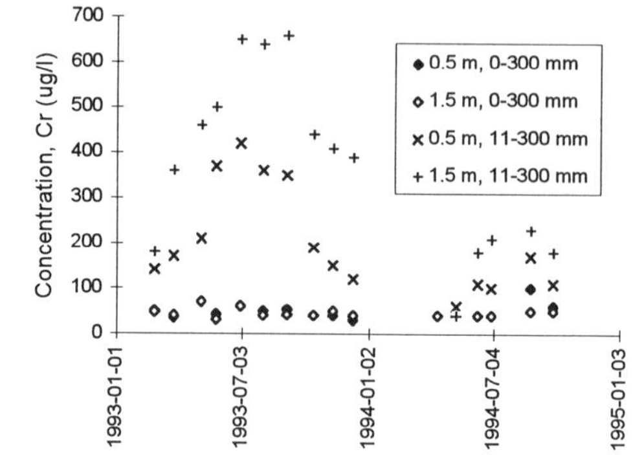

Figure 4 Concentrations of chromium in leachates from the four leachate collection units in the test road.

Differences in leachate composition between the screened and unscreened material were also observed in the test road. Concentrations of chromium in leachates were more than 10 times higher from the screened material than those of the unscreened material. After one year a decrease was observed, but the values from the screened material (see Figure 4) were still clearly higher. The leached amounts of chromium from the screened material (see Figure 5) were clearly separated from those obtained from the unscreened material and from the laboratory column experiments. The majority of the chromium (>75%) in the leachates from the test road were found to be in the oxidised form, Cr(VI). The conclusion was that the higher concentrations of chromium in the leachates from the screened material may be a result of increased oxidation. It was also concluded that the longer retention time, found in the lower leachate collection units, increased concentrations and leached amounts.

An increase was observed in the concentrations of vanadium from the screened part of the test road. In this case, the increase was greater than two orders of magnitude. By plotting the concentrations as a function of pH (see Figure 6) the increased solubility of vanadium upon a decrease in pH was revealed. The column test results were the best examples of this effect where pH decreased with increased L/S. The scatter of the results at the lower pH levels was caused by the depletion of easily leached vanadium with time. The low concentration points at low pH were values from leachates collected during snow melt when retention time was shortened. The release rate decreased with time and seemed to level off at approximately an L/S=0.1. This was not in agreement with the release observed in the column test, see Figure 7. Factors other than a decrease in pH may also control the leaching of vanadium.

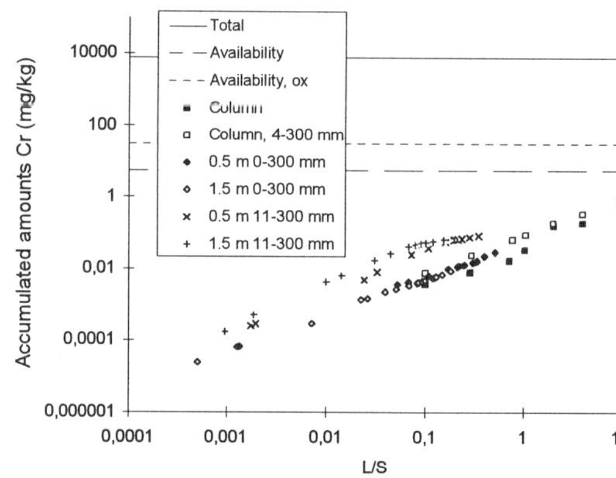

Figure 5 Leached amounts of chromium from the test road and column test in comparison with total content, availability test and column test from Fällman and Hartlén (1994).

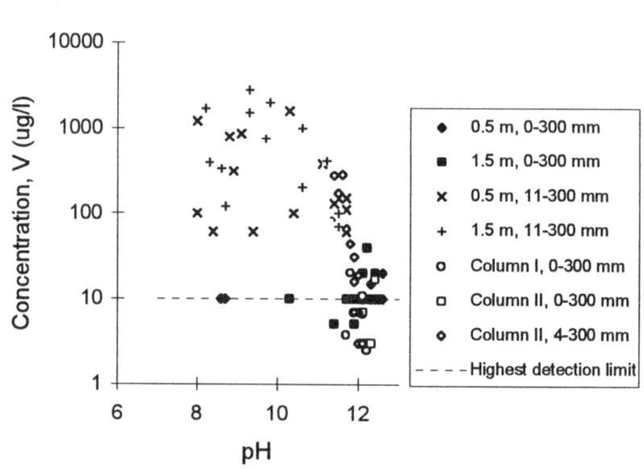

Figure 6 Concentrations of vanadium in leachates from the four leachate collection units in the test road and the columns as a function of pH.

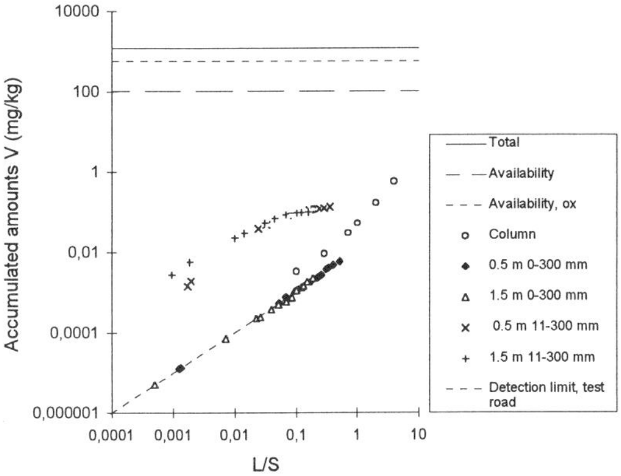

Figure 7 Leached amounts of vanadium from the test road in comparison with total content, availability test and column experiment from Fällman and Hartlén (1994).

The concentration of lead in the leachates from the screened material in the test road were constantly low or below the detection limit. In the unscreened material it varied considerably over time. By plotting the results in relation to pH (see Figure 8) it was shown that high pH increased the concentration of lead in the leachate. Similar but less evident patterns were found also for copper and zinc. All these substances had shown a high leachability in the availability test for the smallest grain size fraction and similar and smaller leachability for the other two fractions. In the screened material low concentrations were expected since there were no fine particles in this material, but no analysis was made on the resulting total contents after the screening.

The pH dependence, including high pH, in the leaching of these substances is known from studies on other

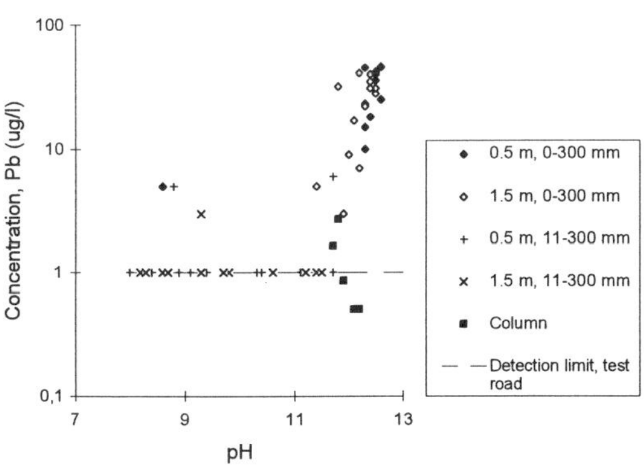

Figure 8 Concentrations of lead in leachates from the four leachate collection units in the test road and column test as a function of pH.

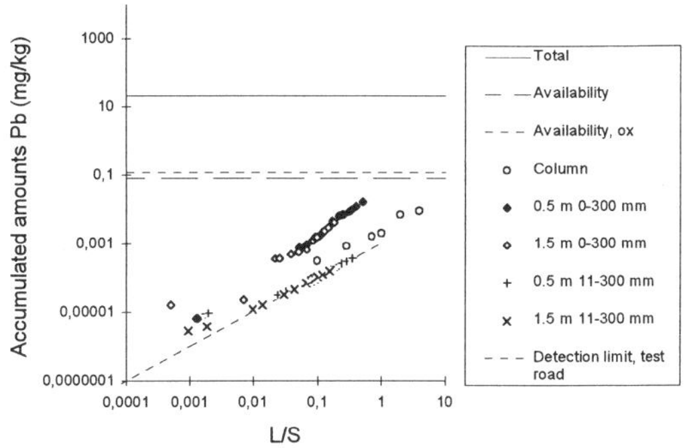

Figure 9 Leached amounts of lead from the test road in comparison with total content, availability test and column experiment from Fällman and Hartlén (1994).

residues (see e.g. de Groot et al. 1989 and van der Sloot et al. 1991) and in modelled systems (e.g. Rai and Zachara 1986).

The accumulated leached amounts of lead from the unscreened material approached the potentially leachable amounts at a higher rate than was obtained in the column test (see Figure 9). There was also a correspondence between the two levels of leachate collecting units, indicating equilibrium conditions.

3.2 *Full scale road construction*

Based on the environmental and physical properties, a permit was granted by the local County Council to use the screened steel slag in a 3.3 km long road for transportation to a public municipal waste disposal site and to an industrial landfill site.

The road has a width of 6.5 m of which 6.0 m has a 0.04 m asphalt pavement. Below this is a 0.08 m bitumen stabilised base course, a 0.08 m base course of crushed rock and a 0.42 m subbase of steel slag. Steel slag is only placed under the pavement area. Underlying subbase consists of crushed rock.

The road has not yet been opened. However, the experiences from the construction period have been good, especially considering the heavy transport of material on the compacted slag surface.

The effect of compaction energy was studied by varying the number of passes and the thickness of the slag layer. A test embankment was initially constructed of slag with a constant slope up to 0.8 m in height, compacted only on the surface. Based on plate load tests and radioactivity density measurements, four passes with a vibratory compaction roller was found to provide optimal compaction. However, based on the plate load tests, the modulus of elasticity from the second load cycle showed lower values than expected from natural aggregates. The performance of the road will be monitored to evaluate the long-term behaviour.

3.3 *Environmental impact assessment*

An environmental impact assessment was performed before approval was granted to build the road as described above. The laboratory and test road leachates showed that the crucial element was chromium and that the steel slag should contain the entire 0-300 mm fraction.

The following conservative assumptions were made for the environmental assessment:

- Slag placement under asphalt cover, resulting after ageing in a net infiltration of 10 mm/year.
- A net infiltration through the embankment slopes of about 400 mm/year from precipitation to which asphalt surface run-off is added. Approximately 10 % of this water comes into contact with the slag.
- Chromium concentration in the leachate of 50 μg/l.

The result of the environmental impact assessment indicated that approximately 27 g/km of chromium would be leached annually. An adsorption of the leached chromium may occur underneath the road especially as the soil contains organic matter, which reduces Cr(VI) to the less mobile Cr(III). Possible impact will be followed in the future with a control programme.

4. CONCLUSIONS

The environmental impact assessment based on the 0-300 mm steel slag gave results acceptable for a permit for utilisation as subbase under pavement in a 3.3 km road. The physical performance of the slag in a test road was acceptable.

The results from the study showed that the environmental improvement of the steel slag, found in the laboratory study by screening off the finest particles, was not reproduced in the test road leachates. The change in the grain size distribution also changed the factors controlling leaching: increased oxidation and decreased pH. Both these were functions of increased air intrusion into the road base. The effect was that leaching of chromium and vanadium from the screened part of the test road was significantly increased. In addition, strong pH dependency was seen for the leaching of vanadium and lead, but the results also indicated that additional factors also may control the leaching of these substances.

ACKNOWLEDGEMENT

This paper was based on work financed by Fundia Special Bar AB in the context of the establishment of a new disposal site and improvement of the company's residues for use as a construction product. The authors wish to express their gratitude for the opportunity to present the data.

REFERENCES

ASTM D 3682-91: Standard Test Method for Major and Minor Elements in Coal and Coke Ash by Atomic Adsorption.

Comans, R.N.J. and Meima, J.A. (1994): Modelling Ca-solubility in MSWI bottom ash leachates. In *Environmental Aspects of Construction with Waste Materials*, Studies in Environmental Science 60, pp 103-110, Elsevier, Amsterdam.

de Groot, G.J., Wijkstra, J., Hoede, D. and van der Sloot, H.A. (1989): Leaching characteristics of selected elements from coal fly ash as a function of the acidity of the contact solution and the liquid/solid ratio. *Environmental Aspects of Stabilization and Solidification of Hazardous and Reactive Waste*, ASTM STP 1033, pp170-183.

Eriksson, L and Höbeda, P. (1992): Utlåtande. Laboratorieärende. Stålslagg. VTI-nr 9224101. Sw. Nat. Road and Transport Inst. Linköping.

FAS Method 231-95. Mineral aggregates. Determination of abrasion value. Different from BS 812. The Swedish Asphalt Pavement Association, Stockholm.

FAS Method 259-95. Mineral aggregates. Determination of abrasion value. The Swedish Asphalt Pavement Association, Stockholm.

Fällman, A-M. (1992): Undersökning av lakegenskaper hos sorterad stålslagg. Fundia Steel AB. SGI-Dnr 2-428/92.

Fällman, A-M. and Aurell, B. (1996): Leaching tests for environmental assessment of inorganic substances in wastes, Sweden. The Science of the Total Environment, 178, pp71-84.

Fällman, A-M. and Hartlén, J. (1994): Leaching of Slags and Ashes - Controlling Factors in Field Experiments versus in Laboratory Tests. In *Environmental Aspects of Construction with Waste Materials*,, Studies in Environmental Science 60, pp 39-54, Elsevier, Amsterdam.

NEN 7341: Leaching characteristics of building and solid waste materials - Leaching tests - Determination of inorganic components for leaching (Availability test). Draft June 1992.

Rai, D. and Zachara, J.M. (1986): Chemical Attenuation Rates, Coefficients and Constants in Leachate Migration. Volume 1: A critical Review. EPRI EA-3356, Volume 1, Electrical Power Research Institute, Palo Alto.

SS 02 81 83 (1986): Metal content of water, sludge and sediments - Determined by flameless atomic absorption spectrometry - Electrothermal atomization in a graphite furnace - General principles and guidelines. Swedish Standards, Stockholm. (In Swedish).

van der Sloot, H.A., Hoede, D. and Bonouvrie, P. (1991): Comparison of different regulatory leaching test procedures for waste materials and construction materials. ECN-C—91-082, Netherlands Energy Research Foundation (ECN), Petten.

Ye, G., Burström, E. and Fällman, A-M. (1995): Utilisation and Stabilisation of Steelmaking Slags. Survey, AFR-Report 57, Stockholm.

Environmental Geotechnics, Kamon (ed.) © 1996 Balkema, Rotterdam. ISBN 90 5410 848 7

Geotechnical investigation on steel slags mixtures for landfills embankments construction

V.N.Ghionna
University of Reggio Calabria, Italy

S.Pedroni
Enel Cris, Milano, Italy

P.Tenani & S.Veggi
Studio Geotecnico Italiano, Milano, Italy

ABSTRACT: This paper reports on an experimental research carried out on steel slags (SS) produced at the ILVA steel mill in Taranto (Italy). The aim of the research was to investigate the possibility of using the SS for landfills construction, alone and mixed with calcareous material (CM) extracted from an adjacent limestone quarry. In the first phase of the research several mixtures of SS and CM were examined and tested according to the following laboratory tests: modified Proctor compaction (MPC) tests, hydraulic conductivity (HC) tests, isotropically consolidated drained triaxial (TX-CID-CL) tests.
In the second phase a real scale field trial embankment of steel converter slags (SCS) was constructed and the compaction procedures were defined through a large number of in situ density measurement (ID) tests and plate loading (PL) tests. Furthermore, the embankment was accurately instrumented to study the deformations and fissures patterns induced in the long period by the swelling that occur in SS constructions.

1. INTRODUCTION

This paper reports on an experimental research carried out on the steel slags (SS) produced at the ILVA steel mill in Taranto (Italy). The aim of this research is to investigate the possibility of using the SS for landfills construction (confinement embankments and top covers), alone or mixed with other materials.

The large quantity of SS produced at the ILVA steel mill in Taranto (approximately 800.000 m^3 per year when the research was started) causes big problems of disposal and elimination.

This research was undertaken in the context of a large project of land reclamation concerning a 30 m deep limestone quarry used to extract the limestone for the steel-casting process at the ILVA steel mill in Taranto.

The project required the construction of several landfills for both urban and industrial wastes for an overall storage capacity of 3.900.000 m^3. All waste fills were confined by huge perimeter embankments having a global volume of 2.000.000 m^3.

Due to the increasing difficulties in finding large volumes of traditional structural fill materials, it was decided to investigate the possibility of using the SS produced at the adjacent steel mill, which had the advantage of being available at a very low cost.

Large quantities of natural cohesive materials were also needed for the construction of the impermeable top covers of the landfills.

The availability of such materials was even more difficult and more expensive than that of the structural materials. Consequently, it was considered very useful to find an alternative solution based on the use of the SS also for the top covers of the landfills.

Studies concerning the use of the steel slags as construction materials have been extensively documented in the literature (Kergoet, 1987; Dron, 1984). The most frequent applications have concerned road subbases, railway ballasts, and foundation bases (Brunelot, 1981; Castan, 1987; Chalancon et al., 1986; Cisti et al., 1984; Fontaine, 1987) but also examples for soil improvement have been reported, such as stone columns (Kernoa and Marchadier, 1986) and heavy tamping (Kernoa and Marchadier, 1981).

However, the studies performed have shown the trend of the SS to undergo a significant swelling in the presence of an increasing pore water content.

The swelling is caused by the hydration of the calcium and magnesium oxides which becomes particularly significant when the water shows an alkaline behaviour (Dron, 1984).

In spite of the process being very slow, the

swelling can result in heavy consequences for the overlying structures (Morales, 1993).

The swelling potential of the SS is directly related to the chemical composition and nature of the minerals but also to the chemical reactivity of the water (pH) and to the fineness and morphological properties of the particles. A possibility to reduce the swelling is given by the use of the steel slags mixed with inert materials such as gravel or sands.

The hydration of the calcium oxide is followed by precipitation and subsequent crystallisation of hydrated calcium silicate. These reactions provide an increase of the strength of the SS in the long term (hardening) which is useful in many applications and represents the main advantage when using the SS.

The hardening processes can be significantly enhanced by even a moderate addition (< 10% by weight with respect of the SS) of some basic additives, such as free lime, phosphogypsum, etc..

According to the aforementioned considerations, preliminary studies were necessary in order to define the most appropriate mixtures of materials to be used for the given problem.

Such studies included:

a) a comprehensive characterisation of both the steel slags and the inert materials (chemical composition, nature of the minerals, physical and mechanical properties);
b) the selection of an appropriate additive and the preparation of several mixtures steel slags / additive / inert materials with different percentages of the components;
c) laboratory tests to determine the mechanical properties of the mixtures both in the short term and long term;
d) an investigation on the compaction procedures of the mixtures, which included both laboratory and trial embankments tests;
e) long term observation of the behaviour of the trial embankments.

For the considered design problem the main concerns related to the use of the SS were a) the internal stability of the landfill embankments, b) the maintenance of the access road running on the top of the embankment, c) the integrity of the lateral impermeable composite liners in spite of the swelling deformations of the embankment, d) impermeable top covers sufficiently flexible to tolerate the differential settlements caused by the consolidation processes of the waste fills.

Another important aspect which had to be taken into account was the request from ILVA to use as much SS as possible both alone and mixed with the inert material extracted from the adjacent limestone quarry.

The experimental research was carried out in two phases. The first one was performed in the laboratory and was addressed to the selection of a suitable mixture to be tested with the field investigation. The second one was carried out on a real scale field trial embankment and was devoted to the definition of the construction procedures and to the observation of the in situ long term behaviour of the mixture.

The research will be continued in order to investigate the possibility of using a mixture of SS and bentonite as construction material for the top covers. At the present moment, only the effects of adding small percentages of bentonite to the calcareous material have been studied.

2. TESTED MATERIALS

Two types of steel slags (SS) were tested, one coming from the blast furnace and quickly refrigerated by hydraulic jets (BFS), and the other coming from the steel converter and slowly refrigerated in air (SCS).

The two types of SS mainly differ for the grading characteristics which correspond to a poorly graded gravel for the SCS and to a rather uniform medium to coarse sand for the BFS (Fig. 1).

Other differences concerning the chemical composition are of minor importance (Table 1).

Both SS are predominantly constituted by CaO, SiO_2 and Al_2O_3; however the SCS have a little higher content of CaO and a little lower content of SiO_2.

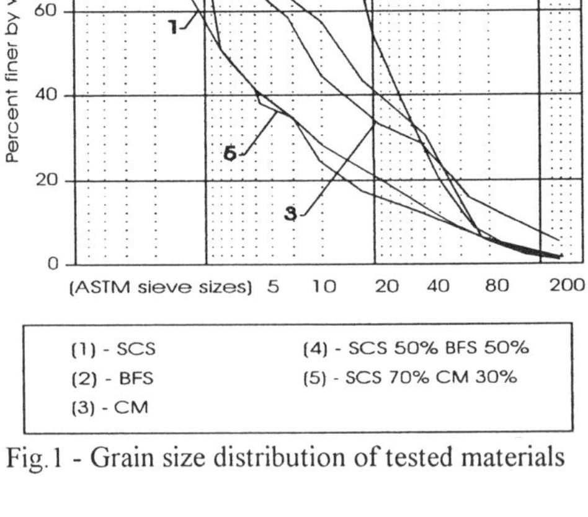

Fig.1 - Grain size distribution of tested materials

Table1 - Chemical composition of the slags

	CaO %	Al_2O_3%	SiO_2 %	MgO %	FeO, MnO
SCS	50 to 60	15 to 25	10 to 15	5 to 10	< 1
BFS	44.1	11.8	35.8	5.4	2.2

Table2 - Tested materials

n°	symbol	designation	description
(1)	—o—	SCS	steel converter slags
(2)	—△—	BFS	blast furnace slags
(3)	—□—	CM	calcareous material
(4)	—●—	SCS50%-BFS50%	mixture of SCS and BFS
(5)	—▲—	SCS70%-CM30%	mixture of SCS and CM
(6)	—■—	CM94%-be6%	mixture of SCS and bentonite

Finally, it is worth mentioning that the BFS have an important commercial use for the cement casting process, while the SCS are generally sent to disposal facilities.

The two SS were tested both alone or mixed with other materials such as a) a calcareous non-uniform coarse grained material (CM) obtained from the milling process of the limestone extracted from the quarry and b) a sodium bentonite (B) in powder.

A complete list of the tested materials is reported in Table 2.

3. LABORATORY TESTS

The laboratory tests were performed to determine the short term behaviour of the tested materials, while the field observations on the trial embankment were meant to study the long term behaviour.

In fact, the hardening processes are significantly influenced by environmental factors such as rainfall, air pollution, temperature, humidity, etc.

The following laboratory tests were performed:

a) modified Proctor compaction (MPC) tests;
b) isotropically consolidated drained triaxial tests sheared at constant confining pressure (σ'_3 = const) (TX-CID-CL);
c) constant head hydraulic conductivity (HC) tests in rigid wall permeameters.

TX and HC tests were conducted on samples compacted at W_{opt} and at 95% of the optimum dry density determined in the MPC tests.

This procedure was adopted to have a reference condition for all the investigated materials and also because it meets the specifications of many compaction works.

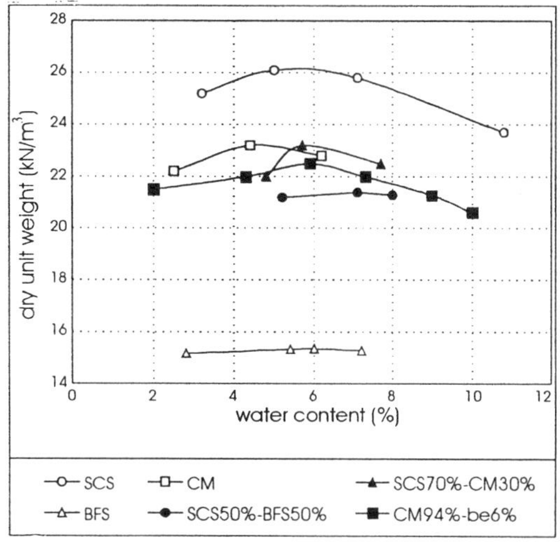

Fig.2.- Compaction curves of tested materials

Samples of large dimensions were used in all the tests listed above, in order to avoid problems due to the presence of coarse particles. In particular, the HC and the MPC tests were carried out on samples having a 30 cm diameter and a 30 cm height, while the TX tests were conducted on samples having a 20 cm diameter and a 40 cm height.

Fig. 2 shows the results of the MPC tests. As it can be seen, the optimum dry density (γ_{opt}) of the SCS is significantly higher than that of the BFS.

Such result reflects the influence of the different mineralogical composition and grain size distribution of the two materials.

Fig. 2 also shows that the optimum dry densities and moisture contents of the mixture lie in the range of the values obtained from the two components.

At the end of the MPC tests, the samples were subjected to HC tests carried out in the same mold used for the compaction tests.

The internal wall of the mold was coated by a 6 mm spongy bitumen sheet glued on it to prevent the side-wall leakage of the permeating water.

The HC tests were performed applying different values of the hydraulic head.

The measured hydraulic conductivity are shown in Fig. 3. As it can be seen, there is significant difference between the hydraulic conductivity (k) of the SCS and that of the BFS. The difference is even more pronounced between the BFS and the CM. However the addition of 30% of CM to SCS is sufficient to lower the k values of the mixture up to those characterising CM.

At the same time the addition of only 6% of bentonite to the CM causes a three order of

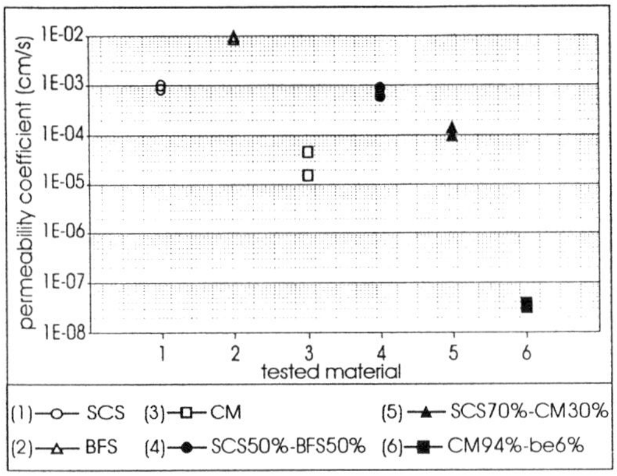

Fig. 3.- Hydraulic conductivity of tested materials

magnitude reduction in the hydraulic conductivity of this material.

Based on the results shown in Fig. 3 the use of the SCS and BFS alone doesn't meet the hydraulic requirements for the construction of the impermeable top covers of the landfills.

Such requirements can be satisfactory met by adding small percentages of bentonite to the CM (or likely to the mixtures of CM and SS) but this part of the research needs more experimental data.

The TX-CID-CL tests were performed on samples previously compacted (at W_{opt} and at 95% of the optimum dry density determined in the MPC tests) in a special mold splitted in two parts. The internal wall of the mold was coated with a 2 mm thick latex membrane kept in place by a small vacuum applied to it. Possible membrane damages caused by tamping were recovered before starting the TX tests. The samples were then saturated by flushing them with de-aired water for a long period.

The TX test apparatus consisted of a 10 t capacity loading frame and a TX cell equipped with an internal pressure cell and an external strain measuring LVDT sensor. The relevant shear strength parameters are listed in Table3.

Table3 - Mohr-Coulomb shear strength parameters

materials	cohesion (kPa)	shearing angle (degree)
(1) SCS	227	45.2
(2) BFS	126	40.9
(3) CM	158	50.0
(4) SCS50%-BFS50%	146	46.9
(5) SCS70%-CM30%	128	49.5
(6) CM94%-be6%	15	47.4

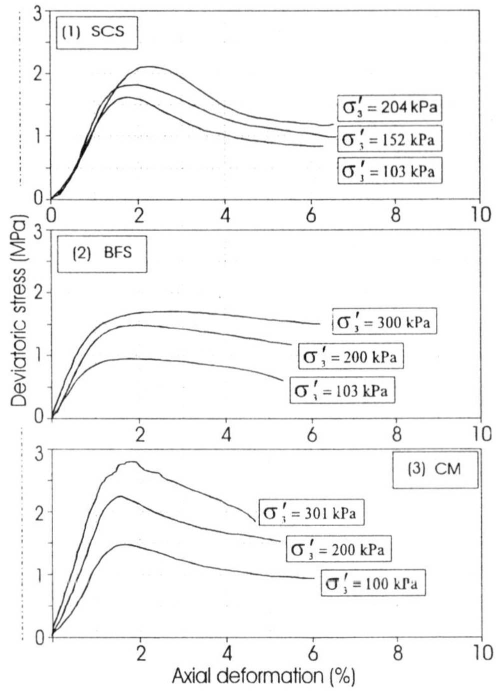

Fig. 4 - Stress-strain curves from triaxial tests

As it can be inferred from Table 3, the mixtures between SS and CM give values of φ' and c' which lie in the range of those separately shown by each component. On the opposite the addition of a small percentage of bentonite to CM causes more pronounced reduction of c'.

Fig. 4 shows some examples of the stress-strain curves obtained from the TX tests. Based on the results obtained from the laboratory tests, and taking also into account economical considerations concerning the other possible commercial uses of the investigated materials, the steel converter slags were considered the most appropriate for the purposes of the present investigation.

4. FIELD TRIAL EMBANKMENT

The construction of the field trial embankment (Fig. 5) was aimed to define the compaction procedures for the selected material (SCS). Such procedures concerned: roller type and weight, travelling speed, thickness of the strata, number of passes, watering system, etc..

The study was performed by constructing a series of preliminary trial compaction fields to verify the

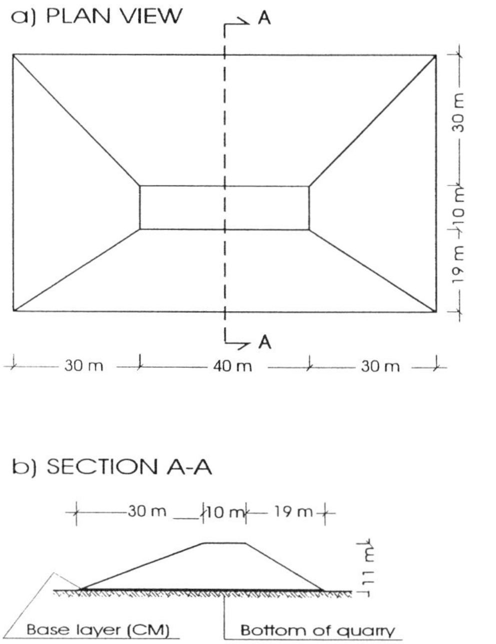

Fig. 5 - Schematic view of the field trial embankment

influence of the several parameters listed above.

The results were checked by performing in situ density measurement (ID) tests and plate loading (PL) tests on the compacted strata. The ID tests were carried out using a water balloon device, while the PL tests were conducted using a 30 cm diameter plate.

The above mentioned studies led to the definition of the following compaction procedure:

- roller: 10 t vibratory roller
- travelling speed: 4 km per hour
- thickness of the strata: 50 cm
- number of passes: 8 passes per stratum
- watering system: road-sprinkler vehicle

This procedure was then used for the completion of the trial embankment up to the final dimensions corresponding to those of the designed landfills embankments.

The suitability of the assumed procedure was checked by 21 ID tests and 13 PL tests performed at different locations within the embankment.

Fig. 6 shows the results of the ID tests carried out at all the investigated locations.

Such results correspond to the following mean values of dry density (γ_d) and moisture content (W):

$\gamma_d = 23.1 \pm 0.17$ kN/m^3

W = 6.0 ±2.4%.

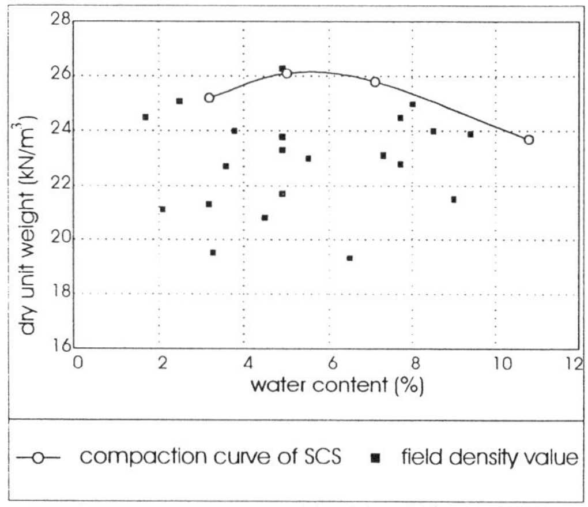

Fig.6 - Field density test results.

The above listed mean values are in quite good agreement with the design specification. However the scatter of the data is too large. Probably this can be attributed to the insufficient size of the rubber balloon device adopted (1000 cm^3) and for some differences in the mineralogical and grading characteristics of the materials coming from the blast furnace process.

An example of the pressure-displacement (p-s) curves obtained from PL test is shown in Fig.7.

The equivalent elastic module (M) of the compacted strata were determined from the PL tests according to the following expression:

$$M = \frac{p_B - p_A}{s_B - s_A} \cdot D$$

being A and B the extreme limits of the assumed calculation range indicated in Table 4 and D the plate

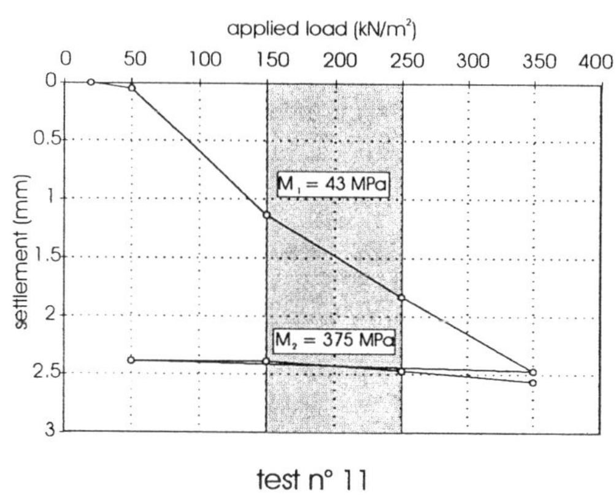

Fig.7 - Pressure-displacement curve of plate load test

Table 4 - Plate loading test results

test n°	embankment elevation (m)	PB - PA (MPa)	M (MPa)
1	0.9	0.15 to 0.25	61
2	0.9	0.15 to 0.25	80
3	1.3	0.15 to 0.25	61
4	3	0.15 to 0.25	72
5	3.6	0.15 to 0.25	56
6	3	0.15 to 0.25	61
7	7.5	0.15 to 0.25	34
8	7.5	0.15 to 0.25	44
9	7.5	0.15 to 0.25	63
10	7.5	0.15 to 0.25	59
11	11	0.15 to 0.25	43
12	11	0.15 to 0.25	41
13	11	0.15 to 0.25	39

diameter. The results obtained are summarised in Table4. They correspond to M values lying in the following interval:

$$M = 55 \pm 13.7 \text{ MPa}$$

It is worth mentioning that the design specifications for the definitive embankments require M values ≥ 40 MPa.

At the end of the construction phase, the field trial embankment was accurately levelled and instrumented in order to monitor the deformations and the fissures patterns induced in the long period by the swelling phenomena associated to the hydration processes of the SCS.

This part of the research is still in progress.

5. CONCLUSIONS

The aim of the research was to investigate the possibility of using the steel slags produced at the ILVA steel mill in Taranto (Italy) for landfill construction (confinement embankments and top covers).

Two possible alternatives were studied, one based on the use of the SS alone and the other based on the use of mixtures between SS and a calcareous material extracted from an adjacent limestone quarry.

Furthermore, the effects of adding small percentages of bentonite to the calcareous material were investigated.

The experimental research was carried out in two phases: the first one in laboratory and the second one on a real scale field-trial embankment. In the first phase, all the considered materials were tested according to the following laboratory tests:

a) modified Proctor compaction tests
b) hydraulic conductivity tests
c) isotropically consolidated drained triaxial tests.

Based on the laboratory tests results and taking also into account economical considerations concerning other commercial uses of the investigated materials, the SCS was considered the most appropriate for the purposes of the present investigation.

The field trial embankment was used to define the compaction procedures of the selected material.

The embankment was extensively investigated for this purpose by means of a large number of in situ density measurement tests and plate loading tests.

Furthermore, at the end of the construction phase the embankment was accurately levelled and instrumented to monitor the long term deformations and fissures pattern induced by the swelling phenomena of the SS which represent the main shortcoming of such materials.

REFERENCES

- Brunelot M.R. (1981). "Le Laitier de Haut-Fourneau Produit a Usinor-Dunkerque et ses Utilisations". Laitiers de Hauts Fourneaux n. 48 - 1/1981 pp. 15 ÷ 36.

- Castan M. (1987). "Les Etudes de Formulation de la Grave-Laitier". Laitiers de Hauts Fourneaux n. 64 - 1/1987 pp. 27 ÷ 43.

- Chalancon J., Georges C. et al. (1986). "Dalle de Foundation en Grave-Laitier - Tout-Laitier Armée de la Nouvelle Batterie de Fours à Coke B6 de la Société Dunkerquoise de Cokéfaction (groupe USINOR). Laitiers de Hauts Fourmeaux n. 62 - 2/1986 pp. 7 ÷ 44.

- Cisti A., Sabo B., Vivier M. (1984). "Grave-Laitier Prete à l'Emploi Activée an Gypsonat - Un Exemple d'Application en Charente-Maritime". Laitiers de Hauts Fourneaux n. 57 - 3/1984 pp. 7 ÷ 18.

- Dron R. (1984). "Le Nouveaux Activants du Laitier". Laitiers de Hauts Fourneaux n. 57 - 3/1984 pp. 7 ÷ 18.

- Fontaine G. (1987). "Le Point de Vue du Laboratoire". Laitiers de Hauts Fourneaux n. 65 - 2/1987 pp. 23 ÷ 26.

- Kernoa M.M. and Marchandier (1981). "Une Utilisation Originale du Laitier de Haut-Fourneau: Fondation d'un Gazometre Humide en Site Tourbeux". Laitiers de Hauts Fourneax n. 48 - 1/1981 pp. 7 ÷ 14.

- Kernoa M.M. and Marchadier (1986). "Un Example d'Utilisation de Laitier de Haut Fourneau pour une Consolidation de Sol Compressible par Colonnes Ballastées". Laitiers de Hauts Fourneax n. 62 - 1/1986 pp. 45 ÷ 50.

- Morales M.M. (1981). "Structural and functional distress due to Slag expansion". Third Intern. Conf. on Case Hystories in Geotechnical Engineering. University of Missouri.

Environmental Geotechnics, Kamon (ed.) © 1996 Balkema, Rotterdam. ISBN 90 5410 848 7

Coal ash as reclamation material and its geotechnical procedures for improvement

Koichi Goto & Koichiro Takezawa
Newjec Inc., Osaka, Japan

Toru Shibata
Department of Civil Engineering, Fukuyama University, Japan

ABSTRACT: This paper describes coal ash as soil materials and a geotechnical approach to the actual ground reclaimed with coal ash. Such ground may show wide irregular variations of soil parameters, which may depend on its kind like chemical components and on its void ratios under sedimentation. The coal ash ground has possibility of liquefaction. Liquefaction resistance of coal ash can be increased with reduction of void ratio which can serve as a design parameter with soil tests to improve the ground. A dynamic compaction with sand piles is one of effective means for improvement of exsisting ground with coal ash.

1 INTRODUCTION

Coal ash is one of industrial waste discharged from coal thermal power stations with energy industrial activity. In Japan, as the annual consumption of coal has been increasing year by year since the oil crisis, it is expected that the amount of coal ash will increase more in future.

The utilization of coal ash has been developed in the many fields. But less than half of coal ash is used as in these fields, or the rest is disposed to reclamation and embankment.

From chemical components, coal ash is near to clay. But, it can be self-hardened by its pozzolanic reaction, therefore its geotechnical characteristics are different from usual soil materials.

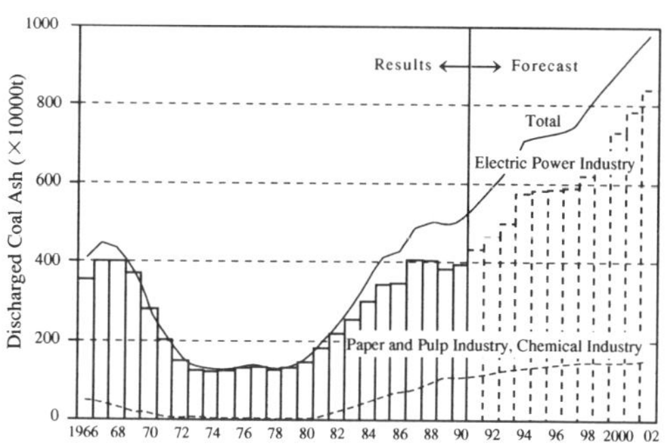

Fig.1 Result and forecast of annual amount of generated coal ash (after JSCE, 1990)

Coal ash as ground, if rolled to compaction, is provided with enough strength in embankment. However, in the case of reclamation in the sea, its strength may not be guaranteed because it is difficult for the ground to be rolled to compaction.

This paper refers briefly to the state of discharge and utilization of coal ash in Japan. And there is indicated a case of soil investigation on the site reclaimed with coal ash to point out the problems in this study. Then, element and field tests are conducted to provide a geotechnical approach how to deal with the coal ash ground.

2 DISCHARGE AND UTILIZATION OF COAL ASH IN JAPAN

2.1 *Discharge and material utilization*

In Japan, coal ash is almost discharged from the electric power industry. A consumption of coal for electric power began to increase from about 1980, and shows the growth more than twice from about 10 to 23 million ton during five years just after then. This is because coal was situated as one of alternatives instead of oil from the oil crisis in 1978 by the energy policy in our country.

Fig.1 shows annual amount of generated coal ash from 1966 to 1990 in result and from the year on forecast (JSCE, 1990). As the discharged coal ash is amounted around 10-20% of coal consumption, the change shows good correspondence to the increase of coal consumption up to the present. And it is

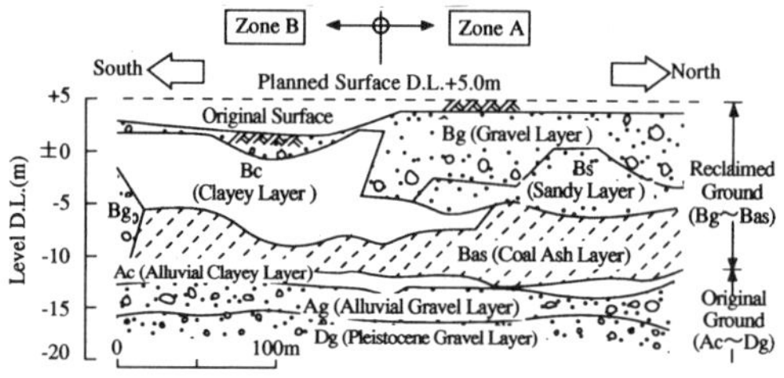

Fig.2 Profile of the reclaimed ground with coal ash

predicted that the amount of discharged coal ash will be about 9 million ton in 2000.

Coal ash is classified roughly into 70-80% of fly ash, 10-15% of clinker ash and 5% of cinder ash.

Fly ash is utilized in the fields of cement, manure and civil engineering works. It is also utilized as admixture material to cement of which technology was already established about 40 years ago. Recently, it has been known that the product has a preventive effect from the alkali-aggregate reaction and it is expected that the utilized amount of fly ash will be increasing for the future.

Table 1 Physical properties of coal ash

Soil Property		Range	Average
Density Of Soil Particles ρ_s (g/cm³)		2.07~2.36	2.18
Grain Size	Gravel Content (%)	0~37	2
	Sand Content (%)	0~65	16
	Silt Content (%)	9~95	67
	Clay Content (%)	0~31	12
Void Ratio e		1.00~2.07	1.47
Bulk Density ρ_t (g/cm³)		1.38~1.58	1.47

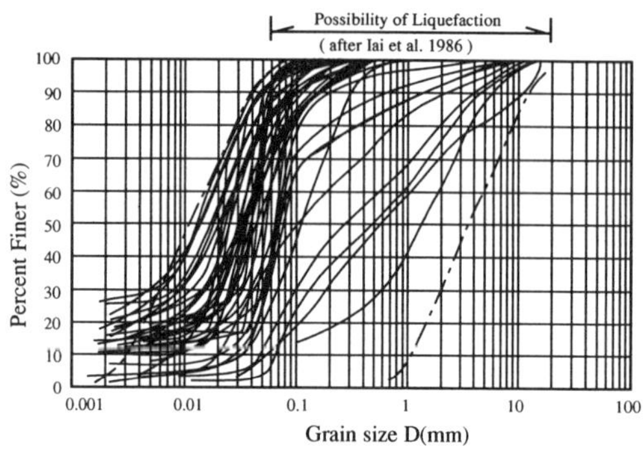

Fig.3 Grading curves of coal ash

2.2 *Disposal for reclamation*

The amount of disposed coal ash for reclamation has been approximately 6.2 million m^3 for 30 years from 1951 to 1980, with 4.0 million m^3 for reclamation in the sea and 2.2 million m^3 for inland reclamation. But, there are few cases of land utilization in which important structures have been constructed at sites reclaimed with coal ash.

The conventional disposal methods are classified into two systems that are the dry disposal system in which coal ash is spread into the sea and the wet disposal system in which coal ash with high water content is spread in low density into the sea.

Sugimoto et al. (1989) suggested the disposal system in which high density slurry of fly ash is placed onto the bottom of the sea ground to make the reclaimed ground of high density and strength.

3 CHARACTERISTICS OF THE RECLAIMED GROUND WITH COAL ASH

3.1 *Profile of the ground*

The ground was reclaimed in the sea pond to dispose coal ash discharged from a thermal power station in 1960's. A typical profile is illustrated in Fig.2. The ground is divided into the original ground and the reclaimed ground. The original ground underlying is almost horizontally layered where there are alluvial clay, alluvial gravel and pleistocene gravel layers, respectively downwards. On the other hand, in the reclaimed ground, coal ash layer, Bas is almost distributed over the whole area on the original ground to form its complex profile that the northern part is mainly reclaimed with sand and gravel Bas (zone A) and the southern part with clay, maybe with dredged clay (zone B).

3.2 *Physical and chemical properties*

The physical properties of the coal ash are shown in Table 1 and its grading curves in Fig.3 The density of particles and the bulk density are smaller than those of natural soil materials. The void ratios show wide variations. The silt contents are around 70%, but the grading curves fall in range with possibility of liquefaction according to the standard of technology of port and harbor in Japan (Iai et al. 1986). Table 2 shows chemical contents of the coal ash where the value of base degree and acidity rate were defined by Kataoka et al. (1991).

Table 2 Chemical contents of coal ash

Sample	SiO_2	Al_2O_3	Fe_2O_3	CaO	MgO	SO_3	Na_2O	K_2O	Base degree	Acidty rate
1	54.0	26.1	4.63	3.69	1.81	1.23	2.01	1.38	0.169	7.907
2	59.1	23.8	5.00	2.15	1.99	0.20	1.91	2.05	0.158	9.070

Base degree= $(Na_2O+K_2O+MgO+CaO+Fe_2O_3)/(SiO_2+Al_2O_3)$

Acidty rate = $(SiO_2+Al_2O_3)/(Fe_2O_3+CaO+MgO)$

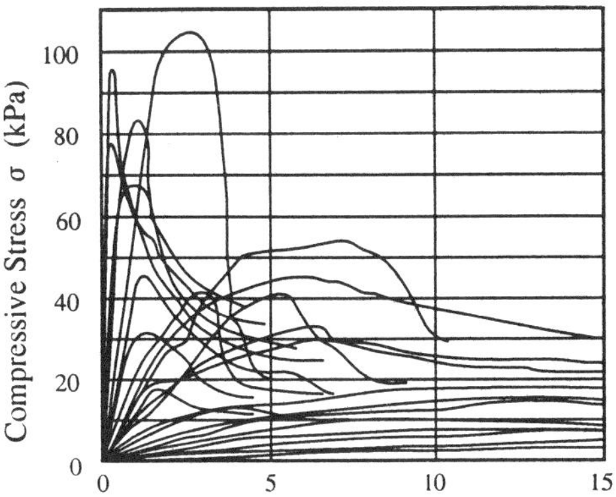

Fig.4 Stress-strain curves of unconfined compression tests on undisturbed coal ash samples

They pointed out that the strength of coal ash gets increasing remarkably when the base degree is more than 0.2 or the acidity rate is less than 10. From the value of base degree and acidity rate, the pozzolanic reaction of the present coal ash is small if anything.

3.3 *Strength characteristics*

Fig.4 shows stress-strain curves of unconfined compression tests on undisturbed coal ash samples. The curves having failure strain smaller than about 4% show considerable strain softening phenomena.

The relations between unconfined compressive strength, qu and effective overburden pressure, po are indicated in Fig.5. The maximum value of qu is about 160kPa, but the minimum is close toward 0, and no tendency of strength increase with augmentation of po is recognized.

Fig.6 is the relations between liquefaction resistance, RL and initial void ratio, eo by cyclic undrained triaxial tests on undisturbed samples of the coal ash and the sandy soil. The RL is defined as the cyclic stress ratio when axial strain reaches 5% double sided amplitude on 20 times numbers of cycles. The RL of the coal ash ranges from 0.20 to 0.45 and shows irregular variations. The RL-eo

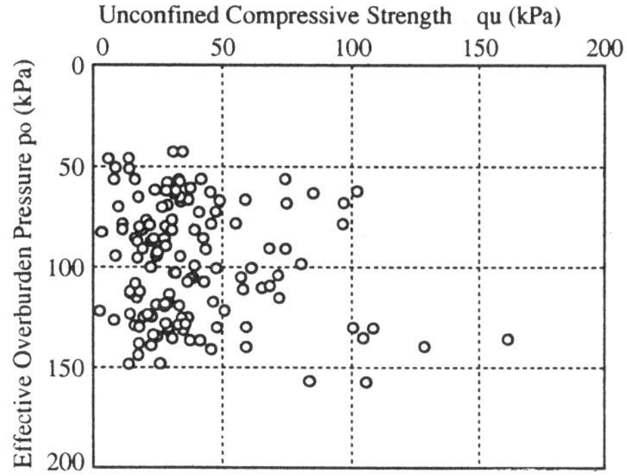

Fig.5 Relations between unconfined compressive strength and overburden pressure of the coal ash

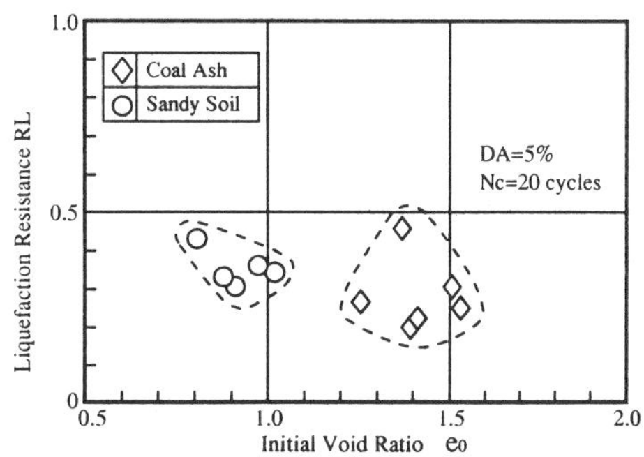

Fig.6 Relations between liquefaction resistance and initial void of the coal ash and sandy soil

relation of the coal ash is not so much clearer than that of sandy soil. As the lower values among RL are near by RL of loose sand, the coal ash layer is potentially liquefiable. On the other hand, the higher values among them may be affected by strength increase with its pozzolanic reaction.

3.4 *Possibility of liquefaction*

The possibility of liquefaction was then investigated by comparing RL with the maximum seismic stress ratio, SS calculated by seismic response analysis using multiple reflection theory considering equivalent linear model. In the seismic response analysis, the base motion was inputted within the pleistocene gavel layer Dg with the peak ground acceleration of 120gal designed reasonably.

Fig.7 shows the result of seismic response analysis and the range of RL given in Fig.6. The distribution of stress ratio generated by several earthquake motions is different from one another but in any case beyond the lower value of strength

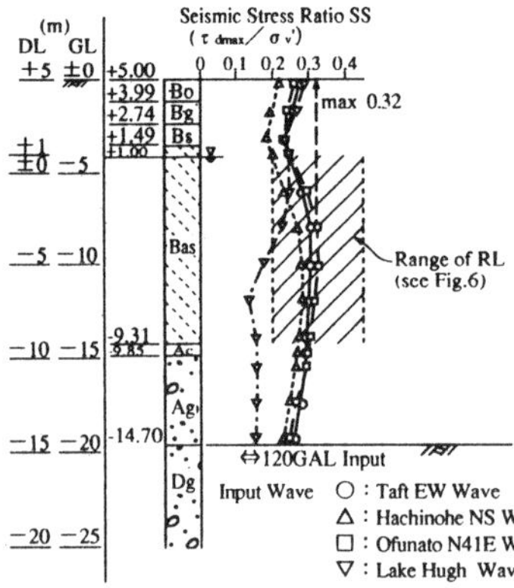

Fig.7 Seismic stress ratio distribution with depth calculated by seismic response analysis.

ratio. This means that the coal ash layer has the possibility of liquefaction.

3.5 *Reclamation process*

As the coal ash employed for reclamation of this ground had been discharged from the same power systems, there was little difference between their chemical contents. On this account, the scattering of soil parameters of the coal ash ground might be caused by other factors than their chemical contents.

In the conventional manner coal ash had been filled by means of wet disposal method with its sedimentation in the sea. Considering self-hardening of coal ash, it is suggested that the scattering of soil parameters may be caused by such wet and slow sedimentation process. Therefore, it is necessary to clarify influence of the process of the reclamation with coal ash.

4 SEDIMENTATION AND CONSOLIDATION TEST

4.1 *Sample and experimental procedures*

The coal ash sample used for the experiments is fly ash adjusted by Japanese Industrial Standard. Physical characteristics of this sample is shown in Table 3 and its grading curves in Fig.8. This sample has more than 70% of silt content. According to the base degree and the acidity rate from its chemical contents shown in Table 4, this sample has higher potential of pozzolanic reaction than the coal ash of Table 2.

Fig.9 illustrates an outline of apparatus employed in consolidation test which can contain a specimen of 9.7cm in diameter and 5.0cm high and it is larger than the standard. The consolidation pressure is loaded by bellofram cylinder.

For sedimentation test, the sample is spread into the water filled in the container comprising the ring and the collar thereon. The experiment is performed with procedures in Table 5.

Table 3 Physical properties of the sample

Soil Property		Sample 1	Sample 2
Density of Soil Particles ρ_s (g/cm³)		2.263	2.258
Grain Size	Maximum Size(mm)	0.425	0.425
	Sand Content (%)	12.0	13.0
	Silt Content (%)	72.0	71.0
	Clay Content (%)	16.0	16.0
	Uniformity Coeff.	6.33	7.73
Maximum and Minimum Density	ρ dmax (g/cm³)	1.275	–
	e min	0.772	–
	ρ dmin (g/cm³)	0.967	–
	e max	1.337	–

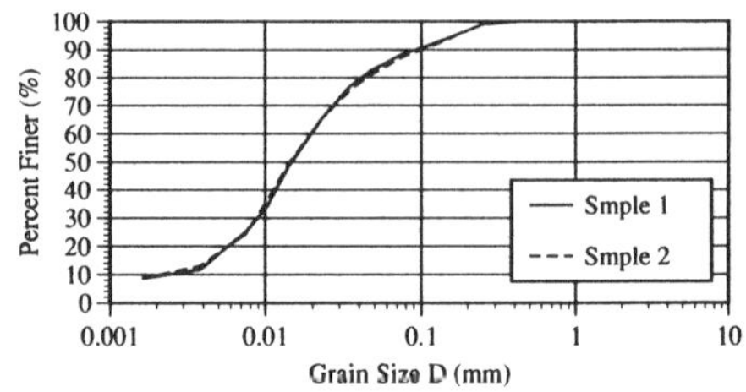

Fig.8 Grading curves of the sample

Table 4 Chemical contents of the sample

SiO_2	Al_2O_3	Fe_2O_3	CaO	MgO	SO_3	Na_2O	K_2O	TiO_2	MnO	P_2O_3	Base degree	Acidty rate
53.2	24.2	5.24	5.67	1.51	0.52	2.36	1.85	1.18	0.02	0.33	0.207	6.232

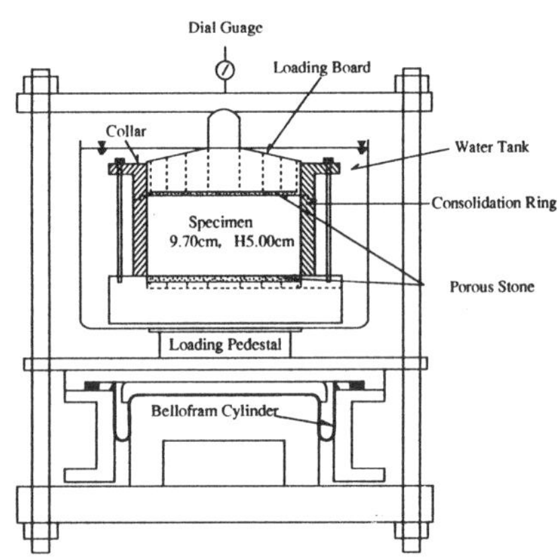

Fig.9 Outline of test apparatus

Table 5 Test condition on sedimentation test

Test No.	Sample State on Spread	Sedimentation
S30	30% water content slurry	Spread at a stretch then left alone 1 day
S50	50% water content slurry	
S70	70% water content slurry	
S90	90% water content slurry	
S110	110% water content slurry	
P1	Dry powder	
P7	Dry powder	7days (1cm high / day)

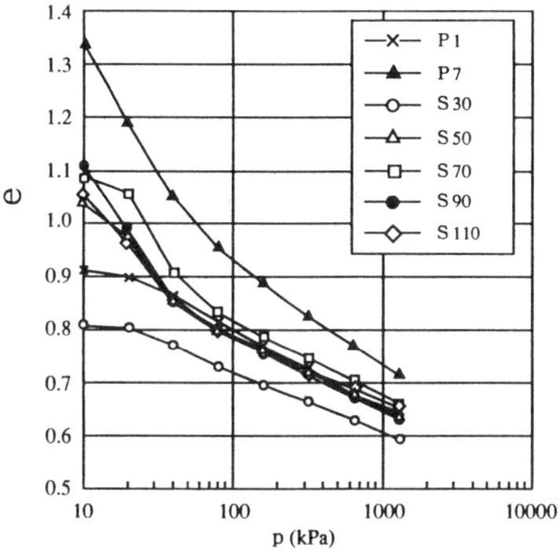

Fig.10 e-log p curves by consolidation tests

After the sedimentation test, the specimens are formed in 5cm high, and then the consolidation test are carried out. The pressures are applied by means of load increment with ratio of 1 within range of 10-1280 kPa, and let to consolidation at each stage of loading for 60 minutes.

4.2 *Results and considerations*

Fig.10 shows *e-log p* curves by the consolidation tests where *e* is the void ratio and *p* is the consolidation pressure. The initial void ratio of S30 is the smallest and the initial void ratio of P7, which is spread by intermittent disposal, is the largest. This means that there is great difference between sedimentation methods. Compressibility becomes larger as initial void ratio steps up. But even if the greatest consolidation pressure of 1280kPa is loaded, the void ratio remains influenced by the initial void ratio and indicating different values. Therefore, it is suggested that the ground remains loose if it is reclaimed loose by coal ash even with consolidation.

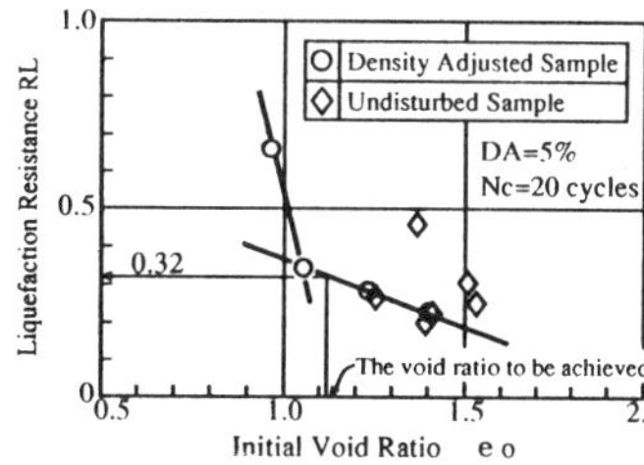

Fig.11 Relations between liquefaction resistance and initial void of the density adjusted samples

5. COUNTERMEASURES

As one of the countermeasures against liquefaction of coal ash grounds, dynamic compaction method with sand piles is considered, and its applicability is evaluated here for the ground shown in chapter 3.

5.1 *Procedures*

On applying the sand compaction pile (SCP) method to the reclaimed ground with coal ash, the geotechnical subjects are as follows:

1) Evaluation of liquefaction resistance, RL with density increase of coal ash.
2) Confirmation of compaction effect in the ground.

For 1), cyclic undrained triaxial tests are carried out using the density adjusted samples prepared by sampling from the reclaimed ground, remolding and reconsolidation. And the design parameter for improvement by SCP is studied from the results of the tests.

For 2), field tests are conducted and improvement effects are confirmed to estimate the design prameter.

5.2 *Cyclic undrained triaxial tests*

The initial void ratios eo of the specimens are adjusted as 0.97, 1.05, 1.24 and 1.41 considering the rate of replacement by SCP. The ratio eo=0.97 corresponds to about 20% rate of replacement.

Fig.11 shows RL-eo relation of the density adjusted samples and the results of undisturbed samples mentioned before are also shown. As the lower group of values of RL among the undisturbed samples are considerably well corresponding to the relation of density adjusted samples, they are equivalent to RL without pozzolanic reaction. And the RL of coal ash increases with decrease of void ratio. As the characteristics are the same as those of

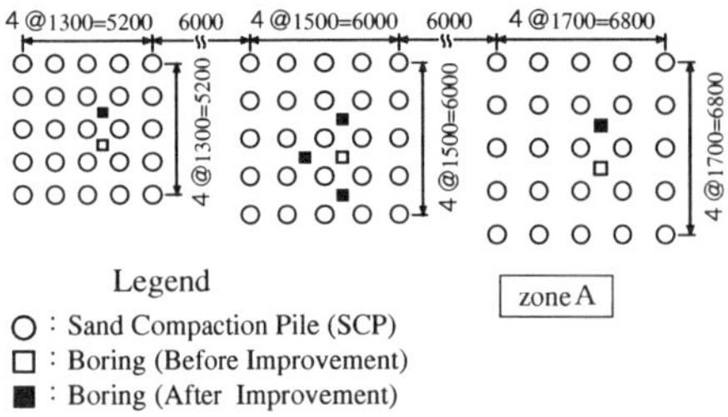

Fig.12 Plan view of SCP arrangements

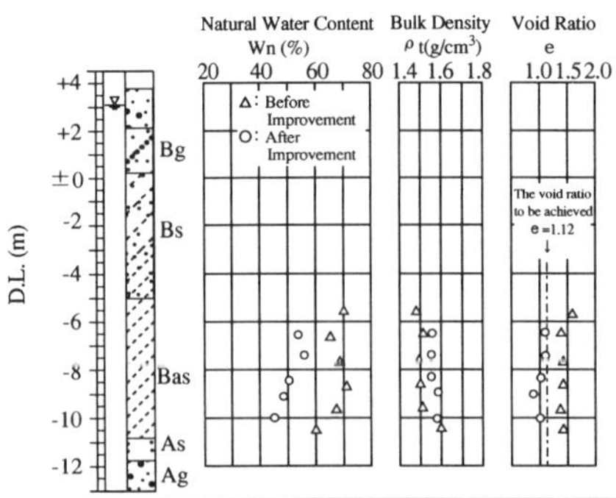

Before Improvement	$\overline{W}_n$=66.6%	$\overline{\rho}_t$=1.51g/cm³	$\overline{e}$=1.434
After Improvement	$\overline{W}_n$=51.0%	$\overline{\rho}_t$=1.56g/cm³	$\overline{e}$=1.019 (e_{max}=1.086)

Fig.13 Results of geotechnical investigation before and after improvement (1.7m square arrangement)

sandy soil, SCP method is considered to be effective countermeasures against liquefaction of the coal ash ground.

The calculated seismic stress ratio, SS is 0.32 as shown in Fig.7, which is suggested for RL necessary to improve the reclaimed ground. Thus, the void ratio can serve as the design parameter and be decided as 1.12 or less from Fig.11 to prevent the coal ash layer from liquefaction.

5.3 *Field test*

The void ratio to be achieved for the coal ash layer improved is e=1.12 or less. As the void ratio before improvement is estimated as eo=1.45, the rate of replacement by SCP will be required to be 13.5%, which is satisfied with square arrangement of sand piles of 70cm in diameter with 1.7m at interval.

The field tests have been operated as three cases of square arrangements with 1.3, 1.5 and 1.7m at intervals of sand piles at zone A shown in Fig.2. Fig.12 illustrates the plan view of arrangements and boring points before and after the test improvement.

The results of geotechnical investigation in the case of 1.7m square arrangement are shown in Fig.13. The changes of void ratios and so on represent increse of density and support this improvement of the coal ash layer. With the void ratios less than the designed value, good effects of improvement have been confirmed. Such effects have also seen in the other cases. Thus, a dynamic compaction such as SCP is one of the effective countermeasures against liquefaction of the reclaimed ground with coal ash.

6. CONCLUSIONS

This paper has presented a geotechnical approach to the actual ground reclaimed with coal ash and brought to the following conclusions:

1) Reclaimed ground with coal ash may show wide irregular variations of soil parameters
2) Such variations may depend on its kind like chemical components and on its void ratios under sedimentation.
3) Reclaimed ground with coal ash has possibility of liquefaction if not improved.
4) Liquefaction resistance of coal ash can be increased with reduction of void ratio which can serve as a design parameter controllable with soil tests to improve coal ash ground .
5) Dynamic compaction with sand piles is one of effective means for improvement of existing ground reclaimed with coal ash.

References

Iai, S., Koizumi, K. and Tsuchida, H. (1986) : A new criterion for assessing liquefaction potential using grain size accumulation curve and N-value. *Rep. of Port and Harbour Res. Inst.* vol. 25, No.3, pp.154. (in Japanese)

JSCE (1990) : Present condition and future view of technology for utilization of coal ash as constructing materials. pp.28-31. (in Japanese)

Kataoka, T., Okamoto, T, and Ogata, N. (1991) : Physical and mechanical properties of reclaimed ground by coal ash - characteristic of strength increase by compaction. *Abiko Res. Lab. Rep. No.* II *91025,* CRIEPI. (in Japanese)

Sugimoto, T., Ohguchi, I. and Horiuchi, S. (1989) : Effective method for underwater disposal of coal fly ash. *Proc. 2nd Int. Symp.* on Environmental Geotechnology.

Environmental Geotechnics, Kamon (ed.) © 1996 Balkema, Rotterdam. ISBN 90 5410 848 7

Use potential of fly ash-residual soil mixture as a dyke material

Daesuk Han
Korea Institute of Geology, Mining and Materials, Taejon, Korea

ABSTRACT: A laboratory investigation was undertaken to appraise the fly ash collected from the Boryong power plant for the utilization as construction material. The fly ash, the residual soil of granite, and the mixtures of the two materials were tested for their specific gravity, grain-size distribution, Atterberg limits, compaction characteristics, permeability, shear strength, unconfined compressive strength, mineralogy, and chemistry.

The hydraulic conductivity of the fly ash-residual soil mixture prepared in the proportion of 1 (residual soil) to 1.5 (fly ash) by dry weight and compacted to the condition slightly lower than its optimum moisture-density ranged from 8.4 to 8.1 x 10^{-9} m/s depending upon the pressures applied to the test specimens while that of the mixture compacted to the condition higher than its maximum dry density was lower than the above-mentioned values; those of the fly ash and residual soil varied from 1.8 to 1.6 x 10^{-7} m/s and from 2.6 x 10^{-10} to 9.5 x 10^{-11} m/s, respectively. The shear strength of the mixture in terms of friction angle and intercept cohesion increased with time due to the pozzolanic action of the fly ash content, that is to say from the values (Ø: 32° and c: 27 kPa) to those (Ø: 38° and c: 40 kPa) for the specimens cured for zero and 7 days, respectively.

The main inorganic constituents of the fly ash were Si and Al; the oxides of these two elements made up 78% of the ash. The results of the leachate quality tests on the fly ash and fly ash-residual soil mixture indicated that both materials were environmentally acceptable construction materials with respect to the U.S. EPA Toxicity Criteria.

Based on the test results, the fly ash alone cannot be used as a dyke material mainly due to its permeability. When the fly ash is blended with a clayey soil, however, its geotechnical characteristics appear to be favorable for a dyke material.

1 INTRODUCTION

The tidal difference along the west and south shore in the Republic of Korea ranges from 3 to 9 m, forming numerous tidal zones. Many of them have been reclaimed for not only agricultural and industrial use but also waste disposal and urbanization. In addition to the two existing landfills, Kimpo and Hwasong at the reclaimed land, several ones are being planned to be constructed along the shore. Since a natural material for dyke construction generally is scarce around the reclaimed land, its substitute appears to be of necessity.

Fortunately, there are five coal-burning power plants, Boryong, Seochon, Kunsan, Honam, and Samchonpo along the shore. They produce about 6,000 metric tons of fly ash every day. A little attention has been paid to the waste products as a potential replacement of conventional material, most of them being disposed of near the power plants.

The Boryong among the five power plants ranks first in ash production, the amount being about 1,800 metric tons per day. Chosen for the study on dyke material were the fly ash collected from the Boryong and the residual soil of Jurassic granite sampled at the Yesan area which is located about 40 km northeast of the Boryong.

A laboratory investigation on the fly ash, the residual soil, and the mixture of the two materials was undertaken during the period between May, 1994 and August, 1995 for the following purposes: (1) to find out their physical, engineering, mineralogical, and chemical characteristics and (2) to evaluate the utilization potential of fly ash-residual soil mixture as a dyke material for landfill facilities.

Table 1. Specific gravity and Atterberg limits for the material Nos. FA-1, FR-1, FR-2, and RS-1.

Material	Mixture, %		Specific	Atterberg limits, %		
No.	Fly ash	Residual soil	gravity	LL	PL	PI
FA-1	100	0	2.33	-	NP	NP
FR-1	80	20	2.39	28.93	NP	NP
FR-2	60	40	2.47	33.05	24.62	8.33
RS-1	0	100	2.68	37.40	21.43	15.97

2 LABORATORY INVESTIGATION

A series of laboratory test was carried out on the Boryong fly ash, the residual soil derived from the weathering of Jurassic granite, and the mixtures of the two materials, including grain-size distribution, specific gravity, Atterberg limits, compaction characteristics, permeability, shearbox strength, unconfined compressive strength, X-ray diffraction analysis, pH, chemical composition, and leachate toxicity.

2.1 Sample preparation

The residual soil of granite was first air dried and then pulverized using a rubber hammer. For most of the tests undertaken, the Boryong fly ash and the pulverized residual soil were oven dried and kept in a desicator until they were used for the preparation of test specimens. Using the two materials, prepared were two kinds of mixtures of 80-20 and 60-40 percent fly ash-residual soil by dry weight.

Throughout this and subsequent chapters, the material Nos. expressed as FA-1, RS-1, FR-1, and FR-2 refer to the Boryong fly ash, the residual soil of granite, the mixture of 80-20 percent fly ash-residual soil, and the mixture of 60-40 percent fly ash-residual soil, respectively.

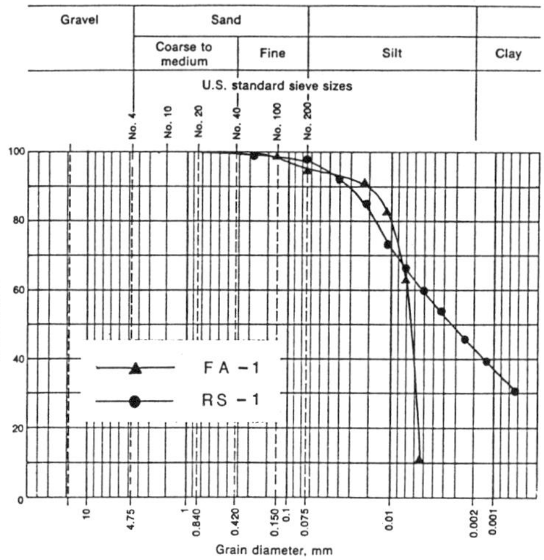

Figure 1. Grain-size distribution curves for the FA-1 and RS-1 materials.

2.2 Physical characteristics

The four materials mentioned in the preceding paragraph were tested for their specific gravity and Atterberg limits in accordance with ASTM D 854 and ASTM D 4318, respectively. The test results are listed in Table 1.

Grain-size analyses were conducted on only two materials, FA-1 and RS-1 in accordance with ASTM D 422. Both materials were first tested following a hydrometer method. After taking the final hydrometer reading, all the ingredients in a mass cylinder were transferred to a No. 200 sieve. The portion retained on the No. 200 sieve was treated following a sieve analysis method. Using the above two test results, grain-size distribution curves were constructed (Figure 1).

2.3 Engineering characteristics

The moisture-density relationships for the materials, FA-1, FR-1, FR-2, and RS-1 were determined following the procedure specified in ASTM D 698, "Standard Methods for Moisture Density Relations of Soils Using 2.5 kg Rammer and 304.8 mm Drop" and using a 101.7 mm diameter mold. Diagrams were constructed illustrating the relationship between dry density and moisture content (Figure 2). The compaction characteristics of the materials obtained from the diagrams are presented in Table 2.

Unconfined compressive strength test was conducted on the materials of FA-1 and FR-2. For each of the materials, 3 sets of specimens (38 mm diameter) were prepared, each set consisting of three specimens. The 1st set was tested right after its preparation; the 2nd and 3rd sets were cured for 7 and 14 days, respectively prior to testing. The strain rate applied to all the test speci-

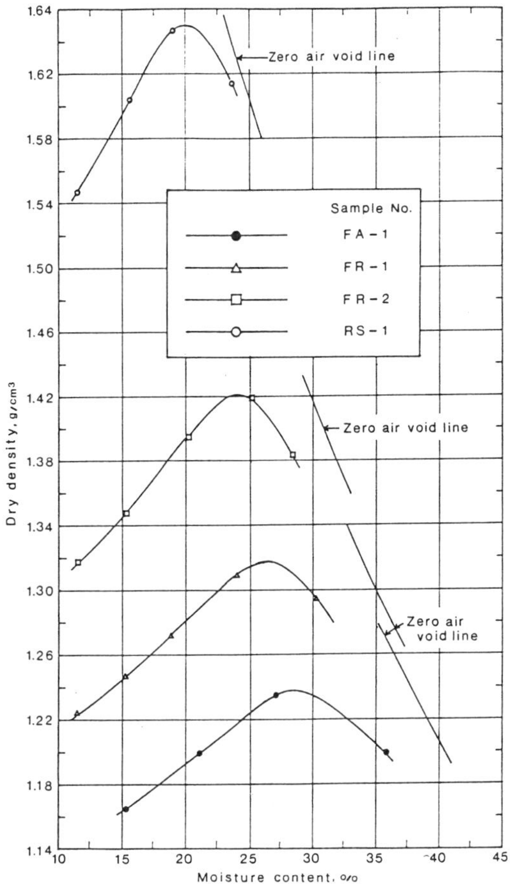

Figure 2. Moisture-density relationships for the materials of FA-1, FR-1, FR-2, and RS-1.

Table 2. Compaction characteristics for the materials of FA-1, FR-1, FR-2, and RS-1.

Material No.	Mixture, %		OMC, %	Max. dry density, g/cm3
	Fly ash	Residual soil		
FA-1	100	0	28.7	1.238
FR-1	80	20	25.5	1.313
FR-2	60	40	24.0	1.421
RS-1	0	100	20.0	1.635

mens was 1.3%/min. The mean values of the test results are presented in Table 3.

For each of the materials, FA-1, FR-1, FR-2, and RS-1, a 101.7 mm compaction mold was used to prepare a sample compacted in close proximity to its maximum dry density listed in Table 2. All the compacted samples

Table 3. Unconfined compressive strength for the materials of FA-1 and FR-2.

Material No.	Specimen No.	MC, %	Dry density, g/cm3	Curing, day	UCS, kPa
FA-1	1	28.0	1.260	0	223
	2	27.8	1.258	7	730
	3	27.6	1.263	14	955
FR-2	1	23.3	1.420	0	205
	2	24.5	1.418	7	603
	3	22.8	1.429	14	704

were used to prepare cutting ring specimens, 50 mm diameter and 20 mm thick. A direct falling head permeability test on the specimens was undertaken right after their preparation using an oedometer consolidation cell provided with a bottom inlet, which is connected to a burett, and an upper overflow outlet; the test was carried out while the specimens were under load. The applied loads were 1, 2, and 4 kg, which correspond to the pressures of 50, 100, and 200 kPa, respectively. The hydraulic conductivities determined at 20° C are presented in Table 4. Besides the above tests, a specimen of the FA-1 material with an initial dry density of 1.229 g/cm^3 was tested after 7-day-curing; the permeability of the specimen with a dry density of 1.232 g/cm^3 under a pressure of 50 kPa was determined at 1.5 x 10^{-7} m/s. Also conducted was the permeability test on a specimen of the FR-2 material compacted to be 1.469 g/cm^3 denser than its maximum dry density presented in Table 2. The determined values also are presented in Table 4 (specimen No. 2 of the FR-2 material).

The measurement of the shear strength on the three materials, FA-1, FR-2, and RS-1 was accomplished by means of direct shearbox test. An ELE motorized machine with a 60 mm shearbox was used to determine the shear strength parameters of five sets of specimens, each set consisting of 3 specimens. The test specimens were prepared by the following method. Firstly, all the materials were compacted into a standard compaction mold using a 2.5 kg rammer and then a thin stainless steel tube, 99 mm diameter and 45 mm high was pushed into the compacted specimens. Three sets of ring specimens were used right after their preparation while two sets were cured for 7 days. A 60 mm square specimen cutter was used for the preparation of test specimens. The applied loading rate was 1.06 mm/min. The shear strength parameters (cohesion and internal friction angle) of five sets of specimens are presented in Table 5.

Table 4. Hydraulic conductivity for the materials of FA-1, FR-1, FR-2, and RS-1.

Material No.	Specimen No.	Mixture, % Fly ash	Residual soil	Initial condition Moisture content, %	Dry density, g/cm^3	During testing Pressure, kPa	Dry den., g/cm^3	k_{20}, m/s
FA-1		100	0	26.8	1.238	50	1.239	1.8×10^{-7}
						100	1.242	1.7×10^{-7}
						200	1.246	1.6×10^{-7}
FR-1		80	20	24.6	1.331	50	1.336	5.4×10^{-8}
						100	1.339	5.3×10^{-8}
						200	1.343	5.1×10^{-8}
FR-2	1	60	40	23.1	1.399	50	1.405	8.4×10^{-9}
						100	1.408	8.3×10^{-9}
						200	1.411	8.1×10^{-9}
	2	60	40	21.4	1.469	50	1.474	4.6×10^{-9}
						100	1.480	4.2×10^{-9}
						200	1.498	3.3×10^{-9}
RS-1		0	100	20.3	1.708	50	1.722	2.6×10^{-10}
						100	1.731	1.1×10^{-10}
						200	1.749	9.5×10^{-11}

Table 5. Results of the shearbox test for the materials of FA-1, FR-2, and RS-1.

Material No.	Specimen No.	Curing time, day	Normal stress, kPa	Max. shear stress at failure, kPa	Dry density at start of shear, g/cm^3	Parameters Int. friction angle, $\emptyset$	Cohesion, kPa
FA-1	1	0	36	29	1.180		
	2	0	62	46	1.184	35°	4
	3	0	88	64	1.182		
	1	7	36	56	1.184		
	2	7	62	84	1.185	44°	28
	3	7	88	106	1.183		
FR-2	1	0	36	50	1.367		
	2	0	62	64	1.360	32°	27
	3	0	88	84	1.362		
	1	7	36	68	1.363		
	2	7	62	92	1.364	38°	40
	3	7	88	115	1.365		
RS-1	1	0	36	64	1.626		
	2	0	62	81	1.624	24°	56
	3	0	88	94	1.625		

2.4 Mineralogy

Petrographic analysis revealed that most of the particles of the FA-1 material (fly ash) are spherical. It also showed that glass is one of the major components of the material. Other components such as mullite and quartz were identified by X-ray analysis. Scanning electron micrograph showed that a lot of acicular compounds had formed on the tiny spheres after the fly ash was mixed with water and cured (Figure 3).

The RS-1 material (residual soil) is composed of quartz, kaolinite, feldspar, muscovite, and smectite according to its X-ray powder diffraction patterns.

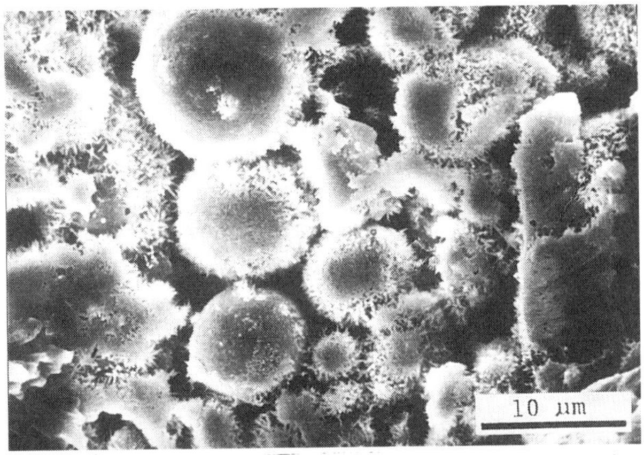

Figure 3. Electron micrograph showing a lot of acicular compounds formed on the tiny spheres of the Boryong fly ash.

2.5 Chemical characteristics

The determination of pH was conducted on the materials of FA-1, FR-1, FR-2, and RS-1 in accordance with the Electrometric Method, BS 1377: 1975, Test 11(A); the pH values were determined at 12.5, 11.8, 11.0, and 4.4, respectively.

Nine chemical constituents were determined for the FA-1 material. The determinations of SiO_2, Al_2O_3, and Fe_2O_3 were made by colorimetric methods, CaO, TiO_2, and MgO by a spectrochemical method, K_2O and Na_2O by flame photometry, and SO_3 by a gravimetric method. The unburnt fuel was determined by loss on ignition. The contents of the chemical constituents are presented in Table 6.

Table 6. Chemical composition of the FA-1 material.

Chemical constituent	Content, %
SiO_2	44.42
Al_2O_3	33.58
CaO	7.13
Fe_2O_3	3.44
TiO_2	1.66
MgO	1.56
K_2O	0.76
Na_2O	0.39
SO_3	0.56
Loss on ignition	4.80
Others (undetermined)	1.70

2.6 Leachate quality

Two materials, FA-1 and FR-2 were used to evaluate the quality of their leachate. The leachate samples were obtained by means of the Toxicity Test Extraction Procedure of the U.S. Environmental Protection Agency. After extraction, the leachate was analyzed with the aid of an iductively coupled plasma atomic emission spectrometer. The concentration of 12 elements in each of the two materials is listed in Table 7. Also presented in the table is the EPA Hazardous Waste Toxicity Criteria.

Table 7. Results of the leachate quality tests on the materials of FA-1 and FR-2.

Element	Concentration, ppm		*Maximum permissible concentration, ppm
	FA-1	FR-2	
Al	2.69	2.21	
As	<0.2	<0.2	5
Ba	0.26	0.66	100
Cd	<0.01	<0.01	1
Cr	0.20	0.089	5
Cu	0.01	0.02	
Fe	0.71	0.33	
Mn	0.66	1.29	
Mo	0.29	0.04	
Ni	0.16	0.12	
Pb	0.081	0.095	5
Zn	0.85	0.16	

*EPA Hazardous Waste Toxicity Criteria (100 times drinking water standards).

3 DISCUSSION

The results of the determination of grain-size distribution reveal that the FA-1 material is mainly composed of silt-size particles, while the RS-1 material comprises 1.9% sand-size particles, 54.6% silt-size particles, and 43.5% clay-size particles. Based on the above gradation and the determination of Atterberg limits, the former and latter are classified as ML and Cl, respectively according to the Unified Soil Classification System; the two mixtures of FR-1 and FR-2 as ML.

As illustrated in the moisture-density relationship (Figure 2), the optimum moisture and maximum dry density of the mixtures decreases and increases, respectively with lower proportion of fly ash.

According to the permeability classification by Terzaghi and Peck (1967), the Boryong fly ash is characterized by low degree of permeability, while the residual soil of granite is practically impermeable when they are compacted in close proximity to their maximum dry densities; the compacted mixtures of the two materials are characterized by very low degree of permeability.

The Boryong fly ash possess a cementitious property what is called pozzolanic action in contact with water. This action must be due to the formation of the acicular compound known as $CaO\text{-}SiO_2\text{-}Al_2O_3\text{-}nH_2O$. The strength characteristics of the mixtures of fly ash-residual soil are enhanced by the pozzolanic action.

Based on the results of the leachate quality tests on the Boryong fly ash and the mixture of fly ash-residual soil, the measured concentrations of the elements are environmentally permissible.

4 CONCLUSION

On the basis of the laboratory investigation, the properties of the mixture of 80-20 percent fly ash-residual soil are dominated by the characteristics of the fly ash, while those of the mixture of 60-40 percent fly ash-residual soil appear to be between the characteristics of the two materials.

The Boryong fly ash cannot be used a dyke material mainly due to its permeability which ranges from 1.8 to 1.6 x 10^{-7} m/s depending on the dgree of density. Its permeability can be decreased by blending it with a clayey soil such as the residual soil of granite. The hydraulic conductivity of the 60-40 percent fly ash-residual soil mixture ranging from 8.4 to 3.3 x 10^{-9} m/s can be decreased when subjected to higher compactive effort, being prctically impervious. At the same time, the strength is more enhanced by not only the increased density but also the pozzolanic action of the fly ash.

The results of the leachate quality test indicate that the above-mentioned mixture is a environmentally acceptable material. It should be emphasized that the tested sample was in a loose state and the concentrations of elements in the leachate from compacted fly ash-residual soil mixtures in dyke structure would be reduced over those obtained by the EPA Toxicity Test Extraction Procedure.

Considering all the available information, it is concluded that the mixture of 60-40 percent fly ash-residual soil by dry weight may be utilized as a dyke material for landfill facilities.

5 ACKNOWLEDGEMENT

The writer wishes to express his sincere appreciation to the Ministry of Science and Technology for its financial support.

REFERENCES

American Scociety for Testing and Materials 1980. Annual Book of ASTM Standards, Section 4 vol. 04.08.

British Standard Institution 1975. Methods of test for soils for civil engineering purposes, BS 1377: 1975.

Han, D. 1989. Geotechnical characteristics of the coal-related wastes in the Republic of Korea. Korea Institute of Energy and Resources Research Report KR-89-(B)-6, p.67-76.

Head, K.H. 1980. Manual of soil laboratory testing vol. 1. London: Pentech.

Head, K.H. 1982. Manual of soil laboratory testing vol. 2. London: Pentech.

McQuade, P.V., et al 1981. Investigation of the use of coal refuse-fly ash compositions as highway base course material. Office of Research and Development, Federal Highway Administration, Department of Transportation, Washington, D.C. Report No. FHWA/RD-801129, p.27-29.

Terzaghi, K. & R.B. Peck 1967. Soil mechanics in engineering practice, p.46-130. New York: Wiley.

U.S. Environmental Protection Agency 1980. Toxicity Test Extraction Procedure, Federal Register vol. 45 no.98.

Environmental Geotechnics, Kamon (ed.) © *1996 Balkema, Rotterdam. ISBN 90 5410 848 7*

Density increase of coal ash reclamation by surfactant addition

Sumio Horiuchi
Institute of Technology, Shimizu Corporation, Tokyo, Japan
Kazuya Yasuhara
Department of Civil Engineering, Ibaraki University, Japan

ABSTRACT: More than 4 million tons of coal ash are discharged annually from coal power plants in Japan, of which 26% is reclaimed in the sea. Sluicing (wet reclamation) is one current method applied in coal ash sea reclamation. However, technical issues that need to be addressed relating to sediment low density have been pointed to (1) an increase in density to prolong the disposal site, and (2) an increase in strength to allow effective use of reclaimed ground. The authors postulated that this low density is due to particle coagulation during sedimentation, and the effect of particle dispersal by surfactant addition on density and strength were experimentally studied. In this paper, the efficiency of surfactant addition is reported with the following results: (1) effectiveness of surfactant addition is dependent on coal ash type and surfactant type, (2) a one percent addition of selected surfactant can increase ground density by 15% on average, (3) the density increase reflects the strength increase by 2-3 times, (4) density increase is aided by an increase in accumulation speed; 15% increase in density can be obtained at accumulation speed of 10cm/day, (5) cement addition increases strength but decreases dry density, and (6) sluicing with both cement and surfactant has little effect on increasing density and strength.

INTRODUCTION

More than 4 million tons of coal ash are discharged annually from coal power plants in Japan. 26% of this coal ash is reclaimed in the sea, and amount of that has increased because of limited land area for reclamation. Conventional sea reclamation methods can be classified into the three types shown in Fig. 1. The slurry reclamation method is the recently developed one in Japan. From the experimental data of reports (Sugimoto *et al.* 1989, Horiuchi *et al.* 1995), average Japanese coal flyash ground properties by the three methods can be summarized as shown in Table 1. Slurry reclamation is the best method in terms of ground use and volume reduction, however, its operation cost is much higher than that of wet reclamation. If the density of wet reclamation could be increased up to that of slurry, this method could be the most efficient.

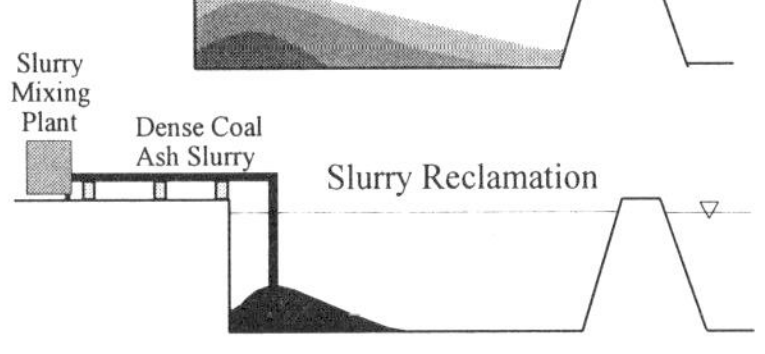

Figure 1 Conventional Sea Reclamation of Coal Ash

Table 1 Properties of Coal Ash Reclaimed Underwater

Properties of Ground reclaimed	Reclamation Method Underwater		
	Wet Type	Dry Type	Slurry Type
Dry Density	0.80 g/cm^3	0.93 g/cm^3	1.25 g/cm^3
28day Compressive Strength	0~4 kPa	0~70 kPa	200 kPa

Properties of coal ash assumed : Maximum dry density=1.33 g/cm^3
Calcium content=4%, pH=12, Mean diameter=0.012mm

One reason for the lower density in wet reclamation could be that the coal ash particles coagulate during sedimentation in the sea, and start to develop the strength as a result of their self-hardening property just after the formation of the loose structure (Horiuchi *et al.* 1995). This means that (1) density should be simultaneously increased during the reclamation, and that (2) higher density will be obtained when the coal ash particles are dispersed by adding surfactant as shown in Fig. 2 Density increase as a result of dispersal is well known in colloid science (Olphen:1977), however, it has rarely been studied in

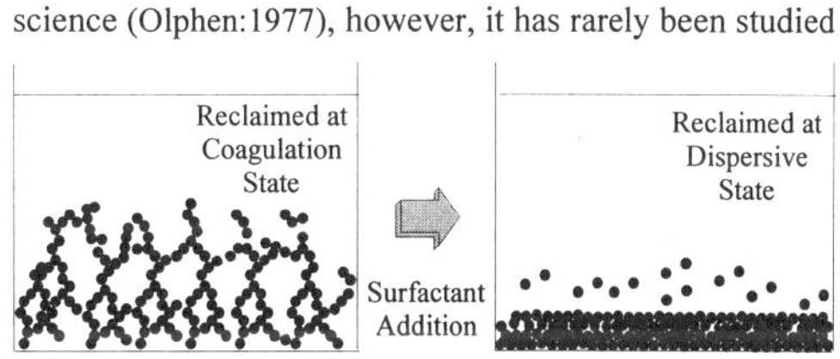

Figure 2 Effect of Surfactant Addition on Sediment

Table 2 Properties of Coal Ashes

Coal Ash		A	G	H	I	L	M	N	V
Specific Gravity		2.23	2.29	2.20	2.34	2.11	2.22	2.44	2.25
Surface Area (cm^2/g)	Blaine	3270	4820	4290	4090	2400	2710	4880	----
	BET	----	29400	39900	3800	4200	7400	19300	34900
Mean Diameter (μm)		41	14	13	9.4	18	19	9.2	17
Max. Dry Density (g/cm^3)		1.37	1.16	1.07	1.10	1.24	1.30	1.33	1.46
Opt. Water Content (%)		20.3	32.4	36.2	34.1	25.5	24.7	26.5	16.8
Ignition Loss (%)		2.40	5.32	8.41	2.33	2.00	5.30	2.78	2.17
Initial pH		----	12.4	11.9	12.1	4.1	11.0	11.7	12.7
Chemical Components (%)	SiO_2	53.0	58.8	49.1	64.6	66.2	58.5	49.4	49.7
	Al_2O_3	27.8	19.7	27.9	24.8	26.8	24.1	27.4	24.1
	Fe_2O_3	5.40	5.56	5.59	3.25	1.00	5.50	10.20	5.09
	CaO	7.60	6.01	3.18	1.62	0.80	2.50	3.98	9.97
	SO_3	0.40	0.75	0.38	0.32	----	----	0.92	0.36
Self-hardening Property		High	High	Low	None	None	None	Low	Mid.

Table 3 Type of Surfactants

A1	Carboxylic Acid
A2	Carboxylic Acid
B3	Naphthalene Sulfonic Acid
C1	Lauryl Sulfonic Acid
D8	Polycarboxylic Acid ether

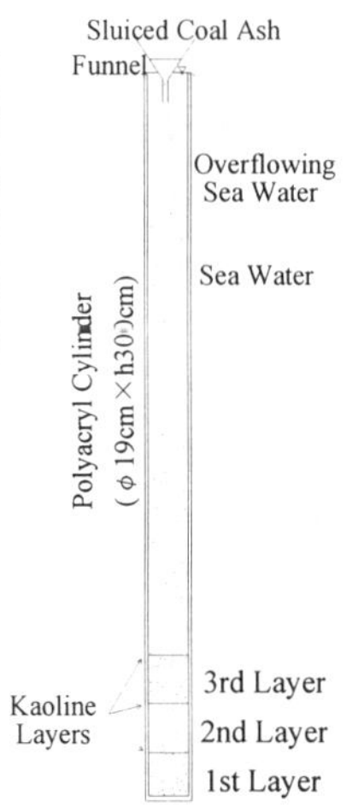

Figure 3 Schematic View of Sedimentation Test

the civil engineering field. Since surfactant addition inhibits the strength development of cement type materials, mechanical properties of the sediment were concerned to be deteriorated by the surfactant addition.

To confirm the efficiency of surfactant addition on density and mechanical properties of the sediment, a series of model sedimentation tests was carried out. This paper reports the merits and limits of surfactant addition in wet reclamation.

EXPERIMENTS

Materials

To simulate sea reclamation, artificial seawater (pH=8) was used throughout this study. Table 2 shows the properties of the seven coal flyashes used in this study. All samples were collected from electric precipitators of pulverized coal power plants, and were kept dry until the sedimentation test. The maximum dry density of the samples ranged from 1.07 g/cm^3 to 1.46 g/cm^3 at 70-78% degree of saturation. Five types of surfactant, listed in Table 3, were selected for the preliminary tests, and applied in larger scale tests.

Test Procedures

Two series of preliminary sedimentation tests were carried out using 1.2-*cm* diameter and 6.5-*cm* diameter cylindrical glass tubes, in which a given amount of flyash was

Table 4 Specs of Sedimentation Tests

Exp. No.	Specs of Coal Ash Sluiced	Diameter of Pipe	Surfactant	Cement	Storage of Coal Ash
A-1	25*kg* × 1*Layer*	19*cm*	Free	Free	> 0.5year
G-1	25*kg* × 1*Layer*	19*cm*	Free	Free	> 0.5year
H-1	20*kg* × 1*Layer*	19*cm*	Free	Free	> 0.5year
H-2	4*kg* × 5*Layers*	19*cm*	Free	Free	> 0.5year
H-3	2*kg* × 10*Layers*	19*cm*	Free	Free	> 0.5year
H-4	1*kg* × 20*Layers*	19*cm*	Free	Free	> 0.5year
I-1	20*kg* × 1*Layer*	19*cm*	Free	Free	> 0.5year
I-2	3.5*kg* × 5*Layers*	17.5*cm*	Free	Free	9 years
I-3	3.5*kg* × 5*Layers*	17.5*cm*	A2; 1%	Free	9 years
N-1	3.5*kg* × 5*Layers*	17.5*cm*	Free	Free	8 years
N-2	3.5*kg* × 5*Layers*	17.5*cm*	D8; 1%	Free	8 years
V-1	17.5*kg* × 1*Layer*	17.5*cm*	Free	Free	> 0.5year
V-2	3.5*kg* × 5*Layers*	17.5*cm*	Free	Free	> 0.5year
V-3	17.5*kg* × 1*Layer*	17.5*cm*	D8; 1%	Free	> 0.5year
V-4	3.5*kg* × 5*Layers*	17.5*cm*	D8; 1%	Free	> 0.5year
V-5	3.5*kg* × 5*Layers*	17.5*cm*	A2; 1%	Free	> 0.5year
V-6	3.5*kg* × 5*Layers*	17.5*cm*	Free	OPC; 3%	> 0.5year
V-7	3.5*kg* × 5*Layers*	17.5*cm*	D8; 1%	OPC; 3%	> 0.5year

directly mixed with the seawater and agitated for 2 *min*, then the density of the sediment was determined from the sediment volume after 24 hours.

Model sedimentation tests were carried out using polyacryl pipes of 17.5-19.0 *cm* diameter and 300 *cm* high. Table 4 and Fig. 3 show the specs and the schematic view of the sedimentation tests. A dilute flyash slurry of 140% water content was poured at a speed of 1 *l/min* into the pipe filled with the seawater. To confirm the effect of accumulation speed, a given quantity of flyash was divided into 5 to 20 pieces, and sluiced every 24 hours.

Surfactant was added to the seawater at a given ratios before preparation of the dilute slurry. In cases V-6 and V-7, ordinary portland cement (OPC) was mixed with flyash before the dilute slurry preparation for the purpose of strength enhancement.

Settlement and consolidation of the flyash sediments was observed through the polyacryl pipe, however, carbon particles contained made precise observation difficult, especially in the surfactant-added tests.

After 90 days from the end of sluicing, block samples were taken from the pipes to determine the dry density. Strength was determined by the vane shear test conducted at 0.05 degree/sec. Salt quantity in the voids was accounted for in determining dry density.

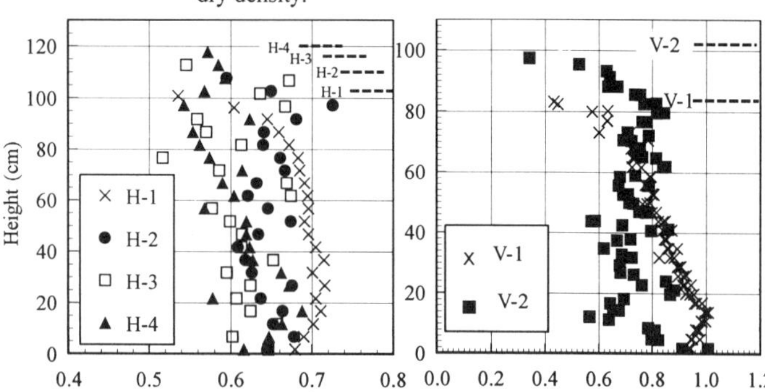

Figure 5 Dry Density and Surface Height of Sediments

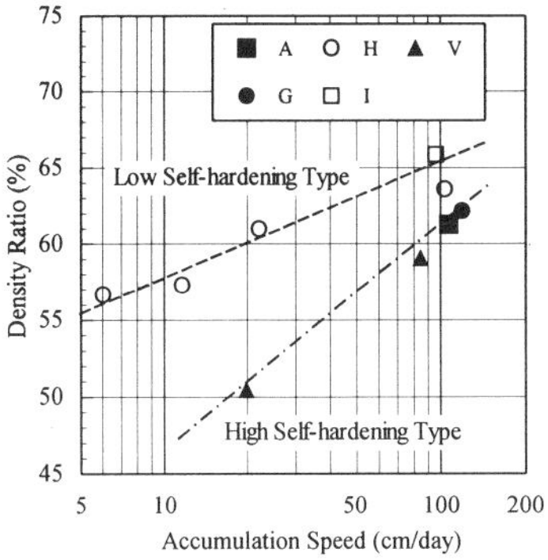

Figure 6 Effect of Accumulation Speed on Sediment's Density

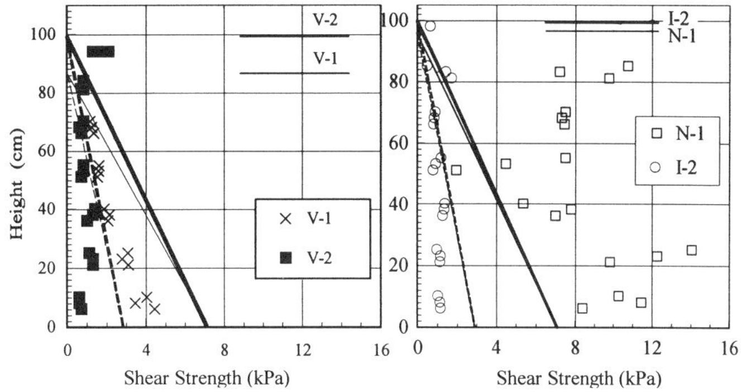

Figure 7 Distribution of 90day Strength of Sediments

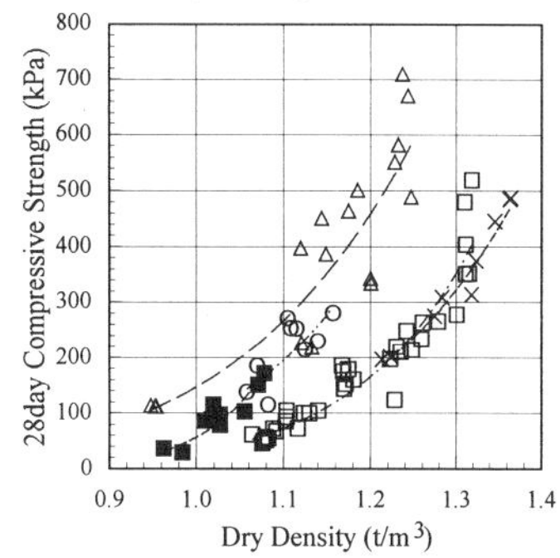

Figure 8 Relation between Strength and Dry density

RESULT

Sediments without Surfactant

The effective stress of the sluiced flyash caused the thickness of the sediment layer to change, but was only detectable within 48 hours. The micro-structure of the sediment, therefore, is stabilized within 48 hours from the sluicing.

Figure 5 shows examples of sediment dry density distribution and total height. Since the upper part of each sediment layer was very loose as typically seen in 20 $kg \times 1$ layer (legend: $\times$), density distribution of the divided sluicing samples shows a zigzag pattern, and that makes the effect of accumulation speed (daily thickness of the sediment) on density vague. The total height, however, shows an obvious trend; the larger the daily quantity of flyash sluiced, the lower the sediment surface height. This shows the importance of accumulation speed for sediment density determination.

Figure 6 shows the relationship between accumulation speed and density ratio, the ratio of density to maximum dry density, where the filled symbols represent the high self-hardening flyashes, and the empty symbols the lower ones. Increased accumulation speed increases density, and this effect is more pronounced in the higher self-hardening ashes. The lower self-hardening type ashes show a higher density ratio due to the following mechanism; (1) loose sediments formed at a lower accumulation speed develop sufficient strength against the lower pressure of the next sluicing, and (2) the sediment consolidates while developing strength as a result of the self-hardening property.

It would be reasonable to assume that the actual accumulation speed is not larger than 10*cm/day*, so the density ratio can be determined from Fig. 6 as 46 % for high self-hardening types, and 58 % for the low self-hardening types. According to the Coal Ash Disposal Manual (EPRI, 1979), dry densities of 0.72 g/cm^3 and 1.12 g/cm^3 are reported for wet reclamation and field compaction, respectively. Provided the density ratio of field compaction is 90%, maximum dry density can be calculated as 1.24 g/cm^3, making the density ratio of the wet reclamation 58 %. This 58 % coincides with the low self-hardening type in Fig. 6.

Strength distribution of the sediments is shown in Fig. 7, where the solid lines represent effective stress. The broken lines represent OCR=1, in which Pc equals 2.5 times shear strength (Yasuhara *et al.*1988). Flyash type significantly affects strength; N shows the greatest strength of 8*kPa*, while the others show only 1~2*kPa*. Excluding type N, strength is closer to the OCR=1 line than the effective stress line. In Fig. 7, strength of V-2 and I-2 are independent of depth, which was also reported by Yasuhara *et al.*(1988). One explanation of this tendency is that there is a strength increase caused by chemical reactions between CO_2 from the air and calcium contained in flyash, that would be concentrated in the upper part of the sediment.

The effect of dry density on strength was reported as shown in Fig. 8 (Horiuchi *et al.* 1995), and the density of the sediments are too low to speculate on the strength of wet reclamation. Since shear strength required for land use would be more than 100 *kPa*, it is necessary to increase density for land use.

Sediments with Surfactant Addition

To determine the optimum quantity of surfactant addition, preliminary sedimentation tests were carried out. Table 5 shows part of the results where 1% of surfactant addition is needed. Though both flyash and surfactant type affect efficiency, a dry density increase of 15% can be achieved on average.

Figure 9 shows dry density distribution of surfactant added sediments, where ● and ○ are surfactant-free and surfactant-added samples, respectively. For three types of flyash, dry

Table 5 Dry density of Surfactant Added Preliminary Sedimentation Test Samples (g/cm^3)

Flyash		G			H			I			L			M			N			V		
Surfactant		0.2%	1.0%	5.0%	0.2%	1.0%	5.0%	0.2%	1.0%	5.0%	0.2%	1.0%	5.0%	0.2%	1.0%	5.0%	0.2%	1.0%	5.0%	0.2%	1.0%	5.0%
30 *ml* test	Free	0.63 (54%)			0.47 (44%)			0.62 (56%)			0.94 (76%)			0.69 (53%)			0.70 (53%)			0.64 (44%)		
	A1	0.76	0.73	0.72	0.50	0.56	0.56	0.72	0.70	0.68	0.95	0.92	0.94	0.75	0.86	0.85	0.78	0.79	0.73	0.77	0.67	0.60
	A2	0.68	0.72	0.72	0.49	0.61	0.73	0.71	0.72	0.74	1.01	1.00	1.05	0.70	0.85	0.90	0.74	0.77	0.78	0.67	0.79	0.98
	B3	0.68	0.75	0.73	0.46	0.63	0.69	0.67	0.77	0.65	0.98	0.97	0.96	0.71	0.93	0.91	0.79	0.81	0.80	0.70	0.78	0.83
	C1	0.66	0.68	0.75	0.45	0.52	0.66	0.64	0.67	0.75	0.96	1.03	1.00	0.70	0.80	0.96	0.71	0.75	0.82	0.63	0.68	0.65
	D8	0.69	0.71	0.71	0.49	0.63	0.70	0.70	0.74	0.76	1.00	1.01	1.04	0.68	0.84	0.88	0.73	0.82	0.84	0.61	0.82	0.95
1 *l* test	Free	0.68 (59%)			0.50 (47%)			0.66 (60%)			1.00 (81%)			0.81 (62%)			0.71 (53%)			0.73 (50%)		
	A1	0.79	0.80	0.77	0.60	0.66	0.69	0.76	0.74	0.73	1.01	1.00	1.00	0.88	0.94	0.93	0.82	0.82	0.77	0.85	0.78	0.63
	A2	0.73	0.77	0.79	0.53	0.65	0.73	0.73	0.76	0.77	1.04	1.05	1.05	0.83	0.92	0.95	0.83	0.82	0.76	0.76	0.82	0.90
	B3	0.73	0.82	0.80	0.51	0.65	0.69	0.72	0.79	0.76	1.03	1.03	1.02	0.82	0.95	0.94	0.76	0.84	0.82	0.79	0.83	0.83
	C1	0.74	0.75	0.83	0.52	0.55	0.71	0.73	0.72	0.79	1.01	1.05	1.05	0.83	0.87	0.98	0.74	0.85	0.85	0.72	0.74	0.79
	D8	0.71	0.77	0.77	0.52	0.66	0.71	0.73	0.76	0.77	1.03	1.04	1.04	0.81	0.90	0.92	0.78	0.82	0.83	0.72	0.83	0.96

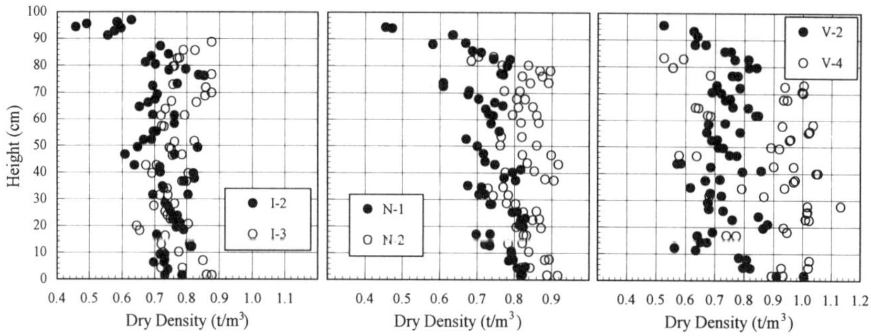

Figure 9 Effect of Surfactant Addition on Dry Density Distribution

density is increased by surfactant addition but is not influenced by effective stress. Since even the lower self-hardening type ashes show density increase by the surfactant addition, the mechanism of the densification is due to the particle dispersal other than the strength inhibition. As seen in Fig. 9, density distribution deviation is increased by surfactant addition. This is because of the existence of a thin lower fraction on the upper part of each sediment layer. Tendencies in Fig. 9 suggest that surfactant addition has little effect on sediment consolidation, but makes the initial micro-structure denser. Comparing the dry density of V-2 (surfactant-free) and V-4 (D8 1% added) at 10～60*cm* height, surfactant addition results in a density increase of 0.12 g/cm^3; from 0.79g/cm^3 to 0.91 g/cm^3. This 0.91 g/cm^3 corresponds to 62% of maximum dry density.

Table 6 shows mean diameter distribution of 3rd layer samples of V-4. It is obvious that the larger flyash particles sedimented faster in line with the Stokes's law. This particle separation suggests flyash particle dispersal by surfactant addition. Particle dispersal can be explained by the increase in surface potential energy. To confirm the electrochemical change of the particle surface, zeta-potential was determined by the Acoustsizer, suitable for dense suspension measurement (O'Brien *et al.* 1993). Examples of the zeta-potential change are shown in Table 7, where the hatched area means that surface charge is too low for the precise measurement. By surfactant addition, especially the anionic surfactant A-2, the negative charge on the particle surface is increased. In this way, flyash particles are dispersed in the seawater.

Table 6 Distribution of Mean Diameter (3rd layer samples of V-4)

Sample height	Mean diameter
40.8 ～ 44.6 *cm*	9.5 μm
37.8 ～ 40.8 *cm*	28 μm
34.8 ～ 37.8 *cm*	37 μm
31.8 ～ 34.8 *cm*	75 μm

Table 7 Zeta-potential Change by Surfactant Addition (mV)

Flyash	I		N		V	
Surfactant	A2	D8	A2	D8	A2	D8
0%	1.4mV	1.3mV	0.8mV	0.6mV	1.3mV	1.0mV
0.1%	1.1mV	1.1mV	0.8mV	0.6mV	0.9mV	1.0mV
0.2%	1.3mV	1.2mV	0.9mV	0.7mV	1.1mV	1.0mV
0.5%	-8.5mV	1.3mV	-5.7mV	0.8mV	-6.9mV	1.1mV
1%	-8.7mV	1.3mV	-7.3mV	0.8mV	-8.4mV	1.1mV
2%	-12.8mV	1.3mV	-9.5mV	-4.1mV	-12.3mV	1.1mV
5%	-12.4mV	1.2mV	-12.0mV	-6.7mV	-12.6mV	1.2mV

It was deduced that changes in flyash surface properties affect its consolidation behavior, so a series of oedometer tests were carried out. Consolidation pressure was exerted at 24 hr, 48 hr, and 672 hr after the sluicing. Three sets of test results are summarized in Fig. 10, where surfactant added samples obviously show a lower void ratio at pressure ranges lower than 0.1*MPa*. The surfactant-free samples, however, show the lower void ratios at higher pressure ranges. This change could be explained by the increase in particle repulsion; *i.e.* surface potential energy is increased by surfactant addition as shown in Table 7, and this higher potential increases repulsion force

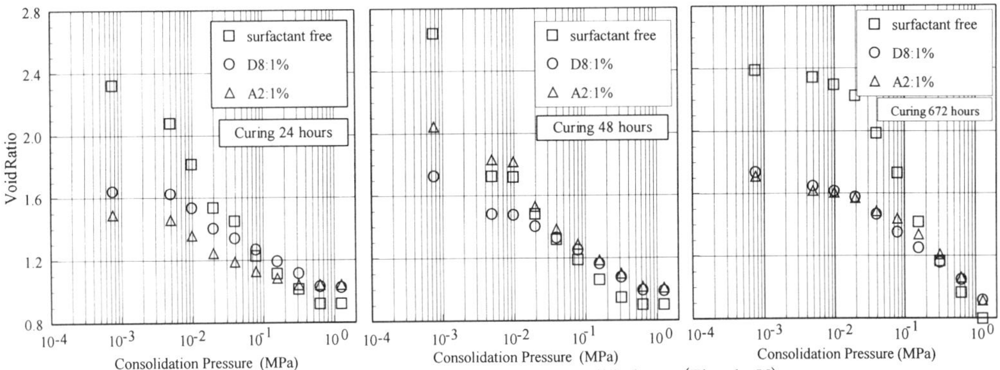

Figure 10 Effect of Surfactant Addition on Consolidation (Flyash: V)

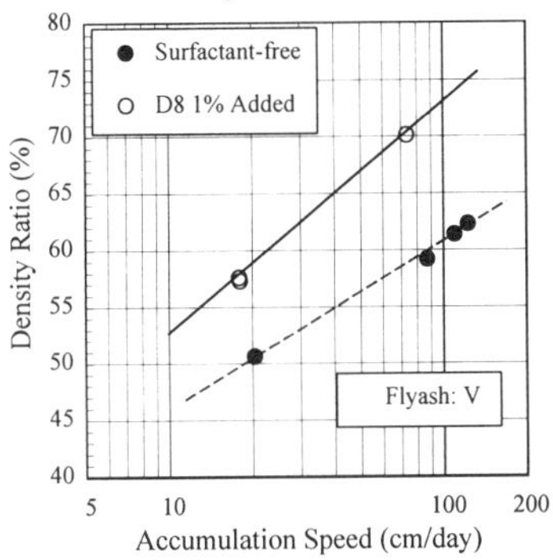

Figure 11 Effect of Accumulation Speed on Surfactant added Sediments' Density

when flyash particles approach the inside of double layer. The tendency seen in Fig. 10 suggests the importance of the circumstances of the ground formation on properties of the natural soil layers. Furthermore, a weak strength development caused by the self-hardening property can be confirmed in Fig.10; the void ratio of the sediments at a given consolidation pressure increases with curing time increasing. This weak strength development prohibits the consolidation of the coal ash sediment.

In Fig.11, the effect of accumulation speed on sediment's density is plotted together with surfactant-free samples. By surfactant addition, this relationship is obviously lifted upwards, and efficiency is increased with higher accumulation speed. At the accumulation speed of 10*cm/day*, the density ratio in surfactant-added wet reclamation can be determined as 53% even for the high self-hardening type. Though a much higher density ratio of 92% can be achieved by slurry reclamation (Horiuchi *et al.*1995), the surfactant addition can increase density by 15% (53/46 =1.15) with only a simple operation.

The effect of surfactant addition on strength distribution is shown in Fig.12. Excluding type I, which shows only a small increase in dry density by surfactant addition, large strength increases of 100-200% can be seen. Figure13 shows the relationship between 28day unconfined compressive strength and dry density of dense flyash slurries, where the solid lines represent surfactant-free samples. Surfactant added samples strength locates below the line, which means strength development caused by the self-hardening property is inhibited by surfactant addition. Therefore, the strength increase by surfactant addition seen in Fig. 12 is caused by the density increase of the sediment.

Evaluating long-term strength from Fig. 12, surfactant added sediments develop shear strength of 20*kPa*. This strength is, however, as the same level as that of soft alluvial silt soils, where it is necessary to improve strength before land use.

Sediments with OPC and Surfactant Addition

Strength of 3% OPC-added slurry are also plotted in Fig. 13. Unconfined compressive strength of 100*kPa*, which may

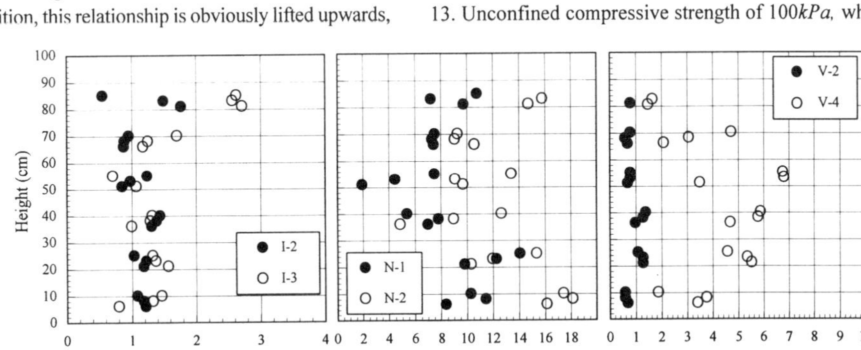

Figure 12 Shear Strength Distribution of Sediments

be sufficient for many land use purposes, is obtained in surfactant-free samples. The points to be confirmed are (1) the degree of strength inhibition caused by surfactant addition at dry density ranges lower than 1.0 g/cm^3, and (2) the density decreasing effect of cement addition on the surfactant added sediment. By a set of sedimentation tests, effects of surfactant and OPC addition were confirmed. Figure 14 shows the distribution of dry density and strength obtained. By adding both surfactant and OPC, a minor increase in dry density and a small decrease in strength is seen. Comparing this figure with Figs. 9 and 12, the following tendencies can be confirmed.

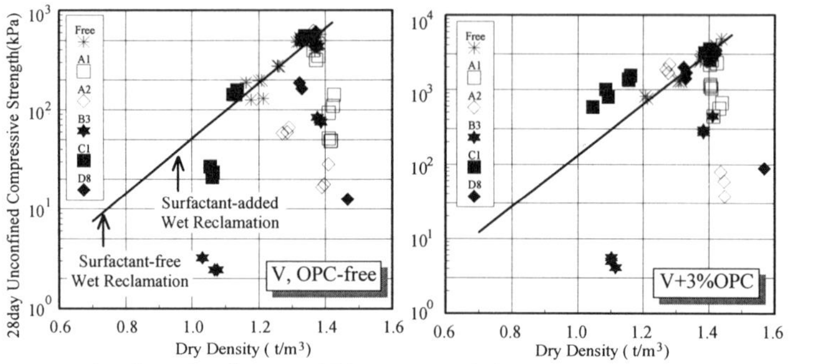

Figure 13 Effect of Surfactant Addition on Strength development of Dense Slurries

(1) Three percent OPC addition decreases dry density of surfactant-free sediment by 0.05 g/cm^3; 0.75 g/cm^3 into 0.7 g/cm^3 by 3% OPC. Sediment strength is increased to 2~3 times that of OPC-free but is the same as that of V-4.

(2) Surfactant addition of OPC-added sediment increases dry density to 0.75g/cm^3, which is same as V-2 sample but lower than V-4 by 0.2g/cm^3. Strength is slightly decreased by surfactant addition, but still larger than V-2 and smaller than V-4.

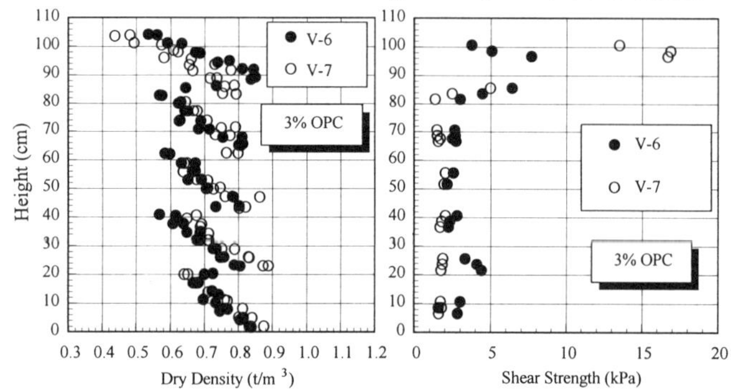

Figure 14 Effect of Surfactant Addition on Density and Strength of OPC Added sediments

According to the results, while cement addition increases strength, it also increases void ratio. However surfactant addition to the cement- added type has little effect on increases in both density and strength. By using a new combination of cement and surfactant, which could control initial strength development of the sediments, effective reclamation increasing both density and strength could be achieved.

CONCLUSIONS

To improve the low density and low strength of wet reclaimed coal ash ground, efficiency of the surfactant addition at sluicing was experimentally studied through a series of sedimentation tests. Based on the results, the following were confirmed.

(1) The effectiveness of surfactant addition is dependent on coal ash type and surfactant type.

(2) One percent addition of selected surfactant can increase sediment density by 15% on average.

(3) Density increasing by surfactant addition increases strength by 2-3 times.

(4) Increased accumulation speed enhances the effectiveness of the density increasing.

(5)At accumulation speed of 10cm/day, density ratio can be easily increased by 15% using a surfactant, however, this density is far below that of slurry reclamation.

(6) Cement addition increases strength but decreases dry density.

(7) Sluicing with both cement and surfactant has little effect on density and strength increase. The appropriate combination of cement and surfactant is needed for the effective reclamation.

References

EPRI (1979): "Coal Ash Disposal Manual (A Report for Electric Power Research Institute, USA)″ EPRI FP-1259.

Horiuchi,S., Tamaoki, K. and Yasuhara, K.(1995):"Coal Ash Slurry for Effective Underwater Disposal" Soils and Foundations (The Japanese Geotechnical Soc.), Vol.35, No.1, pp.1-10.

O'Brien,R.W., Rowlands,W.N. and Hunter,R.J.(1993): ″Determing Charge and Size with the Acoustosizer" NIST Special Publication 856, pp.1-22.

Olphen,H (1977):"An Introduction to Clay Colloid Chemistry (John Wiley & Sons)″ pp.121-125.

Sugimoto,T., Ooguchi,I. and Horiuchi,S.(1989):"Effective Method for Underwater Disposal of Coal Fly Ash" Proc. 2nd Int. Symp. on Environmental Geotechnology, pp.428-437

Yasuhara,K., Hirao,K., Horiuchi,S. and Odawara,T.(1988): "Model Tests on the Use of Fly Ash for Reclaimed Ground" Technical Reports of the Geotechnical Research Institute, Nishinippon Institute of Technology, No.4, pp.17-26 (in Japanese).

Environmental Geotechnics, Kamon (ed.) © *1996 Balkema, Rotterdam. ISBN 90 5410 848 7*

Improvement of a residual soil with industrial by-products

B. Indraratna & R. N. Chowdhury
Department of Civil & Mining Engineering, University of Wollongong, N.S.W., Australia

ABSTRACT: Colluvial residual soils are widely spread in many parts of New South Wales. In hilly terrain, such soils are often subject to considerable erosion and shallow mass movement. Because of the coal mining and steel manufacturing industries, large amounts of by-products such as fly ash and steel slag are produced in New South Wales. This study looks at the benefits of utilising milled slag and fly ash for improving the engineering behaviour of erodible surface deposits (Werombi colluvium), and compares the performance with the conventional lime treatment technique. It is demonstrated that milled slag is very effective in terms of improving the internal friction angle of the treated soil, although it cannot compete with lime with regard to the increase in compressive strength. Fly ash is found to be useful as a void filler if used in substantial quantities, as it does not possess a self hardening (pozzolanic) nature to directly influence the time-dependent increase in strength and stiffness of the blended material.

1 INTRODUCTION

Colluvium, talus and other surface deposits in New South Wales are often characterised by high rates of erosion in hilly terrain. Shallow slides on colluvium and the transport of eroded sediments in selected waterways have been documented (Chowdhury and Young, 1989; Habibi and Sivakumar, 1992). It is well known that about 15% of available land is vulnerable to significant gully and rill erosion, while a further 3% has already been affected by moderate to severe sheet erosion. While the use of lime and cement in soil improvement has been considered in the past, the effectiveness of milled slag and fly ash in surface treatment has not been properly studied in Australia. The purpose of this study was to evaluate and compare the relative merits of slag, fly ash and lime treatment in enhancing the engineering behaviour of a typical colluvium deposit collected from the Werombi region. As this project was concerned with the improvement of erodible surface deposits, the aspects of deep chemical mixing and lime column techniques are not discussed within the scope of this paper.

Approximately, 9 million tonnes of fly ash and 2 million tonnes of steel slag are produced annually in New South Wales. Therefore, it is economically attractive to utilise these waste products for soil improvement rather than employing conventional methods such as lime and cement treatment. Fly ash used in this study is the major by-product of coal combustion plants, where it has been directly collected from the boiler gases with the aid of electrostatic precipitators. Blast furnace slag is a by-product of steel manufacturing industry, and this material was collected from BHP steel works in Wollongong, New South Wales. As neither fly ash nor slag used in this study contained toxic materials or acid forming compounds (eg. pyrites and sulphur), blending with colluvium would not pose a risk of groundwater contamination. Hydrated lime was preferred to quicklime, because the latter affects exposed skin as well as promoting corrosion of farming equipment. Nevertheless, the efficient use of quicklime has been discussed elsewhere by Miura et al. (1986).

2 PROPERTIES OF ERODIBLE COLLUVIUM

Colluvium soil used in this study is originated from residual soils found in Picton area. It can be described as finely textured, reddish brown sandy loam of porous structure, with a small component of angular gravels. The topography of the area is precipitous, with the dominant landform consisting of concave upper slopes and irregular lower slopes with colluvial benches. In many places, highly erodible and unstable shallow deposits exist up to a depth of 150-200 cm, above the fine textured shale beds. Samples were obtained at the upper slopes of Eagle Hill, where symptoms of severe gully erosion are clearly visible. The liquid limit and the plastic limit of the soil were determined to be 37% and 20%, respectively, giving a plasticity index of 17%. The particle size distribution of a typical specimen is shown in Fig. 1, where the gradation of fine particles less than 75 microns was estimated by hydrometer tests. The grain size distributions of milled slag, fly

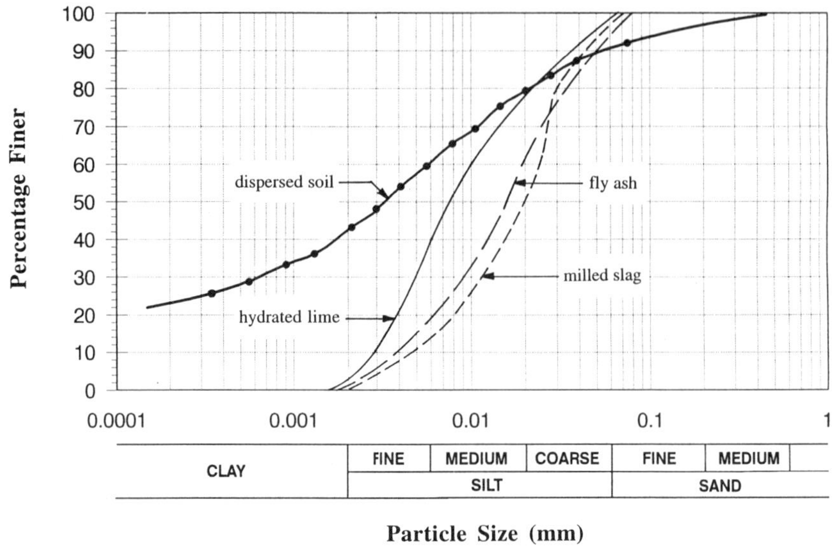

Fig. 1 Particle size distribution of colluvium and additives

ash and hydrated lime as obtained by computerised particle size analyser are also plotted for comparison, which indicates that fly ash and milled slag have very similar distributions. The dispersed soil particles are clearly smaller than the grain sizes of additives which fall within the range of silt size particles. This particular residual soil falls into the category of an inorganic silty clay of low plasticity just above the A-line (CL), and as described by Goldsman et al. (1986), such soils reveal a high degree of erodibility.

The specific gravity of the soil was estimated to be 2.68 which is significantly greater than that of lime (2.21), milled slag (2.32) and fly ash (2.15). Therefore, the blended soil is expected to be light-weight in comparison with the natural soil. Moreover, it was found that the natural soil was acidic (pH @ 4.9-5.1). Consequently, blending with alkaline admixtures such as lime, fly ash and slag would raise the pH levels towards a less acidic material, more suitable for crops.

3 PROPERTIES OF ADDITIVES

Use of lime in ground improvement has been well known for many years (Broms, 1982; Honjo et al., 1991; Yamanouchi et al., 1982; Saitoh et al., 1985). The effectiveness of hydrated lime depends on (a) cation exchange reactions which cause flocculation and (b) pozzolanic reactivity which causes self-hardening of the blended matrix. In contrast, the effectiveness of fly ash and milled slag mainly depends on their free lime contents (F-CaO), which undergoes rapid hydration in the presence of moisture. A small quantity of free lime (2-3%) is adequate to cause significant soil reactions; hence, improvement in strength and deformation properties (Natt and Joshi, 1984). Table 1 gives the chemical composition of fly ash and milled slag used in this study. As the free lime content of this fly ash was less than 0.5%, it could be classified as ASTM class F (non-pozzolanic). The milled slag contained a much higher free lime content (2-3.5%), and can be considered as a low grade cement clinker. Unlike fly ash, milled slag indicated self-hardening properties with time (Fig. 2), where an uniaxial compressive strength of 20 kPa is attained after 3-4 weeks.

4 PROPERTIES OF TREATED COLLUVIUM

The engineering behaviour of colluvium treated with fly ash, milled slag and lime is discussed in this section, with special reference to unconfined compression strength, consolidation and triaxial shear strength.

Table 1 Chemical composition of fly ash and slag

Constitutent (% oxides)	Stabilising Agent	
	Fly Ash	Milled Slag
SiO_2	57-62	14-19
Al_2O_3	24-28	2-5
Fe_2O_3	3-6	16-21
CaO	1.5-4	42-46
MgO	1-2	7-11
MnO	-	5-8
P_2O_5	-	1-3
TiO_2	0.7-1.5	< 0.3
K_2O	1.5-2.5	-
SO_3	0.2-0.4	0.3-0.5
Free Lime (F-CaO)	< 0.5	2.0-3.5

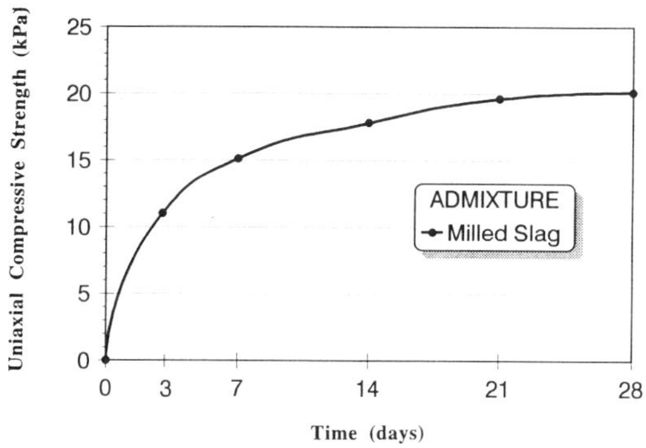

Fig. 2 Increase in uniaxial compressive strength of milled slag with time

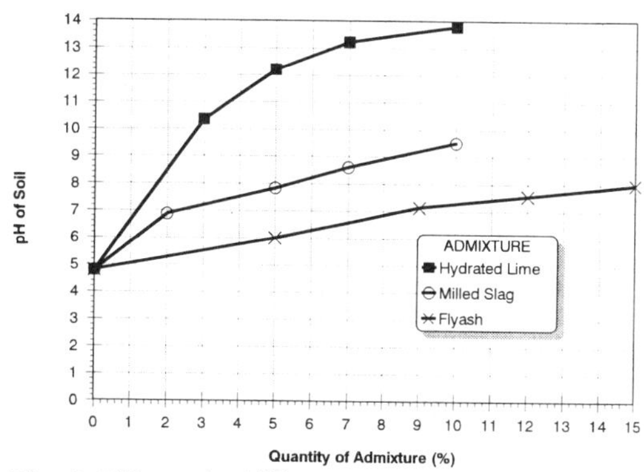

Fig. 3 Effect of additives on pH levels of treated colluvium

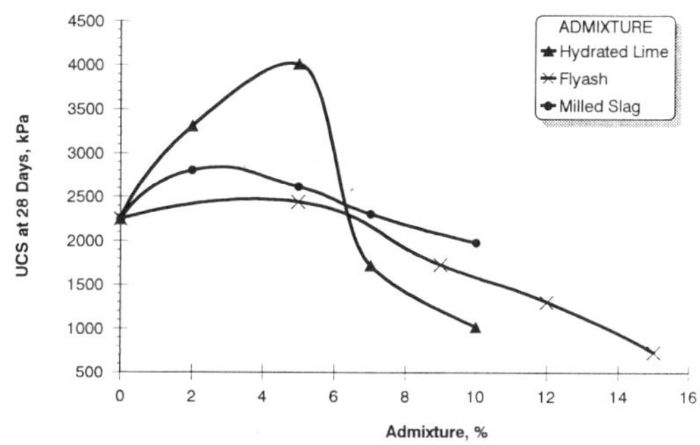

Fig. 4 Effect of additives content on uniaxial compressive strength

4.1 *Effect on soil acidity*

For agricultural purposes, it is relevant to evaluate the effect of admixtures on the soil pH, as the natural soil is acidic (pH 4.9-5.1). Figure 3 indicates that all additives increase the pH value towards an alkaline soil. However, the addition of hydrated lime in small quantities (> 3%) is sufficient to make the blended soil substantially alkaline (pH > 10), which is detrimental to some vegetation as well as being harmful to certain geofabrics used in silt fences. In contrast, fly ash shows the least influence on pH levels, unless a substantial quantity (> 10%) is added. For agricultural purposes, the optimum pH after treatment should be maintained between 6 and 8.5 for most common crops. In this respect, the addition of 2-4% of milled slag seems to be an ideal choice.

4.2 *Effect on Unconfined Compression*

After compacting to standard Proctor condition, blended specimens were tested at various curing periods (7, 14 and 28 days) in unconfined compression, in order to evaluate the time-dependent increase in strength. The blended specimens were air cured at 22° prior to testing, which is not unrealistic considering the field conditions. In all specimens, at least 90% of the 28 day strength was achieved during the first two weeks. Figure 4 illustrates the variation in the maximum unconfined compressive strength (UCS) of treated soil at 28 days curing, as a function of the % of admixture. There is no doubt that neither milled slag nor fly ash can compare with the effectiveness of lime in terms of increasing UCS. Increase in compressive strength due to fly ash is marginal. The optimum lime content is about 5%, which is capable of increasing the strength by a factor of two, whereas the optimum slag content is about 3% associated with an increase in UCS by say 25%.

It is of interest to note that unlike in the case of milled slag, beyond the 5% optimum, the UCS of hydrated lime treated specimens decreases dramatically. This suggests that excessive lime is counter-productive as the available silica in the natural soil is consumed during the hydration phase, an observation also reported by Van Impe (1989). As the current fly ash is non-pozzolanic (ASTM Class F), one cannot expect a significant increase in strength of fly ash treated soil. Nevertheless, fly ash acts as an excellent 'void filler' by producing an overall soil matrix with reduced porosity, hence of slightly increased strength. Excessive fly ash, however, increases the overall porosity once again as well as reducing the intrinsic cohesion of the soil, thereby causing a reduction in strength. In contrast, pozzolanic fly ashes (ASTM Class C) provide significant internal cementation with corresponding increase in strength, as discussed elsewhere by Indraratna et al. (1991).

4.3 *Effect of additives on consolidation*

The use of additives in soil directly affects the consolidation properties due to both physical (compaction) and chemical (hydration) reasons. One-dimensional oedometer consolidation tests were conducted to investigate the nature of compressibility of treated specimens compacted to their standard Proctor densities, and cured for four weeks. Figure 5 illustrates the compressibility curves of the treated specimens, where the consolidation curve of the natural soil is also plotted for comparison. In general, all blended specimens show a decrease in compressibility with the applied pressure, but the rate of void ratio reduction is less than that of the natural soil. Moreover, while hydrated lime tends to increase the initial void ratio (Fig. 5a), fly ash decreases the initial void ratio (Fig. 5c). In the case of the former, the increased porosity can be partially attributed to reduced compaction and hydration induced swelling. In contrast, fly ash acts as a more efficient void filler, and due to its non-pozzolanic nature, hydration induced volume expansion is not expected, even at 15% fly ash content. Similar to fly ash, milled slag also decreases the initial void ratio, except at high contents (10%) which causes significant hydration.

It may be concluded that in lime and slag, the reduction in compressibility is related to hydration and subsequent hardening of the treated soil, whereas in fly ash, compressibility is related to the initial void ratio after blending and compaction. This reduction in initial void ratio has obvious consequences on the resulting permeability, an aspect which has not been discussed in detail in this paper. In the experience of the author, the consolidation settlement of a chemically stabilised specimen can be 5-10 times lower than that of the natural soil, for instance in the case of soft Bangkok clay treated with hydrated lime. In comparison with the other two additives, hydrated lime shows almost an almost flat rebound response upon unloading. Such behaviour can be beneficial for structures such as silos and warehouses subject to changing loads (storage and removal) in minimising foundation cracking.

4.4 *Effect of Additives on Drained Shear Strength*

In order to investigate the effect of additives on the "true" shear strength of treated colluvium, a series of consolidated-drained triaxial compression tests were carried out on compacted specimens (38mm x 76mm) after 28 days of curing. All natural and treated specimens were subject to effective confining pressures of 100, 200 and 400 kPa, and were sheared at small axial strain rates of 0.01-0.03% / min, with no excess pore pressures developing during testing. The stress-strain response was typical of strain-softening nature, and as expected, the peak deviator stresses and the corresponding failure strains increased with the confining pressure. Although most specimens indicated a distinct shear plane at failure, fly ash treated specimens revealed pronounced post-peak brittleness (Fig. 6a) in comparison with the natural soil (Fig. 6b), particularly for fly ash contents exceeding 5%. Irrespective of the confining pressures, these specimens failed within a narrow

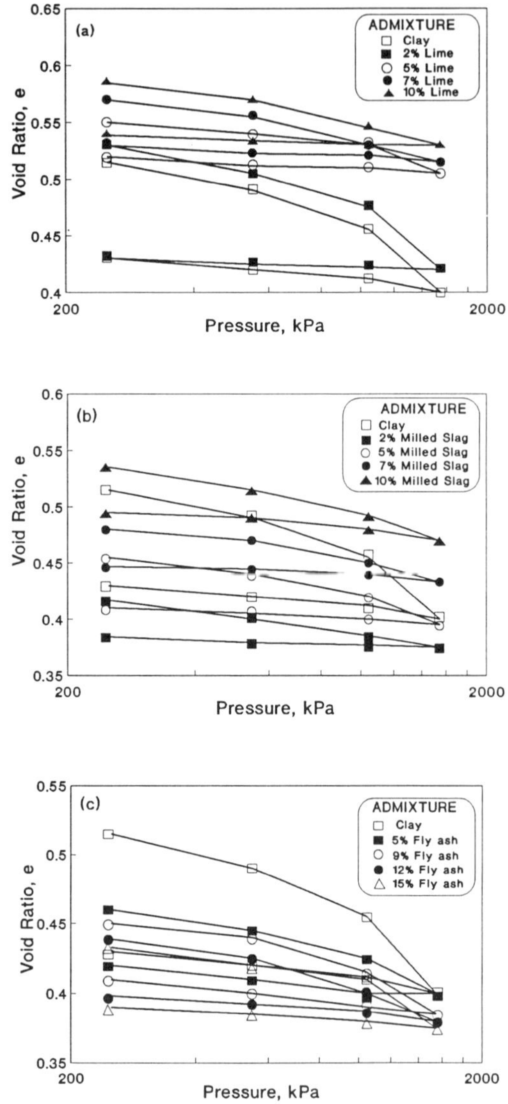

Fig. 5 Consolidation of (a) lime-treated, (b) slag-treated and (c) fly ash-treated soil

range of axial strains (3-5%). Furthermore, they often indicated some tensile cracking prior to the development of distinct shear planes. Specimens stabilised with hydrated lime and milled slag revealed a more gradual strain-softening behaviour, and their failure strains varied from 3% to 9%, with increasing confining pressure, as summarised in Table 2.

The effect of additives is reflected by a significant increase in the apparent friction angle of the treated specimens, while the apparent cohesion did not reveal a specific relationship to the additives content or the

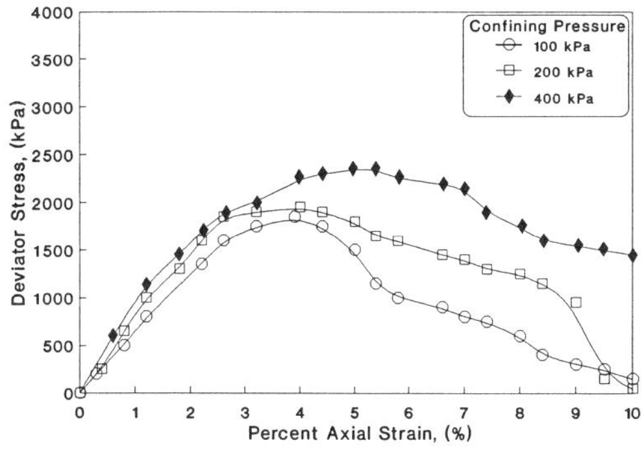

(a)

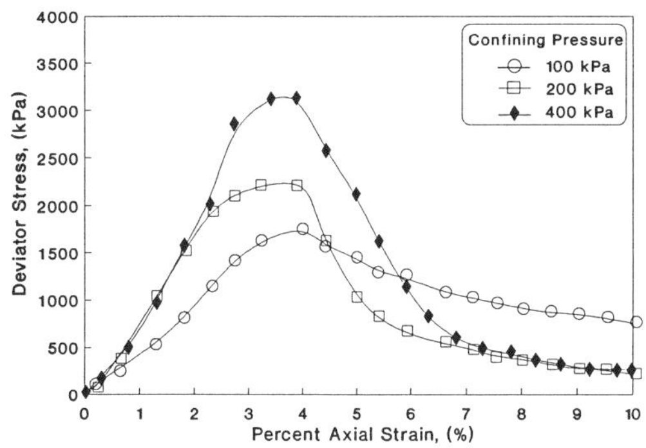

(b)

Fig. 6 Stress-strain behaviour of (a) natural soil and (b) fly ash-treated soil

Table 2 Effect of lime and milled slag on failure strains

Specimen	Failure Strain, %		
	σ'_3=100 (kPa)	σ'_3=200 (kPa)	σ'_3=400 (kPa)
Natural Soil	3.4	3.9	5.2
2% lime	3.6	4.2	5.6
slag	3.8	4.8	6.2
5% lime	3.3	3.9	4.2
slag	2.7	4.3	5.9
7% lime	3.5	4.4	5.4
slag	4.4	5.7	9.0
10% lime	3.8	5.2	6.4
slag	4.4	4.6	6.5

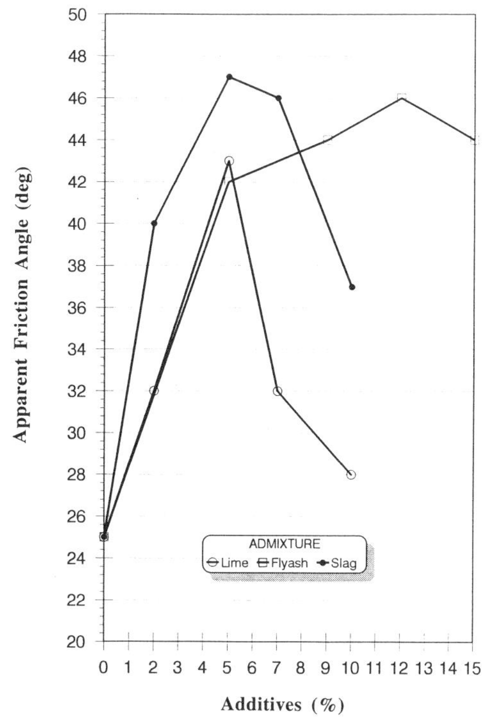

Fig. 7 Effect of additives on apparent friction angle of treated soil

type of additives. The failure envelopes were approximated to the linear Mohr-Coulomb strength criterion, where the shear strength parameters were defined by the drained friction angle, ϕ_d and the drained cohesion intercept, c_d. The variation of the drained friction angle with the content of additives is illustrated in Figure 7. A maximum friction angle of 47° was obtained for the 5% slag-treated specimen, whereas for 5% lime and 5% fly ash-treated specimens, the apparent friction angles were determined to be 43° and 42°, respectively. It is evident from Figure 7 that with respect to the drained friction angle, the optimum lime and slag contents are close to 5%, whereas the optimum fly ash content is considerably greater (12% or so).

In view of the behaviour of stabilised soils studied here, it may be concluded that milled slag is the most effective at relatively low proportions of additives, whereas fly ash is the least effective. The improvement of internal friction by milled slag is the combined effect of (a) cementation bonding and (b) void ratio reduction (except for specimens exceeding 10% slag content). At the same % of additives, the reason why lime does not increase the apparent friction angle as much as slag is perhaps due to the greater consumption of free silica during hydration and the higher porosity of the lime-treated specimens.

5 CONCLUSIONS

Consolidation results confirm that all three additives: hydrated lime, milled slag and fly ash decrease the compressibility of the stabilised soil, but not necessarily its porosity. It is not surprising that hydrated lime and milled slag would decrease compressibility due to self-hardening (hydration and cementation), whereas non-pozzolanic fly ash (ASTM Class C) would only improve soil stiffness by decreasing the void ratio (porosity). Although hydrated lime increases the void ratio of the blended soil, the achievable gain in uniaxial compressive strength (UCS) is considerable. For instance at a 5% (optimum) lime content, UCS of the blended soil increases from 2250 kPa to 4000 kPa. In contrast, fly ash does not significantly increase the UCS, in spite of the fact it decreases the porosity of the blended colluvium.

With respect to drained shear strength, a small percentage (about 5% optimum) of milled slag is sufficient to achieve a high increase in the drained friction angle of the blended specimens (47°, in comparison with the natural soil of 25°). In comparison, at least 12% of fly ash is required in order to obtain a similar gain. From an economic point of view, using milled slag for the stabilisation of this residual colluvial soil is amply justified, although it still cannot compete with more expensive lime products which can quickly achieve a more marked increase in strength, for instance in the case of road construction. As also shown by this study, the use of ASTM Class F fly ash as an additive on its own is not effective, unless co-treated with lime or cement as discussed elsewhere by Natt and Joshi (1984).

ACKNOWLEDGEMENTS

The laboratory assistance of Ken Cunningham (University of Wollongong) and Glen Fordham (TRN Earthmoving Ltd.) is gratefully appreciated.

REFERENCES

Broms, B., 1982. Lime columns in theory and practice, Proc. Int. Conf. Soil Mechanics, Mexico, pp. 149-166.

Chowdhury, R.N. and Young, A.R.M., 1987. Field guide to slope instability in the Wollongong-Picton-Nattai region of New South Wales, Report prepared for the 5th Int Conf and Field Workshop on Landslides, Wollongong, 62p.

Habibi, M. and Sivakumar, M., 1992. Computerisation and application of new sediment transport estimation methods to river systems, 11th AFMC Conference, Hobart, pp. 175-178.

Honjo, Y., Chen, C.H., Lin, D.G., Bergado, D.T. and Okamura, R., 1991. Behaviour ofimproved ground by the.deep mixing method, Proc Kozai Club Seminar, Bangkok, Thailand.

Indraratna, B., Nutalaya, P., Koo, K.S. and Kuganenthira, N., 1991. Engineering behaviour of a low carbon, pozzolanic fly ash and its potential as a construction fill, Canadian Geotechnical Journal, Vol. 28, pp. 542-555.

Miura, N., Koga, Y., and Nishida, H., 1986. Application of deep mixing method with quicklime for the Ariake clay ground, Soils and Foundations, Vol. 34, No.4, pp. 5-11.

Natt, G.S. and Joshi, R.C., 1984. Properties of cement & lime-fly ash stabilized aggregate, Transport Research Record 998, Transportation Research Board, pp. 32-40.

Saitoh, S., Suzuki, Y. and Shirai, K., 1985, Hardening of soil improved by deep mixing method, Proc. 11th Int. Conf. Soil Mech. & Found. Eng., San Francisco, Vol.3, pp. 1985-88.

Serra, M., Robnet, J.C., Mohkam, M. & Daonk, T., 1988. Improvement of dykes by liming, Proc. 8th European Conf. on Soil Mechanics and Foundation Engineering, Vol. 2, pp. 947-950.

Van Impe, W.F., 1989. Soil Improvement Techniques and their evolution, Balkema Publishers, Rotterdam.

Yamanouchi, T., Miura, N., Matsubayashi, N. and Fukuda, N., 1982. Soil improvement with quicklime and filter fabric, J. Geotechnical Engineering, ASCE, Vol. 108, pp. 935-965.

Environmental Geotechnics, Kamon (ed.) © 1996 Balkema, Rotterdam. ISBN 90 5410 848 7

Reuse of industrial wastes as potential liners in landfills

H.I.Jeong & Y.S.Lee
Geotechnical Engineering Division, Korea Institute of Construction Technology, Seoul, Korea

ABSTRACT

Earthen liners have been widely used to contain waste materials in landfills. There is need for cost effective alternatives where clay is scarce. The potentiality of industrial wastes, such as fly ash, crushed waste concrete and clarifier waste sludge, as the althernative liners in landfills is investigated. A series of tests were performed on industrial wastes and admixtures to examine their compaction, compressive strength, hydraulic conductivity characterisitics and the compatibility with representative leachate within landfills. Results from the testing showed that adequate hydraulic conductivity lower than 10^{-7}cm/sec can be achieved with sufficient stabilizer contents and curing. The findings suggest that admixed fly ash, crushed waste concrete, and clarifier waste sludge could serve as an alternative to clay soil in liner construction materials.

INTRODUCTION

Disposal of wastes in an acceptable manner has been of growing concern to the public. One of the most significant concerns has been the possible contamination of ground water from leachates generated by wastes. Recent national regulations require all hazardous waste disposal facilities to be lined with suitable impermeable barriers to protect against contamination.

Polymeric membrane liners, clay, bentonite or cement admixtures and stabilized earth liners have been widely used for this purpose ; however, in many instances such applications are extremely expensive because of high costs of synthetic liners or inadequate because of the lack of suitable clay materials at the disposal site. There is close scrutiny over the durability of liners in contact with various wastes and a need for cost-effective alternatives with superior qualities.

Fly ash, crushed waste concrete, and clarifier clarifier waste sludge present a significant potential as cost-effective liner materials. A series of tests were performed on these industrial waste materials and stabilized admixtures to evaluate the physical, mechanical and hydraulic characteristics pertinent to containment of waste liquids. The influence of construction factors such as water content, compactive effort, and time between mixing water and compacting on density and hydraulic conductivity was considered. Hydraulic conductivity tests using distilled water(DW) and landfill leachate were performed. A summary of the findings of this investigation is presented. The laboratory experiments indicate that admixed fly ash, crushed waste concrete, and clarifier waste sludge have many properties that are attractive as a liner construction material.

BASIC ENGINEERING PROPERTIES

Three types of industrial waste materials, fly ash produced from Boryeong fired power plant, crushed waste concrete generated from demolishing of old apartment, and waste sludge produced from Kwacheon clarifier are used as the major substances for liner materials, and a activated sodium(Na) bentonite is used as an additive material.

The basic properties of the major substances and the additive are shown in Table 1, and the grain size distribution curves are shown in Figure 1. The specific gravity of the bentonite is significantly lower than those of the other materials (Jeong 1995). The coefficient of uniformity (Cu) shows that sludge and bentonite are well graded and the other materials are relatively uniform. Boryeong fly ash and bentonite are composed of fine materials and waste concrete are composed of granular materials which are indicated by the percent of passing No. 200 sieve.

Table 1. The basic engineering properties of the liner materials

Material	Specific gravity	Grain size distribution				Remark
		D_{10} (㎜)	D_{60} (㎜)	Cu	Percent of passing No. 200 sieve	
Boryeong fly ash	2.26	0.003	0.02	6.6	93.0	-
Crushed waste concrete	2.54	0.15	1.1	7.3	0.0	sample passing No.4 sieve
Clarifier waste sludge	1.87	0.4	2.0	5.0	1.1	sample passing No.4 sieve
Activated sodium bentonite	1.92	0.008	0.04	50.0	76.1	-

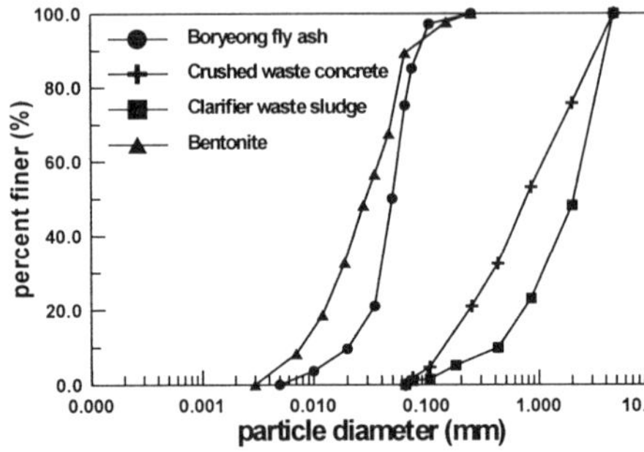

Figure 1. Grain size distribution curves of the liner materials

COMPACTION

When liner materials are built in landfills, they are often compacted to reduce the hydraulic conductivity below the value specified in the regulatory guide and to have the enough strength that can support the weight of wastes. Hence, it is important to test maximum dry density and optimum water content of compacted liner materials (e.g., Goldman 1984).

Compaction characteristics of various materials of major substances and additivies

Table 2. The results of compaction tests (ASTM D698) of the liner materials

Material type (dry weight)	Maximum dry density (g/㎤)	Optimum moisture content (%)
Boryeong fly ash + Bentonite:		
100% + 0%	1.205	21.2
95% + 5%	1.295	22.8
90% + 10%	1.265	24.0
85% + 15%	1.244	26.7
80% + 20%	1.240	28.0
70% + 30%	1.236	31.3
60% + 40%	1.249	32.5
Crushed waste concrete + Bentonite		
100% + 0%	1.618	18.3
95% + 5%	1.605	19.8
90% + 10%	1.570	21.2
85% + 15%	1.526	22.0
80% + 20%	1.490	23.5
70% + 30%	1.489	25.3
60% + 40%	1.408	29.3
Clarifier waste sludge + Bentonite		
90% + 10%	0.903	55.0
80% + 20%	0.935	46.0
70% + 30%	0.985	40.0
60% + 40%	1.026	39.1

were investigated by performing standard Proctor compaction(ASTM D698) on materials using additive contents 0, 5, 10,15, 20, 30, 40 percent.

The bentonite was added on the major substances until the permeabilities of the compacted materials were lower than $1x10^{-7}$ cm/sec(e.g., Bowders 1987). The maximum dry density and optimum moisture content were obtained on admixtures. The results are summarized in Table 2, and the variations of the properties with respect to the percent of bentonite by weight of the sample are shown in Figure 2 and 3.

Crushed waste concrete, having a higher specific gravity, yields higher maximum dry density than clarifier waste sludge, which is a lighter material with a lower specific gravity. Maximum dry density of crushed waste concrete is reduced continuously as the amount of bentonite increases.

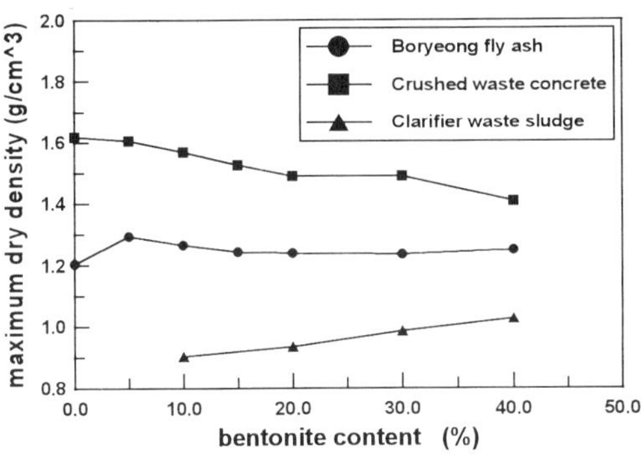

Figure 2. Maximum dry density with bentonite contents

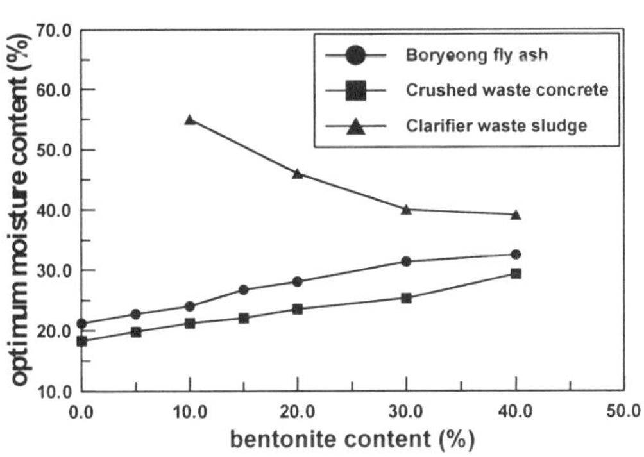

Figure 3. Optimum moistrure content with bentonite contents

This means that the bentonite, which has smaller specific gravity than the other materials, can make the density of admixed materials lower than the density of the materials without bentonite. On the other hand, maximum dry density of fly ash and sludge are increased as the amount of bentonite increases.

The optimum moisture content of fly ash and crushed waste concrete increases, and the optimum water content of sludge decreases as the percent of bentonite increases. The magnitude of optimum moisture content for sludge is higher than that of the other materials(e.g., Cho 1995).

The increase of optimum water content as the increase of the percent of bentonite seems to be caused by the increased hydrophilic tendency of the admixed materials with the addition of bentonite.

COMPRESSIVE STRENGTH

The admixtures included in the compaction study were evaluated for strength by fabricating Proctor-sized specimens, curing them for 7, 14 and 28 days at 25℃ and 100 percent relative humidity, and testing them for unconfined compressive strength immediately after curing. All specimens were compacted at 95 percent of the maximum dry density and 3 percent moist side of the optimum moisture contents established in the compaction study. Compressive strengths of the compacted admixed materials with varying the percent of bentonite are presented in Table 3 and strengths with curing time are plotted in Figure 4. The range of unconfined strength is located in between 100~350 kN/m^2 for all admixtures. The strengths of compacted materials are not quite releted to the curing time and the percent of bentonite for the admixtures.

HYDRAULIC CONDUCTIVITY

Although there is much debate as to the merits of various types of permeameters, the rigid-wall permeameter was used to measure the hydraulic conductivity in this

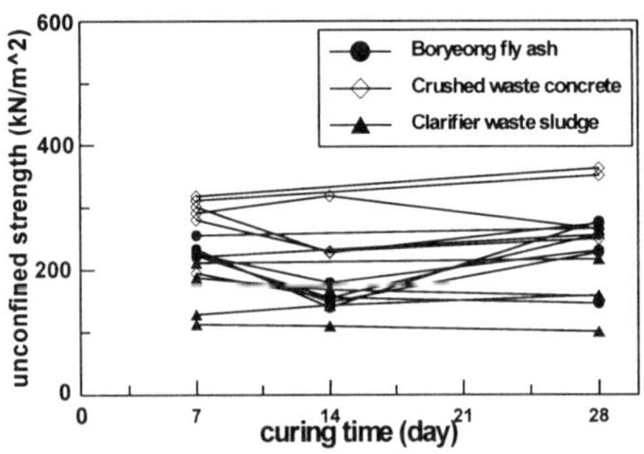

Figure 4. Unconfined strength with curing time

Table 3. The results of unconfined compressive strength of the liner materials.

Material type (dry weight)	Unconfined compressive strength (kN/m^2)		
	7days	14days	28days
Boryeong fly ash + Bentonite			
95% + 5%	223	156	147
90% + 10%	227	149	228
85% + 15%	233	140	277
80% + 20%	226	179	231
70% + 30%	255	-	266
60% + 40%	203	-	256
Crushed waste concrete + Bentonite			
95% + 5%	195	156	256
90% + 10%	291	319	265
85% + 15%	301	228	274
80% + 20%	280	230	250
70% + 30%	312	-	352
60% + 40%	318	-	362
Clarifier waste Sludge + Bentonite			
90% + 10%	113	109	101
80% + 20%	129	144	159
70% + 30%	188	169	158
60% + 40%	211	-	217

study. Experience with various permeameters has shown that the rigid-wall permeameter simulates a worst case scenario due to the potential for side-wall leakage. The specimens for the hydraulic conductivity test are prepared in the same manner as in the compaction test.

The minimum amount of bentonite required to make the hydraulic conductivity of the mixture permeated with distilled water below $1x10^{-7}$cm/sec, which is the required minimum value of K for the liners, is estimated in the hydraulic conductivity test. The minmum amount of bentonite is important for specifying the mixing ratio of bentonite and major substances for producing admixed liners in the field. Hydraulic conductivity results are given in Table 4, and their relation to the percent of the bentonite is shown in Figure 5. A hydraulic gradient of 50 was used to reduce testing time. The dimensions of the specimen are 10 cm the diameter and 8 cm the height.

The results confirm that hydraulic conductivity was reduced in all mixtures

Table 4. The results of hydraulic conductivity of the liner meterials

Material type (dry weight)	Hydraulic conductivity (cm/sec)
Boryeong fly ash+Bentonite	
95% + 5%	2.8×10^{-6}
90% + 10%	1.4×10^{-6}
85% + 15%	6.5×10^{-7}
80% + 20%	3.3×10^{-7}
70% + 30%	3.7×10^{-8}
60% + 40%	2.0×10^{-8}
Crushed concrete+Bentonite	
95% + 5%	1.3×10^{-5}
90% + 10%	9.7×10^{-6}
85% + 15%	5.1×10^{-6}
80% + 20%	8.8×10^{-7}
70% + 30%	4.4×10^{-8}
60% + 40%	1.4×10^{-8}
Waste sludge + Bentonite	
90% + 10%	1.7×10^{-5}
80% + 20%	1.4×10^{-5}
70% + 30%	5.2×10^{-7}
60% + 40%	1.3×10^{-7}

with increasing bentonite content. The cause of the hydraulic conductivity decrease is that the bentonite can swell from 5 to 10 times of its original volume with absorption of water. Admixed materials, depending upon the mix design, can be prepared with coefficient of hydraulic conductivity on the order of 1.0×10^{-7} cm/sec or lower. The percent of bentonite needed to reduce the hydraulic conductivity of the liner materials below $1x10^{-7}$cm/sec is 25% for Boryeong fly ash, 27% for crushed waste concrete and 40% for clarifier waste sludge.

The hydraulic conductivity with respect to the elapsed time for the crushed waste concrete mixtures of 30, 40 percent bentonite is given in Figure 6. This shows hydraulic conductivity of the admixed liner material

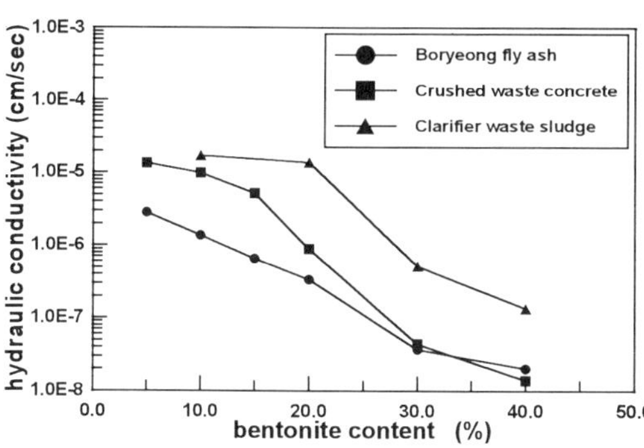

Figure 5. Hydraulic conductivity with bentonite content

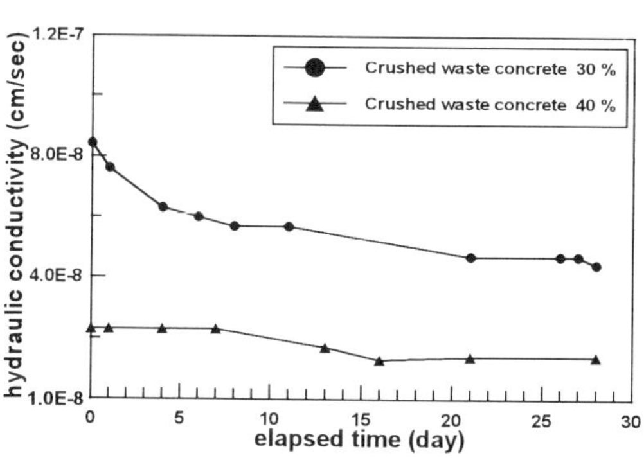

Figure 6. Hydraulic conductivity with elapsed time

reduces also with increasing curing time and onverge to certain values. This is because swelling of the bentonite progresses slowly, and the time for complete swelling compares well with the convergence of the hydraulic conductivity of the liner materials.

COMPATIBILITY TO LEACHATE

The chemical compatibility of the admixtures to the leachate are investigated to examine the reactivity and changes in hydraulic conductivity when exposed to actural landfill environment. The admixed materials used in these tests are admixtures of 30, 40, percent bentonite.

The compatibility results with leachate are shown in Table 5. In this test, the pure leachate was sampled in Kimpo metropolitan municipal waste landfill located in the west seashore in Korea. The properties of the sampled pure leachate is summarized in Table 6. Table 5 shows that the hydraulic conductivity with permeating the leachate increases very slightly than the hydraulic conductivity with permeating the deionized distilled water. Thus, compatibility test shows that municipal waste leachate do not tends to cause and increase in hydraulic conductivity and adversely affect the

Table 5. The result of compatibility to leachate of the liner meterials

Material type (dry weight)	Hydraulic conductivity (cm/sec)	
	water	leachate
Boryeong fly ash + Bentonite		
70% + 30%	3.0×10^{-8}	4.7×10^{-8}
60% + 40%	2.0×10^{-8}	3.9×10^{-8}
Crushed waste Concrete + Bentonite		
70% + 30%	4.4×10^{-8}	4.7×10^{-8}
60% + 40%	1.4×10^{-8}	4.0×10^{-8}
Clarifier waste sludge + Bentonite		
70% + 30%	5.2×10^{-7}	3.6×10^{-7}
60% + 40%	1.3×10^{-8}	2.3×10^{-7}

Table 6 Properties of municipal waste leachate used in compatibility test

Item	Measured value (ppm)
CODmn	627
BOD	2,700
SS	243
TS	11,530

hydraulic conductivity of all mixtures(e.g., Bowders 1985; Daniel 1984; Madsen 1987).

SUMMARY AND CONCLUSIONS

The potentiality of Boryeong fly ash, crushed waste concrete, and clarifier waste sludge as the alternative liner materials in landfills is evaluated by performing the physical, mechanical and hydraulic conductivity test with water and leachate. Based on the results of laboratory experiments, the following conclusions are made.

(1) Compaction studies indicate that the maximum dry density and the optimum moisture content with the addition of bentonite was quite different. This characteristics seems to come mainly from the difference in specific gravity and grain size distributions of the liner materials.

(2) The range of unconfined compressive strength is located in between 1.0~3.5 ㎏/㎠ for all liner materials. The strengths of materials are not quite related to the curing time and the percent of bentonite for the admixtures.

(3) The hydraulic conductivity decreases with increasing bentonite content. The percentage of bentonite needed to make the hydraulic conductivity below $1x10^{-7}$㎝/sec was 25% for Boryeong fly ash, 27% for crushed waste concrete and 40% for clarifier waste sludge.

(4) The effects of the municipal waste leachate on the hydraulic conductivity of the admixtures were very small. From the limited data, the municipal waste leachate may make the increase in hydraulic conductivity smaller, or even negligible.

REFERENCES

Bowders J. J. *The Influence of Various Concentrations of Organic Liquids on the Hydraulic Conductivity of Compacted Clay.* Geotechnical Engineering Dissertation GT85-2, The University of Texas at Austin, 218p., 1985.

Bowders, J. J., Mumtaz A. Usmen, and James S. Gidley, *Stabilized Fly Ash for Use as Low-permeability Barriers.* Geotechnical practice for waste disposal '87, Geotechnical special publication No. 13, pp.320~333, 1987.

Cho, H. J., *A Feasibility Study of the Stabilized Sludge as a Potential Landfill Cover & Liner Materials*, Department of Civil Engineering, Kang Weon National University, 110p., 1995.

Daniel, D. E. and Howard M. Liljestrand. *Effects of Landfill Leachates on Natural Liner Systems.* The university of Texas, Austin, 1984.

Goldman L. J., L. I. Greenfield, A. S. Damle, G. L. Kingsbury, *Design, Construction, and Evalution of Clay Liners for Waste Management Facilities.* US EPA, EPA Contract No. 68-01-7310, 1987.

Jeong, H. I., etc., *Development of Liner Materials for Waste Containment*, Final Report, KICT/95-GE-1202, Korea Institute of Construction Technology, 273p., 1995.

Madsen, F. T. and J. K. Mitchell, *Chemical Effects on Clay Hydraulic Conductivity and their Determination.* Open File Report, No. 13, The Environmental Institute for Waste Management Studies, 1987.

Environmental Geotechnics, Kamon (ed.) © *1996 Balkema, Rotterdam. ISBN 90 5410 848 7*

Utilization of excavated soils from reconstruction of roads in Sapporo City

Mitsuhiko Kamiya & Atsushi Kasahara
Hokkaido Institute of Technology, Sapporo, Japan

Noboru Hoshi & Hiroaki Watanabe
Sapporo Road Maintenance Public Corporation, Japan

ABSTRACT: The waste materials from road construction consist of asphalt mixtures, gravel subbases and subgrade soils. In Sapporo city, they are separately excavated in layers to avoid mixing them with each other, and each is recycled through its own distinct system, taking into consideration the characteristics of the materials and their ultimate end use. The excavated subbase soil is classified into crushed stone, sand and sediment of the settling basin by washing and sieving. A mixture of this crushed stone and sand is supplied as subbase course material and the sand as an anti-frost layer. The fine-grained soil of the settling basin is make up a dewatered filter cake by a pressure dehydrator. The excavated subgrade soils and the dewatered filter cake are placed in a stock pile. The homogeneous soil is used to mix soils. These soils are currently used for embankments and for covering materials of garbage dumping sites.

1 INTRODUCTION

In Sapporo excavated soils of 1.1 million cubic meters come from street reconstruction annually. In the past, excavated soils and gravels have been used as embankment materials for public land, covering soil for reclaimed land or materials for land development.

However, each year it becomes more difficult to secure land in Sapporo suitable for accepting them, so the need to recycle those waste products is pressing. Granular subbase and subgrade soil have often been excavated together and handled together. However, the each has distinct properties, and subbase soil of granular material has considerably higher value. To use these resources most effectively, it is desirable to recycle subgrade soil and granular subbase in differently.

In warm regions, excavated soil containing granular subbase is usually mixed with cement or lime to lower the water content and increase its strength[1)]. It is then used again as road material. However, in cold regions like Sapporo, bearing capacity alone is not sufficient. Subgrade soil and granular subbase must also be non susceptible to frost heave. Therefore, excavated soil from streets can be used as road material only if its gradation is improved to prevent frost heave.

The flow diagram for an excavated soil recycling system in Sapporo is shown in Figure 1. At first, many field tests are conducted for granular subbases and subgrade soils in pavements which will be reconstructed. From those samples, a quantity and a gradation of granular subbase and a nature of the subgrade soil are determined. As a rule we excavate granular subbase and subgrade soil as separate in layers. Granular subbase and subgrade soil with a lot of granular material is recycled at a recycling plant for granular subbase. Other excavated soils are recycled using a stock pile. The important feature of this system is that granular subbase and subgrade soil are recycled separately, taking into account the distinct nature of each.

2 RECYCLING GRANULAR SUBBASE

Granular subbase excavated from streets is recovered by washing with water at a recycling plant, then separating the crushed stone and sand. This way it is possible to recover valuable crushed stone and sand at low cost and with low energy input.

The process flow diagram is shown in Figure 2. Refuse over 150mm diameter (masses of concrete, boulders, garbage, etc.) is removed from the excavated soil by a screen in an initial hopper. Then refuse over 60mm is removed with a

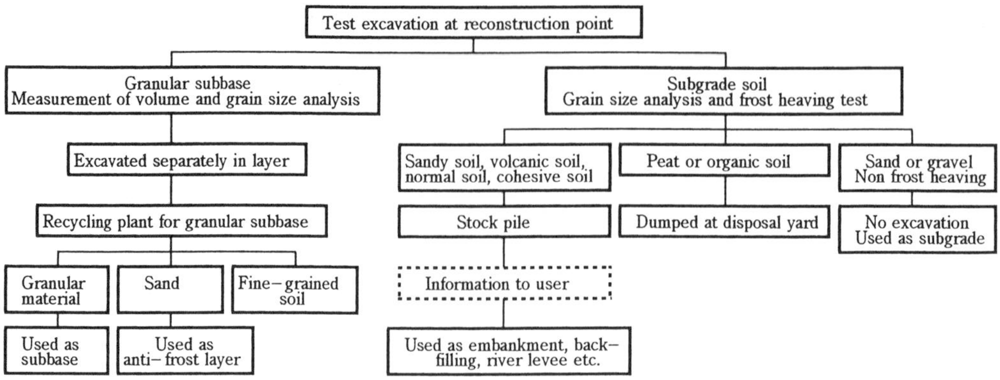

Figure 1 Flow diagram for excavated soil recycling system in Sapporo

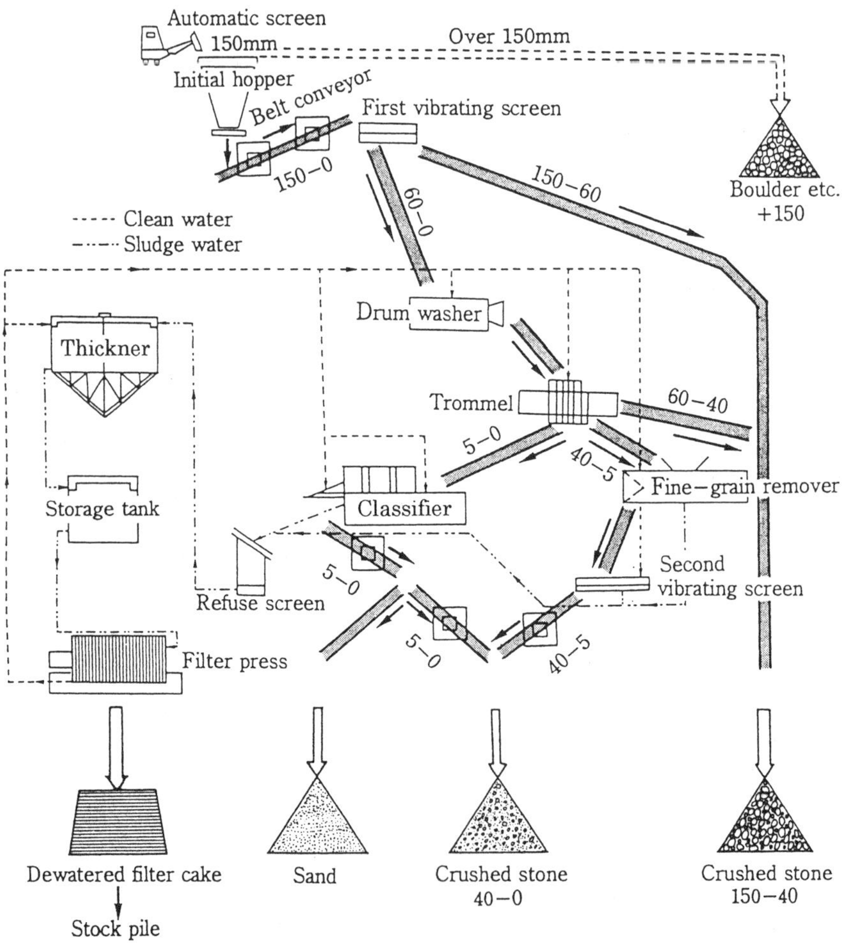

Figure 2 Flow diagram of recycling plant for granular subbase

vibrating screen. Gravel under 60mm is washed with water in a drum washer and then separated by a Trommel into the following sizes: over 40mm, 5-40mm and under 5mm.

Only sand is collected from the under 5mm soil by a classifier. Crushed stone in the 5-40mm range is collected after stirring and washing with pressurized water in a clay bloomer to remove fine grained contents attached to the gravel, and it is then goes through a second vibrating screen. This crushed stone with grain range of 5-40mm has only a small fine grained contents and the gradation is not balanced, so it is mixed with an appropriate amount of sand and piled up as 40-0mm crushed stone.

Silt and clay mixed with water is discharged from each classifier, and goes into a thickener through a refuse screen. The mixture is condensed and settled using chemicals, and becomes a dewatered filter cake after compressing in a filter press. The cake has approximately 40% water content. Dewatered filter cake is mixed with subgrade soil using a stock pile and used for embankment materials, etc.

Well water used for washing is circulated and reused, so polluted water does not flow into rivers during regular operation.

This plant has a capacity of processing about 90 cubic meters of soil per hour. Operating from April to November, it can process about 100,000 cubic meters in a year. The output of recycled crushed stone (grain size: 40-0 mm) and sand (5-0 mm) naturally depends on the properties of the subbase materials excavated from the street. Currently 53,000 cubic meters of crushed stone and 26,000 cubic meters of sand can be output annually, representing 50% and 25% of the input, respectively. Thus 75% of the input can be recycled into products that can be put on the market. Gravel larger than 40mm and fine-graded soil can not be put on the market as products. They account for 10% and 15% of the input, respectively.

Figure 3 shows the grain size distribution curve for the crushed stone in the recycled grunular subbase. The distribution of grain size falls within the standard range specified for civil engineering works by Sapporo City. Between 0.3% and 2.7% of the excavated soil is screened with a 0.075mm screen. This ratio is less than 15% of the standard value specified for subbase materials.

The frequency distribution of the modified CBR is shown in Figure 4. The mean value is 84%, which far exceeds 30%, the granular subbase standard for Sapporo City. Screening the recycled sand at a size of 0.075mm yields an average screening ratio between 0.5 and 4.4 %. Testing has proven that it is not susceptible to frost heave, so it can be used as material of anti-frost layer.

The unit prices of the crushed stone and sand recycled at this plant are almost the same as those of conventional products on the market. Thus, this granular subbase recycling plant is cost effective.

3 RECYCLING SUBGRADE SOIL

The excavated subgrade soil is tested for grain size analysis and then tested for susceptibility frost heave. When the results of the frost heave test indicate that the soil is not susceptible to frost heave and has the prescribed bearing capacity, the soil is not excavated but used as it is for subgrade soil. In this way amounts of excavated soil and newly applied sand for the

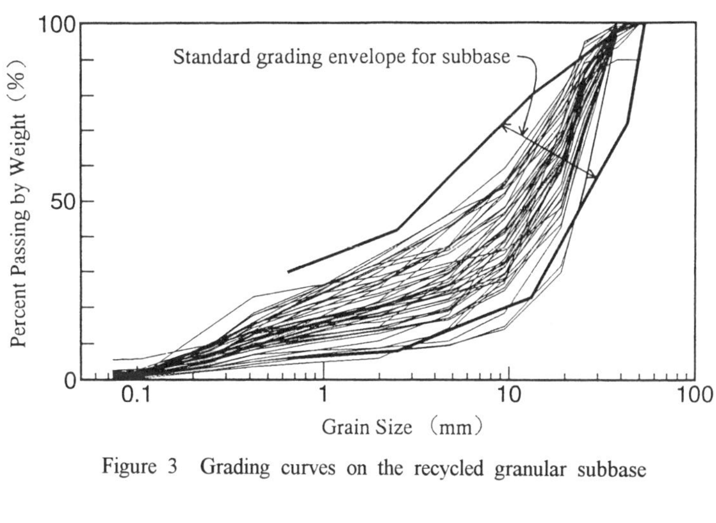

Figure 3 Grading curves on the recycled granular subbase

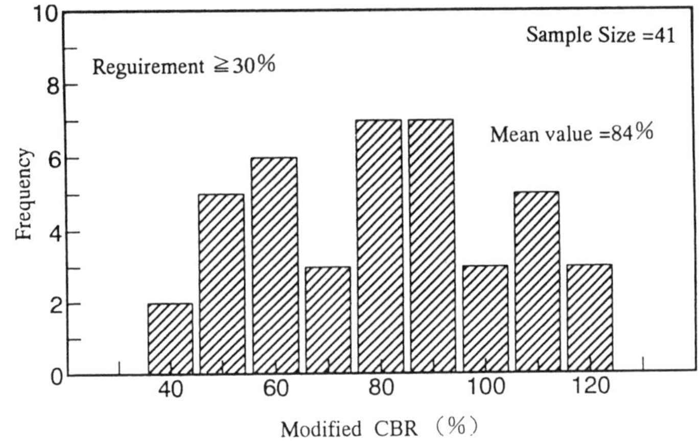

Figure 4 Frequency distribution of modified CBR in the recycled granular subbbase

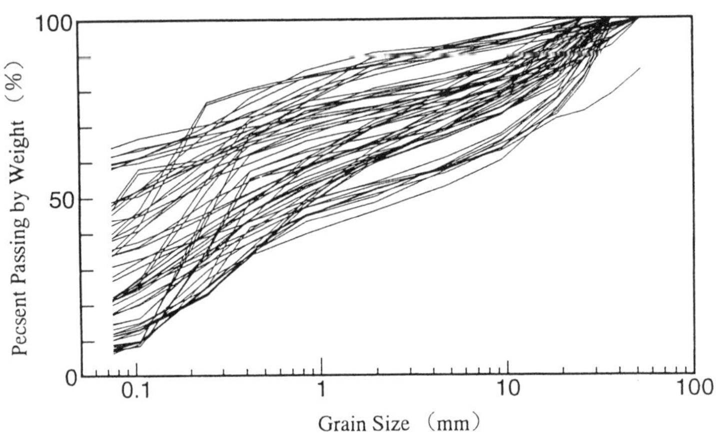

Figure 5 Grading curves on the excavated subgrade soils

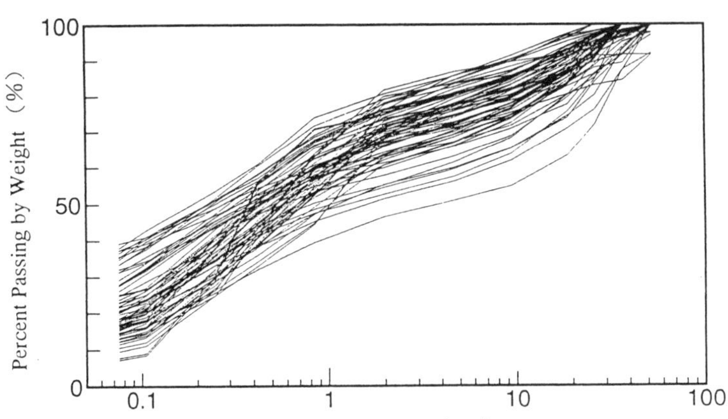

Figure 6 Grading curves on the mixed soils in the stock pile

anti-frost layer can be reduced. However, with the current level of recycling technology it is not possible to utilize subgrade soils consisting of peat or organic soil. These are dumped at disposal yards. On the other hand, soils such as sandy soil, volcanic soil, normal soil and cohesive soil are carried to stock pile yards and classified into these four groups by visual observation. To obtain the prescribed soil, stock piles of these soils are made to form alternating strata. The dewatered filter cake of fine-grained soil from the subbase material recycling plant is also mixed into these stock piles.

The stock pile is used to mix the soils to obtain homogeneous soil keep by mixing soils of different properties. The mixing procedure is as follows: different soils are piled like a sandwich; the piles are cut down by a back hoe and cut-loaded when the soil is carried from the stock pile; at the construction site the soil is unloaded, placed, spread and compacted by rolling.

Stock piles make it possible for a wide range of soils excavated from road construction sites to be recycled by mixing them with good soils. This can be done even for such soils as silt and clay whose usage is limited due to their mechanical properties when used alone. In addition, earth material satisfying prescribed properties can be obtained with the application of appropriate mixing ratios of soils of different properties. The dewatered filter cake produced at the subbase course material recycling plant is also mixed into the stock pile, thus recycling fine-grained soil not used in the recycling process at the plant.

This stock pile method supplies soils of certain properties at appropriate times and amounts from a fixed location. Therefore, many problems inhibiting the recycling of excavated soils from streets are solved.

The grading curve of the excavated subgrade soils taken to the plant and the stock pile are shown in Figure 5 and Figure 6, respectively. The grain size distribution varies because the excavated soils taken to the plant differ according to the area they are taken from. The grain size distributions of mixed soils in the stock pile are almost the same. Sandy or gravelly soils with 12-38% of the grains smaller than 0.075mm were obtained by mixing soils of different properties in the appropriate ratio at the stock pile.

The relationship between the water content as percentage of dry weight and CBR of the mixed soil in the stock pile is shown in Figure 7. As the water content increases, CBR tends to decrease. In addition, CBR is influenced by the percentage of fine-grains smaller than 0.075mm. The relationship between the water content as percentage of dry weight and the percentage of fine-grained soil are classified according to the cone index shown in Figure 8. When the value of the CBR standard for the subgrade and back-filling materials is 6 or more, the area below the solid line in Figure 8 satisfies the standard. The relationship between the water content as percentage of dry weight (W) and the percentage of fine-grained soil content (P) is as follows:

$$W < 17 + 0.15\,P \qquad (1)$$

Homogeneous earth material in terms of strength is produced by using a stock pile and mixing cohesive soil or dewatered filter cake with sandy soil in a ratio that satisfies inequality (1). This mixed soil can be used as the material for

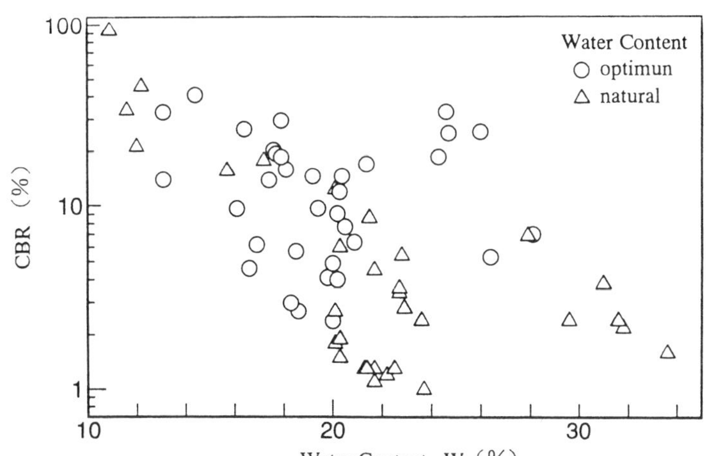

Figure 7 Relationship between water content and CBR of the mixed soils in the stock pile

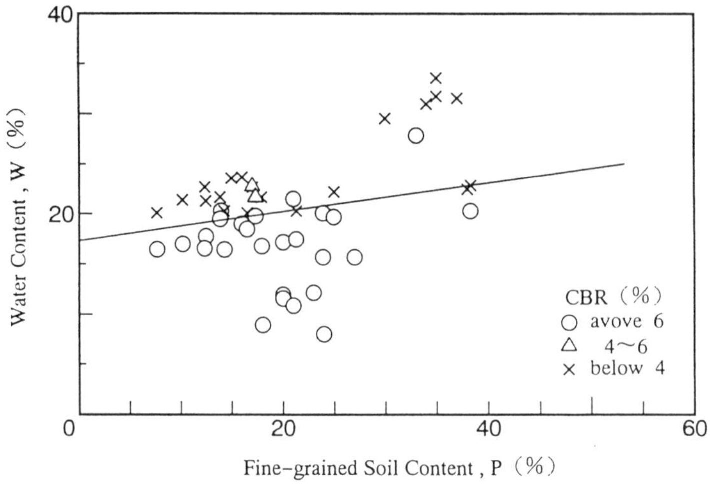

Figure 8 Relationship between water content, fine-grained soil content and CBR

embankment sections, back-filling, river levee and land development.

The capacity of this recycling plant is about 20,000 cubic meters. Depending on the properties of the soil taken to the plant and the demand for recycled soil, this plant is capable of working five cycles annually and treating 100,000 cubic meters of soil, if the improved soil is used regularly.

4 CONCLUSIONS

When excavating the streets, the asphalt mixture, subbase gravel and subgrade soils are excavated in layers to avoid mixing them with each other. Each is recycled through its own distinct system.

The subbase gravel is carried to a newly developed plant and classified into crushed stone and sand by washing and screening. A mixture of this crushed stone and sand is supplied as subbase course material and the sand as material of anti-frost layer. The value of modified CBR of the recycled subbase layer material satisfies applicable standards and the sand is sufficient to prevent frost heave.

The water used to wash the excavated subbase gravel is recycled after being fed into a settling basin to settle the fine-grained content. The sediment of the settling basin is fine-grained soil which is then dredged and pressed by a pressure dehydrator to make a dewatered filter cake with about 40 % water content by dry weight. The excavated subbase soil and the dewatered filter cake are mixed in an appropriate ratio using a stock pile in order to supply soils with a homogeneous property. The resulting soils are sandy or gravely and can be used for back-filling, embankment or land development material by changing their fine-grained content. These soils are currently used for building embankments and covering garbage dumping sites.

Reference

1)Ninomiya,T., Takemoto,R., Hashimoto,K., Takano,T and Tachibana,M. (1989) : Practical application of recycling of excavated surplus soil from road construction, Osaka and its Technology, Osaka Municipal Government, No.14, pp.48-54

Environmental Geotechnics, Kamon (ed.) © 1996 Balkema, Rotterdam. ISBN 90 5410 848 7

Environmental impact and control of alkaline migration by cement stabilized soil

Masashi Kamon & Takeshi Katsumi
DPRI, Kyoto University, Uji, Japan

Sho Oyama
Konoike Construction Co., Osaka, Japan

ABSTRACT: Large amounts of surplus soil are generated from construction works and about half of it is disposed of in waste landfill sites in Japan. To promote effective use of the surplus soil, their properties must sometimes be improved by cement and/or lime stabilization. However, as the soils stabilized by cement or lime have a high pH, the environmental impact of the alkaline migration must be carefully considered. In this study, the methodology and concept to control the environmental impact induced by the use of cement stabilized soil are discussed. The alkaline leachate characteristics from the stabilized soils, buffer ability of filter layer and interaction of these layers were investigated through the experimental studies. Based on the experimental results, the minimum thickness of the filtration layer of the stabilized-soil embankment was estimated through parametric analysis. The buffer ability of the filter layer and the permeability of both stabilized soil and filter layer are important to control the environmental impact of alkaline leachate.

1 INTRODUCTION

Large amounts of wastes and by-products are generated from various industries while waste disposal sites are limited in Japan. Especially, about 437×10^6 m^3 surplus soil was generated from construction works in fiscal 1993, and more than 50 % of it was disposed of in the landfill sites not being reused, according to the investigation by the Ministry of Construction, Japan. Therefore, utilization of these by-products for geotechnical purposes is strongly required. To promote the effective use of the surplus soil, their properties sometimes must be improved by cement and/or lime stabilization (e.g., Kamon and Katsumi 1994a and 1994b). However, as the materials stabilized by cement or lime exhibit a high pH value, the environmental impact of the alkaline migration must be carefully considered.

Importance of the conduction phenomena in soils has recently been emphasized (e.g., Mitchell 1992). Especially, several researchers such as Shackelford and Daniel (1991) and Yong et al. (1992) summarized the migration mechanisms of contaminants in the soil to prevent and control the contamination of the ground. From the viewpoint of environmental compatibility in geotechnical waste utilization, the environmental impact of the alkaline leachate as well as the harmful substances must be discussed. To encourage the recycling of by-products of surplus soils, alkaline leachate migration mechanisms from the stabilized soils have been studied (e.g., Amano et al. 1980; Kitsugi 1989; Mishima et al. 1994) and the clayey soil was found to be a filtration cover of the stabilized soil to minimize the alkaline migration because of the high alkaline buffer ability (Miki et al. 1994; Sano et al. 1993). However, the clayey soils is not always considered to be suitable for the cover because it is difficult to construct the compacted layer by the clay. Besides, alkaline migration and its control from the stabilized materials is strongly affected not only by the individual properties of the stabilized layer and the cover for filtration but also by the combination of these properties. The latter influences have not been sufficiently discussed. Figure 1 shows a typical cross section of the examples of geotechnical waste utilization; application of the stabilized surplus soil to embankment.

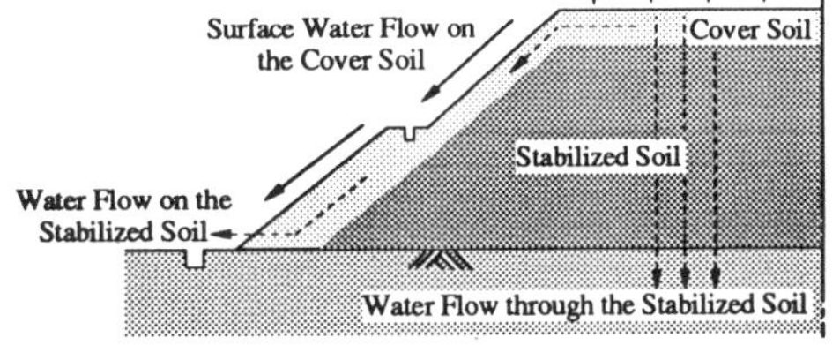

Fig. 1. Water flow around the embankement using stabilized soil

In this study, the methodology to control the environmental impact induced by the application of cement stabilized soil is discussed. Alkaline leachate characteristics from the stabilized materials, buffer ability of filter layer and their interaction are studied. Based on the experimental results, the minimum thickness of filter layer of the stabilized-soil embankment was estimated through the parametric analysis for the example shown in Fig. 1, and the design concept of geotechnical waste utilization is discussed.

2 EXPERIMENTAL STUDY

2.1 *Materials used*

Alluvial clay obtained from Osaka Basin and decomposed granite soil from Ohtsu were used in this study. Table 1 shows their properties. They are the typical soils in the west part of Japan. Alluvial clay was stabilized while decomposed granite was used as filter layer. Mixing proportions of the stabilized soil were determined according to strength development and permeability. Stabilized soil I, the alluvial clay with natural water content (80.4 %) hardened by 10 % ordinary Portland cement in dry weight, has high strength and density, while stabilized soil II is the liquefied alluvial clay with low strength and density; with a 300 % water content mixed with 100 kg/m^3 cement. Table 2 shows their properties.

2.2 *Alkaline neutralization by soil*

Alkaline neutralization by the decomposed granite soil was measured. The soil was mixed in various mixing proportions with the $Ca(OH)_2$ solutions at various concentration levels, and the pH of the suspension was measured 2 hours later. Therefore, the added OH^- can be calculated based on the concentration as well as the mixing proportion of soil and solution. From Fig. 2 indicating the added OH^- and the neutralized per 1 g of soil, most of the added OH^- can be neutralized if the initial OH^- level is lower than 10^{-5} mol/g. However the neutralization is not effective when an OH^- higher than 10^{-4} mol/g is added, which indicates that the neutralization by soil has a limitation.

Alkali neutralization ability, C (mol/g), is defined as the maximum quantity of OH^- that can be neutralized by 1 g of soil. The alkali neutralization ability of the soil used can be determined as 9×10^{-5} mol/g from the curve in Fig. 2. Miki et al. (1994) also measured the alkaline neutralization ability of 10 typical soils by using the saturated solution of $Ca(OH)_2$. The neutralization ability of decomposed granite is similar in value as the one presented here, and is comparatively lower than the clayey soils.

Table 1. Properties of the used materials.

	Alluvial clay	Decomposed granite soil
Particle density (g/cm^3)	2.62	2.64
Natural water content (%)	80.4	7.14
Liquid limit (%)	69.5	-
Plastic limit (%)	31.2	-
Opt. water content (%)	-	18.0
Max. dry density (g/cm^3)	-	1.725
Grain size distribution (%)		
Gravel fraction	0.2	16.2
Sand fraction	2.7	75.4
Silt fraction	59.1	6.8
Clay fraction	38.0	1.6
Permeability (cm/s^2)	-	3.6×10^{-5}
Soil type	CH	S-M
Ignition loss (%)	9.39	-
pH	8.7	5.2
Main mineral	Chrolite, Illite, Smectite	Halloysite, Quartz, Feldspar

Table 2. Properties of the stabilized soils.

	Stabilized soil I		Stabilized soil II	
Aging (d)	7	28	7	28
Compressive strength (MPa)	0.62	1.01	0.032	0.071
Wet density (g/cm^3)	1.53	1.59	1.27	1.27
Dry density (g/cm^3)	0.90	0.90	0.41	0.42
pH	12.8	12.2	12.9	12.8

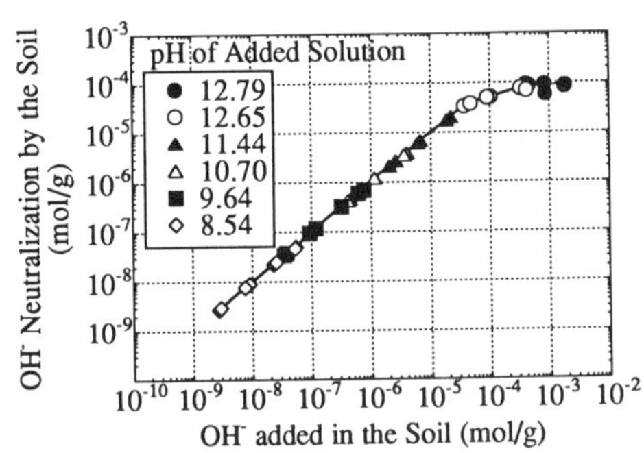

Fig. 2. Alkaline neutralization ability of decomposed granite soil

2.3 *Alkaline migration through stabilized soil*

To express the case in which the water can permeate through the stabilized soil as well as the filter layer, seepage tests were carried out on the specimens as shown in Fig. 3(a). A cylindrical mold was filled with the stabilized soil and the decomposed granite compacted at the optimum water content. After aging, the composite soil specimens were subjected to water head. As shown in Fig. 4 indicating pH change of the discharge, water at a pH higher than 11 was discharged continuously after the seepage exceeded about 650-750 ml. The discharged pH is affected by

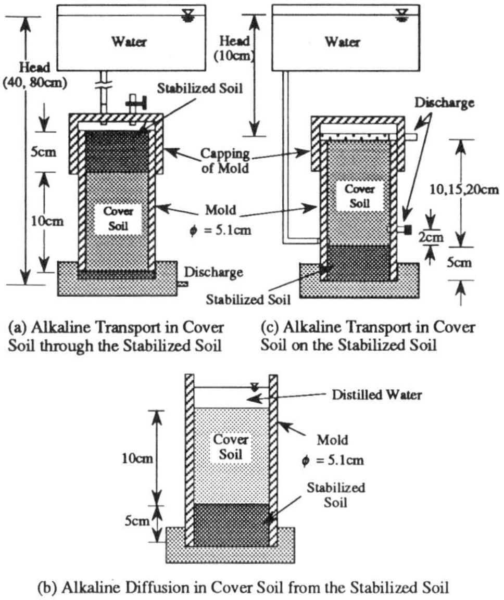

Fig. 3. Experimental equipment on alkaline migration

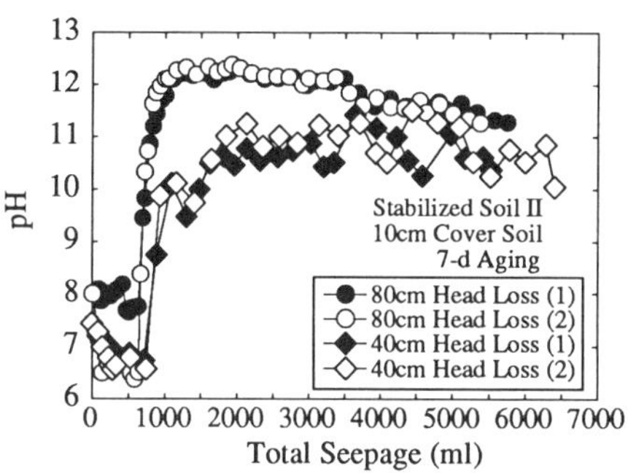

Fig. 4. pH of discharge from the stabilized soil

the water head as shown in Fig. 4, because it may depend on the flow rate. In the case of lower water head (40 cm), the low pH solution was discharged for a longer period and, even after the discharged pH became alkali, the pH was below 11, lower than in the case of higher water head.

Based on the obtained result that neutralization ability of the decomposed granite is 9×10^{-5} mol/g and the assumption that the pH of water immediately after permeating through the stabilized soil is 12.6, the seepage volume required for a pH higher than 8.6 can be calculated to be 1180 ml, which is 1.5-2.0 times larger than the experimental results. The reasons for this difference might be that alkaline seepage penetrates through a selected way in the filter layer and can consequently meet only the selected

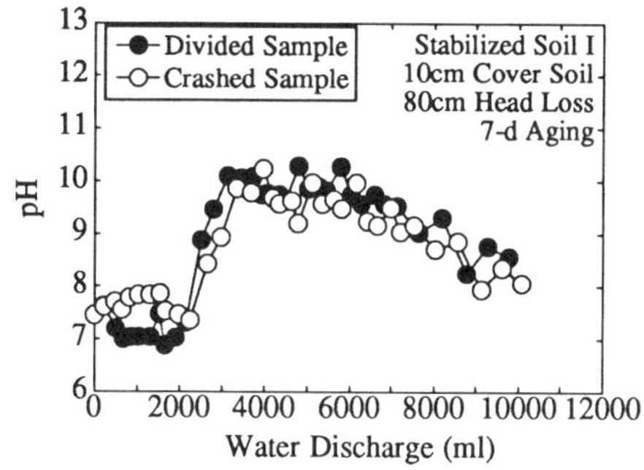

Fig. 5. pH of discharge from the crushed stabilized soil

particle surface for the neutralization, and that alkali neutralization depends on the added OH^-, but not the concentration, in the soil-water systems as discussed in Fig. 2. Flow rate may also affect the neutralization phenomenon.

Similar experiments could not be executed on stabilized soil I because of its low permeability. However, the seepage can permeate through the layer of crushed specimens of stabilized soil I; Stabilized soil was cut into 4 pieces or crushed into pieces 2-5 mm in diameter in these experiments. As shown in Fig. 5, the discharged water became alkaline after the seepage of 2200 ml, but it exhibited a lower pH than in the case of stabilized soil II, and the pH decreased not keeping a high value. One reason is that the seepage water meets only the surface of the divided blocks of stabilized soil, and does not seep into the blocks, and the discharge of alkali from stabilized soil was limited compared to the permeable stabilized soil. Therefore, the permeability of the stabilized soil is an important factor to assess the alkaline migration.

2.4 *Alkaline diffusion from stabilized soil*

To assess alkaline diffusion from the stabilized soil, the experiment shown in Fig. 3(b) was carried out. The bottom of the mold was filled with stabilized soil I. Distilled water was poured after 7 days of curing and left for 94 days. As shown in Fig. 6, only the soil in the location 0-3 cm above the stabilized soil exhibited a higher pH than 10. This shows that the calcium alkali cannot diffuse for a long distance and the pH distribution is kept balanced. According to the diffusion coefficient of ions summarized by Shackelford and Daniel (1991), the diffusion coefficient of hydroxide ion is higher (52.8 m^2/s) than the other ions, but the calcium ion, which is electrically counterbalanced with hydroxide ion here, has a low diffusion coefficient (7.92 m^2/s).

Figure 7 shows the results when the seepage was poured into the filter layer at the point close to the

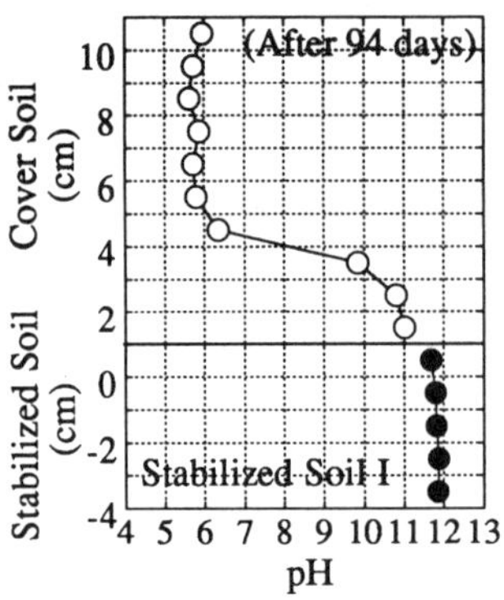

Fig. 6 pH distribution in the stabilized and the cover soil

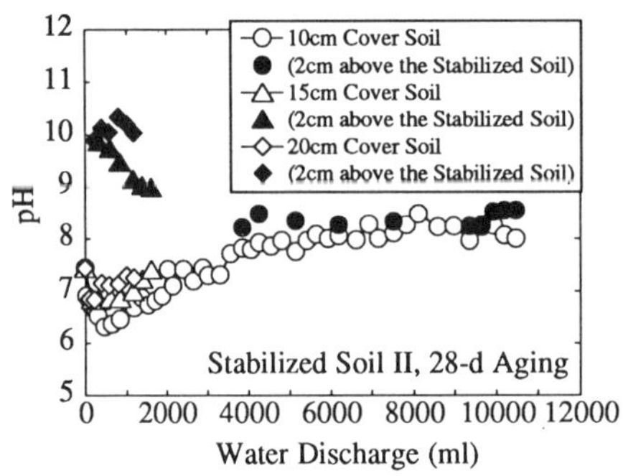

Fig. 7. pH value of discharge from the cover soil

stabilized soil I and discharged from the top of the layer due to the water head, as shown in Fig. 3(c). Therefore, the seepage did not pass through the stabilized soil, and only came in contact with the surface of the stabilized soil. The pH value of the discharge from the top increased as the seepage proceeded, but kept neutral and could not become alkaline. The pH close to (2 cm above) the stabilized soil was high (about 10) at an early stage and decreased as the seepage proceeded. At the stage later than 4000 ml of seepage, the discharge close to the stabilized soil exhibited a similar pH as the discharge from the top of the filter. This tendency indicates that the alkali was distributed and consequently diluted uniformly in the filter layer as the alkali could be supplied from only the surface of the stabilized soil layer.

The tank test was performed to evaluate alkali migration in two dimensional flow. The tank 20 cm in width, 55 cm in height and 250 cm in length, was inclined at 2 %. Stabilized soil I was filled at the bottom of 5 cm thickness except for the 30 cm towards the lower end of the tank. Stabilized soil I was overlain by compacted filter layer (30 cm at the stabilized soil I and 35 cm at the lower end of tank).

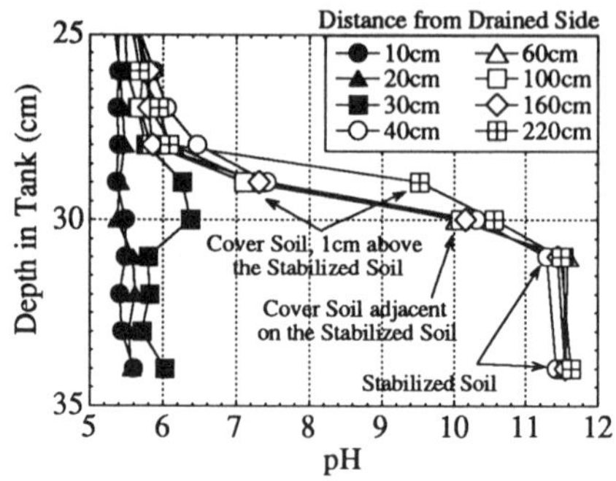

Fig. 8. pH distribution of soils after the tank test

After 7-day curing, the tank was filled with tap water. Water was drained at the bottom of the lower side of the tank for 66 days. Total water discharge was 14,000 ml, which is the estimated amount of seepage in a half month of annual rainfall. One third of the total annual rainfall (1760 mm) is considered to be retained in the soil.

Figure 8 shows the pH distribution in the soil after seepage. While the stabilized soil had a pH higher than 11, only the decomposed granite soil 0 or 1 cm above the stabilized soil was alkali (pH 10-11 and 7-10 respectively), and that more than 2 cm above the stabilized soil maintained the weak acid of decomposed granite soil. On the series of the results of cover soil 1 cm above the stabilized soil, the pH decreased as the measurement point approached the drained side, because the dilution occurred due to the chemical diffusion and water flow.

Therefore, the low permeability of stabilized soil and surrounding cover layer is necessary to control the alkali migration, and the alkaline solution is contained within the cover soil close to the stabilized soil in spite of its low alkaline neutralization ability.

3 PARAMETRIC ANALYSIS

3.1 *Method*

An analytical study was performed to discuss the design concept for alkaline migration control based on the experimental results presented above. For application of the stabilized soil to embankment, the filter layer thickness required for alkaline leachate control was calculated through the parametric analysis.

Typical cross section of the embankment shown in Fig. 1 was simplified as shown in Fig. 9. The high alkaline solution will be generated if the seepage passes through the stabilized soil, and can be neutralized during passage through the filter layer below the stabilized soil. We calculated the thickness

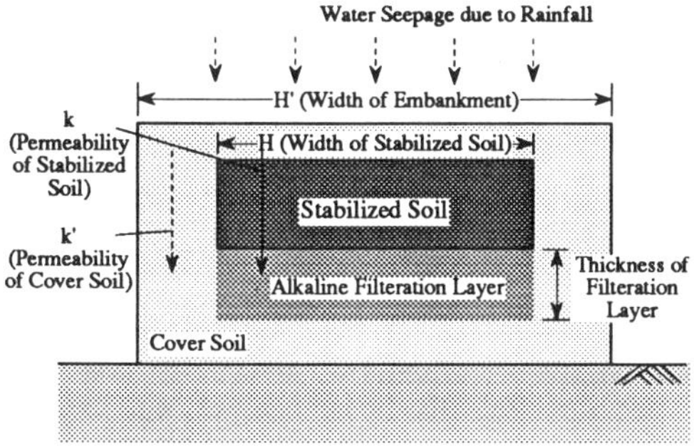

Fig. 9. Simplified cross section for parametric analysis of alkaline migration

of the filter layer that can neutralize the alkaline solution discharged from the stabilized soil. The criterion is a pH lower than 8.6 that is the effluent standard in Japan. The assumptions for the analysis are as follows;

(1) seepage water flows only vertically,
(2) hydraulic gradient in embankment is constantly 1; no influence on the calculated results when the permeability is higher than 10^{-8} cm/s,
(3) alkaline diffusion in cover soil is not considered based on the experimental results,
(4) only the soil below the stabilized soil can neutralize the alkaline, which is indicated as alkaline filtration layer in Fig. 9,
(5) seepage amount into the ground is 1/3 of the average annual rainfall in Japan (1760 mm),
(6) discharged water from the stabilized soil constantly has a high pH; The pH was given parametrically, and
(7) the simplified model section is two dimensions; Axis direction of embankment is not considered.

Seepage quantity, L (mm), is according to;

$$L = 1760/3 \times X/(1+k'/k(H'/H - 1)) \tag{1}$$

where k, k', H and H' are mentioned in Fig. 9 and X is the supposed period (y). H/H' and k/k' are defined as the width ratio and permeability ratio. If the discharge from the stabilized soil is neutralized by alkaline neutralization ability of filter layer, Eq. 2 should be satisfied.

$$C > (10^{(S-14)} - 10^{(8.6-14)}) \times V/W \tag{2}$$

where C (mol/g) is the alkaline neutralization ability of the filtration soil, S and V (l) are pH value and quantity of the discharge for the seepage through the stabilized soil, W (g) is the total mass of the filtration soil under a dry condition. As W and V can be expressed by H, L, thickness of filtration layer (cm), x, and the dry density of filtration soil, ρ_d (g/cm^3), Eq. 2 can be reduced to;

$$x > (10^{(S-14)} - 10^{(8.6-14)}) \times 10^4 L / (\rho_d C) \tag{3}$$

According to the Eqs. 1 and 3, the filtration layer thickness required for the alkaline neutralization was calculated. S, k/k', and H/H' were given parametrically, ρ_d is 1.73 g/cm^3 and C is 9×10^{-5} mol/g based on the experimental results or is 3×10^{-5} mol/g considering the difference between experiment and calculation.

3.2 *Results and discussions*

Figure 10 shows the filter layer thickness required for the alkaline neutralization for 1 year rainfall. The required thickness depends on not only C, but also H/H' and S. Even in the severest condition that S = 12, H/H' = 0.9 and C = 3×10^{-5} mol/g, the required thickness is estimated at 3.6 cm. As the analytical results are strongly affected by S, the decision of initial pH, S, is important to calculate the required thickness. The pH of the stabilized soil, S, was constant in the present analysis, but it might decrease due to carbonation in the field (Kitsugi 1989). After obtaining a more precise relationship between pH change versus time, we can realize a better actual estimation. Considering the influences of width ratio (H/H'), one of the effective options for the alkaline migration control is that more seepage water passes through only the cover soil and less through the stabilized soil due to the dimension of the embankment.

Figure 11 shows the required thickness for 30 years rainfall when H/H' is 0.9, which is the most probable in an actual condition. Therefore, the soil which has high alkaline neutralization ability should be used as the filter soil to control alkaline migration effectively. However, soil that has high neutralization

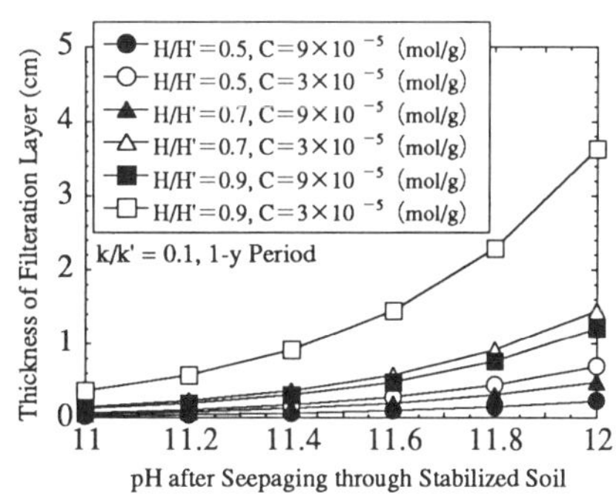

Fig. 10. Required thickness of filteration layer in stabilized soil embankment

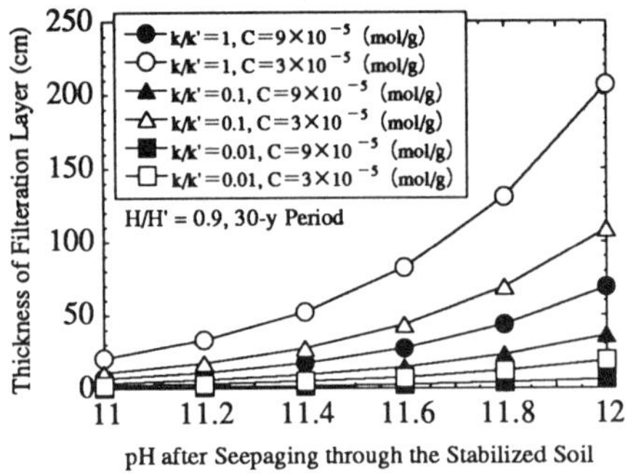

Fig. 11. Required thickness of filteration layer in the stabilized soil embankment

ability, such as clay soil and loam, usually consists of fine-grained particles and exhibits low permeability. These fine soils are not suitable as the cover soil because they are difficult to treat and compact in the field. They should be used as stabilized soil rather than filtration soil. In Fig. 11, when the filtration soil has as low permeability as the stabilized soil (k/k' = 1.0), the layer must be at least 200 cm thick for the alkaline migration control in the severest condition (S = 12.0, C = 3×10^{-5} mol/g). As the permeability ratio decreased, the required thickness decreased markedly. Thus, the permeability of the soils is very important. From the influence of permeability ratio, one of the most effective options was confirmed to be less seepage water passing through the stabilized soil. This means that the stabilized layer should be constructed carefully to become the low-permeable layer, when the environmental influence must be considered.

4 CONCLUSIONS

Alkaline migration mechanisms from stabilized soil have been evaluated and the concept of its control has been discussed through experimental and analytical studies. One of the major problems is the environmental influence of geotechnical waste utilization. The effect of the filtration layer around the stabilized soil for the alkaline migration control has been clarified. The main results can be summarized as follows;

(1) When the quantity of seepage from stabilized soil to filter layer exceeds the criteria, the alkaline solution may leach from the filter. The criteria can be roughly estimated from the alkaline neutralization ability of the soil.
(2) If the stabilized soil has low permeability, seepage water can not penetrate and pass through the stabilized inside and there is little possibility of alkali leachate. If the permeability is high due to the low density or the incidence of crack, seepage water can flow through the stabilized layer and alkali leachate can occur.
(3) If the seepage water flows only in the cover filtration layer, the alkaline will be little diffused and consequently diluted despite its alkaline neutralization ability. Only the filtration layer very close to the stabilized layer will be alkali.
(4) The design concept for the alkaline migration control was discussed for the application to embankment. We must consider not only the neutralization ability but also the permeability of materials and the geometric dimension of soil structure. One of the most effective options is that seepage water does not pass through the stabilized layer.

REFERENCES

Amano, H., & N. Masumoto 1980. Alkaline leachate from cement stabilized soil, *Proc. 34th General Meeting of CAJ:* 472-475 (in Japanese).

Kamon, M., & T. Katsumi 1994a. Utilization of waste slurry from construction works, *Proc. 13th ICSMFE,* 4: 1613-1616.

Kamon, M., & T. Katsumi 1994b. Potential utilization of waste rock powder, *Proc. 1st Int. Cong. on Environ. Geotechnics*: 267-282

Kitsugi, K. 1989. Environmental aspects accompanied with weak soil ground improvement techniques, *Jour. CAJ,* 511: 104-115 (in Japanese).

Miki, H., N. Mori & T. Hurusho 1994. Experimental study on alkaline neutralization ability and alkaline seepage of the soil, *Proc. 49th Ann. Meeting of JSCE,* 3: 1534-1535 (in Japanese).

Mishima, N., K. Hoshino, H. Ishihara & K. Terao 1994. Adherence to pH of soils, *Proc. 29th Ann. Meeting of JSSMFE*: 2327-2328, (in Japanese).

Mitchell, J.K. 1992. *Fundamentals of soil behavior, second edition,* John Wiley & Sons, Inc.

Sano, M., et al. 1993. Experiment results on the prevention effect within cohesive soil ground on immigration of lime-alkalinity adjacent to lime-treated soil mass, *Proc. Ann. Meeting of JSSMFE*: 2585-2588 (in Japanese).

Shackelford, C.D., & D.E. Daniel 1991. Diffusion in saturated soils. I: Background, *Jour. Geotech. Engrg.,* ASCE, 117(3): 467-484.

Yong, R.N., A.M.O. Mohamed & B.P. Warkentin 1992. *Principles of contaminant transport in soils,* Elsevier.

Environmental Geotechnics, Kamon (ed.) © 1996 Balkema, Rotterdam. ISBN 90 5410 848 7

Fly ash and fly ash-soil mixture for embankment construction

Shenbaga R. Kaniraj
Indian Institute of Technology, New Delhi, India

Vasant Havanagi
Central Road Research Institute, New Delhi, India

ABSTRACT: Annually about 60 million tons of fly ash is being produced in India. Fly ash or fly ash-soil mixtures can be used in the construction of embankments. The paper describes the results of an experimental study carried out to determine the geotechnical characteristics of fly ash and fly ash-soil mixtures. The implications on the design of embankments on soft soils are discussed.

1 INTRODUCTION AND SCOPE

There is an accumulation of large quantity of fly ash in India at various thermal power stations and other industrial units using coal as fuel. Less than 5 per cent of this is currently being used for manufacture of brick, pozzolana cement, and other products. Fly ash or fly ash-soil mixtures can be used as a fill material in the construction of embankments. This could result in significant engineering benefits in addition to facilitating mass disposal of fly ash.

Fly ash has been collected from a single electrostatic precipitator of the Rajghat Power Station in New Delhi. Laboratory tests have been carried out as per the Indian Standard Codes of Practice to determine the physical, chemical, and geotechnical characteristics of the fly ash. The soil used in the fly ash-soil mixture is a predominantly silty soil and is generally referred to as Delhi silt. The fly ash is mixed with the Delhi silt in different proportions and the geotechnical characteristics of the mixtures have been studied. The paper describes the results of the experimental study. The implications of the use of fly ash or fly ash-soil mixtures on the stability of embankments on soft soils are discussed.

2 RESULTS OF TESTS ON FLY ASH

2.1 Physical properties

Table 1 shows some of the physical properties of the Rajghat fly ash and the range of values for fly ashes from eight other sources in India namely Bokaro, Bandal, Chandrapur, Durgapur, Harduaganj, Kanpur Madras, and Neyveli.

Table 1. Physical properties of Indian fly ashes

Property	Rajghat	Others
Specific gravity	2.19	-
Per cent finer than 75μ size	85.4	74.1-92.4
Loss on ignition, %	1.4	1.26-15.8
Specific surface area, m^2/N	41	33-69

2.2 Chemical composition

Oxides of silicon (SiO_2 = 61.2%), aluminium (Al_2O_3 = 30.1%), and iron (Fe_2O_3 = 4.2%) together constitute 95 percent (by weight) of the Rajghat fly ash. Oxides of calcium (CaO = 0.1%) and magnesium (MgO = 0.4%) are present in very small quantities. Most of the power stations in India use bituminous or sub-bituminous coal. The calcium content in the fly ash produced from such coals are considered to be small.

2.3 Geotechnical characteristics

The tests carried out include grain size analysis, compaction test, direct shear test, unconfined compression test, drained triaxial shear test, consolidation test, and *CBR* test.

The grain size distribution curve for the Rajghat fly ash is shown in Fig.1.

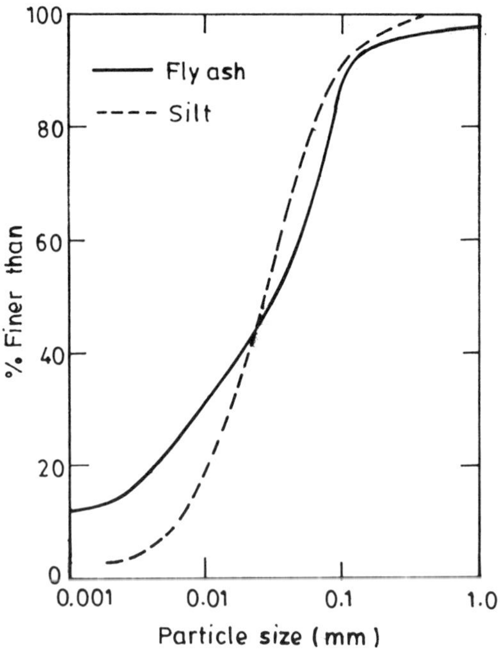

Fig.1 Grain size distribution curves for Rajghat fly ash and Delhi silt

The maximum dry unit weight (*MDD*) and the optimum moisture content (*OMC*) determined from light weight compaction test are 10.72 kN/m^3 and 36.5% respectively. The compaction curve (Fig.2) is relatively flat which implies that the degree of compaction is not sensitive to water content. The degree of saturation at *OMC-MDD* is 80%. de Santayana and Mazo (1994) have also observed the compaction curve to be relatively flat for several types of Spanish fly ash and the degree of saturation at *OMC-MDD* to be less than 85% or even 75%.

Direct shear test is carried out on samples of size 60 x 60 x 20 mm, compacted at *OMC* to *MDD*. The normal stress is varied from 25 to 125 kPa. The values of cohesion intercept c, and angle of shearing resistance ϕ, obtained from the test are 20 kPa and 37.5°, respectively. For the Hanasaari fly ash which is comparable to the Rajghat fly ash, Havukainen (1983) has reported similar values of c and ϕ. The failure state is reached at displacements between 1 and 2 mm. The dilation observed is very small, of the order of 0.8% volumetric strain.

Unconfined compression (*UC*) test is carried out on cylindrical samples of 38 mm diameter and 76 mm height. The samples are statically compacted at *OMC* to *MDD*. The compaction is carried out alternatively from both ends of the mould to ensure uniform compaction of the sample. The unconfined compressive strength q_c, of these samples are in the range of 65-70 kN/m^2. The axial strain at failure is small and ranges from 2-2.5%. The samples fail suddenly with little bulging before failure.

Attempts made to carry out drained triaxial shear test on cylindrical fly ash samples prepared in the manner described for the *UC* tests show that it is not possible to saturate the soil samples by application of back pressure or suction because of the low initial degree of saturation. Saturated samples could not be prepared by increasing the water content at the time of sample preparation, because they collapse due to self weight at high water contents. Therefore, the test is conducted on saturated samples prepared by depositing fly ash through water and compacting the slurry. The dry unit weight of these samples vary from 10.3 to 10.5 kN/m^3. The effective cohesion intercept and angle of shearing resistance of the fly ash are 25 kPa and 38.7°, respectively. Similar observations have been reported by Gray and Ken (1972), and Leonards and Bailey (1982). The measured cohesion appears to be high, possibly due to incomplete dissipation of pore water pressure. More tests are required to be conducted using slower rates of deformation of the sample to confirm this.

The compressibility characteristics of the compacted fly ash is studied in a standard one dimensional consolidation test apparatus. The fly ash is compacted in the consolidation ring at *OMC* to *MDD*. The sample is then allowed to saturate by keeping it in water. The consolidation test is carried out varying the vertical stress from 5 to 12.8 kN/m^2. The consolidation takes place rapidly as has been reported by Havukainen (1983) for the Helsinki fly ashes. The compression index C_c, and the rebound index C_r, of the compacted fly ash are 0.072 and 0.017, respectively.

For the California Bearing Ratio (*CBR*) test, the fly ash is compacted in the *CBR* mould at *OMC* to *MDD*. The specimen is then kept soaked in water for 4 days before testing. The measured *CBR* value of 2.07% is considered to be low compared to the normal soils.

3 RESULTS OF TESTS ON FLY ASH-SOIL MIXTURES

The fly ash may be blended with the local soil before compaction for several reasons. The use of

Table 2. Combinations of fly ash-soil mixture (% by dry weight)

fly ash-soil mixture	fly ash %	soil %	G
A	0	100	2.65
B	25	75	2.53
C	50	50	2.36
D	75	25	2.24
E	100	0	2.19

light weight fly ash in the construction of embankments on soft soils can increase their stability and reduce the settlement. However, if the fly ash were to be imported, mixing it with local soils before use can make the technique cost effective.

The Rajghat fly ash has been mixed with different proportions of the local soil. The local soil is a predominantly silty soil and is generally referred to as Delhi silt. The different combinations of fly ash-soil mixture are identified in Table 2.

3.1 Geotechnical characteristics of fly ash-soil mixtures

The tests carried out to determine the geotechnical characteristics of the fly ash-soil mixtures include, grain size analysis, specific gravity test, compaction test, direct shear test and unconfined compression test.

The grain size distribution curve of the Delhi silt is shown in Fig.1. The particles type present in the Rajghat fly ash and the Delhi silt, according to the Indian standard code, are shown in Table 3. From Fig.1 and Table 3, it is evident that the fly ash and the soil used are predominantly silty material.

The compaction curves of the fly ash-soil mixtures A-E are shown in Fig.2 along with the respective 100% saturation curves. The specific gravity G, values are reported in Table 2. As the fly ash content increases the *MDD* decreases and the *OMC* increases. Tsonis *et al* (1983) have reported similar

Table 3. Particles type in fly ash and Delhi silt

particle type	grain size, mm	fly ash	soil
sand size	0.075-4.75	20%	14%
silt size	0.002-0.075	77%	73%
clay size	< 0.002	3%	13%

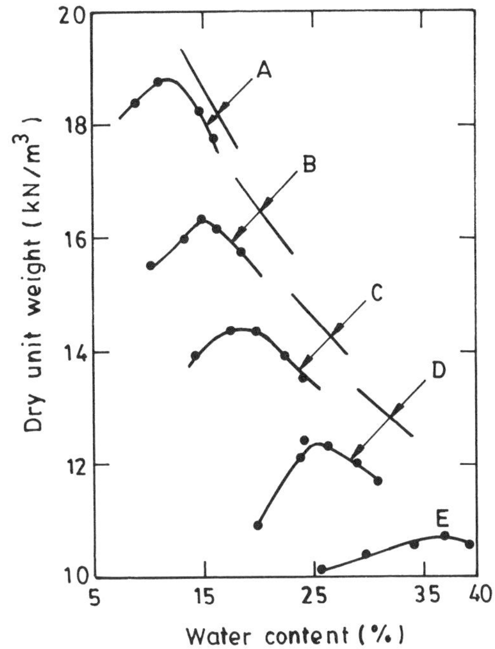

Fig.2 Compaction curves for fly ash-soil mixtures A-E

changes in *MDD* and *OMC* from a study using coal ash-silty clay mixture. The bulk unit weight at *OMC-MDD* also decreases. These variations are shown in Fig.3. The void ratio e, at *OMC-MDD* increases as the fly ash content increases. The change in void ratio and porosity n, are shown in Fig.4. The results suggest that the self weight of the embankment decreases as the fly ash content increases partly due to the lower specific gravity of the fly ash and partly due to the more porous structure of the compacted material. It can be expected that the addition of fly ash to the Delhi silt could increase the coefficient of permeability. But, from the results of consolidation test on fly ash described in section 2.3 it can be inferred that the increase in porosity does not make the compacted material very compressible.

Direct shear tests are conducted on fly ash-soil mixture samples compacted to their respective *MDD*s at the corresponding *OMC*s. Table 4 shows the results of the direct shear tests. With an increase in the fly ash content there is a consistent increase in the angle of shearing resistance. This is perhaps due to the irregular shape of the fly ash particles which are predominant in the Indian fly ashes than the spherical particles (Sharma and Krishnamurthy 1992). The variation of cohesion intercept with fly

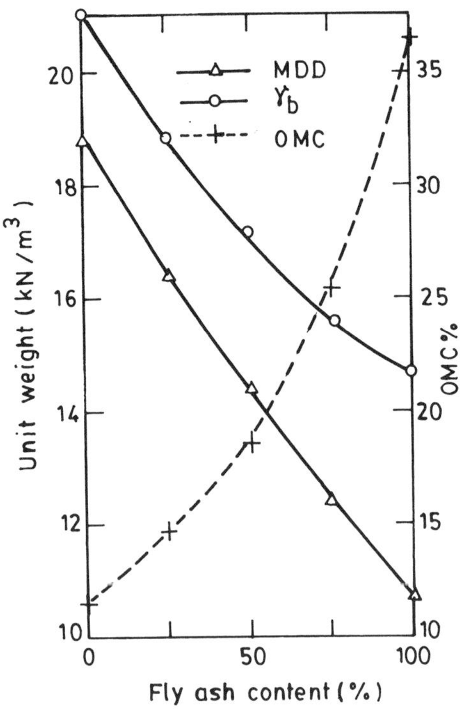

Fig.3 Variation of *MDD*, *OMC*, and γ_b with fly ash content

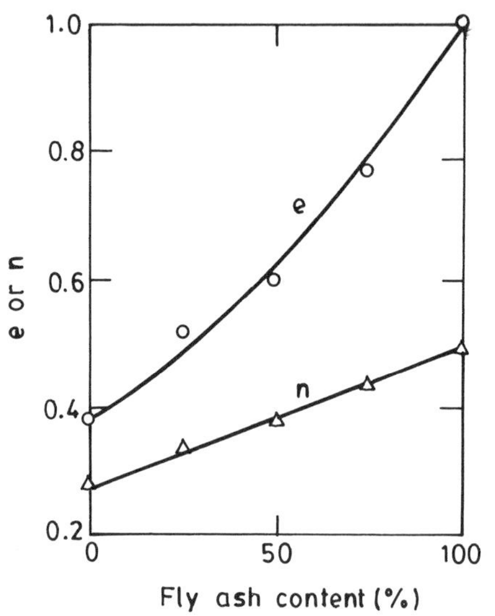

Fig.4 Variation of *e* and *n* with fly ash content

ash content is not consistent. The dilation of the sample during shear could have set up suction or negative pore water pressure in the unsaturated samples. This could have occurred randomly in the direct shear test. More experiments are required to be carried out at slower rates of deformation to confirm the test results.

Unconfined compression tests are conducted on fly ash-soil mixture samples compacted to their respective *MDD*s at the corresponding *OMC*s. The variation of unconfined compressive strength with fly ash content is shown in Fig.5. The unconfined compressive strength is nearly the same till the fly ash content increases to 50%. A further increase in the fly ash content decreases the unconfined compressive strength significantly.

Table 4. Results of direct shear tests on fly ash-soil mixtures

mixture	c kN/m^2	ϕ °
A	38	31
B	36	32
C	51	34
D	40	35
E	20	37.5

4 STABILITY OF FLY ASH-SOIL MIXTURE EMBANKMENTS ON SOFT SOILS

Rotational failure often governs the stability of the embankments constructed on soft soils. Use of light weight material such as fly ash as embankment fill will improve the stability. The rotational stability is usually analyzed using limit equilibrium method and total stress analysis. The failure surface is considered to be an arc of a circle. The improvement in the stability of embankments constructed with fly ash-soil mixtures is demonstrated through an illustrative example.

Figure 6 shows the details of an embankment and the foundation soil. The target factor of safety for a stable embankment is 1.4. The factor of safety of this embankment has been calculated for different fly ash-soil mixtures using the procedure suggested by Kaniraj (1994). The parameters cohesion c, angle of shearing resistance ϕ, and unit weight γ_b, are used in the analysis. It has been noted earlier that suction or negative pore water pressure induced in the sample in the direct shear test could have affected the shear strength parameters. The unconfined compression test on the other hand has shown

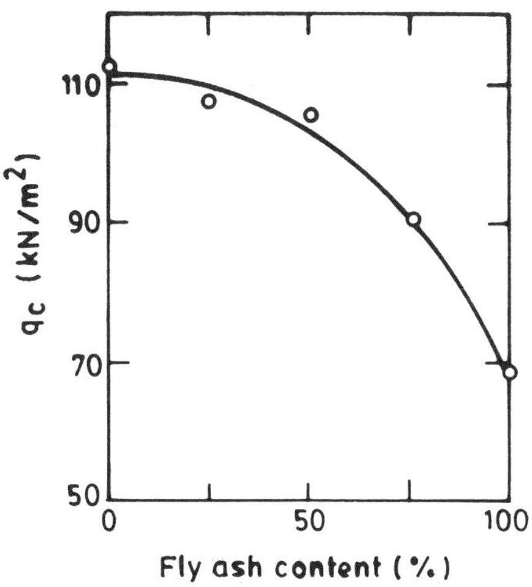

Fig.5 Variation of q_c with fly ash content

consistent results. The analysis, therefore, has been carried out assuming the cohesion to be equal to one half of the unconfined compressive strength. The angle of shearing resistance is neglected. The bulk unit weight corresponding to the *OMC-MDD* condition is used. The values of c, and γ_b used for the different fly ash-soil mixtures are given in Table 5.

Because of the high values of c, a full height tension crack is assumed in the analysis. When a full height tension crack is present, assumptions regarding the cohesion and angle of shearing resistance are of no real consequence as they do not contribute to the resisting force or moment in the calculation for the factor of safety. Table 6 gives the factor of safety of the embankment for the different fly ash-soil mixtures. It is evident that the use of fly ash in embankment construction improves its stability.

From Table 6 it may be observed that the factor of safety in all cases is lower than the target factor of safety 1.4. The stability can be improved by using reinforcement in the form of geotextiles or geogrids close to the base of the embankment. In a single

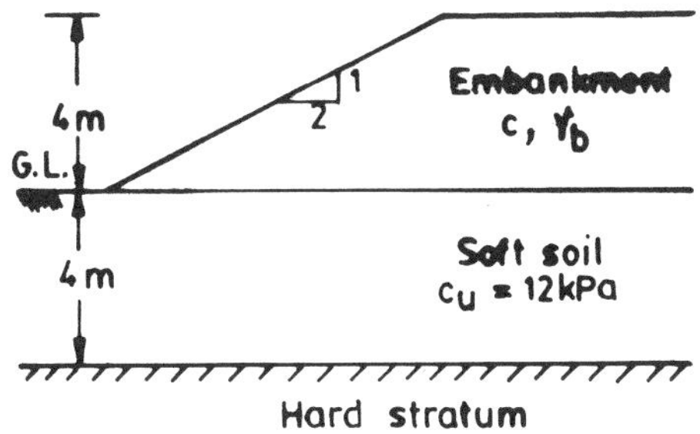

Fig.6 Details of the illustrative example

layer of reinforcement placed at the ground surface, the reinforcement force is assumed to act in the horizontal direction. The maximum working reinforcement force P_{wr}, required to reach the target factor of safety has been calculated in each case using the procedure suggested by Kaniraj (1994). These values are also shown in Table 6. It is apparent that with the increase in fly ash content it becomes possible to improve the stability of the embankment using even low and medium strength geosynthetics as reinforcement.

5 CONCLUSIONS

The following conclusions are made from the study.

1. The Rajghat fly ash collected from electrostatic precipitator is a predominantly silty material and is mainly composed of SiO_2, Al_2O_3, and Fe_2O_3.
2. The degree of compaction of the fly ash is not sensitive to water content. To achieve the required compaction, therefore, precise control of water may not be required during compaction.
3. Increase of fly ash content decreases the maximum dry unit weight and increases the optimum moisture content of fly ash-soil mixtures.
4. When the fly ash content in fly ash-soil mixture becomes more than 50% there is a considerable decrease in the unconfined compressive strength.

Table 5. c and γ_b, used in the illustrative example

fly ash-soil mixture	c kN/m^2	γ_b kN/m^3
A	56	21.00
B	55	18.76
C	51	17.14
D	45	15.56
E	34	14.63

Table 6. Unreinforced factor of safety and working reinforcement force for F = 1.4

fly ash-soil mixture	F	P_{wr} kN/m
A	0.91	-
B	1.02	120
C	1.11	78
D	1.22	42
E	1.30	21

5. Analysis for the rotational stability of an embankment on soft soil shows that the use of fly ash as fill material improves its stability because of the reduction in the self weight.

6. In the case of reinforced embankments on soft soil, by increasing the fly ash content it will be possible to improve the stability of the embankment using even low and medium strength geosynthetics as reinforcement.

REFERENCES

de Santayana, F.P. & Mazo, C.O. (1994). Behaviour of fly ash in experimental embankments, *Proc. XII ICSMFE*, New Delhi, India, 4, 1603-1606.

Gray, D.H. & Ken, Y.K. (1972). Engineering properties of compacted fly ash, *J. Soil Mech. & Found. Eng. Div.*, ASCE, 98

Havukainen, J. (1983). The utilization of compacted coal ash in earth works, *Proc. Eighth European Conf. SM&FE*, Helsinki, 2, 773-776.

Leonards, G.A. & Bailey, B. (1982). Pulverized coal ash as structured fill, *J. Soil Mech. & Found. Eng. Div.*, ASCE, 108, GT4

Kaniraj, S.R. (1994). Rotational stability of unreinforced and reinforced embankments on soft soils, *Geotextiles and Geomembranes*, 13, 707-726.

Sharma, R.C. & Krishnamurthy, S. (1992). Instrumental techniques to classify fly ashes in terms of their reactivity, *Proc. Conf. IL Cemento*, France

Tsonis, P., Christoulas, S. & Kolias, S. (1983). Soil improvement with coal ash in road construction, *Proc. Eighth European Conf. SM&FE*, Helsinki, 2, 961-964.

Environmental Geotechnics, Kamon (ed.) © *1996 Balkema, Rotterdam. ISBN 90 5410 848 7*

Some chemical and strength properties of stored fly ash

Tetsuyuki Kataoka, Hideo Komine & Nobuhide Ogata
Central Research Institute of Electric Power Industry, Chiba, Japan

Toshiro Okamoto
Central Research Institute of Electric Power Industry, Tokyo, Japan

Toshiyuki Yokokura
C.R.S.Corporation, Chiba, Japan

ABSTRACT : The volume of fly ashes discharged from coal fired power stations is increasing. The fly ash has recently attracted increasing attention as a useful material for civil engineering. To use high volume fly ashes as the materials for civil engineering, the system of supplying fly ashes according to demand is needed. This system must be able to stably supply high volume fly ashes of good quality. First, we need to investigate the fundamental properties of fly ashes and the changes of these properties after a long time in order to produce the supplying system described above. This study investigates the concentration of calcium ion eluted from fly ashes, pH and unconfined compressive strength of some kinds of fly ashes stored. From the laboratory test results, it is possible to predict the changes of the concentration of calcium ion eluted from fly ashes and unconfined compressive strength by measuring successively pH of fly ashes.

1. INTRODUCTION

The total volume of fly ashes discharged from coal fired power stations is about four million tons per year at present. About half the total volume of fly ashes is utilized effectively for some way. In the future, the volume of fly ashes discharged from coal fired power station will be increasing more and more, and the development of the utilization of high volume fly ashes is required.

The fly ash has recently attracted increasing attention as a useful material for constructions such as road, embankment, backfilling and grouting. To establish the utilizations described above, it is necessary to develop the system for storing the high volume fly ashes and for supplying the fly ashes according to demand. This system must be able to stably supply high volume fly ashes of good quality.

The storing of fly ashes sprinkled with water on the stock-yard is suitable for the storing method of high volume ashes. However, the quality of fly ashes will change by the chemical reactions in the stock-yard. Therefore, it is very important to understand the changes of properties of fly ashes stored after a long time in order to develop the system for storing and supplying.

In this study, we first measure the concentration of calcium ion eluted from fly ashes, pH and unconfined compressive strength of some kinds of fly ashes stored. From the experimental results, we have considerations concerning the relationship between the strength and the chemical properties.

2. FLY ASHES USED

This section describes the outline of fly ashes used in this study. Table 1 shows the profile of fly ashes. All fly ashes were discharged from coal fired power stations, and have never contacted with water. The fly ash No. G is the ash of coal produced from Japan. The others are the ashes of coals produced from foreign countries. This study also used the samples combining some fly ashes.

Table.1 Chemical and Pysical properties of fly ash

fly ash No.	A	B	C	E	F	G	H
SiO_2 (%)	48.7	51.4	53.8	47.8	53.1	51.2	54.0
Al_2O_3 (%)	17.3	22.5	24.7	22.2	21.1	22.8	22.1
Fe_2O_3 (%)	20.7	5.6	3.2	6.3	5.8	5.6	5.4
CaO (%)	3.8	4.9	6.6	7.1	5.9	10.2	5.6
Ig Loss(%)	4.9	4.6	6.2	7.8	5.8	1.0	2.5
pH	12.6	11.6	11.3	12.7	12.7	12.7	12.5
maximum dry density (Mg/m^3) (m.d.d)	1.39	1.38	1.29	1.14	1.20	1.40	1.40
optimum moisture content (%) (o.m.c)	25.0	20.0	20.0	33.0	29.6	19.5	21.0
fly ash No.	I	J	K	L	M	N	
SiO_2 (%)	66.5	63.9	49.1	59.3	55.7	52.7	
Al_2O_3 (%)	21.1	18.4	31.1	30.1	27.8	25.1	
Fe_2O_3 (%)	5.8	3.3	3.6	2.0	5.3	6.1	
CaO(%)	0.9	5.9	5.5	1.0	1.3	10.0	
Ig Loss(%)	2.5	2.2	4.5	3.7	1.9	2.5	
pH	4.5	12.6	12.0	5.3	4.6	12.6	
maximum dry density (Mg/m^3) (m.d.d)	1.21	1.35	1.21	1.19	1.26	1.34	
optimum moisture content (%) (o.m.c)	32.5	26.1	31.1	32.8	26.9	25.0	

3. STRENGTH AND pH OF FRESH FLY ASHES

This section describes the pH and the unconfined compressive strength of fly ashes.

Almost coal fired power stations recently use the coals imported from foreign countries in Japan. Therefore, the properties of coals diversify and the properties of fly ashes also diversify because the properties of fly ashes influenced by the properties of coals. Almost fly ashes are conventionally alkalinity in Japan. However, there are, recently, the neutral and acid fly ashes as shown in Fig. 1.

It is easy to happen pozzolanic reaction in the fly ashes that highly contain CaO. The unconfined compressive strength at 90 days-curing increases as the content of CaO increases as shown in Fig. 2.

Figure 3 shows the relationship between the pH and the unconfined compressive strength. The unconfined compressive strength decreases as the value of pH decreases because the value of pH varies is correlated with the content of CaO. Figure 3 can be used in order to predict the long-term strength of compacted fly ashes by the value of pH.

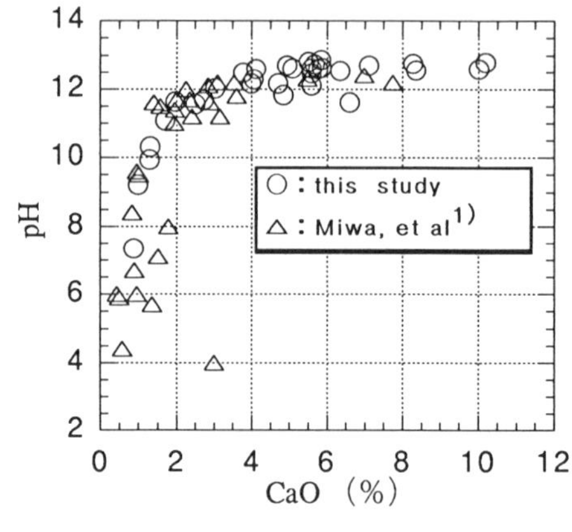

Fig.1 Relation between pH and CaO of fresh fly ashes

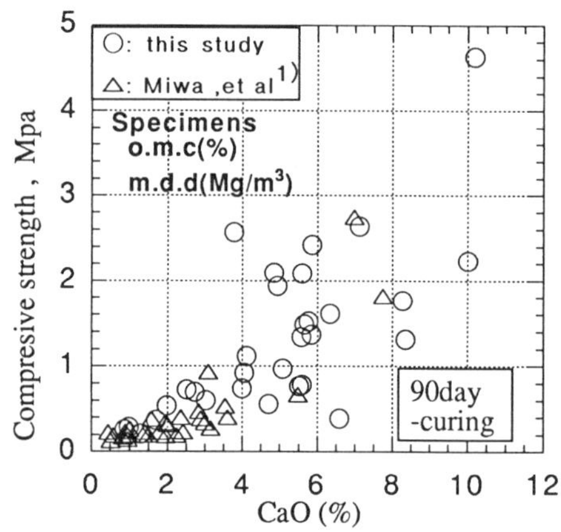

Fig.2 Relation between compresive strength and CaO of compacted fly ashes

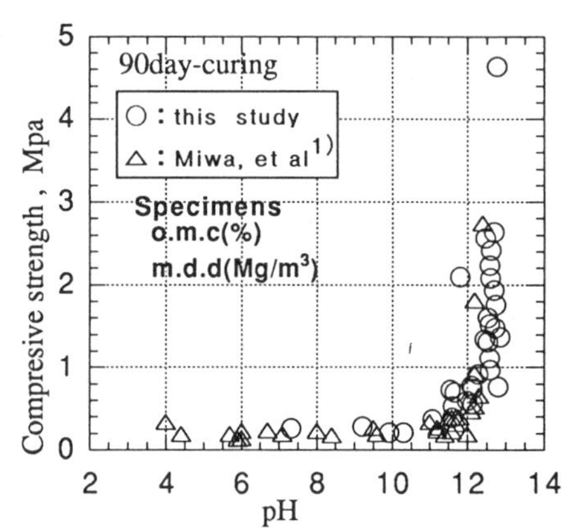

Fig.3 Relation between compresive strength and pH of compacted fly ashes

4. DETERIORATION OF FRESH FLY ASHES STORED

This section describes the deterioration of fresh fly ashes stored.

The experiments described in this section used the fly ash No. G. The fly ashes were stored in the container without a lid and in the container with a lid. Figures 4 and 5 show the measured results of water content, pH and the ion concentration of calcium. The fly ashes kept in a constant temperature (22 °C) and constant humidity (relative humidity : 60%). The changes of water content, pH and the calcium ion concentration of the fly ashes stored in the container without a lid larger than the fly ashes stored in the container with a lid. This experimental result indicates that the changes in quality are strongly dependent on the methods of storing fly ashes.

From the results of thermogravimetric analysis, the thermogravimetric loss of the fly ashes stored for one year in the container with a lid is 0 %, and the thermogravimetric loss of the fly ashes stored for one year in the container without a lid is 1.01 %. Photo 1 shows the observation of the fly ash, which was stored for one year in the container without a lid, by the scanning electron microscope. The micro products can be observed on the surface of the particle of fly ash as shown in Photo 1. Fig. 6 shows the X-ray diffraction plot of the fly ash stored for 1 year in the container without a lid. The diffraction peak of $CaCO_3$ can be observed in the plot shown in Fig. 6. These experimental results indicate that the reduction of pH of the fly ash stored in the container without a lid is mainly caused by the creation of $CaCO_3$ which is refractory and generated by the reaction of calcium ion in the fly ash and CO_2 in the air.

Table 2 shows the unconfined compressive strength of compacted fly ash No. A stored for three years. From the results shown in Table 2, the strength is strongly influenced by the change in chemical characteristics of fly ash.

The results described above point out that we must pay attention to the deterioration of properties of fresh fly ashes when the fly ashes are stored for long-term duration on condition that the fly ashes contact with air directly. Furthermore, it is found that the content of deteriorations of fly ashes can be predicted by the measuring successively the value of pH.

5. DETERIORATION OF CONDITIONED FLY ASHES STORED

This section describes the deterioration of wet fly ashes stored.

The fly ash of which the unconfined compressive strength is less than 0.51 MPa is deprived of the strength by water uptake (Kataoka et al. 1991). Therefore, the strength increasing of fly ash cannot be expected when the unconfined compressive strength is less than 0.51 MPa.

Figure 7 shows the changes of pH and the calcium ion concentration with time of fly ashes No. G and No. F. The fly ash samples (conditioned fly ash) described above were compacted at optimum water content. Figure 8 shows the unconfined compressive strength of the samples above. In the case of fly ash of which ignition loss is large such as No. F, the hydration makes slow progress, and the strength increasing is low in the early step of curing. These figures show that the strength is strongly correlated with the pH and the calcium ion concentration. Table 3 shows the unconfined compressive strength of three kinds of the compacted fly ash stored. The refractory hydrates are being created in the process of drastically decreasing of the calcium ion concentration and the pH (see Fig. 7). Therefore, the components which

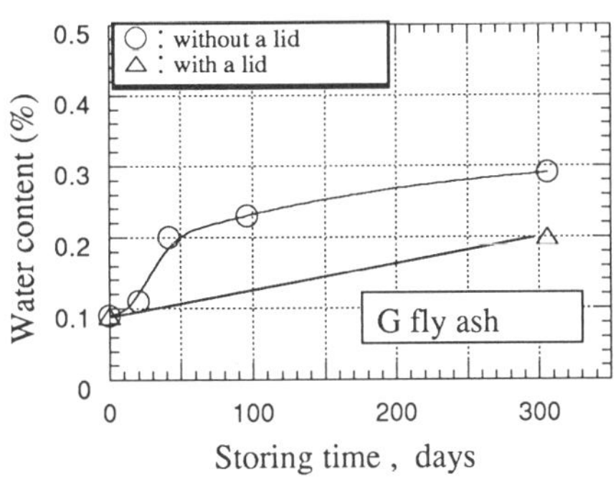

Fig.4 Change in water content of stored fly ash

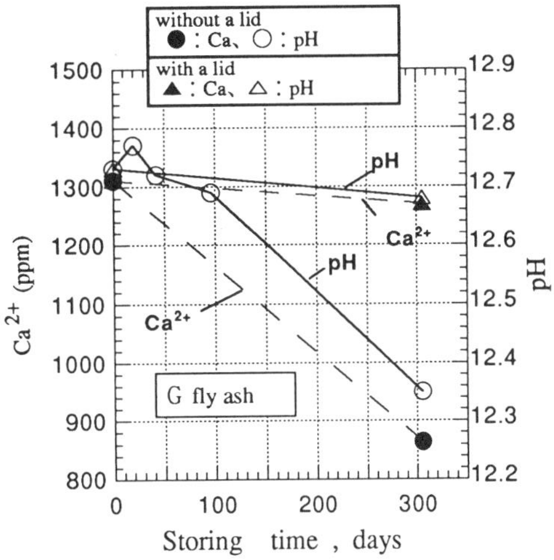

Fig.5 Change in pH and Ca^{2+} of stored fly ash

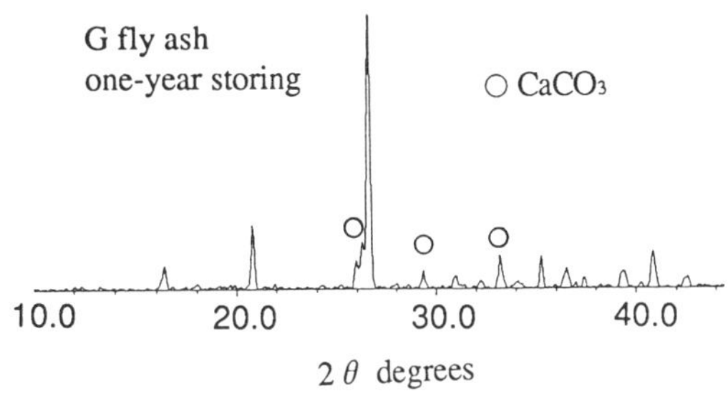

Fig.6 X-ray diffraction diagram of G fly ash stored in the container without a lid

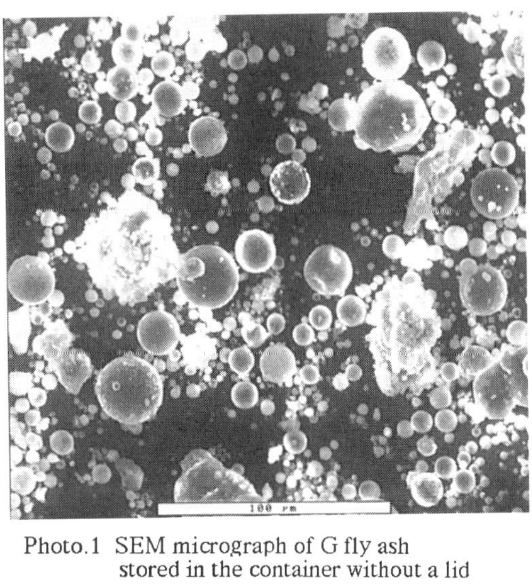

Photo.1 SEM micrograph of G fly ash stored in the container without a lid

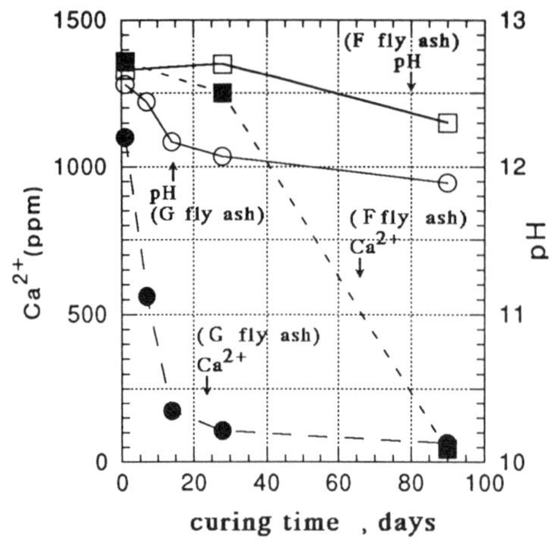

Fig.7 Cange in pH and Ca^{2+} of compacted fly ash with curing time

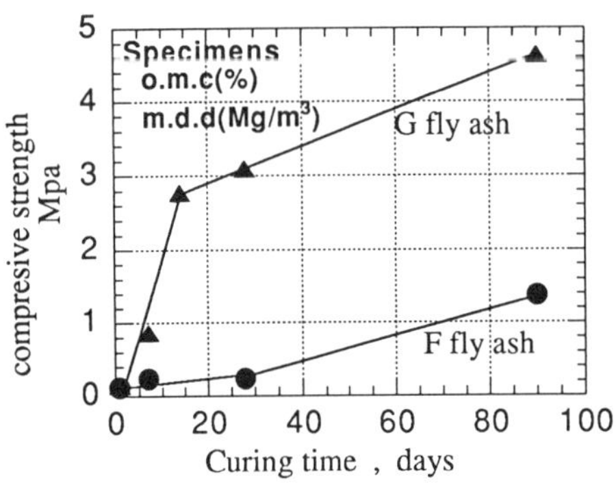

Fig.8 Compresive strength of compacted fly ash with curing time

Table.2 Change in chemical and strength properties of stored A fly ash

Storing time	Water content (%)	CaO (%)	pH	Compresive strength 90day-curing (Mpa)
0day	0.1	3.78	12.6	2.56
3year	0.4	3.65	12.3	1.33

are required for the hydration are remained in the process mentioned above. The data which is shown in Table 3 without shade part are the experimental results in the above process. These experimental results indicate that the strength increasing of fly ashes can be expected in the process of drastically decreasing of the calcium ion concentration and the pH.

Whereas the amount of components which are required for the hydration is little after the process of drastically decreasing of the calcium ion concentration and the pH. The strength of fly ashes is low in the above process as shown in the

Table.3 Change in pH and strength of compacted fly ash

Fresh fly ash		Storage condition		Change in pH of stored fly ash	Compacted fly ash after 90day curing	
No.	pH	water content (%)	Storing time (day)		Compresive strength (Mpa)	pH
G	12.63	17.8	0	—	4.63	11.89
			7	12.50	0.76	12.18
			28	12.32	0.38	12.13
		10.0	7	12.49	0.53	12.13
			28	12.34	0.29	12.20
		5.0	7	12.50	0.76	12.14
			28	12.13	0.32	12.00
A	12.62	25.1	0	—	2.57	11.06
			7	11.48	0.64	no measurement
			28	11.19	0.39	no measurement
			60	11.12	0.31	no measurement
F	12.73	29.6	0	—	1.37	12.28
			7	12.7	2.04	12.28
			28	12.69	2.02	12.22
			60	12.38	1.24	12.20

experimental results which are shown in Table 3 with shade part.

The deterioration and the strength increasing of wet fly ashes stored vary with the kinds of fly ashes. From the consideration described above, the deterioration and the strength increasing of wet fly ashes stored can be obtained by measuring the pH that is strongly correlated with the calcium ion concentration.

6. CONCLUSIONS

The following conclusions were drawn from this experimental study.

1) The strength of fly ashes stored for long-term duration can be predicted by measuring the pH.

2) We must pay attention to the deterioration of properties of fresh fly ashes when the fly ashes are stored for long-term duration on condition that the fly ashes contact with air directly.

3) The deterioration of wet fly ashes which are highly alkalinity begin immediately after storage.

4) The deterioration and the content of strength increasing of fly ashes stored can be evaluated by measuring successively the pH.

REFERENCES

Miwa, K., Sugi, T., Yasuda, M., Tai, F., Hosoda, N. and Nagataki, S. (1992) : Mechanical Properties of Pulverised Fuel Ash by Product, Proceedings of Second Conference on Coal Utilization Technology, pp.58-73.

Kataoka, T., Okamoto, T. and Ogata, N. (1991) : Physical and Mechanical Properties of Reclaimed Ground by Coal ash, CRIEPI report U91025.

Environmental Geotechnics, Kamon (ed.) © *1996 Balkema, Rotterdam. ISBN 90 5410 848 7*

Treatment and utilization of waste sludge/slurry from construction works in Japan

Takeshi Kawachi
Technical Research Institute, Obayashi Corporation, Kiyose, Tokyo, Japan

Takeshi Katsumi
Kyoto University, Uji, Japan

Phi Oanh Tran Duc
Technical Research Institute, Hazama Corporation, Tsukuba, Japan

Masaru Yamada
Osaka City University, Japan

ABSTRACT: Large amounts of waste and by-products have been generated by the Japanese construction industry, while natural resources have been used in large quantities. Proper treatment and effective utilization of the construction wastes have therefore been emphasized. Management of waste sludge/slurry discharged from foundation works in particular is one of the most serious issues in the field of environmental geotechnology. Most of the treated and untreated slurries are only disposed of in landfill sites, and cause environmental problems, such as lack of landfill sites and illegal dumping. Currently, treatment processing and utilization of this waste slurry are under research and development. This paper introduces the present state and the future prospects of the treatment and utilization of the waste or surplus sludge/slurry from construction works in Japan, which are based on the research activities of the JGS Research Committee on Treatment and Potential Use of Industrial Wastes. Some kinds of uses, both already realized and only under research, are summarized or proposed. Required criteria, treatment methods and possible problems are presented for each utilization purpose. Perspectives and proposals on waste slurry utilization are stated in conclusion to mitigate environmental negative impacts due to construction works.

1 INTRODUCTION

Large amounts of wastes or by-products have been generated from the construction industry while natural resources have been used in large quantities in Japan. Proper treatment and effective utilization of the materials have therefore been emphasized (e.g., Honda and Yamada 1990; Kamon et al. 1994). Because most of the by-products, including concrete mass, asphalt-concrete mass and waste sludge, do not contain toxic substances, they can be expected to be reused as construction materials in the same or on other sites. According to the investigation by the Ministry of Construction, although 70 % of the generation of concrete and asphalt wastes is reused as construction materials due to the development of treatment technology, more than 90 % of the sludge/slurry discharged from foundation works is only disposed of and is not utilized. In the fiscal year 1993, about 15000 Gg (= 15 million tons) of the sludge was generated, and only 2 % and 6 % of it was recycled or treated to be reduced in volume for disposal. Such a large amount of sludge disposal sometimes causes environmental problems due to the lack of landfill sites or illegal dumping. Utilization of the waste sludge is therefore strongly required from the viewpoint of environmental geotechnology.

The composition of construction sludge is similar to that of soils. There are many cases in which only the high water content causes technical problems such as fluidity or instability. Dehydration treatment of the sludge has been developed to reduce the volume to be disposed (Kida and Tsuji 1981). The sludge must be properly treated for utilization, such as by dehydration or solidification, for reuse as construction materials. Geotechnical practices for waste sludge utilization have just started, and research and development have been carried out to expand the field for sludge utilization in recent years.

This paper introduces the present state and the future prospects about the treatment and utilization of waste or surplus sludge/slurry from construction works in Japan, and also some proposals to promote waste sludge utilization, which are based on the activities of the JGS Research Committee on Treatment and Potential Use of Industrial Wastes, whose members are shown in Table 1, and mainly performed by the Sub-Committee on Utilization of Construction By-products.

2 BASIC CONCEPT OF SLUDGE UTILIZATION

The characteristics of waste sludge/slurry depend on the construction method, the materials used, and the soil type of the original ground. Table 2 summarizes

Table 1. Members of Research Committee on Treatment and Potential Use of Industrial Wastes (Japanese Geotechnical Society)

Chair: M. Kamon (Kyoto University)

Sub-committee on Utilization of Industrial Wastes		*Sub-committee on Utilization of Construction By-products*		*Sub-committee on Land Use of Waste Reclaimed Site*	
Chief:		*Chief:*		*Chief:*	
Y. Nagasaka	(Geogyro Co., Ltd.)	M.Yamada	(Osaka City University)	K. Onitsuka	(Saga University)
Members:		*Members:*		*Members:*	
K. Fukushima	(Toagosei Co., Ltd.)	T. Kawachi	(Obayashi Corp.)	K. Inoue	(Kansai Soil Research Center)
K. Higaki	(Taisei Corp.)	M. Kamiya	(Hokkaido Inst. of Tech.)	K. Kubota	(Kisojiban Consultants Co., Ltd.)
T. Hirata	(Kagoshima Nat. Col. of Tech.)	K. Kataoka	(Chuken Consultant Co.)	M. Nambu	(Oyo Corp.)
S. Horiuchi	(Shimizu Corp.)	T. Katsumi	(Kyoto University)	J. Okada	(Nakabori Soil Corner Co., Ltd.)
T. Kanda	(Fujita Corp.)	T. Kunito	(Seiko Kogyo Co., Ltd.)	M. Okada	(Toyo Constr. Co., Ltd.)
A. Kasai	(Onoda Chemico Co., Ltd.)	K. Mori	(Kumagai Gumi Co., Ltd.)	S. Ono	(Chuo-Kaihatsu Co., Ltd.)
Y. Nakano	(Konoike Constr. Co., Ltd.)	N. Ogawa	(PWRI, Min. of Constr.)	M. Otsubo	(Kyusyu University)
H. Omori	(Chichibu Onoda Cement Corp.)	P.O. Tran Duc	(Hazama Corp.)	K. Sonoi	(Nikken-Sekkei Ltd.)
H. Sakai	(Railway Technical Res. Inst.)	K. Saitou	(Ohmoto Gumi Co., Ltd.)	K. Terada	(Takenaka Corp.)
T. Yamaguchi	(Sumitomo Constr. Co., Ltd.)	K. Sasaki	(JDC Co.)	M. Tsuchihiro	(Kajima Corp.)
		M. Tokunaga	(Nippon Koei Co., Ltd.)		

the origins and main properties of waste slurry. The most important feature of the slurry is, of course, high water content and difficulty to dehydrate due to the existence of fine soil particles. In particular, waste slurry from slurry excavations usually contains additive materials such as bentonite, polymers, and/or other chemicals besides the soil particles from the original ground, and they can strongly affect not only the engineering properties but also the application of the treatment method and utilization. As waste slurry from ground improvement works has a potential to be hardened by the remained cement components, the application is strongly influenced by the aging period.

More than 10 kinds of practical and possible uses of the waste sludge from construction works have been proposed, and these can be classified into the following groups based on the utilization styles:

1) Utilization as flowable materials: Waste sludge can be recycled as excavation slurry, or used as grouting materials,
2) Utilization as earthen materials: Treated sludge can be used for embankments, reclamation, backfilling, and covering,

Table 2. Properties of waste slurry from construction works.

Origin	Composition	Solid content
Excavation of soft ground	Soil (Clay, Silt)	above 50 %
Slurry excavation method		
Diaphragm wall or piling	Bentonite, Polymer, Dispersant, Soil ...	5-40 %
Shield tunneling	Bentonite, Soil (Clay), Foaming agent ...	5-50 %
Ground improvement (SMW, DMM, Grouting)	Cement, Bentonite, Soil ... (able to harden)	25-50 %
Dredging	Soil, Organic matter	10-30 %

3) Utilization of clay materials: Clays separated from waste sludge can be used as lining material, or as raw materials for cement or ceramic manufacturing, and
4) Utilization of granular materials: Granular materials made from the sludge is proposed to be applied as aggregate, base course material, drain material, or applied for backfilling, and covering.

The waste sludge should be therefore treated through a proper treatment method, to realize the

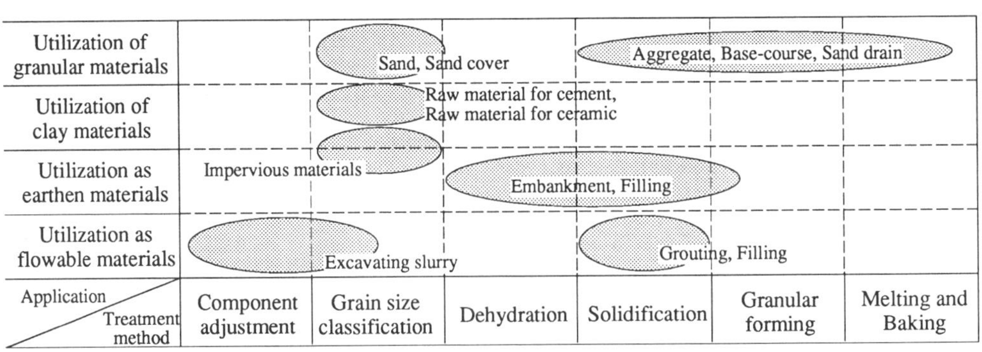

Fig. 1. Concept of relationship between treatment method and application of construction waste sludge/slurry.

above applications. The treatment method must be adopted according to quantity and quality of the sludge, location of generation, application and financial condition. The proposed and realized treatment methods are as follows;

1) Component adjustment by concentration, dilution or addition of other materials: adjustment of grain size distribution and water content,
2) grain size classification by sieving or centrifuging,
3) dehydration for reducing volume and decreasing water content by means of a dehydrator,
4) solidification by the addition of hardening materials such as cement,
5) mechanical treatment for forming granular material by graining or crushing machinery, and
6) thermal treatment: granular materials can be formed by melting or baking processes.

The first step in planning waste slurry utilization is to select the treatment method and the utilization style based on the possible application. Figure 1 shows the relationship between the treatment methods and utilization styles.

3 GEOTECHNICAL PRACTICE FOR WASTE SLUDGE UTILIZATION

Table 3 summarizes the present position concerning waste sludge utilization, its main uses including our proposal, required qualities, treatment methods, satisfactory practices and possible problems. Details of the main uses are presented in the following.

Table 3. Realized and possible uses of waste sludge/slurry from construction works.

Use	Quality required	Treatment method	Satisfactory practice	Possible problems
Excavating slurry	Density, viscosity	- Sieving and addition of clay or chemicals	Partially practiced (for shield tunnel or earth drill)	- Cost for convey - Quality control
Grouting	Strength, gelation time, flowability, bleeding	- Addition of hardening material	Partially practiced (for slurry or soil pressure shields)	- Quality control - Treatment plant
Hardening flowable backfilling	Strength (according to the surroundings)	- Addition of hardening materials	Extensive field application (backfilling for pile)	- Mixing plant
Hydraulic barrier	Permeability, strength, durability	- Dehydration - Solidification	Partially practiced (filling for earth retaining)	- Quality control
Reclamation	(according to the land use)	- Air dry - Dehydration - Solidification	Extensive field application	
Landfill cover	Strength (q_c)	- Dehydration	None	- Adjustment between different governments
Embankment and Subgrade	Compaction, compression, erosion, strength	- Dehydration - and mix with additives or other soils	Partially practiced	- Place for treatment and stock
Soil for vegetation	Grain size, water retention, chemical composition	- Dehydration - and mix with additives - Baking	Experimentally practiced	- Limitation of additives for dehydration
Rock cover for vegetation	Grain size, chemical composition, additives	- Dehydration, chemical addition and spraying	None	- Erosion - Quality control
Stabilized roadbase	Strength (q_u)	- Solidification	Partially practiced (mixed with other granulars)	- Treatment facility
Granular roadbase	Grain size, CBR, stability	- Baking - Melting	Partially practiced (mixed with other granulars)	- Low strength and durability - Grain size distribution
Sand capping	Grain size, chemical composition	- Baking	None	- No criteria - Cost
Sand drain	Grain size, stability	- Baking	Experimentally practiced	- Cost
Backfilling	Grain size, durability	- Baking	Partially practiced	- No criteria - Cost
Aggregate	Grain size, durability, stability, Chemical composition	- Baking - Melting	Experimentally practiced (on the sludge from the water purification plant)	- No criteria - Cost
Filter for water purification	Porosity permeability	- Baking	Experimentally practiced	- Quality achievement and evaluation
Raw material for ceramic manufacturing	Mineral composition grain size	- Dehydration and classification	Experimentally practiced	- Severe quality control - Existing of Cement and Bentonite
Raw material for cement manufacturing	Mineral composition	- Dehydration and classification	Partially practiced (on sewage cake)	- Need of equipment

3.1 *Utilization as flowable materials*

Applications as flowable materials can be divided into two categories; recycling as an excavation slurry and utilization as a slurry which has a potential to harden due to chemical additives.

In the Tokyo Metropolitan area, where many construction works are ongoing, waste slurry discharged from continuous walls, earth drilling and slurry shield works is often re-utilized in kindred work on construction sites located nearby. In such cases, the slurry is treated using an admixture, or concentrated and diluted to reach the required quality with regard to density, viscosity and fluidity. Besides, space is needed to treat or stock the slurry to promote the recycling.

The slurry which has a potential to harden can be used for backfilling and grouting, such as for overbreaking in shield tunneling work, or for filling the driving site around the top of a cast-in-place pile. The waste slurry, to which cement is added, may be filled or grouted into a cavern, and can consequently harden. Control of the fluidity for groutability and choice of the appropriate cement content for strength development are essential. Some methods have been proposed to control the viscosity and maintain the potential of strength development (e.g., Uchiyama and Horiuchi 1994). Figure 2 shows an example of a utilization system for waste slurry in shield tunneling.

Recently, the application of waste slurry for filling the invert section of a shield tunnel has been developed as shown in Fig. 3 (Sukegawa 1994). The filling material for the invert has usually been concrete. To replace it, the filling material must have workability as well as sufficient durability against traffic loading. The hardening flowable mixture to be filled in the invert can be made of the mixture composed of waste slurry and excavated soil with cement and other additives. The waste slurry and the surplus soil excavated from the forward section of the tunnel can be used as the materials for the invert where the excavation has been finished.

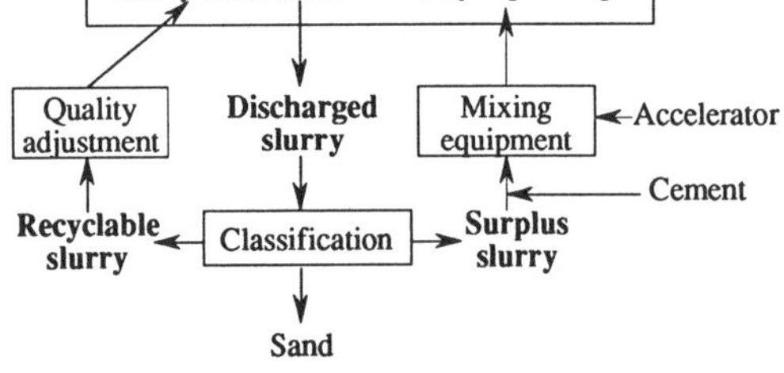

Fig. 2. Utilization system for waste slurry in shield tunneling work.

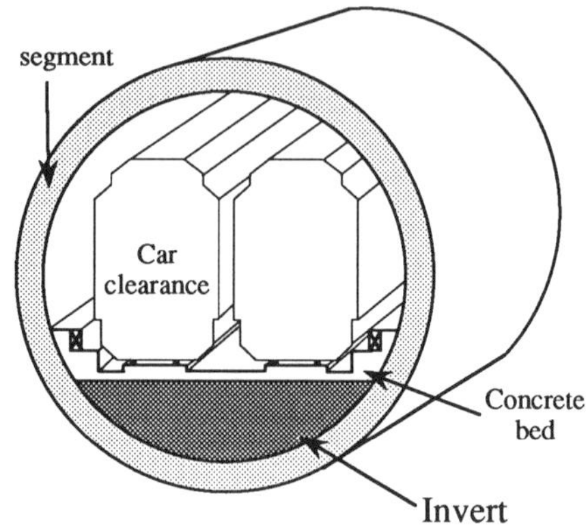

Fig.3. Cross section of shield tunnel.

This kind of waste slurry utilization is usually adopted at the construction site where the slurry is generated. However, utilization through a treatment center is also needed. To treat waste slurry at a treatment center for reuse, development of a method using retarder chemicals to regulate the hardening time has been tried.

3.2 *Utilization of clay materials*

Waste sludge/slurry from construction works is mainly composed of fine particles such as clay and silt. The fine components can be separated from the total slurry by sieving or centrifugal separation methods, and used as the raw clay material for the ceramic industry and cement manufacturing. At present, construction waste slurry has also been used as supplement of insufficient raw material for ceramic products such as tiles. In the case of cement manufacturing, where about 0.3 Mg of clay materials is needed for 1.0 Mg of production, the clay separated from the waste slurry can be used as the supplement for a part of the raw materials; application in this field is being advanced and has been partially practiced. Further applications can be expected due to the large amount of cement production. However, to maintain product quality, the composition of the clay must be strictly prescribed. Consequently the practices will be restricted to the waste slurry from certain regions and from the limited construction methods so as to achieve the quality required and a stable supply.

Separated clay material has occasionally been used for the construction of impermeable walls. In this application, cement and bentonite need to be added to achieve the desired strength and to improve the impermeability respectively.

Table 4. Engineering properties of dehydrated cakes.

Dehydrator		Dehydrated cake	
Type of dehydrator	Working ability	Water content (%)	Cone index (kPa)
Filter-press (ordinary)	500 - 700 kPa	(0.8 - 0.9) x w_L	100 - 1000
Filter-press (high-pressure)	4000 kPa	(0.6 - 0.8) x w_L	1000 - 3000
Belt-press	100 - 150 kPa	1.0 x w_L	0 - 200
Screw decanter	500 - 2000 G	(1.0 - 1.2) x w_L	0

w_L : Liquid limit (%), G : Acceleration of gravity (9.8 m/s^2)

3.3 *Application as earthen materials*

Application of the dehydrated or stabilized waste slurry as earthen material for embankments or filling is the most popular method. The required criteria for the application as earthen material has been prescribed according to purpose by the Ministry of Construction or other agencies. In the manual presented by the Ministry of Construction, the applicable soil is divided into 4 classes based on its engineering properties.

Water-wrung cake which is made of the waste slurry by a dehydrator, such as a filter-press or a roller-press machine, can be used as earthen material. Ordinary filter-presses and roller-presses can press the slurry to pressures of 30-100 kPa and 300-500 kPa respectively. The basic properties of dehydrated cakes, examples of which are presented in Table 4, indicate the possibility of using them as earthen materials. Cement or lime should be added depending on the situation, to obtain the required qualities for construction purposes, such as strength, compressibility and workability. The screw decanter, the dehydrator with the lowest water-wringing ability, can be effectively used in the case of combined dehydration and cement solidification, because the dehydrated cake can be easily mixed with cement due to its high water content. In some cases, the waste sludge is directly solidified because the sludge is difficult to dehydrate due to the remaining bentonite and polymers, or equipment for dehydration is not available. Ogino et al. (1994) presented the application of the cement hardened dredged sludge for embankment construction. Kamon and Katsumi (1994) proposed a utilization system which consists of dehydration and solidification, and emphasized the utilization of other waste materials such as coal ash for dehydration or solidification.

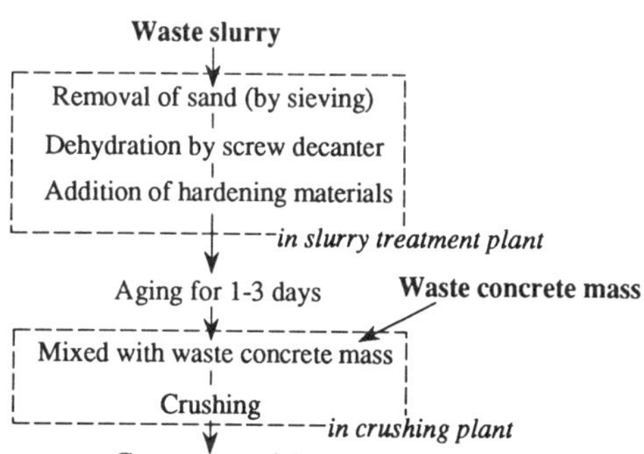

Fig. 4. Treatment system for making coarse materials from waste slurry.

Table 5. Examples of engineering properties of coarse material made from waste slurry.

Index	Result
Particle density (g/cm^3)	2.63
Grain size distribution	
25-30 mm	0.5
13-25 mm	24.4
5-13 mm	27.6
2.5-5 mm	11.0
0.42-2.5 mm	22.6
0.074-0.42 mm	10.4
below 0.074 mm	3.5
Water adsorption (%)	6.7
Volume of abrasion loss (%)	32
Optimum water content (%)	14.0
Maximum dry density (g/cm^3)	1.85
CBR modified (%)	114

Several attempts have been made to use the waste slurry for other purposes, such as for road bases. Figure 4 shows the treatment system for a waste treatment enterprise. In this system, the waste slurry, from which the sand has been removed by sieving, is dehydrated by a centrifugal treatment machine, and the dehydrated cake is mixed with hardening materials. After aging for 1-3 days in a yard, the large mass of hardened mixture is crushed together with the by-product of mass concrete and stocked. The resulting material, whose engineering properties are shown in Table 5, can be used as earthen materials for embankments, coverings and even for road bases.

One of the major problems which prevents waste slurry utilization is that the user of the slurry is different from the generator. As a result, the information about material quality, quantity and time for demand and supply is not sufficient. Besides, because the user, but not the generator, must petition the local government to reuse waste materials under the present legal system, waste utilization tends to be avoided. Therefore, the establishment of an system which offers information about the supply and demand conditions is required.

3.4 *Baking and melting; advanced treatment methods*

Techniques such as baking and melting methods have been developed to produce high strength granular

Table 6. Examples of engineering properties of waste sludge treated through baking process.

Origin	Waste slurry from construction (I)	Waste slurry from construction (II)	Sludge from dam reservoirs
Heating temperature (℃)	1100	1500	1000
Particle density (g/cm^3)	1.72	2.60	1.49 - 1.56
Water adsorption (%)	22.2	0.6	13.1 - 21.1
Crushing strength (MPa)	1.0 - 3.0	13	12.1 - 12.8
Permeability (m/s)	10^{-4} - 10^{-2}	-	-

materials from waste sludge/slurry. Initially, these methods were proposed for the treatment of sewage sludge, and Riley (1951) demonstrated that the composition of the clayey soil is suitable for the production of aggregate material. The treatment process consists of thermal treatment as well as pre-treatment by dehydration, grain-forming and drying and post-treatment by grain size regulating. Dehydration pre-treatment is done by means of a filter-press with about 500 kPa pressure and thermal treatment is performed at 1000-1200 ℃ to melt the particles. The engineering properties, such as strength and durability, of the produced granular materials are excellent, as shown in Table 6 (Kawachi et al. 1994); they can be used in various applications such as for road bases, sand drains and concrete aggregate.

High cost is the most important problem of this method. This method will primarily be adopted for the treatment of slurry which has so significantly poor quality that the slurry cannot be treated by other methods. However, the treatment technique may have advantages because the supply of aggregate of high quality or low-weight is anticipated to be insufficient in the future. The baking method is considered to be more applicable than the melting method because the granular materials produced by this method have properties similar to those of ordinary soil. Application of other wastes as fuel for the method should also be considered in order to reduce the cost.

4 CONCLUSIONS: TO PROMOTE WASTE SLUDGE UTILIZATION

The state of utilization of and the treatment techniques for waste sludge/slurry from construction works in Japan has been presented. Geotechnical practices for waste sludge utilization have only recently started. To promote the waste reuse, many problems need to be solved and the main issues are enumerated as follows;

1) To promote the information exchange about the conditions of waste discharge and possible application.
2) To spread knowledge about the applicable treatment technologies.
3) To have enough space to treat and stock materials.
4) To establish a system to use the waste sludge at the site discharging the waste sludge.
5) To re-estimate authorized standards and design methods.
6) To reduce treatment cost and to link to the recycling of other wastes.

Construction works will continue in order to provide the higher level of infrastructure required, and will continue to have a potential impact on the environment. Environmental compatibility is therefore needed for the construction industry. Re-utilization of waste sludge as earthen material or for other purposes is one of the most effective strategies for mitigating environmental impact from the viewpoint of resource reuse as well as environmental preservation.

REFERENCES

Honda, A. & M. Yamada 1990. *Treatment and reutilization of construction wastes*, Energy-Saving Center (in Japanese).

Kamon, M. et al. (eds.) 1994. *Introduction to environmental geotechnology*, JGS, (in Japanese).

Kamon, M. & T. Katsumi 1994. Utilization of waste slurry from construction works, *Proc. 13th Int. Conf. on Soil Mechanics and Foundation Engrg.*, Vol.4, pp.1613-1616.

Kawachi, T., N. Ogawa & H. Iso 1994. Baking treatment of construction sludge, *Proc. 5th Ann. Conf. of JSWME*, pp.200-202 (in Japanese).

Kida, D. & H. Tsuji 1981. Treatment and utilization of discharged slurry from slurry excavation methods, *Tsuchi-to-Kiso*, JSSMFE, 29(11), pp.57-64 (in Japanese).

Ogino, T., T. Goto, K. Kataoka & M. Kuroda 1994. Utilization of stabilized dredged waste for construction material, *Proc. 1st Int. Cong. on Environ. Geotechnics*, pp.49-56.

Riley, C.M. 1951. Relation of chemical properties to the bloating of clays. *Jour. American Ceramic Society*, 30(4), pp.121-128.

Sukegawa, T. 1994. Development of invert material using excavated soil from slurry shield, *Proc. JSCE*, No.504/VI-25, pp.107-116 (in Japanese).

Uchiyama, N. & S. Horiuchi 1994. In-situ utilization of waste bentonite slurry, *Environmental Aspects of Construction with Waste Materials*, Elsevier, pp.217-226.

Environmental Geotechnics, Kamon (ed.) © 1996 Balkema, Rotterdam. ISBN 90 5410 848 7

Failure pattern and earth pressure of cement treated fly ash

Masaki Kitazume
Port and Harbour Research Institute, Ministry of Transport, Yokosuka, Japan
Tomohiro Ikeda
Third District Port Construction Bureau, Ministry of Transport, Kobe, Japan
Nobuyuki Yamane
TOA Corporation, Yokohama, Japan

ABSTRACT: An experimental investigation was carried out to study the feasibility of using fly ash, the industrial residue of thermo-electric power plants, as a fill material for waterfront retaining structures. In the study, centrifuge model tests were performed on the cement stabilized fly ash ground to investigate the failure pattern and active earth pressure on the retaining wall. As a result of this study it was found that the active earth pressure induced by fly ash mixed with small amount of cement to be lower than conventional sand fill. Consequently, the new application of this waste material will add a new component to cost-effective developments in coastal areas.

1. INTRODUCTION

The world-wide development of coal-fired electric energy production is associated with disposal problem of the waste material as a result of combustion process. Fly ash, removed from flue gas by electrostatic precipitators or filter bag house, is produced in large quantities as by-product of burning coal at these electric power stations. According to Vesperman et al. (1990) more than 70 million metric tons of fly ash is produced annually in the United States; while the production of fly ash from thermo-electric power plants in Japan is estimated to be 7 to 8 million tons per year (Horiuchi et al., 1995); and in Bulgaria 8 million tons (Todorov and Evstatier, 1994).

New environmental regulations in recent years, coupled with the high cost of landfilling and waste disposal, have initiated a new challenge for researchers and engineers to find various applications of different industrial residues to civil engineering structures. In Europe, stabilized coal fly ash was successfully used for light weight landfills, highway embankments, and road subbase construction (Gray and Lin, 1972). In the U.S., the New York State Department of Transportation has published specifications for coal fly ash stabilization mix design (Sgambati, 1985); and Larrimore (1986) completed a study on environmental effects of test sections constructed with stabilized coal fly ash in several highways in the state of Georgia. Edil et al.(1992) investigated the use of fly ash, either by itself or mixed with sand, as a construction material for waste-containment liners and impermeable covers. In Japan, Horiuchi et al.(1992) and Kawasaki et al.(1992) used various kinds of fly ash mixed with cement for construction of a man-made island for Hakucho bridge in Hokkaido area.

In recent years, the construction of many big projects of man-made islands on/off-shore (for example the Haneda airport off-shore expansion project in Tokyo, and the Kansai international airport in Osaka) has gained tremendous attention and priority in Japan. Consequently; the stabilization process of new and cost-effective construction material in port areas should be examined.

In this study, as one of the land development construction methods using industrial residues, the applicability of cement stabilized fly ash for backfill in coastal areas was investigated using geotechnical centrifuge. Although the coal fly ash has self-hardening characteristic, cement stabilized fly ash has relatively high strength and its strength could be well controlled mainly by the amount of cement mixed. Therefore, the stabilized fly ash is expected to reduce the earth pressure acting on the structure.

In this study, centrifuge model tests were performed to investigate the effect of shear strength and geometric conditions of the stabilized fly ash on the failure pattern and earth pressures behind rigid rotating wall. The resulted earth pressure compared with that of conventional sandy soil fill to show applicability to construction of port and harbor facilities.

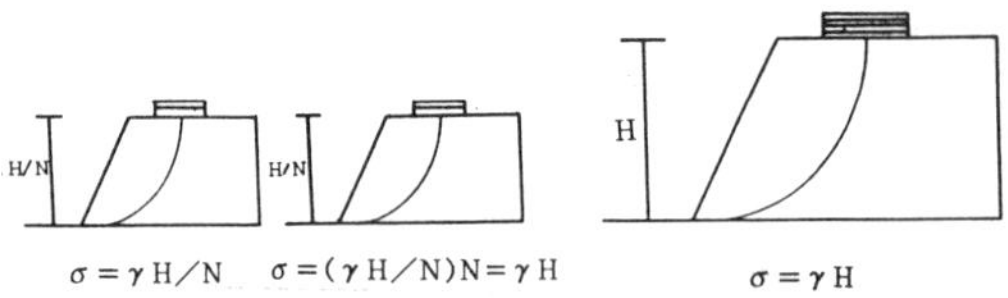

Fig.1 Basic concept of physical modeling.
(a)1-g model, (b)N-g model(centrifuge model),
(c)full scale prototype model

2 .CENTRIFUGE MODELING TECHNIQUE

2.1 *Concept of physical modeling*

The idea of making a small-scale model to study a physical phenomena is common in many fields of engineering. The centrifuge model technique is one of powerful tool being used increasingly by geotechnical engineers to gain insight into the behavior of a physical systems. By placing a small-scale model in an artificially increased gravitational field, to increase the self-weight of the soil, the stress similarity between the model and full-scale prototype can be restored. If the boundary conditions and stress histories correspond in model and full-scale prototype, then the response in the model should be equal to the response in the full-scale prototype, consistent with available scaling relationships and dimensional analysis. The basic concept of centrifuge modeling technique for a slope is shown in Fig.1.

2.2 *PHRI centrifuge facilities*

The centrifuge used in this study was the PHRI (Port and Harbour Research Institute) Mark II gcotechnical centrifuge. The centrifuge has maximum acceleration of 115g, maximum effective radius of 3.8 m and maximum payload of 2710 kg, and is one of the biggest centrifuges operating in the world. The main part of the centrifuge is housed in the underground reinforced concrete pit for safety operation. Two swinging platforms are hinged to the rotating arm via the torsion bar systems to safely deliver the radial force at high acceleration to the end plates at both ends of the arm. The drive unit for the centrifuge of 450 kVA direct current motor is mounted on the underground floor and generates steady acceleration up to 113 g at the surface of the swinging platform whose radius during flight is 3.8m. Figure 2 shows the general view of the centrifuge apparatus. The centrifuge and surrounding equipment are described by Kitazume (1996) in detail.

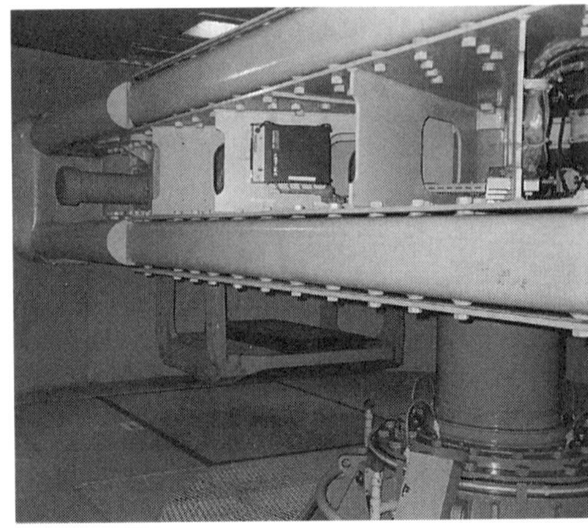

Fig.2 General view of PHRI Mark II geotechnical centrifuge

Table 1. Properties of fly ash

Silica (SiO_2)	55%
Aluminum (Al_2O_3)	21%
Iron Oxide (Fe_2O_3)	3.8%
calcium (CaO)	6.2%
unburned material	6.5%
pH	1.7%

Table 2. Indexes of Toyoura standard sand

Specific gravity	2.65
Maximum void ratio (e_{max})	0.985
Minimum void ratio (e_{min})	0.602
Coefficient of uniformity, Uc	1.34

3. MATERIAL PROPERTIES

The unhumified fly ash used in this investigation was supplied from a dry disposal site of a coal power plant in Kyushu island, located in the western coast of Japan. The typical properties of fly ash are summarized in Table 1. Ordinary portland cement was added to the fly ash slurry in the model tests to increase the strength of the slurry. Toyoura standard sand, which is frequently used for geotechnical engineering in Japan, is used to fill behind the stabilized zone. The major physical properties of Toyoura sand are listed in Table 2.

4. EXPERIMENTAL TECHNIQUE

4.1 *Apparatus and instrumentation*

A rigid specimen box has inside dimensions of 100 cm in length, 32.5 cm in depth and 30 cm in width. The insides of the box are covered by glass plates to decrease friction between the box and the model ground as much as possible. The front side of the box has a glass window of 4.5 cm in thickness to observe the model ground behavior during the test. A rigid model retaining wall was set in the box and fixed during the increasing acceleration to the certain level, and was subjected to rotational displacement about its bottom by means of a motor jack (see Fig. 3).

The retaining wall consisted of three plates, of which the central plate was supported horizontally and hanged by two set of three load cells and two load cells respectively in order to measure horizontal and vertical components of total earth pressures acting on the plate; and ten earth pressure gauges of 7 mm in diameter were installed on the surface of the central plate to measure the horizontal earth pressure distribution along the depth as shown in Fig.4.

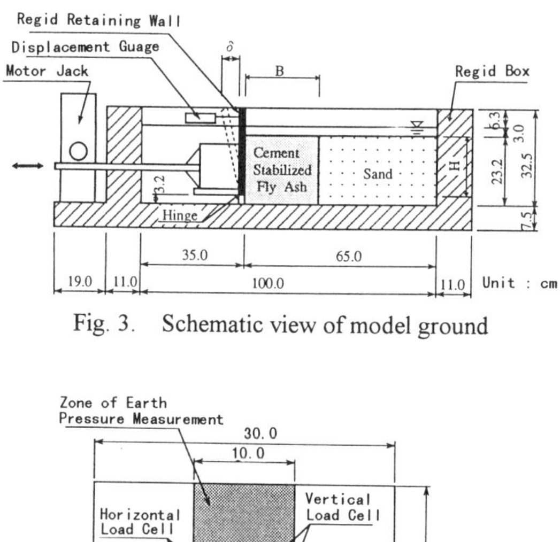

Fig. 3. Schematic view of model ground

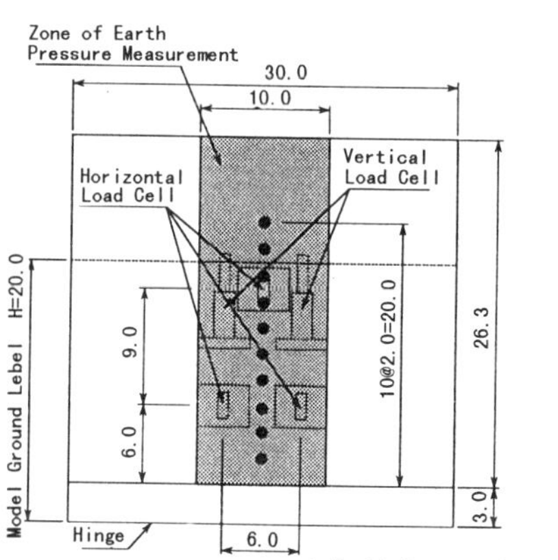

Fig. 4 Instrumentation of model rigid retaining wall

4.2 *Model ground preparation*

The model ground was composed of cement stabilized fly ash ground backfilled with Toyoura sand. Water was added to fly ash powder and mixed for 30 minutes to make a uniform slurry with a water content of 70%. Then, 2~ 4% ordinary portland cement, with respect to dry weight of fly ash, was added to the fly ash slurry and mixed for 10 minutes using an electrical vacuum mixer. The cement fly ash slurry was poured into a part of rigid box whose inside walls were well lubricated with silicone grease to minimize friction between the box and the model ground. After three days curing, the surface of the fly ash ground was leveled and backfilled with Toyoura sand. The box was filled entirely with water to maintain model saturation for a period of one day before the test. The reference specimens for unconfined compression test were manufactured with the same slurry as the model ground.

4.3 *Test procedure*

The model ground thus prepared was brought to an acceleration field of 70 g, which corresponded to a 14 m high retaining wall in a prototype scale. Immediately after reaching 70 g, the rigid retaining wall was allowed to rotate about its bottom at a rotational displacement rate of 0.01 radian per second, in which the model ground could be assumed to fail in an undrained condition. During the loading, the magnitude and the distribution of earth pressure were measured and also model ground deformation was observed. After the centrifuge test the failure pattern of the stabilized fly ash ground was observed in detail and three unconfined compressive tests were conducted on the specimen to estimate the shear strength of the model ground.

5. RESULTS AND DISCUSSIONS

Two series of centrifuge tests were performed as summarized in Tables 3 and 4, to identify the failure pattern and to measure the active earth pressure in cement stabilized fly ash ground.

5.1 *Failure pattern of stabilized fly ash ground*

In the first test series, the stabilized fly ash ground without retaining wall was allowed to fail by increasing acceleration to investigate the failure mode and shape of slip surface. Figure 5 shows a typical cross section of the ground failure obtained in TEST-2, in which an entire failure of the ground took place at

Table 4. Test conditions in test series 1

	Model Scale		prototype scale				
	Width (cm)	Height (cm)	Width (m)	Height (m)	q_u (kN/m^2)	γ_t (Mg/m^3)	Acceleration (g)
TEST-1	37.0	30.0	4.8	3.9	73.5	1.52	13
TEST-2	37.0	30.0	15.5	12.6	98.0	1.51	42
TEST-3	37.0	30.0	20.4	16.5	86.2	1.54	55

Table 5. Test condition in test series 2

	Cement Stabilized Fly ash								Sand	
		Model Scale		Prototype Scale		Cement				
	B/H Ratio	Width (cm)	Height (cm)	Width (m)	Height (m)	content (%)	q_u (kN/m^2)	γ_t (Mg/cm^3)	γ_d (Mg/cm^3)	Acceleration (g)
TEST-10	0	———	———	———	———	———	———	———	1.62	70
TEST- 9	0.5	10.0	20.0	7.0	14.0	2.0	37.2	1.57	1.59	70
TEST- 8	1.0	20.0	20.0	14.0	14.0	2.0	44.1	1.55	1.61	70
TEST-21	2.0	40.0	20.0	28.0	14.0	2.0	50.0	1.51	1.62	70
TEST-11	infinity	infinity	20.0	infinity	14.0	3.0	48.0	1.56	———	70
TEST- 7	2.0	40.0	20.0	28.0	14.0	2.0	111.0	1.57	1.58	70
TEST-22	1.5	30.0	20.0	21.0	14.0	4.0	136.2	1.51	1.60	70
TEST-15	1.0	20.0	20.0	14.0	14.0	4.0	152.9	1.56	1.64	70

the acceleration of 42g. Negligible deformation of the stabilized ground was observed just before the failure. As observed during the test, immediately after development of a vertical tension crack, the model ground failed suddenly. The failure of the stabilized ground was characterized by development of a vertical tension crack and extended to the toe by a slip surface inclined at about 45 degrees with horizontal, as shown in the Fig. 5. The location of the vertical tension crack was within 60% to 75% of the height of the model ground behind the crest of the ground. This failure mode, in which shear failure was dominant, was quite different from that of sandy ground.

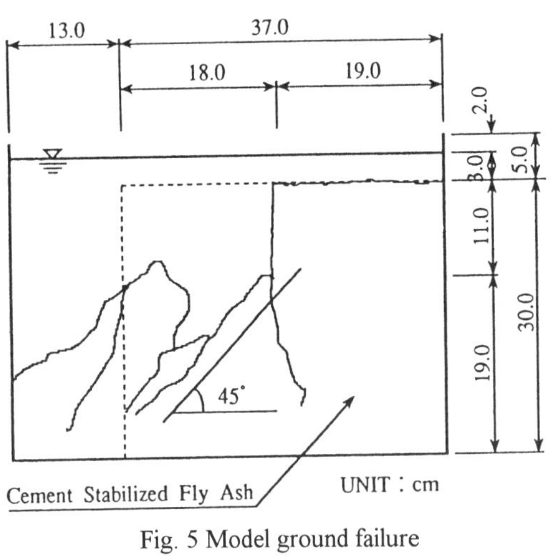

Fig. 5 Model ground failure

5.2 *Active earth pressure tests*

5.2.1 test series

A series of centrifuge model tests were conducted by changing the ratio of width (B) to height (H) of 0.5, 1.0, 2.0 and the infinite in which the stabilized ground extended entirely in the box. And the strength of the stabilized fly ash ground was also changed. The geometric and mixing conditions of each test are summarized in Table 4 together with the strength of the ground. In TEST-10, loading tests on the sandy ground without improved ground was also performed to investigate the earth pressure and compared with that of the stabilized grounds.

5.2.2 earth pressure-displacement

Figure 6 shows the relationships between the rotational displacement of the wall (δ) and the total horizontal earth pressure acting on the retaining wall. In the figure, the displacement is normalized with respect to the wall height (H) and the measured earth pressures are multiplied by the centrifugal acceleration to convert to prototype scale. All the data shown in the figure were obtained in the test cases of TEST-10,-9,-8 and -11 in which the shear strengths of the cement stabilized fly ash grounds were around 50 kN/m^2 with the exception of TEST-10. It can be seen in the figure that at K_0 condition, δ/H=0, highest horizontal earth pressure was obtained in the case of sandy ground, in TEST-10, and earth pressure decreased with increase of the width of stabilized ground. It is widely understood that the earth pressure at rest is mainly dependent on the internal fricion angle, ϕ, of the sandy ground and the shear strength of the stabilized ground. Figure 6 also shows that the earth pressure is also influenced largely by the extension of the stabilized ground behind the rigid retaining wall.

In the case of the sandy ground, the horizontal earth pressure decreases a lot immediately after the loading, and still decreases rapidly with the rotational displacement of the wall reaches a constant value after δ/H=0.01. In the case of B/H ratio of 0.5, TEST-9, the earth pressure decreases rapidly immediately after the loading, but the rate of decrease in earth pressure is smaller than that of the sandy ground. The earth pressure of the stabilized ground decreases with increase of rotational displacement which is similar to the sandy ground. In case of B/H of infinity; TEST-11, the earth pressure at rest is much smaller than that of the sandy ground and decreases immediately and gradually with increase of the rotational displacement. It can be found that the earth pressure decreases faster in larger B/H ground than smaller B/H ground. It can be concluded that the stabilized ground shows smaller earth pressure which is a big advantage of the stabilized ground to be applicable for port and harbor facilities.

Active earth pressure of stabilized fly ash ground with B/H = infinity calculated by Rankine's theory was much smaller than the measured value. It can be emphasized that the earth pressure of stabilized ground should be calculated by taking account the effect of tension crack.

5.2.3 earth pressure -width of stabilized ground

Figure 7 shows the relationships between the width of the stabilized ground and the active earth pressure. In this figure, the filled circles and filled triangles represent the earth pressure at rest (δ/H = 0), and open circles and triangles represent the active earth pressures of the sandy ground either the stabilized grounds of around 50 kN/m^2 or 130 kN/m^2 in strength, respectively. Although the ground strengths are different in each test, the active earth pressures have a tendency to decrease with increase of the B/H ratio, which means that wider improved ground has smaller active earth pressure. However, the stabilized ground having B/H larger than 2, show a constant active earth pressure. As the strength of the stabilized ground increases, the active earth pressure decreases. In this study, it is found that the horizontal earth pressure is dependent on not only the width but also the shear strength of the stabilized ground.

6. PRACTICAL IMPLICATIONS

The utilization of industrial residues as replacement or fill material could be one of the most desirable methods to solve environmental problems posed by the huge amount of wastes produced by electrical power companies. According to an investigation by Peindl et al. (1992), the consumption of coal for power generation is expected to double in the next 10 to 15 years, therefore, the disposal of the by-products will be a major concern.

Figure 8 indicates various types of applications of fill material in waterfront retaining structures. The advantage of this newly developed fly ash is the unit weight of the backfill which is lighter than that of generally available material, and also rapid solidification. Therefore, the induced horizontal earth pressure and bending moment imposed to the retaining system will considerably be decreased. In that case the re-

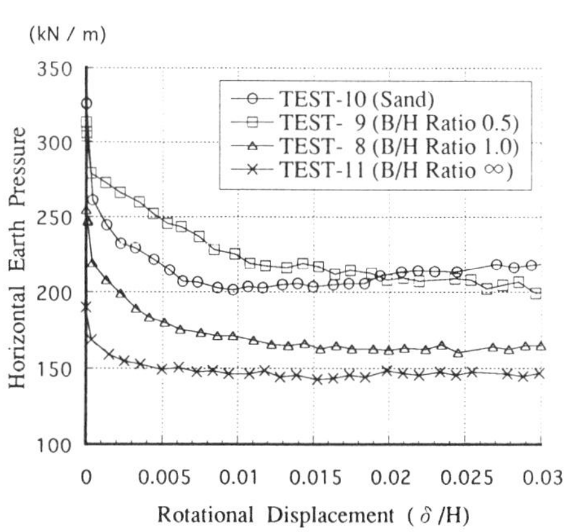

Fig. 6 Rotational displacement and Earth pressure

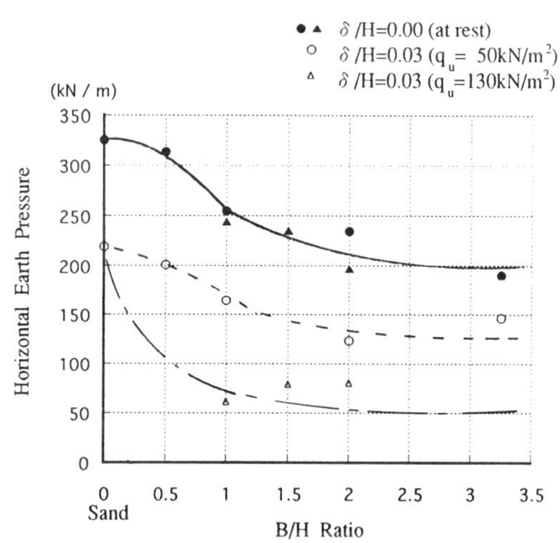

Fig.7 Earth pressure-width of stabilized ground

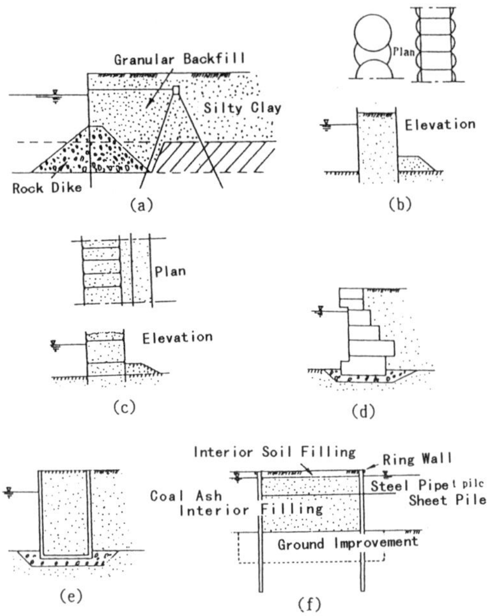

Fig. 8. Different types of waterfront structures (a) anchored bulkhead in rock dike, (b) cellular cofferdam,, (c) double-wall cofferdam, (d) concrete block wall, (e) precast concrete floating box, (f) man-made island

duction in the thickness of the sheet pile quay wall and the total cost of project will provide a more cost-effective means of development in reclamation of coastal areas.

7. CONCLUSIONS

The research reported herein forms a part of an ongoing effort to investigate the applicability of using fly ash in the form of slurry mixed with small amount of cement, as a potential fill material for backfilling water front retaining structures. The specific conclusions derived from this study are as following:

1) The failure behavior of the stabilized ground is characterized by the vertical tension crack and shear failure, which differ from a general sandy ground.

2)The horizontal earth pressure of stabilized ground is affected by the width and the strength of the improved ground.

As a final remark, it is concluded that the use of fly ash as a fill material will not only eliminate the disposal site for solid waste, but also provide low-cost, light-weight, self-hardening, excavatable construction material for various applications in port and harbor areas.

REFERENCES

Edil, T.B., L.K. and Berthouex (1992) Interaction of Inorganic Leachate with Compacted Pozzolanic Geotech. Fly Ash, Journal of Geotech. Eng., ASCE, Vol.118, No.9, pp.1410-1430.

Gray, D.H. and Lin,Y.K. (1972) Engineering Properties of Fly Ash, Journal of SMFE, ASCE, Vol.98, No.1, pp.361-380.

Horiuchi,S.,Taketsuka, M., Odawara, T. & Kawasaki, H. (1992) Fly -ash Slurry Island: I.Theoretical and Experimental Investigations, Journal of Material Eng., ASCE, Vol.4, No.2, pp.117-133.

Horiuchi, S., Tamaoki,K. and Yasuda, K. (1995) Coal ash Slurry for Effective Underwater Disposal, Soils and Foundations, Vol.35, No.1, pp.1-10.

Kawasaki,H., Horiuchi,S., Akathuka,M. and Sano,S. (1992) Fly-ash Slurry Island: II. Construction In Hakucho Ohashi Project, Journal of Material Eng., ASCE, Vol.4, No.2, pp.134-152.

Kitazume, M. and Miyazima ,S.(1996) Development of PHRI Mark II Geotechnical centrifuge. Technical Note of the Port and Harbour Research Institute, 1996 (in print)

Larrimore, C.L. (1986) Ash Utilization in Highways, Georgia Demonstration Project, Research Project 2422-, by Southern Company Services for Electric Power Res. Ins., Palo Alto, California, Apr.

Sgambati, S.A. (1985) Fly Ash Stabilization Mix Design and Specification Development, Technical Report, New York State Dept. of Transportation, New York.

Todorov, K. and Evstatiev, D. (1994) geotechnical study of Lagooned Ashes in Bulgaria, Bulgarian Academy of Sciences, Sofia, 115p.

Vesperman, K.D., Edii, T.B. and Berthouex, R.M. (1985) Permeability of Fly Ash and Fly Ash / Sand Mixtures, Impermeable Barriers for Soil and Rock, Special Technical Publication, No.874, ASTM, Philadelphia, PA, pp.170-180.

Environmental Geotechnics, Kamon (ed.) © *1996 Balkema, Rotterdam. ISBN 90 5410 848 7*

Coal ash utilization as self-hardening slurry in diaphragm wall construction

Hiroshi Kubo & Takeshi Kawachi
Obayashi Corporation, Tokyo, Japan

Shuetsu Fujita, Nobuyuki Hurumichi & Masateru Takahashi
Hokkaido Electric Power Company, Sapporo, Japan

ABSTRACT: In Japan, only 50% of coal ash is effectively recycled while the remaining percentage is dumped. New uses for coal ash need to be developed. The utilization of coal ash as a material in self-hardening slurry (SG) for the construction of an underground diaphragm wall was studied. Ash-SG containing a large quantity of coal ash which was as good as conventional SG in workability and quality was developed.

1 INTRODUCTION

One ton of coal is burned at a thermal power plant to about 130kg of coal ash comprised of approximately 85 to 95% fly ash (Fig.1). In Japan, coal ash presently amounts to 6,500,000 tons per year, and it is estimated that it will amount to ten million tons in the year 2000 (Nagataki 1996). Coal ash is mainly used as a raw material in cement production or concrete admixture and is frequently used as landfill material or back fill material with cement mixing in construction works (Golden 1996, Kawasaki 1992). However, effectively used coal ash is no more than approximately 50% of the total amount. Ineffectively utilized coal ash is disposed of at a dump site. Since the development of such dump sites is becoming increasingly difficult, the promotion of coal ash utilization is a highly important theme.

One of the materials to construct a diaphragm wall beneath the ground is self-hardening slurry (hereafter : SG). SG is made of water, bentonite, cement, etc. In this study, a new utilization of coal ash as bentonite in SGwas studied for the purpose of cost reduction of SG.

2 GENERAL MIX PROPORTION AND FUNCTION OF SG

SG is generally made by mixing together 50 to 60kg of bentonite with 1 m^3 of water, and adding about 200kg of cement (which is high in slag content), and if necessary, 1 to 2kg of retarder. SG keeps a slurry state for several hours and subsequently hardens gradually. In the diaphragm wall method using SG, the trench is excavated while being filled with SG, and the diaphragm wall is formed by the hardening of SG. Since the strength of this diaphragm wall is low, 1/50 to 1/100 of the compressive strength of concrete, this wall is used as a water cut-off wall or temporary earth retaining wall by the insertion of

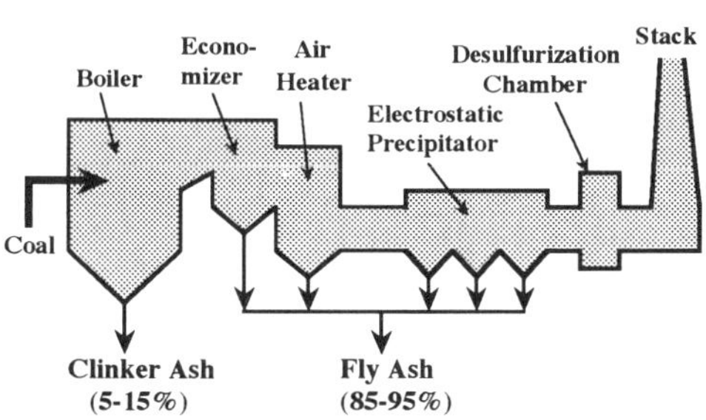

Fig.1 Production Process of Coal Ash

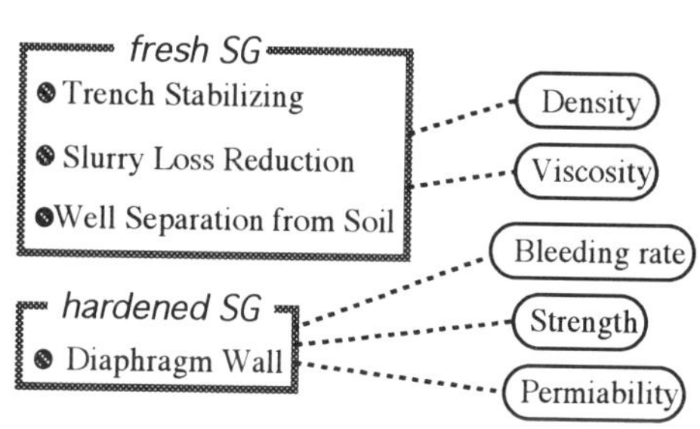

Fig.2 Functions of SG

steel members. As shown in Fig.2, SG used in this way is required to maintain the functions of stabilizing trench walls and preventing slurry penetration into the ground during excavation. It is desirable that the slurry be well separated from the soil. Furthermore, SG is required to keep the designed strength and imperviousness.

3 STUDY ON MIX PROPORTION OF SG USING COAL ASH

3.1 *Properties of Coal Ash Samples*

The properties of coal ash are known to be very dependent on the origin of the coal and the burning conditions. In Japan, since coal is imported from various foreign districts and two or three kinds of coals are frequently mixed and burned, a wide variety of coal ash are produced. Consequently, in the utilization of coal ash for SG, it is necessary to test many types of coal ash on using as SG. If the test method cannot cope with a change in the type of ash, it is not a practical method.

The coal ash samples for laboratory testing consisted of six kinds abbreviated WA, MQ, DD, HS, OB and SR. These samples are all fly ash from the same boiler of a pulverized coal firing type furnace (1300℃). The coal that produced DD and HS is from China, while all other samples were produced by coal from Australia.

The physical and chemical properties of the samples are shown in Table 1. The densities of the particles are 2.1 to 2.3 g/cm³, which are the same as 80 to 85% of particle density in common soil. The samples are mainly comprised of silt fraction, 5 to 75 μm. The relatively strong correlation between the specific surface area (Blaine value) and adsorption value of methylene blue is recognized. The water requirement ratios are obtained from workability test for the mortar with a defined coal ash content.

Table 2 shows the chemical composition of samples and the pH and electrical conductivity of samples suspended in water. SiO_2, Al_2O_3 and Fe_2O_3 contents in DD and HS which originated from Chinese coal were noticeably different than those of the other ash produced from Australian coal. MQ is characteristic of relatively low pH. It was observed using an SEM (scanning electron microscope) that all of these type of coal ash were primarily composed of spherical particles of several to dozens of micrometers in diameter and were strikingly similar in shape. The heavy metal concentrations of the leachate from these coal ash samples did not exceed government criteria.

Table 1. Physical and chemical propeties of coal ash.

	Density of Particles (g/cm³)	Specific Surface Area (cm²/g)	Water require-ment Ratio (%)	Ad. of MB* (mg/g)	Particle Size Distribution(%)		
					< 5 μm	5-75 μm	75 μm -2 mm
WA	2.20	2900	95.3	0.46	23	74	3
MQ	2.08	2560	95.3	0.25	19	77	4
DD	2.37	2970	96.2	0.46	15	81	4
HS	2.30	2840	97.6	0.40	15	81	4
OB	2.25	3230	97.6	0.71	15	83	2
SR	2.05	2640	96.6	0.44	10	83	7

*Adsorption of methylen blue

Table 2. Chemical properties and composition of coal ash.

(%)

	pH	EC*	Ig. loss	SiO_2	Al_2O_3	Fe_2O_3	CaO	MgO	K_2O	Na_2O	SO_3	total
WA	11.2	0.18	1.5	65.7	17.5	7.4	1.3	0.96	1.7	1.1	0.17	97.5
MQ	6.7	0.10	1.4	63.3	22.8	3.6	1.1	0.64	1.4	1.1	0.05	95.4
DD	12.3	0.58	2.1	51.6	20.2	18.2	2.1	0.71	2.7	0.42	0.34	98.4
HS	11.2	0.16	1.4	44.8	41.8	5.4	2.9	0.37	0.76	0.20	0.42	98.1
OB	11.3	0.13	1.9	59.9	29.3	2.8	3.2	0.97	0.65	0.13	0.14	99.0
SR	11.2	0.09	0.8	63.4	27.7	2.8	2.1	1.00	0.61	0.42	0.18	99.0

*Electric conductivity (mS/cm)

3.2 *Bentonite and Cement*

The bentonite used in the test is Na-type, 10.1% in water content, 462% in liquid limit, 18% in plastic limit, 19cm³/2g in degree of swelling and 9.8 in pH. The cement used in the test is the blast furnace cement B-type.

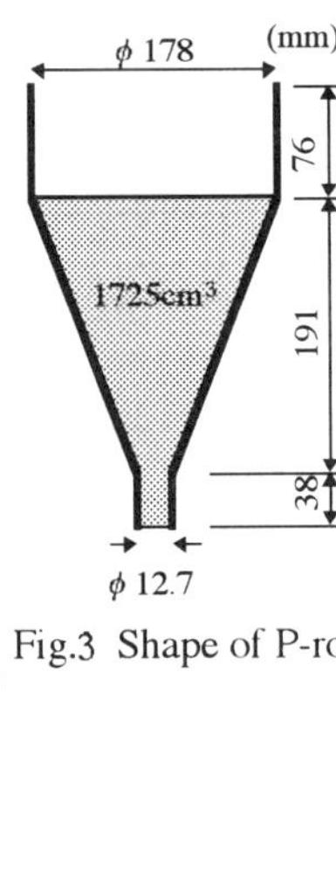

Fig.3 Shape of P-roat

3.3 *Test Method*

The SG added coal ash (hereafter : ash-SG) was made

as follows. Water and and bentonite were mixed for 30 seconds in a high-speed mixer of 4 ℓ capacity then afterwardly letting it sit for 5 minutes. Next, coal ash and cement were further added and the components were mixed together for one minute in the mixer.

In the test of fresh ash-SG, the flow was measured by the time it took for the mixture to flow out from P-roat (Fig.3), and the bleeding rate was measured with a cylinder made of acrylic resin of 5cm in diameter and 30cm in height. In the test of hardened ash-SG, the unconfined compressive strength was measured for samples of 5cm in diameter and 10cm in height, and the permeability was measured under 50 to 200kPa of water pressure.

3.4 *Properties of Slurry Made of Coal Ash and Water*

SG is required to maintain a suitable flow value and a low bleeding rate in the fresh stage. Consequently, to discuss the use of the coal ash for SG, it is first necessary to test the slurry made of ash and water. Fig.4 shows the relation between water content and flow and between water content and bleeding rate for the slurry made of each of the six different kinds of ash and water. With the increase in the slurry water content, the flow value decreased and the bleeding rate increased. The correlation between the flow value and the bleeding rate at 45% of the slurry water content was not high. The slurry showed the greatly varying flows and bleeding rates for the same water content depending on the kind of ash. It also become clear, that the water contents for maintaining a given flow are dependent on the type of ash. These results seem to be signigicantly

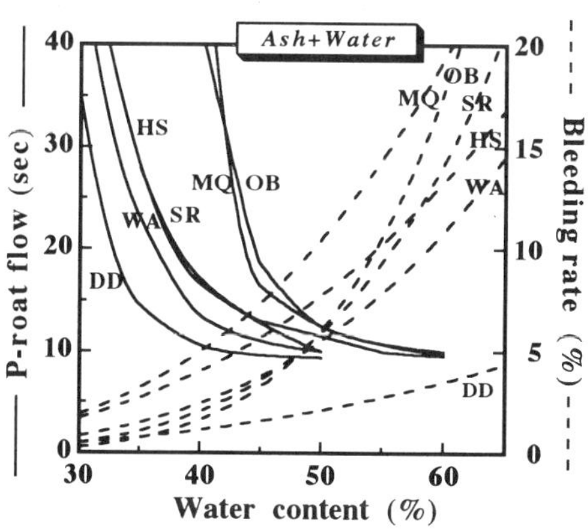

Fig.4 Relations between water content and P-roat flow and between water content and bleeding rate of ash and water slurry.

important in designing the mix proportion of ash-SG.

3.5 *The Mix Proportion of Ash*

The relation between the mix proportion of ash-SG and its properties for ash SG using WA ash is shown in Fig.5. In this figure, the quantity of bentonite is 20kg and that of cement is 175kg in a cubic meter of ash-SG. These quantities are kept constant. The cross axes show the ash added cement content and the approximate water content which change accordingly with supplements.

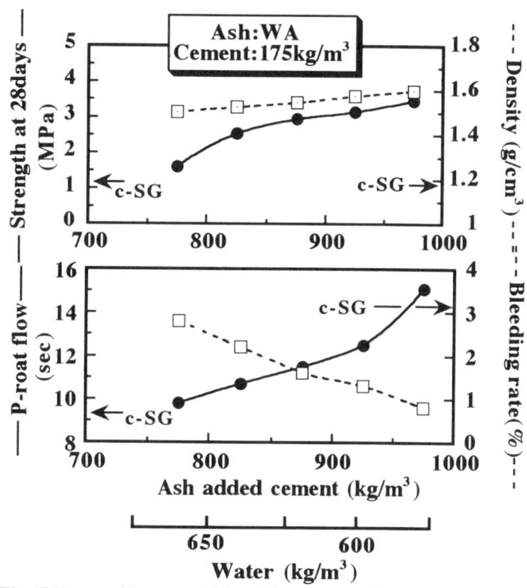

Fig.5 P-roat flow and bleeding rate at fresh stage and strength and density at 28 days for SG with change of ash added cement content.
(Arrows mean the common data of conventional SG.)

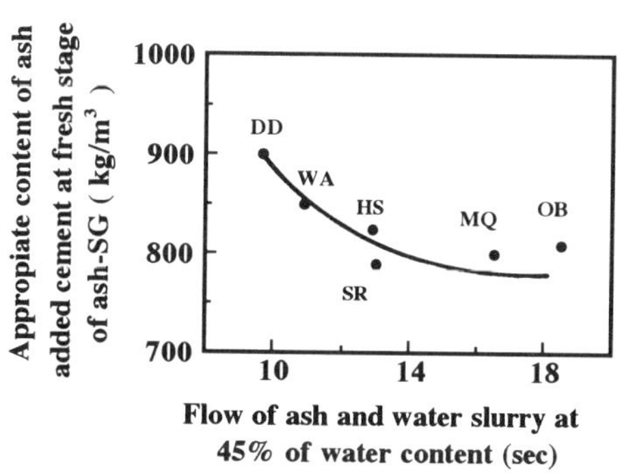

Fig.6 Relation between flow of ash and water slurry at 45% of water content and appropriate content of ash added cement in ash-SG.

Here, the target of the flow was set at <15 seconds and that of the bleeding rate at <3% for the fresh stage of the ash-SG. To attain these two targets, the appropriate content of ash added cement is 725 to 825kg/m^3. As the cement content is kept at a constant 175kg/m^3, this means the appropriate ash content is 550 to 650kg/m^3. The densities of ash-SG were 1.45 to 1.55g/cm^3. Furthermore, the strengths were 1.5 to 3.7MPa at 28 days of curing time.

The mix proportions of ash-SG using the other type of coal ash were tested by the above-mentioned method. Consequently, as shown in Fig.6, the appropriate contents of ash added cement, which were most important in designing the mix proportion of ash-SG, were highly correlated with the flow of the ash and water slurry at 45% of water content (hereafter : FL_{45}). Therefore, in designing the mix proportion of ash-SG, it is possible to cope with the variation in ash types by using FL_{45} gained from simple tests.

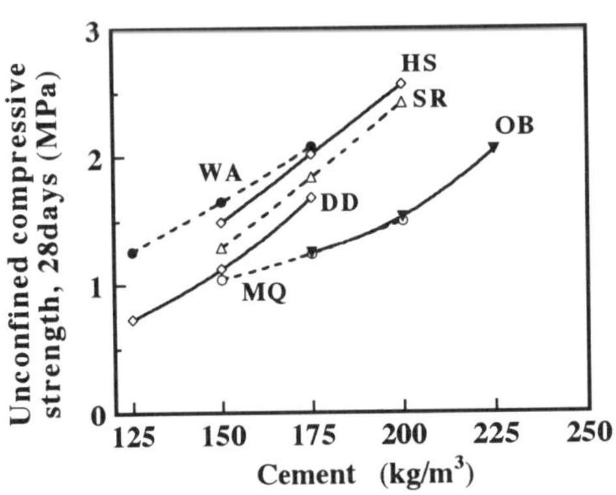

Fig.7 Relation between cement content of ash-SG and strength at 28days of aging.

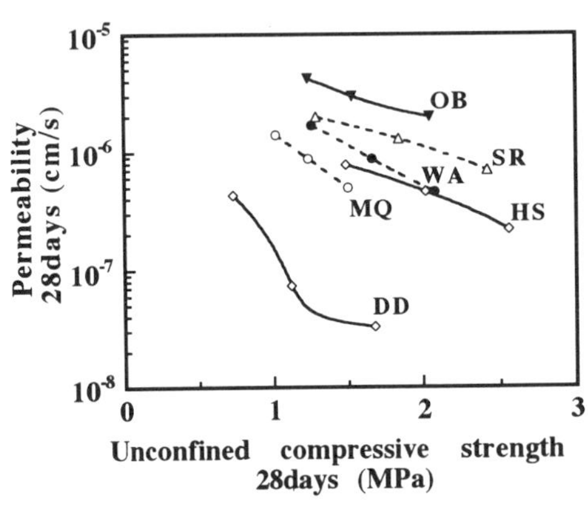

Fig.8 Relation between strength and permeability of ash-SG at 28days aging.

Fig.7 shows the relation between the cement content and strength at a median value of the appropriate content of ash added cement. With an increase in the cement content, the strengths increased. Fig.-8 shows the relation between the strength and permeability for ash-SG using each of six different kinds of ash at 28days of curing time. With the increase in the strength, the permeabilities decreased. Here, the target of the strength was set at >1.8MPa, and that of the permeability at <10^{-5}cm/s. The quality of ash-SG is no less than conventional SG which does not use coal ash. It is possible to use a large amount of coal ash in producing ash-SG, which has the advantage of reducing the bentonite content.

4 CONSTRUCTION OF WATER CUT-OFF WALL USING ASH-SG

4.1 *Overview of Work*

Through the above-mentioned laboratory tests, ash-SG was adopted to construct a vertical water cut-off wall at a dump site for coal ash at the Tomatou-Azuma thermal power station in Tomakommai, Hokkaido by the Hokkaido Electric Power Company.

The dump site was at a swamp near a river. As shown in Fig.9, the stratum was in order from the upper to lower layer a volcanic ash soil (sandy) layer, a peat layer and a clay layer. The permeability were in the order of 10^{-6}cm/s for the clay layer, and as large as approximately 10^{-3}cm/s for the upper two layers. Consequently, construction of the diaphragm wall was needed down to the clay layer.

The method of construction at the coal ash dump site was as follows: First, 0.8m of top soil was removed and paper drains were installed at the place of banking for the purpose of speeding up consolidation settlement by the load deviced from the bank itself. Second, a 5m-high bank around the dump

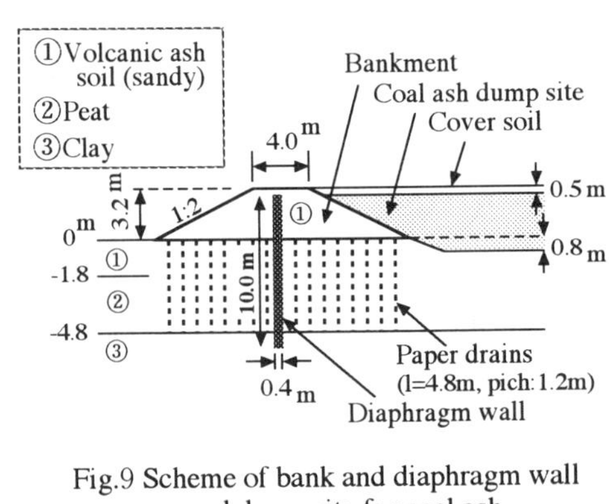

Fig.9 Scheme of bank and diaphragm wall around dump site for coal ash.

site was made using the top soil initially removed. Consolidation settlement primarily occurred in the peat layer and fell off several months after banking. Subsequently, trench excavators were mounted on the bank, whose top was temporarily covered with steel plates, and a trench was excavated to 10m in depth. After that, secondary banking was carried out at a thickness of 0.5m. The dump site was completed with slope protection work and drainage work.

The size of the dump site was 250,000m^2 in area, 1,900m in bank length and 760,000m^3 in dumping capacity. The size of the diaphragm wall was 10m in depth, 0.4m in thickness and 19,000m^2 in area.

4.2 *Execution Method*

Fig.10 shows the flow of work execution with the ash-SG. The coal ash produced at Tomatou-Azuma Electric Power Station was carried into the field stock yard (Photo.1) in moist condition by dump trucks. The ash then loaded into a hopper by a shovel car and conveyed to the SG-plant (Photo.-2) by belt conveyers. Bentonite, the same as that used in the laboratory tests, was mixed with water and changed into bentonite slurry then stored in tanks. The coal ash, cement, bentonite slurry and some needed retarder were put into the mixer in suitable proportions through the respective gages and mixed for 90 to 100 seconds a batch. The ash-SG made in this way was pumped to the agitator and transported to the excavating spot by tank trucks.

Two bucket-type excavators were used for the trench excavation. Operation of the excavators is shown in Photo.3. A total of nine different types of coal ash were used through the work. As most of the

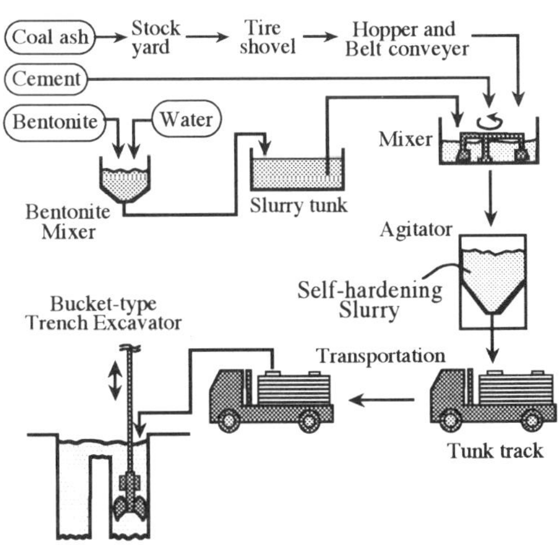

FIg.10 Flow of work excution with ash-SG.

ash types differed from those in the laboratory tests, it was necessary to design the mix proportion of ash-SG early in the field. For this porpose, the above mentioned FL_{45} was applied. Consequently, the mix proportions were dependent on the kinds of coal ash used, and ranged from 610 to 690kg/m^3 of coal ash, 20kg of bentonite, 160 to 210kg/m^3 of cement, 617 to 652kg/m^3 of water and 0 to 2kg/m^3 of retarder.

Photo.1 Coal ash stock yard

Photo.2 SG plant

Photo.3 Operation of excavator

4.3 *Quality of Ash-SG*

The ash-SG made at the field plant was 11 to 14 seconds of flow, 1.47 to 1.52g/cm³ of density and a bleeding rate of <0.5%. Therefore, the ash-SG qualities were suitable for the workability and stability required during trench excavation.

Examples of the quality data of hardened ash-SG are shown in Fig.11. All the strengths exceeded the designed strengh of 0.6Mpa. The relationship between the strengths and permeabilities is shown in Fig.12. The permeabilities also met the required value of $<10^{-5}$cm/s.

5 CONCLUSIONS

For the purpose of developing a new use for coal ash and for cost reduction of self-hadening slurry (SG), ash-SG was studied.

The ash-SG contained 600 to 850kg/m³ of coal ash and maintained suitable properties at the fresh and hardening stage. Ash-SG was successful in reducing the bentonite content by 1/3 to 1/6 as compared with conventional SG not containing coal ash. Although the mix proportion of ash-SG was dependent on the type of coal ash used, the flows of the ash-water slurries with 45% water content were very useful for design of the mix proportions. Ash-SG was found to be comparable to conventional SG in workability and quality when ash-SG was applied to construction a vertical diaphragm wall around the coal ash dump site.

Coal ash is also usable as a self-hardening slurry in the construction of underground walls.

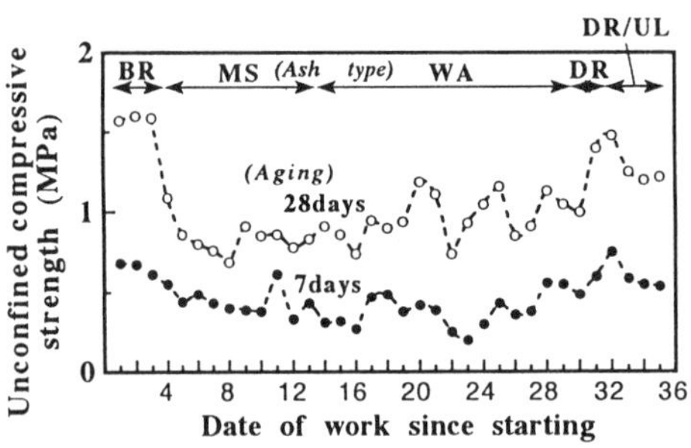

Fig.11 Examples of strength data for hardened SG at 7days and 28days of aging on the work.

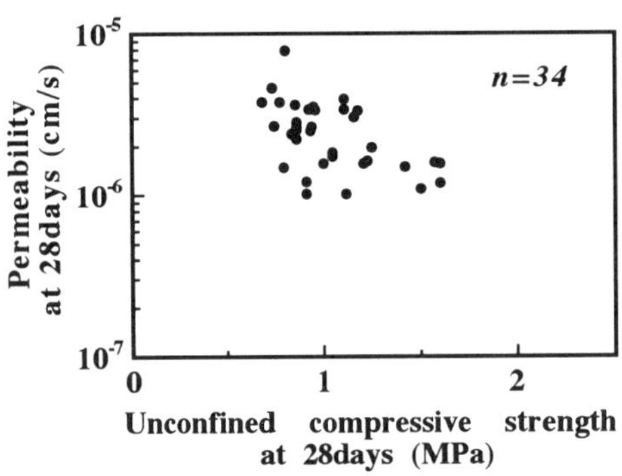

Fig.12 Relation between strength and permeability of ash-SG at 28days aging on the work.

REFERENCES

Nagataki S. 1996. Present Status of Coal Ash Utilization in Japan and Worldwide Trend, Proceedings of International Clean Coal Technology Symposium on Coal Ash Utilization, Tokyo, Supplementary paper, 1-27

Golden D.M. 1996. U.S. Power Industry's Activities to Expand Coal Ash Utilization, Proceedings of International Clean Coal Technology Symposium on Coal Ash Utilization, Tokyo, 25-43

Kawasaki H., Horiuchi S., Akatsuka M. and Sano S. 1992. Fly-Ash Slurry Island:II. Construction in Hakucho Ohashi Project, J.Material Engineering, ASCE, 4(2), 134-152

Environmental Geotechnics, Kamon (ed.) © *1996 Balkema, Rotterdam. ISBN 90 5410 848 7*

Effect of water content on cement stabilization of soft clay

Yorimichi Kuboi
Department of Civil Engineering, Wakayama National College of Technology, Japan

Kazuhiko Nishida
Department of Civil Engineering, Kansai University, Japan

ABSTRACT:
This study investigates the possibility of utilizing the waste rock powder as a supplemental material for cement stabilization of soft clay ground. The water content of the mixture decreases regardless of water-cement ratio when the waste rock powder is mixed with the soft clay. The ratio of designed water content to cement content instead of water-cement ratio is evaluated. The results are summarized as follows:(1) there is an optimum mixed proportion of the waste rock powder for the cement stabilization,(2) the unconfined compressive strength varies in inverse to the ratio of designed water content to cement content,(3) the strength has the highest value in every case when the waste rock powder and cement are mixed with the soft clay at a designed water content around 1.2 times the plastic limit and (4) the corresponding content of waste rock powder and cement depends upon the previously selected value of designed water content adopted above.

INTRODUCTION

Recently, the utilization of resources and preservation of the natural environment have become more important problems in Japan. Considering the increasing amount of various kinds of industrial waste, a by-product of industrial activity, it is necessary to dispose of or utilize them for construction purposes. Crushed stone has to be used instead of natural gravels which is difficult to procure due to environmental preservation. Waste rock powder is an industrial by-product from crusher plants. Since almost no disposal procedure has been developed for the waste rock powder, it causes pollution of the air and water. This waste rock powder is little utilized and is buried in crusher plants.

Each company producing crushed stone has studied utilization of the waste rock powder but can not fully commercialize it at present.[1] The authors have examined how to dispose and how to utilize the waste rock powder as a construction material.[2] If possible, the problems of shortage of resource utilization and environmental pollution can be solved simultaneously. One possible way is to use the rock powder for soil stbilization.

The purpose of this study is focused on clarifing the possibilty of utilizing the rock powder for lime or cement stabilization of clay ground.

Table I. Physical Properties of Samples.

Physical Properties / Samples	Specific gravity (g/cm^3)	Specific surface area $Ss(m^2/g)$	Ig-loss (%)	Liquid limit WL(%)	Plastic limit Wp(%)	Sand (%)	Silt (%)	Clay (%)	Unified soil classification system
Waste rock powder	2.74	14.38	9.787	24.6	20.0	4.1	70.9	25.0	ML
Soft clay	2.59	17.58	5.227	59.3	27.6	18.0	54.3	27.7	CH

Investigation on the relation between the strength of cement stabilized soft clay and the ratio of designed water content to cement content are examined in detail.

SAMPLES AND EXPERIMENTAL PROCEDURE

The sample of waste rock powder used in this experiment is dust collected in cursher plants through the crushing process of stones. The crushed stone contains calcite. The soft clay sample was taken from an agricultural pond. These samples were dried in air and then sieved into particles smaller than 2,000μm. Ignition loss, specific surface area, liquid limit, plastic limit, etc., are determined for these samples.

The physical properties of the samples are shown in Table 1. The waste rock powder can be classified into Silt {ML} with some fine particles.[3] Water content of the soft clay was increased to a desired value(50 %, 60 %, 80 %). These samples were mixed with the various amounts of Portland cement and the waste rock powder. The mass of Portland cement or the waste rock powder were added to the mass of dried soft clay at various proportions. The specimens were prepared in a cylindrical mold with a 50mm diameter, 100mm length and cured for 7 days at a temperature of 20℃ and humidity of 95 %. During the curing period, the water content was kept at the same value as at mixing. Each specimen was wrapped with a membrane to prevent drying. Unconfined compressive tests were carried out on the prepared specimens at a strain rate of 1 %/min.

DISCUSSION OF THE EXPERIMENTAL RESULTS

The unconfined compressive strength has not always close relationship with dry density when the waste rock powder is used.[4] The volume of water necessary for hydration of cement is insufficient when excess amount of waste rock powder is mixed. The simplified model of mixture of soft clay, cement, and the waste rock powder is shown in Fig.1.

the mass of mixture : m

the mass of particle in soft clay : m_S

the mass of water in soft clay: m_W

$$m_W = \frac{w \times m_S}{100}$$

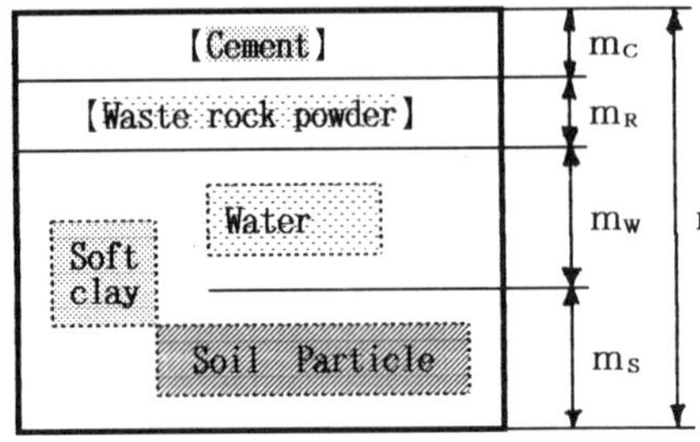

Fig. 1. Model of mixture

the mass of cement: m_C

$$m_C = \frac{a \times m_S}{100}$$

the mass of waste rock powder: m_R

$$m_R = \frac{b \times m_S}{100}$$

water content of soft clay portion: w

$$w = \frac{100 \times m_W}{m_S}$$

cement content : a

waste rock powder content : b

water content of mixture (measured) : w_{NE}

design water content of mixture (calculated from three components) : w_{NC}

From the above model, basic equation (1) can be derived.

$$w_{NC} = \frac{100 \times w}{(100+a+b)} \quad (\%) \qquad (1)$$

The water in specimen may be lost a little through evaporation and reaction of hydration with cement during the curing period. Therefore, a little difference between the water content of specimen and designed water content (W_{NC}) can be recognized for a mixture of cement, waste rock powder, and soft clay. In this case, the water-cement ratio of mixture is kept at the same value even if the amount of waste rock powder is increased. The water content of this mixture decreases when the waste rock powder is mixed with the soft clay. The relationship between the unconfined compressive strength and ratio of designed (W_{NC}) water content to cement content is shown in Fig.2. In this figure, the unconfined compressive strength has a relation for the ratio of designed water content (W_{NC}) to cement content even when water content of the soft clay varies.

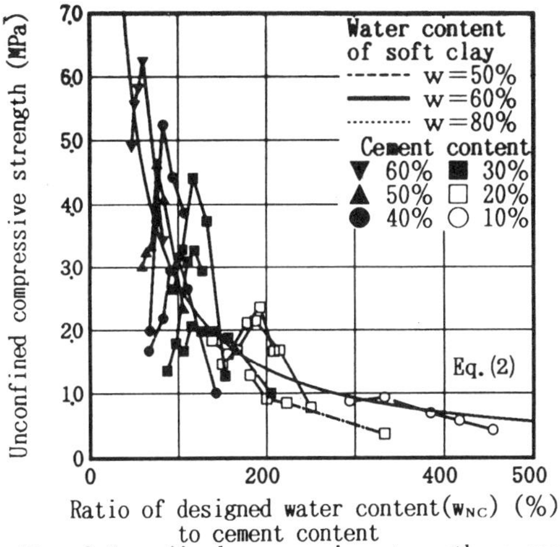

Fig. 2. Unconfined compressive strength versus ratio of designed water content to cement content

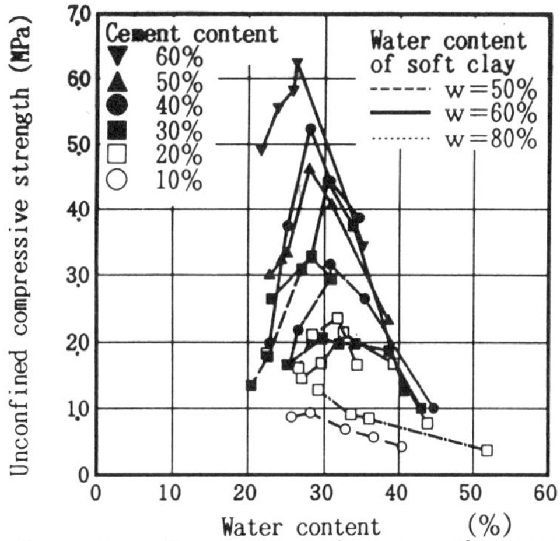

Fig. 3. Unconfined compressive strength versus water content

The unconfined compressive strengths vary in inverse to the ratio of designed water content to cement content.

An experimental equation (2) can be shown in this figure.

$$Y = A \times X^{B} \qquad (2)$$

where, Y is unconfined compressive strength,
X is ratio of designed water content to cement content,
A is coefficient (275 in this case),
B is index (-1 in this case) and
the correlation coefficient is 0.84.

In this equation, the ratio of designed water content to cement content can be determined instead of water-cement ratio for any value of water content of soft clay in terms of the unconfined compressive strength.

The relationship between the unconfined compressive strength and water content is shown in Fig.3. In this figure, the strength curves have peak values around 30 % of water content in every case when the waste rock powder and cement are mixed with the soft clay having various values of water content. This value of water content corresponds to the plastic limit in Table 1.

The relationship between the unconfined compressive strength and designed water content is shown in Fig.4. In this figure, the strength curves have the peak values around the designed water content that is 1.2 times the plastic limit in every case when the waste rock powder and cement are mixed with the soft clay. However

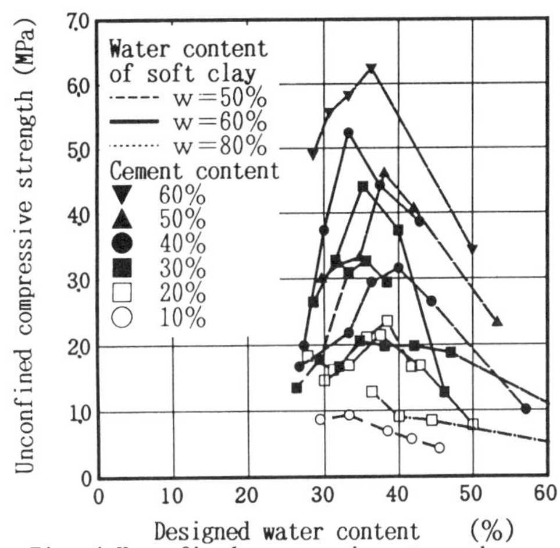

Fig. 4. Unconfined compressive strength versus designed water content

the reason the strength curves show peaks at certain values of water content should be examined. The mass of water necessary to complete reaction of hydration is approximated 35-37 % of the mass of cement. Usually, the proportion of water to cement should be twice the theoretical digestive water.[5]

Japan cement cooperative has adopted the value of 40-70 % for water-cement ratio. According to designing specification of concrete, the amount of designed water should be taken to have 7-10 % of density of concrete and amount of designed cement should be taken to 10-15 % of density for concrete.

Taking these facts into consideration as stated above, the cement content and content of waste rock powder can be estimated. The minimum quantity of water necessary to complete the reaction of hydration for Portland cement is insured if water of 35 % of the amount of cement and the designed water content adjusted to have 35 % of mixture. This value of the experimental result coincides with the designed water content as shown in Fig.4.

1) For the case where 50 % of water content of soft clay; equation (3) can be derived for cement content when substituting 50 % and 35 % for w of the equation (1).

$$a = -b + 42.857 \quad (\%) \qquad (3)$$

The next equation (4) can be derived when cement content corresponds 10 % of the mass of mixture.

$ms \times (100+w+a+b) = 10 \times a \times ms$ as follows

$$a = \frac{(100+w+b)}{9} \quad (\%) \qquad (4)$$

Further we can obtain the equation(5) showing cement content when substituting w=50 % for w of the equation (4).

$$a = \frac{(150+b)}{9} \quad (\%) \qquad (5)$$

The solution of the equations (3),(5) gives cement content: a=19.3 %
content of waste rock powder:b=23.5 %

2) For the case of 60 % of water content in soft clay; Eq. (1),(4) can be as follows

$$a = -b + 71.428 \quad (\%) \qquad (6)$$

$$a = \frac{(160+b)}{9} \quad (\%) \qquad (7)$$

The solution of the equation (6),(7) gives cement content: a=23.3 %
content of waste rock powder:b=48.3 %

3) For the case of 80 % of water content in soft clay; Eq. (1),(4) can be as follows

$$a = -b + 128.5 \quad (\%) \qquad (8)$$

$$a = \frac{(180+b)}{9} \quad (\%) \qquad (9)$$

The solution of the equation (8),(9) gives cement content: a=30.9 %
content of waste rock powder:b=97.7 %

These relations are shown in Fig.5. The values of unconfined compressive strength for each value of cement content and waste rock powder are shown in Fig.5. For the case of 50 %, 60 %, 80 % of water content of soft clay prior to mixing, the values on straight line of Eq.(3), (6), (8) in Fig.5 show content of cement and waste rock powder of designed water content to

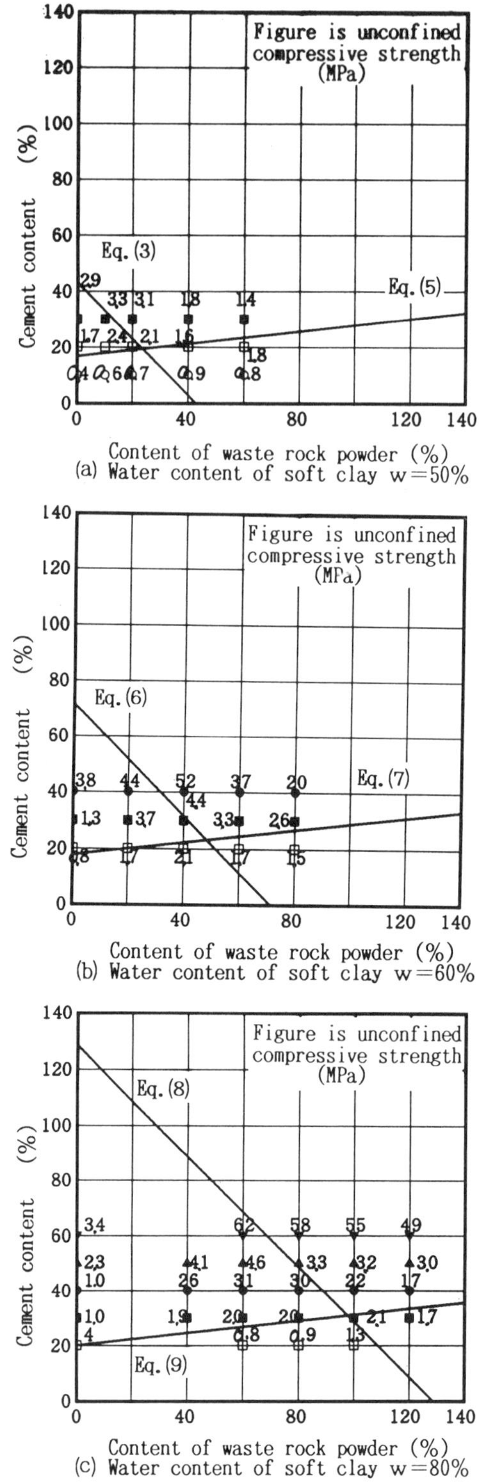

Fig. 5. cement content versus content of waste rock powder

35 % for various water contents of soft clay. The highest values of unconfined compressive strengths can be found on a straight line of Eq.(3),(6),(8) shown in Fig.5 for 50 %, 60 % and 80 % of water content of soft clay, respectively. Therefore, it can be demonstrated that these equations are valuable in examining the results. The unconfined compressive strengths has decreased when the values of cement and waste rock powder are ploted far from the line of these equations. The amount of water required to complete the reaction of hydration for Portland cement becomes insufficient when the design water content decreases to 35 %. The shortage of water may be compensated from the water included in waste rock powder and soft clay. In these cases, the water is adsorbed both on the surface and the inside of waste rock powder and soft clay. The surface cracks and the inside cracks between waste rock power and soft clay make structual defects. Then, unconfined compressive strengths of the specimens give a small value when the particles of waste rock powder and soft clay have some amount of water which can not be involved in the reaction of hydration for Portland cement. If an excess amount of water is added to the specimen, it prevents the combination between cement, waste rock powder and soft clay, and the unconfined compressive strength is small.

The simplified model of the mixture of cement, soil particles, waste rock powder and water is shown in Fig.6. However, following the concrete specification, the amount of designed water for this case should be taken as 7 % of the density for mixture when the amount of cement is 10 % of density for mixture. For this case, the proportion of water to cement is not the double of the theoretical digestive water. This insufficient water must be supplied from the water of waste rock powder and soft clay. If this insufficient water is subtracted from water content of waste rock powder and soft clay, the remainder coincides with the plastic limit in Table 1. For example, the points of intersection of two lines in Fig.5 correspond to the remainder for the case of 50 %, 60 %, 80 % of water content of soft clay. The remainder in water becomes 29.5 % for any values of the water content to waste rock powder and soft clay. This value of remainder in water content is consistent with the water content shown in Fig.5. According to the results, the cement content and waste rock powder can be calculated. It can be insured that the

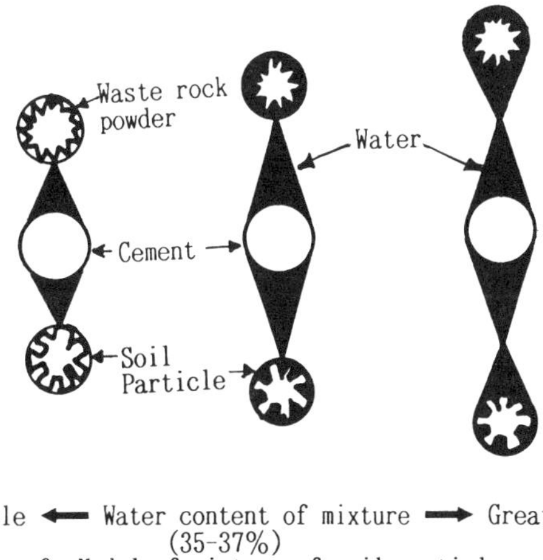

Fig. 6. Model of mixture of soil particle waste rock powder, cement, and water

minimum amount of water required for the complete reaction of hydration for portland cement can be insured if the design water content of the mixture is adjusted to 35 %; this value also corresponds the proportion to cement. This observation is based on the experimental results in Fig.4 and Fig.6.

CONCLUSION

The experimental results on the effect of water content on cement stabilization of soft clay lead to the following conclusions.

(1) The water content of the mixture decreases when the waste rock powder is added to soft clay.

(2) There is an optimum mixing proportion of the waste rock powder for the cement stabilization of soft clay.

(3) The unconfined compressive strength varies inversely to the ratio of water content to cement content.

(4) The strength reaches the peak value in every case when the waste rock powder and cement are mixed with the soft clay at the designed water content around 1.2 times of plastic limit.

(5) The optimum content of waste rock powder and cement can be decided for various values of the water content of clay using the experimental equations.

As stated above, the ratio of designed water content to cement content are determined regardless of water content in soft clay, as far as the unconfined compressive strength is concerned.

This study was to investigate the relation between the strength of the cement stabilized soft clay ground and the ratio of design water content to cement content.

The design method for the cement stabilization of soft clay should be based on the relationship between the unconfined compressive strength and water content, design water content or ratio of design water content to cement content, and the relationship between the unconfined compressive strength and ratio of design water content to cement content.

REFERENCES

[1]Kinki Cooperative Society of Crusher Stones, 1991, A Consulting Report of Market Reseach in 1990, Vision of Utilized Waste Rock Powder

[2]Kazuhiko Nishida,Seiichi Sasaki, Yorimichi Kuboi:1991, Utilization of Waste Rock Powder for Stabilization of Soil, Vol.40, No.459, 1532-1537, The Society of Mterials Science Japan

[3]Kazuhiko Nishida,Yorimichi Kuboi: 1993, Amorphous Materials in Waste Rock Powder and Effect on Stabilizetion, Proceeding of Symposium on Distinction and Classification of Soils,165-170.

[4]Kazuhiko Nishida,Yorimichi Kuboi: 1993, Utilization of Waste Rock Powder for Cement Stabilization of Soft Clay Ground, 1-8, Constructional Material

[5]Masao Shimoda,1984: Method of lime stabilization, 62-81, Kajima Press.

Environmental Geotechnics, Kamon (ed.) © 1996 Balkema, Rotterdam. ISBN 90 5410 848 7

Application of the liquefied stabilized soil method as a soil recycling system

Goro Kuno
Faculty of Science and Engineering, Chuo University, Tokyo, Japan

Hiroshi Miki & Noriyuki Mori
Public Works Research Institute, Ministry of Construction, Tsukuba, Japan

Jotaro Iwabuchi
Central Research Institute for Construction Technology, JARGC, Japan

ABSTRACT In the liquefied stabilized soil (LSS) method, excavated soils are mixed with muddy water (or water) and cementing material, and placed where back-filling is necessary on site. The method is useful for recycling any type of excavated soils, including cohesive soil with a high water content that has conventionally been considered unsuitable for earth works. In addition, the method greatly simplifies the construction sequence since no compaction effort is required during placement.

This paper presents two recent topics: one is a full-scale model experiment for back-filling an old mining tunnel, which is one of adequate uses of LSS, and the other is a field performance test on filling a cavity under the pavement of an urban road.

The full-scale experiment demonstrated the suitable flowing condition of LSS and the adequate placement method for back-filling the tunnel. The field performance test verified the applicability of the combining usage of an agitating truck and a remote plant as well as the ability of the high density liquefied stabilized soil to fill up such a cavity.

1. INTRODUCTION

Liquefied stabilized soil (LSS) is a stabilized soil which is made by mixing excavated soils, muddy water (or water) and cementing materials.

Gravel, gravel-containing soil, sandy soils, or cohesive soils can be used for LSS. Even clay and silt, which have a high water content and are usually considered unsuitable for earth works, can also be used for LSS, so this LSS method should contribute to effective recycling of such excavated soils.

LSS method is being developed under a government-private sector collaborative study as part of the Ministry of Construction's General Technology Development Project "Development of materials for controlling construction by-product generation and methods of utilizing recycled materials (fiscal 1992 - 1996)". This paper presents some of the result of the project : a full-scale model experiment for back-filling an old mining tunnel and a field performance test on filling a cavity under the pavement of an urban road.

2. CHARACTERISTICS AND APPLICATION OF LSS

The major characteristics of LSS are : no compaction is required, narrow spaces are filled well because LSS flows before hardening, and the necessary strength can be obtained by adjusting the mix proportion.

By adjusting the content of cementing material, it is possible to adjust the strength of LSS to suit the purpose, from the strength which permits hand excavation to the equivalent strength of poor mix concrete.

LSS is thus thought to be suitable for following applications:

(1) back-filling for life-line pipes and utility conduits

(2) filling closed spaces with complex shapes like mining tunnel

(3) back-filling for structures such as retaining walls, abutments, culvert boxes, etc.

(4) filling cavities under the pavements of roads

(5) filling spaces between the basement of a structure and the surface of the ground, induced by consolidation of the ground

(6) forming embankments where compaction is difficult.

3. FULL-SCALE MODEL EXPERIMENT FOR BACK-FILLING A MINING TUNNEL

3.1 Overview of the Experiment

One major application of LSS is back-filling or filling old mining tunnels which are in danger of collapse. A full-scale model experiment was therefore conducted in order to determine the necessary flow condition (hereinafter called flow value) at which LSS flows, reaching to the end of a complex tunnel from a remote point of placement, and also to determine a suitable placement arrangement for filling spaces beneath a convex and concave ceiling.

3.2 Experimental Method

3.2.1 Experimental Apparatus

The experiment used three kinds of full-scale models as shown in Figure 1.

Each model was 20 m long, 1.8 m wide, and 1.7 m high. Steel half-cylinders of diameter 0.6 m, called partition cylinders, were mounted at a distance apart beneath a ceiling in order to simulate the convex and concave shapes of a ceiling in a real tunnel.

Note that in Case 3 in Figure 1, the model tunnel branches with straight and elbow parts at the middle. This arrangement was used to compare the cases of LSS flowing into the straight and the elbow parts in terms of flow gradient.

To release the air remaining in the spaces between partition cylinders, Case 1 and 3 used pipes of diameter 5 cm. Case 2 had no pipes but vents in the ceiling of diameter 10 cm. These vents are similar to site boreholes.

3.2.2 Experimental Cases and Method

Figure 1 shows the experimental cases. Soils for LSS were Kanto loam and gravel-containing soil. The mix proportion was set to give the flow values set for the three cases.

(1) Measurement of Flow Gradient

LSS was cast from a pipe at the edge of the model. The height of accumulated LSS was measured from monitoring holes in the ceiling in order to calculate the flow gradient.

Case 1

Back-filling	Flow value	120mm
	Placement	Direct cast
Filling up	Flow value	160mm
	Placement	Pipe Placement
Air releasing pipe for convex part in ceiling	Pipe releasing air	

Case 2

Back-filling	Flow value	160 mm
	Placement	Direct cast
Filling up	Flow value	200 mm
	Placement	Direct cast
Air releasing pipe for convex part in ceiling	Borehole	

Case 3

Back-filling	Flow value	220mm
	Placement	Direct cast
Filling up	Flow value	350mm
	Placement	Pipe Placement
Air releasing pipe for convex part in ceiling	Pipe releasing air	

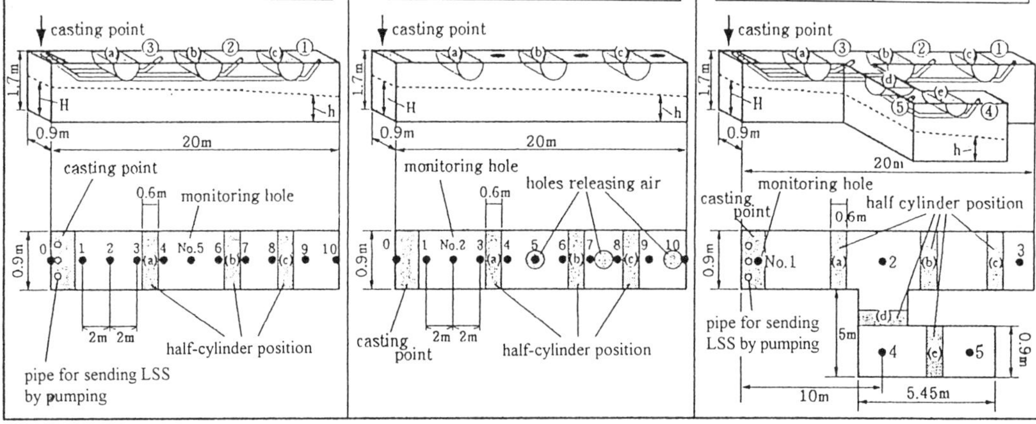

Note. Flow value is an index of the flow condition of LSS; a cylinder 8cm high and diameter 8cm is filled with LSS and lifted up quickly. The diameter to which the LSS spreads out is taken as the flow value.

Fig.1 Experimental Cases and Apparatus

(2) Filling up Spaces under the Ceiling

After obtaining the flow gradient, an experiment of filling a space between a ceiling and the surface of the LSS was conducted. Case 1 and 3 used placement pipes (20 cm diameter) to send LSS at flow values of 160 and 350 mm respective to fill up the spaces. Case 2 used direct placement from the left side of the model at a flow value of 200 mm.

3.3 Experimental Results

3.3.1 Flow Gradient

The lines denoted by —○— in Figure 2 indicate the heights of accumulated LSS for certain amounts of placement. For each case, the flow gradient is shown in Table 1. The flow gradient was obtained by dividing the difference in height by the corresponding distance of the monitoring holes, multiplying by 100, then taking the arithmetic average of the calculated gradient values for each interval.

Table 1 Flow Gradient

Case No.		Flow value	Flow gradient
Case 1		120(mm)	11.3(degrees)
Case 2		160(mm)	2.3 (degrees)
Case 3	Straight	220(mm)	1.9 (degrees)
	Elbow		2.0 (degrees)

At the flow value 120 mm (Case 1), the flow gradient became 11.3%, the highest of all values. In this case, LSS was highly viscous and it accumulated around the inside of the pipe. After placing 5 m^3 of LSS, pumping was no longer possible due to blockage of the pipe.

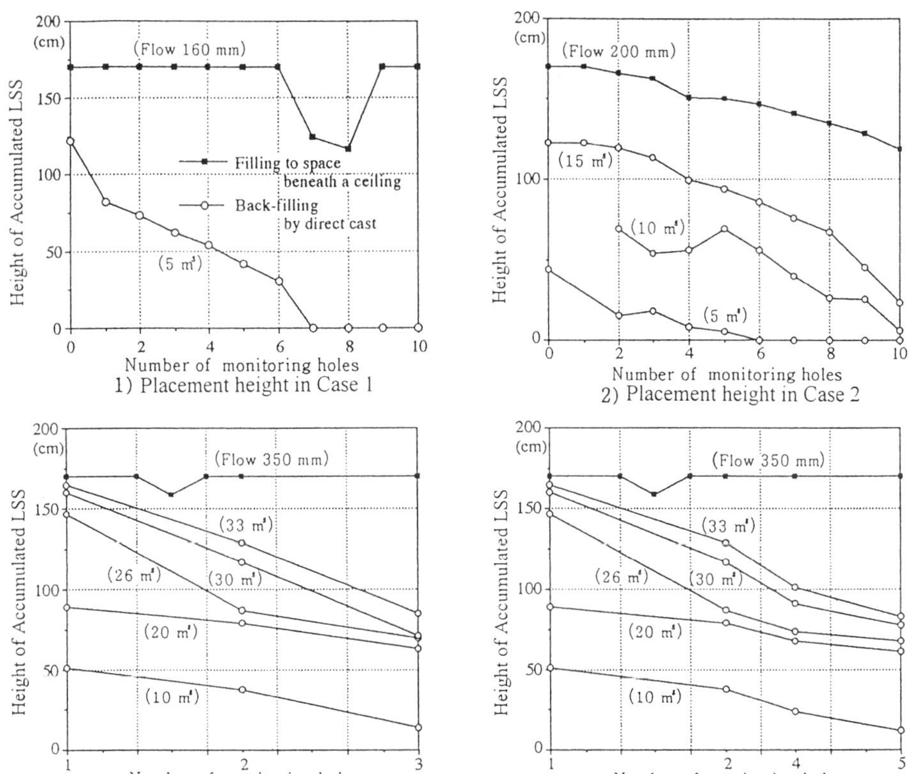

1) Placement height in Case 1

2) Placement height in Case 2

3) Placement height in Case 3 (straight part)

4) Placement height in Case 4 (elbow part)

Fig. 2 Height of Accumulated LSS

On the other hand, at the flow values above 160 mm (Case 2 and 3), the flow gradient was around 2% and LSS could be sent from one end to the other.

Comparing the average flow gradient over straight and elbow parts, no significant differences were observed. Therefore, LSS with a flow gradient of around 2% may flow without being much affected by bends in a tunnel.

3.3.2 Filling up to the Ceiling

The lines denoted by —■— in Figure 2 indicate the heights of LSS after filling up spaces between the ceilings and surfaces of LSS already accumulated. If perfect filling is obtained, the height becomes 170 cm.

The filling state by direct casting (Case 2, flow value 200 mm) is shown in Figure 2-2. In the figure, the height of filled LSS falls steeply near monitoring hole No.3, and filling was not accomplished. This suggests that the partition cylinder (a) around hole No.3 (cf. Figure 1) might act as an obstacle preventing LSS from flowing beyond the partition cylinder.

The filling state by pump casting at the flow value 160 mm (Case 1) is shown in Figure 2-1. In this case, casting was carried out from the pipe end of ① toward the left direction in Figure 1. Twenty-four hours later, filling was resumed at the pipe end of ③. Perfect filling was accomplished in the area from No.9 to 10 by the first filling, but filling was not enough from No.7 to 8, and LSS did not flow into the space on the left side of the partition cylinder (c) near No.9. At the second filling, perfect filling was achieved from No.1 to 7, yet LSS did not go round into the area from No.7 to 8 (between partition cylinders (b) and (c)) and filling was not enough.

The filling state by pump casting at the flow value 350 mm (Case 3) is shown in Figure 2-3 and Figure 2-4. The filling was done sequentially from the pipe ends of ①, ④, ⑤, and ③ in Figure 1. In this sequence, pipe end ② was not used and whether the part between partition (b) and (c) was filled enough could be observed.

In these sequences, almost perfect filling was achieved, including the area between partition (b) and (c). Note that small air voids remained on the upper surface of LSS but adequate filling was confirmed by disassembling the models.

3.4 Summary

① For a flow value exceeding approximately 160 mm, the flow gradient is relatively small, about 2%, and is suitable for back-filling a mining tunnel. In addition, at a flow gradient of about 2%, LSS flows without being affected much by bends in a tunnel.

② To fill up a space between a ceiling and the surface of previously accumulated LSS, LSS having a flow value of around 350 mm, should be pumped through the pipe, regardless of the convex and concave shapes of a ceiling.

4. FIELD PERFORMANCE TEST ON FILLING A CAVITY UNDER THE PAVEMENT OF AN URBAN ROAD

Cavities under the pavements of roads in cities can cause roads to collapse. When such cavities are repaired, it is common to cut the pavement, back-fill the cavities with sand, and cover them. However, in cities where traffic is heavy, cut and cover works are usually restricted to the night time, and the noise and vibration caused by the works are of concern. Field performance on filling a cavity from a hole in the pavement with LSS were therefore conducted.

The field test here used pit sand as the main soil and high density LSS containing a relatively large amount of soil. To reduce the working time at site, LSS was produced at a remote plant, carried by an agitating truck, and then cast on site.

4.1 State of a Cavity under the Pavement

A geological radar and portable scope were used to assess the thickness and area of a cavity. The cavity was found to be just under the pavement (about 40 cm), about 40 to 60 cm thick, 2.5 × 1.5 m wide, with a volume of about 1.0 to 1.3 m^3.

4.2 Mix Proportion of LSS

The mix proportion is shown in Table 2. Pit sand was used as the main material and high density LSS with a large amount of soil was produced. The target flow value was 180mm, which is a relatively low flowing condition, compared to LSS used previously. The utilization ratio R of the main soil was 74%, which is considerably higher than that of LSS used previously (about 50% or less). In addition, the strength of LSS was set at a relatively large value of unconfined compressive strength qu = 980 (kN/m^2).

Table 2 Mix Proportion

Mix(at 1m^3 of LSS)				Density of slurry (kg/m^3)	Mixing ratio of slurry P	Density of LSS (kg/m^3)	Target quality		Utilization rate of main soil R(%)
Slurry Ws		Main soil Wd (pit sand)	Cementation material				Flow value (mm)	qu (28) (kN/m^2)	
Cohesive soil	Water								
126kg	318kg	1269kg	152kg	1100	0.35	1870	180	980	74

Note: $P = Ws/Wd$, $R = Wd/(Ws + Wd) \times 100$

Cohesive soil was Kanto Loam (Density of soil particle 2.775, natural water content 100%).

Cementation material was slag cement Type B.

4.3 Working System

The working system is shown in Figure 3. LSS was produced at a remote plant and transferred to the site by an agitating truck, and placed with a concrete pump. Note that the distance between the plant yard and the construction site was 4 km, taking about 15 minutes.

4.4 Outcome of the Works

① Working Time

About 0.8 m^3 of LSS was cast, and the time spent at the site was 25 minutes from the time of arrival of the agitating truck to the end of filling the cavity. Using a previous system, the plant fixed on the truck was carried to the site, then production of LSS and filling of a cavity was performed. This took about 60 min./m^3 (48 min./0.8 m^3) to complete. Compared with the previous system, the working time at the site could be greatly reduced by the working system shown in Figure 3.

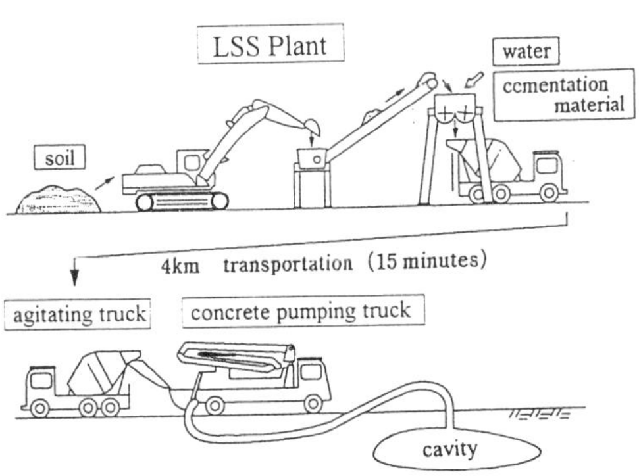

Fig. 3 Working System

② Filling Condition

To detect the cavity, a geological radar was used over an area of 4 m × 7 m in 4 × 7 lattice lines. The results of the primary and post detection operations are shown in Figure 4. Irregular signals were observed during the primary detection, denoting a cavity, whereas after the filling with LSS, these signals vanished, showing that the cavity was steadily filled up.

③ Occupation of Road during the Works

Twenty meters of two lanes were kept for parking the concrete pumping truck and agitating truck, and 30 m of traffic guiding zone were added, giving a total length of road occupation of about 50 m in two lanes.

④ Results of Quality Control Test

For controlling the quality of LSS, the densities of the slurry and LSS, flow value, unconfined compressive strength, and CBR value were measured. The results satisfied the target quality as shown in Table 3.

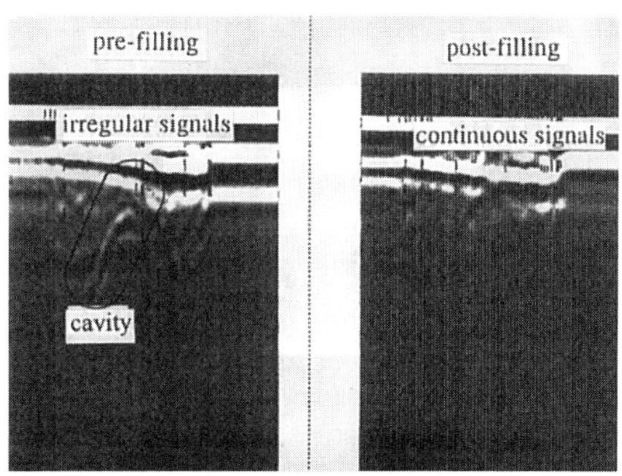

Fig.4 Pre/post Filling Detection Results by Geological Radar

Table 3 Results of Quality Control Tests

Density of slurry (kg/m³)	Density of LSS (kg/m³)	Flow value (mm)		qu(28) (kN/m²)	CBR (%)
		Plant	Site		
1120	1855	174	162	1205	42.8

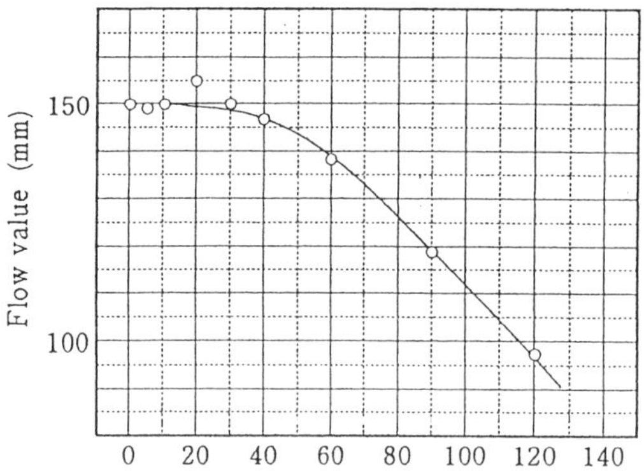

Fig. 5 Decrease in Flow value Over Time

In addition, since the fluidity of LSS decreases during transportation, change in flow value over time was examined. The results are shown in Figure 5. The figure shows that the flow value does not decrease within 30 minutes after production but does decrease thereafter.

4.5 Summary

The test results showed the following:

① The working system of producing LSS at a remote plant, transporting it and placing it on site dramatically reduces the working time at site. With this system, it is possible to carry out two or more operations filling the cavities under a road pavement in one night.

② It is possible to fill up a cavity with high density LSS, even though its fluidity is relatively low.

③ The flow value of LSS starts to decrease gradually about 30 minutes after production. Therefore, when transporting over long distances, the mix design must take into account this drop in flow value.

5. CONCLUSION

The full-scale model experiment for back-filling a tunnel demonstrated the suitable flowing condition of LSS and the adequate placement method. For a flow value exceeding approximately 160 mm, the flow gradient of LSS is relatively small, about 2%, and is suitable for back-filling a tunnel. In addition, at a flow gradient of about 2%, LSS flows without being affected much by bends in a tunnel. To fill up a space between a ceiling and the surface of previously accumulated LSS, LSS having a flow value of around 350 mm should be pumped through the pipe, regardless of the convex and concave shapes of a ceiling.

The field performance test to fill up the cavity under the road pavement verified that the system of producing LSS at a remote plant, transporting it and placing it at site dramatically reduces the working time at site. Also it is possible to fill up a small cavity with high density LSS, even though its fluidity is relatively low. In addition, since the flow value of LSS starts to decrease gradually about 30 minutes after production, the mix design must take into account this drop in flow value when transporting LSS over long distances.

6. ACKNOWLEDGMENT

The full-scale model experiment was undertaken by a collaborative study between the Public Works Research Institute, Ministry of Construction and the Central Research Institute for Construction Technique, JARGC

The field performance test was carried out under the guidance of the "The Detecting Cavities under Pavement of Roads" committee and with the cooperation and advice of the Yokohama National Road Construction Office and the Kanto Technical Office in the Kanto District, Ministry of Construction. The authors sincerely thank all those involved.

References

1) Kuno, G., Miki, H., Mori, N., Iwabuchi, J., Mitsui, T., Ichihara, M., "A Full-Scale Model Experiment for to Back-Filling An Old Mining Tunnel", Proc., 30th Conference of JSSMFE, 1995 (in Japanese)

2) Kuno, G., Miki, H., Koike, K., Mori, N., Iwabuchi, J., Terada, Y., "A Field Performance Test on Filling A Cavity Under Pavement of Road (Part 2nd)", Proc., 50th Conference of JSCE, 1995 (in Japanese)

Environmental Geotechnics, Kamon (ed.) © 1996 Balkema, Rotterdam. ISBN 90 5410 848 7

Characterization of fly ash and other by-products as sorptive subgrades for environmental facility sites

Timothy M. Laumakis
CH2M Hill, Philadelphia, Pa., USA

Joseph P. Martin & Young Chul Kim
Drexel University, Philadelphia, Pa., USA

ABSTRACT: The generation of residual materials (i.e. fly ash or mine spoil) as by-products has continued to grow as a result of both the increased use of natural materials as energy sources (i.e. coal) and the capture of by-products prior to release into the natural environment. However, only a fraction of these soil-textured industrial by-products is productively reused, while the excess is disposed, in landfills, or stored, in lagoons, usually at great expense. Therefore, it is necessary to find alternative applications for these materials in engineering related projects to comply with increasing standards of waste minimization and sustainable development. This paper discusses the use of residual materials as environmental facility subgrades in artificial landfill applications with the emphasis on cost effectiveness, hydraulic control, dimensional stability, and contaminant attenuation.

1 INTRODUCTION

Developments in improved ground modification and geosynthetics have been readily adopted in earthwork practice because they offer solutions to persistent drainage, settlement, and stability problems. However, the onus is on producers of alternative materials, such as fly ash or mine spoils, to demonstrate that construction ease, in-service performance, and cost savings can be obtained by using them as substitutes for traditional materials with established performance histories. One of the most promising potential uses of particulate by-products is as earthfill for traditional uses such as topographic adjustment, dimensionally stable subgrades, etc.. Resistance to the use of unfamiliar fill materials arises from standard practice as well as economics. For example, a roadway or an industrial park require site-specific exploration and testing to determine the properties and load response of the natural subgrade. In contrast, investigations for fill materials are often much less extensive and limited to index properties such as gradation. Experience shows that satisfactory performance is obtained if the borrow is properly compacted and its composition falls inside a specific range of index properties of common soils and will thus have satisfactory engineering properties such as strength, stiffness, and permeability. In some uses, however, it is standard practice to do actual performance tests on fill material candidates.

A potential re-use of residual materials in alternative engineering projects exists when it is necessary to use controlled fill to adjust the elevation and alignment of an environmental facility subgrade for landfills, mining facilities, and transfer stations. There has also been interest in the capability of a subgrade to attenuate contaminant concentrations. The traditional emphasis has been on hydraulic containment by liners and collection systems. However, it is not the seepage that is the concern, but the solutes dissolved in it. Hence, the more recent emphasis is on the attenuation of contaminants by filtration, sorption, neutralization, precipitation, ion exchange and other mechanisms.

Borrow soils have generally been selected on the basis of cost, mechanical properties, and more recently, low permeability. Soil-textured industrial by-products such as mine wastes and fly ash can be economically attractive alternatives to natural soils. Because such materials are available in large quantities and must be disposed of anyway, their use in subgrade, daily cover, and protective cover applications incorporates the sustainable development philosophy into landfilling. In this case, the term "sustainable development" means obtaining as much beneficial usage from a natural resource (i.e. coal), while not degrading the environment as a result of this usage (i.e. air pollution, acid mine drainage, and mounds of non-combustible by-products consuming surface space). The question is whether residual materials from chemical, thermal or physical processing are dimensionally stable and have value in retarding contaminant migration. There is always a basic concern with leaching of contaminants, but is also possible that the processed material can retard contaminants that result from unanticipated releases, not only leachates but, also spills, etc..

Despite demonstration of satisfactory engineering performance and cost savings, regulatory and public relations issues usually arise with the reuse of any industrial waste. Regulations in many states often exempt residual materials from classification as a waste altogether if it is directly reused (Turgeon 1988). However, a favorable regulatory situation will not in itself win acceptance.

2 ENVIRONMENTAL ENGINEERING PROPERTIES

The constituents of industrial by-products are highly variable and are governed by the composition of the source material. Therefore, given that the concentrations of trace metals in fly ash are highly variable and generally greater than background levels the leaching characteristics of a residual material is a major concern when it is being considered for fill to attenuate contaminant migration. Fly ash itself is a highly variable material depending on the source of coal and method of burning. The presence of various heavy metals on the ash particle surface, notably hydrous iron oxides, along with pH, play significant roles in the release of these metals into solution. Therefore, studies have been conducted in an attempt to properly characterize the engineering properties of such residual materials. Through these studies, residual materials have gained acceptance as materials suitable for use in engineering applications. For example, common environmental applications for fly ash include: a component in concrete mixtures, stabilizing road bases (Winterkorn 1975), waste stabilization (Martin et al. 1987), structural fill (Martin 1990), and as a cover material in refuse disposal sites (Mamane & Gottlieb 1992). Another area of fly ash usage is for the treatment of waters contaminated with trace metals (Yadave et al. 1989). In addition, McLaren and DiGioia (1987) concluded that the geotechnical properties of fly ash can be determined using traditional methods. Fly ash, while fine grained, is granular, and can be compacted with vibratory compaction methods similar to those used for coarse soils. The combination of these uses shows promise for constructing low permeability, dimensionally stable and sorbing subgrades. In environmental applications such as liner design for municipal landfills, residual waste piles or saturated and unsaturated contaminant migration modeling, these soil properties provide important data in predicting soil/contamination/pore fluid behavior. Particularly with regard to physical properties and behavior, traditional engineering index property testing (Atterberg Limits and coarse fraction) of soils has been used to determine and predict the behavior of a soil in environmental applications. Traditional engineering index properties and test methods have been used successfully for natural soils in conducting environmental site assessments and evaluating the use of these soils for environmental applications such as low permeability barriers or subgrade layers for solid waste landfills or heap leach pads. However, it is also necessary to assure that the results obtained, when using similar index test methods for residual materials, provide reliable information on how the material will behave under actually field conditions. Properties of these materials can be described in engineering terms such as compaction, strength, compressibility, and permeability. Behavior can thus be predicted. However, analogies with "natural" soils on the basis of simple index tests such as gradation can be exaggerated. For routine use, end users need information on material properties and design details for construction with residual materials similar to that available for current, and thus competing, earthen materials. Utilities must increasingly investigate the properties of residual materials and analyze both the stability and environmental impact of waste depositories (DiGioia 1964). One bonus to designers is that the properties of residual materials from a given source are likely to be more consistent than soils from natural borrow areas. Therefore, a potential geotechnical and geochemical laboratory program could be developed and expanded for the case of screening residual materials. The laboratory screening program would provide guidelines for certain physical and geochemical property

Table 1 Soil Properties that Influence Subgrade Characteristics	
Physical Properties	**Geochemical Properties**
Bulk Density	pH
Surface Area	Redox Potential
Particle Size Distribution	Ion Exchange Capacity
Pore Structure	Organic Matter Content
Permeability (hydraulic conductivity)	Fe and Mn Oxide
	Clay Minerals

requirements of residual materials if they are to be implemented in engineering projects (i.e. compacted subgrade/liner material). The physical and geochemical soil properties that have the most measurable affect on the dimensional stability, retention, mobilization and transport of contaminants within a subgrade are listed in Table 1.

2.1 *Material Investigation*

In two separate investigations (Martin et al. 1987), the properties listed above were investigated for the following three ashes generated by separate power plants located in Eastern Pennsylvania, Delaware, and South Jersey in the United States (USA):

- Freshly generated fly ash from Atlantic Electric and Gas (NJFA)
- Stockpiled Delaware Power & Light fly ash (DSA)
- Stockpiled Pennsylvania Power & Light fly ash (PPL)

For comparative purposes, results of tests for a clean mine spoil generated in Western Pennsylvania are included. All ashes listed originated from burned bituminous coal mined in the Northern Appalachian area (Pennsylvania and West Virginia, USA).

Table 2 shows the gradation indices of the ashes and mine spoil. Column 2 shows that all of the fly ashes are lightweight with the mine spoil resembling a natural soil. It is possible that this fly ash has a high proportion of hollow spheres, but it was not specifically investigated. Column 3 shows that the dry densities of the fly ash, except PPL, are substantially below the values often found with nonplastic soils [14.9-19.6 kN/m^3 (95-125 pcf)]. However, in general, all the fly ashes can be considered a lightweight fill. A well-compacted embankment made with these materials would exert only about 50% of the pressure on a soft subgrade as a fill of equivalent height using coarse granular borrow. In addition, fly ash tends to have a low sensitivity of the compacted density to moisture content. This is a major advantage of fly ash over the fine-grained soils often used in structural fills, where moisture adjustment is necessary and foul weather can seriously retard construction progress. The relatively low permeability of compacted fly ash and mine spoil, as shown in column 4, indicates that seepage will tend to move around a fly-ash/mine spoil body if there is an alternative path of lower resistance. This is favorable for environmental considerations, but also shows a necessity for providing proper diversion of rainfall infiltration. The effective grain size listed in column 5 dominates the hydraulic characteristics (permeability, capillary moisture retention, etc.) of nonplastic soils and can be used as an indicator for permeability behavior.

An index not normally applied to natural soils but commonly applied to coal ashes and mine spoil is the loss-on-ignition (LOI) or unburned-carbon content, ASTM C637. As seen in column 6, the NJFA and DSA have high carbon contents, follow by the mine spoil then PPL fly ash. The low LOI value for the PPL indicates a more efficient burning process compared to that used for the NJFA and DSA fly ash. However, a high LOI value tends to increase organic contaminant attenuation through the increase in the carbon adsorption capacity of the material. This can be seen in column 7. The increase in the distribution coefficient of a soil, an indicator of attenuation,

Table 2 Screening Properties for Suitability of Borrow Source for Compacted Subgrade Composed of Residual Materials								
Sample (1)	Specific gravity Gs (2)	Max. dry density (kN/m^3 (3)	Hydraulic Cond., K (cm/sec) (4)	Effective grain size, d_{10} (mm) (5)	Loss on ignition, LOI (%) (6)	Dist. coeff., K_d (7)	Soil pH (8)	Cation Exchange Capacity, CEC (meq/100g soil) (9)
NJFA	2.16	11.0	1.2e-4	0.011	13.3	1e32	*	1.4
DSA	2.18	10.4	3.2e-5	0.01	14.8	10e9	*	6.0
PPL	2.45	15.6	5.8e-5	0.012	2.3	795.88	4.2	5.0
Mine Spoil	2.69	18.1	3.6e-5	0.05	6.8	122.85	4.2	9.0

* Analysis not performed

increases with the corresponding rise in LOI value. This increase in distribution coefficient is primarily a result of adsorption, either carbon or free oxides, and not to ion exchange as indicated by the CEC values in column 9.

2.2 *Environmental Application*

The rate of migration of contaminants through a liner or within a saturated soil profile is controlled by hydraulic conductivity and attenuation and retardation effects as a result of adsorption or other chemical interactions between the soil and the contaminants in the pore fluid. In environmental applications it is desirable to attenuate or retard contaminants through chemical interactions between the pore fluid and soil. Therefore, the geochemical properties of residual materials have a direct effect on the rate of migration of heavy metals.

The following example will examine the use of the PP&L fly ash as a liner for an environmental battery facility. The scenario consists of a capped facility that is above a liner system composed of a primary liner composed of 0.915 meter (3 feet) of clay underlain by a 1.525 meter (5-foot) thick fly ash secondary liner/subbase compacted to 95% standard proctor [1.51 g/cm^3 (γ_d=94.55 pcf)]. The concern was that the cap failed resulting in a leachate generation rate (v) of 50.8 cm/year (20 inches/year) in the landfill. In addition, the primary liner has failed, exposing the fly ash to the acidic lead leachate. The question of concern is how long will it take for the lead to breakthrough the fly ash subbase to the underlying groundwater table. The calculated porosity (n) of the compacted fly ash subbase, using G_s=2.45, is 0.382, resulting in a seepage velocity (v_s=v/n) equal to 132.10 cm/year (52 in/yr). Next, a linear retardation factor was calculated to give a more conservative estimate of attenuation through the fly ash layer. The linear retardation factor is:

$$R_d = 1+[(\gamma_d)(K_d)/n] \qquad (1.1)$$

where:

γ_d = 1.51g/cm^3 (14.83 kN/m^3)
K_d = 795.88 ml/g
n = 0.382

resulting in R_d = 3,147. Next, the velocity of the reactive solute is calculated by:

$$V_r = V_s/R_d \qquad (1.2)$$

where:

V_s = 132.10 cm/yr
R_d = 3147

resulting in a solute velocity of 0.042 cm/yr (1.66 x 10^{-2} in/yr). Therefore, the time required to achieve breakthrough is then:

$$t = x/V_r \qquad (1.3)$$

where:

$x = 1.525$ meter
$V_r = 0.042$ cm/yr

resulting in 3,631 years required before breakthrough would occur.

In addition, given the same K_d and 1.525 meter thick fly ash liner described above, and an initial lead leachate concentration of 36.97 ppm (also used to generated the above K_d), the mass of lead sorbed can be calculated by,

$$m_{sorb} = C(m_{soil})$$

where:

m_{sorb} = mass of solute sorbed by the residual soil
C = solute concentration : 36.97 mg/l
m_{soil} = unit weight of fly ash layer : (1.525m thick) x (1.51 g/cm^3) = 230.3 g/cm^2

Therefore, the fly ash liner would sorp 8.58 g of lead per square centimeter of liner material.

3 CONCLUSIONS

Under the proper conditions, sediments and soils can act as geochemical barriers to retard the migration of chemical constituents, as well as restrict movement of the solvent (usually water) conveying them. Early in the development of geotechnical practices in waste disposal, Griffen et al. (1976, 1977) quantified the ability of clay-bearing materials to remove heavy metals and other trace elements from seepage solutions resulting from landfill leachate. This use of natural soils to immobilize potential ground-water contaminants and retard their movement is well known. However, residual materials could also provide the same, possibly higher, degree of environmental benefits as natural soils when used in subgrade applications for environmental facilities. Given that the premise of most groundwater standards is based on chemical concentration rather than total mass, attenuation can be a permanent solution in itself. Therefore, the use of properly compacted fly ash would ensure protection to the surrounding environment against contaminant releases while implementing the concepts of waste minimization and sustainable development through the use of this residual material. An additional bonus is that a firm subgrade is also supplied, which will support the structure or impoundment being built.

REFERENCES

Griffin, R.A., Frost, R.R., Au, A.K., Robinson, G.D., and Shimp, N.F. (1977). "Attenuation of Pollutants in Municipal Landfill Leachate by clay Minerals Part 2-Heavy Metal Adsorption," *Environ. Geology Notes*, Illinois State Geological Survey (79), April.

Griffin, R.A., Keros, C., Shrimp, N.F., Steel, J.D., Rutch, R.R., White, W.A., Hughes, G.M. and Gilkeson, R.H. (1976). "Attenuation of Pollutants in Municipal Landfill Leachate by clay Minerals Part 1-Column Leaching and Field Verification," *Environ. Geology Notes*, Illinois State Geological Survey (78), November.

Mamane, Y., and Gottlieb, J. (1992). "Utilization of Coal Ash as a Cover Material in a Municipal Refuse Disposal Site," *Environmental Geotechnology*, Usmen and Acar, eds., 429-435.

Martin, J.P., Felser, A.J., and Vankeuren, E.L. (1987). "Hydrocarbon Refining Waste Stabilization for Landfills," *Geotechnical Practice for Waste Disposal '87*: Geotechnical Special Publication No. 13, R.D. Woods, ed., 668-682.

McLaren, R.J., and DiGioia, A.M. Jr. (1987). "The Typical Engineering Properties of Fly Ash," Geotechnical Practice for Waste Disposal '87: Geotechnical Special Publication No. 13, R.D. Woods, ed., 683-697.

Turgeon, R. (1988). "Fly Ash Fills a Valley" *Civ. Engrg.*, ASCE, 58(12) New York, N.Y., pp. 67-68.

Winterkorn, H.T. (1975). "Soil Stabilization," Foundation Engineering Handbook, H.F. Winterkorn and F.Y. Fang, eds., Van Nostrand Reinhold.

Yadava, K.P., Tyagi, B.S., and Singh, V.N. (1989). "Fly Ash for the Treatment of Water Enriched with Lead (II)," *J. Environ. Health*, A2 4(7), 783-808.

Environmental Geotechnics, Kamon (ed.) © 1996 Balkema, Rotterdam. ISBN 90 5410 848 7

A study of use of H-power ash mixtures as a landfill final cover and a soil stabilizing agent

Maureen T. Lee, Mingxun Ding & Peter G. Nicholson
University of Hawaii, USA

ABSTRACT: Municipal solid waste (MSW) ash is being studied at the University of Hawaii for potential use as an ash landfill final cover and as a soil stabilizing agent. Engineering tests, including permeability, strength, and shrink/swell, were performed on mixtures of ash with local quarry tailings for the landfill final cover. As a soil stabilizer, tests were performed using fly ash mixed with local soils. Tests of the ash-soil mixtures included pH, Atterberg limits, compaction, swell, and strength. Results conclude that the ash mixtures are a viable alternative for a landfill final cover and that the ash may be a suitable stabilizer for certain local soils.

1 BACKGROUND

On the island of Oahu, Hawaii, U.S.A., land is limited. Incineration of Municipal solid waste (MSW) has three major benefits: (1) generation of electricity, (2) recycling of ferrous metals screened in the process, and (3) the decrease in the total volume of the MSW to only 10% of its original volume before it is landfilled. H-Power is a waste-to-energy facility on the island of Oahu. It has a nominal capacity of 544,320 metric tons of waste and produces approximately 108,864 metric tons of ash annually (Jones, et al., 1994). By the reduction of MSW, landfill capacity is conserved.

The City and County of Honolulu have been studying ways to utilize the H-Power ash as a construction material. In 1994, the City and County of Honolulu (City) had closed the Waipahu MSW incinerator. The ash from that incinerator was disposed of in a landfill which needed closure. The City proposed to use the H-Power ash mixed with quarry waste fines as an alternative to a natural soil for the Waipahu ash landfill final cover material. They asked the University of Hawaii to investigate the engineering properties of ash mixtures to determine if they may be a viable alternative. The U.S. Environmental Protection Agency (EPA) will allow the use of an alternate material if it can be proven that it can perform the same or better than a natural material.

Engineering properties investigated included: permeability, compaction, strength, and shrink/swell tests. Identical tests were performed on a local soil that was proposed for the final cover material. Comparisons were made to determine if the ash mixtures were a viable alternative to the local soil for the landfill final cover.

In order to improve poor engineering features of some undesirable local soils, the Federal Highway Administration (FHWA) funded a study to evaluate the potential benefits to engineering properties using soil admixtures, such as fly ash and lime. A series of tests were conducted on mixtures of the H-Power fly ash, lime and less desirable local soils to determine if the ash and lime would be viable stabilizing agents.

2 H-POWER ASH

Coal combustion ash has been studied extensively for use as a construction material and is becoming an accepted practice in the U.S. With MSW ash, there is still the fear that it is a hazardous material and may be detrimental to groundwater and the environment. The utilization of bottom ash is a standard practice in certain European countries. It has been estimated that over 50% of bottom ash generated in the Netherlands, Denmark and Western Germany was utilized in 1989 in road subbase construction, wind barriers, sound barriers, land reclamation, and in asphaltic and portland cement paving applications (Musselman, Eighmy, et al., 1994). Only since the late 1970's, has the U.S. dedicated research to MSW ash utilization. A combination of federal and state regulatory and testing uncertainties, along with local controversy in some places, have delayed the acceptance of ash utilization.

MSW ash can vary with processing procedures, waste stream, and location of facility, i.e., urban, industrial, or rural. In Florida, it took a full scale in situ test to gain acceptance to allow the use of an aggregate product made from bottom ash to be used in road construction in the state (Hooper, 1992). Hawaii, a state of very limited resources, is just beginning to study the utilization of H-Power ash as a construction material.

At H-Power, ash is produced in two forms: bottom and fly ash. Moisture is added to the bottom ash to reduce the temperature. The fly ash is collected in hoppers by electrostatic precipitators in a dry state. Approximately 58% of the total weight of ash produced is comprised of bottom ash. The remaining 42% is fly ash. The ash used in this study was obtained from H-Power separated as "dry" bottom ash and "dry" fly ash. For the landfill cover material, the ash was then oven dried and mixed 60% bottom ash to 40% fly ash by weight to represent typical ash available. The mix was then screened on the No. 4 sieve. The same mixture was used for all the tests for the landfill final cover. Only the fly ash was used in the soil stabilization testing program.

3 COMBINED ASH MIXTURES FOR USE AS A LANDFILL FINAL COVER

3.1 TEST MATERIALS AND OBJECTIVES

The combined ash used in this study was mixed with two different quarry fines at two different mix ratios. The ratios were 80% ash-20% waste fines and 50% ash-50% waste fines. Further testing was performed on a 90% ash-10% waste fines mixture and also 100% ash for comparative purposes. In addition, identical tests were performed on a baseline natural soil that was proposed for the landfill final cover. The soil used in the study was from Mililani and is one of the most abundant and accessible Hawaiian soils, a red-brown lateritic silt, which is currently being excavated throughout Oahu in the development of previous sugar and pineapple fields. The soil has a Unified Soil Classification System (USCS) designation of ML or MH, comprised principally of kaolinite and halloysite (Anderson and Hee, 1995).

The quarry fines used in the study came from two different quarries on the island of Oahu operated by Grace Pacific Corporation (GP), and Ameron HC&D (AM). The Grace Pacific fines are wastes from the sieving of dredged coral, appear tan in color and are granular. The Ameron fines appear gray in color with almost a powder-like fineness. The Ameron fines are by-products from the crushing and grading of quarried "blue rock." Grain size analyses and Atterberg limits were performed on both materials. Results showed the Grace Pacific fines are non-plastic with a USCS classification SP-SM. The Ameron fines were determined to be a fine silt with a plasticity index of 2.6 and a USCS classification of ML.

The intent of these tests was to see if some selected ash mixtures would prove to be a viable alternative to a baseline natural soil for the bottom layer of the Waipahu ash landfill final cover. The major criteria for the final cover material was permeability. The Hawaii Administrative Rules state that the final cover should have a permeability of no greater than $1x10^{-5}$ cm/sec as a guide. The City requested that the material have a permeability of $1x10^{-7}$ cm/sec.

The test plan consisted of initial compaction tests to provide a moisture-density relationship for sample preparation. The Modified Proctor test was used to determine the maximum dry density and optimum moisture content of the materials. All test specimens were prepared at 95% maximum dry density and at optimum moisture content plus 2% which represents typical field compaction specifications. Test specimen preparation consisted of moisture conditioning a measured weight of material and compacting it in equals layers in a sampling ring of a known volume. Other tests performed were falling head permeability for hydraulic conductivity, direct shear for strength, and one-dimensional swell to determine the swell potential.

Test methods used in the analyses were intended to provide conservative results to give a general range of expected field values.

3.2 PERMEABILITY

The falling head permeability test procedure was used as a guideline. While this test method is not an ASTM standard, it is widely used in practice and provides a reasonable estimate of permeabilities achievable. Specimens were compacted in a 2.5 inch (63.5 mm) diameter, 1 inch (25.4 mm) high sampling ring. Tests were performed in an odometer-type apparatus connected to a burette by plastic tubing. The burette was constantly filled with water until the sample was saturated. Once saturated, the burette was filled to a known position and the height was recorded. As the height of the water in the burette fell, time, height and volume change were measured. Measurements were taken at the top of the burette.

The first set of tests were run on mixtures of ash and each of the quarry fines to determine the range of values achievable. Tests were performed for all mixtures with a minimal confining pressure of 1.25 kPa, which was the weight of the top cap. The Mililani soil tests were performed by Hee (1995) with a confining pressure of 9.94 kPa. Results are given below.

The test results showed the Ameron mixture to have the best potential. Therefore, further testing was performed on only mixtures of Ameron fines and ash at a higher confining pressure (22.8 kPa), which was more representative of expected field conditions. Results are given below.

Table 1. Permeability Test Results

Material	Permeability, k (cm/sec)
50% Ash - 50% GP	$6.6x10^{-6}$
80% Ash - 20% GP	$1.1x10^{-6}$
50% Ash - 50% AM	$6.5x10^{-7}$
Mililani Soil	$1.0x10^{-6}$

Table 2. Permeability Test Results

Material	Permeability, k (cm/sec)
80% Ash - 20% AM	$3.0x10^{-8}$
90% Ash - 10% AM	$2.0x10^{-8}$
100% Ash	$6.0x10^{-8}$
Mililani Soil	$2.0x10^{-8}$

All materials tested met the $1x10^{-7}$ cm/sec criteria with the 90% ash-10% Ameron mixture and the Mililani soil being the best. The permeability test results should be viewed as permeabilities readily achievable in the field.

3.3 STRENGTH

Direct shear tests were performed on the ash mixtures to discover any potential problems in material strength and to compare with the strength of the natural soil. The direct shear test was used to determine the soil shear strength parameters: cohesion intercept (c), and friction angle (ϕ). Each mixture was compacted to 95% of the maximum dry density and at optimum water content plus 2%. Samples were compacted directly in the test apparatus sampling ring in five equal layers. Samples were sheared under a constant strain rate of 0.75 mm/min. Shear stress and horizontal displacement were recorded. The maximum shear strength was determined from the plots of the two measurements. This was repeated for three different confining pressures (vertical stresses): 30.7, 61.7, and 92.7 kPa. Results were plotted to determine the friction angle and cohesion intercept for each soil mixture.

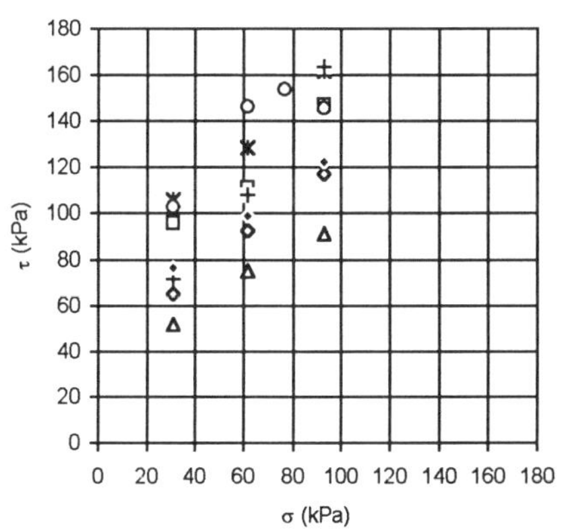

Figure 1. Plot of Direct Shear Test Results

Table 3. Direct Shear Test Results

Material	Shear Strength ϕ (degrees)	Shear Strength c (kPa)
50% Ash - 50% GP	40	40.4
80% Ash - 20% GP	40	66.9
50% Ash - 50% AM	32	33.9
80% Ash - 20% AM	43	75.0
90% Ash - 10% AM	33	92.5
100% Ash	56	23.3
Mililani Soil	36	54.3

Direct shear tests performed gave relative shear strengths of the soil. Trends showed that an increase in percent combined ash led to a higher cohesive shear strength. Results showed the 90% H-Power ash - 10% Ameron waste fines mixture to be the strongest with a nominal cohesive shear strength of 92.5 kPa. The ash alone showed the lowest c value. These results could be due to the strength of the bonds produced in the ash-fines mixtures. The majority of mixtures tested were stronger than the natural Mililani soil.

3.4 SHRINK/SWELL

As a safety precaution, selected potential mixtures and materials were chosen to test for swell and shrinkage potential. The free swell test was based on the ASTM D-4546 Standard Test Methods for One-Dimensional Swell or Settlement Potential of Cohesive Soils, Method A. In this method, each sample was prepared in a consolidometer cell that was 2.5 inches (63.5 mm) in diameter and one inch (25.4 mm) in height: compacted to 95% of the maximum dry density and at optimum water content plus 2% in five equal layers. Once the sample was prepared, a dial gage was placed upon the porous stone cap to measure any swell. The sample was then inundated with water and allowed to swell freely. Measurements were taken for five days. Results are presented below.

This test represented the maximum possible swell of each specimen with zero confining (overburden) stress. While this was not representative of actual expected field conditions, it did provide a conservative indication of potential expansive characteristics. Under these test conditions the swell potential results were considered low, less than 5%. Consequently, in the field, where there would be an overburden pressure due to the topsoil, the swell potential would be

Table 4. Free Swell Test Results

Material	Swell (%)
80% Ash - 20% AM	2.2
90% Ash - 10% AM	4.9
100% Ash	1.4
Mililani Soil	0.44

expected to be insignificant. Therefore, swell was not considered to be a potential problem.

No quantitative shrinkage test was performed; however, the 100% ash and Mililani soil specimens used in the permeability tests were extruded from the sampling rings and observed daily as they air-dried. After three days, the Mililani soil showed substantial vertical shrinkage cracks. The ash samples seemed to expand slightly; however, no quantitative measurements were made and no cracks were noted. Expansion should not be expected in the field due to the confining pressure created by the surrounding soil. However, cracks in the material can create an open flow path, thereby greatly increasing the permeability and infiltration potential of the soil. Since permeability was as low and shrinkage cracking was less, mixtures of ash may prove a better alternative to the natural Mililani soil as a final landfill cover.

4 MSW ASH FOR USE AS A SOIL STABILIZER

4.1 TEST SOIL AND PROGRAM

Unlike the landfill cover test program, this study used only fly ash mixed with a problem local soil. The X-Ray Fluorescence test results showed that fly ash has a significant percentage of CaO (40%) which indicates fly ash itself is a good stabilizing agent.

The soil used in the study was from Halawa Valley, Oahu. It is a silty clay, USCS classification MH, with a high percentage of Halloysite which undergoes irreversible changes in the microstructure and the geotechnical properties upon drying. Therefore, soil specimens need to be prepared using wet methods as described below. This soil has a low pH, low activity, low specific area, and high moisture content.

The test program consisted of pH, Atterberg limits, compaction, free swell, unconfined compression, and direct shear tests.

4.2 SOIL ADMIXTURE REACTIONS

The chemical reactions which cause changes of soil engineering properties include short and long term reactions. Short term reactions consist of hydration and flocculation, which, combined with physical mixing, are responsible for immediate drying effects (10% decrease in moisture content of Halawa soil with 15 % MSW ash). Long term reactions result in the formation of cementitious materials which increase soil strength and stiffness.

4.3 pH

The lime saturation point, where pH reaches a level of 12.3 or 12.4, can be used to indicate whether reactions between soil and admixtures will be complete. Test results given in Table 5 show that 15% MSW ash can only bring up the pH of the Halawa soil mixture to 9.5 while 15% MSW ash and 5% lime can increase the pH of the soil mixture to 12.3. This indicates that the combination of MSW ash and lime is a good additive for a strongly acidic soil, like the Halawa soil, and ash alone may be beneficial for other soils with a higher pH.

4.4 ATTERBERG LIMITS

Wet preparation was used for all tests. The procedure includes washing the soil sample on the No. 40 sieve, and continuing until all the minus No. 40 material has been washed into the pan. Next, the water is allowed to evaporate from the sample until it is without any free water. Finally, the wet sample is mixed with admixtures thoroughly cured, and tested.

Test results in Table 6 showed that with 15% MSW ash or combination of MSW ash and lime, the plasticity index (PI) was decreased about 25%. Both short and long term reactions are responsible for this change. The reduction of plasticity combined with the drying effect due to hydration will significantly ease the workability on the construction site.

4.5 COMPACTION TEST

High natural moisture contents typical of many tropical soils make the compaction and construction in the field difficult. In addition to the drying effect, decrease of the maximum dry density (MDD) and increase of the optimum moisture content (OMC) will be very helpful for ease of compaction of wet soils. This is mainly because there is less of a need for the soil to be dried to a lower moisture content prior to compaction.

Test results shown in Table 7 indicate that for the Halawa soil, there is little change for soil with MSW ash and combination of MSW ash and lime when compared to the original soil. Low pH, low reactivity, and low cation exchange capacity contribute to this effect. Preliminary results from the testing of other soils show that with MSW ash, the MDD and OMC tend to decrease and increase respectively (Nicholson, et al., 1993). All the following tests were performed on soils compacted to 90% relative compaction based on the Modified Proctor test.

Table 5. pH Values

Material	pH
Natural Soil	4.5
15% Ash	9.7
15% Ash + 5% Lime	12.3

Table 6. Atterberg Limits

Material	PL (%)	LL (%)	PI (%)
Natural Soil	61.4	90.5	29.1
15% Ash	53.6	75.6	22.0
15% Ash + 5% Lime	57.6	81.0	23.4

Table 7. Compaction Test Results

Material	MDD (kN/m^3)	OMC (%)
Natural Soil	12.8	35.6
15% Ash	13.1	33.0
15% Ash + 5% Lime	12.5	36.8

4.6 FREE SWELL TEST

The free swell test can be used to measure the maximum magnitude of swell under nominal vertical pressure. Many factors including mineralogy, clay content, the plasticity index and activity influence the swell potential. Test results in Figure 2 show that, when compared to natural soil, there is little initial change in the swell potential with the addition of admixtures, but swell of samples cured for 60 days was reduced to nearly zero. Formation of a cementitious matrix from long term reactions and decrease of the plasticity index make the soil more rigid and more water-resistant. Test results of other soils showed that with fly ash alone the swell potentials were decreased from 30% to 85% (Nicholson, et al., 1993).

4.7 STRENGTH TESTS

In pavement design and construction, choice of design strength is important because selected design strength will have a great effect on the pavement thickness and future pavement performance. Cementation, gel formation and crystallization from long term reactions between soil and admixtures will bind soil particles together which in turn increase strength.

Test results in Table 8 showed that with the addition of 15% ash, the 1-day unconfined compression strength (q_u) of the Halawa soil mixture

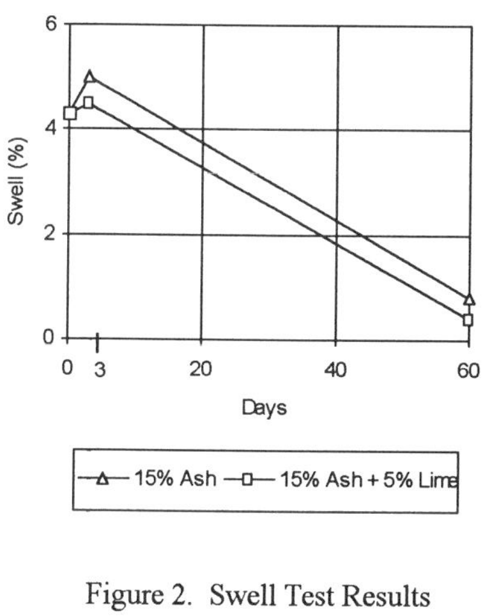

Figure 2. Swell Test Results

Table 8. Strength Test Results

	Natural Soil		15% Ash		15% Ash + 5% Lime	
UCT	q_u (kPa)		q_u (kPa)		q_u (kPa)	
1 day	130		131		-	
7 day	-		85		-	
28 day	-		39		-	
DST	c (kPa)	ϕ	c (kPa)	ϕ	c (kPa)	ϕ
1 day	78	39°	78	44°	80	56°
7 day	-	-	77	35°	100	55°
28 day	-	-	74	45°	128	55°

was unchanged, while 7-day and 28-day strengths were dramatically decreased. Direct shear tests (DST) were conducted because with the addition of admixtures, the soils were altered into less clayey form. Cohesive strength may decline while friction strength is increased, thus making unconfined compression test (UCT) less applicable. Results of direct shear tests showed that the 7-day strength showed little change, but the 28-day strength actually regained which, however, is just slightly higher compared to the natural soil. This indicates that confinement has some influence on soil strength which should be considered in design, and again low pH and low reactivity are responsible for the decrease. Results of the combination of MSW ash and lime showed significant strength increase of soil mixture with time. Again the combination is the most beneficial for strongly acidic soils like the Halawa soil, while MSW ash alone may be a good stabilizing agent for high pH soils.

5 SUMMARY

In conclusion, combined fly ash and bottom ash from MSW when mixed with local quarry fines tested to be a viable construction material. All mixtures and soil tested met the requirements of the Hawaii Administrative Rules for the infiltration layer of the cover material for the Waipahu landfill closure. The 90% combined H-Power ash - 10 % Ameron waste fines performed as good as or better than the natural local soil proposed. The natural soil showed potential for shrinkage cracks which could be detrimental to the permeability, the most prevailing design parameter.

MSW ash and combination of MSW ash and lime, as a soil stabilizer, improved the workability of soil mixtures by the drying effect and decrease of the plasticity index. The swell potential was reduced by increase of soil rigidity and water-resistance. The strength was increased by formation of cementitious materials. Cation exchange capacity, pH, and activity of the natural soil had significant influences on admixture efficiency. The combination of MSW ash and lime was shown to be a good stabilizing agent for highly acidic soil, while MSW ash alone may be beneficial for higher pH soils.

Use of this locally available low-cost, high quality MSW ash can provide a cost effective alternative to designing with undesirable local soils.

REFERENCES

Hee, B.H. (1995). *The Influence of Bentonite Admixtures on the Hydraulic Conductivity of Elastic Silts*. Master's Thesis, University of Hawaii.

Hooper, W.F. (1992). "Processed Ash Demonstration Project." Wheelabrator Environmental System, Inc. Hampton, New Hampshire.

Jones, C.M.; Hartman, R.M.; Kort, D.; and Rapues, N. (1994) "Utilization of Ash From Municipal Solid Waste Combustion, Final Report Phase I." NREL Subcontract No. XAR-3-1322.

Mitchell, J.K. and Sitar, N. (1982). "Engineering Properties of Tropical Residual Soils." *Proceedings of the ASCE Geotechnical Engineering Division Specialty Conference, Engineering and Construction in Tropical and Residual Soils*. American Society of Civil Engineers, New York, NY, pp. 48-49.

Musselman, C.N.; Eighmy, T.T.; Gress, D.L.; Killeen, M.P.; Presher, J.R.; and Sills, M. (1994). "The New Hampshire Bottom Ash Paving Demonstration US Route 3, Laconia, New Hampshire." *National Waste Processing Conference Proceedings*. American Society of Mechanical Engineers.

Nicholson, P.G. and Kashyap, V. (1993). "Fly Ash Stabilization of Tropical Hawaiian Soils." *Fly Ash for Soil Improvement, Geotechnical Special Publication No. 36*. American Society of Civil Engineers, New York, NY, pp. 15-29.

Environmental Geotechnics, Kamon (ed.) © *1996 Balkema, Rotterdam. ISBN 90 5410 848 7*

Utilization of used tyres in civil engineering – The Pneusol 'Tyresoil'

Nguyen-Thanh-Long
Laboratoire Central des Ponts et Chaussées, Paris, France

ABSTRACT : About five hundreds structures have been built with Pneusol-"Tyresoil" (a combination of tyres that cannot be retreated and any type of soil, natural, artificial or wastes!), for many civil engineering applications. This paper presents some principal utilizations of this material.

1 INTRODUCTION

Old tyres constitute a waste that has excellent mechanical properties and is available in quantity in all parts of our countries.

While old tyres do not contribute directly to pollution (unless they are burned in the open air), they affect our environment in the long term because they are not biodegradable. If numerous valorizing processes had been invented (retreads, incineration, farm use, rubber powder, ...) an important quantity of tyres remains nevertheless unused. So as to use a part of those wastes, the Pneusol material was invented by the Laboratoire Central des Ponts et Chaussées.

Pneusol-Tyresoil, a combination of natural, artificial soils or wastes and tyres that cannot be retreated ; they may be whole, partially cut up (one sidewall removed), or completely cut apart (into two sidewalls and one tread) (Long 1980, 1985, 1987).

Today more than 500 structures have been built in France, 12 in Algeria, in the United States of America, in Norway, in England, in Germany, Roumania and even in....Rwanda! covering a wide range of civil engineering applications mainly in order to reinforce earth structures, at a lower cost than conventional technologies.

Our paper deals with the description of some important structures using Pneusol and its different areas of use.

2 AREAS OF USE

One of the qualities of Pneusol is its flexibility, which enables it to withstand large differential settlements. The use of tyres arranged in tiers also gives a better distribution of forces in the reinforced soil mass and on the foundations.

This essential quality makes Pneusol Tyresoil a very good civil engineering material, which has the following advantages :

* standardization and speed of execution ;
* construction intages and in sections ;
* possibility of using a mediocre fill ;
* improvement of the environment through the consumption of an abundant and cumbersone waste.

3 RETAINING STRUCTURES AND PNEUSOL "LIGHT".

The reinforcement of the embankment consists of passenger-car tyres of which both side walls have been cut away. The tyres, tied together by rot-proof straps, are assembled into layers, which are then placed in tiers (0.35 and 0.50 m.apart) on the compacted embankment.

An experimental wall 5 m high and 10 m long, built at Nancy in 1982, has demonstrated how easy it is to place multi-layer Pneusol structures.

The *Bussang Pass*, in 1987, was the first large, major site on the national network to use this material. This site was used for the development, in a real-life situation, of the "technologies" of various potential applications of Pneusol : six hundred meters of Pneusol walls ranging height from 2 to 7 m were built.(Photo n°1)

Photo n°1: Establishing Pneusol with different layers

Photo n°2: Pneusol with concrete and Pneusol facing.

Some of them even used a lightened structure (Pneusol "light") (Long, Valeux 1989). This technique is substantially the same, except that :

* uncut truck tyres are used ;

* they are joined together, if necessary, by metallic parts attached to the sidewalls (for speed of placement) or by ties;

* each layer of tyres is covered with a non-woven geotextile. The voids thus created lighten the embankment, and the linked tyres contribute to its reinforcement. The density of such a material is of the order of 8 to 10 kN/m^3 and depends of course on the thichness of the intermediate layer of fill and also on the volumic weight of the fill itself. For example a Pneusol with clinker (another waste) has only 5 or 6 kN/m^3. Presently many slopes sliding repairs have been made with Pneusol "light" of different kinds.of fills (gravel, clinker, pozzolana,..). Pneusol "light" is extremely cost-competitive.(Bailly and al. 1988, Bricout and al.1992)

In all, 55,000 passenger-car tyres and 2,000 trucks tyres having a combined weight of 500 tons and 1,612 of facing (693 concrete slabs) were used for the widening of the Bussang national road.(Photo n°2)

Repair of the Boulsios landslip in Lozere (1990)

The Boulsios embankment, just outside Millau on the Béziers-Neussargues rail line (southern France), is built on an unstable slope of the right bank of the Tarn. As an indication, the mean settlements of the line have been of the order of 0.50 to 1 cm per month since 1968!. This embankment has therefore often been built back up to raise the line and so compensate for the general slippage of the unstable slope.(Photo n°3)

The SNCF (French National Railways) wished to raise the rails 0.50 m and shift them slightly - about 0.50 m - in the downslope direction to make room for two footpaths along the sides to allow its crews to move about safely.

This solution required adding approximately 7 m^3 of fill per linear meter of line, increasing the loading by about 140 kPa - hardly the thing to promote general stability. The "Mecasol" geotechnical engineering firm had approved only half of this additional loading.

The solution finally chosen to raise the embankment was the following:

- in the part supporting the rail traffic, i.e. under the line itself, with the usual load diffusion angle, ballast is used ;
- elsewhere, the slope of the embankment is stiffened and made lighter by the use of Pneusol with pozzolan; this already has a relatively low density ($10 kN/m^3$), making it possible to produce a lightweight Pneusol having a density of only 5 kN/m^3 (Long 1993).

The work as a whole resulted in a small, even negligible, increase in the surcharge on this unstable slope. This solution is very interesting because the Pneusol can be made very much lighter.

Photo n°3 : Cutaway view of the Boulsios embankment.

4 REDUCING ACTIVE EARTH PRESSURE

First of these structures built in 1986 is a cantilever retaining wall fifty four meters long and five meters high founded on two rows of piles. Before filling behind the wall, the contractor became aware that the shell of the wall was too thin (Long 1990).

"Heavy" Pneusol was used to relieve the active earth pressure and also to eliminate tensile forces on the piles. It was made of treads arranged in layers and tied together by polyester straps. Since the fill inside the treads exerts no active earth pressure and that between two layers is highly confined, the Pneusol mass, provided that the spacing between layers is small, behaves like gravity wall made up of thousands of tiny gabions stacked one to another.

But the structure most used is that of the Fonderie Wall. One hour before work to widen a heavily travelled road was begun, this 18th-Century wall was classified as an historic monument, so it could not be demolished. Luck had it that the whole site lay outside of the right of way of the new pavement. However, a retaining wall near the old foundry, of a very doubtful stability, was right at the base of the embankment of the new road. Being classified, it could not of course, be reinforced from the outside. It is made up of more or less closely spaced blocks.

The historic wall was accordingly lined, inside, with pressure-reducing Pneusol. The Pneusol placed behind the wall consists of layers of truck tyres. Equipped with soil sensors, it gives very good results. (Laréal and Long 1987)

Photo n°4: Reduce active earth pressure with trucks tyres

5 CREATION OF ARCHING

All builders know how to reduce overloading rigid concrete pipes under a large depth : they incorporate a flexible material with a lower modulus than the one of the fill.

So the French authorities have conducted some experiments to relieve the stress concentration on underground culverts and quantify the arching created by the use of Pneusol. (Long 1990)

The Monistrol work in Haute-Loire, built in 1985 which is 137 m long, as a span of 5.10, and buried under 13 m is the first application of this type. The work consists of an ogival vault 22 cm thick supported by two sills resting on micropiles. Runners at regular intervals provide transverse stiffening.

To avoid overloading the work, in particular its foundation, pressure relief using the Pneusol technique was chosen because of the consistency of its geotechnical characteristics. A thichness of about 2.2 m of Pneusol, made using six layers of truck tyres, was placed above the work, over a zone wider than the work itself. To test the efficacy of the process, seven pressure sensors were placed in each of two profiles.(Fig.n°1)

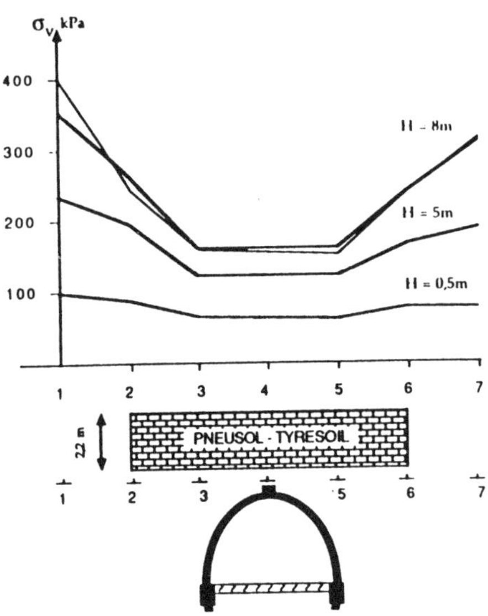

Figure 1 : Culvert of Monistrol-sur-Loire

The pressure curves show that the desired effect appears immediately with the first placement of fill above the Pneusol. The ratio of the mean vertical pressure on the work to the vertical earth pressure is about 0.8 here, whereas the theoretical value for conventional fill is 1.6.

The calculation of this thickness of "soft" Pneusol to create the arching effect has, strangely, never been approached by any researcher (to our knowledge).

In a first stage, we opted for simplicity in the calculations, to encourage widespread use of Pneusol by civil engineering contractors.

To avoid the Marston effect, it suffices to cause a settlement of fill above the top generating line of pipes, a settlement larger than that of surrounding engineering fill. This is the principle of "pressure relieving placement", which is to stage the earthworks and the modulus of the fill in such a way that the plane of equal settlement is located as close as possible to the top generating line of the structure. In this way, the weight of earth will not not be increased.

Let E_1 and H_1 be the modulus and height, respectively, of the engineered fill on the sides between the foundation level and the plane of equal settlement.

Let E_2 and H_2 be the corresponding values for the Pneusol fill above the pipe, between the top generating line and the plane of equal settlement.

Let σ_v be the uniform stress exerted above this plane. Assuming that the pipe is very rigid (not deformable), then by the definition of the plane of equal settlement, one has approximately (Fig.n°2):

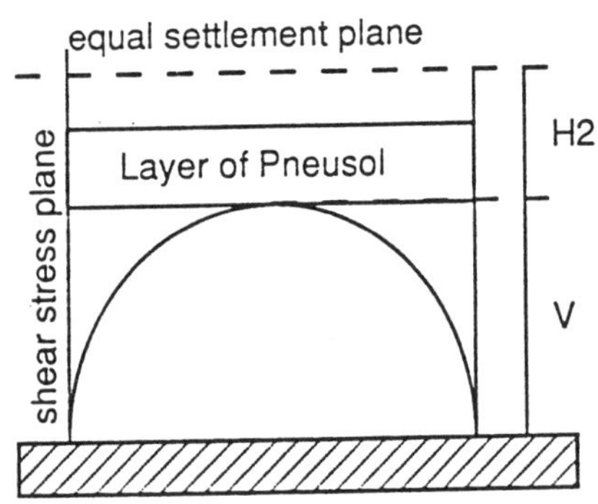

Figure 2 : Thickness calculation

$$(H_1/E_1) = (H_2/E_2)$$

If the sag of the structure is V :

$$H_2 = V\,(E_2/(E_1 - E_2))$$

The numerical modeling for analysing the elastic and elasto-plastic behaviour was carried out thanks to a general-purpose finite element software (CESAR-LCPC). The numerical results are obtained in fair agreement with the experimental observations.

Since then, more than 200 other structures using this technique were built in France, twelve in Algeria, in Spain.

6 ENERGY ABSORPTION

The idea of Pneusol as an "energy absorption" springs from a common place observation of daily life. One often sees tyres stored casually against garage walls. And tyres are used at tricky corners on motor racing circuits to slow the occasional car that spins off the track. The test results showed that the energy restitution coefficient of Pneusol is very low, about 0,10 (this is the ratio of rebound height H of a falling weight to height H_0 from which it was dropped).

A real test ist performed on a three layers'Pneusol built on a concrete roof (about 25 m^2)supported by twelve poles equipped with strain gages. Experimental results show that the Pneusol has a very good absorption power.

7 OTHERS APPLICATIONS

Pneusol is a combination of two components, tyres and soils. So it is possible to built any structure you wants depending on the destination of your structure. For example, a Pneusol "light" is a good material for the subgrade of a road in cold countries again frost because it has many voids (Québec). (Long 1990, 1993).

It is sometimes difficult to ensure the stability of structures built in water, because the soil must be held in place without disturbing the flow of water along the embakment, since this may lead to erosion of the banks.

We present here the investigations and works of reinforcement of a dike built between 1852 and 1859, using Pneusol to protect the banks.

The dike is 950 m long and reaches a maximum height of 11 m. It has always been the site of leaks, the extent of which increases with the height of water.

Its upstream slope has a gradient of 1 in 2.7 and covered with a layer of dry stones, of which the portion out of the water has been repointed.

Its dowstream slope, which is steeper 1 in 1.67, is covered with brambles, bushes and many trees. At the low point of the site, the dike has a masonry drainage aqueduct, the downstream head of which shows signs of fracturing, overturning and extensive seepage.

This dike was instrumented with three piezometers tubes and a clinometer at the south head of the drainage aqueduct. In addition, cores were taken and penetrometer, shear and permeability tests were carried out.

It was found that in the tallest cross-section of the dike, at the aqueduct, body of the dike consists of a very loose ground that is waterlogged and shows local signs of internal erosion. There is also water pressure on the wing walls of the downstream head of the aqueduct and extensive fracturing.

Because of the wooded character of the downstream slope and the width of the dike, it was decided to improve the overall watertightness of the dike by waterproofing the upstream slope.

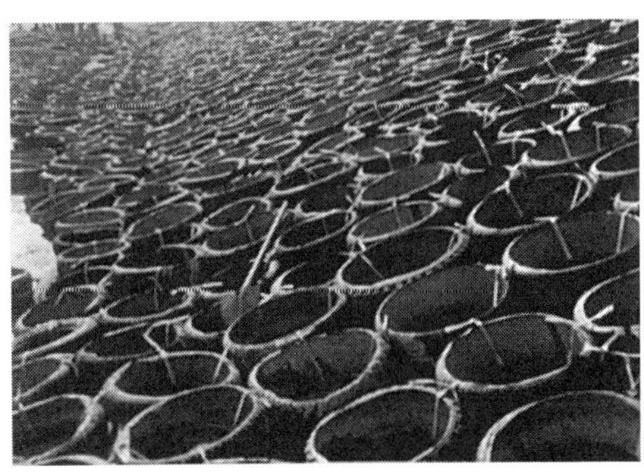

Photo n°5 : Protection of slope at Etang du Puits

The approach chosen included the following stages :

* application of a coasting of mortar to the dry stone surfacing;

* placement of a GSUV PVC geomembrane 1.5 mm thick, protected on both sides by MT 300 woven and needle-bonded polypropylene felt to provide both drainage and protection against punching;

* vertical waterproofing of the foot of the dike, to a height of 2 to 2.5 m, by continuous cement grout wall;

* mechanical protection of the geomembrane by Pneusol-Tyresoil including, at the top, a tier of tyres tied together after removal of the upper sidewall; this system is then covered with a layer of levelled sand; at the top over a length of 3.5 m, there is a second tier of tyres holding 80/160-mm loose stones to protect the slope from impacts by motorboats.(Photo n°5)

8 CONCLUSIONS

This papier touches on pratically all facets of the use of Pneusol-Tyresoil as a civil engineering material.

Basing upon ten years of experience (the first real stucture was built in 1984) some significant findings may be pointed out :

*Tensile tests of tyre parts and soil-tyre adherence have yielded good results.

*The construction Pneusol "light" cost found at different sites is quite attractive and compares favourably to polystyrene foam or expanded clay. In some cases (in particular for embankments on unstable slopes), it is possible to compensate for the difference in mean density by doing additional earthwork and using more Pneusol light.

* Globally, the Pneusol embankment behaves as an embankment that is reinforced and therefore has cohesion. This cohesion enables it to withstand much larger differential settlements than conventional embankments.

*The presence of the tyres and the voids they create gives the Pneusol embankment good anti-vibration, anti-frost and perhaps anti-sismic properties.

* Finally, among the things we are sure of, we may mention the long life of the tyres, their ability to withstand agression of all kinds (in particular chemical), the simplicity of the process, its univerticality - the basic materials are available practically on the spot in all countries, and, of course, its ecological character - wastes are not only eliminated but also made useful in process.

The promising research results, the good behaviour of different structures built (500), ease of use and excellent long-term performance make Pneusol-Tyresoil a good civil engineering material.

REFERENCES

Nguyen-Thanh-Long 1980 *Le Pneusol* Coll. International "Routes et Développement" Institut des Sciences et des Techniques de l'Equipement et de l'Environnement pour le Développement, Mai, Paris.

Nguyen-Thanh-Long 1985 *Le Pneusol* Rapport des Laboratoires n.7 Juillet.

P.Laréal, Nguyen Thanh Long 1987 *Le Pneusol réducteur de poussée* 4ème Colloque Franco-polonais, Grenoble.

Nguyen-Thanh-Long 1987 *Domaines d'utilisation du Pneusol* 4ème Colloque Franco-polonaise, Grenoble.

J.C.Bailly, B.Bescond., Y Guillard, Nguyen-Thanh Long, D.Schmutz, P.Ursat, J.C.Valeux, E.Waschkowski, 1988 *Le Pneusol : Recherches et réalisations* Congrès International de la Route de l'An 2000, Berlin, Septembre.

Nguyen-Thanh-Long, D.Schmutz, P.Ursat.1989 *Lutzelhouse Landslide Repair using Pneusol* 2nd International Symposium on Environmental Geotechnicology, Tongji University, Shangai, May

Nguyen-Thanh-Long, J.C.Valeux.1989 *Le Pneusol léger*, Rapport des Laboratoires GT n.37

Nguyen-Thanh-Long 1990 *The Pneusol.* Rapport des Laboratoires des Ponts et Chaussées, Série Géotechnique GT 44.

D.Bricout, J.L.Dupressoir, Nguyen-Thanh-Long 1992 *Réparation d'un glissement par le Pneusol léger*. 6th International Symposium on Landslides. Christchurch, New Zealand

Nguyen-Thanh-Long 1993 *Pneusol : Recherches-Réalisations-Perspectives* - Thèse de Doctorat, Insa de Lyon - Décembre 1993 - 200 p.

Environmental Geotechnics, Kamon (ed.) © *1996 Balkema, Rotterdam. ISBN 90 5410 848 7*

Utilization of bottom ash as material to form a stabilized base course

Yuji Maeno & Tokio Hirata
Kagoshima National College of Technology, Japan

Masataka Hanashima & Ryuji Yanase
Fukuoka University, Japan

ABSTRACT: This paper is concerned with the effective use of bottom ash mixed with either hydrated lime or cement as a subgrade or a base course material. The result of this research clearly points out that bottom ash is indeed dynamically compatible for use as a subgrade and a base course material. We can however state that the strength of stabilized bottom ash is peculiar and should be examined with care.

1. INTRODUCTION

The disposal of wastes in big cities has become more difficult because of the ever decreasing lifespan of the final disposal sites appropriated for these wastes. We are considering the possibility that wastes should not only be used as a landfill material but should also be utilized for other purposes whenever possible and in the process significantly decrease the amount of wastes being disposed. With the large amount of bottom ash resulting from the burning of general wastes, the disposal of bottom ash itself presents yet another problem. Lately the analysis of bottom ash and assessment of its possible effective use have begun but the implementation stage remains to be realized. We have been involved with the dynamic test of waste materials ever since and as a result of its properties being made clear, we are now considering the possibility of bottom ash utilized as a subgrade and subbase material. We have used the bottom ash taken from one of K-City's incineration facilities in our test.

2. SAMPLE CHARACTERISTICS

2.1. Physical properties

Table 1 specifies the physical properties of the bottom ash sample used in the test. The density of it is 2.49g/cm^3 and in comparison with Toyoura sand's (considered as a standard in Japan) 2.5 g/cm^3(the method and explanation of soil testing, 1991) , it can therefore be stated as having a density of ordinary sand. The combustible material comprised is 28.5%, roughly 22% of which is iron and the rest being glass, porcelain and the like. The ignition loss is 3.4% based on the ignition loss test (3 hours exposure under 600 degrees Celsius).

Table 1 Physical properties

Density of soil particle		2.49 g/cm^3	
Plasticity index(particle passing 425 μm sieve)		NP	
Composition of bottom ash			
Ash	Combustible material		
	Total	Iron	The rest
71.5%	28.5%	6.3%	22.2%
Ignition loss			
Total		Ash	The rest
3.4%		4.2%	1.1%

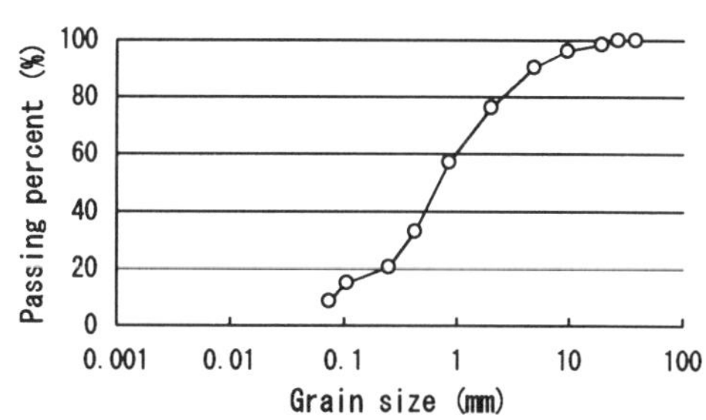

Fig.1 Grain size distribution curve

Since the density of typical bottom ash ranges from 1.8 to 2.6 g/cm^3 and the ignition loss is 0.6–37%(M. HANASHIMA, 1988), the ignition loss of this sample is fairly small and then, we can say that this sample is sufficiently incinerated and the ignition loss is in the lower range values. Furthermore, the moisture content at the time of extraction from the incineration plant is between 35–40%.

Table 2 The result of the dissolution

	Environmental water quality standard	Dissolution value
alkylmercury	ND	ND
Total mercury	0.0005mg/lit.	ND
Cadmium	0.01mg/lit.	ND
Lead	0.1 mg/lit.	0.19mg/lit.
Organophosphorus	ND	ND
Hexavalent chromium	0.05mg/lit.	ND
cyanogen	ND	ND
Polychlorobipheny l	ND	ND

Fig. 1 represents the grain size distribution curve of the bottom ash. The uniformity coefficient U(c) is 11.25 and the curvature coefficient U(c') is 2.22 which means that the sample is well grading.

2.2. Chemical properties

Table 2 lists the results of the dissolution test conducted on the toxic materials included in the sample. It can be seen that except for lead, the amount of all the other toxic matter included satisfies the environmental water quality standard. Clearly the 0.1mg/lit. obtained from lead does not pass the environmental water quality standard for lead which is 0.19mg/lit. These results are roughly the same with those of previous test (A. KIDA, 1988).

It has become clear that if liquid chelate is mixed at 0.3–3% of the total quantity of bottom ash, the dissolution of heavy metal particle can be prevented.

It is also a fact that if this chemical chelate is added on a certain cement mixture, the strength of the cement is unaffected(T. HONGOH, 1994).

The problem of heavy metal dissolution can then be solved with the proper utilization of this chemical chelate.

3. COMPACTED PROPERTIES

Fig. 2 and 3 show the compacted characteristics of bottom ash samples mixed with only cement, and then only with hydrated lime (hereon to be referred to as lime). Fig.2 represents the sample compacted with "E" compaction process (rammer weight is 44.1N, mold diameter is 15cm, it compacted to 3 layers and 92 blows to each layers and its compaction energy is 245cm·N/cm^3) and fig. 3 shows the same sample compacted with compaction process(24.5N , 10 cm, 3 layers and 25 blows, Its compaction energy is 55cm·N/cm^3), both graph in relation to the corresponding dry density and moisture content. Furthermore, tests were conducted on 2, 4, and 6% cement , and then lime samples. With the "E" compaction process, the optimum moisture content and the maximum dry density for the cement mixture samples are 12% and 1.64–1.65g/cm^3, respectively. Those of the lime mixture samples are 18% and 1.58–1.61 g/cm^3. The maximum dry densities of the cement mixture are therefore slightly greater than those of the lime mixture's. On the other hand, with the "A" compaction process, the maximum dry densities for the cement mixture samples corresponding to 2, 4, and 6% mixture quantities are 1.250, 1.277, and 1.293 g/cm^3, which means that the maximum dry density increases with the amount of cement additives.

The 2, 4 and 6% lime mixture samples gave 1.351, 1.343, 1.334g/cm^3 values for maximum dry density which means that the latter decreases even as the amount of lime increases, but the values of the maximum dry density for the lime mixture sample are clearly greater than those for the cement mixture sample.

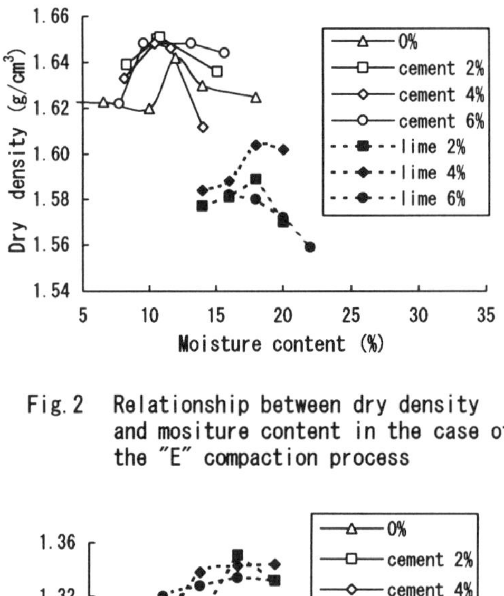

Fig.2 Relationship between dry density and mositure content in the case of the "E" compaction process

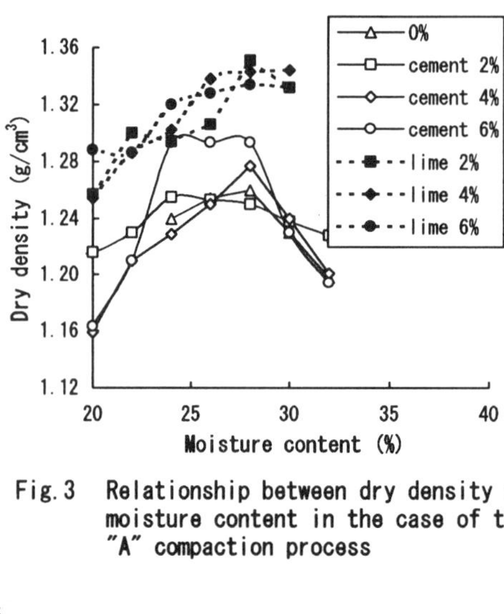

Fig.3 Relationship between dry density and moisture content in the case of the "A" compaction process

Table 3 Dry density and CBR value of bottom

Passing sieve size	2mm	5mm	37.5mm
Dry density(g/cm^3)	1.61	1.61	1.62
CBR	0.7%	0.7%	1.0%

4. CBR VALUE

We also carried out CBR tests to determine if bottom ash can be used as a subgrade material. We used bottom ashes sieved through 2, 5, and 37.5mm sieves and the results are in table 3. The values of CBR for the 2 and 5mm sieve bottom ashes are the same at 0.7% and that for the 37.5 mm sieve bottom ash is 1.0 %, all being small CBR values. It is clear that bottom ash sieved through a 37.5 mm sieve exhibited the greatest CBR value but still is unfit to be used as a subgrade material in its pure state.

4.1. Bottom ash samples stabilized by cement

We then conducted CBR tests on bottom ash mixed with cement at 2, 4, and 6%, which compacted with "E" compaction process. Fig. 4 shows these cement mixture samples graphed against their corresponding moisture contents and CBR values.

The peak values of CBR appear at the moisture content (12%, see fig. 2) for those samples and the CBR values obtained for the 4%, 6% samples are 19.1% and 20.3% which are fairly large values.

Fig. 5 shows the relationship between the amount of cement additives and the CBR values.

The values of the design CBR were obtained from samples that were made into 3 layers and compacted 67 times for each layer. This method of compaction is stated in the Manual for Asphalt Paving (Japan Road Association, 1989) and is used as the standard against which other materials will be gauged to determine whether it is suitable to be used as a subgrade material. For the 2, 4, and 6% samples, 11.9%, 19.1%, and 20.3% CBR values were obtained (see fig. 4) while the corresponding design CBR values are 9.3%, 10.5%, and 15.2%. It is further stated in the same manual that the design CBR values must be higher than 3% value for it to be used as a subgrade material. From the above values we can state that even the 2% sample fulfills this condition and can therefore be used as a subgrade material.

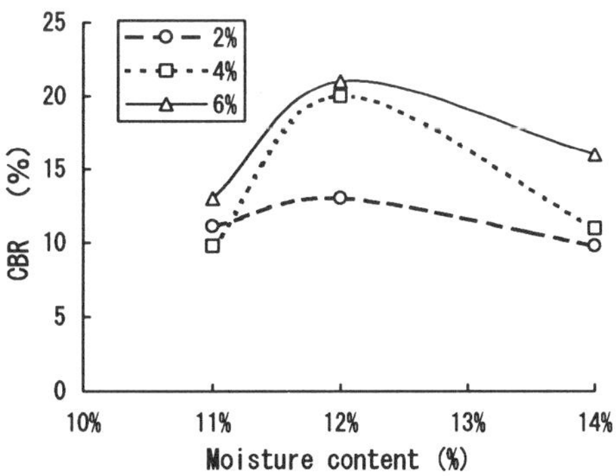

Fig. 4 Relationship between CBR value and moisture content of stabilized bottom ashes mixed with cement

4.2. Bottom ash samples stabilized by lime

Fig. 6 is the graph for the lime–bottom ash samples having an 18% moisture content. With the addition of lime, it can be seen that the design CBR values as well as the CBR values of samples compacted with the "E" compaction process abruptly increases.

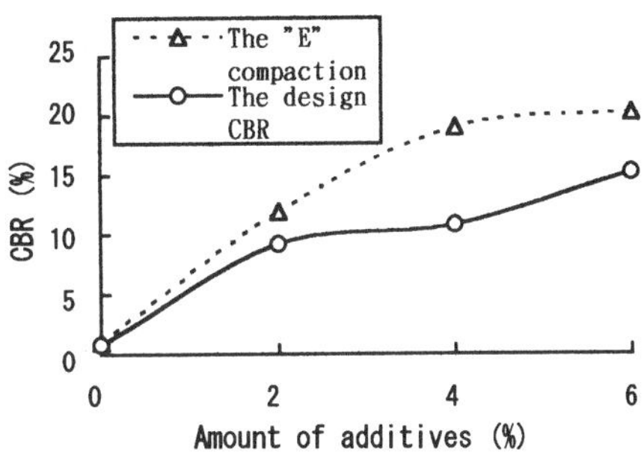

Fig. 5 Relationship between CBR value and amount of cement additives (moisture content = 12%)

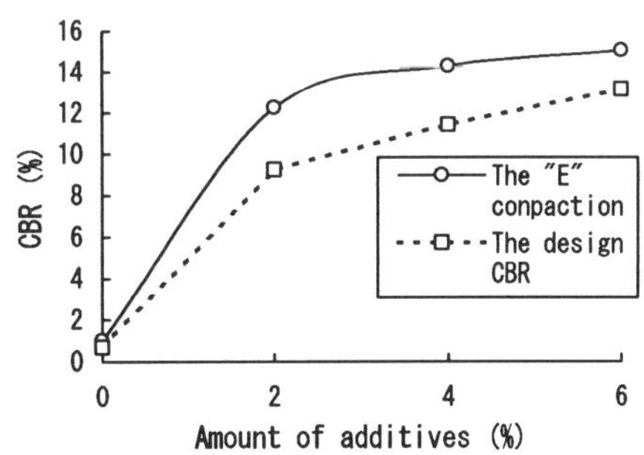

Fig. 6 Relashonships between CBR value and amout of hydrated lime additives (moisture content = 18%)

Furthermore, the design CBR values of the samples are more than 3% values and therefore can be used as a subgrade material. But compared to the CBR values of the cement bottom ash samples, these CBR values are generally lower.

We next tried to check whether bottom ash freshly taken from the incineration facility will be strong enough or not, without regulating its moisture content. We then added lime to this "unregulated" sample and performed CBR tests, the results of which are presented in fig.7. The lime was added at 2, 4, 6, and 8% quantities. It can be seen from the graph that the CBR of the bottom ash itself was a low 0.45% but suddenly increased upon the addition of lime. It can also be seen that even if the quantity of lime is increased beyond 4%, the CBR does not increase anymore. Furthermore, the behavior of the curve is roughly the same with that of the design CBR curve. Even the 2% sample does fulfill the required CBR value, which from the graph is 4%.

5. UNCONFINED COMPRESSIVE STRENGTH

Further to verify the subbase material suitability of bottom ash mixed with cement, and bottom ash mixed with lime, we also performed unconfined compression tests of samples compacted with the "A" compaction process.

5.1. Bottom ash samples stabilized by cement

We conducted the unconfined compression tests on 2, 4, 6, 8, 10, 12, 14, and 18% samples with a 7 day curing period. The optimum moisture content for each of the samples was 28% and had their maximum unconfined compressive strength values. The results are represented in fig. 8. Because bottom ash itself is a mixture of many different materials, the unconfined compressive strength values aren't always constant. From this graph, we can say that as the amount of cement additives is increased, the unconfined compressive strength values of the sample increase in accordance to the locus of a straight line. We have derived the following equation based on these data:

$$q(u) = 0.2035X - 0.193$$

where q(u): unconfined compressive strength(MPa)
X: amount of cement additives(%)

Also, the coefficient of correlation for this equation is 0.84.

Fig.9 shows the corresponding strain and compressive stress for each specific amount of cement additives. The value of the deformation modulus, which is the slope of the curve for the 2% sample is 2.7MPa, 8.8MPa for the 4% sample and then 12.0 MPa for the 6% sample, which means that the deformation modulus increases with the increase in cement additives. For the 12–18% samples, when the strain is between 0 to 2%, the value of

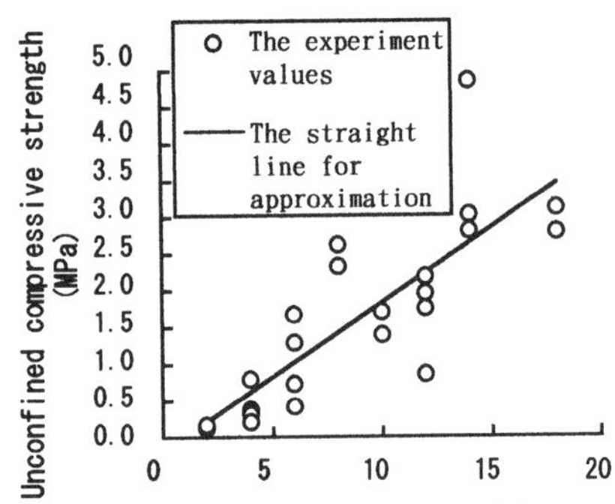

Fig. 8 Relationship between unconfined compressive strength and amout of additive

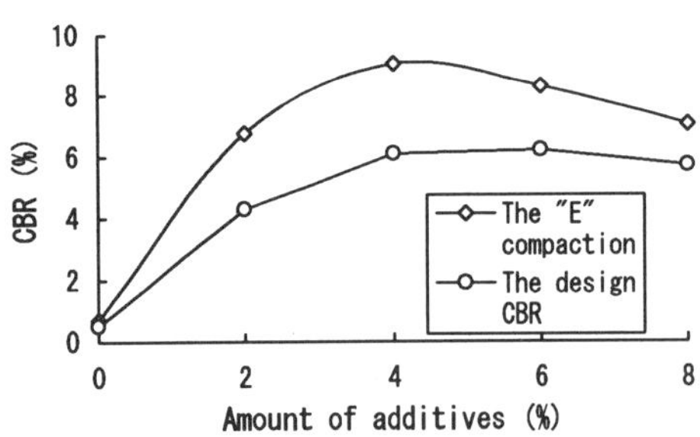

Fig. 7 Relationship between CBR values and amount of hydrated lime additives

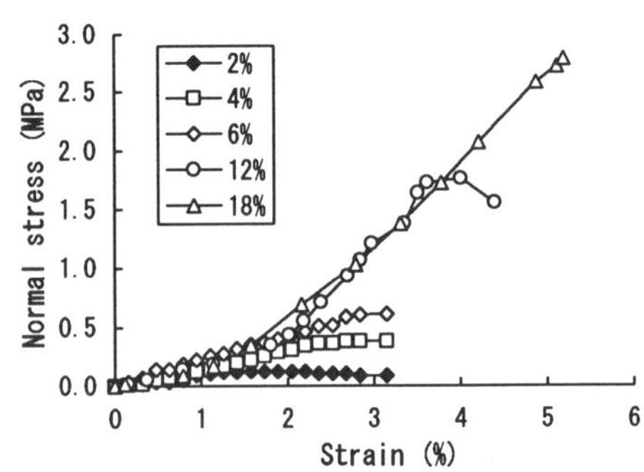

Fig. 9 Relationship between normal stress and strain

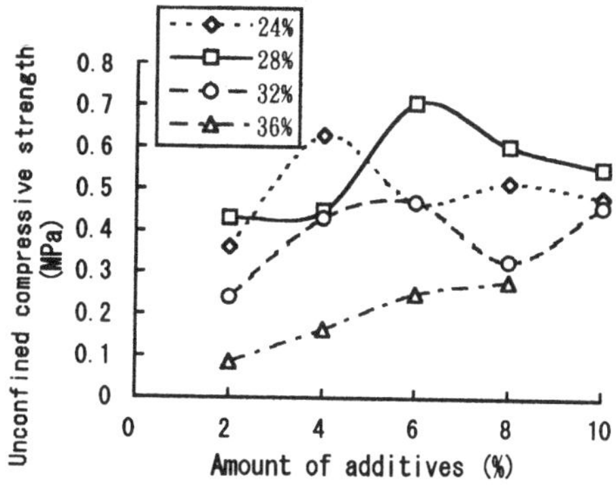

Fig. 10 Relationship between amount of hydrated lime additives and unconfined compressive strength

deformation modulus is 15.0MPa and is 73.0MPa when the value of the strain goes beyond 2%. When 4% cement is added to ordinary soil, the unconfined compressive strength rises to 0.8–2.5 MPa which means that the unconfined compressive strength of the bottom ash-cement samples is greater(The method and explanation of soil testing, 1991).

5.2. Bottom ash samples stabilized by lime

Fig. 10 shows the relationship between the amount of lime additives and unconfined compressive strength. The lime was added at 2, 4, 6, 8, and 10% proportions with a curing period of 10 days. Since the sample would not always yield a maximum unconfined compressive strength even if the moisture content was adjusted to its optimum value, we took note of the value of the moisture content for each sample. As can be seen in the curve, for the 2–6% samples the unconfined compressive strength increased but failed to do so for the 6–10% samples.

Even if the moisture content was adjusted to its natural moisture content value of 36%, the unconfined compressive strength likewise increased for the 2–6% samples and again did not increase for the 6–8% samples but these samples of 36% moisture content are all below 0.3MPa. The greatest unconfined compressive strength obtained is then 0.7 MPa for the 6% sample with an optimum moisture content of 28%.

We can now say that the unconfined compressive strength is not always directly proportional to the amount of lime additives.

6. CONCLUSION

The main results of this paper may be summerized as follows:

1) Lime or cement added to bottom ash abruptly enhances the CBR of the latter and can therefore be used as a subgrade material. Also, bottom ash obtained directly from the incineration facility on its natural, moisture content is dynamically capable of being used as a subgrade material if lime is added at a 2% proportion.

2) It is stated in the asphalt paving manual that for a material to be used as a subbase course, it must have an unconfined compressive strength of 0.98MPa and 2.9MPa if it is to be used as a base course. Bottom ash mixed with cement at 8% proportion and a 7 day curing period has an unconfined compressive strength of above 0.98MPa and if the cement proportion is 18%, the unconfined compressive strength is above 2.9MPa, which therefore makes it suitable for use as a subbase course or base course. 3) If lime within a 10% proportion is added to bottom ash, the maximum obtainable unconfined compressive strength is 0.7MPa and even if the proportion of lime is increased beyond that, the unconfined compressive strength will cease to increase. Therefore bottom ash mixed with lime is difficult to be used as a base course material.

4) The stabilization of bottom ash with the use of cement additives has become clear but because the incombustible portion comprises up to 4.4%, how its strength will change over a long period of time needs to be scrutinized. There is also a need for long term research regarding the dissolution possibility of the poisonous compound lead.

REFERENCES

Japan road association (1988) : Manual for asphalt pavement, pp. 8

The Japanese society of soil mechanics and foundation engineering (1991) : the method and explanation of soil testing, pp. 46

The Japanese society of soil mechanics and foundation engineering(1991) :The method and explanation of soil testing, pp. 244

Akiko KIDA and Yukio NOMA(1988) :Metal and anion leaching properties by column extraction method, Vol. 41, Journal of Japan waste management association, No. 164, pp. 223–228

Masataka HANASHIMA, Yasushi MATUFUJI, Syuji NAGANO, Ryuji YANASE and Tokio HIRATA (1988):Soil mechanics characteristics of solid waste, Journal of Japan waste management association, Vol. 41, No. 163, pp. 148–167

Teruo HONGOH, Toshitaka YAMASAKI, Mamoru TANAKA and Takuma ORIHIRA (1994): Manufacturing of interlocking block from incineration residue, Journal of Japan waste management association, Vol. 47, No. 227, pp. 227–281

Environmental Geotechnics, Kamon (ed.) © 1996 Balkema, Rotterdam. ISBN 90 5410 848 7

Studies on reclamation using slag red mud mixture

T. Mitsunari
Toda Corporation, Japan

M. Hanada
Yoshizawa Lime Industry Co., Ltd, Japan

ABSTRACT: The paper describes the results of experiments conducted for the purpose of estimating the shear strength of slag red mud mixture. The relation between strength and the cement water ratio of slag red mud mixture is shown, and from it, the shear strength of its mixture is predicted on the basis of the determination of its self-weight consolidation history via the slice method.

1 INTRODUCTION

In the reclamation management of alumina refinement waste known as red mud(RM), two problems are faced: to predict its available storage, and to improve an extremely soft RM layer. An approach to solve the former problem was presented last year(Mitsunari, 1995). To use the RM pond as land, requires soil its improvement. With conventional practices, RM had to be left intact for a long period of time. Further, its improvement incurred tremendous cost. This study focuses on one of the preparatory treatments of red mud slurry with blast furnance slag, in which the slag red mud mixture (SRM) is deposited in a pond enclosed by a dike. SRM is consolidated by self-weight and hardens via pozzolanic reaction. This method therefore equivalent to soil improvement, and differs from the usual practice with respect to its route toward red mud land.

This paper describes the experimental results of SRM and makes a prediction of the shear strength of its layer if applied in the RM pond.

2 PROPERTIES OF RED MUD SLURRY

Red mud is discharged at a high alkali slurry state of low concentration of 30% to 40%, and its equilibrium concentration registers about 45%. Its specific gravity is 3.3, and its red color due to the bauxite's ferricoxide that it contains, has led to the substance being called "red mud". The diameter of its particles is almost less than 5 μ. A plasticity index is about 20% with liquidity index near to one. RM is extremely soft due to the mutual contact resistance of particles and lacks the strength of soil skeleton.

Table 1 Chemical analysis of reclamation materials

	SiO_2	Al_2O_3	Fe_2O_3	TiO_2	Na_2O	CaO	MaO
Virgin red mud	16.2	20.4	32.0	5.2	9.3	4.9	–
Blast furnance slag	35	15	1.5	–	–	41.5	5.5

The slag in use is a sandy-glassy material produced by refrigerating blast furnance slag with water and has hydraulic properties. The chemical composition of the two mixing materials is shown in Table 1.

3 LABORATORY TESTING

Mold tests were carried out with a slag red mud mixture (SRM) from high alkali red mud slurry which was cured in the mold, in the manner of concrete test specimens until their predeterminated times, after which a series of soil tests were undertaken with the samples. Field testing was also done at Tomakomai in Hokkaido. Results and conditions of the determination of slag content (s), water content (w), cement water ratio (c_w), age (t), and unconfined compressive strength (q_u), are shown in Table 2.

Table 2 Experimental conditions and results

<table>
<tr><th></th><th colspan="2">Grain size of slag</th><th>Slag content s (%)</th><th>Water content w (%)</th><th>Cement water ratio c_w</th><th>Age t (months)</th><th colspan="2">Compressive strength q_u (kPa)</th><th colspan="2">Symbols shown in Fig.2</th></tr>
<tr><td rowspan="9">Laboratory tests</td><td colspan="2" rowspan="3"></td><td>10</td><td>60</td><td>0.12</td><td>6</td><td colspan="2">140</td><td colspan="2">①</td></tr>
<tr><td>20</td><td>52</td><td>0.27</td><td>6</td><td>330</td><td>1200</td><td>②</td><td>②c</td></tr>
<tr><td>30</td><td>47</td><td>0.45</td><td>6</td><td colspan="2">730</td><td colspan="2">③</td></tr>
<tr><td rowspan="6">Less than 10 mm</td><td rowspan="6">Less than 0.6 mm</td><td rowspan="2">10</td><td rowspan="2">111</td><td rowspan="2">0.06</td><td>6</td><td>0</td><td>60</td><td rowspan="2">④</td><td rowspan="2">⑤</td></tr>
<tr><td>17</td><td>30</td><td>170</td></tr>
<tr><td rowspan="2">18.2</td><td rowspan="2">100</td><td rowspan="2">0.13</td><td>6</td><td>140</td><td>260</td><td rowspan="2">⑥</td><td rowspan="2">⑦</td></tr>
<tr><td>17</td><td>240</td><td>390</td></tr>
<tr><td rowspan="2">25</td><td rowspan="2">91.7</td><td rowspan="2">0.19</td><td>6</td><td>270</td><td>360</td><td rowspan="2">⑧</td><td rowspan="2">⑨</td></tr>
<tr><td>17</td><td>510</td><td>1770</td></tr>
<tr><td rowspan="5">Field test</td><td colspan="2" rowspan="5">Less than 10 mm</td><td rowspan="5">7.6</td><td rowspan="5">70 ~ 80</td><td rowspan="5">0.06 ~ 0.07</td><td>6</td><td colspan="2">0 ~ 20</td><td colspan="2" rowspan="5">⑩</td></tr>
<tr><td>12</td><td colspan="2">10 ~ 80</td></tr>
<tr><td>17</td><td colspan="2">10 ~ 120</td></tr>
<tr><td>23</td><td colspan="2">10 ~ 170</td></tr>
<tr><td>37</td><td colspan="2">20 ~ 180</td></tr>
</table>

The relation between age and unconfined compressive strength of SRM is shown in Fig. 1. The greater the slag content and the smaller the curing water content, the higher the strength of the samples. In the case of high temperature (:20℃) during curing, the sample strength increased early (cf. ②^{c} sample). Owing to pozzolanic reaction, the long-term strength of SRM has been revealed.

Fig. 2 shows the unconfined compressive strength of samples with various degrees of slag content (s [%]), or the dry weight percentage of slag to that of the samples. A s value of ten percent is sufficient to attain the strength of reclaimed ground equal to 100 kPa.

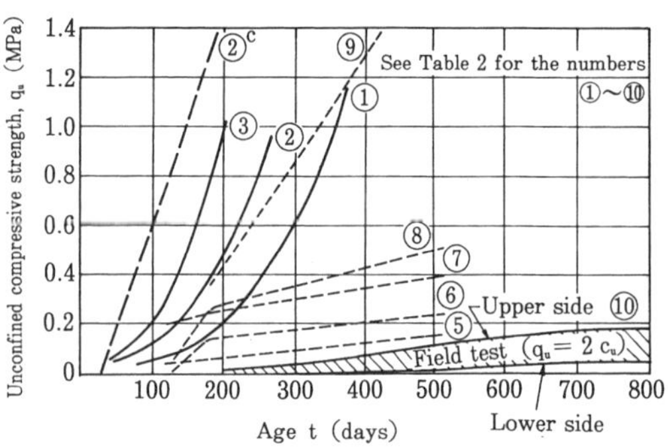

Fig. 1 Relation between age and compressive strength of SRM

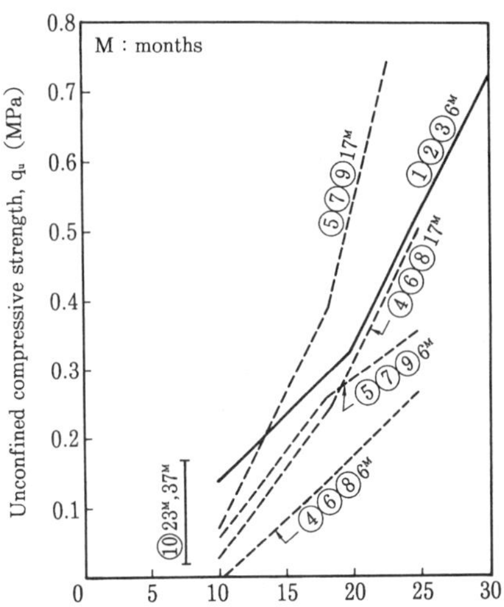

Fig. 2 Relation between slag content and strength

4 FIELD TESTING

SRM slurry was sedimented into a pond having a depth of 3m and filling 240 m³ of its volume, as shown in Photo 1. The concentration of red mud slurry was 40% with a pH value of 12 to 13. Slag content of grain size less than 10 mm was 7.6%. Grain size distribution of the reclamation material is shown in Fig. 3. The surface of the pond was filled by soil 1m thick, then cured for 3 years. In situ vane shear testing was conducted at predetermined ages.

Fig. 4 shows the observed time-settlement curve after filling the surface. Consolidation of the layer finished within 2 months and its settlement became about 1m. The water content of the layer was 139 % at the outset, while that of the consolidated layer decreased from 80% to 70%, and thereafter remained constant.

Photo 1 Deposition of SRM slurry

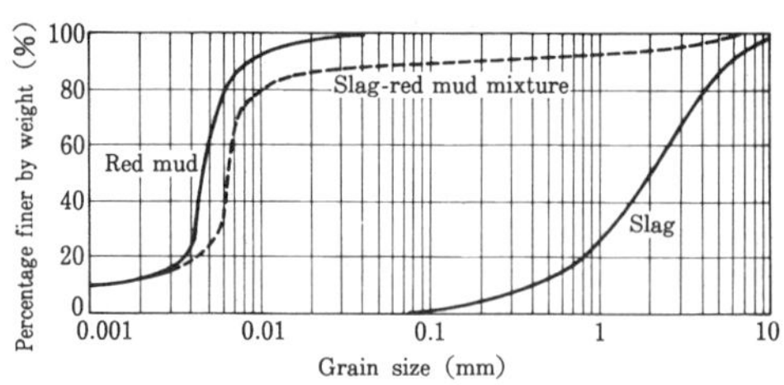

Fig. 3 Grain size distribution of materials

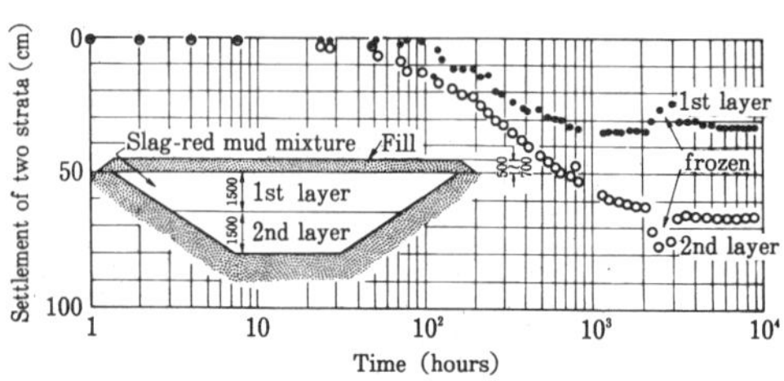

Fig. 4 Observed time settlement of two strata of the pond in the field test

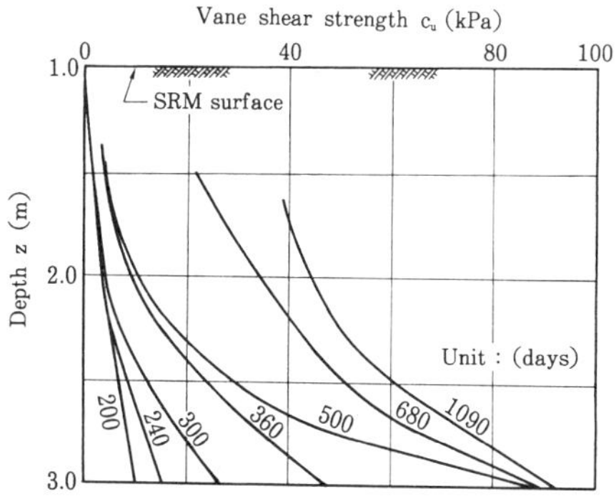

Fig. 5 Shear strength distribution

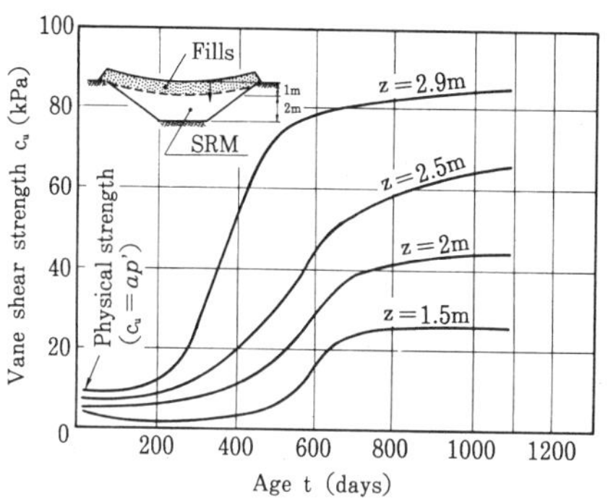

Fig. 6 SRM shear strength curves

The modified distribution curves of vane shear strength are presented in Fig. 5, and show that the strength of the SRM layer increased not only at the lower side, but also at the upper side of the pond. Based on the results, the shear strength revelation curves as shown in Fig. 6 were calculated from $q_u = 2c_u$, as previously illustrated in Fig. 1.

5 RELATION BETWEEN STRENGTH AND CEMENT-WATER RATIO

The chemical oomposition of slag is almost the same of cement ; thus the ratio (x) of slag lime content to that of cement can be defined, with the cement content of slag assumed to be xs. The ratio (c_w) of xs to water content (w) during curing can be called "cement-water ratio" as in the case of concrete. Therefore, cement-water ratio (c_w) is expressed by the Eq. (1):

$$c_w = xs/w \tag{1}$$

c_w was previously shown in Table 1. These data and the data of red mud cement mortar (Yano, 1974) were plotted on a logarithmic diagram, as shown in Fig. 7.

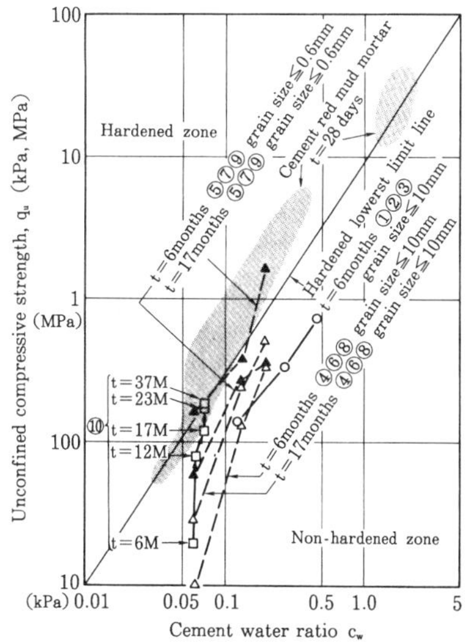

Fig. 7 Relationship of unconfined compressive strength and cement water ratio

Final long term-strength and c_w show a good correlation. A border line with hardened and non-hardened zones could be drawn as shown in Fig. 7. The strength of the sample of one age lies parallel with the border line. Given a longer period, pozzolanic reaction causes the strength to increase toward the hardened zone.

The curing temperature and grain size of slag affect its speed in reaching the hardened zone. The results of field testing, in which c_w was less than 0.1, showed that the greater the age, the closer it came to the border line of the hardened zone. In the case of c_w exceeding 1.0, its strength became nearly that of concrete. Even with c_w value very small (0.06), a strength of 100 kPa was obtained. It is concluded that final SRM strength has a constant relation with c_w, and its factors are, slag content (s), cement water ratio (c_w), and age (t).

6 PREDICTION OF SHEAR STRENGTH OF SRM POND

If water content of the SRM layer increasing in thickness with time as shown in Fig. 8 could be predicted, final shear strength could also be predicted through the q_u-c_w relation shown in Fig. 7. Thus, c_w is, so to speak, an index linking physical consolidation with chemical reaction. When SRM is secularly sedimented in a pond, hardened strength will be revealed according to the time history of the SRM layer with its water content. Self-weight consolidation of red mud slurry progresses sooner at the lower side of the pond.

The red mud layer reaches an almost stationary state except for a depth of 2m or 3m from its surface (Mitsunari, 1995).

At the early stage of self-weight consolidation, physical phenomena can be observed owing to the late effect of slag; thus, an increase of the c_w value due to consolidation is not prevented. After the SRM layer is fully consolidated, its strength will develop.

The water content of SRM layer has been determined using the slice method, as suggested by the author. There is no great difference in distribution of specific volume using this method as shown in Fig. 10 when data obtained from oedometer tests are used.

6.1 Slice method

Self-weight consolidation of a simultaneously sedimented slice progresses until the next slice sediments according to the relation between sediment depth and time, as shown in Fig. 8.

The effective stress of the *i* th slice at the *n* th stage sedimentation can be estimated from the expression:

$$ {}_np_i = 0.5({}_{n-1}\gamma'_{ti} \cdot {}_{n-1}\Delta H_i + \gamma'_{t0}\Delta H_0) + \sum_{i=2}^{n-1} {}_{n-1}\gamma_{ti} \cdot {}_{n-1}\Delta H_i \qquad (2) $$

where γ_t and ΔH are the unit weight and thickness of the slice, respectively; suffix means the number of slices, and the zero refers to the original state.

c_v values are obtained from log c_v-log *p* relation, and drainage length is determined by the equivalent-depth method using c_v value and the thickness of each slice. Time factor (T_v), with respect to degree of consolidation of sediment (U_n) can be calculated by the Eq. (3):

$$ T_v = \Delta t c_{v0} / (H_e/2)^2 \qquad (3) $$

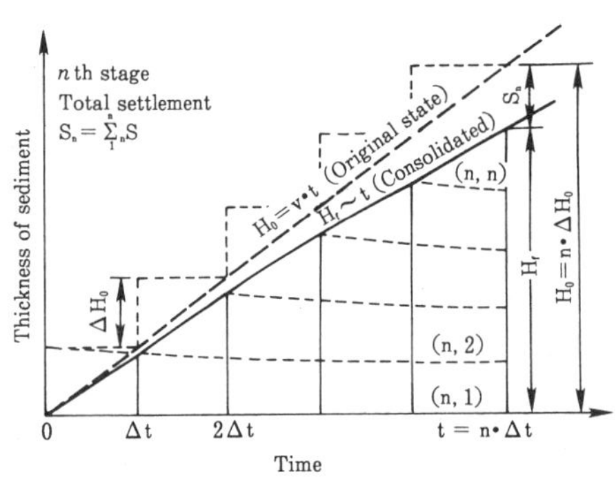

Fig. 8 Relation between sediment and time

where Δt : the period until the subsequent slice sediments, as shown in Fig. 8; c_{v0} : standard value of c_v ; H_e = equivalent depth.

The void ratio of *i* th slice at *n* th stage can be obtained from the Eq. (4) :

$$ {}_ne_i = {}_{n-1}e_i - U_n\{{}_{n-1}e_i - {}_n(e_f)_i\} \qquad (4) $$

where i=1, n : when i=n, ${}_{n-1}e_i = e_0$.

The sediment settlement (${}_nS$) for Δt at the *n* th stage will then be given by the Eq. (5) :

$$ {}_nS = \sum_{i=1}^{n}\left(\frac{{}_{n-1}e_i - {}_ne_i}{1 + {}_{n-1}e_i} \cdot {}_{n-1}H_i\right) \qquad (5) $$

Here, i=1, n : when i=n, ${}_{n-1}H_i = \Delta H_0$, ΔH_0=the slice thickness at original state of red mud.

Total settlement is expressed by the Eq. (6) :

$$ S_n = \sum_i^n {}_nS \qquad (6) $$

The total thickness of red mud layer after settling is :

$$ H_{f(n)} = {}_n\Delta H_0 - S_n \qquad (7) $$

When deposition ceases, the degree of consolidation U is expressed by the Eq. (8) :

$$ U = S_n / \Sigma(s_{fi})_n \qquad (8) $$

where $(s_{fi})_n$: the final settlement of the *i* slice at the *n* th stage.

6.2 Prediction of SRM layer shear strength

RM slurry, not SRM, was deposited into a pond enclosed by a dike 15m high and with base area of 5,500 m², as shown in Photo 2 and Fig. 9, in the same district where the field test was conducted.

Photo 2 RM pond enclosed by a dike at Tomakomai

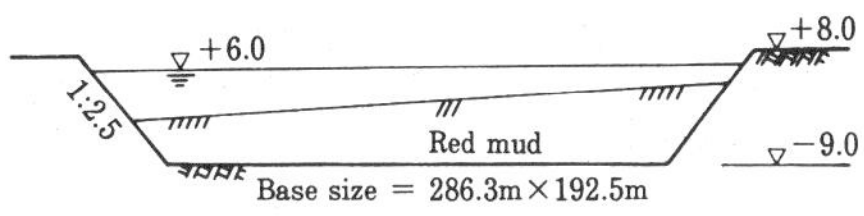

Fig. 9 Size of red mud pond

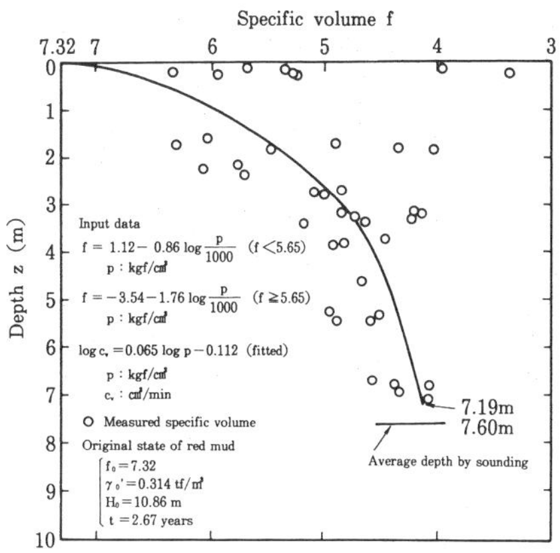

Fig. 10 Predicted specific volume distribution of red mud layer by slice method

If SRM is deposited into the same pond, SRM layer shear strength can be predicted based on the following conditions.

① Original state of SRM

Slag content : $s = 7.6\%$
Specific gravity : G $= 3.17$ ($G_{RM} = 3.3$)
Original water content : $w_0 = 177\%$ ($w_{0RM} = 191\%$)
Original Specific volume : $f_0 = 6.60$ ($f_{0RM} = 7.12$)
Submerged unit weight : $\gamma_0' = 0.329$ tf/m³
($\gamma'_{0RM} = 0.314$ tf/m³)

② f $-\log p$ and $\log c_v - \log p$

$$f = 1.16 - 0.84 \log \frac{p}{1000} \qquad p \text{ ; kgf/cm}^2$$

$$\log c_v = 0.665 \log p - 0.112 \qquad c_v \text{ ; cm}^2\text{/min}$$

③ SRM storage speed

SRM slurry having a slag content of $s = 7.6\%$ can be converted by both original water content and amount of dry weight of red mud as shown in Table 3.

Table 3 SRM slurry storage speed

Time (year)	Dry weight of RM (t)	Deposit volume V (m³)	Rate of deposition v (m³/year)	Original depth H_0 (m)
0.67	74,000	167,000	249,000	2.86
2.67	337,000	760,000	297,000	10.98
4.00	553,000	1,249,000	367,000	16.27

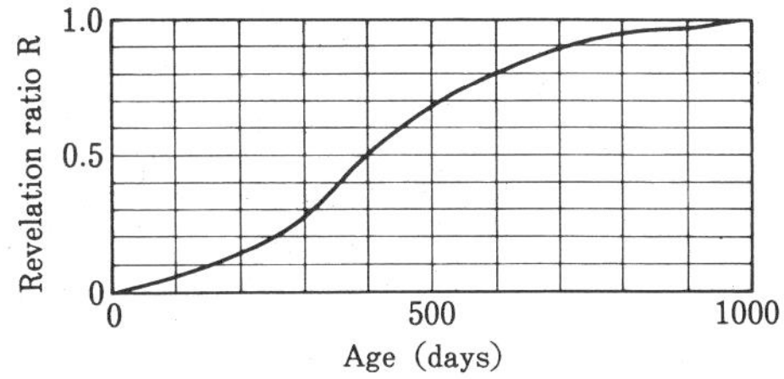

Fig. 11 Revelation ratio of SRM

④ Shear strength revelation curve

The nomalized curve can be expressed as shown Fig. 11 on the basis of the shear strength revelation curves of SRM as presented in Fig. 6.

The calculated sedimentation process of SRM via the slice method according to the above conditions is shown in Fig. 12, which indicates the consolidation history of each slice simultaneously sedimented.

Based on this result, the shear strength was predicted by the following procedure.

i) Calculation of the mean void ratio (e_m) of each slice:
$e_m = (e_i + e_t)/2$, where e_i : initial value after Δt years, e_t : after t years. $w_m = e_m / G$, where w_m : water content corresponding to e_m.

ii) Calculation of c_w : $c_w = sx / w_m$, where $s = 0.076$, $x = 0.4$

iii) Determination of the final strength (q_{uf}) using the border line shown in Fig. 7.

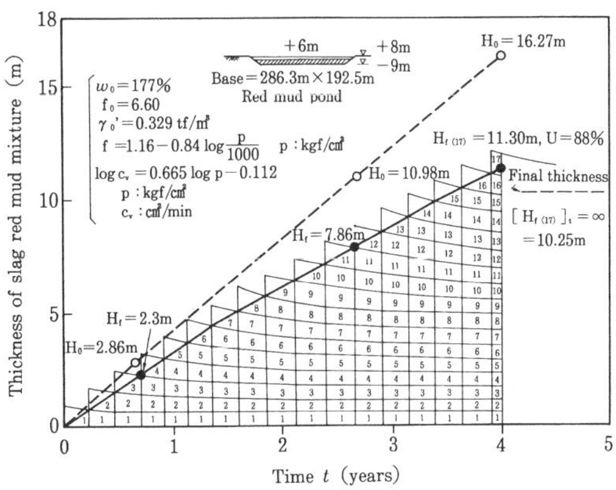

Fig. 12 Predicted thickness of SRM with no drainage

iv) Determination of the revealed strength (q_{ut}) of t_i age after reading both sedimentation age (t) in Fig. 12 and revelation ratio(R) in Fig. 11: $q_{ut} = R\, q_{uf}$, where $t_i = t - \Delta t$.

v) Determination of the consolidated shear strength of SRM, q_{u0} : $q_{u0} = \alpha p_i$, where α : rate of strength increases, p_i : effective stress of ith slice, which can be figured from the f-log p relation.

vi) Computation of the shear strength of each slice using the following expression, q_u : $q_u = q_{u0} + q_{ut}$

6.3 *Results of establishing shear strength*

The predicted shear strength of the SRM layer is illustrated in Fig. 13 after $t = 2.67$ at 4.0 years and when deposition ceases after 4.0 years. In fact, RM slurry was deposited in the pond shown in Photo 2 according to the conditions of RM original data shown previously in ① and the storage speed of equivalent slurry volume of dry weight of RM as shown in Table 3.

Depth sounding, sampling and in situ cone penetration tests were carried out after 2.67 years. These data and calculated curves of RM layer shear strength via the slice method also appear in Fig. 13, showing good agreement in these distribution curves. The pond had no seepage from its bottom. In case of the SRM layer, pozzolanic reaction developed for self-weight consolidation periods, and strength revealed from the bottom of the pond after curing more than 100 days. When deposition ceases after 4 years and the pond is left as is, strength revelation will spring up since time passes at a constant cement-water ratio, and finally in the stationary state after consoldation, strength distribution will show a concave function curve. However, the surface of the SRM pond is in a state of suspension since it can not be consolidated by self-weight, thus no strength is detected. For this reason, it is not easier to fill the surface of the pond than in the case of RM alone.

Once filled with soil, even if the slag content is small in value (7.6%), it has a shear strength of nearly 100 kPa. This phenomenon makes preparatory soil improvement of the RM pond possible. It is suggested that if the slag content (s), is increased, stronger ground can be obtained.

The foregoing represents studies on handling high alkali red mud slurry and, therefore, with a view toward influence on the enviroment, SRM slurry should be deposited into an enclosed pond. The pozzolanic reaction is accelerated by alkali state and makes the needle-like particles (ettringite) and crystal needles of calcium aluminate hydrate. If red mud slurry neutralized by sulfuric acid is

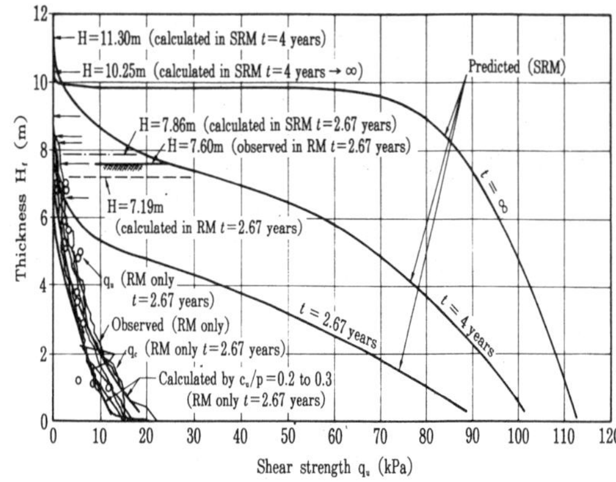

Fig. 13 Predicted shear strength distribution of SRM

used, it was testified that the above hardening failed to occur for two years in SRM.

7 CONCLUSION

The results of both laboratory and field tests indicated that the unconfined shear strength of slag red mud mixture (SRM) is dependent on its slag content (s), its cement water ratio (c_w), and its age (t), and even with a c_w value of 0.06, an in situ shear strength of 100 kPa was obtained after three years at Tomakomai in Hokkaido.

Since the progress of self-weight consolidation of red mud layer has been verified by a symplified procedure (the slice method), the shear strength of SRM layer was predicted based on the consolidation history of each slice and of the $q_u - c_w$ relation in the same pond of RM. Strength distribution displays a concave function curve toward its value of nearly 100 kPa after 2.67 years. This strength equals seven times that of the RM layer, which makes preparatory soil improvement of the RM pond possible.

Slag is appropriate to red mud as a reclamation material.

REFERENCES

Mitsunari, T. 1995: Self-weight consolidation settlement of red mud layer. *Proc. of the International Sympsium on Compression and Consolidation of Clayey Soils* IS-Hiroshima 95/Japan/10−12 May 1995 Vol.1 pp. 715−718.

Yano, M. 1974: Studies on Strenght of Cement-Red Mud Mortar. *Toda Technical Research Report* Vol.1 (in Japanese).

Environmental Geotechnics, Kamon (ed.) © 1996 Balkema, Rotterdam. ISBN 90 5410 848 7

Effects of organic decomposition on paper mill sludges used as landfill cover material

Horace K. Moo-Young
Morgan State University, Baltimore, Md., USA

Thomas F. Zimmie
Rensselaer Polytechnic Institute, Troy, N.Y., USA

ABSTRACT: Paper mill sludges are being substituted as the hydraulic barrier in landfill covers. In spite of high water contents and high compressibility, paper sludges can provide an adequate barrier layer. Paper mill sludges are composed of 40-60% organics. A linear relationship exists between the organic content and the specific gravity of paper sludges. A linear relationship exists between the consolidation parameters and the organic content. The magnitude of permeability decreased with time as the organic decomposition. Computerized respirometer tests were conducted on various sludges to determine the effects of organic decomposition.

1. INTRODUCTION

Compacted clays have been widely used as the impermeable barrier layer in landfill covers. However, when an abundant source of clay is not readily available, the cost of landfill closure is greatly increased. The elevated cost of waste disposal may be reduced by the use of unconventional material in the construction of landfills. The high price of disposal has sparked interest in the development of alternative uses for waste sludges (paper mill sludges and water treatment plant sludges). Numerous studies have been conducted to find alternative uses for paper mill sludges (NCASI, 1992a; NCASI, 1992b; Kraus and Benson, 1994; Floess et al., 1995; Moo-Young 1995; Moo-Young and Zimmie, 1995; Zimmie and Moo-Young, 1995; Moo-Young, 1992).

Paper mill sludges, in spite of high water contents and low solids contents in comparison to clays, can be compacted to low permeabilities and can substitute for clays in landfill covers. Moreover, since paper mill sludges are considered a waste product, they are provided to the landfill owner at little or no cost. This may reduce the cost of construction by $20,000 to $50,000 per acre. Since 1975, paper mill sludges have been used to cap landfills in Wisconsin (Stoffel and Ham, 1979; Pepin, 1985).

Seven sludges were used in this study. Sludge A is a waste water treatment plant sludge from a deinking recycling paper mill. Sludge B is a blended sludge from a waste water treatment plant that receives its effluent from a recycling paper mill and the neighboring community. Sludge C is a blended sludge from an integrated paper mill. Sludge C was mined from a sludge monofill that was in operation since 1973. Samples were collected from different sections of the monofill to represent different sludge ages: one week (C1), 2-4 years (C2), and 10-14 years (C3). Sludge D is a primary waste water treatment plant sludge from a recycling paper mill. Sludge E is a primary waste water treatment plant sludge from a non-integrated paper mill that uses titanium oxide as the primary filler. The purpose of this paper is to describe the effects of organic decomposition on paper mill sludges used in landfill covers.

2. ORGANIC CONTENT

Paper mill sludges are composed of kaolinite clay, organic fibers and tissue, and water. Organic content tests were performed on the sludges according to American Society for Testing and

Materials (ASTM) procedure D2974 method C for geotechnical classification. A muffle furnace was used to burn off the organics at a temperature of 440° C. At this temperature, the organic matter is burned off, and the mineral constituent, kaolinite or titanium oxide, forms an ash. Sludges with greater organic contents tend to have higher water contents and void ratios, and will be more compressible (LaPlante, 1993). This relationship was also found to be true for peats (MacFarlane, 1966).

The organic content results are summarized in Table 1. Paper mill sludges are characterized by a high organic content. In comparison, typical clays have organic contents of less than 1% (Lambe and Whitman, 1969). Table 1 shows that for sludge C, there is a decrease in organic content with an increase in age. This decrease in organic content results from the biological decomposition of the organic fibers and tissue with time.

3. SPECIFIC GRAVITY

Specific gravity tests were performed on the sludges according to ASTM procedure D854. Slight modifications were made to apply the procedure to paper sludge (Moo-Young, 1995; LaPlante, 1993). Upon drying, the sludge samples formed flocs, developed a coarse texture, and were not easily pulverized. Thus, the sludge samples were taken at their natural water content and soaked in water for an hour before pulverization. This behavior was noted by Wang et al. (1991) for a water plant sludge and by Feustel and Byers (1930) for various peat types. Specific gravity results are summarized in Table 1.

The data in Table 1 indicates a direct relationship between specific gravity and organic content. Paper mill sludges containing lower percentages of organics have higher percentages of kaolinite. Since the kaolinite mineral has a specific gravity of 2.60 to 2.65, and the fibers are estimated to have a specific gravity of 1.5, a higher kaolinite content in the sludge will increase the overall specific gravity. Figure 1 shows the relationship between the specific gravity and organic content for all sludges. As the organic content decreases, the clay content increases, and the specific gravity increases.

Table 1. Ranges of Organic Content and Specific Gravity for Sludges

Sludge	Organic Content (%)	Specific Gravity
A	45-50	1.88-1.96
B	55-70	1.83-1.85
C1	54-56	1.80-1.84
C2	40-50	1.90-1.93
C3	30-40	1.96-2.15
D	44	1.93-1.95
E	35-40	1.96-2.08

4. CONSOLIDATION CHARACTERISTICS

One dimensional consolidation tests were performed on the seven sludges to obtain a relationship between compressibility and organic content using ASTM procedure D2435 methods A and B. Twenty-two consolidation tests were conducted to obtain a relationship between organic content and compressibility. The sludges were tested at an initial water content ranging from 109% to 224%. The correlation coefficients between the organic content and the compression index and the organic content and coefficient of compressibility are 0.47 and 0.53, respectively, which indicate that there are positive linear correlations between the variables.

Figure 2 plots the organic content versus the compression index for the various sludges. The solid line represents the first order linear regression model that passes through the origin. Figure 2 also shows the ±95% prediction interval which illustrates upper and lower range values with respect to the first order linear regression on which a future observed response for the organic content with respect to the compression index may be predicted.

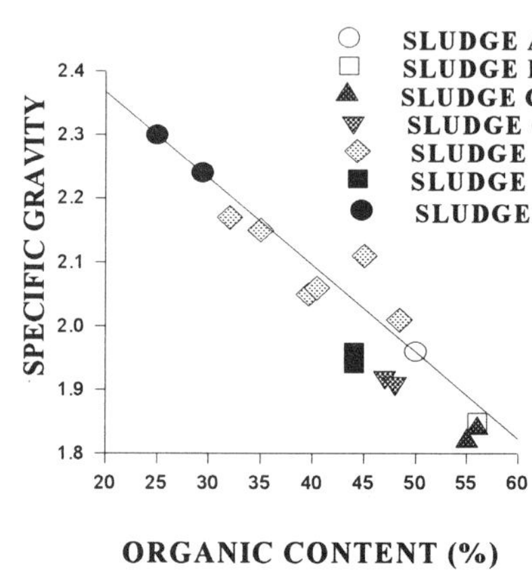

Figure 1. Specific Gravity and Organic Content Relationship

The relationship between the compression index, C_c, and the organic content, O_c (%), from Figure 2 is as follows:

$$C_c = 0.027\ O_c \qquad (4.1)$$

Figure 3 plots the coefficient of compressibility, a_v, and the organic content, O_c, for the various sludges. (In this study, the coefficient of compressibility represents the average slope for the coefficient of compressibility along the arithmetic plot of the void ratio and vertical pressure plot between pressures of 24 kPa (0.25 tsf) and 96 kPa (1 tsf).) The coefficient of compressibility and the organic content have a first order linear regression (solid line) which passes through the origin. The upper and lower 95% prediction intervals (dotted lines) are also plotted. The relationship for the coefficient of compressibility, a_v (m^2/kN) and the organic content, O_c (%), is as follows:

$$a_v = 0.000274\ O_c \qquad (4.2)$$

There is little or no data in the literature relating the organic content to compressibility for sludges.

5. PERMEABILITY CHARACTERISTICS

The flexible wall permeameter was used to determine the one dimensional permeability of various remolded and Shelby tube paper sludge samples following the American Society of Testing and Materials procedure D5084-90. Figure 4 illustrates the relationship between the organic content and permeability plotted on a semi-logarithmic scale. The regression correlation coefficient between the organic content and permeability is 1.19 that indicates a linear association between the organic content and permeability with a positive slope when plotted on a semilogarithmic scale. The first order regression model for the average organic content, O_c, (%) and permeability, k (cm/sec), relationship is as follows:

$$O_c = -5.5 \log k \qquad (5.1)$$

Figure 4 also shows the ± 95% prediction intervals that illustrate an upper and lower range of values with respect to the first order regression model on which a future observed response for the organic content with respect to the permeability may be predicted.

The organic contents ranged from 25% to 73%. For the average permeability line, the hydraulic conductivity ranged from about 2×10^{-8} to 2×10^{-7} cm/sec. The upper prediction interval ranges from 4×10^{-7} to 1×10^{-5} cm/sec, and the lower prediction interval ranges from 1×10^{-9} to 2×10^{-8} cm/sec. As the organic content decreases, there is a decrease in permeability. Obviously, this permeability reduction due to biodegradation has important practical implications relative to landfill cover design.

To evaluate the effect of organic decomposition on permeability, sludge C1 (1 week), sludge C2 (2-4 years), and sludge C3 (10-14 years) were remolded at a water content of 250%, and the permeability of the sludges was measured at an effective stress of

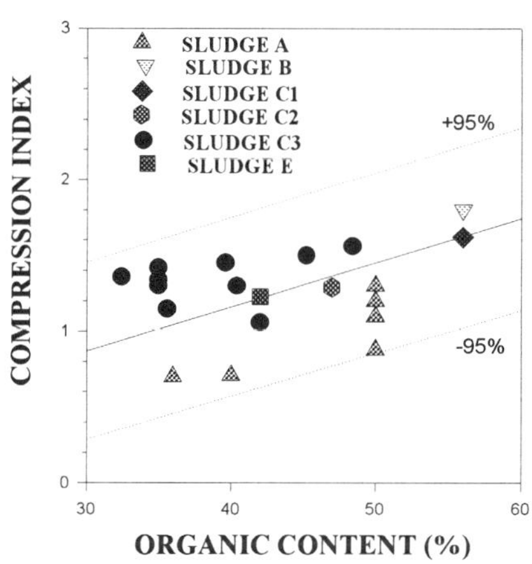

Figure 2. Compression Index and Organic Content Relationship

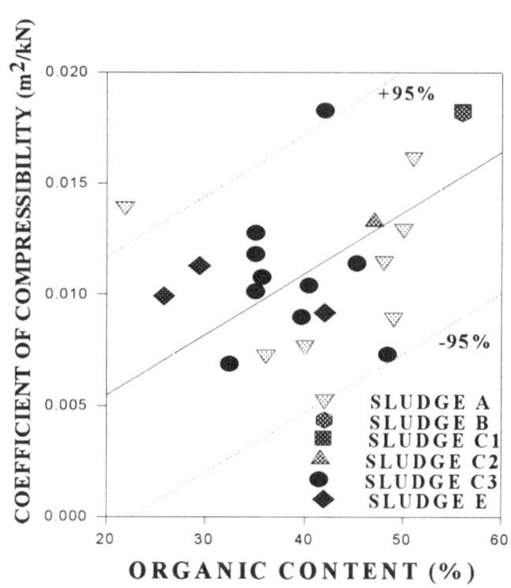

Figure 3. Coefficient of Compressibility vs. Organic Content Relationship

34.5 kPa (5 psi). Sludge C1 had a permeability of 1×10^{-6} cm/sec and an organic content of 55%. Sludge C2 had a permeability of 3.5×10^{-7} cm/sec and an organic content of 45%. Sludge C3 had a permeability of 1×10^{-7} cm/sec and an organic content of 40%. Since the sludges were molded at the same water content, this indicates that the permeability decreased as the organic content decreased. The reduction in permeability may be attributed to the increase in the percentage of kaolin clay in the sludge matrix as the organics decompose with time (Moo-Young, 1995; LaPlante, 1993). The older sludge (C3) appeared more clayey in texture than the new sludge (C1).

6. BIOLOGICAL DECOMPOSITION OF THE ORGANICS IN PAPER SLUDGES

Wood provides the main source of cellulose fibers for paper. Wood consists of approximately 50% cellulose fiber, 30% lignin, and 20% hemicellulose, pentosane, and sugar. Lignin holds the fibers together and is removed during the paper making process to secure the separation of the cellulose fibers. The organics in paper sludges are generally composed of cellulose fibers and lignin which are recovered in the waste water treatment process.

Biochemical oxygen demand (BOD) is the most commonly used parameter to define the strength of a municipal or organic industrial waste water. BOD is also used to quantify the biodegradability of contaminated soils. In this portion of the study, the natural biological decomposition of the cellulose organics within sludges A, C1, and C3 was assessed using respirometry to determine the oxygen uptake rate.

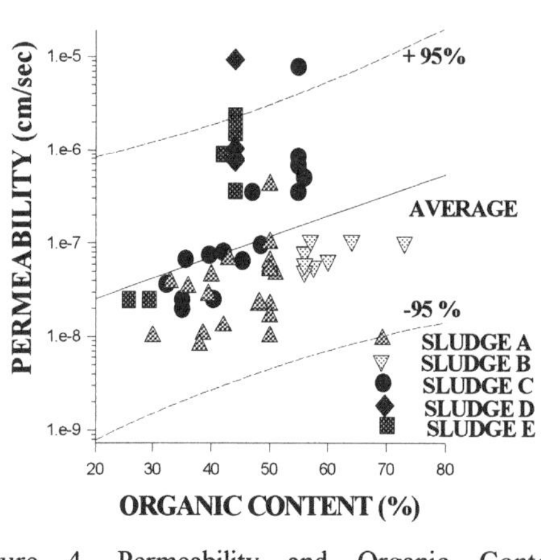

Figure 4. Permeability and Organic Content Relationship

6.1 Testing Methods

Respirometry is the measurement of oxygen uptake (BOD) exerted by a microbial population. When an aerobic condition exists, the microbial population exerts an oxygen demand proportional to the concentration of organic matter and biomass in the sample. The Comput-ox Computerized Respirometer (NCON Model WB512) was used in this study to supply pure oxygen to the reactors in response to the pressure differences created by oxygen consumption. Carbon dioxide generated during respiration is absorbed by potassium hydroxide pellets fixed at the top of each reactor. Numerous respirometer bench scale studies have been conducted for waste water treatment and bioremediation investigations (Graves et al., 1991; Santiago et al., 1991; Sun et al., 1994; Christodoulatous et al., 1995).

The mineral media used in this study consisted of the following compounds: magnesium sulfate, calcium chloride, ferric chloride, and potassium dihydrogen phosphate. Two sets of organic decomposition tests were conducted on paper sludges A, C1, and C3. Sludge A represents a one week old recycled paper sludge that has been used as the capping material in two landfill covers in Massachusetts. Sludge C1 represents a one week old integrated paper sludge. Sludge C3 represents a 10-14 year old integrated paper sludge used as the capping material in Corinth, New York. Sludges C1 and C3 were used to compare the biological decomposition of new paper sludge to older paper sludges. Test one was conducted without microbial enhancement for 72 hours. A paper sludge concentration of 400 mg/l was placed into the 500 ml reactor. Reactor one tested the reagent water. Each test was conducted at 20° C with an initial pH of 7.

Test two was conducted with microbial enhancement for 240 hours on sludge A, C1, and C3. The microorganisms used in set two were from the waste water treatment system of a recycling paper plant. A paper sludge concentration of 400 mg/l was placed into the 500 ml reactor with a seed volume of 10 ml. Tests were conducted at 20° C with an initial pH of 7. Reactor one tested the reagent water and the microorganisms (10 ml).

6.2 *Respirometer Results*

Figure 5 plots the oxygen uptake (BOD) versus time for test 1 and 2. For test 1, the blank reactor (the nutrient water) had little oxygen uptake as indicated in Figure 5. For sludge A in test 1, there was a lag time of approximately 1 day (24 hr) before the oxygen uptake increased steadily at an average rate of 1.2 mg/l-hr (Figure 5). For sludge C1 in test 1, there was a lag time of approximately 20 hours before the oxygen uptake increased sharply over the next five hours. After this sharp increase, the oxygen uptake for sludge C1 increased steadily at an average rate of 0.6 mg/l-hr (Figure 5). For sludge C3 in test 1, there was a lag time of approximately 1 day before the oxygen increased at an average rate of 0.9 mg/l-hr (Figure 5). These tests indicate that the paper sludges decompose naturally, since the oxygen uptake increases with time. Moreover, natural biological decomposition is apparent because the oxygen uptake is proportional to the paper sludge concentration (Jacquez et al., 1991).

For test 2, the blank reactor (the nutrient water and the microorganisms) had little oxygen uptake as indicated by Figure 5. For sludge A in test 2, there was a lag time of one day before the oxygen uptake began to increase at an average rate of 3.7 mg/l-hr (Figure 5). In comparison to test 1, the rate of oxygen uptake is three times greater for test 2 which would be expected, since the microorganisms used in this portion of the study were acclimated to sludge A. For sludge C1 in test 2, there was a lag time of approximately 1.5 day (36 hr) before the oxygen uptake increased at an average rate of 0.82 mg/l-hr from 36 to 100 hours. After a sharp increase in the oxygen uptake (at t = 101 hours), the oxygen consumption increased at an average rate of 1.1 mg/l-hr (Figure 5). For sludge C3 in test 2, there was a lag time of 110 hours before the oxygen consumption increased at an average rate of 1.1 mg/l-hr (Figure 5). The long lag times indicate that the microorganisms were unacclimated to sludges C1 and C3 (Jacquez et al, 1991). Thus, the rates of oxygen uptake for sludges C1 and C3 in test 2 were similar to the values for test 1. This error may have also resulted from insufficient seed or the presence of inhibiting substances.

This study represents the biodegradation of paper sludges in an aerobic environment. When used as a landfill cover, paper sludges will be covered with a vegetation layer and drainage layer, and paper sludges will be in an anaerobic environment. To obtain a greater understanding of the biodegradation

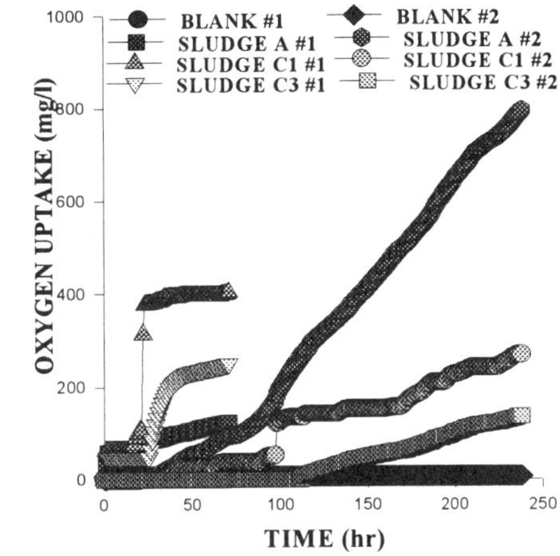

Figure 5 Oxygen Uptake vs. Time

of paper sludges in landfill covers, more testing is needed to compare the aerobic and anaerobic biodegradational rates to other parameters such as the chemical oxygen demand and the total organic carbon. This is an area for future research.

7. SUMMARY AND CONCLUSIONS

The following conclusion can be drawn with respect to the organic decomposition of paper sludges.

1. Paper mill sludges are characterized by a high organic content in comparison to clay soils.
2. As the organic content increases, the specific gravity of paper sludges decreases.
3. As the organic content of paper sludges increases, the compressibility increases.
4. As the organic content decreases, the permeability of paper sludges decreases.
5. Paper sludges exhibit natural biological decomposition at a rate ranging from 0.6 to 1.2 mg/l-hr.

ACKNOWLEDGMENT

Support for this research was received from the following organizations: Erving Paper Company, Erving, MA; International Paper Company, Corinth, NY; Marcal Paper Mill Inc., Elmwood Park, NJ; Mead Specialty Paper Division, Lee, MA; Clough Harbour and Assoc., Albany, NY, and the Army Research Office, Research Triangle Park, NC.

8. REFERENCES

Christodoulatous, C., N. Pal, & R. Vemuri 1995. Biodegradation of PAHs in soil-water slurries with enhanced bioavailability. *GeoEnvironmental 2000: ASCE Geotechnical Special Publication No. 46.* Y.B. Acar and D.E. Daniel (eds.), 2:1493-1505. New York: ASCE.

Feustel, I.C. & H.G Byers 1930. The physical and chemical characteristics of certain American peat profiles. *Technical Bulletin 214*, Washington, D.C.: US. Department of Agriculture, Bureau of Chemicals and Soils

Floess, C.H., R.F.J. Smith, & R.H. Hitchcock 1995. Capping with fiber clay. *Civil Engineering: ASCE*, August: 62-63.

Graves, D.A., C.A Lang., & M.E. Leavitt 1991. Respirometric analysis of the biodegradation of organic contaminants in soil and water. *Applied Biochemistry and Biotechnology.* 28:813-826

Jacquez, R.B., M.H. Li, W.H Zachritz,., & M.I. Beach, 1991. *Minimization of hazardous waste generation: preliminary investigation of waste management properties of a solvent substitute for TCE.* Larchmont: N-Con Systems.

Kraus, J.F. & C.H. Benson 1994. *Effect of freeze-thaw on the hydraulic conductivity of three paper mill sludges: laboratory and field evaluation.* Environmental Geotechnic Report 94-6, University of Wisconsin-Madison.

Lambe, T.W. & R.V. Whitman 1969. *Soil Mechanics.* New York: John Wiley and Sons.

LaPlante, K. 1993. *Geotechnical investigation of several paper mill sludges for use in landfill covers.* Masters of Science Thesis, Rensselaer Polytechnic Institute, Troy, NY.

MacFarlane, I.C. 1966. Peat structure as a basis of classification. *Proceedings of the Eleventh Muskeg Research Conference*, NRC ACSSM Tech. Memo. 87: 26-31.

Moo-Young, H.K. & T.F. Zimmie 1995. Design of landfill covers using paper mill sludge. *Research Transformed into Practice: Implementations of NSF Research.* J. Colville and A.M. Amde (eds), 14-28. New York: ASCE.

Moo-Young, H.K. 1995. *Evaluation of paper mill sludges for use as landfill covers.* Ph.D. Thesis, Rensselaer Polytechnic Institute. Troy, NY.

Moo-Young, H.K. 1992. *Evaluation of the geotechnical properties of a paper mill sludge for use in landfill covers.* Masters of Science Thesis, Rensselaer Polytechnic Institute. Troy, NY.

National Council of the Paper Industry for Air and Stream Improvement Inc. (NCASI) 1992a. *Progress in reducing water use and waste water loads in the U.S. paper industry.* NCASI Technical Bulletin No. 603. New York: NCASI.

NCASI 1992b. *Solid waste management and disposal practices in the U.S. paper industry.* NCASI Technical Bulletin No. 641. New York: NCASI.

Pepin, R.G. 1985. The use of paper mill sludge as a landfill cap. *Proceedings of the 1984 NCASI Northeast Regional Meeting*, Special Report No. 85-02. New York: NCASI.

Santiago, I., F. Cadena, & M. Beach 1991. *Modification of aerobic respirometry to assess biodegradability of compounds in anaerobic processes.* Larchmont: N-Con Systems.

Stoffel, C.M. & R.K. Ham 1979. *Testing of high ash paper mill sludge for use in sanitary landfill construction.* Prepared for the City of Eau Clairc, WI, by Owen Ayers and Associates, Inc., Eau Claire, WI.

Sun B., N. Nirmalakhandan, E. Hall, X.H. Wang, J., Prakash & R. Maynes 1994. Estimating toxicity of organic chemicals to activated sludge microorganisms. *Journal of Environmental Engineering.* 120:1459-1469.

Wang, M.C., J.Q. He, & M. Joa 1991. *Stabilization of water plant sludge for possible utilization as embankment material.* The Pennsylvania State University, State College, PA.

Zimmie, T.F. & H.K. Moo-Young 1995. Hydraulic conductivity of paper sludges used for landfill covers. *GeoEnvironmental 2000: ASCE Geotechnical Special Publication No. 46.* Y.B. Acar and D.E. Daniel (eds.), 2:932-946. New York: ASCE.

Environmental Geotechnics, Kamon (ed.) © *1996 Balkema, Rotterdam. ISBN 90 5410 848 7*

Properties and immobilization from slag municipal waste

R. Nüesch, P. Amann & A. Mendoza
Division of Geotechnical Engineering, IGT, ETH Zürich, Switzerland

ABSTRACT: Slag and fly ash commonly are washed prior to disposal to decrease their soluble salts and chemical reactivity. But the present data suggest that, to the contrary, washing may increase reactivity. It is desirable to reuse the residues from waste incineration. This could be done by a more comprehensive preparation of the residues than is possible today. These reflections will also be applied to existing and future waste disposal sites.

1 INTRODUCTION

The aim of the Swiss waste management program is to produce either recyclable waste products, or waste with a composition similar to that of soil or naturally-occurring materials so that the waste can be disposed of without harm to the environment (Baccini, 1993)(Fig. 1).
In Switzerland, municipal waste incinerators produce 760,000 tons of slag and 60,000 tons of fly ash per year. The air and water regulations existing at present, as well as the new technologies employed in waste incineration, guarantee that waste gas and leachate from a waste incineration installation be of relatively good quality. However, cleaner waste gas and leachate decrease the quality of the solid products. These wastes have to be deposited in landfills because they contain leachable heavy metals. Presently the incinerated waste does not

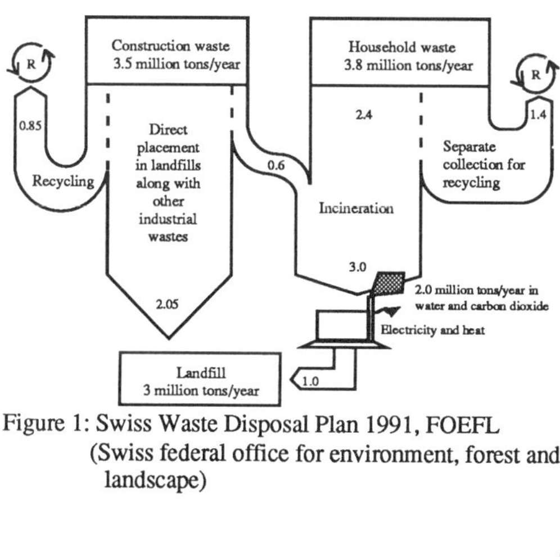

Figure 1: Swiss Waste Disposal Plan 1991, FOEFL (Swiss federal office for environment, forest and landscape)

meet the required standards because it is chemically unstable and it is too high in heavy metal and organic content. Studies of slag show a high reactivity, including heat production (80° C) and changes in composition over time (Lichtensteiger & Zeltner, 1993). The actual practice of slag storage is contrary to the principles of supply and long time effectivity. In the short-term the liberation of heavy metal is very slight; in the long-term, however, the reaction and contaminant potential of the waste incineration residues may influence the groundwater and the soil in a very negative way (project EKESA, 1992). Recently, efforts have been made to improve the residue quality by improving the incineration method (e.g. Belevi, 1993), or by adding a secondary thermal treatment of the residue, e.g. melting (Baccini et al. 1990). The composition of slag and fly ash in contact with water determines the chemistry of leachates and therefore also affects its interaction with other materials when it is used as a recycled construction material or when it interacts with landfill barrier systems. The quality of leachates and the mobilization of heavy metals from incinerated residue have been the main focus of investigations lasting several years. Generally , slag leachate has high alkalinity (pH 11) and a high concentration of salt and organic compounds. The leachates are dominated by ions such as sodium, potassium, cloride and sulfate (Johnson, 1993). Admixtures with concrete and clay were investigated to test the mobility of heavy metals of fly ash and to increase fly ash stability (Pluess 1993). Residues were washed in water before deposition in a landfill to leach soluble salts, as is the common practice. Finally many studies were concerned with possible interaction of leachates with clayey sediments used as liners in landfills (e.g. Schwaighofer et al., 1995). The adoption of the Environment-Protection-Law enabled the federal authorities to mandate and support environmental research projects. Thus, in the framework of the research programme "Environmental technologies and research (SPP)", - which is funded by the Swiss National Science Foundation -, different solutions to environmental problems having a national and global impact will be studied. The Institute of Geotechnical Engineering of the ETH Zurich is taking part in the joint project "Environmental technologies: Disposal and Recycling" which treats the immobilisation of municipal waste residues installations and the possible use of these residues as construction materials.

In this paper the physical and chemical consequences of washing slag and fly ash are discussed and soil mechanics data of slag and slag-clay mixtures are presented.

2 CHARACTERISATION OF THE RESIDUES

2.1 Mineralogical and chemical investigations

Four materials from a Swiss incinerator plant were used: unwashed (SU) and prewashed (SW) slag from which the metal iron parts were magnetically removed, and unwashed fly ash (FU) and washed fly ash (FW) residue. The fly ash, about 9% of the total incinerated waste, came from the waste gas cleaning system of the incinerator plant. Opalinus clay, which is an industrially ground Jurassic clay from Switzerland, was added to the slag as a binder to investigate soil mechanics behavior. The slag and fly ash were characterized mineralogically, physically and chemically. The organic content was determined by TG/MS (Thermogravimetry Mass Spectroscopy) the other elements by AAS (Atomic Adsorption Spectroscopy) and ICP/MS (Ionized Coupled Plasma Mass Spectrometry).

2.1.1 Mineralogical composition and physical properties

The mineralogical composition and physical properties of slag and fly ash are summarized in Table 1. The washing process creates new mineral phases in slag and fly ash. The calcite content increases and a change in physical properties such as specific surface area (by BET), water content and grain size distribution also occurs. The grain size distribution in Table 1 was determined by sedimentation in water. Fly ash shows a significant loss of material, which is interpreted mainly as dissolution of soluble salts. The disolved material is higher in the washed fly ash than in the unwashed.

Physical properties of different grain sizes obtained by dry sieving are shown in Table 2. The amount of small grain sizes decreases for washed slag but increases for washed fly ash. The specific surface area of washed slag and fly ash increases for all grain sizes compared to unwashed samples. Unwashed fly ash shows very high specific surface area for grain size >0.5mm, which is attributed to organic matter. The adsorbed water content in washed slag and fly ash increases with smaller grain sizes. The water content of the whole washed slag decreases slightly from 0.57 to 0.52% (see Table 1), because the smaller grain size distribution represents a minor part of the whole material.

The porosity distribution of slag and fly ash

Table 1: Mineralogical composition and physical properties of materials

	unwashed slag SU	washed slag SW	unwashed fly ash FU	washed fly ash FW	Opalinus clay OP	Montigel clay MO
minerals	quartz calcite 5% hydroxyapatite feldspar chromite aluminium Fe-oxide	quartz calcite 12% hydroxyapatite feldspar chromite aluminium Fe-oxide ankerite mirabelite melanterite masanite	anhydrite sylvite halite quartz calcite 9% hematite goethite rutile basanite	quartz sylvite halite anhydrite calcite 10% hematite goethite rutile gypsum	quartz 30% ml i/s 20% illite 15% chlorite 10% kaolinite 10% carbonate 10% accessories 5%	montmori-llonite 66% quartz 8.3% mica 12-15% feldspar 2-4% carbonate 3.8 % kaolinite 2% accessories 3%
BET m2/g	9.5	10.7	10.7	19.5	21	490
CEC meq/100g	-	-	-	-	12	62
γ_s kN/m3	28.52	27.76	26.36	26.71	26.50	28.1
water wt%	0.57	0.52	1.45	5.82	2.7	3
>63micron wt%	53	52	17	11	2	1.4
<20micron wt%	38	43	60	59	89	87.9
<2micron wt%	9	12	1	1	57	77.6
dissolved wt%	-	-	15	19	-	-

Table 2: Grain size obtained by dry sieving versus grain size distribution, specific surface area and water content of unwashed and washed slag and fly ash

	grain size mm	>2	0.5-2	0.25-0.5	0.063-0.25	<0.063
wt % of grain size	SU	46	23	11	16	4
	SW	50	24	11	13	2
	FU	1	3	8	80	8
	FW				55	45
specific surface m2/g	SU	5	6.2	8	11.7	15
	SW	4.5	7.7	10.5	15.1	15.7
	FU	208	93	19.3	15.7	6.9
	FW				16	27.1
water content wt %	SU	0.37	0.56	0.65	0.9	1.28
	SW	0.23	0.41	1.13	1.19	1.48
	FU	2.47	1.63	0.71	1.51	1.38
	FW				5.41	6.31

measured with high pressure mercury intrusion, shows smaller pore sizes after the washing process. Washed slag shows a median pore size of 363 nm compared with 661nm for an unwashed sample. Washed fly ash has a smaller size (182nm) and more uniform pore size distribution compared to that of unwashed fly ash (741 nm).

2.1.2 Chemical composition

The chemical composition of washed and unwashed slag and fly ash, divided into major elements (Si, Ca, Fe Al), soluble salts (i.e. Na and K), reactive elements (C, S and P) and some toxic heavy metals, are summarized in Table 3 according to the washing procedure.

The elements Si, Ca, Fe and Al are potential sources for new mineral phases and influence the chemistry of leachates (e.g. Ca which forms lime und calcite). The major elements in slag, except for Fe, show no change in concentration due to the washing process. All major elements in washed fly ash show higher concentrations than in unwashed fly ash. Elements which are often in soluble salts, such as K and Na, show no concentration change in washed slag. Therefore is it assumed that slag does not contain much K and Na from soluble salts. But fly ash shows leaching of K and Na from washed material compared to unwashed. The chosen heavy metals show no systematic trend as a result of washing.

Table 3: Concentration (wt%) of major elements, salt ions reactive elements and some toxic heavy metals

Elements	SU	SW	FU	FW
major elements				
Si	20.0	20.0	14.7	15.4
Ca	13.2	13.3	12.6	14.1
Fe	9.6	7.3	2.9	3.0
Al	5.5	5.6	5.8	7.1
elements in soluble salts				
Na	1.9	1.9	2.5	0.9
K	0.8	0.8	3.9	1.2
reactive elements				
C	1.3	4	4	7.3
P	1.76	1.06	0.95	1.02
S	0.6	0.8	2.7	2.8
toxic heavy metals				
Cd	0	0	0.05	0.05
Cr	0.05	0.05	0.07	0.08
Cu	0.26	0.23	0.12	0.15
Pb	0.16	0.12	0.86	0.99
Zn	0.22	0.30	3.02	3.47
Sn	0.02	0.02	0.15	0.18

2.1.3 Discussion of the results

The appearance of new mineral phases induced by the washing process indicates that the slag and fly ash react with water. The carbonate content increases after the washing and generally exerts a major influence on the chemistry of the leachates and on particle cementation. The total amount of calcium does not change for slag and therefore an external Ca-source, such the washing water, can be excluded. The specific surface area of washed slag and fly ash is higher compared with unwashed material and shows the largest increase in smaller grains (<0.063mm). The water content increases for small fraction as well, and washed fly ash has double the amount of water as the Opalinus clay. The higher surface area and higher water content can cause a long term chemical instability and reactivity for washed slag and fly ash. Negative long term interaction with clay due to high alkalinity and chemical reactivity is increased due to the washing process of slag and fly ash.

Washed fly ash is even more reactive than washed slag, as was observed during exposure to water during sedimentation tests in the lab. The smaller grain sizes of washed fly ash show a higher amount of dissolved material during the sedimentation test in water compared to unwashed fly ash. The washed fly ash has a larger surface area and therefore dissolves more than unwashed fly ash. The removal of soluble salts is linked with higher surface area and water content, which enhance the reactivity of slag and fly ash . An unwanted, longterm interaction with clay due to high alkalinity and chemical reactivity may be increase due to the washing of slag and fly ash.

2.2 Results obtained by other participants in the joint SPP project (Environmental Investigations)

In order to determine the environmental compatibility of the reference material, the Institute of Plant Biology at the University of Zurich carried out toxicity tests. The investigated waste gas residues presented an eco-toxicology more critical than that of the waste incineration slag, but they are nevertheless to be considered slightly toxic. At present the microbiological aspect of extracted metals from waste gas cleaning residues is being examined. At the Professorship for Biocompatible Materials at the ETH Zurich tests were made to find out the toxic impact of the residues on animal cells. *Fibroblaster* cells proved to be sensitive indicators in relation to zyto-toxicology of treated as well as untreated residues. The *Fibroblaster* cells tests showed that as a rule the zyto-toxicity decreases in the following sequence: untreated waste gas cleaning residues/treated waste gas cleaning residues, waste incineration slag, vitrified products. The Chemtech Department at SULZER's carried out (alkaline) tests on the reference material using concentrated acids. The washing out of the material with concentrated acids at low pH-values of the leachate leads to a swelling of the material and filtering was not possible any more. The tests have shown that a combination of acids with complex-forming agents and oxidation agents is necessary to wash out the various metals in quantitative parts.

3 GEOTECHNICAL INVESTIGATIONS

Based on the results of the characterization of the residues up to the present time, separation of the materials into building material and waste material could not be made, because the heavy metals exist evenly in all grain size fractions. The question of a possible use of these residues as construction materials is therefore still open. The quality of today's slag and deposit practice shows that in a long view the elutriation of the heavy metals out of the deposit body with a drop of the pH has to be expected (Lichtensteiger, 1993). For this reason, in the second phase of the project the waste disposal safety will be investigated. The basic concept for these tests will be the multi-barrier system. With a view to use residues as a leachate barrier, the immobilisation of incineration residues with clay will be investigated. At the same time the soil mechanical properties (soil classification, compactability, permeability, leaching) of the slag and the slag-clay mixture will be determined to permit the best possible use of these materials in waste disposal sites. The aim of the project is to use the disposal site as a raw material source with a possibility to recover raw materials.

3.1 Research Program

For the first investigations unwashed slag (SU), Montigel (unactivated Calcium-Bentonit, MO) and Opalinus clay (OP) were used. The data of the Opalinus clay and the Montigel can be seen in Table 1. The composition of the investigated samples are presented in Table 4.

The classification of the materials was carried out

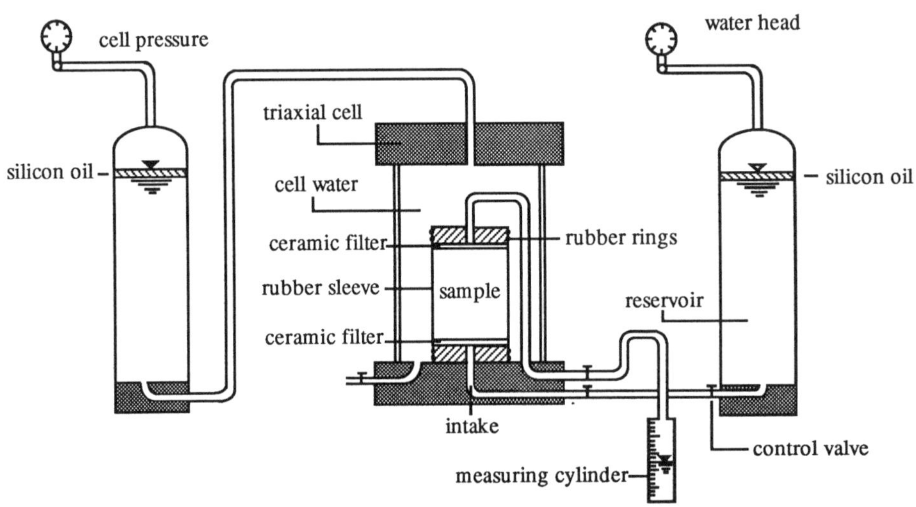

Figure 2: Testing equipment for the determination of permeability in the triaxial cell

Table 4: Composition of the mixtures

	Slag %	Opalinus-clay %	Montigel %
Slag	100	0	0
Mixture 1	80	20	0
Mixture 2	80	15	5

in accordance with the Swiss Standard 670 810c (USCS). The mixing process was done manually and corresponds almost to the mixing process in a concret-mixer (Plüss 1993). Immediately after the mixing process the mixture was placed into the Standard Proctor apparatus. After compaction the samples were transferred to a triaxial pressure system (triaxial cell) (Fig. 2). Tap water was then allowed to flow through the samples. The tests were carried out with a lateral pressure of 0.65 bar and a pressure gradient of i = 8.0 to i = 8.4 .The leaching behaviour of the samples was tested by means of percolation cells (Plüss, 1993). The samples were built in the cell and slowly run trough with the eluate six times altogether. The eluate used was destilled water with a pH of 6.02. The quantitative analysis of the percolates resulted from the AAS.

3.2 Results

The results of the soil classification showed that the slag can be designated as a silty gravel with fine particles (GM). The investigated slag shows no plasticity and has a specific unit weight of 29 kN/m^3. The particle size distribution is indicated in Fig. 3. No particle crushing of the material due to compaction was observed. The particle size distribution of Mixtures 1 and 2 are also indicated in Fig.3. The results of the Proctor tests are shown in Fig. 4.

The permeability of a pure slag sample could be measured only during a short period of time, because this sample was very porous and the capacity of the sampling container was limited. The permeability coefficient of the pure slag is within the range 1x10^{-5} and 1x10^{-6} m/s. The results of the permeability tests of the mixtures are indicated in Fig. 5.

Figure 6 presents the eluted quantities of Zn and Cu as a cumulative curve. Pb and Cd did not show any changement in concentration.

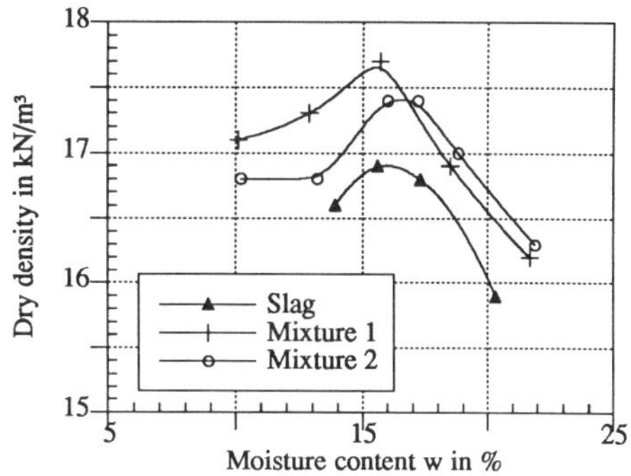

Figure 4: Proctor curve of investigated samples

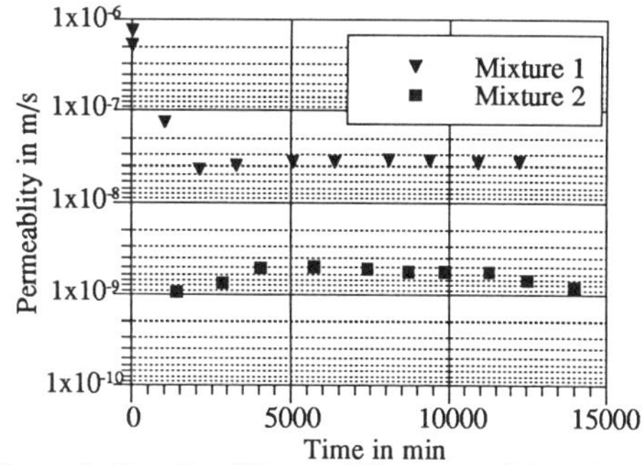

Figure 5: Results of Permeability tests of the mixtures

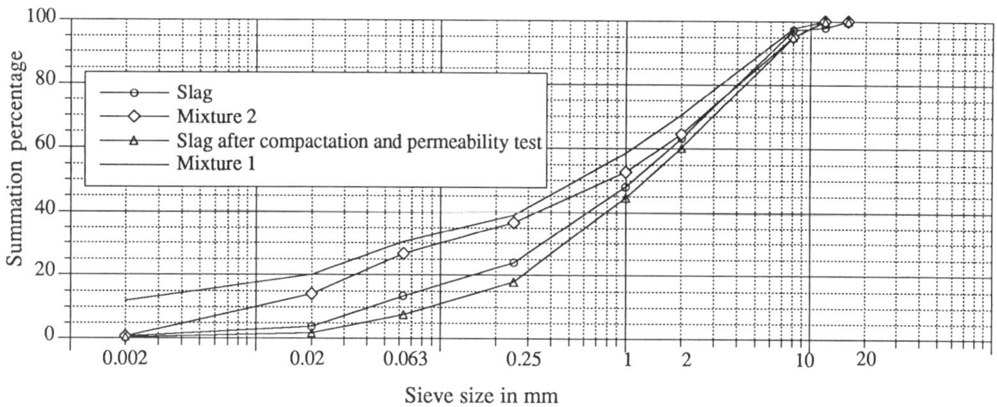

Figure 3: Particle size distribution of the material investigated

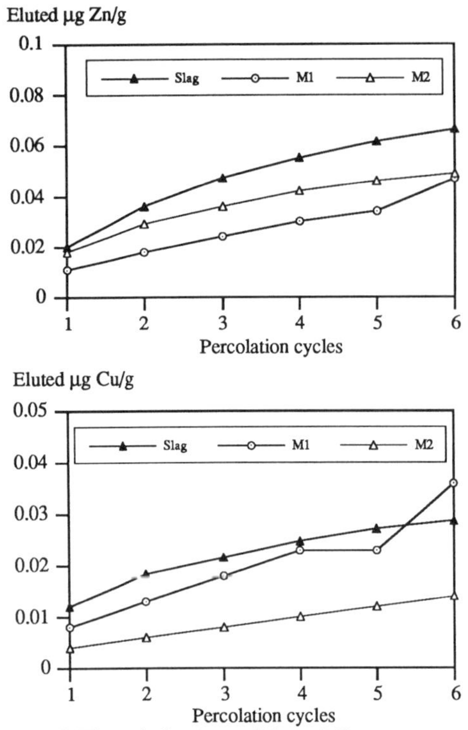

Figure 6: Eluate behaviour of Zn and Cu

With additional clay there are only 80 weight per cent of slag in the mixture, and therefore the original concentration of Zn and Cu without additional clay had to be diminished by 20% in order to compare these values with the measured concentration after additional clay. The sixth percolation from mixture 1 shows an increase of eluted cations, which is possibly due to the slow velocity of flow-trough.

3.3 Discussion of the Results

The slag investigated is a coarse-grained material with approx. 14% fine particles. During preparation of the samples no special preparation conditions were observed. The material is easily mixable and workable.

The Proctor curve of the investigated samples shows that the samples with a percentage of water $w = w_{opt} \pm 3\%$ can still be compacted up till 97% of $\gamma_{d_{max}}$. The dry density of very wet samples decreased vigorously.

The permeability tests have shown that the slag is a very porous material. With the addition of Opalinus clay (Mixture 1) the permeability coefficient could be decreased by two orders of magnitude, whereas the addition of both Opalinus clay and Montigel (Mixture 2) decreased it by approx. three orders of magnitude.

Even if only small quantities of Zn and Cu could be eluted, the total of the eluted quantities of Zn up till 45% and Cu till 56% could be diminished by adding clay.

4 CONCLUSIONS

Slag is a very reactive material. The washing process of the residues, which is a usual practice, augments the reactivity even more. The possible washing out of salts and heavy metals prevents the direct use of incineration residues as construction material, although from a geotechnical point of view it would be suitable for this purpose. The addition of clay minerals has decreased the permeability of the slag by up to three orders of magnitude and has diminished the possibility of elution of the heavy metals Cu and Zn significantly.

The lower permeability allows that the penetration of atmospheric water into the deposit body is reduced, which reduces directly the quantity of eluate, i.e. less water for treatment and smaller charge of the effluent and ground water. The eluate behaviour of the heavy metals has to be confirmed by other eluate agents, e.g. compounds similar to eluate.

The results of the investigations carried out are encouraging, but they require more precise clearings in the lab (short-term) and conversion of the lab investigations in the field (long-term).

REFERENCE

Amt für Gewässerschutz und Wasserbau des Kantons Zürich (1992): Projekt EKESA (Emissionsabschätzung für Kehrichtschlacke).

Baccini, P.,1993: Warum neue Qualitätsanforderungen in der Abfallwirtschaft. Hrsg. P.Baccini und B. Gamper, vdf, ETH Zürich.

Belevi, H., 1993: Wie wird Schlacke erzeugt?In: Deponierung fester Rückstände aus der Abfallwirtschaft. Vdf, ETH Zürich.

Johnson, A., 1993: Chemische Eigenschaften und Langzeitverhalten der Müllschlacke. In: Deponierung fester Rückstände aus der Abfallwirtschaft. Vdf, ETH Zürich.

Lichtensteiger T., Zeltner, Ch., 1993: Wie lassen sich Stoffqualitäten beurteilen?, Seminar: Endlagerqualität am Beispiel Kehrichtschlacke, Ittingen, CH .

Pluess A., Ferrell R.E. jr, 1991: Characterization of lead and other heavy metals in fly ash from municipal waste incinerators, Hazardous waste & hazardous materials. VOL.8/4, Mary Ann Liebert, Inc., Publishers, 275-292.

Plüss, A., 1993: Immobilization and Stabilization of fly ash with clay minerals. Proc. Joint CSCE-ASCE Nat. Conf. of Env. Eng. Montreal, Quebec, Canada, 813-820.

Schwaighofer, B., Knaak, B., Ottner, F., Müller, H.W., 1995: Observation on Austrian clay rocks and their barrier functions for waste disposals-influence of the time factor. *Appl Clay Sci.*, 10, 199-208.

Environmental Geotechnics, Kamon (ed.) © *1996 Balkema, Rotterdam. ISBN 90 5410 848 7*

Status on utilization and disposal of industrial waste in Japan

Yuji Nagasaka
GEOGYRO Corporation, Tokyo, Japan

Sumio Horiuchi
Shimizu Corporation, Tokyo, Japan

Kanji Higaki
Taisei Corporation, Yokohama, Japan

ABSTRACT More than 398 million tons of industrial wastes are discharged annually in Japan. To minimize the environmental impact of industrial wastes, many efforts have been made for increasing of utilization and volume reduction of disposal. Utilization for geo-materials and reclamation with latest technologies are effective solution for decreasing disposal quantity, because of a huge amount of wastes can be consumed in such construction projects as structural fill or man-made island. The technical issues such as chemical and mechanical properties have not necessarily been evaluated, however. To promote the waste utilization as for a stable ground material, "the research committee for industrial waste disposal and utilization" of the Japanese Goetechnical Society has studied and provided the useful information in Japan. In this paper, technical issues including quantity, characteristics, utilization as effective geo-materials are reviewed for the following seven different kinds of industrial wastes: coal ash, paper sludge ash, MSW ash, steel slag, sewage ash, waste plastics, and waste tire.

1. INTRODUCTION

More than 398 million tons of industrial wastes are discharged annually in Japan. Because of the vast types of industrial wastes are discharged from different industries, the first step of the sub-committee on utilization of industrial wastes in the Research Committee on Treatment and Potential Use of Industrial Wastes for industrial waste (The Japanese Goetechnical Society), shown in Table 1, was choice of major substance to be studied regarding to (1) annual quantity, (2) less environmental impact at utilization, and (3) members' interests. Table 2 shows the substances chosen in this sub-committee. Excepting steel slag and waste tire, amount of utilization is less than disposal. The Law for Promotion of Resources Recycling, being enacted from 1991 in Japan, has gradually promoted the utilization, more effort is needed to minimize environmental impact.

Geo-materials, which have been used for stable earth structures such as structural fill, back-fill and road, is expected to be the most suitable target to increase waste utilization, because an enormous quantity of materials could be used in such civil engineering works. One issue that need to be addressed relating to the low utilization rates would be a lack of technical information exchange especially of the chemical and mechanical properties in actual application.

In this paper, the latest technical information of the following seven kinds of wastes are reviewed focusing on geo-

Table 1 Members of Research Committee on Treatment and Potential Use of Industrial Wastes (The Japanese Geotechnical Society)

Chair: M. Kamon (Kyoto University)					
Sub-committee on Utilization of Industrial Wastes		*Sub-committee on Utilization of Construction By-products*		*Sub-committee on Land Use of Waste Reclaimed Site*	
Chief:		*Chief:*		*Chief:*	
Y. Nagasaka	(Geogyro Co.,Ltd.)	M.Yamada	(Osaka City University)	K. Onitsuka	(Saga University)
Members:		*Members:*		*Members:*	
K. Fukushima	(Toagosei Co.,Ltd.)	T. Kawachi	(Obayashi Corp.)	K. Inoue	(Kansai Soil Research Center
K. Higaki	(Taisei Corp.)	M. Kamiya	(Hokkaido Inst.of Tech.)	K. Kubota	(Kisojiban Consultants Co.,Ltd.)
T. Hirata	(Kagoshima Nat. Col. of Tech.)	K. Kataoka	(Chuken Consultant Co.)	M. Nambu	(Oyo Corp.)
S. Horiuchi	(Shimizu Corp.)	T. Katsumi	(Kyoto University)	J. Okada	(Nakabori Soil Corner Co.,Ltd.
T. Kanda	(Fujita Corp.)	T. Kunito	(Seiko Kougyo Co.,Ltd.)	M. Okada	(Toyo Constr.Co.Ltd.)
A. Kasai	(Onoda Chemico Co.,Ltd.)	K. Mori	(Kumagai Gumi Co.,Ltd.)	S. Ono	(Chuo-Kaihatsu Co.,Ltd.)
Y. Nakano	(Konoike Constr.Co.,Ltd.)	N. Ogawa	(Min.of Constr.)	M. Otsubo	(Kyusyu University)
H. Omori	(Chichibu Onoda Cement Corp.)	P.O. Tran Duc	(Hazama Corp.)	K. Sonoi	(Nikken-Sekkei Ltd.)
H. Sakai	(Railway Technical Res. Inst.)	K. Saitou	(Ohmoto Gumi Co.,Ltd.)	K. Terada	(Takenaka Corp.)
T. Yamaguchi	(Sumitomo Constr.Co.,Ltd.)	K. Sasaki	(JDC Co.)	M. Tsuchihiro	(Kajima Corp.)
		M. Tokunaga	(Nippon Koei Co.,Ltd.)		

Table 2 Industrial Wastes Studied in the Sub-committee

Industrial Wastes		Quantity(Mt/y)		Properties			Chemical Components (%)				
Name	Subclass	Util.	Disp.	Gs	Mean Dia.	pH	SiO_2	Al_2O_3	Fe_2O_3	CaO	SO_4
Pulverized Coal Ash	Fly Ash	1.5	2	2.2	0.02mm	11	41-65	20-36	1.9-13	0.5-9	0.1-1
	Bottom Ash	0.7	0	2.2	2mm	8	〃	〃	〃	〃	〃
Steel Slag	Blast Furnace	21.8	0	1.6	2-20mm	9	30-35	5-15	0-2	40-50	1-5
	Converter 〃	8.7	1	2.0-2.4	----	----	10-15	1-2	10-20	40-50	0-1
	Electric 〃	1.7	0.8	>2.5	30-300mm	----	15-30	5-10	1-20	3.5-5.5	0-1
Sewerage Sludge Ash	Lime-treated	0.02	0.065	2.5-3.7	0.01-0.8mm	12-13	10-30	5-20	5-20	15-35	0-3
	Polymer- 〃	0.025	0.14	2.3-2.8	0.01-0.2mm	9	35-50	10-20	5-20	1-10	0-3
	Combined- 〃	----	0.08	2.8-3.0	0.02-1mm	7-13	20-40	10-15	5-20	5-40	0-3
Paper Sludge Ash		0.08	0.41	2.0-3.3	0.005-0.4mm	8-13	25-55	25-54	1-4	2-11	----
MSW Ash	Fly Ash	0.02	6	2.4-2.9	<0.1mm	8-12	15-25	4-15	4-8	15-30	----
	Bottom Ash	----	----	----	5mm	----	40-45	12-16	3-9	14-20	----
Waste Plastics		0.6	4.3	1	----	----	----	----	----	----	----
Waste Tire		0.6	0.07	1.6	----	----	----	----	----	----	----

material utilization; (1)coal ash, (2)paper sludge ash, (3)MSW ash, (4)steel slag, (5)sewage ash, (6)waste plastics, and (7)waste tire.

2. COAL ASH

In Japan, 10% of the total electric power is generated by pulverized coal combustion and is expected to increase to 20% by 2001. As shown in Table 2, a half of pulverized coal ash discharged from coal power plants are utilized, in which 70% are for the cement industry. Because of the limitation of cement production, the other usage is demanded for the future power plans. Its use as geomaterials have been believed to be effective. In resent, comprehensive studies are carried out for the use in harbor and air-port construction projects.

Coal Ash for Reclamation

The reclamation of coal ash is of significant from the both points of waste disposal and creating new land area. Conventional reclamation methods can be classified into three types as shown in Fig.1, in which the slurry reclamation is recently developed in Japan. From the experimental data (Sugimoto *et al.* 1989, Horiuchi *et al.* 1995), properties of the reclaimed ground can be summarized as in Table 3 for the typical Japanese coal fly ash. It can be concluded that the slurry reclamation is the best method for ground use and for disposal volume reduction, its running cost is higher than the others, however. For density increasing of the wet reclamation, addition of surfactants are studied by Horiuchi *et al.* (1994), and 15% increase was confirmed in lab tests.

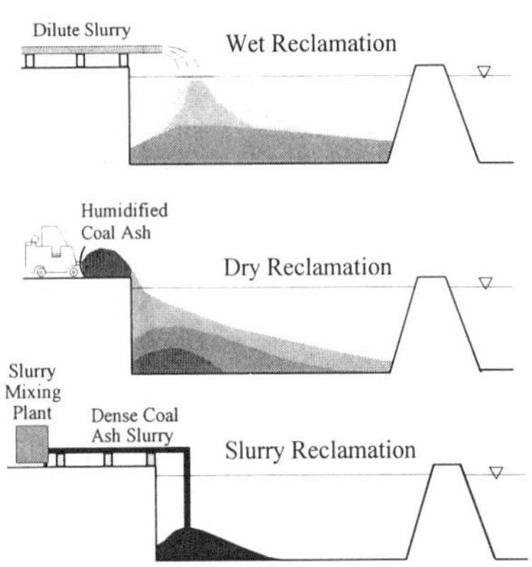

Fig.1 Methods of Coal Ash Reclamation

Table 3 Comparison of Coal Fly Ash at Sea Reclamation

Properties of Ground reclaimed	Underwatr Reclamation Method		
	Wet Type	Dry Type	Slurry Type
Dry Density	0.80 g/cm^3	0.93 g/cm^3	1.25 g/cm^3
28 day Compressive Strength Underwter	0～4kPa	0～70kPa	200kPa

Properties of coal ash assumed

Maximum dry density : 1.33 g/cm^3, Calcium content : 4%, pH : 12
Mean diameter : 0.012mm

Coal Ash Mixtures for Geo-Materials

Many usage has been reported on efficiency of coal fly ash as a partial substitute of cement and lime through lab tests and field applications. The following advantages of coal ash usage were confirmed; (i) suitable strength, and (ii) and higher fluidity of the slurry. Outstanding examples are as follows.

(A) Slurry composed of coal flyash, volcanic ash, cement and seawater was used for two man-made islands of 50,000m^3 each (Horiuchi *et al.*1992, Kawasaki *et al.*1992). Advantages of the method confirmed are, (a) lower lateral pressure to the cofferdam, and (b) sufficient strength for ground use. Figure 2 shows the slurry plant and cofferdam used in the project, and 90,000 tons of coal flyash were utilized.

(B) Lightweight air mortar, in which a large part of cement was substituted by fly ash, was used for structural fill surrounding a high-rise building to prevent subsidence and sliding failure of the low strength alluvial layer (Shioya *et al.* 1992).

(C) Flash-setting grouts composed of fly ash, cement, bentonite and water glass were developed through lab tests,

Fig.2 Coal Ash Slurry applied in the Hakucho Ohashi Project

and then were applied as back-fills in several projects (Horiuchi *et al.*1991, Konno *et al.*1989, Higaki *et al.* 1994).

(D) High strength ground materials was developed by mixing aluminum carbonate compound with fluidized-bed combustion coal ash (Kamon *et al.* 1991).

Coal Ash for Special Cement

New approaches of coal fly ash usage have been challenged in many projects, *i.e.* special cements using coal fly ashes combined with other chemicals were developed for soil stabilization and waste treatment.

(1) Special cement composed of coal ash, cement and aluminum carbonate compound was effective for the stabilization of dredged soil (Kamon *et al.* 1987).

(2) Applicability of special mixtures provided by mixing lime with coal fly ash for stabilization of dispersive soils were studied through a series of lab tests. The stabilized soils were concluded to be suitable material for the earth-filled dam construction (Nontananandh *et al.* 1994).

(3) Mixture of fluidized-bed combustion coal ash and portland cement showed high potentials for stabilization of municipal waste incineration fly ash both on strength development and controlling leachates (Kamon *et al.* 1994).

(4) New synthesized cement using coal ash and lime showed a high potential as a soil improvement cement developing lower strength (Sekiguchi *et al.* 1991).

(5) Potential of mixtures composed of coal ash, shellfish incineration ash and FGD sludge was confirmed on soil stabilization. Appropriate mixtures were proposed through the lab tests (Takahashi *et al.* 1994).

3. STEEL SLAG

An enormous efforts have been paid for the utilization of steel slag, and two kinds of JIS (Japanese Industrial Standard) are issued for civil engineering applications; JIS A 5012 of "Granulated Blast Furnace Slag Fine Aggregate for Concrete", and JIS A 5015 of "Iron and Steel Slag for Road Construction". As shown in Table 2, a large part of the steel slag is used in Japan, and many technical issues were published (Japan Steel Association, 1982, Nagai,1991, Steel Slag Association,1988, Ikeda, 1995). In this paper, therefore, only an outline of utilization in Japan is described.

Blast Furnace Slag

Almost 100% of this slag is utilized, in which 70% is for cement and concrete industries, and 20% is for road and railway constructions. Because of its high calcium content, reclaimed aggregate develops a sufficient strength for land use.

Converter Furnace Slag

Forty percent of this slag was disposed of until 1980, but more than 90% is now utilized for steel production process (17%), road and railway construction (9%), and civil engineering works (46%).

Electric Furnace Slag

Different from the other types of steel slag, more than 30% is disposed of, and this ratio has being constant from 1980's. This is because of bad site location and lack of supply system. Utilization for civil engineering usage is still less than 40% in 1992.

4. SEWAGE SLUDGE ASH

Result of the recent spread of sewage system in Japan, amount of the sludge discharged from the water treatment plants is increasing every year. The volume of this sewage sludge is reduced by such processes as thickening, dewatering, and incineration before filling or reclamation. Because of the shortage of disposal sites, system has being moved toward application of incineration.

As shown in Table 2, sewage sludge incineration ash is classified into three types by the additives for flocculation; lime-treated, polymer-treated, and combined-treated ash. The polymer treatment type has being increased in Japan for the final volume reduction.

Lime- treated Sewage Sludge Ash

Since the lime-treated ash contains 20-50% of calcium, it is expected that the ash might be usable as soil stabilizer. A corporate study conducted between Nagoya City Authorities and Ministry of Construction has confirmed its applicability and prepared a tentative manual. Further studies, being carried out in the field, show a high potential of this idea; (1) improved soil can be easily compacted than sandy soils, (2) sufficient initial strength can be attained, and (3) strength is not so large as difficulty in re-excavation. Based on the results obtained, Nagoya City Authorities plans to make an actual recycle plant system shown in Fig.3 from 1996.

Polymer-treated Sewage Sludge Ash

Because of the added polymer is decomposed during incineration, chemical compositions of the polymer-treated ash resembles to natural soils. This type of ash, therefore, has been studied for the following application; (1) alternative raw materials in ceramic industry, such as bricks, con-

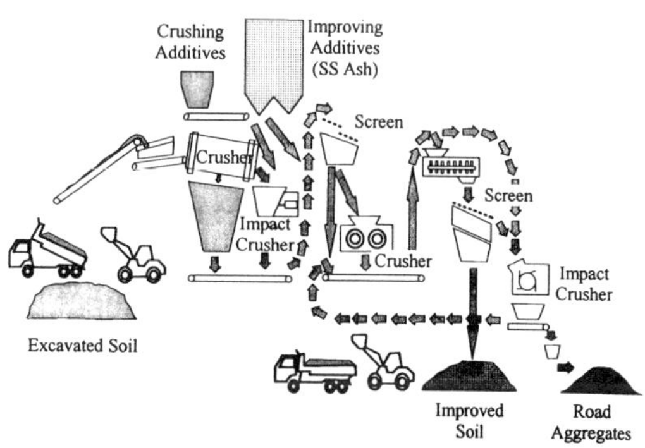

Fig.3 Soil Recycle Plant in Nagoya

crete blocks, and lightweight aggregate, (2) porous aggregates for garden soils, and (3) cement added slurries for back-fill.

In recent, new attempt of this ash has studied for a raw materials of cement production; eco-cement.

5. PAPER SLUDGE ASH (PS ASH)

Properties of paper sludge ash are dependent on the process; sludge from virgin chips process contains wooden fiber, however recycle paper process contaminates clay powders and pigments.

The bulk density of the ash is 0.25 to 0.50 g/cm^3, which seems to be suitable for lightweight geo-material. Specific gravity varies widely from 2.4 to 3.3, but this value is similar to ordinary geo-materials except the special paper ash that contains ferrotitanium, for the reason of its chemical components. Grain size ranges $5\text{-}425 \times 10^{-3} mm$ in general, however depends on the incinerator type.

According to leaching test results, the ashes meet to environmental quality standards excepted pH which ranges widely from 8 to more than 12.0.

Geotechnical properties

The following data are reported for compacted PS ashes (PS Ash Committee,1994);

(1) Compaction Characteristics: Figure 4 shows compaction curves on various kinds of PS ashes. Comparing with general Japanese soils, effect of water content on dry density is extremely small, and density itself is very small.

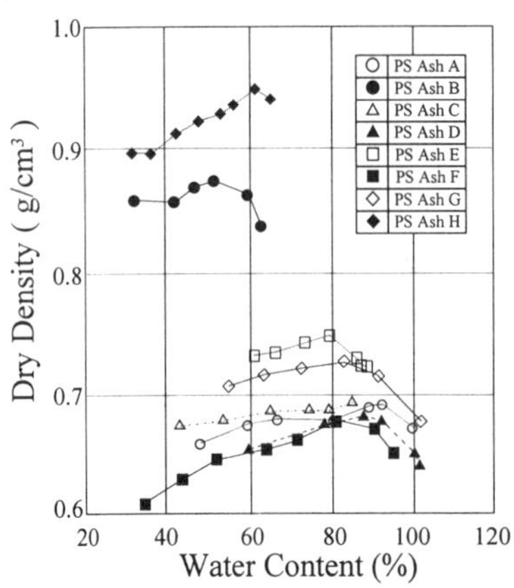

Fig. 4 Compaction Curves of PS Ashes

Table 4 Static Properties of Compacted PS Ashes

	A	B	C	D	E	F	G	H
Consolidation Yield Stress (MPa)	1.52	1.96	1.70	1.68	1.70	1.27	2.06	1.58
Compression Index	0.98	0.51	1.16	1.34	0.65	1.23	1.01	0.97
Effective Cohesion (MPa)	0.02	0.04	0.02	0.01	0.03	0.02	0.04	0.02
Effective Frictional Angle (°)	39.6	39.8	40.9	43.7	41.8	42.5	38.0	41.8

(2) Static Properties; Table 4 shows tri-axial and consolidation characteristics. Cohesion of the compacted ash appears to be zero, and particle degradation was observed during compression process.

Geotechnical Utilization

According to a report by Fuji Paper Manufacturing Guild, PS ash are utilized as follows;(1) raw materials for blocks and soil stabilizing agents, and (2) an artificial soil for planting slopes. Now in order to reuse a large quantity of the ash, the research on ordinary geo-materials for embankment, filling and so on, has been executed.

Though many studies have being made for geo-material utilization, only 20% is used as shown in Table 2. One reason of this low rate may be due to a severe legal restriction in Japan.

6. MSW ASH

In Japan, more than 50million tons municipal solid waste is discharged annually and approximately 73% of the waste is incinerated. Annually, about 6 million tons municipal solid waste incineration ash (MSW ash) is consequently produced, the utilization of MSW ash is rather limited, however. It is important to establish the excellent utilization techniques considering the shortage of disposal sites.

MSW ash is basically classified into bottom ash and fly ash, and its properties are dependent on combustion furnace type, *e.g.* stoker or fluidized-bed. Fly ash is generally more toxic than bottom ash because of adsorption of heavy metals. After the revision of the Law in 1991, processes of melting, cementation, chelation and dissolving processes were designated as intermediate treatment for the fly ash.

Bottom ash contains considerable amount of carbon and particle diameter is about 5 *mm*. Fly ash is much finer and has a higher specific surface area ranges from 2,000 to 10,000 cm^2/g. According to typical leaching test results, heavy metals and chlorides are released in the leachates.

Geotechnical Utilization

Though much effort has been paid for the utilization, less quantity of MSW ash is actually utilized. The following three processes are believed to be essential and effective for the civil engineering utilization.

1) Slag process: Most effective for both reduction of volume and toxicity, but needs highest cost. Properties of slag depend on types of melting furnace. Concentrations of leachates are negligible in general. The slag could be applicable for structural fill or pavement.

2)Cement solidification: Stabilization process by cement addition is the most economical and widely applied for disposal. Since some problems, however, remain unsolved on leachates, and this process is not used for geo-technical utilization. Such applications as cover material, light weight filling and concrete admixture are being investigated steadily.

3)Alinite cement: Alinite rich cement can be made from MSW ash. Production techniques are being investigated using pilot plant through a national project sponsored by the Ministry of International Trade and Industry. This process and material draws considerable attention in its high utility value.

7. WASTE PLASTICS

In Japan, annual production of all the plastics is 12 million tons, and 5 million tons are disposed as wastes, in which 44% is from industries and the rest is in the municipal wastes. In plastic products, additives such as pigments or softening agents are usually mixed with, and it would be suspected that some of the materials would be changed to a toxic substance at an extreme conditions.

Major utilization of waste plastics is reuse as economic plastics and for oil production through cracking process. Because plastic is significant resource, the description of plastic type and additive mixed is essential for the reuse.

Geotechnical Utilization

Plastic Blocks: Large plastic blocks of 1.0~3.0m^3 size are produced by compression at 15-20*Mpa,* and used as retaining walls, covers of slope, etc. Surface of the block is reinforced by wire mesh and coated by mortar or asphalt. Density of the block is 1.2-1.4 g/cm^3.

Lightweight Geomaterial: Expanded polystyrene (EPS) is widely used for cushions and container, and a large part of the annual production of 5 million m^3 are disposed. A lightweight geomaterial using this waste EPS (SLS) has been studied and used in many construction projects; chipped EPS (finer than 10mm diameter) is mixed with excavated soil, portland cement and water. Two kinds of lightweight mixture are developed; (1) less water content type (optimum water content +5 ~10%), and (2) slurry type (water content of 1.5-2.5 times of liquid limit). Density of this new geo-material is 0.9~1.3 g/cm^3 as shown in Fig.5. The less water content type is used for road base, back-fill, etc. The slurry type is used for under water fill, back-fill, and partial sea reclamation. Strength and density can be controlled by the composition. The mechanical properties could be determined by the common soil tests using waste materials to be mixed.

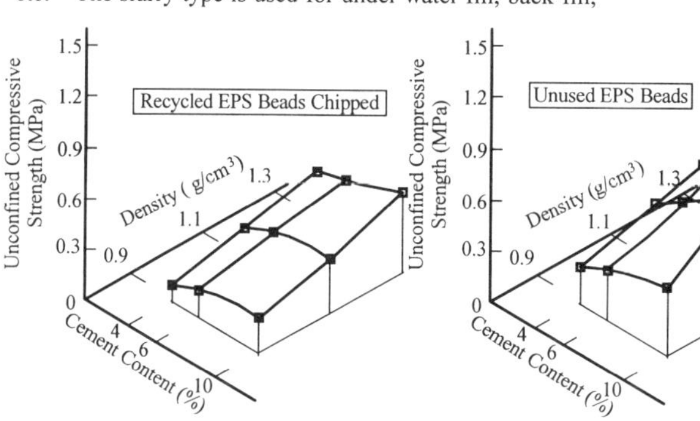

Fig.5 Density of Lightweight Geomaterial using EPS Beads (Cured 7 days)

8. WASTE TIRE

The Japanese production of automobile tire has been increasing year by year. It is very difficult to estimate the amount of waste tire accurately but has also been increasing gradually. From a crude estimation, recycle ratio of waste tire is around 90%, and the recycle was classified into the following three types; (1) thermal energy resource (35%), (2)renewed tire production (34%), and (3) trading as used tire (17%). The recycle for energy resources is introduced in thermal processes in cement, paper and metal industries, however, much more cost benefit will be needed to promote it because additional investments are necessary to clean the gas discharged after the combustion.

Geotechnical Utilization

(1)Aggregate for asphalt concrete: Mixtures of chipped tire (finer than 5 *mm* diameter) and asphalt are used for special pavement in the cold area. This type of pavement is able to improve both tire grip and noise reduction in winter, and also effective to prevent from ice adhesion onto frozen pavement. The same idea has been utilized in Sweden and USA since 1970.

(2) Back-filling for Dual-purpose channel: To prevent frost heaving of irrigation and drainage channel in the cold area, a new back-filling idea using tire chips, which shows lower heat transference than conventional soil back-fill, has been utilized (Kubo, 1995). The concept is direct placement of tire chips, finer than 50 *mm* diameter, behind channel, as shown in Fig.6. Properties of the placed tire chip layer are as follows;

Mass Density : 0.61 g/cm^3
Permeability: 7-11 $\times 10^{-2}$ *cm/sec*
Compressive stress at 20% strain; 130 *kPa*
Metal Leachates : Not detected except iron.

Results obtained in a field test shows that this new back-fill keeps the ground temperature more than 2°C higher than the conventional one.

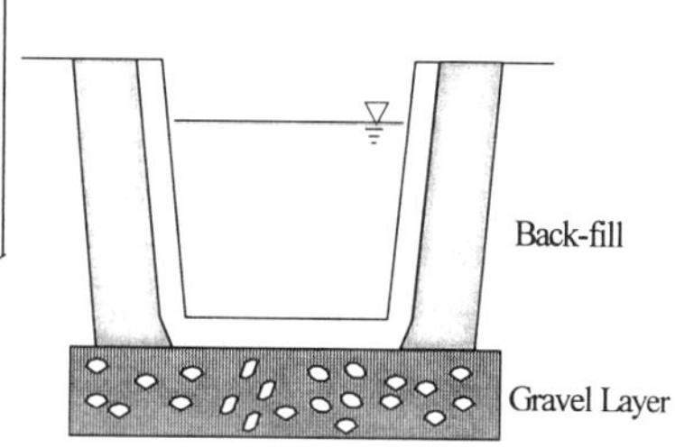

Fig.6 Cross section of New Back-filled Channel using Waste Tire Chip

9. CONCLUSION

In many construction projects, vast kinds of industrial wastes are actually utilized. Pulverized coal flyash and steel slag are the typical examples. Utilization of wastes itself is of course an important economical issue, however we have to keep in mind that the utilization is an action for environmental protection. We, therefore, should take environmental conditions into account carefully together with waste properties. "Prevention is better than cure" and "Spilt water cannot be gathered again."

References

Pulverized Coal Ash

Higaki, K. *et al.* (1994): Proc. First National Symp. on Environmental Geotechnology (The Japanese Geotechnical Soc.), pp.195-200 (in Japanese).

Horiuchi,S. *et al.*(1992): J. Material Engineering, ASCE, 4(2), pp.117-133.

Horiuchi,S. *et al.* : Waste Materials in Construction (Elsevier Science Publishers), pp.545-552.

Horiuchi,S. *et al.* (1995). Soils and Foundations(The Japanese Geotechnical Soc.), Vol.35, No.1, pp.1-10.

Kamon, M. *et al.* (1987): Proc. of the 41st JCA General Meeting (Japan Cement Association), pp.98-101(in Japanese).

Kamon, M. *et al.* (1991): Proc. Symp. on Land Use of Waste Landfill (The Japanese Geotechnical Soc.) , pp.89-94 (in Japanese).

Kamon, M. *et al.* (1994):Proc. First National Symp. on Environmental Geotechnology (The Japanese Geotechnical Soc.), pp.231-222 (in Japanese).

Kawasaki,H. *et al.* (1992): J. Material Engineering, ASCE, 4(2), pp.134-152.

Nontananandh, S. *et al.* (1994) :Proc. First National Symp. on Environmental Geotechnology (The Japanese Geotechnical Soc.), pp.121-128.

Sekiguchi,M. *et al.* (1994): Proc. First National Symp. on Environmental Geotechnology (The Japanese Geotechnical Soc.), pp.213-216 (in Japanese).

Shioya, H. *et al.* (1992) : Tsuchi-to-Kiso(The Japanese Geotechnical Soc.), Vol.40, No.12, pp.29-34 (in Japanese).

Sugimoto,T. *et al.* (1989): Proc. 2nd Int. Symp. Environmental Geotechnology, Vol.1, pp.428-437.

Takahashi, M. *et al.* (1994):Proc. First National Symp. on Environmental Geotechnology (The Japanese Geotechnical Soc.), pp.217-222 (in Japanese).

Steel Slag

Ikeda (1995): Iron and Steel, Vol. 81, 4, pp60 (in Japanese).

Japan Steel Association(1982): Properties and Utilization of Steel Slag (in Japanese).

Nagai (1991): Resource and Material, Vol.107, 2, pp78 (in Japanese).

Steel Slag Association(1988) : Characteristics and Utilization of Steel Slag, 1 (in Japanese).

Sewage Sludge Ash

Japan Institute of Wastewater Engr. Tech.(1994): Final report on "The Use of Sewage Sludge as a Cement Processing Material and Application Technologies" (in Japanese).

Japan Sewage Works Association(1994): Statistics of Sewage Work in1992 (in Japanese).

Sakurai (1987): Civil Engineering Journal, Vol.129, No.10, pp32 (in Japanese).

Matukawa *et al.* (1993) : Papers of Sewage Works Association, Vol.30, No.357, pp85 (in Japanese).

Paper Sludge Ash

Ikuta,K.(1991): Clean Japan, Vol.90, pp11~15 (in Japanese).

Iio,M. *et al.*(1994): Proc. 5th Conference on Waste Management Research, pp203~206 (in Japanese).

PS Ash Comittee (1994): Research & Development Committee of PS Ash, p.1-46 (in Japanese).

Nikkei Construction(1992): Recycling of Civil Engineering, No. 9.25, pp51~53 (in Japanese).

Shimono, K. *et al.* (1982): Polymer Working, Vol.31, No.3, pp5~14, 1982 (in Japanese).

Shimono, K. *et al.* (1980):Research Report of Paper Industry in Shizuoka Prefecture, pp109-112 (in Japanese).

Tominaga,M.(1991): Clean Japan, Vol. 90, pp16~24 (in Japanese).

MSW Ash

Abe : PPM, Vol.25, No.1, pp18~31 (in Japanese).

Act of Ministry for Health and Welfare (1992): Disposal or Reutilization of Special Controlled General Waste and Special Controlled Industrial Waste, No.194, Notification by Ministry of Health and Welfare, July 3 (in Japanese).

Andoh *et al.* (1994) : Waste Management Research, Vol.5, No.1, pp18~31 (in Japanese).

Hada (1993): Industrial Pollution, Vol.29, No.8, pp47~61 (in Japanese).

Hiraoka *et al.* (1994) : Waste Management Research, Vol.5, No.1, pp3~17 (in Japanese).

Kamon : PPM, Vol.25, No.1, pp62~66 (in Japanese).

Kanda *et al.* (1993) : Proc. 48th Conference on Japan Society of Civil Engineering, session3, pp183~184, 1993 (in Japanese).

Kimura *et al.* : PPM, Vol.25, No.1, pp32~38 (in Japanese).

Shibata and *et al.* : Proc. the 5th Conference on Waste Management Research, pp339~342 (in Japanese).

Shimaoka *et al.* (1994) : Waste Management Research, Vol.5, No.1, pp32~45 (in Japanese).

Tashiro *et al.* (1994): Technical Research Report of JDC Corporation, No.13, pp107~111 (in Japanese).

The Japanese Environmental White Paper (1993), pp23 (in Japanese).

Waste Plastics

Nagasaka, Y. *et al.* (1985): Proc. of International Conf. of Ocean Space Utilization.

Pradan, T. B. S. *et al.* (1993): Proc. 3rd International Offshore and Polar Eng. Conf.

Pradan, T. B. S. *et al.* (1994): Proc. 1st International Congress on Environmental Geotechnics.

Waste Tire

Kubo,H.: J. JSIDRE, Vol.63,No.5, pp.467-470 (in Japanese).

Environmental Geotechnics, Kamon (ed.) © *1996 Balkema, Rotterdam. ISBN 90 5410 848 7*

Utility effect by ark shell ash for improving surplus soft clay construction work

Masaru Nanri
Saga Prefectural Government, Japan

Katsutada Onitsuka
Saga University, Japan

ABSTRACT: Incinerated ark shell (ark shell ash) has a characteristic of moisture absorption. The decrease in water content for surplus soft clay with high water content can be expected. Then the strength increase caused by pozzolanic reaction which occurs between ark shell ash, water and clays was also studied. As a result, it is clear that the utility effect by ark shell ash for improving is better, at least not more than that by quicklime.

1 INTRODUCTION

Japan has many soft grounds. In particular, extreme soft clay called Ariake clay has widely deposited at the coast of the Ariake sea in Kyushu Island. When such clays are produced by construction work, it will be impossible to re-utilize them without improvement. Hence the surplus clay is often dumped. It is clear in the future, however, that disposal area of soft clay produced by construction work will be limited. Effective treatment methods for problematic soft clay should be studied. Chemical treatment methods and physical dehydration are currently influential. Cement, Lime and Interface active agent are often used as chemical improving materials.

Ariake sea is rich in marine products. In particular a lot of ark shell is gotten to be canned and consumed. However left ark shell is a waste. It is embarrassed to dispose of it. The main component of ark shell is carbonic calcium ($CaCO_3$). Ark shell is incinerated and dissolved by heat more than 1000 ℃. As this incinerated ark shell is superior with a characteristic of moisture absorption. The decrease of water content can be expected if this will be mixed with clay at a high water content. In addition, the strength increase caused by pozzolanic reaction which occurs between incinerated ark shell, water and clays was also studied.

When Ariake clay produced by construction work has extremely high water content, it is not surprised to treat it as industrial waste (polluted sludge by construction). This paper also describes the characteristic of improved soil with incinerated ark shell for the purpose of effective utility of waste.

2 THE CHARACTERISTIC OF USED MATERIAL AND EXPERIMENTAL METHOD

2.1 *The characteristic of Ariake clay as soft clay by construction work*

Ariake clay which has a high water content and a high clay content is usually impossible to use effectively as a soil material. However, because its disposal is difficult, Ariake clay has been somewhat treated by quicklime or cement. Table 1 indicates the characteristic of Ariake clay. It is found by Table 1 that Ariake clay has a high clay content in physical and a large amount of silica and alumina component in chemical. It can be predicted that digestion, water absorption and pozzolanic reaction occur between Ariake clay and quicklime.

2.2 *The characteristic of incinerated ark shell*

Hideshima et al.(1992) have studied the effect of incinerated ark shell (ark shell ash) as a desiccant on laver. The preservation test by ark shell ash on

Table 1 Soil characteristic of experimenntal sample

Name of soil			Ariake clay
Natural water content (%)			160 ~ 180
Density of soil particle (g/cm^3)			2.617
Consistency	W_L	(%)	133.2
	W_P	(%)	46.7
	I_P		86.5
Grain size distribution	Gravel	(%)	0.0
	Sand	(%)	0.8
	Silt	(%)	31.5
	Clay	(%)	67.7
Chemical component	SiO_2	(%)	59.24
	Al_2O_3	(%)	17.51
	CaO	(%)	1.21
	the others	(%)	22.04

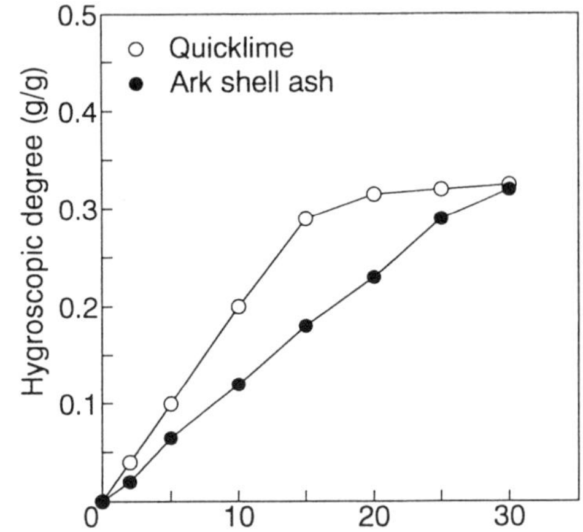

Fig.1 Relation between Hygroscopic degree and Time

Table 2 Chemical component of Ark shell ash and Quicklime

	Ark shell ash	Quicklime
CaO (%)	95.50	96.34
MgO (%)	0.37	—
Na_2O (%)	1.26	—
MnO (%)	0.01	—
SiO (%)	0.36	—
Al_2O_3 (%)	0.12	—
Ig.loss (%)	1.10	—
the others (%)	1.28	3.66

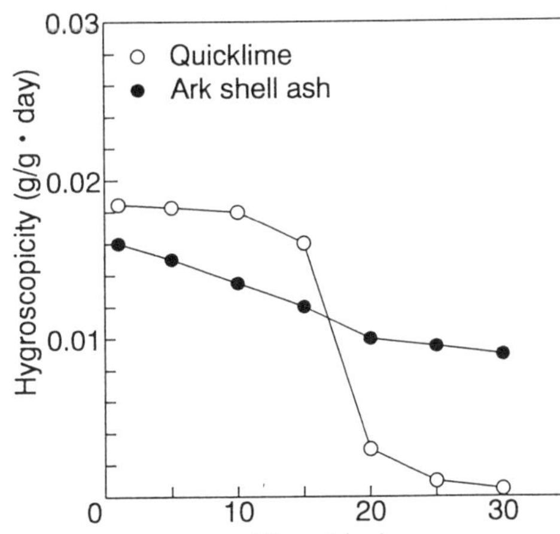

Fig.2 Relation between Hygroscopicity and Time

laver was performed by Hideshima et al. As aresult, it was clear that ark shell ash can be used as a desiccant for long term preservation. In this investigation, hygroscopic degree (g/g) and hygroscopicity (g/g·day) were examined to confirm the effectiveness as a desiccant. Fig.1 and Fig.2 indicate the change. Hygroscopic degree of ark shell ash is inferior to quicklime at an early time, but equal to quicklime after four weeks. Hygroscopicity of quicklime falls after two weeks, but that of ark shell ash is nearly the same for a month.

In this study, ark shell was incinerated at 1050 ℃ for an hour. Table 2 indicates the chemical component of ark shell ash and quicklime. It can bebe seen in this table that the chemical component of ark shell ash is not different from that of quicklime.

2.3 *Experimental method*

To estimate the effect of the improvement on the unconfined compressive strength, Table 3 shows experimental condition. The additive content of ark shell ash and quicklime is expressed in terms of the dry weight of clay. After added, the treated soils

Table 3 Experimental condition

Test	Unconfined compression test	
Mixing method	Electric mixer for 10 minutes	
Additive content	Ark shell ash	10, 20, 30%
	Quicklime	10, 30%
Compaction method	Dynamic compaction by rammer Static compaction	
Curing time	1, 3, 7, 28 days	

are cured for 1 day in (a room with) the temperature of 20 ± 3 ℃ to make it ispossible that they can have some strength to be compacted. It is supposed that rolling compaction is necessary for efficient use of improved soft soil. Hence, compaction after mixing additive is employed to make a specimen. And two methods of compaction are employed. One is the dynamic compaction which consists of filling three layers of soil mixed with additive in the mold (a diameter of 5.0cm, a height of 10.0cm) and hitting 25 times each layer by a rammer (a weight of 1.25kg, a drop of 11.7cm, a compaction energy by Standard Proctor). The other is the static compaction, its spesimens were adjusted the soil weight to be equal to the dry density of a specimen made by the dynamic compaction and compacted slowly by a equipment. They are called dynamic specimen and static specimen respectively. The both specimen were stored in moist box at a fixed temperature after compaction.

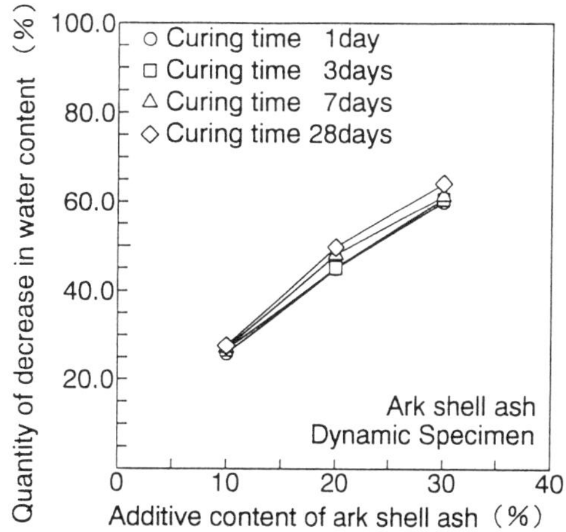

Fig.3 Relation between decrease in water content and additive content of ark shell ash

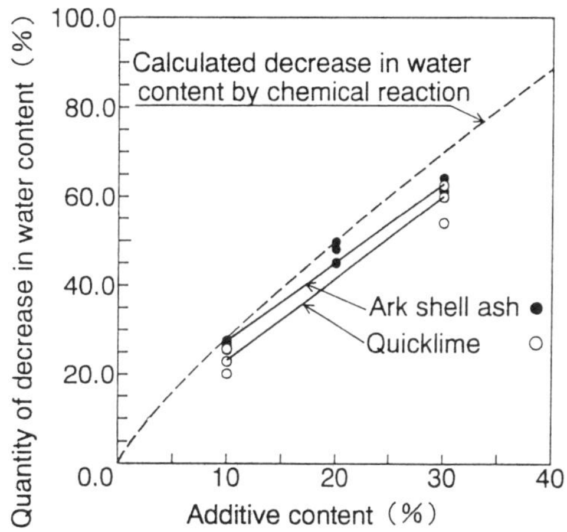

Fig.4 Relation between decrease in water content and additive content

3 THE EFFECT OF DECREASE IN WATER CONTENT FOR SURPLUS SOFT CLAY BY CONSTRUCTION WORK

Decreasing high water content is necessary for utilizing effectively surplus soft clay produced by construction work. It is clear by early investigation (Nanri, et al., 1994) and practical construction that decrease in water content is caused by digestion reaction when quicklime is used. The ark shell ash with a similar chemical component to quicklime was used for improving the surplus soft Ariake clays. The effect on the decrease in water content was investigated in laboratory tests.

Fig.3 shows the relation between the decrease in water content and the additive content of ark shell ash in dynamic specimen. It is seen that the water content decreases with the increase in the additive content. In addition, to examine the effectiveness in the decrease in water content by quicklime and ark shell ash, they were compared with the calculated line based on the chemical reaction by quicklime. This calculated line is obtained on the assumption that the water is vanished by the chemical change with quicklime in the weight corresponding to 79% of the quicklime.

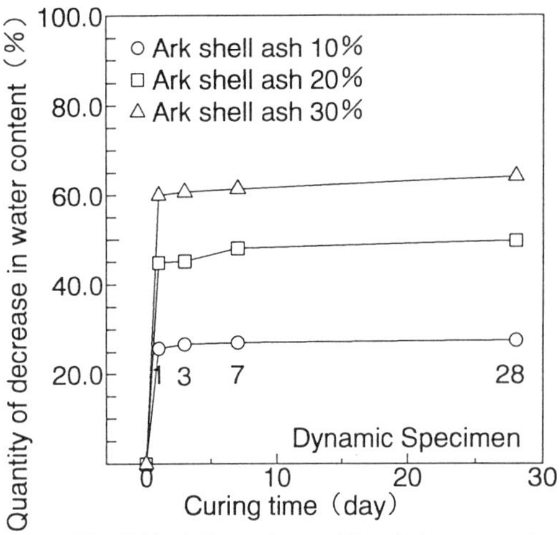

Fig.5 Variation of quantity of decrease in water content with curing time

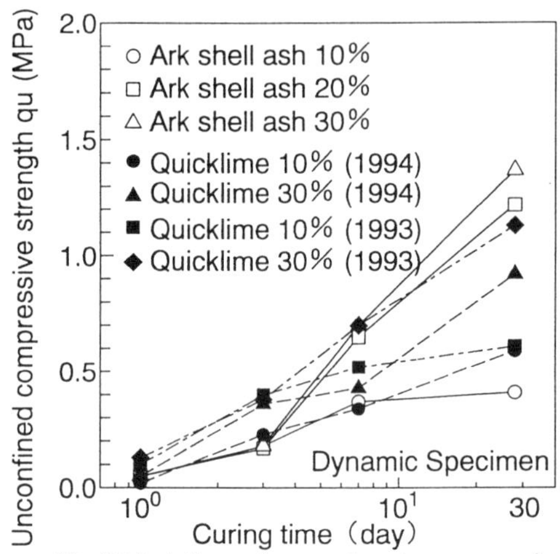

Fig.6 Variation of unconfined compressive strength with curing time

Fig.4 shows that the value in ark shell ash is equal to that in quicklime at the additive content of 10% ~ 30%. And it issupposed that ark shellash has a similar line to the calculated line for quicklime. As a result, it is cofirmed that the ark shell ash has a similar effectiveness on the decrease in water content as quicklime.

Fig.5 shows the decrease in the water content with the curing time up to 28days (in the case of dynamic specimen). It can be considered that a rapid decrease occurs within 1day. After the curing of 1day, the water content shows little change. Ark shell ash shows a similar behavior to that of quicklime for both dynamic and static specimens.

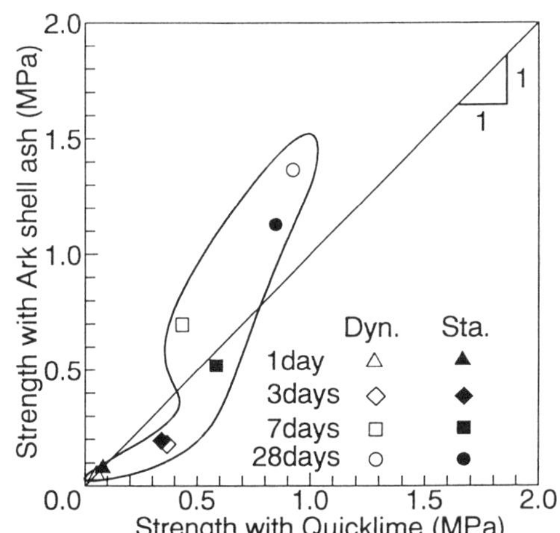

Fig.7 Comparison between Quicklime and Ark shell ash

4 THE STRENGTH INCREASE OF SURPLUS SOFT CLAY PRODUCED BY CONSTRUCTION WORK

4.1 *The strength increase with curing time*

Fig.6 shows the variation of unconfined compressive strength with the curing time (logarithmic expression) for dynamic specimens. It can be seen that strength increases lightly at early time (1,3days) for the specimen with ark shell ash. However the strength increases rapidly after early time at additive contents of 20% and 30%. On the other hand, the strength increases linearly from 1day to 28days for the specimen treated with quicklime. Fig.7 shows the comparison between unconfined compressive strength of the specimens treated with quicklime and those with ark shell ash. It can be seen that the strength of the specimen treated with quicklime is larger than that with ark shell ash at early time (1day to 3days). But the strength of the specimen with ark shell ash is larger than that with quicklime afterafter 7days.

The used quicklime was a product which has some grain size more than 2mm. On the other hand, ark shell was incinerated and crushed finely. The short term reaction with quicklime is larger than that with ark shell ash because quicklime has some of particle which has a big surface area. However, the long term reaction with ark shell ash is larger than

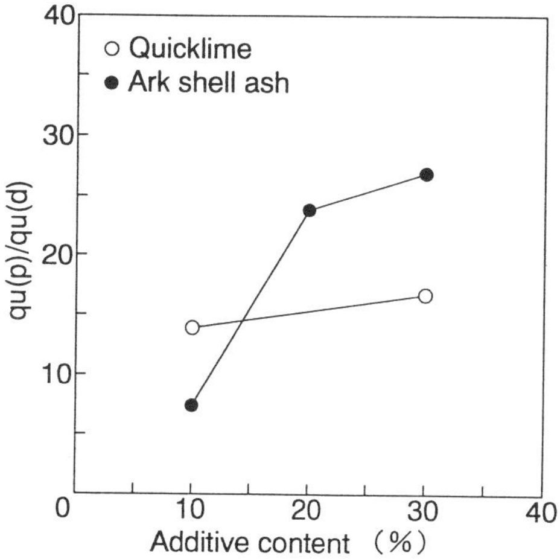

Fig.8 Relation between qu(p)/qu(d) and additive content

that with quicklime because ark shell ash has plenty of particle which has a small surface area. It is clear that ark shell ash is superior to quicklime for long term strength.

4.2 *The developing degree of pozzolanic reaction*

It seems that the difference in the strength increase caused by different types of additive material depends on the developing degree of pozzolanic reaction. The unconfined compressive strength was divided into the strength qu(p) due to pozzolanic reaction and the strength qu(d) due to digestion reaction (Nanri et al., 1995). And the strength qu(p) was obtained by subtracting the strength qu(d) from the overall qu measured by experiment at 28days. Then qu(p)/qu(d) was calculated. This value can be considered to be the developing degree of pozzolanic reaction. Fig.8 shows the relation between qu(p)/qu(d) and additive content of both quicklime and ark shell ash in the dynamic specimens. The qu(p)/qu(d) of quicklime is larger than that of ark shell ash at additive content of 10%, but ark shell ash is larger than quicklime at additive content of more than 20%. It can be expected that ark shell ash which has plenty of fine particle has larger pozzolanic reaction than quicklime at additive content of more than 20%. The static specimen has a similar behavior to the dynamic specimen. This phenomenon can be explained by the following two factors. The strength due to digestion reaction is small on ark shell ash because the hygienically degree of ark shell ash at early time is inferior to quicklime. Afterward the pozzolanic reaction develops greatly untill 28days. It is clear that effectiveness of ark shell ash for improving is equal or more than that of quicklime.

4.3 *Study on the compaction method*

Two compaction methods are adopted in this study to make clear their difference of effectiveness for improving. The writers (Onitsuka et al., 1995) have described that there is no difference in improved strength by the compaction method if dry density of specimen is less than 0.66g/cm ^{3}. The dry density with ark shell ash is 0.48g/cm ^{3} (additive content of 10%), 0.55g/cm ^{3} (additive content of 20%), 0.60g/cm ^{3} (additive content of 30%). The dry density at every content are all less than 0.65 g/cm ^{3}. It seems that there is no difference between two compaction methods in strength increase by ark shell ash. It is considered that the difference in strength by two compaction methods can be estimated by the dry density of ark shell ash which is similar to that for quicklime and pozzolanic materials.

5 CONCLUSIONS

The effectiveness of improving surplus soft clay by construction work by incinerated ash of ark shell was investigated in this study. The results can be summarized as follows.

1) When ark shell ash is used as a desiccant, Hygroscopic degree and Hygroscopicity of ark shell ash is inferior to that of quicklime at an early term, but equal to that of quicklime after four weeks.

2) The behavior of decrease in water content by ark shell ash is almost the same as the calculated line based on chemical reaction by quicklime. It is confirmed that ark shell ash has a similar effectiveness to quicklime on the decrease in water content.

3) It is found that the strength with ark shell ash increases rapidly at additive content of more than 20% after early term (3days). It is guessed that ark shell ash which has plenty of fine particle is superior to quicklime on long term effectiveness for improving.

4) It can be considered that qu(p)/qu(d) indicates the developing degree of pozzolanic reaction. The

qu(p)/qu(d) of quicklime is larger than that of ark shell ash at additive content of 10%, but ark shell ash is larger than quicklime at additive content of more than 20%. It is guessed that the influence of grain size in ark shell ash grows large to the developing degree of pozzolanic reaction as an additive content of ark shell ash increases.

5) It is considered that the difference in improving characteristic by two compaction methods can be estimated by the dry density of ark shell ash which is similar to quicklime.

REFERENCE

Hideshima,Y. and Tanaka,N.(1992) : The study on development of a desiccant by Ark shell ash, Saga Prefectural Industrial Research Institute Report, pp.13 ~ 17 (in Japanese).

Nanri,M. and Onitsuka,K.(1994) : Decrease in Water Content and Some Change of Properties by Quicklime Stabilization of Soft Surplus Clays, First National Symposium on Environment Geotechnogy, Japanese Society of Soil Mechanics and Foundation Engineering, pp.253 ~ 260 (in Japanese).

Nanri,M. and Onitsuka,K.(1995) : Improvement of Soft Clays of Construction Debris by Using Quicklime and the Mechanism of Hardening, Tsuchi-to-Kiso, Vol.43, No.8, pp.27 ~ 30 (in Japanese).

Onitsuka,K. and Nanri,M.(1995) : Characteristic of Strength Increase for Improved Surplus Clays in Construction Works with Quicklime and Pozzolanic Materials, submitted to JSCE (in Japanese).

Environmental Geotechnics, Kamon (ed.) © *1996 Balkema, Rotterdam. ISBN 90 5410 848 7*

Recycling of excavated surplus soils in Osaka City

Toshiaki Ninomiya
Osaka City Government, Japan

Takashi Takano
Public Works Division, Public Works Bureau, Osaka City Government, Japan

Takashi Yoshida & Keiichiro Muramatsu
Administration Division, Public Works Bureau, Osaka City Government, Japan

ABSTRACT: Owing to the rapid and extensive urbanization, waste soils have been increasingly discharged from various construction works in Osaka City. Recently, a serious urban problem has been surfaced regarding how to provide dumping yards for those waste soils in and around the city which has been highly urbanized. Consequently, a study on the system of effectively re-using waste soils discharged from road construction projects was commenced by the Osaka City Government in 1976. The study concluded that an "improved soil" consisting of excavated surplus soils mixed with about one percent of quicklime can be reused for subbase course of road pavement structure. Based on this conclusion, the first soil improvement plant in Japan was constructed by the Osaka City Government in January 1983. Since then, recycling of waste soils has contributed to the effective use of natural resources, the extension of existing spoil-bank's life, the stable supply of subbase course materials, saving in disposal costs of surplus soils, the mitigation of traffic congestion as a result of shortened conveyance distance of surplus soils, and the prevention of environmental demolition caused by excavation of pit soil/gravel to be used for subbase course.

1. HISTORICAL AND TECHNICAL BACKGROUND OF THE SOIL IMPROVEMENT PLANT

In order to prevent excavated surplus soils from illegal dumping, the Osaka City Government had enforced a system of appointing a specific place for dumping yards. However, the limited capacity of appointed dumping yards requested the Government to investigate the recycling of excavated surplus soils discharged from road construction works. Following the establishment of the Surplus Soils Recycle Project Team in the Public Works Bureau of the Osaka City Government (hereinafter referred to as the "Public Works Bureau") in 1976, the "Committee of investigating on the recycling of excavated surplus soils discharged from road construction works" chaired by Dr. Tadashi Mise (present post : Honorary Professor of the Osaka City University) was organized in 1978. The investigation had

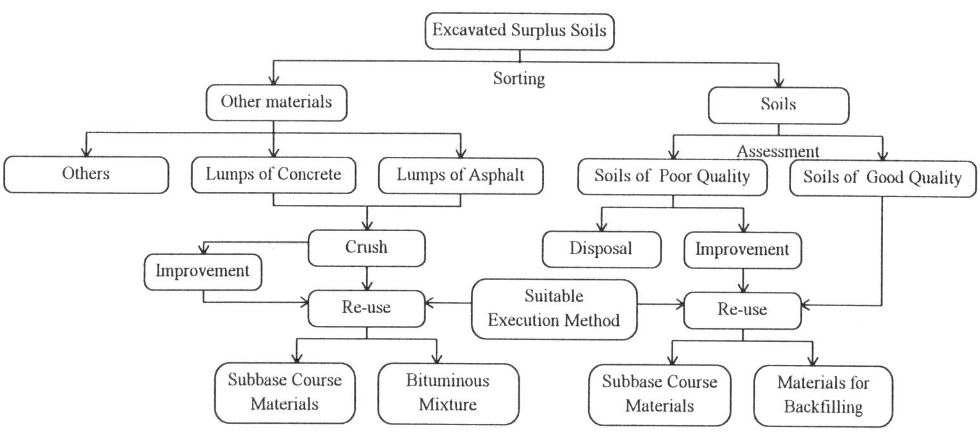

Figure 1-1 RE-USE OF EXCAVATED SURPLUS SOILS

been undertaken based on the conception shown in Figure 1-1 and was completed in 1981.

Table 1-1 shows a total volume of soils excavated/filled in various construction works in Osaka City in the fiscal year 1977. Of the excavation volume, approximately 1,360,000 m^3 (excluding lumps of asphalt and concrete), only 246,000 m^3 or 18% were reused for backfilling and the rest (1,114,000 m^3) were dumped for reclamation. On the other hand, about 644,000 m^3 or 72% of the total backfilling materials were purchased and conveyed long ways from mountainous areas located in the outskirts of the City.

Table 1-1 Volume of Soils Excavated/Filled in 1977

		(in 1,000 m^3)
Excavation	Soil	1,069
	Crashed Stone	291
	Asphalt/Concrete	189
	Total	1,549
	Disposal	1,303
Backfilling	Re-used Materials	246
	Purchased Materials	644
	Total	890

Results of soil tests conducted for the abovementioned excavated soils are presented in Table 1-2. The natural moisture content is slightly higher than the optimum moisture content. Soils having the ratio of fine of 25% or less (screen mesh of 0.074 mm or less) account for about 60% of the total excavated soils. Those soils are mostly composed of sandy soil and/or sand mixed with fine.

Table 1-2 Results of Soil Tests conducted for Excavated Soils in 1977

	Ratio of Fine	Natural Moisture Content
No. of Specimens	436	436
Maximum (%)	45.0	25.1
Minimum (%)	8.1	4.7
Average (%)	22.9	15.3
Standard Deviation	4.9	2.9

Properties of soils stabilized with quicklime (hereinafter referred to as "improved soil") were compared in different ratio of added quicklime with those of excavated soils (hereinafter referred to as "raw soil") as shown in Table 1-3. Regardless of the ratio of added quicklime, the ratio of fine decreases as the raw soils are improved. Quite large modified CBR values prove that improved soils can provide greater bearing capacity. The following factors contribute to these results:

(1) The grain sizes of raw soils are made even through the use of sieve with 25 mm meshes; and

(2) Soil particles are well interlocked because of fine particles grained by mixing raw soils with quicklime.

Since improved soils are not always reused in construction works immediately after the stabilization, those are often stocked in the open yard for some time. The stocked soils which are left under the conditions of repeated dry and wet weather could change in their quality. Therefore, the deterioration of improved soils with the lapse of time was investigated for one year after the soils were improved. As shown in Figure 1-2, it was found that improved soils statically stocked for one year still had enough quality.

Table 1-3 Properties of Raw and Improved Soils (%)

Ratio of Added Quicklime		1.0	1.0	2.0	2.0	3.0	5.0
Raw Soil	Ratio of Fine	23.3	39.0	24.7	19.1	28.5	34.5
	Moisture Content	7.6	14.0	14.9	13.9	14.1	17.8
	CBR	9.1	14.0	4.9	6.9	5.0	2.5
Improved Soil	Ratio of Fine	12.6	30.7	19.6	16.9	24.8	20.9
	Moisture Content	6.6	13.7	12.7	10.9	9.8	15.1
	Non-immersion CBR	118.0	73.0	68.0	91.6	78.8	76.4
	Immersion CBR	97.0	81.0	97.5	108.0	108.0	96.1
	Modified CBR	82.0	88.0	80.0	86.0	63.5	83.0

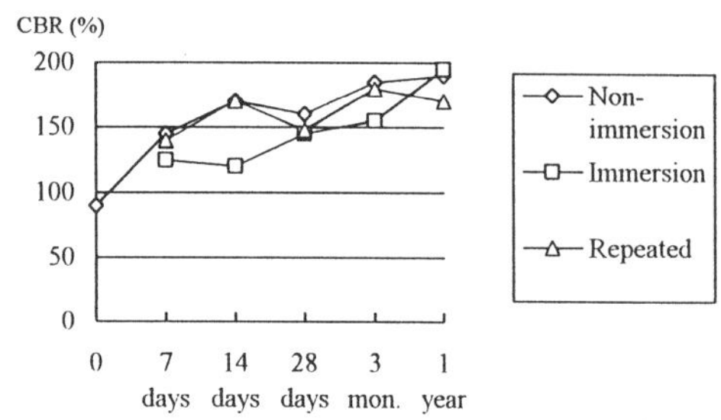

Figure 1-2 Changes in CBR with the lapse of time

2. PLANT FACILITIES AND PROCESS OF SOIL IMPROVEMENT

Based on the findings and recommendation of the abovementioned investigation, the soil improvement plant was constructed in January 1983. The location and layout of the plant are shown in Figures 2-1 and 2-2 respectively.

2.1 Outline of the Soil Improvement Plant

The outline of the soil improvement plant is described as follows:

- Location : Nanko-minami 4-chome, Suminoe-ku, Osaka City (at the western foot of Kamome Ohashi Bridge)
- Area of the Plant : 8,066 m^2 (2,000 m^2 of materials yard, 2,000 m^2 of improved soils yard and 3,180 m^2 of additional stock yard)

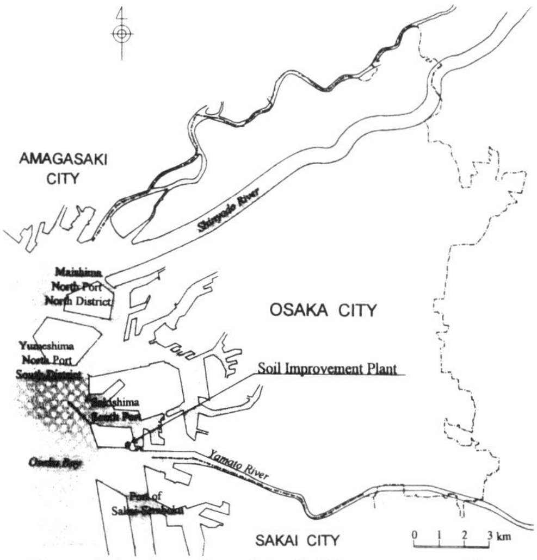

Figure 2-1 Location of the Soil Improvement Plant

- Capacity : 1,000 kN/hr. (average 3,300 kN/day, maximum 10,000 kN/day)
- Completion Date : January 1983
- Cost : Approximately 390 million yen (including 310 million yen for plant construction)

According to the findings of the investigation, all excavated surplus soils discharged from road construction works implemented by not only the Public Works Bureau but also other agencies, can be improved for recycling, if three units of the plant with a capacity of 3,000 kN/hour and two units of the plant with a capacity of 3,900 kN/hour are installed. However, it was decided that the plant improving the surplus soils excavated in road construction works of the Public Works Bureau only should be constructed as an experiment.

The annual consumption of subbase and backfilling materials for road construction works implemented by the Public Works Bureau had then totaled to about 100,000 m^3. Therefore, the capacity of the plant for improving soils was proposed to be about 100,000 m^3/year or 1,000 kN/hour.

2.2 Process of Soil Improvement

Because the plant was proposed to improve excavated surplus soils discharged only from the road construction works in the city, properties of raw soils were not expected to be remarkably dispersed. Therefore, the system could be simplified as shown in Figure 2-3. Although the time required for soil improvement depends on the detection frequency of metallic substances, it usually took the plant about 4-5 minutes to improve soils in their quality since those had been dumped into the supply hopper. If the metal detector detects reinforcing bars or other metallic substances in raw soils, belt conveyors are stopped to remove them by hand or using an electromagnet.

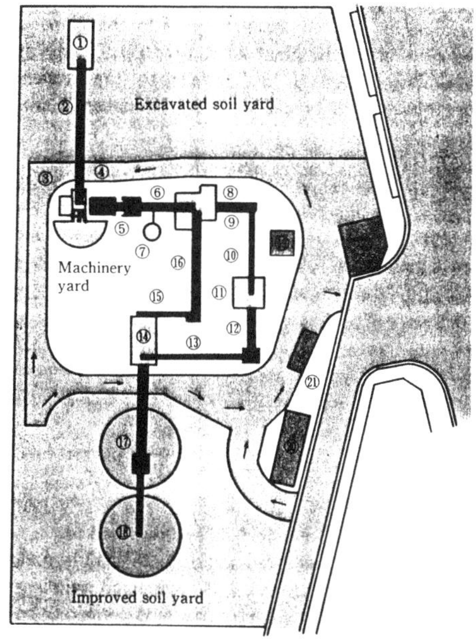

① Supply hopper
② Belt conveyor B-1
③ Raw material hopper
④ Feeder of fixed quantity lots
⑤ Picking floor
⑥ Belt conveyor B-2
⑦ Quick lime tank
⑧ Double roll crusher
⑨ Belt conveyor B-3
⑩ Belt conveyor B-4
⑪ Crushing mixer
⑫ Belt conveyor B-5
⑬ Belt conveyor B-6
⑭ Vibration screen
⑮ Belt conveyor B-7
⑯ Belt conveyor B-8
⑰ Belt conveyor B-9
⑱ Belt conveyor B-10
⑲ Control room
⑳ Office
㉑ Truck scale
㉒ Power receiving station

Figure 2-2 Layout of the Soil Improvement Plant

3. QUALITY OF IMPROVED SOIL

The stabilization with ten-odd percent of quicklime mixed at site was frequently applied in the improvement of weak subgrade and excavated surplus soils to be used as lower subbase course materials. However, this conventional improvement method has caused various problems such as heterogeneous quality due to insufficient mixture and smoking at sites.

As for the proposed soil improvement method which requires to mix only a few percentage of quicklime, on the other hand, its performance has been proved to sufficiently meet the specifications for subbase course materials through the various field and laboratory tests as well as follow-up tests for the accomplished works.

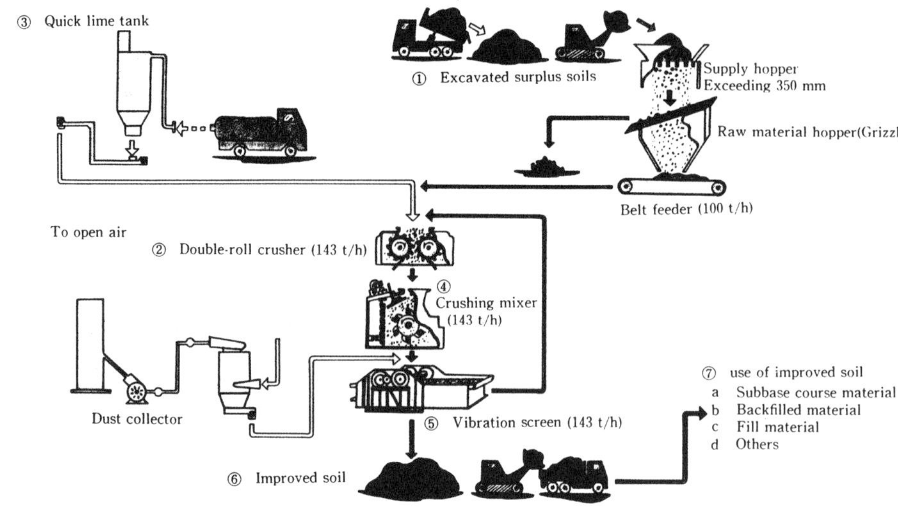

Figure 2-3 Work Flow of Soil Improvement Plant

3.1 Physical Properties of Improved Soils

The improved soils are composed of excavated surplus soils and fragments of concrete/asphalt articles which are mixed with some one percent of quicklime and crushed to make grain size even (25 mm or less) in the centralized in-place mixing plant. Since physical properties of the improved soils could be changed with the passage of time, the properties have been investigated for fresh soils immediately after improvement as well as aged improved soils.

(1) Fresh Improved Soils

Various quality control tests were carried out in the plant. Results of the tests are summarized in Table 3-1.

(2) Aged Improved Soils (after reused for pavement)

In the years 1985 and 1986, CBR tests (non-immersion) and unconfined compression tests were carried out for undisturbed samples picked out from subbase course materials of eight roads which were constructed using improved soils in 1983. Test results are presented in Table 3-2.

In-situ CBR values had slightly increased after 2-3 years of construction and then became almost steady. As for confined compression strength of soils which was about 200 kN/m^2 immediately after they were improved, it increased to 1,450 kN/m^2, on the average, for 2 years and to 2,260 kN/m^2 for 3 years.

Table 3-1 Results of Quality Control Tests (April 1983 - March 1995)

		Raw Soils					Improved Soils				
		n	X	X_{max}	X_{min}	σ	n	X	X_{max}	X_{min}	σ
Moisture Content (%)		2,139	9.98	47.2	1.0	2.726	2,287	8.10	27.80	0.70	1.999
Unit Weight (kN/m^3)							299	13.71	15.61	11.93	0.65
Grain	2.36 mm sieve (%)	1,068	59.3	84.1	32.7	8.207	2,287	53.2	83.3	25.1	6.577
Size	0.075 mm sieve (%)	1,068	15.9	55.1	4.3	5.134	2,287	13.2	41.1	2.7	3.880
Plastic Index		67	N.P.				67	N.P.			
Abrasion (%)							116	23	36	17	2.903
Optimum Moisture Content (%)		212	8.4	35.5	5.8	2.322	610	8.7	13.1	5.4	0.948
Maximum Dry Density (kN/m^3)		212	20.41	24.50	12.02	0.93	610	20.06	22.15	17.79	0.43
CBR	Non-immersion (%)	1,002	128.0	278	2.5	42.62	2,287	98.1	233	9	27.56
	Immersion (%)	581	73.5	211	2	45.19	613	128	276	21	39.10
	6 days in the air & 4 days immersed						382	155	365	28	49.47
Modified CBR (%)		212	54.7	99	16	16.12	610	83	188	24	22.79
Unconfined Compression Strength (kN/m^2)							73	194.9	700.0	70.0	79.9

Note, n : Number of samples, X : Average, X_{max} : Maximum value, X_{min} : Minimum value, σ : Standard deviation

Table 3-2 Properties of Reused Soils

Road	Non-immersion CBR (%)		Unconfined Compression Strength (kN/m^2)	
	1985	1986	1985	1986
A	116	85	1,140	1,460
B	178	160	2,340	3,480
C	68	51	1,090	1,380
D	65	122	1,070	2,030
E	142	157	1,460	2,280
F	90	97	1,070	1,450
G	125	135	1,770	3,470
H	130	154	1,690	2,530
Average	114	120	1,450	2,260
σ	38.4	39.4	460	860

3.2 Estimate of Layer Coefficient

The layer coefficient of a layer material is defined as a coefficient of structural strength of the material against that of asphalt concrete. The coefficients of various layer materials were determined by AASHTO based on theoretical and statistical analyses of data obtained from a number of experimental pavement construction works. As it is a great trouble to determine the layer coefficient of a new material in the same way as AASHTO, several methods have been proposed to estimate them simply. Among the simplified methods are the methods of estimating them based on the elastic modulus of each material, the methods of estimating them in comparison with materials having known coefficients, and others.

The Public Works Bureau constructed experimental pavement and carried out various field and laboratory tests on the experimental pavement to obtain the data necessary for estimating the coefficients of layers made of the improved soils. The experimental pavement was constructed in three segments of a certain secondary arterial road in the Osaka City. Although the three segments had the same cross-section of pavement structure, upper subbase course of each segment was composed of a material different from other segments. They were improved soil, crushed stone and blast-furnace slag. Comparing with properties of pavement structure composed of materials having known coefficients, the layer coefficients of upper and lower subbase courses in the three segments were estimated.

The layer coefficients of the improved soil layer estimated by various methods are presented in Table 3-3.

As a result, the following values were tentatively determined as layer coefficients of the improved soil layers to be used in the design of pavement structure:

- Upper subbase course : 0.45
- Lower subbase course : 0.35

Appropriateness of these values have been verified through the follow-up surveys on the experimental pavement.

3.3 Quality of Improved Soils

As a result of the preceding discussions, quality of improved soils are summarized as follows:

(1) Fresh improved soils

Although the improved soil contained much fine soil particles than graded grain materials, it had values of modified CBR of more than 80% and met the specifications of granular materials for upper subbase course materials. However, since only a few quicklime (1% more or less) was added, unconfined compression strength of improved soil was as low as about 200 kN/m^2, and did not meet the specifications of not only hydraulic graded slag but also soils stabilized with lime.

(2) Improved soils during and soon after construction

No problems regarding the construction activities and management were found. The pavement had enough bearing capacity soon after the construction. The deflection of the pavement in the early stage had the same tendency as pavement structures constructed using other conventional materials, while the pavement with improved soils had smaller deflection after one month of the construction. This fact indicated that the improved soils were stabilized as a layer in an early stage.

(3) Aged soils after 2 or 3 years of the construction

Unconfined compression strength of improved soils had increased as time goes by. It became 1,450 kN/m^2 after 2 years and 2,260 kN/m^2 after 3 years. The CBR

Table 3-3 Elastic Modulus and Layer Coefficient

		Methods based on Elastic Modulus			Methods based on the Chart		Method	Integrated Results
		Two-layer theory	Unconfined compression tests	CBR tests	Deflection	Surface roughness	based on Wheel trucking tests	
Layer Coefficient	Max.	2.08	0.58	0.67	-	-	0.46	0.67
	Min.	0.43	0.26	0.36	-	-	0.37	0.37
	Average	0.95	0.40	0.49	0.63	0.55	0.42	0.48
Remarks		10 routes (n = 56)	10 routes (n = 56); Average of values estimated by formulas of Dr. Takeshita and Public Works Research Institute	10 routes (n = 56)	1 route (n = 3)	1 route (n = 3)	n = 20	Note, excluding the results from Two-layer theory

values have also increased slightly.

(4) Layer Coefficient

The strength of the improved soil was evaluated to be equivalent to or greater than that of hydraulic graded slag. Therefore, layer coefficients of the improved soil to be used for the pavement design under the direction of the Public Works Bureau were tentatively determined at 0.45 for upper subbase course and 0.35 for lower subbase course.

4. ACCOMPLISHMENT

During the period from 1983 to 1994, the total volume of excavated surplus soils annually discharged from the construction works of the Public Works Bureau as well as that of soils annually conveyed into/improved in the soil improvement plant have varied as shown in Figure 4-1.

As number and scale of public works have increased year after year, excavated surplus soils discharged from those construction works have also gone on increasing. However, volume of produced improved soils has remained about 1,000,000 kN per year which was the original target volume, because of limited capacity of soil yards, increased nocturnal construction works of which surplus soils could not be accepted by the plant, increased poor soils which had difficulties in their improvement, and other causes. Therefore, it is proposed that each of these causes shall be resolved and a more efficient operation system shall be established.

5. CONCLUSIONS

This paper outlines the Osaka City's soil improvement plant and its products, or improved soils. Reclamation areas which have been constructed along the shore of Osaka Bay consuming surplus soils are not currently expected to be expanded as rapidly as before. On the other hand, volume of excavated surplus soils will undoubtedly increase. Therefore, the construction of the soil improvement plant with larger capacity of improving surplus soils has been proposed.

A new plant must have its capacity as large as it can accept all surplus soils discharged from not only construction works of the Public Works Bureau but also all other public works in Osaka City. Since it is quite difficult to secure lands required for a such large scale plant in the city which has been highly urbanized, a possible location will be found in the western reclamation area. How to use or what shall be constructed in such reclamation areas has been discussed and became one of the current topics even though they have not been constructed. It must be time that a soil improvement plant is one of the answers to those discussions and topics. It should be our mission to be fulfilled for the 21st century to re-use excavated surplus soils as useful resources, because we are the ones who are pursuing "the creation of a comfortable town".

6. REFERENCES

1) Japan Road Association: 1992, Manual for Asphalt Pavement.
2) Y. Takemoto, T. Takano and T. Ogawa: 1981, A Study on the Applicability of Improved Soil as Subbase Course Materials (First Report), Proceedings of the 35th Annual Conference of the Japan Society of Civil Engineers, Volume 5.
3) N. Sakai, M. Takaoka, K. Yoshikawa and T. Takano: 1982, A Study on the Applicability of Improved Soil as Subbase Course Materials (Second Report), Proceedings of the 14th Japan Road Conference, Volume 4.
4) T. Ninomiya, H. Saeki, Y. Takemoto and T. Takano: 1987, Properties of Improved Soil (First Report), Proceedings of the 42th Annual Conference of the Japan Society of Civil Engineers, Volume 5.
5) T. Ninomiya, M. Yamada and M. Tachibana: 1988, Application of Soil improved in the Soil Improvement Plant to Subbase Course, Pavement Vol. 23, No. 10.
6) T. Ninomiya, T. Takano and M. Tachibana: 1989, Properties of Improved Soil (Second Report), Proceedings of the 44th Annual Conference of the Japan Society of Civil Engineers, Volume 5.

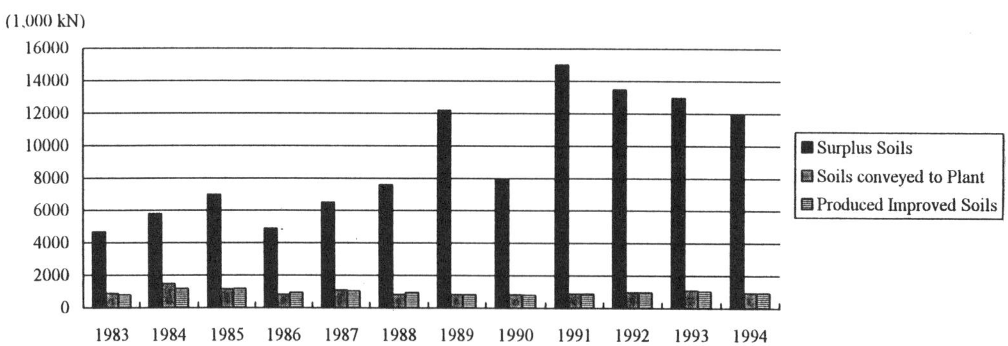

Figure 4-1 Volume of Excavated Surplus Soils, Soils conveyed to Plant and Produced Improved Soils

Environmental Geotechnics, Kamon (ed.) © 1996 Balkema, Rotterdam. ISBN 90 5410 848 7

Hydration mechanisms of fly ash stabilized by lime

Supakij Nontananandh
Department of Civil Engineering, Kasetsart University, Bangkok, Thailand

Masashi Kamon
Disaster Prevention Research Institute, Kyoto University, Japan

ABSTRACT: Many current researches have recognized that coal fly ashes stabilized by cement and/or lime steadily gain strengths over time due to the pozzolanic reaction. The major attraction in using this material in pavement or grouting work is that the structure will be progressively stronger over its design life. Understanding of certain reaction mechanisms significantly provides improvement on fly ash stabilization to a satisfactorily high level. The experimental studies presented in this paper have been devoted to elucidate hydration mechanism and major reaction products of lime-stabilized fly ash mixtures. Evaluation of hardening effects was performed based on the compressive strength test and by means of the physico-chemical tests. It is found that hardening mechanism can be improved when a certain proportion of chemical additives such as aluminum sulfate and sodium carbonate are blended into the hydrated lime. Consequent improvement on strength is due to the enhancement of hydration by a through-solution mechanism. Strength development is closely related to chemical compositions of fly ash, appropriate type and amount of stabilizer, types, degree of crystallization and morphology of reaction products, and changes in their microstructures during the course of reactions.

1 INTRODUCTION

Coal fly ashes (fly ashes) are presently generated in large quantities worldwide due to increasing demand for energy consumption. Fly ash utilization as construction materials can concomitantly compromise concerns for technical and economical feasibility along with the environmental control. Potential uses for this material include a design of lime-cement-fly ash for runway pavement (Jefts, 1986), remedial works on the leachate problems by fly ash-cement grouting, materials for concrete dam and recently for roller compacted concrete (RCC) dam, and fly ash-bentonite containment walls (Cabrera et al., 1994).

Concerning technical advantages, fly ashes accelerate the early strength of cement, reduce heat of hydration, improve workability of concrete, and improve resistance to sulfate attack and chloride ion penetration. In the presence of hydrated lime, coal fly ashes contribute to long-term strength due to the carbonatious and pozzolanic reactions.

For hydration mechanisms of the lime-pozzolana mixtures, several hypotheses have been proposed. Uchikawa and Uchida (1980, 1986) found that the initial rate of hydration of the C_3A-lime-gypsum mixtures was activated by pozzolanic material. Chemical reactivity was enhanced by its characteristics such as specific surface area, exchange capacity of cation, and glass phase contents. Enhancement of hydration in this case is basically due to a topochemical mechanism.

Preliminary study on the stabilization of lime-fly ash mixtures by Kamon et al. (1988) indicated that changes in microstructures were due to certain hardening effects. Significant improvement on strengths could be observed

Table 1a Chemical compositions of fly ashes

Materials	Compositions (%)								
	SiO_2	Al_2O_3	Fe_2O_3	CaO	MgO	Na_2O	K_2O	TiO_2	SO_3
Fly Ash I	49.3	22.0	4.40	5.40	1.10	2.01	1.20	1.23	0.70
Fly Ash II	56.9	32.7	1.70	0.50	0.30	1.12	0.11	1.74	<0.1

Table 1b Selected national standard specifications for fly ashes used in concrete and properties of fly ashes used in this study.

Requirement	Fly ash		Australia	Canada		India	Japan	UK	USA	
	I	II		Class C	Class F				Class C	Class F
Ignition loss, max%	1.1	5.1	5.0	12.0	6.0	12.0	5.0	7.0	12.0	6.0
SO_3, max %	0.7	0.1	-	5.0	5.0	3.0	-	5.0	5.0	5.0
MgO, max %	1.1	0.3	5.0	-	-	5.0	5.0	5.0	5.0	5.0
Alkali metal as Na_2O, max%	2.0	1.1	-	-	-	1.5	-	1.5	1.5	1.5
SiO_2, min %	49.3	56.3	-	-	-	35.0	45.0	-	-	40.0
$SiO_2+Al_2O_3+Fe_2O_3$, min%	75.7	91.3	-	-	-	70.0	70.0	70.0	-	50.0
Specific surface area, m^2/kg	310	330	-	-	-	320	270	-	-	-

when lime was mixed with some chemical additive such as sodium carbonate and aluminum sulfate. However, conceptual views on how strengths are improved, especially in the presence of this type of additives, are unclear and thus need to be further substantiated. It is, therefore, our main objective in this study to elucidate hydration mechanism of coal fly ashes stabilized by lime, in order to achieve strengths applicable as road subbase and base materials.

2 EXPERIMENTAL PROCEDURE

2.1 *Materials*

Two types of fly ashes, designated as fly ash I and fly ash II, were used. According to ASTM C-618, these fly ashes are classified as low-lime (class F) fly ashes. Both types of fly ash consist mostly of glassy, spherical particles as observed by scanning electron microscopic observation. X-ray diffraction patterns of the untreated ashes clearly reflected peaks corresponding to quartz (SiO_2) and mullite ($3Al_2O_3 \cdot 2SiO_2$). Material compositional analysis revealed that Si, Al, and Ca are identified as dominant elements for fly ash I, while Si and Al predominate in fly ash II. Chemical compositions of fly ashes are given in Table 1a. As illustrated in Table 1b, the properties of fly ashes used are consistent with the standard specifications of other countries.

2.2 *Specimen preparation and test*

Specimens were prepared by mixing fly ashes with hydrated lime and lime-based stabilizer, using mix proportions of 10% and 20% by total weight. Lime-based stabilizer used as a hardening material consists of 70% $Ca(OH)_2$, 18% $Al_2(SO_4)_3$ and 12% Na_2CO_3. In order to obtain a uniform consistency of freshly mixed fly ash with workability (flow value) within a range of 150-190 mm, it was decided to control the mixtures with a mixing water of 60% by total weight. Mixing conditions are summarized in Table 2.

Mixing and preparation of specimens were carried out according to the method of making and curing noncompacted - stabilized soil specimens (JSF T-821). Fly ash fresh pastes were put in a mold for obtaining a cylindrical specimen of 5 cm in diameter and 10 cm in length. After approximately 18-24 hours, the specimens were demoulded and cured in a constant-temperature room (20°C, 80% relative humidity) until testing. Compressive strength tests were carried out after the curing periods of 3, 7, 28, and 90 days. Some were available for

Table 2 Strength of the fly ash-stabilized mixtures

Mixtures	Materials & Contents		Compressive Strength (MPa)				
Symbol	fly ash	stabilizer	3 days	7 days	28 days	90 days	2.75 years
MIX-1	90% ash I	10% lime	0.16	0.27	0.48	2.63	6.40
MIX-2	80% ash I	20% lime	0.08	0.18	1.10	5.15	*
MIX-3	90% ash I	10% LBS	0.23	0.65	7.14	7.17	*
MIX-4	80% ash I	20% LBS	0.49	1.07	10.11	11.01	*
MIX-5	90% ash II	10% lime	0.12	0.13	0.28	0.28	2.48
MIX-6	80% ash II	20% lime	0.06	0.09	0.46	0.57	*
MIX-7	90% ash II	10% LBS	0.07	0.17	0.84	2.79	*
MIX-8	80% ash II	20% LBS	0.08	0.31	4.38	5.15	12.30

Note : LBS stands for lime-based stabilizer which is composed of 70% lime, 18% $Al_2(SO_4)_3$, and 12% Na_2CO_3
* stands for no observed data

the test almost after 3 years. In order to elucidate hydration mechanisms and major reaction products, X-ray diffraction analysis (XRD), scanning electron microscopic observation (SEM) and energy dispersive X-ray microanalyzer (EDX) were performed on the failure specimens after strength test.

3 RESULTS AND DISCUSSION

3.1 *Strengths of fly ash mixtures*

Table 2 shows comprehensive strengths of all mixtures. It is most likely that strengths of lime-stabilized mixtures were developed slowly in the early age and steadily increased as curing time increased. For such mixtures as MIX-1, MIX-5 and MIX-6, strength development appears to be more pronounced after 33 months. Actually, a relatively high strength in long term elucidates the potential for utilization of the stabilized fly ash as subbase for road.

For the same mixing condition, the mixtures of fly ash I gave considerably higher strengths than those of fly ash II in all respects. It is, however, can be noticed that at long term the fly ash II mixtures can develop strengths comparable to those of fly ash I mixtures. As shown in Table 2, the 2.75-year specimen of MIX-5 and MIX-8 developed strengths to a certain level of the 90-day strengths of MIX-1 and MIX-4, respectively.

As a material for road base purpose, it is recommended that a mixture should gain 10-day strength of 0.7 MPa for a lower base and 1.0 MPa for upper base, without yielding a loss in strength. It is found that the strengths were

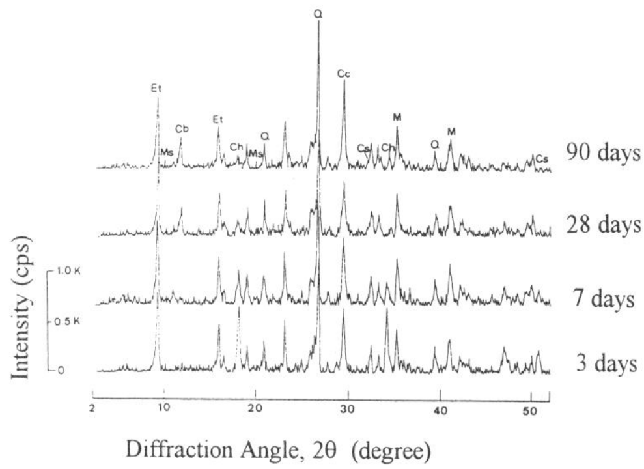

Figure 1 XRD patterns of 20% lime-based-stabilized fly ash I at various curing periods.

markedly improved in agreeable with the recommendation when lime-based stabilizer was used. For example, the 28-day strength of fly ash I, MIX-1 and MIX-3, are 0.48 and 7.14 MPa, while, those of fly ash II, MIX-6 and MIX-8, are 0.46 and 4.38 MPa, respectively. In this study, the stabilizer was found more active in the fly ash I mixtures than in the fly ash II mixtures. This indicates that chemical compositions of the original ash affect the hardening in the resultant mixtures.

The evidence of strength development thus brings to light some essential roles by which aluminum sulfate and sodium carbonate in combination with lime contributes to strength of fly ash mixtures. It is believed that lime-based stabilizer accelerates hydration of ash through

Figure 2 Typical SEM micrographs of 20% lime-stabilized fly ash I at specified curing periods of 3, 7, 28, and 90 days.

Figure 4 Typical SEM micrographs of 10% lime-stabilized fly ash II at specified curing periods of 3, 7, 28, and 90 days.

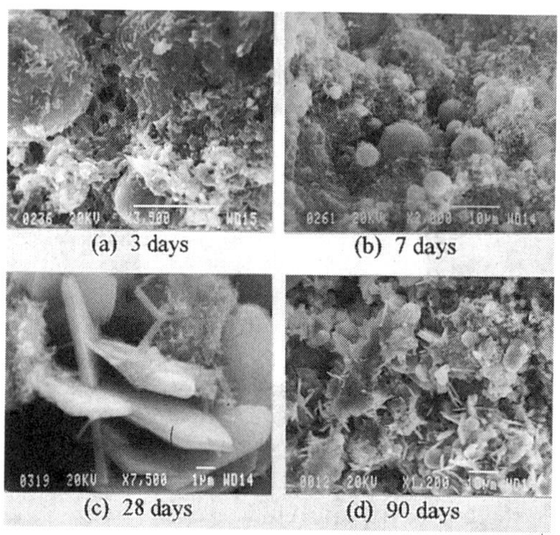

Figure 3 Typical SEM micrographs of 20% lime-based-stabilized fly ash I at specified curing periods of 3, 7, 28, and 90 days.

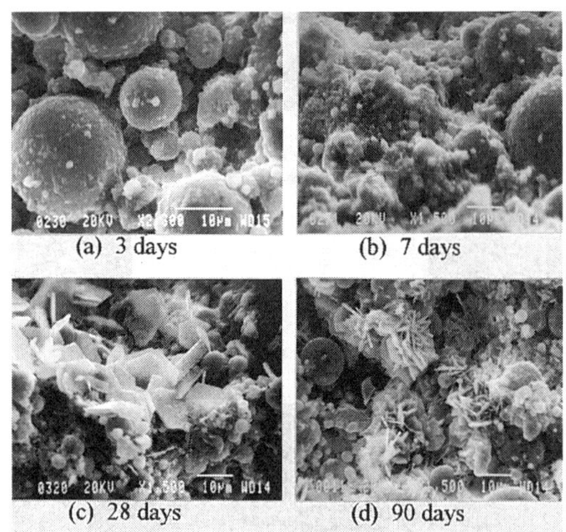

Figure 5 Typical SEM micrographs of 20% lime-based-stabilized fly ash II at specified curing periods of 3, 7, 28, and 90 days.

particular mechanisms. Further discussion is presented to substantiate this assumption.

3.2 *Changes in mineral composition and micro-structures in relation to strength development*

Typical XRD patterns of the fly ash I mixed with lime-based stabilizer (Figure 1) show reduction of the crystalline components of fly ash and hydrated lime. They converted into reaction products which can be identified as calcium silicate hydrate (CSH), calcium aluminate hydrate (CAH), ettringite, and calcium carbonate ($CaCO_3$). Mixtures having superior strengths exhibit larger amounts of these major reaction products, while those with low strength exhibit weak diffractions of such cementitous materials but strong peaks of hydrated lime. For lime-fly ash I, no marked changes in ash particles and the overall microstructures could be seen even at long term, as illustrated in Figures 2a - 2d. The reactions in this case seemingly proceed in a slow rate. Conversely, for lime-based stabilizer, the reaction products were abundantly produced on the fly ash surface, filling up the voids which were previously filled with pore water (Figure 3).

Conversion of ettringite crystals into hexagonal cardboard-shape monosulfates could be observed (Figures 3b-3c). Accordingly, fly ash particles virtually changed into smaller in sizes after 28 days. Figures 3c and 3d reveal that fly ash particles act as spherical cores where upon reactions products are formed extensively. The combination of reaction products such as this resulted in a higher rigidity.

In agreement with the strength test results, there were no marked changes for fly ash II when mixed with lime (Figures 4a-4d). Some remained unreacted even at long time. When lime-based stabilizer was used, changes in microstructures could be observed (Figures 5a-5d). However, there are some distinct differences between the developed microstructures of fly ash I and fly ash II mixtures.

First, monosulfate phases formed in fly ash II mixtures are the hexagonal rosette features (Figure 5d), while, in fly ash I mixtures the hexagonal cardboard features are formed (Figure 3c). Second, some inert fly ash II particles remain unreact and inherently impede in a dense structure (Figure 5d).

It can be concluded that fly ash I consists of more active alumino-silicate particles than fly ash II. In addition, substantial changes in microstructure into a denser texture contribute to strength development of the lime-fly ash stabilized materials.

3.3 *Hydration mechanisms of fly ash mixtures*

Based on investigations of changes in microstructures and their morphology in terms of hydration time, the conceptual pictures regarding the reaction mechanisms can be highlighted. The essential roles of sodium carbonate and aluminum sulfate as a chemical additive for lime can be asserted using the reacting solid particle models as illustrated in Figure 6. Based on the results of strength test and microscopic observations, reactivities of the

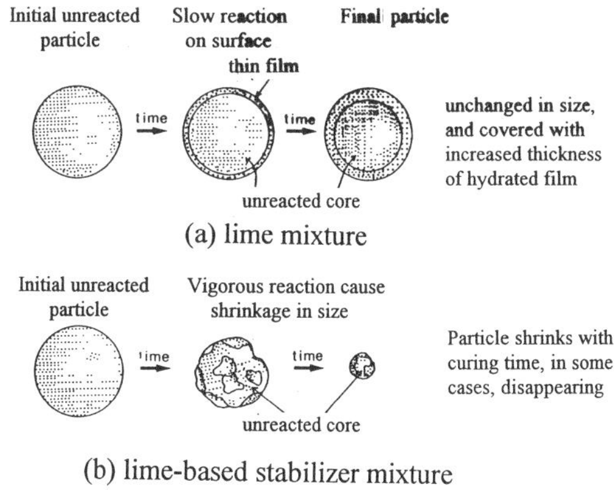

Figure 6 Reaction mechanisms of reacting fly ash particles for: (a) in lime mixtures, and (b) in lime-based stabilized mixtures.

mixtures can be categorized by the following order; lime-based stabilized ash >> lime-stabilized ash > hydrated ash. In all cases for the same mixing condition, fly ash I gains superior reactivity than fly ash II.

Beneficial effects concerning an addition of additives into lime are attributed to three main reasons.

First, the chemicals added improve reaction by increasing the rate of solubility of fly ash particles, thus, increase the rate of initial hydration and pozzolanic reaction in long term.

Second, in addition to this, major reaction products such as CSH, ettringite and monosulfate are richly formed and gradually strengthened structures as reaction progresses.

Third, cementation is activated due to the presence of sodium carbonate. Formation of $CaCO_3$ substantially fills the voids and thus provides additional cementation of the phases.

In the lime-fly ash system, ash particles initially react slowly with hydrated lime. Thin hydrate films, presumedly a CSH phase, are formed, covering the unreacted cores. These protective layers then retard further reaction of fly ash particles.

However, as curing time increases, thickness of the hydrate film increases slightly, probably due to a diffusion of Ca^{2+} , OH^- , and H^+ into the inner cores (Figure 6a). Through such reaction mechanisms, the particles sizes are mostly unchanged as previously illustrated by SEM micrographs.

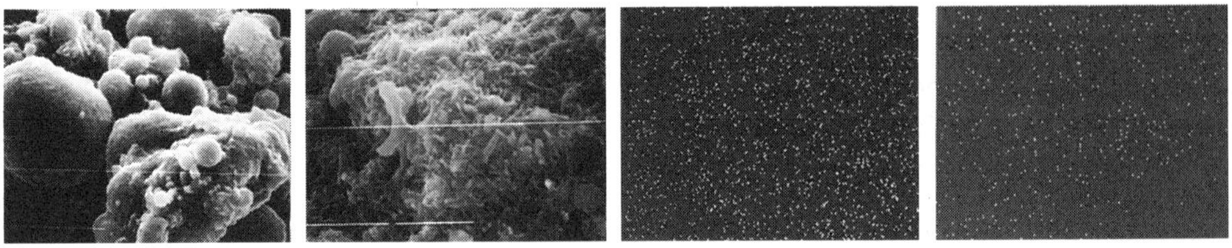

(a) (b) (c) (d)

Figure 7(a) Dissolution and agglomeration of fly ash in lime-based mixtures.
7(b) SEM micrographs of 10% lime-based stabilized fly ash at 3 days.
7(c) X-ray mapping of Ca element on the scanned view of 7(b).
7(d) X-ray mapping of Si element on the scanned view of 7(b).

Conversely, the particles seem to dissolve vigorously at the very beginning when sodium carbonate and aluminum sulfate are present in the mixtures (Figure 6b).

As clearly seen in Figure 7a, surfaces of the spherical ashes are leached out and changed into irregular shapes, while some grains are agglomerated into a cluster. Dissolution of fly ash in this case can be further substantiated by an investigation of the constitutive elements on the surface. It is noted that fly ash is the only source of silicon. Therefore, as illustrated in Figures 7b-7d, the distribution of Si and Ca over the whole surface will assert to the fact that silicate as well as aluminate minerals dissolve into the pore solution, and thus activate hydration of the lime-based-fly ash system, resulting in the development of strength. As can be expected, such observations were not found in the lime-fly ash systems.

4 CONCLUSIONS

Based on the experimental study, the main results are concluded as follows:

1. Strength development of the coal fly ash is substantially influenced by several factors, namely, reactivity of original fly ash, type and amount of stabilizer and mixing condition.

2. Strengths have improved in relation to the types of hydration products and their morphological characteristics.

3. The major reaction products which contribute to strengths are calcium silicate hydrate, calcium aluminate hydrate, ettringite and monosulfate and calcium carbonate.

4. Certain proportion of lime, sodium carbonate and aluminum sulfate markedly improve the reactions of the lime-fly ash mixtures by an increase in dissolution of ashes and starting materials, acceleration of the hydration and pozzolanic reactions, simultaneous activation of the carbonation.

5. Hypothesis in relation to strength developing mechanisms is proposed. Strength of the fly ash mixtures can be improved when hydrations are enhanced by the through-solution mechanism.

6. Strengths of the successful mixtures of lime-stabilized fly ash show promise for as geomaterials such as road base and subbase.

REFERENCE

Cabrera, J.G. and Woolley, G.R. (1994); "Fly Ash Utilisation in Civil Engineering", Environmental Aspects of Construction with Waste Materials, edited by Goumans J.J.J.M. et al., pp.345-356.

Jefts, A.R. (1986); "New Runway Takes Off", Civil Engineering, ASCE, pp.47-49.

Kamon, M., Nontananandh, S. and Tomohisa, S. (1988); "On Lime-Stabilized Fly Ashes by X-ray Microanalyzer in relation to Strength Development", Procs. of the 43th JSCE Conference, Hiroshima, Japan, pp.168-169.

Uchikawa, H. and Uchida, S. (1980); "Influence of Pozzolana on the Hydration of C_3A", the 7th International Congress on Chemistry of Cement, Sub-theme IV, Paris, pp.24-29.

Uchikawa, H. and Uchida, S. (1986); "Effect of the Character of Glass Phase in Bleeding Components on their Reactivity in Calcium Hydroxide Mixture" the 8th Int.Congress on Chemistry of Cement, Rio de Janeiro, Sub-theme 3.2, pp.245-250.

Environmental Geotechnics, Kamon (ed.) © *1996 Balkema, Rotterdam. ISBN 90 5410 848 7*

Oyster shell-sand mixture as material of sand compaction piles

T. Okumura
Okayama University, Japan

M. Kobayashi
DRAM Engineering Inc., Tokyo, Japan

ABSTRACT: In Miyagi Prefecture, 75% of oyster shell is either disposed of or stocked without any eventual use in mind. For the purpose of effective usage, geotechnical properties of crushed oyster shells and shell-sand mixtures were investigated, and then the latter were utilized successfully as the material of sand compaction piles to improve the soft clay layer under a breakwater.

1. INTRODUCTION

In Miyagi prefecture, northeastern Japan, 4 thousand tons of oyster are harvested annually and 30 thousand tons of oyster shell are produced. Only 25% of oyster shell is utilized, and the remaining 75% is either disposed of or stocked without any eventual use in mind. Stocked shell in Ishinomaki City, Miyagi Prefecture, for example, amounts 140 thousand tons.

Oyster shell is said to be composed mainly of calcium carbonate, containing organic matter of several percent. The shell is 8-14 cm in length, 4-7 cm in width, 2-4 cm in thickness, having inside pores, rather thin but well graded when crushed. Bulk density of shells is 0.3 (uncrushed) -1.0 (crushed) t/m^3, comparatively small, coefficient of permeability is 10^{-3}-10^{2} cm/s, toxic substances are not dissolved from them, and mixability with soil is good.

Making use of such good characteristics, large amount of oyster shell can be utilized in port and harbor constructions for back filling of revetments and quay walls, sand or gravel drains and sand compaction piles for soft ground under the breakwaters and quay walls, littoral nourishment, sea bottom covering, etc.

For the purpose of effective usage, however, such geotechnical properties of oyster shells and shell-sand mixtures as bulk density, shear strength and compressibility should be clarified. In 1991 fiscal year, laboratory tests of oyster shells, crushed finely and coarsely, and shell-sand mixtures were carried out.[1] In 1992, trial usage of oyster shell-sand mixtures to 81 sand compaction piles for improving soft clay layer under the breakwater in Ishinomaki port, and some relevant laboratory and field tests were performed.[2] In 1993, practical usage to 420 sand compaction piles and relevant investigations were performed.[3]

2. LABORATORY TESTS

Samples for the laboratory tests are tabulated in Table 1, and details of the tests in Table 2. Sand G

Table－1 Kinds of samples

Year	Symbol	Detail
1991	A	crude oyster shell
	B_1	crushed shell(max. dia. 30mm)
	C	crushed shell(max. dia. 5mm)
	D	powder shell(incl. impurities)
	D'	powder shell(not incl. impurities)
	E_1	B_1 50% + G_1 50%(in weight)
	F	B_1 25% + G_1 75%(in weight)
	G_1	sand(50% crude + 50% washed)
1992	B_2	crushed shell(max. dia. 30mm)
	H	B_2 33% + I 33% + J 33%(in volume)
	I	crude sand incl. impurities
	J	washed sand
1993	B_3	crushed shell(max. dia. 30mm)
	E_3	B_3 50% + G_3 50%(in volume)
	K	B_3 40% + G_3 60%(in volume)
	G_3	sand(50% crude + 50% washed in vol.)

Table－2 Laboratoy tests

Year	Soil test	Samples	Remarks
1991	Grain Density	A, B_1×3, C, D, D', G_1	
	Grain Size Distribution	D, D', G_1 Max. Density(B_1, C)×4, Direct Shear(A~G_1)×18, Compaction(B_1, C)×8	before & after tests
	Max., Min. Density	A, B_1, C, E_1 F, G_1	
	Moisture Content	A, B_1, C, D, D', E_1, F, G_1	
	Permeability	B_1, C	Dr 80%
	Direct Shear	A, B_1, C, E_1, F, G_1	5 pressures, Dr 80%
	Compaction	B_1, C	3 energies
	Modified CBR	B_1, C	
1992	Grain Density	B_2, H, I, J	
	Grain Size Distribution	B_2, H, I, J	
	Max., Min. Density	B_2, H, I, J	
	Moisure Content	B_2, H, I, J	
	Direct Shear	H	
	Bulk Density	B_2, H, I, J	
1993	Grain Density	B_3, E_3, K, G_3	
	Grain Size Distribution	B_3, E_3, K, G_3	
	Max., Min. Density	B_3, E_3, K, G_3	
	Direct Shear	E_3	

Table－3 Laboratory test results

year		grain density g/cm³	50% grain size mm	moisture content %	dry density max. g/cm³	dry density min. g/cm³	compression index	bulk density g/cm³
1991	A	1.737	59	33.9	0.251	0.182	7.000	
	B_1	2.174	13	27.6	0.617	0.447	0.770	
	C	2.609	3.1	45.7	0.730	0.546	0.460	
	D	2.629	0.49	12.0				
	D'	2.662	0.57	26.8				
	E_1	2.420	0.95	19.0	1.092	0.889	0.325	
	F	2.543	0.59	15.8	1.364	1.078	0.175	
	G_1	2.666	0.50	12.4	1.606	1.278	0.044	
1992	B_2	2.702	2.5	24.0	1.081	0.804		0.907
	H	2.699	0.63	14.6	1.452	1.108		1.159
	I	2.698	0.56	15.2	1.551	1.263		1.281
	J	2.697	0.61	9.1	1.686	1.384		1.310
1993	B_3	2.674	4.0		0.873	0.675		
	E_3	2.668	0.66		1.416	1.098		
	K	2.666	0.62		1.481	1.187		
	G_3	2.664	0.51		1.507	1.237		

		coeff. of permeability cm/s	optimum moisture content %	max. dry density g/cm³	modified CBR 90%	CBR 95%
1991	B_1	3.18×10^{0}	34.5	1.130	14.8	16.5
	C	8.07×10^{1}	30.5	1.307	16.2	20.8

is the mixture of crude sand I 50% and washed sand J 50% in volume. Containers for testing grain density, maximum and minimum density, and direct shear behavior of samples including large size shell grains such as A, B, C, etc. are specially big.

Laboratory test results are tabulated in Table 3. Grain density of shell samples (A-D) is larger with smaller grain size, because inside pores of smaller grain will be less than those of larger grain. Maximum and minimum density of shell samples are smaller than those of sand, and larger with smaller grain. Void ratios of shell samples are larger than those of sand at both maximum and minimum density, and hence coefficient of permeability is larger.

Constant pressure drained direct shear test results of the first year are shown in Fig. 1. Shear resistance of samples including large size shell tends not to increase at a high pressure. It will be due to effect of grain crushing during shear. Angle of internal friction also tends to decrease at a high pressure, but for samples C and G it is constant over the pressure of 100kPa, showing no influence of crushing for finer grains.

Crushing of shell grains takes place during shear, consolidation, and compaction. Fig. 2 shows grain size distributions before and after direct shear test, as an example. Larger size grain crushes more.

From the above laboratory test results, characteristics of oyster shell can be concluded as follows :

a) Grain density and maximum & minimum density of the shell is smaller than those of sand.
b) Compressibility is larger than that of sand.
c) Angle of internal friction (shear strength) is comparable to sand. However, affected by grain crushing, internal friction of large size shell decreases at a high pressure.
d) Permeability is larger than that of sand.
e) CBR is comparable to sand.
f) Crushed shell is well graded.
g) Characteristics of shell-sand mixtures are in between shell and sand.
h) There exist some variations in the properties of shell-sand mixtures depending on the producing place and the crushing method.

3. STANDARD PENETRATION TEST ON MODEL GROUND

Fig. 3 shows model ground where standard penetration tests (SPT) were carried out for the shell-sand mixture H of various densities. Test

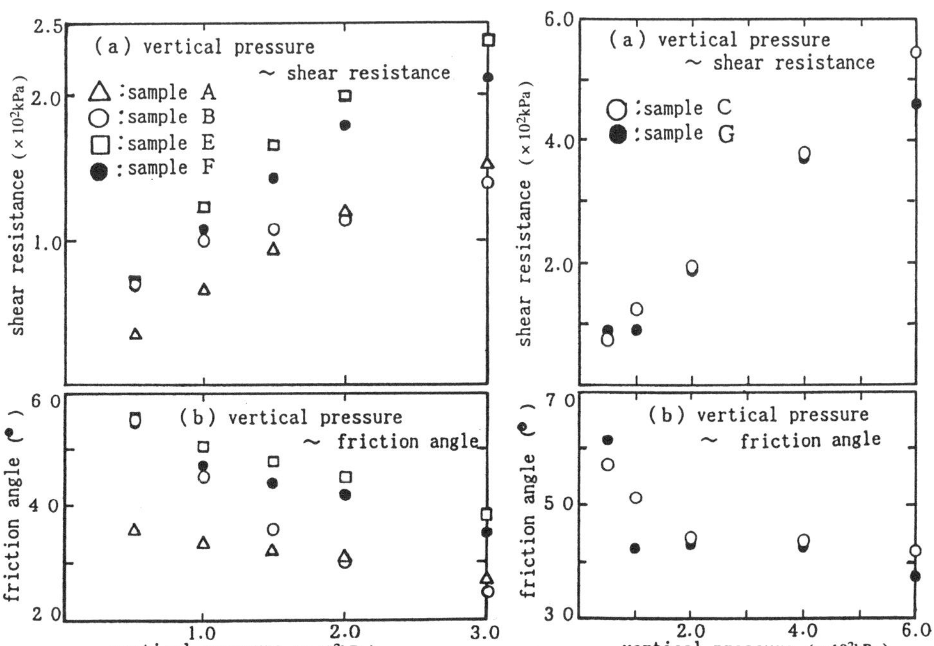

(a) large scale (1991)　　(b) Mikasa type

Fig. 1 Direct shear test result

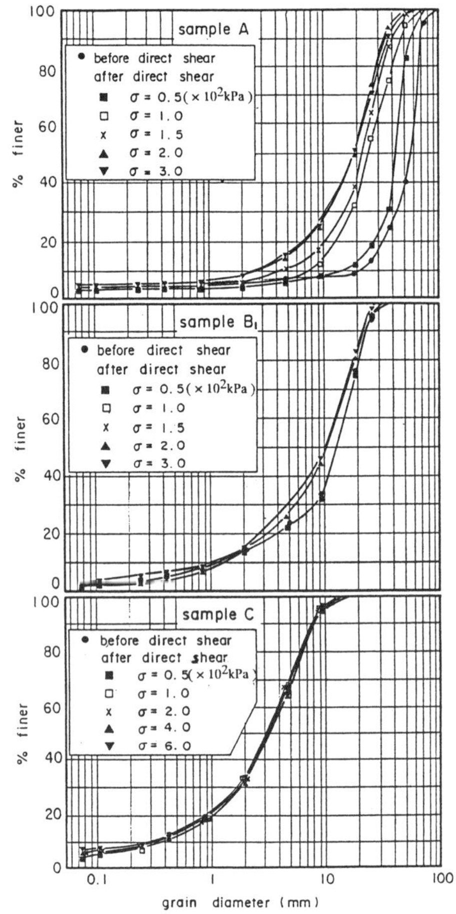

Fig. 2 Grain size distribution before and after direct shear test

results are tabulated in Table 4. The greater the dry density or relative density is, the higher the N value or equivalent N value is.[4)]

4. SAND COMPACTION PILES AND THEIR EVALUATION

In 1992 fiscal year, 81 sand compaction piles (SCP) of H sample were driven experimentally as indicated by ◎ in Fig. 4. Mixing of shell and sand was made naturally on the way of transportation from the stock yard to the working vessel. Homogeneity is confirmed by examining grain size distribution of SPT samples taken from the driven SCP.

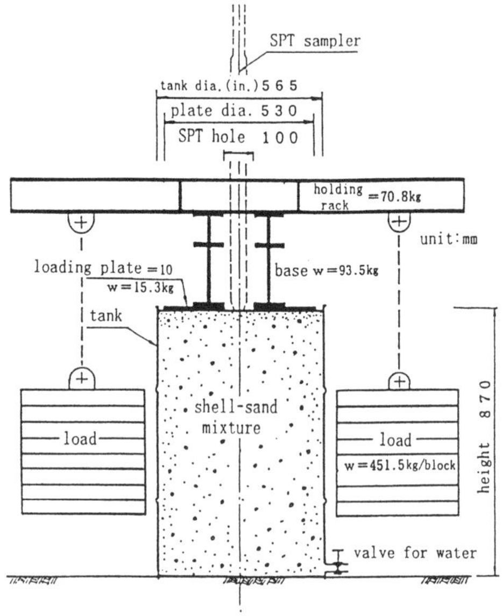

Fig. 3 Model ground for SPT

Shear test of undisturbed samples of the driven SCP is practically impossible. Therefore, shear strength (angle of internal friction) of SCP was estimated from field and model N values as follows :

a) Make diagram correlating dry density to equivalent N value from Table 4.

b) Make diagram correlating dry density to angle of internal friction with parameter of overburden pressure from the direct shear test results.

c) Compute equivalent N value from field N value and assumed bulk density of SCP. Repeat computation until the correlation a) is satisfied within an error of 5%, and finally decide the field bulk density of SCP.

d) Angle of internal friction is estimated from the correlation b) .

Estimated angle is 48° in average and greater than the designed value of 35°. This high value of angle of internal friction may be attributed by well gradation due to mixing of sand and crushed shell.

Settlement of the soft ground improved by shell-sand compaction piles was also estimated from the results of direct shear test. Estimated amount of settlement was greater but not so much than that of sand.

In 1993, 420 shell-sand compaction piles (volume of crushed shell is 10 thousand m^3) were driven into the same site as 1992 under the breakwater as a regular construction. Field SPT and other investigations showed sufficient quality of shell-sand

Table－4 Results of model SPT

No. of compaction	0	5	20	40	100
N value	0	4.5	9	12	19
dry density t/m^3	1.359	1.523	1.610	1.661	1.719
relative density %	78	115	132	141	150
overburden pressure tf/m^3	1.99	7.93			
equivalent No. 66	—	4.0	8.3	11.2	17.8

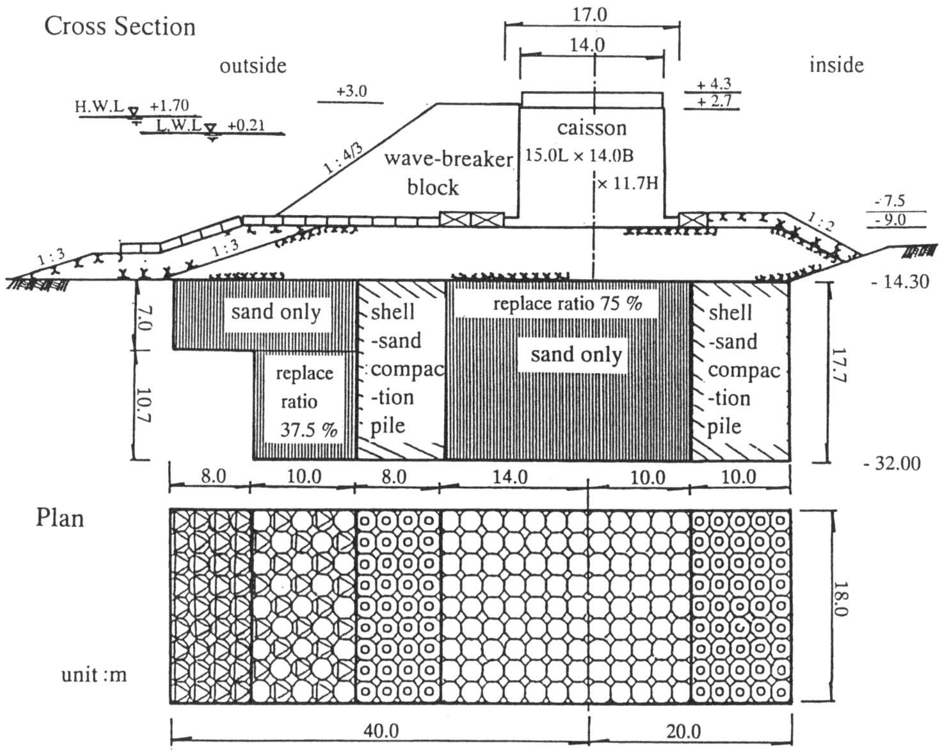

Fig. 4 Experimental shell-sand compaction piles under a composite breakwater

compaction piles (average angle 47°). In 1994 and thereafter 10 thousand m^3 of crushed shell are being utlized annually as the material of sand compaction piles under the breakwater.

5. CONCLUDING REMARKS

For the purpose of recycling, geotechnical properties of oyster shells and shell-sand mixtures were investigated in the laboratory, in the model ground, and in the field. Making use of advantageous characteristics, oyster shell-sand mixtures were utilized successfully as the material of sand compaction piles to improve experimentally and practically the soft clay layer under a breakwater.

These investigations ware carried out by Shiogama Port construction office (Head ; Mr. Y. Hashidate), 2nd District Port Construction Bureau, Ministry of Transport, under the instruction of " Investigation Committee for Recycling Construction Materials in Ishinomaki Port (Chairman : Prof. E. Yanagisawa) " and with cooperation of Ishinomaki City, Miyagi Prefecture, and Miyagi Prefectural Federation of Fishermen's Cooperative Associations. The authors are much obliged to the above persons.

REFERENCES

1) Hashidate, H., Fukuda, T., Okumura, T. and Kobayashi, M. (1993): Engineering properties of oyster shell-sand mixtures, Proc. 28th Annual Meeting of Japanese Soc. of Soil Mech. & Foundation Eng., Vol. 1, pp. 869-872 (in Japanese)
2) Hashidate, H., Fukuda, T., Okumura, T. and Kobayashi, M. (1994): Engineering properties of oyster shell-sand mixtures and their application to sand compaction piles, Proc. 29th Annual Meeting of JSSMFE, Vol. 1, pp. 717-720 (in Japanese)
3) Okumura, T. and Kobayashi, M. (1995) : Geotechnical properties of oyster shell-sand mixtures, Proc. 30th Annual Meeting of JSSMFE, Vol. 1, pp. 733-734 (in Japanese)
4) Iai, S., Koizumi, K. and Tsuchida, H. (1986) : A new criterion for assessing liquefaction using gradation and N value, Rept. of Port & Harbour Research Inst., Vol. 25, No. 3, pp. 125-234 (in Japanese)

Environmental Geotechnics, Kamon (ed.) © 1996 Balkema, Rotterdam. ISBN 90 5410 848 7

Geotechnical properties of fresh coal ashes from thermal power plants

T.Okumura – *Okayama University, Japan*
K.Takahashi – *Port and Harbor Research Institute, Japan*
M.Tezuka – *Center for Coal Utilization, Japan*
Y.Umehara – *Coastal Development Institute of Technology, Japan*
M.Kobayashi – *DRAM Engineering Inc., Tokyo, Japan*

ABSTRACT: Geotechnical properties of fresh fly and clinker ashes from thermal power plants all over Japan were investigated and roughly concluded: Grain density is smaller than that of soil; Fly ash is classified into silt and clinker ash into sand; Variations in properties of coal ash are rather large; Consolidation and shearing characteristics of fly ash are equivalent to those of silt ; Shearing resistance of clinker ash is equivalent to that of sand; Reactivity of fly ash is high.

1. INTRODUCTION

For the purpose of recycling coal ash from thermal power plants as material for port and air port constructions, geotechnical properties and their variations were investigated. Fresh fly ashes were sampled twice (Nov. and Jan.) from all plants (22) in Japan, and 12 times (Nov. through Feb.) from two representative plants. Fresh clinker ashes were sampled 6 times from the same two plants. Both ashes were tested in the laboratory as shown in Table 1.[1)2)]

2. PHYSICAL PROPERTIES

2.1 Grain density

Grain density of fly ash is 2.19 ± 0.11(standard deviation)(g/cm^3) as a whole (Table 2), which is smaller than that of common soils. There is rather large difference in the mean values of grain density between two plants. This difference of 0.17 is much larger than the standard deviation of each plant (0.08, 0.07). It means that there exists clear difference between two plants owing to kind of coal, method and temperature of burning, sampling place, etc. Coefficient of variation is 4.8 % in among all plants but smaller value of 3.5 % in time dependent variation at one plant. This tendency of variation is common in most properties as described later on. Comparing with that of soils in harbor area of 1.9 %, the variation of fly ash is much higher.

Grain density of clinker ash is 2.26 ± 0.12, which is a little larger than that of fly ash . Also in this case, difference between two plants is large. The range of grain density data of fly and clinker ash is not different from that of literatures.

Table-1. Investigations

test	ash	nos.
1) Time dependency		
① grain density	FA	24
	CA	12
② grain size	FA	24
	CA	12
③ max. & min. density	FA	24
	CA	12
④ consistency	FA	24
⑤ compaction	FA	24
⑥ consolidation	FA	24
⑦ triaxial compression	FA	24
⑧ direct shear	CA	12
⑨ chemical component(9item)	FA	24
	CA	12
2) All plants		
① grain density	FA	40
② grain size	FA	40
③ max. & min. density	FA	40
④ consistency	FA	20
⑤ compaction	FA	20
⑥ consolidation	FA	20
⑦ triaxial compression	FA	20
⑧ chemical component(9item)	FA	40

FA : fly ash CA : clinker ash

Correlations exist and are proportional between grain density and maximum density of fly ash, and maximum and minimum density of clinker ash.

Table-2 Grain density

	mean	standard deviation	coeff.of variation	nos.
Fly ash (whole)	2.190	0.107	0.0489	64
time dependency	2.184	0.115	0.0527	24
plant No.7	2.269	0.076	0.0335	12
plant No.16	2.098	0.073	0.0348	12
all plants	2.193	0.105	0.0479	44
Nov.	2.195	0.094	0.0428	22
Jan.	2.192	0.115	0.0524	22
Clinker ash	2.260	0.120	0.0531	12
plant No.7	2.334	0.125	0.0536	6
plant No.16	2.186	0.053	0.0242	6

Table-3(a) Grain size (clay content) (%)

	mean	standard deviation	coeff.of variation	nos.
Fly ash (whole)	14.54	11.71	0.8054	53
time dependency	14.77	4.87	0.3295	13
plant No.7	19.33	2.66	0.1375	6
plant No.16	10.86	1.68	0.1543	7
all plants	14.48	12.80	0.8840	43
Nov.	14.62	12.42	0.8495	21
Jan.	14.35	13.45	0.9373	22

Table-3(b) Grain size (silt content) (%)

	mean	standard deviation	coeff.of variation	nos.
Fly ash (whole)	73.03	11.21	0.1535	53
time dependency	69.62	4.87	0.0700	13
plant No.7	73.17	2.64	0.0361	6
plant No.16	66.57	4.28	0.0642	7
all plants	73.99	12.10	0.1635	43
Nov.	74.47	12.58	0.1689	21
Jan.	73.53	11.89	0.1617	22
Clinker ash	8.17	5.24	0.6421	12
plant No.7	9.67	6.56	0.6789	6
plant No.16	6.67	3.45	0.5167	6

Table-3(c) Grain size (sand content) (%)

	mean	standard deviation	coeff.of variation	nos.
Fly ash (whole)	12.57	10.55	0.8393	64
time dependency	14.67	6.83	0.4652	24
plant No.7	9.08	2.78	0.3058	12
plant No.16	20.25	4.65	0.2298	12
all plants	11.58	11.84	1.0225	44
Nov.	11.19	12.57	1.1233	22
Jan.	11.97	11.34	0.9474	22
Clinker ash	57.67	12.94	0.2244	12
plant No.7	59.00	6.753	0.1145	6
plant No.16	56.33	17.84	0.3167	6

2.2 Grain size distribution

Content of fines of fly ash is 88 % in average, and silt content is 73 ± 11 (%) (Table 3). Therefore, it is classified into silt (M) of fine (F), according to the Japanese Unified Soil Classification System. Difference in each content between two plants is larger than the standard deviation of each plant. Coefficient of variation is {clay cotent:88%, silt:16%, sand:102%} in among all plants, and {clay:15%, silt:6%, sand:31%} in time dependent variation, which is much smaller. Uniformity coefficient is 5.9 ± 3.1 (Table 4), and is poorly

Table-4 Uniformity coefficient, Uc

	mean	standard deviation	coeff.of variation	nos.
Fly ash (whole)	5.876	3.107	0.5288	54
time dependency	8.843	2.590	0.2929	14
plant No.7	9.229	3.287	0.3562	7
plant No.16	8.457	1.838	0.2173	7
all plants	5.070	2.658	0.5243	43
Nov.	5.057	2.711	0.5361	21
Jan.	5.082	2.671	0.5256	22
Clinker ash	14.110	7.780	0.5514	9
plant No.7	15.150	2.240	0.1597	4
plant No.16	13.280	10.710	0.8065	5

Table-5(a) Maximum density, ρ_{dmax} (g/cm^3)

	mean	standard deviation	coeff.of variation	nos.
Fly ash (whole)	1.125	0.177	0.1574	64
time dependency	1.232	0.118	0.0961	24
plant No.7	1.279	0.134	0.1048	12
plant No.16	1.186	0.082	0.0688	12
all plants	1.073	0.177	0.1646	44
Nov.	1.061	0.146	0.1377	22
Jan.	1.086	0.205	0.1891	22
Clinker ash	0.825	0.167	0.2021	12
plant No.7	0.933	0.130	0.1398	6
plant No.16	0.718	0.128	0.1785	6

Table-5(b) Minimum density, ρ_{dmin} (g/cm^3)

	mean	standard deviation	coeff.of variation	nos.
Fly ash (whole)	0.778	0.140	0.1796	64
time dependency	0.860	0.089	0.1037	24
plant No.7	0.873	0.096	0.1101	12
plant No.16	0.848	0.084	0.0992	12
all plants	0.735	0.140	0.1901	44
Nov.	0.720	0.116	0.1609	22
Jan.	0.751	0.162	0.2151	22
Clinker ash	0.665	0.131	0.1963	12
plant No.7	0.739	0.092	0.1249	6
plant No.16	0.591	0.126	0.2128	6

graded. Also in this case, variation in among all plants is larger than that in time.

Average gradation of clinker ash is fine content 8 %, sand content 58 %, gravel content 34 % (Table 3), and it is classified into sand (S). Uniformity coefficient is 14.1 ± 7.8, and is well graded. Difference in two plants is not clear.

Consistency tests clarified that both ashes were all non-plastic.

2.3 Maximum and minimum density

According to the Japanese Geotechnical Society Standard, JGS-T-161, maximum density of fly ash is 1.13 ± 0.18 (g/cm^3), and minimum density is 0.78 ± 0.14 (g/cm^3) (Table 5). Difference between two plants is within the standard deviation of each plant. Coefficient of variation in among all plants is twice that in time of each plant.

Maximum density of clinker ash is 0.83 ± 0.17, and minimum density is 0.67 ± 0.13 (g/cm^3). Difference between two plants is large, but is inconclusive because of few numbers of data.

Correlation between maximum and minimum density is clear as follows,

Fly ash : $\rho_{dmax} = 1.17\ \rho_{dmin} + 0.21$(g/cm^3),

Clinker ash : $\rho_{dmax} = 1.22\ \rho_{dmin} + 0.01$ (g/cm^3).

Some weak correlation is recognized between each density and SiO_2 content of fly ash.

3. CHEMICAL CHARACTERISTICS

Chemical components of fly ash are SiO_2: 63 ± 6 (%), Al_2O_3: 24 ± 4 (%), Fe_2O_3: 4.1 ± 2.3(%), CaO: 2.8 ± 2.7 (%), MgO: 1.0 ± 0.6 (%), etc. Unburned component (carbon, etc.) is 3.2 ± 2.3 (%) and pH is 10.9 ± 2.2. As for clinker ash , SiO_2: 61 ± 5 (%), Al_2O_3: 21 ± 5 (%), Fe_2O_3: 5.0 ± 2.6 (%), CaO: 2.6 ± 2.0 (%), MgO: 1.0 ± 0.5 (%), unburned component: 5.1 ± 5.8 (%), pH: 9.5 ± 1.3 are not much different from those of fly ash.

As for SiO_2, as main component, time dependent variation at each plant is 7 -10 %, and that among all plants at a time is 10 - 11 %, not so different from each other (Table 6). Same tendencies exist in the case of Al_2O_3 and others.

Calcium oxide, having large effect of hardening, has large variation of about 100 %, and magnesium oxide, having negative effect, about 60 %. Therefore, self hardening properties of fly ash will have large variation.

Rather weak correlations exist between CaO • MaO - pH, Fe_2O_3 - ρ_s, SiO_2 - ρ_{dmax} • ρ_{dmin}, etc.

4. MECHANICAL PROPERTIES

4.1 Compaction characteristics

Examples of compaction tests of fly ash are shown in Fig. 1. Test results are summarized in Table 7. Maximum dry density is 1.25 ± 0.13 (g/cm^3) and optimum moisture content is 25.8 ± 5.8 (%).

Differences between two plants in maximum dry density and optimum moisture content are larger than standard deviation of each plant. Coefficient of variation of maximum dry density is 11 % among all plants, but 6 or 7 % in each plant. That of optimum moisture content is 25 % between all plants, but 15 or 12 % in each plant.

Table-6 SiO_2 content (%)

	mean	standard deviation	coeff.of variation	nos.
Fly ash (whole)	62.7	6.16	0.098	64
time dependency	61.9	5.36	0.087	24
plant No.7	60.9	5.93	0.097	12
plant No.16	62.9	4.77	0.076	12
all plants	63.0	6.45	0.102	44
Nov.	63.1	6.26	0.099	22
Jam.	63.0	6.78	0.108	22
Clinker ash	61.1	5.24	0.086	12
plant No.7	58.9	5.44	0.092	6
plant No.16	63.3	4.32	0.068	6

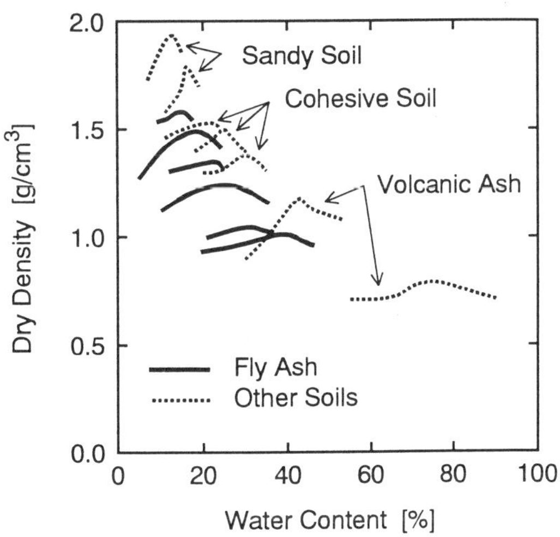

Fig. 1 Compaction curves of fly ash and soil

Table-7-(a) Mximum dry density, ρ_{dmax} (g/cm³)

	mean	standard deviation	coeff.of variation	nos.
Fly ash (whole)	1.252	0.132	0.105	44
time dependency	1.277	0.119	0.093	24
plant No.7	1.362	0.093	0.068	12
plant No.16	1.192	0.073	0.062	12
all plants	1.218	0.138	0.113	22

Table-7-(b) Optimum moisture content, w_{opt} (%)

	mean	standard deviation	coeff.of variation	nos.
Fly ash (whole)	25.77	5.80	0.225	44
time dependency	24.10	3.77	0.156	24
plant No.7	22.15	3.30	0.149	12
plant No.16	26.06	3.25	0.125	12
many plants	27.77	6.86	0.247	22

Table-8 Initial void ratio, e_0

	mean	standard deviation	coeff.of variation	nos.
Fly ash (whole)	1.164	0.403	0.346	44
time dependency	0.857	0.137	0.160	24
plant No.7	0.851	0.155	0.182	12
plant No.16	0.862	0.123	0.142	12
all plants	1.490	0.307	0.206	22

Table-9 Coefficient of consolidation, c_v (cm²/day)

	mean	standard deviation	coeff.of variation	nos.
fly ash (whole)	3,254	1,734	0.533	44
time dependency	4,585	1,143	0.249	24
plant No.7	4,892	1,208	0.247	12
plant No.16	4,278	1,033	0.242	12
all plants	1,997	1,260	0.631	22

Table-10-(a) Coefficient of volume compressibility, m_{v100} (10^{-4}/kPa)

	mean	standard deviation	coeff.of variation	nos.
fly ash (whole)	4.33	8.75	2.024	44
time dependency	4.39	11.79	2.684	24
plant No.7	7.27	16.40	0.443	12
plant No.16	1.52	1.95	1.286	12
all plants	4.17	2.15	0.516	22

Table-10-(b) Coefficient of volume compressibility, m_{v1000} (10^{-5}/kPa)

	mean	standard deviation	coeff.of variation	nos.
Fly ash (whole)	3.424	1.874	0.544	44
time dependency	2.095	1.160	0.554	24
plant No.7	2.261	1.507	0.667	12
plant No.16	1.928	6.922	0.359	12
all plants	4.775	1.377	0.288	22

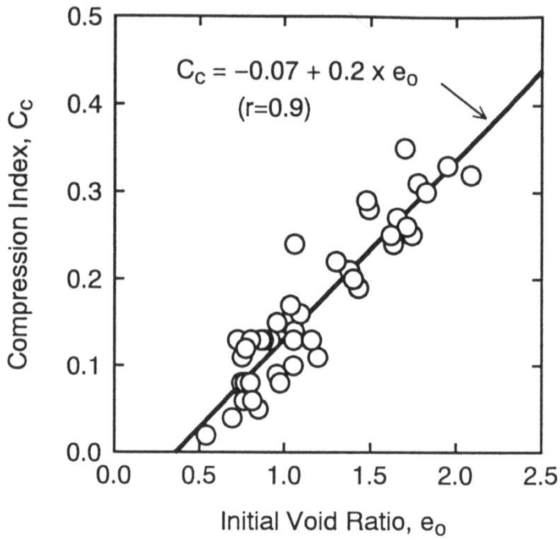

Fig. 2 $c_c \sim e_0$ relation of fly ash

As in the case of common soil, maximum dry density of fly ash increases inversely with optimum moisture content.

4.2 Consolidation characteristics

Fresh fly ash was prepared to be the moisture content of 10 %, compacted satically (JGS-T-812) to be the relative density of 95 %, soaked in water for 28 days, and then supplied for consolidation test. Initial void ratio of the sample is 1.16 ± 0.40 (Table 8). Relative density is about 80 % because of expansion during soaking. Variation of initial void ratio among all plants, 21 % , is larger than that of each plant, 18 or 14 %.

Coefficient of consolidation of fly ash is 3.25 ± 1.73 (× 10³cm²/day) (Table 9), and equivalent to that of silt. Compression index is 0.16 ± 0.09, coefficient of volume compressibility is 4.3 ± 8.8 (×10^{-4}/kPa) at the pressure of 100kPa and 3.4 ± 1.9 (×10^{-5}/kPa) at 1MPa (Table 10). Coefficient of permeability is 8.9 ± 6.8 (× 10^{-7}cm/s) at 100kPa and 1.92 ± 2.73 (×10^{-7} cm/s) at 1 MPa. As a whole, consolidation characteristics of fly ash are equivalent to those of silt.

There exists a correlation between compression index and initial void ratio as shown in Fig. 2.

4.3 Triaxial compression characteristics

Fly ash samples prepared as in the case of consolidation test were consolidated isotropically and tested undrained. Initial void ratio is 1.07 ±

0.34 (relative density is about 90 %).

As shown in Table 11, cohesion is 0.64 ± 1.12 (×10^2kPa), angle of internal friction is 19.9 ± 6.0 (°), and hence, triaxial comoression characteristics of fly ash are equivalent to those of silt.

4.4 Direct shear characteristics

Fresh clinker ash prepared as in the case of consolidation test was supplied for direct shear test. Initial void ratio is 2.06 ± 0.51 (relative density 65 %).

Shearing resistance of clinker ash does not increase proportionally with consolidation pressure, and angle of internal friction decreases with pressure. It will be effect of grain crushing of clinker ash tested. At the consolidation pressure of 100kPa, cohesion is 2.06 ± 0.51 (×10^2kPa) and angle of internal friction is 46.6 ± 8.5 (°) (Table 12).

Correlation between angle of internal friction and

Table-11-(a) Cohesion, Ccu (100kPa)

	mean	standard deviation	coeff.of variation	nos.
Fly ash (whole)	0.644	1.123	1.743	44
time dependency	1.010	1.425	1.411	24
plant No.7	1.256	1.887	1.502	12
plant No.16	0.763	0.744	0.975	12
all plants	0.274	0.270	0.985	22

Table-11-(b) Angle of internal friction, ϕ cu(°)

	mean	standard deviation	coeff.of variation	nos.
Fly ash (whole)	19.94	5.99	0.301	44
time dependency	21.19	6.11	0.288	24
plant No.7	17.93	6.23	0.347	12
plant No.16	24.45	4.01	0.164	12
all plants	18.46	5.41	0.293	22

Table-12 Direct shear test (clinker ash)

	mean	standard deviation	coeff.of variation	nos.
Angle of internal friction (°)	46.58	8.45	0.181	12
plant No.7	49.83	5.64	0.113	6
plant No.16	43.33	9.99	0.231	6
Cohesion (10^2kPa)	2.058	0.507	0.247	12
plant No.7	1.757	0.237	0.135	6
plant No.16	2.359	0.541	0.229	6

Table-13 Density and moisture content (fly ash)

cement content	wet density (g/cm^3)	moisture content (%)	dry density (g/cm^3)
0 %	1.691	37.16	1.233
1 %	1.622	36.84	1.185
2 %	1.635	38.35	1.182

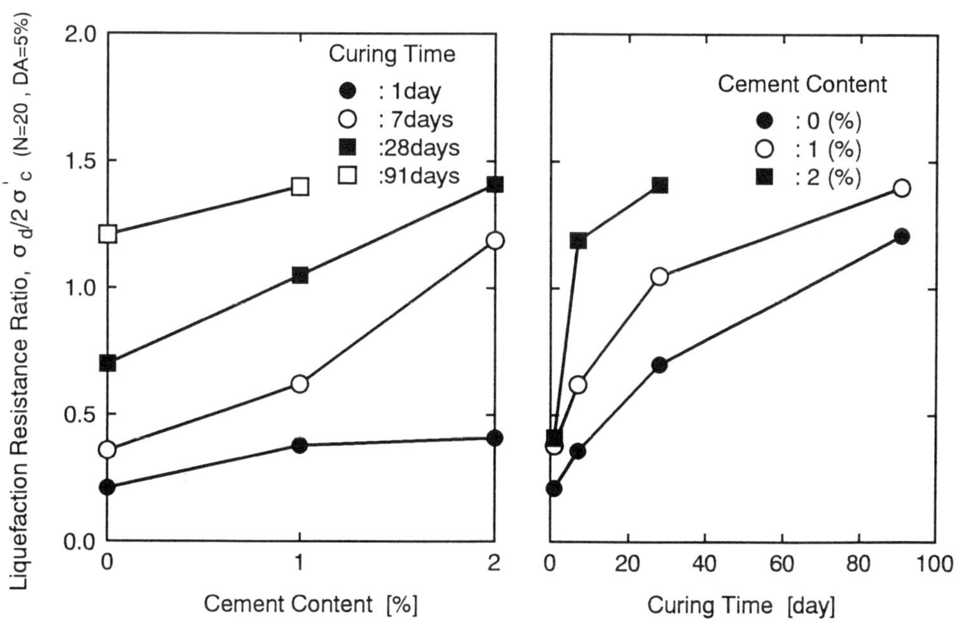

Fig. 3 Liquefaction test results (fly ash)

initial void ratio exists,

$\rho d = 70.6 - 11.56e_0$.

5. PORTLAND CEMENT REACTIVITY

Fly ash sampled at No.16 plant on November was supplied for Portland cement reactivity test such as unconfined, triaxial and liquefaction test. The ratio of water to fly ash and cement is 40 %, and preparation is accorded to JGS-T-821. Density and moisture content are tabulated in Table 13.

Unconfined cmpressive strength and total shearing resistance in triaxial compression test are greater with increasing cement content and curing time period.

Examples of liquefaction test results are shown in Fig. 3, where number of cycles, N, is 20, strain amplitude, DA,is 5 %, $\sigma_d/2$ is shear stress amplitude, and $\sigma'c$ is consolidation pressure. From the figure it is clear that Portland cement reactivity of fly ash is very high.

6. CONCLUDING REMARKS

Geotechnical properties of fresh fly and clinker ashes from all the thermal power plants in Japan were investigated in the laboratory and summarized as follows:

a. Grain density of coal ash is about 2.2 g/cm^3 and smaller than that of common soils.

b. Fly ash is classified into silt, and clinker ash into sand.

c. Variations in most geotecnical properties among all plants at a time are greater than the time dependent variations of one plant. The latter is still greater than those of marine soils in harbor area.

d. Consolidation and shearing characteristics of fly ash are equivalent to those of silt.

e. Shearing resistance of clinker ash is equivalent to sand, but the angle of internal friction decreases at high confining pressure owing to grain crushing.

f. Portland cement reactivity of the selected fly ash is very high.

REFERENCES

1) Takahashi,K. et al. (1995): Engineering properties of fly ash from thermal power plants (Part 1)---Physical and chemical properties---. Proc. 30th Annual Meeting of Japanese Society for Soil Mechanics and Foundation Engineering, pp.757-760 (in Japanese)

2) Takahashi, K. et al. (1995): Ditto (Part 2)---Mechanical properties and reactivity---, Ditto, pp.761-764 (in Japanese)

Environmental Geotechnics, Kamon (ed.) © 1996 Balkema, Rotterdam. ISBN 90 5410 848 7

Effect of plastic wastes in improving cement-treated soils

K.Omine, H.Ochiai, N.Yasufuku & T.Kato
Faculty of Engineering, Kyushu University, Fukuoka, Japan

ABSTRACT: Recently, there has been an increasing amount of plastic wastes and it has become necessary to make use of them. As one of the methods for reuse of plastic wastes in geotechnical engineering, mixing these plastic wastes within the shallow part of soft ground is expected to make soft ground stable and improve its trafficability. The purpose of this study is to investigate the effect of the shape of plastic pieces and cement content on the strength-deformation properties of cement-treated soils. The improvement, brought about by mixing with plastic pieces, is demonstrated by the results of unconfined compression test on the treated soils containing plastic pieces. The plastic pieces were obtained by cutting plastic sheets. The same results are also confirmed for the treated soils mixed with PET-bottle pieces which are ordinary plastic wastes.

1 INTRODUCTION

The production of plastics in Japan has increased gradually and so has the amount of the plastic wastes. The plastic wastes are often classified as a trash unsuitable for burning, because of problems such as a damage to the chimney due to extreme heat and the formation of hydrogen potassium gas. Moreover, considering the difficulty in obtaining reclaimed land for the waste disposal, developing a technique for reusing plastic wastes is one of the important problems in environmental and geotechnical engineering.

There have been some studies on reusing plastic wastes, from a geotechnical engineering point of view. T. Hirata and Y. Maeno (1992) have investigated the mechanical properties of mixtures with trash ash and hard or soft plastic wastes and have discussed their effects on the compaction degree, compressibility and shear strength of the mixtures. T.B. Edil and P.J. Bosscher (1994) have researched the engineering properties of tire chips and soil mixtures for reusing shredded scrap tires as a construction material for light-weight fill material. These methods for the reuse of plastic wastes would not only provide a means of disposing them but would also help to solve difficult economical and technical problems.

As a new method for reusing plastic wastes, mixing PET-bottle pieces within the shallow part of soft ground is expected to make soft ground stable and improve its trafficability (T. Kato et al., 1995). In order to investigate the mechanical properties of cement-treated soils mixed with plastics, unconfined compression tests are conducted. The plastic pieces were obtained by cutting plastic sheets. The effects of slenderness of plastic pieces and cement content on the strength-deformation properties of treated soils are discussed based on the test results. Furthermore, the improvement effects are also confirmed for the treated soils mixed with PET-bottle pieces.

2 UNCONFINED COMPRESSION TEST ON CEMENT-TREATED SOILS WITH PLASTIC PIECES

2.1 *Specimens and Testing Method*

Three kinds of plastic pieces with different lengths L and widths B are used as mixing materials, which are obtained by cutting plastic sheets of 0.04cm thickness. As shown in Table 1, the area (L× B) of these plastic pieces is uniform (=1cm^2). The ratio of length and width L/B of plastic pieces is one of the parameters which represents the shape of the plastic piece and is herein referred as slenderness value. First, Kaolin clay (w_L=50.6%, I_P=19.6, ρ_s=2.70g/cm^3) in a slurry of water content of 100% is mixed adequately with portland cement in water-cement ratio of 1. The mixture samples are made by mixing with plastic pieces. There are four types of mixtures with different plastic contents by volume,

Table 1 Size of plastic pieces and slenderness value

L/B	L(cm)	B(cm)
1	1	1
16	4	0.25
64	8	0.125

Table 2 Results of the unconfined compression test

Cement Content (kg/m^3)	L/B	Plastic Content (%)	q_u (kPa)	ε_f (%)
50	16	0	51.8	0.502
		2.5	59.2	0.553
		5.0	61.2	0.695
		7.5	62.3	0.836
100	16	0	85.8	1.075
		2.5	107.9	0.796
		5.0	115.5	0.743
		7.5	121.6	0.900
150	1	0	167.8	0.698
		2.5	173.8	0.776
		5.0	161.1	0.689
		7.5	151.1	0.866
150	16	0	180.5	0.539
		2.5	209.0	0.607
		5.0	222.4	0.769
		7.5	223.2	0.585
150	64	0	167.6	0.889
		2.5	183.6	0.642
		5.0	213.2	0.688
		7.5	269.7	1.071
300	16	0	588.6	0.733
		2.5	510.4	0.733
		5.0	464.1	0.783
		7.5	421.8	0.618

i.e., 0, 2.5, 5.0 and 7.5%.

The specimen size is 10cm×10cm×20cm. These specimens are wrapped in moisture proof bags and are cured for 7 days in a humid room at the temperature of 20±3℃. The undrained compression tests of the treated soils were conducted under the axial strain rate of 1%/min.

2.2 *Test Results*

The results of unconfined compression test of cement-treated soils mixed with plastic pieces are shown in Table 2. As seen in this table, the unconfined compressive strength of the treated soils is influenced by the plastic content and the slenderness of plastic piece, while the axial strain at failure is in the range of 0.5 to 1.0%.

The typical stress-strain relationship of the treated soil with plastic pieces is shown in Fig.1. As seen in Fig.1 (a) and (b), the plastic pieces of the slenderness value L/B=16 contribute significantly to the increase in peak strength and residual strength of the treated soils at cement content of 100kg/m^3, whereas the values at cement content of 300kg/m^3 decreases with increase in plastic content. Thus, the stress-strain behavior of the treated soil with plastic pieces depends on not only the content of plastic pieces but also the cement content.

Figure 2 shows that the relationship between deformation modulus E_{50} and unconfined compressive strength q_u of the treated soils with different plastic content, where E_{50} indicates secant moduli at a point of $q_u/2$ in the stress-strain curve. As shown in the figure, E_{50} is almost proportional to q_u and is independent of the cement and plastic contents. On the other hand, the strain energy of the treated soil under a residual state can be used to represent the degree of resistance.

Therefore, The influences of the slenderness

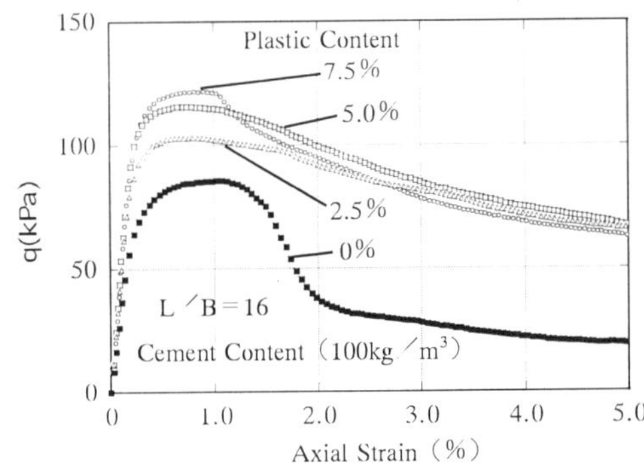

(a) Cement content 100(kg/m^3)

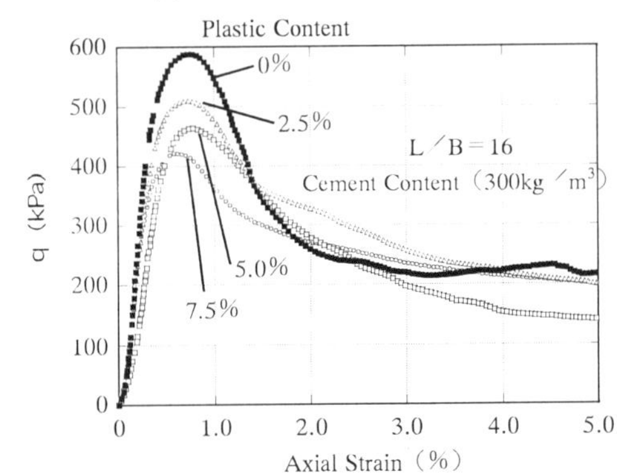

(b) Cement content 300(kg/m^3)

Fig.1 Typical stress-strain relationship of cement-treated soils with plastic pieces

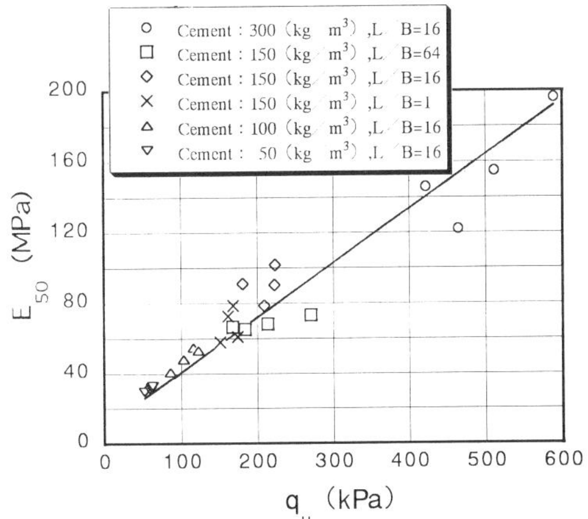

Fig.2 Relationship between deformation modulus and unconfined compressive strength of cement-treated soils with plastic pieces

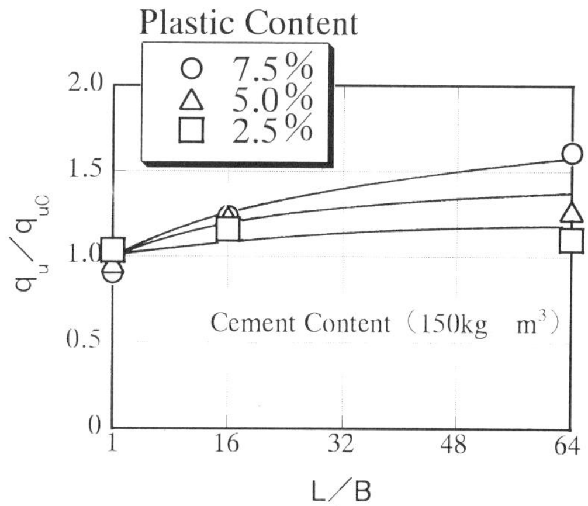

Fig.3 Influence of the slenderness value L/B on the unconfined compressive strength ratio q_u/q_{u0} of the treated soils

value of plastic piece and the cement content are discussed by focusing on the unconfined compressive strength and the strain energy of the treated soil in the study.

2.3 *Influence of Slenderness of Plastic Pieces on Improved Properties*

Strength of composite materials mixed with fibers is represented as the function of fiber properties and fiber-matrix interaction properties (C.L. Victor, 1993). It is, therefore, considered that the shape of plastic pieces is a very important factor in evaluating mechanical properties of the mixtures. From this point of view, the influences of the slenderness of plastic pieces on the strength-deformation properties of the treated soils are discussed.

Figure 3 shows the influence of the slenderness value L/B on the unconfined compressive strength ratio q_u/q_{u0} of the treated soils with cement content of 150kg/m^3, where q_u and q_{u0} represent the unconfined compressive strengths of the treated soils with and without plastics respectively. The value of q_u/q_{u0} represents a degree of improvement by mixing with plastic pieces. As seen in this figure, more slender plastic pieces contribute to the increase of the unconfined compressive strength of the treated soils and the increase is more pronounced as plastic content increases. These facts indicate that the slender the plastic pieces and the more its content, the better the treated soil is in preventing the development of cracks.

On the other hand, the strain energy required to deform the treated soil up to axial strain of 5% is defined as follows.

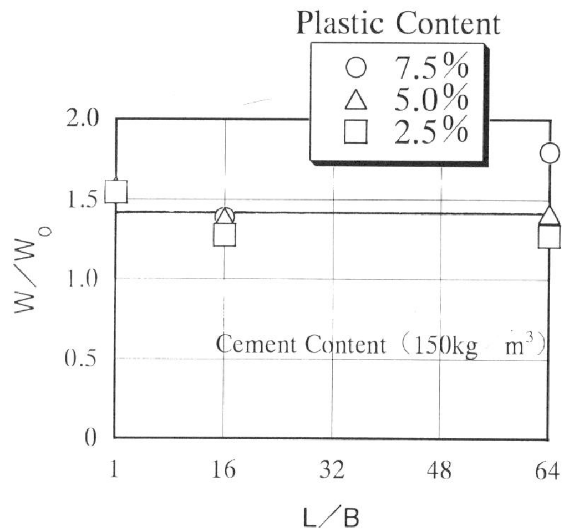

Fig.4 Relationship between the energy ratio W/W_0 and the slenderness of the plastic pieces

$$W=\int_0^{0.05} \sigma_1 d\varepsilon_1$$

where σ_1 is an axial stress and $d\varepsilon_1$ is an incremental axial strain. The energy W of the treated soils is one of the parameters for evaluating improvements in properties and represents a degree of resistance for deformation of the treated soils under a residual state. The axial strain of 5% is decided by considering that the almost treated soils are in a residual state at the axial strain of 5%. Figure 4 shows the relationship between W/W_0 and the slenderness value of the treated soils with cement content of 150kg/m^3, where W_0 indicates the strain energy of the treated soil without plastic pieces. The value of W/W_0 represents a degree of improvement by mixing with plastics and is positive when W/W_0 is more than 1. As shown in this figure, the values of W/W_0 is independent of the slenderness value L/B and is almost constant (about 1.4) even for different plastic contents. Therefore, it can be said that large strain energy is needed to deform the treated soils when plastic content is more than 2.5% at least.

2.4 *Influence of Cement Content on Improved Properties*

In the previous section, it is mentioned that the strength of the treated soils with cement content of 150kg/m^3 is influenced by the slenderness value L/B of plastic pieces, while the strain energy needed to deform the treated soils to residual state is independent of the slenderness value. In this section, the influence of cement content on the strength-deformation properties of the treated soils mixed with plastic pieces are discussed.

The relationship between the unconfined compressive strength ratio q_u/q_{u0} of the treated soils and cement content in the case of the slenderness value L/B=16 is shown in Fig.5. The obvious peak of the strength ratio is indicated and the cement content corresponding to this value is about 100kg/m^3 for the treated soils with any plastic content. It is also found that the peak value of the strength ratio increases as plastic content increases. On the other hand, q_u/q_{u0} of the treated soils with cement content of 50 to 150kg/m^3 is larger than 1, whereas q_u/q_{u0} at cement content of 300kg/m^3 becomes less than 1. At a cement content of 300kg/m^3, soil grains are closely bound together and since the friction between plastic pieces and the grains is less than that between grains, sliding easily occur at the plastic pieces-grain surface. This

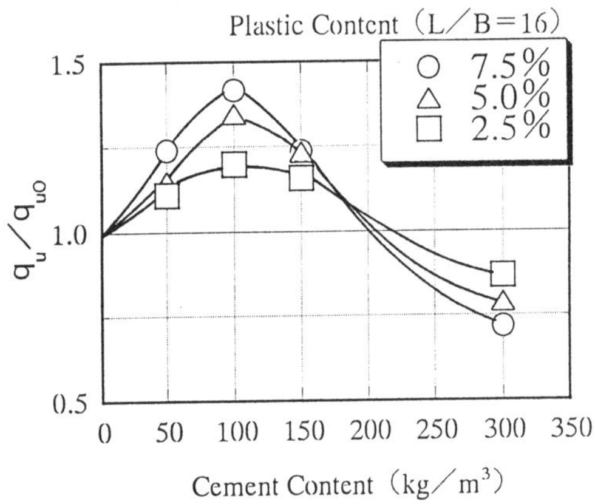

Fig.5 Relationship between q_u/q_{u0} of the treated soils and cement content

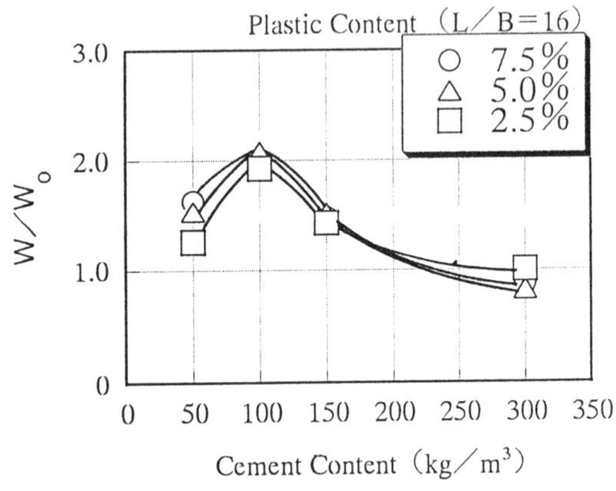

Fig.6 Relationship between the energy ratio W/W_0 of the treated soils and cement content

explains the reduction in strength at high cement contents. This fact means that the improvement by mixing with plastics is not expected at all when relatively soft plastic pieces are included in the cement-treated soil with high strength.

Figure 6 shows the relationship between the energy ratio W/W_0 of the treated soils and cement content in the case of the slenderness value L/B=16. As seen in this figure, the value of W/W_0 becomes a maximum at cement content of 100kg/m^3 and decreases gradually when cement content is more than 100kg/m^3. This tendency is similar to the relationship between q_u/q_{u0} and cement content.

Thus, it is concluded that the maximum improvement is obtained at about 100kg/m^3 for both strength and strain energy. Although this result is obtained for the treated soils in the case of the slenderness value L/B=16, it is considered that similar property will be obtained for the treated soils with plastic pieces with other slenderness values.

3. IMPROVEMENT ON TREATED SOILS WITH PET-BOTTLE PIECES

3.1 *Comparison between PET-bottle Pieces and Plastic Pieces*

In order to see whether similar improvement effects to the treated soils with plastic pieces are obtained for PET-bottle, which is an ordinary plastic waste, a series of unconfined compression tests of the treated soils mixed with PET-bottle pieces was conducted. Improvements on treated soils with PET-bottle pieces are discussed based on the test results.

Figure 7 shows the stress-strain relationship for the treated soils with PET-bottle pieces at cement content of 100kg/m^3, where the used PET-bottle pieces were of a mean thickness of 0.08cm and a slenderness value L/B=16 (L=4cm, B=0.25cm). The stress-strain curve of the treated soils with PET-bottle pieces is similar to that with plastic pieces in the same conditions and a big improvement for the deformation property by mixing with PET-bottle pieces is indicated.

Figure 8 shows the comparison of the strength ratio q_u/q_{u0} for the treated soils with PET-bottle pieces and those with plastic pieces. As shown in this figure, q_u/q_{u0} of the treated soils increases as plastic content increases and this tendency of the treated soils with PET-bottle pieces is almost equivalent to that with plastic pieces at the same plastic content.

The relationship between the enrgy ratio W/W_0 of the treated soils and plastic content is shown in

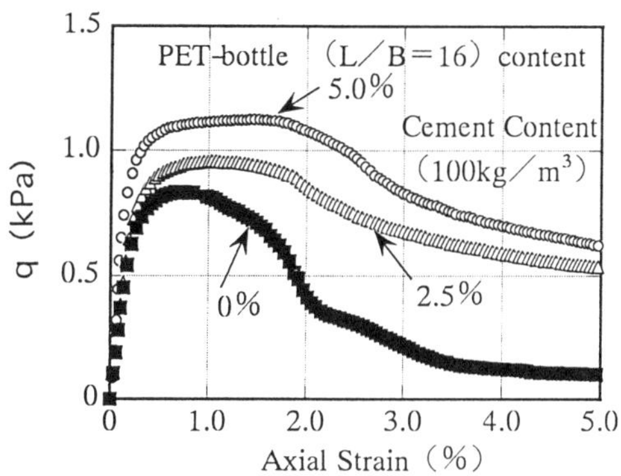

Fig.7 Stress-strain relationship of treated soils with PET-bottle pieces

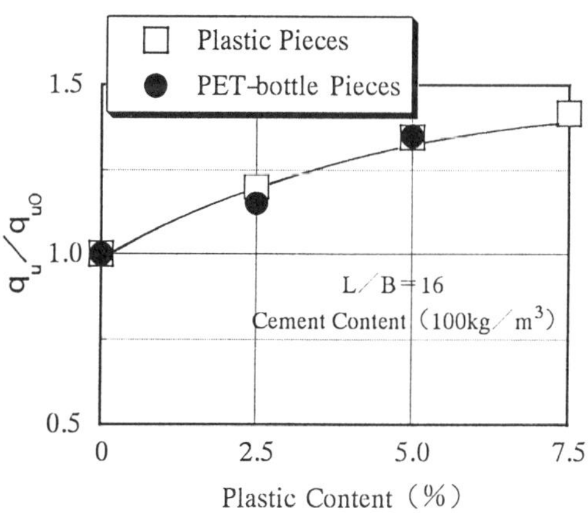

Fig.8 Comparison of the strength ratio q_u/q_{u0} for the treated soils with PET-bottle pieces and plastic pieces

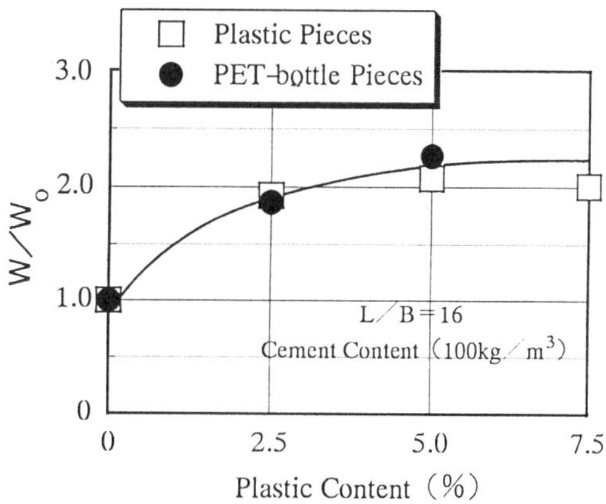

Fig.9 Comparison of the energy ratio W/W_0 for the treated soils with PET-bottle pieces and plastic pieces

Fig.9. The value of W/W_0 increases gradually and becomes constant at plastic content of about 5%. The increase of the energy for the treated soils with PET-bottle pieces is almost equivalent to that of plastic pieces at the same plastic content.

Thus, it is concluded that the improvement on properties of the treated soils with PET-bottle pieces is nearly equal to that with plastic pieces.

3.2 *Unconfined Compression Test of Treated Soils with Large Size PET-bottle Pieces*

The improvement of properties of the treated soils mixed with small PET-bottle pieces is mentioned in the previous section. However, in considering the practical use of PET-bottle, it is necessary to investigate the effect in the case of large size PET-bottle pieces. From the point of view, the unconfined compression test of the relatively large size specimen is performed. The specimen size is 25cm×25cm×50cm and PET-bottle pieces are L=24cm and B=1.5cm (L/B=16). The method of specimen preparation is the same as the one described in the section 2.1.

The stress-strain relationship of the treated soils with large size PET-bottle pieces is shown in Fig.10. As seen in the figure, the treated soil with large size of PET-bottle pieces has significant increase in unconfined compressive strength comparing with the treated soil without PET-bottle pieces. The improvement on strength is almost the same as treated soil with small size of PET-bottle pieces. Although the strength of the treated soil with PET-bottle pieces decreases remarkably after indicating a peak value, the residual strength is larger than that of the treated soil without PET-bottle pieces.

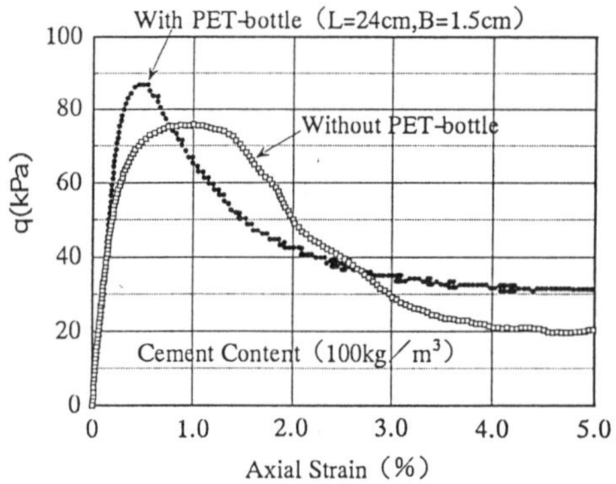

Fig.10 Stress-strain relationship of treated soils with large size PET-bottle pieces

Thus, the same improvement effects are confirmed for the treated soil with large size PET-bottle pieces.

CONCLUSIONS

In this study, the following conclusions are obtained from the unconfined compression test results on cement-treated soils mixed with plastic pieces.

1) The stress-strain property of the treated soil with plastic pieces depends on not only the content of plastic pieces but also the cement content.

2) More slender plastic pieces contribute to the increase of the unconfined compressive strength of the treated soils and the increase is more pronounced as plastic content increases.

3) Large strain energy is needed to deform the treated soils when plastic content is more than 2.5% at least.

4) The maximum improvement is obtained at cement content of $100kg/m^3$ for both strength and strain energy.

5) The improvement by mixing with plastics is not expected at all when relatively soft plastic pieces are included in the cement-treated soil with high strength.

6) The improvement on properties of the treated soils with PET-bottle pieces which are ordinary plastic wastes is nearly equal to that with plastic pieces.

7) The same improvement effects are confirmed for the treated soil with large size PET-bottle pieces.

REFERENCES

Edil, T. B. and Bosscher, P. J. (1994) : Engineering properties of Tire Chips and soil mixtures, Geotechnical Testing Journal, GTJODJ, Vol.17, No.4, pp.453-464.

Hirata , T. and Maeno , Y. (1992) : Mechanical properties of municipal waste and their improvement, TSUCHI-TO-KISO JSSMFE Ser.No.413, Vol.40, No.6, pp.29-34 (in Japanese).

Kato, T., Ochiai, H., Yasufuku, N., Omine, K. and Ohno, S. (1995) : Effects of improvement of cement-treated soils mixed with plastic pieces, Technology Reports of Kyushu University, Vol.68, No.6 (in Japanese).

Victor, C.Li. (1993) : From micromechanics to structural engineering-The design of cementitious composites for civil engineering applications, Structural Eng./Earthquake Eng. Vol.10, No.2, pp.37s-48s.

Environmental Geotechnics, Kamon (ed.) © *1996 Balkema, Rotterdam. ISBN 90 5410 848 7*

Use of ground tires for organic compound containment in the soil-bentonite (SB) slurry cutoff wall

Jae K. Park, Jae Y. Kim & Tuncer B. Edil
University of Wisconsin-Madison, Wis., USA

Curtis D. Madsen
RMT, Inc., Madison, Wis., USA

ABSTRACT: Soil-bentonite (SB) slurry cutoff walls have been used to contain contaminants; however, such systems are susceptible to organic compound breakthrough. Using the capability of ground tires to sorb organic compounds, it may be possible to retard the movement of organic compounds through a SB slurry cutoff wall. Mathematical modeling is used to evaluate the transport of organic compounds through SB slurry cutoff walls with and without ground tire amendment under various seepage velocities. When 10% by weight of ground tires are added into a 1 m thick SB slurry cutoff wall, the times for methylene chloride, trichloroethylene, and *m*-xylene to break through are delayed by 1.8, 16, and 70 times, respectively, for a seepage velocity of 2.2×10^{-6} cm/sec.

1 INTRODUCTION

Since the soil-bentonite (SB) slurry cutoff wall was first used in the United States in the early 1940's, the technology has been used as one of the lowest-cost alternatives to reduce the spread of groundwater contamination (D'Appolonia 1980; Ryan 1985). Several studies have provided considerable evidence that organic compounds have high mobility through engineered containment systems (Johnson et al. 1989; Edil et al. 1994; Park et al. 1995a). Park et al. (1995a; 1995b) showed that automobile tires could sorb a significant amount of volatile organic compounds (VOCs) and proposed its use as a supplement to the engineered waste containment systems, e.g., solid waste landfill liner systems and slurry cutoff walls, in order to retard organic compound migration. Column tests showed that the hydraulic conductivity of the SB slurry cutoff wall backfill material is not affected by the addition of ground tires (Park et al. 1995a). According to the study, *m*-xylene did not break through a ground tire amended SB slurry cutoff wall backfill material in 450 days; however, in the case of a conventional SB slurry cutoff wall backfill material, *m*-xylene broke through the specimen in 180 days to 10% of the incoming concentration.

This study is based on the concept of placing a SB slurry trench around a leaking hazardous waste site to control off-site migration of contaminated groundwater during remediation activities. Although the conventional SB cutoff wall is effective in reducing groundwater flow, the movement of organic compounds would not successfully be retarded due to the lower sorption capacity of the SB slurry cutoff wall material and the diffusivity of organic compounds. Thus, to retard the off-site migration of the organic contaminants in the groundwater, it is proposed to mix ground tires in the SB slurry cutoff wall backfill material to provide a higher sorption capacity for organic compounds, thus converting the SB slurry cutoff wall into a more effective reactive barrier.

The performance of the SB slurry cutoff wall material amended with ground tires is compared with the conventional SB slurry cutoff wall under various conditions and the effect of the ground tire content on the organic compound retardation of the SB slurry cutoff wall is evaluated by mathematical modeling.

2 MASS TRANSPORT MODEL

Under the isotropic and homogeneous conditions of a porous medium, the one-direction (z direction) mass transport of a non-decaying and non-reactive compound by advection and dispersion through the medium can be written as follows:

$$\frac{\partial C}{\partial t} = \frac{D_h}{R_f}\frac{\partial^2 C}{\partial z^2} - \frac{v_z}{R_f}\frac{\partial C}{\partial z} \qquad (1)$$

where C = concentration of a target compound [M/L^3], t = elapsed time [T], D_h = hydrodynamic dispersion coefficient [L^2/T], z = distance [L], v_z = seepage velocity [L/T], and R_f = dimensionless retardation factor used to describe the attenuation of a target compound during the transport process through the porous medium.

When the partition of the target compound between the solid and the solution is assumed to be linear, the retardation factor can be expressed using the partition coefficient, particle density, and total porosity of the medium (Hashimoto et al. 1964):

$$R_f = 1 + \left(\frac{1 - n_t}{n_t}\right) \cdot \rho_p \cdot K_p \tag{2}$$

where n_t = total porosity of the medium, ρ_p = particle dry density [M/L^3], and K_p = partition coefficient [L^3/M].

Eq. 1 has been used to express the movement of contaminants in groundwater. The solution of Eq. 1 depends on the initial and boundary conditions. Using the following initial and boundary conditions:

$$C(0,t) = C_0 \quad ; \quad t \geq 0$$
$$C(z,0) = 0 \quad ; \quad z > 0$$
$$C(\infty,t) = 0 \quad ; \quad t \geq 0$$

an analytical solution was obtained (Ogata and Banks 1961):

$$\frac{C(z,t)}{C_0} = \frac{1}{2} \cdot \mathrm{erfc}\left(\frac{R_f \cdot z - v_z \cdot t}{2\sqrt{R_f \cdot D_h \cdot t}}\right) + \frac{1}{2} \cdot \exp\left(\frac{v_z \cdot z}{D_h}\right) \cdot \mathrm{erfc}\left(\frac{R_f \cdot z + v_z \cdot t}{2\sqrt{R_f \cdot D_h \cdot t}}\right) \tag{3}$$

In this study, Eq. 3 was used to estimate the concentration of a target contaminant at the outer end of the SB slurry cutoff wall. The concentration of a target contaminant at the outer end of the SB slurry cutoff wall can be expressed as follows:

$$C_{out,t} = \frac{C_0}{2} \cdot \mathrm{erfc}\left(\frac{R_f \cdot L - v_z \cdot t}{2\sqrt{R_f \cdot D_h \cdot t}}\right) + \frac{C_0}{2} \cdot \exp\left(\frac{v_z \cdot L}{D_h}\right) \cdot \mathrm{erfc}\left(\frac{R_f \cdot L + v_z \cdot t}{2\sqrt{R_f \cdot D_h \cdot t}}\right) \tag{4}$$

where $C_{out,t}$ = concentration of a target contaminant at the outer end of the SB slurry cutoff wall at a certain time [M/L^3], and L = width of the SB slurry cutoff wall [L].

The seepage velocity, v_z, can be estimated from the hydraulic conductivity, the hydraulic gradient, and the effective porosity of the SB slurry cutoff wall:

$$v_z = \frac{K_h \cdot i}{n_e} \tag{5}$$

where K_h = hydraulic conductivity [L/T], i = hydraulic gradient, and n_e = effective porosity of the medium.

When the SB material is mixed with ground tires, the overall partition coefficient and particle density can be calculated as follows using the partition coefficient and particle density of each material:

$$K_p = \sum_{i=1}^{n} K_{p,i} \cdot \theta_i \tag{6}$$

$$\rho_p = \sum_{i=1}^{n} \rho_{p,i} \cdot \theta_i \tag{7}$$

where n = number of components, and θ = weight fraction.

3 PARAMETERS USED FOR MODELING

Before conducting mathematical modeling of the contaminant transport through a SB slurry cutoff wall, the parameters required were selected. Most of the parameters were determined as closely as possible to the field condition.

The typical width of the SB slurry cutoff wall is between 0.5 to 1.5 m (D'Appolonia 1980; Evans 1994). The width of 1 m was selected for the purpose of modeling. Even though the filter cake plays an important role in the control of the hydraulic conductivity of the overall SB slurry cutoff wall, the thickness and the sorption capacity of the filter cake were ignored in the modeling. The thickness of the filter cake is generally less than 3 mm (D'Appolonia 1980).

For the estimation of the seepage velocity of groundwater flow, the hydraulic gradient and effective porosity are required. The hydraulic conductivity of the overall SB slurry cutoff wall is a function of both the filter cake and the backfill materials. For a wide variety of practical applications, the hydraulic conductivity of the filter cake is in the magnitude of 10^{-9} cm/sec. The hydraulic conductivity of the backfill material is more site dependent. It is affected significantly by the characteristics of the native soil and the bentonite mixing content. The typical value of the hydraulic conductivity of the backfill widely ranges between 10^{-5} and 10^{-7} cm/sec. In most SB slurry cutoff walls, the hydraulic conductivity of 1×10^{-6} cm/sec can be achieved (D'Appolonia 1980). The typical hydraulic conductivity of a SB slurry cutoff wall is ranged between 1×10^{-7} and 1×10^{-8} cm/sec (Evans 1994).

The hydraulic gradient is another site dependent factor. The hydraulic gradient may vary during the service life of a SB slurry cutoff wall. Based on column tests (Park et al. 1995a), the effective porosity of a SB slurry cutoff wall backfill material was estimated to be ranged from 70 to 100% of the total porosity by a tracer test. The total porosity of the specimen was approximately 0.45. In this study, seepage velocity was simplified to be constant and different levels of seepage velocity values were tested. The total porosity of 0.45 was applied for modeling.

Shimizu et al. (1992) derived an empirical relationship between the soil organic carbon partition coefficient and the octanol-water partition coefficient of organic compounds based on 150 data reported by various investigators. Accordingly, the soil-water partition coefficient of organic compounds can be estimated using Eqs. 8 and 9.

$$\log K_{oc} = 0.98 \cdot \log K_{ow} - 0.26 \quad (8)$$

$$K_p = f_{oc} \cdot K_{oc} \quad (9)$$

where K_{oc} = organic carbon partition coefficient, K_{ow} = octanol-water partition coefficient, and f_{oc} = soil organic carbon content.

The organic carbon content in SB material was assumed to be 1% by weight. In the case of ground tire, another empirical equation, Eq. 10, was used to estimate the partition coefficients of organic compounds (Kim et al. 1995):

$$\log K_p = 0.978 \cdot \log K_{ow} - 0.124 \quad (10)$$

The hydrodynamic dispersion includes the molecular diffusion and the mechanical dispersion (Domenico and Schwartz 1990). Due to the relatively slow groundwater flow through the SB slurry cutoff wall system (in most cases), the mechanical dispersion can be considered negligible. The hydrodynamic dispersion coefficient of contaminant was estimated by multiplying the free solution diffusion coefficient by 0.707 (Carman 1939; Perkins and Johnston 1963).

Three different levels of ground tire content were tested: 0, 5, and 10% by weight. The values of the specific gravity of soil and ground tires used were 2.70 and 1.15, respectively (Edil and Bosscher 1994).

Using Eqs. 8,9, and 10, the retardation factors of organic compounds by the ground tire amended SB slurry cutoff wall were computed. Figure 1 shows the effect of the ground tire content on the retardation factor for a wide range of octanol-water partition coefficients.

As the ground tire content increases, the retardation factor also increases. Furthermore, the greater octanol-water partition coefficient of an organic compound has, the more significantly affected is the retardation factor by the ground tire content.

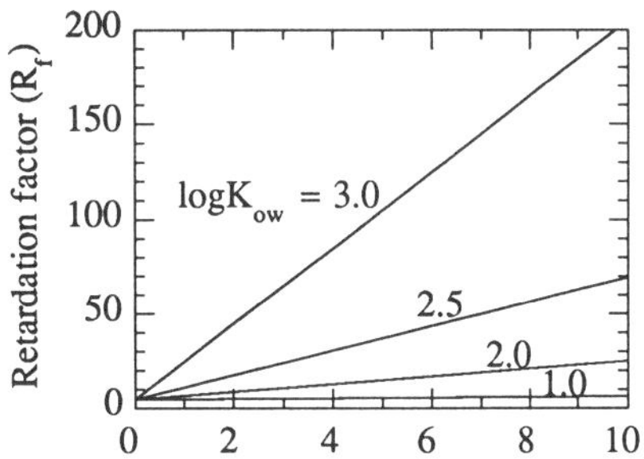

Figure 1. Effect of the ground tire content on the retardation factor of the ground tire amended SB slurry cutoff wall.

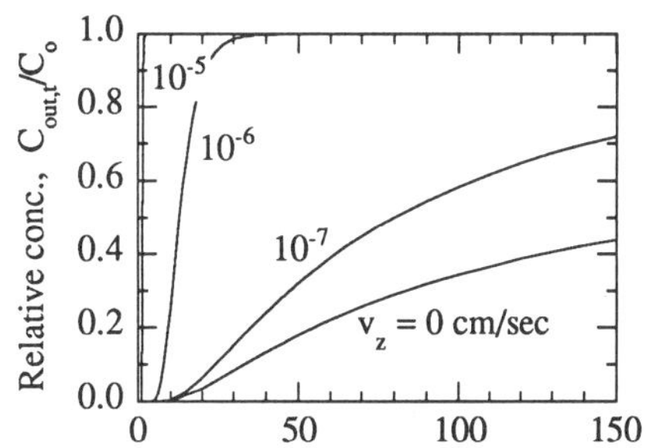

Figure 2. Methylene chloride breakthrough curves of the conventional SB slurry cutoff wall with different seepage velocities.

4. MODELING RESULTS

Methylene chloride is one of the most widely and frequently detected organic compounds in the hazardous solid waste landfill sites (Plumb and Pitchford 1985). Figure 2 is the breakthrough curves of methylene chloride with various seepage velocity conditions in the conventional SB slurry cutoff wall which contains no ground tires. The seepage velocity of 0, 10^{-7}, 10^{-6}, and 10^{-5} cm/sec were applied. The values of the octanol-water partition coefficient and the free solution diffusion coefficient were adopted from U.S. EPA. (1990) and Yaws (1995).

As the seepage velocity increases, methylene chloride breaks through the SB slurry cutoff wall faster. If the seepage velocity of the SB slurry wall is kept low enough, the mass transport of contaminants can be effectively retarded. The seepage velocity can be controlled by hydraulic gradient or hydraulic conductivity. However, the hydraulic conductivity of the SB slurry cutoff wall can be increased by the loss of fine particles (Marcotte et al. 1994). The filter cake can be lost due to the low hydraulic conductivity and small thickness or damaged by the backfill process (D'Appolonia 1980). The loss of filter cake can significantly increase the overall hydraulic conductivity of the cutoff wall. The exposure to contaminants can also affect the hydraulic conductivity. According to the column tests, the hydraulic conductivity of a SB slurry specimen increased up to 40 times in a period of a year (Gipson 1985).

The addition of ground tires to a SB slurry cutoff wall can compensate for the increased mass transport of contaminant caused by the deterioration of the cutoff wall. Figure 3 shows the breakthrough curves of methylene chloride in the ground tire amended SB slurry cutoff walls at the seepage velocity of 1×10^{-5} cm/sec. When a SB slurry cutoff wall is amended with 10% of ground tires, the initial and complete breakthrough times of methylene chloride are delayed by approximately 1.7 times.

In the case of more hydrophobic organic compounds, the breakthrough is more significantly affected by the addition of ground tires. Figure 4 shows the trichloroethylene (TCE) breakthrough curves. TCE can break through the conventional SB slurry cutoff wall in two years. However, the complete breakthrough of TCE is delayed by approximately 16 times with addition of 10% ground tire.

The mass transport of three organic compounds, i.e., methylene chloride, TCE, and *m*-xylene are simulated under the following conditions: (1) the hydraulic conductivity of the overall SB slurry cutoff wall is 1×10^{-7} cm/sec; (2) the hydraulic gradient is 10; and (3) the effective porosity is 0.45, identical to the total porosity. The resulting seepage velocity was 2.2×10^{-6} cm/sec.

The times for 10% and 50% breakthrough of the target compounds were computed by the model using the above parameter values and the results are summarized in Table 1. It appears that a conventional SB slurry cutoff wall cannot effectively retard the target compounds under the testing conditions. All the target compounds reach 10% and 50% of the incoming concentrations in five and seven years, respectively, without ground tires. With 10% ground tire additions, however, breakthrough times were delayed by 16 times for TCE and 70 times for *m*-xylene, respectively.

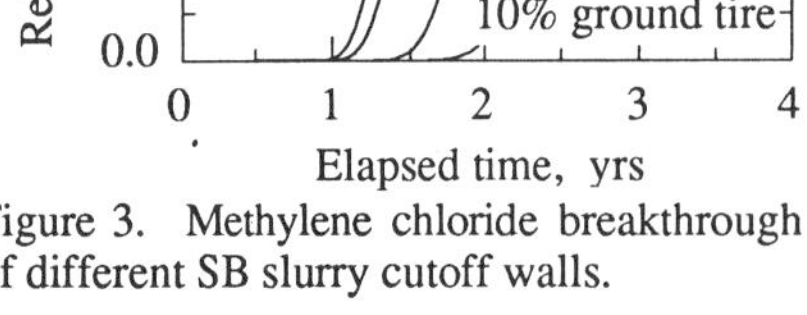

Figure 3. Methylene chloride breakthrough curves of different SB slurry cutoff walls.

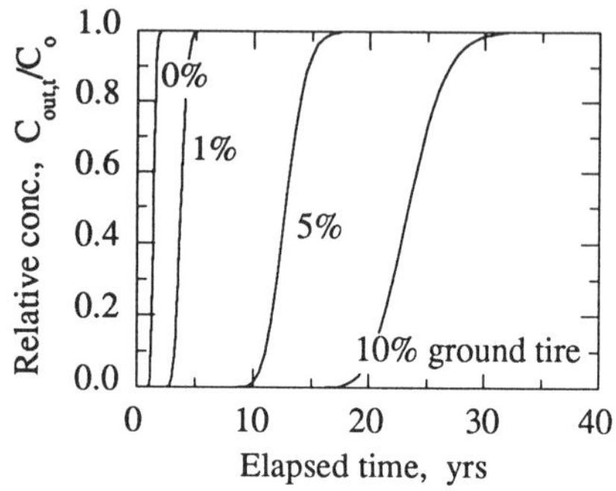

Figure 4. TCE breakthrough curves of different SB slurry cutoff walls.

The times for 10% breakthrough of the incoming concentration are compared for three different ground tire contents and a range of hydraulic gradients (Figures 5, 6 and 7). Under the pure diffusion condition, i.e., zero hydraulic gradient, the time for 10% breakthrough of methylene chloride is delayed by 25 years with the addition of 10% ground tire by weight (Figure 5). As the seepage velocity increases, the difference between the conventional and the ground tire amended SB slurry cutoff walls becomes less significant, e.g., 3.4 years delay with 10% ground tire addition. It is caused by the high diffusivity and hydrophilicity of methylene chloride. More hydrophobic compounds such as TCE and *m*-xylene have longer breakthrough times than methylene chloride, even under a higher seepage velocity condition (Figures 6 and 7). When the hydraulic gradient was 10, the times for 10% breakthrough of TCE and *m*-xylene with 10% ground tire addition were 16 and 70 times longer than those in a conventional SB slurry cutoff wall. The 10% breakthrough time of *m*-xylene was delayed from 4.9 to 348.5 years when 10% of ground tires were added.

Table 1. Required times (years) for 10 and 50% breakthrough at different ground tire contents.

	10% Breakthrough			50% Breakthrough		
	Ground tire content			Ground tire content		
Organics	0%	5%	10%	0%	5%	10%
Methylene chloride	4.3	6.1	7.7	6.0	8.5	10.7
TCE	4.7	42.5	77.9	6.4	57.2	105.1
m-Xylene	4.9	182.0	348.5	6.5	237.7	455.2

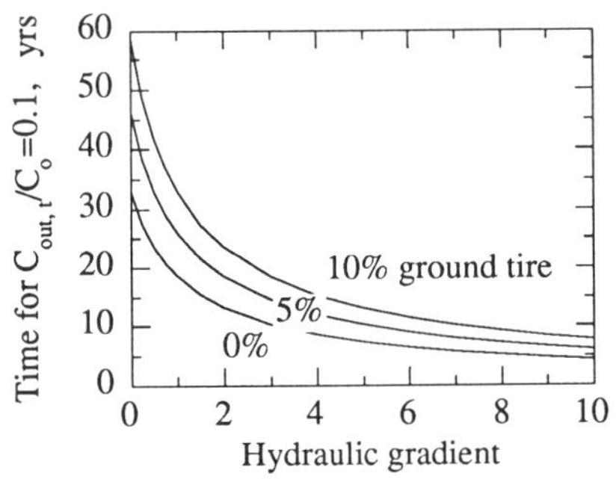

Figure 5. Effect of ground tire content on the methylene chloride 10% breakthrough time.

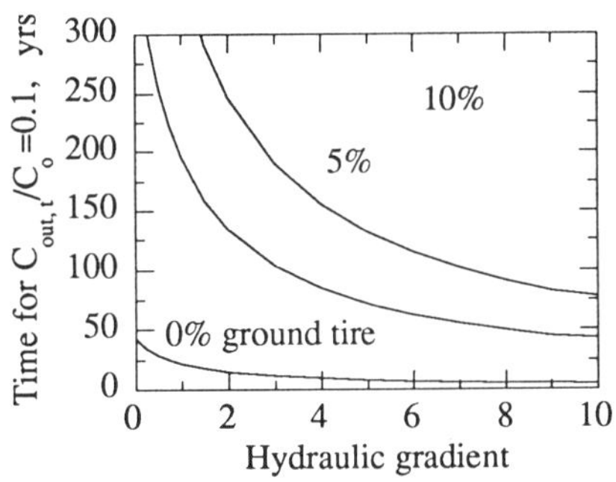

Figure 6. Effect of ground tire content on the TCE 10% breakthrough time.

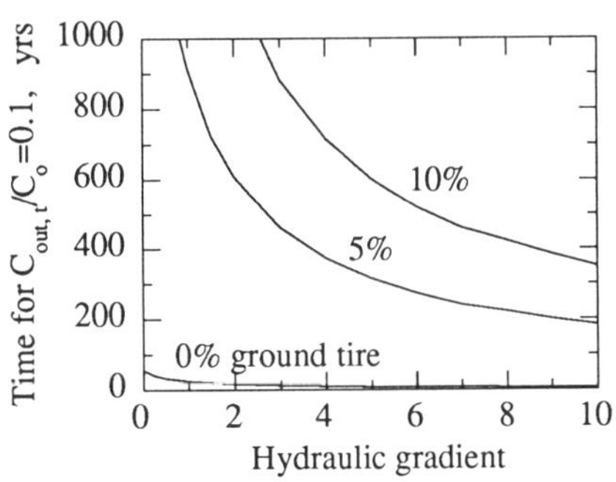

Figure 7. Effect of ground tire content on the *m*-xylene 10% breakthrough time.

5 CONCLUSIONS

Addition of ground tires into a conventional SB slurry cutoff wall leads to a significant increase in the retardation factor. This in turn results in a significant delay in organic compound breakthrough times. Addition of 10% ground tires to a 1 m thick SB slurry cutoff wall delays the breakthrough times of methylene chloride, TCE and *m*-xylene by 1.8, 16, and 70 times, respectively at the seepage velocity of 2.2×10^{-6} cm/sec. Addition of ground tires can significantly improve the organic compound retardation of the SB slurry cutoff wall with minimal additional construction costs.

ACKNOWLEDGMENTS

This study was based on the work financially supported in part by the University of Wisconsin System Ground Water Research Advisory Council and the Wisconsin Department of Natural Resources Waste Tire Management or Recovery Grant Program.

REFERENCES

Carman, P. C. 1939. Permeability of saturated sands, soil and clays. *Journal of Agricultural Science*, 29(2): 262-273.

D'Appolonia, D. J. 1980. Soil-bentonite slurry trench cutoffs. *Journal of Geotechnical Engineering*, 106(4): 399-417.

Domenico, P. A. and Schwartz, F. W. 1990. *Physical and Chemical Hydrogeology*, John Wiley & Sons, New York, N.Y.

Edil, T. B. and Bosscher, P. J. 1994. Engineering properties of tire chips and soil mixtures. *Geotechnical Testing Journal*, 17(4): 453-464.

Edil, T. B., Park, J. K., and Heim, D. P. 1994. Large-size test for transport of organics through clay liners. *Hydraulic Conductivity and Waste Contaminant Transport in Soil, ASTM STP 1142*: 353-374.

Evans, J. C. 1994. Hydraulic conductivity of vertical cutoff walls. *Hydraulic Conductivity and Waste Contaminant Transport in Soil, ASTM STP 1142*: 79-94.

Gipson, A. H. Jr. 1985. Permeability testing on clayey soil and silty sand-bentonite mixture using acid liquor. *Hydraulic Barriers in Soil and Rock, ASTM STP 874*: 140-154.

Hashimoto, I., Deshpande, K. B., and Thomas, H. C. 1964. Peclet numbers and retardation factors for ion exchange columns. *Industrial & Engineering Chemistry Fundamentals*, 3(3): 213-218.

Johnson, R. L., Cherry, J. A., and Pankow, J. F. 1989. Diffusive contaminant transport in natural clay: a field example and implications of clay-lined waste disposal sites. *Environmental Science & Technology*, 23(3): 340-349.

Kim, J. Y., Park, J. K., and Edil, T. B. 1995. Sorption of organic compounds onto ground tires in the aqueous phase, submitted to *Journal of Environmental Engineering*.

Marcotte, D., Marron, J. C., and Fafard, M. 1994. Washing of bentonite in laboratory hydraulic conductivity tests, *Journal of Environmental Engineering*, 120(3): 691-698.

Ogata, A. and Banks, R. B. 1961. A solution of the differential equation of longitudinal dispersion in

porous media. *Geological Survey Professional Paper 411-A.*
Park, J. K., Madsen, C. D., and Kim, J. Y. 1995a. Retardation of volatile organic compound movement by a soil-bentonite slurry cut-off wall amended with ground tires. *Proc. of the WEF 68th Annual Conf. & Exp., Miami Beach, Fla.*: 433-444.
Park, J. K., Kim, J. Y., and Edil, T. B. 1995b. Mitigation of organic compound movement in landfills by shredded tires. to appear in *Water Environment Research.*
Perkins, T. K. and Johnston, O. C. 1963. A review of diffusion and dispersion in porous media. *Society of Petroleum Engineers Journal*, 3(1): 70-84.
Plumb, R. H. and Pitchford, A. M. 1985. Volatile organic scans: implications for ground water monitoring. *Proc. of NWWA/API Conf. on Petroleum Hydrocarbons and Organic Chemicals in Ground Water - Prevention, Detection and Restoration, Houston, TX*: 207-222.
Ryan, C. R. 1985. Slurry cutoff walls: applications in the control of hazardous wastes. *Hydraulic Barriers in Soil and Rock, ASTM STP 874*: 9-23.
Shimizu, Y., Takei, N., and Terashima, Y. 1992. Sorption of organic pollutants from vapor phase: the effects of natural solid characteristics and moisture content. *Water Science and Technology*, 26(1-2): 79-87.
U.S. EPA. 1990. *WERL Treatability Data Base*, Risk Reduction Engineering Laboratory, U.S. EPA., Cincinnati, OH.
Yaws, C. L. 1995. *Handbook of Transport Property Data*, Gulf Publishing, Houston, TX.

Environmental Geotechnics, Kamon (ed.) © 1996 Balkema, Rotterdam. ISBN 90 5410 848 7

Use of pyrolized carbon black from scrap tires in hot mix asphalt design

Taesoon Park
Indiana Department of Transportation, Division of Research, West Lafayette, Ind., USA (Currently: Korea National Housing Corporation, Division of Research, Seoul, Korea)

C.W. Lovell
School of Civil Engineering, Purdue University, West Lafayette, Ind., USA

ABSTRACT: Various concepts and techniques have been developed for efficient and economical disposal and utilization of scrap tires. One of these is pyrolysis of scrap tires. Pyrolysis is a method of decomposing tires by a cooking process in order to break down the rubber into salable byproducts. Tire pyrolysis yields approximately 25% carbon black (CB) which is called pyrolized carbon black (CBp) in this study. Due to the impurity of CBp (75% carbon black and 25% other elements), CBp can not be used for industrial purposes without special processing. The main objective of this study is to investigate the viability of using CBp as an additive in hot mix asphalt concrete. The analyses of test results showed that CBp modified asphalt generally performed equal or better than commercial CB modified asphalt. The optimum dosage of CBp for improving the performance of hot mix asphalt is 10 to 15 percent of the asphalt binder used.

1. INTRODUCTION

It is estimated that over 242 million waste tires are generated each year in the United States. In addition, it is reported that approximately 2 billion waste tires are accumulated in stockpiles or uncontrolled tire dumps across the country. (EPA, 1991). In 1991 approximately five million scrap tires were used in various civil engineering applications and another five million were used in asphalt paving applications. The interest of the highway construction industry is increasing in identifying an effective and economical application of waste tires to help solve the waste tire problems.

Tires contains five different types of carbon black. The carbon black is added to tires to increase these durability. These carbon blacks can be extracted from the tire pyrolysis process and are called pyrolized carbon black (CBp) in this study. Due to the pyrolysis, the structure and contents of carbon black are modified. The pyrolized carbon black typically contains 75% carbon black, 9% ash, 4% sulfur and 12% butadine copolymer.

More pure commercial carbon black have been used to improve the performance of asphalt mixtures. The inclusion of the commercial carbon black increases the rutting resistance, the durability, and reduces the temperature susceptibility of asphalt mixtures(Yao and Monismith, 1986, Vallerga and Gridley, 1980)

This study investigates using pyrolized carbon black in lieu of commercial carbon black as reinforcing agent for asphalt mixture. Due the high contents of pure carbon black in pyrolized carbon black, it is expected that it will enhance the performance of asphalt mixture. Pyrolized carbon black is obtainable as a byproduct of tire pyrolysis, and is a relatively inexpensive raw material. Furthermore, tire pyrolysis could be a major remedy for the mass disposal problem of waste tires.

2. TIRE PYROLYSIS

Pyrolysis is a method of decomposing tires by a cooking process in order to break down the rubber into salable by-products. Tire pyrolysis yields approximately 55 % oil, 25 % carbon black, 9 % steel, 5 % fiber and 6 % gas (Roy et al., 1990). The yield can be varied depending on the operation conditions, such as temperature and pressure. High temperature (i.e., 900°C), pyrolysis yields larger quantities of residues (CBp), steel and ashes. Lower temperature pyrolysis yields larger quantities of oils.

Destructive distillation and carbon black recovery are the two main operations in the pyrolysis of tires. As many as 5000 to 8000 waste tires can be processed in a day by a single facility (Wolf Industires, 1994) The products recovered in pyrolysis can be reprocessed for the manufacturing industry and for the construction industry.

3. MATERIALS AND EXPERIMENTAL WORK

3.1 *Materials Used*

The AC-10 asphalt was used as asphalt binder. The gradation of aggregate was controlled after the initial round of the Marshall method. Two types of additives were used, commercial carbon black and pyrolized carbon black, to compare the test results and the performance of CB mixtures and CBp mixtures. Pyrolized carbon black(CBp) was provided by Wolf Industries, Brazil, Indiana. According to the test data provided by Wolf Indusries, the CBp is environmentally safe. The CBp is blended with asphalt binder as received. Table 1 shows general properties of carbon black produced during vacuum pyrolysis of waste tires.

Commercial carbon black was purchased form CABOT Industries, Boston, Massachusetts. REGAL 300R, high structure high abrasion furnace (HAF) type carbon black was used. The reason for choosing this type carbon black is that Rostler et al. (1977) reported that the HAF type produced the best performance among other carbon blacks. The analytical specifications of REGAL 300R carbon black from the CATBOT industries are shown in Table 2.

Table 1 General properties of carbon black produced during vacuum pyrolysis of waste tires (After Roy et al., 1990)

Iodine index (mg/g)	144.2 - 151.4
DPB absorption (cc/100g)	84.6 - 93.0
Heat loss at 105 °C (%)	57.1 - 60.6
Tint strength (% ITRB)	15.5 - 17.0
Ash (%)	3.3 - 4.9
Volatile matter	2.5 - 3.0

Table 2 Analytical Specifications for REGAL 300R Carbon Black (After CABOT Industries,1995)

Property	Test Method	Spec.
Iodine Index (mg/g)	ASTM D1510	76 ± 5
Density (Kg/m^3)	ASTM D1513	200 ± 48
Ash (%)	ASTM D1506	1.0 max
DPB Absorption (cc/100g)	ASTM D2414	85 ± 5
Tint Strength (% ITRB)	ASTM D3265	113 ± 3.0

3.2 *Preparation of binder*

Heated CBp and CB (145°C to 150°C) were blended separately with heated asphalt cement. Different ratios of CBp and CB (5%, 10%, 15% and 20% by weight of the asphalt cement used) were blended with AC-10 asphalt cement before the aggregate was mixed with the binder.

3.3 *Experimental work*

Laboratory tests were carried out to evaluate the reinforcing effects of the CBp. The laboratory tests included the Marshall method, Gyratory Testing Machine (GTM), and Dynamic Creep Test (Confined). The tests were performed in accordance with the ASTM and Indiana Department of Transportation (INDOT) specifications.

The Marshall method was used to determine the optimum binder content, and the characteristics of mixtures in terms of Marshall stability, flow, air-voids, and voids in mineral aggregate (VMA).

The U.S. Army Corps of Engineer gyratory testing machine was used to investigate the stress and strain relationship of the mixtures. The compactibility, stability, and variation of unit weight of mixtures were evaluated.

The dynamic creep test (confined) was carried out to investigate the rutting potential of each mixture. The creep test is not yet standardized. Gabrielson's (1992) recommendation was used. Tests were carried out on a Material Testing System (MTS 810), with a feedback control hydraulic tester and with a temperature controlled environmental chamber. The specimens were reproduced with optimum binder content determined from the Marshall method. They were conditioned at 50°C in an oven for 2.5 hours. A 138 Kpa axial stress was applied in the environmental chamber for 30 minutes, and then a 827 Kpa axial stress was applied. After 10 seconds, the deviatoric stress of 689 Kpa was applied for 3,600 cycles, including a 15 minute rebound period. The duration of a repeated loading cycle was 1.0 second, which consisted of 0.1 second loading and 0.9 second unloading duration. The results were analyzed in terms of mix stiffness and corrected cumulative creep.

Relative comparison were made to evaluate the characteristics and performance of CBp mixtures with respect to CB mixtures and AC-10 mixture.

4. TESTS RESULTS AND DISCUSSION

4.1 *The Marshall method*

The Marshall method provides the optimum binder contents and the mechanical properties of the mixtures. While the Marshall method is standardized in ASTM D1559, the criteria of acceptance vary from agency to agency. Indiana DOT criteria were used in this study. Figure 1 shows the optimum binder contents. The optimum binder content increases with increasing CBp and CB.

Figure 2 shows the relationship of the air voids, and the VMA versus CBp contents. The air voids increase almost linearly with increasing CBp content. The CB mixtures show the same trend except showing slightly higher air voids than the CBp mixtures. The VMA increases with increasing CB content, however, a general trend is not maintained for the inclusion of CBp. A particular observation is that 20 percent mixtures produce sudden decrease of the VMA. The VMA is affected by the aggregate gradation, surface texture, and shape. Therefore, it is assumed from the test results that coarser portions of CBp acts as a mineral filler contribution to a denser gradation.

The Marshall stability increases with increasing CBp and CB content up to 15 percent in 5 to 5.5 percent binder, which is around the optimum binder contents, as can be seen in Figure 3.

However, the Marshall stability begins to decrease after 15 percent CBp and CB. In general, CBp shows similar effects as CB in stability. The improvement of flow is not significant due to the inclusion of either CBp or CB. The flow tends to increase with increasing CBp and CB contents. A small improvement is observed in the vicinity of the optimum binder contents for the mixtures.

The Marshall method shows that the mechanical properties of the CBp mixtures are almost the same as those of the CB mixtures and the AC-10 mixture.

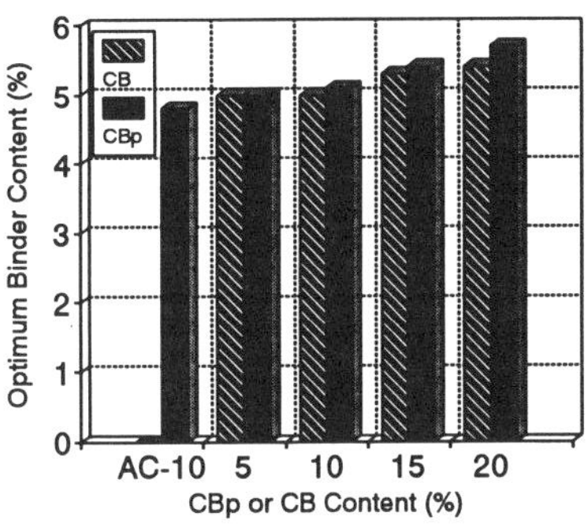

Figure 1. The optimum binder content

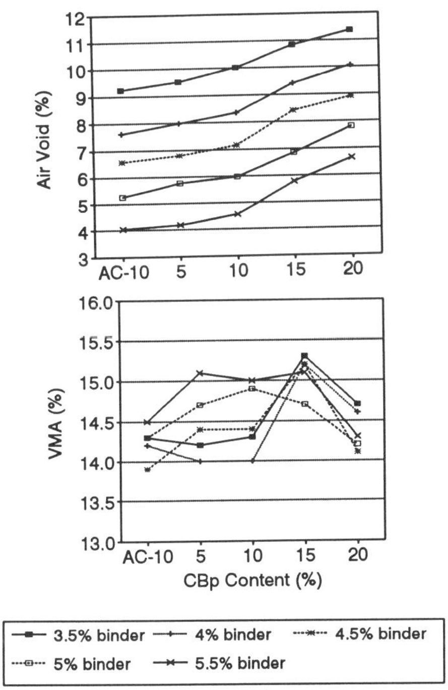

Figure 2. Air voids and VMA vs. CBp content

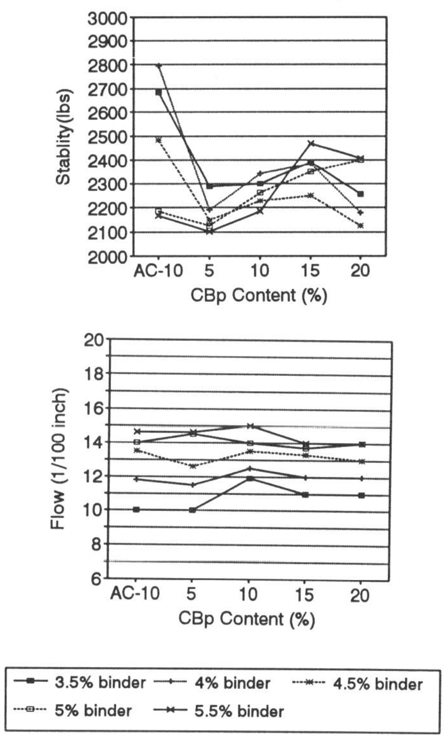

Figure 3. Marshall stability and flow vs. CBp contents

The most significant improvement is that the rate of reduction of the Marshall stability with decreasing air voids is not as severe as that observed in the AC-10 mixture.

4.2 *Gyratory testing machine*

The Gyratory testing machine provides stress-strain relationship and predicts performance of the mixtures under the traffic loading. Figure 4 shows the relationship between air voids and GTM revolutions for CBp and CB mixtures. The variation of air voids of CBp mixtures is less severe than that of CB mixtures. The air voids in the mixture is an important parameter since it permits the physical properties and performance of the mixture to be predicted for the service life of the pavement. The effect of the inclusion of CBp is beneficial for controling air voids.

The Gyratory compactibilty index (GCI) is an indicator of the compactibility of the mix. The closer this value to the unity, the easier the mix is to compact. The GCI of the AC-10 mixture is 0.96, that of CBp mixture is 0.96, and that of the CB mixture is 0.95. Therefore, the compactibility of all mixtures are about the same, with the CB mixture requring only slightly higher compaction efforts.

Table 3 summarizes Gyratory stability indices (GSI) and Gyratory shear factors (GSF) at GTM 250 revolutions. At GTM 250 revolutions, the ultimate compaction was achieved. The value of the unit weight is almost constant (less than 157 N/m^3 per 100 revolutions) at GTM 250 revolutions, which is assumed to represent the ultimate compaction in the pavement. The GSI is related to the plastic deformation of pavement. As the GSI value is closer to unity, the mixture become more stable and plastic deformation less likely to occur. The GSF is a factor of safety type of index. When the GSF value is less than unity, there is an inadequate shear strength for the anticipated maximum shear in pavement. The GSF can be obtained from the following relationship:

$GSF = Sg/\tau max$,

where, τmax = vertical pressure/π = 263.4 KPa

As shown in Table 3, the AC-10 mixture has satisfactory performance in terms of GSI and GSF. The CBp mixtures also show good performance except for the 20% mixture. While the stability for the CBp mixtures decreases with increasing CBp contents. the GSF for the CBp mixtures remains almost constant. The CB mixtures indicates similar results for stability, however, the GSF for CB mixtures is significantly decreased with increasing CB contents. The CB mixtures show significantly less shear stress than the CBp mixtures and AC-10 mixture.

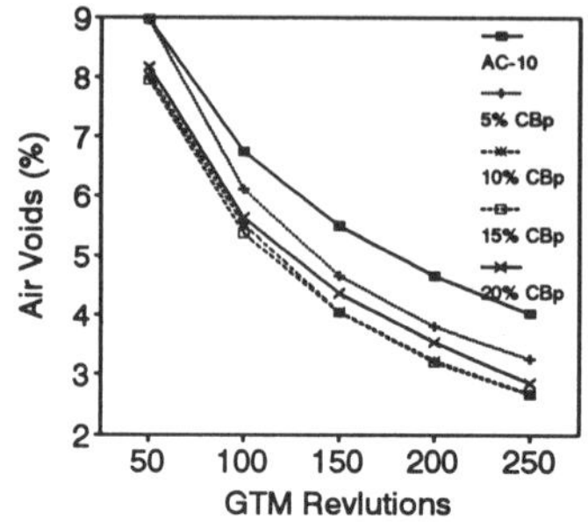

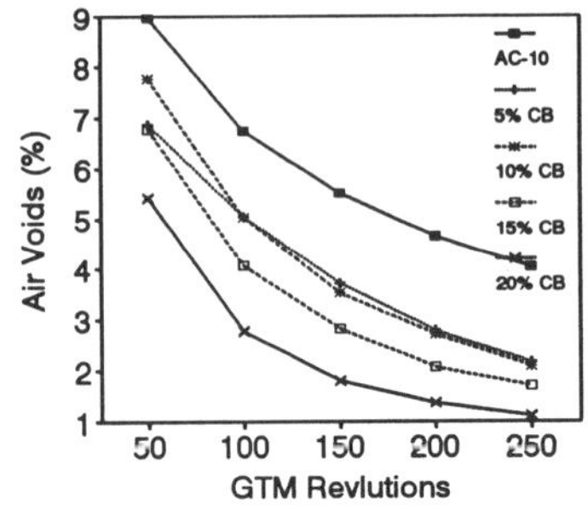

Figure 4 Relationship between air voids and GTM revolution

Table 3 GSI and GSF at GTM 250 revolutions.

	AC10+CBp		AC10+CB	
Mixture	GSI	GSF	GSI	GSF
AC10	1.12	1.07	1.12	1.07
5 %	1.16	1.09	1.16	0.81
10 %	1.19	1.09	1.14	0.70
15 %	1.21	1.09	1.28	0.45
20 %	1.25	1.08	1.45	0.18

Note : Design criteria (GSI≤1.2, GSF≥1.0)

The test results of the GTM reveal that the inclusion of CBp in the asphalt mixture enhances the shear resistance and reduces plastic deformation. The optimum CBp contents for this purpose are between 10 and 15 percent. The CBp mixtures are equal to the AC-10 mixture, and better than the CB mixtures.

4.3 Dynamic creep test

The total permanent deformations at the end of each rest period were obtained through the creep test. The influence of the inclusions of CBp and CB were analyzed through preliminary analysis of creep data. The analyses produced values for the mix stiffness and for corrected cumulative creep.

The mix stiffness modulus was defined as follows (Hill et al, 1974):

$$Smix\ (T,t) = \frac{\sigma}{\varepsilon_t}$$

where *Smix (T,t)* = mix stiffness at a specified temperature (T) and time of loading (t)
σ = applied stress and
ε_t = axial strain at $t = \Delta h/h$, where Δh is change in height of specimen, and h is height of specimen

Figure 5 shows mix stiffness of the CBp mixtures and the CB mixtures. As mix stiffness increases, the rutting resistance of asphalt mixtures increases. The inclusion of CBp and CB indicates an improvement in the rutting resistance. Both CBp and CB mixtures show almost the same effects for mix. Therefore, it is expected that the use of CBp can reduce the rutting problem in asphalt pavement.

Figure 6 shows the corrected cumulative creep for CBp and CB mixtures. The slope for creep potential decreases as the CBp and CB content increase. The cumulative creep behavior also shows the same influence by the inclusion CBp and CB.

The CBp mixtures indicate equal or better reinforcing effects than the CB and AC-10 mixtures. Although the reinforcing mechanism of CBp and CB is not fully understood. Vallerga and Gridley (1980), Yao and Monismith (1986) have pointed out the importance of the dispersing effects of CB in asphalt cement. When CB is properly dispersed, the microsize and the hydrophobic property of CB particles preferentially saturated by the asphalt cement makes CB a part of the asphalt cement. This is called a "microfiller action of CB", and is in contrast to conventional mineral fillers. Therefore, Vallerga and Gridley (1980) recommended the use of pelletized CB to improve the dispersing effect and high speed mixing (approximately 15,000 rpm) in order to achieve the full development of the dispersing effects in asphalt mixtures. As mentioned earlier, loose power type CB and regular speed (800 rpm) mixing were used in this study have the same testing conditions as the CBp and also to simulate the pug-mill operation in the field. On the other hand, the CBp may not require the dispersing effects that the CB needs, because, as a result of pyrolysis, the structure and microfiller action different. It is assumed by the authors that the coarse and the fine particles of CBp acts as the fraction of the fine aggregate gradation and also micro fine particles act as a microfiller in the asphalt binder. However, further study of the surface chemistry is needed to understand the reinforcing mechanism and to verify the assumptions of the action of CBp and the CB in the asphalt mixture. These may provide the potential reasons that the performance of the CB mixtures not being as significant as for pervious studies, and showing lesser reinforcing effects with respect to the CBp mixtures and the AC-10 mixture.

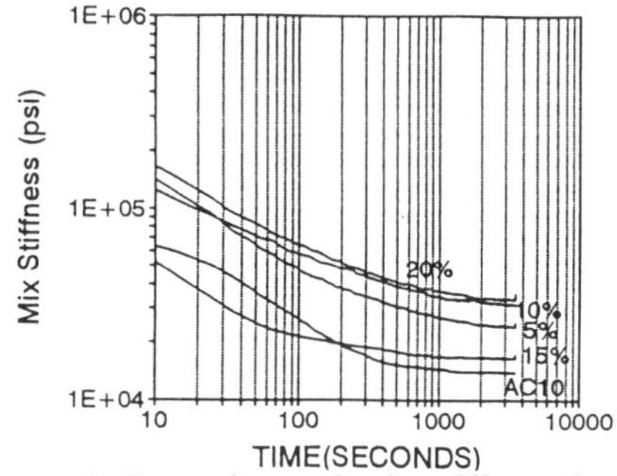

Figure 5 Comparison of mix stiffness for CBp mixture

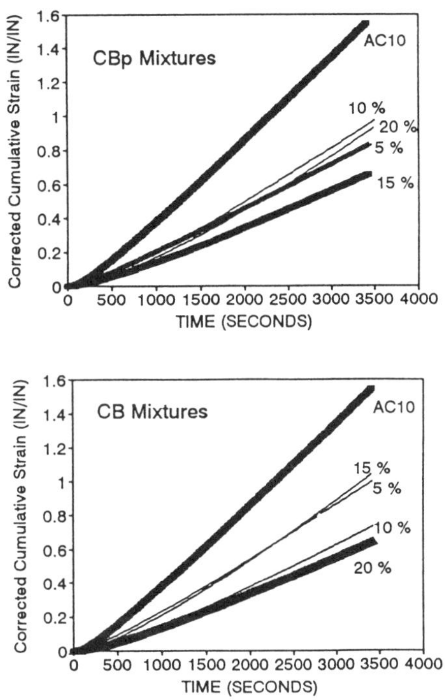

Figure 6 Comparison of corrected cumulative creep

5. MATERIAL COST AND CONCERNS

The raw material cost of the CB was 32 cents/Kg (CABOT, 1995), while the cost of the CBp was estimated at 7 cents/Kg (Wolf Industires, 1995). If a 8 Km, two lane road is overlaid to depth of 37 mm with asphalt concrete of an optimum binder content of 5 %, approximately 1818 Mg of asphalt cement are needed. Therefore, 18 Mg of CBp and CB are required for a 10% CBp or CB mixture. The increase of raw material cost over the conventional, unmodified mixture is 6.3% for CBp, and 35% for CB. The additional cost to use CBp can be potentially be compensated by extending the service life of the pavement and through a better performance of the pavement. However, this can be only be verified through field trials.

For a consistent use of CBp in pavement construction, quality control of the CBp is needed. The pelletized form of CBp is highly desirable to reduce handling problems in the field.

6. CONCLUSIONS AND RECOMMENDATIONS

The following conclusions and recommendations are based on the laboratory tests of this study.

1) Pyrolized carbon black is useful as a reinforcing agent to improve the performance of asphalt mixture.
2) The typical performance of CBp mixtures is improved with respect to CB mixture and AC-10 mixture.
3) A pyrolized carbon black content of 10 % to 15 % by weight of asphalt is recommended for the improvement of asphalt mixtures.
4) Rheological study of CBp modified asphalt binders should provide fundamental understanding of the effects of the CBp behavior in the asphalt mixture.

7. Acknowledgment

The authors appreciate the financial support of .the Indiana Department of Transportation (INDOT) and the Federal Highway Administration for this research.

REFERENCES

CABOT Industries (1995), "Carbon black for specialty applications", CABOT Corporation Special Blacks Division, Catalogue, Massachusetts.

EPA (1990), "Markets for Scrap Tires," Report No. PA/530-SW-90-074A, United States Environmental Protection Agency, Washington D.C., 115p

Gabrielson R.J. (1992), "Evaluation of hot mix asphalt static creep and repeated load tests," Ph.D. Dissertation, Auburn Univ., Alabama, 219 pp..

Hills, J., Brein, D and Van de Loo, P. (1974),"The correlation of rutting and creep tests on asphalt mixes," paper IP 74001, Institute of Petroleum, London, England, January.

Rostler, F., White, R., and Dannenberg, E. (1977),"Carbon black as a reinforcing agent for asphalt", AAPT, vol.46, pp. 376-401

Roy, C. Labrecque, B. and Caumia de, B. (1990),"Recycling of scrap tires to oil and carbon black by vacuum pyrolysis," Resources Conservation and Recycling, No.4., pp.203 - 213.

Vallerga, B. and Gridley, P. (1980), "Carbon black reinforcement of asphalt in paving mixtures:, Asphalt Pavement Construction, New Materials and Techniques, ASTM STP 724, American Society for Testing and Materials, pp.110-128.

Wolf Industries (1994), "Product specification data and material safety data sheet," Personal letter from Mr. Foster.

Environmental Geotechnics, Kamon (ed.) © 1996 Balkema, Rotterdam. ISBN 90 5410 848 7

The engineering properties of phosphogypsum for stack closure

Shailesh K. Patel & Wayne A. Ericson
Bromwell & Carrier, Inc., Lakeland, Fla., USA

Steven G. Richardson
Florida Institute of Phosphate Research, Barton, Fla., USA

Jerry B. Sartain
University of Florida, Gainesville, Fla., USA

ABSTRACT: Phosphogypsum (PG) is a by-product of the phosphoric acid production process. Low-level radioactivity in PG, mostly from radium-226, has restricted its use to research and development purposes. Therefore, most PG will remain stockpiled in unlined above ground stacks.

Recent regulations by the Florida Department of Environmental Protection regarding stack closure require that a final cover consisting of a barrier layer covered with soil or amended PG capable of sustaining vegetation be installed at closure to preclude rain water infiltration and reduce leachate generation. A low-permeability soil or synthetic membrane may be used as a barrier layer, but the cost associated with either component could be enormous.

Therefore, a study to demonstrate the potential for a cost-effective final cover system using compacted PG and vegetation was conducted. The objectives of the study were to determine the influence of PG compaction on vegetation establishment and stack hydrology. Field lysimeter and runoff plots consisting of varying degrees of compaction and slopes were constructed on a PG stack. The plots were vegetated with bermudagrass (*Cynodon dactylon [L.] Pers.*). Field tests performed to determine the horizontal permeability and water infiltration and laboratory tests to determine vertical permeability with additional compactive efforts were performed.

Additional compaction efforts over typical modified Proctor reduced permeability of PG from 10^{-5} to 10^{-6} cm sec^{-1}. Field tests indicated that for all lysimeter plots, regardless of the initial degree of compaction, the horizontal permeability was 10^{-2} cm sec^{-1}, suggesting that with time, uncompacted PG tends to re-compact. However, horizontal permeability was three orders of magnitude greater than vertical permeability. This should result in longer residence time for the water and increase the potential for its loss via evapotranspiration, thus reducing infiltration. This was further supported by negligible infiltration collection from lysimeter plots. Similarly, lysimeter infiltration rates were 10^{-4} cm sec^{-1} regardless of compaction or vegetation, further suggesting that PG tends to re-compact upon repeated wetting and drying. Initially, runoff increased with slope, but varied with rainfall duration. Compaction hindered seed germination and vegetation establishment, but this was rectified using simple tillage operations. Vegetation also improved water quality to satisfy state surface water standards. This study demonstrates that PG can be compacted to reduce infiltration and has potential as an alternate to a compacted soil or synthetic membranes as barrier layers. Use of PG as a final cover component would also reduce impacts or surrounding soil borrow areas and result in substantial cost savings to the phosphate chemical industry.

1 INTRODUCTION

For each metric ton (mt) of phosphoric acid produced from phosphate rock, there is a co-production of approximately five mt of phosphogypsum (PG) as a by-product. The standard practice for the Florida based chemical fertilizer plants is to pump PG, using process water, as a slurry into an impoundment where a dragline is used to move some of the settled PG to raise the dike height, resulting in a "pile" or "stack" up to 60 meters (m) in height. Currently, stacks in Florida range from 20 hectares (ha) to 200 ha and occupy over 2,000 ha of land. Chemical processing results in 30 million mt of PG annually (May and Sweeney, 1980), with over 700 million mt already accumulated. Low-level radioactivity present in PG, mostly from radium-226, has resulted in restrictions on the use of PG (Federal Register, 1992). With this final ruling on PG removal and uses, it is a foregone conclusion that most PG will remain stockpiled in approximately the 20 stacks within Florida. When these stacks cease usefulness as deposition sites, they will need to be reclaimed in an environmentally acceptable manner.

A concern for potential ground water contamination by the low pH (<2) pond water and its metallic ion content is promulgated and addressed by the Florida Department of Environmental Protection (FDEP) under Chapter 62-673, phosphogypsum stack management (Florida Administrative Code, 1993). This rule requires that all PG stacks constructed after March 25, 1993 must have a composite liner and leachate collection system in place. Since most of the stacks in Florida were constructed prior to this date, only two PG stack in the state meets these requirements. The remaining stacks that do not meet these requirements are expected to begin mandatory closure by March 25, 2001, unless by this date the stack owner or operator demonstrates that the stack system was not or is not causing any violations of the ground water standards within the permitted zone of discharge.

The provision for an impermeable, final stack cover is to reduce infiltration and promote lowering of the internal water mound. As part of the closure plan for unlined stacks, the rules require construction of a final cover consisting of a barrier soil layer at least 45 centimeters (cm) thick, having a maximum permeability of 1×10^{-7} cm sec^{-1}. The barrier layer will need to be covered with an additional 45 cm of soil or amended PG capable of sustaining vegetation. Alternatively, synthetic membranes, covered with 60 cm of soil or amended PG may be utilized. Both these methods are very expensive.

Therefore, the objective of this paper is to present results from laboratory and field infiltration and moisture migration testing to demonstrate the extent to which PG alone can be compacted, enabling its use as an alternative to a barrier soil layer or synthetic membrane component of the final stack cover.

2 MATERIALS AND METHODS

Field Test Plot Installation

The test plot program consisted of 12 plots constructed on an inactive PG stack. Each plot was three m^2. An additional area of 39 m^2 surrounding each plot was also vegetated with grass to provide conditions similar to a grass covered stack. Eight out of the 12 plots were installed with basin lysimeters to collect infiltrating rainwater and are described in Table 1. The remaining four plots were equipped with runoff collection troughs and are described in Table 2. Each lysimeter was 1.2 m wide by 2.4 m long by 0.3 m deep, constructed of pressure treated wood, lined with 20 mil PVC and filled with pea gravel. The surface of the lysimeter was covered with filter fabric to prevent the overlying PG from migrating onto the gravel drainage medium. The bottom of each lysimeter was located approximately 1.2 m below the test plot surface. The lysimeter bottom sloped at approximately 16H:1V which drained to a five cm diameter PVC pipe connecting to an external water sampling and collection system. The PG was compacted using a vibratory roller. A dry density of 95 percent of the American Society for Testing and Materials (ASTM) D1557 maximum dry density was the goal for field compaction. Adequate dolomitic limestone and fertilizer was added to raise the pH above 5.0 and provide nutrients for the annual rye grass (*Lolium perenne [L.]) Pers.* and bermuda grass (*Cynodon dactylon [L.] Pers.*) which were established from seed.

The runoff plots were of similar dimensions except open at the bottom. A 30 cm diameter PVC half trough collected water which drained similarly to the sampling and collection

Table 1. Basin Lysimeter Test Plot Specifications and Descriptions

Plot Number	Size (m x m)	Plot Profile Description
1A and 2A	1.2 x 2.4	Uncompacted, unvegetated
1B and 2B	1.2 x 2.4	Uncompacted, vegetated
1C and 2C	1.2 x 2.4	60 cm compacted, vegetated*
1D and 2D	1.2 x 2.4	15 cm uncompacted cover, 60 cm compacted, vegetated

* Subsequently surface rototilled to help grass seed germination.

Table 2. Runoff Plot Specification, Slopes, and Descriptions

Plot Number	Size (m x m)	Percent Slope	Plot Profile Description
E	1.2 x 2.4	1%	60 cm compacted, vegetated *
F	1.2 x 2.4	3%	15 cm uncompacted cover, 60 cm compacted, vegetated
G	1.2 x 2.4	25%	uncompacted, vegetated
H	1.2 x 2.4	50%	uncompacted, vegetated

* Subsequently surface rototilled to help grass seed germination.

system. The water collection system consisted of 208 liter plastic drums for infiltration and runoff quantitative measurements.

Permeability Tests

Phosphogypsum permeability was determined by using various compactive efforts, beginning with the standard Proctor (ASTM D698) and progressively increasing the input energy to more than the modified Proctor (ASTM D1557) test. Falling head permeability tests were then run on the moisture-density relationship test samples to determine the hydraulic conductivity.

In situ Horizontal Flow and Vertical Infiltration Test

Horizontal flow and vertical infiltration tests were conducted on all lysimeter test plots for comparative purposes. A falling-head slug test (ASTM method D4044) was used to determine horizontal permeability, whereas a double-ring infiltration test (ASTM method D3385) was used to determine the vertical infiltration.

Quantitative and Qualitative Monitoring of Vegetation, Infiltration and Runoff

During the initial stages of vegetation establishment, visual observations were made to determine any deficiency, toxicity, or water stress symptoms. Infiltration and runoff measurements were scheduled periodically depending on rainfall events. Similarly, qualitative sampling was conducted immediately after each heavy rainfall event and sent for determination of various water quality parameters.

3 RESULTS AND DISCUSSION

Laboratory permeability tests were conducted on several compacted samples of PG with varying degrees of compaction. Preliminary results (Table 3) indicate that additional

Table 3. Laboratory Permeability of Phosphogypsum With Compaction

Soil Composition	Compaction Effort	Moisture Content (%)	Dry Density ($kg\ m^{-3}$)	Permeability ($cm\ sec^{-1}$)	Permeant Fluid
Phosphogypsum	25 blows/5 layers *	17.7	1467	1.89E-05	Distilled Water
Phosphogypsum	25 blows/5 layers *	18.9	1485	1.22E-05	Distilled Water
Phosphogypsum	25 blows/5 layers *	15.8	1495	1.77E-05	Gypsum Saturated Water
Phosphogypsum	100 blows/5 layers **	18.7	1507	7.56E-06	Distilled Water

*Typical of modified Proctor Test (ASTM D1557).
**Extra compactive effort.

compactive effort has beneficial effects on reducing the permeability. From these tests, field permeability of the phosphogypsum will be on the order of 7×10^{-6} to 1×10^{-5} cm sec^{-1}. Data from Chang and Mantell (1990) indicated that the falling head permeability of several phosphogypsum sources in Florida was about 3×10^{-5} cm sec^{-1} which is consistent with our results.

The horizontal permeability of the phosphogypsum plots was measured between 0.3 and 0.6 m from the surface. These data indicate the horizontal permeability of all the plots is of the same order of magnitude (Table 4). This suggests that with time, phosphogypsum reverts to a relatively compacted form. However, Plots 1A and 2A have a relatively lower average horizontal permeability. The lower permeability in these plots is attributed to the compaction caused by the excavation and reconstruction of plots. Plots C are compacted to a depth of 60 cm from the surface and therefore explain the lower permeability. Plots D have 15 cm of uncompacted phosphogypsum underlain by 60 cm of compacted material and, therefore, should also have shown lower permeability, but did not. The bermuda grass roots may have influenced the near-surface permeability on these plots.

Comparison of data in Tables 3 and 4 show that the horizontal permeability is greater than vertical permeability by three orders of magnitude. Also, as the compaction effort is increased, the data indicate a four order of magnitude difference between the horizontal and vertical permeability. The additional compactive effort also led to one order of magnitude decrease in vertical permeability. These differences in permeability could affect the residence time of the water between the compacted and uncompacted layers and also affect the quantity of water permeating the stack and that which may be available for

Table 4. In situ Horizontal Permeability of Phosphogypsum Stack Lysimeter Test plots

Plot	Horizontal Flow* (cm sec⁻¹)
1A and 2A	1.55E-02
1B and 2B	1.90E-02
1C and 2C	1.21E-02
1D and 2D	1.96E-02

* Measurement is an average for the two plots.

Table 5. In situ Infiltration in Phosphogypsum Stack Lysimeter Test Plots

Plot	Average Infiltration* (cm sec⁻¹)	Lowest Infiltration Value* (cm sec⁻¹)
2A	2.67E-04	7.66E-05
1B and 2B	1.84E-04	1.04E-04
1C and 2C	2.46E-04	1.64E-04
1D and 2D	3.85E-04	1.92E-04

* Measurement is an average for the two plots, except for plot 2A.

additional loss via evapotranspiration. Upon excavation of the uncompacted 1A plot, roots were found to penetrate through the fiber material around the plot perimeter, which was placed at a depth of approximately 1.2 m from the surface. Roots at such depths should effectively increase the zone of water loss via transpiration and also reduce the amount of water available for infiltration.

Vertical infiltration test results using the double-ring method are shown in Table 5. Measurement on Plot 1A was not possible due to excavation and reconstruction of this plot prior to testing. The vertical infiltration from all plots is generally of the same order of magnitude. These data suggest that the presence of vegetation does not significantly affect infiltration and that the uncompacted phosphogypsum tends to recompact or crust at the surface upon repeated wetting and drying cycles. Such a surface would likely lead to larger residence time and therefore greater water loss via evaporation.

Initially, subsidence and resulting fractures in the lysimeter plots hindered measurement of representative water infiltration data. However, after the repairs to the plots, the data indicated no measurable infiltration, generally less than one percent of the total rainfall for the measurement period. Water quality parameters were not determined for water from the lysimeter plots due to the insufficient accumulation of representative water after any rainfall event.

The overall quantity of runoff was about 20 percent of the measured rainfall, but was between 32 and 96 percent during the rain months. The data suggest several phases in runoff quantity. As expected, runoff increased with slope when plots were devoid of vegetative cover (data not shown). As the vegetation increased, the uncompacted plots (G and H) showed slightly lower runoff than the lesser vegetated, compacted plots (E and F). However, as the vegetation increased on the compacted plots, the runoff was similar for all plots.

Compaction at the surface of the tests plots hindered seed germination. However, loosening the surface resulted in a good ground cover with adequate liming and fertilizer application. Similar observations and finding have been reported by Patel et al., (1995) and Richardson (1994).

Runoff water quality was also observed to improve with increasing vegetative cover. Once the grass cover was established, runoff water quality parameters improved to satisfy state surface water standards.

4 CONCLUSION

According to Chapter 62-673 (FAC, 1993), a final cover system on phosphogypsum should be designed to promote drainage off the stack, minimize ponding, erosion and infiltration into the stack, and function with little or no maintenance.

As demonstrated by this study, PG can be compacted to reduce permeability and also reverts to a relatively compacted form at the surface which should reduce infiltration rates. Vegetation can be directly established on PG provided adequate liming and proper fertilizer nutrients are applied. Vegetation would also improve the quality of runoff water from the stacks. A gentle slope on the top surface of the stack would promote drainage and should prevent ponding. Additionally, these results indicated that the use of compacted PG has potential as a barrier layer and should be less costly than a barrier soil layer or synthetic membrane.

REFERENCES

American Society for Testing and Materials. 1995. Annual Book of ASTM Standard, Section 4, Construction. Volume 04.08. Soil and Rock (I): D420 - D4914.

Chang, W. F. and M. I. Mantell. 1990. Engineering properties and construction applications of phosphogypsum. Phosphate Research Institute, University of Miami, Coral Gables, Florida.

Chapter 62-673, Florida Administrative Code, 1993. Phosphogypsum Management. Vol. 9A p. 856-6A-856-6K.

Federal Register. 1992. 57(107), 23305-23320 (June 3, 1992). EPA 40 CFR Part 61. National emission standards for hazardous air pollutants: national emission standards for radon emission from phosphogypsum stack.

May, A. and J. W. Sweeney. 1980. Assessment of environmental impacts associated with phosphogypsum in Florida, p. 481. In D.P. Borris and P.W. Moody (eds.). Phosphogypsum: Proceedings of the International Symposium on Phosphogypsum. Florida Institute of Phosphate Research, Bartow, FL.

Patel, S.K., S.G. Richardson, J.B. Sartain, and F.M. Rhoads. 1995. Influence of tillage and amendments on vegetation establishment in phosphogypsum. In National Meetings of the American Society of Surface Mining and Reclamation, Gillette, WY. June 5-8, 1995.

Richardson, S.G. 1994. Establishing vegetative cover on phosphogypsum. 1994 Progress Report No. 89-01-086. Florida Institute of Phosphate Research, Bartow, FL.

Environmental Geotechnics, Kamon (ed.) © *1996 Balkema, Rotterdam. ISBN 90 5410 848 7*

Geotechnical properties of a mixed light soil using wood chips and loam

Tej B.S. Pradhan
Yokohama National University, Japan

T. Sato
Kanto Landscaping Co-operative Association, Tokyo, Japan

R. Kozuka
Yokohama City Office, Japan

ABSTRACT: The main objective of the present paper is to examine the possibility of reutilization of the waste wood extracted from the tree branches and demolished concrete blocks, as a new geotechnical material. Laboratory tests were carried out to clarify the geotechnical properties such as compaction, permeability, strength and deformation characteristics of the mixed composite soil fabricated by mixing wood chips, crushed concrete material with soft loam. A series of compaction tests, permeability tests, compression tests were performed under different fabrication conditions and the results are discussed.

It was found that (a) the density of the mixed soil can be controlled between 1.0 to 1.5 Mg/m^3, (b) coefficient of permeability can be controlled between 10^{-3} to 10^{-6} cm/sec and (c) compressive strength can be controlled between 130 to 360 kPa according to the mixing ratio and the compaction method.

1 INTRODUCTION

Regulations by trimming of branches of trees in parks, gardens or along the roads are carried out periodically as one part of environmental glorification of cities. As one of the municipal wastes, the treatment of such tree branches is becoming a great environmental problem. On the other hand, a great problem is arising on the treatment of construction wastes like soft loam, crushed concrete etc. Recently, an extensive research work has been carried out to clarify the mechanical properties of a newly developed light weight mixed soil fabricated by mixing waste soil and expanded polystyrene beads with a little amount of cement (Pradhan, et al., 1993, Pradhan et al., 1994, Pradhan et al., 1995).

Many attempts for the reutilization of municipal as well as construction wastes like old tyres, rock powder, plastic bottles etc, in geotechnical field could be observed as a contribution to solve some environmental problems.

In this study, an attempt has been made for the utilization of (a) tree branches pruned from gardens, parks, roadsides tress etc, (b) concrete crush from building demolition and (c) soft soil loam, as a new geotechnical material. A series of laboratory tests has been performed in order to clarify the mechanical properties of the new mixed composite soil.

2 PHYSICAL PROPERTIES OF TESTED MATERIALS

Wood Chips: Tree branches were crushed to make wood chips and used in tests after passing through a sieve of 4.76mm. The bulk density of the wood chip was 0.5 to 0.6 Mg/m^3.

Concrete Crush: Concrete crushed particles were passed through a sieve of 2mm. The mean diameter was 0.55mm and the fines content below 75 μ m was almost zero. The particle density of this material was 2.71 Mg/m^3.

Loam: The loam used is called ‘Kanto Loam’, which is a deposit of volcanic ash, and is widely distributed in eastern part of Japan. The loam was extracted from surface of a deposited place and the water content was about 20% in airdried condition. This loam when mixed with water becomes very soft and leads to many geotechnical problems in the field and hence is hardly used in construction sites. The sand fraction was about 50% and the fines content below 75 μ m was 50%. The particle density was 2.8 Mg/m^3.

3 FABRICATION OF SPECIMENS

In order to determine the mixing ratio of the ingredients to fabricate the specimen, a series of preliminary compaction tests were carried out. The compaction method was based on Japanese standard JIS1210 with the compaction energy of 560 $m{\cdot}kN/m^3$. Firstly, compaction tests on the mixture of loam and concrete crush under different mixing ratios was carried out. Fig.1 shows the compaction curves for different mixing ratios. It could be noticed from the figure that with the addition of concrete crush, the optimum water content (w_{opt}) decreases and the maximum dry density increases. This w_{opt} was used to prepare the specimen with wood chips.

Wood chips were mixed with the mixture of loam and concrete crush with different ratios in terms of percentage by weight. Two types of compaction energies (560 and 280 $m{\cdot}kN/m^3$) were chosen. The low compaction energy was specified by reducing to half the impact number. The relationship between wet density ρ_t and mixing ratio of wood chips relative to the total weight of the loam and concrete crush, is shown in Fig.2. It could be observed that by increasing the mixing ratio of the wood chips, the density of this new material can be reduced to nearly 1 Mg/m^3. Wet density depends on the compaction energy and also slightly on the mixing ratio of loam and concrete crush. Another important point to be noticed is that as the mixing ratio of the wood chips exceeds 30%, the decrease in ρ_t is significantly low. Based on the results shown in Fig.2, three different values of densities of the specimen is specified from the practical point of view.

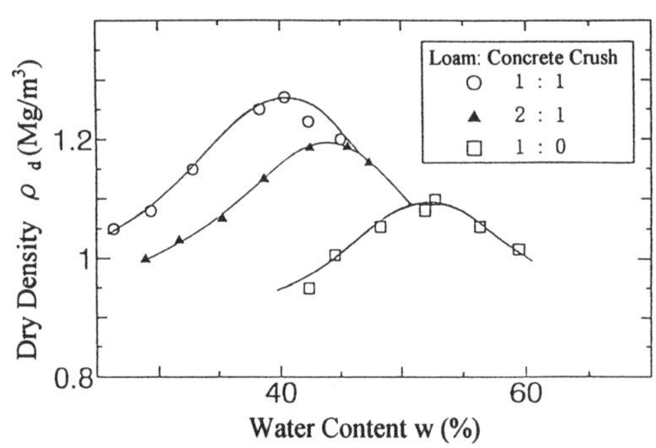

Fig.1 Compaction curves

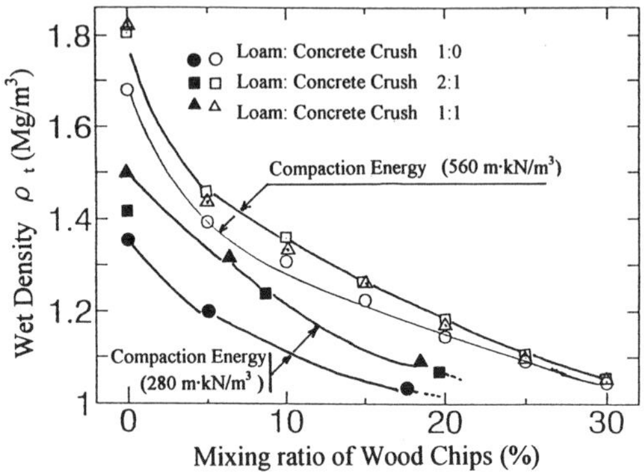

Fig.2 Mixing ratio against density of specimen

Specimens were fabricated by the mixing ratio as shown in Table-1. A plastic mold of inner diameter 5cm and height 12cm with collar was used to prepare the specimen. The wood chips, concrete crush and loam was mixed well with the specified optimum water content w_{opt} and put into the mold and compacted in three layers with the specified energy. The mold was then wrapped by thin plastic film and cured for 1 day in a constant temperature room (20 ° c). The weight and the specimen dimensions were measured after it is taken out from the mold. The specimen is 5cm in diameter and 10cm in height. A picture of a specimen is shown in Fig.3. The density uniformity of the specimen was checked by cutting in three pieces and measuring their unit weights. The variation was utmost 3%.

Table 1. Specimen conditions

Loam: Concrete Crush	Density *1 (Mg/m^3)	Density *2 (Mg/m^3)	M_1 (g)	M_2 (g)	M_3 (g)
1:0	1.70	1.35	1000	0	0
	1.40	1.19			50
	1.20	1.03			175
2:1	1.80	1.43	667	333	0
	1.40	1.25			80
	1.20	1.07			195
1:1	1.80	1.50	500	500	0
	1.40	1.33			65
	1.20	1.09			180

*1: At High Compaction Energy (560 $m{\cdot}kN/m^3$) M_1: Dry Mass of Kanto Loam
*2: At Low Compaction Energy (280 $m{\cdot}kN/m^3$) M_2: Dry Mass of Concrete Crush
M_3: Mass of Wood Chips

Fig.3 Photograph of a specimen

4 TESTING METHOD

The specimen was set in a triaxial cell with a latex membrane of thickness 0.3mm. A vacuum pressure of 10kPa was applied for 30 minutes and deaired water was percolated for 2 hours. After that the vacuum was replaced by a small confining pressure of 20 kPa. This small confining pressure of 20 kPa was chosen because the compressive strength was found almost equal to the uniaxial compressive strength for this loam. The triaxial system was modified to be capable of conducting falling head permeability test. Each specimen was subjected to permeability test prior to the axial compression test under undrained condition and with constant axial strain rate of 0.5%/min. It was confirmed from preliminary experiments that no effect was found in stress strain behavior due to the prior execution of permeability test.

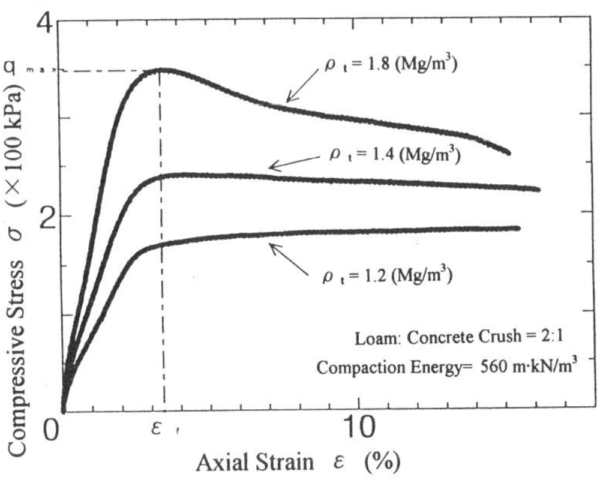

Fig.4 Typical stress strain curves

5 TEST RESULTS AND DISCUSSIONS

Typical compressive stress and axial strain curves for different densities are shown in Fig.4 for the case of compaction energy 560 m·kN/m^3 and loam and concrete crush ratio of 2 to 1. The stress strain curve is quite similar to a stiff clay, however the initial portion seems more linear and elastic. From these curves, compressive strength q_{max} and axial strain at failure ε_f was measured. Under the identical specimen condition, at least two tests were performed in order to see the fluctuation in the data and the repeatability of the fabrication. The fluctuation in strength, deformation and permeability values were utmost 5% .

6 STRENGTH CHARACTERISTICS

Compressive strength in terms of maximum deviator stress q_{max} at a confining pressure of 20 kPa is plotted against density ρ_t in Fig.5 (a), (b) for different compaction energies. Following points could be noticed from the figures:

(1) By the addition of wood chips the strength decreases, however the degree of decrement is significant in case of high compaction energy. Addition of 18% (ρ_t = 1.2 Mg/m^3) of wood chips reduces the strength of loam to about 50%. The reason for this may be due to the reduction of cohesion by the presence of wood chips in soil.

(2) Presence of concrete crush increases the compressive strength. However, the increase is not significant when the mixing ratio of loam and concrete crush is in the range of 1 : 0.5 and 1 : 1.

It can be concluded from the figures that the compressive strength at confining pressure of 20 kPa can be controlled to be between 130 to 190 kPa for high compaction energy and between 50 and 100 kPa for low compaction energy specimens at a wet density of 1.2 Mg/m^3. These strength values are somewhat smaller than an usual clay, however due to the addition of small amount of cement a rather higher strength could be obtained if needed. The author believes that those strength values are enough since this new material is mostly expected to be used in the backfill of retaining walls, slope of embankments, parks etc. with vegetation purposes.

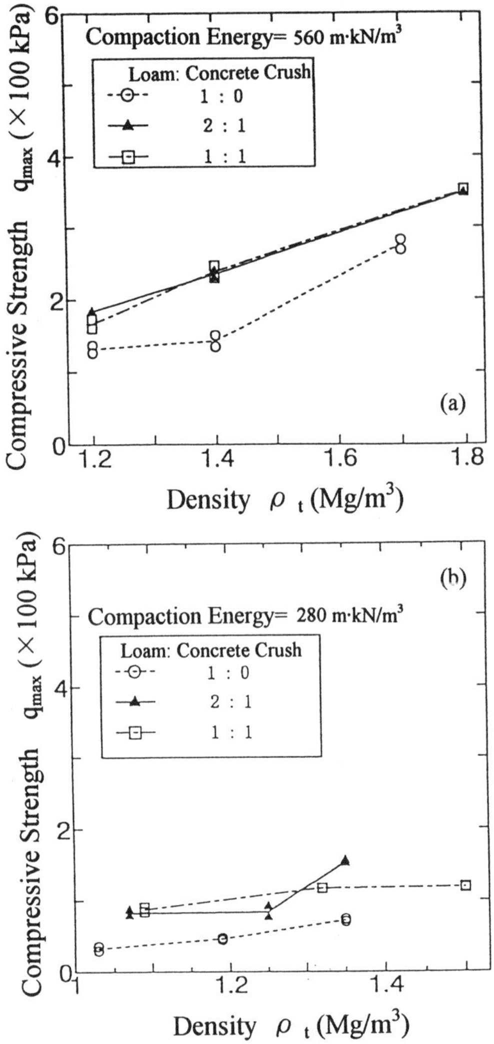

Fig.5 Compressive strength vs density

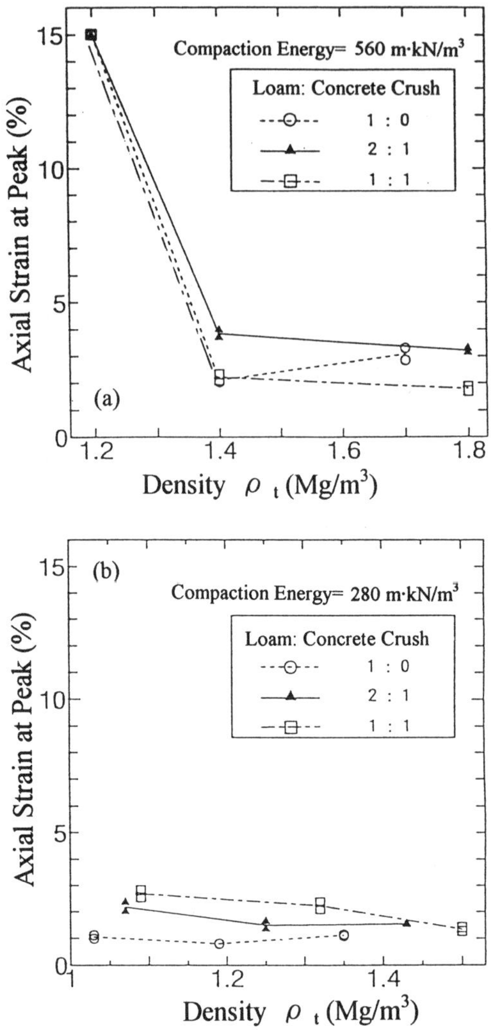

Fig.6 Axial strain at peak vs density

7 DEFORMATION CHARACTERISTICS

Axial strain ε_p observed at the peak compressive strength for high and low compaction energies are shown in Figs 6(a) and (b) against density. ε_p was taken as 15% in the case where no sharp peak was observed. Following points could be noticed:

(1) Due to the addition of wood chips (lower density), axial strain at peak becomes higher. This tendency is predominant in case of higher compaction energy. This implies that the material shows ductile behavior at lower density for higher compaction energy. On the other hand, in case of low compaction energy, the material showed somewhat brittle behavior. However, it should be stated here clearly that even though ε_p was smaller, the stress reduction after the peak was very small and hence the material is not perfectly brittle in behavior. The residual strength is almost 95% of the peak strength. The reason for this tendency may be due to some kind of unstable structure formed by low compaction energy.

(2) ε_p is almost unaffected when the density is greater than 1.4 Mg/m^3.

8 PERMEABILITY CHARACTERISTICS

The coefficient of permeability k is plotted against density for different compaction energies in Figs 7 (a), (b). Following points could be noticed:

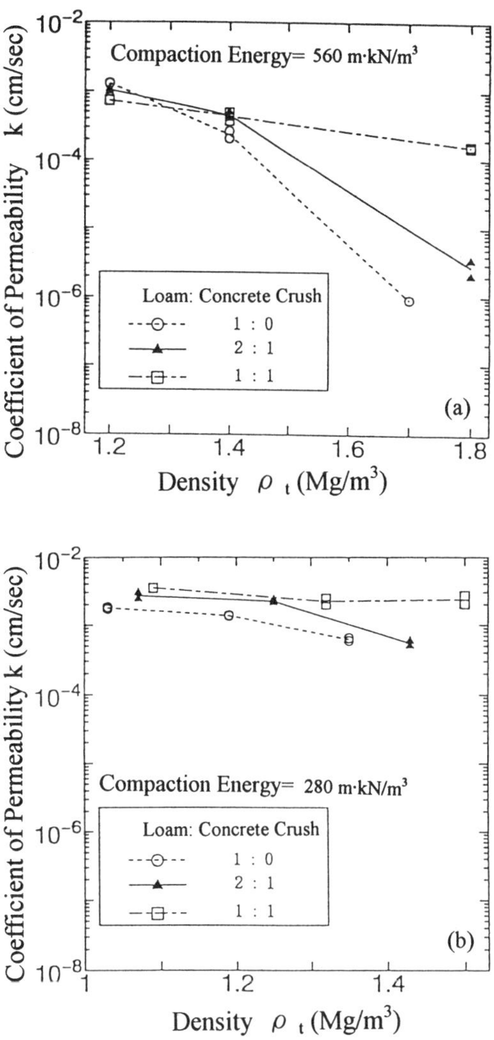

Fig.7 Coefficient of permeability vs density

Due to the addition of wood chips, the permeability increases at lower density. This behavior is quite understandable since wood chips are fibrous material with higher permeability as compared to soil. This tendency is predominant in case of high compaction energy. On the other hand, at low compaction energy, the permeability is not significantly affected by the presence of wood chips. This may be due to the reason that the void ratio of the material itself is quite high and hence the presence of wood chips is not so effective. The results show that the coefficient of permeability lies in the range of 2×10^{-3} to 10^{-4} cm/sec for the density less than 1.4 Mg/m^3.

It is said that the adequate coefficient of permeability of soil for good vegetation is around 10^{-3} to 10^{-4} cm/sec, however it differs according to the type of trees. Standing on this point of view, it can be concluded that this new light soil has a great possibility of using it as a material for vegetation and hence can be used in parks, slopes of embankment etc. Kita et al. (1992) has also shown that a mixture of Kanto loam and expanded polystyrene beads can also be used as a material for vegetation.

9 EFFECT OF CURING PERIOD

It seems quite necessary to know how the wood chips behaves in side a clay in long term. Specimens were cured for different periods and subjected to permeability and compression tests. The coefficient of permeability against curing period is shown in Fig.8. Although the maximum curing period was 90 days, no significant change in permeability was found within this period.

Compressive strength against curing period is shown in Fig.9. A slight increase in the strength was noticed however not so significant. It can be surmised that the physical property of wood chips do not change abruptly inside the soil at least till 90 days. Further research will be needed to clarify the effect due to biodegradation of wood chips in long term.

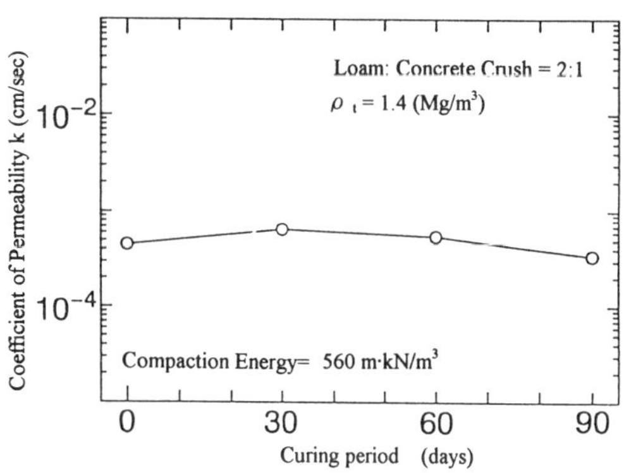

Fig.8 Coefficient of permeability vs curing period

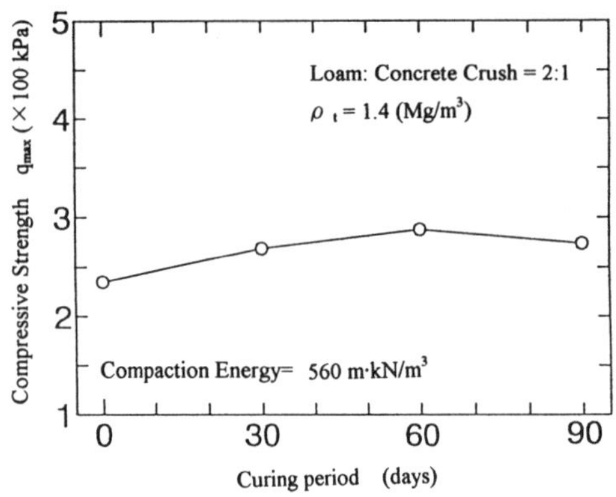

Fig.9 Compressive strength vs curing period

10 CONCLUSIONS

From a limited number of tests on a new light material fabricated from wood chips, concrete crush and loam, following conclusions could be drawn.

1. Density of the new material can be adjusted upto 1.0 Mg/m^3 by changing the mixing ratio of wood chips and loam.
2. Undrained compressive strength of the material lies in the range of 130 to 190 kPa under compaction energy of 560 m·kN/m^3 and at density of 1.2 Mg/m^3.
3. The material shows ductile behavior with the addition of wood chips.
4. The coefficient of permeability of this material is in a range of 10^{-3} to 10^{-4} cm/sec for density less than 1.4 Mg/m^3 and hence can be used as a material adequate for vegetation.

REFERENCES

Pradhan, T., Imai, G., Hamano, M. and Nagasaka, Y. 1993. Failure criterion of a new light geotechnical material SLS. Proc. 3rd International Offshore and Polar Engineering Conference, Singapore, Vol.1 :652-659.

Pradhan, T., Imai, G. and Uchiyama, M. 1994. Mechanical properties of stabilized light soil using expanded polysterene. Proc. 1st International Conference on Environmental Geotechnics, Edmonton, Canada, Vol.1: 755-761.

Pradhan, T., Hirano, Y., Tanabe, T. and Natori, H. 1995. Stabilized light soil using vinylidene chloride foam. Proc. of the 11th European Conference on Soil Mechanics and Foundation Engineering, Copenhagen, Vol.2 :121-126.

Kita, D., Shiota, K. and Sugimoto, H. 1992. Characteristics of light soil compaction for artificial ground. Proc. 27th annual meeting of JGS (in Japanese).

Environmental Geotechnics, Kamon (ed.) © *1996 Balkema, Rotterdam. ISBN 90 5410 848 7*

Research into the effective re-use of concrete waste

Hideaki Shibata
Kokushikan University, Tokyo, Japan

Kenji Saito
Nishimatsu Construction Co., Ltd, Tokyo, Japan

ABSTRACT: The disposal of construction waste material that is designated as industrial wastes, is causing major social problems such as noise and dust, in urban areas. Also, supplies of fine aggregates for use in highway construction, particularly river sand, are becoming exhausted, and the procurement of good quality aggregates has become difficult. This research was carried out to determine to what extent a particular construction waste material, namely, concrete waste, could be re-used as an aggregate substitute in concrete mixtures, or for the base course or subgrade of highways.

1 INTRODUCTION

In recent years, the amount of concrete waste arising through the rehabilitation of roads, concrete building demolition and as surplus during construction work, has been increasing. As this concrete waste is disposed of as industrial waste, its disposal is restricted to such places as harbor landfill or industrial waste disposal sites. Therefore the amount that can be disposed of at any one time is extremely small. In addition, the noise and dust produced and the resulting defilement of the natural environment are newly emerging social problems. On the other hand, supplies of fine aggregates for use in road construction, particularly river sand, are becoming exhausted, causing problems in their procurement. These various problems are steadily becoming serious in urban areas. As a result, recently, seminars have been held at the Ministry of Construction, and specialist committees have been set up by the Japanese Geotechnical Society (JGS); both public and private sectors (ACTEC 1993, Committee on Construction By-products Countermeasures 1994, Honda and Yamada 1994, NTS 1994) have become actively involved in research relating to industrial by-products. This particular research, as a means of effective reuse of construction waste material (Shibata and Saito 1993, 1994, 1995), was focussed on the possibility of re-using concrete waste and asphalt concrete waste as recycled aggregate.

2 CONSTITUENTS AND PROPORTIONS OF CONSTRUCTION WASTE MATERIAL

In Figure 1 is shown the constituents and proportions of mixed construction waste material ($\phi \leq 10$mm) obtained from an intermediate disposal plant. As can be seen from the figure, when building structures are destroyed, apart from concrete and plasterboard rubble, there is a large proportion of fine soil included. In addition, about 12% is impurities such as glass or wood that cannot be completely removed at the treatment plant.

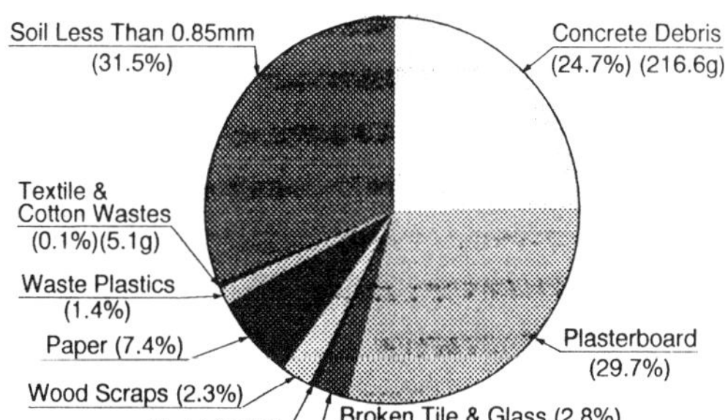

Figure 1. Constituents and proportions of mixed construction waste material

3 PHYSICAL PROPERTIES OF THE RECYCLED AGGREGATE

In Table 1 are shown the specific gravity and coefficient of water absorption for the recycled aggregate used in the experiments. From the table it can be seen that the value of specific gravity is 2.573, a little less than regular sand or gravel. However, the coefficient of water absorption of 6-10%, is very high compared to regular sand. Also, from Photograph 1, in the case of concrete waste, apart from the quartz and feldspar found in regular aggregate, calcium crystals can also be seen to exist. In addition, crystals present in regular concrete, such as calcium hydroxide, calcium carbonate, gypsum, and ettringite were also found. From visual inspection, the presence of considerable air voids was also confirmed.

4 RESULTS OF GRAIN CRUSHING AND ABRASION TESTS

The grain reduction ratio of the recycled aggregate, determined from the impact compaction test prescribed in the California Bearing Ratio (CBR) test, and the Los Angeles Abrasion Test, is indicated in Figures 2 and 3. From Figure 2, for the case of concrete waste it was found that even for a low energy level, (17 impacts), about 12% grain crushing occurred, and that further energy caused considerable crushing. It can be seen from Figure 3

Table 1. Concrete waste absorption coefficient and specific gravity

	Asphalt Concrete Waste		Concrete Waste	
	Coarse Aggregate	Fine Aggregate	Coarse Aggregate	Fine Aggregate
Absorption (%)	6.257	7.967	6.760	10880
Specific Gravity (gm/cm^3)	2.361		2.573	

Photograph 1 Concrete Waste (× 5000)

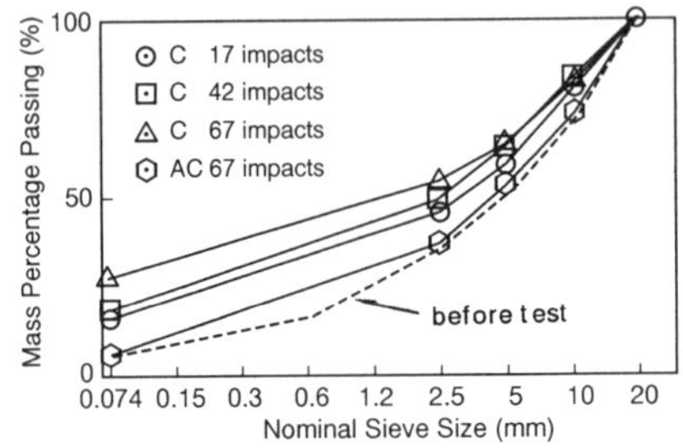

Figure 2. Particle size distribution with successive impacts

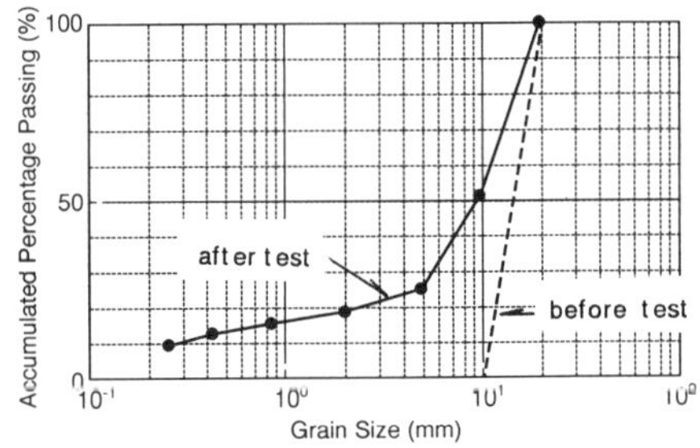

Figure 3. Grain size accumulation curve

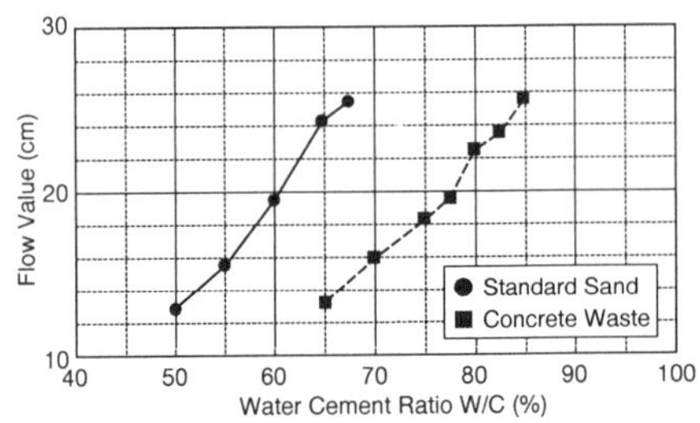

Figure 4. Variation of flow with W/C Ratio

that the abrasion test resulted in about 51% grain crushing. However, in the case of asphalt concrete waste, grain crushing was extremely small, being only a few percent. From these results it can be seen that when using concrete waste as recycled aggregate, it is necessary to take into consideration the degree of grain crushing that will occur.

5 THE USE OF RECYCLED AGGREGATE IN CONCRETE MORTAR

5.1 *Flow test*

The results of flow tests to determine the flow properties of cement mortar made using the recycled aggregate, can be seen in Figure 4. The figure shows the curves for cement mortar made using standard sand and for that using the recycled aggregate. In

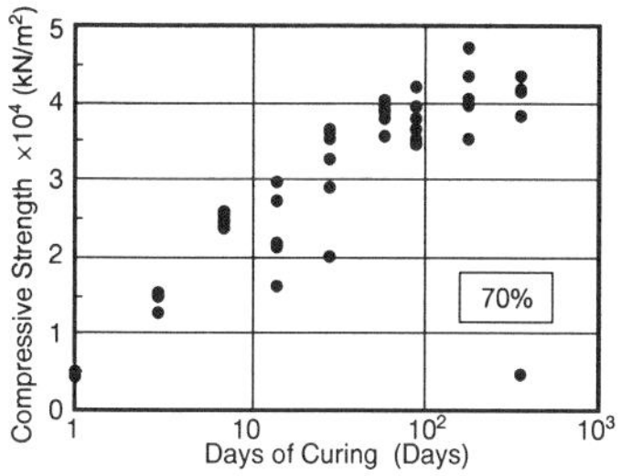

Figure 5. Variation of compressive strength with days of curing

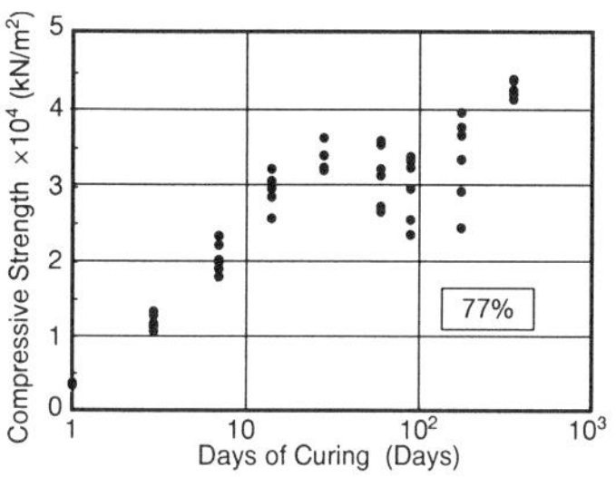

Figure 6. Variation of compressive strength with days of curing

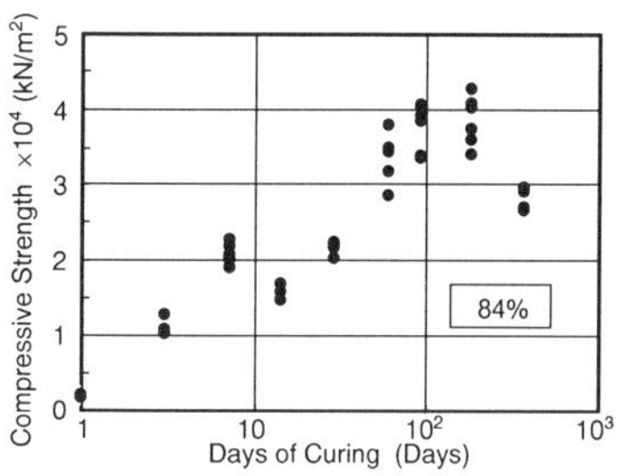

Figure 7. Variation of compressive strength with days of curing

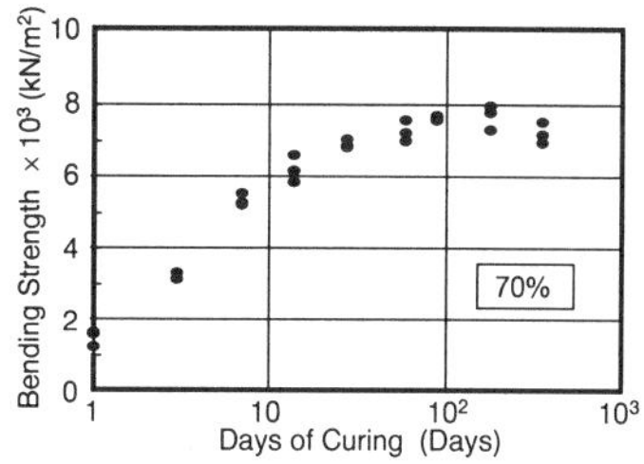

Figure 8. Variation of bending Strength with Days of Curing

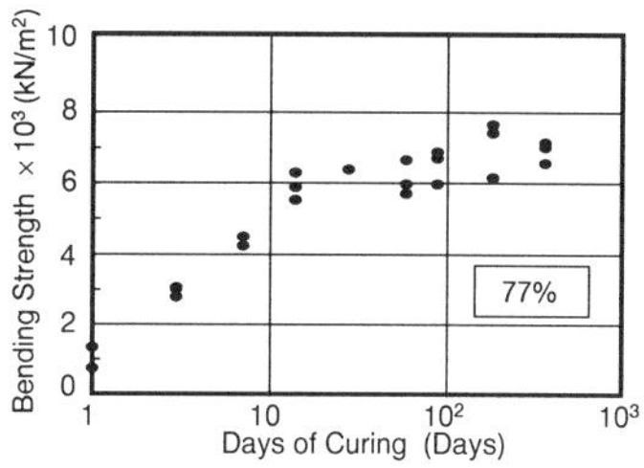

Figure 9. Variation of bending strength with days of curing

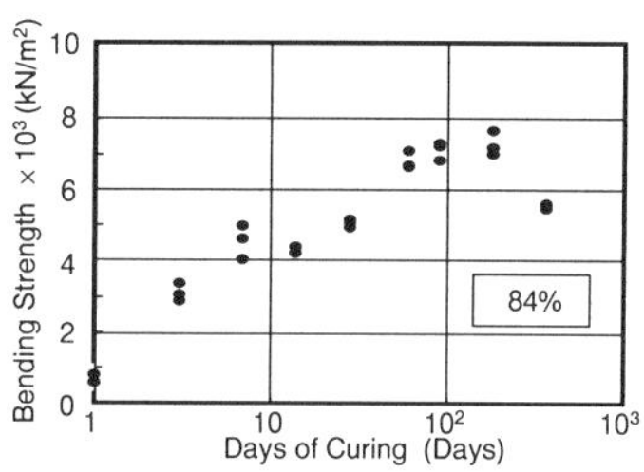

Figure 10. Variation of bending strength with days of curing

both cases it can be seen that the flow value increases with increasing W/C ratio, and a comparison of the two shows that for the same flow value, the W/C ratio is about 18% higher in the case of recycled aggregate. This is the result of the high coefficient of water absorption due to the considerable internal porosity of the recycled aggregate which consequently leads to the higher W/C ratio.

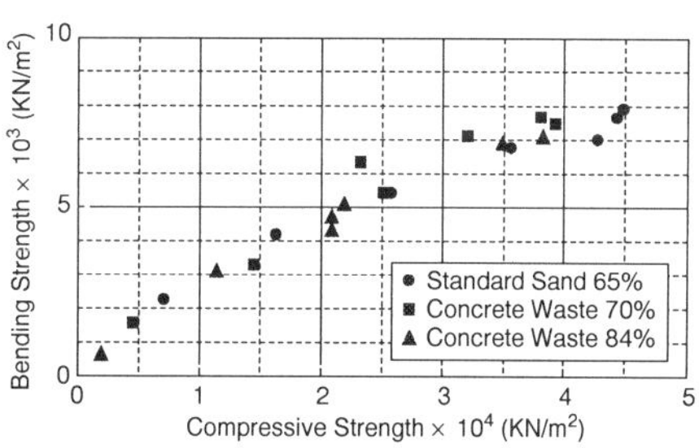

Figure 11. Compressive and bending strength relation

5.2 *Strength test*

The increase in compressive strength with number of days curing for the cement mortar made using the recycled aggregate, can be seen in Figures 5, 6, and 7, and the bending strength in Figures 8, 9, and 10. The figures within the graphs are the values of W/C ratio. From Figures 5, 6, and 7, the compressive strength when using concrete waste, for a W/C of 70%, is about 2.94×10^4 kN/m^2 after 60 days, and reaches a maximum value of 4.41×10^4 kN/m^2 at 120 days. Although there is a slight drop in strength after that, the strength exceeds the design strength of 2.94×10^4 kN/m^2. It was found that up to W/C = 77%, adequate strength was maintained, but that when the W/C becomes as large as 84%, the reduction in strength is also large, falling to about the design strength value. From Figures 8, 9, and 10, it can be seen that the same tendencies are present in the bending strength. The relation of compressive and bending strength are shown plotted in Figure 11, from which it is clear that whether concrete waste or standard sand are used, the values fall on the same curve. From the above it can be seen that, providing adequate attention is paid to the value of W/C ratio, the recycled aggregate can be successfully used for cement mortar.

6 THE USE OF RECYCLED AGGREGATE FOR ROADWAY SUBGRADE

The CBR test was used to investigate the stability of subgrade mixed with the recycled aggregate. In Figure 1 it was shown that concrete waste contains a certain percentage of impurities such as glass or plastic. The California bearing ratio (CBR) was evaluated for recycled aggregate containing these impurities. The values of corrected CBR for the case where concrete waste was used, are indicated in Figure 12. The same method was also used to determine the corrected CBR for various other materials. The values of CBR for these various materials are indicated in Table 2. From this table it can be seen that the value of corrected CBR for concrete waste containing a quantity of fine particles is 10.5% greater than the value for concrete waste alone. Also, even when 10% plastic is included, the CBR, at 21%, is seen to be fairly stable. On the other hand, the inclusion of about 10% glass makes evaluation impossible, indicating that the application of glass to roadway subgrade would present difficulties. This is due to the considerable grain crushing that occurs with the glass.

From the above it was concluded that concrete waste not containing impurities could be effectively used for roadbed subgrade, but that the inclusion of impurities such as glass, that exhibit considerable grain crushing, would not be suitable for this application.

7 CONCLUSIONS

From the results of the above experiments, the possibility of using concrete waste as recycled aggregate and the various problems involved may be summarized as follows:

i) If cement mortar is made using the level of concrete waste of these experiments, a compressive strength of over 3×10^4 kN/m^2 can be achieved, indicating the possibility of application, for example, for secondary concrete products.
ii) Concrete waste that does not contain impurities can be used as recycled aggregate in roadway subgrade. However, the presence of impurities such as glass, makes the material unsuitable.

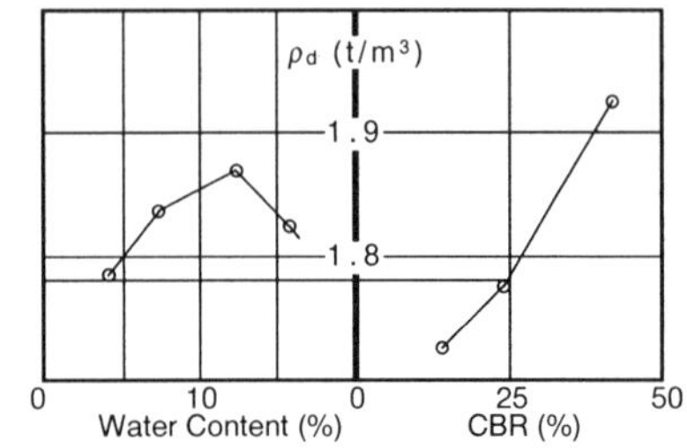

Figure 12. Corrected CBR values

Table 2. Corrected CBR values for various materials

Material	CBR
Concrete Waste	14.5
Concrete Waste (with 10% smaller than 2.5 mm inclusions)	25.0
Glass 10% added to mix	Measurement impossible
Glass 5% added to mix	23.5
Plastic 10% added to mix	21.0

iii) Concrete waste is a material that is highly susceptible to grain crushing and having a high absorption coefficient. It remains necessary to carry out further research into the influence of the characteristics of this material in order to increase its scope of application as a recycled aggregate.

REFERENCES

Advanced Construction Technology Center (ACTEC) (ed) 1993. *Guidelines on the proper treatment of construction by-products.* Taisei Publishing.

Committee on Construction By-products Countermeasures, Economic Affairs Bureau, Ministry of Construction (ed) 1994. *Coping with construction by-products and case studies.*

Honda, A., and M. Yamada 1994. *The recycle of construction by-products and waste materials.* Energy Conservation Center.

NTS 1994. Technological data on the treatment and use of construction by-products and waste materials. *Lecture records.*

Shibata, H., and K.Saito 1993. Study on the effect of concrete wastes. *Proc. 28th Meeting of the Japanese Society of Soil Mechanics and Foundation Engineering (JSSMFE)*: 9-10.

Shibata, H., and K.Saito 1994. Study on the effect of concrete wastes (No.2). *Proc. 29th Meeting of the Japanese Society of Soil Mechanics and Foundation Engineering (JSSMFE)*: 29-30.

Shibata, H., and K. Saito 1995. Study on the effect of concrete wastes (No.3). *Proc. 30th Meeting of the Japanese Society of Soil Mechanics and Foundation Engineering (JSSMFE)*: 27-28.

Environmental Geotechnics, Kamon (ed.) © 1996 Balkema, Rotterdam. ISBN 90 5410 848 7

Treatment and utilization of waste soils from construction sites

Hiroshi Sogabe & Atsuo Aihara
Tokyo Metropolitan Government, Bureau of Sewerage, Japan

Kiyoomi Yamada
Nihon University, Tokyo, Japan

ABSTRACT: The authors developed a high pressure dewatering method of cement-mixed soft soil for increasing strength and decreasing volume of soft cohesive soil, together with mud and slurry derived from urban construction works. The dewatering characteristics of cement-mixed soft soil and the strength characteristics of dewatered cakes made from soft cohesive soils and the various properties of those cakes became clear, and it was proved that crushed cakes of classified small-gains size by a crusher could be used as materials for roadbed or refill works instead of natural gravel and sand. In order to put the high pressure dewatering method to practical use, we constructed the full scale experiment plant and succeeded in producing the granular improved soil of good quality continuously.

1 INTRODUCTION

Waste soil, which is non-used soil and sludge, of 375.360 million m^3 is annually generated from construction works mainly in urban areas in Japan, and the disposal of such soil has become one of the social problems. For solving this problem, it is necessary to reduce the total volume of waste soil or to promote the utilization of the soil for new purposes other than the conventional land reclamation. Particularly, soft cohesive soils generated from construction sites have difficulty to handle or utilize. To cope with this problem, the author proposed here a new treatment technology 'High Pressure Dewatering Method of Cement Mixed Soft Soil'. Using the high pressure dewatering method, the volume of waste soils can be reduced, and the treated soft cohesive soils can be used as materials for earthwork instead of natural sand and gravel.

In this paper, the pressuring and dewatering characteristics of the cement-mixed soft soil, determined by a large-scale consolidation test apparatus, are clarified, and the strength characteristics of dewatered cake, which is the compressed and dewatered cement-mixed soft soil, and the material characteristics (specific gravity, coefficient of water absorption, abrasion ratio and slaking) of granular improved soil which was made by crushing dewatered cakes into predetermined particle size were measured in a laboratory.

2 METHODS OF EXPERIMENTS

2.1 *MATERIALS USED*

Soil samples used in the testing are shown in Table 1. Cohesive soils in the soil samples are the Kawasaki soil (A), Kinshicho Soil (B), and Ukima Soil (C), and sandy soil is the Ohtemachi Soil (D). Ordinary Portland cement (OPC) was used for mixing with soil.

2.2 *EXPERIMENTAL EQUIPMENT*

Fig. 1 shows a large-scale consolidation test apparatus with consolidometer 30 cm in diameter. A compression equipment is of hydraulic type capable of loading 70 tons maximum. Compression load was measured with load cells, and the measurement of vertical displacement was performed with displacement gauge with the accuracy of 1/1,000 mm.

Table 1. Basic properties of soil samples

Property		A	B	C	D
Natural water content W(%)		102.3	69.4	83.0	36.2
Density of ρ_s (g/cm^3)		2.633	2.620	2.631	2.660
Liquid limit w_L (%)		71.8	47.5	82.7	38.5
Plastic limit w_P (%)		34.4	29.1	39.9	23.3
Plasticity index I_P		37.4	18.4	42.8	15.2
Grain size	Sand(%)	4	11	3.1	66
	Silt (%)	45	55	46.6	15
	Clay (%)	51	34	50.3	19

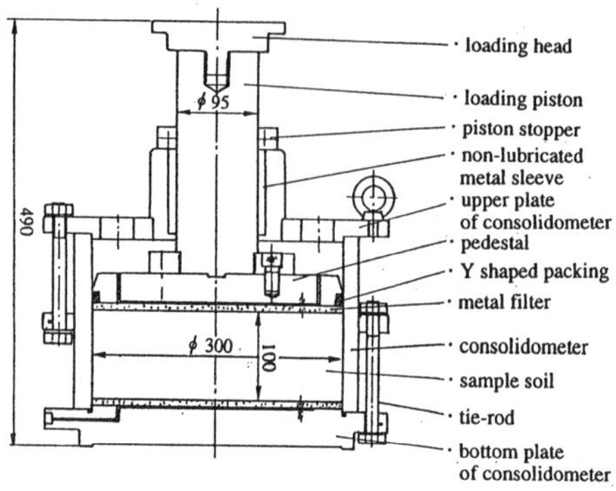

Fig.1. Large-scale consolidmeter

2.3 *PREPARATION OF CEMENT-MIXED SOFT SOIL AND PRESSURING AND DEWATERING TEST METHOD*

In the preparation of cement-mixed soft soil, the cement content (aw) was specified as 0%, 5%, 10% and 20%, and soil slurry was mixed with OPC of water-cement ratio (W/C) 60% for about 5 minutes.

The soil slurry (cement-mixed soft soil) was placed in the large-scale consolidometer (specified initial sample height : h_0 = 7.5 cm) shown in Fig. 1 and the loading pressures P were 0, 5, 10, 20, 40, 60, 80, and 100 $\times 10^2$ KPa. At the same time, vertical displacement (S) was measured by the displacement gauge. Using the value of h_0 and S, the rate of volume decrease ($\varepsilon_v = S/h_0$) was calculated [1].

3 RESULTS AND DISCUSSION

3.1 *PRESSURING AND DEWATERING CHARACTERISTICS*

Fig. 2 shows the relation between the rate of volume decrease (ε_v) and loading time (t) of the Kawasaki Soil under different loading pressures (P).

Fig. 3 shows the relations between the dewatering completion time (T) and the loading pressure (P) for soil samples.

Fig. 4 shows the relation between the water content(W) and loading pressure for soil samples after pressuring and dewatering. The dewatered soil sample is called dewatered cake.

From these figures, the following results can be obtained:

(1) As the cement content increases, the pressuring and dewatering time is shortened.

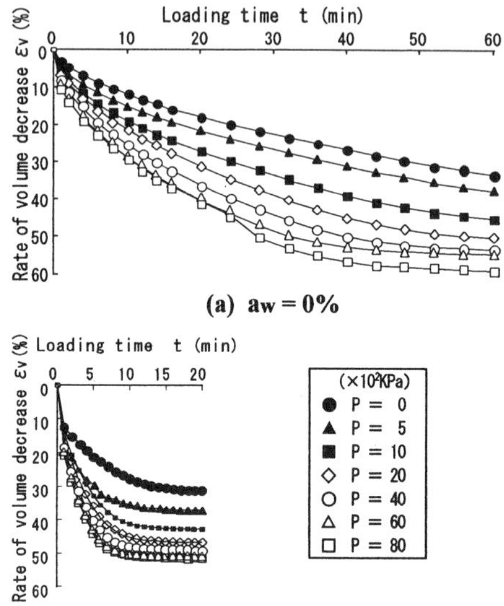

Fig.2. Relation between S, ε_v ,t and P

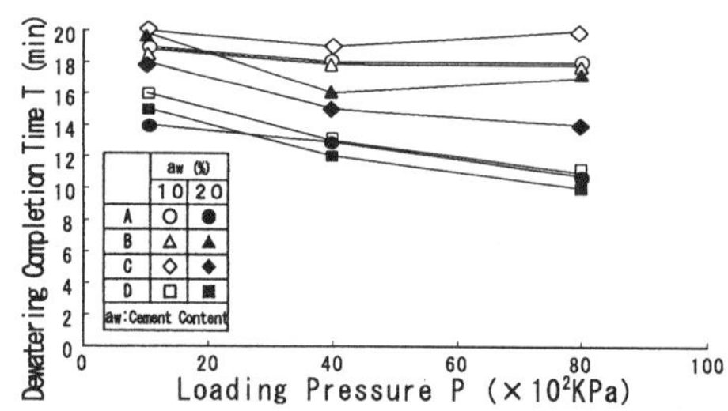

Fig.3. Relation between the dewatering completion time T and loading pressure P

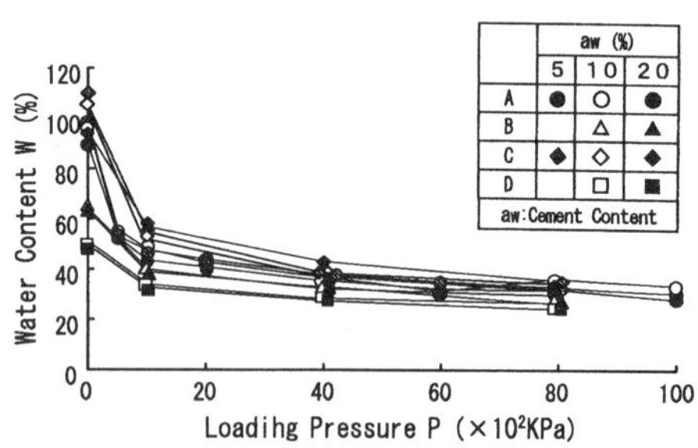

Fig.4. Relation between the water content W and loading pressure P

(2) From the viewpoint of the rate of volume decrease, the practical loading pressure can be

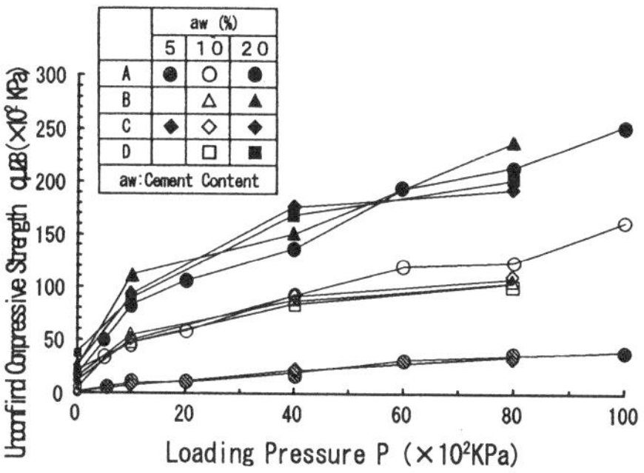

Fig.5. Relation between unconfined compressive q_u and loading pressure P

approximately 40 (× 10^2 KPa) ,and the rate of volume decrease εv at that time is 20 to 40%.

(3) The Ohtemachi Soil with a large sand content has a shorter dewatering time compared to other soil samples.

(4) The water content greatly varies in the range of P = 10 to 20 (× 10^2 KPa).

3.2 *STRENGTH CHARACTERISTICS OF DEWATERD CAKE*

At first, water content and wet density of the dewaterd cake after the consolidation test were measured, then core samples (diameter of 2 cm and height of 4 cm) were taken in the lateral direction from the dewatered cake [2]. The core samples were cured for the predetermined period. An unconfined compression test was conducted for these samples at the curing age of 28 days.

Fig. 5 shows the relation between the unconfined compressive strength (q_u) and the loading pressure (P).

From this figure, the following results can be obtained:

(1) As the loading pressure increases, the unconfined compressive strength q_u of dewatered cake increases remarkably.

(2) q_u can be roughly predicted by the cement content and loading pressure regardless of the kind of soil. For instance, when the loading pressure is 40 (× 10^2 KPa), q_u is 20 (× 10^2 KPa) at a_w = 10%, and 130 to 170 (× 10^2 KPa) at a_w = 20%, respectively.

3.3 *MATERIAL CHARACTERISTICS OF GRANULAR IMPROVED SOILS*

The dewatered cake was crushed into 20 to 40 mm by a crusher and by hand. This is called granular

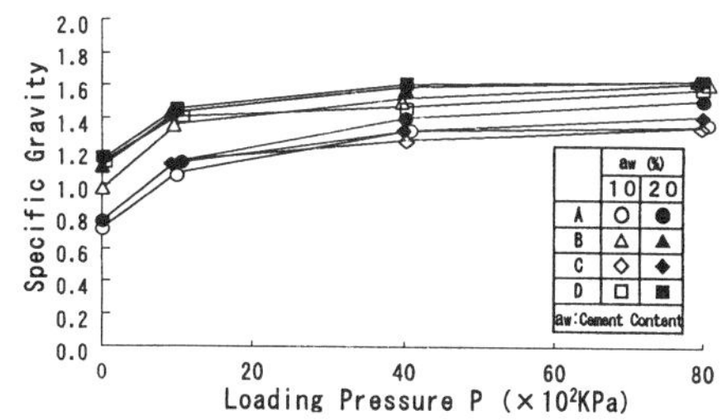

Fig.6. Relation between specific gravity in absolute dry condition and loading pressure P

improved soil. For this granular improved soil, we conducted several kinds of tests for determining the specific gravity in absolute dry condition, coefficient of water absorption ratio, and crushing and slaking characteristics. These tests were conducted in accordance with the standards of Japanese Industrial Standard (JIS) and the standards prepared by the Japan Road Association and the Japan Highway Public Corporation.

Fig. 6 shows the relation between the specific gravity in absolute dry condition and loading pressure of the granular improved soil [3]. The specific gravity increases with the loading pressure. Comparing with crushed stone being used as the material for earthwork, the specific gravity in absolute dry condition of the granular improved soil is relatively smaller than the values of crushed stone. For example, if the loading pressure is 40 (× 10^2 KPa), then the specific gravity in absolute dry condition of granular improved soil is 1.27 to 1.61. This value is almost the same as the specific gravity of artificial lightweight aggregate of 1.00 to 1.60.

Fig. 7 shows the relation between the water absorption ratio and loading pressure of the granular improved soil. The water absorption ratio decreases with the increase of loading pressure and is 27% to 54% at a_w = 10%, and 25% to 39% at a_w = 20%, respectively. These values are close to about 37% to 48% for the value of the natural lightweight aggregate.

Fig. 8 indicates the relation between abrasion ratio and loading pressure. The granular improved soil made from sandy soil at the cement content a_w of 20% fulfills the abrasion ratio standards for public construction work in Japan which value is less than 30% to 40%.

Fig. 9 shows the relation between the percentage of crushing and loading pressure and Fig. 10 the relation unconfined compressive strength. According to these results, the percentage of crushing decreases as the loading pressure increases for both aw of 10% and 20%, but its rate of decrease becomes slightly small when the loading pressure exceeds 40 ($\times 10^2$ KPa). Moreover, the granular improved soil with unconfined compressive strength more than 150 ($\times 10^2$ KPa) will permit to reduce the percentage of crushing to less than 20%.

Fig. 11 shows the particle size distribution of the granular improved soil of the Kawasaki Soil and Ohtemachi Soil after crushing tests. For avoiding the generation of fine granule smaller than 2 mm, the cement content aw of 20% and the loading pressure P more than 40 ($\times 10^2$ KPa) is needed.

Fig. 12 indicates the results of slaking test which was plotted on a rock compressibility evaluation diagram specified by the Japan Highway Public Corporation [4]. The granular improved soil is within C zone in the diagram where the slaking does not occur easily.

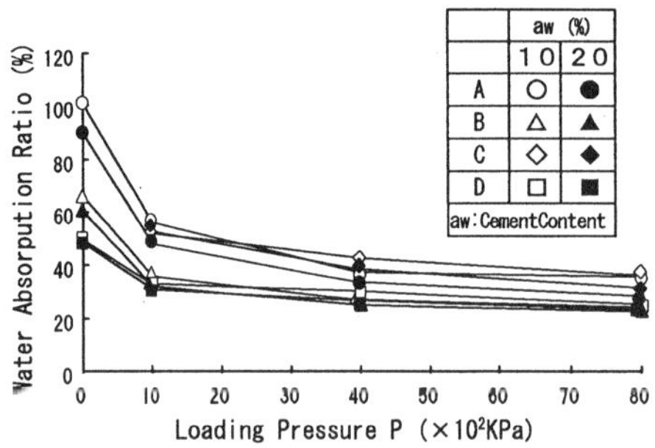

Fig.7. Relation between the water absorption ratio and loading pressure

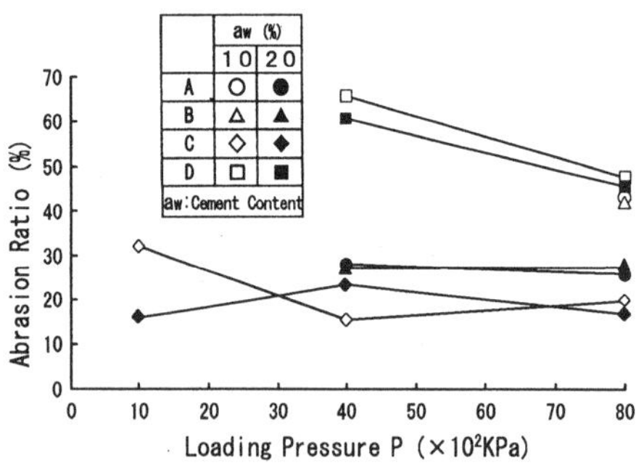

Fig.8. Relation between abraision ratio and loading pressure P

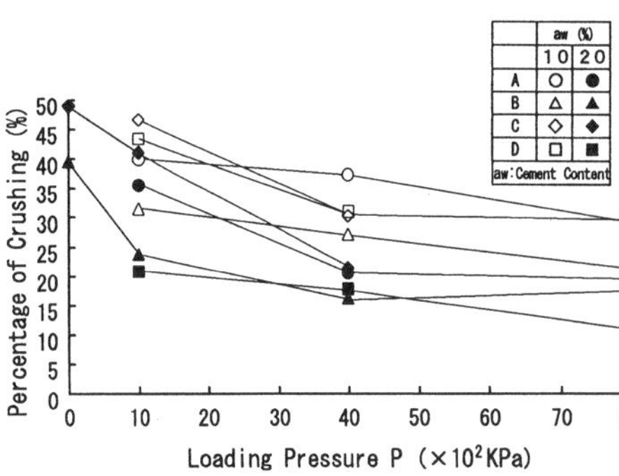

Fig.9. Relation between the percentage of crushing and loading pressure P

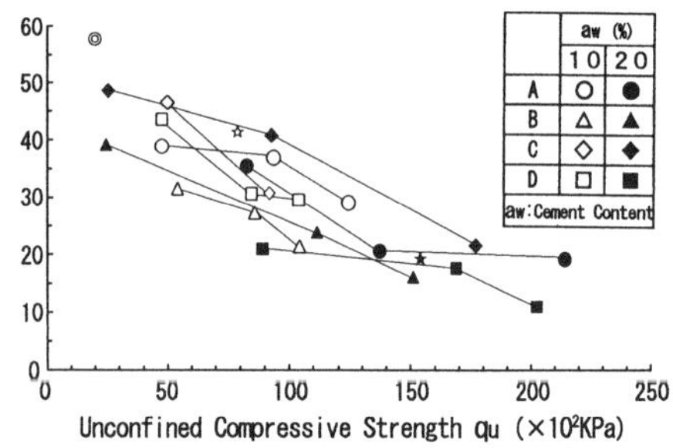

Fig.10. Relation between the percentage of crushing and unconfined compressive strength q_u

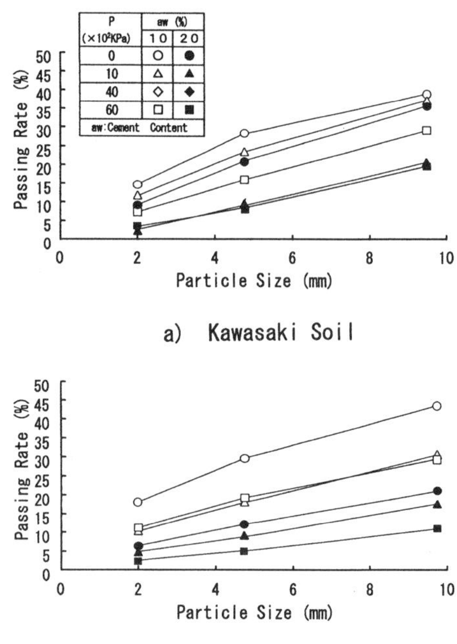

Fig.11. Relation between passing rate and particle size

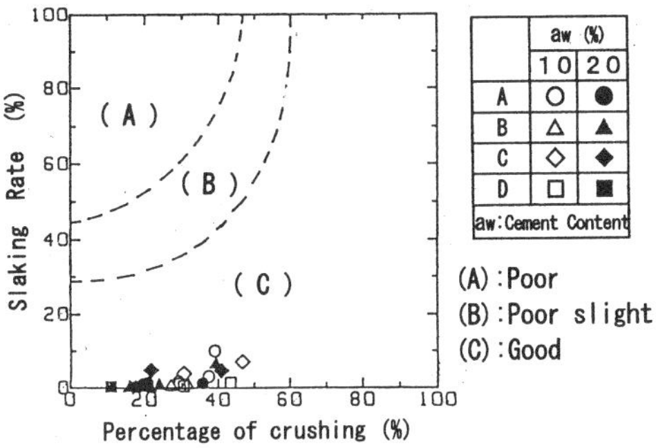

Fig.12. Relation between slaking rate and percentage of crushing

Fig. 13 indicates the relation between the slaking rate of the granular improved soil and loading pressure. If the cement content aw is more than 10 to 20% and the loading pressure is more than 40 ($\times 10^2$ KPa), then the slaking rate of the granular improved soil can be less than 5%.

4 FULL SCALE EXPERIMENT PLANT

As the results of basic experiments in a laboratory described above, it was proved that the granular improved soil could be used as a useful material for earthwork. Therefore, in order to put the high pressure dewatering method to practical use, we constructed the full scale experiment plant and conducted the technology development to produce the granular improved soil continuously [5].

4.1 *Process flow*

Figure 14 shows the process flow of the full scale experimental plant. This plant is comprised of four processes.

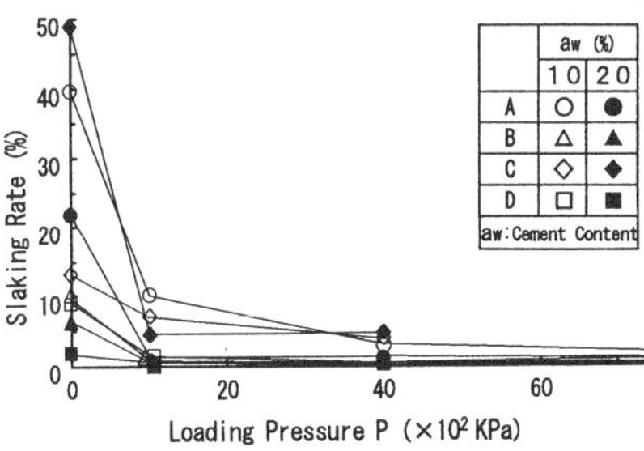

Fig.13. Relation between slaking rate and loading pressure P

① Conditioning process : gravel is removed from soft cohesive soil and the soft cohesive soil is made into slurry of approximately 200 % water content.

② Mixing process : the slurry is mixed with the cement slurry.

③ Dewatering process : cement-mixed soft soil is dewatered using the 40 ($\times 10^2$ KPa) high-pressure dewatering machine to dewatered cake like a board.

④ Crushing process : Dewatered cake is crushed to granular.

Table 2. shows equipements of the experimental plant.

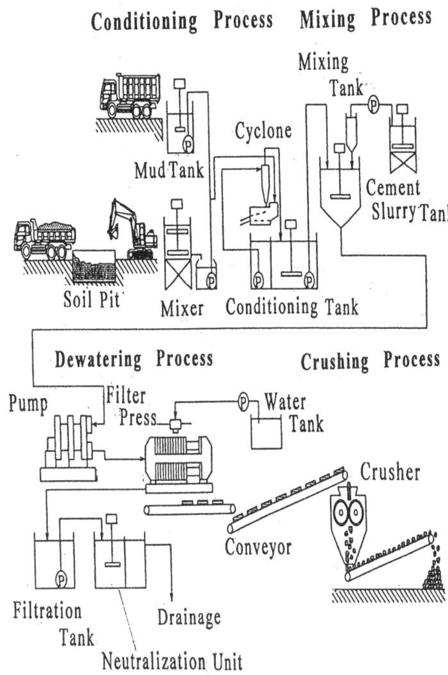

Fig.14. Process Flow

Table 2. List of equipments

Equipment	Specification
Soil Pit	$12m^3$
Mud Tank	$10m^3$
Soil Mixer	$1m^3$
Conditioning Tank	$8m^3$
Mixture Tank	$0.75m^3$
High Pressure Pump	40×10^2KPa, $6m^3$/hr
Filter Press	14 Chambers-$0.174m^3$
Filtration Tank	$1m^3$
Neutralization Tank	$5m^3$
Double Roll Crusher	ϕ 550mm × w750mm

4.2 *GRANULAR IMPROVED SOIL*

The soil samples used in the full scale experiment are shown in Table 3. As the cement-mixed soft soil was to be transported to the dewatering machine by a high pressure pump, its water content was conditioned to 200%. The dewatering machine was operated under the condition that the operating pressure was 40 ($\times 10^2$KPa) and the cement dosage rate aw is 20 %. Unconfined compressive strength of 90 ($\times 10^2$KPa) on average could be obtained for the dewatered cake at the age of 28 days.

Fig.15 shows the dewatered cake which is crushed to granular less than 40 mm, that is, the granular soil.

As a result of CBR test, the granular improved soil had CBR value of almost 60 % and was proved to be a good material for earthwork.

The full scale experiment plant could produce the granular improved soil of 30 m^3 during the four month operation without any trouble.

Table 3 . Basic properties of soil samples

Property		E	F	G
Natural water content W(%)		46.7	68.4	60.8
Density of ρ_s (g/cm^3)		2.699	2.672	2.694
Liquid limit w_L (%)		47.7	66.5	59.3
Plastic limit w_P (%)		22.4	28.9	26.4
Plasticity index I_P		25.3	37.6	32.9
Grain size	Sand(%)	28.7	11.7	9.6
	Silt (%)	44.2	52.0	55.5
	Clay (%)	27.1	36.3	34.9

Fig.15. Granular improved soil

5 CONCLUSIONS

(1) Pressuring and dewatering completion time for the cement-mixed soil is very short compared to non cement-mixed soil.

(2) As pressuring proceeded, the unconfined compressive strength qu of dewatered cake increases remarkably.

(3) The specific gravity in absolute dry condition of the granular improved soil is almost the same as that of lightweight aggregates.

(4) In the case of the granular improved soil made from cohesive soil, the abrasion ratio is less than 30% when the cement content is increased.

(5) Strength and slaking ratio of the granular improved soil with cement are considerably improved.

(6) Utilizing the granular improved soil as materials for earthwork, it is to be used as the subgrade roadbed material., In this study, we proved that the granular improved soil made from cohesive soil fulfilled the requirements of material standards in Japan.

(7) We developed the high pressure dewatering process that could produce the granular improved soil.

6 REFERENCES

1) Yamada, K., Nakano, T., SAI, K., Dekura, M. and Saito, S. (1991): "Use of excavated soil improved by adding cement and applying pressure," Proc. of The 46th Annual Conference of The Japan Society of Civil Engineers, 3, pp. 1092-1093.

2) Morita, Y., Yamada, K., Sai, K., Aihara ,A. and Saito ,S. (1992): "Use of excavated soil improved by adding cement and applying pressure -aggregate tests-," Proc. of The 47th Annual Conference of The Japan Society of Civil Engineers, 3, pp. 1001-1002.

3) Yamada, K., Morita, Y., Aihara ,A. Saito, S. and Minegishi, D. (1992): "Use of excavated soil improved by adding cement and applying pressure -durability tests-," Proc. of The 47th Annual Conference of The Japan Society of Civil Engineers, 3, pp. 1002-1003.

4) Japan Highway Public Corporation Laboratory (1978): Technical Report No. 212, pp. 1-30.

5) Yamada, K., Sogabe, H., Aihara ,A., Uryuu, T. and Fujii, Y. (1994): " Utilization of Soft Cohesive Soils from Construction Sites, " Proc. of The 49th Annual Conference of The Japan Society of Civil Engineers, 6, pp. 504-505.

Environmental Geotechnics, Kamon (ed.) © 1996 Balkema, Rotterdam. ISBN 90 5410 848 7

Properties and usage of iron and steel slag

Kunio Takahashi
Port and Harbour Research Institute, Yokosuka, Japan

Takao Komuro & Yoshio Horinouchi
Nippon Slag Association, Tokyo, Japan

ABSTRACT : Iron and steel slag is produced 33 million ton a year in Japan. Utilizing the great amount of slag is significantly important. The paper presents the properties and usage of the three types of slag:granulated blast furnace slag,air-cooled blast furnace slag and steel slag. The granulated blast furnace slag has especially useful properties such as light weight,high permeability and latent hydraulic property.Because of the useful properties, the granulated blast furnace slag was utilized in port and harbour construction of three types,which is introduced as an example.The steel slag is comparatively difficult to use because of its properties,and is disposed in special site.But steel slag has useful properties such as heavy weight. The authors are making some efforts to utilize steel slag in port and harbour construction of three types,which is also introduced.In order to focus on discussion about the granulated blast furnace slag and steel slag,only a brief description is made for the air-cooled blast furnace slag in this paper.

1. INTRODUCTION - OUTLINE OF IRON AND STEEL SLAG

The iron and steel slag, industrial waste in steel industry, consists of three categories namely blast furnace slag(BF slag), converter slag(C slag) and electric furnace slag(EF slag). Typical generation process of those slags are shown in the Fig.1. The BF slag is generated during the process producing pig iron from iron ore. The steel slag(S slag) is generated during converting pig iron to steel and the EF slag is generated during recycling steel scrap. Both the C slag and the EF slag are often called as S slag. The BF slag consists of two types, the air-cooled blast furnace slag (ABF slag) and the granulated blast furnace slag (GBF slag) quenched with water. Table 1 shows typical chemical composition of iron and steel slag . The BF and the S slag are composed of lime and silica, and similar to ordinary sedimentary rock and Portland cement. The ABF slag and the S slag are coarse like gravel while the GBF slag is sandy. Hydraulicity is common feature of all slags. The released water from all slags are alkaline due to contained calcium. The past assays reports that no ingredients in the released water is beyond the upper limit stated in the Japanese environmental standard .The S slag expands because contained free-lime reacts with water. This is a marked feature of the S slag.

Fig.2 shows the production and consumption by uses for the BF slag and S slag in Japan in 1994. Total consumption of the ABF slag is about 7.2 million ton and it is mainly utilized for road construction, cement material and earthwork material, while 14.4 million ton of the GBF slag for cement and earthwork material, and 11 million ton of the S slag for temporary work, recovered metal recycling , recycling in steelworks and road construction.Total production quantity of both BF slag and S slag is 32.6 million ton. Almost all BF slag generated in Japan is effectively consumed. However ,effective consumption ratio of S slag is only 61 per cent.

The ABF slag have been used as subgrade material since long time ago because it has similar features as ordinary crushed stone. The Japanese Industrial Standard (JIS) states the required quality for the ABF slag and the S slag as subgrade materials. This JIS states that the S slag can be used for subgrade

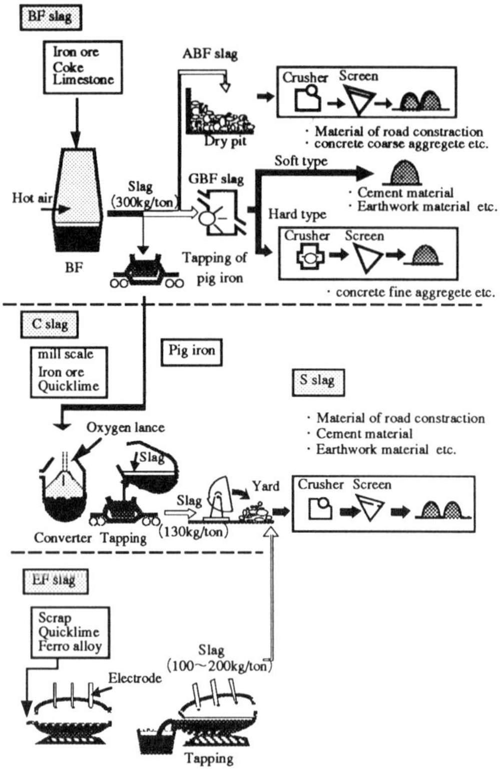

Fig.1 Generation process of iron and steel slag

Table 1 Typical chemical composition of iron and steel slag

(Unit; %)

	BF slag	C slag	EF slag	Andesite	Ordinary Portland cement
SiO_2	33.8	13.8	19.0	59.6	22.0
CaO	42.0	44.3	38.0	5.8	64.2
Al_2O_3	14.4	1.5	7.0	17.3	5.5
T-Fe	0.3	17.5	15.2	3.1	3.0 *
MgO	6.7	6.4	6.0	2.8	1.5
S	0.84	0.07	0.38	-	2.0 **
MnO	0.3	5.3	6.0	0.2	-
TiO_2	1.0	1.5	0.7	0.8	-

(Note) *as Fe_2O_3 , **as SO_3

material, if its self-expanding ratio is below 1.5 per cent by accelerating reaction with air and water. Utilization of slag for subgrade is about 5.72 million ton in 1994, and this is about 17 per cent of total quantity of slag. This figure says that the road

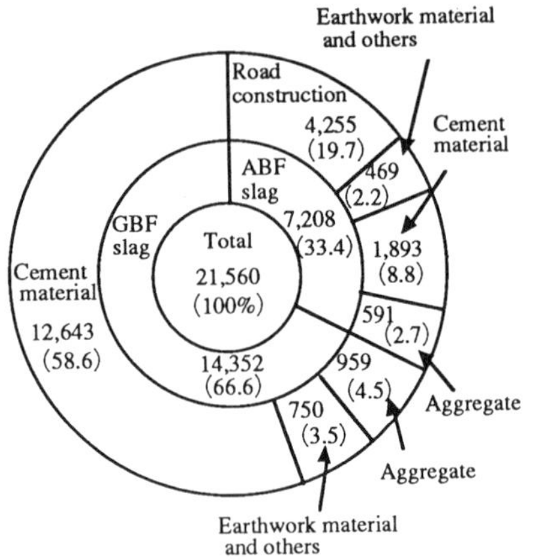

(a) BF slag

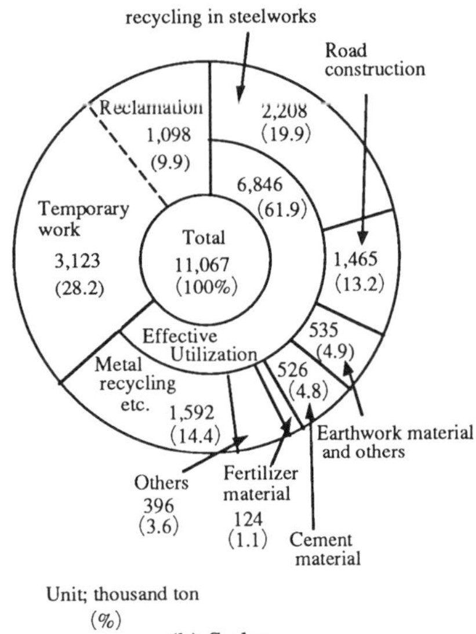

(b) S slag

Fig.2 Production and consumption by uses for the BF slag and S slag

construction is one of the major field for slag utilization.From the following section, this paper focuses on the utilization for earthwork material and others shown in the Fig.2.

2. THE GBF SLAG - CHARACTERISTICS AS SOIL AND ITS APPLICATION

2.1 Characteristics and application in port and harbour construction

Table 2 shows physical properties of the GBF slag.

Table 2 Physical properties of GBF slag and ordinary sand

Physical property	GBF slag	Ordinary sand
Unit weight (ton/m^3)	1.21-1.64	1.8
Internal friction angle (degree)	35	30
Permeability (cm/sec)	1-0.1	1 - 0.01

The GBF slag has lighter in weight and larger internal friction angle and permeability comparing with the ordinary sand. Due to this feature, the following utilization can be made advantageously.

- backfilling material for behind revetment and wharf
- trafficability improvement
- ground stabilization
- embankment material

The authors of this paper had published a guidebook for the GBF slag utilization.

2.2 Application to the wharf in Okidate area in Aomori Port

Fig.3 shows the typical section of the wharf construction in Okidate area in Aomori Port. As this wharf was constructed in soft ground , the GBF slag was used for backfilling material in order to decrease the soil pressure. The area 27 - 36 m wide and 9.8 m deep behind the wharf was backfilled by the slag. Total volume used for this backfilling was 59 thousand ton.

2.3 Application to the soil disposal area in Kanda Port

Fig.4 shows the typical section of the cell revetment for the soil disposal area of Kanda port in Fukuoka prefecture. In this disposal area, a high dike had to be constructed on the soft ground with high ground water level. Therefore stability of the ground and good workability were important factors in design and construction.The characteristics of GBF slag, light weight and latent hydraulic,were quite suitable as the material of this dike. Approximately 465 thousand ton of the GBF slag was used and no problem was found during construction. The

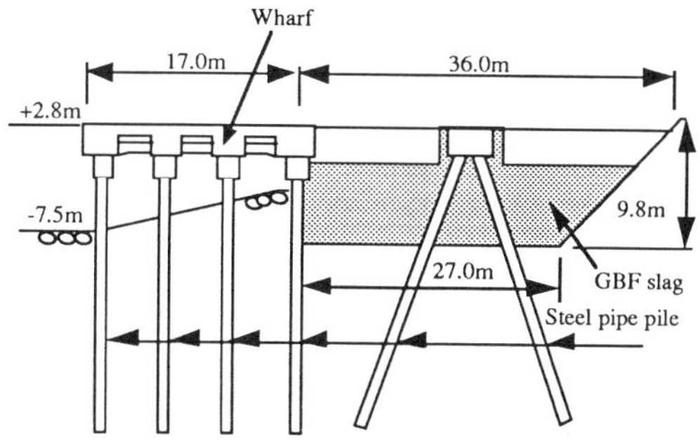

Fig.3 Application to the wharf in Aomori Port

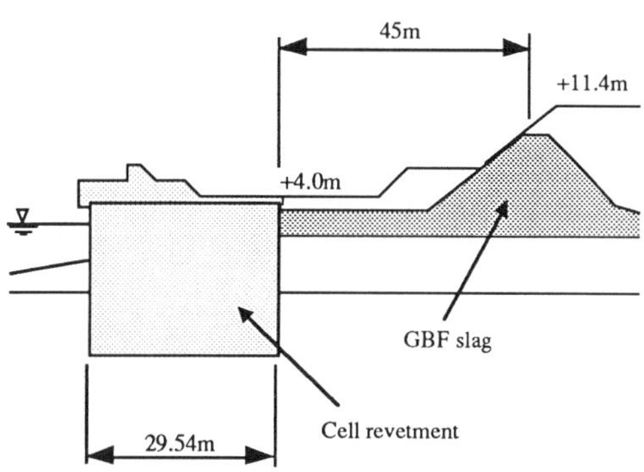

Fig.4 Application to the soil disposal area in Kanda Port

Swedish weight sounding (SWS) was carried out six month after the construction completion,and the test result reports that the SWS value of GBF slag was increased to 6 to 28 due to its hydraulic,while the SWS value of sand was 3 to 4.

2.4 Application to the wharf of the Rokko Island in Kobe Port

Fig.5 shows the typical section of the wharf of the Rokko Island in Kobe Port. The caissons were seriously moved due to the Hyogo Prefecture Earthquake (M7.2) occurred in January 1995. After putting back those caissons, the soil improvement by sand compaction piles(SCP) and the soil pressure mitigation by light weight backfilling material were done to improve quake proof of this wharf. The GBF slag was utilized for the light weight material in the area 19 m wide behind the wharf. Approximately one million ton of the GBF slag is expected to be used eventually for this purpose.

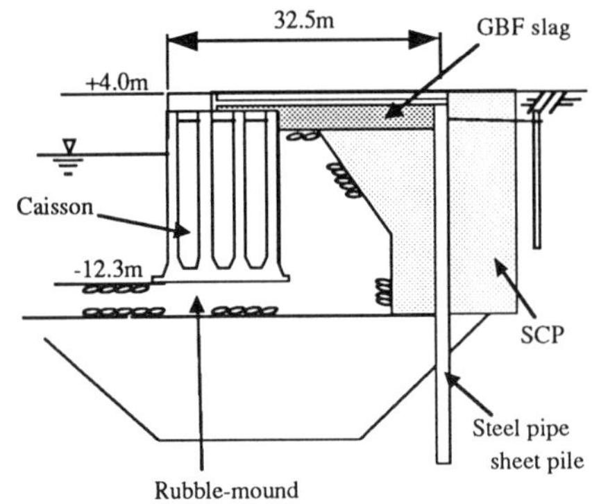

Fig.5 Application to the wharf in Kobe Port

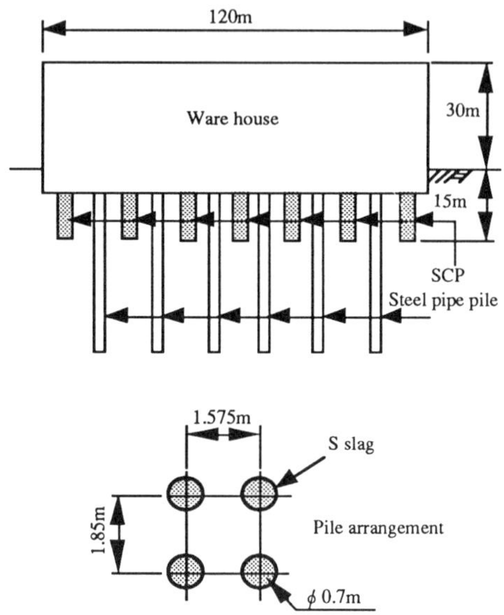

Fig.6 Application to the steelworks foundation in Chiba

3. THE S SLAG - CHARACTERISTICS AS SOIL AND ITS APPLICATION

3.1 Characteristics and applicatin in port and harbour construction

Table 3 shows physical properties of the S slag. Unit weight of the S slag is 1.87-2.39 ton/m^3 and greater than the ordinary sand. It was regarded as disadvantage of the S slag to expands while later because contained free-lime reacts with water. But the aging technology to complete its expansion prior to utilization has been developed now.

The advantage of large unit weight was brought out in port construction, when it was used for material of sand drain(SD) and SCP, surcharge material for soil improvement and fill material for cell revetment. Besides,utilization as replacement soil and fill material for caisson and double sheet pile wall are concluded to be expectable in accordance with our recent study.

Table 3 Physical properties of S slag and ordinary sand

Physical property	S slag	Ordinary sand
Unit weight (ton/m^3)	1.87-2.39	1.8
Internal friction angle (degree)	40	30
Permeability (cm/sec)	1 - 0.01	1 - 0.01

3.2 Application to soil improvement for steelworks foundation in Chiba

Fig.6 shows the application to soil improvement for warehouse of steelworks in Chiba. Since the ground is loose sand, soil investigation was necessary in order to prevent liquefaction due to earthquake. SCP method using S slag was adopted for this improvement. The sand pile was designed to be 700 mm diameter and 15 m long, and overall pile length was 42,474 meters. Total quantity of used S slag was 42,800 ton. The average N value of the soil between the sand piles was increased from 12 to 22. Consequently safety against liquefaction was improved.The N value of the sand pile was 48 at the end of driving, and it was drastically increased to 111 after two years. Because hydraulicity and self-expansion, typical features of the S slag, are effective to soil improvement, S slag is more suitable material for this purpose than ordinary sand. When the Chiba-ken Toh-ho-oki Earthquake occured in 1987, many liquefaction were found in the reclaimed land in this region.But no liquefaction was found the area improved by the S slag.

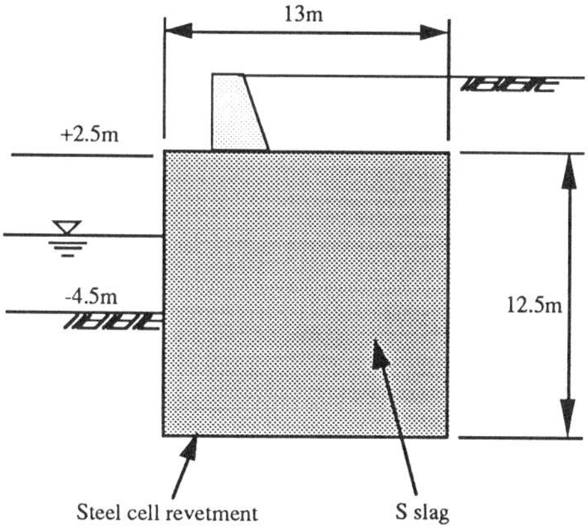

Fig.7 Application to the revetment in Yokkaichi Port

3.3 Application to fill material for steel cell revetment in Yokkaichi Port

Fig.7 shows the application in Yokkaichi Port. Large unit weight of the S slag was expected to effective to stability of the cell. The cell was 13 m outer diameter and 12.5 m high. Steel slag fill was carried out for one cell only because the application was experimental purpose in this case. Used quantity was 1,800 ton. PH of the sea water around the cell was about 9.3 during construction and it was recovered to its stationary value of 8.4 after two hours. The turbidity caused by S slag was less than the ordinary sand, together with smaller diffusion area and shorter convergence time. This shows that the advantage of S slag in this application. The follow-up survey to investigate variation of inclination, diameter and top elevation of the cell is being done. After five months , no significant problem was found due to expansion of the S slag.

3.4 Application to pre-loading mound material in Honmoku Wharf

Fig.8 shows the application in Honmoku Wharf. SCP were driven behind the wharf to improve stability of the soft ground. Slag was used for pre-loading mound and the slag was left partially even after the pre-loading was completed in order to use as subgrade of pavement.
The slag was mixture of S slag and ABF slag in the ratio of 4 to 1. Two different height of the pre-loading mounds were constructed to apply the loads

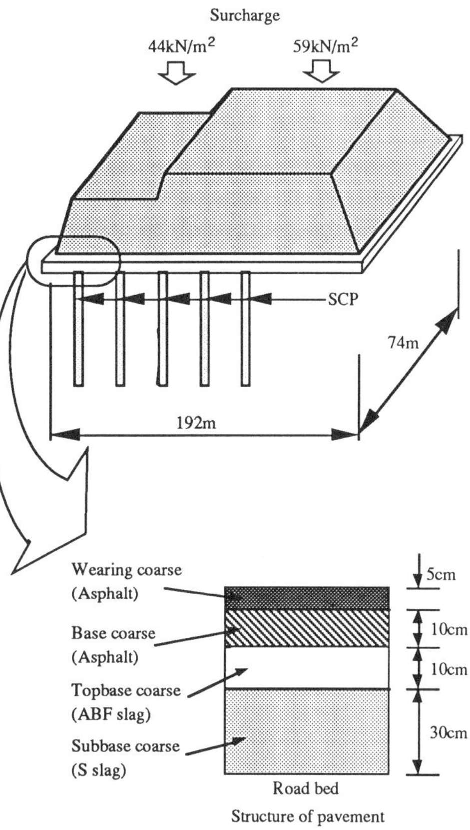

Fig.8 Application in Honmoku Wharf

of 44 kN/m^2 and 59 kN/m^2. After the pre-loading was completed the mounds were removed to a certain elevation so that the remaining slag could be used for subbase coarse. On the subbase coarse,BF slag was filled and compacted for the upper base coarse and then asphalt pavement was executed. The 250 thousand ton of S slag was used for this work. Application of S slag contributed to save both time and cost, because the required quantity of soil for the mound was reduced due to large unit weight of the S slag and the mound was appropriated to the subbase.Quantity of material could be decreased to 70 to 80 per cent of the original quantity.

4. CONCLUSION

Past conventional utilization of the slag in port and harbour construction is mainly for subgrade material.

However utilization for revetment and wharf construction is increasing recently as described in the previous section.Utilization of slag,taking advantages of light weight of the GBF slag and larger unit weight of the S slag, leads to construction of the stable and economical structure. In addition, utilization of slag is not only an effective recycle of industrial waste but also a prevention of environmental disruption because it is an alternative material for natural rock and sand.
The waterfront area is especially important to Japan because this country is surrounded by sea. This is the reason for significance of waterfront development including port and harbour facilities. In order to expand the usage of slag in this filed, further research and development, collection of application data and publicity for advantage of slag have to be pursued. Especially for S slag,the authors will proceed study on the utilization of following five purposes which are expected to be advantageous to its characteristics . The utilization manual will be provided by collecting technical data including insitu full scale test results.

- Replacement material
- Surcharge material
- SD
- SCP
- Fill material for cell,caisson and double sheet pile wall

At present slag is classified as "an industrial waste" , but we hope the slag will be called "a valuable product from steelmakers" in the near future.

REFERENCES

1) T. Kimura et al., Soil Improvement Using Converter Slag Compaction Pile Method, Kawasaki Steel Technical Report 20 (1988) 1, pp 69-75

2) Y. Yamaguchi et al., Application of Converter Slag to Pavement for the No.4 Pier D in Honmoku Wharf, Proc. of No.16 Road Conference Papers, pp 279-280

3) K. Takahashi, Effective Utilization of Recycled Materials in Port Construction

4) S. Wada, Basic Concept of Kobe Port Rehabilitation, Doboku-Seko, 36 (1995) 8, pp 33-37

5) Nippon Slag Association, Granulated Blast Furnace Slag for Earthworks, 1995

6) Nippon Slag Association, Characteristics and Values of Steel Slag, 1994

7) Ministry of Transport , Construction method of the inner bulkhead of the soil disposal area in Kanda Port,1987

8) Coastal Development Institute of Technology, Reseach on behavior of temporary steel cell revetment in Yokkaichi Port, 1996

Environmental Geotechnics, Kamon (ed.) © 1996 Balkema, Rotterdam. ISBN 90 5410 848 7

Mechanical properties of ash generated in sewage sludge incineration and its effective use

Enao Takayanagi
Osaka City Government, Japan

Tomio Tamano
Osaka Sangyo University, Japan

Kiyoshi Matsui
Japan Sewerage Works Agency, Tokyo, Japan

Satoshi Ono
Chuo Kaihatsu Co., Osaka, Japan

ABSTRACT: In most major cities of Japan, a great percentage of land area is now being served by sewage works, causing increases in the volume of sewage sludge, and thus making the securement of sludge disposal site increasingly difficult. In order to cope with the situation, efforts have been made to reduce the volume of sludge by incineration. Nevertheless, the acreage that can be secured for disposal site is extremely limited, particularly in major cities. In order to cope with the potential increase in the volume of sludge caused by the progress of sewerage system construction, the adoption of advanced treatment in the water-treatment process and the construction of combined sewage works, and in order also to achieve the long-term use of such limited disposal sites, attaining the effective use of ash resulting from the sludge incineration process not only as a landfill material but as other types of construction material is a matter of urgency.

Incinerator ash, a residual product of sludge incineration, can be used either as a construction material or part or whole of primary material. In the case of the former, incinerator ash is used for backfilling, as well as a subgrade or base course material. Meanwhile in the case of the latter, fired blocks, bricks, tiles and lightweight aggregate are possible uses.

In the light of the existing circumstances, this paper first considers the fundamental soil mechanics characteristics of incinerator ash when used as a construction material. It then discusses the mechanical properties of incinerator ash based on the results obtained from various field tests carried out to validate the potential uses of incinerator ash as landfill, backfill, as well as subgrade and base course materials. Finally, this paper cites a typical case of incinerator ash application to a pedestrian road construction where incinerator ash is being used as a primary material of permeable fired blocks.

1 INTRODUCTION

In Japan, the construction of sewage works has been furthered with the view to achieving such goals as flood prevention, the enhancement of living environments, and the conservation of the water quality of public water bodies. The percentage of land area served by sewage works, which was 28% in 1979, has increased in 1994 to 51% (*Sewage works in Japan* 1995).

In major cities confronted with two major difficulties, namely, increases in the volume of sewage sludge and the shortage of sludge disposal site brought about by the increasing percentage of coverage by sewage works, efforts have been made to reduce the volume of sludge by dehydrated sludge incineration. Nevertheless, the acreage that can be secured for disposal site is extremely limited, particularly in major cities. In order to cope with potential increases in the volume of sludge caused by the progress of sewerage system construction, the adoption of advanced treatment in the water-treatment process and the construction of combined sewage works, and in order also to achieve the long-term use of such limited disposal sites, attaining the effective use of incinerated ash not only as a landfill material but as other types of construction material is a matter of urgency.

Sludge comes from sewage works in the following different forms: 1) dehydrated sludge, 2) composted sludge, 3) ash, and 4) liquid slag. The use of sludge has to be considered for each form. Of all sludge forms, ash is considered to have the greatest potential as a construction material in terms of both quality and quantity. Ash can be used either directly as a construction material or indirectly as part or whole of primary material. In the case of the former, ash is used for backfilling, as well as a subgrade or base course material. Meanwhile in the case of the latter, possible uses are fired blocks, bricks, tiles and lightweight aggregate which come from different manufacturing processes.

In the light of the existing circumstances, this paper first considers the fundamental soil mechanics characteristics of incinerated ash when used as a construction material. It then discusses the mechanical properties of incinerated ash based on the results obtained from various field tests carried out to validate the potential uses of incinerated ash for landfilling, backfilling, as well as a subgrade or base course construction material. Finally, this paper cites a typical case of incinerated ash application to a pedestrian road construction where incinerated ash is being used as a primary material of permeable fired blocks.

2 PRESENT STATE OF EFFECTIVE USE OF SEWAGE SLUDGE

The state of sewage sludge disposal as of 1994 is presented in Table 1. As can be seen from the Table, 63% of the entire volume of sludge is subjected to landfill disposal, 24% is being used effectively, and the remaining 13% is subjected to other forms of disposal. Table 2 shows the existing state of the effective use of sewage sludge. A large percentage, 92%, of sewage sludge is being used in open space and farm land. 8% is used as a construction material, a large proportion of which is taken up by incinerated ash (*Sewage works in Japan* 1995). The figures given in the abovementioned Tables are aggregates obtained from various parts of the country. In urban communities, incinerated ash accounts for large proportions. In Osaka City for example, incinerated ash accounts for nearly 100% of landfill proposal.

Possible uses of ash as a construction material are listed in Figure 1. The use of ash for different purposes, such as soil improvement, subgrade and base course construction, concrete secondary materials, asphalt filler, tiles, bricks, and permeable blocks, is currently under consideration. Yet, research and development efforts are necessary to establish a system that can satisfy both quality and quantity requirements. Possible items that require further considerations include: 1) the improvement of quality, 2) building up of a market, 3) cost reduction and 4) stabilizing the mechanical properties of incinerated ash.

Table 1. Disposal of sewage sludge (Sewage works in Japan, 1995) (April 1, 1993 to March 31, 1994)

(Unit: thousand m^3/year)

Type of sludge	Type of disposal				
	Land reclamation	Sea side land reclamation	Effective use	Others	Total (%)
Dehydrated sludge	863	312	462	91	1,728 (75)
Incinerated ash	108	125	46	17	296 (13)
Dried sludge	28	0	65	0	93 (4)
Digested sludge, concentrated sludge	10	0	1	179	190 (8)
Total (%)	1,009 (44)	437 (19)	574 (24)	287 (12)	2,307 (100)

Note: Dried sludge includes composted sludge. Incinerated ash includes liquid slag. Volume ratio: dehydrated sludge : incinerated ash : dried sludge : digested sludge, concentrated sludge : 1:11:4:0.1

Table 2. Effective use of sewage sludge (Sewage works in Japan, 1995) (April 1, 1993 to March 31, 1994)

(Unit: thousand m^3/year)

Type of use		Type of sludge					
		Dehydrated sludge	Incinerated ash	Dried sludge	Composted sludge	Digested sludge	Total
Open space/farm land	Undertaken by municipalities	44	2	9	24	0	79
	Delivered to fertilizer manufacturers	406	9	3	29	1	448
	Subtotal	450	11	12	53	1	527
Construction material		12	35	0	0	0	47
Total		462	46	12	53	1	574

Note: A large percentage of dehydrated sludge delivered to fertilizer manufacturers are composted for use in open space and farm land.

3 SOIL MECHANICS CHARACTERISTICS

In order to obtain the general knowledge of sludge, its soil mechanics characteristics in particular, soil tests were carried out, covering an extensive range of form variation. Since the soil mechanics characteristics of sludge are affected by its final form, or the type of the final disposal process (digestion, dehydration, incineration, or composting), as well as the type and the dosage of a coagulant, seven types of sludge were selected for the tests, as shown in Table 3, to gain a clear insight into the unknown mechanism. The first characters of the notations that appear in Table 3 indicate the types of sludge (or the types of final disposal process), and the second character, the types of coagulants added.

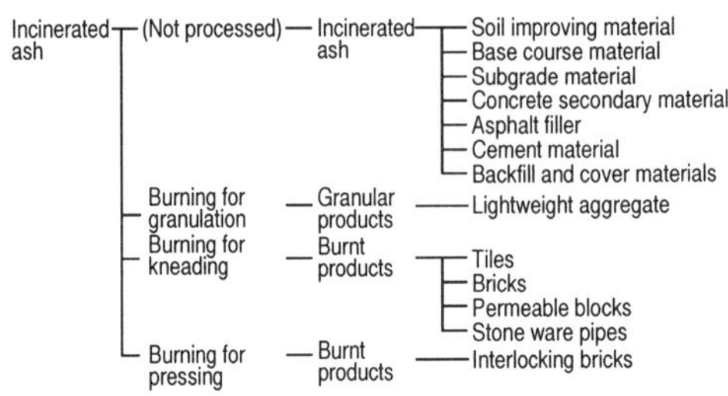

Figure 1. Uses of incinerated ash

Table 3. Type variation of investigated sludge

Symbol	Type	Remarks
Ap − 1	Incinerated ash	Burnt Dp-1 (Digested sludge with polymer added)
Aℓ − 1	Incinerated ash	Burnt Rℓ-1 (Raw sludge with lime added)
Aℓ − 2	Incinerated ash	
Aℓ − 3	Incinerated ash	
Dp − 1	Digested and dehydrated sludge	
Rℓ − 1	Raw and dehydrated sludge	
Cℓ − 1	Aerobic fermentable sludge	Composted Rℓ-1 (raw sludge with lime added)

A: Incinerated ash D: Digested sludge R: Raw sludge
C: Composted sludge ℓ: Lime added p: Polymer added

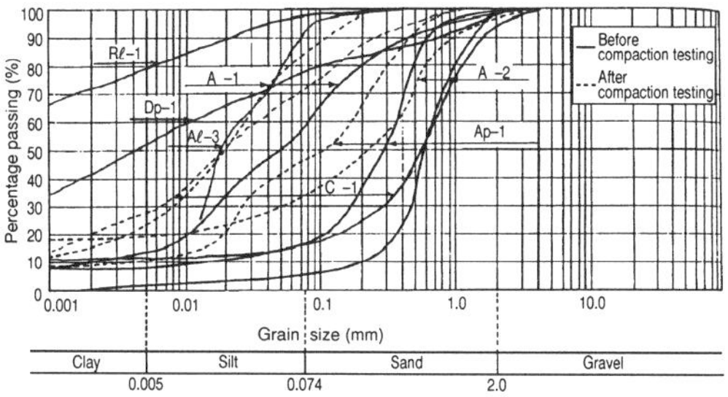

Figure 2. Grain size distribution curves

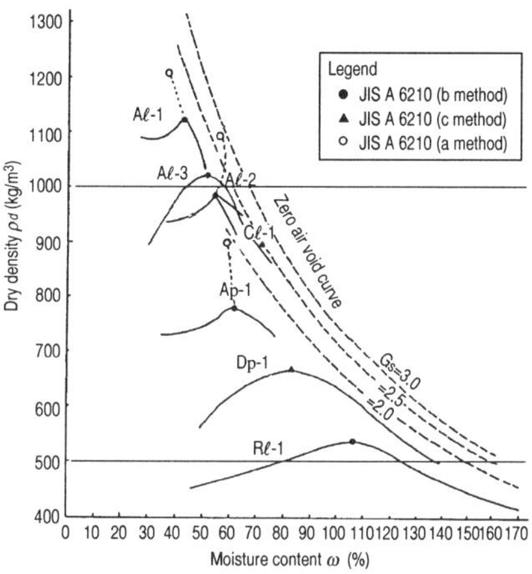

Figure 3. Moisture-density curve and maximum dry density of each sludge type

Grain size accumulation curves are shown in Figure 2. As can be seen from the Figure, the dehydrated sludge is predominantly fine grained. However, its grain diameter increases to a certain extent after being subjected to the composting or incineration process. From the results shown in Figure 2, that is, the proportion of fine grains in the sludge increases after being subjected to compaction testing, it was became clear that the grains in the dehydrated sludge are not closely bonded to each other and are susceptible to breakage.

In Figure 3, moisture-density curves for different sludge types (ρ_d-ω) and the relationship between the maximum dry density and optimum moisture content are presented. The relationship shows a tendency that the greater the maximum dry density becomes, the smaller the optimum moisture content becomes. This tendency is common among other general soil types. The maximum dry density of sewage sludge is 0.8 to 0.4 times as small as that of the general soil (1600 to 1900 kg/m^3 or so). The maximum dry density of calcareous ash, high-molecular ash, calcareous composted sludge and dehydrated sludge is 900 to 1100 kg/m^3, 800 to 1000 kg/m^3, 900 kg/m^3 and less than 700 kg/m^3, respectively.

Presented in Figure 4 is a diagram showing

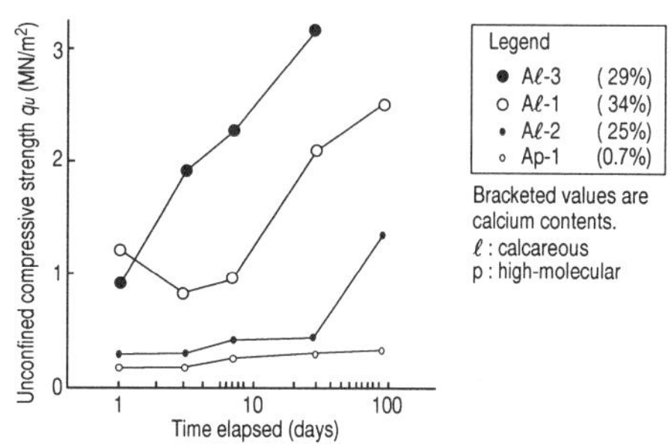

Figure 4. Variations with time in strength at the optimum moisture content (incinerated ash)

Table 4. Permeability coefficient of each sludge type

Type of sludge	Permeability coefficient (cm/sec)	Permeability of soil
Incinerated ash	$10^{-4} \sim 10^{-6}$	Low
Composted sludge	$10^{-8} \sim 10^{-10}$	Practically impermeable
Dehydrated sludge	Less than 10^{-10}	Impermeable

variations with time in the unconfined compressive strength of incinerated ash when the optimum moisture content is reached. The unconfined compressive strength of sewage sludge varies depending on its form and the type of the coagulants added. In the case of the calcareous ash, a considerable increase was caused under the influence of quick lime and iron chloride contained in the ash. Taking the unconfined compressive strength immediately after specimen fabrication as a basis of comparison, a threefold increase occurred in four weeks and more than a fivefold increase in 12 weeks. Whereas in the case of the high-molecular ash, no marked increase was observed. When viewed from the relationship with the calcium content, a qualitative relationship can be observed in which the strength increases with increases in the calcium content. The deformation at failure is as small as 0.3 to 1.1%, exhibiting failure characteristics inherent in the brittle substances.

The modified California bearing ratio (CBR) is a commonly used indicator of the strength of base course materials (Japan Road Association 1992). The modified CBR of calcareous ash is greater than 60%, and that of calcareous composted sludge is 10% or so. The swelling rate is as small as less than 0.1% in the case of ash, and less than 1.8% in the case of composted sludge. Both were found to be a viable earthwork material.

Table 4 shows the permeability of each sludge type. Composted sludge and dehydrated sludge, when compacted up to nearly the levels of the optimum moisture content and the maximum dry density, can be determined impermeable. Meanwhile ash falls under the category of low permeability.

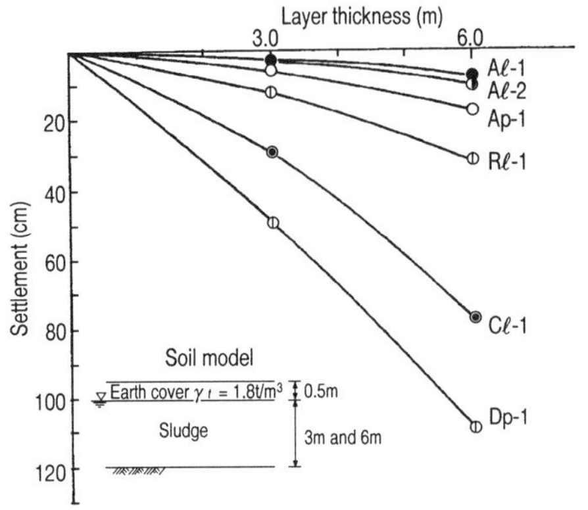

Figure 5. Settlement of a landfill caused by the self weight of ash

4 EXAMPLES OF EFFECTIVE USE OF SLUDGE AND ITS MECHANICAL PROPERTIES

4.1 *Landfilling and backfilling*

The test results given in the preceding section revealed that the soil mechanics characteristics of composted sludge and ash are well suited for a filling material. From the point of workability, moisture content equivalent to that of composted sludge is desirable. Ash in an absolute dry condition requires hydration. Adequate strength can be expected as long as proper control of the optimum water content is ensured.

When incinerated ash is used for landfilling, there is no problem in terms of both load bearing capacity and settlement. In the case of composted sludge as well, there is no difficulty in achieving required bearing capacity. However, in some cases where compacting is impossible, substantial settlements may result. Whereas in the case of dehydrated sludge, settlements of great magnitude are most likely, and not much can be expected thereof in terms of load bearing capacity. So, soil improvement and stabilization measures of some kind will be necessary according to the type of a overlying structure.

Figure 5 shows the settlement of each sludge type caused by its self weight when used for a landfill with a 50 cm earth cover. In the case of incinerated ash, the amount of settlement is so small and the settlement progresses at a rapid rate, it can be considered that settlement will cease by the end of the filling operation.

4.2 *Rolling compaction testing*

Rolling compaction testing was carried out on incinerated ash with the aim to investigating its compaction and strength characteristics at the filling

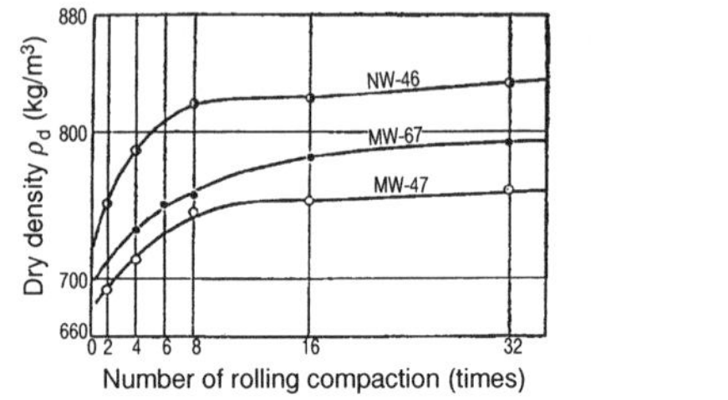

Figure 6. Number of rolling compaction - Dry density relationship

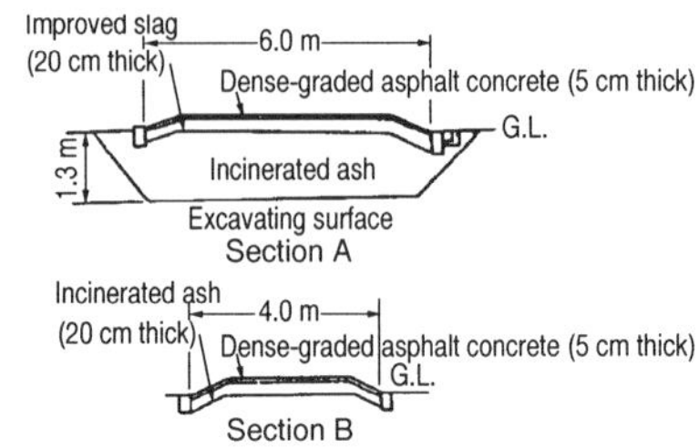

Figure 7. Pavement profiles

site. The testing was carried out on incinerated ash (A1-3) under the following three different moisture states: 1) 46%, the natural moisture content of the ash left as it was piled at the site (hereinafter referred to as 'NW-46' for short); 2) 47% (MW-47) and 3) 67% (MW-67) attained by mixing the ash having the initial moisture content of about 30% with water in a slowly rotating mixer. Crusher-run was laid at the testing site, which was then rolled and graded. 135 kN- and 80 kN- tired rollers were used in compacting operation.

The relationship between the number of rolling compaction and dry density is presented in Figure 6. In all moisture states, dry density values settled after around 8th compaction. The level of dry density achieved was only 70 to 80% of the maximum dry density (1000 to 1200 kg/m³) obtained from the compaction testing of the fill soils. Although the moisture content of NW-46 and that of MW-47 are almost identical, enhanced particle size distribution and thus enhanced compressibility have been attained in NW-46 due presumably to the fact that ash particles became well mixed with water while it was left at the filling site over an extended period of time and that grain diameter increased under the influence of lime.

4.3 *Experimental paving*

In order to grasp the behavior of incinerated ash when used for subgrade and subcourse construction, experimental paving was carried out. Figure 7 shows the profile of the pavement. Section A was set up to conduct tests on incinerated ash when used to fill the voids around sewer pipes. Experimental subgrade

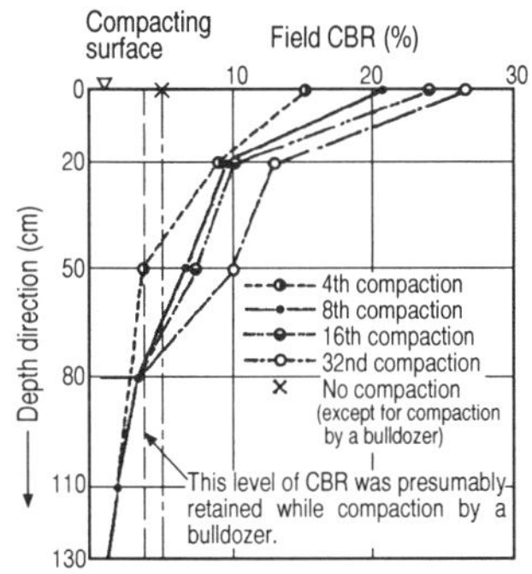

Figure 8. Number of rolling compaction - Field CBR in depth direction relationship

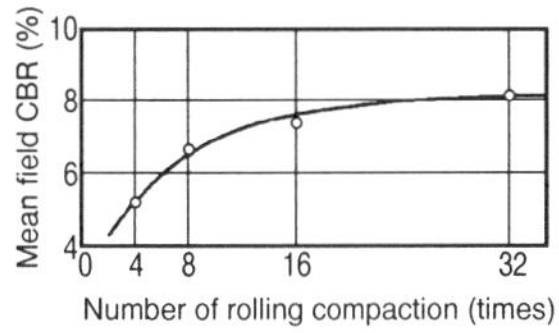

Figure 9. Number of rolling compaction - Average field CBR relationship

Photograph 1. Field CBR testing in process

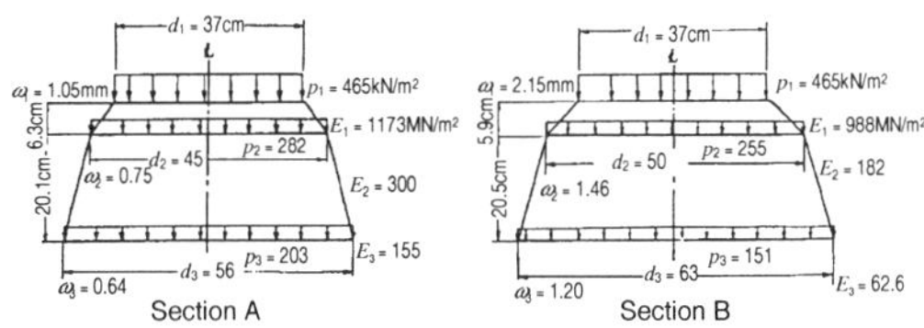

Figure 10. Load distributions

construction was carried out in such a way that first the existing ground was excavated to a depth of 1.3 m, which was then filled with incinerated ash. Meanwhile Section B was set up to carry out the experimental paving of light traffic streets or an experimental base course construction for cycle tracks. The incinerated ash used was one piled at a temporary storage yard and left there for one month. The moisture content of the ash was as high as about 70%. The lime content was 12 to 15%.

In Section A, excavations were filled with the incinerated ash, which was then spread over by a 2-ton bulldozer. The filling and spreading operation was repeated three times on a lay-by-layer basis. Throughout the operation, the dry density of the ash in the direction of depth was maintained nearly uniform, 600 kg/m³. After completion of the filling and spreading operation, compaction was repeatedly carried out thirty two times for subsequent subcourse construction.

In the case of Section B, surface course construction was accomplished by scraping and grading the existing ground, which was followed by the spreading of the incinerated ash and finally by thirty two times of compaction (ρ_d up to 720 kg/m³ or so).

Presented in Figure 8 is a diagram illustrating the relationship between the number of rolling compaction and variations in the field CBR in the direction of depth. In Section A, a layer as thick as 1.3 m was compacted in a single operation. The field CBR at the compacting surface at the time of the 32th compaction was as large as 26%. However, the field CBR markedly decreased in the direction of depth, and at around 20 cm below the compacting surface, it decreased to half that at the compacting surface. Compaction effect attained was up to 50 cm or so.

A diagram showing the relationship between the field CBR (mean CBR values in the depth direction) and the number of rolling compaction is presented in Figure 9. Sharp increases can be observed in the field CBR up to the 8th compaction. The averaged field CBR at the 32th compaction is 8.1%. Photograph 1 shows field CBR measurement in process.

In Section B, field CBR testing and Benkelman beam testing were carried out on the existing subgrade. The field CBR and deflection were found to be 15 to 20% and 1.2 to 2.1 mm, respectively. In cases where random paving is planned for conventional gravel roads, and its deflection is 2 mm, a pavement thickness 12 to 13 cm is adequate to bear light traffic flow. In cases where incinerated ash is to be used as a base course material, a 25 cm thick pavement is sufficient.

Using deflection values obtained through the abovementioned tests carried out both in A and B Sections, layer configurations were analyzed by means of the Nacimento method. Figure 10 shows load distributions. The deflection of the subgrade surface is about 1.0 to 1.1 mm and its modulus of deformation is 155 MN/m². Moduli of deformation within a range between 150 and 300 MN/m² are large enough for a subgrade, and it can be said from this that incinerated ash has better strength characteristics compared to other subgrade materials, such as crusher-run 100 MN/m² or pit sand 35 MN/m² (Ueshita et al. 1973).

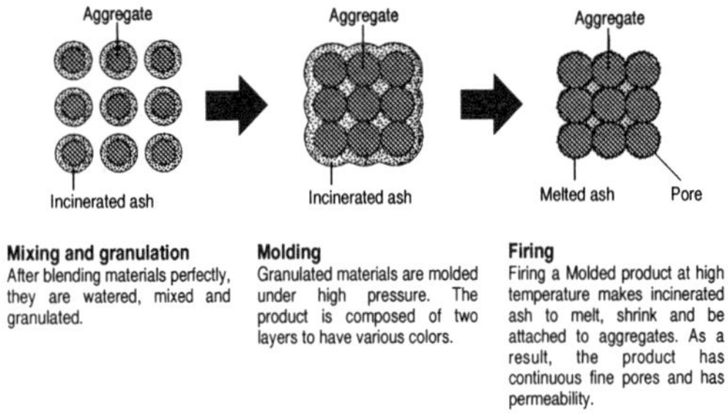

Figure 11. Mechanism of fired block manufacturing

Table 5. Characteristics of the fired block

Dimensions	198 × 98 × 60 mm
Number of pieces per $1m^2$	50 pieces/m^2
Weight	2.0 kg/piece
Bending strength	3 MN/m^2 or more
Permeability coefficient	1×10^{-2} cm/sec or more
Wear tolerance	0.1 kg or less (JIS A5209)
Surface hardness	6~7
Sliding resistance coeffcient	0.7 or more (JIS A1407)
Absorption capability	250 cc of water/piece

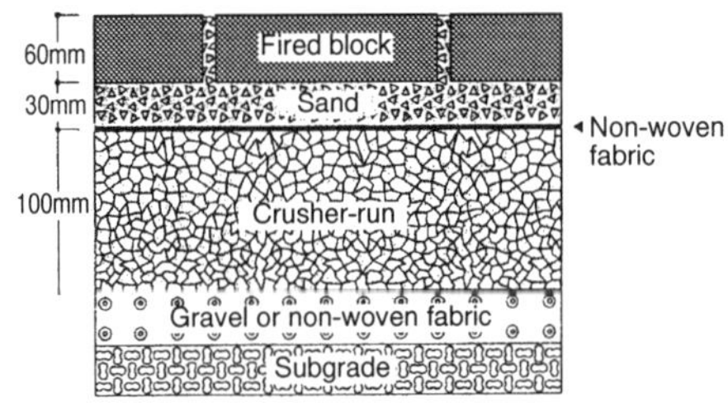

Figure 12. Typical construction of permeable block pavement

Photograph 2. Use of fired blocks for a pavement

The amount of deflection of the subgrade in B Section constructed with incinerated ash was found to be 1.5 mm or so, although a considerable fluctuation was caused by the disuniformity of the strength of the existing base course. The modulus of deformation of the subgrade was found to be 182MN/m^2, which is about half to one third that, 400 to 500 MN/m^2 of the usual base course provided with lime stabilization.

4.4 *Use of incinerated ash as a construction material*

In some cases incinerated ash is being used as part of a primary material in the process of fabricating construction materials, such as bricks, tiles and stone ware pipes. One of typical examples of successful practical use of incinerated ash as a primary material can be found in the case of fired block manufacture in Osaka City. Figure 11 illustrates the mechanism of fired block manufacture. The fired block has two layers, of which upper layer is provided with color coating. The fired block has properties shown in Table 5. The fired block has stormwater runoff control effects because of its improved permeability and moisture retaining capacity. The fired block is used for pavements, as shown in Photograph 2, and many other purposes. The cross section of the pavement using permeable fired blocks is presented in Figure 12.

5 CONCLUSION

The contents of this paper can be summarized as follows.

(1) The soil mechanics properties of incinerated ash indicate that incinerated ash is a viable subgrade material. The properties vary from calcareous ash to high-molecular ash. The ash with an increased lime content can make a subgrade material of improved quality.
(2) The experimental paving that used incinerated ash as subgrade and base course materials revealed the compaction and strength characteristics of incinerated ash and proved that incinerated ash is a most likely candidate for a pavement component.
(3) Incinerated ash can also be used for permeable fired blocks. The manufacturing process of the fired blocks and its typical application are presented.

REFERENCES

Japan Road Association 1992. *Guidelines on asphalt paving*:pp. 20-36.
Sewage Works in Japan 1995. pp. 224-232.
Ueshita, K., T. Yoshikame and T. Tamano 1973. Analysis of pavement structures measured by core boring and the Benkelman beam test. *Proc. of JSCE*. No. 214: pp. 17-25 (in Japanese).

Environmental Geotechnics, Kamon (ed.) © 1996 Balkema, Rotterdam. ISBN 90 5410 848 7

Recycling of sludge and mud generated at construction sites

Yukihiro Tsukada & Nobuyoshi Ogawa
Public Works Research Institute, Ministry of Construction, Tsukuba, Japan

ABSTRACT : This report outlines the recent trends of construction by-products and concept of recycling in Japan. The paper also reports on the progress of research and development that have been undertaken with the aim of reducing the amount of construction-generated sludge (sludge and mud generated at construction sites) and developing the technology to recycle them.

1 INTRODUCTION

Disposal of wastes such as surplus soils, asphalt concrete fragments, cement concrete fragments, sludge and mud, wood, and their mixtures generated at construction sites (which are collectively called "construction by-products") has become a pressing problem in Japan as it is becoming increasingly difficult to secure dumping sites.

To minimize the generation of construction by-products, and to promote the recycling , the Ministry of Construction in 1992 launched a five-year comprehensive technological development project entitled the "Development of Technology to Minimize and Recycle Construction By-products."

A cooperative research has been undertaken by the PWRI, Advanced Construction Technology Center (ACTEC), and 22 private enterprises with the aim of reducing the amount of construction-generated sludge (hereafter called "construction sludge") and developing the technology to recycle them (hereafter called "the Cooperative Research").

This report outlines the recent trends of construction by-products and concept of recycling. The paper also reports on the progress of the Cooperative Research.

2 RECENT TRENDS AND R&D ON CONSTRUCTION BY-PRODUCTS

The construction industry accepts some kinds of by-product from other industries such as fly ash, granulated slag. This is indicated by the fact that the 1981-86 project of the Ministry of Construction entitled the "Development of Technology to Utilize Wastes" placed emphasis on utilization of by-product generated by other industrial activities as construction materials.

In 1990, a nationwide statistical survey to collect information on the treatment and disposal of construction by-products was conducted for the first time in Japan. This was in response to the growing need for new methods for treatment and recycling of wastes generated as a result of construction work.

In 1991, the Law for Promotion of Resources Recycling was enacted. Thus, the concept of recycling was legally established, and public attitudes toward wastes in general began to change. The image of construction by-products has also changed from one of difficult-to-treat or difficult-to-dispose-of materials to one of valuable and reusable resources.

The year 1991 saw the establishment of the Liaison Councils for Reduction and Recycling of Construction By-products by local contracting agencies. Since then, various efforts have been undertaken to reduce the amount of construction by-products, and recycle and treat them properly. In 1994, each council set its own specific goals for each type of construction by-product for achievement by the year 2000, and the goals thus set were published in the form of the Action Plan for Reduction and Recycling of Construction By-products (popularly called "Recycling Plan 21"). Various efforts are now undertaken to achieve these goals.

In 1992, a five-year comprehensive technological development project entitled the "Development of Technology to Minimize and Recycle Construction By-products" was launched. This research project focuses on concrete fragments, surplus soils, and construction sludge and mud. The purpose of this project is technological development not only of "hardware" for reduction and recycling of construction wastes, but also of "software" including formulation of technical standards for recycling and guidelines for selecting planning/design methods and construction methods, and development of techniques for evaluating the economy of recycling.

In 1993, a series of model works for increased recycling was began in order to increase recycling mainly in public works projects directly executed by the Ministry of Construction and explore possible applications of newly developed recycling technologies.

3 WHAT IS CONSTRUCTION SLUDGE?

In Japan, the term "construction sludge" refers collectively to construction by-products which cannot be used for filling or other purposes without some kind of treatment, such as waste bentonite slurry and fine, high-water-content excavated mud. Typical examples of construction sludge are shown in Table 1.

Construction sludge are defined as sludge and mud which cannot be heaped on hauling vehicles, on which humans cannot walk, or which becomes fluid during haulage. To express conditions like this in terms of geotechnical indicators, the above condition roughly corresponds to cone indexes of 2 kgf/cm^2{200 kPa} or less and unconfined compressive strengths of 0.5 kgf/cm^2 {50kPa} or less. The relationship between construction sludge and construction-generated soils is understood as shown in Fig.1 (Construction By-Products Recycling Publicity Council 1994).

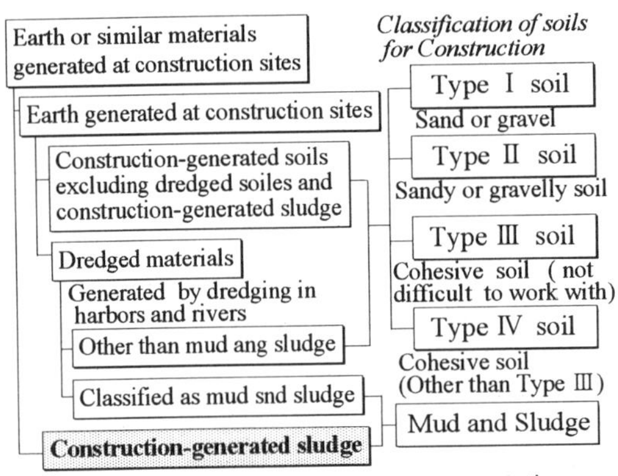

Fig.1 The relationship between construction-sludge and soils

Construction sludge and mud consist mainly of naturally occurring soils and are in most cases harmless. Therefore, it is possible to recycle by relatively simple treatment such as dewatering and stabilization.

In Japan, however, construction sludge, like sludge produced by wastewater treatment, are officially classified as industrial wastes. Their handling, therefore, is under legal restrictions, and so recycling of construction sludge is being severely hampered.

4 GENERATION AND DISPOSAL OF CONSTRUCTION SLUDGE AND GOALS OF RECYCLING

Figs.2 and 3 show the percentages of generated and disposed of construction by-products excepting soil of in Japan, which have been obtained from a 1993 survey of construction by-products.

The annual yield of construction sludge is estimated at 15,000,000 tons, which accounts for about 20 percent of the total yield of construction by-products excepting soil. But only 6 percent is saved by reduction and 2 percent by recycling.

The percentage of reduction and recycling combined, which is an indicator of recycling, is only 8 percent, which is the lowest value among the five principal categories of construction by-products.

The Recycling Plan 21 has set the national target value of the percentage of reduction and recycling of construction sludge at 35 percent.

In addition to efforts on the part of those directly involved in construction work, research and development efforts to reduce and recycle construction sludge are needed.

Table 1 Typical Examples of Construction Sludge and Mud

Type of Construction Work	Generation Process
・Diaphragm Wall Method ・Reverse Circulation Pile Method ・Earth-Pressure-Balance Shield Method	Construction sludge is a deteriorated and therefore unusable slurry or mud water, or that was left unused after completion of construction work. In many cases, excavated materials produced under these construction method become construction sludge.
・Column-Type Diaphragm Wall Method ・High-Pressure Deep Mixing Method	Mixtures of cement slurry discharged into the original ground and earth in the original ground become construction sludge.
・Cut-and-Cover Method	High-water-content Soils excavated from soft cohesive ground become construction sludge.

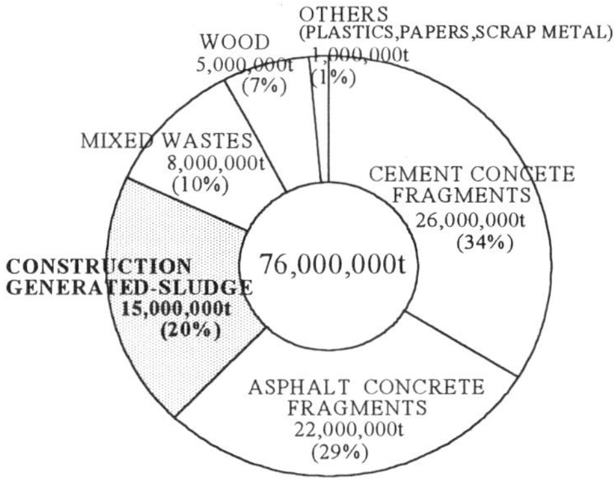

Fig.2 Annual Quantity of Construction By-product (1993) Generation in Japan

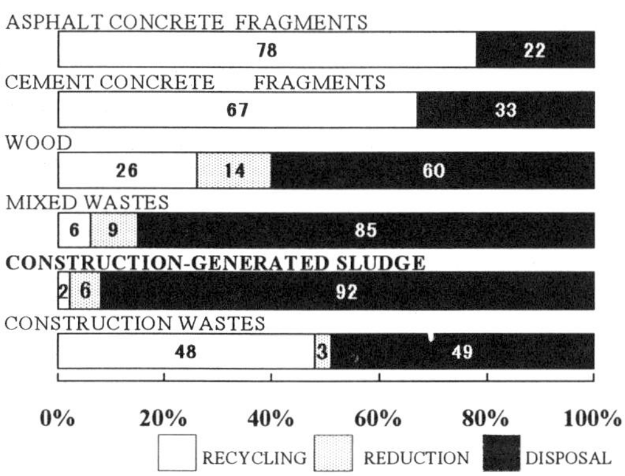

Fig.3 Recycling and disposal of Construction By-products in Japan

5 CLASSIFICATION OF RECYCLING

5.1 *Classification from the Viewpoint of Relationship with Processing*

Generally, in order to recycle waste, it is necessary to process or reprocess it though the degree of processing varies. From the viewpoint of this processing or reprocessing, recycling technology can be classified into two types:

1) Processing Technology

Technology to make used materials reusable

2) Manufacturing Technology

Technology to facilitate the above processing by altering used materials

It is also possible to further classify "Processing Technology" by the method or degree of processing.

In trying to establish recycling technology, it is important to coordinate research and development efforts so that the above two technologies are balanced and complement each other.

5.2 *Classification by Uses*

Another classification is one from the viewpoint of the relationship between the uses of the original materials and those of recycled materials. Thus, recycling can be classified as one of the three types listed below or a combination thereof. The concepts of these types of recycling are illustrated in Figure 4.

1) Primary Recycling

For the same uses as those of the original materials

2) Creative Recycling

For different uses from those of the original materials

3) Return-to-Nature Recycling

For returning used materials to nature without exerting adverse effects on the natural world.

In the field of recycling-related technological development, disproportionately much attention tends to be directed to creative recycling. Needless to say, however, the most important type of recycling is primary recycling which often makes it possible to achieve the same goal through relatively simple and easy processing.

Technologies for return-to-nature recycling, which aims to return used materials to nature without exerting adverse effects on the natural environment, are expected to become increasingly important. In this approach, the major part of used materials will be returned to nature in the form of soil, and smaller parts will be restored as water or air. These are important subjects of "environmental geotechnics."

5.3 *Classification of Construction Sludge Recycling*

Construction sludge is classified the following three categories:

1) Recycled as Slurry or Mud Water

Used slurry or mud water which has become construction sludge can be recycled by restoring its original properties or functions. Since this type of recycling reduces the quantity of slurry or mud water that needs to be prepared newly, the quantity of slurry or mud water to be disposed of can be reduced. This is, therefore, a type of recycling which helps to reduce construction sludge.

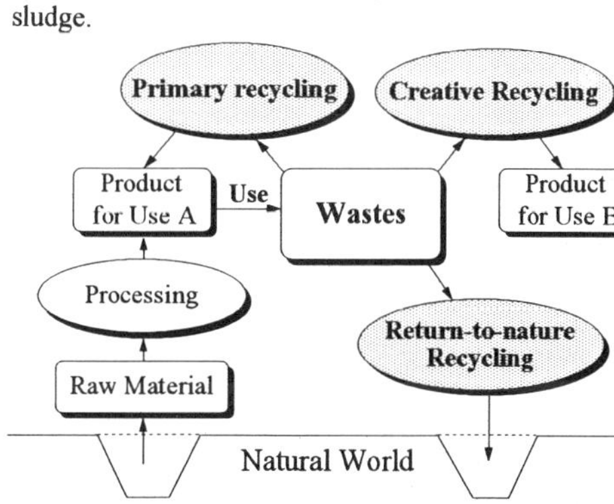

Fig.4 Concepts of Major Categories of Recycling

It is to be noted, however, that because slurry or mud water is mostly water, transportation of slurry or mud water should be minimized so as to prevent inefficiency of recycling and adverse effects on the environment.

2) Recycled as Products or Raw Materials

Construction sludge can be recycled as materials with relatively high values added as compared with conventional materials for earthwork applications. For example, sludge and mud can be used as artificial aggregate and raw materials for cement etc.

To satisfy quality standards for particular applications, the physical and chemical properties of sludge or mud are often specified.

3) Recycled as Dewatered or Stabilized Materials

Excess water contained in construction sludge can be removed to obtain materials which have the same properties as those of ordinary cohesive soils.

Sludge can also be stabilized by adding cement, lime or other stabilizers to obtain the strength required for such applications as fill, backfill and reclamation materials.

Dewatering is also important to reduce the volume of sludge by separating water and nonwater components.

6 EXAMPLE OF RECYCLING

Typical examples of the categories of the above classification are given below..

6.1 Primary Recycling

In primary recycling, used materials are reused for the same purposes as those of the original materials.

An example of primary recycling is seen at treatment plants where bentonite slurry is recycled by removing coarser grains with cyclones, neutralization with carbon dioxide, and adding bentonite. Another example is seen in systems for controlling the properties of slurry or mud water used at construction sites.

6.2 Creative Recycling

Examples of research and development and other activities classified as creative recycling in which used materials are reused as products or raw materials are given in Table 2 (Tsukada and Ogawa 1995).

Particular attention has been given to some advanced plants in Shizuoka and Osaka which have already gone in operation on private-sector projects. Other examples include use of reprocessed materials as backfill materials behind shield segments or as ingredients of invert concrete in railway shield tunnels.

Now being developed in the Cooperative Research is the technology to manufacture relatively small (2.5～15 mm or so) grains and use them as materials for gravel drains. To achieve the required quality and economy, the research is now focusing on granulation and baking methods and the types of construction sludge and mud that can be used for these applications.

Photo 1 Granuiar Material Baked in a Roataly Kiln at about 1,100℃

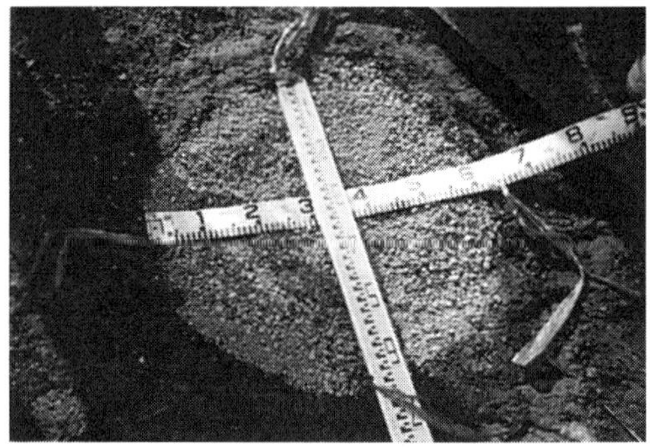

Photo 2 Head of a Drain Formed in the Field

As an attempt at recycling construction sludge, sludge obtained by dewatering slurry used at a diaphragm wall construction site was granulated and used to form gravel drains. The granular material was obtained by crushing and granulating dried sludge and baking the grains in a rotary kiln at about 1,100℃. The granular material thus obtained did not crumble during placement and proved to be easy to place because it has a higher fluidity to fill out drain holes than conventional single-grained crushed stone does.

The granular material and the head of a gravel drain in the field are shown in Photos 1 and 2.

6.3 Return-to-Nature Recycling

Relatively few examples have been reported of return-to-nature recycling. This type of recycling, however, can be effected with relative ease by combining existing techniques, such as dewatering and stabilization.

Construction sludge may be affected by the components of earth being excavated or may contain admixtures. In recycling sludge or mud which was recovered from a contaminated ground or that contains admixtures. Therefore, it is necessary to make sure the collected material is safe before it is reused.

The problem of alkaline solutions released from stabilized soils can be solved by, for example, covering the stabilized soil with a layer of natural soil, as has been well documented (e.g.,Kitsugi 1989).

Photo 3 shows a 3,700m^3 embankment which was formed experimentally, using a material obtained by improving construction sludge, along the bank of the Tone River (Ogawa 1995). Planting was planned at the site of this experimental construction in order to study the long-term effect of planting.

Photo 3 A View of an Experimental Embankment along the Tone River

Table 2 Ongoing Activities of Creative Recycling of Construction Wastes in Japan

Use	Type of Waste	Treatment Method	Research Institution Duration Stage of Development
·Backfill , etc.	·Clay ·Waste Slurry	Sediments obtained by the coagulating sedimentation method from clay left after sand and gravel have been removed from excavated material or from slurry used in reverse circulation piles are dewatered using a filter press. The dewatered material is granulated with an extrusion-type granulator and baked in a rotary kiln to produce 20-30 mm grains.	·Private enterprise ·Start-up of plant: Oct.1992 (Shizuoka Pref.) ·Practical Use
·Base course, etc.	·Sludge ·Concrete fragments	Mixtures of crushed concrete and sludge or mud are stabilized with a cement stabilizer, and hardened mixtures are broken up.	·Private enterprise ·Start-up of plant: 1990 (Osaka Pref.) ·Practical Use
·Lightweight aggregate	·Sludge	Powder of spontaneous combustibles is added to sludge or mud, and the mixture is let to burn in a vertical or other type of kiln. Since the combustibles will burn away, the mixture will become a porous material usable as lightweight aggregate.	·Ministry of International Trade and Industry Japan Contractor Management Association ·From 1992 ·Prototype
·Lightweight aggregate ·Horticultural soil ·Filter material	·Sand pit sludge ·Sludge from stone cutting ·Purification plant sludge ·Bottom sediment	A foaming agent is added to a mixture of sand pit sludge and sludge produced by stone cutting or other materials to form 3-30 mm grains. The grains are dried, baked in a rotary kiln and cooled in a rotary cooler. Lightweight aggregate can be obtained by sieving the material thus obtained.	·Ceramics Laboratory, Ibaraki Industrial Technology Center ·From 1990 ·Prototype
·Lightweight aggregate ·Horticultural soil ·Filter material	·Quarry sludge	Dewatered cakes of quarry sludge (sandstone, shale) and purification plant sludge are crushed and re-formed into 0.6-20 mm grains. These grains are baked and foamed in a rotary kiln while the grains are being dried. The baked and foamed grains are then cooled in a rotary cooler and sieved to produce a material usable as lightweight aggregate, etc.	·Clean Japan Center ·From 1990 ·Mass Production Experiment
·Lightweight aggregate ·Horticultural soil ·Filter material	·Wastes containing cement	Dewatered cakes of cement-containing waste liquid are dried and pulverized. Aluminum sludge and calcareous stone powder are added to the pulverized material. After the mixture is baked and rapidly cooled, it is pulverized again to obtain this hydraulic material.	· Applied Material Division, Mie Industrial Technology Center ·1992 to 1995 ·Prototype
·Molten slag	·Sludge	Dried sludge and 15% or so steel dust are mixed and molten in a furnace. The molten mixture is put into alkaline hot water, and the solids thus obtained are crushed with an impact crusher to obtain slag aggregate.	·Japan Machinery Federation Engineering Advancement Association of Japan ·1990 to 1992 ·Prototype
·Base course , etc.	·Self-hardening sludge (cement-containing)	Porous and highly absorptive sand-like material manufactured by baking old paper ash is added to and mixed with self-hardening (cement-containing) sludge. The mixture is then improved to obtain a material usable as base course material, etc.	·Private enterprise ·From 1993 ·Prototype

7 TECHNIQUES FOR DEWATERING

Recently developed techniques for dewatering construction sludge include 6 techniques which were evaluated, under the Construction Technology Appraisal Program, in a 1992 project entitled the "Development of High-Efficiency Law-Water-Content-Type Dewaterer and Dewatering Methods for Construction Sludge".

In the Cooperative Research, a new dewaterer, "High Pressure Thin Layer filter press dewaterer" has been developed. The newly developed dewaterer is essentially a squeeze type pump, which is usually used to push newly mixed concrete through a pipeline, equipped with a pressure control system.

High dewatering pressure (15kgf/cm^2 {1,500KPa} under normal conditions), which has been achieved economically, reduces the water content of the dewatered cake, thus increasing the strength of the dewatered soil (dewatered cake).

The pressure control system activates the pump to restore dewatering pressure only when it has fallen below the predetermined level. The pump is run intermittently once the filter chamber has been filled with slurry. Since energy is not wasted in restoring the dewatering pressure as in conventional centrifugal pumps, energy efficiency is improved.

Furthermore, the thickness of the dewatered cake has been reduced, to the extent that automatic removal of the dewatered cake is not hampered, so that the time required for dewatering is shortened.

Fig.5 compares the newly developed dewaterer with the conventional filter press dewaterer.

Experiments were carried out with dredged sludge from five locations in Japan. Both conventional dewatering methods and the High Pressure Thin Layer Filter Press Dewaterer were used. The addition of short fibers was also studied. In this figure can be seen the relationship between the water content and cone index for the dewatered cake.

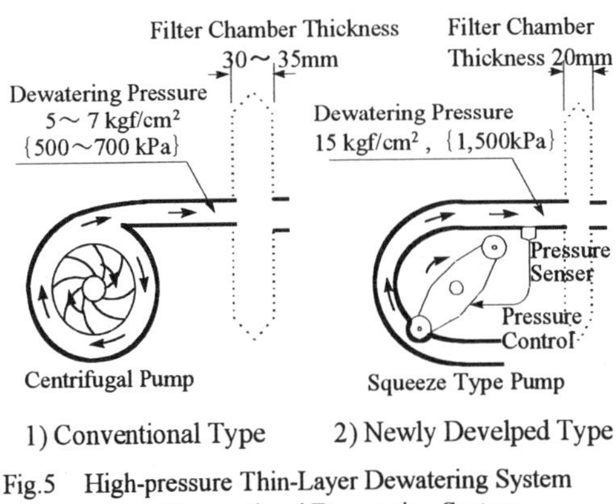

Fig.5 High-pressure Thin-Layer Dewatering System vs. Conventional Dewatering System

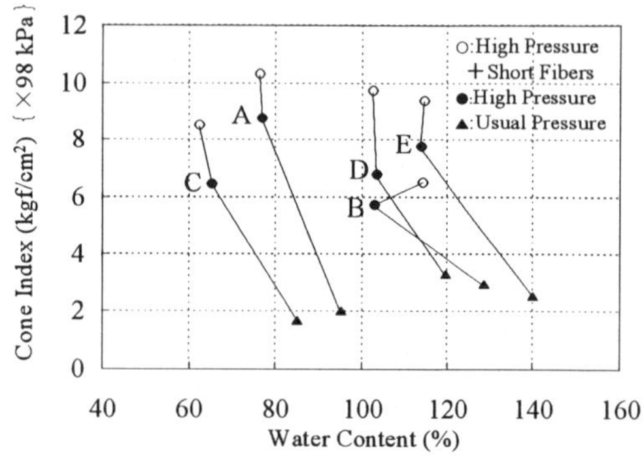

Fig.6 Relationship between the water content and cone index for the dewatered cake

By high pressure thin layer dewatering, the water content has been reduced to 20~30%. Also, the strength has been increased to give cone index values in the range 4~6 kgf/cm^2 {400～600kPa}. Furthermore, by the addition of short fibers, the cone index was increased by a further 2~3kgf/cm^2 {200～300kPa}.

8 CONCLUDING REMARKS

This paper has reported on the recent trends of construction by-products and concept of recycling.

The paper also reports on the technologies being developed for reduction and recycling of construction sludge.

We are going to present the "Construction Sludge Recycling Manual (provisional title)" based on the results of the Cooperative Research.

REFERENCES

Construction By-Products Recycling Publicity Council. 1994: Integrated Reduction and Recycling of Construction By-products.

Civil Engineering Research Center. July 1994 Construction-Generated Soil Utilization Manual.

Kitsugi, K. 1989. Soil Improvement Technology and Environmental Problems: pH Control in Soil Improvement, Cement and Concrete, No. 511.

Japan Society of Soil Science and Plant Nutrition. 1981. Adsorption of Soil: Fundamentals and Applications.

Ogawa, N. 1995. Recycling of Sludge and Mud Produced by Foundation Work, The Foundation Engineering & Equipment.

Tsukada, Y. and Ogawa, N. 1995. Recycling Techniques for Construction-Generated Sludge, JCM Monthly Report.

Ogawa, N. et al. 1995. Treatment and Effective Use of Dredged Sludge, 6th International Conference on the Conservation and Management of Lakes-Kasumigaura

Environmental Geotechnics, Kamon (ed.) © 1996 Balkema, Rotterdam. ISBN 90 5410 848 7

Reuse of stabilized petroleum drilling wastes as sub-bases

Ahmet Tuncan, Mustafa Tuncan & Hakan Koyuncu
Anadolu University, Eskisehir, Turkey

ABSTRACT: Most petroleum drilling wastes contain varying amounts of petroleum hydrocarbons, organic and inorganic chemicals and metals. The presence of these pollutants affects the physical and chemical process in soil and water after the release is completed. Petroleum drilling wastes are disposed of in open pits known as "mud-pit." These wastes are stabilized in the laboratory to find economical proportions of drilling waste, fly ash, lime and cement. After laboratory investigations were completed, the field study was conducted as a joint effort by researchers at Anadolu University and Turkish Petroleum Corporation (TPAO). In this study, petroleum drilling wastes were stabilized with a pozzolanic lime, fly ash and cement. This stabilization produced chemically and physically stable and mechanically handable new products. These products were then used as sub-base material in the pilot field study. The parameters that are involved in such a determination are strength, permeability, durability, leachate, pH, oil and total metal content of stabilized mixtures.

1 INTRODUCTION

Oil well drilling activities produce large quantities of wastes in the form of drilling muds and cuttings. These wastes are disposed of in open pits (mud-pits) and large quantities of unstabilized wastes remain at a number of sites. To reduce the high risks of environmental impacts, effective and economical alternatives are to stabilize the wastes and reuse them as sub-base material in road construction. Anadolu University and Turkish Petroleum Corporation have been cooperating to determine the feasibility of reuse of petroleum drilling wastes as sub-base after stabilizing with pozzolanic materials. Turkish Petroleum Corporation constructs roads between petroleum tower and the nearest village. Therefore, stabilized mixtures can be economically and effectively used for the road construction.

An economical and fast solution is to stabilize petroleum contaminated drilling wastes by mixing it with additives. Stabilization is the improvement of physical and mechanical properties of wastes, encapsulation of pollutants and reduction of solubility of toxic substances. The stabilization method is applied to petroleum drilling wastes to bind the oil and metals in a structure formed by cementing pozzolanic materials to produce chemically and physically stable and mechanically handable mixtures. The stabilization of petroleum drilling wastes is accomplished by pozzolanic reactions which include both chemical and physical interactions between the wastes and additives. Stabilization with lime, fly ash and cement is the simplest and inexpensive technique. Lime, fly ash and cement are used to increase pH, to contribute surface area and to retain hydrocarbons, respectively.

The main purpose of stabilization is to increase the strength of sub-base material to support road traffic, to maintain integrity of the material during repeated freezing/thawing, to reduce the leachability of the waste constituents. Therefore, it is provided to have a substantial strength and environmental durability. This paper reports the results of a comprehensive field study on physical, mechanical and chemical properties of stabilized mixtures in order to use as sub-base material in the road construction.

2 MATERIALS

2.1 Petroleum drilling waste (PDW)

Petroleum drilling waste contains drilling muds and cuttings which consist of both liquid and solid phases. Drilling muds are normally classified as oil-base and water-base mud (Rogers, 1963). Oil-base mud contains a continuous oil phase consisting of usually diesel oil, bentonite clay and chemicals. Water-base mud contains fresh or salt water phase, bentonite clay and chemicals. Water-base mud is preferred for its low cost. However, certain situations require an oil-base mud for drilling operations in formations sensitive to water-base fluids. Drilling cuttings are the solid cuttings which are brought to the surface from the well during borings. In this study, water-base mud was used. This mud consists of high chloride, barite, bentonite clay, caustic and carboxy methly cellulose (CMC). In addition to these, wastes also consist of diesel oil and rock fragments. Constituents of drilling waste are given in Table 1.

Table 1. Some properties of petroleum drilling wastes.

Oil Content (%)	: 2.25-2.50
Moisture Content (%)	: 50-80
Organic Matter Content (%)	: 1.80-2.40
pH	: 6.50-7.00
Cation Exchange Capacity (meq/100g mixture)	: 15-20
Grain Size Analysis	
Gravel (%)	: ----
Sand (%)	: 18
Silt (%)	: 74
Clay (%)	: 8

2.2 Additives

One of the most popular pozzolans is fly ash. Most fly ash is a by-product of coal burning power plants. Fly ash is a fine-grained dust and is primarily composed of silica, alumina and various oxides and alkalies. It is pozzolanic in nature and can react with hydrated lime to produce cement-type products. For this reason, lime/fly ash mixtures can be used for stabilization of highway bases and sub-bases. Fly ash acts as a conditioner in the waste to absorb liquid and to coat waste globules.

Lime has been used as a practical soil stabilizing agent for a long time. The most common alkali used in the stabilization is lime. The hydrated form is more common because it is easier and safer to store and handle. It decreases the plasticity and oil content of wastes (Van Keuren et al.,1987). It increases workability and improves the strength and deformation properties of wastes. If the lime is mixed with iron and aluminum bearing minerals, it will form a crude form of cement. Many types of coal power plant fly ash contain at least some amounts of iron and aluminum. If fly ash is mixed with lime, then the mixture can react with the wastes in a manner analogous to portland cement. Lime and lime/fly ash mixtures tend to be less expensive than cement. Lime can be used for the treatment of acidic wastes due to its high acid buffering capacity. As a result, the pH of the stabilized product remains high for a longer period of time minimizing the mobility of metals.

Portland cement was used to improve the interparticle binding and mechanical properties of wastes. It reacts with water and binds to the aggregate within a short time. It is a mixture of a number of minerals such as calcium silicates, calcium aluminates and iron oxides. Cement is used as a stabilizing material for the construction of highways. It increases the strength of soil and its strength increases with curing time. Some properties of additives are given in Table 2.

3 TESTING

A testing program was conducted to investigate how much improvement could be accomplished in the physical, mechanical and chemical properties of petroleum drilling wastes by hydrated lime, fly ash and cement treatment.

Petroleum drilling wastes were excavated with conventional construction equipment from the mud-pit. Specimens were prepared by mixing the petroleum drilling wastes with water at the predetermined optimum moisture content of the mixed system before adding additives. Additives were then mixed by workers to form homogeneous mixtures. The mixtures were divided into 3 sets. Each set was treated with combination of additives at different percentages. The amount of the additive was measured as weight percentage of petroleum drilling wastes. The proportions of these additives were selected as:
Fly Ash (1 unit)+Lime(1/3 unit)+Cement (1/12 unit)
Fly Ash (1 unit) + Lime (1/3 unit) +Cement (1/6 unit)
Sand (1 unit) + Fly Ash (1/10 unit) + Lime (1/3 unit) + Cement (1/12 unit) by weight of the petroleum drilling wastes. Final mixtures were compacted by using a drum roller in the field. Roads were 2.60 m by 6.00 m in cross section and 30 cm in thickness and 1.00 m by 3.50 m in cross section and 30 cm thickness. The schematic presentation of roads is shown in Figure 1.

A pilot field study is needed to determine the construction feasibility and the most appropriate mixing technique for road construction. The laboratory study indicated that the following mixtures were appropriate:

Mixture 1: Petroleum drilling waste + Fly ash (40%) + Lime (13.33%) + Cement(3.33%)
Mixture 2: Petroleum drilling waste + Fly ash (50%) + Lime (20%) + Cement(5%)
Mixture 3: Petroleum drilling waste + Sand(50%) + Fly ash (5%) + Lime (16.7%)+Cement(4.17%)

3.1 Determination of oil content

Stabilized and unstabilized specimens were dried at 40 °C for 24 hours. After drying, 20 gr of specimens were placed into the chamber of the soxhlet extractor and extracted with 175 ml of methylene chloride. The solvent reservoir was heated to 60 °C by electrical heater for 2 hours. Methylene chloride is relatively

Table 2. Some properties of the additives.

Fly Ash	
Source	: Coal burning power plants
Specific Gravity	: 2.31
Cu	: 4.1
Cc	: 1.20
Sand(%)	: 85
Silt(%)	: 15
Lime[$Ca(OH)_2$]	: Hydrated
Cement	: Portland
Sand (mm)	: 0.05-1.50

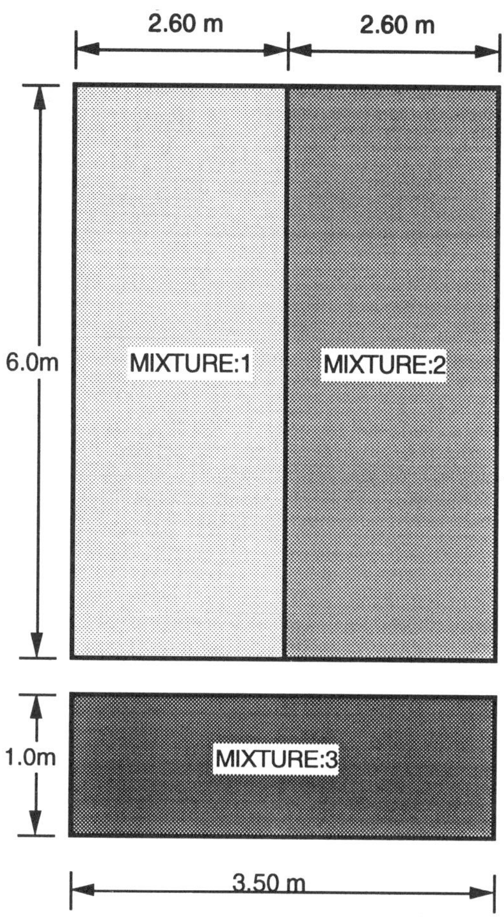

Figure 1. The schematic presentation of field study.

inexpensive, not particularly hazardous to use, extracts oil effectively and has a sufficiently low boiling point to be conveniently used in a soxhlet extraction apparatus (Mc Gill and Rowell, 1980).

3.2 Determination of total metal contents

Stabilized and unstabilized specimens were dried at 105 °C for 24 hours and grounded to pass through No. 100 sieve. 10 gr of specimens were then digested in 3:1 concentrated nitric acid and hydrochloric acid mixture for 2 hours at 80 °C. The diluted digests were then filtered through pore size of 0.2 μm cellulose nitrat filter. Copper, iron, manganese, zinc and lead were determined by atomic absorption spectrophotometer. Chromium, cadmium, nickel, barium and arsenic were determined by inductively coupled plasma spectroscopy.

3.3 Determination of cation exchange capacity (CEC)

CEC was determined by sodium saturation method (Chapman, 1965). First, specimens were saturated with sodium acetate to replace the cations held on the surface and in the interlayer positions. Secondly, specimens were washed with isoprophyl alcohol to remove the excess saturating salt. Thirdly, specimens were washed with ammonium acetate, by which the absorbed sodium was released into the solution. The amount of sodium was determined by atomic absorption spectrophotometer. All CEC measurements were made at pH 7 solution of the specimens.

3.4 Determination of moisture content and ignitable organic matter content

The moisture content was measured by placing the mixtures in an oven at 105 °C. The organic matter content of the mixtures was determined by the ignition loss method. Stabilized and unstabilized specimens were pre-dried at 105 °C for 24 hours. These specimens were then burned at 550 °C for 2 hours. The loss of weight of the specimen as the result of the second heating was assumed to be primarily due to organic matter ignition.

4 EXPERIMENTAL RESULTS

The following tests and analysis were conducted on the stabilized specimens: unit weight, moisture content, cation exchange capacity, oil content, total metal contents, unconfined compressive strength, permeability (hydraulic conductivity), freeze/thaw durability and leachate analysis. At least 3 replicate samples were run for each test and the results were averaged.

4.1 Unit weigth, moisture content, oil content and cation exchange capacity (CEC)

The variation of unit weight, moisture content, oil content and CEC properties of stabilized mixtures are given in Table 3. Addition of lime/fly ash/cement reduces oil content of petroleum drilling wastes.

Table 3. Some properties of stabilized mixtures

Mixture	Unit Weight (g/cm^3)	Moisture Content (%)	Oil Content (%)	CEC (meq/100g)
PDW	2.43	86	3.20	15.20
Mixture:1	2.05	78	2.45	18.45
Mixture:2	1.88	73	1.25	15.20
Mixture:3	2.15	79	0.85	19.55

Addition of lime/fly ash/cement increases the CEC of petroleum drilling wastes due to the existence of fly ash/cement in the mixture. Cation exchange capacity is an indication of the relative capacity of a soil to retain metal ions. Most metals in petroleum industry wastes exist as compounds that are not very water soluble.

4.2 Unconfined compressive strength (UCS)

The unconfined compressive strength is the most popular indicator of mechanical stabilization, especially for the cement-base techniques. The variation of unconfined compressive strength of stabilized mixtures is shown in Table 4.

The unconfined compressive strengths of mixture:1 and mixture:2 are significantly lower than mixture:3, as expected. When the petroleum drilling wastes are treated with sand, there is an increase in the strength values for fresh, 7 and 28 day constant humidity cured specimens, because petroleum drilling waste consists of mainly silt size material. Therefore, addition of sand significantly increases the strength of stabilized mixtures, although the cement content of mixture is around 4.17%. Addition of lime also increases the strength of mixtures due to the flocculation and agglomeration effect of calcium ions. UCS provides insight to the ongoing cementation and pozzolanic reactions during the period of curing. Unconfined compressive strengths of lime stabilized natural soils range from 80 psi to 1100 psi depending on the amount of pozzolanic material present and curing period (Pamukçu and Winterkon, 1989).

4.3 Permeability (hydraulic conductivity)

Stabilized mixture permeability is used as a measure of its capacity to physically isolate contaminants as well as an indication of restricted internal fluid movement. Most solidification processes result in material permeability of 10^{-5} to 10^{-6} cm/sec (Pamukçu et al., 1989).

The permeability test results are shown in Table 5.

Table 4. The variation of unconfined compressive strength of stabilized mixtures.

Mixture	Curing Period (day)	Unit Weight (g/cm^3)	UCS (kg/cm^2)
Mixture:1	Fresh	1.895	1.492
	7	1.991	1.650
	28	1.950	3.165
Mixture:2	Fresh	1.856	1.500
	7	1.991	1.721
	28	1.906	3.510
Mixture:3	Fresh	1.889	3.011
	7	2.022	6.410
	28	2.000	8.535

It can be seen that addition of sand decreases the permeability of stabilized mixture, because it creates large pores for water to move. On the other hand, addition of lime and cement decreases the permeability values of mixtures 1 and 2. Addition of lime and cement not only increases the strength but also decreases the permeability of stabilized mixtures. Curing period is also important for mixtures 1 and 2.

4.4 Total metal contents

Total metal concentrations of stabilized mixtures are shown in Table 6.

Lead occurs in soils, rocks and minerals. It is also found in substantial amounts in cement and lime. Petroleum industry wastes contain both inorganic and organic lead compounds. Allowable lead concentration in the petroleum drilling waste is 100 mg/kg (Mucsy et al., 1984). If the pH of mixture increases, soluble lead concentration decreases. Lead concentrations of 3 different mixtures are significantly decreased under the allowable limit after stabilization.

Drilling muds contain both soluble and difficult soluble chromates for corrosion control and chromium lignosulfonates. Allowable chromium concentration in the petroleum drilling waste is 100 mg/kg (Mucsy et al., 1984). Chromium concentration in petroleum drilling waste is around the allowable

Table 5. The variation of permeability test results of stabilized mixtures.

Mixture	Curing period (day)	Permeability (cm/sec)
Mixture:1	Fresh	1.77×10^{-5}
	28	4.77×10^{-6}
Mixture:2	Fresh	------------
	28	8.26×10^{-6}
Mixture:3	Fresh	1.28×10^{-5}
	28	1.19×10^{-5}

Table 6. Total metal concentrations of stabilized mixtures.

Metals (mg/kg)	PDW	Mixture:1	Mixture:2	Mixture:3
Pb	184	57	57	100
Mn	1696	403	301	426
Zn	112	69	50	42
Cr	101	44	44	40
Cu	112	123	58	42
Cd	<0.5	<0.5	<0.5	<0.5
Ni	67	51	67	43
As	<5	<5	<5	<5
Ba	5891	1354	1547	1449
Fe	23027	7384	6565	6637

limit. Chromium concentrations of 3 different mixtures are significantly decreased under the allowable limit after stabilization.

Cadmium and its compounds are highly toxic. The primary treatment method is precipitation with alkali, usually lime. The pH of mixture is the most important factor controlling cadmium solubility. Solubility of cadmium decreases as pH increases. Between pH 8 and 11 about 99% of the cadmium is speciated as the solid carbonate (Conner, 1990). Allowable cadmium concentration in the petroleum drilling waste is 5 mg/kg (Mucsy, et al., 1984). Cadmium concentration in petroleum drilling waste is under the allowable limit.

Zinc is an aquatic toxin according to EPA. Zinc is found in all natural waters and soils, and is an important nutrient in plant and animal life. Allowable zinc concentration in the petroleum drilling waste is 300 mg/kg (Mucsy et al., 1984). Zinc concentration in petroleum drilling waste is under the allowable limit. Zinc concentrations of 3 different mixtures are significantly decreased under the allowable limit after stabilization.

Barium is classified as an alkali earth metal. It is almost in the form of barite. Barite is used in oil well drilling as an additive to drilling mud. Allowable barium concentration in the petroleum drilling waste is 100 mg/l (Morgan and Novoa, 1984). Barium concentration can not be decreased under the allowable limit even after stabilization.

Arsenic is classified as a nonmetal, although it is grouped with the metals for most environmental purposes. Most of arsenic compounds are highly toxic. Allowable arsenic concentration in the petroleum drilling waste is 5 mg/l (Morgan and Novoa, 1984). Arsenic concentration in petroleum drilling waste is under the allowable limit.

Copper is an aquatic toxin according to EPA. Copper is abundant in minerals and soils. Allowable copper concentration in the petroleum drilling waste is 100 mg/kg (Mucsy et al., 1984). Copper concentration in petroleum drilling waste is around the allowable limit. Copper concentrations of 3 different mixtures are significantly decreased under the allowable limit after stabilization.

Allowable nickel concentration in the contaminated soil is 100 mg/kg (Saarela, 1992). Nickel concentration in petroleum drilling waste is under the allowable limit.

Allowable manganese concentration in the contaminated soil is 50 mg/kg (Saarela, 1992). Manganese concentration can not be decreased under the allowable limit even after stabilization.

Allowable iron concentration in the contaminated soil is 100 mg/kg (Saarela, 1992). Iron concentration can not be decreased under the allowable limit even after stabilization.

4.5 Freeze/Thaw durability

Stabilized mixtures are subjected to repeated freezing/thawing in their typical service environment as highway sub-base. Over time these environmental stresses cause cracking, peeling and crumbling of stabilized mixtures. Therefore, environmental durability is one of the most important parameters for stabilized mixtures. Freeze/thaw durability test results are shown in Table 7. The average cumulative mass loss was least for the mixture 2 which survived 7 cycles of freeze/thaw prior to failure.

Table 7. Freeze/Thaw durability test results of stabilized mixtures.

Mixture	Moisture Content (%)	Number of Cycles	Mass Loss (%)
Mixture:1	28	2	total deterioration
Mixture:2	29	7	52
Mixture:3	20	6	68

4.6 Leachate analysis

The basic indicator of immobilization is the leachate obtained in the laboratory. Distilled water was used as a permeant. The leachates collected from permeability apparatus test of four different samples were analyzed for total metals, pH, sodium and electrical conductivity. These leachates were collected after 24 hours of steady state water permeation through the saturated matrix of each sample. Table 8 shows the results of pH, sodium and electrical conductivity of the stabilized mixtures. It can be seen that after 24 hours, addition of lime/fly ash/cement to the petroleum drilling waste significantly increases pH values of stabilized mixtures, as expected. pH is an important parameter which affects the immobilization of metals (Tuncan et al., 1995). With proper pH management, which is provided in this case, metals are effectively immobilized. High pH is desirable because metal hydroxides have minimum solubility in the range of pH 7.5-11. Sodium content and electrical conductivity of stabilized mixtures are increased due to the additives in the mixtures.

Table 8. Leachate test results of stabilized mixtures after 24 hours.

Mixture	pH	Sodium (ppm)	Electrical Conductivity (mSimens/cm)
PDW	6.23	840	26.1
Mixture:1	13.06	1680	35.3
Mixture:2	13.35	1900	43.6
Mixture:3	13.15	960	28.6

Table 9. Total metal concentrations of leachate test after 24 hours.

Metals (ppm)	PDW	Mixture:1	Mixture:2	Mixture:3
Pb	1.04	1.61	0.47	1.15
Mn	64.4	0.02	0.00	0.00
Zn	2.42	2.07	0.48	1.47
Cr	0.50	0.20	0.20	0.20
Cu	1.56	1.40	1.28	0.61
Cd	<0.05	<0.05	<0.05	<0.05
Ni	0.20	0.08	0.09	0.07
As	<0.5	<0.5	<0.5	<0.5
Ba	0.30	0.04	0.27	0.60
Fe	4.58	0.28	0.16	0.22

Total metal concentrations of stabilized mixtures are shown in Table 9. U.S. EPA toxicity limits for lead , chromium, cadmium, arsenic and barium are 5.0 mg/l, 5.0 mg/l, 1.0 mg/l, 5.0 mg/l and 100 mg/l, respectively. The concentrations of total metals are under the limits before and after stabilization, in general.

5 CONCLUSIONS

The stabilized mixtures showed significant improvement in their physical, mechanical and chemical properties with respect to those of unstabilized petroleum drilling waste. The following conclusions have been drawn from this study.

1. Additions of lime/fly ash/cement and lime/fly ash/cement/sand significantly increases pH, however decreases oil content of petroleum drilling wastes.
2. After stabilization, concentrations of lead, chromium, zinc, nickel and copper are reduced under the allowable limits. However, concentrations of manganese, barium and iron could not be reduced under the limits.
3. Additions of lime/fly ash/cement and lime/fly ash/cement/sand produce improvement in the strength and durability of the stabilized mixtures, in general. Samples of stabilized mixtures, cured for 28 days, showed that unconfined compressive strength doubled the strength of mixtures.
4. Permeability is reduced after stabilization. The total metal concentration in the leachate of the stabilized mixtures is reduced under the EPA toxicity limits.

The best result in all of the tests conducted is obtained with petroleum drilling waste stabilized with Sand (%50) + Lime (%16.7) + Fly ash (%5) + Cement (%4.17). This mixture can be effectively and safely used as sub-base material for the road construction. This is due to the combined action of the pozzolanic, cementing and sorbent reactions of each additive.

ACKNOWLEDGMENTS

This research was supported by the Scientific and Technical Council of Turkey (TUBITAK), Turkish Petroleum Corporation (TPAO) and Anadolu University, Eskisehir, Turkey.

REFERENCES

Chapman, H.D. 1965. Cation exchange capacity. In Methods of Soil Analysis, by Black, C.A. Editor, Part 2 pp. 891-901, No. 9, in the Series Agronomy; Am. Inst. Agronomy, Madison, Wisconsin.

Connor, J.R. 1990. Chemical fixation and solidification of hazardous wastes. Van Nostran Reinhold Publishing Co., New York.

McGILL, W.B. and Rowell, M.J. 1980. Determination of oil content of oil contaminated soil. The Science of the Total Environment. Vol.14. pp. 245-253, Elsevier Scientific Publishing Co., Amsterdam, Nedherland.

Morgan, D.S. and Novoa, J.I. 1984. Oil sludge solidification using cement kiln dust. Journal of Environmental Engineering, Vol. 110, No. 5, pp.935-948.

Mucsy, G., Kranicz-Pap, E. and Urbanyi, G. 1984. Disposal of oil-containing sludges on farmlands. Hazardous and Industrial Waste Management and Testing, Third Symposium, ASTM STP 851, pp. 135-151, Philadelphia, USA.

Pamukcu, S. and Winterkon, H.F. 1989. Soil stabilization and grouting, Ch. 9 in Foundation Engineering Handbook 2nd ed., H.Y. Fang, ed., Van Nostrand Reinhold Co, New York, N.Y.

Pamukcu, S., Lynn, J.B. and Kugelman, I.J. 1989. Solidification and re-use of steel industry sludge waste, Proceedings of the 21st Mid Atlantic Industrial Waste Conference, Technomic Press, Lancaster, PA., pp.3-15.

Rogers, F.W. 1963. Composition of properties of oil well drilling fluids. Gulf Publishing Co., Houstan, Texas.

Saarela, F.W. 1992, Effects of Tailings spread around three mines in Finland, Proceedings of the Mediterranean Conference on Environmental Geotechnology, A.A. Balkema Publishing Co., pp. 483-489, Çeşme, Turkey.

Tuncan, A., Tuncan, M. and Koyuncu, H. 1995, Physical and chemical stabilization of petroleum drilling wastes, Proc. of the Second International Conference on Mediterranean Coastal Environment, MED COAST 95, pp.1483-1494 Tarragona, Spain.

Van Keuren, E., Martin, J. and De Falco, A. 1987, Pilot field study of hydrocarbon waste stabilization, Proc. of the 19th Mid-Atlantic Industrial Waste Conference, pp. 330-341, Technomic Publishing Company.

Environmental Geotechnics, Kamon (ed.) © 1996 Balkema, Rotterdam. ISBN 90 5410 848 7

Fundamental study on development of valuable reuses of Ohya stone dust

K.Ueno, Y.Yokoyama & S.Imaizumi
Utsunomiya University, Japan

H.Yagi
Central Consultant Co. Ltd, Yokohama, Japan

ABSTRACT: Ohya stone dust, which is a kind of industrial waste and has been produced at least 60 kt in a year, was treated with calcium hydroxide and evaluated its engineering properties to find valuable reuses as a road construction material. The authors conducted a set of index tests, compaction tests, unconfined compression tests and CBR tests in this study. Ohya stone dust was graded well, and well compacted the dust fulfilled the specification for subgrades required for Japanese road construction. Furthermore, stabilization with 10 % of addition of calcium hydroxide in dry weight brought satisfactory amount of strength required for base courses for Japanese highways.

1 INTRODUCTION

Ohya stone, which is widely used as building stone, has been produced in Ohya district located about 100 km north of Tokyo since the period of ancient burial mounds. The quarrying has become an important industry in Ohya district. Fig.1 shows progress of annual production of Ohya stone. Currently the estimated an annual production is ranging from 200 to 300 *kt* approximately. According to Organization of Ohya Stone Producers, the amount of stone dust currently equals 30 % of the production of Ohya stone: 60 kt par year approximately.

A certain part of the stone dust are used to fertilize soil or processed to make into a kind of artificial building stone, most amount of the stone dust, however, has been thrown into dead pits and/or thrown to be tailing mounds without processing. To find valuable reuses of such amount of the stone dust is a significant problem to reduce industrial waste. On the other hand, Kanto loam, a volcanic clayey soil, widely covering the ground surface of Kanto area, is so sensitive and its moisture content is so high that this soil is not appropriate for back filling, especially in road construction. In addition, Kanto loam is likely to lead frost heaving. Therefore, we believe that success to find valuable reuses of Ohya stone dust, especially in road construction, becomes an advantage in the fields of environmental and civil engineering.

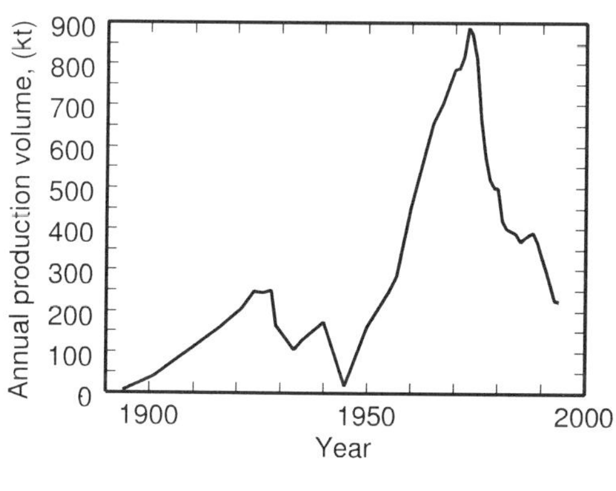

Fig. 1: Progress of annual production of Ohya stone

Unconfined compression tests and California Bearing Ratio tests (CBR tests) were employed to investigate strength characteristics of compacted Ohya stone dust in the presented study. Effects of lime stabilization and influence of submersion were also examined in this study.

2 MATERIALS

2.1 *Ohya stone dust*

Table 1 shows physical properties of Ohya stone dust. They were evaluated by means of the procedures specified by Japanese Society of Soil Mechanics and Foundation Engineering (1990).

The value of 2.40 (g/cm^3) of specific density of Ohya stone dust, as listed in Table 1, is similar to attapulgite, rather than ordinary soils and rocks. Ohya stone, which is a kind of tuff, consists of crynoptillolite having a pipe-shaped void structure with about 0.3 nm diameter. The diameter of the pipe-shaped void is so small that water cannot percolate the void to saturate it. According to Tohgo (1989), evaluated the specific density of Ohya stone, therefore, tends to be smaller than that of ordinary soils and rocks. The specific surface defined by surface area par gram ranges from 30 to 150 m^2/g. This indicates that particles comprising Ohya stone have a large number of micro void structures.

Fig.2 shows the grain size distribution of Ohya stone dust. Ohya stone dust is well graded and classified into volcanic sandy soil (SV) according to the fraction of its components. The range of grain size of the soil that would be likely to cause frost heaving is also presented in Fig.2 (Kawakami 1992). It can be expected that Ohya stone dust prevents frost heaving. Atterberg limits of the fraction finer than 420μm is also tabulated in Table 1. These indicate that fine Ohya stone dust is classified into volcanic cohesive soil type I (VH_1). Coarse particles in Ohya stone dust are crushable, and seem to be aggregates of the particles whose grain size is finer than 10 μm, as shown in Fig.3. Therefore, Ohya stone dust should be regarded as a fragment of a heavily overconsolidated volcanic clayey soil.

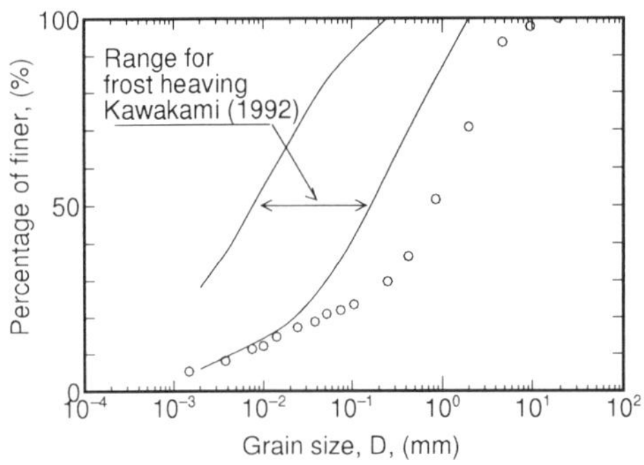

Fig. 2: Grain size distribution of Ohya stone dust

Fig. 3: A microscopic photograph of Ohya stone dust (×3000, 20kV SEM)

Table 1: Index of Ohya stone dust

Specific density	ρ_s	2.40	(g/cm^3)
Moisture Content	w_n	24.5	(%)
Specific surface		30-150	(m^2/g)
Gravel fraction		29.2	(%)
Sand fraction		47.6	(%)
Silt fraction		14.0	(%)
Clay fraction		9.2	(%)
Uniformity coefficient	U_c	230	
Liquid limit†	w_L	50.9	(%)
Plastic limit†	w_P	33.3	(%)
Index of plasticity†	I_P	17.6	(%)

† for the fraction finer than 420μm

2.2 *Lime treated Ohya stone dust*

Calcium hydroxide was employed as an additive in this study. Changes in pH and Atterberg limits of the stone dust, treated with calcium hydroxide with certain addition ratios R are tabulated in Table 2. The ratio of addition was defined as;

$$R(\%) = 100\frac{m_a}{m_d} \quad (1)$$

where R is the ratio of addition in percent; m_a, the mass of the additive; m_d, the mass of the dried Ohya stone dust. The value of pH for lime treated Ohya stone dust with five percents of addition suddenly increased, then the values of pH with further addition were kept almost constant. Although no noticeable change was observed in liquid limit w_L, plastic limit of treated Ohya stone dust became greater and was distinguished from that of non-treated one. Because no changes in the values of pH and I_P with the ratio of addition more than R = 5 % were observed, at least 5 % of additive in dry mass is required to expect pozzolanic reaction in Ohya stone dust.

3 EXPERIMENTS

Three kinds of mechanical tests, compaction tests, unconfined compression tests and California Bearing Ratio tests were conducted to evaluate adaptability in back filling of Ohya stone dust.

Table 2: Changes in pH and Atterberg limits of mixtures with addition of calcium hydroxide

$R(\%)$	0	5	10	15	20	25
pH	8.9	11.5	11.6	11.7	11.7	11.7
$w_L(\%)$	50.9	51.9	49.2	49.4	49.7	49.5
$w_P(\%)$	33.3	44.7	40.2	43.8	42.4	40.9
$I_P(\%)$	17.6	7.2	9.0	5.6	7.3	8.6

3.1 *Compaction tests*

Two series of compaction tests, termed Compaction tests with Non-treated samples (CN tests) and Compaction tests with Lime-treated samples (CL tests), were conducted by means of JSF T 711 (Japanese Society of Soil Mechanics and Foundation Engineering 1990). The stone dust used in the compaction tests in category A were cured into a cylindrical steel mold with an inner diameter of 100 mm and a height of 127 mm, and layered each a third of the height. Each of these three layers were blown 25 times with a rammer having a mass of 2.5 kg and a dropping height of 300 mm. The value of compaction energy par volume became 0.55 $MJ \cdot m^{-3}$. On the other in category E, a 2.48 $MJ \cdot m^{-3}$ of compaction effort were provided to specimens in a mold with an inner diameter of 125 mm and a height of 150 mm by using a rammer having a mass of 4.5 kg and a dropping height of 450 mm.

Test names, addition ratios and categories of compaction tests are listed in Table 3. The maximum dry density ρ_{dmax} and the optimum moisture content w_{opt} obtained these compaction tests are also presented.

3.2 *Unconfined compression tests*

In order to investigate strength characteristics of Ohya stone dust, unconfined compression tests, termed Unconfined Compression tests with Non-treated Ohya stone dust (UCN tests) and Unconfined Compression tests with Lime treated Ohya stone dust (UCL tests), were carried out. Table 4 shows the test conditions employed. Not only the effects of lime stabilization but the influence of submersion was also taken into account. In UCL tests, we had three series having different addition ratios R and curing conditions: UCLa series with 10% of addition and air curing; UCLb, 10% of addition and submersion; and UCLc, 5% addition and submersion.

The compaction procedure, described in JSF T 711 category A, was employed to remold non-treated and lime treated Ohya stone dust and to make cylindrical specimens having dimensions of 100 mm in diameter and 127 mm in height. After having been compacted under the optimum moisture content obtained from CN-1 test (w_{opt} = 33.3%), the specimens used in UCN tests were displaced from the mold, then immediately examined their unconfined compression characteristics. In UCL tests, after Ohya stone dust and calcium hydroxide were well mixed and their moisture content were adjusted to the values of the corresponding optimum moisture content listed in Table 3, remolding by means of compaction were performed as same as in UCN tests. Pozzolanic reaction probably required certain days of curing. The specimens used in UCL tests were, therefore, wrapped with filter paper, paraffin and plastic film in order to retain their moisture conditions as soon as they were displaced from the mold. After having been cured in the air for nine days, the specimens exposed themselves to the air or water for required successive curing periods. The temperature and the humidity of the room in which all of the specimens were cured were regulated at 20 ± 3 °C and 50%, respectively. Finally, unconfined compression tests with the cylindrical specimens, thus prepared, having dimensions of 100 mm in diameter and 127 mm in height, were conducted with a 1%/min of a loading rate.

Table 3: Compaction tests

Name	Ratio R(%)	Category	ρ_{dmax} (g/cm^3)	w_{opt} (%)
CN-1	0	A	1.252	33.3
CN-2	0	E	1.323	28.5
CL-1	5	A	1.221	34.6
CL-2	10	A	1.206	34.7
CL-3	15	A	1.196	34.8
CL-4	10	E	1.309	30.1

3.3 *CBR tests*

Effects of lime treatment and influence of submersion were examined in CBR test programmes termed CBRN tests and CBRL tests: the former for non-treated specimens, while the latter for lime treated specimens.

The value of addition ratio employed in CBRL

Table 4: Unconfined compression tests

Name	Ratio R (%)	Density ρ_d (g/cm^3)	Curing period (days)	Num. of tests
UCN	0	1.252	0	3
UCLa-2	10	1.143,1.172	7†	2
UCLa-3	10	1.160-1.187	14†	3
UCLa-4	10	1.149-1.210	28†	3
UCLb-1	10	1.142-1.188	1	4
UCLb-2	10	1.160-1.181	7	4
UCLb-3	10	1.157,1.172	14	2
UCLb-4	10	1.143-1.163	28	4
UCLc-1	5	1.160-1.174	1	3
UCLc-2	5	1.154-1.215	7	3
UCLc-3	5	1.167-1.182	14	3
UCLc-4	5	1.168-1.180	28	3

†cured in air

Table 5: CBR tests

Name	R (%)	Density ρ_d (g/cm^3)	Submersion period (days)	Num. of Tests
CBRN-1	0	1.147-1.327	0	3×2
CBRN-2	0	1.151-1.320	4	3×2
CBRN-3	0	1.170-1.290	7	3×1
CBRN-4	0	1.157-1.320	14	3×2
CBRN-5	0	1.198-1.270	28	2×1
CBRL-1	10	1.148-1.251	0†	2×1
CBRL-2	10	1.124-1.232	4†	2×2
CBRL-3	10	1.129-1.257	7†	2×2
CBRL-4	10	1.135-1.234	14†	2×1
CBRL-5	10	1.142-1.238	28†	2×1

†after six days air curing

tests was R = 10 %. In CBRL tests, compacted specimens were cured for six days in the air, of which temperature and humidity were regulated at 20±3 ℃ and 50%, respectively. Test names, addition ratios R, submersion periods, the values of dry density and the number of tests are listed in Table 5. These series of CBR tests were conducted according to KODAN A 1211 (Japan Highway Public Co. 1985), except shortage of a compaction energy in some cases. Values of compaction energy of 0.458, 1.13 and 2.48 MJ·m^{-3} (blow counts were 17, 42 and 92 for each three layers) for CBRN tests, 0.458 and 1.13 MJ·m^{-3} (17 and 42 blowing) for CBRL tests were employed in this study.

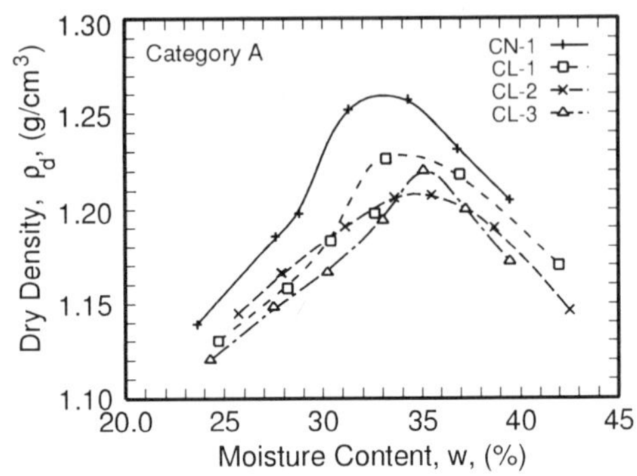

Fig. 4: Compaction test results (category A)

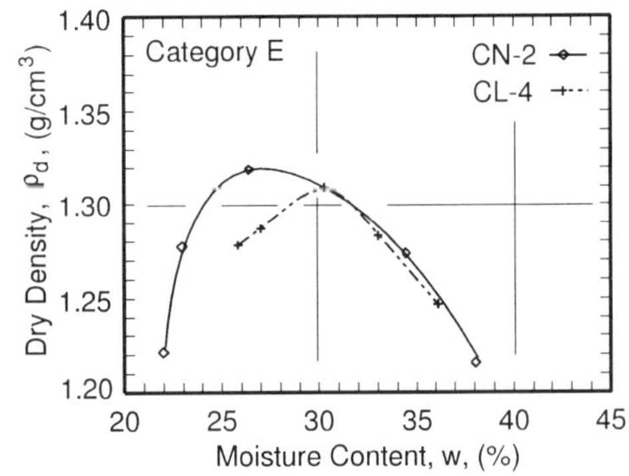

Fig. 5: Compaction test results (category E)

4 RESULTS AND DISCUSSIONS

4.1 *Compaction tests*

Fig.4 and Fig.5 show compaction curves obtained from non-treated and treated Ohya stone dust specimens. Lime treated Ohya stone dust shows smaller maximum dry density ρ_{dmax} and higher optimum moisture content w_{opt} than these values for non-treated one. It can be expected that increase of finer content of samples tested, brought by addition of calcium hydroxide, resulted in smaller maximum dry density ρ_d and higher optimum moisture content w_{opt}.

4.2 *Unconfined compression tests*

Fig.6 directly shows comparison between test results obtained in UCN-1 tests and UCLc-1 tests. A noticeable effect of lime treatment was observed among these test results. This implies that pozzolanic reaction, which is a chemical reaction between calcium hydroxide and silica (SiO_2) or alumina (Al_2O_3), occured in the specimens treated with 5 % of addition.

The values of unconfined compression strength q_u obtained from UCN-1 tests ranges from 179kPa to 198kPa, while the values from UCLc-1 tests from 836kPa to 1.00MPa which are 4.7 to 5.1 times greater than those values of non-treated Ohya stone dust. The effect of lime treatment also appeared in secant modulus E_{50}. The values for lime treated specimens were at least E_{50}=100MPa, which were about 10 times larger than the highest value of E_{50} obtained from non-treated one.

The values of unconfined compression strength q_u obtained in UCN-1 tests, UCLb-1 tests and UCLc-1 tests are plotted against addition ratios R with hollow diamonds in Fig.7, respectively. The values which coincide to 95 % of degrees of compaction are also indicated and connected with two straight lines.

The values of 686kPa and 980 kPa of unconfined compression strength q_u of lime stabilized road materials are required for subbases and base courses by Japanese highway administrator. In the case of Ohya stone dust, the values of 5% and 10% of addition ratios R provide satisfactory unconfined compression strength for subbases and base courses, respectively.

The influence of submersion of specimens was also

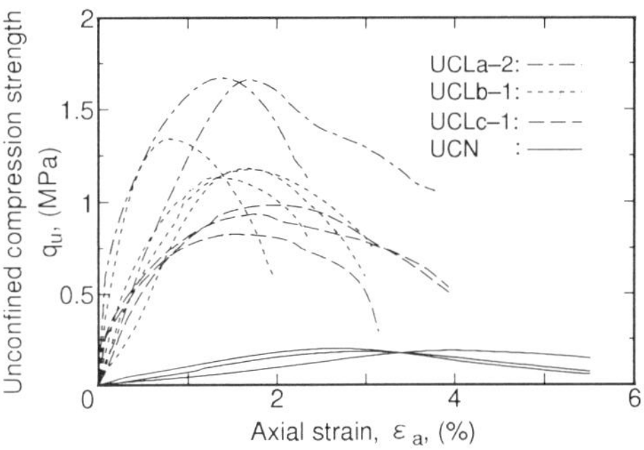

Fig. 6: Comparison between non-treated and lime treated Ohya stone dust

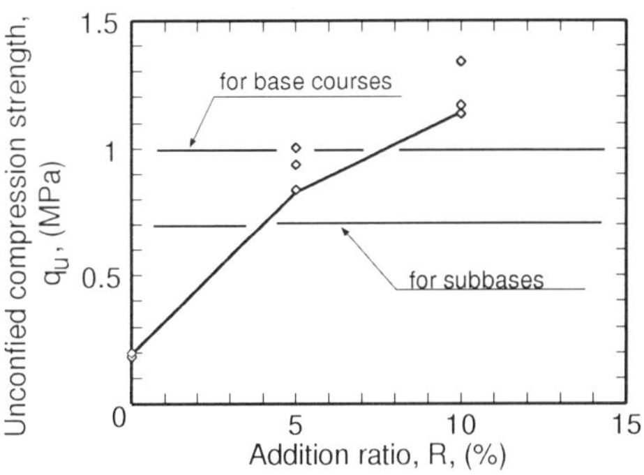

Fig. 7: Relationship between unconfined compression strength q_u and addition ratio R

investigated in this study. One day's submersion of non-treated specimens in UCN tests destroyed all of the specimens because of dissipation of apparent cohesion made from capillary forces. On the other hand, no lime treated specimen was destroyed even after 28 days' submersion. This fact implies that pozzolanic reaction initiated from calcium hydroxide produced effective cementation among particles of Ohya stone dust.

Fig.8 shows the relationship between unconfined compression strength q_u and curing and/or submersion periods: symbols indicate test results; lines, the values obtained by means of interpolation at 96.2% of degrees of compaction, as described in this figure. In UCLa tests, unconfined compression strength q_u of air cured specimens linearly increase with increasing curing period. After 28 days curing, the interpolated value of q_u for a specimen with 96.2% of degree of compaction reaches to 2.84MPa. On the other hand, the values of unconfined compression strength of submerged specimens in UCLb tests show almost constant values until 14 days' submersion, then rapidly increase and reach to 2.75MPa after 28 days submersion. In UCLc tests, unconfined compression strength

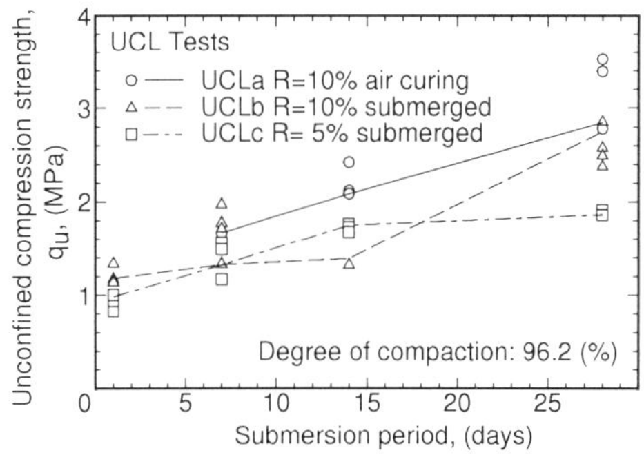

Fig. 8: Influence of submersion on unconfined compression strength q_u

q_u of submerged specimens with an addition ratio of $R = 5\%$ increase with increasing submersion period until 14 days submersion, then become constant at 1.86MPa. Although pozzolanic reaction of submerged specimens seems to be delayed due to dilution of calcium hydroxide with sucked water, sufficient curing provides satisfactory strength for submerged lime treated specimens.

4.3 *CBR tests*

The values of CBR were determined at 2.5 mm of penetration as:

$$CBR(\%) = 100\frac{q_{2.5mm}(\mathrm{MPa})}{6.86(\mathrm{MPa})} \tag{2}$$

where CBR is the California Bearing Ratio in percent; $q_{2.5mm}$, the load intensity in MPa at a penetration of 2.5 mm. The denominator of 6.86 MPa in Equ.2 is the standard load intensity employed when the value of the penetration is 2.5 mm.

The relationship between CBR and the submersion period is illustrated in Fig.9. Hollow diamonds in this figure represent the results of the non-treated specimens in CBRN tests (R=0%), while crosses indicate the results of the lime treated specimens with an addition ratio of $R = 10\%$ in CBRL tests. Since the values of the dry density ρ_d of the specimens examined were rather scattered as shown in Table 5, the values of CBR corresponding to 90, 93 or 95% of degrees of compaction were determined then illustrated with lines in Fig.9.

The values of CBR for non-treated Ohya stone dust tend to decrease with increasing submersion period. The values corresponding with a degree of compaction of 95% dropped from 48.0% to 30.0% during 14 days submersion. Contrarily, the values for lime treated Ohya stone dust were made almost constant at 130% during seven days submersion, and

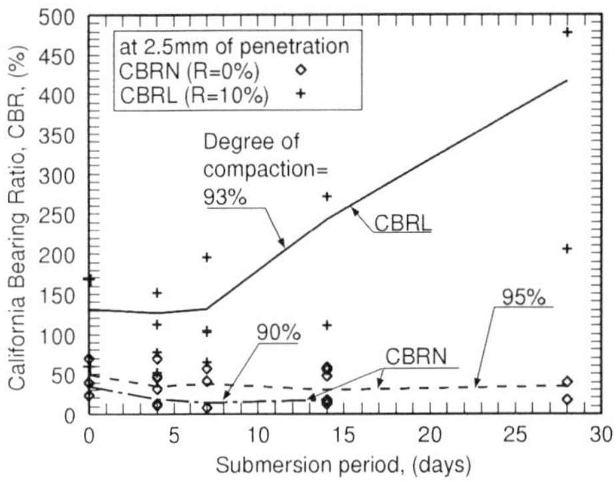

Fig. 9: CBR test results

increased during the successive submersion period. Furthermore, the value for lime treated Ohya stone dust at 93% of degrees of compaction were about 2.7 times larger than that for non-treated one.

Japanese specifications for highway construction requires values of 80% and 30% of CBR for base courses and subbases with a compaction quality with 93% of degrees of compaction, respectively. Also the values of 5% and 10% of CBR are specified for lower and upper subgrades. Because the test results show that non-treated Ohya stone dust with a proper compaction effort can be used for both upper and lower subgrades, the stone dust might be used for back filling in subgrades. Note further, lime treated Ohya stone dust with 10% of addition ratio and a proper compaction fulfills the recommendation for both base courses and subbases. Lime treatment with 10% of addition is worthwhile for an utilization of Ohya stone dust in road construction.

5 CONCLUSION

Adaptability of Ohya stone dust as a road construction material was evaluated in this study. Samples of Ohya stone dust were either compacted or stabilized with calcium hydroxide and examined their unconfined compression strength q_u and CBR.

Ohya stone dust was graded well. Non-treated stone dust with proper compaction work fulfilled the specification required for subgrades according to CBR tests results. Furthermore treated stone dust with addition of calcium hydroxide of 5 % and 10 % in dry weight fully satisfied the specifications for subbases and base courses for Japanese highways, respectively. Performance of lime treated Ohya stone dust under *in situ* situation will be presented in a future publication.

6 ACKNOWLEDGEMENT

The writers wish to express thanks to Mr. Takehito Chikabayashi and Mr. Takae Mizunuma, a technician and a graduate student in Utsunomiya University, for their efforts to conduct laboratory testing. The study presented here is a part of the outcome of an industrial joint research program between Yoshizawa Lime Industry Co. and Utsunomiya University. The writers are also grateful to Dr. Mitsuo Hanada, an executive of Yoshizawa Lime Industry Co., for his support and advice.

REFERENCES

Japan Highway Public Co. (1985). *Testing Procedures in Japan Highway Public Co.* Japan Highway Public Co. (in Japanese).

Japanese Society of Soil Mechanics and Foundation Engineering (1990). *Procedures and interpretations of soil testing.* Japanese Society of Soil Mechanics and Foundation Engineering. (in Japanese).

Kawakami, F. (1992). *Soil Mechanics.* Morikita Publisher Co. Ltd. (in Japanese).

Tohgo, S. (1989). *Estimation of absorption performance by means of liquid phase absorption method — especially Ohya stone and volcanic ash —.* Doctor thesis, Waseda University. (in Japanese).

Environmental Geotechnics, Kamon (ed.) © 1996 Balkema, Rotterdam. ISBN 90 5410 848 7

Consolidation behavior of a water treatment sludge

M.C.Wang
The Pennsylvania State University, Pa., USA

W.Tseng
Hazama Corporation, Tokyo, Japan

ABSTRACT: The consolidation behavior involving compressibility and consolidation rate of a water treatment alum sludge was investigated. The test sludge was both untreated and treated separately with 60% by weight of a slaked lime and a class C fly ash. The treated sludge was cured for one month. The test specimens for untreated sludge were both uncured and cured separately for one and two months. The study was conducted using the conventional consolidation test, and the consolidation behavior was analyzed using both the traditional small-strain theory and the finite-strain theory of consolidation.

Results of the study indicate that the small-strain coefficient of consolidation decreases with increasing consolidation pressure for both untreated and treated sludges. The magnitude of small-strain consolidation coefficient is greater for the lime-treated, followed by the fly ash-treated, then the untreated sludge. The finite-strain coefficient of consolidation is more than two orders of magnitude smaller than that of the small-strain coefficient of consolidation. Both small-strain compression index and swelling index are considerably greater for the untreated than the treated sludges. The ratio between swelling and compression indexes is much lower than that of natural soils. These compressibility and consolidation data are essential sludge properties for predictions of sludge landfill settlement.

INTRODUCTION

The continuous treatment of drinking water to meet the daily domestic and industrial consumptions has produced an enormous amount of residuals. The rate of residuals accumulation is being augmented by population growth. In the meantime, governmental regulations for environmental protection are becoming more stringent. Thus, the disposal of the ever increasing volume of water treatment residuals in compliance with the stringent regulations is a challenging task and has received increasing attention nationwide.

Of the various possible disposal methods, landfill methods are often preferred for the disposal of water treatment residuals. Possible reasons for this option include that water treatment residuals are typically inert and nonhazardous, that the volume of residuals disposed per unit area can be many times greater than other land disposal options, and that, if properly designed, a landfill may eventually be utilized for other purposes such as for supporting roadways, parking lots, recreation parks, buildings, and other engineering structures.

Under its own weight, the landfill will undergo consolidation. As a result, the residuals will densify creating more air space for further deposition. To effectively utilize the available land, the increased air space due to landfill consolidation must be known in terms of the total volume and the time rate of volume increase. The determination of these two quantities requires knowledge of the consolidation behavior of the residuals.

Available information on the consolidation behavior of water treatment residuals is quite limited. Using the method of dewatering under varying pressure differentials, Knocke and Wakeland (1983) reported an average coefficient of compressibility varying between 0.94 and 0.97 for two alum sludges and about 0.79 for a lime sludge. Raghu et al. (1987) conducted a conventional consolidation test on a lime/alum/polyamine coagulant sludge and presented a void ratio vs pressure curve of which the void ratio ranged between 1.2 and 1.5 and the compression index was approximately 0.3. They also reported a dial gage reading vs square-root-of-time curve with a coefficient of consolidation equal to 0.18 mm^2/s. It should be noted, in comparison of these available data, that the relation between the coefficient of compressibility and the compression index is not yet available.

Most recently, Wang et al. (1992a) presented some data on the consolidation behavior of two alum and one iron water treatment sludges. They reported that the compression indexes of the two alum sludges -- one from Chesapeake, VA and the other from Durham, NC, were 6.69 and 5.29, respectively, and that of the iron sludge (from Pittsburgh, PA) was 1.99. They also presented the swelling index of 0.17 for the Chesapeake alum sludge, 0.15 for the Durham alum sludge, and 0.03 for the Pittsburgh iron sludge. Elsewhere, Wang et al. (1992b) reported the compression index of the Chesapeake alum sludge admixed with soil, lime, and fly ash, separately. The lime and fly ash treated sludges were cured for 2 weeks and 4 weeks. The compression indexes were 4.04 for the soil-treated; 3.56 and 4.34 for the lime-treated sludge cured for 2 weeks and 4 weeks, respectively; and 3.70 and 3.93 for the fly ash-treated sludge cured for 2 weeks and 4 weeks, respectively. These available data are useful for the determination of the magnitude of landfill settlement. For settlement rate determination, other properties are needed. This paper presents some of the needed properties obtained to date.

TEST SLUDGE

The test sludge was an alum sludge obtained from Chesapeake, VA. The sludge contains approximately 20% (by weight) clay size and 46% silt size particles with a median size of 0.013 mm. Having 550% liquid limit and 311 plasticity index, the test sludge is classified as CH according to the Unified classification system.

The test sludge was also treated by admixing separately with a slaked lime and a class C type fly ash. The additive content was arbitrarily chosen at 60% by dry weight of sludge. For the treated sludges, the test specimens were cured for 1 month. For the untreated sludge, three sets of specimens were tested -- uncured, 1 month cured, and 2 months cured specimens.

CONSOLIDATION RATE

The needed consolidation behavior was obtained by performing the conventional consolidation test, ATM D-2435 (Annual, 1992). Consolidation pressures ranging from 1.72 kPa (0.25 psi) to 541.28 kPa (78.56 psi) were applied in 9 steps. The test results were analyzed using the conventional (small-strain) consolidation theory proposed by Terzaghi and the finite-strain theory of Gibson et al. (1967 & 1981). When the consolidation pressures are 8.48 kPa (1.23 psi) and lower, the settlement vs logarithm-of-time (log t) curves do not exhibit the typical inverse S shape. Thus, it is difficult to find the traditional (small-strain) coefficient of consolidation (c_v) for low consolidation pressures using the logarithm-of-time (log t) function. For higher consolidation pressures, the coefficients of consolidation (c_v) determined from the log t functions are slightly smaller than those obtained from the square-root-of-time ($\sqrt{t}$) functions. Such a difference has also been reported for cohesive soils by Tavenas et al. (1983), Mikasa and Takada (1985), Olson (1985), and others. They also showed that the c_v determined from the $\sqrt{t}$ function is more accurate. Mainly for this reason, the c_v values obtained from the $\sqrt{t}$ functions are used in the following analysis.

For the untreated residuals, the c_v values are shown graphically as a function of logarithm-of-consolidation pressure (log p) in Figure 1 (A); they are the best fit curves of the data points. Although the data points are scattered, the trend is clear that c_v decreases with increasing log p at a rate which is very slow in the beginning, intensifies in the middle range, and then gradually diminishes at the later stage. The shape of the curves resembles that of a silty clay reported by Lambe (1951). The decrease in c_v with increasing consolidation pressure can be attributed to the reduction in void ratio and coefficient of permeability under higher consolidation pressures, since c_v is directly proportional to these two quantities.

The figure also shows that curing results in an increase in c_v, and that the amount of increase is greater under low consolidation pressures and becomes insignficicant when the pressures are higher. The effect of curing on c_v may be related to thixotropic hardening which has been elaborated by Wang et al. (1992a). Due to thixotropic hardening, the cured specimens are less compressible and, therefore, under the same consolidation pressure both void ratio and permeability are greater and hence the values of c_v are higher than those of uncured specimens. Furthermore, because the rate of hardening decreases with curing time, the rate of increase in c_v is greater in the first month than in the second month as shown in Figure 1.

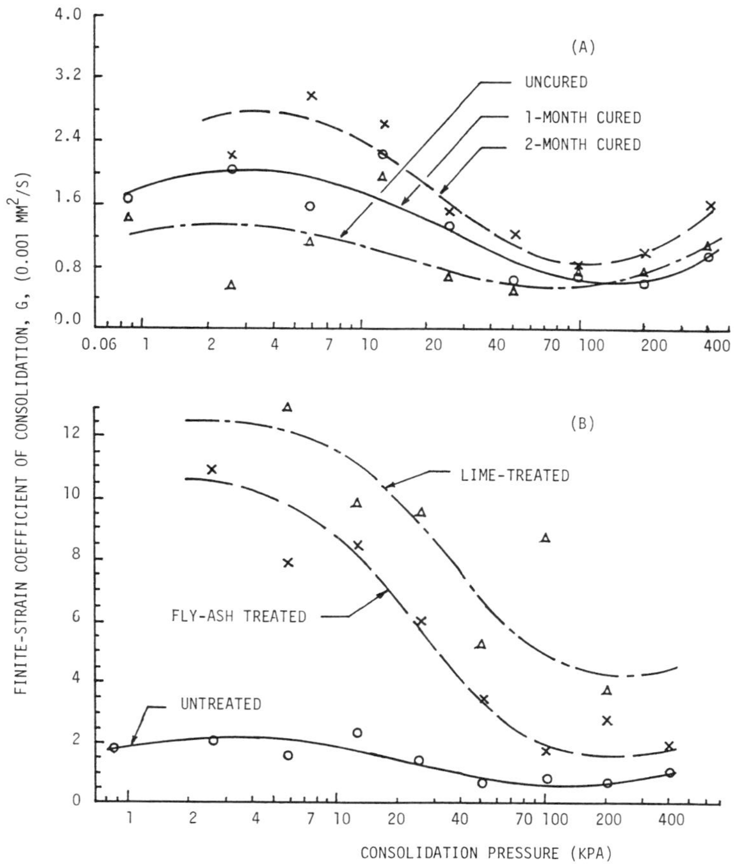

Figure 2. Variation of Finite-Strain Coefficient of Consolidation with Consolidation Pressure for (A) Untreated and (B) 1-Month Cured Specimens

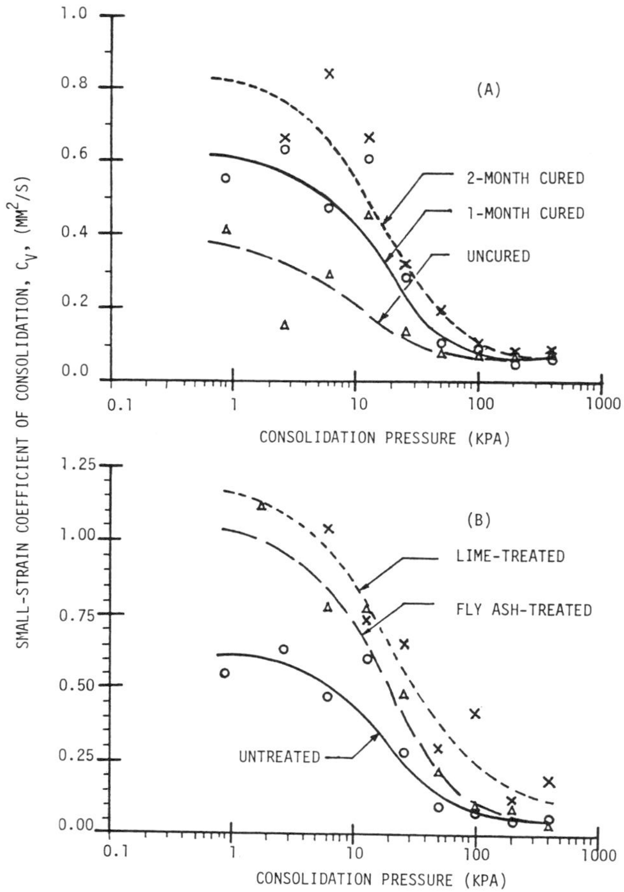

Figure 1. Variation of Small-Strain Coefficient of Consolidation with Consolidation Pressure for (A) Untreated and (B) 1-Month Cured Specimens

For the residuals treated with lime and fly ash, the c_v values are plotted against log p in Figure 1 (B) together with the data of one-month cured untreated residuals. The shape of curves for the treated specimens resembles that of the untreated. Thus, the effect of consolidation pressure on c_v is about the same regardless of whether or not the residuals are treated. For the treated residuals, however, the c_v values are considerably greater than that of the untreated one. Furthermore, the lime-treated residuals appear to have somewhat higher values of c_v than the fly ash-treated residuals. The difference in c_v among the three different conditions can be related to the changes in both strength and permeability resulting from admixture treatment. According to Wang et al. (1992b), under the same curing time, the undrained shear strength is greatest for the lime-treated, intermediate for the fly ash-treated , and smallest for the untreated residuals. Meanwhile, Wang and Tseng (1993) showed that, under the same void ratio, the permeability also decreases in the same order of lime-treated, fly ash-treated and untreated sludges. Therefore, as reasoned before, the c_v values are highest for the lime-treated, intermediate for the fly ash-treated, and lowest for the untreated residuals.

From Figure 1, the values of c_v for the untreated residuals vary from approximately 0.043 to 0.828 mm^2/s depending on consolidation pressure and curing time. For the treated residuals, the c_v values range between about 0.043 and 1.129 mm^2/s. When compared with the c_v values of some industrial wastes, these values are considerably higher than 0.0054 to 0.0322 mm^2/s of an acidic petroleum refining waste sludge treated with clay, fly ash, cement, and lime (Biehl and Browning, 1990). For the phosphatic clay studied by Keshian et al. (1977), the c_v values fluctuate around 0.02 mm^2/s which also is considerably lower than the values of this study. However, the c_v values of a red mud (Somogyi and Gray, 1977) ranging from 0.03 to 6.9 mm^2/s have an upper limit much higher than the values shown in Figure 1.

According to Gibson, et al. (1981), the finite-strain coefficient of consolidation (g) can be determined, as a reasonable approximation, from the small-strain coefficient of consolidation (c_v) using the relation $g = c_v/(1 + \bar{e})^2$, in which $\bar{e}$ = average void ratio for each load increment. The computed values of finite-strain coefficient of consolidation are related with the consolidation pressure (*p*) in Fig. 2(A) and (B). It is seen that the data points are more scattered than those in Fig.1, and that the data points scatter is much greater for treated than untreated residuals. Despite data points scatter, it is seen that the difference in g values between the untreated and treated sludges is considerably greater than that of c_v values. Such a difference can be attributed to a much smaller $\bar{e}$ value for the treated sludges. For the untreated sludge both uncured and cured, the values of g vary between 0.5 x 10^{-3} mm^2/s and 3.0 x 10^{-3} mm^2/s depending on curing time and consolidation pressure. For the lime-treated and fly ash-treated residuals with 1 month curing, the values of g vary from 2.0 x 10^{-3} to 11.0 x 10^{-3} mm^2/s and from 4.0 x 10^{-3} to 13.0 x 10^{-3} mm^2/s, respectively. For comparisons, the values of g vary from about 0.1 x 10^{-3} to 0.5 x 10^{-3} mm^2/s for Canaveral Harbor material, from about 0.2 x 10^{-3} to 1.0 x 10^{-3} mm^2/s for Craney Island material (Cargill, 1984), and from 2.3 x 10^{-3} to 4.4 x 10^{-3} mm^2/s for Osaka Harbor mud (Gibson et al., 1981).

COMPRESSIBILITY

The void ratio (e) vs logarithm-of-consolidation pressure (log p) curves for the untreated and treated residuals both cured for 1 month were similar to those presented by Wang et al. (1992b). The curves for the uncured and 2-month cured untreated residuals, which are not presented, also show the same shape. Each of the loading portion of the e-log p curves exhibits a slight concave segment revealing the sensitive nature of the test residuals.

The compression index and swelling index of both untreated and treated residuals with different curing conditions each obtained from the average of two data are summarized in Table 1. Note that some of the data presented in Table 1 were obtained previously (Wang et al., 1992 a & b); they are included for comparison and discussions. It is seen that the compression and swelling indexes are considerably greater for the untreated than those for the treated residuals. The data also reveal that except for the 1-month cured specimens, the lime-treated residuals have slightly smaller compression and swelling indexes than the fly ash-treated residuals. The smaller index values for the lime-treated residuals may possibly be attributed to the greater strength increase than the fly ash-treated residuals as discussed earlier. Since the higher strength specimens will be compressed less under the same pressure, the compression index and, consequently, the swelling index will be lower than those of the lower strength specimens.

Table 1. Compression and Swelling Indexes

Test Residuals	Compression Index (C_c)	Swelling Index (C_s)	$\frac{C_s}{C_c}$	Finite-Strain Compression		
				e_{00}	e_∞	λ
Untreated						
Uncured	6.69[a]	0.17[a]	0.025[a]	16.2	6.9	0.079
1-month cured	6.68	0.17	0.025	17.5	6.7	0.084
2-month cured	6.59	0.16	0.024	17.0	6.0	0.071
Lime Treated						
2-week cured	3.56[b]	0.04	0.011	-	-	-
1-month cured	4.34[b]	0.08	0.018	8.5	3.7	0.053
Fly Ash Treated						
2-week cured	3.70[b]	0.06	0.016	-	-	-
1-month cured	3.93[b]	0.09	0.023	9.5	3.4	0.057

[a] from Wang et al. (1992a)
[b] from Wang et al. (1992b)

For both untreated and treated residuals, the compression and swelling indexes vary with curing time. The data seem to reveal a trend that longer curing results in greater index values for the treated but smaller values for the untreated residuals. Additional data are needed to verify this trend, however. The data also show that the ratios as between swelling and compression indexes (C_s/C_c) are smaller for the treated than for the untreated residuals. Meanwhile, all of these values are considerably lower than those of natural soils. Such characteristics of water treatment residuals have been pointed out earlier (Wang et al., 1992a).

Based on the approximate solution of Cargill (1984), the coefficient, λ, which is a measure of material compressibility, can be computed from $\lambda = -(1/p) \ln [(e - e_\infty)/(e_{00}-e_\infty)]$ in which e_{00} and e_∞ are the void ratios at zero and maximum consolidation pressures, respectively. To determine λ value requires fitting the void ratio vs consolidation pressure (e vs log p) curve. According to Cargill (1984), the best curve fit can be obtained by matching the generated curve with the experimental curve at about 20% of the maximum consolidation pressure. Following this method, the values of coefficient λ together with e_{00} and e_∞ are computed and included in Table 1. The data show that the treated sludges have considerably smaller e_{00}, e_∞ and λ values than those of the untreated sludge. For the untreated sludge, little effect of curing on these values is seen.

SUMMARY AND CONCLUSIONS

The consolidation behavior of a water treatment residuals both untreated and treated separately with a slaked lime and a fly ash was investigated using the conventional consolidation test. The test residuals were cured to different durations up to two months. The consolidation behavior was analyzed using both the conventional small-strain theory and the finite-strain theory. The effect of consolidation pressure, curing time, and treatment on consolidation behavior was presented.

Results of the study indicate that the small-strain coefficient of consolidation decreases with increasing pressure and is greater for the treated than for the untreated residuals. For the untreated residuals, curing results in a greater coefficient of consolidation. For each test residual, the finite-strain coefficient of consolidation is more than two orders of magnitude smaller than that of the small -strain coefficient of consolidation. Furthermore, both compression and swelling indexes are greater for the untreated than the treated residuals. Also, the ratios between small-strain swelling and compression indexes are considerably smaller than those of natural soils. With these compressibility and consolidation data, it will be possible to predict the amounts and time rate of consolidation settlement of sludge landfills. Thus, the results of this study should provide a database useful for effective planning of disposal landfill for water treatment residuals.

ACKNOWLEDGMENT

The water treatment residuals used for the study were provided by Chesapeake Water Treatment Plant. The cooperation of F. Saunders, Plant Manager and C. Maples is greatly appreciated.

APPENDIX - REFERENCES

Biehl, F. J. & J. S. Browning, III. 1990. Evaluation and Analysis for Subsidence of Stabilized Sludge. Hazardous and Industrial Wastes, Proceedings of the 22nd Mid-Atlantic Industrial Waste Conference, pp. 594-609.

Cargill, K. W. 1984. Prediction of Consolidation of Very Soft Soil. Journal of the Geotechnical Engineering Division, ASCE, Vol. 110, No. 6, June, pp. 775-795.

Gibson, R. E., G. L. England, & M. J. L. Hussey. 1967. The Theory of One-Dimensional Consolidation of Saturated Clays. I. Finite Non-Linear Consolidation of Thin Homogenous Layers. Geotechnique, Vol. 17, No. 3, pp. 261-273.

Gibson, R. E., R. L. Schiffman, & K. W. Cargill. 1981. The Theory of One-Dimensional Consolidation of Saturated Clays. II. Finite Non-Linear Consolidation of Thick Homogeneous Layers. Canadian Geotechnical Journal., Vol. 18, No. 2, pp. 280-293.

Keshian, B. Jr., C. C. Ladd, & R. E. Olson. 1977. Sedimentation-Consolidation Behavior of Phosphatic Clays, Geotechnical Practice for Disposal of Solid Waste Materials. Proceedings of Specialty Conference of the Geotechnical Engineering Division, ASCE, Ann Arbor, Michigan, pp. 188-209.

Knocke, W. R. and D. L. Wakeland. 1983. Fundamental Characteristics of Water Treatment Plant Sludges. J. AWWA, 75 (10), 516-523.

Lambe, T. W. 1951. Soil Testing for Engineers. John Wiley & Sons, Inc., New York.

Mikasa, M. & N. Takada. 1985. Determination of Coefficient of Consolidation (c_v) for Large Strain and Variable c_v Values. Consolidation of Soils; Testing and Evaluation, ASTM STP 892; pp. 526-547.

Olson, R. E. 1985. State of the Art: Consolidation Testing. Consolidation of Soils; Testing and Evaluations, ASTM, STP 892, pp. 7-70.

Raghu, D., H. N. Hsieh, T. Neilan, & C. T. Yih. 1987. Water Treatment Plant Sludge as Landfill Liner. Geotechnical Practice for Waste Disposal '87; Geotechnical Special publication 13, ASCE, New York, NY, 744-757.

Somogyi, F., & D. H. Gray. 1977. Engineering Properties Affecting Disposal of Red Muds. Geotechnical Practice for Disposal of Solid Waste Materials, Proceedings of Specialty Conference of the Geotechnical Engineering Division, ASCE, Ann Arbor, Michigan, pp. 1-22.

Tavenas, F., P. Leblond, P. Jean, & S. Leroueil.1983. The Permeability of Natural Soft Clays., Part I: Method of Laboratory Measurement. Canadian Geotechnical Journal, Vol. 20, pp. 629-644.

Wang, M. C., J. Q. Hull, M. Jao, B. A. Dempsey, & D. A. Cornwell. 1992a. Engineering Behavior of Water Treatment Sludge. Journal of Environmental Engineering, ASCE, Vol. 118, No. 6, November/December, pp. 648-664.

Wang, M. C., J. Q. Hull, & M. Jao. 1992b. Stabilization of Water Plant Sludge for Possible Utilization as Embankment Material. Transportation Research Record No. 1345, Transportation Research Board, Washington, D.C., pp. 36-43.

Wang, M. C. & W. Tseng. 1993. Permeability Behavior of a Water Treatment Sludge. Journal of Geotechnical Engineering, ASCE, vol. 199, No. 10, Oct. 1672-1677.

Environmental Geotechnics, Kamon (ed.) © 1996 Balkema, Rotterdam. ISBN 90 5410 848 7

Bearing capacity of reclaimed coal fly ash

Kazuya Yasuhara
Department of Urban and Civil Engineering, Ibaraki University, Hitachi, Japan

Sumio Horiuchi
Institute of Technology, Shimizu Construction Co. Ltd, Tokyo, Japan

Kazutoshi Hirao
Department of Civil Engineering, Nishinippon Institute of Technology, Kanda, Fukuoka, Japan

Noriyuki Yasufuku
Department of Civil Engineering, Kyushu University, Fukuoka, Japan

ABSTRACT : Model tests have been carried out on a coal based pulverised fuel ash (PFA) with a view to its use for land reclamation. The bearing capacity characteristics were investigated by plate loading and cone penetration tests on model reclaimed PFA. An analysis of the bearing capacity of the PFA deposit was performed using cavity expansion theory combined with results from undrained triaxial and oedometer tests. The results of the analysis were compared with those observed in plate loading tests. The investigation has highlighted the following:
1) An increase in the bearing capacity of PFA is caused by both increased dry density and enhanced or added cementation. In this respect, the slurry disposal method (HAS) for PFA is best suited to producing stable reclaimed land.
2) Terzaghi's classical theory of bearing capacity is sufficient for predicting the bearing capacity of PFA in 1g laboratory model tests under low confining pressures and with a small diameter loaded area. However, Vesic's cavity expansion theory which takes compressibility into consideration, is necessary for predicting the in-situ bearing capacity under higher confining pressures and larger loaded areas.

1 INTRODUCTION

The currently used in-situ placement methods for reclamation with pulverised fuel ash (PFA) can be classified into two types: dry and wet disposal systems, as shown in Fig. 1. In the former method, fly ash is usually compacted and then reclaimed while in the latter PFA is sluiced into a lake or a sea. Although the wet disposal system is widely used in Japan, PFA landfills reclaimed by this system sometimes do not have strength enough to support buildings or earth structures, and may be difficult and otherwise costly to use after closing the site. On the other hand, PFA landfills reclaimed by the dry disposal system have sufficient strength due to the self-hardening properties of PFA, and it is possible to utilize such land for buildings and structures.

The low density of fly ash given by the wet disposal system results from the free sedimentation of fly ash particle aggregates (Yasuhara et al., 1987). Such free sedimentation, however, can be avoided through the use of underwater placement techniques which give the fly ash a higher density (Horiuchi et al., 1985 : 1995, Yasuhara and Horiuchi, 1996). This material has been named high density ash slurry (HAS), detail of whose placement method is shown in Fig. 1. Horiuchi et al. (1985) indicated that PFA placement, improved by this slurry disposal method, provided great potential for producing stable reclaimed land.

Following on the work of Horiuchi et al. (1985), the present paper aims at demonstrating the advantages of the HAS disposal method from the viewpoint of the improvement of strength and bearing capacity and hence the stablility of reclaimed landfill.

	System Process	Appearance
Wet Disposal System	① Mixing Fly Ash with a large amount of sea water. ② Pumping the low density FA slurry to the ash pond. ③ Spreading the FA slurry into the ash pond.	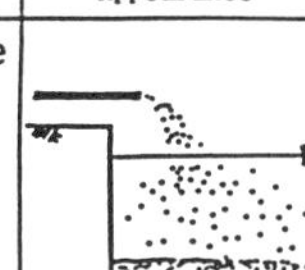
Dry Disposal System	① Mixing with sea water at around the optimum water content of Fly Ash. ② Transfering the humidified FA to the disposal site by belt-conveyor or dump trucks. ③ Spread the humidified FA into the ash pond.	
Slurry Disposal System	① Mixing with sea water by the double mixing method to make the high density Fly Ash slurry. ② Pumping the high density slurry to the disposal site ③ Placing the high density slurry onto the bottom of the ash pond.	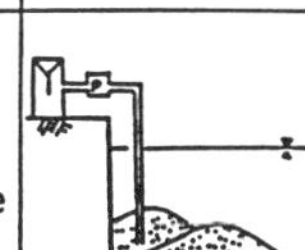

Fig. 1 Disposal methods for coal fly ash

2 LABORATORY MODEL TESTS FOR INVESTIGATING THE BEARING CAPACITY

To investigate the bearing capacity characteristics of PFA, both plate loading tests and cone penetration tests were carried out using a PFA laboratory model in a poly-acryl soil tank (1.0m length, 0.8m width and 0.5m depth).

Two kinds of fly ash with a calcium oxide (CAO) content of 4.5% (labeled A) and 2.0% (labeled B), respectively, were used for plate loading tests. Portland cement or gypsum was added as a cementing material in some tests. The details for all types of test are listed in Table 1.

Static cone penetration tests and portable cone penetration tests ere conducted to investigate the bearing capacity characteristics of three PFA models: dry disposal PFA with 1% and 0% cement added, called DAS, and the slurry PFA. The rates of penetration are 2 to 3 mm/sec and 10 mm/sec respectively for the static cone penetration tests and portable cone penetration tests.

3 BEARING CAPACITY CHARACTERISTICS FROM LABORATORY MODEL TESTS

3.1 Bearing capacity from cone penetration tests

The variations of the cone resistance, q_c, obtained from cone penetration tests on PFA-B are plotted against depth in the model, in Fig. 3. The results in this figure indicate the following:

1) the variations of dry density with depth and the cone resistance strength are almost constant except for DAS with the addition of cement;
2) possibly due to a non-uniform distribution of cement, the dry density and cone resistance strength of DAS with cement varies with depth;
3) in the case of HAS, the tendency is different; the dry density of HAS is largest among all the tests;

1 Loading plate 2 Load cell
3 Oil cyclinder 4 Oil tank
5 Motor 6 Relief valve
7 Switch box 8 Sampling mould

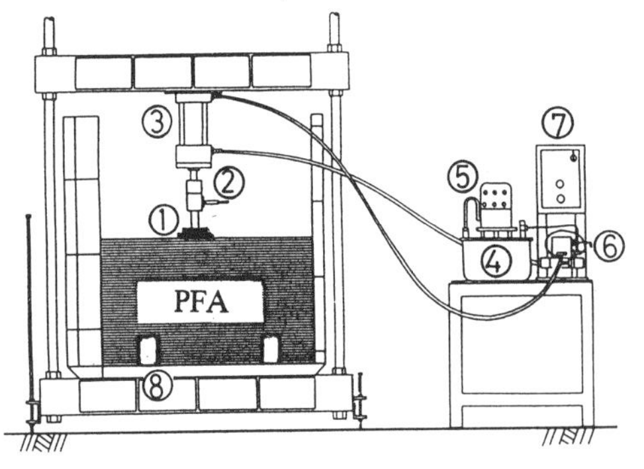

Fig. 2 A layout of model footing testing equipment

Table 1 Test conditions and CaO content

Fly ash	No.	Reclamation method	Combination (%) FA:Cement:Gypsum	CaO (%)
A	A1	DAS	100: 0: 0	4.9
	A2	DAS	96: 2: 2	7.0
	A3	DAS	98: 2: 0	6.7
B	B1	DAS	100: 0: 0	2.0
	B2	DAS	99: 1: 0	2.6
	B3	HAS	100: 0: 0	2.0

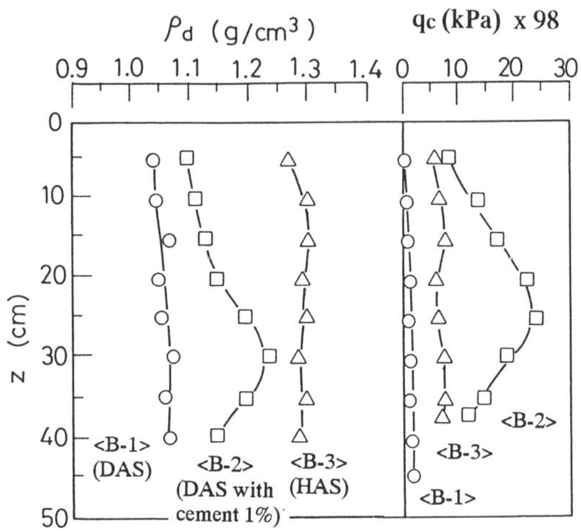

Fig. 3 Distribution of ρ_d and q_c against depth (Series B tests)

however, the cone penetration strength of HAS is smaller than that of DAS with 1% cement added.

Fig. 4 plots the variations of cone penetration resistance against the dry density for both kinds of PFA. These results indicate the following:

1) it can be seen from Figs. 4a and 4b that the CPT strength increases with the addition of cement; judging from the fact that the calcium oxide (CAO) content increases with the cement content, it is thought that CAO must contribute to the increased bearing capacity of PFA;

2) when we compare the q_c versus ρ_d relations for HAS and DAS shown in Fig. 4b, the cone resistance of PFA increases with increasing the dry density; but this effect is less significant than that of cementation;
3) it is therefore concluded from the above-mentioned findings that the cone resistance of PFA relies upon cementation and dry density.

3.2 Results from Plate Loading Tests

Figs. 5a and 5b illustrate the results from plate

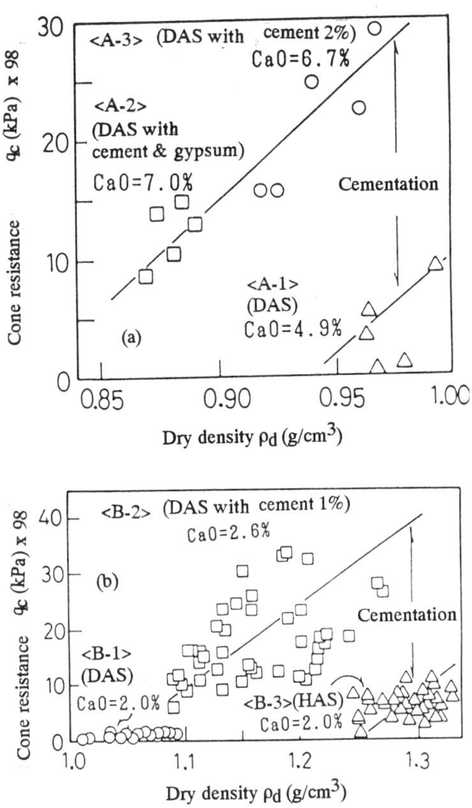

Fig. 4 Relations between q_c and ρ_d ((a) fly ash-A, (b) fly ash-B)

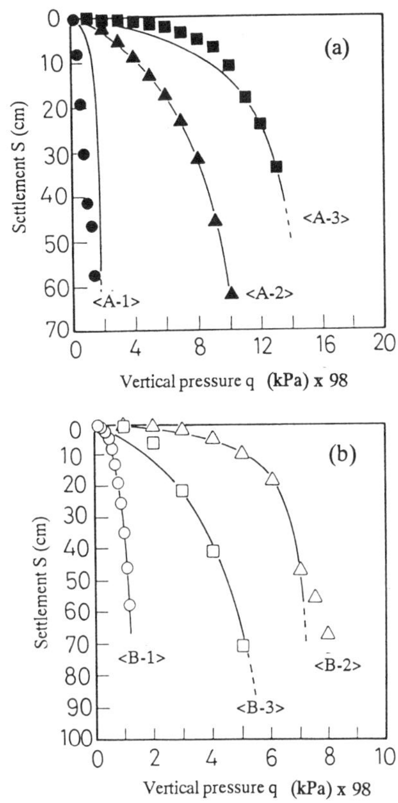

Fig. 5 Hyperbolic approximation for load versus settlement curves ((a) fly ash-A, (b) fly ash-B)

loading tests on the PFA model using the dry disposal method. The bearing capacity in these figures is defined by the inflection point, that is, yielding point in the settlement versus vertical stress curves. The results from Fig. 5a indicate that the bearing capacities of PFA with 2% cement and 2% gypsum added (test-A2) and with only 2% cement added (test-A3) increase from 100 kPa up to 750 kPa and 950 kPa, repectively, while the bearing capacity of PFA with no additives (test-A1) remains small (approximately 100 kPa). This difference is simply because of the increased CAO content which contributes to the self-hardening effect of the PFA.

Fig. 5b show the results from plate loading tests on PFA-B in which the amount of CAO content is smaller than that in PFA-A. The results in Fig. 5b, showing that the bearing capacity of PFA-B is larger than that of PFA-A, reflects this difference in CAO content. When we compare the results in Figs. 5a and 5b, the slurry fly ash is endowed with a higher bearing capacity than the dry fly ash but a smaller one than the dry fly ash with cement.

In order to further interpret the results from plate loading tests on each type of PFA, we have attempted to approximate the settlement versus load curves in Fig. 5a and 5b using the hyperbolic formulation, following the proposals by Yamaguchi (1977) and Kusakabe and Kawai (1989) given by:

$$S = \frac{p/K_i}{1 - p/p_f} \tag{1}$$

where p and p_f are current and ultimate vertical stress, and K_i is the coefficient of ground reaction and corresponds with the modulus of stiffness in Fig. 6 determined from the results of undrained triaxial tests. Although the results are limited, we obtain From Fig. 6 the following simple relation:

$$K_i = 0.25\ G \tag{2}$$

3.3 Analysis of the Results from plate loading tests

In analysing the results from plate loading tests on

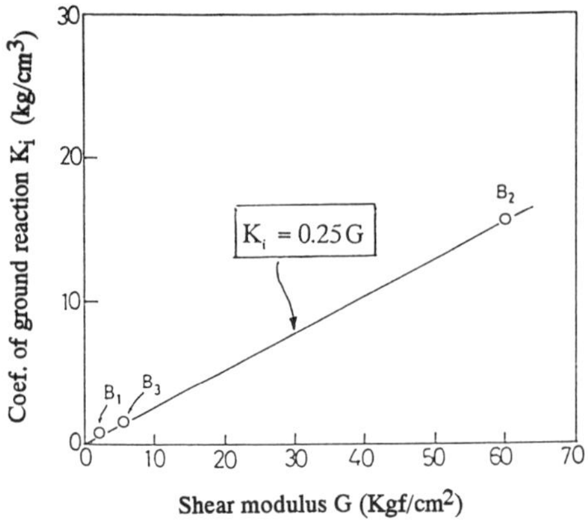

Fig. 6 Relation between coefficient of K_i and G

soils, we should consider the compressibility of soils. However the effect of compressibility is not considered in the conventionally used Terzaghi theory of bearing capacity because the soil is assumed to behave as a rigid-plastic body. To eliminate this deffect and to take into account the effect of compressibility in evaluating bearing capacity Vesic (1972, 1973) extended the Terzaghi theory into:

$$q_d = cN_c\zeta_c\zeta_{cc} + qN_q\zeta_q\zeta_{qc} + 0.5\gamma BN_\gamma\zeta_r\zeta_{rc} \tag{3}$$

where ζ_{cc}, ζ_{qc}, and ζ_{rc} are coefficients relating the compressibility of soils given by:

$$\zeta_{qc} = \exp\left[\left\{-4.4+0.6\frac{B}{L}\right\}\tan\phi' + \left\{\frac{(3.07\sin\phi')(\log 2I_r)}{1+\sin\phi'}\right\}\right] \tag{4a}$$

$$\zeta_{cc} = \zeta_{qc} - \frac{1-\zeta_{qc}}{N_q \tan\phi'} \tag{4b}$$

$$\zeta_{rc} = \zeta_{qc} \qquad (\leq 1) \tag{4c}$$

In Eq. (4a) I_r is a rigidity index included in the Vesic (1972) cavity expansion theory and is given by:

$$I_r = \frac{G}{c' + p\tan\phi'} \tag{5a}$$

where G is shear modulus, c' is cohesion ϕ' is internal friction angle and $p = (1+2K_o)\sigma_v/3$ (K_o is coefficient of earth pressure at rest; σ_v is effectice overburden pressure). When we consider the compressibilty of ground into consideration, the following relation instead of Eq. (5a) should be used:

$$I_{rr} = \frac{I_r}{1+\Delta_{av}} \tag{5b}$$

When we adopt Eq. (3) for evaluating the bearing capacity of PFA, we need the following parameters : (i) strength parameters in terms of the effective stress, c' and ϕ' : (ii) shear modulus, G, (iii) a constant related to the compressibity and (iv) unit volume weight, γ'. Among the above, the shear modulus and compressibility constants are included in the following relations:

$$G = \alpha_1\left(\frac{p}{\sigma_o}\right)^{\beta_1} \tag{6}$$

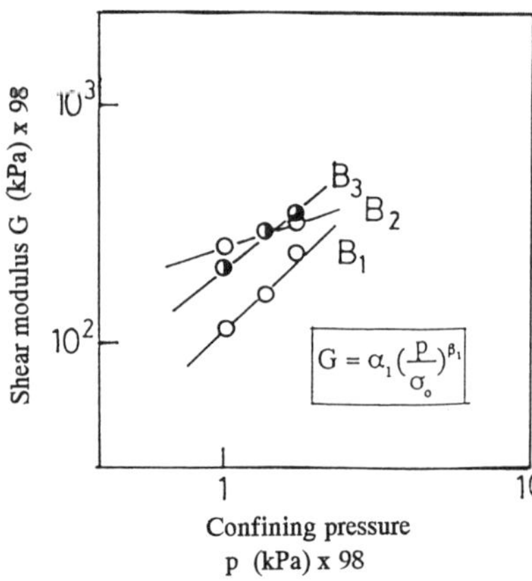

Fig. 7 Shear modulus versus confining pressure relations obtained from undrained triaxial tests

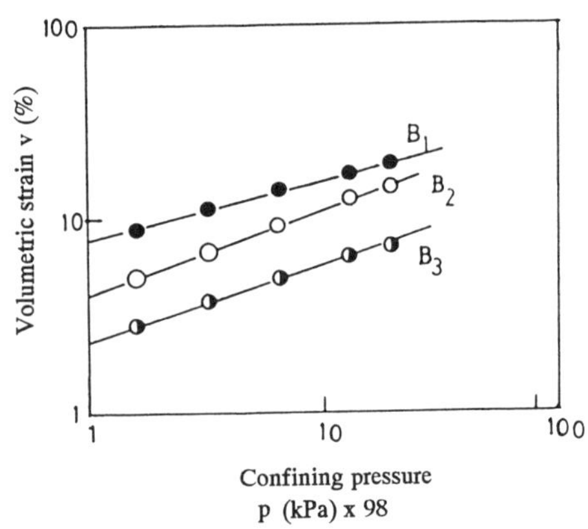

Fig. 8 Volumetric strain versus confining pressure relations in oedometer tests

$$\Delta_{av} = c_o(F_m - 1)(\frac{p}{\sigma})^m \tag{7}$$

where :

$$F_m = \frac{3(1+\sin\phi')}{3+(3-4m)\sin\phi'} \tag{8}$$

Experimental constants, a_1, b_1, c_o and m included in both equations are determined from Fig.7 and Fig. 8, using the results of consolidation and triaxial tests, respectively. Each value used for the present analysis is summarized in Table 2.

Bearing capacities for three cases calculated using Terzaghi equation and Vesic's Eq. (3) are summarized in Table 3 and compared with those observed in plate loading tests. Although the diameter of the loading area for the plate loading tests in the laboratory was restricted to only one of 15cm, the calculated results include the variations of bearing capacity with the diameter. The upper values in Table 2 show the results calculated using Terzaghi's equation which does not consider the soil compressibility, while the lower value shows the results calculated using Vesic's equation which takes it into consideration. In comparing both values, it is pointed out that in the case of the Vesic's value being larger than the Terzaghi one, that the PFA ground should be regarded as the non-compressible and can be evaluated using Terzaghi's conventional bearing capacity theory. The small value with the underline in both columns was adopted for prediction. Thus the calculated bearing capacity are plotted against the diameter of loading plate in Fig. 9. The following is also indicated from Table 3 and Fig. 9:

Table 2 Material parameters in analysis

	B1	B2	B3
ϕ' (°)	17.2	21.5	36.5
c' (kPa) x 98	0.08	0.40	0.08
α_1	111.0	252.0	205.0
β_1	0.946	0.332	0.787
c_0	0.078	0.040	0.023
m	0.307	0.441	0.413
γ (g/cm^3)	1.50	1.66	1.74

Table 3 Predicted and experimental results of limited bearing capacity

Diameter of loading plate B(m)	Experimental results (kPa) x 98 B1	B2	B3	Calculated results (kPa) x 98 B1	B2	B3
0.15	1.47	7.76	7.02	<u>1.4</u>*	<u>9.6</u>	<u>8.3</u>
				3.8**	37.7	14.6
1.00	-	-	-	<u>1.7</u>	<u>10.2</u>	<u>13.1</u>
				3.9	28.1	17.2
5.00	-	-	-	<u>3.2</u>	<u>12.9</u>	<u>35.6</u>
				6.0	25.8	36.0
10.00	-	-	-	<u>4.9</u>	<u>16.3</u>	63.7
				8.6	28.2	<u>57.3</u>
20.00	-	-	-	<u>8.5</u>	<u>23.0</u>	119.9
				13.7	34.4	<u>95.8</u>

* : Limited bearing capacity calculated by Terzaghi's formula

** : Limited bearing capacity calculated by Vesíc's formula

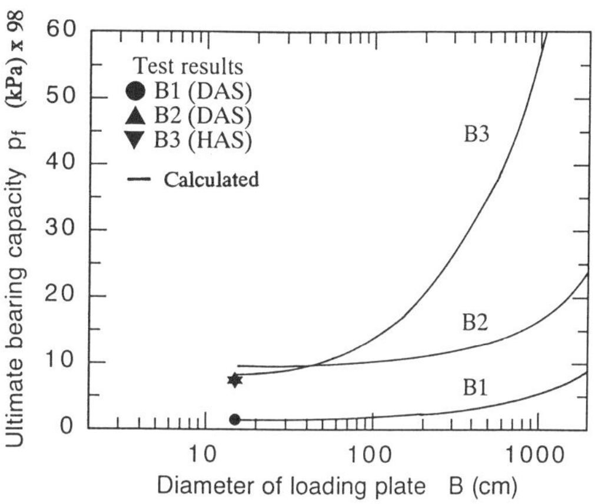

Fig. 9 Comparison between calculated and observed results of utlimate bearing capacity

i) the bearing capacity calculated using Terzaghi's equation agrees well with that observed in plate loading tests;

ii) it can be said from the calculated results for bearing capacity of PFA that not much of an increase in bearing capacity for B1 (DAS) and B2 (DAS with cement) with increasing diameter of loading plate should be expected and on the other hand, the bearing capacity of HAS increases considerably with increases in the diamater of loaded area, B;

iii) when the diameter of the loading plate is beyond 5.0 m, the bearing capacity from Terzaghi equation is larger than that from Vesic equation by Eq. (3). This suggests that it is necessary to consider the compressibility when evaluating the bearing capacity in the case of larger diameters of loaded plate as may occur in the field.

CONCLUSIONS

1) An increase in bearing capacity of PFA is caused by both increased dry density and enhanced or added

cementation.

2) In this respect, the slurry disposal method (HAS) for PFA is best suited for producing stable reclaimed land.

3) Terzaghi's classical theory of bearing capacity is sufficient for predicting the bearing capacity of PFA for 1g laboratory model tests under low confining pressures and with a small diamater of loaded area. However, Vesic theory which takes material compressibility into consideration is necessary for predicting the bearing capacity in the field under high confining pressures and larger loaded areas.

REFERENCES

Horiuchi, S. et al. (1985) : Rheological and mechanical properties of high density fly ash slurry, Proc. 7th Intn'l Ash Utilization Symp., pp. 907 - 917.

Horiuchi, S., Tamaoki, K. and Yasuhara, K. (1995) : Coal fly ash slurry for effective underwater disposal, Soils and Foundations, Vol. 35, No. 1, pp. 1 - 10.

Kusakabe, O. and Kawai, N. (1989) : Reduction in bearing capacity and increase in settlement of a footing due to upward seepage flow, Journal of JSSMFE (Tsuchi to Kiso), 37-6 (No. 377), pp. 57 - 62 (in Japanese).

Terzaghi, K. (1943) : Theoretical Soil Mechanics, John Wiley & Sons Inc., New York.

Vesic, A. S. (1972) : Expansion of a cavity in an infinite soil mass, J. SMFD, Proc. ASCE, Vol. 98, No. SM3, pp. 265 - 290.

Vesic, A. S. (1973) : Analysis of ultimate loads of shallow foundations, J. SMFD, Vol. 99, No. SM1, pp. 45 - 73.

Yamaguchi, H. (1977) : Panel Discussion, Proc. 9th ICSMFE, Vol. 3, pp. 382 - 384.

Yasufuku, N. (1990) : Yielding characteristics and elasto-plastic constitutive equation for anisotropically consolidated sand with a wide stress range, Ph D Thesis, Kyushu University, pp. 210 - 238 (in Japanese)).

Yasuhara, K., Hirao, K., Hyodo, M. and Horiuchi, S. (1990) : Fundamental research on the utilization of pluviated coal fly ash for reclamation, Proc. 34th Japan Geotechnical Eng. Symp., pp. 341 - 348 (in Japanese).

Yasuhara, K. et al. (1994) : Liquefaction characteristics of coal fly ash as a reclamation material, Proc. 1st Intl. Conf. Environmental Geotechnology, Vol. 1, Edmonton, Canada.

Yasuhara, K. and Horiuchi, S. (1996) : Effective use of flyash slurry as fill material, Proc. Intn'l Symp. Clean Coal Tech. Symp. on Coal Ash Utilization, Vol. 1, pp. 81 - 95.

Environmental Geotechnics, Kamon (ed.) © 1996 Balkema, Rotterdam. ISBN 90 5410 848 7

Strength characteristics of surplus quarry soil with slag added

Kinutada Yokota, Norio Yagi & Ryuichi Yatabe
Ehime University, Faculty of Engineering, Matsuyama, Japan

ABSTRACT: Nowadays, industrial wastes are produced in Japan where is difficult to dump easily. Every product needs to be utilized to protect the environment. A slurry produced from the crushed stone business is increased with a development of industrial action. But, the slurry is not utilized effectively as a recycle material. We tried to make materials of a high compressive strength by adding Portland-cement, slag and gypsum to the slurries.
In this study, the authors used three kinds of slurry made from sandstone, andesite and decomposed granite soil, which are widely distributed in Shikoku Island. The generation of strength was different depending on the kind of slurry because the pozzolanic reaction is different depending on the mineral components of the slurry. The authors checked the influence of the mineral components on the generation of strength. The results show that the presence of mica results in a decrease in strength while the presence of montmorillonite makes results in an increase in strength of slag with and gypsum added.

1 INTRODUCTION

On account of increasing in quantity of discharged industrial wastes, distraction of environment and increasing of cost for treatment and disposition are in full swing. Therefore, Low for the Promotion of Utilization of Recyclable Resources (1991) and Waste Disposal and Public Cleaning Low (1992) were put into operation. Constructive wastes are larger percentage of the quantity of wastes and as a type of industry appointed according to the law, construction industry is powerfully wanted to use actively recycling natural resource and to reduce the quantity of wastes.
It was tried that the slurries produced from crushed stones are changed to useful materials adding cement, slag and gypsum. Slag and gypsum are also by-product materials. The materials made by this method are expected to have high strength.

2 SITE OF SAMPLING PLACES

Shikoku Island is more than 80% covered with mountains. Therefore, crushed stones for aggregate of concrete are produced at many places. Three kinds of slurry generated during the production of crushed stone at three places in Ehime Prefecture show in Fig. 1 were gathered.
The first, second and third samples which consist of decomposed granite soil, sandstone and andesite, respectively, were gathered at Hojo, Kawauchi and Kuma, respectively. These were called the Hojo, Kawauchi and Kuma samples, respectively.

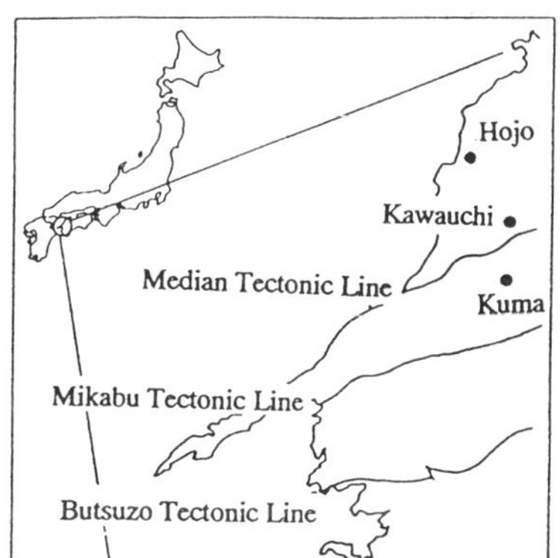

Fig. 1 Sampling site.

3 PHYSICAL AND GEOLOGICAL PROPERTIES OF SLURRIES AND ADDED AGENTS

The physical properties of each slurry are shown in Table 1. The liquid limit and the plastic limit are the same as those of a general clayey soil. The slurry at each site was like cake of which water content is about 30%. The specific gravity of each sample is slightly larger than that of the general soil. The grain size distribution curves of the slurries are shown in Fig. 2. About 35~40% has a grain size smaller than 5 μ corresponding to clay particle. From Fig. 2 these slurries are classified as clays or silt clays in terms of soil engineering.

From X-ray of analyse it was found that the Hojo sample is composed of quartz, feldspar and mica, the Kawauchi sample is composed of quartz and feldspar and the Kuma sample is composed of montmorillonite, feldspar and mica.

Portland-cement, hydrated slag, gypsum slag and gypsum were used as additive agents.

Slag and gypsum are also industrial wastes. The pozzolanic reaction due to Ca and Al which is effective for strengthening a material such as sludge, occurs in these materials.

Table 1 Physical properties of specimens.

Clay	Gs	L.L.(%)	P.L.(%)	Ip
Hojo	2.778	56.5	28.9	27.6
Kuma	2.739	52.6	30.8	21.8
Kawauchi	2.689	31.2	18.7	12.5

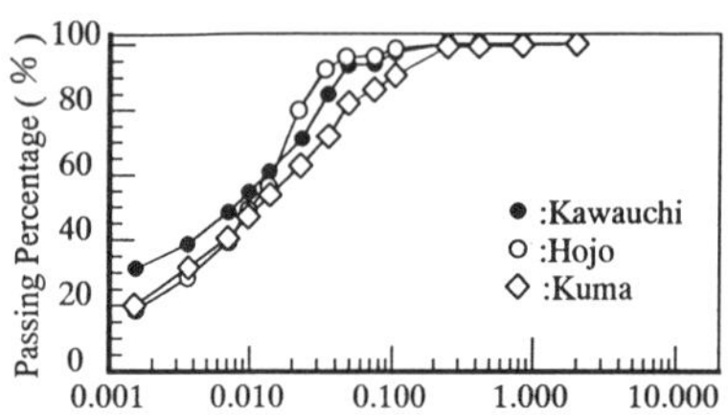

Fig. 2 Grain size distribution curve.

4 MECHANICAL PROPERTIES OF THE SLURRIES

A high strength slurry material can be made by consolidating it under high pressure. Then we tried consolidation tests for the slurries under very high pressure using the oedometer. The relationship between void ratio and consolidation pressure, that is the $e \sim \log p$ curve for the Kawauchi sample was

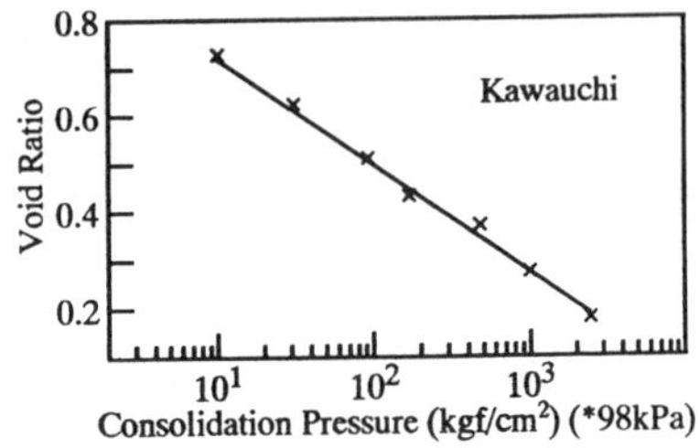

Fig. 3 e - log p curve.

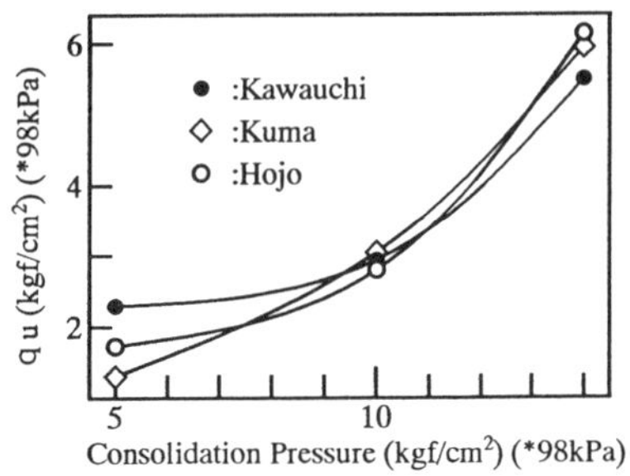

Fig. 4 Unconfined compressive strength of soil samples.

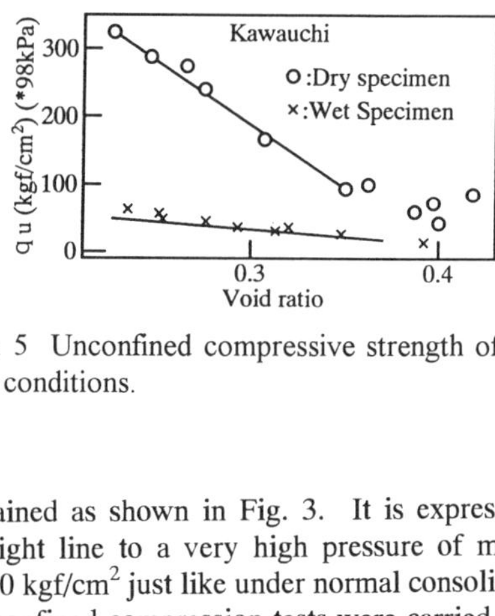

Fig. 5 Unconfined compressive strength of wet and dry conditions.

obtained as shown in Fig. 3. It is expressed by a straight line to a very high pressure of more than 1000 kgf/cm^2 just like under normal consolidation.

Unconfined compression tests were carried out after consolidating the slurries under high pressure. The increasing of the unconfined compressive strength q_u with the consolidation pressure p_c less than 14 kgf/cm^2 is shown in Fig. 4. The ratio q_u/p_c generally decreases with increase of the p_c in clay soil but vice versa in the slurry. Also the relationship between q_u and void ratio e compressed under very high pressure is shown in Fig. 5. In this figure the unconfined compressive strengths q_u of the specimens dried in the air after consolidation under various pressures are also plotted. It can be seen from Figs. 3,4 and 5 that the q_u/p_c of the specimens consolidated under less than 20 kgf/cm^2 is relatively large, like 0.3 ~ 0.4, and are while those,consolidated under more than 100kgf/cm^2 are small, like less than 0.1. So high strength can be obtained by drying the specimens after consolidat under very high pressure. But this procedure is considered to be very expensive, therefore we tried to get high strength slurries by mixing agents in them.

5. UNCONFINED COMPRESSIVE STRENGTH OF THE SLURRIES WITH ADDED AGENTS

5.1 Method of making specimens

The method of making the slurry specimens with some agents added is as follows.

1) The agents, that is cement, slag and slag plus gypsum,are added to the slurries in various percentages by weight of the solid part.
2) These samples are mixed at water contents a little over the liquid limits.
3) The samples are compressed under the appropriate pressure of about 13 kgf/cm2 to make the specimens.
4) The specimens are cured for various periods.
5) The specimens are cut to the required diameter and height.
6) The dillerent kinds of specimen are shown in Table 2.

5.2 Comments on the unconfined compressive strengths

The increase in unconfined compressive strength q_u of each specimen with increase in the percentage of cement content is shown in Fig. 6. The Kawauchi and Kuma specimns have nearly same q_u which is greatter than that of the Hojo specimen.

The relationship between the q_u and the slag content is shown in Fig. 7. The addition of slag is not effective for increasing the q_u of the Kuma and the Hojo specimens, although it has nearly the same or more effect as cement for the Kawauchi specimen. But when the slag and the gypsum are added to the slurries, the q_u values of the Kawauchi specimens

Table 2 Rate of addition of slag, cement, gypsum.

slag %(W)	cement %(W)	gypsum %(W)	remark each sample
3	3	0.3	
5	5	0.3	ϕ=3.0cm
10	10	0.3	H=6.0cm

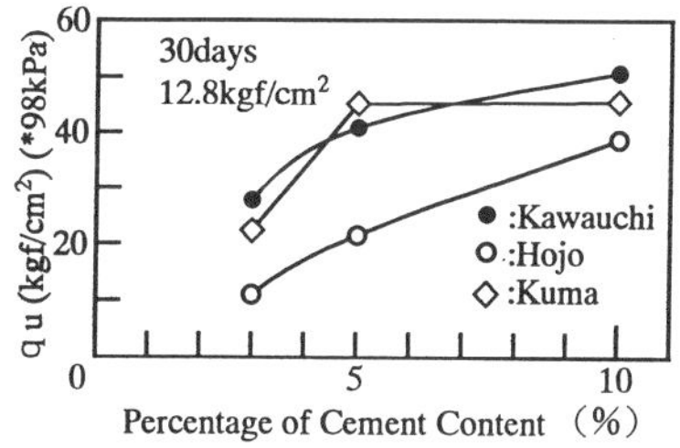

Fig. 6 The difference of unconfined compressive strength by adding of cement.

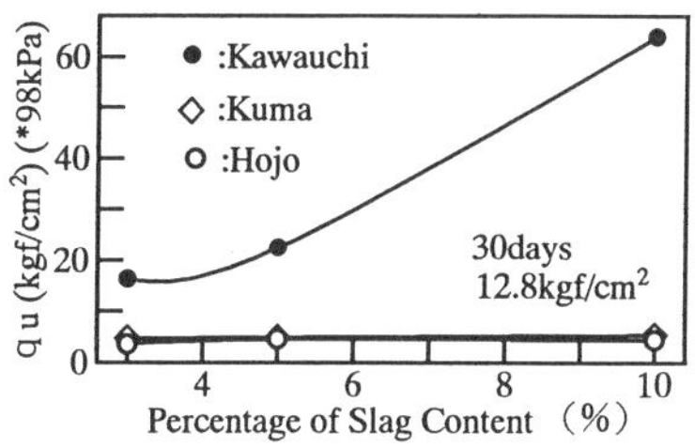

Fig. 7 The difference of unconfined compressive strength by adding of slag.

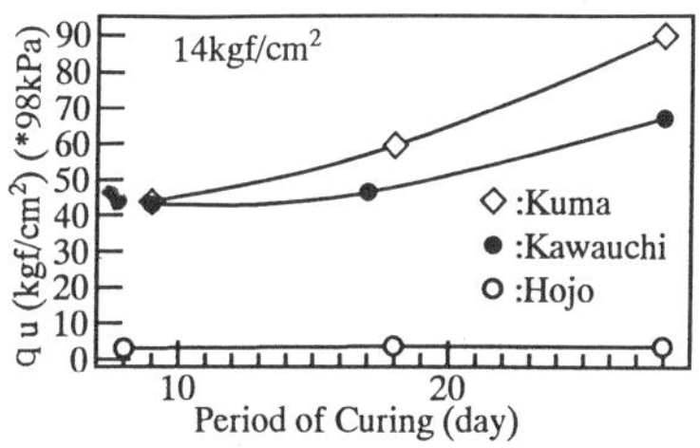

Fig. 8 The difference of unconfined compressive strength by adding of slag and CaO

increase, although there is still no change in q_u values of the Hojo specimen as shown in Fig. 8. The ratio of the slag and the gypsum by weight is nine to one.

Changes of water content during curing are shown in Fig. 9 for the specimens to which the slag is added. The water content of the The Kawauchi sample specimen decreases a little with curing period, though there are no changes in the water content of the Kuma and Hojo specimens. This may be one of the reason why the q_u of only the The Kawauchi sample specimen increases when adding the slag.

It is considered that one of the reasons for the differences between the unconfined compressive strengths when adding the agents is the presence of minerals like mica and montmorillonite contained in the samples. Therefore, the unconfined compression tests were carried out on Kawauchi specimens containing slag and gypsum to which different amount of mica and montomorillonite were added. The Kawauchi sample sample was used. The results of the unconfined compressive tests are shown in Fig. 10. From the results it is estimated that mica reduced the q_u value while montmorillonite increases it when slag and gypsum are the added strengthening agents.

The mineral composition of these slurries was investigated by X-ray analyses. The results of the X-ray analyses are shown in Fig. 11 from which the minerals contained in each sample are given in Table 3.

Increase in unconfined compressive strength q_u is expected by the hydaulicity and the pozzolanic reaction. The pozzolanic reaction is promoted in an alkali material. The slurry samples have pH values ranging from 7.3 to 8.95. The Kawauchi and the Kuma samples have high pH values. Therefore, it is considered that the Kawauchi and the Kuma specimens to which cement, slag and gypsum have been added as strengthening agents can result in high q_u values. The reason that the Kawauchi sample specimen exhibits high q_u with every agent, is because the minerals of quartz and feldspar are only contained in it. On the other hand, the Kuma and the Hojo samples contain montomorillonite and mica, respectively as well as quartz and feldspar. Fig. 10 shows that montomorillonite is effective in increasing the q_u valuce while mica is less effective. Therefore, adding the agents slag or slag and gypsum to the Hojo specimen does not increase its q_u value.

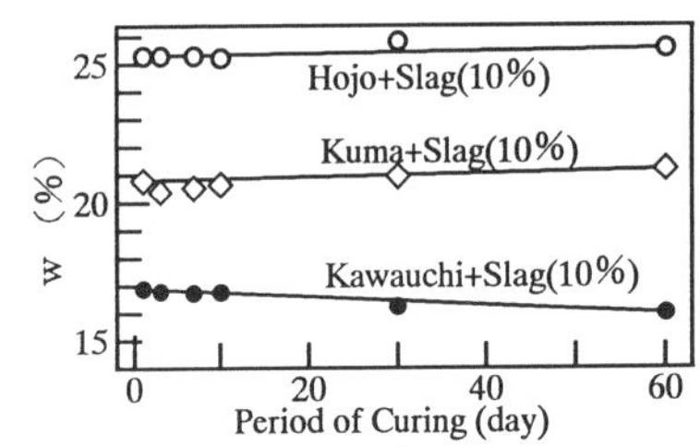

Fig. 9 The Change of moisture content.

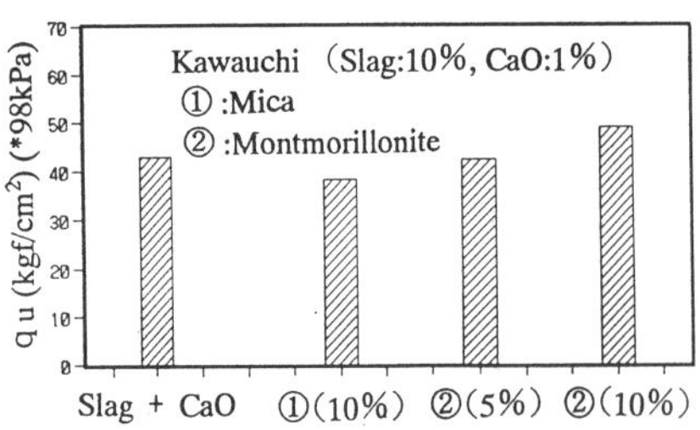

Fig. 10 The difference of unconfined compressive strength by adding of mica and montmorillonite.

Table 3 Mineral component.

Sample	mineral
Kawauchi	quartz > feldspar >
Kuma	quartz > feldspar > montomorillonite
Hojo	quartz > mica > feldspare

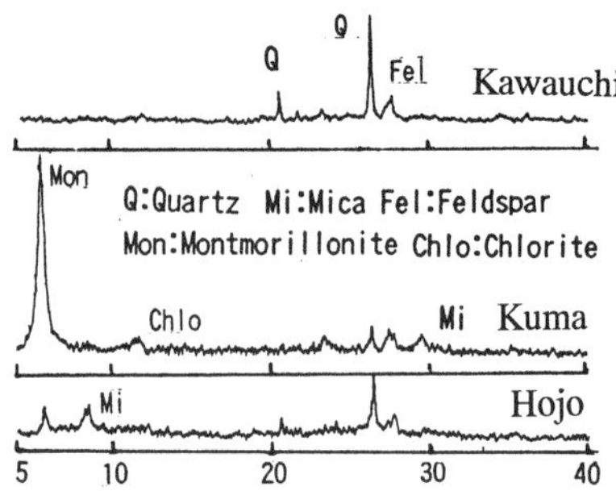

Fig. 11 The chart of X-ray.

6. CONCLUSIONS

The following results were obtained:

1. Pozzolanic reaction by the addition of a slag is notable in the Kawauchi and the Kuma specimens but is not notable in the Hojo specimen. The Kuma and the Kawauchi samples are alkali soils but the Hojo sample is not alkaline.
2. The water content and void ratio of the Kawauchi specimen are the smallest of the three samples and the Kawauchi and the Kuma specimens are stiffened by the products which are generated by the pozzolanic reaction.
3. The unconfined compressive strength of the Kuma specimen is increased by the addition of montmorillonite, because the pores of the specimen are filled by the montmorillonite which is swelled by Ca ions from the slaked lime.
4. The unconfined compressive strength of the Hojo specimen is reduced by the addition of mica.

References

G. Kuno. , M. Sato. 1984. Hydraulic hardening of blast furnace slags by means of curing procedures.No.19 Annual proceeding in domestic meeting of JSFE. 1649-1650.

K. Hiroi. , K. Komanobe. , K. Segawa. 1994. Study of effective utilization for crushed stone powder. No.29. Annual proceedings in domestic meeting of JSFE. 2237-2238.

S. Seike. 1986. Dynamics of Material. 143-1 48, Kyoritsu Co.

Environmental Geotechnics, Kamon (ed.) © *1996 Balkema, Rotterdam. ISBN 90 5410 848 7*

Characterization of fly ash as construction material

Yudhbir
Indian Institute of Technology, Kanpur, India

S.K. Das
Formerly: Indian Institute of Technology, Kanpur, India

ABSTRACT : For 3 low calcium and high carbon and one high calcium and low carbon Indian flyashes, Electron Dispersive X-ray (XRD) and Xray Diffraction (XRD) analyses have been used to evaluate chemical and mineralogical characterictics of both the total fly ash and individual grains. Indirect methods based on XRD patterns are used to determine nature and amount of glass content of fly ashes. A relationship between chemical composition (k/A ratio) and glass content for low calcium fly ashes is suggested as a useful tool for estimation of glass content. Free lime content and carbon content of fly ashes are shown to govern their gain in strength with time pattern and pozzolonic reactivity.

1 INTRODUCTION

Large quantities of poor quality low calcium, high carbon - fly ashes are produced annually in India. Limited reuse of these poor fly ashes and storage practices - open dumping and discharge as slurry in ash ponds pose serious economic and environmental problems. Nontananandh (1990) and Yudhbir and Honjo (1991) have stressed the need to use this poor quality industrial waste as a resource construction material. Systematic characterization of fly ashes in respect of physical chemical and mineralogical composition is necessary to categorise fly ash in terms of their use as construction materials.

In this paper results of chemical and mineralogical characterization of four Indian fly ashes are presented. Methods to estimate quantity and nature of glass content of fly ash are discussed. Self hardening potential and pozzolonic reactivity of fly ashes are explained in terms of chemical, mineralogical and carbon content of fly ashes.

2 FLY ASHES INVESTIGATED

The fly ashes were obtained directly from hoppers from four power plants : Choudwar (Orrisa), Parichha and Panki (U.P.), (all fired by high carbon anthracite coal) and Neyveli (Tamilnadu) (fired by lignite). Choudwar, Parichha and Panki fly ashes are low calcium, high carbon variety with little reuse whereas Neyveli is low carbon, high calcium fly ash which has extensive reuse as substitute for cement.

3 CHARACTERIZATION METHODS

Das (1992) studied grannulometry, chemistry and mineralogy of these fly ashes. Electron Dispersive X-Ray (EDX) was used for chemical analysis and X-Ray Diffraction (XRD) was used for mineral analysis and determination of glass phase. Details of these studies on bulk and individual grains and floaters along with morphological characteristics evaluated by Scanning Electron Microscope (SEM) are reported by Das (1992).

4 TEST RESULTS AND DISCUSSION

Fig.1 shows particle size gradation from hydrometer test. Coulter counter method using Methanol to prevent grain aggregation of high calcium fly ash gave fines content (<0.02 mm size) 15-20% higher for Neyveli sample (see Das 1992). EDX spectra and RDX patterns are shown in Figs. 2 and 3 (a) respectively. Chemical and mineralogical composition of coarse and fine fractions of Neyveli ashes is shown in Fig. 4. XRD patterns and location of hump or diffused band for fly ashes with different CaO content are depicted in Fig.5.

Choudwar, Parichha and Panki are characterized by 60-65% Silica (SiO_2), 18-28% Alumina (Al_2O_3), 4-8% iron (Fe_2O_3), 1-2.6% CaO, 2-8% carbon and traces of SO_3, TiO_2, MgO, Na_2O, K_2O. Neyveli on the other

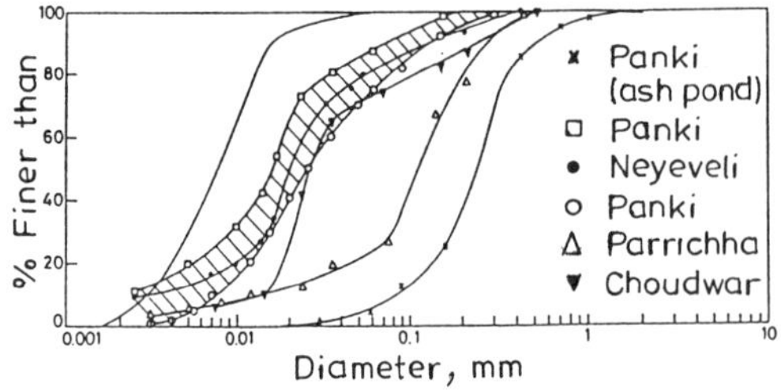

Fig. 1 Particle size distribution

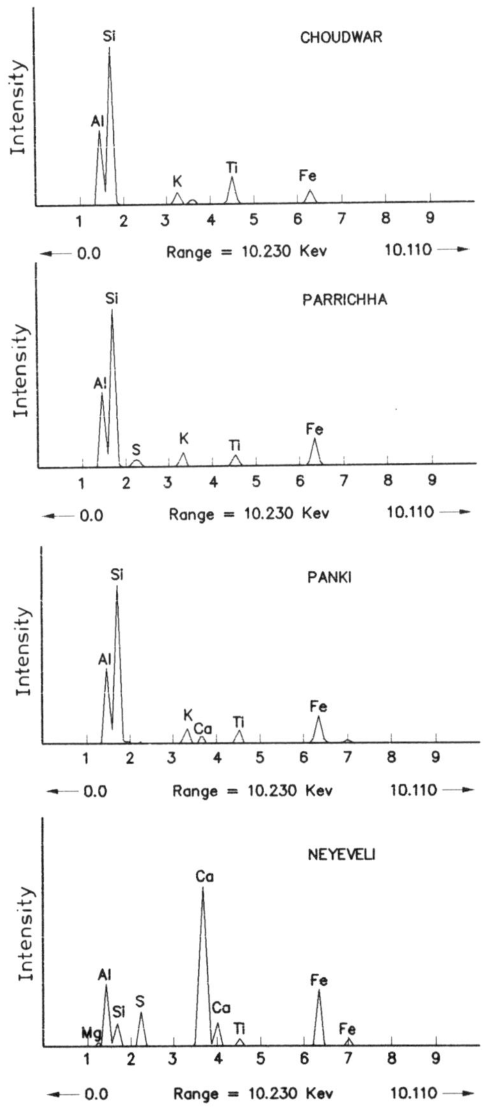

Fig. 2 EDX spectra

hand has 25-28% SiO_2, 23-29% Al_2O_3, 24-28% Fe_2O_3, 16-18% CaO, 1.5% carbon, 0.95% SO_3, and traces of MgO, TiO_2, etc. This gross chemical composition is based on chemical extraction analysis, whereas the EDX spectra (Fig.2) helps to investigate chemistry of individual grains, floaters and coatings on these (Nontananandh 1990, Das 1992)., Fig. 2 shows Choudwar has no calcium whereas Neyveli has very high calcium content and Panki and Parichha have intermediate values. Choudwar, Panki and Parichha have similar portions of Si and Al as compared to much smaller Si Value for Neyveli. XRD patterns (Fig. 3 (a)) reveal that low calcium, high carbon fly ashes are composed of predominantly inert crystalline minerals - quartz, mullite, magnetite and hematite - whereas high calcium, low carbon Neyveli ash contains reactive minerals such as calcium aluminate (C_3A), calcium silicate (CS) and anhydrite ($C\bar{S}$) (See Mehta 1979, 1985 and Lüxan et al (1988). Differences in chemical and mineralogical composition betwen coarse (>75μ m) and fine (< 45μ m) fractions of Neyveli are highlighted in Fig. 4. The coarse fraction is rich in Si and Fe (Fig. 4 (a) and inert minerals quartz and hematite (Fig. 4 (b) and the fine fraction is rich in Ca and Al (Fig. 4 (a)) and reactive minerals C_3A, CS and $C\bar{S}$ (Fig. 4 (b)). Similar studies for Choudwar, Panki and Parichha ashes did not show any significant differences between coarse and fine fractions. These results confirm earlier findings (see Das 1992) and bring out an important point that fineness of a low calcium, high carbon fly ash is no indicator of its increased reactivity. Reactivity of high calcium Neyveli is further illustrated by the increase in amount of ettringite (responsible for gain in strength with time on curing) with curing period of hydrated sample as shown in Fig. 3 (b).

Quantity and nature of glass content of fly ashes is known to be an important indicator of their reactivity (Yudhbir and Honjo 1991). XRD patterns are employed in determination of glass phase (Mehta 1979, Diamond 1983, Das 1992) and its proportion (Watt and Thorn 1965, Mehta 1985, Hubbard et al 1985, VanRoode et al 1987) in fly ashes. 2θ values for the fly ashes studied, their floaters and lime treated Panki fly ash (reburnt at ≥ 1100°C with 10% and 20% lime) are shown in Figs. 5 and 6. Results show good agreement with those reported for other fly ashes. The hump location for low calcium fly ashes varies linearly between 23° - 27° with analytical lime content and the shift may be due to structural arrangement in the siliceous glass as CaO content is increased. For high calcium fly ashes position of hump shifts to 32° due to the presence of calcium aluminate type of glass (See Diamond 1983). 2θ values are 21° for Choudwar (~

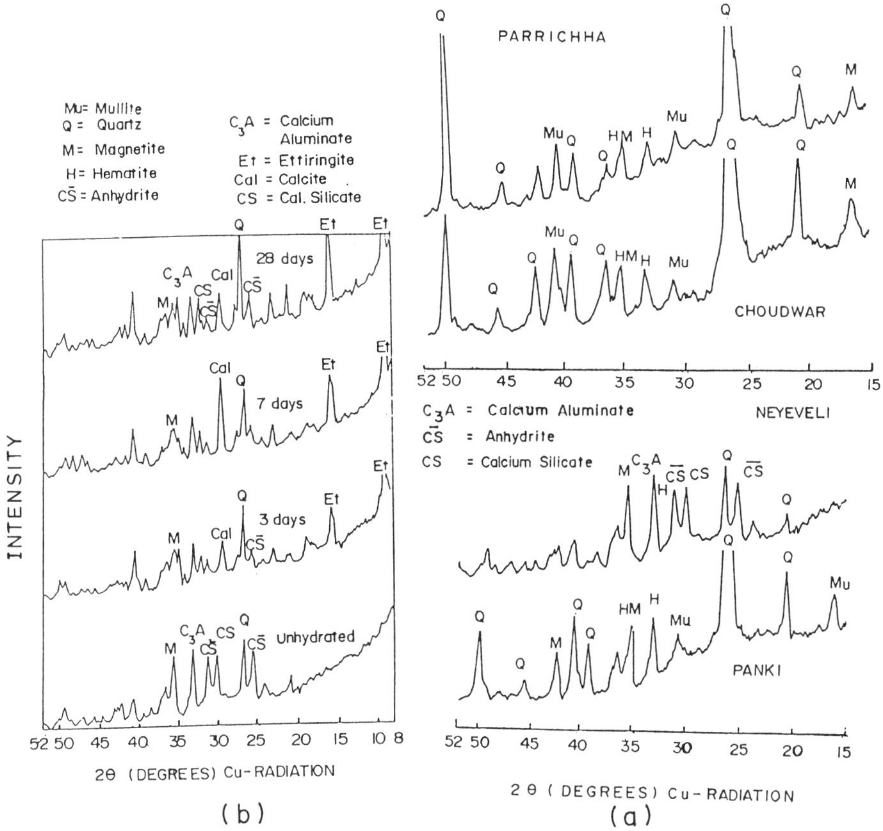

Fig. 3 XRD patterns for fly ashes (a) investigated & (b) hydrated Neyveli fly ash

nil lime), 23° for Panki and Parichha (2.5 and 1.5% lime) and 28° for Neyveli (16% lime). 2θ values for floaters are 25°-26° for Choudwar, Panki and Parichha and 32° for Neyveli. Das (1992) using EDX spectra of these floaters confirms that the Neyveli floaters are composed of calcium aluminate as compared to floaters of Choudwar, Panki and Parichha which are primarily made of alumino-silicate. The jump in 2θ values between whole fly ash and the floaters is thus indicative of major change in the quality of glass present.

The % of glass content is obtained by quantitative XRD indirect method by subtracting from 100 the sum of the % ages of crystalline minerals and loss on ignition. For low calcium fly ashes glass phase can be directly determined by dissolving fly ash with flourosilicic acid (See Valent et al 1988). Following Hubbard et al (1985) a relationship between ratio (K/A) x10 and total glass content for low calcium fly ashes is presented in Fig. 7; where K = K_2O/M.W and A = Al_2O_3/M.W, M.W being molecular weight. This relationship seems promising and may be used for estimating glass content (± 12%) in a fly ash on the basis of ratio K/A obtained from chemical composition.

5 REACTIVITY OF FLY ASHES

Characterization of a fly ash in terms of its chemistry, mineralogy, carbon content, CaO, amount and nature of glass present is very valuable in determining its self - hardening potential and pozzolonic reactivity. Figs. 8 and 9 depict self - hardening characteristics and pozzolonic reactivity of some fly ashes reviewed by Yudhbir and Honjo (1991). They proposed a linear relationship between 28 day self-hardened strength and free lime content for three categories of fly ashes as shown in Fig. 10. The categories are based on CaO and carbon content : I - CaO = 2-16%, Loss on Ignition (LOI) = 0.4 -2%; II-CaO = 1-4%, LOI = 3-7% and III - CaO = 2-4%, LDI = 10-24%. This relationship suggests that for a given free lime content 28 day

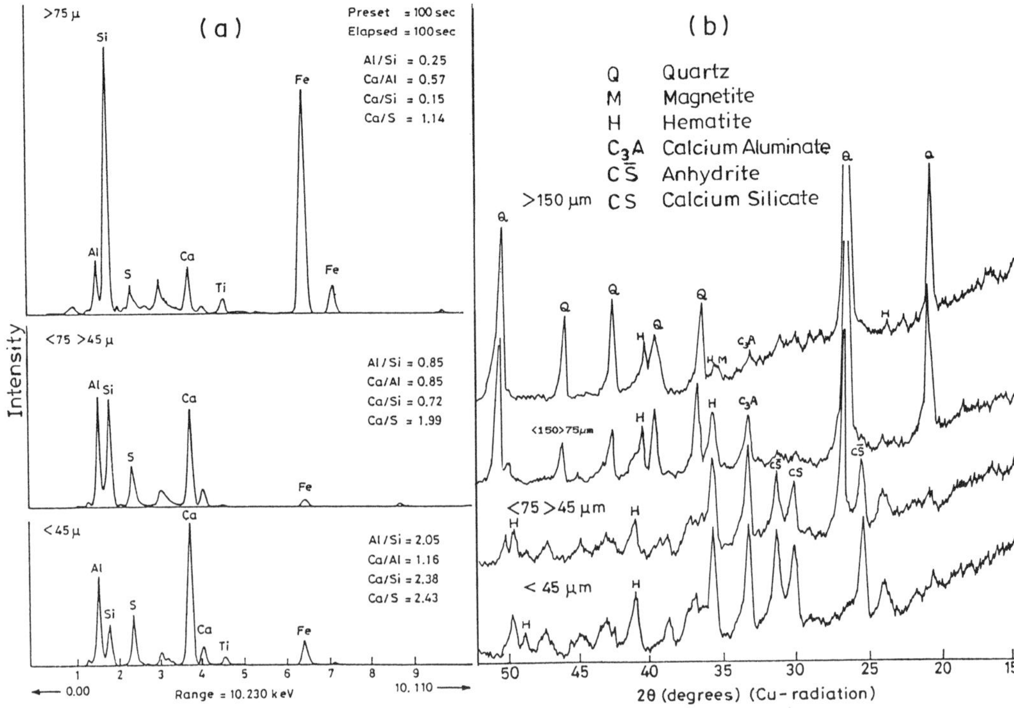

Fig. 4 (a) EDX spectra & (b) XRD patterns of different fractions of Neyveli fly ash

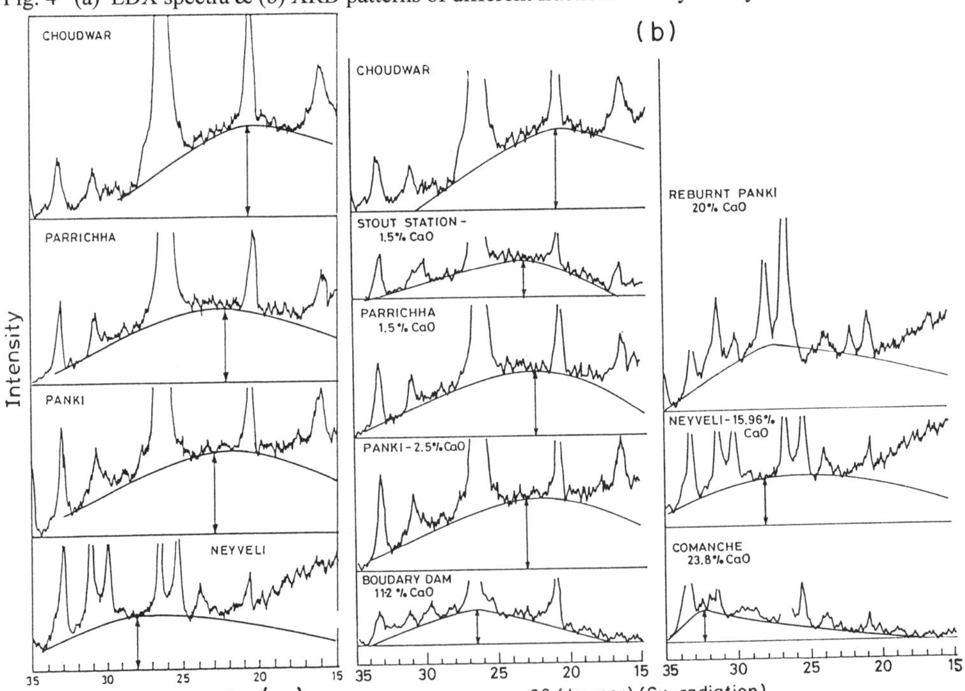

Fig. 5 (a) XRD patterns & location of hump for fly ashes investigated & (b) Comparative study of hump position for different fly ashes

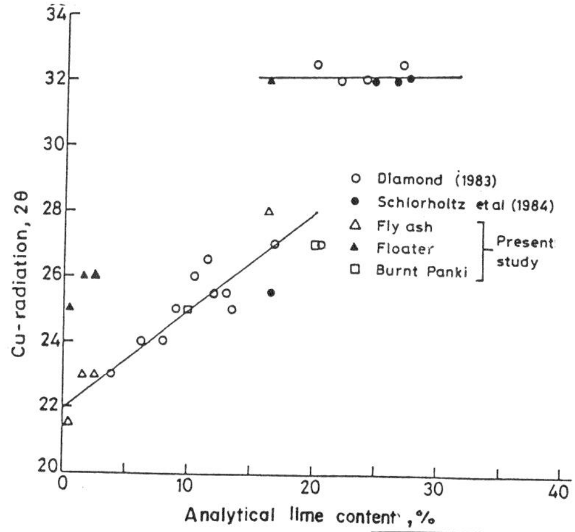

Fig. 6 Relationship between hump position (2θ Value) and analytical lime content

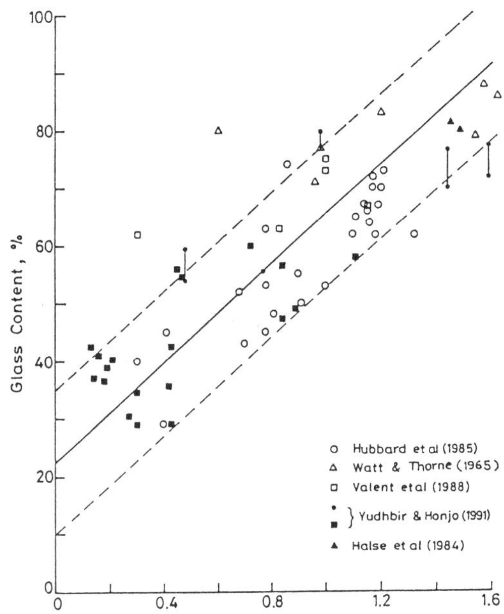

Fig. 7 Relationship between glass content and K/A ratio

strength decreases with increasing carbon content. Based on free lime content and LOI, gain in strength with time (Fig. 8) and pozzolonic reactivity (Fig.9) of different fly ashes can be evaluated.

6 CONCLUSIONS

EDX and XRD techniques to characterize fly ashes in respect of chemical and mineralogical composition are shown to be powerful tools. Nature of minerals present, chemical composition of bulk fly ashes and individual grains, and nature and amount of glass

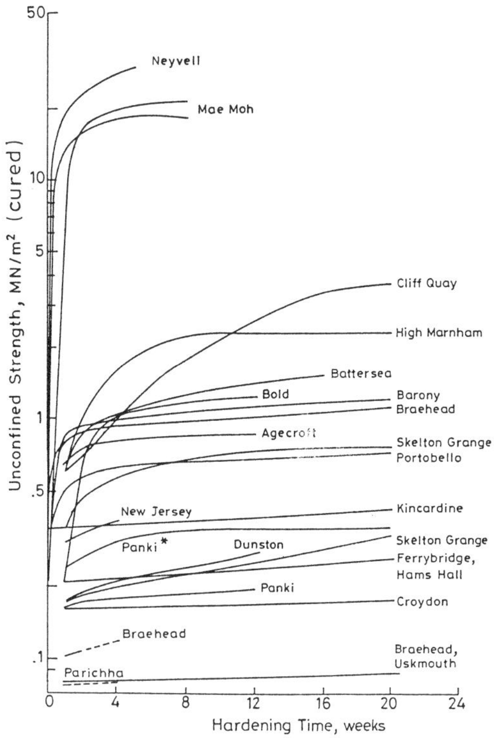

Fig. 8 Gain in compressive strength with time

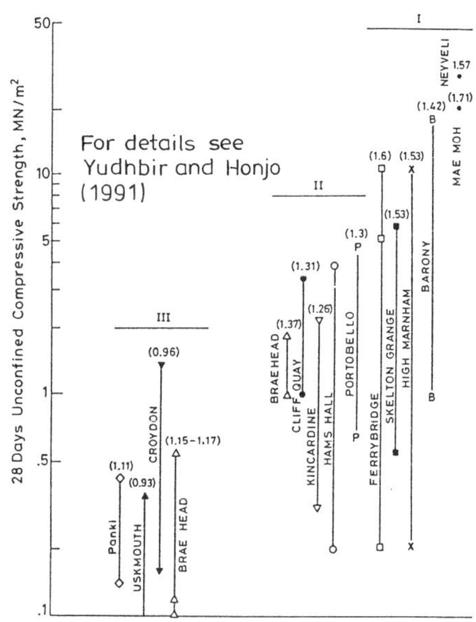

Fig. 9 Gain in compressive strength after treatment with lime

content-which strongly govern the reactivity of fly ashes-can be quantified so as to categorise fly ashes in terms of their use as construction materials. Relationship between chemical composition (K/A) and glass content is suggested as a useful aid to estimate glass content for low calcium fly ashes. Gain in strength with time and pozzolonic reactivity of fly ashes is shown to be governed by free lime content and LOI.

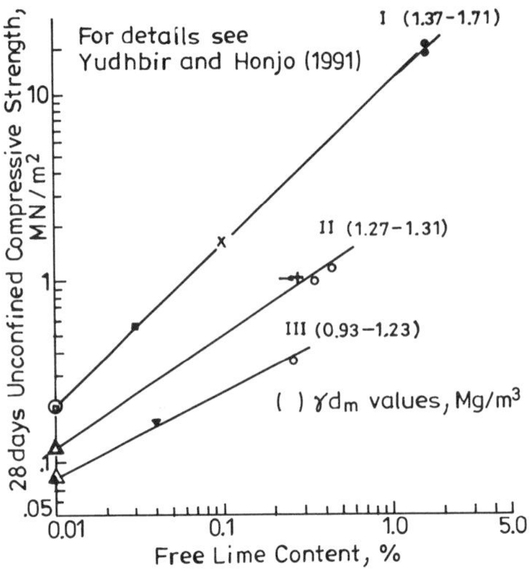

Fig. 10 Relationship between compressive strength, free lime and carbon content

REFERENCES

Das, S.K. 1992. Morphological, chemical and mineralogical characterization of some Indian fly ashes. M. Tech. thesis, civil engineering department, I.I.T. Kampur

Diamond, S. 1983. On the glass present in low and high calcium fly ashes. *Cement and Concrete Research,* U.S.A. 13:459-464

Halse, Y. et al 1984. Development of microstructure and other properties of fly ash - OPC system. Cement and Concrete Research, U.S.A. 14:491-498

Hubbard, F.H. et al 1985. Pulverized fuel ash for concrete : compositional chareaterization of United Kingdom PFA. *Cement and Concrete Research,* U.S.A. 15:534-544

Lüxan, M.P. et al 1988. Investigation of the fly ash - calcium hydroxide reactions - a critical review. *Cement and Concrete Research,* U.S.A. 19:69-80

Mehta, P.K. 1979. Pozzolanic and cementitious by - products as mineral admixtures for concrete - a critical review. *Fly ash, Silica, Fume, Slag and other mineral by - products in concrete,* 1, special publication.

Mehta, P.K. 1985 Influence of fly ash characteristics on the strength of Portland - fly ash mixture. *Cement and Concrete Research,* U.S.A. 15:669-674

Nontananandh, S. 1990. Industrial waste utilization as construction materials by chemical stabilization. Ph.D. thesis, Kyoto University, Kyoto, Japan.

Schlorholtz, S. et al 1984. An examination of the ASTM pozzolonic activity test for low calcium fly ashes. *Cement and Concrete Research,* U.S.A. 14:499-504

Valent, G.L. et al 1988. Influence of chemical and physical properties of Italian fly ashes, on reactivity towards lime phospho gypsum and water. *Cement and Concrete Research,* U.S.A. 18:91-102

Van Roode, M. et al 1987. X - ray diffraction measurement of glass content of fly ashes and slags. *Cement and Concrete Research,* U.S.A. 17:183-197

Watt, J.D. and Thorne, D.J. 1965. Compaction and pozzolonic properties of pulverized fuel ashes. Part I & II, *Journal of Applied chemistry,* London. 15:585-594; 595-604

Yudhbir and Honjo, Y. 1991. Application of geotechnical engineering to environmental control. 9th Asian Regional Conference on Soil Mechanics and Foundation Engineering 2. Bangkok, Thailand.

Environmental Geotechnics, Kamon (ed.) © 1996 Balkema, Rotterdam. ISBN 90 5410 848 7

Effect of pyrolyzed carbon black on asphalt binder

Y.D.Zeng & C.W.Lovell
School of Civil Engineering, Purdue University, USA
C.Brian
Research Division, Indiana Department of Transportation, Ind., USA

ABSTRACT: In this paper, research is carried out to investigate the potential use of tire-derived pyrolyzed carbon black as an asphalt modifier. The pyrolyzed carbon black was combined with the asphalt cements in different percentages. Tests such as penetration, softening point were performed to determine the temperature susceptibility of the modified binders as altered by the pyrolyzed carbon black. The results illustrate the viability of the pyrolyzed carbon black as an asphalt additive.

1 INTRODUCTION

There are approximately 240 million waste tires generated annually in the United States [S. N. Amirkhanian, 1993]. While some of these tires are recapped or ground into crumbs for special uses, most are dumped in rural areas or in landfills, which will pose serious environmental problems or use space needed for municipal solid wastes.

Scrap tires, can on the other hand, represent a source of energy and chemicals. By thermal decomposition, it is possible to recover useful products. Typical yields include light hydrocarbons, carbon black, gas, steel, sludge and water. The carbon black derived from used tires contains several impurities, which makes it difficult to be reused commercially.

The commercial Carbon black has been used as a reinforcement in tires to increase the life of tires. A similar concept of applying carbon black to reinforce asphalt cement was first reported in the early 1960's [Allioti 1962]. Since then, several research studies have been undertaken. The results show some positive effects, and no adverse effects due to the use of carbon black have been reported. But the carbon black used in research so far is comprised largely of pure carbon in pelletized or microfil form. The pyrolyzed carbon black from scrap automobile tires is comprised of about 75% pure carbon. Therefore, laboratory studies are necessary to verify the technical viability of the pyrolyzed carbon black as a modifier in asphalt binder.

2 MATERIAL PROPERTIES

The materials used in the study include commercial carbon black (CB), pyrolyzed carbon black (CB_p) and asphalt cement (AC10 and AC20). The general properties of these materials are described as fellows:

2.1 *Carbon Black*

To provide a comparison for the effect of pyrolyzed carbon black on asphalt properties, commercial carbon black is also utilized in this study. The typical characteristics of this carbon black are tabulated in Table 1:

Table 1. Characteristics of carbon black (CB) used (After Cabot, 1994)

Properties	Specification
Iodine Index (mg/g)	76±5
DBP absorption	85±5
Ash (%)	1.0 max
Density (kg/m^3)	200.0
Particle size (nm)	100~500
Surface area (m^2/mg)	15~100

2.2 *Pyrolyzed Carbon Black*

Pyrolysis is a process to decompose the scrap waste tires. The waste tires are broken down at high temperature into saleable products. The pyrolyzed carbon balck used in this research is derived from the reductive pyrolysis process. This pyrolyzed carbon black is a grey to black powder. Its particle size is much larger than that of commercial carbon black. It has a higher specific gravity than commercial carbon black. It is insoluble in water. Partial clods can be observed, but these are easily broken down.

The general properties of the pyrolyzed carbon black, as studied by Roy, et al, 1990, were listed in Table 2. Some significant differences were found compared with the properties of carbon black.

2.3 *Asphalt Cements*

The most commonly used asphalt cements for pavement construction in the United States are AC10 and AC20. Consequently, these two kinds of asphalt cements were used.

3 TEST RESULTS

3.1 *Design of the Tests*

To study the effect of carbon black and pyrolyzed carbon black on the properties of asphalt, consistency tests should be performed on the asphalt cements modified by commercial carbon black and pyrolyzed carbon black. Penetration tests, ring and ball softening point tests were conducted. All these tests were run based on the standard ASTM testing procedures with a few modifications due to the specific problems encountered.

The settlement of carbon black in asphalt cement during the cooling period was found to cause large deviations in penetration results. Three practical methods were used in trying to solve the problem:

(1) The penetration box containing the specimen is immersed in ice water to reduce the time needed to cool the sample.

(2) The box is set into a machine which rotates at 60 revolutions per minute to prevent the settlement of the carbon black to the bottom of the box.

(3) In order to observe the settlement of carbon black in asphalt cement, a penetration box, open at both ends is employed in the test so that the penetration can be performed on both ends of the box.

Table 2. Characteristics of pyrolyzed carbon black (CB_p) used (After Roy, et al, 1990)

Properties	Specification
Iodine Index (mg/g)	144.2~151.4
DBP absorption	84.6~93.0
Heat Loss at 105°C (%)	0.4~1.0
Ash (%)	15.5~17.0
Volatile Matter (%)	4.0~3.3
Sulfur (%)	2.5~3.0

3.2 *Effect of Commercial Carbon Black on Asphalt Cement*

The commercial carbon black was used as an additive to AC10 and AC20 in this research for comparison. The percentages of the carbon black added are 5%, 10%, 15% and 20% of the total weight of the asphalt binder.

Figure 1 shows the penetration of AC10 and AC20 versus percentage of carbon black. From the Figure, at a temperature of 25°C, the penetration of AC10 and AC20 decreases steadily from 101 to 68 and 65 to 48, respectively, when the percentage of the carbon black added increases from 0% to 20%; at 15°C, the penetration decreases from 40 to 31 and 27 to 19, respectively, with the increase of carbon black added.

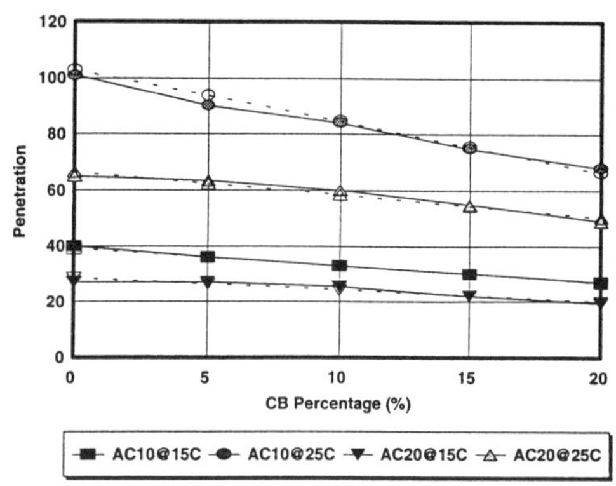

Figure 1. Penetration of AC10 and AC20 modified by carbon black (Dotted lines are regression lines)

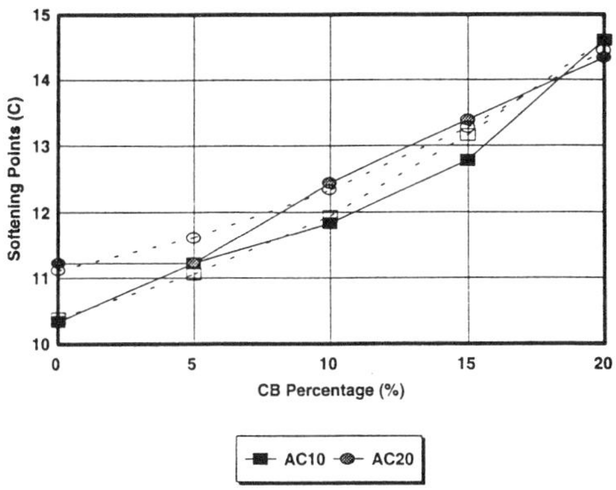

Figure 2. Softening points of AC10 and AC20 modified by carbon black

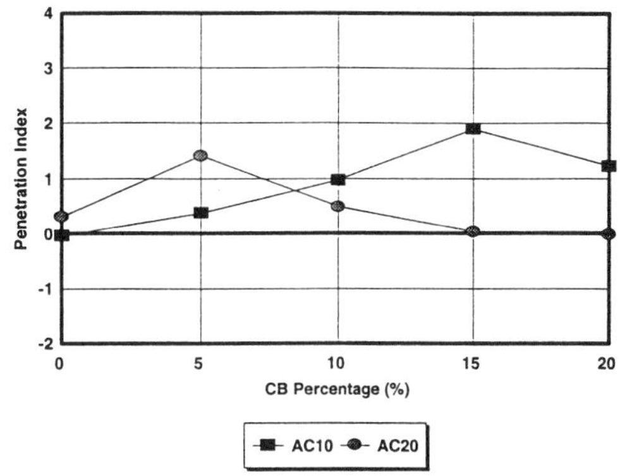

Figure 3. Temperature susceptibility of AC10 and AC20 modified by carbon black

Figure 2 presents the softening point test results.The softening point of AC10 increases from 50.6°C to58.3°C when the carbon black percentage increases from 0% to 20%; the softening point of AC20 increases from 52.2°C to 57.8°C when the carbon black percentage increases.

The consistency of asphalt changes with temperature. Temperature susceptibility is the rate at which the consistency of an asphalt cement changes with a variation in temperature, and is a very important parameter of asphalt cement. The penetration index is used as a measure of temperature susceptibility. The lower the penetration index value, the higher its temperature susceptibility. Most paving asphalt cements have a penetration index between ±2.0. When penetration index is less than -2.0, the asphalt cement is highly temperature susceptible, usually exhibits brittleness at low temperature and is very prone to transverse cracking in cold regions. The penetration index (PI) is given by:

$$PI=\frac{20-500A}{1+50A}$$

here

$$A=\frac{\log P_1-\log P_2}{T_1-T_2}$$

In the above equations, T_1, T_2 are different testing temperatures, P_1, P_2 are penetrations at T_1 and T_2 respectively.

Figure 3 shows the penetration index (PI) of AC10 and AC20 versus percentage of carbon black added. For AC10, PI tends to increases slightly from 0.0 to about 1.8 and then drops to 1.2. For AC20, penetration index does not show much increase. The value increases from 0.38 to about 1.4 and then drops to 0.0.

3.3 *Effect of Pyrolyzed Carbon Black on Asphalt Cement*

The same tests procedures were used to investigate the effect of pyrolyzed carbon black on asphalt cements. The results demonstrate some special phenomena due to the unique characteristics of the pyrolyzed carbon black. The density of the pyrolyzed carbon black is greater than that of commercial carbon black and thus the separation problem is more serious.

Figure 4 illustrates the penetrations at 15°C and 25°C on both ends of the penetration box for AC10 and AC20. The trends are essentially the same as those for commercial carbon black on asphalt cements; the penetrations at both temperatures decrease when the percentage of carbon black added increases.

Figure 5 shows the softening point of AC10 and AC20 against the percentage of pyrolyzed carbon black. The softening point of AC10 increases from 50.6°C to 55.0°C, and increases from 52.2°C to 55.6°C for AC20.

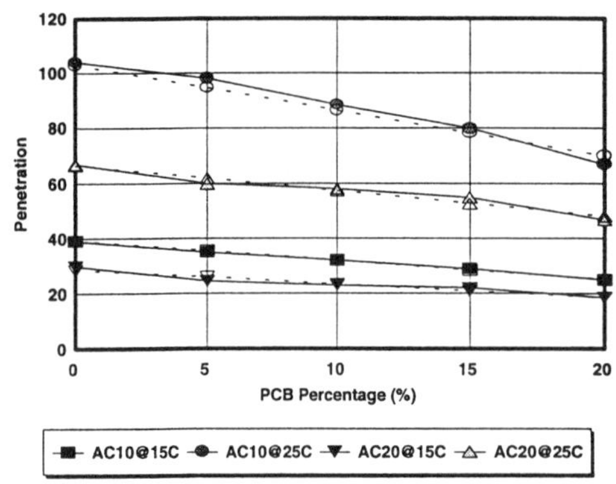

Figure 4. Penetration of AC10 and AC20 modified by pyrolyzed carbon black

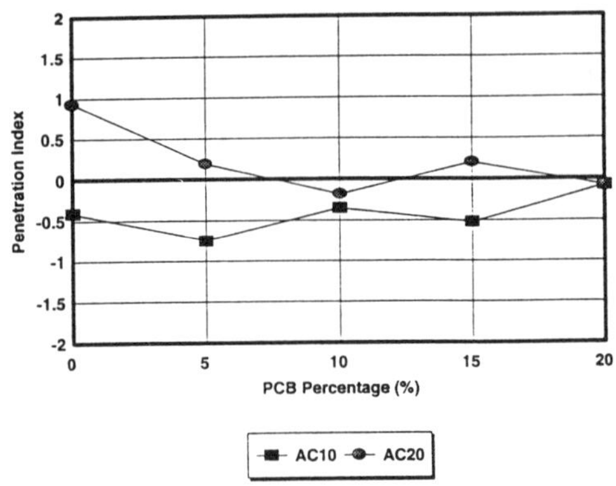

Figure 6. Temperature susceptibility of AC10 and AC20 modified by pyrolyzed carbon black

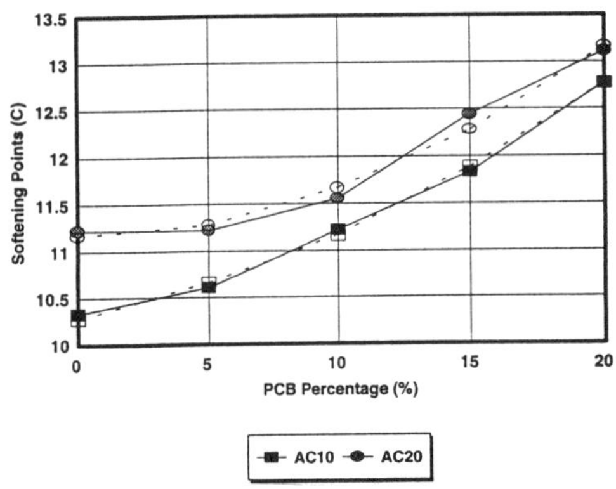

Figure 5. Softening points of AC10 and AC20 modified by pyrolyzed carbon black

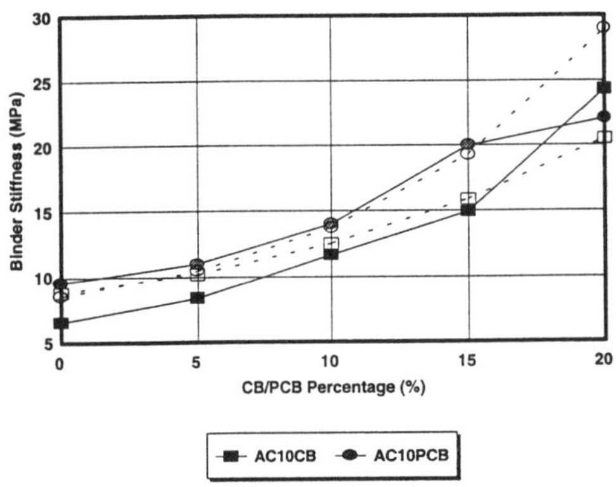

Figure 7. Binder stiffness of AC10 modified by Carbon black and pyrolyzed carbon black

As shown in Figure 6, the penetration index of AC10 modified by pyrolyzed carbon black is approximately a straight line with a slight increase. The plot also indicates a decrease of penetration index.

Previous researches indicated a linear relationship of the penetration of asphalt binders against carbon black percentage [Rostler et al, 1977]. Figure 1 and Figure 4 illustrate the regression curves of penetration of asphalt at 25°C and 15°C against the percentage of carbon black and pyrolyzed carbon black added. The difference between regression and raw data is about 0-5% of the test results.

Many experiments have shown a parabolic relation between the softening point of asphalt and carbon black. As such, the regression curves in Figure 2 and Figure 4 for asphalt cements modified by carbon black and pyrolyzed carbon black indicates a small deviation of test results. The raw data fit very well with the parabolic regression data.

Asphalt cement is inherently a visco-elastic material. The deformation potential of asphalt binder at certain temperature can be evaluated using binder stiffness, which is analogous to elastic modulus. A regression equation was provided by Ullidtz (1979) for binder stiffness as:

$$S_b(MPa)=1.157\times10^{-7}t^{-0.368}e^{-P.I}(T_{RB}-T)^5$$

where, S_b is binder stiffness (MPa); t is the time of loading (Second); PI is the penetration index; T_{rb} is the result of ring and ball softening point (°C) and T is the ambient, or material testing temperature (°C).

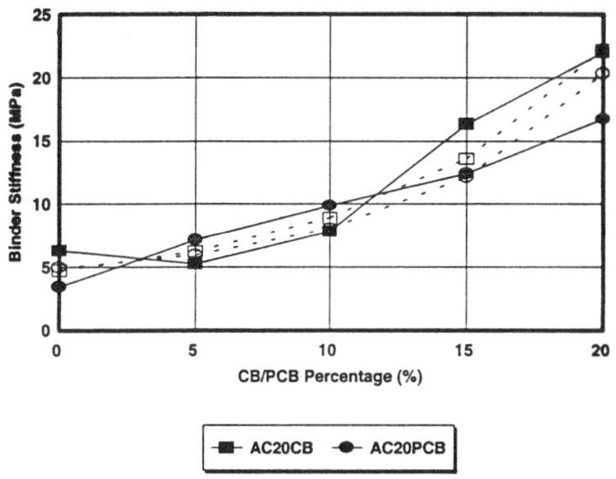

Figure 8. Binder stiffness of AC20 modified by Carbon black and pyrolyzed carbon black

Figure 7 and Figure 8 show the raw and regression curves of binder stiffness of AC10 and AC20 modified by carbon black and pyrolyzed carbon black respectively. The Figures indicate:

(1) Binder stiffness increases with an increase of carbon black or pyrolyzed carbon black percentage in asphalt binder.

(2) The binder stiffness of those modified by pyrolyzed carbon black is generally larger than those modified by carbon black at the same testing temperature and additive percentage.

4 CONCLUSIONS

The use of pyrolyzed carbon black as an additive of hot mix asphalt cements produces the following benefits:

(1) Disposal of scrap automobile tires. This might be a most desirable techniques from an economic and environmental standpoint.

(2) The inclusion of pyrolyzed carbon black demonstrates no detrimental effect on the temperature susceptibility of asphalt cements, and hence the service life of hot mix asphalt pavements.

The test results in Figure 1 to Figure 8 indicate that:

(a) Penetrations of the asphalt cements modified by both carbon black and pyrolyzed carbon black decrease steadily with temperature.

(b) Softening point of the asphalt cements mixed with carbon black or pyrolyzed carbon black increases when the quantity of carbon black added increases. The increase due to the addition of pyrolyzed carbon black is more significant.

(c) Though the results do not indicate much improvement of temperature susceptibility, the addition of pyrolyzed carbon black will not have adverse effect on asphalt properties.

(d) The prize of the pyrolyzed carbon black is currently about U. S. $0.16 per pound. Compared with a relatively higher prize of carbon black at U. S. $0.70 per pound the pyrolyzed carbon black use has a cost advantage over that of commercial carbon black.

REFERENCES

Alliotti A. G. (1962) “Carbon Black - It’s Nature and Possible Effects on the Characteristics of Bituminous Road Binder”, *Proceedings Australian Road Research Board*, Vol 1, Part 1, 912pp.

Amirkhanian S. N. (1993) “Utilization of Scrap Tires in Flexible Pavements--Review of Existing Technology”, *ASTM STP 1193*.

Cabot Industries (1994) “Carbon Blacks for Specialty Applications”, Cabot Corporation Special Blacks Division, Catalogue.

Rostler F. S, White R. M, Dannenberg E. M. (1977), “Carbon Black as a Reinforcing Agent for Asphalt”, *Association of Asphalt Paving Technologists*, Vol. 46, pp376-401.

Roy C., Labrecque B., Caumia de B.(1990) “Recycling of Scrap Tires to Oil and Carbon Black by Vacuum Pyrolysis”, *Conservation and Recycling*, No. 4., pp203-213.

Ullidtz P.(1979) “A Fundamental Method for Prediction of Roughness, Rutting and Cracking of Pavements”, *Association of Asphalt Paving Technologists*, Vol. 48.

Remediation of contaminated ground

Environmental Geotechnics, Kamon (ed.) © 1996 Balkema, Rotterdam. ISBN 90 5410 848 7

On-site treatment of soil contaminated with chlorinated compounds

Jacques Benoit
AGRA Earth & Environmental, Calgary, Alb., Canada

ABSTRACT: Low Temperature Thermal Desorption (LTTD) is a term applied to a number of different ex-situ processes that use either direct or indirect heat to raise the temperature of contaminated soil and volatilize the organic contaminants into a gaseous state. The gases can then be either combusted in an afterburner or condensed and collected. Where the soil is heated directly in an oxidative environment and the gases are combusted in an afterburner, the technology is suitable for non-halogenated hydrocarbon contaminated soils. Conversely, where the soil is heated indirectly in a limited oxygen environment and the gases are condensed, the technology can be suitable for the remediation of halogenated and chlorinated hydrocarbon contaminated soils.

This paper will present the results of a study that utilized 'thermal phase separation' for the treatment of 7,000 tonnes of PCB contaminated soils. It is essential to keep in mind that thermal phase separation is **not** an incineration or direct fired process. Thermal Phase Separation is a process method whereby heat generated by the combustion of propane is transferred to the host solids matrix via indirect means; the fuel gas flame does not come into contact with the contaminated soil. Heat is transferred from the firebox indirectly, through the soil extraction chamber shell, to the contaminated soil so that hydrocarbon contaminants are volatilized from the soil.

1 TECHNOLOGY OVERVIEW

The primary process factors affecting thermal desorption performance and applicability are the maximum soil temperature achieved and the heating method (direct or indirect). The soil treatment temperature required to treat a specific hydrocarbon in a LTTD system can be estimated to be slightly higher than the original distillation temperature of the virgin hydrocarbon. Figure 1 presents a range of temperatures at which selected contaminants vaporize.. The information in Figure 1 is related to the evaporation of a liquid contaminant at pure state under the relevant operating pressure.

This concept can be presented in another format. If the effectiveness of thermal desorption is dependent upon the distillation temperature of the contaminant then the removal efficiency must also be dependent upon the temperature of the soil. Figure 2 shows this relationship for the removal efficiency of PCB's in relation to the soil temperature at the exit of the Thermal Phase Separator (TPS) during actual field operations at several tonnes per hour throughput. It is clear that, in order to obtain the remediation objective for a particular contaminant given the feedstock characteristics, the removal efficiency must be calculated. Once the removal efficiency has been calculated the operating conditions can then be accurately determined.

This methodology applies to both direct and indirect thermal desorption units. Only the treatment of off gases is different.

2 THERMAL PHASE SEPARATION TECHNOLOGY

TPS technology is a cost effective technology for the remediation of sites contaminated with halogenated and polyaromatic hydrocarbon compounds. It is well-suited to removing either chlorinated or non-chlorinated hydrocarbons with boiling points up to 650 ^{0}C. Because the waste is indirectly heated and not subject to combustion, the only gaseous emissions are associated with the burning of the fuel gas. Since products of incomplete combustion or chlorinated organics are neither produced nor emitted, this unit is permitted to treat hazardous materials. The TPS technology is typically the preferred technology when:

- the contaminants are halogenated compounds;
- particulate, acid gases and contaminant emissions are of concern;
- products of incomplete combustion (PICs) are of concern; or
- the contaminants have very high boiling point characteristics.

Figure 3 presents a schematic of the indirect heat transfer process utilized in the TPS.

The thermal phase separation process is a two-staged process. The first consists of removing the organics from the soil through the application of heat in the process described above. This is the phase separation process. The second stage involves cooling the gases and vapours and condensing them into liquid form. The condensate is then separated into oil, water, and solids fractions. The gases are first cooled to about 28 ^{0}C in a quench process and then further cooled to a few degrees above ambient by a condenser.

3 EQUIPMENT DESCRIPTION

The thermal phase separation system consists of five distinct sub-systems as described below.

3.1 *Feed Handling*

Feed soil, screened and crushed to less than 25 mm, is deposited into a feed hopper by a small front end loader, from which it is moved away by a horizontal conveyor belt. Material such as large rocks, metal and debris are manually removed. The soil is then discharged into a clay lump-breaker and, in turn, the fragmented soil is transferred to an inclined conveyor for delivery to the extraction chamber.

3.2 *Extraction Chamber/Thermal Unit*

The soil is discharged from the inclined conveyor to a small hopper, which directs the material to two rotary airlock valves. Upon passing through the airlock the soil drops into the extraction chamber. Two parallel screw augers move the material through the heat treating zones. See Figure 4

The extraction chamber is physically separated from the combustion system by a steel plate and is completely enclosed by the firebox. The firebox derives its heat by combusting propane in a series of burners along the entire length of the extraction chamber so as to achieve even heat distribution. Heat is transferred from the combustion firebox to the soil via conduction through the steel plate.

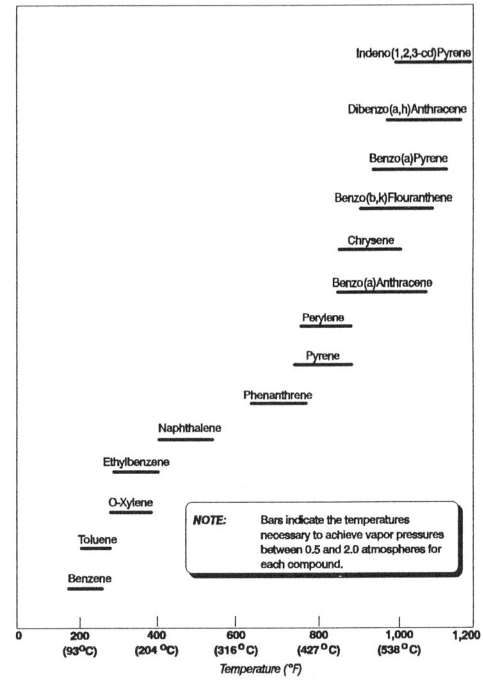

Figure 1. Vaporization Temperature of Hydrocarbon Constituents

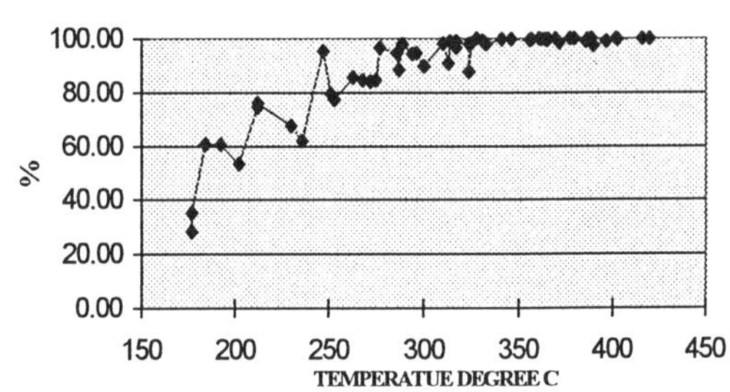

PCB REMOVAL EFFICIENCY OF THE TPS

Figure 2

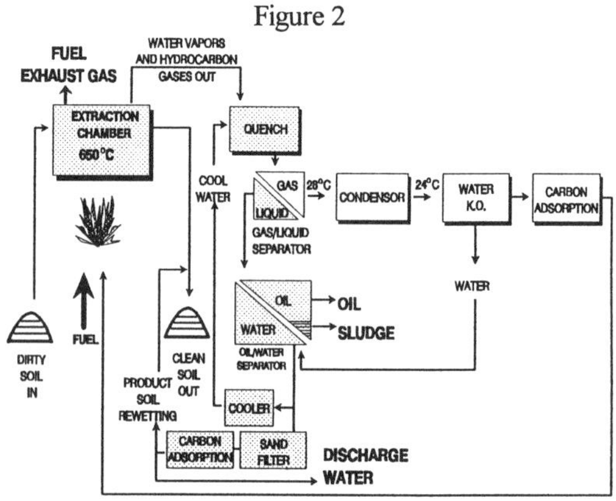

THERMAL PHASE SEPARATOR FLOW DIAGRAM

Figure 3

3.3 *Treated Soil Handling*

The two extraction chamber augers direct the treated soil to a discharge auger. The hot processed soil exits the discharge auger through a heat resistant paddle wheel airlock valve, which in turn directs the treated material to a pugmill. Water is sprayed into the pugmill to achieve cooling and re-wetting of the soil to a nominal 10% moisture content. The spray also serves to prevent fugitive dust emissions.

3.4 *Gas Treatment*

The water vapour and gaseous hydrocarbons originating in the extraction chamber are subjected to three sequential treatment stages.

First, the gas stream is cooled to approximately 28 ^{0}C via direct contact water sprays in a quench chamber.

Second, the gas stream is cooled to a temperature only a few degrees above ambient, by a fin fan cooler. The resulting water content of the saturated gas stream is estimated to be between one and five percent.

Third, the relatively dry gas stream of non-condensable gases (essentially representing the volume of system leakage air) is subjected to final polishing in a mist eliminator (for fine aerosol removal) and carbon adsorption beds. A positive displacement induced draft fan ensures movement of gaseous products through the system, from the extraction system to the condensing components.

3.5 *Water/Condensate Treatment*

During treatment operations water is recirculated through the quench auger to achieve cooling of the gas stream. This volume of recirculated water, together with products condensed from the gas phase are routed through a sludge settling chamber and a three phase oil/water/solids separator.

Any excess water generated by the treatment process is typically treated using a mobile wastewater treatment system. The mobile wastewater treatment system incorporates both physical and chemical processes to remove organic and inorganic contaminants from wastewater. It involves a combination of processes which can include the addition of chemical reagents, pH adjustment, coarse filtration, ceramic ultrafilters, ion exchange, and carbon adsorption.

The TPS comes on six mobile trailers and takes up a small footprint of only 25 m x 30 m. The footprint can be smaller depending upon the need for additional water treatment and fuel storage tanks. The TPS generally takes 5 to 7 days for mobilization and processes soil at 2 - 5 tonnes per hour. The TPS can accommodate both propane or natural gas fuels and the consumption ranges from 5 to 7 MM Btu/hr.

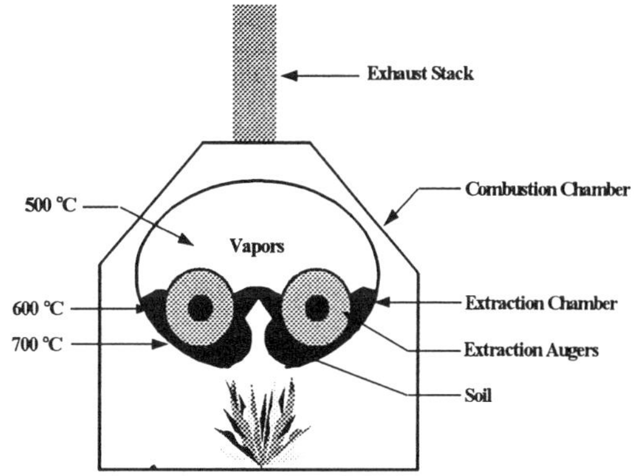

Figure 4
EXTRACTION CHAMBER

4 THERMAL PHASE SEPARATION PROJECTS

The mobile TPS technology is new in the treatment of contaminated soils in Canada. In 1993, soils containing pentachlorophenols and polyaromatic hydrocarbons in British Columbia and chlorinated solvents such as orthodichlorobenzene in Alberta were successfully treated. In 1994 & 1995, 7000 tonnes of PCB contaminated soil were processed in central Alberta and Newfoundland.

A summary of project remediation results are shown in Table 1.

Table 1
PROJECT REMEDIATION RESULTS

	Feed Soil Conc. mg/kg	Treated Soil Conc. mg/kg	Removal Efficiency %
TPH	11179	59.3	99.4
Oil & Grease	1013	<40	>96.1
PAHs	6169.7	21.98	99.6
Chlorophenols	85.5	.1	99.7
PCBs 1242	521	1.71	99.67
Dioxins & Furans (TEQ)*	209.8	<2.7	>98.7
PCBs 1260	169	<0.066	>99.96

*Expressed as ng/kg. TEQ Toxicity Equivalence Factor

Engineering improvements and design modifications to the TPS technology have continued to ensure remedial objectives are achieved. This is particularly evident in the most recent project contaminated with Aroclor 1260 PCBs which are the most difficult PCBs to desorb. The TPS demonstrated excellent removal efficiencies of greater than 99.96%. Further, in the last two projects, stack emissions were measured and 99.9999% removal efficiency was also demonstrated. Of particular interest were the very low levels of dioxin and furan stack emissions which were measured at 0.013 pg/Nm^3.

5 DESRT - THERMAL PHASE SEPARATION OF PCBS DEMONSTRATION RESULTS

Most recently, under Development and Demonstration of Site Remediation Technology (DESRT) program, the TPS has undergone rigorous testing to meet new regulations and concerns from various regulatory agencies. Unfortunately, at this time very few TPS type technologies exist in Canada and therefore limited data has been accumulated. In the absence of regulations specific to indirectly heated thermal desorbers, the TPS in general is not well understood and it is typically forced to follow fixed facility "incinerator" type regulations which are developed for a very different process. Despite this obstacle, it was agreed to perform a test under the existing "incinerator" regulations, with a target destruction efficiency of one milligram of PCB in the stack per one kilogram of PCB in the feed. Note that the TPS does not destroy the contaminants but rather condenses them. Concerns over the production and emission of dioxins and furans in the TPS were also raised and therefore rigorously tested.

The demonstrations took place in July and October, 1994 at sites contaminated with PCBs. The samples from the triplicate test were collected from feed soil, treated soil, oil, sludge, treated water and stack. Soil and water samples were taken every half hour and composited for each test.

The averaged results from the performance test done in triplicate are as follows.

5.1 *Soil Results*

The TPS has demonstrated over 99.9% removal efficiency with inlet PCB concentrations of 700 mg/kg during normal operating conditions (Table 2). However, during the DESRT tests, the average inlet PCB concentration was 169 mg/kg. The TPS was very successful in removing PCBs from the soil. The PCB analysis of the treated soils, at a method detec tion limit of 0.002 mg/kg, produced non-detectable concentrations for tests 1 and 3. Test 2 contained PCB concentrations of 0.14 mg/kg. This represents an average removal efficiency of PCBs from the soil of > 99.9%.

Table 2
RESULTS FROM THE TESTS CONDUCTED UNDER THE DESRT PROGRAM

PARAMETER	RESULTS	LIMIT
STACK		
PCB Concentrations $ng/(Nm^3)$ c/o 11% O_2	130.5	
Stack Gas PCB Mass Loading (mg/hr)	0.3758	
HCL mg/Rm^3	<0.002	75
Particulate mg/Rm^3	14.28	200
Dioxins and Furans ng/Rm^3 (as TEQ)	**0.0130**	**12**
mg PCB Stack/kg PCB Feed	**0.5721**	**<1**
DRE (%)	**>99.9999**	**99.9999**
FEED SOILS		
Feed Rate (tonnes/hr) dry	**4.432**	
Soil Moisture Content %	12.5	
PCB Content (mg/kg) wet	148	
Dioxins and Furans (as TEQ) (μg/kg)	0.0150	
TREATED SOILS		
Dioxins and Furans (as TEQ) (μg/kg)	**0.003**	**<1**
PCB (mg/kg) wet	**0.058**	**<5**
PCB Removal Efficiency (%)	**>99.96**	

The soil was characterized as 36% silt and clay with the remaining material as coarse to very fine sand. In general, a difficult matrix for thermal desorption.

The soil feed rate into the TPS was on average 3.9 tonnes per hour.

5.2 *Soil Results - Dioxins and Furans Breakdown*

Regulators raised concerns over the issue of PCDDs and PCDFs. In the event the TPS generated excessive amounts of dioxins and furans, the quality of treated soil may be adversely affected by a rise in dioxins and furans. Although the soil contained small amounts of PCDDs and PCDFs, the quality of the treated soil improved by removing additional dioxins and furans by 80.4%. Based on these results, it is believed that the TPS has direct applications for the remediation of dioxins and furans contaminated soils.

5.3 *Water & Oil Results*

If the soil contains moisture of greater than 12%, the TPS will have a net gain of water which must be treated. This treated water is required to meet strict discharge criteria.

The total PCB analysis of the treated water shows non-detectable values and therefore easily meets the criteria of 5 μg/L. The evaluation criteria for dioxin/furan concentrations of the treated water was stated as 0.25 nanograms per litre for each congener group. The sum of all the dioxin and furan congeners are 0.0538 ng/L and 0.5679 ng/L respectively. A more detailed review of the water data indicates that all dioxin/furan congener groups to be less than this criteria with the exception of Tetrafuran which marginally exceeded the criteria by 0.120 ppt. However, a new criteria based on TEQ factors, indicates that the dioxins and furans criteria are easily met.

Due to very low levels of extractable hydrocarbons in the soil, no significant oil was removed from the soil. In general, the hydrocarbons are found in the sludge.

5.4 *Sludge Results*

As expected, the removal of PCBs from soil was concentrated in the sludge with a concentration of 7,523 mg/kg. The volume of sludge is approximately 1% of the total original volume of soil. It is believed that based on the sludge moisture content of 75.9%, sludge volumes can be further reduced by using dewatering technologies. The small amounts of native dioxin and furan (total concentration of 5.4 ppb) found in the contaminated soil desorbed by the TPS are concentrated in the sludge. Similar to the dioxin and furan results in the soil, the sludge results demonstrate that the TPS did not preferentially desorb specific congeners.

5.5 *Stack Emission Results*

Despite the average low soil feed rate of 0.655 kg/hr, the TPS was able to meet the required ratio of less than one mg of PCB in the stack per one kg of PCB feed on all individual three tests. The average measured ratio was 0.572 which equates to greater than "six nines" removal efficiency.

During the inception of this project, concern was raised over the potential of releasing excessive levels of dioxins and furans. The performance test concluded that the TPS technology is more than capable of handling the criteria limit of 12 ng/Rm^3 (TEQ). The dioxins and furans levels were three orders of magnitude lower.

The measured flue gas HCl (<0.002 mg/Rm^3) and particulate (14.28 mg/Rm^3) are well below the evaluation criteria of 75 mg/Rm^3 and 200 mg/Rm^3 respectively. The measured particulate concentration of the flue gas is of the same order of magnitude as that which might be expected from the combustion of natural gas.

6 UNIT COSTS

Unit costs range from $450 to $600 per tonne which excludes mobilization. The unit costs are dependent upon:

- Total volume
- Clean up criteria
- Type and concentration of contaminant
- Host matrix conditions such as moisture and particle size distribution
- Site conditions

7 BENEFITS

- Proven technology;
- Excellent removal efficiency - in most cases soil can be placed back on site;
- Costs are less expensive compared to other remediation options;
- The site can be remediated in a short period of time.

8 FURTHER DEVELOPMENTS

Based on the experience with this TPS, a second unit has been designed and built which features:

- Design soil throughput of 14 tonnes per hour based on 12% moisture
- Requires only a 18.3 m x 33.6 m footprint
- Contains a microprocessor based process control system
- Skid mounted ensuring mobility and maximum stability
- Enhanced condensation system maintaining an uncondensed off-gas temperature to 7 °C.

9 CONCLUSION

The application of Low Temperature Thermal Desorption for the treatment of soils contaminated with hazardous compounds has been demonstrated by the development of the Thermal Phase Separator. The Thermal Phase Separator, which is not an incineration process, has treated over 7,000 tonnes of soils contaminated with PCBs. The soils were treated to levels well below accepted standards and at a substantially lower cost than incineration. The TPS reduces the amount of contaminated soil to less than 1% of the original volume therefore allowing for most of the soil to be replaced on-site. This technology has now been evaluated and tested by a number of regulatory agencies in Canada and is now ready for full commercialization.

Environmental Geotechnics, Kamon (ed.) © 1996 Balkema, Rotterdam. ISBN 90 5410 848 7

Pollutant-trapping materials – Two case histories

A. Bouchelaghem, D. Gouvenot & G. Raillard
Solétanche Entreprise, Nanterre, France

ABSTRACT
Pollutant-trapping materials are produced by mixing pollutant-trapping substances in a cementitious slurry. They retain pollutants by various mechanisms. One case history concerns the containment of chiefly organic chemical waste with a diaphragm wall incorporating activated charcoal. The other describes a pollutant-trapping drainage trench providing in situ treatment of an area seriously polluted with hexavalent chrome.

1 INTRODUCTION

Extensive research over the last ten years into the behaviour of dissolved pollutants has culminated in a range of materials capable of slowing or preventing their migration. These patented pollutant-trapping materials are marketed under the name Ecosol.

Two recent projects illustrate the advantages of these materials for protecting the environment against pollution.

The first example concerns the containment of an old chemicals dump by means of a diaphragm wall built with a slurry containing activated charcoal. The pollutants were mostly organic chemicals although there were also some heavy metals present.

The second case history concerns a pollutant-trapping drainage trench backfilled with porous material to prevent hexavalent chrome from contaminating the water table.

These projects took place in 1994 and 1995 and used site-specific Ecosol pollutant-trapping materials tailored to the problems in hand.

2 POLLUTANT-TRAPPING MATERIALS

2.1. Concept

The self-hardening pollutant-trapping materials have a water/cement mix base.

Various other components can be added while the mix is still a slurry, such as special clays, siliceous materials and trapping agents. The resulting mix hardens over time to form a stable, durable microstructure of low permeability with a skeleton consisting of insoluble crystals of hydrated calcium aluminates and silicates.

The specific trapping agents are selected with reference to the properties required in the final material. Developing the proper formulation is done in the laboratory. After hardening, the material is tested to assess its trap efficiency under actual conditions of use.

Formulation development is based on knowledge of the physical and chemical mechanisms liable to affect the mobility of the dissolved pollutants. (Bouchelahem et al, 1993 and Gouvenot, 1993)

2.2 Trapping Mechanisms

There are many physical and chemical mechanisms that can be used in connection with dissolved chemical species : precipitation, oxidation, complexation, adsorption, etc.

These mechanisms involve the specific properties of certain materials, such as filtration-adsorption on activated charcoal.

All these mechanisms operate within the pores of the pollutant-trapping material, whose trap efficiency increases as the specific area is greater. To obtain this property, the Ecosol mix makes use of minerals with very high specific surface values, such as clays (100 m^2/g) and activated charcoal (500-1000 m^2/g).

Trap efficiency is governed by two complementary mechanisms : precipitation and adsorption.

2.2.1 Precipitation

Chemical precipitation is used, among other things, for treating acid industrial wastes containing dissolved ions of metals such as iron, nickel, copper, lead, trivalent chrome, chrome VI, etc.

These cations are mainly precipitated in the form of metallic hydroxides in chemical reactions controlled by the pH.

In cement-based materials, lime solubility dictates a pH of 12-12.5. However, this reserve of alkali in cements, although large, is not infinite, and alkaline buffers are incorporated in Ecosol materials to extend their trapping performance in time.

2.2.2 Adsorption

Adsorption is a physical/chemical mechanism based on the properties of porous materials with high specific surface.

On these surfaces, there are sites to which certain dissolved chemical species are attracted and bound by various mechanisms:

- by physical bonds, in pores of similar sizes to the molecules,
- by van der Waals forces, due to polarisation of certain molecules at the surface,
- by ion exchange: clays can exchange their compensating ions with ions from the solution,
- surface affinity: hydrophobic organic molecules have affinities for sites on the activated charcoal.

3 SLUDGE LAGOON CONTAINMENT BY ECOSOL WALL

3.1 Remediation Project

The waste disposal yard at the Spolana (Czech Republic) carbon factory lies beside the Elbe river. It is a former borrow area that has been used for decades as a waste dump by the nearby factory.

It is split into several separate waste disposal areas surrounded by embankments 9 metres high above original ground line. The many different types of waste are classified according to their toxicity, the most highly toxic being confined to two areas 175 metres wide by 300 metres long (Fig. 1).

To increase the holding capacity of these areas while improving safety, it was decided to provide perimeter containment and arrange a new dumping area for toxic waste above the old one by raising the existing embankments.

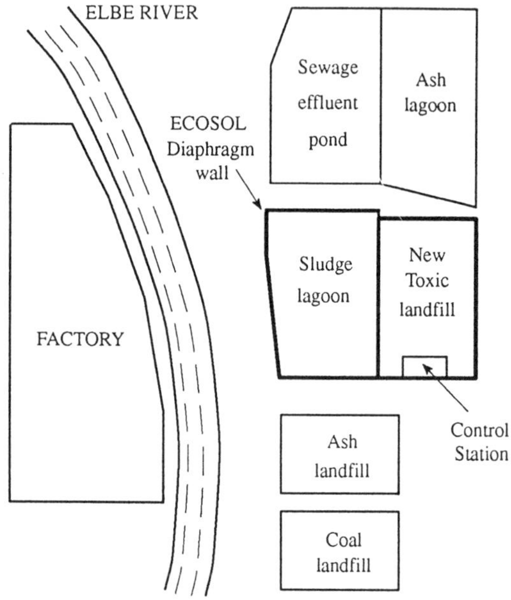

Fig. 1 - Site and Waste Dumps

3.2 Site Characteristics

Underlying a variable thickness of waste (between 7 and 9 metres), are sandy clays and sand with some gravel. Beyond 14-15 metres depth, there is a bed of indurated sandy marl.

There are a variety of pollutants present but they are mostly organic pesticides, PCB's, hydrocarbons, cyanides and sundry organochlorine compounds. Mineral pollutants are also present, chiefly mercury.

3.3 Engineering Design

Containment of the old chemicals dump is provided by a vertical diaphragm wall socketed into the marl, with a horizontal capping incorporating a geomembrane.

3.3.1 Cap

The whole 53 hectares area of the dump was dynamically compacted to provide a stable seat for the geomembrane and prevent tears due to settlement of the waste. The compaction reduced the thickness of the waste by approximately 60cm (Fig. 2).

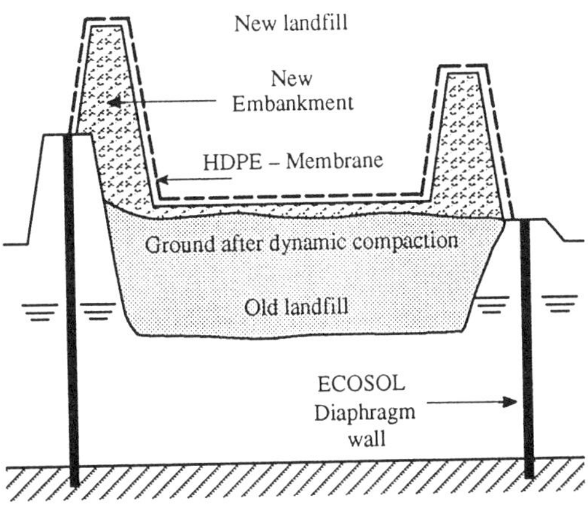

Fig. 2 - Section through Containment Zone

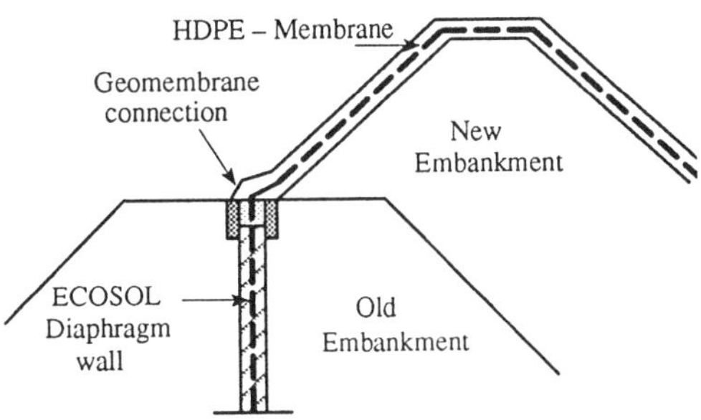

Fig. 3 - Diaphragm Wall and Geomembrane

Continuity between the horizontal and vertical containment was effected by joining the horizontal geomembrane to the membrane in the diaphragm wall (Fig. 3).

3.3.2 Ecosol Perimeter Wall

The perimeter wall provides vertical containment. It is therefore in permanent contact with the organic pollutants in the dump, and it was accordingly attractive to use a trapping material to slow down the migration of pollutants through the wall, when driven by a hydraulic and/or concentration gradient. The trapping substances selected with reference to the organic nature of most of the pollution were activated charcoal and special clays (Cherel, 1991).

The diaphragm trench was dug with a KS 3000 hydraulic bucket on a telescoping kelly enabling it to be guided as it moved between the trench and the spoil trucks. By carefully controlling the rheological properties, the specially-formulated slurry was able to act as shoring mud to stabilise the trench during digging (Bouchelaghem et al., 1989).

The Ecosol diaphragm wall built in this way was 1680 metres long, 11 metres to 20 metres deep, and socketed one metre into the sandy marl. It is 0.60m thick with a total area of 30,000 m^2.

4 TREATMENT OF CHROME POLLUTION BY POLLUTANT-TRAPPING DRAINAGE TRENCH

The principle of the pollutant-trapping drainage trench is to intercept polluted groundwater and treat it in situ. The trench does this without maintenance, staff or energy input.

The method was applied to solve a chromate-polluted groundwater problem. (Bouchelaghem et al, 1994)

4.1 Pollution Source and Type

The pollution was caused by materials containing hexavalent chrome being used in the construction of the A22 motorway in northern France, near Lille.

The toxicity of the material, known at the time of construction, prompted the incorporation of engineered safeguards in the fill. Unfortunately, the design of the central platform did not completely control seepage. Soon after the road was opened, there were indications of chrome VI in the seepage water at some spots.

In 1987, a company located below the motorway, (i.e. down-gradient of groundwater flow) reported hexavalent chrome in its drainage water, and this pollution was accompanied by bare patches in the vegetation of the motorway bank.

4.2 Site Characteristics

The ground under the slip road embankment consists of 4-5 metres of sandy silt overlying Flanders clay (Fig. 4). The water in this silt flows gently towards a bank slope some ten metres down-gradient of the fill, directly overlooking the car park.

In the course of time, rain infiltrating into the fill leached the soil. This water, contaminated with chrome VI, had access to the aquifer despite the layer of rolled shale at the base of the fill. Water in the toe ditch below the bank had the characteristic lemon yellow colouring of chromate pollution. In dry weather, patches of yellowish efflorescence appeared on the slope. Seeps of polluted water that killed off the vegetation could be seen on the slope above the company car park.

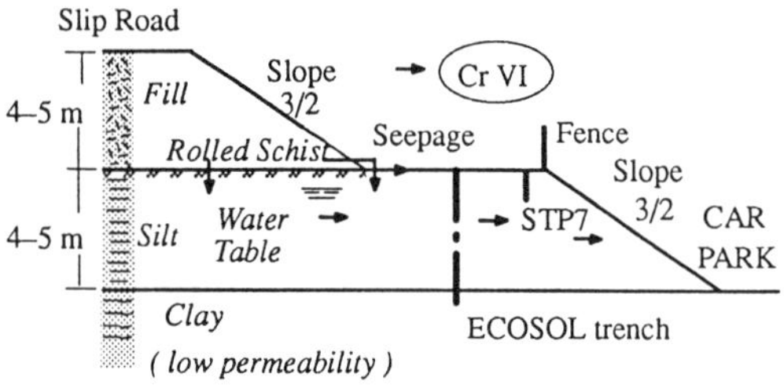

Fig. 4 - Sketch of Site Configuration

The site was surveyed to explore the extent of the pollution. Auger holes were sunk for soil samples. Chemical analysis revealed serious soil pollution, with chrome VI present in concentrations of up to 1000 mg/kg, but disappearing beyond a depth of 5-6m because of the impervious Flanders clay bed. Groundwater samples taken from piezometers yielded very high chrome VI contents, practically up to 2000 mg/l. Chrome VI is soluble and highly toxic, and the maximum permitted level in drinking water is 0.05 mg/l.

4.3 Engineering Design

The proposed solution consists of intercepting the polluted groundwater with an Ecosol drainage trench designed to trap chrome VI. More precisely, a trench is dug and backfilled with a pervious, porous, pollutant-trapping material.

The required material is obtained by mixing sorted gravel with a special slurry from the Ecosol range. These slurries are special formulations developed and patented by Soletanche to suit the exact nature of the pollutants in question: radioactive elements, heavy metals, organic compounds.

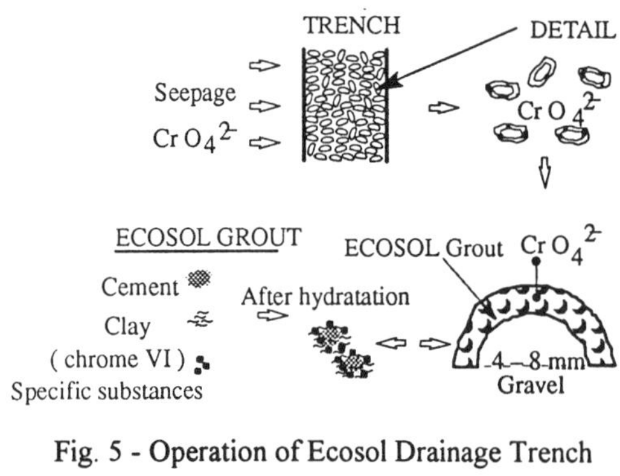

Fig. 5 - Operation of Ecosol Drainage Trench

The gravel grains coated with pollutant-trapping slurry bond together so that, when the mix hardens, they form a rigid, pervious structure through which water readily circulates (Fig. 5). As the water seeps through, the chemical pollutants are trapped on the film of hardened Ecosol formulation adhering to the grains of gravel.

4.3.1 Preparation of Ecosol Material

Several crucial parameters must be scrupulously controlled when preparing the Ecosol material.

- *Slurry/gravel weight ratio*:
This ration must not be too high, or the structure will not be sufficiently porous. Nor must it be too low, or the gravel grains will not be completely coated.

- *Gravel size*:
For any given ratio between a unit volume of slurry and the weight of gravel, a finer gravel size will increase the active surface area of the structure and reduce the thickness of the film of hardened slurry coating the gravel grains. This speeds up the trapping reactions. On the other hand, it reduces the permeability of the structure, with the attendant risk of the pores becoming clogged.

- *Selection of gravel and slurry*:
The bond between the film of hardened slurry and the gravel skeleton must be sound and durable. The use of cement slurry and compatible gravels ensures durability provided the same precautions are used as when designing standard concrete mixes.

- *Rheological properties of slurry*:
The rheology of the freshly-mixed slurry paste must be designed to form a stable film coating the gravel. Therefore, apart from the specific pollutant-trapping substances, the mix must be so designed that the initial stiffness of the paste when the gravel is incorporated is not too great, nor too low.

-*Slurry formulation*:
All types of Ecosol slurry can be used provided their rheological properties are suitable. The exact formulation of the slurry depends on pollutant types and concentrations and required trap capacity.

The mix design is done in the laboratory. In this instance, the cumulative quantity of chrome VI trapped per cubic meter of material could be determined.

The Ecosol material is prepared by adding the various reagents in the composition at the mixer. This slurry is then used for coating the gravel. This is done in a truck mixer, so that the sorted gravel is delivered, the slurry added, the mix blended and the material delivered to the point of use all in one vehicle.

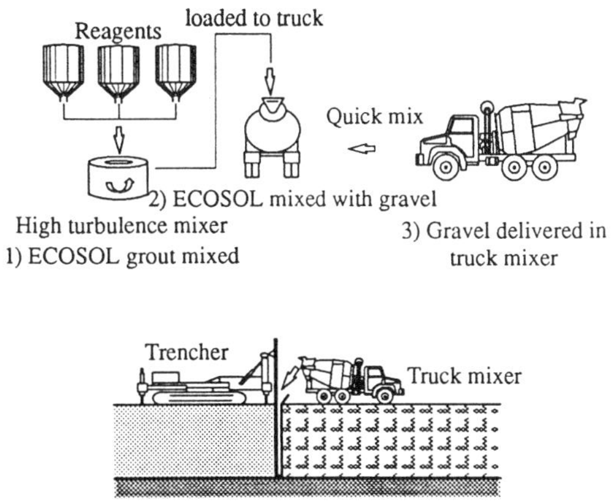

Fig. 6.Preparation and Deposition of Ecosol Material

4.3.2 Trenching

On site, the trench is dug by a trenching machine, capable of digging to a depth of six meters. The mobile frame carries a chain and square-sides chute guiding the backfill material from the top hopper into the trench. The Ecosol-coated gravel is brought in by the truck mixer to the trench hopper and the blended material flows like concrete (Fig. 6). The end result is a trench backfilled with Ecosol-coated gravel.

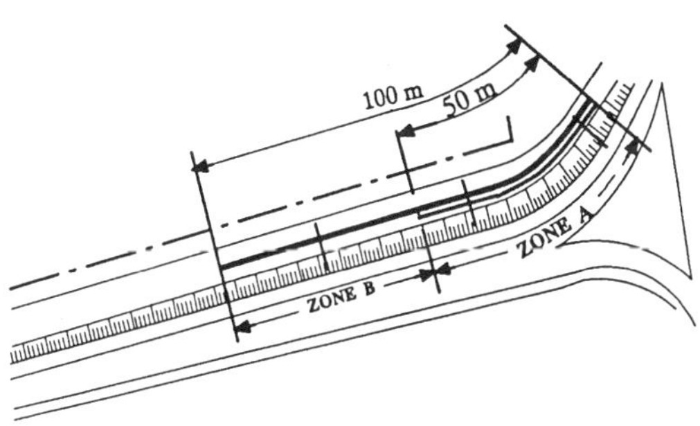

Fig. 7 - Trench Layout

There were two areas (Fig. 7) :
- Zone (A) where groundwater is intercepted by two trenches, because this area has the highest chrome VI concentrations.
- Zone (B), where the flow is intercepted by a single trench (Fig. 8).

The 0.25m wide drainage trenches were widened at the top. This extra thickness of trap potential deals with surface water contaminated with chrome VI liable to seep into the aquifer.

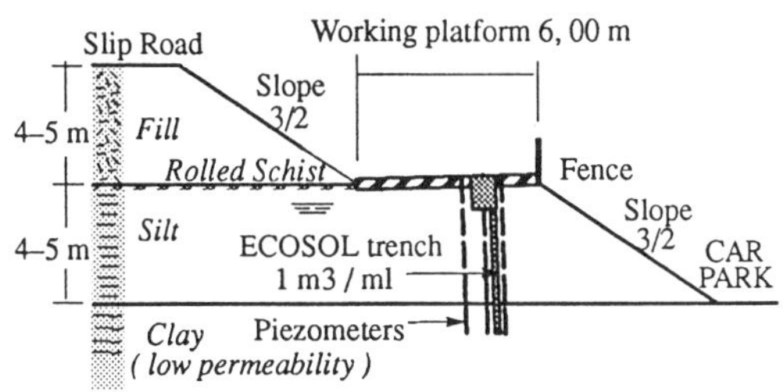

Fig. 8 : Completed Ecosol Drainage Trench

4.4 Trench Performance Monitoring

Three piezometer profiles perpendicular to the trench monitor the trend in chrome VI concentration up- and down-gradient of the trench. Eleven piezometers were installed to monitor trench performance.

The Ecosol material confers a high trap potential on the trench, around 15 kg chrome VI per linear meter of trench. In the absence of data on chrome VI abundance up-gradient of the trench, one of two scenarios may be applicable.

In the first scenario, trench trap capacity is sufficient to intercept all the chrome VI present up-gradient. Since the chrome VI is irreversibly fixed, the pollution problem can be considered solved. If desired, the trench backfill could be removed and disposed of at an approved site. The coating of Ecosol grout on which the chrome VI is fixed can be separated mechanically from the gravel, and the gravel can be re-used.

In the alternative scenario, trench trap capacity is insufficient to intercept all the chrome VI present up-gradient. In this case, a second trench could be built to duplicate or replace the first.

The upstream and downstream piezometers monitor ground water levels and provide regular samples for chemical analysis. These data are used to quantify the volum of chrome VI on the upstream side and check the functioning and dregree of saturation of the trench. The first results are successful.

5. CONCLUSION

These two successful projects are indicative of the wide range of uses for Ecosol pollution-trapping materials :

- To trap essentially organic pollutants in a diaphragm wall.
- To trap chrome VI in a drainage trench.

The second technique, treating pollution with an Ecosol-filled drainage trench, can be generalised to other pollutants: heavy metals such as lead, cadmium, copper, nickel, zinc, chrome, mercury and arsenic, and organic compounds like aromatics, volatile halogens, pesticides, PCB's, etc.

The Ecosol drainage trench is an economical alternative to pumping and treatment plant. It dispenses with the need for the small treatment station which needs maintenance, energy and staff to operate. The slurry film containing the specific pollutant-trapping substances is only active in contact with pollutants, and the material automatically adapts to changes in pollution levels.

The technique can be used for prevention as well as cure, to protect a delicate ecosystem against accidental pollution. The slurry containing the specific trapping substances is only active when in contact with the pollutants for which it is specifically designed. There might, for example, be a preventive trench between a factory and a river, whose length would be limited to the factory's impact area, and running perpendicular to the direction of flow towards the river. Under normal conditions, the trench would remain passive until an overflow release, leaking tank, fire or other accident allowed pollutants to reach the water table.

6 REFERENCES

Bouchelaghem A., Cherel A & Gouvenot D, 1989. Les techniques spéciales de génie civil appliquées à la protection de l'environnement. *L'eau, l'industrie, les nuisances - Juin 1989.*

Bouchelaghem A. & Gouvenot D, 1993. Barrières antipollution souterraines. *Géoconfine, Montpellier (France), 8-11 juin 1993.*

Bouchelaghem A. & Gouvenot D., 1994. Barrières ouvragées pour la protection des eaux souterraines. *XIIIème ICSMFE, New Delhi, Inde, Janvier 1994.*

Cherel, A & Gouvenot D. 1991. Techniques de pointe de travaux spéciaux dans le sol pour la protection des eaux souterraines contre les polluants minéraux et organiques, *XXIème journées de l'hydraulique, Sophia-Antipolis, Janvier 1991.*

Gouvenot D, 1993. L'Innovation Géotechnique au service de la protection de l'Environnement. *Colloque International Environnement et Géotechnique, Paris, 6-8 Avril 1993.*

Environmental Geotechnics, Kamon (ed.) © 1996 Balkema, Rotterdam. ISBN 90 5410 848 7

An innovative approach to insitu remediation of metal-contamination aquifer

S.C.H.Cheung – *Vanbo Environmental Engineering Consultants Inc., Montreal, Que., Canada*

C.P.Chen – *Foci Consulting Inc., Westerville, Ohio, USA*

Tim Law – *Civil & Environmental Engineering Department, Carleton University, Ottawa, Ont., Canada*

Y.Hirata – *Taisei Kiso Sekkei Co., Tokyo, Japan*

M.Fukue – *Ocean Civil Engineering Department, Tokai University, Japan*

ABSTRACT

Heavy metals are commonly found in the aquifer at many of the contaminated sites. The conventional pump-and-treat method is among the most widely used system for groundwater remediation. However, this method is often questioned as it requires decades of costly operations to achieve the desired levels of cleanup because of problems with slow tailings due to physical and/or chemical cause. This paper focusses on the development of an innovative approach using a combination of the chemical enhancement and resin extraction technology for implementation into this method for slow tailings caused by chemical causes. For physical cause, an innovative in situ underground water flow meter will be presented. This instrument is attractive for use to study the groundwater flow field which is required in the design of wells system to overcome the problem of slow tailings arising from different hydrogeological flow rates.

1. INTRODUCTION

Many waste disposal sites are contaminated with heavy metals (EPA 1990). A large amount of such metals have migrated into the aquifer. Because the heavy metals are highly toxic to humans, animals, and aquatic life, the contaminated aquifer must be remediated in order to protect the environment. A common method for groundwater remediation, known as pump-and-treat removal method, is to withdraw the contaminated water from the aquifer and treat it on-site as shown in Figure 1. However, this method is very often questioned as it will require decades of costly operations to achieve the desired levels of clean-up. The lengthy period is related to the residual contaminants, hereto referred as slow tailing, caused by the physical and/or chemical processes in the contaminated aquifer as shown in Figure 2 (Keely, 1989; Palmer and Fish,1992)

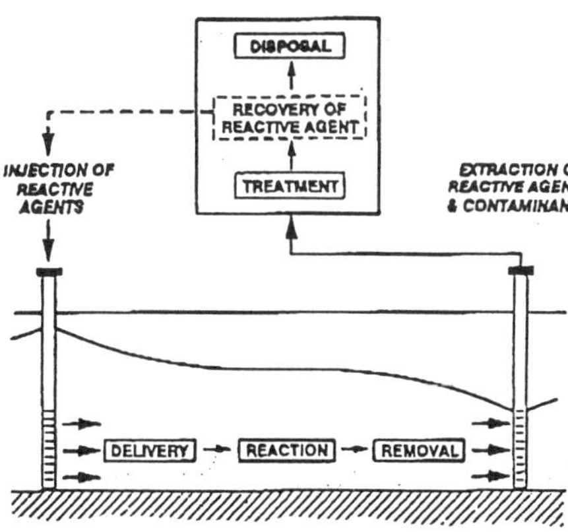

Figure 1. Schematic representation of chemical enhancement of a pump-and-treat operation.

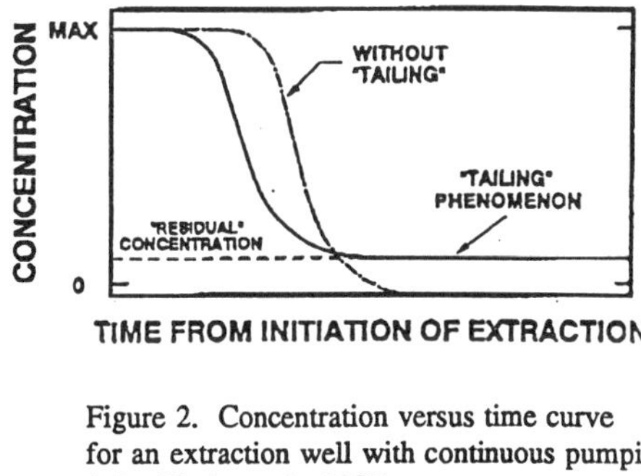

Figure 2. Concentration versus time curve for an extraction well with continuous pumping (after Keely et al., 1987).

The cause of the slow tailing is expected to be site specific. For instance chemical cause may be due to the slow release of contaminants from the soil constituents and contaminant precipitates. The extent of contaminant release depends largely on the compositions of soil constituents and groundwater which vary from site to site. Slow release due to chemical cause may be remediated by chemical reagents. Physical cause may be due to the slow transfer of contaminants arising from different hydrogeological flow rates to the extraction well and the heterogeneous permeabilities of soil masses within the contaminated zone (Palmer and Fish,1992). Chemical reagents may be used to modify the various flow properties arising from heterogeneous soil masses to such a level that the potential natural physical cause of slow tailing can be greatly reduced. The problem of hydrogeological flow rates can be amended by the design of the wells system from a knowledge of the groundwater flow field.

Other problems associated with the conventional pump-and-treat method are the disposal of the large volume of secondary wastes which are often generated by the recovery of heavy metals from extraction solution and the costly remediation process. Because of these problems, the evaluation of the effectiveness of the pump-and-treat remediations and the improvement of this effectiveness have become the current issues confronting site remediation decision makers.

Innovative technologies must be developed to overcome the above-mentioned problems. The technologies must be able to concentrate contaminants prior to extraction to minimize the volume of water recovered through extraction well prior to above ground treatment and to reduce the lengthy periods of time required for contaminant reduction. Chemical enhancement may be used to increase mobilization of metals in groundwater and subsequently recover the metals using an extraction well. Such technologies must be able to demonstrate the delivery and uniform dispersal of the required reagents to the zone of contamination as well as recovery of metals and unconsumed reagents and metals in the soil, and the effects of reactions of the reagents with the soil or contaminants on the physical properties of the saturated sediments. Innovative technologies are also needed to recover the heavy metals from extraction solution without generating a large volume of secondary wastes which require additional costly disposal. This paper will present an innovative underground flow meter to be used for hydrogeologic flow field and discuss the approach leading to the development of a technological system which will enable the clean-up of heavy metals to the desired level in a cost effective and practical manner as follows.

2. UNDERGROUND WATER FLOW VELOCITY METER

We have developed an in situ device which can directly measure the groundwater flow rate and direction simultaneously at any desired location in the field as shown in Figure 3. The device is consisted of three parts: (1) a data acquisition system,(2) a pressure control unit and (3) a cylindrical probe. The first two parts stay on the ground surface while the probe is lowered to the bottom of a cased borehole where measurements are made. The probe has a porous chamber filled with glass beads to allow the groundwater to flow through in the same rate and direction as in the natural soil. The flowrate in the chamber, however, is different from that in the surrounding soil because of the difference in permeability coefficient between the glass beads and the soil. This difference can be measured in the laboratory and calibrated for use in the field. The primary function of the flowmeter therefore is to measure the chamber velocity from which the natural groundwater velocity can be obtained in the field. The chamber has a central electrode and twelve other electrodes lining its circumference. The groundwater flow rate and direction are measured with the aid of distilled water serving as a tracer. Released from the central electrode, the distilled water merges with the prevalent natural groundwater flow and changes the electrical conductivity between the central and the circumferential electrodes. Monitoring all these changes and their associated locations, the data acquisition system will yield the speed and the direction of the groundwater flow.

Figure 4 shows a typical set of test results for a sand aquifier at an industrial site in Japan. The three highest voltage change-time curves from among the 12 electrodes are shown here. The rest are clustering near the time axis and are not shown in the figure. As can be seen, the maximum curve is associated with electrode No. 6 which is used to evaluate the groundwater flow rate and direction. The seepage velocity is calculated to be $4x10^{-5}$ m/s, a value compatible with the hydraulic gradient at the site and the type of sand. Because electrode No. 1 is usually placed towards the north direction, the groundwater is flowing along a direction south 15° east.

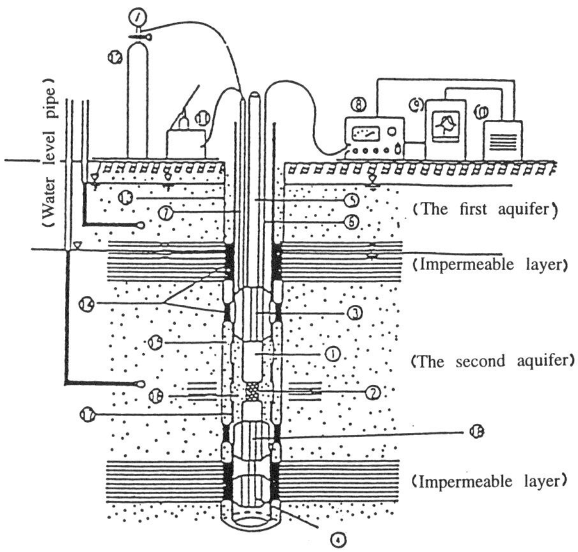

1. Measuring probe
2. Electrode section (including glass balls)
3. Upper rubber air packer
4. Lower rubber air packer
5. Guide rods (boring rods)
6. 16-core type cable
7. Connecting air or water pressure vinyl pipes
8. Data logger including indicator
9. Data printer
10. Power supply
11. Manual water pump
12. Gas cylinder
13. Bore hole
14. Water interception material
15. Filter material
16. Sand to fill the space between probe and casing pipe
17. Screen
18. Packer to prevent the sand from falling

Figure 3. Schematic Diagram of Groundwater Flow Meter System

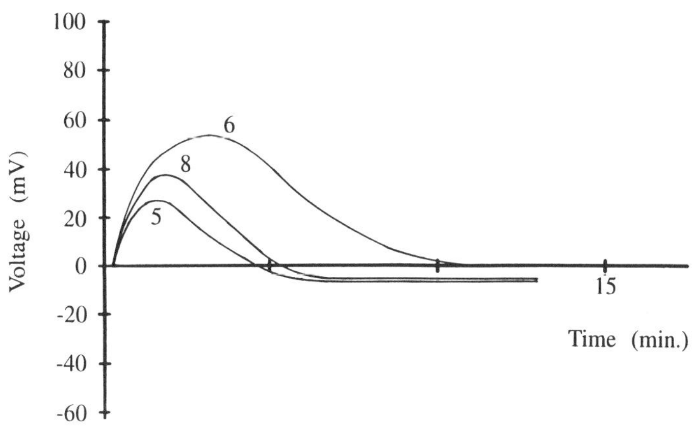

Figure 4. Voltage-time curve for various electrodes for an industrial site

3. IDENTIFICATION OF REAGENTS FOR CHEMICAL ENHANCEMENT TECHNOLOGY

To identify the potential chemical reagents for use, it is fundamental to know how the heavy metals are held by soil constituents and the effects of chemical reagents on the release of heavy metals. Heavy metals are retained by various soil constituents through the processes of adsorption/absorption and coprecipitation. The relative amount of which is dependent on soil type, contaminant type and groundwater pH as shown in Figure 5. The problem

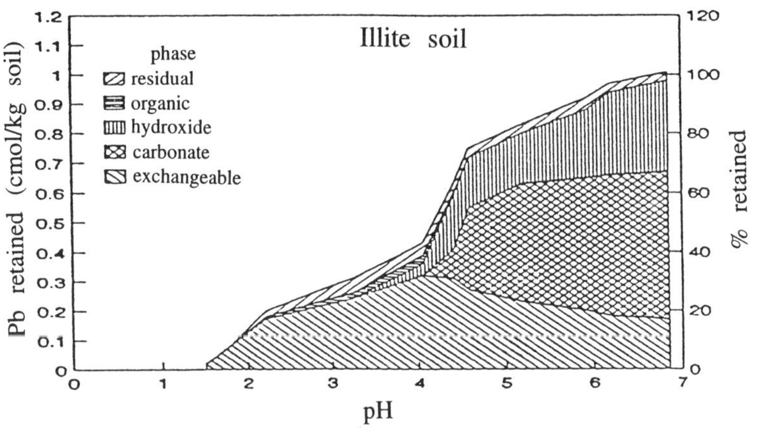

Figure 5a. Amounts of Pb retained by the various constituents and mechanisms as a function of pH for an illite soil.

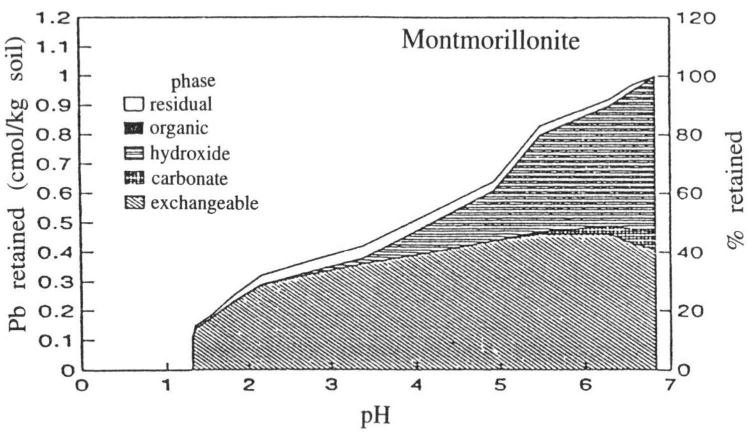

Figure 5b. Amounts of Pb retained by the various constituents and mechanisms for montmorillonite as a function of pH.

of tailing is associated with the slow release of heavy metals from the adsorption sites and precipitates. To enhance the release, chemical reagents may be chosen to compete with the contaminants for the adsorption sites, complex the contaminant, change the redox state of the contaminants, change the solvation properties of the groundwater, ionize the contaminants in a precipitate. However, the chemical reagent may be hazardous and the unconsumed chemical reagent may not be easily recovered from the aquifer. The chemical reagent may induce precipitates and substantiate the growth of microorganisms in the aquifer which may plug the injection and extraction wells (Palmer and Fish,1992). This condition is undesirable as it will impair the performance of the wells system. The chemical reagent may influence the permeability and mechanical properties of soil significantly to have impact on its surrounding environment such as settlement of the building (Yong et al., 1992; Fukue et al., 1992). The chemical may affect the treatment process (Haas and Vamos,1995). The use of chemical reagent may also be limited if the slow tailing is largely due to physical process such as that caused by the natural heterogeneous soils in the aquifer. For chemical enhancement method to be effective, the chemical reagents is desirable to possess the potential ability to minimize this natural physical cause of tailing. This may be achieved by using a chemical reagent solution which minimizes the effect of heterogeneous clay minerals on the soil permeability. A good candidate for this may be saline solution which can eliminate the mineral surface force effects by depressing the ion layers resulting in almost the same permeabilities in heterogeneous soils. (Cheung et al.,1987)

From the above studies, it seems feasible to identify the potential chemical reagents from the knowledge of chemical reaction in the soil matrix and the effect of soil permeability with the chemical reagent in the field. Because the constituent and groundwater composition vary in the field, a range of chemical reagents that can enhance the continuous release/transfer from a specific process is required to formulate the composition and sequence of application. It is recognised that even if the chemical reagents are non-hazardous, the composition of the chemical reagents must also be optimized for effective performance in the field condition with due compliance to the government regulation.

In this approach , it is first necessary to identify the potential chemical reagents that are able to enhance the heavy metals from the soil matrix in the field. One of these chemical reagents will be selected to possess the ability to modify the permeability of clay soil for uniform transfer of contaminant in heterogeneous soil masses. The potential chemical reagents will then be screened by considering its toxicity and its potential impacts on the aquifer and the on-site treatment process. Chemical reagents of hazardous nature will be eliminated. After the potential chemicals reagents are identified, it is required to select its chemical composition for various soil types, heavy metals type and concentration, and groundwater compositions to be expected in the field. This can be achieved with the help of a geochemical model. One common geochemical model presently in use is GEOCHEM (Sposito and Mattigold, 1979). This model sets up the thermodynamic relationship for the vaious reactions, processes, and species and solves the relationships simultaneously to give equililbrated mass distribution of the various species dissolved , sorbed, and precipitated. This model will indicate the effect of chemical reagents on groundwater such as precipitates which have the potential of plugging the wells system and affecting the rate of contaminant removal. This information gives guidance under what groundwater conditions the chemical reagent is not applicable. Moreover, it will show the relative effect of the chemical reagents on the individual chemical processes and their interdependence and also the relative amount of the contaminants held by the soil constituents (Yong et.al.,1992)

It should be noted that this model assumes equilibrium condition and is valid only for instantaneous reaction which is generally the case for heavy metals. To use this model for prediction, dynamic equilibrium must be achieved, depending on reaction and flow rate of the chemical reagents to be assessed experimentally. After the composition of the chemical reagent is identified, it is desirable to have an estimate of how much the chemical reagent is required as well as how much it will be consumed. This can be achieved using the GEOCHEM model with the knowledge of soil properties such the sorption isotherm, ion exchange and pH buffering capacity of the soil through laboratory testing. Finally, the sequence of the chemical reagents application for uniform dispersion and effective performance remains to be identified. This can be achieved by considering the individual chemical/physical processes and the flow rate of chemical reagents(Cheung and Chan 1988).

4. RESIN/MEMBRANE SYSTEM FOR CHEMICAL EXTRACTION

It is desirable to develop a resin/membrane system that can extract only heavy metals contained in the chemical reagents. The use of resin/membrane systems is innovative in that it allows continuous on-site removal of heavy metals as compared to the intermittent conventional precipitation method. The process is continuous until the resin is depleted. After the resin is depleted, it can be regenerated and reused. The use of resin to remove or to extract metals from polluted solution have been studied for different applications (Kingston et al.,1975 and Sengupta and Sengupta 1993). Much work remains to identify resin/membrane systems for extraction on-site of heavy metals. This work should be able to address the following questions.

(1) What are the advantages and disadvantages of using this method as compared to the conventional chemical process such as precipitation/flocculation?

(2) Are there resin/membrane systems commercially available or is it worthwhile to have this developed?

(3) What are the most effective method to recover heavy metals?

(4) What are the most effective methods to regenerate and reuse the resin?

(5) What are the secondary waste,if any, generated from the process of recovering metals and regenerating resins?

(6) Can chemical reagents be separated and reused?

5. CONCLUSION

We have developed an innovative in-situ groundwater flow meter. This device is very attractive for use in studying groundwater flow pattern for contamination remediation. An innovative approcah has also been developed to enhance chemical release from the soil matrix and to extract the heavy metals from the pumped water on-site. This approach is directed to improve the efficiency and practicality of the conventional pump-and -treat technology.

REFERENCES

Cheung, S.C.H. and T. Chan, "Parameter-Sensitivity Analysis of Near-Field Radionuclide Transport through Engineered Barriers" International Journal of Energy Systems, Vol 8,No.1, p37,1988.

EPA. "Handbook on In Situ Treatment of Hazardous Waste Contaminated Soils", EPA/540/2-90/002, Risk Reduction Engineering Laboratory, Office of Research and Development,U.S> Environmental Protection Agency, Cincinnati, Ohio, January 1990.

Fukue, M., T. Nakamura, T. Negota, and S.C.H. Cheung "Cementation of Marine Soils due to Calcium Carbonate", Proc. of 1992 CSCE Annual Conference, Quebec City, May 1992.

Haas, C.N. and R.J. Vamos,"Hazardous and Industrial Waste Treatment", Prentice Hall, New Jersey,1995.

Keely, J.F., "Performance Evaluations of Pump-and-treat Remediations", EPA/540/4-89/005, Office of Research and Development, U.S. Environmental Protection Agency, 1989.

Kingston, H.M., I. Barnes, T.J. Brady, and T.C. Rains,"Separation of Eight Transition Elements from Alkali and Alkaline Earth Elements in Estuarien and Sea Water with Chelating Resin and Their Determination by Graphite Furnace Atomic Absorption Spectrometry", Analytical Chemistry,Vol.50, No.14, p.2084-2070, 1975.

Palmer, C.D. and W. Fish, "Chemical Enhancements to Pump-and-Treat Remediation" EPA/540/S-92/001, Office of Research and Development, U.S. Environmental Protection Agency,January 1992.

Sengupta, S. and A.K. Sengupta," Characterizing a New Class of Sorptive/Desorptive Ion Exchange Membrane for Decontamination of Heavy-Metal-Laden Sludges", Environmental Sci. Technology, Vol.27, No.10, pp.2133-2140,1993.

Sposito, G. and Mattigold, "GEOCHEM: A Computer Program for Calculation of Chemical Equilibria in Soil Solutions and other Natural Water System", University of California, Riverside,California, 1979.

Yong, R., A.M.O. Mohammed, and B.P. Warkentin, "Principles of Contaminant Transport in Soils", Elsevier Science Publisher, Netherlands, 1992.

Environmental Geotechnics, Kamon (ed.) © *1996 Balkema, Rotterdam. ISBN 90 5410 848 7*

Slope stabilisation and hydrocarbon remediation at a fuel service station

R.G.Clark
CL Associates, Halesowen, Birmingham, UK

ABSTRACT: Cracking and structural displacement occurred at the Nene Service Station in the years following its construction. Investigation revealed that the promontory of fill supporting the Service Station had been built out over a periglacial landslip which had been reactivated by the loading due to the fill. Details are presented of the stabilising measures carried out to restrict future movement and the remediation of hydrocarbon contamination in the flood plain below the Service Station which was necessary before the stabilising measures could start.

1. INTRODUCTION

The present Service Station at this site was built some eight years ago. Following construction, the forecourt slab and the brick sales building both cracked due to ground movement. Displacements of up to 50 mm horizontal and 40 mm vertical occurred. A previous service station at the same site also cracked and the fuel tanks rotated in the ground. This paper discusses the reasons for the movement and presents details of the stabilising remedial measures that were carried out, together with the extent of hydrocarbon contamination that required clean up before the remedial measures could be implemented.

2. THE SITE AND GEOLOGY

The Service Station site is located immediately to the south of the A47 Trunk Road about 1 km to the east of Wansford in Cambridgeshire (see Fig 1). Here the A47 runs on a low embankment which has been built across a natural slope and the Service Station is constructed on a promontory of fill extending out from the A47 over the slope.

The natural slope falls by about 9 m to the south at an inclination of about 15° to the horizontal from the road level to the flood plain of the River Nene. The strip of land forming the flood plain is about 45 m wide and is at an elevation of approximately 10 m AOD. The River Nene flows from West to East.

The slope comprises alluvium over Terrace Gravels and Northampton Sand directly overlying Upper Lias clay. There is some alluvium on the flood plain, which also overlies the Upper Lias clay. Periglacial features are known to occur in the Lias Clay and to cause landslips (Weeks, 1969).

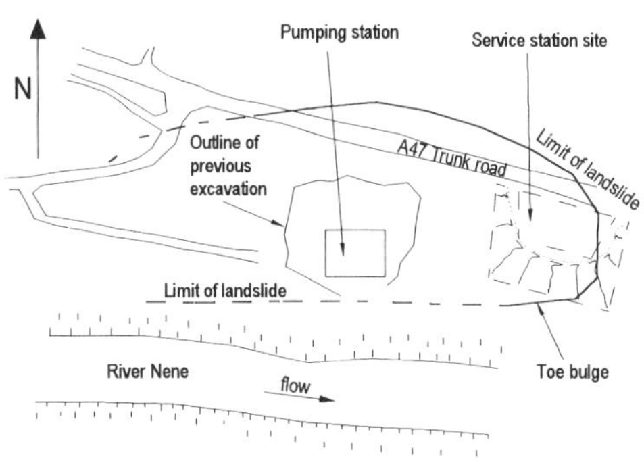

FIG 1. PLAN OF LANDSLIDE

3. DESK STUDY

Inspection of the site revealed a very evident bulge in the natural ground surface at the toe of the fill slope approximately due south of the Service Station. It was also noted that trees growing at the top of the fill slope had characteristically bowed trunks indicating that movement at the site had been taking place for some time. The trees were about eight to ten years old.

To the east of the Service Station between the A47 and the river there is open pastureland. No development has taken place on this land and the ground profile is likely to represent the slope in its natural state. The slope topography is typical of previously slipped ground, with numerous low undulations.

At the toe of the slope a spring line is clearly marked by patches of reeds. Between this spring line and the River Nene the ground is marshy and an attempt has been made in the past to drain this area by excavating shallow ditches draining into the river.

About 100 m to the west of the Service Station and adjacent to the river there is an Anglian Water Pumping Station which experienced stability problems during construction as reported by Chandler, 1979. During excavation for the foundations of the Pumping Station in 1972, major slips occurred as a result of the excavations. Investigations revealed a major landslip, the extent of which is shown on Fig 1. It can be seen that the area of landslip included a section of the A47 and the site of the Service Station. Slip surfaces were found in a soliflucted layer beneath the Terrace Gravels and in the underlying weathered Upper Lias Clay.

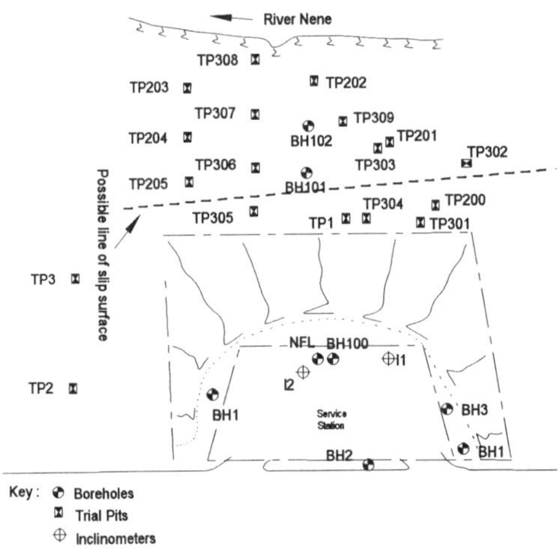

FIG 2. SITE PLAN SHOWING EXPLORATORY HOLE POSITIONS

4. GROUND INVESTIGATIONS

Following the incidence of movement and cracking at the Service Station, a ground investigation was carried out in 1990 comprising soft ground boring, rotary drilling, trial pits and piezometer installation. The investigation positions are shown on Fig 2. Where the promontory of fill is present the investigation revealed Made Ground over Terrace Gravels over Northampton Sand over Upper Lias Clay. In the boreholes potential zones of slippage were found comprising extremely closely spaced discontinuities infilled with very soft clay in an otherwise firm to very stiff clay stratum.

Trial pits were excavated at the toe of the promontory of fill due south of the Service Station and at the side of the promontory of fill on natural ground to the east. These showed a layer of soliflucted Lias Clay below the Terrace Gravels with distinct shear planes and gleyed surfaces. A pit excavated into the toe bulge found that the near surface strata had been folded such that the shallower layers had moved towards the river relative to the deeper layers as shown in Fig 3 (Trial Pit 1) which indicates a geological section on a north-south line through the Service Station. The folding could be clearly seen because of coloured marker layers within the clay.

The groundwater level in the flood plain was about 1 m below ground surface. The regional groundwater flow was generally southwards, ie downslope under the Service Station towards the River.

5. MECHANISM OF FAILURE

The results of the investigation indicated that the promontory of fill had moved as a result of slippage along pre-existing shear surfaces within the solifluction layer. This landslipping would have been reactivated by the placing of the fill with greatest movement occurring during wet periods. There was some evidence to suggest that the toe bulge had increased in size as the movement proceeded. The line of intersection of the toe of the slip with the ground surface is shown on Fig 2.

6. SLOPE STABILITY REMEDIAL WORKS

In order to stabilise the slip and allow the Service Station to continue operating a variety of possible remedial measures were considered. These included drainage, toe weighting, anchors and piles (Hutchinson, 1977). On the basis of practicality and cost a solution involving a combination of toe weighting and drainage was adopted.

To determine the required size of the toe weighting a slope stability back analysis was carried out in order to determine appropriate material parameters (in combination with laboratory test results) for a factor of safety of unity (Chandler, 1977). The analyses were for circular (Bishop, 1955) and non circular (Morgenstern and Price, 1965) slips.

The actual slip surface was likely to have been non circular. The resultant parameters were then used

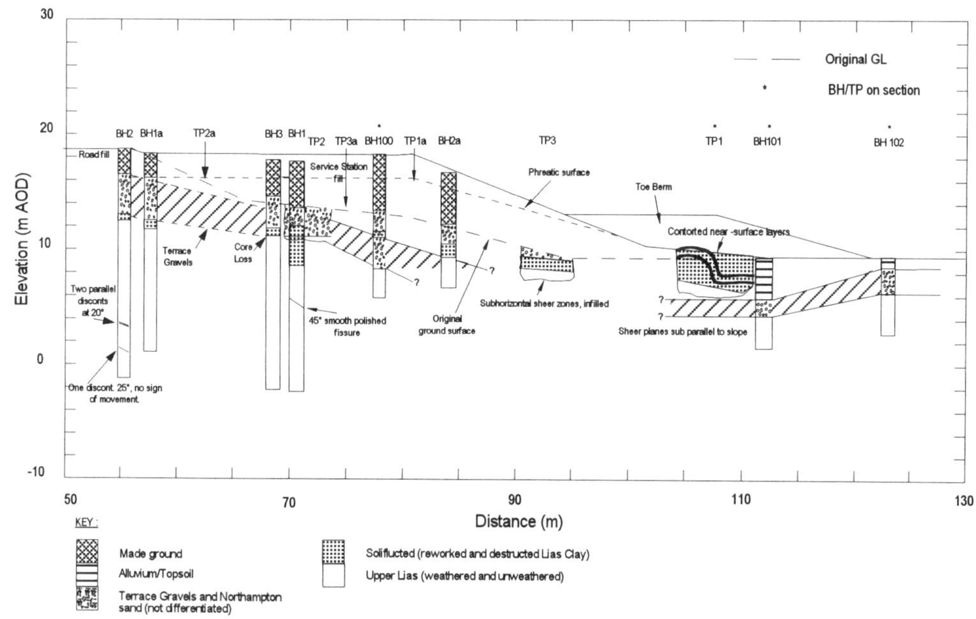

FIG 3. GEOLOGICAL CROSS SECTION.

in the design of the toe weighting. The geometry of the designed toe berm, to give a factor of safety of 1.3, is shown in Fig 3. The following parameters for the natural materials were used in the analysis.

Material	c'(kN/m^2)	ϕ'
Terrace Gravels	0	35
Solifluction Layer	0	14
Weathered Upper Lias	0.5	19

The scheme also included improved drainage under the forecourt, in the fill and in the flood plain.

As part of the remedial works two inclinometers to a depth of 25 m were installed from the top of the fill slope. To date these have shown no movement.

7. HYDROCARBON CONTAMINATION

Hydrocarbon contamination was found in the gravels on the flood plain whilst investigating the cause of instability at the site. A specific investigation comprising both trial pits and probeholes was therefore carried out to investigate the extent and degree of this contamination.

The method of probing that was used obtained small 'undisturbed' samples from the last penetrated cohesive material in each hole. These samples were subjectively evaluated for visual and olfactory evidence of hydrocarbon contamination. Soil gas concentrations were measured in the probeholes using a portable photoionization analyser (PID) immediately after the completion of every probe. The soil atmosphere was sampled for total volatile hydrocarbon (TVH) fractions at a depth of 0.25 m bgl after temporarily sealing the top of the hole to prevent dilution of the soil atmosphere. Soil and water samples were sent to the laboratory for quantitative gas chromatography analyses to determine total petroleum hydrocarbon (TPH) concentrations. The probehole locations and concentration contours of total hydrocarbons in soil gas (expressed as ppm Benzene equivalent) are shown in Fig 4. The locations of the trial pits are shown in Fig 2 (300 series).

The PID was used to enable rapid detection. This type of instrument provides an indication of the total concentration of volatiles in the soil gas but provides no information on the specific type of contamination present or its composition and nature. Nevertheless, the method enabled the distribution and overall extent of the contamination to be determined.

The laboratory analyses on samples of the clay below the gravels showed TPH concentrations below the level of detection. The maximum TPH concentration in the groundwater in the gravels was 390 mg/l in Trial Pit 303.

Petroleum behaviour in the ground is difficult to predict because of its low relative density, its tendency to float on the water table and the fact that it forms "mounds" beneath surface spills or underground leaks. Floating petroleum can often move in directions other than the regional hydraulic gradient. For this reason, a large number of probes were put down to enable a large area to be tested for volatile aromatic hydrocarbons in the soil gas. It can be seen from Fig 4 that there was a well defined plume of petroleum contamination in the flood plain generally following the direction of flow in the river. The plot also indicates where the petroleum contamination was exiting from the toe of the fill slope. The laboratory results confirmed the visual evidence that the hydrocarbon contamination was primarily confined to the groundwater within the alluvial gravel horizon. Although the plot suggests that some migration of hydrocarbon contamination was taking place towards the river, none was found in the river water.

8. HYDROCARBON REMEDIATION

Before the stabilising works could be constructed it was essential that the hydrocarbon contamination in the flood plain was cleaned up. There were two reasons for this. Firstly, the scheme included the provision of drainage in the flood plain discharging to the River, which would simply have provided a direct flow path to the River, even though interceptors were to be included in this drainage. Secondly, the construction of the toe berm would have made it virtually impossible to carry out remediation of the hydrocarbon contamination at a later date.

The clean up method involved a combination of the use of absorbent pads and circulation of contaminated groundwater through a granular activated carbon filter. This combination was chosen because the contamination was substantially held within a relatively shallow granular layer in the floodplain above low permeability clay. There was therefore no need for deep remediation.

A pit (see Figs 4 and 5) approximately 2.5 m deep and with plan dimensions of 7.5 x 7.5 m at the surface was excavated at the location where the highest concentrations of petroleum product had been found (Recovery Pit). Another pit approximately 2 x 1 m in plan and deep enough to penetrate the surface of the gravel was excavated further downstream of the Recovery Pit at the eastern edge of the contamination plume (Recharge Pit).

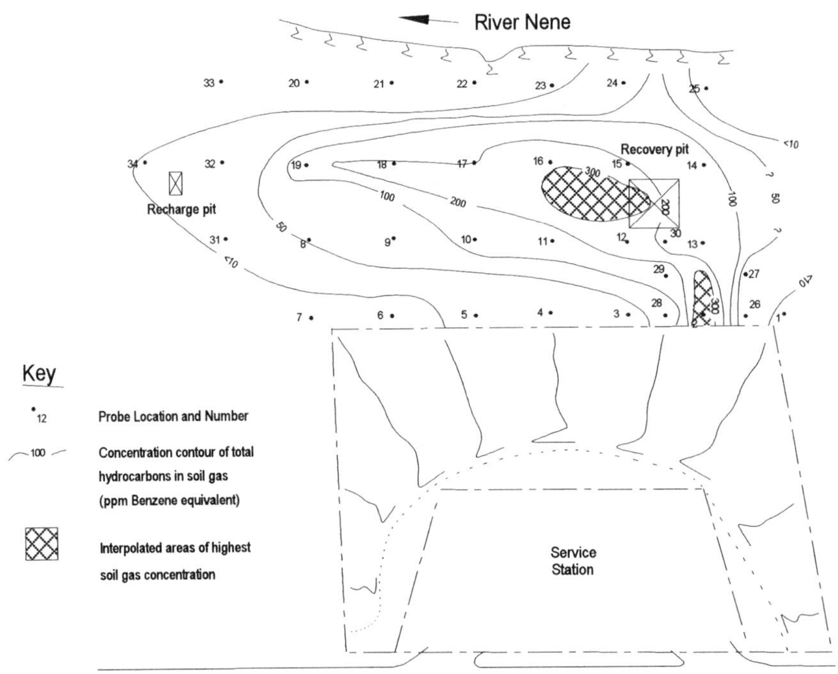

FIG 4. HYDROCARBON PLUME

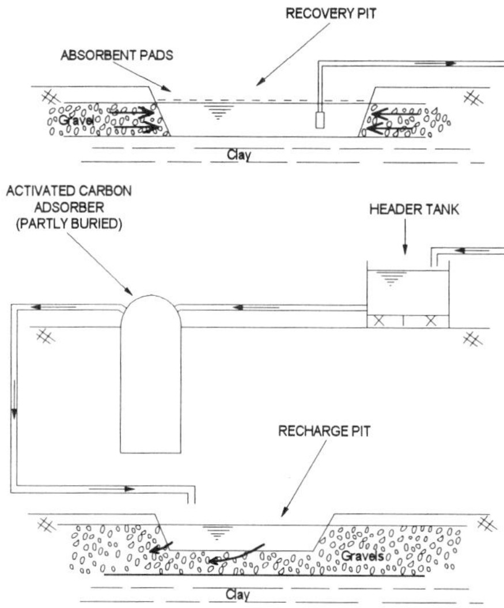

FIG 5. SCHEMATIC DIAGRAM SHOWING HYDROCARBON REMEDIATION.

The excavation of the Recovery Pit caused a considerable inflow of hydrocarbon contamination into the pit. This was effectively removed by the use of the absorbent pads. These were 450 by 450 mm and were supplied in rolls with a total length of about 45 m and placed on the surface of the water within the pit. The pads float and were therefore placed, tied together, in strips. They were replaced as necessary during the subsequent circulation of groundwater and the used pads were sent to a licensed facility for disposal.

Additionally, a submersible pump (capacity approximately 5m^3/h, 230 volt, 0.5 kw) was installed in the Recovery Pit just below the surface water level. The pump was connected to a float switch to ensure that free hydrocarbon product floating on the surface of the water was not drawn into the pump.

The pump discharge was directed to a drum type granular activated carbon adsorber (diameter 1100mm, overall height 1700mm and capacity 1000 litres) via a header tank. The pump discharge from the Recovery Pit entered the drum filter through a nozzle, percolated downward through the carbon bed (flow rate 6.6 m^3/h) and was then pumped by an internal pump back to the exit nozzle at the top of the adsorber. The treated water was then discharged via a pipeline into the Recharge Pit as relatively clean, clear water with the appearance of fresh drinking water.

Activated carbon adsorption is a proven technology that is capable of removing a wide range of organic compounds to produce a very high-quality discharge. The method is based on the principle that certain organic constituents preferentially adsorb to organic carbon. It is an efficient process that is most effective in removing high molecular weight compounds and moderately effective in adsorption of chlorinated solvents. The contaminated liquid moves upward or downward by gravity or under pressure through columns of carbon. The granulated carbon is `activated' by being placed in suspension under the upward gradient of the pressurised fluid thus allowing an optimum surface area of the carbon particles to come into contact with the contaminated liquid for more efficient adsorption.

During the decontamination extra trial pits were excavated in the flood plain to ensure that the Recovery Pit was located in the centre of the contamination plume and that at the end of the decontamination process the water in the Recharge Pit was representative of the surrounding groundwater.

The majority of the hydrocarbon contamination was removed by the absorbent pads in the Recovery Pit and by volatilisation of the free product from the surface of the Recovery Pit prior to the start of circulation through the activated carbon adsorber. Before any remediation was carried out TPH levels in the groundwater were up to 390 mg/l. During circulation of groundwater through the activated carbon adsorber, BTEX (Benzene, Toluene, Ethylbenzene and Xylenes) levels were reduced from a pre-clean up value of about 1850 µg/l to a post clean up value of about 5 µg/l. The remediation period was six weeks.

In addition to the above, water samples were also analysed for methyl-tertiary butyl ether (MTBE). Concentrations varied considerably during the remedial works from below the detection limit of 0.1 µg/l to 130 µg/l. This was considered to be due to the very high solubility and mobility of MTBE in water which makes it unsuitable for adsorption by granular activated carbon. However, removal of MTBE had not been intended as a specific requirement of the project.

9. CONCLUSIONS

The increased loading on the slope due to the promontory of fill caused a reactivation of a periglacial landslide.

This slide was an extension of the one which occurred in 1972 at the adjacent Pumping Station as reported by Chandler, 1979.

The slope at the Service Station has been stabilised by means of toe weighting and drainage.

Inclinometers have shown that no further movement has taken place.

Hydrocarbon contamination in the flood plain below the Service Station has been remediated by a combination of absorbent pads and a granular activated carbon adsorber to achieve of the order of 99% clean up.

Simple earthworks and drainage in the form of a stabilising berm, together with low technology remediation measures, have been used to reinstate the Service Station and to allow it to continue operating.

ACKNOWLEDGEMENTS

The author wishes to thank Heron Garage Properties and Snax 24 together with the Directors of C L Associates for permission to publish this paper. The consulting structural engineers for the project were Cameron Taylor Bedford and the civil works were carried out by Weldon Construction. The absorbent pads were supplied by Oil Pollution Environmental Control Limited and the activated carbon adsorber was manufactured by Chemviron Carbon Limited.

REFERENCES

Bishop AW : 1955 : The use of the slip circle in the stability analysis of slopes. Geotechnique, Vol 5 No 1.

Chandler RJ : 1977 : Back analysis techniques for slope stabilization works : a case record. Geotechnique, Vol 27 No 4.

Chandler RJ : 1979 : Stability of a structure constructed in a landslide : selection of soil strength parameters. Proc 7th Euro Conf. SM and FE, Brighton, Vol 3.

Hutchinson JN : 1977 : Assessment of the effectiveness of corrective measures in relation to geological conditions and types of slope movement. Bull Int Ass Eng Geol. No 16.

Morgenstern NR & Price VE : 1965 : The analysis of the stablity of general slip surfaces. Geotechnique, Vol 15

Weeks AG : 1969 : The stability of natural slopes in south east England as affected by periglacial activity. Quarterly Journal of Engineering Geology, Vol 2.

Environmental Geotechnics, Kamon (ed.) © *1996 Balkema, Rotterdam. ISBN 90 5410 848 7*

Depollution of oil-contaminated soils by electro-injection

Jean-Marie Fleureau & Siba Kheirbek-Saoud
Laboratoire de Mécanique, Ecole Centrale Paris & CNRS URA, Châtenay-Malabry, France

ABSTRACT: The paper presents the results of an experimental research on the use of electric fields to decontaminate permeable soils polluted by hydrocarbon wastes. Several models of soils (permeability = 10^{-8} to 10^{-6} m/s; porosity = 15 to 40%) were tested in the laboratory, ranging from sands to clayey consolidated soils, as well as oils of different polarities and acidities. The method used consists in simultaneously or alternately injecting water and electric current in the medium and measuring the oil recovery as a function of time. The interpretation of the results focuses on electrochemical and electrokinetic phenomena.

The conclusion of this research is that electro-injection can be an efficient depollution method in the case of oil-contaminated permeable soils, but its use requires the knowledge and/or control of many parameters, including the intensity of the electric field and the conditions of its application, the pH and normality of the pore water, the type of oil, the clay content of the soil, etc.

1. INTRODUCTION

In the last 30 years, lots of research have been devoted to the study of the effect of an electric field or an electric current in soils and rocks either to consolidate the soil (Casagrande et al. 1961, Bjerrum et al. 1967, Gray and Mitchell 1967), to enhance oil recovery in petroleum reservoirs (Tikhomolova 1964, 1969, Chilingar et al. 1970) or, more recently, to clean polluted soils (Shapiro et al. 1989, Hamed et al. 1991, Hatfield et al. 1993, Yeung & Mitchell 1993).

The analyses of the electrokinetic phenomena taking place under the effect of an electric field in the case of monophasic flows are all based on the assumption of the existence of a double or diffuse layer in the vicinity of the solid surfaces and on the fact that the mobile ions of the diffuse layer are driven by the electrical forces (Mitchell 1976). Helmoltz-Schmoluchowski theory is widely used to estimate the flow of water in these conditions. It consists in writing the equilibrium of an element of fluid under the combined effets of the electric driving force, of pressure forces when there is an hydraulic head gradient and of the viscous drag, and leads to the following expression of the velocity of the fluid:

$$u = \frac{D\zeta E}{4\pi\eta}$$

where D is the dielectric permittivity, ζ, the Zeta-Potential of the solid surface, E, the electric field and η, the viscosity of the fluid. Many authors have pointed out the shortcomings of this approach (Haydon 1964) but it remains globally valid and useful.

On the other hand, when there are several types of fluid in the porous medium, interfacial phenomena must be taken into account. A few possible effects of an electric field on the interfacial phenomena have been investigated in recent years: these effects are related either to the presence of surfactants migrating in the electric field and accumulating at the interface (electro-adsorption: Watanabe et al. 1978, Dupeyrat & Nakache 1980) or, in the case of very high electric fields ($\geq$ 1 kV/cm), to an electrostatic modification of the water-oil menisci (Joffre et al 1984, 1986). Moreover, the possibility of electrokinetic phenomena also happening in polyphasic media was mentioned by several researchers.

Many experiments were performed at the Ecole Centrale de Paris in order to assess the efficiency of an electric field, in addition to the injection of brine, to increase the oil recovery from polluted soils or rocks (Fleureau, 1979). Several models of soils were studied as well as several concentrations of brine, and various electrical parameters. At the same time, a study of the influence of the current on the interfacial parameters of the system was made to better understand the driving mechanisms of the tests. Finally, a theoretical model is proposed to express the simultaneous flows of both fluids in the samples.

2. ELECTRO-INJECTION EXPERIMENTS

The tests were carried out in a triaxial cell, modified to allow the use of the sintered bronze porous filters as electrodes, under a small confining pressure (0.2 MPa), on several materials: (a) Fontainebleau sand; (b) unconsolidated clayey sandstone; (c), (d) and (e) consolidated clayey sandstones; (f) consolidated limestone, with porosities ranging from 15 to 40% and monophasic permeabilities, from 10^{-8} to 10^{-6} m/s (Table 1).

The samples (D= 6-10 cm, H= 16-20 cm) were first saturated with brine (with concentrations in Sodium Chloride between 5 and 35 g/l), then the water was displaced by oil (ESSO Bayol 52), until the irreducible water saturation was reached. The first phase of the tests consists in injecting brine through one end of the sample, at a constant flow rate, and measuring the quantity of water and oil recovered at the other end. When no more oil flowed from the sample, the electric field (limited to a few Volts per centimeter) was added in the direction of the water flow, either in the same or in the opposite direction.

Examples of the results obtained with the sand (a) and the unconsolidated sandstone (b) are shown in Figs. 1 and 2, as the oil recovery (i.e. the ratio of the volume of oil recovered to the initial oil volume) versus the ratio of the volume of water injected to the pore volume, V_P. A significant increase in oil recovery is observed each time the electric field is applied, but the volume of oil tends to stabilize after some time, corresponding to approximately ten times the pore volume.

Globally, when considering all the tests, the initial (hydraulical) recoveries were comprised between 50 and 70%, while the additional recoveries (with the electric field) ranged from a few percents to 30%, depending on the type of material and on the electrical parameters. No clear correlation with the mineralogy of the materials was observed, but it was noted that, in the case of fissured or unconsolidated samples yielding the lowest initial oil recoveries, the electric field was the most efficient. On the other hand, in the case of compact limestone, very homogenous, the electrical recovery did not exceed 5%.

It was also remarked that, the higher the salinity of water, the higher the oil recovery by the electric field.

The values of the interfacial parameters play a major part in the mechanisms of displacement of oil by water: a significant decrease in the water-oil interfacial tension may reduce the entrapment of oil.

Moreover, electroosmosis of water may result in

Table 1. Description of the materials used in the tests

	Composition Mineralogy	Porosity %	Permeability m/s
(a)	Pure silica sand	30-40%	10^{-6}
(b)	Clayey sandstone Clay content: 10% (montmorillonite; kaolinite)	25%	8.10^{-7}
(c)	Sandstone with clayey cement	25%	6.10^{-8}
(d)	Sandstone with clayey cement	15-18%	10^{-6}
(e)	Sandstone Clay content: 6% (kaolinite)	17%	6.10^{-8}
(f)	Limestone	15%	10^{-8}

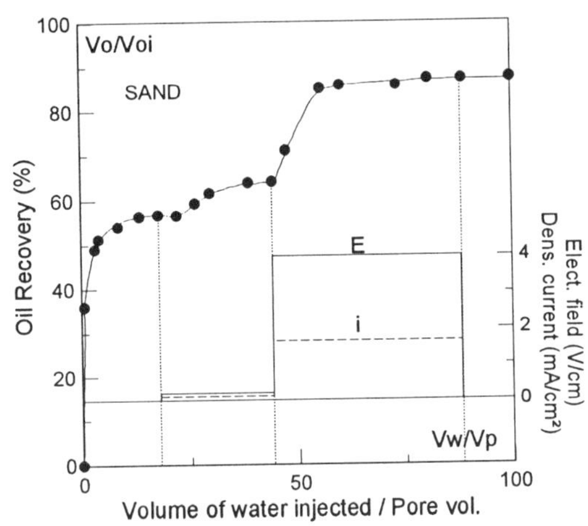

Fig. 1. Oil recovered from Fontainebleau sand vs. volume of water injected during several phases of injection of current

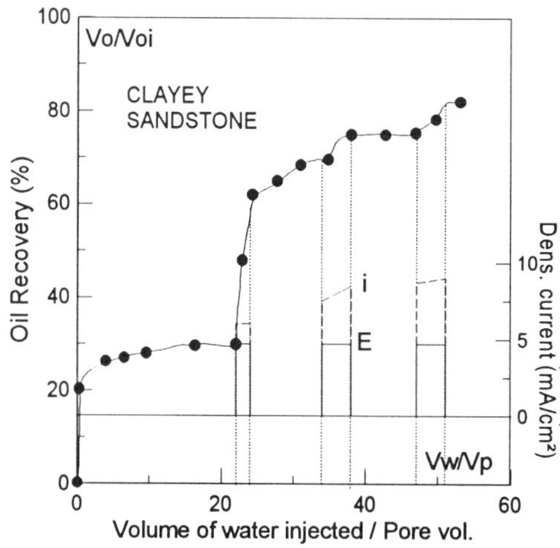

Fig. 2. Oil recovered from fissured sandstone vs. volume of injected water during several phases of injection of current

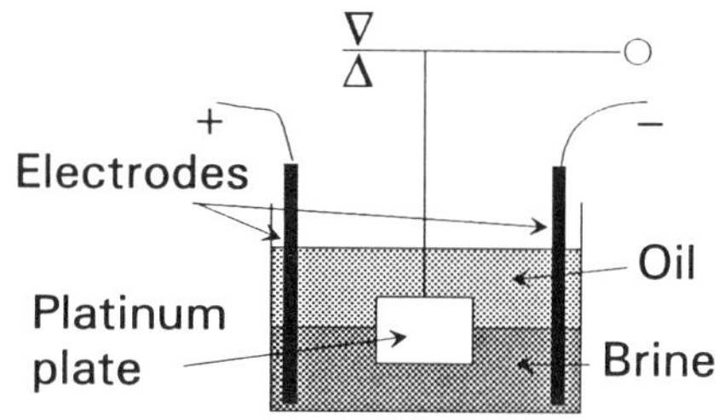

Fig. 3. Experimental device for the measurement of interfacial tensions and the electrolysis of brine

the displacement of the oil occupying the center of the channels in a water-wet porous medium. On the other hand, direct electrophoresis of oil globules by the electric field is possible if surfactant molecules are adsorbed at the water-oil interface and provide a kind of ionization of oil. As, in our experiments, surfactants were not directly added, we had to examine the possibility of in-situ formation of surfactant species through electrolysis reactions.

3. INFLUENCE OF AN ELECTRICAL FIELD ON THE INTERFACIAL PARAMETERS

Considering the relatively low values of the electric field used in the experiments, a direct electrostatic effect on the interfacial tension is not possible. The only way to change the interfacial parameters is therefore through the products formed by the electrolysis of water and their reaction on oil.

The interfacial tension was determined using an equilibrium method, by measuring the force exerted by the interface on a platinum plate suspended to a precision scale (Fig. 3). The oil and brine were placed in a beaker containing approximately 50 cm^3 of each liquid and the electric field was applied by 2 vertical electrodes crossing the water-oil interface. Three types of electrodes (graphite, copper and iron) were used. Once electrolized, the solutions were agitated for 20 hours, in order to reach the partition equilibrium of the surfactant species present in the system.

The results obtained with the graphite electrodes are shown in Fig. 4 as the changes in interfacial tension γ_I versus the quantity of electricity Q, for 4 concentrations in Sodium Chloride. They show a rather important decrease in interfacial tension, from 20 to 6 mJ/m^2 for the 0.7 N solution and from 20 to 3 mJ/m^2 for the 2.6 N solution, followed in both cases by a return to the initial value when the quantity of electricity increases. The pH of the aqueous solution regularly increases during the electrolysis, to reach 10 or 11 at the end of the test at most concentrations. Similar results were obtained with the copper electrodes, though with more limited changes in interfacial tension (Fig. 5); on the other hand, there was practically no change in γ_I when using iron electrodes.

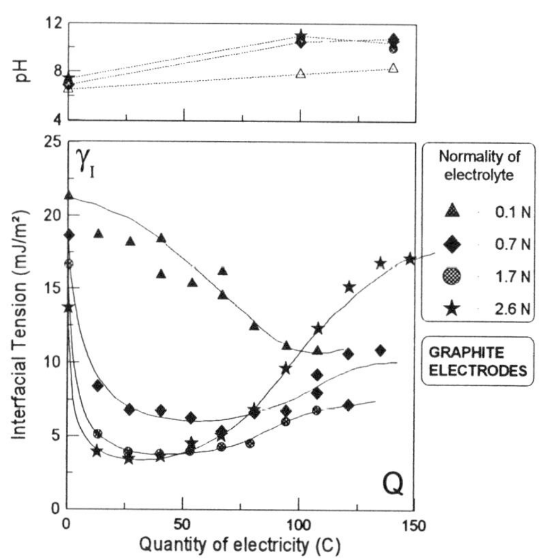

Fig. 4. Water-oil interfacial tension and pH of the aqueous phase during electrolysis of brine with graphite electrodes

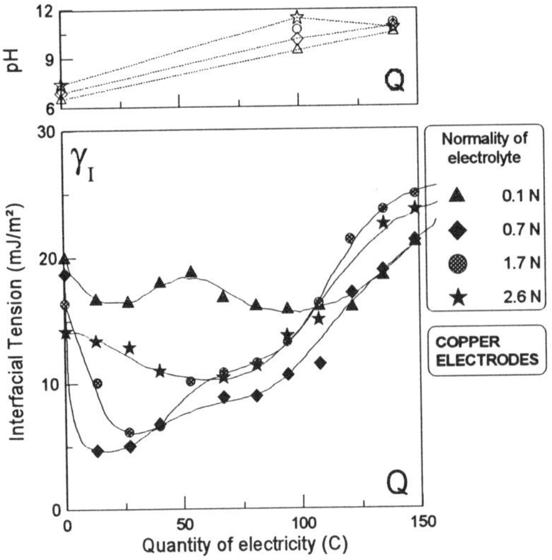

Fig. 5. Water-oil interfacial tension and pH of the aqueous phase during electrolysis of brine with copper electrodes

With graphite electrodes, the electrolysis reactions (in a neutral or slightly basic medium) lead to the formation of hydroxile ions OH^- and the emission of hydrogen at the cathode, and to the formation of H_3O^+ ions and the emission of oxygen and chlore at the anode, following the reactions shown below:

At the cathode, in a neutral or slightly basic medium and in normally oxygenated water:

$$H_20 + 1/2\ O_2 + 2e^- \rightarrow 2\ OH^-$$

or, in water with little oxygen:

$$2\ H_2O + 2e^- \rightarrow 2\ OH^- + H_2 \nearrow$$

At the anode, in a basic medium:

$$2\ OH^- \rightarrow 1/2\ O_2 \nearrow + H_2O + 2e^- \qquad (V_0 = 1.36\ V)$$

$$2\ Cl^- \rightarrow Cl_2 \nearrow + 2e^-$$

and in an acid medium:

$$3\ H_2O \rightarrow 1/2\ O_2 \nearrow + 2\ H_30^+ + 2\ e^- \qquad (V_0 = 1.23\ V)$$

Both reactions have similar reference potentials but the decomposition of water is highly dependent on pH and decreases when the medium becomes basic. The second reaction will be therefore more likely to happen but, globally, the pH will increase as there will be more OH^- ions produced at the cathode than H_3O^+ ions at the anode, which is confirmed by experimental results. The decrease in interfacial tension shows that the hydroxyle ions can react on the acid components of the oil to form carboxylates which accumulate at the water-oil interface. The same decrease can be observed when Sodium Hydroxide is directly added to the aqueous phase, as shown in Fig. 6. When the OH^- ions produced at the cathode pass through the sample as in our experiments, the same phenomenon can happen and lead to a kind of ionization of oil droplets, making their displacement possible by the electric field.

With copper or iron electrodes, additional reactions lead, respectively, to the formation of copper and Cu^{++} ions, or to the formation of hydrated oxides FeO_3 and Fe_3O_4. In the latter case, the OH^- ions produced at the cathode disappear in the reactions, the solution becomes slightly acid and the interfacial tension remains nearly constant.

These results were confirmed by another study on Fontainebleau sandstone, with several types of oil (Decane, Soltrol, North Sea crude oil), which showed that the effect of the electolysis products was all the more important that the polarity and acidity of the oil were high (Fleureau & Dupeyrat 1988). This study also showed that the wettability of the solid surface could be affected in the same conditions as for the interfacial tension, i.e. strongly with polar crude oils and very little in the case of refined and purified oils. It was proved that the effect of the electrolysis products, through their reactions on the acid impurities of oil, was similar to that of an anionic surfactant (Sodium Dodecyl Benzene Sulfonate).

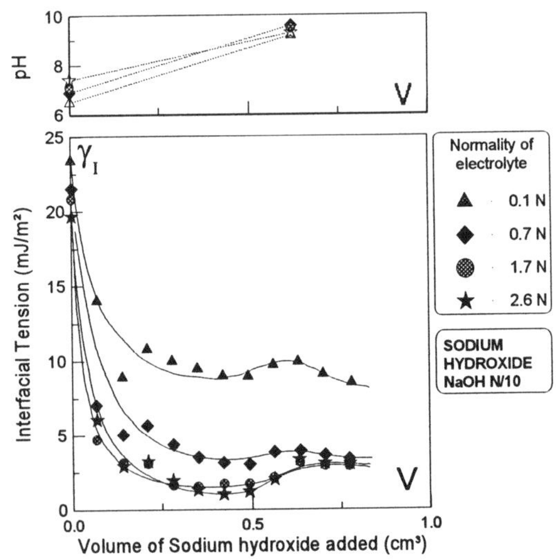

Fig. 6. Changes in water-oil interfacial tension and pH of the aqueous phase when Sodium Hydroxide N/10 is directly added to brine

4. MODEL OF DIPHASIC FLOW THROUGH A POROUS MEDIUM WITH AN ELECTRIC FIELD

In order to assess the role of the electrokinetic phenomena in the displacement of oil, a model based on Helmholtz-Schmoluchowski theory was developed, where the pores are represented as capillaries. A basic assumption of the model is that, in the pores containing oil, a continuous film of ionized water still exists near the wall of the capillary. The driving force of the displacement of oil is the viscous drag exerted by the layer of water displaced by the electric field. The calculations are first made for a single horizontal capillary subjected to axial electric field (E) and hydraulic pressure gradient (dp/dx); they will be later generalized to an ideal porous medium. We suppose that the diffuse layer is assimilated to an Helmholtz distribution, which is equivalent to replacing the diffuse water layer by a charged layer at the water oil interface with the surface charge:

$$\sigma = \frac{D\ \zeta_e}{4\pi\ \delta_e}$$

where D is the dielectric permittivity of the medium, δ_e, the equivalent thickness of the "double-layer" and ζ_e, the corresponding Zeta potential (Fig. 7). To take into account Stern's layer of fixed ions, the effective radius of the capillary is taken equal to $R_1 = R - \delta_0$. The radius of the part of the capillary containing the oil is R_2.

The basic equilibrium equations between pressure and viscosity forces are, for each fluid (with the subscripts w for water and o for oil):

$$\frac{dp}{dx} = \frac{\eta}{r}\frac{\partial}{\partial r}\left(r\frac{\partial u}{\partial r}\right) = -a$$

where u is the Darcyan velocity of the fluid, η, its viscosity, and p, its pressure. By integrating these equations in the boundary conditions of the capillary, it is possible to derive the expressions of the flows of water (q_w) and oil (q_o) in the single capillary. On the equivalent charged layer, there is continuity of the velocities, but not of their derivatives.We must write the equilibrium of this layer under the effect of the viscosity and electrical forces, i.e.:

$$-\eta_w \frac{\partial u_1}{\partial r}(R_2) + \eta_0 \frac{\partial u_2}{\partial r}(R_2) + |\sigma| E = 0$$

Finally, we obtain:

$$q_o \approx \frac{\pi}{8\eta_0} R_2^4 \frac{dp}{dx} - \frac{D|\zeta_e|}{4\eta_w} R_2^2 E$$

and:

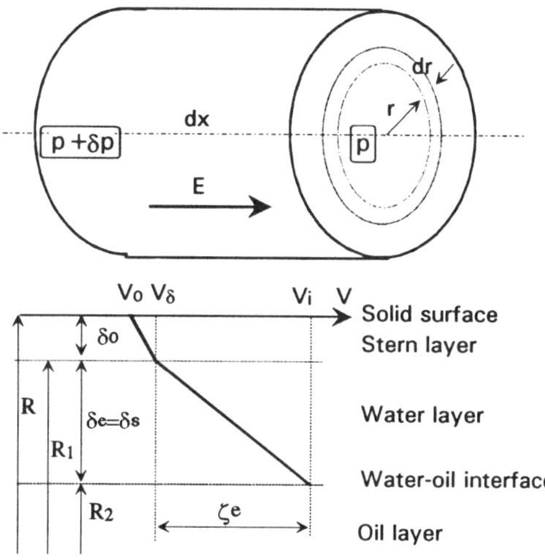

Fig. 7. Schematic representation of a capillary with diphasic flow of water and oil and definition of the corresponding parameters

$$q_w \approx \frac{\pi}{2\eta_w} \delta_e^2 R_1^2 \frac{dp}{dx} - \frac{D|\zeta_e|}{4\eta_w} \delta_e R_2 E \approx 0$$

When considering a porous medium with porosity n, a cross-surface area Σ, and S_w and S_o, the degrees of saturation in water and oil, the flows of water and oil become, respectively, when dp/dx = 0:

$$Q_w = \frac{DE|\zeta_e|}{4\pi\eta_w} S_w n\Sigma \quad \text{and} \quad Q_o = \frac{DE|\zeta_e|}{4\pi\eta_w} S_o n\Sigma$$

This model was applied to some of our experiments and a good agreement was observed between the calculated and measured volumes of oil recovered (Fig. 8). The prediction concerning the flow of water was not so good but, in that case, the real repartition of water in the pores must be taken into account, which is not done in the model.

Therefore, it was concluded that the mechanisms considered in the model probably gave a reasonably accurate picture of the mechanisms involved in the displacement of oil, at least in the case of materials containing a certain amount of clay. Of course, one may be aware of the shortcomings of this approach and of the discrepancy between the description of the diffuse layer given above and reality. In fact, its interest is mainly to highlight the possibility of such a displacement and to give a means of computing an order of magnitude of the flows of water and oil, though other electrokinetic mechanisms, like electrophoresis, may take part in the phenomenon as well. It

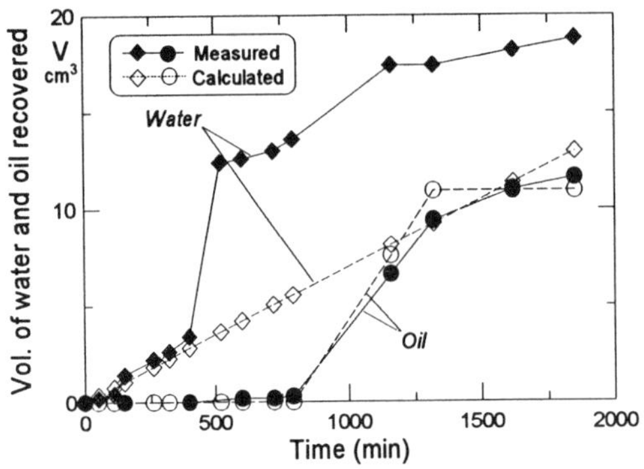

Fig. 8. Comparison between calculated and measured electroosmotic oil and water recoveries

must be considered therefore as a first approach to a very complex problem.

5. CONCLUSION

The experiments presented in the paper and others performed at the Ecole Centrale Paris show that electro-injection, i.e. the application of an electric field or current simultaneously or alternately with the injection of water, can noticeably improve the quantity of oil and other non-polar liquids recovered from polluted soils and rocks, independently of their clay content. The influence of some parameters, like the salinity of the injection water or the heterogeneity of the porous medium, is highlighted.

Several possible mechanisms have been identified. Among these, the reactions of the electrolysis products of water on the acid impurities of the oil are shown to play a major part in forming surfactants which adsorb at the water-oil interface or on the surface of the substrate. The presence of surfactants results both in the decrease in the water-oil interfacial tension and in the creation of a surface charge on the oil droplets which, in turn, makes electrophoresis possible under the effect of the electric field.

On the other hand, electroosmotic phenomena can also happen in polyphasic media. A simplified model, based on Helmholtz-Schmoluchowski theory, was developed to show that these phenomena could explain the displacement of oil, as the displacement of water by the electric field results in a shear stress exerted on the external surface of the oil, and to give at least a qualitative estimate of the flows of water and oil in these conditions. A general agreement was observed in most cases between the predictions of the model and the experimental data.

6. REFERENCES

Bjerrum, L., Moum, J. & Eide, O. 1967. Application of electroosmosis to a foundation problem in a Norwegian quick clay, *Geotechnique* 17: 214-235.

Casagrande, L. 1962. Electroosmosis and related phenomena, *Ingenería*, April: 51-62.

Chilingar, G.V., El Nassir, A. & Stevens, R.G. 1970. Effect of direct electrical current on permeability of sandstone cores, *J. Pet. Techn.* July: 830-836.

Dupeyrat, M. & Nakache, E. 1980. Electrocapillarity and electroadsorption, *J. Coll. Interf. Sci.* 73, 2: 332-344.

Fleureau, J.M. 1979. Influence d'un champ thermique ou électrique sur les phénomènes d'interaction solide-liquide dans les milieux poreux, *Doctorate Thesis*, Ecole Centrale Paris.

Fleureau, J.M. & Dupeyrat, M. 1988. Influence of an electric field on the interfacial parameters of a water/oil/rock system, *J. Coll. Interf. Sci.* 123, 1: 249-258.

Gray, D.H. & Mitchell, J.K. 1967. Fundamental aspects of electroosmosis in soils, *J. Soil Mech; Found. Div.* ASCE 93, SM6: 209-235.

Hamed, J., Acar, Y.B., Gale, R.J. 1991. Pb(II) removal from kaolinite by electrokinetics, *J. Geotechn. Eng;* 117, 2: 241-271.

Hatfield, J.H., Thompson, J.C., Reed, B.E., Mitchell, T.B. 1993. Lab-scale investigation of in-situ electrokinetics, *Proc. 25th Mid-Atlantic Ind. Waste Conf.*, A.P. Davis, ed., Technomic Pub. Co., Lancaster, Basel: 55-64.

Haydon, D.A. 1964. The electrical double-layer and electrokinetic phenomena, in *Recent Progress in Surface Science*, New-York: Academic Press 1: 94-158.

Joffre, G.H. 1984. Calcul par une méthode variationnelle de la forme de ménisques électrisés, *J. Mécan. Théor. Appl.* 3, 4.

Joffre, G.H., Cloupeau, M. 1986. Characteristic form of electrified menisci emitting charges, *J. Electrost.* 18: 147-161.

Mitchell, J.K. 1976. Fundamentals of soil behavior, New York: J. Wiley and Sons.

Shapiro, A.P. & Probstein, R.F. 1993. *Environm. Sci. Techn.*

Tikhomolova, K.P. 1964. *Kolloidnyi Z.* 26, 4: 513-517; 1969. Joint electroosmosis of 2 immiscible liquids, *Kolloid Z.* 31, 4: 588-593.

Watanabe, A., Higashitsuji, K. & Nishizawa, K. 1978. Studies on electrocapillary emulsification, *J. Coll. Interf. Sci.* 64, 2: 278-289.

Yeung, A.T. & Mitchell, J.K. 1993. Coupled fluid, electrical and chemical flows in soil, *Geotechnique* 43, 1: 121-134.

Environmental Geotechnics, Kamon (ed.) © 1996 Balkema, Rotterdam. ISBN 90 5410 848 7

Experimental evaluation of in-situ chemical displacement of Trichloroethene (TCE) from aquifer media using ethanol

D.G.Grubb
Georgia Institute of Technology, Atlanta, Ga., USA

N.Sitar
University of California, Berkeley, Calif., USA

ABSTRACT: In-situ chemical displacement using alcohols is currently being evaluated as a possible environmental restoration technique for cleanup on DNAPL contaminated sites. Horizontal flow experiments in a two-dimensional apparatus were conducted under simulated aquifer conditions in homogeneous, layered and heterogeneous clean sands to evaluate the phenomena associated with the mobilization of TCE by ethanol in water-wet porous media. The experimental results show that the mobilization of TCE by ethanol can lead to uncontrolled migration of TCE.

1 INTRODUCTION

The problems associated with pump-and-treat groundwater restoration techniques for cleanup of sites contaminated with dense non-aqueous phase liquids (DNAPLs) have served as an impetus to develop technologies such as in-situ chemical displacement using alcohols to accelerate DNAPL remediation. The ability of ethanol to displace TCE from coarse and fine sands was therefore investigated experimentally. Ethanol was chosen because it is mutually soluble with TCE and water and can therefore significantly enhance the aqueous solubility of TCE. Ethanol also reduces the interfacial tension between the DNAPL and water thereby lowering the capillary pressures required for mobilization.

The experiments were conducted in homogeneous, layered and heterogeneous sandy soils using a vertically-oriented, two-dimensional apparatus with a flow geometry and velocities that simulated typical aquifer conditions (Grubb, 1995). The apparatus contained sand samples that were 51.5 cm long with a height of 10 cm and width of 2 cm. An injection manifold enabled uniform injection along the entire vertical height of the sand sample. Tensiometers and liquid extraction ports were installed at various locations on the apparatus.

Six ethanol-water displacement experiments were conducted in uncontaminated porous media to establish a base-line performance for ethanol flooding. The baseline performance of the cosolvent front propagation in porous media was needed to evaluate how the stability of the cosolvent front is affected by the interfacial forces in NAPL contaminated porous media. The observed inclination angles of the propagating ethanol interface in homogeneous sands compared well with the inclination angles predicted by the Gravity Number, the ratio of the gravity forces to viscous forces (Grubb, 1995). For flow velocities less than 5 m/day, the time required for the propagating ethanol interface to reach its stable configuration compared well with predictions based on a model of gravity segregation of miscible liquids in a no-flow domain (Gardner et al., 1962).

After the ethanol was completely flushed out of each soil sample, a small quantity (5-7 ml) of pure TCE was injected into the center of the sample and the ethanol displacement was repeated under the same flow conditions. The focus of the five TCE mobilization experiments was to evaluate the phenomena associated with the simultaneous displacement of a DNAPL (TCE) and water from porous media using a cosolvent (ethanol). Since liquid properties such as density and viscosity and interfacial tension can change significantly during displacement, flow instabilities arising from such changes were of interest, not only in terms of coupled phenomena, but also in terms of assessing the efficacy of cosolvent displacement as a potential DNAPL cleanup technique for use in highly contaminated soils, i.e., DNAPL source zone areas. Only the results of the TCE mobilization experiments in layered and heterogeneous sand samples are presented in this paper. A more detailed discussion of the entire experimental program is given by Grubb (1995).

2 TWO LAYER MODEL

In these experiments, the lower third of the apparatus was packed with a #60 fine crystal silica sand. A #30 coarse crystal silica sand was used in the upper two-thirds of the apparatus, as shown in Figure 1. The hydraulic conductivities of the fine and coarse sand were approximately 0.0196 and 0.12 cm/s, respectively. Approximately 5.0 ml of TCE was injected into the coarse sand, its initial distribution being indicated by the darkest-shaded areas with black borders shown in Figure 1. The horizontal spreading of TCE reached a maximum width of approximately 7.0 cm from Stations 7.5 to 10.25. The lowest (visible) extremity of the TCE was at the interface the coarse and fine sand, approximately 3.2 cm above the bottom of the apparatus. Ganglia of TCE were observed betweens Stations 8 and 10, and directly below sampling port S1 (Station 11) along the interface between the sands. The approximate location of the propagating ethanol front at 2 to 5 minute intervals is also shown, the flow direction being from left to right. The flow rate was approximately 4 m/day with a total head drop of 4.2 cm H_20.

In the first 15 minutes (0.20 PV), the ethanol front shapes and locations were essentially identical to those observed in uncontaminated media suggesting that the TCE-affected zone has very little influence on the flow field. The ethanol front first encountered the TCE-affected zone at approximately 14 minutes. Rapid migration of TCE was observed along the ethanol interface, the movement of the TCE accelerating as more TCE was contacted by the ethanol. As the displacement progressed, a bulb of TCE formed at the leading edge of the ethanol front and TCE continued to drain from the upper portion of the coarse sand along the ethanol front.

At this stage, the inclination angle of the interface steepened from approximately 28 degrees just before ethanol reached the TCE to a maximum of 40 degrees six minutes later (t=20 min.; 0.26 PV). Also, after most of the TCE drained from the coarse sand and accumulated on top of the fine grain sand, the bottom portion of the ethanol flood front stalled and the front quickly began to override both the TCE and water.

At 28 minutes (0.36 PV), TCE was distributed along the ethanol-water interface from Station 5.75 to 11.75. TCE was perched above the fine sand between Stations 5.75 and 8, its vertically-downward migration arrested by the higher capillary forces required to enter the smaller pore throats of the fine grained media. A liquid sample collected as the ethanol front arrived at S1 had a TCE concentration of 105,000 mg/l (10.5%). At approximately 33 minutes (0.43 PV) into the experiment and 19 minutes after the TCE-affected zone was first contacted by ethanol, the TCE penetrated the fine sand at Station 5.5. Once in the fine sand, TCE continued to migrate downward and upgradient to Station 4.

Overall, the passage of the ethanol front through the TCE-affected zone mostly redistributed the TCE in the sample. The propagation and stability of the ethanol front was only temporarily affected as ethanol swept the TCE-affected zone. In addition, no increase in total head was measured when the ethanol front initially contacted the TCE, indicating that the TCE and interfacial forces had no measurable effect (>1 cm H_20) on the pressure regime in the #30 sand.

After breakthrough, the TCE that was perched on top of the fine sand was eventually detached from the TCE in the fine sand and was advected downgradient along the coarse/fine sand interface. Interestingly, no TCE was detected in sampling ports S4 and S7 which were embedded in the fine sand even though TCE was directly above these locations. The last visible trace of separate phase TCE appeared to have been entirely flushed from the coarse sand by 85 min (1.11 PV). In the fine sand, a ganglion of TCE trapped near Station 3.75 in the fine sand persisted until at least 66 minutes (0.86 PV) into the displacement. The presence of separate phase TCE in the fine sand was not discernible by 76 minutes (0.99 PV). By 85 minutes, the last traces of dissolved TCE appeared to have been removed from the fine sand.

At times the effluent from the sample contained two phases. An effluent sample of the non-polar phase had a TCE concentration of approximately 590,000 mg/l or 59%. Interestingly, this sample contained very little ethanol, indicating that even by the time the separate phase reached the effluent collection manifold, some 75 minutes after being initially contacted by ethanol, appreciable quantities of ethanol still did not accumulate in the non-polar phase. This is evidence that the mobilization of TCE by cosolvents, even pure cosolvents, is mass transfer-limited and that the assumption of miscibility during cosolvent displacement is inadequate.

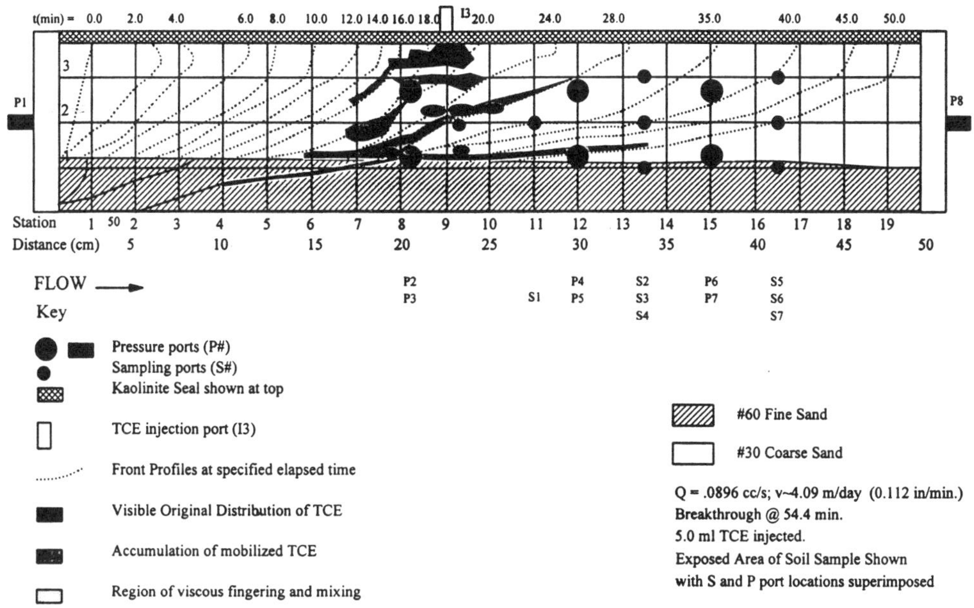

Figure 1 Ethanol front profiles for TCE mobilzation experiment in layered #60 fine and #30 coarse sands.

3 HETEROGENEOUS MODEL

Since the TCE had entered the fine sand in the layered sample, the purpose this experiment was to evaluate whether TCE could penetrate and traverse lenses of fine sand, or if preferential flow of TCE in the coarse sand would occur around the fine sand lenses during cosolvent displacement. The position of the fine sand lenses are shown in Figure 2. Approximately 5.0 ml of TCE was injected into the sample, and the initial distribution of TCE and the location and shape of the propagating ethanol front at 2 to 5 minute intervals are also shown in Figure 2.

Lateral spreading of the TCE was evident along the fine sand layer between Stations 6.75 and 10. Although the sample was packed horizontally in layers and the coarse sand had the general appearance of being homogeneous, TCE responded to subtle heterogeneities by accumulating in the bedding planes within the coarse sand between Stations 10.5 and 12.5 at elevations corresponding to the upper and lower boundaries of the fine sand lens. The lowest observed extremity of the TCE was at an approximate elevation of 7 cm. The flow velocity for this experiment was approximately 4.12 m/day with a total head drop of 3.3 cm H_2O.

The ethanol front propagated horizontally through the coarse and fine sand lens, but the observed profile in the fine sand lens was influenced by the buoyancy of ethanol. Upward migration of ethanol through the bottom of the fine sand lens gave the appearance of gravity underride by ethanol, as shown in Figure 2 by the ethanol profile at Station 3 at 10 minutes. The upward migration of ethanol occurred at multiple locations, leading to trapping of water in the fine sand. The ethanol front profile in the fine sand lagged considerably behind the front in the coarse media.

The ethanol front encountered the TCE-affected zone at approximately 14.25 minutes (.19 PV), rapidly mobilizing TCE in the vicinity of Station 6. Based on visual observations, the first penetration of TCE into the fine sand lens occurred at Station 5.25 at 16 minutes. The distribution of TCE and ethanol at 18 minutes (t_1, see Figure 2) indicates that the TCE penetrated the fine sand lens between Station 5.25 and 6.0. The distribution of TCE in the fine sand indicates upgradient and downward migration of TCE.

At 20.9 minutes (0.28 PV), the TCE attained its maximum upgradient position at Station 4.75, approximately in the center of the fine sand lens. The upgradient migration of TCE was essentially halted at this location because breakthrough of TCE occurred through the bottom of the fine sand lens in the vicinity of Station 5.25. Assuming vertical movement only, the computed travel time across the

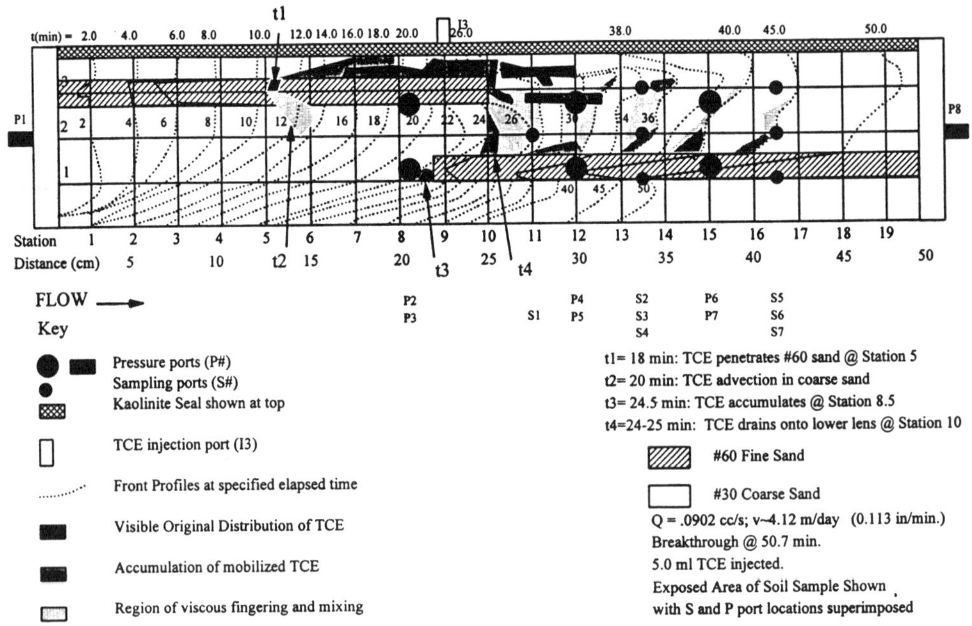

Figure 2 Ethanol front profiles for TCE mobilzation experiment in heterogeneous #60 fine and #30 coarse sands.

1.3 cm thick fine sand lens is 3.9 minutes. This travel time corresponds to a TCE migration velocity of 7.65×10^{-3} cm/s, or 6.6 m/day. After TCE broke through the underside of the upper fine sand lens, it was advected downgradient as shown by the plume between Stations 5.25 and 6.0, see t_2 in Figure 2.

As the leading edge of the ethanol front propagated from Station 8.5 (Figure 7.28) to the end of the upper fine sand lens (Station 10) at approximately 24 minutes (0.32 PV), a complex distribution of ethanol and TCE evolved. TCE leaked through the bottom of the fine sand at multiple locations. TCE continued to seep through the fine sand while the available mass and driving force persisted. Eventually, downward flow of TCE through the fine sand slowed and/or halted because it became detached from its source as the TCE in the upper coarse sand began to be displaced downgradient by the ethanol. Also, the prevailing horizontal gradient in the upper fine sand lens resulted in the smearing of TCE in the downgradient direction within the lens. At other locations, the buoyant ethanol traversed the upper fine sand lens. This combination of events resulted in the trapping of water within the upper fine sand at multiple locations.

Between approximately 22.5 and 24 minutes (0.30 to 0.32 PV), TCE was visually detected in the vicinity of Station 8.5 to 9 at an elevation of 5 cm. The exact source of this TCE is not known, but it is quite possible that it was part of the initial mass of TCE that was not visible. At about 24.5 minutes (0.33 PV), the TCE located at Station 8.5 migrated vertically-downward by gravitational forces to the leading edge of the lower fine sand lens (see t_3). At 30 minutes (0.40 PV) this mass of TCE was horizontally displaced downgradient to Station 9.5 below the lower fine sand lens.

At 24 minutes (0.32 PV), the leading edge of the ethanol front reached Station 10, the downgradient edge of the upper fine sand lens. When the ethanol contacted the TCE situated directly above it, rapid vertically-downward migration of TCE occurred. At the same time, the buoyant ethanol was moving vertically upwards. The concurrent flow of TCE and ethanol resulted in the "raining" of TCE through the column of the ethanol and water saturated coarse sand, as indicated by t_4 in Figure 2. TCE began accumulating on top of the lower fine sand lens at Station 10, and exhibited a tendency to migrate upgradient. However, the flow of ethanol between the two fine sand lenses appeared to limit upgradient migration of TCE to Station 9.5. Additionally, the bulk migration of TCE between 24 to 30 minutes was aided by the ongoing horizontal displacement of TCE by the ethanol in the coarse sand situated above the upper fine sand lens which appeared to force TCE over the edge of the upper fine sand lens.

At about 28 minutes (0.38 PV), the upward migrating ethanol encountered the two ganglia of TCE situated in the bedding planes within the coarse sand at elevations of 7 and 8.3 cm. Once contacted, the ganglia of TCE drained giving further rise to the "raining" effect of TCE before they were ultimately engulfed by the propagating ethanol front at 38 minutes (0.51 PV). During this time period, vertically-oriented multipore stringers/streaks of TCE were observed between Stations 10.25 and 12.5. TCE then accumulated on top of the lower fine sand lens and eventually, over time, migrated into the lower fine sand lens as a dissolved phase with the ethanol.

As the experiment progressed, the trapping of water at multiple locations was evident. At 40 minutes (0.54 PV) water was still trapped in the upper fine sand lens. Also, a pocket of water was trapped behind the leading edge of the ethanol front in the vicinity of S2 and S3 between Stations 13 and 15. At no time was water trapped in the lower fine sand lens owing to the stable configuration of the fluids (no buoyancy driven flow).

As in the layered model, an increase in pressure was not associated with the displacement, indicating that the TCE and interfacial forces had negligible effects (>1 cm H_2O) on the pressure regime in the sample, and thus the ethanol front. Moreover, the shapes and locations of the ethanol fronts in the uncontaminated and contaminated soil heterogeneous sample experiments were essentially identical and the breakthrough times were 52.7 and 50.7 minutes, respectively (3.8% difference).

In this experiment TCE was detected in sampling ports S4 and S7 which were located along the underside of the lower fine sand lens whereas in the layered model it was not. The measured TCE concentrations at these locations were on the order of non-detect to 50,000 mg/l. While it was not visually evident that TCE migrated into the lower fine sand lens (at least as a pure phase), the dissolved concentrations of TCE collected from S4 and S7 confirm that TCE migrated through the lower fine sand lens and/or around the lower fine sand lens to S4 and S7 from locations upgradient of the leading edge of the lower fine sand lens. Regardless of the exact pathway, the results show that TCE reached the bottom of the apparatus.

4 CONCLUSIONS

The experimental results show that the ethanol displacement of TCE in layered and heterogeneous sand samples can produce rapid downward mobilization of TCE at the leading edge of the ethanol front, indicating that interfacial tension reduction is more decisive than NAPL dissolution as a transport mechanism under these conditions. In addition, TCE was observed to migrate *upgradient* along the sloping ethanol-water interface.

Moreover, heterogeneities promoted a variety of bypassing and liquid mixing phenomena that resulted in multiple instabilities, which include: physical displacement of TCE over the edges of discrete fine-grained lenses due to localized confined flow conditions; upward migration of ethanol through finer-grained soils due to buoyancy; rapid downward mobilization of TCE; *upgradient* mobilization of TCE around heterogeneities; migration of TCE into *fine-grained* soils at locations both *upgradient* and downgradient of the initial emplacement of TCE; and enhanced TCE dissolution. In coarse (#30) sands, the bypassing and density-driven flows were not accompanied by increases in total head larger than 1 cm H_2O.

The experimental results indicate significant complications could be encountered when applying (pure) cosolvent displacement for *in-situ* remediation of DNAPLs. Therefore, this methodology is likely to be applicable only in very limited situations in which excellent control over the flow field can be exercised.

ACKNOWLEDGEMENTS

This research was sponsored in party by a National Science Foundation Creativity Award in Engineering; USEPA Cooperative Agreement CR-818956; a Jane Lewis Fellowship from U.C. Berkeley; and a research grant provided by GeoSyntec Consultants.

REFERENCES

Gardner G.H.F., J. Downie, and H.A., Kendall, 1962. Gravity segregation of miscible fluids in linear models, *SPE Journal*, 95-104, June.

Grubb, D.G., 1995. Experimental evaluation of in-situ chemical displacement of Trichloroethene (TCE) from Aquifer Media using Ethanol, *Ph.D. dissertation*, University of California, Berkeley, pp. 330.

Environmental Geotechnics, Kamon (ed.) © 1996 Balkema, Rotterdam. ISBN 90 5410 848 7

Geotechnical systems of groundwater protection

J. Hulla, E. Bednárová & D. Grambličková
Slovak Technical University, Bratislava, Slovakia

M.T. Rahman
Ain Shams University, Cairo, Egypt

ABSTRACT: In accidental penetration of pollutants into an unprotected aquifer, it is possible to predict the range of pollutant transport. If self purification processes are not sufficient to eliminate the pollution, it is possible for thick aquifers to use pumping wells. For thinner aquifers, sealing walls can be considered. They intensify the self purification capacity or fully close the polluted space.

1. INTRODUCTION

Acute contamination is considered the most dangerous one. Accidental spillovers or oil and chemical spillage penetrating into the ground-water causing their contamination are rather frequent. The most unfavourable conditions are induced in aquifers where highly permeable gravel soils are not covered with less permeable soil layers, and also where the groundwater level is close to the surface of territory.

In such cases prompt decisionmaking must consider protective measures to reduce the damage to the quality of the groundwater. Rather quickly proceeding pollutants, interve-ning into considerable distance, thus deterio-rating the groundwater resources.

The purpose of this paper is to give informa-tion about orientation predicting the accidental transport of pollutants into groundwater, on possibilities pumping, sealing, or eventually combined protective systems, together with some experiences from the Slovak Republic.

2. PREDICTION OF TRANSPORT

Along with pollutant transport, a whole series of processes takes place between them and the water, porous or fissured medium (dispersion, diffusion, adsorption, decay of radioactive mate-rials, degradation of organic matter, etc.) resulting in decreasing pollutant concentration. Bear, 1972; Bear and Verruijt, 1987; Klotz and Moser, 1979; Seiler, 1985; Lenda and Zuber, 1970, published important works dealing with this topic.

Results obtained in our studies have been published in the contribution worked out by Hulla, Plško, and Rahman, 1994. Here only the simple equation is presented

$$\frac{c_{\max}}{c_o} = \frac{V_o}{2A\, n_{ef}\sqrt{\pi\,\alpha_L\, x\, R_D}} \exp(-0.693\frac{x R_D}{vT}) \qquad (1)$$

which enable to estimate the maximum concentration of the pollutant ($c_{\max}$) at the distance x; c_o = initial pollutant concentration when penetrating the groundwater; V_o = volume of pollutant; A = area perpendicular to the transport direction; n_{ef} = effective porosity of the medium; R_D = retardation factor; T = half-time of change; v = real groundwater velocity; α_L = the longitudinal dispersivity.

On the bases of numerous of experiments carried out in Slovakia and abroad using gravel soil, the following relationship has been developed for longitudinal dispersivity at distance x measured from the source of pollution

$$\alpha_L = \exp(-6.908 + \sqrt{(11.513)^2 - (\ln x - 11.513)^2}\,) \qquad (2)$$

For conservative pollutants (stable, non adsorbed - $R_D = 1$, do not undergo any decay or degradation - $T = \infty$), the graphical relation in the distance x (m) has been developed using Eqs.

(1) and (2) under transport conditions existing over experiments given by the relation

$$\frac{V_o}{A\,n_{ef}} = \frac{0.834\,V_o}{x} \tag{3}$$

with various values of the volumes V_o (m^3), demonstrated in Fig. 1. By means of this relationship, it is possible to determine quickly the ratio c_{max} / c_o and consequently the maximum concentration for the distance x.

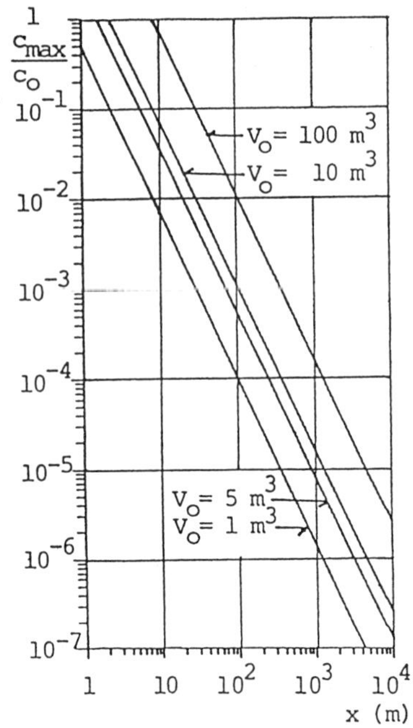

Fig 1 Orientation transport prediction

In designing protective measures, the topmost priority is placed upon the maximum concentration of pollutants in locations close to drinking water or service water resources.

If the permissible concentration is not exceeded to a great extent (approximately $c_{max} \leq 10\ c_{perm.}$), it would be possible to consider temporary cessation of the water resources operations. The contaminated water would not be conveyed into the network and would be retained into the storage reservoirs until its treatment or dilution. Then it would be possible to convey it into the surface streams. Drinking or surface water supply could be restored again if the pollution concentration in the pumped water is less then the permissible value ($c_{max} < c_{perm.}$).

According to the pollution prediction, if the maximum pollutant concentration in local resources is considerably higher than the permissible values, it would be necessary to take into account supplementary pumping, or sealing protective measures.

3. PUMPING SYSTEMS

If a drilled well is constructed to remove pollutants from the groundwater, or if a suitable well creating part of the exploited water resources is used, the flow field in its surroundings will undergo changes due to water pumping, (Fig. 2).

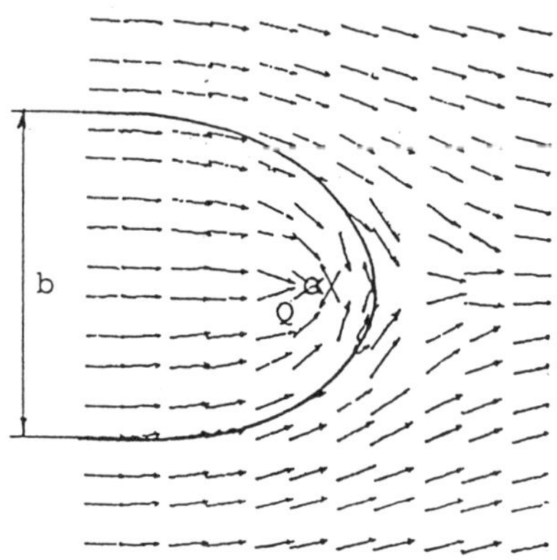

Fig. 2 Flow field in a well surroundings

From the specified area having a width b, the pollutants may be transmitted into the well. The capture zone width from which the water is drained into the well depends mainly on the pumped amount Q. Results of our parametric study are presented in Fig. 3. They are based on the simple Dupuit relationship for a hydraulically perfect well into which water with unconfined groundwater level is flowing (water depth = 10.0 m, well diameter = 0.3 m).

Under certain conditions, actual basic values of b_o and Q_o are specified over the pumping test, it is feasible by means of graph in Fig 3 to determine by adequate accuracy also the values of b and Q for other conditions.

At water flowing through gravel soils approximately in a horizontal direction, the pollutant spreading occurs due to hydrodynamic dispersion at an angle up to 50°. This value was identified during several experiments performed in quaternary alluvial deposits of the rivers Váh and Danube in the Slovak Republic.

For the removal of pollutants after a long period of penetration into the soil, or at some distant allocation of existing pumping wells, one well is usually not sufficient. Using the finite element method, it was realized that it is more advantageous to construct several wells in the flow direction in a row. That enables the withdrawal of the pollutants more reliably from a larger area than in cases when wells are constructed beside each other perpendicular to the pollutant direction transport.

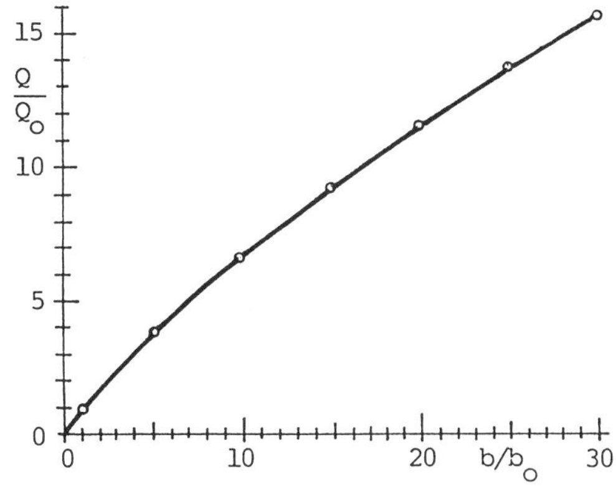

Fig. 3 Dependence between the relative pumped amount Q/Q_o and the relative width of capture zone b/b_o

A pumping system creating a hydraulic screen has been providing, for a rather long period, the protection of high quality drinking water resources of the Danube Island downstream of the petrochemical refinery Slovnaft, situated close to Bratislava, (Fig. 4).

Due to lack of financial support, the refinery was not provided with technology for preventing leakage of petroleum substances into groundwater during transport or damage to equipment in technological processes, having prevalently a continuous character. According to the information published by Pazourek, Puna, and Žila, (1995), petroleum substances had alredy been identified in some wells near the village of Podunajské Biskupice by 1965. In 1966 practically all wells were contaminated and the first traces of petroleum substances were found in the important drinking water resources supplying a population of 400 000 in Bratislava, having a capacity of nearly 1 $m^3.s^{-1}$. These resources had to be put out of operation and substitute resources had to be promptly provided.

Fig. 4 Layout of the hydraulically protected territory

The hydraulic screen, preventing the transport of pollutants to more remote areas, is established by means of 21 hydraulically imperfect wells (the gravel deposit extends down to more than 100 m). Large amounts of water are continuously pumped out from the wells, (all together about 1 $m^3.s^{-1}$). Some wells of the polluted water resources are also used for the protection of this territory. The flow of groundwater and the transport of pollutants within the area of the refinery and its surroundings have been monitored on an area of about 100 km^2 with the aid of 600 observation wells. The extent of pollution by petroleum substances in 1974 and 1993 is illustrated in Fig. 5, showing satisfactory efficiency of the hydraulic screen.

Contaminated water from the hydraulic screen and also wastewater from the refinery are treated in an advanced wastewater treatment plant delivered by the Japanese firm Kurita.

The technology consists of mechanical, chemical and biological processes. Excellent efficiency of the treatment process is documented by actual results (Klátik, 1995):

- non-polar extractives in the polluted water 436 - 860 $mg.l^{-1}$, after treatment 5 $mg.l^{-1}$,
- biological oxygen demand before treatment 1068 $mg.l^{-1}$, after treatment < 20 $mg.l^{-1}$,
- phenol before treatment 40 $mg.l^{-1}$, after treatment < 0.1 $mg.l^{-1}$,
- solids in the contaminated water 90 - 150 $mg.l^{-1}$ after treatment < 30 $mg.l^{-1}$,
- pH before treatment 8.8 -10.2, after treatment 6.5 - 8.5.

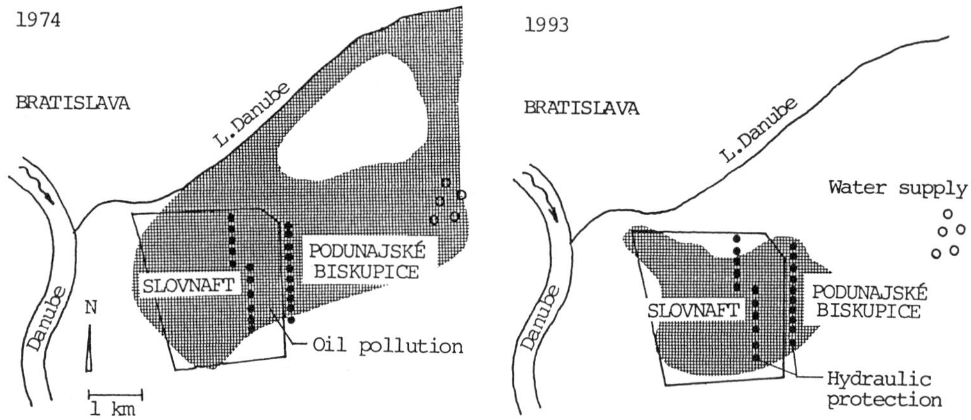

Fig. 5 Extent of pollution by petroleum substances in 1974 and 1993 (Pazourek, Puna and Žila , 1995)

Thus, it may stated that water of better quality is being discharged into the Danube than is being withdrawn from the river. The cost of the operation and the power demands of the pumping system is considered as a disadvantage.

Recycling systems can be developed by means of two wells applying two methods:

1 - The contaminated water from the pumping well is conveyed through a pipe into a recharging well in the counter current direction. The contaminated water is closed and circulates between two wells. A certain risk occurs at the boundary between the contaminated and clean water, which can not remain sharp for a long time. Due to diffusion, a small amount of pollutants escape into the clean water zone. This method is suitable when pollutants are degradable or may undergo decay. It can be also applied when the treatment processes are time consuming, or when large amounts of water have to be treated.

2 - The contaminated water is conveyed from the pumping well into the recharging well situated in the flow direction. A protected area with clean water is created between the wells, however the contaminated water continues to flow behind the recharging well. This method can be used if there is enough free space behind the recharging well enabling the development of self purification processes.

It is also possible to develop recycling systems by means of several pumping and recharging wells.

It is not possible to develop pumping and recycling systems in the case of thin aquifers, when it is not feasible to attain a significant water level decrease in wells and confined areas from which the contaminated water may be drained, corresponding to low pumping amounts. Thus, drain trenches or sealing walls can be considered.

4. SEALING SYSTEMS

Protection against accidental emergency pollutant losses is possible, under certain conditions, using sealing protection systems. It is evident that there would be no time to develop them if the spillage already has occurred. However, they may be constructed as preventive measures for the intensification of self purification processes, especially in locations where traffic roads are close to important water resources.

Sealing systems are constructed as diaphragm walls of various thickness and made using various technologies. By means of grabs and hydrocutters, walls of about 40 cm thickness are constructed for this purpose. Thin walls formed by vibrated steel profiles with classic or jet grouting are 10 to 40 cm thick. Walls, several centimetres thick can be constructed using jet grouting.

Sealing walls are suitable in locations where they could be embedded into a relatively impermeable layer. Anisotropy is a favorable factor in gravel soils reaching down to deep layers. Also in such conditions shallow sealing walls may represent an effective protective measure.

Application of sealing diaphragm walls can provide a more intensive self purification capacity or completly confining the space containing the contaminated groundwater.

The principle of the sealing wall effect on self purification processes is illustrated in Fig. 6. A relatively non-permeable wall forces the polluted water to bypass it. The flow lines get longer, and thus intensify self purification processes (especially dispersion and adsorption). The efficiency of such measures is illustrated in Fig. 1.

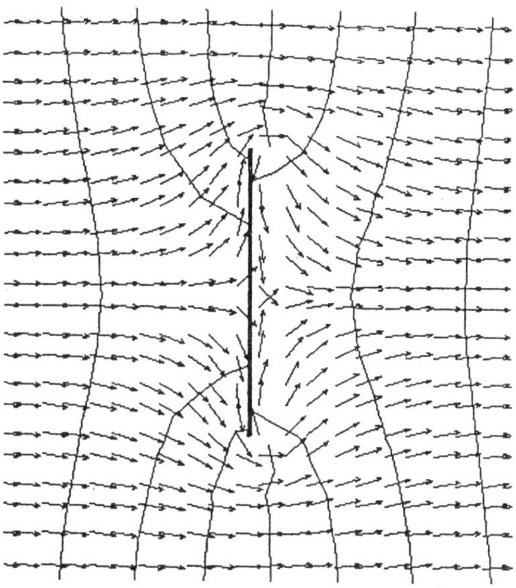

Fig. 6 Flow lines bypassing the sealing wall.

From Fig. 6 we can deduce that, when the groundwater flows in a perpendicular direction to the sealing wall, the extension of the pathway is a function of its position. The greatest elongation of the pathway is identified by streamlines lying close to the sealing wall center. There the track is elongated on both sides approximately by the whole wall length l_w.

Track extension for stream lines existing at the wall edge is negligible. Streamlines lying between the edge and half of the wall are usually extended by half of the wall length l_w.

Adequate tightness of the wall is the precondition of self purification capacity intensification. Our performed analysis using the finite element method showed that, a wall of permeability $k_w = 10^{-7}$ m.s^{-1} (approximately a thousand less than the soil permeability) is sufficient and adequate for gravel soils. In the case of more permeable walls, some of the contaminated water would leak through them which in the case of a wall permeability $k_w = 10^{-6}$ m.s^{-1} would be equivalent to the wall shortening to half its length.

When the sealing walls have fully confined the space with contaminated water, it is necessary to pay attention to the fact that the water level in the confined space should always be lower than the clean water level in the surrounding area. However, the amount of pumped water is dependent to a great extent upon the wall permeability.

Over a period of five years, we had the opportunity to monitor the development of sealing wall permeability. The walls were made of a self hardening suspension of 60 cm thickness and 44 to 48 m depth. They were used at the foundation of the hydropower plant and the navigation locks in the quaternary gravel sediments of the Danube (reaching down to a depth of 400 m) approximately 60 km southeast of Bratislava. By means of the monitoring system it was possible to determine from the total pumped amount, the proportion of leakage through the sealing walls and to calculate the values of permeability coefficients. The results are presented in Fig. 7.

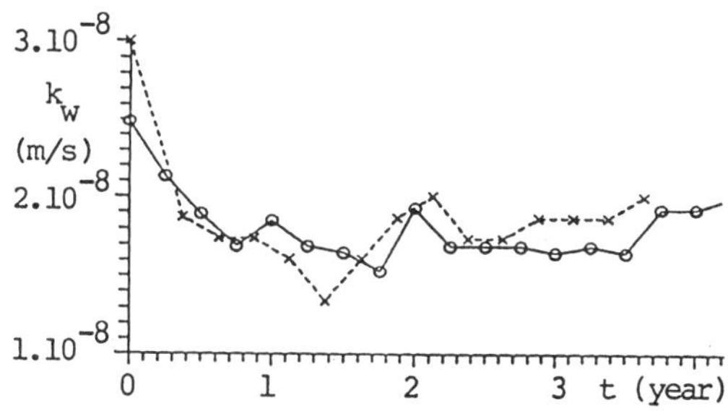

Fig. 7 Development of permeability of sealing walls made of a self hardening suspension (o - navigation lock, x - hydropower plant)

The permeability coefficients were in the range of 1.10^{-8} to 3.10^{-8} m.s^{-1}. After two years, their values showed a slightly increasing trend, how- ever, that did not cause any problem to the construction works.

The sealing walls must provide long-term protection for groundwater quality, stability of hydraulic structures and of the adjacent porous or fissured medium. Thus, a long-term development of sealing wall permeability is considered as a topical problem and we are creating conditions for its solution.

The effect of pollutants on the development of sealing wall permeability is also an important problem in cases of protected waste disposal sites, sanitary landfills, dumps, or sludge beds from waste materials.

Pavilonsky, (1985), published interesting conclusions resulting from experimental research. The passage of various pollutants through clayey (mainly bentonite and kaolinite) thin sealing layers was observed over 200 to 1100 days. The results were compared in the form of relative permeability coefficients for contaminated and clean water. For 0.36 to 7% solution of acids (sulphuric, hydrochloric, and nitric) the relative permeability coefficients were in the range of 5.6 to 11.6. When industrial wastewater was used, containing 70 various organic and inorganic substances, the obtained ratios were 3 to 924. The industrial wastewater had an alkaline reaction, the relative permeability coefficients were 150 to 480, in extreme cases even to 1000.

Thus, when contaminated water infiltrated, the values of the permeability coefficients were 10 to 1000 times higher as compared with clean water leakage. It is evident that more attention has to be given to this problem when thin walls are used for long-term sealing of the contaminated medium.

Traditionally the design of earthen burriers has been based on the assumption that the hydraulic conductivity controls the rate of leachate migration. However, recent field studies have indicated that diffusion is the controlling mechanism of solute transport of many fine grained soils.

5. CONCLUSIONS

In the case of accidental emergency pollutant losses, penetrating into an unprotected medium, it is possible to draft an orientation prediction of transport or spreading by means of the graph in Fig. 1.

If natural self-purification processes are not sufficient for the elimination of accidental pollution, then protective measures must be used.

In locations with thick aquifers it is possible to use existing drilled wells or to construct new ones for withdrawing contaminated water . The relationship between the width of the capture zone and the amount of pumped water for practical use is presented in Fig. 3.

For thinner aquifers, sealing walls may be considered as potential protective measures. By means of sealing walls, self purification processes can be enhanced, (Fig. 6), or the contaminated space can be fully confined. Priority should be given to the negative impact of pollutants on the sealing walls permeability.

REFERENCES

Bear, J. 1972. *Dynamics of fluids in porous media.* New York: American Elsevier.

Bear, J. & Verruijt A. 1987. *Modeling groundwater flow and pollution.* Dordrecht: D. Reidel Publishing Company.

Hulla, J., Plško, J. & Taha, M. 1994. Influence of nuclear power stations on the groundwater. In W.D.Carrier III. (ed.), *First International Congress on Environmental Geotechnics*: 867-872. Richmond: BiTech.

Klátik, J. 1995. The mechanical - chemical - biological water cleaning plant in Slovnaft. *Ropa a uhlie.* 37: 29-31.

Klotz D. & Moser, H. 1974. Hydrodynamic dispersion as aquifer characteristic. In *Isotope techniques in groundwater hydrology*: 341-355. Vienna: IAEA.

Lenda, A. & Zuber, A. 1970. Tracer dispersion in groundwater experiments. In *Isotope hydrology*: 619-641. Vienna: IAEA.

Pavilonsky, V.M. 1985. Varying permeability of clayey soils linings. In *Proc. of 11th ICOSMFE - vol.3*: 1213-1216. Rotterdam: Balkema.

Pazourek, J., Puna, M. & Žila, J. 1995. The groundwater protection - its history and present state. *Ropa a uhlie.* 37: 15-18.

Seiler, K.P. 1985. Results of field experiments on hydrodynamic dispersion in quaternary gravels of southern Germany. In *Scientific Basis for Water Resources Management.* Jerusalem: IAHS.

Environmental Geotechnics, Kamon (ed.) © 1996 Balkema, Rotterdam. ISBN 90 5410 848 7

Remediation of VOC contaminated soils by deep vacuum

Kenneth H. Kastman
Woodward-Clyde Consultants, Chicago, Ill., USA

Saad Ghalib
Woodward-Clyde Consultants, Detroit, Mich., USA

James Husted
Woodward-Clyde Consultants, Philadelphia, Pa., USA

Rosemary Lewis
Woodward-Clyde Consultants, Santa Ana, Calif., USA

ABSTRACT: Remediation of soils contaminated by volatile organic compound (VOC) constituents using soil vapor extraction (SVE) techniques is commonplace in the United States. Some soil formations require the use of a high or deep vacuum to effectively remove target constituents. Use of a deep vacuum not only removes vapor phase constituents, but also removes the water phase from the soil structure, giving rise to the technology description of dual-phase or multi-phase extraction (MPE). This paper summarizes the use of MPE at several sites with differing subsurface conditions and differing constituents. The impact of geology and soil permeability, vacuum requirements, the rates of soil vapor phase and water phase removal, and the effectiveness of MPE systems are presented in this paper.

INTRODUCTION

Groundwater remediation using conventional pump-and-treat technologies in medium-to-low-permeability soils general suffers from low groundwater removal rates and thereby low effective VOC removal efficiencies. The presence of constituents which are sorbed onto the aquifer media, and nonaqueous phase liquids (NAPLs) which are immiscible in water, provide a constant source of constituents which slowly bleed into the aquifer. The low removal rates and presence of sorbed constituents result in extended periods of cleanup, and high operation and maintenance costs.

Applying a vacuum pressure to the soil formation has been demonstrated to be an effective way to remove constituents and speed the cleanup process. The vacuum pressure removes constituents from both the water and the vapor phase of the soil. The multi-phase extraction process, which is also known as 2-PHASE Extraction (a patented process by the Xerox Corporation) addresses the limitations of conventional pump and treat operations for certain soil conditions.

TECHNOLOGY DESCRIPTION

Conventional groundwater extraction by pump-and-treat technology depends on gravity drainage of water to the pumping well. Extraction of groundwater alone will directly influence removal of the dissolved constituent mass, while the sorbed constituent mass and NAPLs remain. Constituent mass sorbed to the aquifer media in contact with the plume may be on the same order or greater than the dissolved constituent mass. Low permeability formations may provide a significant source of sorbed mass as the porosity of such soils may be as great as the adjacent aquifer materials.

MPE was developed as a result of observations of vapor extraction technology effectiveness on constituent removal in unsaturated soil. Vapor extraction takes advantage of the volatile nature of petroleum hydrocarbons and halogenated hydrocarbons by inducing air flow through the soils. Vapor phase constituents are entrained in the extracted air and removed from the subsurface. As the constituent mass is removed in the vapor phase, mass which is sorbed onto the soil matrix is desorbed and entrained in the extracted air flow.

MPE takes the theory of vapor extraction and applies it to groundwater extraction MPE uses high vacuum (generally 10 to 60 CM. of Hg) applied to a wellhead to increase groundwater removal rates (and consequently the dissolved phase mass) and to volatilize and extract that portion of constituents from the sorbed or free product (NAPL) phases. Vacuum lift of water is not a limiting factor in the application of the technology as a mixed vapor/liquid column is extracted from the well. The technology uses a conventional well casing and a patented "straw-like" tube within the well casing to apply the vacuum, and withdraw the mixed vapor/liquid column.

MPE produces an enlarged drawdown condition when compared to conventional pump-and-treat systems (Figure 1). MPE increases the hydraulic gradient toward the extraction well, which increases groundwater well yield. Concurrently, air is also pulled through the subsurface toward the well through the unsaturated and dewatered areas. Volatilized constituents are entrained in the extracted air stream and are removed with the water, thus enhancing the total yield of constituents from the well.

MPE is optimized when it is applied to formations which can sustain a dewatered condition. This takes advantage of the vapor phase removal component. Although dewatering of high conductivity soil can occur with a sufficient number of wells, experience has shown that soils with a hydraulic conductivity of less than approximately 1 gpd/ft^2 are good candidates for MPE or permeability of less than $1x10^{-4}$ cm/s.

PILOT TESTING

The design of a MPE system is typically based on a pilot test. Equipment related to operational performance is usually selected based on the pilot test results. In particular, blower or vacuum pump selection is dependent on the vacuum developed and the volume and rate of soil gas that can be withdrawn. There are approximate correlations between the soil permeability and both the vacuum that can be developed and the volume of soil gas that can be withdrawn. Table 1 summarizes representative pilot test data and indicates the soil formation, permeability and pilot test parameters.

Figures 2 and 3 indicate the correlations between pilot test parameters There is a direct relationship between soil permeability and soil gas removal rate (e.g., the lower the permeability, the lower the soil gas removal rate). There is an inverse relationship between soil permeability and vacuum (e.g., the lower the permeability, the higher the vacuum required).

Soil with a permeability of 10^{-7} cm/s or less is a limiting factor to use of MPE. The high vacuum requirements and low soil gas volume withdrawn at this permeability, renders the technology infeasible.

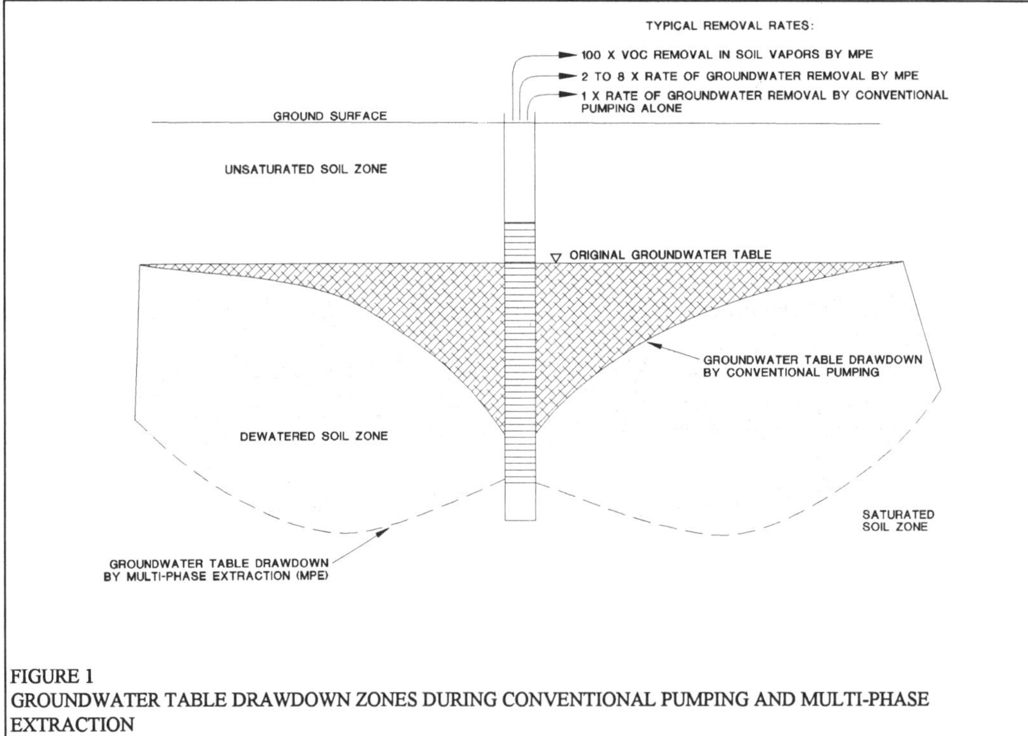

FIGURE 1
GROUNDWATER TABLE DRAWDOWN ZONES DURING CONVENTIONAL PUMPING AND MULTI-PHASE EXTRACTION

TABLE 1
REPRESENTATIVE MULTI-PHASE EXTRACTION PILOT TEST RESULTS

MPE Case No.	Formation Description at Extraction Well	Hydraulic Conductivity of Formation (cm/s)	Extraction Well Casing Vacuum Pressure (cm of Hg)	Soil Gas Extraction Rate (SCFM)[1]	Groundwater Extraction Rate (lpm)[2]	Site (Well)	Constituent Removal Rate-GW only (kg/day)	Constituent Removal Rate-Total (kg/day)	Ratio of Total to GW Only
1	Silty clay glacial till	$8.7x10^{-6}$	55	1.5	0.03	A(MW1)	0.0004	0.091	220
2	Calcareous shale bedrock	$1.0x10^{-6}$	55	25	0.03	B(19)	0.0018	4.86	2700
3	Clay residuum, weathered lim and limestone bedrock	$1.0x10^{-5}$	50	4.7	0.005	C(PR-3)	NA[3]	NA[3]	---
4	Clay residuum, weathered lim and limestone bedrock	$1.0x10^{-5}$	47	10	0.35	C(PR-6)	0.018	0.55	30
5	Fine sand to silty sand	$1.0x10^{-4}$	13	71	1.06	D	0.82	124	150
6	Silty clay, clayey silt, silt	$1.0x10^{-4}$	38	28	0.05	E(MW7)	0.0009	0.32	350
7	Medium sandy clay to clay	$1.0x10^{-6}$	53	10	0.05	F(EXT-1)	0.015	1.57	105
8	Medium sandy clay to clay	$1.0x10^{-6}$	56	8	0.02	F(MW-2)	0.008	0.19	23
9	Gravelly clay	$1.0x10^{-5}$	35	10.6	0.04	G(TW-)	0.57	72.3	127
10	Medium to fine sand	$1.0x10^{-3}$	1	92	0.09	H(RW-2)	0.012	25.5	2200
11	Fine silty, clayey sand	$1.0x10^{-4}$	34	19	0.34	H(RW-1)	0.25	35.3	1450

[1] SCFM = Standard Cubic Feet per Minute; multiply by 28.3 to convert to liters per minute
[2] lpm = Liters per minute
[3] NA = Not available

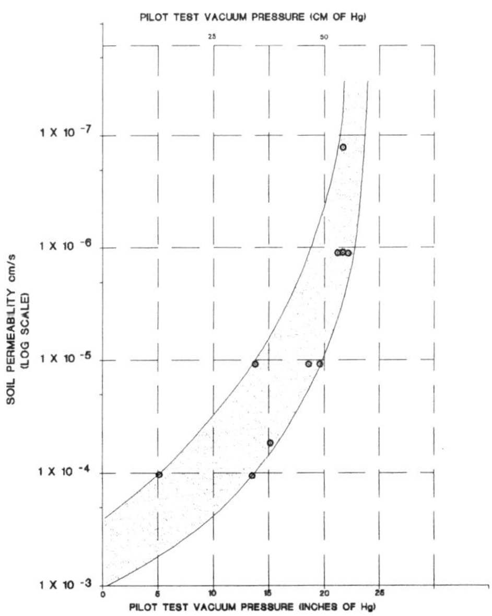

FIGURE 2
MULTI-PHASE EXTRACTION PILOT TEST RESULTS SOIL PERMEABILITY Vs. VACUUM PRESSURE IN EXTRACTION WELL

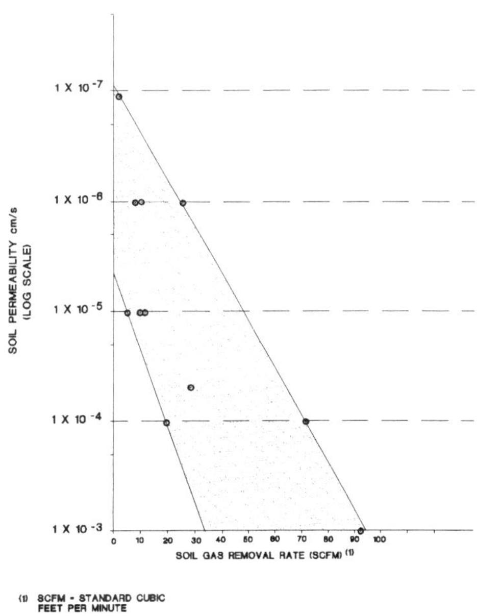

FIGURE 3
MULTI-PHASE EXTRACTION PILOT TEST RESULTS SOIL PERMEABILITY Vs. SOIL GAS REMOVAL RATE

Pilot test results generally indicate a significantly greater rate of constituent removal for the combined soil vapor and soil water phases when compared to just the water phase removal rate. The ratio of removal for the combined phases is typically tens to hundreds of times greater than the water-only phase rate (Table 2). The ratios measured during pilot testing are generally higher than measured in long term performance. However, there is always a significantly higher rate of removal for the combined soil vapor phase and water phase than the water-only phase.

IMPACT OF COMPLEX GEOLOGY

It is often difficult to accurately correlate the measured permeability of soil formations with the performance of a cleanup technology. This is especially true for sites with complex geology or mixtures of soil types. The Xerox site in Irvine, California, USA (Irvine site) is such an example. The Irvine site was extensively investigated with many borings and wells. The site is comprised of mixed zones of weathered bedrock residual soils overlain by water deposited bay soils. Soils range from sands to clays within the same lithologic unit. The measured permeability of soils at the Irvine site ranged from $1x10^{-3}$ cm/s to $3x10^{-7}$ cm/s. Table 2 indicates the range of soil type and permeability.

Results of pump tests in sandy zones at the Irvine site (e.g., permeabilities of 10^{-3} cm/s) suggested that pump-and-treat technology may be potentially feasible fro cleanup of chlorinated constituents at the site. A conventional pump-and-treat groundwater extraction system was installed. The results proved essentially futile. Over 50 groundwater extraction wells were installed and operated for up to 48 months, with the result of extraction of a total of approximately 40 kg (90 lbs) of VOCs.

The difficulty achieving the desired results by the conventional pump-and-treat system led to the development of the Xerox 2-PHASE Extraction Process. A 2-PHASE Extraction System was installed. The system utilized 70 wells that were operated for up to 30 months. The result was extraction of a total of approximately 1150 kg (2500 lbs) of VOCs. The site has received closure from the state regulatory agency.

It is estimated that use of the 2-PHASE Extraction technology reduced cleanup time by a factor of 20 times.

TABLE 2
GENERAL LITHOLOGIC SUBDIVISIONS OF XEROX IRVINE FACILITY

Lithologic Unit	Depth Of Unit (Meters)	Description Of Unit	Permeability Testing	Measured Permeability Range cm/s	Measured Permeability Average cm/s
1	0-5	Areally and lithologically variable unit; lithology may vary from sand to clay, usually brown or reddish; caliche common..	Clayey zones (4 slug tests)	$3x10^{-5}$ to $4x10^{-6}$	$1x10^{-5}$
			Sandy zones (1 pump test)	$3x10^{4}$	—
2	5-12	Clay, silty clay, and clayey sand usually olive or gray, grades from tan/brown downward to gray/olive; clay commonly contains large, clear euhedral gypsum crystals and gypsum-filled fractures; small amounts of organic material, generally roots; fossils (shells) rare; clayey sand occurs near base.	Clayey zones (7 slug tests)	$3x10^{-5}$ to $3x10^{-7}$	$6x10^{-6}$
			Sandy zones (1 pump test)	$3x10^{4}$	$7x10^{4}$
3	12-16	Clay and sand; clay is similar to Unit 2 clay; sand is usually gray, but occasionally tan; abundant fossils (shells).	Mixed zone (2 pump tests)	$7x10^{4}$ to $2x10^{4}$	$4x10^{4}$
Overall Site Average				$1x10^{-3}$ to $3x10^{-7}$	$4x10^{4}$

LONG-TERM PERFORMANCE

Predicting full-scale performance of VOC constituent removal and time of cleanup can be achieved by evaluating long term performance of full scale sites. A short-term pilot test removes easy-to-recover constituents close to the recovery well location, resulting in typically high rates of recovery. This phenomenon can be evident even for week long pilot tests. Gradually, however, the recovery rate declines due to the difficulty of removal of more tightly sorbed VOC constituents at further distances from the recovery well. As with pump-and-treat systems, MPE systems typically reach an asymptotic level of performance. That is, the MPE system reaches a point where continued operation does not result in substantial additional recovery of target chemicals. Typically, pilot tests cannot fully predict the recovery rate decline or asymptotic level of performance.

Test results from several MPE systems were reviewed to assess the long term performance. The three sites had the Xerox 2-PHASE systems installed. The sites were located in different parts of the USA (Irvine, CA; Atlanta, GA; and Blauvelt, NY) and were in differing geologic conditions.

Table 3 summarizes the long-term performance results for the three full-scale sites. Figure 4 shows the removal of constituents (on a normalized total system basis) versus time. Figure 4 demonstrates the decline in removal rate and the approach of an asymptotic limit of total VOC constituent recovery for each site. The asymptotic limit for constituents in individual wells was below or near the site-specific cleanup objectives for each site.

The VOC constituent removal rate for the MPE system was greater than the pump-and-treat removal rate at each of the three sites. The ratio of total mass of constituents removed by the MPE systems versus a pump-and-treat system is summarized on Table 3. From 6 to over 500 times the removal rate was achieved for the MPE systems. The application of a MPE system was estimated to reduce the total remediation time to less than 5 years at each site, a reduction of from 7 to 20 times the time for a conventional pump-and-treat system.

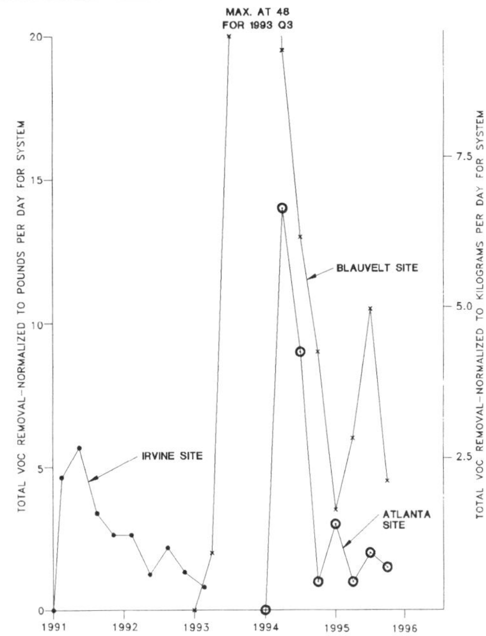

FIGURE 4
HISTORY OF TOTAL VOC REMOVAL AT THREE FULL SCALE MULTI-PHASE EXTRACTION PROJECTS

TABLE 3
COMPARISON OF TOTAL VOC RECOVERY BY GROUNDWATER EXTRACTION WELLS TO MULTI-PHASE EXTRACTION AT THREE FULL SCALE SITES

Project	Remediation System	Months of Operation	Number of Wells	Average Daily Recovery of Total VOCs		Ratio of MPE Recovery to GW Well Recovery
				kg/well	lb/well	
Xerox-Atlanta	GW Extraction by Wells	11	17	0.02	0.05	14
	Multi-Phase Extraction	21	6	0.28	0.72	
Xerox-Blauvelt	GW Extraction by Wells	36	10	0.73	1.61	16
	Multi-Phase Extraction	31	22	11.8	26.1	
Xerox-Irvine	GW Extraction by Wells	48	52	0.02	0.04	30
	Multi-Phase Extraction	30	70	0.54	1.19	

CONCLUSIONS

Multi-phase Extraction is most applicable in the following situations:

- Moderate to low permeability soils,
- Mixed permeability soil formations, and
- Formations where conventional pump-and-treat operations are marginal.

The benefits of MPE are:

- Shortens the time for cleanup, and
- Reduced cost.

REFERENCES

Basel, Michael D., Clare, Jay B., and Mihalovich, Robert D., 'Hydraulic Containment in Low Permeability Soils Using Vacuum-Enhanced Groundwater Extraction", Presented at the 1994 Petroleum Hydrocarbons and Organic Chemicals in Ground Water: Protection, Detection, and Remediation Conference, American Petroleum Institute and National Ground Water Association, November 2-4, 1994.

Brown, Richard A. et.al., 'Dual Phase Extraction: Tool or Panacea", Proceedings of Twenty-Sixth Mid-Atlantic Industrial Waste Conference, University of Delaware, August 7-10, 1994.

Falotico, Robert J. et.al., 'Dual Phase Vacuum Extraction Systems-Design and Utilization", Proceedings of Twenty-Sixth Mid-Atlantic Industrial Waste Conference, University of Delaware, August 7-10, 1994.

Ghalib, Saad, and Kastman, Kenneth H., 'DNAPL Remediation Using Dual-Phase Extraction", Proceedings of the Sixteenth Superfund Conference, November 6-8, 1995.

Environmental Geotechnics, Kamon (ed.) © 1996 Balkema, Rotterdam. ISBN 90 5410 848 7

Migration of contaminants in contaminated soil on electro-kinetic remediation

Takeshi Kawachi & Hiroshi Kubo
Obayashi Corporation Technical Research Institute, Tokyo, Japan

ABSTRACT: In Japan, the treatment of contaminated soil or ground has become serious problem in redevelopment of old industrial area. In these cases, in-situ remediation is preferable, but a reliable technique has not been established in case of heavy metal contaminated soil yet. So, the experimental study viewing to elecro-kinetic remediation has been carried out. In this paper, migration behaviour of some heavy metal in artificial contaminated soil and outline of out-door experiment to decontaminate Cr^{6+} in real contaminated soil have been described. It has been observed that heavy metal in form of anionic compound is concentrated in anode zone and eliminated efficiently by dewatering. So, electro-kinetic method is supposed to be hopeful in-situ decontamination technique especially in contaminated ground by anionic contaminants.

1. INTRODUCTION

In Japan, contaminated ground has become serious hazard to live or develope region because of groundwater pollution and difficulty of treatment of excavated soil and so on. Soil or ground contaminated by heavy metal is sometimes found in the site of old industrial region and reclaimed land near an estuary. Contaminant in these ground is not decomposed and destroied, therefore insolubilization of contaminant is often adopted to treat these soil in construction work. Even after insolubilization these soils have to be contained in controlled landfill in Japan by regulation. It is not so easy to find controlled landfill for these purposes near site. So, technique to remove contaminant from contaminated soil has been expected.

Among the techniques to remove contaminant from soil, in case of ground contaminated by heavy metal, electro-kinetic method has been examined by some different groups in Europe and America. (Lageman R. 1993, Acar, Y. B. et al 1993) At present, this method is thought being developed to make clear optimum condition and limit of application. In Japan, it is not clear whether this method is applicable to Japanese ground condition, environment and regulation or not. So, basic phenomena have been examined using artificial contaminated soil by laboratory study. On the basis of laboratory experiment outdoor experiment to remove Cr^{6+} from contaminated soil has been carried out. In this paper the process and results of these experiments are described.

By the way, in Japan ground improvement using electro-osmotic dewatering was researched and applied about 40 years ago on some projects but replaced by new methods such as paper drainage method and sand drainage method. (S. Mise 1959)

Recently, construction methods of cut-off wall by electro-phoresis were developed to prevent diffusion of contaminant. (T. Kawachi 1994)

2. EXPERIMENTAL METHOD

2.1 Materials and condition

Soil was artificially prepared by mixing silica sand, bentonite and kaolinitic clay and saturated

Table-1 Properties of artificial soil

component	silica sand bentonite kaolinitic clay	100 part 5 part 15 part
water content	about30%	
hydraulic conductivity	1.2×10^{-7} cm/sec	

by adding water. Soil condition is shown in Table-1. To this soil contaminant was added in form of reagent such as $CuCl_2$, PbO, $K_2Cr_2O_7$ and mixed by mortar mixer.

2.2 Experimental apparatus and condition

Direct current was charged using apparatus shown in Figure-1. Aluminum plate was used as cathode and graphite carbon was used as anode. Applied potential gradient was about 0.6V/cm and charged period was from 7days to 26days as shown in Table-2.

Migrated water by electro-osmosis was drained from cathode cotinuously in all case and only in case 3 water close to anode was drained also periodically. Water was supplied to keep saturated condition.

2.3 Measurement

Electric current was measured and, pH and concentration of contaminant of drained water were analysed periodically during charging electric potential. After charging, distribution of contaminant, pH in soil was analysed. Chemical analysis of contaminant was performed by atomic absorption method or ICP method.

3. EXPERIMENTAL RESULTS

3.1 Phenomena during electro-kinetic process

In all cases water was continuously drained through tube incerted to cathode. For example, Fig.-2 shows the result in case 1. Increase of out-flow water from cathode was nearly proportional to charged time and pH of this water is very high in spite of relatively neutral pH value of pore water at initial. This trend was obserbed in all cases. These phenomena were attributable to electro-osmosis and electrolysis of water caused by direct current. In these cases, the electric current supplied was about 20mA to 50mA. Generally, the electric current increased gradually for one or two weeks, and then became to decrease rapidly. Finally, the electric current of the final state was lower than that of the initial state.

3.2 Distribution of contaminant in cationic form

Cu and Pb are thought to behave mainly as cationic form. The results for these metals are shown in

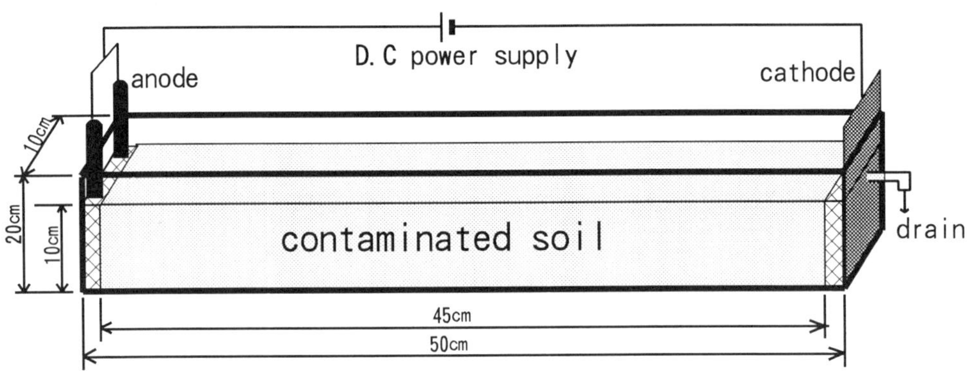

Figure-1 Apparatus for experiment

Table-2 Experimental condition

case	contaminant	concentration mg/kg	electric charge Volt	charged period days	drained water m ℓ
1	Cu	97	30	21	1600(contineous)
2	Pb	199	30	26	7100(contineous)
3	Cr^{6+}	500	25	7	2100(once a day)

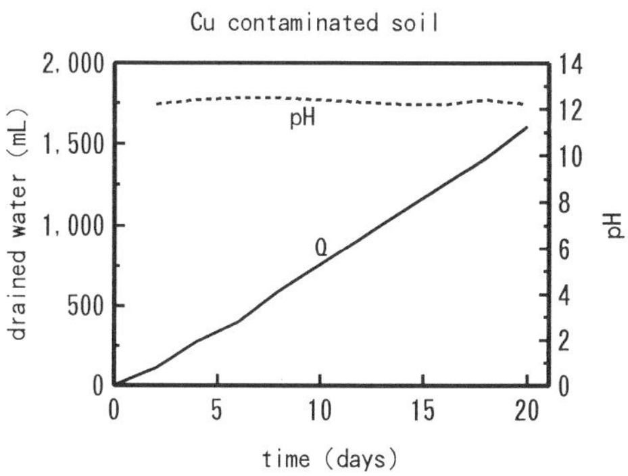

Figure-2 Examle of quantity of drained water and pH

cathode did not get higher than initial level and finally Cu was not detected in out-flow. On the other hand, pH value of soil got lower than initial near anode and higher near cathode due to electrolisis of pore water. This change of pH value probably caused the change of the form of Cu. Namely, Cu ion added migrates to cathodeand precipitates as hydro-oxide near cathode due to high alkaline condition. Because of low solubility of precipitated $Cu(OH)_2$ in water, Cu concentration of out-flow did not increase. So, Cu has accumulated near cathode as hydro-oxide.

Then, distribution of Pb in case 2 is shown in Figure-4. Distribution pattern of Pb and pH is similar to that of case 1. That is to say, concentration of Pb gets lower near anode and gets

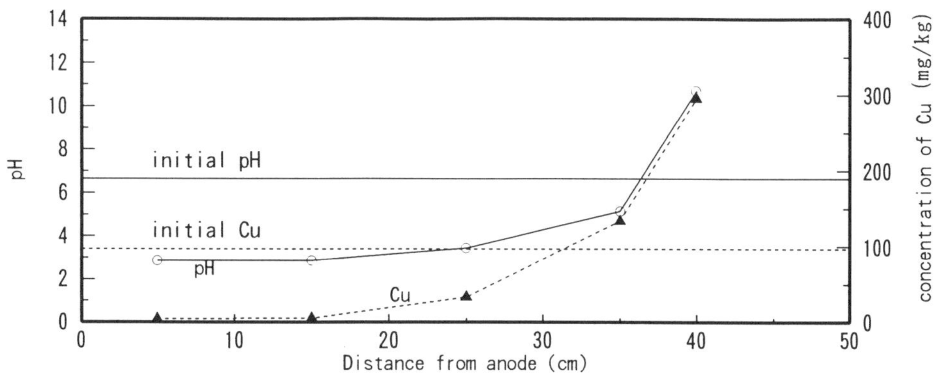

Figure-3 Distribution of Cu and pH

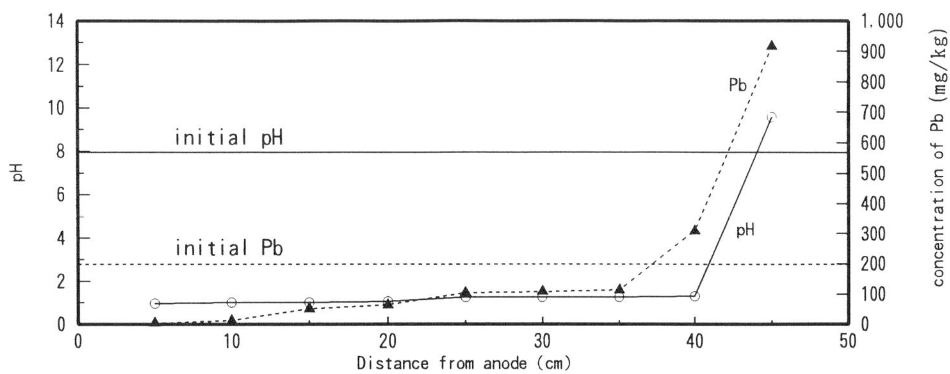

Figure-4 Distrinbution of Pb and pH

Figures-3 and 4. Figure-3 shows distributin of Cu ion and pH value in soil after 21 days. Concentration of Cu decreased drastically near anode and increased near cathode. This phenomena mean migration of Cu ion from anode to cathode by direct current. But, Cu concentration of out-flow from higher near cathode. In spite of lower pH value Pb remains in about a half concentration near anode and intermediate zone. Pb is said to exist in the form of Pb^{2+}, Pb, PbO_3^{2-} dipendingly to pH value. In the region of pH higher than 10.0, anionic Pb such as PbO_2^{2-} is dominant (W.L. Lindsay 1979). So, Pb concentrated as Pb cation near cathode by electrophoresis and transformed to PbO_2^{2-} due to high pH value in this regeon.

3.3 Distribution of contaminant in anionic form

In case 3, Cr^{6+} ($K_2Cr_2O_7$) was added as contaminant to artificial soil and analyzed to identify the form of Cr. Analytical result showed fairly portion of added Cr was reduced to Cr^{3+} in soil after a few days. Experiment was carried out using this increased near cathode drastically. So, accumulation of Cr^{3+} in intermediatezone is refered to formation of hydro-oxide. In case of Cr^{3+}, it was not clear why Cr^{3+} decreased near cathode working attractive force.

Then, mass balance of Cr was computed as shown in Table-3. Before charging direct current about 2900 mg of Cr was contained in soil totally. During charging about 30% of Cr^{6+} was removed in drained water in 7days. This value is a total of Cr^{6+} contained in out-flow water. On the other hand, Cr^{3+}

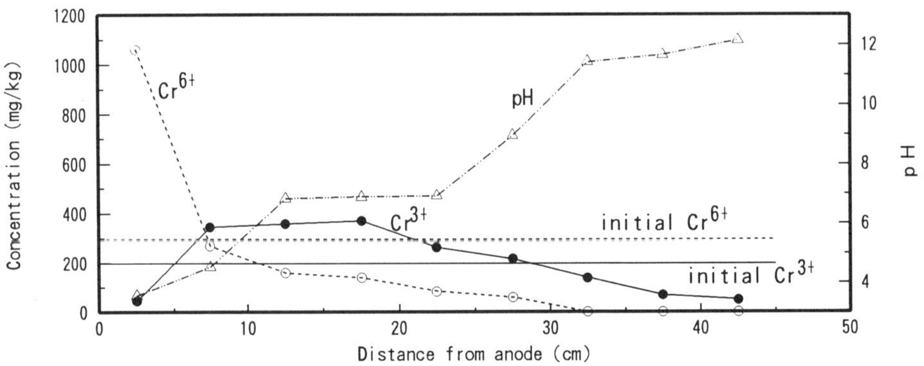

Figure-5 Distribution of Cr^{6+}, Cr^{3+} after 7days

soil containing Cr as Cr^{6+} and Cr^{3+}. Concentration of Cr^{6+} and Cr^{3+} is 490 and 200ppm respectively. Distributions of Cr^{6+} and Cr^{3+} and pH are shown in Figure-5 after 7days of electric charging. Cr^{6+} decreased near cathode, while concentrated near anode. This behaviour of Cr^{6+} is reflected by migration characteristics of anionic $Cr_2O_7^{2-}$. On the other hand, Cr^{3+} decreased near cathode and anode but accumulated slightly in intermediate zone. Cr^{3+} is thought to behave as cation form and precipitate as hydro-oxide in neutral to alkaline zone. In this case, pH decreased near anode and was not removed in drained water. As a result, about 95% of Cr^{3+} remained in soil after charging. So, electro-kinetic method is not suitable to remove Cr^{3+} in this way.

Now, no matter how this case was finished after 7days, main root or path of Cr^{6+} removal was certified to be through out-flow of anode. Concentration of Cr^{6+} in out-flow remained stable from the begining to the end. So, it is expected that efficiency of Cr^{6+} removal does not decrease so markedly near the end.

Table-3 Mass balance of added Cr after 7days

	contained in soil (initial)	remained in soil (after charge)	Removed in drainage water anode	cathode
Cr^{6+}	1711	1202	455	47
Cr^{3+}	1193	1128	33	3
total	2904	2330	488	50

(expressed in mg)

4.DISCUSSION

Phenomena happening incidentally to electro-kinetic remediation has been shown by laboratory

To establish electro-kinetic remediation method understanding and control of these complex phenomena lead to sucsess. In Figure-6, these concepts are shown schematically.

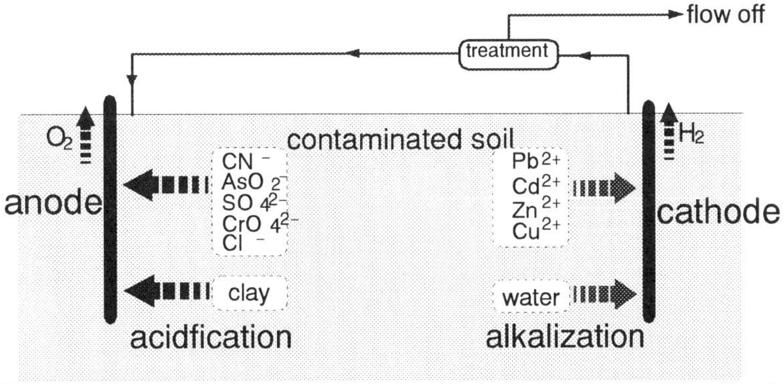

Figure-6 Schematic expression of chemical reaction induced by direct current

experiment using artificial soil. It can be said that mainly three reaction happens in soil at electric field. These are as follows.

① Migration of cation and anion induced by electro-phoresis
② Migration of water to cathode induced by electro-osmosis
③ pH change and gas evolution induced by electrolysis of water

Actually, charged particle such as clay or colloid migrates as well as ion in suspension. In this experiment, Cu and Pb migrated to cathode and Cr^{6+} migrated to anode. This reaction is the main mechanism to remove ionic contaminant. Especially, Cr^{6+} was removed efficiently.

Water migration caused continuous out-flow from cathodic drainage pipe. But, contaminant was not so much detected in out-flow water. So, removement through cathodic drainage water attributes little to remove heavy metal as it is. But, if heavy metal can be transformed to anionic form using chelating reagent, metal could be removed by anodic out-flow water (D.C.Guzman 1990).

Alkalization by electrolysis is remarkable near cathode and acidification occured near anode. These are shown in following chemical reaction.

$2H_2O + 2e \qquad H_2 + 2OH^-$ (cathode)

$2H_2O \qquad O_2 + 4H^+ + 4e$ (anode)

Cathodic reaction causes hydrolysis of heavy metal ion. As a result, insoluble precipitates such as $Cu(OH)_2$, $Pb(OH)_2$, $Cr(OH)_3$ are formed near cathode and these are not removed from sistem. So, it is nessesery to prevent precipitate formation to remove contaminant to perfect level as reported (A.P. Shapiro 1989).

On the basis of these experiments, outdoor experiment has been carried out to make clear applicability and efficiency of electro-kinetic method using real Cr^{6+}contaminated soil. Scale and layout of electrode are shown in Figure-7. In this experiment, surrounding water of anode has been pumped up periodically to remove Cr^{6+}. Removed Cr^{6+} was nearly proportional to charged period and total electricity consumed as shown in Figure-8.

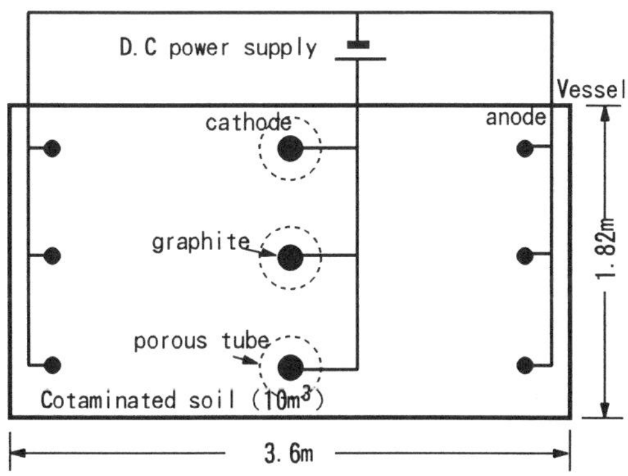

Figure-7 Lay-out of electrode

Phenomena observed in laboratory experiment such as migration of water due to electro-osmosis and gas evolution by electrolysis are observed in this case remarkablly. These facts suggest applicability of this method. This experiment is to being performed about 6 to 8 months to clean up soil perfectly as to soluble Cr^{6+}. Finally the efficiency and economy of electro-kinetic remediation will be discussed to some extent. Soil properties after

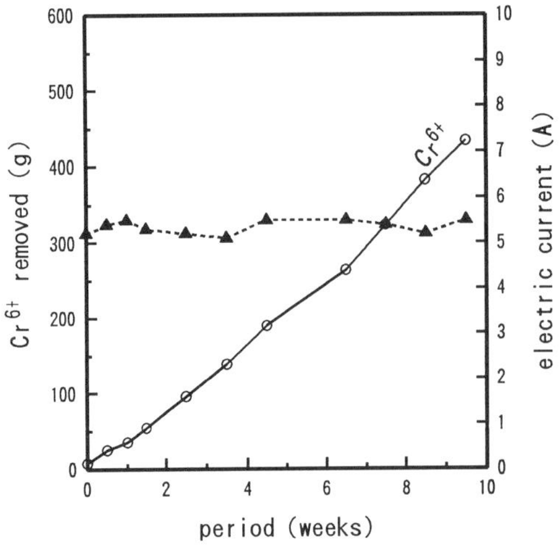

Figure-8 Accumulated Cr^{6+} removed and chared current in outdoor experiment

Photo-1

remediation will be also tested. The state of experiment of early stage is shown in Photograph-1.

5. CONCLUSIONS

Basic phenomena have been confirmed by laboratory experiment to make clear fiesibility of electro-kinetic remediation. In electro-kinetic remediation three reactions occur in parallel as follows.

①electro-phoresis of ion or charged particle
②electro-osmosis
③electrolysis of water

In this paper, artificially contaminated soil with Cu, Pb, Cr^{6+} was tested by charging direct current. As a result, contaminants such as Cu and Pb migrate to cathode reflecting cationic properties. On the other hand, Cr^{6+} migrates to anode efficiently in form of chromate anion ($Cr_2O_7^{2-}$). In the case of cationic migration, metal cation makes hydro-oxide and stops to migrate in alkaline region induced by electrolysis near cathode. So, countermeasure to prevent hydrolysis of metal ion is necessary. In this stage, contaminant taking anionic form is thought more suitable to electro-kinetic method.

Although, electro-kinetic method is promising in-situ remediation technique, but many problems are remained to be designed and applied plactically.

REFERENCE

1) Acar Y.B. and Alshawabkeh A.N. (1993): Principles of electro-kinetic remediation. Environ. Sci. Technol. Vol.27, 2638-2647

2) D.C.Guzman et al (1990): The use of electrokinetics for hazardous waste site. J.Air Waste Manage.AssocVol.40, No.12, 1670-1676

3) T.Kawachi (1994): Developement of cut-off wall using electrophoresis. Proceeding of First simposium on ground improvement, 151-156 (in Japanese)

3) Lageman R. (1993): Electroremediation: applications in the Netherlands. Environ. Sci. TechnolVol.27, 2638 - 2647

4) W.L.Lindsay (1979): Chemical equilibria in soils, 338-339

5) S.Mise (1959): On the dehydration mechanism by electroosmosis, Proceeding of JSCE, No.69, 73-78 (in Japanese)

6) A.P.Shapiro et al (1989): Preliminary studies on the removal of chemical species from saturated porous media by electroosmosis, PhysicoChemical Hydrodynamics, Vol 11, No.5/6, 785-802

Environmental Geotechnics, Kamon (ed.) © 1996 Balkema, Rotterdam. ISBN 90 5410 848 7

Attenuation modeling and source area mitigation

J.R. MacPherson, J.S. Andrilenas, F.T. Schelby & K.V. Lew
AGRA Earth & Environmental Group, Inc., West Virginia, Phoenix, Ariz., USA

ABSTRACT: Attenuation modeling used in conjunction with source area mitigation was proposed and accepted by the New Mexico Environment Department in a remedial action plan (RAP) prepared for a leaking underground storage tank (UST) facility. This approach was unusual in that it involved no capture of contaminated groundwater and no hydraulic control. The approach reduced "source" strength primarily by volatilization, and relied on natural biodegradation to remove the plume. The RAP has been implemented and contaminant concentrations at the site are now below regulatory levels.

Source area mitigation consisted of bio-sparging along with a soil vapor extraction (SVE) system. After estimating the source mitigation rate based on a bench-scale biodegradation rate study, attenuation modeling was used to predict the evolution of the dissolved contaminant plume over time. This evolution involved contaminant transport times, distributions, and concentrations at compliance points as affected by decay (biodegradation) and retardation (sorption). Input parameters to the model included source strength, source duration, source size, decay rates, groundwater velocities, aquifer effective porosity, retardation coefficients, aquifer dispersivities, and aquifer thickness.

Two general solution methods were considered: analytical and numerical. Application of the analytical model "SOLUTE" developed by Milovan S. Beljin and distributed by the International Groundwater Modeling Center (IGWMC) was selected based on more robust sensitivity analyses available through its application. This model allowed thorough analysis of input parameters. SOLUTE was run 34 times varying source strength, duration, shape, degradation rates, retardation coefficients, longitudinal and transverse dispersivities, and aquifer thickness. Results of the preliminary SOLUTE runs were compared to known historical events at the site and physical and chemical data. Later simulations were then fit tested based on the data available for the site.

Results of attenuation modeling indicate that source area mitigation through bio-sparging and SVE along with natural biodegradation should result in decreased concentrations of benzene in off-site compliance wells. Results from field testing after initiation of the source area mitigation indicate that benzene concentrations have indeed decreased, both on- and off-site.

1. INTRODUCTION

Keeping costs of remediation low is a primary goal in site clean-ups. Toward this end, an approach was employed to avoid costly pump and treat by relying on natural biodegradation occurring in site soils. The contaminant concentrations in the source area were too high (>3000 parts per million (ppm)) to allow significant biodegradation in that area, and would continue to yield contamination to groundwater flowing through the source area. Thus, source mitigation was necessary if the microbes were to successfully remediate the overall plume. The source mitigation technique chosen was a combination of bio-sparging and a SVE system, for both volatilization and increased oxygen content.

1.1 *Site background*

The leaking UST facility consisted of three 22,800 L tanks reportedly operating from 1961 through 1991. Gasoline contaminated soils were reported during tank removal operations within the UST excavation located west of the on-site building (Figure 1). Site characterization studies followed, with the highest levels of gasoline contamination to soils detected in the vicinity of MW-2, just west of the pump islands. Apparently, a product line was the primary release point. Gasoline had migrated down to the groundwater surface, 14 m below surface grade (bsg). From there, the gasoline hydrocarbons migrated northward with groundwater flow. Dissolved phase contamination migrated beneath U.S. Highway 66 as evidenced by low volatile aromatic hydrocarbon and 1,2-dichloroethane (EDC) concentrations in groundwater sampled from MW-7, a compliance point 60 m downgradient of the UST cavity.

Soils underlying the site consist predominantly of sandy and silty clays. The clay encountered in the top 5 to 9 m contained more sand than silt. Between 5 to 12 m bsg, silt content increased, and the sand content decreased. At depths of approximately 12 m beneath the site, sand and silt contents decreased to trace amounts. A clayey, gravelly sand was encountered at 15 m bsg, within the saturated zone.

Groundwater occurs at 14.1 to 14.8 m bsg under semi-confined conditions. The piezometric gradient averages approximately 0.0078 m/m. The condition of having the piezometric head up into a low permeability clay, with a moderate permeability sand less than 1 m below, suggested that sparging would have a large radius of influence and may be an effective remediation technology at the site. This lead to the concept of using source mitigation if natural biodegradation proved sufficiently active and attenuation modeling confirmed this remediation approach would prove effective.

1.2 *Feasibility Testing*

Feasibility testing consisted of slug tests, SVE testing, sparge testing, and bio-feasibility testing. Attenuation modeling applied the results from these tests to estimate contaminant transport times, distributions, and expected concentrations at compliance points effected by decay (biodegradation) and retardation (sorption).

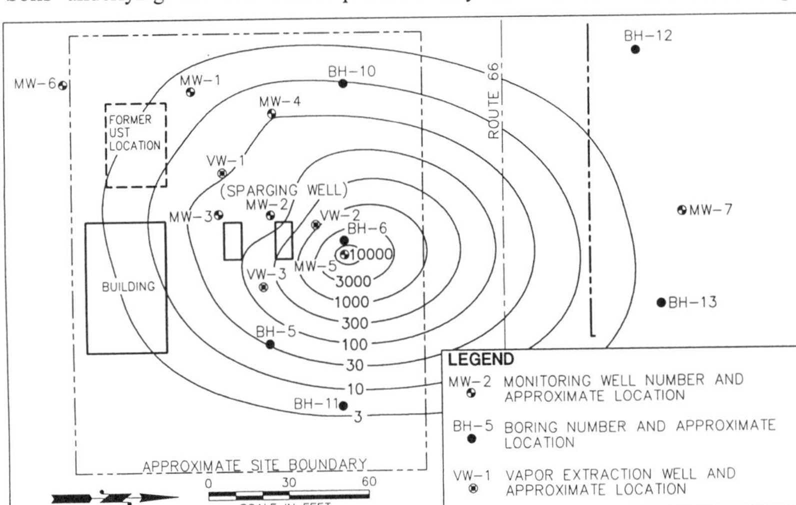

FIGURE 1 – BTEX CONCENTRATION CONTOURS (in ppb) FOR MARCH 1994

1.3 *Attenuation Modeling*

Attenuation modeling required several input parameters. These included: source strength, source duration, source size, decay rates, groundwater velocities, aquifer effective porosity, contaminant retardation coefficients, aquifer dispersivities, and aquifer thickness.

The feasibility testing provided initial and boundary conditions, and the primary fate and transport parameters. Attenuation modeling involved selection between the general solution methods: analytical and numerical. Though analytical solutions are limited due to the implicit assumptions of constant source, decay, flow, retardation, and dispersion, the amount of field data for this site were also limited to a degree that time variable functions for these conditions and parameters were not known. Therefore, a numerical solution only extended modeling capability by allowing inclusion of speculative values for these conditions and parameters. Application of an analytical model allowed for a more comprehensive sensitivity analysis of decay and retardation effects. Based on this, the program SOLUTE was applied (Beljin 1991). SOLUTE calculates two-dimensional advection-dispersion with decay and retardation. The governing equation is:

$$R\frac{\partial C}{\partial t} = \frac{\partial}{\partial x_i}\left(D_{ij}\frac{\partial C}{\partial x_j}\right) - \frac{\partial}{\partial x_i}\left(C\overline{v}_i\right) - \lambda RC - \frac{C'W^*}{n} \quad (1)$$

where R = retardation coefficient; C = contaminant concentration; x_i = distance in the x_i direction tensor; x_j = distance in the x_j direction tensor; D_{ij} = dispersion coefficient (mechanical dispersion and molecular diffusion); $\overline{v}_i$ = average pore velocity; λ = decay; C' = solute concentration in source/sink fluid; W^* = source/sink flow rate; and n = total porosity. SOLUTE was run thirty-four times varying source strength, source duration, source shape, degradation rate, retardation coefficient, and longitudinal and transverse dispersivities.

1.3.1 *Source Strength, Duration, and Size*

The exact date the gasoline release began, the duration of the release, and the rate of release were unknowns. To assure that the remedial design would succeed, "worse case" values for the release were assumed. These values included: a) the duration of release was 10 years; and b) the rate of release reaching the groundwater was constant at 0.0035 kg/day/m^2 over 23 m^2, or a total rate of 9.5 kg/day (12.5 L/day). Over ten years, this amounts to a total release volume of 34,900 kg (46,400 L for gasoline density of 0.75 kg/L). The 23 m^2 source area at the water table was estimated based on soil and groundwater analytical results. Additionally, this contact area is assumed to be centered just north of Service Island No. 2 where the release appears to have contacted the water table.

Since the release primarily occurred from a product line, the release rate was assumed to vary. This pulsed input typically favors natural remediation processes (volatilization through the ground surface and biodegradation) over the duration of the release compared to a continuous or catastrophic release. This effectively allows for an actual source somewhat larger than the 34,900 kg modeled.

During sensitivity analysis, source strengths used were 21, 9.5, and decreasing from 9.5 to 1.4 kg/day. Source durations were 1, 5, 10, and 20 years. The majority of simulations had a source release duration of 10 years and a constant source. One and five year source durations were also used to evaluate the effect of remediation efforts. Source configuration was square and octagonal.

1.3.2 *Decay (Biodegradation)*

A site-specific biodegradation rate half-life ($t_{1/2}$) can be estimated using the concentration of the contaminant when the source was terminated and the contaminant concentration found in the investigation. It is assumed that free product gasoline reached the groundwater table and, thus, soil concentrations reached residual saturation.

The highest benzene groundwater concentration in 1993 was 1.4 ppm from MW-5, located slightly downgradient (north) of the release area. The ratio of benzene to TPH is expected to be on the order of 1:30 to 1:70 based on benzene's more rapid volatilization loss but higher water solubility relative to other gasoline compounds, and based on field experience. Given 1.4 ppm benzene, and the above ratios, the high TPH concentration should have been between 42 and 98 ppm.

Assuming soil in the source area was at residual saturation (100,000 to 300,000 ppm, or 10 to 30% of an assumed total porosity of 45%), the data indicate that the natural biodegradation half-life is 200 to 255 days (300,000 decreasing to 42 ppm and 100,000 decreasing to 98 ppm, respectively). It is probable oxygen diffusion became the rate limiting step as D.O. levels were zero at MW-5. Any increase in D.O. should accelerate biodegradation and shorten the half-life to less than 200-255 days.

During sensitivity analysis scenarios used half-lives of either 0, 6, 60, 365, or 600 days. The $t_{1/2}$ in the biofeasibility study was 3 days, and represents a $t_{1/2}$ if nutrients were added to groundwater. A value of 6 days, as opposed to 3 days, was chosen since laboratory results are often difficult to recreate in the field. The 60 day value represents the effect of oxygen addition through sparging without nutrient amendment. The 365 days was lower than the half-lives derived from the site-specific rates discussed above for conservative bias (e.g., no action). A $t_{1/2}$ of 600 days was chosen for further evaluation of model sensitivity to $t_{1/2}$.

1.3.3 *Retardation*

Retardation coefficient (R) values for gasoline components vary widely. Simulations were run with multiple values. "Worst case" simulations used the R for benzene. In saturated media, R is:

$$R = 1 + \frac{\rho_b}{n} K_d \qquad (2)$$

where ρ_b = soil bulk density (1.8 g/mL; assumed); n = total soil porosity (0.45, assumed); and K_d = contaminant partition coefficient.

The important variable in this equation is K_d where $K_d = K_{oc} \times f_{oc}$. Benzene has an equilibrium partition coefficient to organic carbon (K_{oc}) of 65, toluene 300, and naphthalene 940 (Mabey *et al.* 1982). K_{oc}s for gasoline components are well characterized. The largest uncertainty in calculating R is the f_{oc}. An f_{oc} value of 0.4% (4 g/kg) was considered conservative since this is a common value for sands and also because sorption to site clay mineral surfaces occurs and was not included in this calculation. Thus, the R values for benzene, toluene, and naphthalene were 2, 5.6, and 15, respectively.

Sensitivity simulations used Rs of 1 (no retardation), 2, 4, and 10. The value of 1 was used to evaluate the effect of biodegradation alone. When R=2, the scenario represented benzene transport. Reported concentrations under these scenarios do not truly depict benzene concentrations because they assume the source is a single component, while in reality the source gasoline probably contained 1 to 5% benzene (Stavinoha 1973 and 1975).

1.3.4 *Dispersivity*

The longitudinal and transverse dispersivities for this site were initially estimated, modeled, and fitted to observed data. Initial values were estimated at 2 m for longitudinal coefficient of dispersion (σ_L) and 0.2 m for transverse coefficient of dispersion (σ_T) based on values reported by Sudicky *et al.* (1983), Sudicky *et al.* (1985), Freyberg (1986), and Sudicky (1986). It is assumed the aquifer is 3 m thick. Darcy's law is assumed appropriate for calculating transport in the saturated zone at this site.

Sensitivity simulations were run to evaluate the dispersion coefficient. During these runs, longitudinal and transverse dispersivities were 1.2, 2, 2.4, and 4.6 m and 0.12, 0.2, 0.24, 0.3, and 0.5 m, respectively.

2. RESULTS

2.1 *Feasibility Tests*

Slug testing at wells MW-5, MW-6, and MW-7 yielded hydraulic conductivies of 1.6, 2.1, and 0.04 m/day, respectively. A high gradient of 0.0078 m/m combined with a median hydraulic conductivity (K) of 1.6 m/day (mean of 1.2 m/day) and an assumed effective porosity (n_{eff}) of 0.3 yields an average linear groundwater flow velocity of 0.042 m/day. This velocity value was used in attenuation modeling to give the fastest probable transport rate ("worst case" value).

Results of SVE testing indicated a theoretical radius of influence of 14 m at approximately 2.4 m^3/min. A design radius of 10 m was chosen for the site. Results of sparge testing indicated a theoretical 30 m radius at 0.99 m^3/min at a pressure of 1.34 atm (absolute). Dissolved oxygen levels at the contaminant plume center were zero and low (<2 mg/L) out to an approximate 15 m radius of the center. The plume center was estimated between MW-2 and MW-5 based on the zero dissolved oxygen (D.O.) reading observed at MW-5 early in the sparge test.

Results of bioremediation studies performed on groundwater collected from the site indicated the presence of indigenous hydrocarbon degrading bacteria including *Pseudomonas spp.* Subsequent biodegradation rate analyses of contaminated site groundwater indicated a half-life on the order of three days, under optimum laboratory conditions.

2.2 *Attenuation Modeling*

Preliminary modeling simulations generated a best fit source shape, source strength, and dispersivity values. The best fits were an octagonal source shape, a source strength of 9.5 kg/day for 10 years, and σ_L and σ_T of 2.4 and 0.24 m, respectively.

Subsequent runs used these best fits, varying biodegradation, retardation, time, and the source duration. EDC was considered to represent a nearly conservative tracer (very low sorption and decay rate).

EDC was detected at the site at 98 ppb in MW-5 and 6.2 ppb downgradient in MW-7, a factor of 16 drop. Though this concentration gradient was less than that simulated (e.g., a factor of 50), the EDC source strength undoubtedly decreased with time due to its high water solubility (8,690 mg/L at 20C; Verschueren 1983) while the model used a constant source. Assuming EDC was a nearly conservative tracer confirms the choices for input values used here (Figure 2).

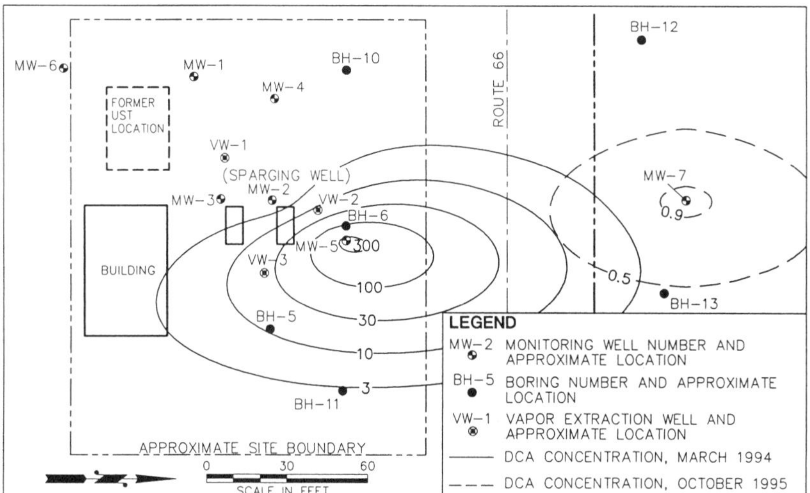

FIGURE 2 – EDC CONCENTRATION CONTOURS (in ppb) IN MARCH 1994 AND OCTOBER 1995

A $t_{1/2}$ of 600 days had little effect on plume size. The biodegradation half-life of 6 days degraded the plume too fast. Assuming biodegradation rates increase linearly with increased D.O., increasing from 2 mg/L to 8 mg/L (a factor of four) would change the observed site half-life from 200-255 days to about 60 days or less. This was the assumed half-life that bio-sparging would cause.

A drastic reduction in plume size occurred when R was increased from 2 to 10. Benzene has an R of 2 and was observed in MW-7, while the xylenes have an R of about 10 and were not observed in MW-7.

Six or more years had past since the gasoline input to the system was terminated. During that time biodegradation was active. Based on 1993 concentrations compared to model results, it was apparent that the plume either decreased in size or grew at a much slower rate than modeled, affirming the "worst case" input values chosen. These results indicated source mitigation would be beneficial.

It was estimated that bio-sparging would decrease the half-life to 60 days by increasing the D.O. Sparging also removed gasoline components by volatilization. A simulation with a conservatively decreasing source function was run to mimic the removal of residual hydrocarbons from the source area by volatilization. The resultant modeled plume indicated that sparging alone would result in substantial reduction of plume size and that contaminant concentrations at the compliance well (MW-7) would decrease. Combined with feasibility testing, a final remedial design consisted of SVE at four wells and one bio-sparging well (MW-2).

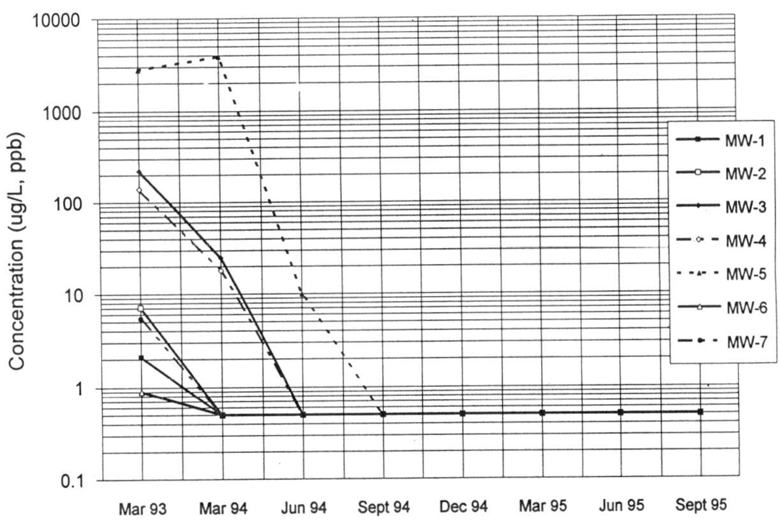

Figure 3. Benzene concentrations (ug/L) over time.

The entire BTEX plume has been removed (Figure 1 and Figure 3). Benzene concentrations at MW-5 near the source dropped from 3,900 ppb to not detected (ND) within 6 months. Benzene at the downgradient compliance point MW-7 dropped from 5.4 ppb to ND. EDC at MW-5 dropped from 450 ppb to ND in 9 months. EDC at MW-7 has fluctuated between ND and 4.3 ppb, although over the last 9 months it has averaged 1.7 ±0.1 ppb. Recovery of total gasoline volatiles by the SVE system have decreased a factor of 4 since shortly after startup.

3. CONCLUSIONS

Field and laboratory feasibility testing with attenuation modeling to compare the effects of biodegradation, retardation, and source function indicated that the plume size was most dependent on source function. Bio-sparging reduced the source function through both volatilization and enhanced biodegradation. Bio-sparging alone was a cost-effective groundwater treatment technology at this site.

REFERENCES

Beljin, M.S. 1991. SOLUTE, A Program Packag of Analytical Models for Solute Transport in Groundwater. *IGWMC*. Colorado School of Mines, Golden, CO.

Freyberg, D.L. 1986. A Natural Gradient Experiment on Solute Transport in a Sand Aquifer 2. Spatial Moments and the Advection and Dispersion of Non-reactive Tracers. *Water Resources Research*. 22(13):2031-2046.

Mabey, W.R., J.H. Smith, R.T. Podoll, H.L. Johnson, T. Mill, T.-W. Chou, J. Gates, I. Waight Partridge, H. Jaber, & D. Vandenberg 1982. *Aquatic Fate Process Data for Organic Priority Pollutants*. EPA: 440/4-81-014.

Stavinoha, L.L. 1973. Internal Standards for the Isolation and Determination of Aromatics in Motor Gasoline. *J. Chrom. Sci.* 11:515-520.

Stavinoha, L.L. 1975. Hydrocarbon-Type Analysis of Gasoline Using Stabilized Olefin Absorption and Gas Chromatography. *J. Chrom. Sci.* 13:72-78.

Sudicky, E.A., J.A, Cherry, & E.O. Frind 1983. Migration of Contaminants in Groundwater at a Landfill: A Case Study. *J. Hydrology*. 63:81-108.

Sudicky, E.A., R.W. Gillham, & E.O. Frind 1985. Experimental Investigation of Solute Transport in Stratified Porous Media 1. The Nonreactive Case. *Water Resources Research*. 21(7):1035-1041.

Sudicky, E.A. 1986. A Natural Gradient Experiment on Solute Transport in a Sand Aquifer: Spatial Variability of Hydraulic Conductivity and Its Role in the Dispersion Process. *Water Resources Research*. 22(13):2069-2082.

Verschueren, K. 1983. *Handbook of Environmental Data on Organic Chemicals*. 2nd Ed. New York: Van Nostrand Reinhold.

Environmental Geotechnics, Kamon (ed.) © *1996 Balkema, Rotterdam. ISBN 90 5410 848 7*

Degradation of petroleum oil by white rot fungus *Phanerochaete chrysosporium*

Kaori Maeda & Y.Yamashita
Konoike Construction Co. Ltd, Osaka, Japan
H.Ozaki, Y.Terashima, F.Wu & K.Ono
Kyoto University, Japan

ABSTRACT: Degradation of petroleum oil by *Phanerochaete chrysosporium* was investigated. At initial concentration of 1,000 mg/liter, about 60% of heavy oil was degraded during 22 days of incubation in liquid cultures. For the purpose of learning the ability of *P.chrysosporium* to degrade polynuclear aromatic hydrocarbons (PAHs), degradation of phenanthrene in liquid culture was also investigated. At initial concentration of 10 mg/liter, about 60% of phenanthrene was degraded during 7 days of incubation. The activities of extracellular lignin degrading enzymes (ligninases) synthesized by *P.chrysosporium* were measured, and it seems that ligninases might relate to degradation. *P.chrysosporium* immobilized on wood chips was mixed with soil contaminated by heavy oil at initial concentration of 5,000 mg/kg-soil, and about 20% of heavy oil was degraded during 14 days of incubation.

1. INTRODUCTION

Petroleum oil is the most important resources of energy and used widely. In recent years, the contamination of soil and groundwater by wastes and wastewater from the petroleum industry has been made apparent. Fractionation products divided from crude oil, such as gasoline, diesel oil, heavy oil, *etc.*, contain polynuclear aromatic hydrocarbons (PAHs). The crude-oil source and fractionation process have an effect on the PAHs content of the final fuel products, and higher concentrations of PAHs are associated with the higher-boiling-point distillation products[1]. PAHs are relatively stable and less easy to degrade than many other organic compounds are. Many PAHs, or their metabolites possess toxic, mutagenic and carcinogenic properties. In recent studies, certain bacteria and fungi have been shown to metabolize various pollutants, and white rot fungi is presently receiving much attention. The ability of white rot fungi to degrade pollutants is evidently connected with its capacity to degrade lignin, a plant polymer of randomly linked phenyl-propane subunits that resists breakdown by most organisms. The basidiomycete *Phanerochaete chrysosporium* is one of the best-characterized white rot fungi. Morphology of *P.chrysosporium* is shown in Fig. 1. Ligninolysis by *P.chrysosporium* is catalyzed in part by extracellular lignin degrading enzymes (ligninases), and they have been considered for various applications in pollutants degradation. The major components of its ligninases are two families

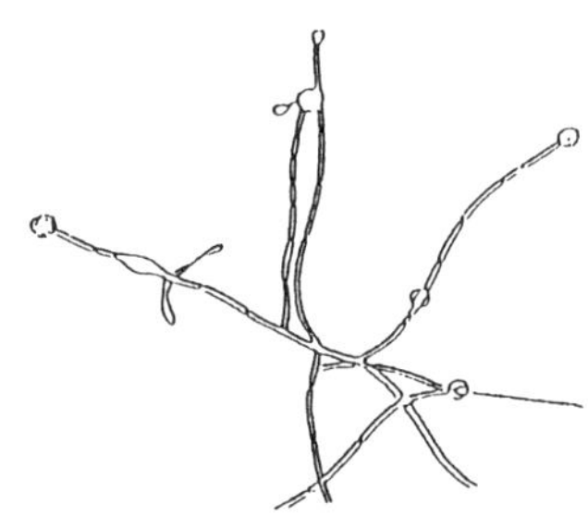

Fig. 1 Morphology of *P.chrysosporium* based on the photography by microscope (×400)

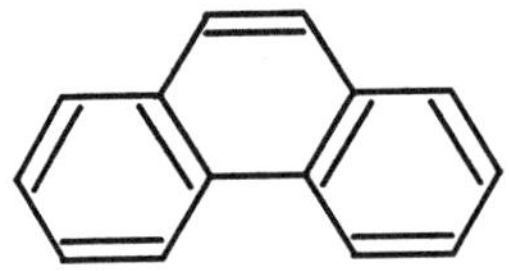

Fig. 2 The chemical structure of phenanthrene

of isoenzymes which are designated lignin peroxidase (LiP) and manganese-dependent peroxidase (MnP). These enzymes have been said to be triggered by nutrient limitation. In the previous studies, it has been shown that both LiP and MnP were activated under carbon-limited condition[2]. Moreover, LiP activity has been shown to be increased by the addition of veratryl alcohol, the secondary metabolite of glucose. This fungus has been reported about its ability to degrade various organic materials, such as organochlorine pesticides, azo dyes, *etc*., which are known to be persistent for microorganisms[3].

In this study, the ability of *P.chrysosporium* to degrade petroleum oil in water and soil was investigated. As the substrate, heavy oil and phenanthrene were selected, because heavy oil is the residual oil of petroleum refinery which is distilled at the higher boiling point, and phenanthrene is one of the major PAHs found in petroleum-contaminated sites which consists of three fused benzene rings (Fig. 2).

2. MATERIALS AND METHODS

2.1 *Strain and medium condition*

P.chrysosporium Burds BKM-F-1767 was obtained from the Wood Research Institute, Kyoto University, Japan. The fungus was incubated at 39°C on 2% malt extract agar slants and maintained at 4°C. A spore suspension was prepared by washing the spores of the fungus into 5 mL of sterile distilled water. Liquid culture was grown by inoculating 100 μL of spore suspension (*ca.* 1×10^6 spores/mL) in 20 mL of sterile carbon-limited culture medium contained in 200-mL Erlenmeyer flasks. This culture medium contained 0.2% glucose, 12 mM nitrogen, 0.1% Tween 80, 10 mM Na-aconitate buffer (pH = 4.3), 5 mg/liter thiamine-HCl, and trace elements. Veratryl alcohol (2 mM) was added after 2 days of inoculation. The flasks were flushed with O_2 for 1.5 minutes at the time of inoculation and then every three days. For measuring enzymes activities, 1 mL of samples were collected and measured as followed. Soil inoculum consisted of aspen (*Populus tremuloides*) pulpwood chips (*ca.* 1.5 by 0.5 by 0.2 cm) grown with *P.chrysosporium* and prepared as follows. The chips were placed in 200-mL Erlenmeyer flasks with a proper amount of distilled water and sterilized by autoclaving at 120°C for 2 hours. Sterile chips were inoculated by incubating with pieces of malt agar from *P.chrysosporium* slants in 200-mL Erlenmeyer flasks at 39°C for 10 days. The flasks were flushed with O_2 for 1.5 minutes at the time of inoculation and then every three days.

2.2 *Enzyme Activities*

LiP activity was measured by monitoring the oxidation of veratryl alcohol to veratraldehyde[4]. The assay mixture contained 90 mM sodium tartrate (pH = 3.0), 0.45 mM H_2O_2, 1.2 mM veratryl alcohol, and 200 μL of extracellular fluid in a total volume of 3.3 mL. LiP activity was indicated by the linear increase in A_{310} and 1U was defined as 1 μmol of veratryl alcohol oxidization per minute. MnP activity was also measured by monitoring the oxidation of vanillylacetone[5]. The assay mixture contained 100 mM sodium tartrate (pH = 5.0), 0.1 mM H_2O_2, 0.1 mM vanillylacetone, 0.1 mM $MnSO_4$, and 200 μl of extracellular fluid in a total volume of 3.0 mL. MnP activity was indicated by the linear decrease in A_{336} and 1U was defined as 1 μmol of vanillylacetone oxidization per minute.

2.3 *Degradation of heavy oil in liquid culture*

Heavy oil was added to 5-day-old culture medium at initial concentration of 1,000 mg/liter and incubated at 39°C. At regular intervals, each sample was mixed with 50 mL of chlorofluorocarbon for 1 hour and analyzed by oil concentration analyzer (HORIBA OCMA-300). Control cultures contained sterile liquid medium and heavy oil.

2.4 *Degradation of phenanthrene in liquid culture*

Degradation of phenanthrene by *P.chrysosporium* was investigated as described above at initial concentration of 10 mg/liter, while enzymes activities were monitored at the same time. Phenanthrene was extracted by 20 mL of acetonitrile, and analyzed at 250 nm by high performance liquid chromatography (HPLC, YANACO L-5000) with UV spectrophotometer. Used column was a reversed phase column (YANAPAC ODS-A), and were eluted with a 750 : 250 : 1 mixture of methanol : water : acetic acid, flow rate was 1.5 mL/min, and column temperature was fixed at 50°C. Ligninases activities were measured as described above.

2.5 *Degradation of heavy oil in soil*

A horticultural red earth (red soil) and sandy soil for filtration media (sand) were used in this study. Red soil (10g) or sand (20g) was placed in 200-mL Erlenmeyer flasks. Then, heavy oil dissolved in ethanol was added to the soil at initial concentration of 5,000 mg/kg-soil. The ethanol was allowed to evaporate, and the soil was then mixed with 5 pieces of aspen chip inoculated with *P.chrysosporium.* The moisture content of the soil was adjusted (*ca.* 40%, wt/wt) by adding sterile water and incubated at 39°C. Control cultures contained 5 pieces of sterile aspen chip, instead of inoculated aspen chip.

3. RESULTS AND DISCUSSION

3.1 *Degradation of heavy oil in liquid culture*

Degradation rate of heavy oil in liquid culture is shown in Fig. 3. Degradation rate was obtained with the formula:

$$R = \frac{C_C - C_P}{C_C} \times 100$$

where R is the degradation rate of heavy oil, C_C is the concentration of heavy oil in control culture, and C_P is the concentration of heavy oil in inoculated culture. At the day 0, LiP activity was 0.285 U/liter, and MnP activity was 0.0629 U/liter. Compared with the previous study [2], these values were considered enough for degradation. *P.chrysosporium* degraded about 60% of heavy oil during 22 days of incubation.

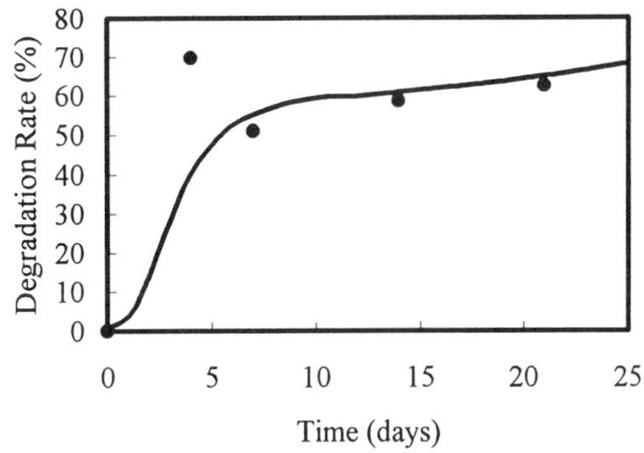

Fig. 3 Degradation rate of heavy oil in liquid culture

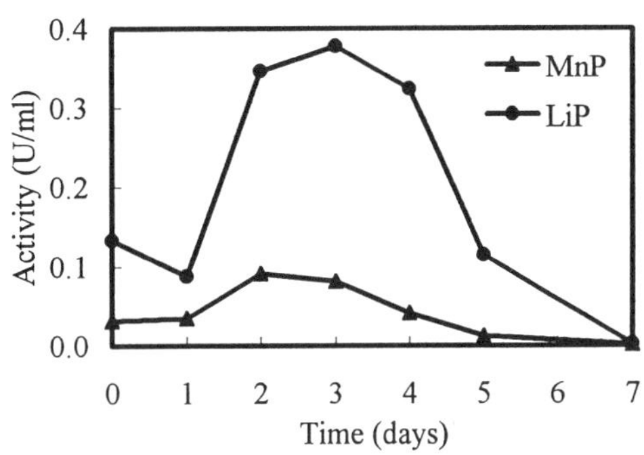

Fig. 4 Time series of ligninases activities

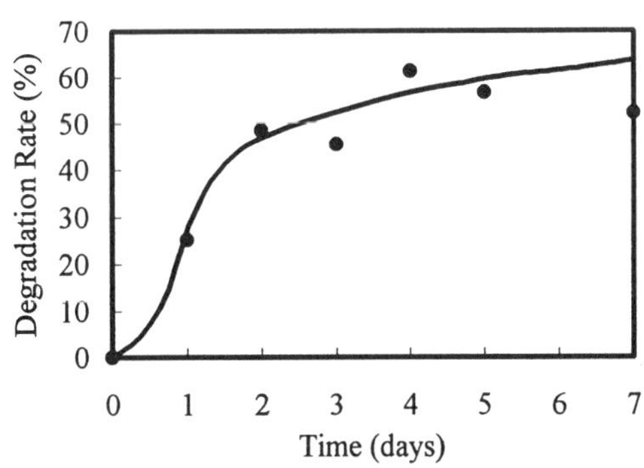

Fig. 5 Degradation rate of phenanthrene in liquid culture

The degradation rates increased rapidly during the first 7 days, but slowly after 7 days. This result indicates that ligninases were inactivated during the first few days. Further, heavy oil contains many kinds of hydrocarbons, so that *P.chrysosporium* might degrade the lower-molecular-weight hydrocarbons in heavy oil. Therefore, further investigation on the degradation characteristics is required.

3.2 *Degradation of phenanthrene in liquid culture*
Time series of ligninases activities is shown in Fig. 4 and degradation rates of phenanthrene in liquid culture are shown in Fig. 5. The degradation rates were similar to those described above. *P.chrysosporium* degraded about 60% of phenanthrene during 7 days of incubation. The degradation rates increased rapidly during the first 3 days with an increase of ligninases activities, but slowly with a decrease of activities. This result indicates that degradation of phenanthrene is related, at least, to ligninases activities and might be caused by radical reaction.

3.3 *Degradation of heavy oil in soil*
Degradation rate of heavy oil in soil is shown in Fig. 6. The degradation rates were similar to those described above. *P.chrysosporium* degraded less than 10% of heavy oil during 7 days of incubation, but more than 20% during 14 days. This result indicates that *P.chrysosporium* can degrade heavy oil in soil, but it proceeds slowly compared with the degradation rates in liquid culture. Further investigation on the optimal conditions for degradation is required to raise growing rate and ligninases activities of *P.chrysosporium*.

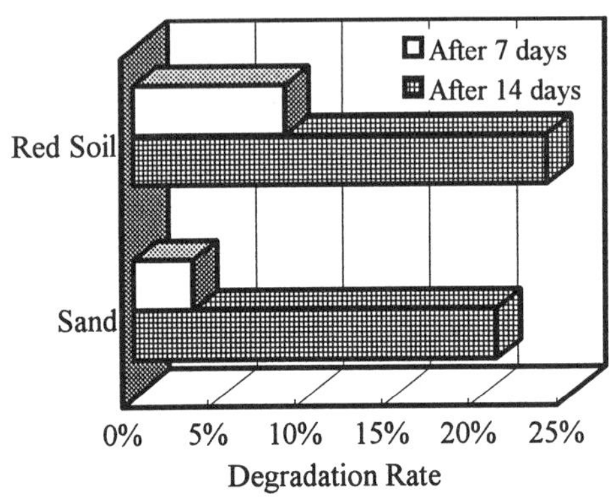

Fig. 6 Degradation rate of heavy oil in soil

4. CONCLUSIONS

1) *P.chrysosporium* could degrade heavy oil and phenanthrene under carbon-limited condition. This reaction is considered to be related to LiP and MnP produced by *P.chrysosporium*.
2) *P.chrysosporium* could also degrade heavy oil in soil. Both the growth of *P.chrysosporium* and the degradation rates were, however, slower than those in liquid culture.

We are grateful to Professor Mikio Shimada and Doctor Toshiaki Umezawa for the generous providing of the strain of *P.chrysosporium* BKM-F-1767 and for their valuable comments to our studies.

REFERENCES

1)Wilson, S. C., et. al. 1993. Bioremediation of soil contaminated with polynuclear aromatic hydrocarbons (PAHs). *Environmental Pollution.* 81:229-249
2) Ozaki, H., et. al. 1995. Factors Influencing the activities of ligninolytic enzymes and decolorization of azo dyes by the crude enzymes. *Environmental Engineering Research.* 32:111-119
3)Ozaki, H., et. al. 1994. Factors influencing the growth characteristics of the white rot fungus, *Phanerochaete chrysosporium* and its recalcitrant substance degradability. *Environmental Engineering Research.* 31:359-367
4)Tien, M., et. al. 1988. Lignin peroxidase of *Phanerochaete chrysosporium. Methods in Enzymology.* 161:238-249
5)Paszczynski, A., et. al. 1985. Enzymatic activities of an extracellular manganese-dependent peroxidase from *Phanerochaete chrysosporium. FEMS Microbiology Letters.* 29:37-41

Environmental Geotechnics, Kamon (ed.) © *1996 Balkema, Rotterdam. ISBN 90 5410 848 7*

Experimental and modeling studies on electro-kinetic extraction of lead from Georgia kaolinite

Rajendra M. Menon, Cheng-non Hsu & Albert T. Yeung
Department of Civil Engineering, Texas A&M University, College Station, Tex., USA

ABSTRACT: Electro-kinetic extraction has been proven to be a viable technique to remove contaminants from fine-grained soils. However, many electrochemical reactions associated with the technique are not fully understood. In the paper, results of experimental and modeling studies on electro-kinetic extraction of lead from Georgia kaolinite are presented. The developed numerical model can predict the non-linear voltage distribution, pH distribution, and concentration distributions of chemical species along the length of the specimen as a function of time. The good agreement between the experimentally measured results and the numerically simulated results indicate that a better understanding of the physics and chemistry involved in the innovative *in-situ* hazardous waste site remediation technology has been developed.

1. INTRODUCTION

The adverse effects of lead in the environment on public health are of great concern. U. S. EPA has ranked lead as the number one priority hazardous substance. Soil is contaminated by lead from various sources including: (a) flaking lead paint; (b) leaded gasoline; (c) improper waste disposal; and (d) mine tailings, etc. Remedial measures normally implemented are dig-and-haul or solidification and stabilization followed by an engineered cover. Dig-and-haul may be very expensive if the extent of contamination is large. Immobilization of the contaminant *in-situ* renders the site unsuitable for future use.

The combined actions of electro-osmosis and ionic migration induced by a direct-current electrical potential in a wet soil mass results in the possibility of electro-kinetic extraction (Yeung 1990; Mitchell and Yeung 1991; Yeung 1994). The technique has been proven to be effective in removing spiked heavy metals from saturated Georgia kaolinite (Acar et al. 1994; Eykholt and Daniel 1994). However, the physics and chemistry involved are not fully understood yet. In this study, electro-kinetic experiments were conducted on lead(II) contaminated Georgia kaolinite to investigate the impact of various physicochemical factors on the cleanup efficiency of the technique. A plausible approach to model various processes involved was also developed to predict the variations of some of the important parameters along the length of the specimen as a function of time. These parameters include: (1) voltage distribution; (2) pH distribution; and (3) concentration distributions of anions, cations and non-ionic species This paper will present the results of these studies.

2. EXPERIMENTAL STUDIES

Approximately 1000 ml of 12.06 mol/m^3 lead nitrate solution (2500 ppm of lead) was prepared in an Erlenmeyer flask. The solution was acidified to dissolve the lead nitrate crystals completely. Approximately 1 kg of Georgia kaolinite was weighed and placed in a clean bucket. The lead nitrate solution was added to and mixed thoroughly with the soil to obtain a slurry. The soil slurry was then consolidated to a uniformly contaminated saturated specimen of 76.2 mm in diameter and 152.4 mm in length by a static loading system consisting of a piston, a hanger, and a series of weights.

The prepared specimen was then mounted onto the electro-kinetic cell. A schematic of the experimental setup is shown in Fig. 1. Major components of the electro-kinetic cell are: a specimen cylinder, two end flanges housing the reservoirs, two perforated graphite electrodes, two gas volume measuring devices, two flow volume measuring devices, and five voltage

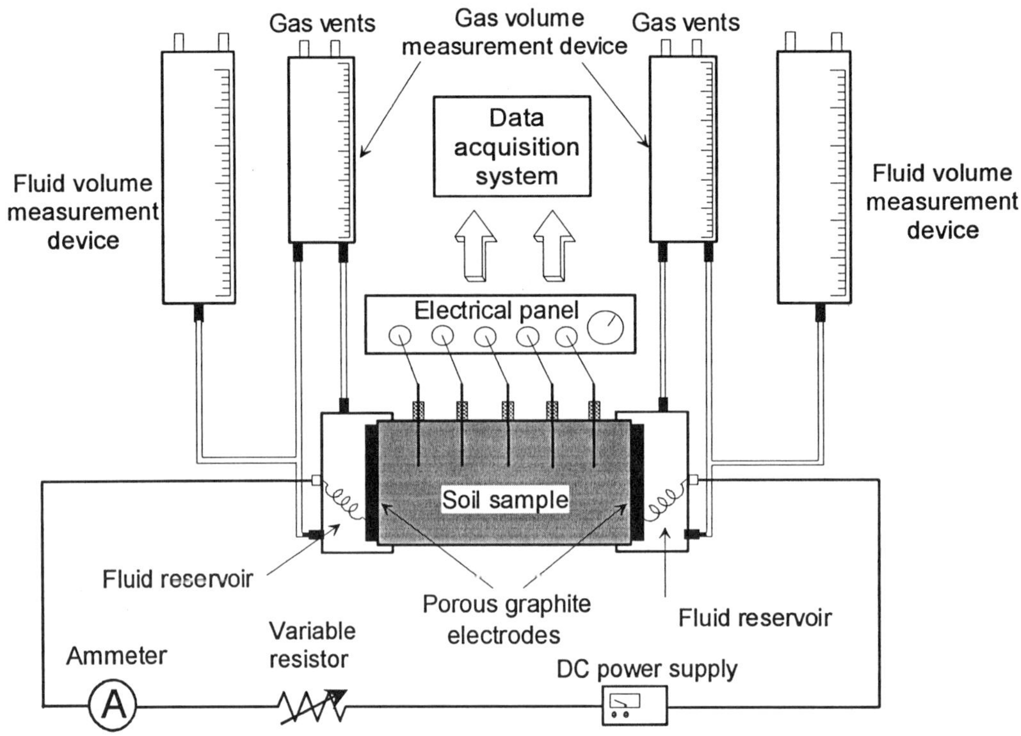

Fig. 1. A Schematic of the Experimental Setup

measurement probes. The gas and flow volume measuring devices are graduated cylinders filled by deionized water. The water levels in these cylinders were read at regular intervals during the test to determine the amount of gas generated and the electro-osmotic flow volume induced. The electrical current passing through the specimen and the voltages at seven different locations along the length of the specimen were also measured. After a constant electrical potential had been maintained on the specimen for five days, the specimen was disassembled and cut transversely into fifteen disc shaped sections. Each section was divided into four sub-sections for water content, total lead concentration, pH, and electrical conductivity measurements using standard procedures. Thus, the final distribution of water content, pH and total lead concentration, and electrical conductivity along the length of the specimen were determined.

3. MODELING STUDIES

The numerical model developed simulates: (1) transport of different lead species, hydrogen ions, and hydroxide ions induced by electro-osmosis, ionic migration, and diffusion; (2) chemical speciation of lead at different pHs; and (3) soil-lead interactions during the electro-kinetic extraction process.

The transport equations were formulated using the formalism of non-equilibrium thermodynamics by Yeung (1990) for the coupled flows of chemicals, electricity and water under the combined influences of chemical, electrical, and hydraulic gradients. The advantage of this approach is that coupling effects on the migration of each species can be considered (Datla and Yeung 1994; Yeung and Datla 1995). The equations for the flow of cations and anions relative to the soil are:

$$J_c = (L_{31}+C_cL_{11})\gamma_w\nabla(-H) + (L_{32}+C_cL_{12})\nabla(-E) + (L_{33}+C_cL_{13})\frac{RT}{C_c}\nabla(-C_c) + (L_{34}+C_cL_{14})\frac{RT}{C_a}\nabla(-C_a) \quad (1)$$

$$J_a = (L_{41}+C_aL_{11})\gamma_w\nabla(-H) + (L_{42}+C_aL_{12})\nabla(-E) + (L_{43}+C_aL_{13})\frac{RT}{C_c}\nabla(-C_c) + (L_{44}+C_aL_{14})\frac{RT}{C_a}\nabla(-C_a) \quad (2)$$

where L_{ij} = coupling coefficients defined elsewhere (Mitchell 1991, 1993; Yeung 1990; Datla and Yeung 1994; Yeung and Datla 1995); J_c = flux of cations; J_a = flux of anions; C_c = concentration of cation; C_a = concentration of anion; $\nabla(-H)$ = hydraulic gradient; $\nabla(-E)$ = electrical gradient; $\nabla(-C_c)$ = chemical gradient of cation; $\nabla(-C_a)$ = chemical gradient of anion; R = universal gas constant; T = absolute temperature; and γ_w = unit weight of water.

The determination of pH along the length of the specimen is very important in determining the behavior of reactive contaminants. The numerical model developed determines the amount of H^+ and OH^- generated by electrolysis of water at the anode and the cathode, respectively. pH buffering in the reservoirs is also considered. The transport of H^+ and OH^- into the soil are modeled by Eqs. (1) and (2). The pH is then determined as a function of time and space from the total hydrogen balance after the acid/base buffer capacity of the soil has been properly considered.

In order to model the physicochemical reactions of lead, it is important to study its chemical speciation in the pore fluid. Only three lead species are considered since the concentrations of the others are negligible (i.e. lower than 1×10^{-5} mol/m^3). The three species considered in this study were Pb^{2+}, $PbNO_3^+$, and $PbOH^+$. Sorption edge tests were performed to determine the proportion of total lead sorbed on the soil particle surface at different pHs. Hence, the amount of mobile lead at any time can be determined. A system of non-linear equations is then formulated using the mass balance, mass action, and chemical equilibrium equations for the mobile lead and nitrate in the dissolved phase. The solutions to these non-linear equations give the concentrations of different lead and nitrate species as a function of time and space.

The electrical voltage distribution is determined on the assumption that the non-linearity is caused by the removal of conducting contaminants. The electrical conduction through soil solid particles is assumed to be negligible. However, the electrical conductivity of the diffuse double layer is taken into account. The total electrical conductivity at any location is taken to be the sum of electrical conductivity of the pore fluid and that of diffuse double layer. Initially, the voltage distribution was linear due to the uniform distribution of ions resulting in a soil specimen of homogeneous electrical resistance. As conducting contaminants are being removed from the ends of the specimen, the electrical resistances at these locations increase. These heterogeneous distribution of resistances cause more rapid voltage changes at the ends of the specimen. Goodisman (1987) presented a relationship between electrical conductivity and ion concentrations for electrolyte solutions,

$$k = F\sum_i C_i u_i z_i \quad (3)$$

where k = electrical conductivity of the solution (S/m); F = Faraday's constant (96500 C/mol); C_i = concentration of species i (mol/m^3); u_i = ionic mobility of species i (m^2/V-s); and z_i = charge of species i.

If k_o is the electrical conductivity of the soil-water mixture without any contaminant, the total or bulk electrical conductivity of a contaminated soil k_T is given by

$$k_T = k + k_o \quad (4)$$

The resistance of each control volume is then determined using the relationship between electrical conductivity, dimensions of the control volume, and resistance. The total resistance of the specimen is thus the sum of these individual resistances as they are connected in series. The voltage at any point within the specimen can be determined by Ohm's law as the voltage difference across the specimen is known.

The transport equations and the non-linear physicochemical reaction equations are solved alternatively to simulate the electro-kinetic extraction process.

4. RESULTS AND DISCUSSIONS

An electro-kinetic experiment was performed on a saturated Georgia kaolinite specimen uniformly spiked with lead nitrate. The constant direct-current voltage maintained across the specimen was 48 volts. The initial concentration of lead solution added to the specimen was 12.06 mol/m^3. The initial pH of the soil specimen was 3.45. The cation exchange capacity of the soil was 20 mmol/kg of dry soil. The duration of the test was 5 days.

The final concentration distributions of total lead in

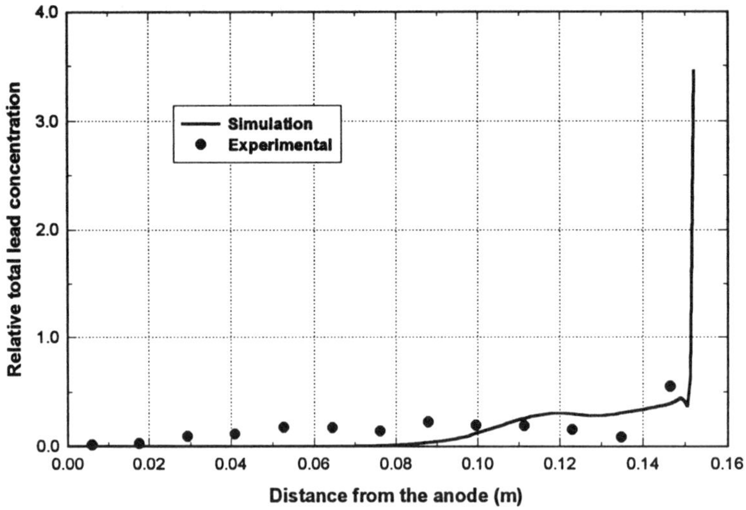

FIG. 2 Final Total Lead Distribution after Treatment

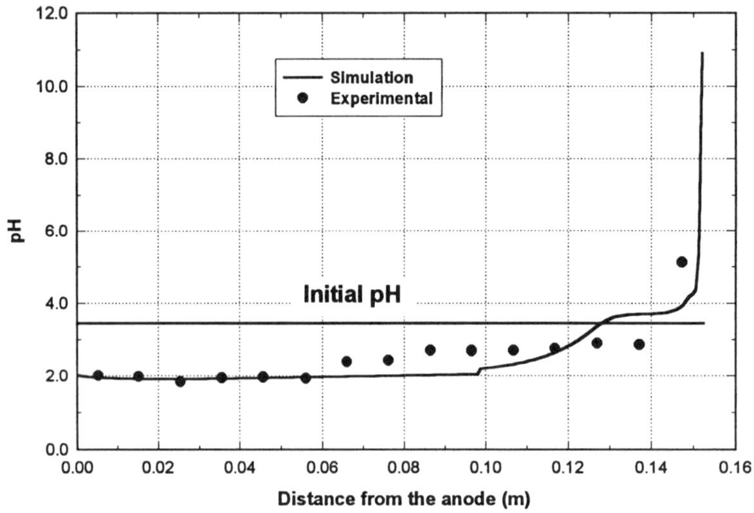

FIG. 3 Final pH Distribution after Treatment

the specimen relative to the initial total lead concentration as measured by the experiment and simulated by the numerical model are presented in Fig. 2. Generally, they are in good agreement except in the close vicinity of the cathode where the model predicts a very high concentration of total lead. Such lead accumulation was not observed in the experiment. However, it may be attributed to the lack of experimental data points in the close vicinity of the cathode.

A comparison of the final pH distribution between simulated and measured results is shown in Fig. 3. They are in close agreement indicating that the assumptions made in the development of the model are reasonable. It can also be observed that the values of pH in more than three-quarters of the soil specimen are lower than 3. As most lead species exist in the dissolved phase under such an acidic environment, it is possible to obtain a high cleanup efficiency as demonstrated in this study. Moreover, Georgia kaolinite has a relatively low acid/base buffer capacity and it is not able to maintain its initial pH during the process. Hence, it is important to note that the treatment efficiencies will be soil specific.

The comparison of voltage distribution between model predictions and experimental results is shown in Fig. 4. They are generally in good agreement. The voltage changes in the model are caused by the removal of lead and nitrate at the anode and the cathode, respectively. However, the electrical gradient near the anode was lower than the initial value owing to the greater influx of H^+ ions than the removal of lead ions resulting in the increase in electrical conductivity. However, the model underpredicts the voltage near the cathode. There may be other factors that causes the rapid voltage change at the cathode that have not been accounted for in this model.

The model predictions on chemical speciation of lead along the length of the specimen after 5 days of treatment are presented in Fig. 5. As a result of the low pH in most parts of the specimen, the speciation of lead was dominated in the free metal form, i.e Pb^{2+}, which was migrated effectively toward the cathode. At the cathode, the soil pH increased and the proportion of Pb^{2+} decreased. Moreover, the proportion of lead sorbed on the soil particle surface also increased with a rise in pH (Yeung et al. 1996) resulting in a high concentration of final total lead in the vicinity of the cathode as shown in Fig. 2.

5. CONCLUSIONS

This study presents a plausible approach to model the electro-kinetic extraction of reactive contaminants such as lead from fine-grained soils. The model predicts the variations of many important parameters as a function of time satisfactorily.

The experimental results indicate that the technique is effective in removing lead from a soil with a low buffer capacity. As the initial pH of the soil was low and the influx of H^+ from the anode made the pH even lower, the lead in the soil existed in its free metal form which could be removed effectively by an applied direct-current electric field. However, the results also demonstrate the importance of the physicochemical soil-contaminant interactions on the efficiency of the technique.

6. ACKNOWLEDGEMENTS

This material is based in part upon work supported by the Texas Advanced Technology Program under Grant No. 999903-258. This project has also been funded in part with Federal Funds as part of the program of the Gulf Coast Hazardous Substance Research Center which is supported under cooperative agreement R815197 with the United States Environmental Protection Agency and in part with funds from the State of Texas as part of the program of the Texas Hazardous Waste Research Center. The project is also supported by a 1995 Clay Minerals Society Student Research Grant awarded to the second author. The support is gratefully acknowledged. The contents do not necessarily reflect the views and policies of the U. S. EPA, the State of Texas, or the Clay Minerals Society nor does the mention of trade names or commercial product constitute endorsement or recommendation for use.

REFERENCES

Acar, Y. B., Hamed, J. T., Alshwabkeh, A., and Gale, R. J. (1994). "Cd(II) removal from saturated kaolinite by application of electrical current." *Géotechnique*, 44(3), 239-254.

Datla, S., and Yeung, A. T. (1994). "Subsurface migration of contaminants under the coupled influences of hydraulic, electrical, and chemical gradients." *Proc. 8th Int'l. Conf. for Computer Methods and Advances in Geomechanics*, Morgantown, West Virginia, 2, 1043-1048.

Eykholt, G. R., and Daniel, D. E. (1994). "Impact of system chemistry on electroosmosis in contaminated soil." *Journal of Geotechnical Engineering*, ASCE, 120(5), 797-815.

Goodisman, J. (1987). *Electrochemistry: theoretical foundations*, John Wiley & Sons, New York, New York.

Mitchell, J. K. (1991). "Conduction phenomena: from theory to geotechnical practice." *Géotechnique*, 41(3), 299-340.

Mitchell, J. K. (1993). *Fundamentals of soil behavior*, 2nd Edition, John Wiley & Sons, New York, New York.

Mitchell, J. K., and Yeung, A. T. (1991). "Electrokinetic flow barriers in compacted clay." *Geotechnical Engineering 1990*, Transportation Research Record 1288, Transportation Research Board, Washington, D.C., 1-9.

Yeung, A. T., (1990). "Coupled flow equations for water, electricity and ionic contaminants through

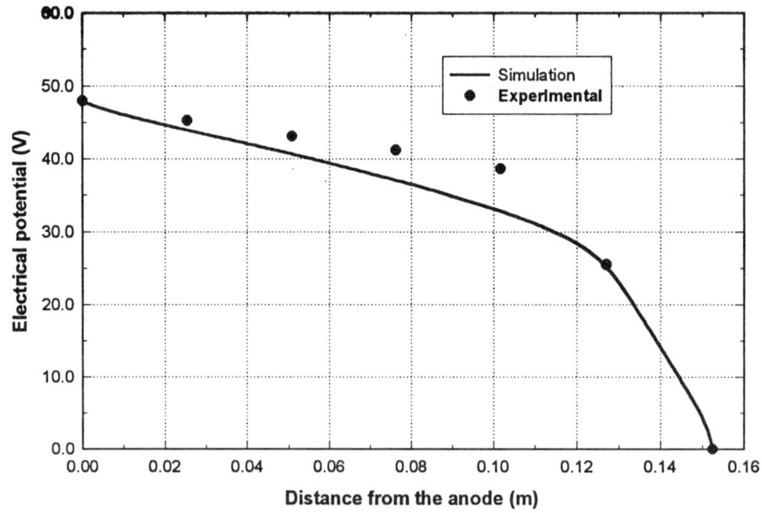

FIG. 4 Voltage Distribution after 5 Days of Treatment

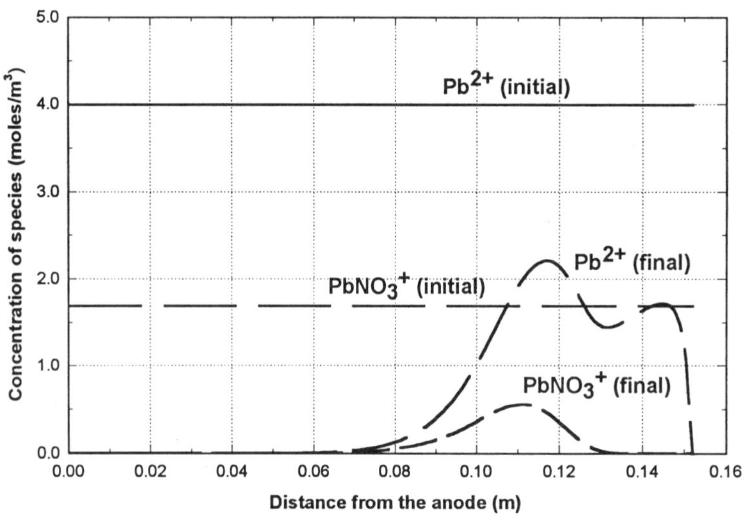

FIG 5. Speciation of Lead in Dissolved Phase

clayey soils under hydraulic, electrical and chemical gradients." *Journal of Non-Equilibrium Thermodynamics,* 15(3), 247-267.

Yeung, A. T. (1994). "Electrokinetic flow processes in porous media and their applications." *Advances in porous media*, Elsevier, Amsterdam, The Netherlands, 2, 309-395.

Yeung, A. T., and Datla, S. (1995). "Fundamental formulation of electrokinetic extraction of contaminants from soil." *Canadian Geotechnical Journal,* 32(4), 569-583.

Yeung, A. T., Hsu, C., and Menon, R. M. (1996). "EDTA-enhanced electro-kinetic extraction of lead." *Journal of Geotechnical Engineering*, ASCE, 122(8), in press.

Environmental Geotechnics, Kamon (ed.) © *1996 Balkema, Rotterdam. ISBN 90 5410 848 7*

Evaluation methodology *BESAL* for the selection of remedial action technologies

Th. Neteler
Schaub GmbH, Dorsten, Germany

ABSTRACT: Within the framework of an interdisciplinary research project “development and assessment of remediation concepts for contaminated sites and abandoned polluted deposits“ the evaluation methodology *BESAL* for the selection of remedial treatment technologies was developed at the Institut of Soil Mechanics and Foundation Engineering at the Ruhr-University of Bochum. It was ordered by the Environmental Agency of North Rhine-Westphalia, headed by Prof. Dr.-Ing. H.L. Jessberger and coordinated by Dr.-Ing. Th. Neteler. This methodical guidance is expected to facilitate the choice of the most appropriate and available treatment technology.

1 INTRODUCTION

During the investigation in the field of remedial actions on contaminated land, an appropriate concept is worked out, which is expected to determine also the technology or rather the combination of technologies that could be used for remediation. Prior to the selection of the most suitable technologies applicable for the distinct site, an evaluation of the existing alternative remedial techniques is necessary. The comparative evaluation has to consider all facts relevant to the decision in order to find an adequate solution covering all demands of a remedial action.

The investigation for remediation will take place after evaluating the contaminated site risks. The investigation to choose the most appropriate remediation concept can have different occasions:

- prevention of the danger from hazardous potential
- precaution for damages
- preparation for re-use
- rehabilitation of land
- preparation for development
- setting priorities between several sites with similar contamination

Different soil cleaning treatment technologies (e.g. thermal treatment, biological treatment, soil washing technology) and saveguarding technologies as solidification or encapsulation are available. The selection of the most appropriate remediation concept requires many individual decisions, which result from the motives on the basis of the investigation and from the site-specific requirements:

- decisions of cost-effectiveness
- decision between different re-use concepts
- delimitation of different remediation areas
- decision between different remedial actions e.g. decontamination of contaminated soil and encapsulation

The German expert group on environment ("Sachverständigenrat für Umweltfragen" [SRU 89]) and other authors [e. g. Salzwedel 86] take the view that the remediation of contaminated sites cannot only be guided by prevention. Instead, remedial actions should be rather part of a systematic concept, in which several projecting levels and instances are to be passed through.

Landscape planning, municipial development, land-use control and also projects for the utilization of groundwater have to be considered when establishing concrete use-specific quality standards for a remedial action. Therefore a methodical guidance has to provide the respective information.

Due to the numerous site-specific factors influencing the technical applicability of the different remediation technologies, and the case-specific importance of the single evaluation criteria, the evaluation method has to be adaptable to variable circumstances.

The present report gives a first overview of the evaluation methodology leading the officials in charge to a systematic and comprehensible procedure during the process of selecting suitable remedial technologies. It is developed in order to assist the authorities, the consultants and other persons involved to take the necessary decisions. The evaluation method will be published as a guideline; in addition, a computer programme is in preparation.

2 THE EVALUATION METHODOLOGY *BESAL*

The evaluation methodology *BESAL* intends a fair comparison of the decontamination and encapsulation technologies in consideration of site specific restrictions [Jessberger & Neteler 93]. The result of this comparison facilitates the decision to be taken concerning the available technologies and taking into account all important objectives.

The evaluation methodology consists of various phases (see fig. 1), which will be shortly presented in the following.

2.1 *Characterization of the initial situation for the selection of suitable remedial technologies*

Input data for the evaluation methodology consist of all results that have been obtained during the risk assessment and the investigation on remedial actions to be taken. In addition to these results the site specific definition of the objectives is to be determined by the official in charge before deciding on a specific treatment technology. Re-use specific trigger values for soil cleaning treatment technologies (e.g. thermal treatment, biological treatment) and the required degree of reduction of contamination mobility for technologies as solidification or encapsulation must be established during the investigations on remediation.

2.2 *Pre-evaluation*

The next steps of the methodology provide a pre-evaluation intending to exclude those technologies from further investigations which are not applicable for the remediation of the special site.

- Pre-evaluation of appropriate cleaning technologies
  Treatment technologies which are not applicable to the site specific soil or pollution are excluded from further investigation.
- Pre-evaluation of appropriate encapsulation technologies (long term isolation technologies)
  Each single techniology is based on specific criteria limitating the possibilities of respective realization; these criteria are supposed to make possible decisions on the realization in single cases.
- Pre evaluation of appropriate combinations of different specific remediation technologies
  It is intended to prove the applicability of combinations of treatment technologies and encapsulation technologies.
- Feasibility test
  The applicablility of techniques and combinations has to be proved by "feasibility criteria", which often lead to exclusions of single technologies (e.g. stage of technical development)

2.3 *Development of site-specific remedial scenarios*

Planning site specific concepts on remedial actions, the official in charge has to develop "remedial scenarios" using the technologies and combinations, which are proved to be applicable to the site specific requirements in "pre-evaluation". E.g. identifying thermal treatment to be applicable in the special case the user of the specific methodology has to consider, that

- the contaminated soil has to be excavated and transported to the on/off-site treatment plant,
- if necessary, areas for interim storage must be available,
- space for the treatment plant must be provided,
- the treated soil has to be suitable for re-use.

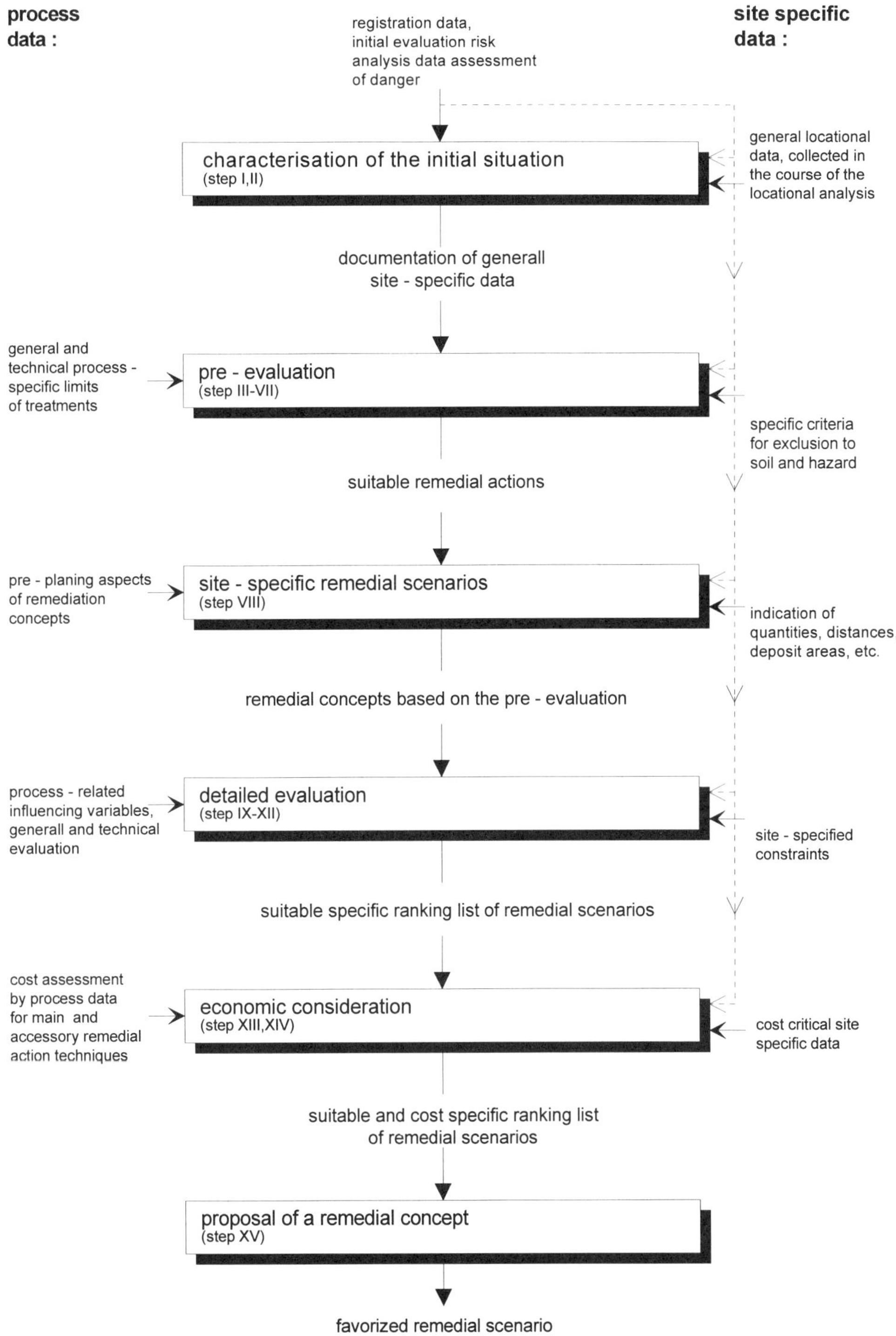

Fig. 1: Methodic procedure for the selection of an appropriate remediation concept [Neteler 95]

2.4 *Determination of the objectives importance for the evaluation*

The evaluation methodology *BESAL* provides a detailed evaluation with respect to site specific requirements. Remediation objectives are e.g. the decontamination of the soil following particular trigger values, reaching high safety standards, and avoiding new environmental impacts. Figure 2 shows an objective system comprehending general remediation objectives, which usually underlie the investigations.

The site specific importance of the general objectives results by considering all available site data in each single case.

2.5 *Detailed evaluation of the remedial scenarios*

The presented methodology demands a site specific comparative evaluation of the different technologies with respect to the remediation objectives. Two alternative procedures in rating the remediation scenarios are provided:

- rating and allocation of points
- rating by verbal reasoning.

In both procedures the remediation scenarios are compared whith each other in accordance with an approach applied to analyse the utility value. The utility value of remedial actions is defined in terms of relation between reached objectives and the determined objective sytem. The user of the methodology will gain a ranking system of the applicability of the technologies in the special case.

2.6 *Economic considerations*

Detailled cost estimates are necessary to get an economical overview. The methodology provides cost estimates with a high accuracy using a special questionnaire, which was developed in the course of the research project. Present Worth Analysis allow to compare the cost of remedial actions which occur over different time periods (like for instance investment costs, operating costs etc.) [LAWA 90].

The result attained after the qualified evaluation (utility values) will be related to the costs amounting for the respective remedial scenarios. It will then be possible to make some cost effectiveness considerations. The definition of limits (maximum cost limit, minimum utility value limit) and further instructions (minimization of costs, if necessary, considering different profit margins) leads to a limitation of decision alternatives. The remaining measures are considered as optimal for the purposes of the evaluation methodology.

2.7 *Proposal of a remedial concept*

After final considerations, a remedial concept is proposed to the responsible authorities. This proposal will prepare the decision on the way and the dimension of measures to be taken in single cases. With this proposal a complete documentation of the evaluation steps will be given.

3 CONCLUSIONS

According to the risk assessment the decision on whether a site has to be cleaned up or not has to be made. In case that remediation is necessary, a choice out of various possible measures/ technologies has to be made to transform the contaminated site into an area bearing no further risks for men and environment. The authorities and other persons involved to take the necessary decisions have to prove a great variability of aspects concerning the differences of the physical, chemical or biological treatment technologies and encapsulation technologies. Numerous objectives and planning intentions are also to be taken into account.

With the help of the presented evaluation methodology *BESAL*, it is possible to choose the most suitable remediation technology concept for a specific site. The decision-making prozess will gett more transparent and retraceble.

Furthermore, the computer aided model will make it possible to collect results of taken remedial actions, to evaluate them systematically and prepare them for the responsible authorities as a decision basis for future cases [see Schmalenbach & Jessberger 95].

REFERENCES

Länderarbeitsgemeinschaft Wasser LAWA (1990): Leitlinien zur Durchführung von Kostenvergleichsrechnungen, Gebr. Parkus KG, München.

main objectives	secondary objectives
1 reduction of existing risks	1.1 pollution specific applicability
	1.2 soil specific applicability
	1.3 good site specific adaptability
	1.4 reilable and simple monitoring- and control possibilities
2 availability	2.1 stage of development
	2.2 short duration of the remedial action
	2.3 minimal requirements on infra - structure
	2.4 minimal limitations of use / applicability
3 minimisation of new enviro-mental impacts	3.1 minimal impact by emission
	3.2 minimal energy and ressource consumption
	3.3 saving of soil specific characteristics
	3.4 minimal consequences of groundwater lowering
	3.5 minor residual products
4 safety standards	4.1 safe soil excavation
	4.2 safe soil treatment
	4.3 safe additional measures

Fig. 2: System of objectives for the site specific selection of concepts for remedial technologies

Jessberger, H.L. & Neteler, Th. (1993): „Leitfaden zur Bewertung und Auswahl geeigneter Sanierungsverfahren *BESAL*“. In: H.L. Jessberger, Abschlußbericht zum Forschungsvorhaben „Entwicklung und Beurteilung von Sanierungskonzepten für Altstandorte und Altlasten“, Bochum (unveröffentlicht).

Neteler, Th. (1995): Bewertungsmodell für die nutzungsbezogene Auswahl von Verfahren zur Altlastensanierung. Dissertation, Fakultät für Bauingenieurwesen der Ruhr Universität Bochum; zugleich Heft 23 der Schriftenreihe des Instituts für Grundbau

Salzwedel, J. (1986): Wasserrechtliche Instrumente zur Durchführung einer Sanierung und zur Begrenzung des Sanierungsaufwandes; in: VDG (Hrsg.): Altlastensanierung aus der Sicht des Gewässerschutzes; Schriftenreihe der VDG 52, Bonn.

Schmalenbach, U. & Jessberger, H.L. (1995): Evaluation and selection of remedial measures by means of computer-based desision support system. In: W.J. van den Brink, R. Bosmann & F. Arents (eds.): Contaminated Soil `95. Netherlands, Kluver Academic Publischers, pp 1478 - 1495.

SRU, Rat von Sachverständigen für Umweltfragen (1989): Sondergutachten "Altlasten"; Metzler-Poeschel-Verlag, Stuttgart.

Environmental Geotechnics, Kamon (ed.) © 1996 Balkema, Rotterdam. ISBN 90 5410 848 7

Immobilization of oil contaminated soils with cement and oil-stabilizing admixtures

Masataka Okumura & Tetsuji Yamada
Konoike Co. Ltd, Japan

Koichi Ono
Kyoto University, Japan

ABSTRACT : Treatment of wastes that is to be landfilled must be done in such a way that leaching test results of each hazardous substance complies with the regulations. Application of stabilizing admixtures to individual waste is still under investigation. In practice the laboratory test has to be done for each type of waste to find out the suitable admixture and required dosage.

In this study oil-contaminated soil was used to verify whether stabilizing admixtures would immobilize the oils in the soil.

Tests on leaching of oils out of the stabilized substance and unconfined compressive strength were carried out.

This paper proposes stabilizing admixtures suitable for each different type of oil-contaminated soils and the minimum required dosage.

1. INTRODUCTION

Today there are a number of contaminated sites that have soils contaminated with oils, organic solvents and heavy metals. Often these soils must be landfilled and must be treated to meet the requirements of the landfill. These requirements may include an oil-slick analysis, hexane extraction or other analytical testing.

Solidification by stabilizing admixtures is one of the ways that oils can be immobilized. To date stabilizing admixtures suitable for oil-contaminated soil have not been investigated clearly since the oils tend to obstruct the cement-water reaction. [2),3),4)]

In this study, using oil-contaminated soil, three types of experiments were carried out to examine if stabilizing admixtures can immobilize the oils and what kind of stabilizing admixtures are suitable for treatment of oil-contaminated soil.

Experiment ① deals with the relationship between oil concentration and unconfined compressive strength.

Experiments ② and ③ deal with the immobilization of the oils by stabilizing admixtures.

2. *EXPERIMENT ①*

2.1 MATERIAL AND METHOD

The objective of this experiment was to investigate the effect of the oils on the ability of the cement to stabilize the oils in the soil matrix. The measured parameters in this experiment were as follows.

Relationships between concentration of the oils and unconfined compressive strength

Influence of water-stabilizing admixtures ratio

Immobilization of the oils

Initial soil conditions are shown in Table-1.

Table-1 Physical test data

Characteristics	Value
pH	8.2
Conductivity	750 μ s／m
Hexane-extraction	110mg/l
Ignition loss	10.0%
Particle size	
Gravel (2～75mm)	18.4%
Sand (75μm～2mm)	63.8%
Silt, Clay (<75μm)	17.8%
Classification of soil	sand with gravel

In this experiment, the soil was sifted using a 5 millimeter mesh size. The type of stabilizing admixture used here was for organic soils. Three soil samples were made by adding of heavy oil. Initial oil concentrations are shown in Table-2.

Table-2 Initial soil analysis

Soil	Hexane extraction	Carbon tetrachloride extraction	Moisture content
SOIL-①	6500 mg/kg	<2 mg/l	18.6%
SOIL-②	17000 mg/kg	2.7 mg/l	12.9%
SOIL-③	28000 mg/kg	2.8 mg/l	10.7%

The stabilizing admixture was mixed in to the soil in a slurry. [5] The ratio of water to stabilizing admixture for each soil was .5, .8 and 1 respectively. The admixture dosage was 10 % by weight. Each test sample size was 50 millimeter in diameter and 100 millimeter in height. Samples were cured in 100% moisture condition at 20 degrees until the test-age was reached. Three test ages were used; 14 days, 28 days, and 90 days. Three test methods were used; unconfined compressive strength test, oil slick analysis and carbon tetrachloride extraction.

Figure-1 shows the results of the unconfined compressive strength test at the 14 days. Figure-2 is for 28 days.

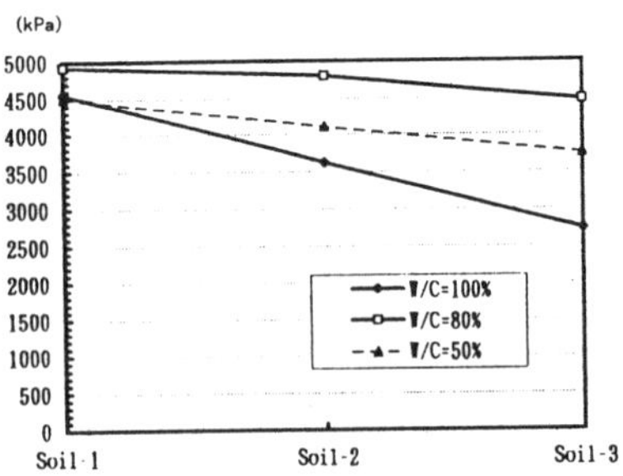

Figure-1 Unconfined compressive strength test results at the age of 14days

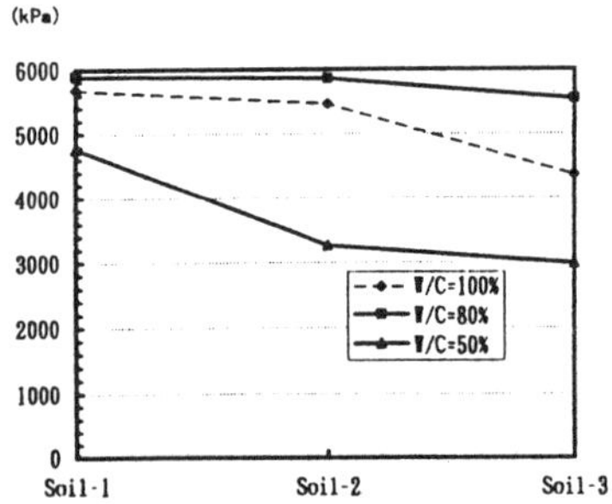

Figure-2 Unconfined compressive strength test results at the age of 28days

《Concentration of oils : SOIL ①<SOIL ②<SOIL ③》

Table-3 indicates the carbon tetrachloride extraction results. At 14 days, the oil concentration, by carbon tetrachloride-extraction, for all samples was less than 2mg/l.

Table-3 Results of oil concentration by carbon tetrachloride extraction

SOIL	Initial	14 days	28 days
SOIL-①	<2mg/l	<2mg/l	<2mg/l
SOIL-②	2.7mg/l	<2mg/l	<2mg/l
SOIL-③	2.8mg/l	<2mg/l	<2mg/l

At 28 days, an oil slick was not detected for all three samples.

2.2 DISCUSSION

a) The unconfined compressive strength is lowered as the concentration of the oils increase. It would appear that the oil controls the hardening of the cement.

b) The optimum water to stabilizing admixture ratio was .8.

c) Based on the carbon tetrachloride extraction and oil slick analysis results, the oils appear to be immobilized by stabilizing admixtures at the age of 28 days.

3. *EXPERIMENT ②*

3.1 MATERIAL AND METHOD

In this experiment, three types of stabilizing admixtures were used. Type-A was for an uncontaminated low organic soil, type-B was for an organic soil, and type-C was for a contaminated soil. The parameters measured were as follows.

Immobilization of the oils

Relationships between unconfined compressive strength and concentration of stabilizing admixture

Three different concentrations of oil were used in the soil samples. The soil was sifted using a 5 millimeter mesh size. Initial oil concentrations are shown in Table-4.

Table-4 Initial soil analysis

Soil	Hexane extraction		Moisture content
SOIL-A	4900 mg/kg	<2 mg/l	7.7%
SOIL-B	13000 mg/kg	<2 mg/l	8.1%
SOIL-C	24000 mg/kg	<2 mg/l	9.4%

The stabilizing admixture was mixed into the soil in a slurry with a water to stabilizing admixture ratio of .8. The concentrations of admixture dosage were; 100 kg/m^3, 150 kg/m^3, and 200 kg/m^3. Each test sample size was 50 millimeter in diameter and 100 millimeter in height. Samples were cured in 100% moisture condition at 20 degrees until the test-age was reached. Three test ages were used; 14 days, 28 days, and 90 days. Three test methods were used; unconfined compressive strength test, oil slick analysis and carbon tetrachloride extraction.

Figure-3 shows the results of the unconfined compression strength test after 14 days for SOIL-C. Figure-4 shows the results after 28 days.

No oil slick was detected in all samples after 14 days. Also oil concentrations were not detected by carbon tetrachloride extraction in all samples.

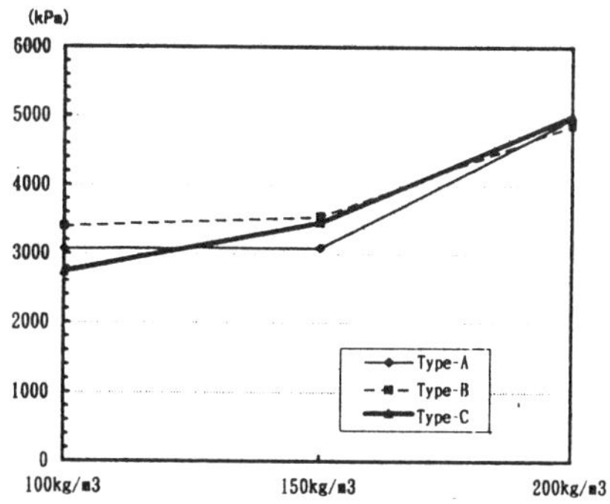

Figure-3 Unconfined compressive test results at the age of 14days

3.2 DISCUSSION

a) The admixture used for the contaminated soil produced the highest unconfined compressive strength. (Figure-3,4)

b) Increasing oil concentrations decreased the unconfined compressive strength. It would appear once again that the oil controls the hardening of the cement(Figure-5)

c) Based on the oil slick analysis, the stabilizing admixture used is effective in immobilizing the oils.

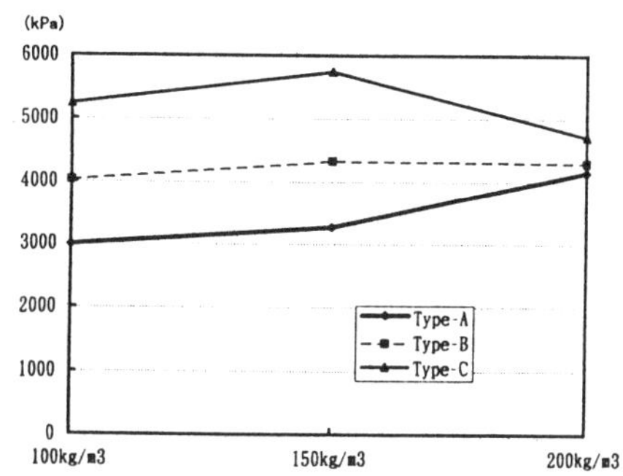

Figure-4 Unconfined compressive test results at the age of 28days(Soil-C)

4. *EXPERIMENT ③*

4.1 MATERIAL AND METHOD

In this experiment, we examined the effect of the stabilizing admixture on high water content soil (water content 37%). Parameters measured were as follows.

Immobilization of the oils

Relationship between unconfined compressive strength and concentration of stabilizing admixture

The type of stabilizing admixture was for the solidification of organic soil. Admixture dosages were; 5%, 10%, and 15% by weight.

The test methods used were; hexane-extraction and unconfined compressive strength test. Test ages were; 28, 90 and 180 days.

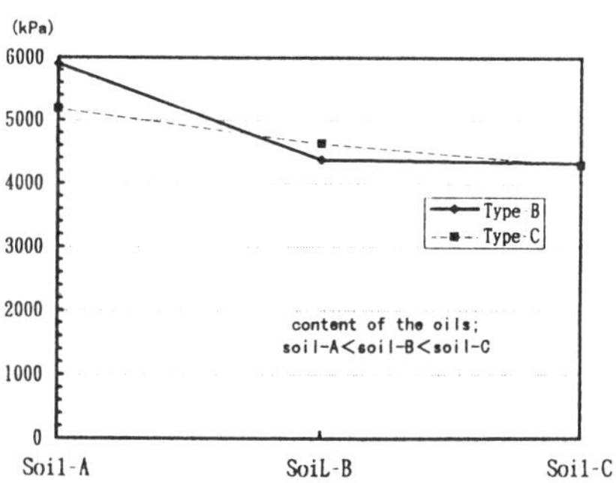

Figure-5 Unconfined compressive test results of three types of soils

Figure-6 shows the unconfined compressive strength test results.

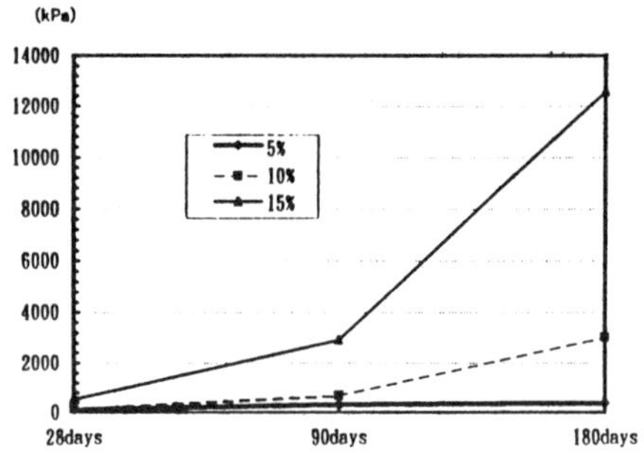

Figure-6 Unconfined compressive strength test results

Table-5 gives the pH and conductivity values measured on samples at stabilizing admixture dosage of 15%, Table-6 shows hexane-extraction data.

Table-5 pH and conductivity data at stabilizing admixture dosage 15%

Characteristics	initial	28days	90days	180days
pH	8.2	12.3	11.7	11.4
conductivity	750μs/cm	3300	1400	1400

Table-6 Hexane-extraction data

admixture dosage	initial	28 days	90 days	180 days
5%	110mg/l	38mg/l	2.6mg/l	7.2mg/l
10%	−	47mg/l	15mg/1	7.7mg/l
15%	−	26mg/l	26mg/l	6.1mg/l

4.2 DISCUSSION

1) Both Conductivity and pH were the highest after 28 days. It would appear that the hydration reaction is the most significant at that time.
2) Between 90 and 180 days, the conductivity and pH are stable. It appears that the hydration reaction at 180 days is as active as that at 90 days.
3) The oil concentration by hexane-extraction decreases with time. It can be assumed that the hydration reaction immobilized the oils.
4) Though normally the unconfined compressive strength is almost fixed after 28 days, in this experiment, the unconfined compressive strength continued to increase well after that time. The reason is assumed to be that the hydration reaction was retarded by the oils.

5. CONCLUSION

1) The tendency of oils to be immobilized by the use of stabilizing admixtures was confirmed by these three experiments.
2) The oils reduce the unconfined compressive strength by oils retarding the hardening of the cement.
3) This study used oil as a contaminant. In the future, it will be necessary to carry out experiments using other contaminants to test their leachability under similar conditions.

The authors would like to acknowledge the support of Mr. Isizaki and Mr. Araki (Sumitomo Osaka Cement. Co. Ltd.) for this work.

6. REFERENCES

1)Ministry Of Health and Welfare 1995. *Guide Book Of Wastes Treatment.*

2)Soil Quality Institution 1994. *Environmental Ground Society Primer : 203-208.*

3)Soil Quality Institution 1993. *Treatment Of Poor Ground : 211-216.*

4)T.Murata 1993. *Easy Chemistry of Wastes II : 158-172*

5)Sediment treatment incorporate association 1995. *Manual of Sediment Experiment And Test : 390-410*

6)N.Suzuki 1996. *Treatment Of Contaminated Soil With Geo-Cement*

Environmental Geotechnics, Kamon (ed.) © 1996 Balkema, Rotterdam. ISBN 90 5410 848 7

Centrifuge modelling of the removal of heavy metal pollutants using electrokinetics

M. Penn & C. Savvidou
University of Cambridge, UK
E. E. Hellawell
W.S. Atkins Consultants Ltd, UK (Formerly: University of Cambridge, UK)

ABSTRACT: Centrifuge modelling of contaminant transport and soil remediation requires knowledge of scaling relations between model and prototype. This paper presents the derivation of scaling laws specific to the centrifuge modelling of electrokinetics. A centrifuge test, which considers the electrokinetic remediation of a kaolin clay model initially polluted with a 0.1 molar copper sulphate solution, is also described and discussed qualitatively.

1 CENTRIFUGE MODELLING OF ELECTROKINETICS

The electrokinetic remediation technique employs a direct electric current to mobilise contaminants in fine grained deposits. The processes governing the removal include ionic mobility and electroosmosis. Ionic mobility is the movement of an ion towards the oppositely charged electrode under the influence of an electric potential gradient. Electroosmosis is a phenomenon caused by the diffuse double layer present at the particle-fluid interface (Mitchell, 1993). The mobile ions within this layer migrate towards the oppositely charged electrode (generally the cathode) producing a drag force on the surrounding fluid, and hence movement of the pore fluid. As electroosmosis is generally towards the cathode, it inhibits the migration of negative ions, and aids the migration of positive ions.

Centrifuge modelling offers a technique for replicating the stress distribution of a prototype, within a reduced scale model. The logarithm of the stress is related to voids ratio, and hence to the porosity and tortuosity of the soil. The governing processes of electrokinetics, electroosmosis and ionic migration, are affected by the porosity and tortuosity of a given soil.

During the electrokinetic process, changes in pore water pressures have been observed due to non-uniform voltage gradients, which produce non-linear electroosmotic flow through the clay (Eykholt & Daniel 1994). These differential pore water pressures cause contaminant movement as they dissipate. It is important that in a model the initial pore water pressure corresponds to the prototype value so that the differential pore water pressures induced by electroosmosis are superimposed on the actual flow regime and soil conditions encountered in the field. This ensures that the soil structure and the contaminant migration are correctly simulated. The centrifuge model tests replicate prototype stresses and hence initial pore water pressure distributions. Therefore by simulating the prototype changes in stress and therefore voids ratio, porosity, toruosity, and pore water pressure, centrifuge modelling provides the correct environment for observing electrokinetic remediation within a scaled model of a prototype.

Reduced scale modelling enables simulation of prototype problems such as different electrode configurations and depths, various pollutants and soil types to be investigated within a controlled environment. The three-dimensional model also allows areas around and below the electrodes to be studied, thereby complementing the observations made using one-dimensional electrokinetic flow cells.

1.1 *Scaling Laws for Transport Processes*

The aim of centrifuge modelling is to simulate the correct prototype stress conditions at points in the model, which geometrically correspond to points in the prototype. This is achieved by making a scale model, reduced by a factor N, and subsequently testing the model at N gravities (Schofield 1980).

The pore water pressure differential between

two points in the model is the same as the differential observed between two corresponding points in the prototype, but the linear distance is reduced in the model. This means that the hydraulic gradient will be increased by a factor of N. As the fluid velocity is proportional to the hydraulic gradient, the pore fluid velocity in the model is also increased by a factor of N;

$$\bar{v}_m = N\bar{v}_p \tag{1}$$

where the subscripts p and m indicate the prototype and model respectively. Since model velocities are N times faster, and model distances are N times smaller than the corresponding prototype values, it follows that the model advection times (defined as distance divided by velocity), are N^2 times shorter than the prototype times;

$$t_m = \frac{1}{N^2} t_p \tag{2}$$

Scaling laws for centrifuge modelling of contaminant transport due to hydraulic and concentration gradients have been presented by Arulanandran et al (1988) and Culligan-Hensley & Savvidou (1995). The scaling laws for electrokinetic transport processes are derived below.

The one-dimensional equation governing contaminant migration due to electrical, concentration and hydraulic gradients for a particular ionic species (Alshawabkeh & Acar, 1992) is:

$$nR_d \frac{dc}{dt} = D_d * \frac{d^2c}{dx^2} - \bar{v}\frac{dc}{dx} \tag{3}$$

where n is porosity, R_d is the retardation factor, c is concentration of the ionic species (kg/m^3), t is time (s) and D_d* is the effective diffusion coefficient (m^2/s) defined as:

$$D_d* = D_d \tau n \tag{4}$$

where τ is the tortuosity of the soil. $\bar{v}$ is the average velocity of an ion (m/s) defined below:

$$\bar{v} = -(u* + K_e)\frac{dE}{dx} - K_h\frac{dh}{dx} \tag{5}$$

where $u*$ is the effective ionic mobility (m^2/Vs), K_e is the coefficient of electroosmotic permeability (m^2/Vs), E is the electric potential (V), K_h is the hydraulic permeability (m/s) and h is the hydraulic potential (m).

The contaminant transport equation (eqn. 3) ignores the effect of mechanical dispersion. The authors have modified the equation to incorporate both diffusion and mechanical dispersion by substituting the coefficient of longitudinal hydrodynamic dispersion, D_{hl}, for D_d*. D_{hl} is defined below;

$$D_{hl} = D_d* + \alpha_l \bar{v} \tag{6}$$

where α_l is the longitudinal dispersivity of the soil.

If the pollutant, pore fluid and homogeneous soil are the same in both model and prototype, and the microscopic pore and particle geometry is the same, it follows that;

$$(u*)_m = (u*)_p \tag{7a}$$

$$(D_d*)_m = (D_d*)_p \tag{7b}$$

$$(\alpha_l)_m = (\alpha_l)_p \tag{7c}$$

If the model velocities are small, D_{hl} is dominated by diffusion, rather than dispersion, and the contaminant transport is correctly scaled (Boyce, 1994). It can be assumed that for small velocities;

$$(D_{hl})_m = (D_{hl})_p \tag{8}$$

The coefficient of electroosmotic permeability is defined as:

$$K_e = \frac{n\varepsilon\zeta}{\eta} \tag{9}$$

where ε is the electrical permittivity of the medium (F/m), ζ is the zeta-potential (V) and μ is the dynamic viscosity of the pore fluid (kg/ms). Assuming that the soil porosity and the characteristics of the double layer are the same at prototype and model scale, the electroosmotic permeability is unchanged by centrifuge gravities.

$$(K_e)_m = (K_e)_p \tag{10}$$

If the electric potential applied to both prototype and model is the same;

$$E_m = E_p \tag{11}$$

and the model is at reduced scale;

$$x_m = \frac{1}{N} x_p \quad (12)$$

then the one-dimensional equation for velocity in the model can be written for zero hydraulic gradient as:

$$\bar{v}_m = -[(u^*)_m + (K_e)_m]\frac{dE_m}{dx_m} \quad (13)$$

Substituting (7) - (12) into (13) the following expression is obtained.

$$\frac{1}{N}\bar{v}_m = -[(u^*)_p + (K_e)_p]\frac{dE_p}{dx_p} \quad (14)$$

Applying the one-dimensional equation at prototype scale yields the following;

$$\bar{v}_p = -[(u^*)_p + (K_e)_p]\frac{dE_p}{dx_p} \quad (15)$$

Comparing (14) and (15) gives;

$$\bar{v}_m = N\bar{v}_p \quad (16)$$

Therefore, the model velocities associated with electroosmotic flow and ionic migration are N times greater than in the corresponding prototype. The scaling for pore fluid velocities due to electrical gradients are the same as those presented for hydraulic gradients (eqn. 1).

The modified one-dimensional equation for ion movement (eqn. 3) due to electric and concentration gradient can be written for the model;

$$n_m(R_d)_m\frac{dc_m}{dt_m} = (D_{hl})_m\frac{d^2c_m}{dx_m^{\ 2}} - \bar{v}_m\frac{dc_m}{dx_m} \quad (17)$$

Equations (7) - (12) and (16), can be substituted into (17) to yield;

$$\frac{1}{N^2}n_p(R_d)_p\frac{dc_p}{dt_m} = (D_{hl})_p\frac{d^2c_p}{dx_p^{\ 2}} - \bar{v}_p\frac{dc_p}{dx_p} \quad (18)$$

The one-dimensional equation can be written for prototype scale;

$$n_p(R_d)_p\frac{dc_p}{dt_p} = (D_{hl})_p\frac{d^2c_p}{dx_p^{\ 2}} - \bar{v}_p\frac{dc_p}{dx_p} \quad (19)$$

Comparing equations (18) and (19) yields the scaling law for time;

$$t_m = \frac{1}{N^2} t_p \quad (20)$$

Therefore migration related to electroosmosis and ionic mobility takes place N^2 times faster in the model than in the prototype. The scaling for migration times is consistent with the scaling for flows due to hydraulic gradients (eqn. 2).

Therefore, equations (16) & (20) indicate that if the same electrical potential is applied across both model and prototype, then there is consistency between the scaling of contaminant transport due to hydraulic and electrical gradients.

1.2 *Scaling Law for Electric Current*

If the soil and pore fluid are the same for the model and prototype, then it follows that;

$$\rho_m = \rho_p \quad (21)$$

where ρ is the resistivity of the soil (Ωm) and is defined as;

$$\rho = \frac{RA}{l} \quad (22)$$

where R is resistance (Ω), A is area (m^2) and l is length (m). The resistivity of the model can be written as;

$$\rho_m = \frac{R_m A_m}{l_m} \quad (23)$$

Using (11), (12) & (21) to substitute into (23);

$$N\rho_p = \frac{R_m A_p}{l_p} \quad (24)$$

This can be compared to the resistivity of the prototype to give the scaling for resistance;

$$R_m = NR_p \quad (25)$$

The scaling law for current, I (Amps), can be found using Ohm's law:

$$E = IR \quad (26)$$

The scaling for E and R, given in equations (11) and (25) produce the following scaling for current and power (P):

$$I_m = \frac{1}{N} I_p \tag{27}$$

$$P_m = \frac{1}{N} P_p \tag{28}$$

This means that the current flow within the model is N times smaller than the current flow within a corresponding prototype.

2 EXPERIMENTAL PROCEDURE

Centrifuge tests were carried out using the Cambridge University 10m diameter beam centrifuge. The cylindrical models (420 mm diameter x 300 mm depth) were made from an E-grade kaolin clay mixed with copper sulphate solution (0.1 molar) and were tested at 50g. A constant electrical potential of 40 V was applied across the clay by inserting rectangular platinum electrodes, 120 mm x 65 mm deep, at a depth of 65 mm and a separation of 200 mm, which corresponds to a prototype separation of 10m, into the clay within porous plastic wells. Contaminants that migrated to the wells during the process were removed by flushing the wells with deionised water.

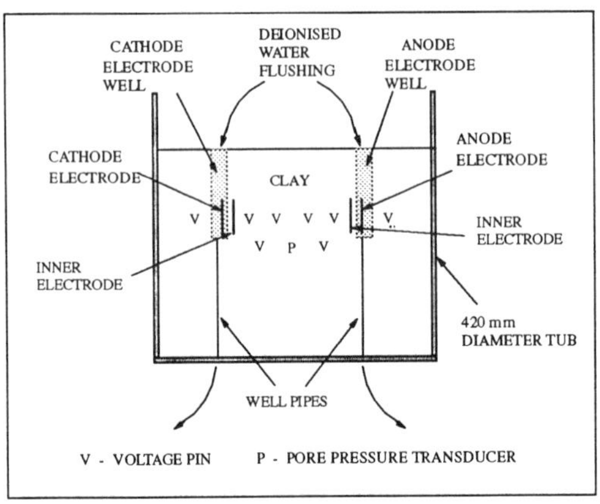

Fig. 1 Cross-section through model showing positioning of instrumentation.

The voltage field, temperature and pore water pressures within the model were monitored by voltage pins, thermocouples and pore pressure transducers respectively. A vertical section of the model and instrumentation is presented in Fig. 1.

The test duration was 22 hours. After the test, sampling was carried out to determine the final concentration of copper and sulphate ions in the pore fluid. The model was divided into sections by cutting the clay into cubes. The pore fluid was extracted from the clay and tested for copper, using atomic absorption spectrophotometer, and sulphate using a turbidity method with a spectrophotometer. The initial copper and sulphate concentrations were found by sampling the clay removed during the model-making, prior to testing. The initial concentrations were assumed to be constant throughout the soil mass.

3 DISCUSSION OF EXPERIMENTAL RESULTS

The initial and final voltage profiles for a vertical cross-section through the model are presented in Figs. 2 (a) & (b). The initial voltage field is uniform indicating the uniform conductivity of the medium, Fig. 2(a). In Fig. 2(b) the voltage field is distorted indicating a change in conductivity of the soil with time. The area close to the cathode, and to a lesser extent the area around the anode, become more resistive. These are indicated by the close spacing of equipotentials and therefore, higher voltage gradients across those regions.

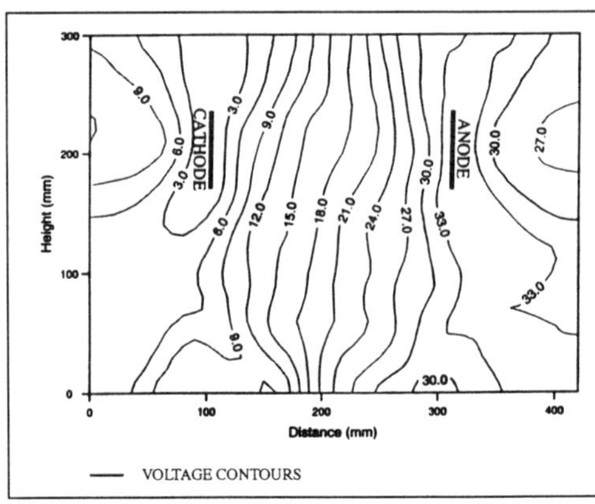

Fig. 2 (a) Initial voltage field for a vertical cross-section through the model.

Electrolysis of water at the two electrodes produces hydroxide ions and hydrogen gas at the cathode, whilst oxygen and hydrogen ions are produced at the anode. The primary electrolysis reactions are:

$4H_2O + 4e^- \rightarrow 2H_2\uparrow + 4OH^-$ Cathode

$2H_2O - 4e^- \rightarrow O_2\uparrow + 4H^+$ Anode

The resultant ions from the electrolysis reactions migrate towards the oppositely charged electrode, forming acid and alkali fronts. (Alshawabkeh & Acar, 1992). The acid front migrates from the anode to the cathode, whilst the alkali front migrates from the cathode to the anode. The copper ions, which move towards the cathode, meet the hydroxide front close to the cathode and precipitate out of solution in the high pH environment forming copper hydroxide;

$Cu^{2+} + 2OH^- \rightarrow Cu(OH)_2$.

This precipitate is not conductive and causes a high resistive band to build up close to the cathode.

The wells are flushed with deionised water, which is drawn into the soil around the anode due to electroosmosis. This high resistance fluid acts as a resistive band around the anode. The distorted final voltage field (fig. 2(b)) shows these areas of high resistance. Fig. 3, which is the voltage profile between the electrodes, shows the high resistance areas close to both electrodes as large voltage gradients.

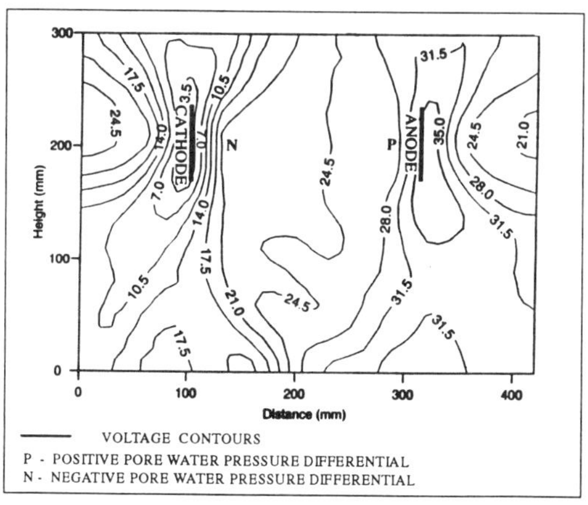

Fig. 2 (b) Voltage field at 22 hours for a vertical cross-section through the model.

The model can be idealised conceptually as a number of resistors in series, with a constant electric potential being applied across the circuit. The clay can be divided into different sections between the electrodes, and each can be considered to be a resistor with a certain resistance. As the total resistance of the circuit is the sum of the individual resistors, just one small area of clay with a large resistance can effectively control the circuit, reducing the current in the whole system. This means that the efficiency of the process, when using a constant voltage system, is vulnerable to small areas of high resistance. These areas require high voltage gradients to keep the current flowing through them, hence reducing the driving voltage gradient over the majority of the sample and limiting effectiveness of the total supply voltage.

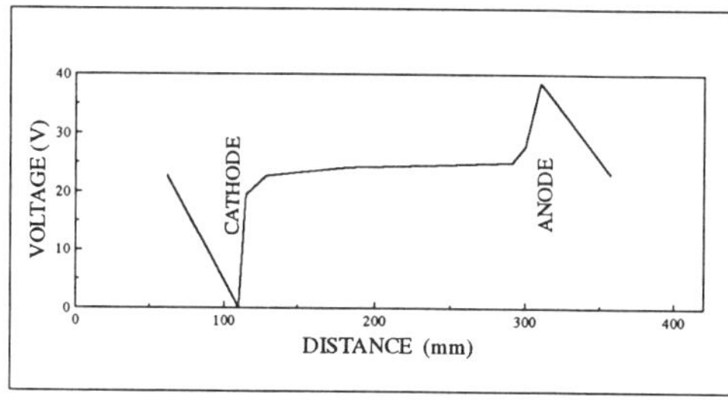

Fig. 3 Voltage profile between the electrodes at depth of 90 mm, at 22 hours.

The final concentrations for a vertical cross-section through the model show some reduction in concentration for both ions, mainly around the electrodes (Fig. 4a & b). As expected, there is a greater reduction in copper ions at the anode, as the positive copper ions migrated towards the cathode. There is also a reduction in sulphate ions at the cathode as the negative sulphate ions migrated towards the anode.

In the test, the ion movement was affected by the non-linear voltage gradient. Charge balance throughout the model dictates that the mass flux of ions through each section of the model must be constant. This means that with a reduced concentration close to either electrode, the voltage gradient must increase so that within that section, fewer ions are moving faster to maintain the ion mass flux. In the central part of the model, a greater number of ions are moving more slowly under a smaller driving gradient. The concentration profiles at the end of the test, both show reduced concentrations around the electrodes, which correspond to the areas with large voltage gradients, whilst the central section has a higher concentration and smaller electrical driving gradient, Fig. 2 (b).

Electroosmotic flow is dependent on the zeta-potential of the soil, which is influenced by pH. As pH gradients are formed by electrolysis reactions, it is reasonable to expect non-uniformities in electroosmotic flow due to pH changes through the

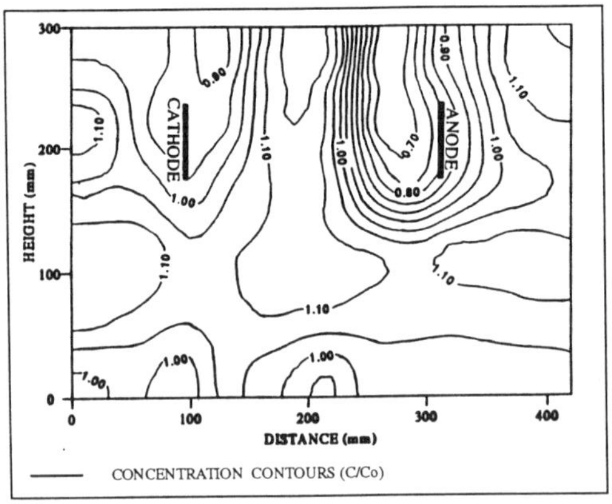

Fig. 4 (a) Normalised final copper concentrations for a vertical cross-section through the model.

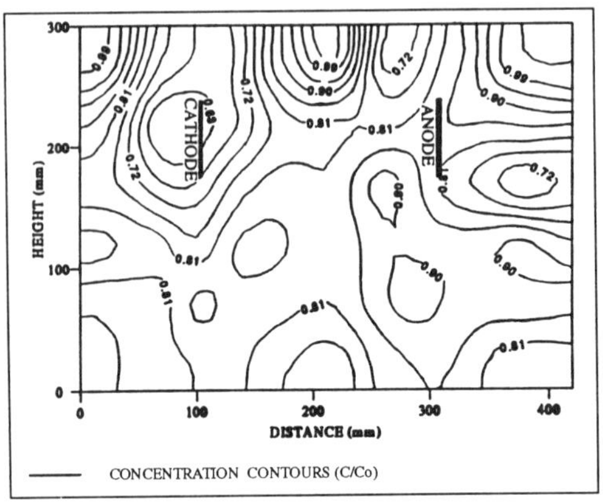

Fig. 4 (b) Normalised final sulphate concentrations for a vertical cross-section through the model.

model (Eykholt & Daniel, 1994). Electroosmotic flow is also proportional to the voltage gradient, and so a non-linear voltage gradient will produce non-uniform electroosmotic flows and will generate differential pore water pressures across the model. The electroosmotic flow will be high in areas of large voltage gradients in Fig. 2(b) and (3), close to the electrodes. As the electroosmotic flow is from the anode to cathode, and is non-linear across the model, differential pore water pressures will develop close to the changes in gradient in the voltage profile. Positive differential pore water pressures will develop at the change in voltage gradient, a few centimetres from the anode, whilst negative differential pore water pressures will develop just before the change in voltage gradient, in the model a few centimetres from the cathode. These areas are marked on Fig. 2(b) by a P and N respectively. As these pore water pressures dissipate, they will cause contaminant transport. For accurate simulation of contaminant migration due to these resultant pore water pressures, it is necessary to superimpose them on the actual pressures encountered in a prototype.

The post-test concentration readings are influenced by the movement of pore water as the clay model swells after testing. The effect is difficult to quantify, and this highlights the need to develop sensitive in-situ instrumentation for transient pH and ionic concentration measurements.

4 CONCLUSIONS

Centrifuge testing has been demonstrated to be a useful technique for studying remediation of contaminated soil using electrokinetics. The scaling laws show that there is no distortion between hydraulic and electrically induced contaminant movement. The centrifuge test that simulated a three dimensional prototype showed some contaminant removal close to the electrodes and highlighted the fact that electrokinetic remediation is susceptible to small areas of high resistance between the electrodes, which limit the process by reducing the driving gradient across the majority of the contaminated soil.

REFERENCES

Alshawabkeh, A., & Acar, Y.B. (1992) Removal of contaminants from soils by electrokinetics: A theoretical treatise. *J. Envir. Sci. & Health.* A27(7) : 1836-1861.

Arulanandran, K. et al (1988) Centrifuge modelling of transport processes of pollutants in soils. *J. Geot. Eng. ASCE,* Vol. 114, No. 2 : 185 - 205

Boyce, E. R. (1994) Modelling of transport processes in heterogeneous soils. PhD thesis, Cambridge University.

Culligan-Hensley, P.J. & Savvidou, C. (1995) 'Environmental geomechanics and transport processes' in *Geotechnical Centrifuge Technology,* ed. Taylor R.N. , Blackie Academic & Professional.

Eykholt, G.R. & Daniel, D.E. (1994) Impact of system chemistry on electroosmosis in contaminated soil. *J. Geot. Eng. ASCE,* Vol. 120, No. 5 : 797-815.

Mitchell, J.K. (1993) *Fundamentals of soil behaviour.* 2nd Ed., John Wiley & Sons.

Schofield, A.N. (1980) Cambridge geotechnical centrifuge operations *Geotechnique,* Vol. 20.

Environmental Geotechnics, Kamon (ed.) © 1996 Balkema, Rotterdam. ISBN 90 5410 848 7

The behaviour of light, nonaqueous phase liquids under hydraulic flushing

S. Ratnam, P.J. Culligan-Hensley & J.T. Germaine
Massachusetts Institute of Technology, Cambridge, Mass., USA

ABSTRACT: Centrifugal modelling techniques were used to investigate the behaviour of a light nonaqueous phase liquid (LNAPL) under upward hydraulic flushing over a wide range of conditions. The work demonstrates the influence of buoyancy and viscous pressure forces on NAPL entrapment. Data from experiments conducted using two different sands are reported in a phase-diagram describing the percentage of LNAPL trapped in contaminated soil following one pore-volume injection of water, as a function of a combined capillary and Bond number (N_{CB}). This form of diagram may be used to predict the amount of LNAPL trapped behind the passage of a wetting front for any value of N_{CB}.

1. INTRODUCTION

In the United States, it is estimated that more than 300,000 sites may have contaminated soil or groundwater requiring some form of remediation (NRC, 1994). Over the past decade, the technological response to demands for clean-up at such sites has exclusively been the application of so-called "pump-and-treat" technology. In essence, pump-and-treat technology attempts to "flush out" the contaminants, by physical displacement to an extraction well where the recovered effluent can be treated with established above ground technologies.

Many of the prevalent chemicals found at contaminated sites are organic compounds. Such chemicals typically enter the subsurface as organic liquids that are sparingly soluble with water. When the extraction of these nonaqueous phase liquids (NAPLs) is undertaken by hydraulic flushing (water flooding), discontinuous "blobs" of NAPL can become entrapped in the porous medium behind the wetting (water) front. The entrapment of NAPLs under hydraulic flushing is pertinent to many aspects of site remediation. Entrapment mechanisms determine what proportion of NAPL can be removed using traditional pump-and-treat technology, and consequently, how much NAPL remains for possible recovery by alternative methods, such as bioremediation, *in-situ* air sparging or surfactant flushing.

This paper describes a research program that is being undertaken to investigate LNAPL (Light Nonaqueous Phase Liquid) entrapment under hydraulic flushing. Geotechnical centrifuge modelling is being used as an experimental tool for the investigation.

1.1 Mechanisms for NAPL entrapment

When a wetting fluid (such as water) displaces a nonwetting fluid (such as oil) in a porous medium, a proportion of the nonwetting fluid will remain entrapped in the medium pores once the wetting front has passed. In general, the nonwetting fluid is trapped in the form of disconnected "blobs" or "ganglia," whose complicated morphology and distribution is dependent upon (i) the geometry of the pore space; (ii) fluid-fluid properties such as interfacial tension, density contrast, and viscosity ratio; (iii) the fluid-solid interfacial properties effecting wetting behaviour; (iv) the applied pressure gradient; and (v) gravity (Morrow and Songkran, 1981).

The program described here is investigating the effects of applied pressure gradient and gravity on residual nonwetting phase saturation for systems in which (i) there is complete wetting, (ii) the fluid-fluid properties are held constant, (iii) the porous media are geometrically similar and (iv) the immiscible displacement is stable.

During stable displacements, nonwetting phase (NAPL) entrapment is caused by capillary forces and overcome by viscous pressure forces and/or buoyancy forces (Mayer and Miller, 1992). The ratio of viscous pressure to capillary forces and gravity to capillary forces are usually expressed as dimensionless groups, known as the capillary number (C_a) and the Bond number (B_o), respectively.

$$C_a = \frac{u\mu}{\sigma} \qquad B_o = \frac{\Delta\rho g R^2}{\sigma} \tag{1}$$

where u is the displacing (D'arcy) fluid velocity, μ is the displacing fluid viscosity, σ is the interfacial tension, Δρ is the fluid density contrast, *g* is the acceleration due to gravity, and R is the average particle radius of the porous medium.

Above a certain upper, critical capillary or Bond number, no capillary trapping occurs, and the residual nonwetting phase saturation is close to zero. Conversely, below a certain lower, critical capillary or Bond number, the trapping is dominated by capillary forces, and the residual nonwetting phase saturation reaches a constant maximum value.

Between these two extremes, the residual nonwetting phase saturation depends upon the combined effects of capillary, viscous pressure and gravity forces.

The mapping of these three domains in the plane C_a - B_o, constitutes the "phase-diagram" for a given system.

This research is using geotechnical centrifuge modelling to establish experimental phase-diagrams for three uniform porous media systems, under both vertical and horizontal displacement conditions.

2. APPROACH

The experimental work is being conducted using the geotechnical centrifuge facility in the Department of Civil and Environmental Engineering at Massachusetts Institute of Technology; the reader is referred to Pahwa (1987) for a detailed description of this facility.

Centrifuge modelling is an experimental method traditionally used by geotechnical engineers to obtain soil-stress conditions that are homologous in model and prototype (Schofield, 1980). This is achieved by subjecting a scale soil model, where all linear dimensions are reduced by a factor n, to a centrifugal acceleration of n gravities (n*g*), Figure 1.

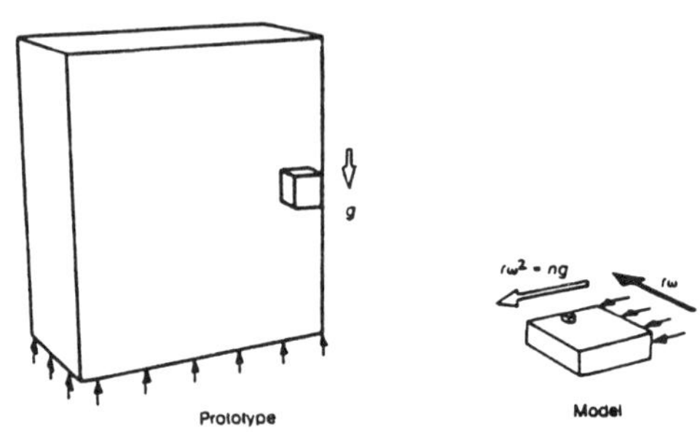

Figure 1. Principle of centrifuge modelling (after Schofield, 1980)

For the particular case of nonwetting phase entrapment, centrifuge modelling offers a simple means of varying the Bond number characterizing a particular system: this is achieved by simply varying the gravitational force (g-level) acting upon that system (refer to Equation 1). No other experimental technique can offer the same versatility as centrifuge modelling in this respect.

2.1 Vertical displacement tests

Figures 2 and 3 describe the model form that is being used to investigate LNAPL entrapment under vertical hydraulic flushing.

All experimental models are constructed from uniform silica sand fully saturated with an aliphatic oil (Soltrol-220). During each experiment, the nonwetting phase fluid (Soltrol-220) is displaced upward at a constant velocity by water, which is acting as the wetting phase fluid. The wetting phase displacement velocity is fully controlled by a constant rate displacement pump. The degree of entrapped nonwetting phase remaining after a single passage (one pore volume) of the wetting front is determined from appropriate pre- and post-test weight and volumetric measurements.

The effective sum of the capillary and Bond numbers is varied between tests, by changing the wetting phase displacement velocity through the model and/or the gravitational force acting on the model.

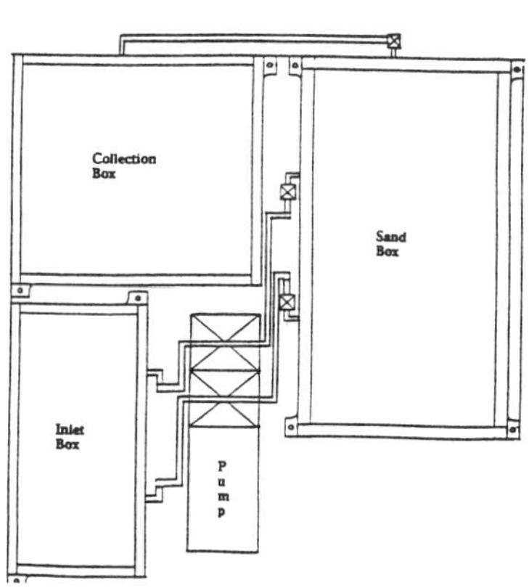

Figure 2. Plan of Experimental Set-up.

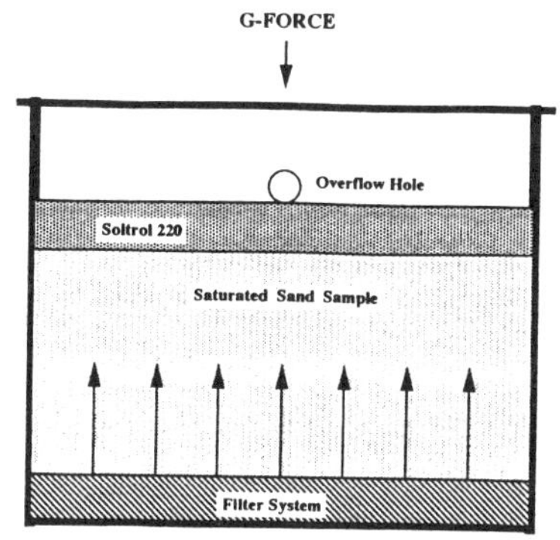

Figure 3. Section Through Centrifuge Model.

Three porous medium systems are being used during the investigation, Figure 4. The first two systems (referred to as 90P and 100P, respectively) comprise two uniform silica sands having similar average particle radii, but each having a slightly different particle-size distribution. The third system (40P) comprises a uniform silica sand having the same particle-size distribution as the first system, but having a larger average particle radius. All sands model a uniform, high permeability aquifer system, typical of formations where pump-and-treat technologies have proven to be most effective (NRC, 1994).

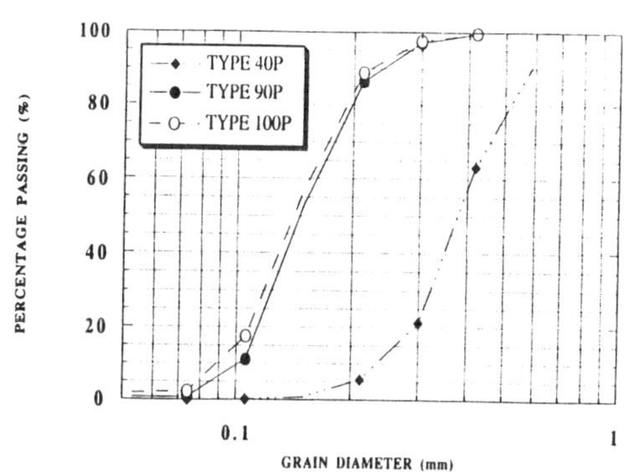

Figure 4. Particle size distribution curves.

The material properties of the three porous medium systems, together with those of the Light Nonaqueous Phase Liquid (Soltrol-220) are described in Table 1.

Table 1. Material Properties.

Silica Sands

Soil Type	D_{50} (mm)	Specific Gravity	Intrinsic Perm. (m^2)
40P	0.35	2.66	$6x10^{-11}$
90P	0.17	2.66	$2x10^{-11}$
100P	0.17	2.65	$1x10^{-11}$

Soltrol-220

Viscosity (Pa s)	$4.12x10^{-3}$
Density (g/cm^3)	0.79
Surface tension (dyn/cm)	25.30
Flash point (deg. C)	99
Boiling point (deg. C)	233

3. EXPERIMENTAL DATA AND DISCUSSION

To date, 15 vertical displacement tests have been conducted in two of the porous medium systems. Table 2 summarizes results for tests conducted in the coarse type 40P sand, while Table 3 summaries results for tests conducted in the finer type 90P sand. During both test series, the variation in the sample void ratio was kept to a minimum between experiments.

Table 2. Data from tests using 40P sand.

Test	*g*-level	C_a x 10^6	B_ox10^2	Trapped residual (%)
Sang4	1.0	3.11	0.24	9.4
Sang7	1.0	3.11	0.24	9.3
Sang5	7.25	3.95	1.77	4.4
Sang1	14.02	3.70	3.42	3.4
Sang6	35.83	4.19	8.7	2.9
Sang2	37.18	3.46	9.1	2.8
Sang3	57.98	3.23	14.1	2.3

Table 3. Data from tests using 90P sand.

Test	*g*-level	C_a x 10^6	B_ox10^2	Trapped residual (%)
Sang10	1.0	3.47	0.06	8.2
Sang11	12.8	3.95	0.74	4.9
Sang15	24.07	3.94	1.40	3.15
Sang9	35.60	3.87	2.10	2.67
Sang12	55.48	3.96	3.19	2.49
Sang14	77.32	3.82	4.4	2.31

In the case of the upward displacement of an LNAPL by water, viscous pressure and buoyancy forces act in the same direction, and are thus additive in overcoming the mechanism of capillary entrapment. When this occurs, it is possible to combine the individual effects of the capillary and the Bond number in the following dimensionless group (Morrow and Songkran, 1981).

$$N_{CB} = C_a + k_{wf} A B_o \quad (2)$$

where k_{wf} is the relative permeability of the wetting phase at the flood front, and A is a constant presumed to link the intrinsic permeability of a porous medium to its average particle radius, vis

$$k = AR^2 \quad (3)$$

Based on previously reported work (Naar et al., 1962), an average relative permeability k_{wf} = 0.63 was assumed at residual LNAPL saturation for all sands, while the A value for each sand type was computed using the data presented in Table 1.

The percentage of LNAPL entrapment versus N_{CB} is illustrated for all 15 vertical displacement tests in Figure 5. The data clearly shown the influence of buoyancy and viscous pressure forces on capillary entrapment. At higher values of N_{CB}, the supplemental pressure provided by viscous pressure and/or buoyancy acts to reduce the proportion of LNAPL remaining entrapped in the sand pores once the wetting front has passed.

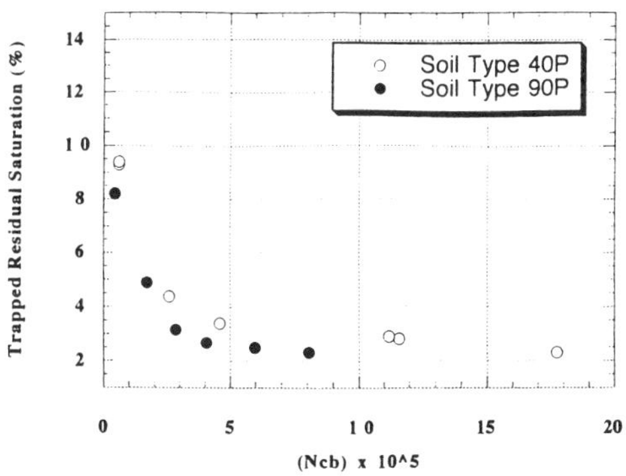

Figure 5. Trapped LNAPL residual vs. N_{CB}

The phase-diagram presented in Figure 5 suggests an upper, critical value of $N_{CB} \approx 15 \times 10^{-5}$ for the system under investigation, above which the trapped residual saturation reduces to a minimum of approximately 2.3%. In addition, the diagram demonstrates a lower, critical value of $N_{CB} \approx 1 \times 10^{-5}$, below which the trapped residual reaches a maximum. Between these two extremes, the diagram can be used to predict the amount of LNAPL trapped behind a single passage of the wetting front for any magnitude of N_{CB}. The behaviour reported in this diagram lies within the range of behaviour reported for nonwetting phase entrapment of Soltrol-130 in packings of close-sized glass beads (Morrow and Songkran, 1981).

The construction of phase-diagrams, such as that given in Figure 5, is not original and has been reported elsewhere in the literature (see for example Lenormand et al., 1988). However, the use of centrifuge modelling as an experimental tool to conduct this form of investigation is new, and the work presented here demonstrates the potential offered by centrifuge modelling in this area.

Finally, it should be noted that the problem under investigation here concerns contaminated sands that are initially fully oil-saturated. In reality, oil contaminated aquifers would likely be pre-wet with water: this would preclude their being fully oil-saturated, even under heavy contaminant loading. This fact should be accounted for in the interpretation of saturation phase-diagrams.

4. CONCLUSIONS

The behaviour of a light nonaqueous phase liquid under vertical hydraulic flushing was investigated over a wide range of conditions using centrifugal modelling techniques. The work clearly demonstrates the influence of buoyancy and viscous pressure forces on capillary entrapment. Data from a series of experiments conducted in two porous medium systems were used to construct a phase-diagram describing the percentage of residual LNAPL entrapped following one pore volume injection of water, as a function of a combined capillary and Bond number (N_{CB}). This diagram can be used to predict the amount of LNAPL trapped behind a single passage of the wetting front for any magnitude of N_{CB}

An understanding of the behaviour of NAPLs under hydraulic flushing is important when planning the effective site remediation of organic-contaminated soils. The construction of phase diagrams, as described in this paper, can contribute to this understanding. In addition, such diagrams can be used to predict what proportion of nonaqueous phase contamination may be realistically removed under hydraulic flushing and, consequently, how much remains for recovery by tertiary methods.

The work described here is currently being extended to investigate nonwetting phase entrapment under horizontal displacement conditions, where the effects of the capillary and Bond number are no longer additive. Future plans also include the effects of surfactants and dynamic behaviour, in the form of capillary, viscous or gravity fingering onset, on phase-diagram behaviour.

ACKNOWLEDGMENTS

This work was supported primarily by the MRSEC Program of the National Science Foundation under Award Number DMR-9400333. The authors would like to express their gratitude for this support.

REFERENCES

Lenormand, R., E. Touboul & C. Zarcone 1988. Numerical models and experiments on immiscible displacements in porous media. *J. Fluid Mech.*, 189: 165-187.

Mayer, A.S. & C.T. Miller 1992. The influence of porous media characteristics and measurement scale on pore-scale distributions of residual nonaqueous phase liquids. *J. Cont. Hydrology.* 11:189-213.

Morrow, N.R. & B. Songkran 1981. Effect of viscous and buoyancy forces on nonwetting phase trapping in porous media, In D.O. Shah(ed), *Surface Phenomena in Enhanced Oil Recovery* : 387-411. Plenum Press: N.Y.

Naar, J., R.J. Wygal, & J.H. Henderson 1962. Imbibition relative permeability in unconsolidated porous media, *Soc. Pet. Eng. J.,* 2(13): 225.

National Research Council (NRC) 1994. *Alternatives for groundwater cleanup.* National Research Council Committee on Groundwater Cleanup Alternatives, TD426.A48, National Academy Press: Washington.

Pahwa, A. 1987. Quick release tests on centrifuge of wall retaining saturated sand backfill. Master of Science Thesis. Department of Civil & Environmental Engineering, M.I.T.

Schofield, A.N. 1980. Cambridge geotechnical centrifuge operations. *Geotechnique.* 30(3):227-268.

Environmental Geotechnics, Kamon (ed.) © 1996 Balkema, Rotterdam. ISBN 90 5410 848 7

Anisotropy of gasoline-contaminated sands grouted with microfine cement

Lois G. Schwarz & Raymond J. Krizek
Department of Civil Engineering, Northwestern University, Evanston, Ill., USA

ABSTRACT: Determining the effectiveness of grouting a contaminated soil depends in large part on the interactions of (a) the contaminant and the grout and (b) the contaminant and the soil. The presence of gasoline strongly affects both the tendency of a cement grout to bleed and the resulting microstructure of the hardened grout; therefore, it might be expected to exert a strong influence on the degree of anisotropy manifested in the permeability and strength of the grouted sand. Among other factors, the permeability and strength of a grouted sand will depend on (a) the pre-grouting and post-grouting contaminant residuals which, in turn, are a function of the fineness of the soil and the viscosity of the injected grout, (b) the degree to which the original pores are filled with grout, (c) the behavior of the grout prior to setting, and (d) the resulting interconnectivity of the grouted pores. Rheological variances in grout mixes with the same composition and solid concentration are considered to be contributory causes to the measured degree of anisotropy

1 INTRODUCTION

Concern about organic contaminants originating from non-aqueous phase liquids (NAPLs) stems primarily from their persistence in the subsurface and their ability to contaminate large volumes of water because of the generally poor attenuating capability of the soil. Petroleum hydrocarbons, termed LNAPLs (light non-aqueous phase liquids), have densities less than water and are usually found near the top of the saturated zone; however, water table fluctuations can effectively trap them below the water table during periods of depression. Because capillary forces make the removal of LNAPLs in the subsurface environment very difficult, ground modification by permeation grouting may provide a viable technique for controlling LNAPL contaminant migration.

Natural soil deposits often manifest high degrees of anisotropy in their permeability and strength behavior, with the actual magnitude of the directional differences depending on the shape, orientation, and concentration of the particles which form the media (Rice et al., 1970; Dullien, 1992). At the micro-level this anisotropy is caused primarily by grain shape and orientation, whereas at the macro-level it may be attributed in large part to the presence of alternating thin layers with different properties. The permeability in the horizontal direction may sometimes be ten to a hundred times that in the vertical direction, but strength anisotropy is usually much less. Permeation grouting involves filling the soil voids with a grout and allowing it to gel in place. If a suspension grout is used, the possibility of sedimentation (or bleed) prior to set might contribute to the anisotropy of the grouted mass. The amount of potential bleed is controlled by (a) the permeability of the grout to its own pore water, (b) the distance such water must travel to reach a drainage surface, and (c) the hardening time of the grout (Jefferis, 1994).

2 OBJECTIVE

The primary objective of this study is to demonstrate and quantify the effect of gasoline contamination on the resulting anisotropy in the permeability and strength of different sands injected with a microfine cement grout at different water:cement ratios.

3 EXPERIMENTAL INVESTIGATION

Sand specimens saturated with either water or gasoline were injected with a microfine cement grout at a water:cement ratio of either 2:1 or 3:1 and cured for 28 days in either a vertical or a horizontal orientation. Then, they were tested in the direction of their long axis to determine the effect of the resident pore fluid prior to grouting on the permeability and unconfined compressive strength after grouting.

4 MATERIALS USED

The microfine cement used in this study was MC-500, a blended slag-Portland cement manufactured by Onoda in Japan and marketed in the United States by Geochemical Corporation of Ridgewood, New Jersey. One percent (by dry weight) NS-200, a napthalene formaldehyde condensate superplasticizer, was added to the grout to enhance full wetting of the cement particles during mixing. Three different gradations of sand, obtained from U.S. Silica Company in Ottawa, Illinois, were grouted. Ottawa 20-30, F-42, and F-62 sands had approximate grain sizes ranging from 850 to 600 μm (with a mean of 725 μm), 600 to 50 μm (with a mean of about 375 μm), and 425 to 75 μm (with a mean of 225 μm), respectively. The contaminant was a regular unleaded gasoline (octane rating of 87) refined by Mobile Oil Company and obtained from a local service station.

5 PREPARATION OF SPECIMENS

Grouts with a water:cement ratio of either 2:1 or 3:1 were prepared in a high energy mixer (estimated free-running rotational rating of 17,000 rpm) with a volume capacity of 0.65 liter. The NS-200 dispersant was added to the water and mixed thoroughly for about ten seconds, after which the microfine cement was added and mixed for one minute. Cylindrical specimens were prepared by placing the sand at a density of 1.81 ± 0.02 g/cm^3 (same for each of the three different sands) in PVC tubes with an inside diameter of 51 mm and a length of 152 mm. Water-saturated specimens were permeated by about 2500 ml of tap water at a pressure of 15 kPa, and gasoline-contaminated specimens were saturated by flowing 400 ml of gasoline through them at a similar pressure. Details of the injection system and procedures have been reported by Schwarz and Krizek (1995a).

6 GROUTED SANDS

The specimens were grouted from the bottom by injecting two void volumes of uncontaminated grout into either water-saturated or gasoline-saturated sand. Grouting pressures ranged from about 15 to 30 kPa for the 20-30 sand, 20 to 50 for the F-42 sand, and 35 to about 200 kPa for the F-62 sand, but they were held relatively constant for each gradation of sand, initial soil condition, and water:cement ratio of the grout. All specimens were grouted with their long axis vertical, but, immediately after injection, half of them were rotated 90° to make their long axis horizontal. Both "vertical" and "horizontal" specimens were cured in these orientations at room temperature for 28 days, after which they were tested in the direction of their long axis.

7 VISCOSITY AND BLEED OF GROUTS

During the course of this study, grouts were prepared from seven different shipments of microfine cement. Although not initially intended to be a variable, rheology and sedimentation tests conducted on the cement in each shipment indicated variations which may influence the macro-behavior (permeability and strength) of the grouted sands. For all practical purposes these seven cements can be grouped into three categories according to their viscosity and bleed, and given in Table 1 are average values for the apparent viscosity (determined at a shear rate of 7 sec^{-1}) and bleed for each of these three groupings

Table 1. Viscosity and Bleed of Grouts

Cement Group	Water: Cement Ratio			
	2:1		3:1	
	Viscosity (mPa·sec)	Bleed (%)	Viscosity (mPa·sec)	Bleed (%)
A	70	8.5	12	23
B	52	9.5	11	21
C	41	11	8	21

Viscosity was determined using a shear rate of 7 sec^{-1}.

(designated as A, B, and C). The viscosity and bleed varied by more than 70% and 25%, respectively, for 2:1 water:cement ratio mixes and about 50% and 10% for 3:1 mixes. These differences lend support to the hypothesis that grout rheology contributes to the degree of anisotropy manifested in the engineering properties of grouted sand, but the effect of rheology can be appreciated more strongly in the subsequent comparison with data from an earlier study where the differences are larger. Because higher viscosity grouts undergo less bleed, sands grouted with them would be expected to exhibit a lower permeability than sands grouted with lower viscosity grouts. In the case of the contaminated specimens, more viscous grouts would tend to (a) displace any pre-grouting contaminant residual more effectively and (b) provide a post-grouting contaminant residual that is less interconnected. In general, the presence of some degree of gasoline contamination will decrease the tendency of the cement particles to settle prior to hardening in the void spaces of the sand; this will result in a more homogeneous, albeit less dense, distribution of the hydration products within the grout matrix and an associated decrease in the accumulation of bleed water in the upper portion of the void spaces. Schwarz and Krizek (1995a) showed that the addition of as little as one percent gasoline (based on total volume) significantly reduced the bleed capacity of the grout and resulted in a "honeycomb" structure in the region where the bleed water would normally have accumulated.

8 PRESENTATION OF RESULTS

Table 2 summarizes the permeability and strength data obtained in this test program. In general, the independent variables are the type of sand (20-30, F-42, and F-62), the water:cement ratio of the grout (2:1 and 3:1), and the initial condition of the sand (water-saturated or gasoline-saturated). Addressed first will be the effect of sand type and grout concentration on the vertical permeability of water-saturated specimens; next will be the influence of gasoline contamination on the vertical permeability; and last will be comparisons of horizontal permeabilities to corresponding vertical permeabilities. Then, an interpretation of the strength data will follow a similar pattern. Following the foregoing presentations will be a comparison of data from this study with similar data from a previous study (Krizek and Helal, 1992) on a water-saturated Ottawa 20-30 sand.

Table 2. Permeability and Strength Data

Sand	Water: Cement Ratio	Vertical Permeability (10^{-6} cm/sec)		Horizontal Permeability (10^{-6} cm/sec)		Horizontal to Vertical Permeability		Vertical Strength (MPa)		Horizontal Strength (MPa)		Horizontal to Vertical Strength	
		Water	Gas	Water	Gas	Water	Gas	Water	Gas	Water	Gas	Water	Gas
20-30	2:1	<0.1(C)	5(B)	0.3(B)	1000(C)	3+	200	5.0	4.1	7.1	4.9	1.39	1.20
	3:1	8(C)	60(B)	20(C)	1000 (B)	2.5	17	1.9	1.8	2.0	1.9	1.05	1.06
F-42	2:1	<0.1(C)	0.4(A)	<0.1(B)	10(C)	?	25	6.6	7.0	6.0	7.1	0.91	1.01
	3:1	<0.1(C)	7(B)	1(C)	20(B)	4^+	3	3.3	3.1	3.5	2.9	1.06	0.94
F-62	2:1	<0.1(C)	0.6(B)	<0.1(B)	20(C)	?	33	11.4	10.9	13.5	12.0	1.18	1.10
	3:1	<0.1(C)	2(A)	<0.1(C)	40(B)	?	20	6.2	6.9	7.0	7.3	1.13	1.06

Note: The letter (A, B, or C) after each measured permeability or strength value designates the cement group from which the injected grout for that particular specimen was prepared.

9 VERTICAL PERMEABILITY

After curing, permeability tests (using deionized water as the permeant) were conducted at a gradient of 15 to 18 with no back-pressure for 2 to 5 days, depending on the permeability of the specimen. Each contaminated specimen had its own permeant supply source to minimize contamination among the specimens due to possible solubility of the contaminant.

9.1 Effect of sand gradation and grout concentration

As seen in Table 2, with the exception of the 20-30 sand grouted with a 3:1 mix, the vertical permeability of all three water-saturated grouted sands was less than 10^{-7} cm/sec (the limit that the testing procedure was capable of measuring with a reasonable degree of accuracy); comparable measurements for ungrouted sands were 0.114, 0.044, and 0.020 cm/sec for the 20-30, F-42, and F-62 sands, respectively. These data show that both the 2:1 and 3:1 grouts and their resultant hydration products did an excellent job of filling the voids in all of these soils, except for the case of the less viscous 3:1 grout in the larger voids of the 20-30 sand; in this latter situation the bleed prior to set obviously left some interconnected void spaces through the sand.

9.2 Effect of gasoline contamination

The data in Table 2 show that the permeability values for the gasoline-saturated grouted sands were significantly higher (up to almost two orders of magnitude) than the corresponding values for the water-saturated grouted sands, with the former generally decreasing with the fineness of the sand and increasing with the water:cement ratio of the grout. This increase in permeability is probably due to preferential flow through the interconnected porous regions caused by the contaminant residual in the interfacial zone of the gasoline-saturated specimens; these interfacial microstructures were observed by Schwarz and Krizek (1995a). The permeability of the contaminated 20-30 sand was distinctly higher (an order of magnitude) than that of the contaminated F-42 sand, whereas there were only minor differences in the permeabilities of the contaminated F-42 and F-62 sands. This is attributable to the fact that the coarser 20-30 sand has larger grains and correspondingly larger pores, with the latter being more difficult to "fill," and therefore the residual is not effectively disconnected. Lower water:cement ratio grouts will exhibit greater stability and are more viscous with correspondingly more cement particles; hence, they will be more effective in (a) decreasing pore interconnectivity, (b) expelling the residual contaminant as the injection front advances, and (c) encapsulating the residual which is not mobilized. Schwarz and Krizek (1995b) estimated that contaminant pre-grouting residuals were about 38%, 36%, and 26% for the F-62, F-42 and 20-30 sands, respectively.

10 PERMEABILITY ANISOTROPY

In general, the measured horizontal permeability values given in Table 2 are at least several times and up to one or two orders of magnitude higher than the corresponding vertical permeability values, with the contaminated sands exhibiting by far the greater differences. The ratio of horizontal to vertical permeability for the water-saturated sands must be viewed with considerable uncertainty because five of the six vertical permeabilities and three of the six horizontal permeabilities were determined to be less than 10^{-7} cm/sec; grouting 20-30 sand with a 3:1 mix was the only case where a well defined permeability ratio was obtained (the vertical permeability was about 40% of the horizontal). As with the vertical permeability values, the contaminated specimens manifested much higher (generally about two orders of magnitude) horizontal permeabilities than their water-saturated counterparts. With regard to anisotropy, the gasoline-saturated sands gave horizontal permeability values on the order of 20 to 30 times the vertical values (with two exceptions -- one higher and one lower). It is interesting to note in Table 2 that, in four of the six cases for gasoline-saturated sand, the horizontally cured specimens were grouted with a less viscous grout than the vertical specimens (exceptions being the 20-30 and F-42 sands which were grouted with a 3:1 mix from the same shipment), and this is consistent with the relatively high ratios of horizontal-to-vertical permeability in these cases.

The resulting anisotropy in the permeability of the grouted sand may be explained in large part by the bleed that occurs in the injected grout prior to set (Helal and Krizek, 1992). The suspended particles settle in the pore spaces between the sand grains and either the bleed water (in the case of water-saturated sands) or a relatively porous "honeycomb" structure (in the case of gasoline-saturated sands) accumulates at the tops of the pores. Accordingly, the primary reason for the increased horizontal permeability relative to the vertical is the presence of these interconnected spaces at the tops of the grouted soil voids. This being the case, it is logical to expect some concentration of cement particles at which the ratio between horizontal and vertical permeability values is the maximum. For lower water:cement ratio grouts, very little bleed occurs; hence, the soil voids that were filled at the time of injection are essentially filled with hydration products after set, and the permeability, albeit very low and difficult to measure accurately, is approximately the same in both directions for all practical purposes. For higher water:cement ratio grouts, most of the hydration products are concentrated at the contact points and on the surfaces of the sand grains; consequently, a complete or almost complete blockage of the voids does not occur and the permeability is high and more or less the same in both directions.

11 VERTICAL COMPRESSIVE STRENGTH

Subsequent to permeability testing, the grouted sand specimens were removed from their PVC tubes, trimmed to a length of 102 mm, sulfur-capped, and tested in unconfined compression with a time-to-failure between 30 and 75 seconds; summarized in Table 2 are the results from these unconfined compression tests. In general, water-saturated grouted sands tested in the vertical direction were slightly stronger (about 5% on average) than their gasoline-saturated counterparts. This is attributable to the weaker interfacial bonds caused by the gasoline contaminant on the surfaces of the sand grains. In both cases the strength increased as the sand became finer and as the water:cement ratio of the grout decreased. In the case of the finer sands the strength increases because there are more particle contacts to be bonded together, and in the case of the grout concentration the strength increases because the grout is stronger and the soil voids are filled more completely due to less bleed.

12 STRENGTH ANISOTROPY

In general, the values of strength in the horizontal direction were about 5% to 20% higher than those in the vertical direction, with the F-42 sand providing the primary exceptions to this generality. These relatively modest degrees of strength anisotropy may be explained by the fact that the specimens tested in the horizontal direction have "columns" of sand grains and cementitious material (the hydration products from the settled cement in the soil voids) to transmit the load, thereby enhancing somewhat the load-carrying capacity in the horizontal direction relative to the vertical.

13 PERMEABILITY -STRENGTH CORRELATION

Notwithstanding the scatter in the data normally collected in any experimental program of this type, there is a reasonably strong inverse correlation between permeability and strength for three-quarters of the test values. For the remaining one-quarter of the data, the evidence supporting this correlation is moot because the lowest permeability values measured with a reasonable degree of accuracy were 10^{-7} cm/sec.

14 COMPARISONS WITH PREVIOUS STUDY

Collaborative evidence supportive of the findings in this experimental program have been reported by Krizek and Helal (1992). Summarized in Table 3 are the viscosity and bleed values for the 2:1 and 3:1 mixes used to grout water-saturated 20-30 sand and the resulting 28-day directional permeability and strength values. Although the same type of cement and essentially the same experimental technique were used to prepare the grouts in this earlier study, a cursory comparison with the grout properties in Table 1 will show that the viscosities of the grouts were much lower (over 200%) and the bleed values were much higher (up to 150%). Measured vertical and horizontal permeability values in the Krizek-Helal study were about three orders of magnitude higher than those measured in this study, and corresponding strengths ranged from 40% to more than 200% higher. However, there were several important differences between the two studies. The first important difference was in the density of the 20-30 sand and, therefore, the volume of voids available to be "filled" with grout. In the previous study, the sand was packed into the test column to a density of 1.65 ± 0.02 gm/cm^3 with a porosity of 0.375, whereas in this study the density was 10% higher with the resulting porosity being about 16% smaller. Consequently, more cement particles may be retained in the larger

Table 3. Properties of Grouts and Grouted 20-30 Sand (Krizek and Helal, 1992)

Water: Cement Ratio	Grout Properties		Grouted Sand Properties			Ratio of Horizontal to Vertical	
	Viscosity* (MPa·sec)	Bleed (%)	Orientation	Permeability (cm/sec)	Strength (MPa)	Permeability	Strength
2:1	15.7	24	Vertical Horizontal	$2x10^{-5}$ $1x10^{-3}$	8.3 10.1	50	1.22
3:1	6.0	45	Vertical Horizontal	$5x10^{-2}$ $3x10^{-2}$	4.1 6.3	0.6	1.54

*Viscosity was determined using a shear rate of 7 sec.$^{-1}$

volume of void space and thereby increase the strength. Two other differences pertain to the specimen preparation and handling procedures. In the Krizek-Helal study, the injection pressure was about 70 kPa (possibly influencing the mass of cement particles captured by the sand) and the specimens were cured in a high humidity room (possibly affecting the hydration rate of the cement particles and the associated interconnectivity of the voids); this is in contrast to injection pressures ranging from 15 to 30 kPa and curing being done in sealed columns for this study.

15 CONCLUSIONS

Based on the results of this study with its inherent limitations in the materials tested and the ranges of the variables investigated, the following conclusions may be advanced.

1. Grouted gasoline-contaminated sands are more permeable (up to two orders of magnitude) and slightly weaker (5% on the average) than their non-contaminated counterparts. Relative to the grouted water-saturated sands, the permeability anisotropy of the grouted gasoline-saturated sands is considerably higher for the 20-30 sand and essentially undetermined (although possibly equal) for the other two sands and the strength anisotropy is approximately the same to slightly lower. Increased permeability is attributed to preferential pathways for water migration in the interfacial zone around the sand particles and decreased strength is explained by weaker interfacial sand/grout bonds, both of which are due to the presence of a contaminant residual on the granular surfaces.

2. Relative to the 20-30 sand, the finer F-42 and F-62 grouted sands, both contaminated and non-contaminated, are generally less permeable (an order of magnitude) and stronger (ranging from 1.5 to 4 times). Since the finer sands have smaller pore sizes and somewhat rougher grain surfaces, there are more locations for grout particles to attach prior to hardening, and sedimentation and self-consolidation of the grout are impeded. However, these conditions are conducive to more homogeneously grouted void spaces and the finer sands exhibit somewhat less anisotropy.

3. As the water:cement ratio of the grout suspension decreases, the cement particle volume and the viscosity increase and the bleed decreases. Although the presence of gasoline contamination noticeably reduces the tendency of a cement grout to bleed, the post-grouting contaminant residual due to an adsorbed layer of gasoline on the granular surfaces is apparently more influential in controlling the permeability and strength of the grouted soil.

REFERENCES

Dullien, F.A.L. 1992. *Porous Media Fluid Transport and Pore Structure*, Academic Press, New York, pp. 303-306.

Helal, M. and Krizek, R.J. 1992. Preferred Orientation of Pore Structure in Cement-Grouted Sand. *Specialty Conference on Grouting, Soil Improvement and Geosynthetics*, Geotechnical Special Publication No. 30, American Society of Civil Engineers, New Orleans, Louisiana, pp. 526-540.

Jefferis, S.A. 1994. Fresh Properties of Portland Cement Grouts. *Structural Grouts*, P.L.J. Domone and S.A. Jefferis, eds., Chapman and Hall, Inc., New York, pp. 28-57.

Krizek, R.J. and Helal, M. 1992. Anisotropic Behavior of Cement Grouted Sand. *Specialty Conference on Grouting, Soil Improvement and Geosynthetics*, Geotechnical Special Publication No. 30, American Society of Civil Engineers, New Orleans, Louisiana, pp. 541-550.

Rice P.A., Fontugne D.J., Latini, R.G., and Barduhn, A.J. 1970. Anisotropic Permeability in Porous Media. *Flow Through Porous Media*, American Chemical Society, Washington, D.C., pp. 48-56.

Schwarz, L.G. and Krizek, R.J. 1995a. Grouting Gasoline-Contaminated Sand with Microfine Cement. *Geo-environment 2000*, American Society of Civil Engineers, New Orleans, Louisiana, pp. 1366-1380.

Schwarz, L.G. and Krizek, R.J. 1995b. Effects of Gasoline on Cement-Grouted Sands. *Xth Pan-American Conference on Soil Mechanics and Foundation Engineering*, International Society for Soil Mechanics and Foundation Engineering, Guadalajara, Mexico, pp. 685-698.

Environmental Geotechnics, Kamon (ed.) © 1996 Balkema, Rotterdam. ISBN 90 5410 848 7

Case studies – Removal of pesticides buried in the ground

M. Shimizu
Faculty of Engineering, Tottori University, Japan

ABSTRACT: In Tottori prefecture, Japan, several kinds of pesticides, more than 200 tons, that had been banned were buried in the ground in 1972. The prefectural government judged that they may be a source of contamination of sites and the groundwater and bagan to remove them from 1994 at five among 45 sites. The soil profiles are different from site to site: sandy or cohesive soils under groundwater table or over it. The procedures for the investigation made before and after the removal work and the methods of taking soil samples to be chemically analyzed and monitoring the groundwater quality are described.

1. INTRODUCTION

In 1972, in Tottori Prefecture, Japan, pesticides that had been or would be banned were buried in the subsoil (Fig. 1). The prefectural government started, in 1994, to remove them from the subsoil because they may be the sources of the degeneration of environments.

This paper deals with:
1) why and how pesticides were disposed in the ground;
2) how they were removed; and
3) how subsoil and groundwater were investigated to get information about their eventual contamination.

Fig. 1: The location of Tottori Prefecture.

Geotechnical methods involved in the procedures employed, including methods of the soil sampling for chemical analyses and geophysical exploration techniques, are described with the discussion about their effectiveness.

2. HISTORY

The Japanese government set, in 1968 to 1970, the standard values for the maximum permissible concentration of residues of nine chemical substances used as pesticides in 14 farm products: the nine substances are BHC, DDT, alderin, dieldrin, endrin, parathyon, EPN, arsenic, and lead arsenate. The first five are organic chlorine compounds and the sixth and seventh are organic phosphorous compounds. Every one of these has acute or chronic toxicity. By 1978, these substances and mercury compounds were banned, except EPN. In the following, we call all these substances pesticides.

In August 1970, the dieldrin in excess of the standard value was detected in cucumbers produced in Tottori. The detection shocked people concerning farm plants because a main industry of Tottori Prefecture is agriculture; they stopped immediately the shipping of cucumbers. The prefectural and municipal authorities and Agricultural Cooperation Association started to campaign for the proper use of pesticides, and they recovered, by the end of 1970, all of the above mentioned nine pesticides and mercury compounds from farmers. They also recovered envelopes on which DDT was applied. The pesticides that were recovered amounted to about 250 tons; the envelopes of 52 million were burned. The recovery cost about 75 million Japanese yen (about 750 thousand US dollars in current rate).

The recovered pesticides were buried, by June 1972, at 45 sites within the prefectural area, according to the rule that had been made by the prefectural authority. The rule was based on an ordinance and a notice of the national government: the ordinance expresses the methods for the control of poisons and deadly violent substances; and the notice for the use of organic chlorine pesticides and the disposal of banned pesticides. The rule orders to bury the pesticides at the depth more than 1m below the ground surface and to cover the disposal with cement concrete thicker than 10 cm, or to envelope it by a sheet of vinyl chloride or polyethylene thicker than 0.3 mm.

The actual ways that were employed were different from site to site. They can be classified into five patterns (see Fig.2):

a) the disposal was put into envelopes or enveloped by sheets without cover (26 sites);

b) the disposal was put into envelopes or enveloped by sheets and covered by cement concrete (5 sites);

c) the disposal was put into envelopes or enveloped by sheets and put on a cement concrete base plate (one site);

d) the disposal was put into a cement concrete capsule (11 sites);

e) the disposal was put in a cave in the side of a cliff and the entrance of the cave was closed with cement concrete (2 sites);

In July, 1984, a member of the prefecture assembly posed a question if the buried pesticides are still safe and do not pollute the areas, especially the groundwater, around the sites. The question triggered a fact-finding survey: the prefectural authority organized a committee, composed of specialists working for national institutes and universities, and investigated, by October this year, the actual conditions of all the sites where pesticides had been buried; and chemically analyzed, at representative eleven sites, the soils and water in the rivers near the sites. The survey showed, as the results, that pesticides had not flowed out and they had not degenerated the soils and the water. After the survey, the prefectural authority decided to examine regularly the conditions of the sites.

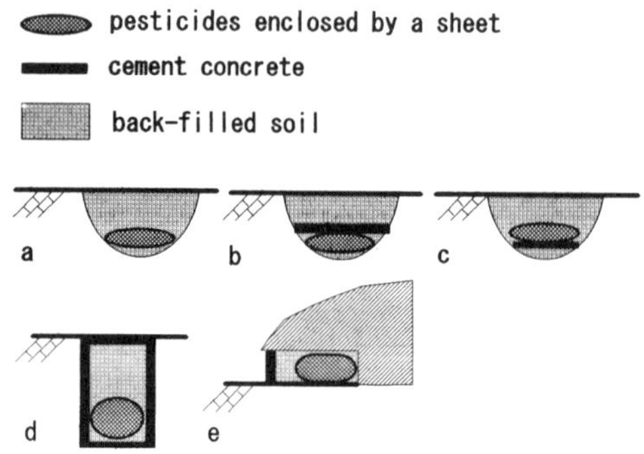

Fig. 2: Sketches of configurations of buried pesticides

In 1993, a non-government group of environmentalists posed questions about the environmental safety in areas near the sites where pesticides were buried. The prefectural government considered that they might be a source of environmental degeneration; the government organized a committee of specialists in various fields as did in 1984. The government and the committee decided to remove all the pesticides and to examine the environmental quality of the areas near the sites. Five sites were selected in order to establish the procedures for removing buried pesticides and for investigating the environmental quality of subsoil near the sites. The details of the procedures are described in the next section.

3. PROCEDURES FOR THE REMOVAL OF PESTICIDES

The features of five selected sites are shown in Table 1.

The site No.1

This site is located within the area of Tottori Dune, which is a part of San-in Seaside National Park and a representative tourist resort. The pesticides were buried in the sand of the dune. In this area, man activities are forbidden by a law. After some legal procedures for the permission, the following procedures were taken: The location could not be identified exactly because the scene around the site had changed since the pesticides were buried in 1972. The exact location had to be found. A ground probing radar using electromagnetic waves and a seis-

Table 1: Features of the sites.

site	amount of buried pesticides (kg)	soil	others
No. 1	3,700	sand	in Tottori Dune
No. 2	8,500	-	in an old cement-concrete tank
No. 3	2,000	-	in an old cement-concrete silo
No. 4	3,000	volcanic soils	on a hill
No. 5	5,300	silt	in a lowland area

mic method using the Raleigh wave were used; at the same time hand-operated augers were used.

Trees were cut. Sheet piles were driven to surround the area where the pesticides were buried, because the excavation of sand would cause instability of the sand slope without sheet piles. Overburdening sand was carefully removed by a power shovel and man-operation (see Fig. 3).

The pesticides were found dry and the sheet enveloping them did not alter, which indicates that the sheet was very effective to prevent water from flowing into. The sheet is the same kind to those usually

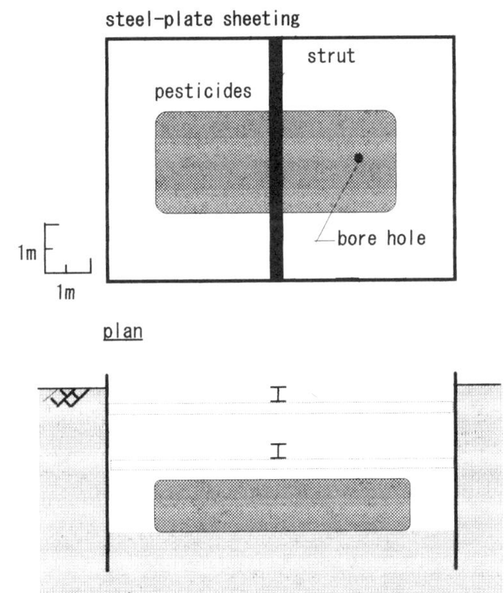

Fig. 3: Key sketch of the site No. 1.

used in construction sites in Japan.

The pesticides were separated, in the field, according to their chemical components and removed up in a cage. The removed pesticides were put in drums.

Soil samples were taken at 27 points; hand augers and a boring machine were used. The chemical-analysis data from the samples taken using hand augers would not be reliable because the zone near the ground surface would be possibly contaminated. This is the reason why the boring machine was used. The same problem, however, would occur even with a boring machine. A technique which can prevent the contaminated surface soil from entering into the lower zone was used (Hoffman and Dresen, 1990; Hoffman,1993).

The pesticides such as BHC, DDT, alderin and dieldrin and mercury were detected in the zone shallower than 50 cm below the bottom of the excavation; the detected values were so low that they are nearly the limit of quantitative analysis. It seems that a very bit of substances that may have fallen during the removing work caused their detection. For the future safety, the sand from which pesticides were detected was removed: the sand shallower than 10 cm in the area surrounded by the steel plate sheeting; and the soil shallower than 20 cm in the area where the mercury was detected.

After the removal of pesticides, the area was repaired to the original state.

The site No. 2

Pesticides were disposed into a cement concrete tank which had been used to salt farm products before. The tank was covered by a cement-concrete plate, of which the thickness was not known; the inside could not be seen from the outside. People said that the inside of the tank was partitioned off by a vertical wall. See Fig.4, in which the map near the tank is sketched.

Two holes of 30 cm diameter was carefully opened on the top plate by a core boring machine to observe the inside. No rock pneumatic machine was used because it would cause some destruction of the tank and the flow out of the pesticides. The top plate, which was found 23 cm thick, was removed after cutting its circumference by a plate cutter.

The inside of a cell was found to be filled with water that seemed to have entered through a crack in the top plate. The pesticide in this cell was muddy. The pesticides in the other cell was dry.

After the removal of all the pesticides, the inside surfaces of the tank were carefully observed to in-

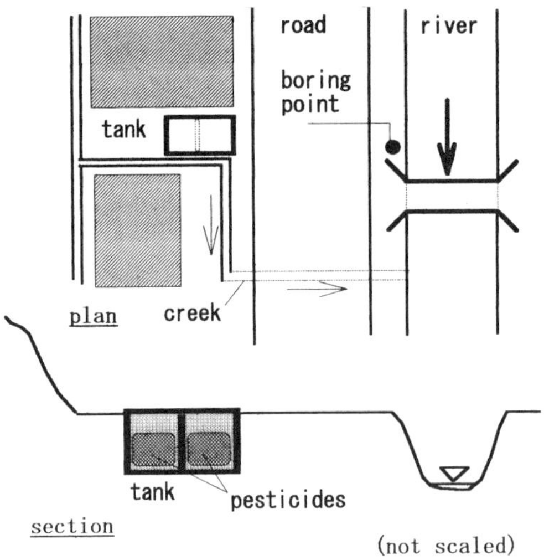

Fig.4: Sketch of the site No.2.

vestigate whether there were some cracks or not. Fortunately no crack was seen although the surface mortar had been crumbled.

Chemical analyses were carried out for the water filling inside the tank and for the water flowing in a creek of which the width is about 50 cm and the depth also 50 cm. For the water filling the tank, BHC and arsenic were detected in excess of the standard values; for the water in the creek no pesticide was detected.

A boring was made, at a point 20m apart from the tank, to take soil samples from various depths, ranging from 50 to 425 cm below the ground surface. The same technique to that employed at the site No.1 was employed here, too. From all the soil samples no pesticide was detected. It was concluded that no seepage-out of pesticide and water occurred and the environment had not been degenerated.

The site No. 3

The pesticides had been put into a cement concrete silo used for storing farm products. As at the site No. 2, the top plate covering the pesticides was carefully cut away by a core-boring machine and a plate cutter.

The disposed pesticides were found to be covered with soil. The pesticides were wet and could not be separated. At the bottom of the silo there was about 15 liters of water. However, the cement concrete of the silo was very intact and no crack was found, which shows that almost no seepage occurred through the wall of the silo. No chemical survey was carried out.

The site No. 4

The site is located on a hill. In this area, a group of tumuluses have been found. In a de facto investigation of the tumuluses carried out a few years before, the sheet enveloping pesticides was partly broken by a power shovel and pesticides spread a little.

Before the removal of pesticides, samples were taken for chemical analyses at nine points from the depth shallower than 2m below the ground surface. Two samples showed a high concentration of BHC, but other samples showed nothing.

The pesticides were removed as at the site No.1, but without steel plate sheeting. The pesticides were found in about 30 envelopes of vinyl, rather thick. Some envelopes had been broken. The pesticides were wet, but they could be separated.

Chemical survey was made after the removal. As the results, BHC was detected almost everywhere over the area in which the pesticides were buried but the depth was very close to the surface. The soil shallower than 2m, which is the depth at which BHC was detected by the preliminary survey, was removed.

The site No. 5

The site is located in a lowland area. The soil profile, shown in Fig. 5, shows that a very soft layer of silt with sand exists from 2.0 to 3.6 m below the surface and that the upper level of the soft soil layer is under the water table. It was supposed that the seepage of groundwater would make it difficult to remove pesticides.

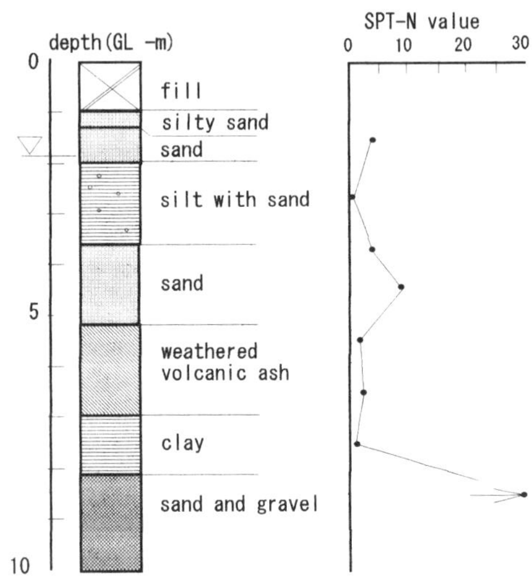

Fig.5: Soil profile at the site No.5.

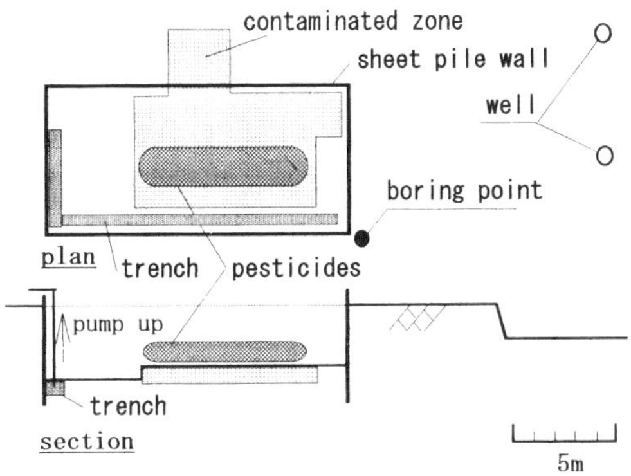

Fig.6: Sketch of the site No.5.

Also at this site, the exact location of pesticides was not clear. The surveying was carried out by using a ground probing radar.

The block of pesticides, being covered separately by 14 envelopes of vinyl chloride, was carefully exposed by a power shovel and man-work. A trench was dug to partly surround the block (fig. 6). The water seeping into the trench was pumped up. The pumped water was stored in septic tanks to be purified; the purified water was poured into a river.

The pesticides were partly wet and partly dry. Two thirds could be separated and the rest not.

The groundwater and soils were served to chemical analyses both before and after the removal of pesticides. The analyses before it showed that, from two soil samples, BHC was detected, from one the mercury and from one another the lead. Those soil samples from which pesticides and others were detected were taken from the depth of 2m below the ground surface. On the other hand, the groundwater sample, which is the water seeping in the borehole to 2m below the ground surface, showed BHC and arsenic.

The survey was made, after the removal of pesticides, for the soil under the bed of the block of pesticides. At the zone shallower than 10 cm below the bed, BHC, mercury, arsenic and lead were detected in excess of the standard values for the environment. The spread of thc contamination was worried about; the groundwater, taken from two wells that are located at 14 and 18 meters apart from the site, was analyzed. The results showed that the groundwater was not contaminated.

By considering the results of chemical surveys, the soil that seemed to be contaminated was removed, i.e., the soil covering the area in which the pesticides had been buried.

4. DISCUSSION

At the sites No.1 and No.5, the exact location of the pesticides was not known before work. A ground probing radar was used, but it was not very effective to specify the boundary of the area where the pesticides were buried. This technique is based on the principle that the materials that possess different values of density have different characteristics of electromagnetic wave reflection. Pesticides are produced by mixing chemicals with clay powder or by remolding the mixed powder with water. The pesticides produced in this a way are clays in the sense of geotechnical engineering. Therefore, the electromagnetic radar is not very effective to find buried pesticides.

In the investigations of which procedures have been described in this paper, it was emphasized that results of chemical analyses are strongly influenced by the quality of soil samples. If even a particle of pesticides falls into the sample, it will cause incorrect results of quantitative chemical analysis.

Vertical sampling of soil samples for chemical analyses should be taken with special techniques: one way is to use multiple bore holes in each of which sampling is made from different depth; or , if single borehole is used, the contamination of the lower soil by the upper contaminated soil has to be avoided. It is possible with the technique used at the sites No.1 and No.2: the sampler used for sampling to a certain depth is washed completely and it can be used to a next depth. By such a way, the contamination of the lower soil can be avoided. More refined technique is introduced by Hoffman and Deresen (1990) or Hoffman (1993).

5. CONCLUDING REMARKS

The pesticides of which the use was forbidden were disposed also in the prefectures other than Tottori. Even de facto investigation has not yet been done in some prefectures. In the future, the problem to which Tottori Prefecture is facing might occur in other prefectures. The Author hopes that the experiences stated in this paper would be helpful for a solution to problems occurring in other areas.

Acknowledgments
The author is a member of the committee organized, in 1994, by Tottori Prefecture Government. The Author wishes to thank the government for letting

him have very valuable experiences and for allowing him to use many data. He also appreciates the members of the committee: Dr. T. Nakamura, chairman and former professor of Shimane University and Tottori University; Dr. S. Kuwatsuka, emeritus professor of Nagoya University; Dr. O. Nakasugi, chief researcher of National Institute for Environment; Dr. M. Tanaka, director of National Institute for Public Health; and Dr. S. Okada, professor of Tottori University.

References

Hoffman, F. and Dresen, M. D. (1990): A method to evaluate the vertical distribution of VOC's in ground water in a single borehole, *Ground Water Monitoring Review*, Vol.10, No.2, pp.95-100.

Hoffman, F. (1993): Characterization of ground water contamination at the LLNL main site: a case study, *Technologies for Environmental Cleanup: Soil and Groundwater*, Kluwer Academic Publishers, pp.115-133.

Environmental Geotechnics, Kamon (ed.) © 1996 Balkema, Rotterdam. ISBN 90 5410 848 7

Transport and biodegradation phenomena of volatile chlorinated hydrocarbon

Masanori Shimomura & Satoshi Imamura
Technology Research Center of Taisei Corporation, Yokohama, Japan
Tetsuo Nagatou & Takashi Motoyama
Taisei Corporation, Tokyo, Japan

ABSTRACT: Soil and groundwater contamination with volatile chlorinated hydrocarbons (VCHs) has become a serious environmental problem in Japan. These pollutants are degraded under anaerobic condition in the field and understanding this behavior is important. This paper presents the results of a numerical calculations of biological degradation of VCHs and comparison with field data. The results of the investigation show that half-life of tetrachloroethylene (PCE) and trichloroethylene (TCE) are longer than those reported in previous studies.

1 INTRODUCTION

Soil and groundwater contamination with volatile chlorinated hydrocarbons such as tetrachloroethylene (PCE) and trichloroethylene (TCE) has become a serious environmental problem in Japan because of the health hazard they pose to man. Many studies have been conducted in the past to understand the behavior of these organic solvents in groundwater. In the beginning, these organic solvents were considered not to be biodegradable but later studies showed these solvents were transformed to other hazardous materials by underground microbes.

Ninomiya, et al., (1991) studied the biological transformation of PCE and TCE in the laboratory using contaminated soil in Yokohama, Japan. He reported that the half-life of these organic compounds were 0.8 - 2.5 days for PCE and 0.5 - 1.1 days for TCE. Yagi and Uchiyama (1989) reported half-lives for TCE of 12 days, 26 days and 27 days by using soil from various places. Using an in-situ recharge test of waste water, Barrio-Lage, et al., (1986), reported that the biodegradation constants of both PCE and TCE are approximately 0.003 per day. Based on their results the half-lives of PCE and TCE are 230 days. Arihara, et al., (1994), reported a half-life of approximately 3.8 years by measuring PCE distribution in unsaturated soil. Previous studies have reported vastly different VCHs half-lives depending on type of soil and/or test method but one thing they had in common was that they did not consider transport phenomena in the groundwater.

In this paper, VCHs half-lives are investigated under actual aquifer environments by numerical simulation which can treat the advection, dispersion, sorption and biodegradation.

2 Biodegradation pathway and Properties of volatile chlorinated hydrocarbons

Volatile chlorinated hydrocarbons such as PCE and TCE are man-made compounds most frequently used as a solvent for degreasing, cleaning, and extraction in many industries. Dichloroethylene (DCE) has three kinds of isomers such as 1,1- , trans-1,2- and cis-1,2-DCE. These isomers are different in combination of chlorine, and trans- and cis- 1,2- DCE has not been produced or consumed for the commercial use.

Biodegradation of VCHs in anaerobic environments is considered to occur by sequential reductive dehalogenation. Under anaerobic conditions, VCHs undergo a sequential reductive dehalogenation from PCE to TCE then to 1,1-trans-1,2- and cis-1,2-DCE and finally to vinyl chloride (VC), (Fig.1).

A contaminated site in which only PCE was dumped reported traces of TCE and DCE in observation wells dug at the site. The reports of TCE and DCE at the site in which these chemicals were not used provide evidence of biological transformation under anaerobic environment.

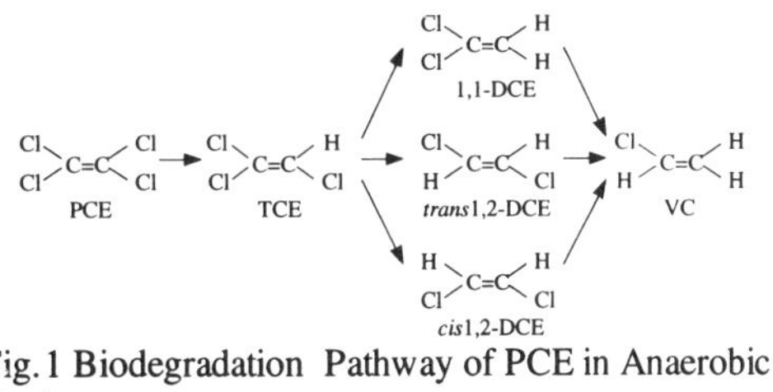

Fig.1 Biodegradation Pathway of PCE in Anaerobic Environments.

3 Description of the Site and Results of the Field Surveys

Soil profiles of a site located on a terraced deposit are shown in Fig. 2 in which an aquifer lies below a silt layer. The aquifer consists of laminated layers of sand, gravel and silt. A dominant gravel aquifer is located at a depth of 10 m and its permeability is approximately 10^{-2} cm/sec. A section through the aquifer section and layout of observation wells are, also, shown in Fig. 2.

The mechanism of groundwater contamination of PCE in this site is considered to be as outlined below.

1. PCE leaked from the waste tank or drain as a non-aqueous phase liquid (NAPL).
2. PCE penetrated through the sand seam or pore of silt as non-aqueous and aqueous phase liquids.
3. PCE dissolved into the groundwater and was transported in the aquifer along with groundwater flow.

Symbols W1 - W5 in Fig. 2 represent observation wells in which groundwater level and concentration of organic solvents were measured. As a result of a month of monitoring, direction of the groundwater flow was established to be from W1 to W4 and velocity of groundwater was approximately 10 m/year.

In order to analyze concentration of contaminants in the groundwater, the head space method was adopted and the concentration of head space gas was analyzed by gas chromatography with a photo ionization detective (PID) lamp. The observations made at wells W1 - W5 indicate various relationships between PCE and TCE, TCE and DCE as shown in Fig. 3 and Fig. 4.

Concentration of PCE at wells W1, W2 and W5 were approximately 20 mg/L and relatively higher than at the other wells. W1, W2 and W5 are located around the source of the contamination and specifically W1 and W5 are sources of contamination while W2 is a little apart from the source of contamination. Consequently, soil concentration in top soil borings at W1 and W5 were over 100 mg/kg of dry soil and only 1 mg/kg of dry soil at W2. This difference can be attributed to the TCE concentrations at W1, W2 and W5.

Figs. 3 and 4 show that there is a difference of about 10 times in the concentrations at the observation wells. The large difference is considered to arise

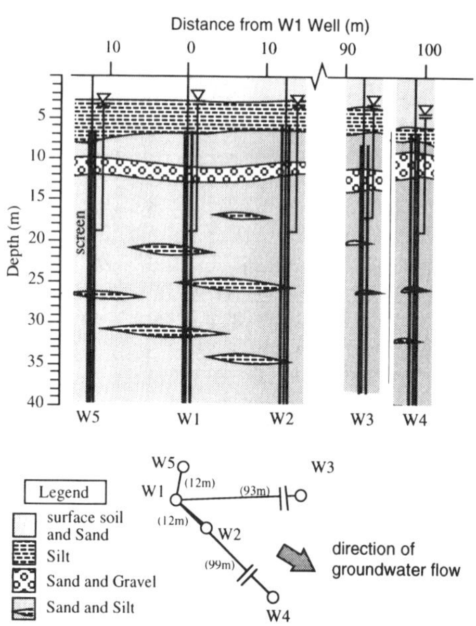

Fig. 2 Aquifer Section and Layout of Observation Wells.

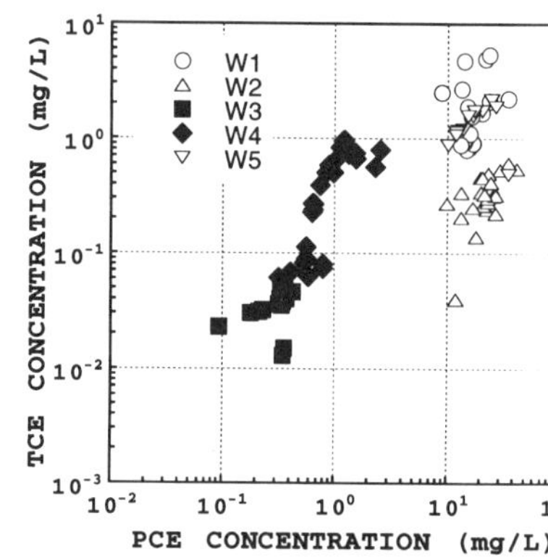

Fig. 3 In-situ Relationships Between PCE and TCE Concentrations in Groundwater

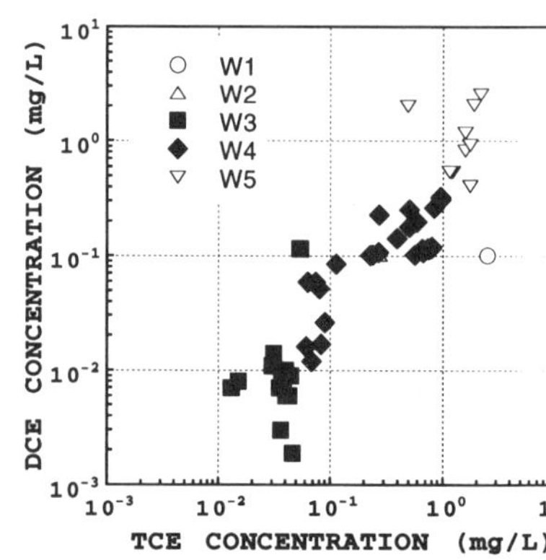

Fig. 4 In-situ Relationships Between TCE and DCE Concentrations in Groundwater

from the effects of environmental conditions such as temperature and humidity on the flow of VCHs.
In this study, numerical calculations are carried out using average values of concentration.

4 Overview of Numerical Simulation

4.1 Governing Equation

The governing equation of advection, dispersion, sorption and degradation is given by Eq. (1) in which the half-life is represented by the degradation constant λ_k [= ln 2 / $(t_{1/2})_k$] (Huyakorn 1987).
In this model transport of organic solvents is not the result of only the parent organic component but also daughter organic components which are produced by biological transformation.

$$\frac{\partial}{\partial x_i}\left(\phi D_{ij}\frac{\partial C_k^w}{\partial x_j}\right)-\frac{\partial}{\partial x_i}\left(V_i C_k^w\right)-\frac{\partial}{\partial t}\left(\phi R_k C_k^w\right)-\phi R_k \lambda_k C_k^w+\sum_{m=1}^{M}\phi R_m \xi_{km}\lambda_m C_m^w=0 \quad (1)$$

where Dij = hydrodynamic dispersion tensor (cm²/sec); Vi = velocity of groundwater (cm/sec); C_k^w = concentration in the water of chlorinated organic component k (e.g. mg/L) ; ξ_{km} = ratio of parent component m transforming into chlorinated organic component k; λ_k = degradation constant of chlorinated organic component k (1/sec); ϕ = porosity ; ρ_s = specific gravity of chlorinated organic component k (-); R_k = retardation coefficient;

$$\left(=\frac{1-\phi}{\phi}\rho_s K_{dk}+1\right)$$

K_{dk} = distribution coefficient ; M=number of chlorinated organic component transforming into component k.

4.2 Simulation Model

The behavior of organic solvents is usually performed by a three dimensional analysis. The conditions in the site are dictated by a high permeability layer situated in the aquifer which leads to a horizontal transport of organic solvents. Therefore, a two dimensional analysis was carried out. The finite element model is shown in Fig. 5. The simulated area is 500 x 1000 m and is modeled by 1581 nodes and 1500 elements. This area includes W1, W2 and W5 wells, which as described earlier are the sources of PCE contamination and the boundary conditions of the area is taken to be as one of constant concentration. Also, the flow of groundwater is taken to be uniform because of the negligible difference in the groundwater usage in the modeled area.

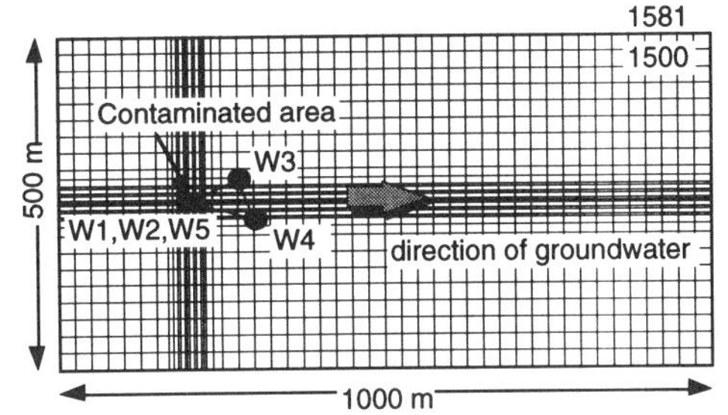

Fig.5 Numerical Model for Advection, Dispersion and Biodegradation.

Table 1 Parameters of Numerical Analysis.

Flow velocity of groundwater (m / year)	Porosity ϕ	Dispersivities DL (m)	Dispersivities DT (m)	Distribution coefficient Kd
10.0	0.24	10.0	1.0	0.289 (PCE) 0.071 (TCE)

The parameters used in the numerical simulation are shown in Table 1.
The velocity of groundwater is determined from the groundwater level and permeability as mentioned above. The dispersivity in the field is not easily defined and it is correlated to migration distance of the contaminant, Anderson (1984). As migration distance at the site is approximately 300 meters, longitudinal dispersivity is determined from the relationships between distance and dispersivity up to a distance of 10 m.
At this site, PCE has been dumped since 20 years ago and for the purpose of this study the simulated time was taken to be 50 years and a constant concentration of contaminant source is assumed as the boundary condition.

5 Results and Discussion of Numerical Simulation

Parametric studies were carried out by changing the degradation constants of PCE and TCE. The numerically calculated distribution of PCE and TCE are comparable to their respective in-situ measured values when the half-lives of PCE and TCE are 1600 days and 200 days. Concentration contours of PCE and TCE in groundwater are shown in Figs. 6 and 7. The change in PCE and TCE concentrations with time at wells W1 - W5 are shown in Fig. 8. The calculated concentrations show a fairly good agreement with the average of observed data.

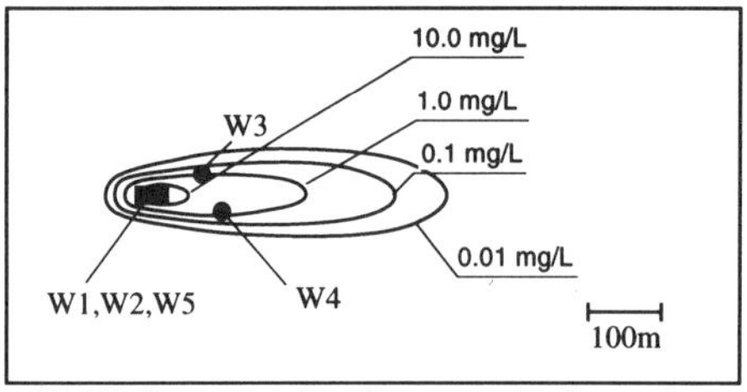

Fig.6 PCE Concentration Contours in Groundwater.

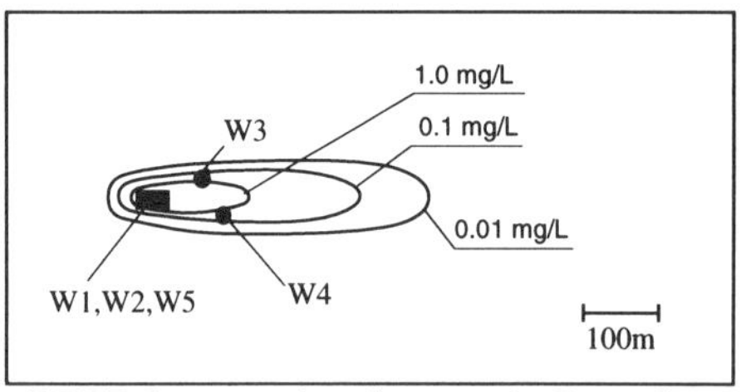

Fig.7 TCE Concentration Contours in Groundwater.

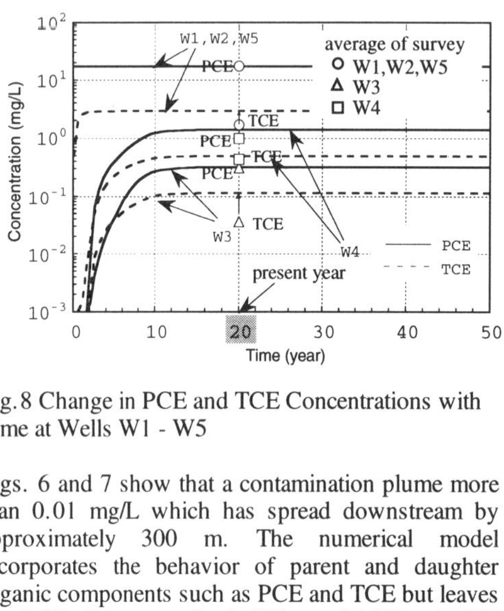

Fig.8 Change in PCE and TCE Concentrations with Time at Wells W1 - W5

Figs. 6 and 7 show that a contamination plume more than 0.01 mg/L which has spread downstream by approximately 300 m. The numerical model incorporates the behavior of parent and daughter organic components such as PCE and TCE but leaves out DCE. However, the half-lives of PCE and TCE can still be determined from the distribution of these organic solvents in the site because the degradation of TCE is considered in the analysis.

In Fig. 8, it is understood that concentrations of PCE and TCE have reached their steady state values. If degradation is not considered in the simulation, a contaminant plume is seen to spread over the entire region. In the numerical model, biodegradation was accounted for and the spread of the contamination is found to finally converge when the supply of contaminant becomes equal to the degradation of contaminants. Also, in the field biological degradation actually occurs therefore the model incorporating the degradation is more realistic.

6 Conclusions

In this paper, numerical simulations of biological degradation of volatile chlorinated hydrocarbons were performed and the results compared to data gathered from in-situ measurements. Results of this study are:

(1) The half-lives of tetrachloroethylene (PCE) and trichloroethylene (TCE) 1600 days and 200 days, respectively. Both are longer than those reported in previous studies.

(2) The numerical model accounted for biological degradation and therefore the actual behavior of contaminants in the field can be understood.

In the future, the influence of soil and other environmental conditions need to be investigated at various fields. This study represents a first step in this direction.

REFERENCES

Anderson,M.P. 1984. Movement of Contaminants in Groundwater: Groundwater Transport - Advection and Dispersion, in Groundwater Contamination, J.D.Bredehoeft et al. eds., National Academy Press, 37-45

Arihara,N. et al. 1994. Study on behavior of volatile chlorinated hydrocarbons in he soil and groundwater , Proceedings of the 3rd Symposium on Groundwater/Soil Pollutions and Their Management, 293-298, in Japanese.

Barrio-Lage,G. and Persons,F.Z. 1986. Sequential Dehalogenation of chlorinated ethens : *Environ.Sci.Tecknol. 20* , 96-99.

Huyakorn,P.S. and Pinder,G.B.1987. Computational Methods in Subsurface Flow , 183-185.

Ninomiya,K. and Sakai,M. 1991. Groundwater Contamination with Volatile Chlorinated Hydrocarbons (II) - A Consecutive First Order Reaction of Tetrachloroethylene - , 15th Yokohama Environmental Res. Inc. Report, 151-156, in Japanese.

Yagi,O. and Uchiyama,H. 1989. Biodegradation of Volatile Chlorinated Aliphatic Hydrocarbons , Microorganism Vol.5 No.6 , 13-22.

Environmental Geotechnics, Kamon (ed.) © *1996 Balkema, Rotterdam. ISBN 90 5410 848 7*

Remediation of Hg contaminated soil

I.A.Sobolev, A.S.Barinov, L.B.Prozorov & V.M.Kuptsov
Department of Engineering Supply, Government of Moscow, Scientific and Industrial Association (SIA) 'Radon', Russia

ABSTRACT: The sorption and desorption processes had been studied for moraine and surface clays and for granulometric fractions of the sand from north part Moscow region. We may decontaminate by the electrokinetic methods from the sand practically no more 50% of mercury, if use 1 N solution of hydrochloric acid in order to transfer of the mercury in the solution.

1 INTRODUCTION

The mercury belongs to the group of heavy metals (Hg, Pb, Cd, As, Zn), distinguished by high anthropogenic background. These metals are very toxic, suppressing ability to lives of the animals and plants [1].

Anthropogenic mercury discharged into environment comes from several global sources: fossil fuel combustion, chloralkali factories, mining industries.

Mercury and its compounds are actively used in an electro-technical industry, pharmacology, medicine, vacuum technology, scientific researches. Therefore the territories around large industrial centres are strongly polluted by mercury.

Electrokinetic soil processing using low-level direct current (on the order of amperes per square meter of area) is perhaps one of the most promising in situ decontamination processes capable of removing heavy metals and radionuclides from soil. The significance of the technology appears to be in its projected low operation cost anditspotential applicability to a wide range of contamination situation [2].

The procedure of the electrokinetic decontamination is realized in three stages: 1) transformation of a polluting impurity in the soluble forms (anions, cations or neutral atoms), 2)movings of an impurity by electrical current in the anode-cathode space, 3) accumulation of polluting impurity in the collector of metals.

In practice we must establish connection of the medium parameters (porosity, content pore water with introduced desorbed solution) and the parameters of equipment (the polution collector construction, electric current, voltage and distance between the electrodes, time of decontamination).

In principle, key of the all this problems decision consist in the polution forms detection, the content and the concentration of desorption solution. We try to resolve this in the present research.

2 PROCESS OF DESORPTION

The adsorption of Hg on the surface of the solid can be described by the exchange reaction of two species (*Hg* and *X*) between their aqueous solution (*aq*) anl solid (*s*) [3]:

$$Hg(aq) + X(s) \Leftrightarrow X(aq) + Hg(s)$$

The rate of change of the concentration of *Hg* is:

$$\frac{dHg(aq)}{dt} = -K_1 Hg(aq)\,X(s) + K_{-1} Hg(s)X(aq),$$

where $Hg(aq)$, $Hg(s)$, $X(aq)$, $X(s)$ - concentration of species dissolved or suspended in aqueous solution (moles/ml);
K_1 and K_{-1} forward and backword reaction rate constants respectively.
Assuming a first order reaction kinetics for a steady state we have:

$$Kap = \frac{K_1}{K_{-1}} = \frac{Hg(s)X(aq)}{Hg(aq)X(s)}$$

Kap - apparent equilibrium constant.

For real systems we must use the thermodynamic activities instead of the concentration, using the conventional relationship between activity, concentration and activity coefficient.

The activity of the dissolved species is $a(aq) = C(aq)\gamma$, where γ is the molal activity coefficient, $C(aq)$ is molal concentration. Activity of the adsorbed species is $a(s) = c(s)\lambda$, where $c(s)$ is mole fraction, λ-is the activity coefficient in the adsorbed state. The activity coefficients of ions in solution γ can be estimated from Debye-Huckel [4] and related equations, but for the activity coefficients of adsorbed species λ there is no comparable method and the equilibrium constants Kap are known only for several specific system [4].

It is generaly difficult to evaluate how several dissolved substances in natural waters would distribute themselves between a solution of given composition and a mixture of several solid phases. Experimental measurements are certainly the most realible way to find out the type of exchange or adsorption that take place in a given solution - solid system under given condition. In reality things are not always so simple: the species may not be exchanged rapidly especially with those not in surface sorption sites. Sorption may involve oxidation and precipitation, adsorption on colloides, then coagulation on larger particles.

Therefore in practice ordinary was used coefficient *Kd*:

$$Kd = \frac{Hg(s)}{Hg(aq)Cp},$$

Kd - distribution coefficient at steady state [ml/g]
Cp - concentration of solid in aqueous solution [g/ml].

Kd (opposit from *Kap*) is not on equilibrium constant and can vary as pH, solution composition, physical and chemical properties of suspended particles changes.

It is common practice to express the desorption results of metals in terms of the "% desorbed".

For the best understanding of processes we studed as desorption , as sorption.

3 METHODS OF RESEARCHES AND MERCURY ANALYSES

The known amounts (ordinary 2 g) of contaminated (desorption experiments), or uncontaminated (sorption experiments) soils was placed in the 0.5 l glass bottle. Than was added 100 ml the different content solution. The solution was contained 50 µg of mercury in the form of nitric acid salt in sorption experiments. Six solution types were studied: distilled water, natural water from the place of sampling, solutions of hydrochloric acid. Ca^{2+} (79%), Mg^{2+} (14%), (Na^{+}+K^{+}) (7%) are basic cations of natural water; HCO_3^- (84%), SO_4^{2-}(14%), Cl^- (2%) are basic anions. Total mineralization is 230 mg/l.

The solution with contaminated soil were mixed constantly during 4 hours. After this solution was filtered and concentration of mercury was measured on the filter and in the solution. The results are presented in term of h (for desorption) and g (for sorption):

$$h = \frac{C(aq)}{C_o(s)} 100\% , \quad g = \frac{C(s)}{Co(aq)} , \text{ where}$$

$C(aq)$ mercury content in filtrate after experiment,
$C_o(s)$ mercury content in solid before experiment,
C(s) mercury content in solid after experiment,
$C_o(aq)$ mercury content in solution before experiment.

The quantitative chemical analysis for the mercury was performed by atomic absorption spectrophotometric measurement.

Mercury was extracted by tioether from 0.5 l of water, after filtration tioether was diluted by nitric acid, limit of detection is 0,0005 µg/ml. Mercury from 2 g of solid samples was extracted by treatment of the sample by nitric acid, limit detection is 0.2 µg/g .

4 MATERIAL FOR RESEARCH

The soil types were studied: sands, moraine clays and surface clays. They was sampling in Sergiev Posad branch SIA "Radon" , located in 90 km north-east from Moscow. All soil types are most extend to north part Moscow region and present the worked by ice material of Scandinavian sheet.

The desorption processes had been studied only for the four different sand fractions. The mercury in the form of nitric acid salt had been introduced in the sand lens on the territory of branch two years ago. In results of natural proc esses (wash by atmospheric precipitation, change of the forms, diffusion, sorption) stationary distribution of mercury in the different size fractions had been formed. The granulometric content (w_i) and the mercury distribution in different fractions (η_i) are shown in Fig.1.

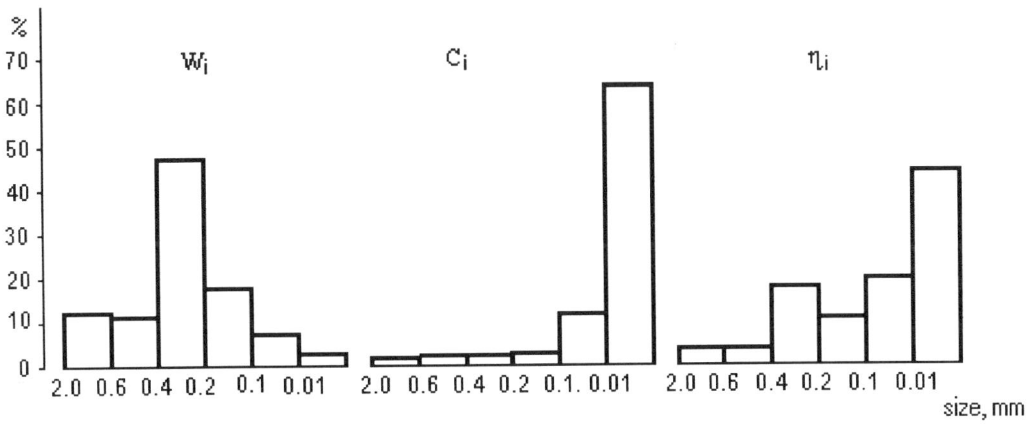

Fig.1. The granulometric composition *(w_i)*, the mercury concentration *(c_i)* and the mercury distribution *(η_i)* for the different granulometric sand fractions.

$$\eta_i = \frac{w_i c_i}{\Sigma w_i c_i} 100\,\%, \quad \text{where}$$

w_i - % weight fraction of i, c_i - mercury concentration, µg/g,
η_i - mercury distribution in the different fractions, %.

The size 0.2-0.4 mm fraction have composed 48 weight %, but the mercury content have in forty time lesser, than fraction <0.01 mm which have only near 3% weight. Just therefore fraction with size <0.01 mm has contained 44% of mercury.

5 SORPTION PROCESSES

The results of experiments are listed in Table 1. The most level of sorption have surface clays and the 0.1-0.01 mm sand fraction. It may be explained as being due to high concentration of organic matter in surface clay and the high sorption surface of fraction.

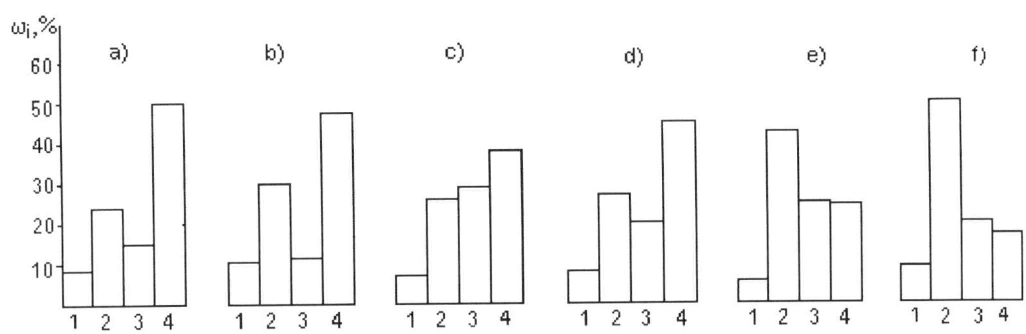

Fig.2 Distribution of the ω parameters for different sand fraction (1-0.6-0.4; 2- 0.2-0.4; 3 - 0.2-0.1; 4 - 0.1-0.01 mm) and solutions (a - distiled water; b- natural water; c - 0.01 N acid; d - 0.1 N acid; e - 0.5 N acid; f - 1.0 N acid).

Table 1. *g*-parameters for different soils and solutions.

soils / *solutions*	*morain clay*	*surface clay*	*sand* all	0.6-0.4	0.4-0.2	0.2-0.1	0.1-0.01
				fractions, mm			
distil. water	20.8	71.3	14.9	11.7	8.7	12.3	83.4
natural water	12.9	30.5	11.9	8.2	7.9	12.2	68.2
acid 0.01N	10.0	23.0	8.9	4.7	4.3	8.0	31.4
0.1 N	6.3	16.1	6.2	3.5	2.7	4.9	24.0
0.5 N	4.4	13.6	4.8	2.4	3.3	4.8	9.2
1.0 N	2.7	10.2	3.2	2.5	2.9	3.3	4.9

The surface clay differ from moraine clay in particular high concentration of organic matter.

The sorption is considerably lower in the natural water than in the distilled water, owing to exchange reactions of the Ca, Mg ions with the mercury ions. The sorption have been decreased sharp with increase of the hydrochloric acid concentration.

The contribution of the each sand fraction (ω_i) in the all sorption is given in Fig.2.

$$\omega_i = \frac{g_i w_i}{\Sigma g_i w_i} 100\,\%$$

One may see, that in the distiled and natural waters main part of mercury have been sorbed on the small size fraction, which compose near 10% weight. The mercury content is practically in two time less in the fraction 0.2-0.4 mm, which compose 48% weight.

The influence of the small size fraction have been decreased with increace the hydrochloric acid content and for 1 N acid the ω_i and w_i distributions is practically the same.

6 DESORPTION PROCESSES

The experimental results are presented in Table 2.

The *h* sand parameter has been no more 3% for the distilled and natural waters and has been determined by exchange reactions. This parameter increase with growth acid content and near 50% of mercury has transitioned in the solution for 1 N acid, all mercury has located in solution for 10 N acid.

Table 2. *h*-parameter for the different solutions and fractions.

soil / *solution*	*sand* all	0.6-0.4	0.4-0.2	0.2-0.1	0.1-0.01
		fractions, mm			
distilled water	2.45	-	-	-	-
natural water	2.76	-	-	-	-
acid 0.01 N	3.64	-	-	-	-
0.1 N	7.75	-	12.9	-	8.8
0.5 N	27.4	19.9	18.7	18.6	18.7
1.0 N	48.4	44.1	62.8	41.0	29.9

The contribution of the each sand fraction (λ_i) in the all desorption had been given in Fig.3 for the 0.5 N and 1.0 N acid concentration:

$$\lambda_i = \frac{h_i w_i}{\Sigma h_i w_i} 100\,\%$$

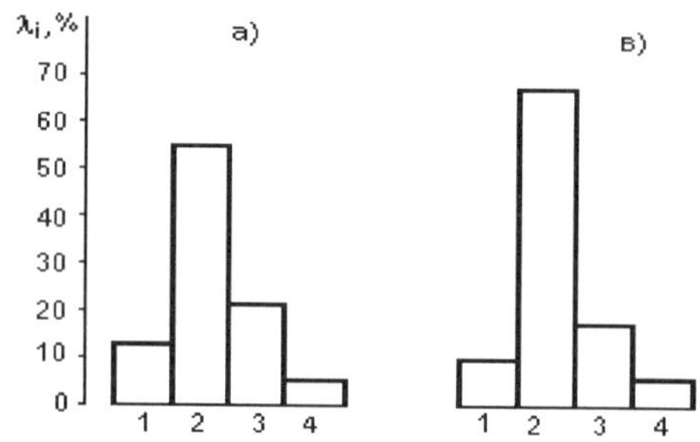

Fig. 3 Distribution of the λ parameters for the different sand fractions (see Fig. 2) and silutions: a - 0.5 N acid; b - 1 N acid.

Main part of mercury has transitioned in solution from the 0.2-0.4 mm fraction. In order to draw on mercury from the small size fraction necessary use the more acid concentrations.

7 THE INFLUENCE OF THE CONTACT TIME AND THE SOLUTION VOLUME

We had carried out several experiments in order to study influence the contact time and the solution volume on the g and h parameters. It had been used 2 g of the fraction 0.1-0.01 mm, which had high concentrations of the mercury and the solution of distilled water. The experimental results are shown in Fig. 4, 5. The desorption processes don`t have the dependence from the time and the solution volume (in parameters of experiments).

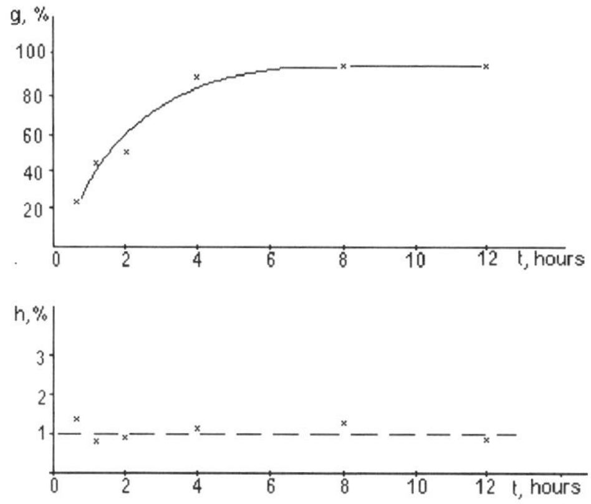

Fig. 4 Dependenced of *g, h* parameters from the contact time.

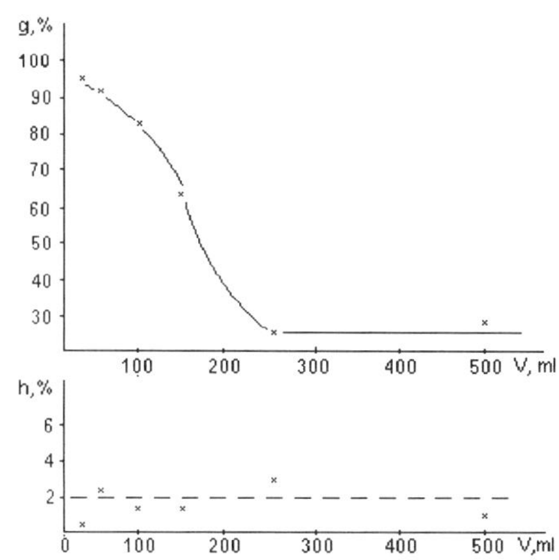

Fig. 5 Dependences of the *g, h* parameters from the solution volume (V).

As opposed, sorption processes in our experiments have been dependent both from the time and the solution volume. In the first 4 hours sorption was increased in 4 time and has reached of the saturation. It showed, that in the sorption processes take part the more deeper layer of the solids.

The sorption have been strongly dependent from the solution volume, but only for the laboratory experiments. The solution volume is very small in the real soil systems.

8 CONCLUSION

1. It had been determined the mercury sorption properties of basic types of soils in the north part of Moscow region.

2. Main part of mercury in the sand had been sorbed on the small size fraction from the natural or distilled water, but in the acid solution (0.5 N and 1.0 N) distribution of the granulometric fractions and distribution of the sorbed mercury have appeared practically the same.

3. The mercury desorption is very small in the distilled and natural water (~3%), hydrochloric acid solution (1N) had desorbed near 50% of mercury, mainly from the more widespread of the 0.2-0.4 mm fraction. In order to desorbed mercury from the fraction small size necessary use more concentrate acid solution (8-10 N).

4. Practically the metal collector may work using current 1 A. It means, that for 100 days exposition we may collect 90 gramme-equvalent of ions. If we use 1 N solution, may decontaminate of ~50% mercury only from 90 l of soil solution. Use the 10 N acid solution is practically unreally.

5. The sorption and desorption processes are established very quickly (<0,5 hour) for the 12 hours time interval in the real soil systems.

ACKNOWLEDGEMENTS

The author wish to recognize the contribution made to the project by the following individuals:

A.V.Timofeeva and A.A.Kostrova - analytical chemists,

V.N.Titov - the help in the work,

Y.G.Tkachenko - the useful discussion,

Y.G.Beltyukova - for help in preparation of article.

REFERENCES

1. V.B.Ilin. Heavy metal in system soil plant (in russian). Novosibirsk, Nauka, 1981, 151 p.

2. J.K.Wittle, S.Pamukcu. Electrokinetic treatment of contaminated soil, sludges and lagoons. Final report. Contract № 02112406. Argonne National Laboratory. Department of Energy. p.61.

3. Y.H.Li, L.Burkhardt, M.Buchholtz, P.O`Hara and P.Santschi. Partition of radiotracers between suspended particles and seawater. Gecochim. et cosmochim Acta.1987, vol.48, p.2011-2019.

4. L.I.Antropov. Theoretical electrochemistry (in russian). Moscow, Highschool, 1984, p.518.

Environmental Geotechnics, Kamon (ed.) © 1996 Balkema, Rotterdam. ISBN 90 5410 848 7

Experiments of applying electromigration to soil contaminated with ^{137}Cs

I.A.Sobolev, L.B.Prozorov, A.S.Barinov & V.V.Martyanov
Government of Moscow, Engineering Ensuring Department, Moscow Research and Production Association 'Radon', Russia

ABSTRACT: The essence of electro-migration technology includes three basic stages:

- transformation of contaminant elements in a mobile condition (as ions) by activators;
- migration under action of an electrical field from the anode to cathod or on the contrary;
- accumulation of contaminant elements in electrode zones and their subsequent withdrawal from these zones.

In the given work results of researches on three field experiments are presented.

1 INTRODUCTION

During recent five years specialists from Mos RPA (Moscow Research-and-Production Association) "Radon" (Russia) have been conducting extensive investigation on utilization of electro-migration processes for decontamination of soils from radionuclides and heavy metals.

Contamination with radionuclides and heavy metals, as a rule is characteristic of soils containing clay particles. Those particles possessing high sorption properties, are capable to retain contaminants on their surface, and even to include them in their mineral structure. Strength of sorption links is defined by physical-and-chemical properties both for separate varieties of clay particles and for the elements-contaminants. Under field conditions most common the thickness of soil layer with increased content of elements-contaminants is limited to 1m.

Field experiments on remediation of ^{137}Cs contaminated soil by Electro-Migration Technology were conducted on flood-lands of the stream in 1993-1995.

During experiments two types of electro-migration (3 tests) were applied to soil contaminated with ^{137}Cs:

- test 1 and 2 - removing contaminants vertically downwards in clay having by high sorption properties (figs.1-5);
- test 3 - moving contaminants in a horizontal direction (fig.6) and their accumulation in near-electrode (cathod) space (with the next removing and disposal of them in repositories).

2 TEST 1

Total length of chosen site was about 15 m., width and depth were 0.4-1.0m.

Radiometric investigations have determined that main contamination is observed in clay formations to a depth of 40 cm and peak depth of dose intensity has been identified at the distance of 10-20 cm from the surface. Total dose intensity on the surface of the site reached 200-300 mR/h and peak of gamma-radiation on the soil has reached 400 mR/h. Total ^{137}Cs activity at the test site has exceeded 2 x10^7 Bq., and the total mass of contaminated soil was 4.5t. Main purpose of the activities consisted in moving of activity peak downward along soil cross section and in decreasing of its value at the expense of summary dissipation and sorption of covering loams and morain sediments by clay particles. The site was divided into 5 blocks, and 34 electrode and observation holes have been drilled on it. Electrode arrangement allowed to provide for vertical component of mass transfer in the soil. Anode electrodes were located on the surface of soil, and cathode ones in electrode holes up to a depth of 1 m. As desorbing agents (within block boundaries) solutions of potassium and ammonium nitrate salts have been used and besides a number of special additives preliminary tested in laboratories.

Total durability of the test made up about 70 days, operating period of decontamination made up 40 days. It should be noted that more than half of the period was characterized by unfavorable weather conditions, at average temperature of -5 C.

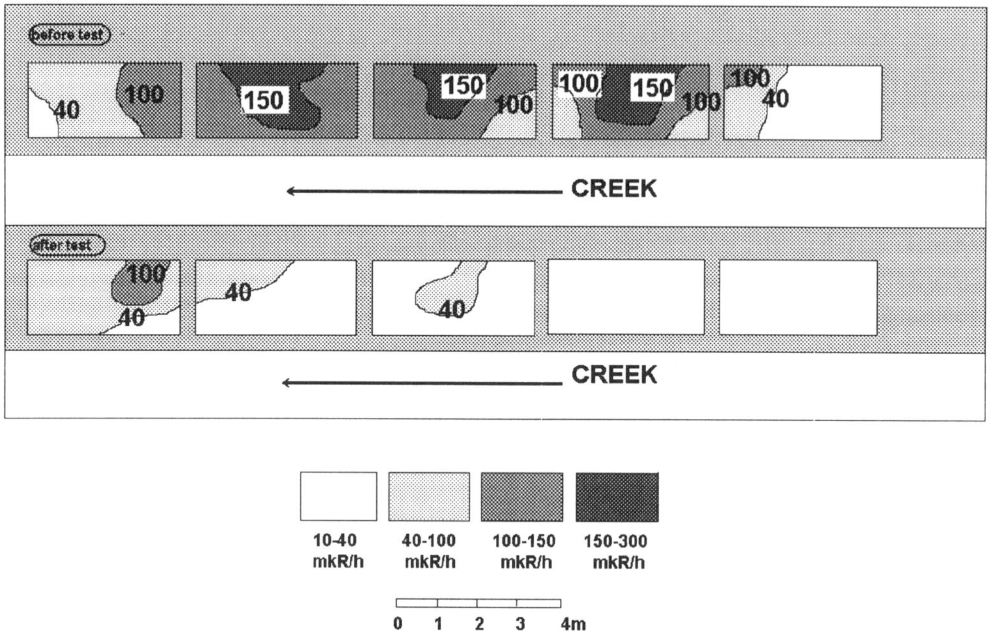

Fig.1. Gamma dose distribution on the surface (Site 1).

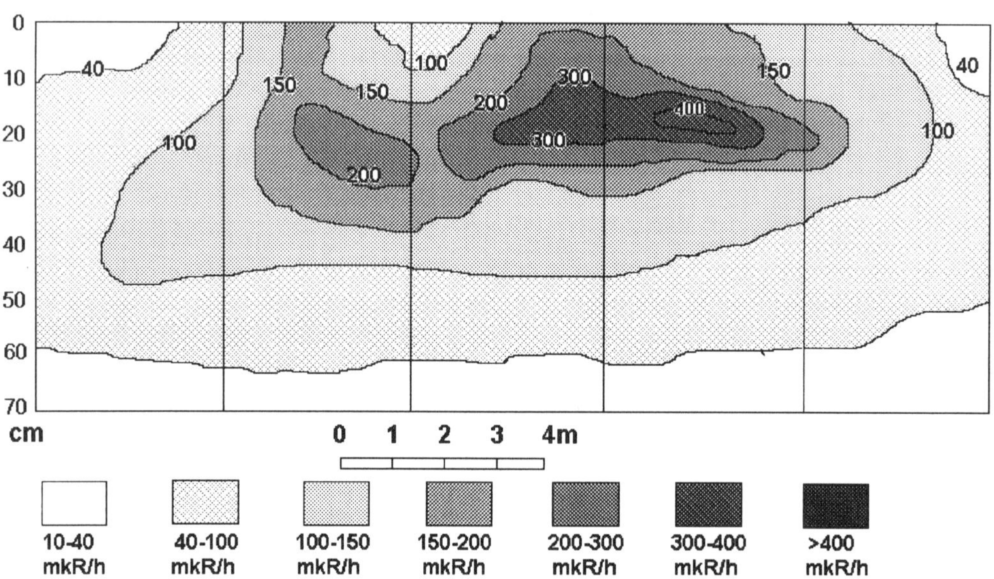

Fig.2. Radiometrical profile (Site 1, before test).

It was determined that (figs.1,2 and 3):

- maximum degree of decontamination has reached 63%;
- on the surface the intensity of gamma-radiation has not exceeded twofold natural background (40 mR/h);
- activity peak moved 10-20 cm downward;
- for decontamination of 1 m^2 of surface has been spent about 200 kWh of electric power, specific energy consumption for desorption of 1 Bq of ^{137}Cs were within boundaries of 0.2-2 Watt-hour.

3 TEST 2

Total square of chosen site was 13 m^2. The contamination was located at the opposite banks of the stream. Initial gamma dose rate at the surface of the site was 110-120 mkR/h. The site was divided into 2 blocks and 13 electrode and 11 monitoring wells (holes) have been drilled on it.

The were 2 types of contamination along the depth: the first one - at the depth 15-20 cm (more than 600 mkR/h) and the second - at the depth of 50-70 cm (less then 200 mkR/h). Summary depth of contamination was 80 cm.(fig.4) Specific activity of the soil - $6.9*10^3$ - $2*10^4$ Bq/kg. As desorbing agent solutions of NH_4NO_3 + HCl have been used.

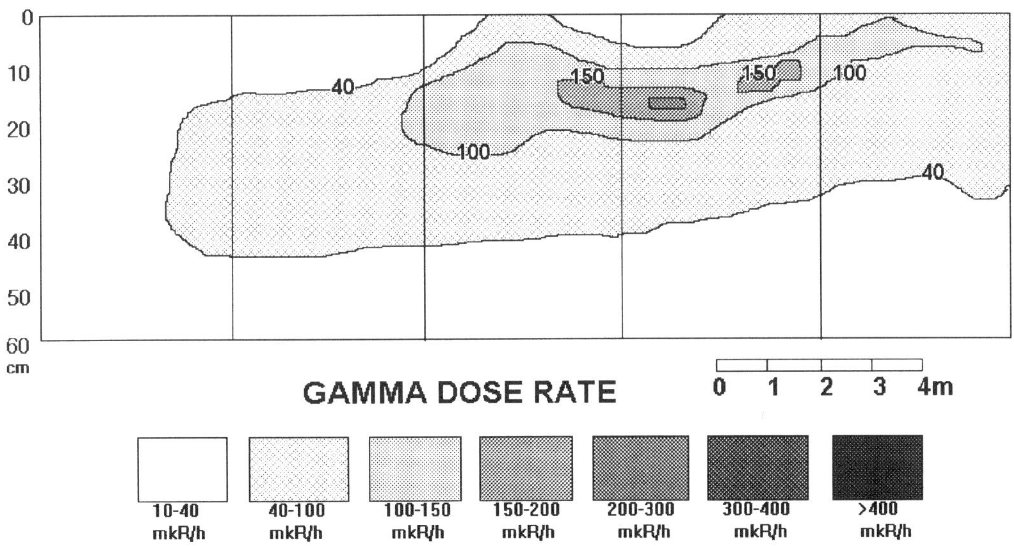

Fig.3. Radiometrical profile (Site 1, after test).

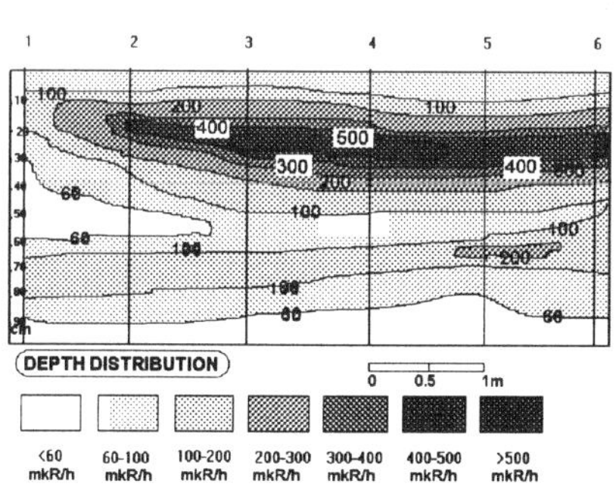

Fig.4. Radiometrical profile. Site 2 before test.

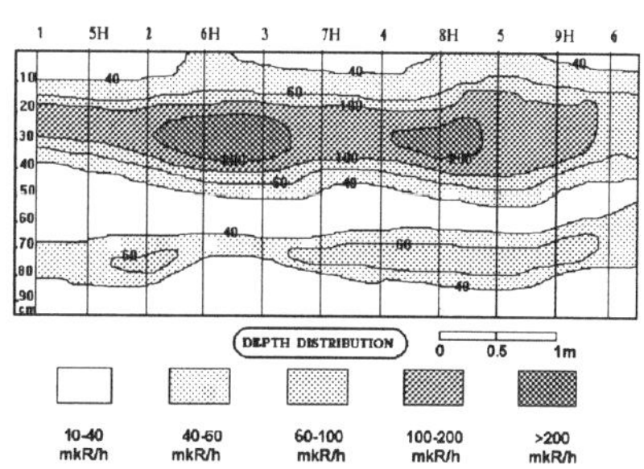

Fig.5. Radiometrical profile. Site 2 after test.

Total durability of the soil decontamination made up about 45 days.

It was determined that (fig.5):

- maximum degree of decontamination has reached 67.2%;
- the surface gamma dose rate has not exceeded 20-35 mkR/h;

4 TEST 3

Total square of chosen site was 60 m^2. The soil was contaminated by ^{137}Cs. A depth of contamination was 1 m. Gamma-radiation peak has been identified at the distance of 15-25 cm from the surface. Total dose intensity on the surface of the site reached 400 mkR/h (background - 190 mkR/h), in the depth - more then 1000 mkR/h. Total ^{137}Cs activity at the site has exeeded 0.02 Ci.
The site was divided with special electrodes in holes (cathodes and anodes), monitoring holes (for gamma-dose rate), power suply (for voltage drop 1 V/cm., and I=9A) and other methods. The "cathode" holes (wells) contained special powder for sorbtion ^{137}Cs. During the experience the solutions of Na^+ and K^+ were added into the "anode" holes and to the surface of the site for desorbtion ^{137}Cs.
Total durability of the experience made up about 2.5 monthes. The experience results are follows (fig.6):

- the oreol of ^{137}Cs-contamination removed from "anode" holes to "cathode" holes (to the middle of the site). It was determinated, that gamma-dose rate reduced to 40%;
- secondary oreols of ^{137}Cs-contamination were formed around "cathode" holes as a result of current-and-chemical treatment. Gamma-dose rate in these areas increased from 300 - 500 mkR/h up to 600-940 mkR/h;
- the contamination of ^{137}Cs was located on the contact of elecrode columns constructions with the soil, but not on the special powder;
- the special powder (sorbent) was destroyed as a result of current-and-chemical treatment;
- the current and reagent conditions of this experiance are good enough satisfactory for desorbtion and migration of ^{137}Cs in the soil.

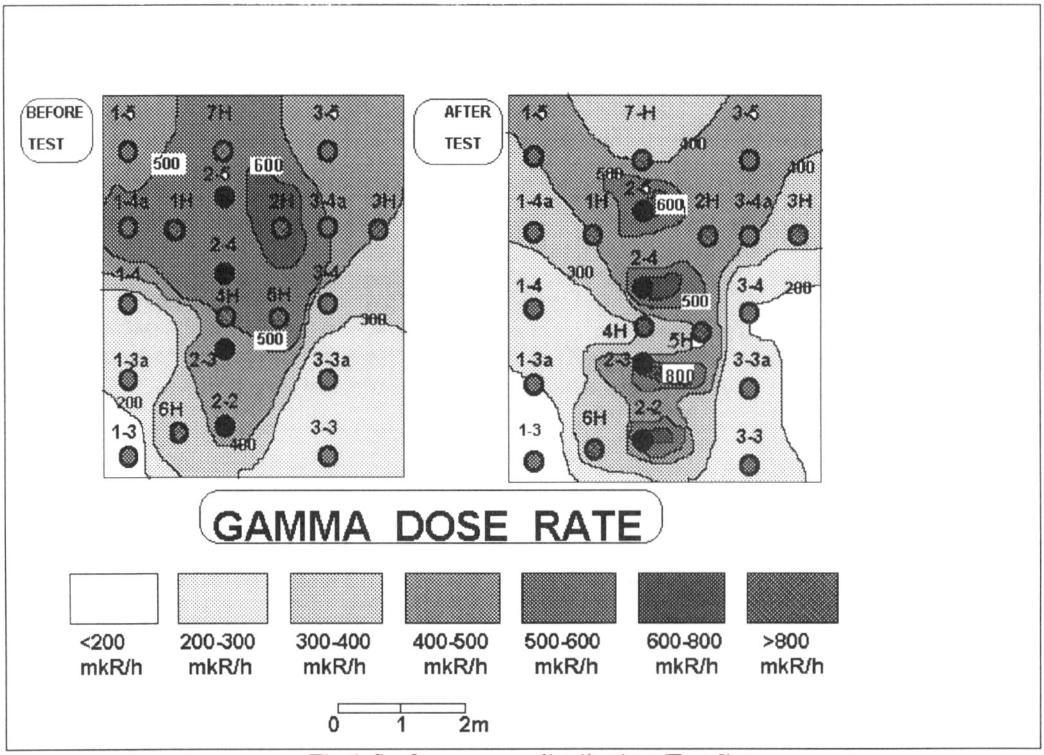

Fig.6. Surface gamma distribution (Test 3).

5 CONCLUSION

Preliminary assessment of the results of experiments on soil decontamination permits to consider electromigration technology as scientific and technical trend for surface rehabilitation from radionuclides and heavy metals. In this case the best effect can be expected while decontaminating of local areas with high contamination levels. Further efficiency of proposed technology is connected with the development of ecologically safe reagents for transferring contaminants in mobile state and in constructive improvements of electrodes.

The experiences of soil decontamination will continue till 1997 and we hope to recieve new results.

REFERENCES

Reiniers R.S.& Krenkel P.A. 1974. Kinetics of mercury adsorption in sediments, J. Water Polut. Control Fed., 46(2): 215.

Hamed, J., Y.B. Acar & R.J. Gale 1991. Pb(II) removal from kaolinite by electrokinetics, J. Geotech. Eng., ASCE, 117(2): 241-271.

Environmental Geotechnics, Kamon (ed.) © 1996 Balkema, Rotterdam. ISBN 90 5410 848 7

In situ bioventing of oil contaminated soil in cold climates

M. Sparrevik & G. D. Breedveld
Norwegian Geotechnical Institute, Ullevaal Hageby, Oslo, Norway

ABSTRACT: Pilot scale experiments were performed to assess the feasibility of stimulating in situ biodegradation of oil contaminated soil by venting. Large undisturbed soil samples (1 m^3) were taken from a diesel polluted field site and remediation was conducted for 1 year at a normal Scandinavian soil temperature (8 °C). Highest degradation rates were obtained when aeration was combined with circulation of a nutrient solution of nitrate and meta-phosphate. Without nutrient addition almost no biodegradation was observed. With nutrient addition the oil content decreased from initial values of 2 000 - 5 000 mg TPH / kg soil, to 600 mg TPH / kg soil at the end of the remediation period. This corresponds to a degradation rate of 4,5 mg oil / kg soil / day.

1. INTRODUCTION

In Situ bioventing has the potential to remediate oil contaminated soil, also in cold climates like Norway (Leeson et al., 1992, Ong et al., 1994). The technique is based on the ability of micro-organisms to transform hydrocarbons to biomass, carbon dioxide and water. Degradation takes place under aerobic conditions, requiring addition of oxygen and nutrients (nitrogen and phosphorus). In general straight chained hydrocarbons and mono aromatic compounds are easily degradable whereas branched hydrocarbons are more persistent (Atlas, 1984). Oxygen may be supplied to the soil by vacuum extraction of soil-gas or by injecting air directly below the water table (air sparging) (Eyk and Vreeken, 1988, Kampbell et al., 1993). Nutrients may be supplied by irrigation of nitrogen / phosphorus solutions (Hinchee et al., 1991).

Pilot scale experiments have been performed for 1 year to assess the feasibility of the technique as a remediation method for oil contaminated soil in the unsaturated soil zone.

2. EXPERIMENTAL SETUP

Undisturbed soil samples, in the form of soil cores were taken by vibrating a 200 cm diameter by 500 cm high steel cylinder in to the soil. Four cylinders were removed, containing 10 to 30 year old diesel contaminated soil with a concentration of 2 000 to 5 000 mg oil / kg soil. The soil was classified as glaciofluvial sand with a low content of organic matter.

The cylinders were placed in a rock cavern and maintained at normal Scandinavian soil temperature (8 °C). Equipment was installed to study the effects of aeration and nutrient addition to the soil (fig. 1).

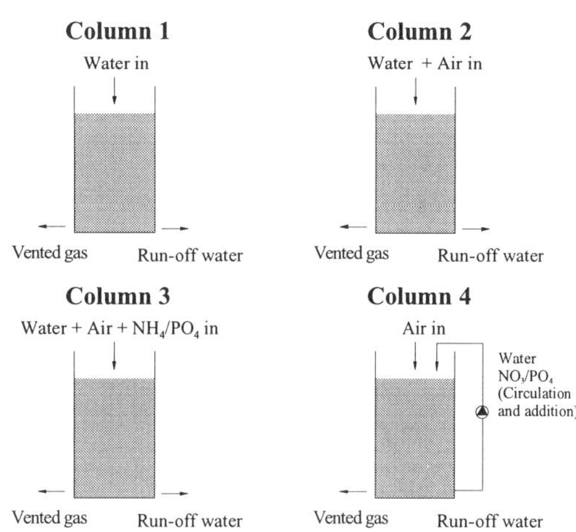

Fig. 1. Experimental setup

Sample points were installed on 3 levels for gas sampling of the unsaturated soil. Column 1 was used as an untreated reference column. In column 2, 3 and 4, an air volume of twice the calculated soil pore volume was ventilated through the soil daily. Every week oxygen and carbondioxide were measured in the gas sampling points.

All columns were supplied with equal amounts of water (34 ℓ / week).

Column 3 was supplied weekly with an ammonium (200 mg N / ℓ) and ortho-phosphate solution (50 mg P / ℓ). In column 4 nitrate (200 mg N / ℓ) and meta-phosphate (50 mg P / ℓ) were circulated as a nutrient solution.

During the remediation period the run-off water was analyzed for oil content (TPH-GC), total organic carbon (TOC) and nutrients (NH_4^+, NO_2^- NO_3^-, total-P and O-PO_4).

After one year of remediation soil samples were taken from the columns by drilling holes on 3 levels in the cylinder wall and taking horizontal cores (diam. 4 cm). The soil was analyzed for total carbon (TC), total organic carbon (TOC) and total petroleum hydrocarbons (TPH).

3. RESULTS AND DISCUSSION

Degradation

During the remediation period the production of CO_2 and the consumption of O_2 was used as an indicator for biological degradation (Sharab and Bartha, 1993, Wood et al., 1993). The accumulated production of CO_2 indicated degradation in all the aerated columns but a significant enhancement in degradation rates was observed when nutrients were added.

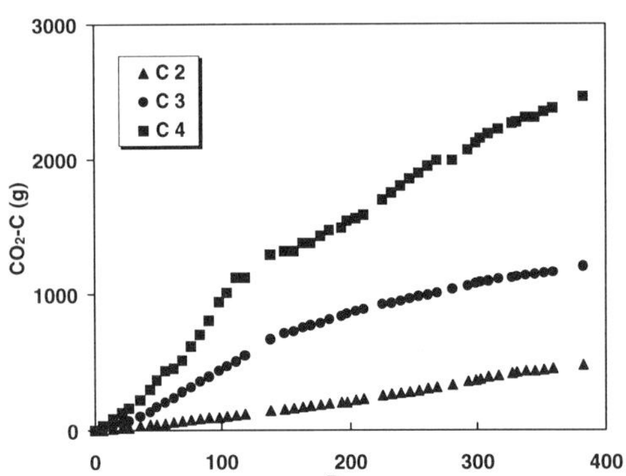

Fig. 2. Accumulated CO_2 production

Circulation of nitrate and meta-phosphate (column 4) yielded the highest degradation rates in the experiment (fig. 2).

Results from the CO_2 analysis at different levels in columns 3 and 4 indicated variations in the degradation pattern between the columns. Weekly addition of ammonium to column 3 resulted in a degradation front moving downward in the column with time (fig. 3). The biological activity for the glaciofluvial soil depends on the availability of nitrogen. Strong adsorption of ammonium by the soil matrix retards the nitrogen transport. The slow movement of ammonium through the soil profile will therefore determine the velocity of the degradation front.

Uniform degradation was achieved when nitrate was circulated in the soil (fig. 4). As a result of low adsorption by the soil, nitrogen would be easily available and therefore rapidly transported in the column (Follet, 1989).

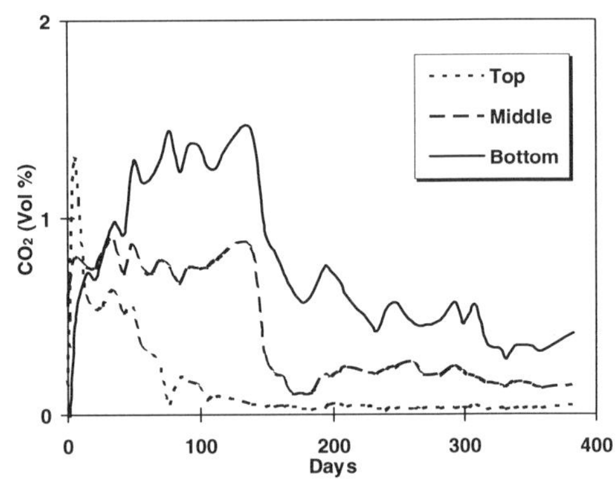

Fig. 3. CO_2 production in column 3

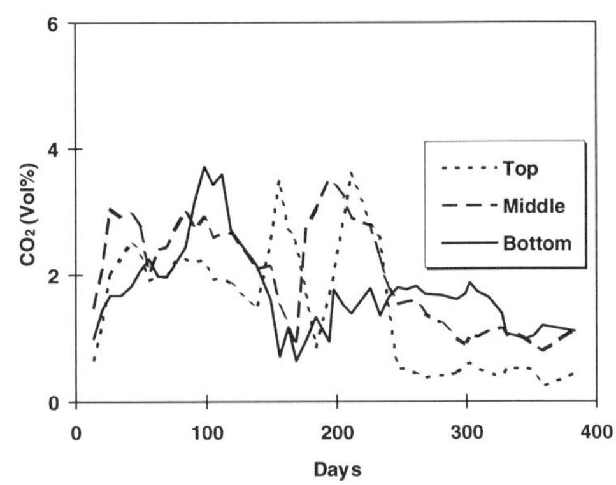

Fig. 4. CO_2 production in column 4

Nutrients in the water

Nitrate was found in the run-off water of both column 3 and column 4 (fig. 5).

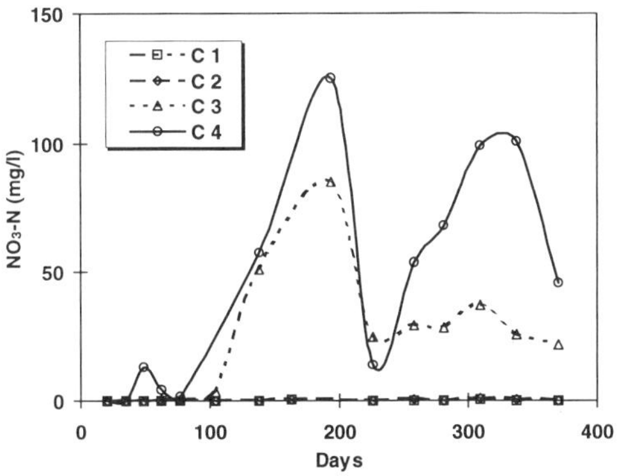

Fig. 5. Nitrate in run-off water

The presence of nitrate in column 3, to which ammonium was added is a result of nitrification. The nitrification of ammonium by micro-organisms in the soil, can be described with the following formula (Paul and Clark, 1989):

$NH_4^+ + 1\frac{1}{2} O_2 \rightarrow NO_2^- + 2H^+ + H_2O$ Step 1

NO_2 is then rapidly oxidized to NO_3 as follows:

$NO_2^- + \frac{1}{2} O_2 \rightarrow NO_3^-$ Step 2

High concentrations of nitrite were detected in the run-off water from column 3 (fig. 6).

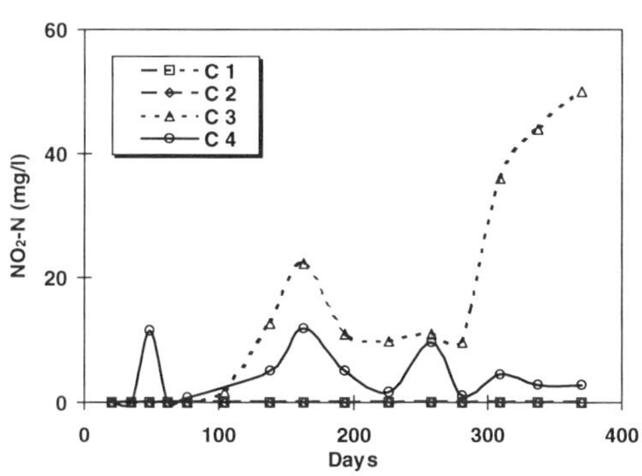

Fig. 6. Nitrite in run-off water

This was unexpected as nitrite normally is rapidly oxidized to nitrate. The high levels of nitrite may result from a slow rate during the second step in the nitrification process. The micro-organisms responsible for this second step (Nitrobacter) may be negatively affected by high ammonium concentrations (Schlegel, 1985). The production of nitrite in column 3 may therefore be determined by the amount of ammonium in the soil.

Nutrients in the soil

After 1 year of remediation soil samples were taken and easy available nutrients were extracted from the soil with $CaCl_2$ solution.

Nitrate was found in moderate concentrations in all soil layers (fig. 7).

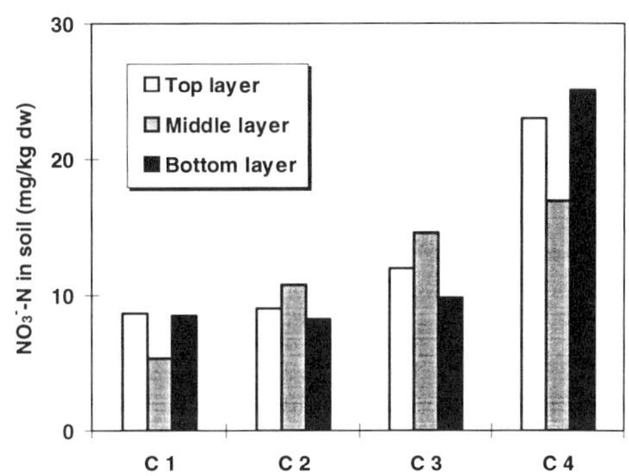

Fig. 7. Extractable nitrate in soil

The major differences in nitrate concentration are observed in the water phase whereas the nitrate concentration in the soil is relatively similar between the columns. The soil is therefore believed to have a limited storage capacity for nitrate.

Ammonium was only found in the top and middle layers of column 3 (fig. 8). This illustrates the strong retardation of ammonium in the glaciofluvial soil. The results correspond with the observation of the CO_2 levels in column 3 (fig. 3). The nitrogen source in the form of ammonium becomes available over time as the nitrification process proceeds.

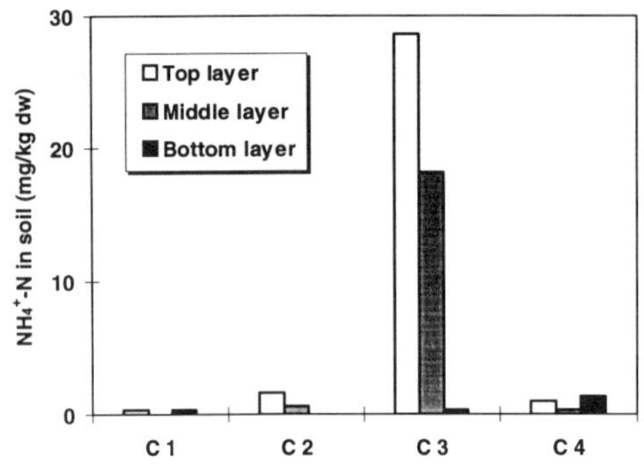

Fig. 8. Extractable ammonium in soil

Organic content in run-off water

The run off-water from the columns was analyzed for total petroleum hydrocarbons (TPH) and the total organic carbon content (TOC). TPH analyses indicated a two phase leakage pattern. An initial period with high TPH concentrations was followed by a stable leakage level of approximately 2 mg TPH / ℓ in the rest of the period (fig. 9). TPH analyses indicated changes of the oil composition in the water during the experimental period. A gradual transformation in the composition of the oil, from shorter to longer chained hydrocarbons also was observed.

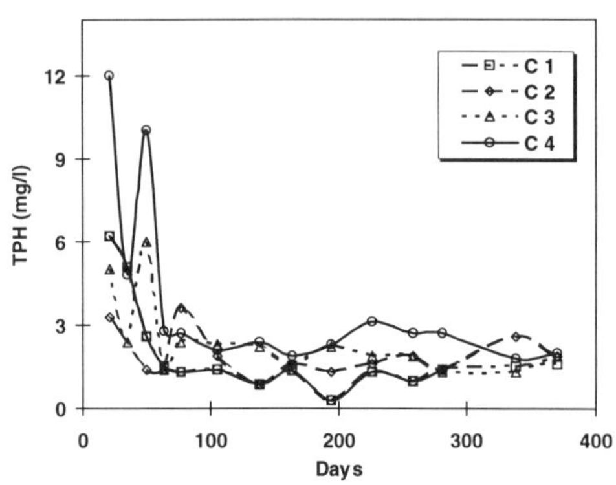

Fig. 9. TPH in run-off water

Results from TOC analysis of the run-off water yielded higher values for column 4 compared to the other columns. The TOC value reflects the content of all organic material in the water. TOC in the run-off water consists of approximately up to 10 % oil components. The organic content in the glaciofluvial soil is low (0,2 %) and is believed to give only a small contribution to TOC. It can therefore be assumed that the TOC value largely indicates the content of biomass and microbial products in the run-off water. The large TOC values in column 4 may therefore be a sign of high biological activity (fig. 10).

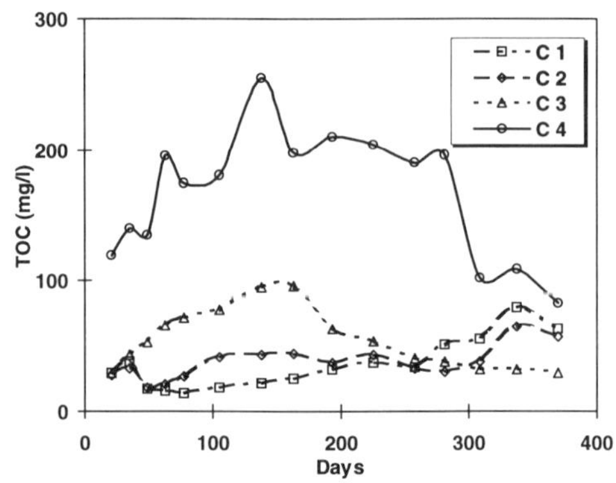

Fig. 10. TOC in run-off water

Organic content in the soil

Analyses of TOC reflect the low organic matter content of the soil which is typical of glaciofluvial nutrient poor soils. The difference between TC (fig. 11) and TOC (fig. 12) is relatively small thus indicating generally low carbonate levels.

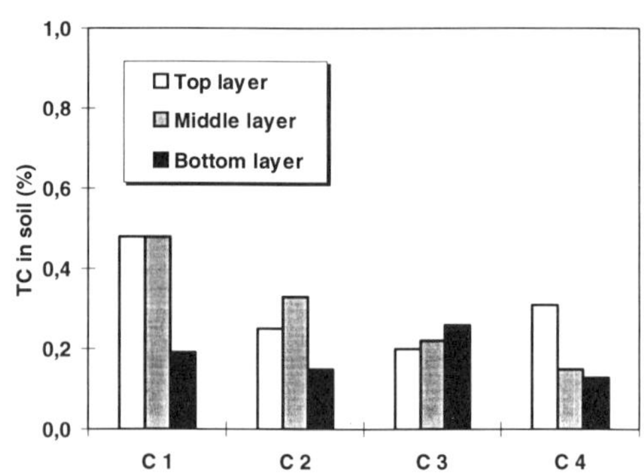

Fig. 11. TC in soil

The slightly larger difference between TC and TOC in column 1 may be a consequence of carbonate precipitation as a result of high CO_2 values in the non ventilated column (Eyk and Vreeken, 1988).

The TOC values do not indicate any build up of biomass in the soil of the columns with nutrient addition.

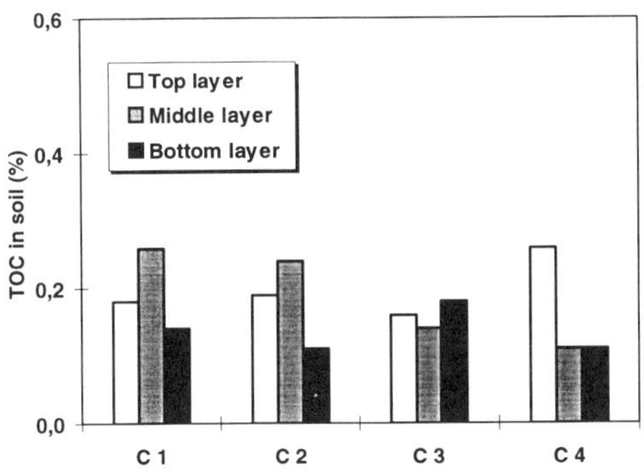

Fig. 12. TOC in soil

The TPH values confirm the observation of CO_2 production in column 3 and 4. In column 3 the TPH values increased towards the bottom of the column, indicating a higher degradation efficiency in the top soil. In column 4 similar TPH values are observed throughout the whole soil profile indicating uniform degradation rates (fig. 13).

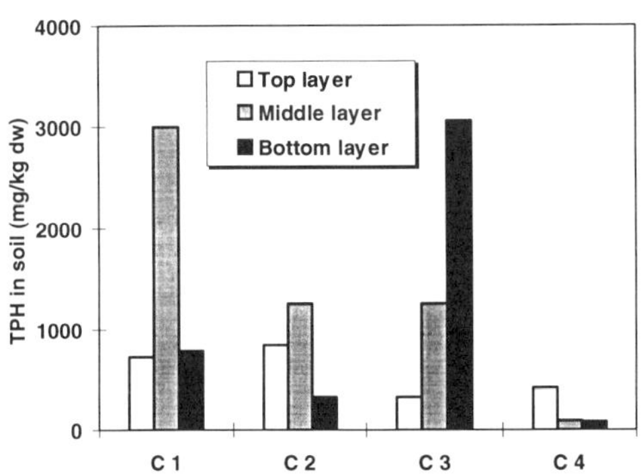

Fig. 13. TPH in soil

Based on the measurements of soil, water and air, a mass balance for the hydrocarbon removal can be made. For column 4 the mass balance indicates dominant removal of hydrocarbons by biodegradation. In this column the degradation rate during the first year of remediation was 4.5 mg oil / kg of soil / day. This is approximately 1/3 of the degradation rate found for a soil polluted by fresh fuel oil (Breedveld et al, 1994) The transport of hydrocarbons to the environment was dominated by leakage of free phase oil (table 1). The contaminant in the soil (weathered diesel) yielded low concentration of hydrocarbons in the vented gas.

Table 1. Mass balance for hydrocarbon removal in column 4.

Source	Removal (%)
Biodegradation as produced CO_2	87
Hydrocarbons in vented gas	1
Free phase oil	9
Soluble hydrocarbons in water	0.1
Biomass and organics in water	3

4. CONCLUSIONS

In situ bioventing has proven to be an efficient method to remediate oil pollution in unsaturated glaciofluvial soil under Scandinavian climatic conditions.

In the pilot scale experiments the highest degradation efficiency was observed when aeration was combined with circulation of nitrate and metaphosphate. The removal rate was significantly higher compared to the use of ammonium as a nitrogen source. This is a consequence of the low mobility of ammonium. It was found that addition of nutrients was necessary to achieve degradation in the nutrient poor glaciofluvial soil.

The experiment confirmed that biodegradation was the most dominant way that the contaminant (weathered diesel oil) was removed from the soil.

5. ACKNOWLEDGEMENT

The research projects discussed in this article were financed by the Norwegian Research Council, the State Pollution Control Authorities (SFT), the Norwegian Defence Service (NODCS), Noteby and Sintef SI. The author would like to thank all personnel involved in the research projects.

REFERENCES

Atlas, R. M. (1984)
Petroleum microbiology
Macmillan Publishing Company, New York. 685 p.

Breedveld G. D., Olstad G., Briseid T., and Hauge A. (1995)
Nutrient demand in bioventing of fuel oil pollution
In: Hinchee, Miller and Johnsson (eds.), In Situ Aeration: Air Sparging, Bioventing, and Related Remediation Processes. Batelle press, Columbus Richland. pp.391-400.

Eyk, J. and C. Vreeken (1988)
Venting-mediated removal of hydrocarbons from subsurface soilstrata as a result of stimulated evaporation and enhanced biodegradation.
Med. Fac. Landbouww. Rijksuniv. Gent, vol. 53, no. 4b. pp. 1873-1884.

Follett, R.F. (ed.) (1989)
Nitrogen management and groundwater protection.
Developments in agricultural and management forest ecology 21.
Elsevier Science Publ. Amsterdam. 393 p.

Hinchee, R.E., D.C. Downey, R.R. Dupont, P. Aggarwal and R.E. Miller (1991)
Enhancing biodegradation of petroleum hydrocarbon through soil venting.
J. of Hazardous Materials vol 28 no. 3.

Kampbell, D.H., C.J. Griffin and F.A. Blaha (1993)
Comparison of bioventing and air sparging for in situ bioremediation of fuels.
EPA report no. 600/A-93/178, U.S. Environmental Protection Agency, Washington. 8 p.

Leeson, A., R.E. Hinchee, J. Kittel, G.D. Sayles, C.M. Vogel and R.N. Miller (1992)
Optimizing bioventing in shallow vadose zones and cold climates.
NATO/CCMS, International Conference, Budapest 18-23 october 1992. 9 p

Ong, S.K., A. Leeson, R.E. Hinchee, J. Kittel, C.M. Vogel, G.D Sayles and R.N. Miller (1994)
Cold climate applications of bioventing.
In: Hinchee, Alleman, Hoeppel and Miller (eds.), Hydrocarbon Bioremediation. Lewis Publishers, Boca Raton. pp. 444-453.

Paul, E.A. and F.E. Clark (1989)
Soil microbiology and biochemistry.
Academic Press, San Diego. 273 p.

Schlegel, H.G. (1985)
Allgemeine Mikrobiologie 6.
Thieme verlag, Stuttgart. 571 p.

Sharab, N. E. and R. Bartha (1993)
Testing of Some Assumptions about biodegradability in soil as measured by carbon dioxide evolution
Applied and Environmental Microbiology, vol. 59, pp. 1201-1205.

Wood, B.D., C.K. Keller and D.L. Johnstone (1993)
In situ measurement of microbial activity and controls on microbial CO2 production in the unsaturated zone.
Water Resources Research, vol. 29, no. 3, pp. 647-659.

Environmental Geotechnics, Kamon (ed.) © 1996 Balkema, Rotterdam. ISBN 90 5410 848 7

Electrokinetic soil decontamination – Effect of local resistivity variations

Douglas I. Stewart
Department of Civil Engineering, University of Leeds, UK

L. Jared West
Department of Earth Sciences, University of Leeds, UK

ABSTRACT: The electrokinetic decontamination of reconstituted kaolinite contaminated with lead nitrate has been investigated in laboratory column experiments. During these experiments the voltage was measured at points along the specimen. The reported experiments show that both the degree of decontamination, and the locality where it occurs depend on (i) the distribution of the contaminant between the sorbed and free aqueous phases, (ii) the relative importance of diffuse layer and the free pore fluid ions as a current carriers, and (iii) the influence of high resistivity zones.

1 INTRODUCTION

Electrokinetic processing is a technique for transporting chemical species through fine grained soils insitu. It has potential for both removal of contaminants and injection of nutrients/fixing compounds and is particularly suited to fine grained soils where hydraulic flushing/injection is inappropriate (see for example, Acar and Alshawabkeh, 1993 or Lageman, 1993). Other potential uses of electrokinetic processing include providing a barrier to contaminant leakage through landfill liners (Yeung, 1992), injection of stabilising chemical grouts, diversion of contaminant plumes (Renaud and Probstein, 1987), and dewatering and decontamination of wastewater sludges (West and Stewart, 1995a).

This paper presents data from seven electrokinetic column experiments on kaolinite contaminated with lead nitrate at levels above and below the cation exchange capacity (CEC).

2 MATERIALS

The materials used in this study were Speswhite kaolinite supplied by ECC International, analytical grade lead (II) nitrate supplied by Merck Ltd, and distilled water. Speswhite kaolinite is a weathering product of Cornish granite which is used commercially in paper making. Its properties are listed in Table 1, and the adsorption isotherm for lead (II) ions at the natural clay pH of about 4.5 is shown in Figure 1. For details of the isotherm determination see West and Stewart (1995b).

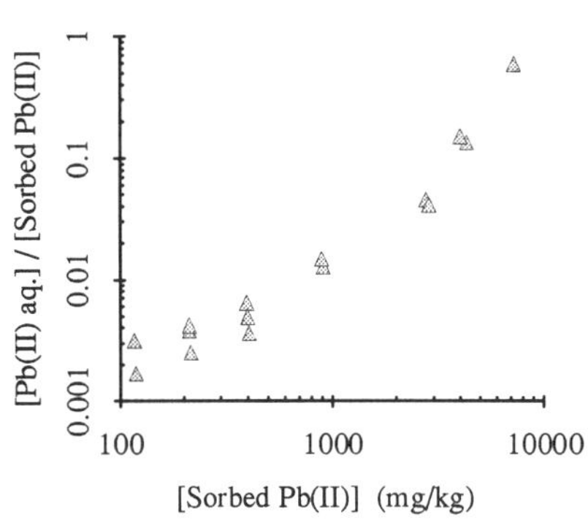

Figure 1. Adsorption Isotherm

3 COLUMN EXPERIMENTS

The kaolinite powder was mixed at 100% moisture content with solutions of 1000 ppm and 10,000 ppm lead (II) nitrate. The slurry was vacuum-de-aired for approximately one hour, and then consolidated directly into a 200mm long, 90mm internal diameter acrylic tube using an extension tube. The final consolidation stress was 19.4kN/m^2, which resulted in a moisture contents of about 67%. Sufficient clay was used to ensure a consolidated specimen length of approximately 240mm. After consolidation, the column extension was removed and the specimens trimmed, and a sample was retained for analysis. Acrylic end-caps containing graphite power supply electrodes and porous stones were placed on each end of the tube (see Figure 2). A gap of 3mm between the porous stones and the power electrodes allowed fluid to be circulated between the end caps and external reservoirs by peristaltic pumps. This maintained electrical contact between the specimens and the power electrodes and prevented drying of the specimens. Flushing of the electrode surfaces also reduces power electrode polarisation and removes gas generated by electrolysis reactions. Specimen details are given in Table 2.

During testing a constant voltage of 7.5V was applied between the power electrodes, and reservoir pH, anode reservoir mass, electric current, and the voltages at each of seven monitoring electrodes along the front of the column were recorded at regular intervals[1]. Test durations varied from 1 to 8 weeks and are detailed in Table 2. After testing, the clay specimens were extruded and sliced into sections, and each slice was homogenised. The pH of each slice was measured by inserting a Reagacon

Table 1. Properties of Speswhite Kaolinite

Characteristic	Data
Mineralogical Composition⊕	
Kaolinite	94%
Mica	4%
Montmorillonite	1%
Feldspar and quartz	1%
Specific Gravity	2.6
Surface area m^2/g	14.0
Percent of particles <2μm+	80%
Index properties+	
Plastic limit	38%
Liquid limit	80%
CEC meq/100g at pH7	4 - 5
Hydraulic conductivity (e = 1.6), m/s	2- $4x10^{-9}$
Initial clay pH (glass spear electrode)	4.1
pH of supernatant*	4.5

⊕ McGuffog, ECC International Ltd. (pers. comm.)
+ Al-Tabbaa (1987)
* from washing 2.5g clay with 25ml water

[1] The full array of seven electrodes were only present in test H4 onwards

Table 2. Column experiments

Test	Duration days	Lead content of clay ppm	Initial Resist. Ω.m	K_e cm^2/Vs	Total Energy kJ
H2	8	886	12.9	$2.8x10^{-5}$	33
H4	14	865	13.9	$2.1x10^{-5}$	40
H1	15	825	13.0	-*	42
H9	15	727	12.1	-*	47
H3	54	921	13.3	-*	70
H5	13	7650	2.7	$1.2x10^{-5}$	65
H10	14	7775	2.1	$1.4x10^{-5}$	51

* accurate values were not determined due to water leakage from the reservoir.

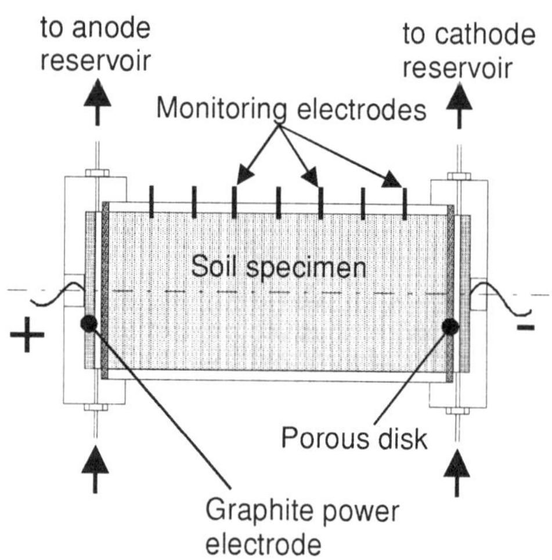

Figure 2. Schematic diagram of the column apparatus (monitoring electrodes are shown rotated from the horizontal into the vertical plane for clarity).

GCPFC11 glass pH combination penetration electrode directly into the clay, and moisture contents were determined by oven drying at 105°C. The oven dried samples were analysed for lead content using X-ray Fluoresence Spectroscopy (this technique gives total levels of an element, but does not give chemical speciation). Most specimens were sliced into 20mm thick sections, but two repeat test specimens (H9 and H10) were divided into six 33mm thick slices. These thicker slices provided less resolution in the lead and pH profiles, but yielded sufficient clay for a pore fluid sample to be extracted by centrifuging at an acceleration equivalent to 5,000 times earth's gravity for 15 minutes. The pore fluid pH was measured using the pH electrode mentioned above, and the pore fluid conductivity was measured using a Jenway PCM121 conductivity electrode.

4 RESULTS

Pore water flow was from the anode towards the cathode, and was significantly greater in the less contaminated specimens. This is characteristic behaviour for Speswhite Kaolinite contaminated with lead to levels below and above the CEC, and results from the effect of lead ions on the zeta potential of the clay (West and Stewart, 1995b).

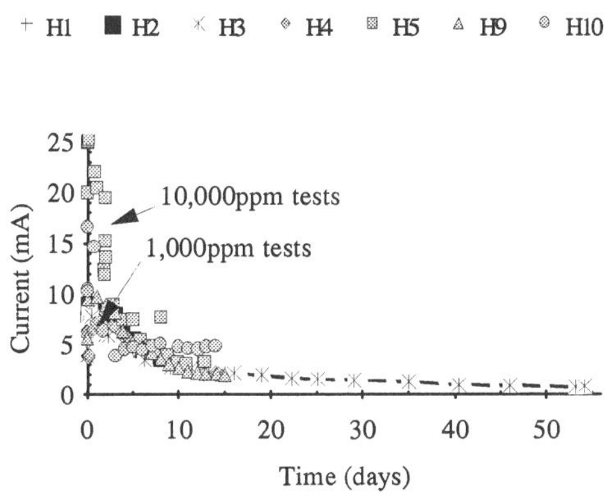

Figure 3. Electrical current data.

Figure 3 shows the electrical current during testing. In all tests, the electrical current increased over the first hour as the conductivity of the electrode reservoirs increased, and then decreased with increasing test duration. Current in the 1000ppm specimens showed similar behaviour in all the tests, whereas that in the 10,000ppm tests was more susceptible to small variations in test conditions (such as interruptions in the applied voltage for the purpose of geophysical data collection not reported here), although the current was usually greater than that in 1000ppm specimens for equivalent test duration.

Figures 4 (a), (b) and (c) show the voltage profiles in the column experiments after 2, 5 and 14 days, respectively (the voltage gradients were initially uniform). In the 10,000ppm experiments there was localisation in the voltage profile after two days, with most of the applied voltage dropped across the last 25mm of the specimen near the cathode (25mm is the distance between the last monitoring electrode and the end of the specimen). This voltage drop endured to the end of the tests. In the 1,000ppm experiments there was little variation in the voltage gradient along the specimens after 2 days. Later, the two 10,000ppm specimens behaved slightly differently, with varying amounts of localisation in the cathode half of the specimens. The zone of localisation in the 1,000ppm specimens was at least twice as thick as in the 10,000ppm specimens.

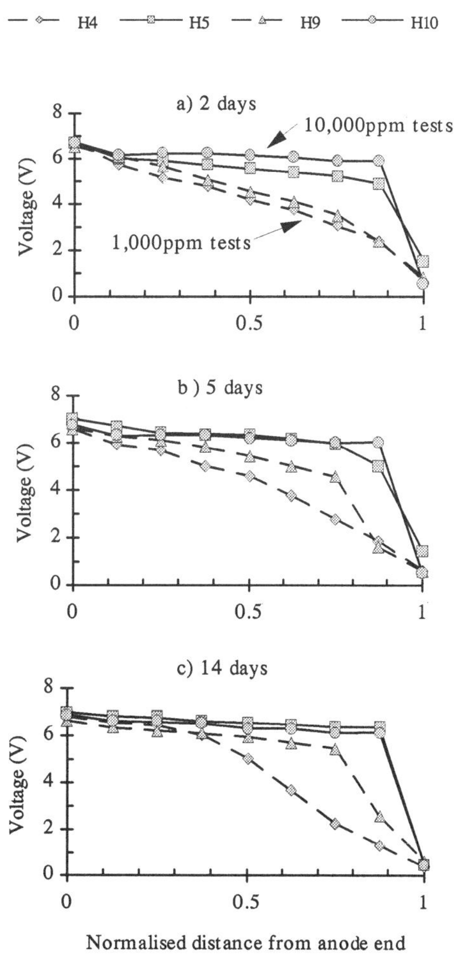

Figure 4. Voltage profiles after 2, 5 and 14 days.

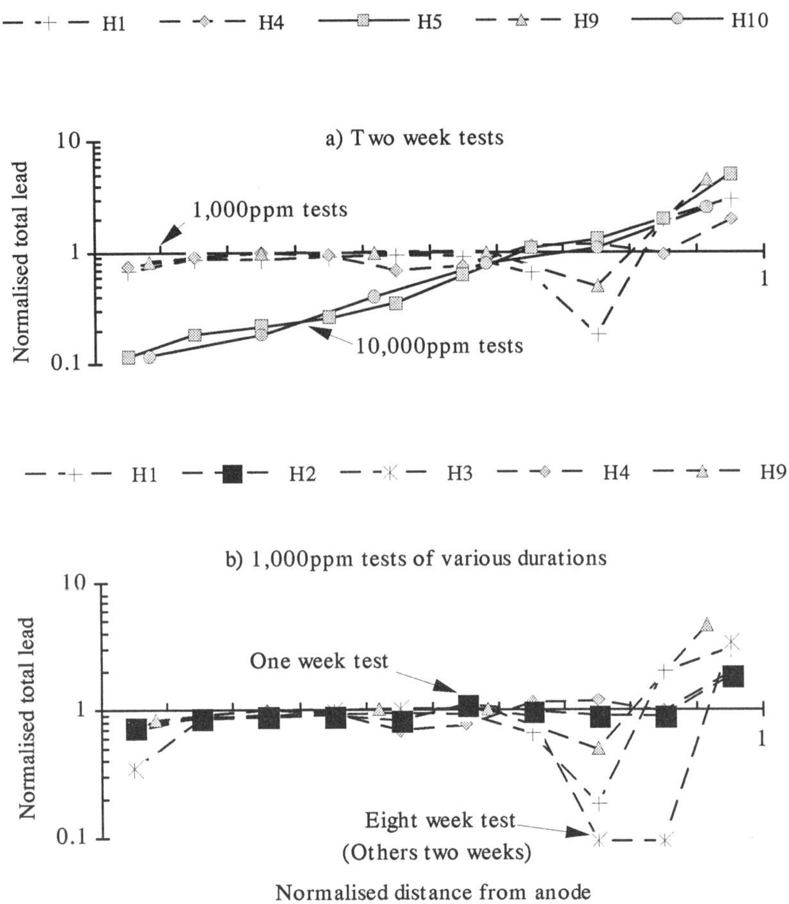

Figure 5. Normalised lead profiles a) two week tests b) 1,000ppm tests

Post-test observation of the 10,000ppm specimens revealed an approximately 5mm thick zone immediately adjacent to the cathode, where the soil was stiff, friable and discoloured with occasional black spots (with the exception of this last 5mm, the clay remained plastic). Post-test observation of the 1000ppm specimens indicated that they became slightly firmer from anode to cathode with the exception of the last 10mm adjacent to the cathode end, which was very soft.

The initial pH of the clay was 4.1 and 3.3 in the 1,000ppm and 10,000ppm specimens respectively. As a result of transport of electrolysis products, the pH of all specimens increased near the cathode and decreased over the remainder of the specimens.

Figure 5a shows the normalised lead profiles for the specimens tested for 14 days. Significant lead removal occurred over two thirds of the 10,000ppm specimens (greatest near the anode), whereas there was little depletion over the anode half the 1000ppm specimens, with lead depletion occurring instead about half to two thirds the way along the column. Lead accumulated adjacent to the cathode in all specimens, although the absolute lead levels in this region were much higher in the 10,000ppm specimens.

Figure 5b shows the normalised lead profiles for the 1000ppm specimens tested for one, two and eight weeks. There was little decontamination over the bulk of the specimen even after eight weeks of current application, although a distinct trough in the

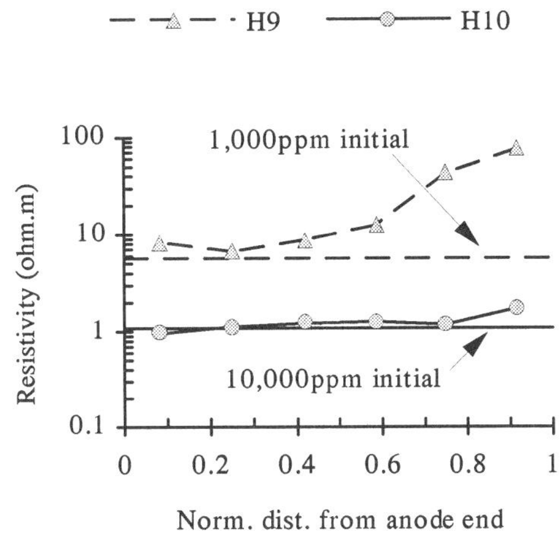

Figure 6. Pore fluid resistivity

profile had formed. This began as a slight depression near the middle after one week of testing, and progressively migrated towards the cathode and deepened. The three specimens tested for two weeks differed in the size and the position of this depleted zone, which indicates that decontamination behaviour is sensitive to minor variations in the test conditions.

Figure 6 shows the pore fluid resistivity in tests H9 (1,000ppm) and H10 (10,000ppm). The initial pore fluid resistivity of the 1,000ppm specimen was about five times larger than the 10,000ppm specimen. The 1,000ppm specimen showed a marked pore fluid resistivity increase along the entire specimen but particularly in the cathode third. This corresponds with the steep portion of the voltage profile and the trough in the lead profile. The 10,000ppm specimen showed a slight pore fluid resistivity decrease near the anode, with a slight increase over the bulk of the specimen, largest in the cathode slice. It should be noted that the value given for the cathode slice is a bulk average value for the pore fluid in the last sixth of the specimen (33.3mm) and is not representative of the pore fluid in the final 5mm of the soil, where the soil was noticeably stiff (this narrow region is believed to be responsible for most of the resistance increase).

5 Discussion

The differences in lead transport behaviour between the 1000 and 10,000ppm specimens is related to the distribution of lead ions between the free aqueous phase and the electrical double layer. In the 10,000ppm tests, a large proportion of the lead was in the free aqueous phase, and migrated across the specimens on application of the electrical field until precipitated adjacent to the cathode. In the 1000ppm tests, over 95% of the lead was initially in the electrical double layer (see Figure 1). It is proposed that lead transport within the double layer was significant in these specimens in regions where the pore fluid was resistive. This resulted in depletion of double layer lead in a region near the cathode (see Figure 5b).

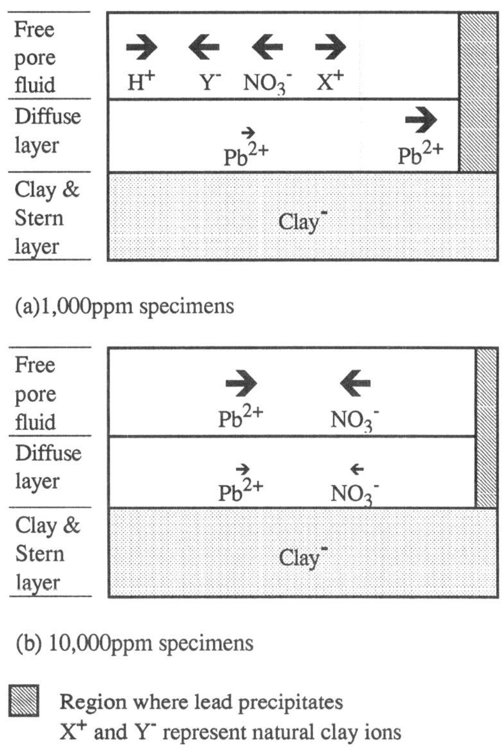

Figure 7. Idealised distribution of major ions after about two weeks of voltage application

The specimen resistance increased during the tests (Figure 3). The voltage profiles indicate that development of local zones within each specimen were the cause. In the 1000ppm tests, the voltage gradient was steepened across a broad region in the cathode half of the specimen. The pore fluid resistivity also showed a marked increase in this region (see Figure 6), indicating ion depletion. Ion depletion in this region probably resulted from neutralisation/precipitation reactions involving ions derived from the clay. In contrast, increased resistivity in the last 5mm of the 10,000ppm specimen probably resulted from precipitation of lead[2].

The schematic diagrams in Figure 7a and 7b show the proposed distribution of the major ions present and their relative velocities in the 1,000 and 10,000ppm specimens respectively (minor current carriers are omitted for clarity). In Figure 7a, fastest transport of lead occurs in the region where the pore fluid is most resistive, because more current is forced through the double layer. In figure 7b, double layer lead transport is not significant because the region of high pore fluid resistivity is restricted to very near the cathode, where the pH is too high for lead transport.

In summary, electromigration within the free pore fluid is the major current carrying mechanism at the start of all tests. In the 10,000ppm specimen, lead and nitrate ions were the dominant species and significant lead transport occurred. In the 1,000ppm specimens, the natural clay cations and nitrate anions were the dominant species in the pore fluid, as most of the lead was within the double layer (see Figure 1). Therefore, little lead transport occurred, until pore fluid ions became depleted.

6 Conclusions

- Localisation of the voltage gradient can occur during electrokinetic decontamination of specimens contaminated both above and below the exchange capacity.
- When the cathode reservoir is not buffered, precipitation of hydroxides will increase resistivity adjacent to the cathode, and most of the electrical energy will be dissipated within a region where there is no mobile contaminant.
- When the contaminant is in the electrical double layer, decontamination will only occur when the pore fluid becomes resistive.
- Lead depletion patterns reflect local pH and the importance of double layer ions as charge carriers.

[2] The pH in the cathode reservoir in all test was 11-12, which is well above the precipitation pH range for lead hydroxides of pH 5-8 (Pourbaix, 1966).

References

Acar Y.B. and Alshawabkeh A.N. (1993). Principles of electrokinetic remediation. Environ. Sci. Technol., Vol. 27, 2638-2647.

Al-Tabbaa A. (1987). Permeability and stress strain response of Speswhite Kaolinite. Ph.D. thesis, University of Cambridge.

Lageman R. (1993). Electroremediation: applications in the Netherlands. Environ. Sci. Technol. Vol. 27, 2648-2650.

Porbaix, M. (1966). Atlas of electrochemical equilibria in aqueous solutions. Pergamon Press, London. 644p.

Renaud, P.C. and Probstein, R.F (1987). Electro-osmotic control of hazardous waste. Physico-chemical Hydrodynamics, Vol. 9, 345-360.

West, L.J. and Stewart, D.I. (1995a). Electrokinetic processing of sewage sludge contaminated with phytotoxic metals. Symposium on Natural hazardous and Environmental Geotechniques, Bangkok, 27-36.

West L. J. and Stewart D. I. (1995b) Electrokinetic decontamination: the effect of zeta potential. Geoenvironment 2000. American Society of Civil Engineers Special Publication No. 46, 1535 - 1549.

Yeung, A.T. (1992). Environmental engineering applications of electrokinetic flow processes: state of the art. Hong Kong Engineer, November 1992, 25-30.

Environmental Geotechnics, Kamon (ed.) © 1996 Balkema, Rotterdam. ISBN 90 5410 848 7

Treatment of oil contaminated soil with geo-cement

Noriyasu Suzuki, Toshiyoshi Goto & Takuya Ogino
Chuken Consultant Co. Ltd, Osaka, Japan

Tomohiko Chaen
Estec Co. Ltd, Osaka, Japan

ABSTRACT: This paper discusses the use of geo-cement to stabilize and detoxify oil contaminated soil at an industrial plant.

For site characterization, test-pit excavation and power auger boring was carried out to estimate the extent and volume of the contaminated soil. To evaluate the level of contamination, oil content tests, carbon tetrachloride extracts tests and elution tests of hazardous materials(11 items based on land-filling regulations) were performed on samples obtained.

As the content of oil was found to be about two times greater than the criterion and as utilization of the treated ground was expected , addition of pure soil and stabilization with geo-cement were considered. After several treatments were tested in the laboratory to determine the most appropriate method the following procedure was selected:

1. Contaminated soil is mixed with pure soil in the ratio of 67:33.
2. 300kg of geo-cement is added to 1 m^3 of soil in the slurry state(w/c=80%).

To mix the geo-cement slurry with soil evenly, a special attachment for the backhoe was prepared. After the treatment: unconfined compressive tests, oil content tests and extracts tests were carried out every 75－266 m^3 of stabilized soil. The results were within the criteria and the treated area is now used as a driving range for the employees of the factory.

1. INTRODUCTION

In an industrial plant, a large volume of sludge which contained high concentrations of oil, was placed in a temporary storage yard for evaporation.

To leave it in that state might have caused the contamination to disperse into the surrounding soil and water system. It was also undesirable from the view point of landscaping and utilization of the ground.

At this site, treatment with geo-cement was employed for stabilization of the oil contaminated soil and also to increase the baring capacity of the ground for utilization. The flow of the whole treatment work is shown in Fig.1.

In this paper, site characterization, laboratory tests, execution work and quality control tests are described.

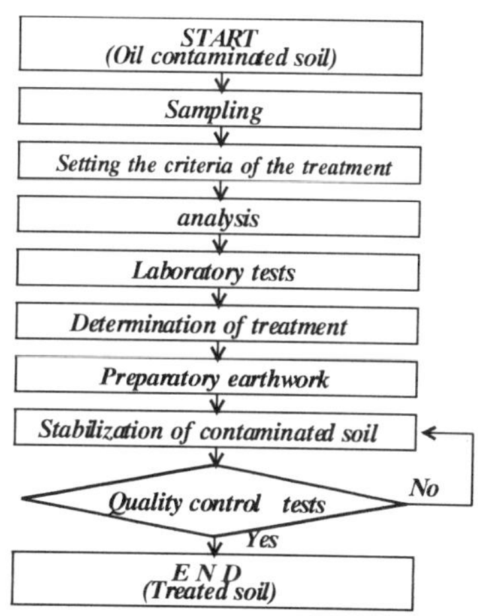

Fig.1 The flow chart of the treatment procedure

2. SITE CHARACTERIZATION

2.1.SAMPLING

In order to evaluate the oil contaminated area, the depth of contamination and the oil content, power auger boring and backhoe excavation were carried out at 16 points and 4 points respectively. Fig.2 shows the tested points and a cross section of the site. Soil samples were divided into two: the contaminated and uncontaminated soil by observation and odor. Four contaminated soil samples and one uncontaminated soil sample underwent laboratory tests.

2.2.CRITERIA

To evaluate the level of contamination, criterion values based on the law needed to be determined. For this site, the "Waste Disposal and Public Cleansing Law" was used as reference. For the burying treatment, the criterion values were as follows:

(1) Criteria for oil component
  1) oil content(n-hexane extracts): less than 5%
  2) carbon tetrachloride extracts: less than 100 ml/1 liter of sample solution
(2) Criteria for hazardous materials (metals etc.)

The criterion values are shown in Table 1.

Table 1. Criteria of hazardous materials for landfill disposal

hazardous materials.	criteria (elution test) (mg in 1litter of sample solution)
alkyl mercury compound	ND
mercury or its compounds	less than 0.005
cadmium or its compounds	less than 0.3
lead or its compound	less than 3
organophoshorus compounds	less than 1
chromium(VI) compound	less than 1.5
arsenic or its compounds	less than 1.5
cyanide	less than 1
PCB	less than 0.003
trichloroethylene	less than 0.3
tetrachloroethylene	less than 0.1

Table 2 Physical properties of soil samples

Sampling point	wet density (g/cm^3)	water content (%)	ignition loss (%)	classification of soil*
A2	1.581	97.6	10.8	contaminated
C2	1.468	104.8	12.3	contaminated
C4	1.599	92.8	+3.7	contaminated
C12	1.318	131.5	24.1	contaminated
A1	1.528	81.6	-	uncontaminated
pure soil	1.510	77.2	-	-

*by observation(view and odor)

2.3.TEST AND ANALYSIS

The results of the tests regarding the physical properties of the soil are shown in Table 2.

Soil samples were loam and those classified as contaminated by observation have a high water content and are very soft.

The soil samples were also analyzed to

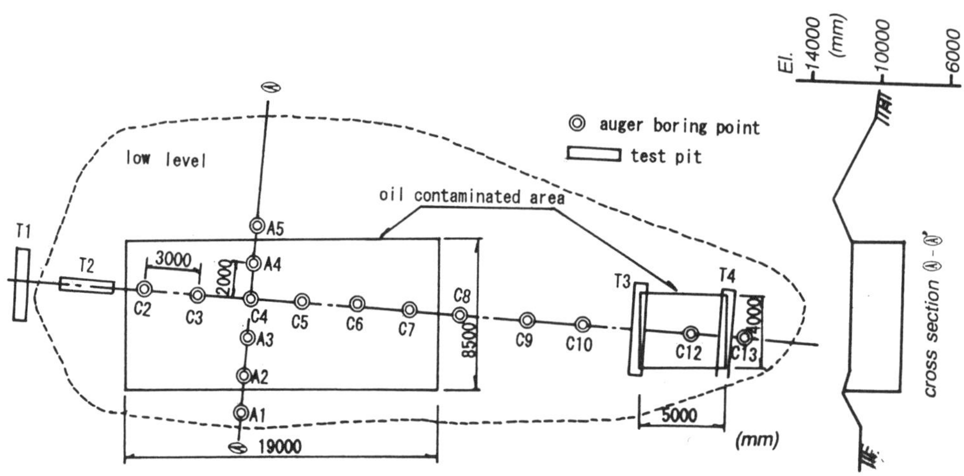

Fig.2 The site and sampling point

obtain the oil content and extracts by Sediment Monitoring Methods(n-hexane extracts test) and the methods specified by the Environment Agency(carbon tetrachloride extracts test). The results are shown in Table 3 .

Table 3. Results of chemical analysis

soil samples	n-hexane extracts (%)	carbon tetrachloride extracts (mg/l)
A2	5.6	19
C2	5.5	58
C4	4.3	20
C12	14.0	62
A1	2.3	24

(shaded) :Value exceeded the criterion

The soil samples(A2,C2,C12) classified as contaminated by observation exceed the criterion. Though C4 dose not exceeded the criterion, the value was two times larger than that of uncontaminated and nearly equal to the criterion value. Consequently, all the four samples classified as contaminated by observation decided to be treated.

By the sampling and the tests, two contaminated areas and the depth were fixed. The total amount of contaminated soil was estimated as 700m^3 approximately.

3. LABORATORY TESTS

To determine which treatment satisfied the criteria, laboratory tests were carried out. Unconfined compressive strength criterion of 220 kPa was introduced to assure the bearing capacity of 98 kPa for future utilization of the treated ground.

3.1.TREATMENT AND METHOD

Pure soil mixing ratio and amount of added stabilizing agent were selected as the test parameter. The treatments of nine cases are shown in Table 4.

As a stabilizing agent, geo-cement was selected, which was developed for soil stabilization. The chemical components of geo-cement are shown in Table 5 together with those of ordinary portland cement.

Table 4. Conditions of laboratory tests

Geo-Cement (kg/m^3)	200,250,300
water-cement ratio (%)	80
mixing ratio of soil (m^3) A : oil contaminated soil B : pure soil	[1]: A : B = 1 : 0 [2]: A : B = 0.67 : 0.33 [3]: A : B = 0.50 : 0.50

Table 5 Chemical components of cements

type	ig.loss	SiO_2	CaO	Al_2O_3	SO_3
GEC	0.7	21.7	59.0	6.3	6.1
OPC	1.0	21.6	64.4	5.5	1.8

Geo-cement was used in the slurry state to avoid the powder getting into the atmosphere. The contaminated soil, pure soil and geo-cement slurry were mixed in the soil mixer. Part of the mixtures were stored in plastic envelopes for analysis and were placed in molds(ϕ 5 × 10cm) for compressive tests.

3.2.OIL CONTENTS AND EXTRACTS

At 28 days, oil contents and extracts were analyzed and unconfined compressive tests were carried out. Treated soil was crushed and analyzed. Results are shown in Table 6. As expected, oil contents decreased as the mixing ratio of pure soil and amount of added geo-cement were increased.

3.3.SPECIMEN IMMERSION TEST

To estimate the influence of the treated soil on the permeated water or ground water, an immersion test of a treated soil specimen was carried out. The specimen (ϕ 5× 10 cm; 270g) was immersed in pure water of about 2700g. After 7days and 28days, n-hexane extracts of the immersed water was analyzed according to the method set by the Environmental Agency and JIS K 0102 (industrial effluent testing method).

Results are shown in Table 7.

The n-hexane extracts of the water was below the detection limit at each immersion period and below the criteria of industrial effluent(5.0mg/l). From this result it is expected that the dispersion of oil contamination has been prevented.

Table 6. Properties of treated soils

case No.	soil No.	Geo-Cement (kg/m^3)	unconfined compressive strength (kPa)	oil contents (%)	oil extracts (mg/l)
I	[1]	200	77.5	8.4	58
II		250	204	7.9	52
III		300	482	7.5	62
IV	[2]	200	137	5.8	17
V		250	284	5.2	19
VI		300	537	4.6	22
VII	[3]	200	208	3.8	13
VIII		250	423	3.7	11
IX		300	792	3.1	18

(shaded) :Value unsatisfied the criteria.

Table 7 Results of immersion test

immersed period	n-hexane extracts (mg/l)
7days	below detection limit(0.5)
28days	below detection limit(0.5)

3.4.COMPRESSIVE STRENGTH

Fig.3 shows the relationship between the amount of geo-cement and unconfined compressive strengths.

The strength of the treated soil becomes high as the amount of geo-cement and mixing ratio of pure soil increases.

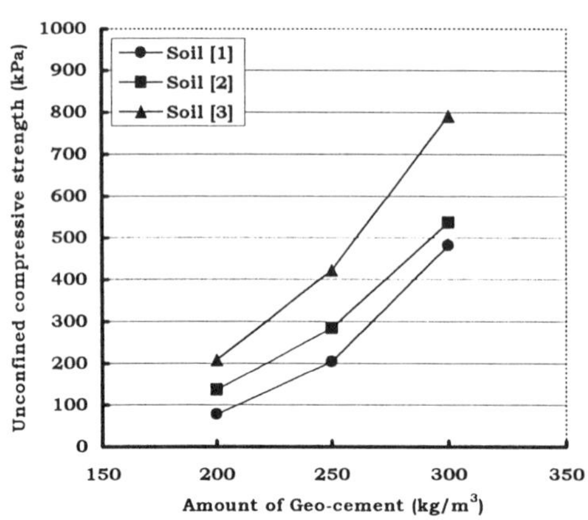

Fig.3 Relationship between amount of geo-cement and unconfined compressive strength

3.5.DETERMINATION OF TREATMENT

As shown in Table 6, we found that three cases satisfied every criteria. Among these, the treatment VI was chosen with due consideration given to the needed amount of pure soil and geo-cement. Treatment VI is summarized as follows.

1. Contaminated soil is mixed with pure soil in the ratio of 67:33.
2. w/c of the slurry is 80%.
3. 300kg of geo-cement is added to 1 m^3 of mixed soil in the slurry state.

Treated soil underwent the elution tests for hazardous material. No sample exceeded the criteria.

4. EXCUTION WORK

4.1.PREPARATORY EARTH WORK

Prior to the main treatment, soil mixing work was executed. At first, contaminated soil was completely removed from two sites and placed temporarily near the experiment site. Then a pit was excavated as shown in Fig.4. The contaminated soil and pure soil were mixed in the ratio of 2:1 and stored in the pit.

4.2.TREATMENT FOR STABILIZATION

A 0.7 m^3 bucket class backhoe was used for stabilization. The backhoe was equipped with a special attachment for mixing instead of the usual bucket(see Fig.5).

Fig.6 shows the equipment used. Geo-cement slurry made in the mixing plant is pumped to the backhoe through the flexible hose and is sprinkled on to the soil mixture from the nozzles of the attachment. The flow rate of the slurry was 190 l/min and was controlled using a flow meter. The soil and

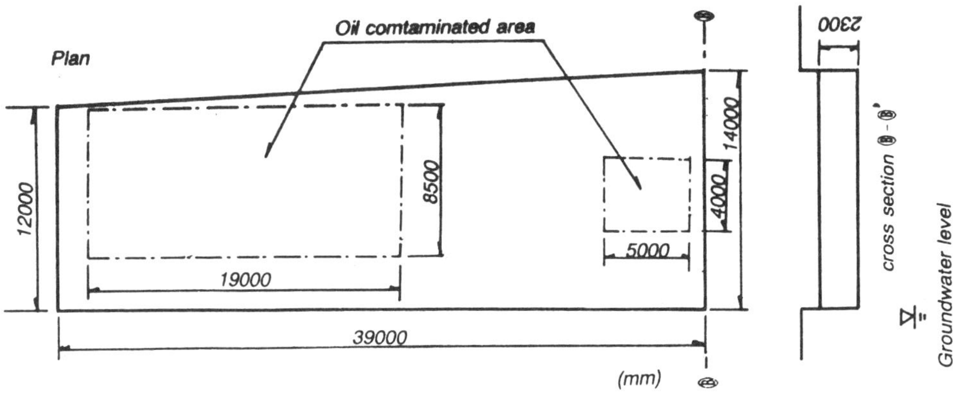

Fig.4 The shape of a pit for the treatment

the slurry were mixed until the color of the soil became uniform. 75-266 m^3 of soil was treated in one day. Treatment of 1000 m^3 of soil mixture was completed in six days. Treated and hardened soil was covered with pure soil and a lawn was planted to prevent flashing by rain. An imaginary completed view of the site is shown in Fig.7.

4.3.QUALITY CONTROL

Treated soil was quality sampled once a day. Oil contents and extracts and unconfined compressive strength were measured. The results are shown in Table 8. All values satisfied the criteria.

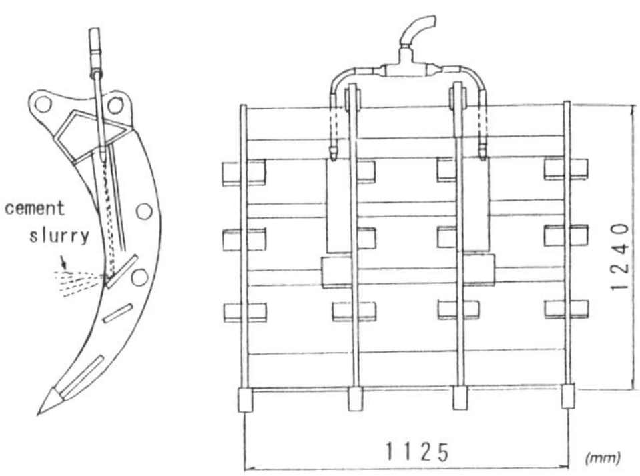

Fig.5 Special attachment for mixing work

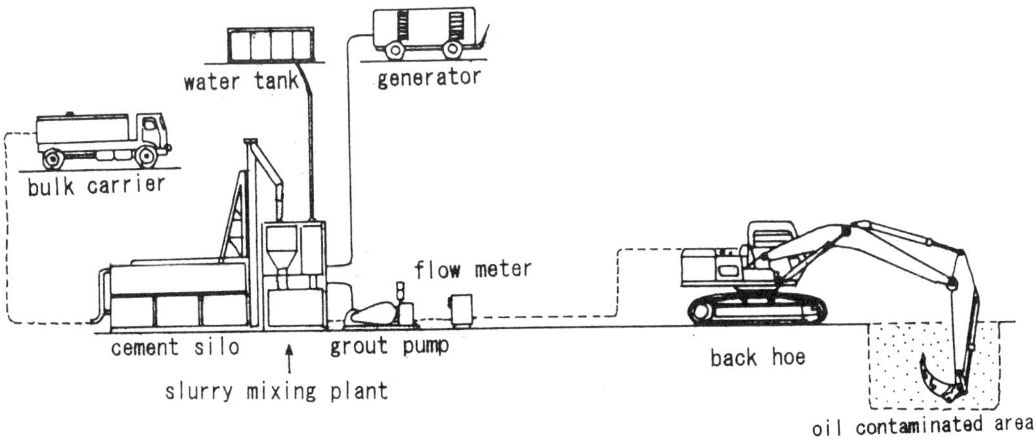

Fig.6 The execution system

Table 8 Results of quality control tests

oil contents (%)	oil extracts (mg/l)	unconfined compressive strength (kPa)
0.9	8	796
1.5	13	921
1.9	12	923
1.2	12	1120
1.3	16	1430
0.7	5	1230

4.4.ASSIGNMENT

Some problems were noted concerning methods of analysis and quality control testing.

1) Test methods for burying treatment in regard to Waste Disposal and Public Cleansing Law are severe, however extracts from the hardened specimen was below the detection limit.
2) There is no guideline for the frequency of sampling and the probability of criteria clearance.
3) It took 3 weeks to obtain the test results.

Establishment and the development of suitable tests for in situ stabilization and a quick test method and newly devised measuring instruments are expected.

5. CONCLUSION

The treated ground is now effectively used as a driving range for the employees of the factory. Treatment of the oil contaminated soil with geo-cement slurry contributed to the utilization of the contaminated soil without resorting to its disposal as waste. The method of treatment of contaminated soil presented here seems to be applicable to usual sites.

Recently, the numbers of waste treatment areas have reduced on the other hand, the regulations governing the waste treatment process have become more severe. The authors hope this method will contribute to waste treatment.

ACKNOWLEDGMENT

The authors would like to express their appreciation to the many people who gave their advice and support. Grateful thanks are also expressed to Dr. Shidawara for the support we received when writing this paper.

REFERENCES

[1] M. Okumura(1996) Immobilization of Hazardous Substances with Stabilizing Admixtures, Second International Congress on Environmental Geotechnics .

Fig.7 Imaginary completed view of the site

Environmental Geotechnics, Kamon (ed.) © 1996 Balkema, Rotterdam. ISBN 90 5410 848 7

Solubilization of TNT in surfactant solutions

M. Raihan Taha
Universiti Kebangsaan Malaysia, Malaysia
I. Henry Soewarto, Yalcin B. Acar & Robert J. Gale
Louisiana State University, Baton Rouge, La., USA
Mark E. Zappi
US Army Engineer Waterways Experiment Station, Vicksburg, Miss., USA

ABSTRACT: Three different types of surfactants were investigated to solubilize TNT in order to assist its extraction from soils. A soil from a site contaminated with TNT was used in this study. The soil was shaken with the surfactants with concentration ranging between 0.2 to 10%. TNT analyses were conducted using High Performance Liquid Chromatography (HPLC). The soil had TNT concentration of 16,500 mg/kg. Batch solubilization test results indicated that the two anionic surfactants produced solubilization higher than the aqueous control solution (water), at surfactant concentrations ranging up to 10%. The nonionic and the cationic surfactants did not improve solubilization above that achieved with water at concentrations 1% and below. The negatively charged soil surface may have competed for the nonionic and cationic surfactant molecules thus reducing the effectiveness of the surfactants to enhance TNT solubilization.

1 INTRODUCTION

Trinitrotoluene or TNT is toxic and carcinogenic. TNT in its purest state is nonpolar, nonvolatile, has low water solubility and it is yellowish in color. TNT contamination is a consequence of uncontrolled disposal of explosive by-products from munitions factories. Various soil decontamination methods such as incineration, alkaline hydrolysis, soil washing, enhanced pumping, bioremediation and electrokinetic (EK) remediation have been and are currently investigated in the USA to remove TNT from the ground (Garg et al. 1991, and Taha et al. 1995a). Consideration of the solubility of TNT is very important in the success of decontamination methods such as soil washing, surfactant enhanced pumping and micellar EK remediation. Solubility of TNT in water ranges only from 120 to 160 mg/L (0.12 to 0.16%), however many contaminated sites have concentrations up to 20,000 mg/kg or more. Therefore, solubility enhancers (surfactants) are required to release the organics from the soil surface and to increase its availability in the solution phase for the efficiency of the removal process. This paper will present the results of the experimental work and the implications of the results on extraction and remediation of TNT from soils.

2 MATERIALS AND METHODS

Soil samples were obtained from a former military site in Nebraska, USA, where TNT contamination is widespread (Zappi et al. 1993). The basic properties of the tested soil are given in Table 1. Grain size analysis (Taha et al 1995b) indicated that more than 85 % of the soil particles were retained on the ASTM No. 200 (74 μm size opening) standard sieve. Attempts to evaluate clay size particles by hydrometer analysis were unsuccessful due to the emergence of bubbles which made hydrometer readings impossible. The soil is classified in the CL or OL region (inorganic clay, or organic silts of low plasticity) in the Unified Soil Classification System (USCS). Soil pH indicated a neutral condition. Brady (1974) reports the range of cation exchange capacity (CEC) to be between 20-26 mmol charge per 100 g of dry soil in neutral conditions. The value of 24.6 obtained in this study matched this range.

SDS and DOWFAX 8390 were selected as the anionic (negatively charged) surfactants, Tween 80 and Brij 35 were the nonionic and Cetyltrimethylammoniumbromide (CTAB) and Cetyltrimetylammoniumchloride (CTAC) were the cationic (positively charged) surfactants. Some selected properties of the surfactants are shown in Table 2.

Table 1. Basic properties of soil used in for all the surfactants used experiments

Properties	Values/Observation
Natural moisture content	16.8 %
Specific gravity, G_s	2.53
Liquid limit, LL	43.9
Plastic limit, PL	21.1
pH	7.16
CEC (mmol charge/100g of dry soil)	24.57

Table 2. Selected properties of surfactants used in this study

Name	Formula/Structure	CMC* (mg/L)
SDS	$C_{12}H_{25}O(SO_3)^{-}Na^{+}$	2100
DOWFAX 8390	$C_{16}H_{33}C_{12}H_7O(SO_3)_2^{-}Na_2^{+}$	2000
Tween 80	$C_{64}H_{124}O_{26}$	13
Brij 35	$C_{58}H_{118}O_{24}$	74
CTAB	$CH_3(CH_2)_{15}N(CH_3)^{+}Br^{-}$	361
CTAC	$CH_3(CH_2)_{15}N(CH_3)^{+}Cl^{-}$	41

*CMC is the critical micelle concentration. At the CMC, an agglomeration of surfactant molecules known as a micelle is formed. The changes in many physico-chemical characteristics of a surfactant solution, such as solubility of organic nonpolar molecules, conductivity, etc., occur at or above the CMC.

All surfactant solutions were prepared using deionized, deaired water with pH ranging between 6.5-7.5. In desorption experiments, 3 g of dry soil was first shaken with 10 mL of various concentrations (ranging up to 10%) of surfactants solutions for 18 hours. The soil-surfactant mixtures were then centrifuged for 1 hour at 15,000 rpm. The solutions were diluted subsequent to centrifuging. Initially, trends in TNT concentrations obtained from HPLC analysis were quite erratic and the results were seldom repeatable. It was hypothesized that colloid particles in the soil-surfactant mixtures interfered with the dilution and analysis even after extensive centrifuging and filtration using a 0.5 μm pore size filter. Reliable results were obtained only after filtering the solution twice before dilution. First, the solution was passed through a 25 mm diameter 2.7 μm pore size and then through a 1.0 μm pore size glass fiber filter (Whatman, England). The trends in TNT concentrations became more consistent and repeatable, showing that the procedure was reliable and that accurate TNT analyses could be carried out.

3 RESULTS AND DISCUSSION

In the first series of tests, dilute surfactant concentrations were used. The concentration of surfactants tested ranged from 0.1 to 1% (w:v). This range of concentration is at least at or above the critical micelle concentration, CMC. The CMC is the concentration at which micelles begin to form and it is dependent on the surfactant structure, composition, temperature, ionic strength, and the presence of type of organic additives in the solution (Rosen, 1978). Zappi *et al.* (1994) demonstrated that surfactant concentrations much higher than the CMC were required to have significant desorption of TNT from the contaminated soil. Edwards *et al.* (1994a) reported that surface tension reduction levels off at surfactant doses greater than that needed to attain aqueous CMC for nonionic surfactants. This means that in order to promote enhanced desorption, the concentration of surfactant required is at least the sum of CMC and the amount of surfactant expected to be sorbed on the soil surface. Therefore, for surfactant molecules that have high affinity to the soil surfaces (e.g. cationic and nonionic surfactants) the concentration required for enhanced desorption is much higher than its CMC.

The effects of surfactant concentrations (up to 1% solution) on the amount of TNT desorbed from the contaminated soil are shown in Figure 1. All surfactants were first tested up to 1% concentration, except for CTAB (up to 0.5%) because it reaches its aqueous solubility limit below 1%. Results for the desorption of TNT by water are also shown and the value (135 mg/kg) matches the reported solubility of TNT in water, which ranges from 120 mg/L to 150 mg/L depending on the temperature of the solution (Freeman and Colitti, 1981; and Yinon, 1990). This value provides the datum for all enhancement comparisons.

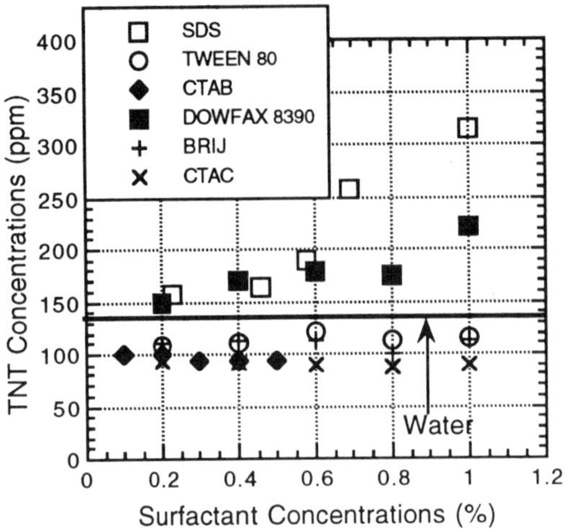

Figure 1. Extraction of TNT in contaminated soil using surfactant concentration ranging between 0-1%.

From Figure 1, both anionic surfactants showed better desorption of TNT than the aqueous control, i.e. water. SDS, however gave a higher TNT equilibrium concentration than DOWFAX 8390. Cationic and nonionic surfactants did not provide good desorption at concentrations of 1% and below. Indeed, the solubilities of TNT at these concentrations are below that obtained in water, indicating that water is a better solubilizing agent than these solutions. The inefficiency to extract TNT with the cationic and nonionic surfactants at these concentrations is possibly due to competition by the negatively charged soil surfaces for the positive and nonionic surfactant molecules. Theoretically, cationic surfactants will mostly suffer losses by adsorption from coulombic attraction (ion pairing mechanism) in which compounds of opposite (positive) charge are taken up by the negatively charged soil particles. Adsorption also takes place due to ion exchange mechanism which involves replacement of counterions adsorbed onto the soil surface by similarly charged surfactant ions (Rosen, 1978). These may be the fundamental reasons why cationic surfactants are not commonly used when soils are involved. In some cases, a substantial fraction of nonionic surfactant in a soil-aqueous phase system can sorb onto soil (Liu *et al.*, 1992). This is possible through hydrophobic interaction in which nonpolar organic compounds are attracted onto the soil surface possibly by hydrogen bonding between the soil and the surfactant (Edwards *et al.*, 1994b). Sorbed surfactant can enhance the capacity of the soil to act as a sorbent for organics. The negatively charged anionic surfactant molecules would suffer less sorption problems and their micelles could desorb the TNT molecules from the soil surfaces unimpeded. Therefore a relatively high desorption (compared to that of water) was achieved with SDS and DOWFAX 8390. Although soil minerals are usually electronegative and thus repel anions, it has been reported that soils high in amorphous iron oxide-hydroxide coatings can have a high attraction for anions (Fink *et al.*, 1970).

The results provides a preliminary assessment of the potential candidates among surfactants to enhance the desorption of TNT from the soil. However, it must be realized that the amount of TNT in the soil used is relatively high (16,500mg/kg). One percent SDS solution desorbed only around 3.2 mg of the approximately 49.5 mg TNT available in the soil sample used in all batch tests that was only 2 g. This corresponds to 6.5% of TNT being removed from the soil. Although the percentage of TNT removed from the soil surface is small, it is hypothesized that continuous supply of the surfactant over extended periods of time could result in higher removal efficiencies as shown by Zappi *et al.* (1995) for Tween 80.

Although the concentrations of all surfactants were above the CMC, there was not an improvement in TNT desorption and/or solubility using the cationic and nonionic surfactants at doses exceeding 1%. It was expected that enhanced desorption (and solubilization) would start to occur at the CMC. However, the CMC concept was developed for the case of a solute-surfactant phase system and a soil-solute-surfactant phase mixture is a more complex one. It was hypothesized that if enough surfactant molecules are made available to achieve equilibrium with the soil surfaces, the excess molecules will begin to form micelles and desorb TNT from the soil. A greater solubilization than what was observed with low solution concentrations would then be possible. Surfactant concentrations up to 10% were then evaluated. Concentrations beyond

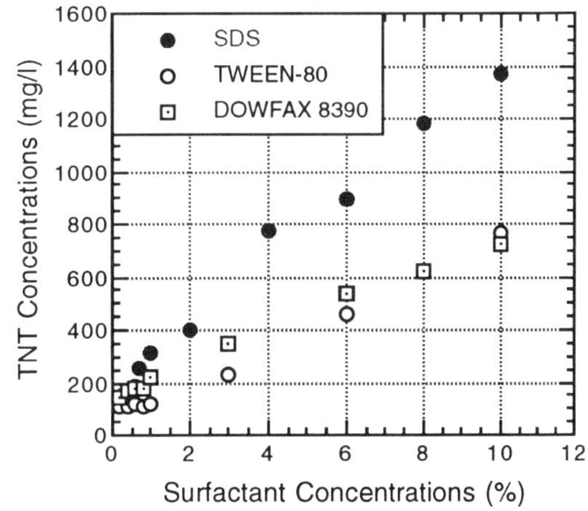

Figure 2. Extraction of TNT from contaminated soil at surfactant concentration up to 10%.

10% were not tested because the solution became very thick and viscous. Mixing and handling becomes tedious and probably the cost factor will also hinder the use of surfactants at these levels. No cationic surfactants were tested because they were quite difficult to mix at high concentrations due to their limited aqueous solubility.

The enhancement achieved in solubilization using the higher surfactant doses is illustrated in Figure 2. SDS gives the best results among the surfactants tested. Approximately 1360 mg/L of TNT was desorbed by the 10% SDS solution which represents about 27 % of TNT in the soil. TNT desorption from soil is enhanced significantly above that achieved at 1% for Tween 80. The improvement in extraction with Tween 80 is from about 3% to 23% removal of TNT at 1% and 10% surfactant solutions, respectively. The results of this study may be used to assist decontamination techniques such as EK soil processing and other remediation techniques such as hydraulic recovery, soil washing, etc. which rely upon efficient desorption and solubilization of species for the recovery nonpolar organics from soils. and 10% surfactant solutions, respectively. There does not seem to be an optimum surfactant concentration in the range tested. The extent of extraction increases with the increase in surfactant concentration.

4 CONCLUSIONS

An investigation was conducted to assess the increase in solubility of TNT and its desorption from the soil using surfactants. Increasing the solubility of TNT in the soil through desorption by surfactants will enhance its removal using conventional and emerging soil remediation technologies.

At low surfactant concentrations (0.1-1%), SDS gave the best desorption results. Desorption of TNT using nonionic and cationic surfactants at these concentrations were relatively constant and below the values obtained using water. At concentrations beyond 1 % and up to 10 %, SDS displayed a steadily increasing trend in TNT extraction. Tween 80 also showed significant improvement, indicating that sufficient amounts of surfactant molecules were available for enhanced desorption to occur. The results of this study may be used to assist decontamination techniques such as EK soil processing, hydraulic recovery, soil washing, etc. which rely upon efficient desorption and solubilization of species for the recovery nonpolar organics from soils.

REFERENCES

Brady, N.C. 1974. *The Nature and Properties of Soils*, 8th edition, MacMillan Pub. Co.,Inc., New York, 639 p.

Edwards, D.A., Liu, Z., and Luthy, R.G. 1994a. *J. Of Environmental Engineering*, Vol. 120, No. 1, pp. 5-22.

Edwards, D.A., Adeel, Z., and Luthy, R.G. 1994b. *Environmental Science and Technology*, 28, pp. 1550-1560.

Fink, D.H., Thomas, G.W., and Meyer, W.J. 1970. *J. Water Pollution Control Federation*, Vol. 42, No.2, Part 1, pp. 265-271.

Freeman D.J., and Colitti, O.A. 1981. *36th Industrial Waste Conf.*, Butterworths, Boston, pp. 383-394.

Garg, R., Grasso, D. and Hoag, G. 1991. *Hazardous Waste and Hazardous Materials*, Vol. 8, No. 4, pp. 319-340.

Liu, Z, Edwards, D.A., and Luthy, R.G. 1992. *Wat. Res.*, Vol. 26, No. 10, Pergamon Press Ltd, Great Britain, pp. 1337-1345.

Rosen, M.J. 1978. *Surfactants and Interfacial Phenomena*, John Wiley & Sons, New York, 304 p.

Taha, M.R., Acar, Y.B., Soewarto, I.H., Gale, R.J., and Zappi. 1995a. *J. Haz. Mat.* (Submitted for publication) Elsevier:Netherlands.

Taha, M.R., Soewarto, I.H., Acar, Y.B., Gale, R.J., and Zappi. 1995b. *Water, Air and Soil Poll.* (Submitted for publication). Kluwer:Netherlands.

Yinon, Y. 1990. *Toxicity and Metabolism of Explosives*, CRC Press Inc., Boca Raton Florida, 176 p.

Zappi, M.E., Gunnison, D., Pennington, J., Teeter, C., Coyle, C., and Rope, C. 1994. *HMCRI Superfund Conference*, Washington DC.

Zappi, M.E., Gunnison, D., Pennington, J., Teeter, C., Coyle, C., and Rope, C. 1993. *Evaluation of Bioslurry Systems for Treating Explosives Contaminated Soils From the Hastings East Industrial Park*," USAE Waterways Experiment Station, Vicksburg, MS.

Environmental Geotechnics, Kamon (ed.) © 1996 Balkema, Rotterdam. ISBN 90 5410 848 7

Monitoring a soil flushing remedial design using neutron moderation

Michael A. Williams & Craig A. Willis
Black & Veatch Special Projects, Corporation, Overland Park, Kans., USA

ABSTRACT: An insitu soil flushing remedial design was selected to remove chromium-contaminated pore water from deep, unsaturated soils beneath a wood treating facility in California. During almost 25 years of pressure injection operations, excess wood treating solutions spilled below the pressure injection chambers and percolated through the soil into a complex alluvial aquifer system.

Following the remedial investigation and feasibility study, a 6-month pilot study was conducted in 1991. The pilot study consisted of soil flushing injection wells, groundwater extraction wells, tension lysimeters, and monitoring tubes. The monitoring tubes were constructed using an innovative design capable of maximizing neutron penetration through unsaturated soils and allowing sampling of groundwater. During the pilot study, water was injected through a gravity-feed system and used to mobilize hexavalent chromium in the unsaturated soils.

An integral part of the design consisted of monitoring moisture content increases using the neutron moderation method. Neutron moderation is a down-hole method involving a weak radioactive source capable of emitting high velocity neutrons The neutrons are approximately the same mass as hydrogen atoms; therefore, the fast neutrons collide with hydrogen atoms and are slowed or moderated. A source detector is used to measure the slower velocity neutrons. The neutron counts can then be correlated with subsurface moisture contents.

This system helped determine advancing moisture fronts, preferential flow pathways, and spacing of injection wells. The full-scale design has been in operation for 3 years. Site closure will likely involve additional moisture profile monitoring using the neutron moderation method.

1 PROJECT HISTORY

For almost 25 years, a 121,406-m^2 manufacturing facility injected chromated copper arsenate (CCA) and acid copper chromate preservative solutions into precut lumber (Figure 1). During the pressure injection process, excess wood treating solutions spilled below the pressure injection chambers, percolated through the soil, and ultimately migrated into a complex alluvial aquifer system. Based on soil and groundwater assessment programs at the site, subsurface soils and groundwater beneath the retort pit area are known to have been contaminated by the wood treating activities at the site. In 1991, wood treating operations were discontinued and soil flushing remediation efforts were initiated.

BLACK & VEATCH Special Projects Corp. (Special Projects) personnel and others have conducted several assessment programs at the facility since 1984. Groundwater extraction and onsite treatment were initiated as interim remedial actions in 1987. An onsite pond closure was completed in 1990. In 1991 and 1992, groundwater extraction and treatment systems were expanded and in situ soil flushing was implemented as required by the California Environmental Protection Agency's Remedial Action Plan for the facility.

The Soil Flushing Project has progressed from a pilot study to full-scale implementation. The purpose is to flush highly mobile and soluble hexavalent chromium from unsaturated soils to the groundwater. Groundwater is currently pumped and treated through an onsite treatment

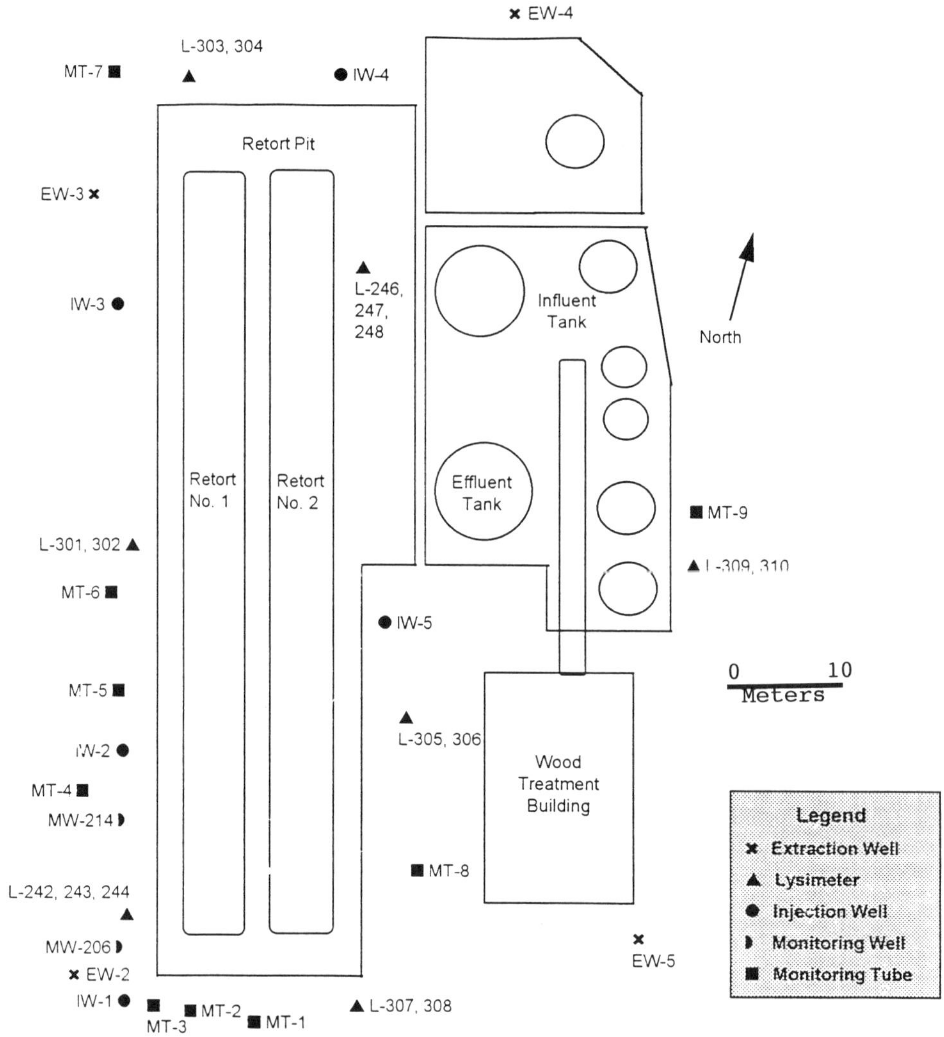

Figure 1. Plan view of the full-scale soil flushing system

system and reinjected as flushing water or discharged in accordance with a National Pollution Discharge Ellimination System permit.

Neutron probe monitoring was selected to verify the success of flushing unsaturated pore water from approximately 6 to 24 meters bls. The method is inexpensive, non-destructive, and repeatable.

2 PROJECT SETTING

Unconsolidated strata underlying the site consist of discontinuous lenses of clays, silts, sands, and gravel. Groundwater at the site is currently approximately 7 meters bls after several favorable years of regional groundwater recharge. Groundwater flow seasonally varies at the site from south-southwest to south-southeast because of agricultural pumping.

Based on shallow aquifer testing and grain size analyses, shallow aquifer hydraulic conductivity below the site is known to vary between $1.1x10^{-2}$ and $1.3x10^{-2}$ cm/sec. In addition, the average linear velocity under natural gradient conditions varies between approximately 171 m/year and 205 m/year.

Concentrations of arsenic and copper in the soils directly beneath the retort pit attenuate with depth and are less than the site-specific action levels by approximately 5 meters bls. Total chromium (predominately hexavalent chromium) concentrations have been detected as high as 4,270 milligrams per kilogram (mg/kg) at 0.3 meters bls.

3 SYSTEM COMPONENTS

Components of the full-scale soil flushing system include the following (Figure 1):

- Shallow injection wells (IW-1 through IW-5).
- Shallow extraction wells (EW-2 and EW-3).
- Intermediate extraction well (MW-214).
- Shallow lysimeters (L-242, typical).
- Neutron probe monitoring tubes (MT-1 to MT-9).
- Influent water holding tank (reused existing tank).
- Reinjection of treated groundwater from onsite treatment systems with potable water backup supply.
- Influent and effluent flow meters at all injection and extraction well locations.
- Extracted water holding tank (reused existing tank).
- Several additional existing extraction wells, monitoring wells, and piezometers in the vicinity of the site (not all shown on Figure 1).

4 NEUTRON MODERATION

Neutron logging is a proven and inexpensive method for monitoring the moisture content of unsaturated soils. It has been used in both the oil industry to determine porosity and by agronomists on a smaller scale to measure soil moisture. The geophysical logging tool used in the oil industry has the benefit of monitoring a greater distance from the source and it can be submersed below the water. However, the lower operating cost, portability, and low source strength make the agronomist's tool the preferred choice for most environmental applications (Kramer et.al.,1990).

The neutron probe source consists of a low-level encapsulated nuclear source, usually Americium-Beryllium. Alpha particles bombard the Beryllium and emit fast neutrons (neutrons with an excitation energy of 4.5 to 11 MeV). The fast neutrons collide with atoms in the soil and the resulting collisions are either elastic or inelastic depending upon their masses. The mass of most common atoms in the soil and pore water is much greater than the mass of the fast neutrons and the collisions are essentially elastic. The mass of Hydrogen (H^+) atoms is nearly identical to the mass of fast neutrons and the resulting collisions cause the neutrons to loose energy in the form of heat. A source detector located within the probe measures the number of thermalized or moderated neutrons. Since H^+ atoms are primarily found in pore water, a calibrated relationship can be established between the moisture content and the neutron count.

5 SYSTEM OPERATIONS

Operation of the full-scale soil flushing system has been continuous since June 1992, with all five injection wells injecting water at the site. Effluent from existing water treatment systems at the facility has been the primary water supply source for the soil flushing. An existing tank at the former wood treating site has been used as an influent holding tank for the gravity-fed injection system. Hydraulic control of groundwater below the site has been maintained by cycled groundwater extraction from an existing intermediate depth groundwater monitoring well (MW-214). After being temporarily held in an existing effluent holding tank, extracted groundwater is transferred to existing onsite water treatment systems at the facility.

6 DESIGN

The monitoring network was designed to maximize neutron penetration by minimizing borehole size. It is also important to log the borehole and to collect soil physical and chemical samples. In some geologic environments hole integrity may be a concern.

Our design consisted of using the dual wall percussion hammer drilling technique, with 14 cm outside diameter and 8 cm inside diameter steel tubes. The percussion hammer uses a powerful diesel hammer to drive dual steel tubes into the ground. Compressed air is forced through the dual tubes, and borehole materials are removed through the center of the tubes and collected at a discharge "cyclone." The drilling can be extremely quick and efficient. Borehole size is crucial in an effective design on neutron monitoring. Keller et al., 1990, demonstrated that although polyvinyl chloride (PVC) had a greater masking effect than stainless steel, the greatest inhibiting factor for neutron penetration was large diameter wells. In addition, Amoozegar et al., 1989, found that slurry backfills yielded neutron count rates as good as access tubes tightly fitted to the borehole; however, increasing access hole size resulted in decreased sensitivity.

Our monitoring tube design consisted of 6.4 cm nominal, Schedule 80 aluminum pipe with approximately 3 m of PVC well screen placed at the bottom of each monitoring tube. The monitoring tube served as both a monitoring well and neutron probe access tube. The monitoring tube was constructed much like a typical monitoring well; however, the water-to-cement ratio used in the backfill was carefully controlled to adhere to previous laboratory results by Kramer et al., 1990. That is, each mix of grout contained 19 liters of water per 42.6 kilogram sack of cement, with approximately 5 percent sodium bentonite powder. According to the results presented by Kramer et al., this mix should require approximately seven days for proper maturation.

7 CONCLUSIONS

The soil flushing system has been successfully used to mobilize hexavalent chromium for groundwater extraction and treatment. Neutron moderation was also successfully used to verify lateral moisture migration from vertical injection wells. Monitoring moisture profiles allowed for a better understanding of preferential pathways Approximate lateral migration rates were inferred from peak neutron counts at monitoring tubes located at discrete distances from the injection well.

Coarse-grained lithologies showed the greatest increase in moisture content, based upon neutron counts. This is because pre-flushing moisture contents were low in the sandy horizons and preferential flow probably occurred through these highly permeable zones. Clay horizons showed limited success in flushing, as evidenced by gradual increases in neutron counts. Consequently, we assume that limited flushing is occurring in these zones (Figure 2).

Although we had hoped that a saturated wedge or mound would develop near the injection well, water level elevations remained

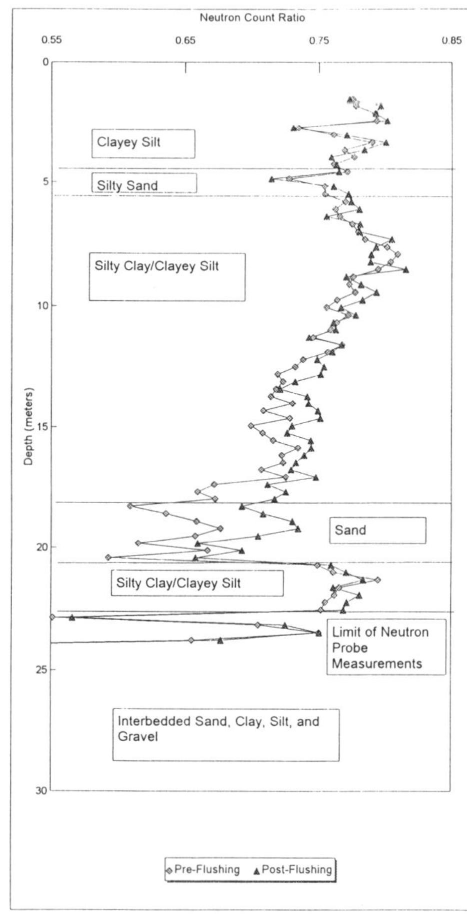

Figure 2. Soil flushing neutron logs at MT-4

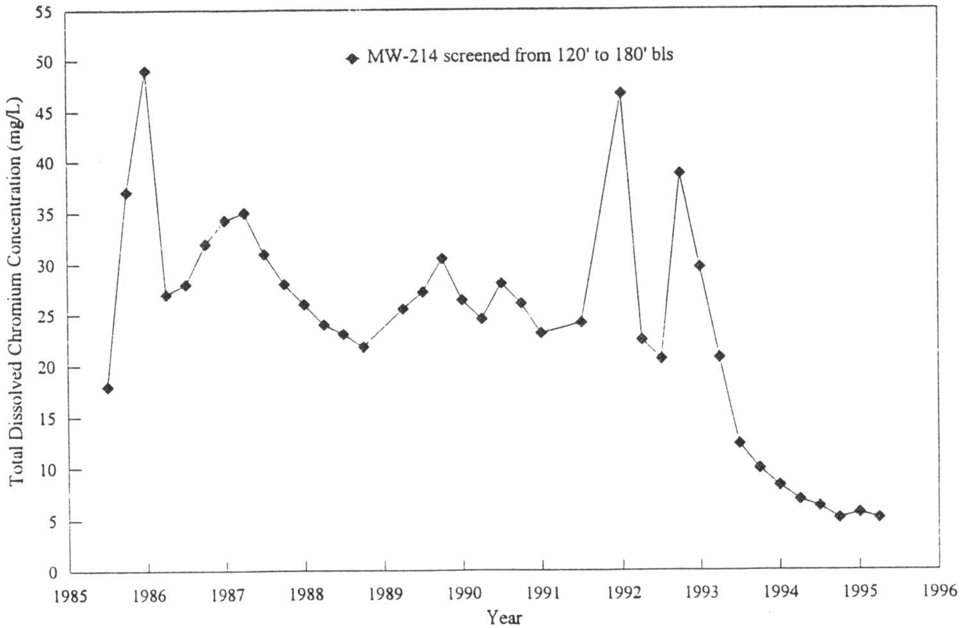

Figure 3. Total dissolved chromium detected in groundwater at MW-214

nearly the same. This may indicate that moisture fronts were advancing, but matric potentials (negative pressure heads) were still present, much like an induced capillary fringe. In addition, regional drawdown of the ground water surface may have inhibited the development of a mound. A system of nested monitoring wells would have failed to observe the moisture increases, because a positive pressure head was not established.

The success of soil flushing can be seen in the groundwater concentrations of chromium at MW-214 (Figure 3). Following initial flushing in June of 1992, concentrations increased. After continuous flushing, concentrations have decreased to their present levels.

REFERENCES

Amoozegar, A., K.C. Martin and M.T. Hoover. 1989. Effect of Access Hole Properties on Soil Water Content Determination by Neutron Thermalization. Soil Sci. Soc. Am. J., v. 53, pp. 330-335.

Hillel, D. 1982. Introduction to Soil Physics. Academic Press, London, England, 364 p.

Keller, B.R., L.G. Everett, and R.J. Marks. 1990. Effects of Access Tube Material and Grout on Neutron Probe Measurements in the Vadose Zone. Ground Water Monitoring Review, v. 10, no. 5, pp. 96-100.

Kramer, J.H., L.G. Everett, and L.A. Eccles. 1990. Effects of Well Construction Materials on Neutron Probe Readings with Implications for Vadose Zone Monitoring Strategies. In Proceedings of the Fourth Annual Outdoor Action Conference on Aquifer Restoration, Ground Water Monitoring and Geophysical Methods, presented by NWWA and the U.S. EPA EMSL.

Environmental Geotechnics, Kamon (ed.) © 1996 Balkema, Rotterdam. ISBN 90 5410 848 7

Effect of soil-water content on adsorption of volatile organic chemicals in soils

T. Yamaguchi & R. I. Farcasanu
Hiroshima University, Higashi-Hiroshima, Japan

P. Moldrup
Aalborg University, Denmark

L.W. Petersen
Danish Ministry of Agriculture, Tjele, Denmark

D.E. Rolston
University of California, Davis, Calif., USA

ABSTRACT: The purpose of this study is to evaluate the effect of low soil-water content on VOC retardation (adsorption) in soils under soil-venting conditions, i.e., under rapid convective-dispersive gas transport. Retardation of 1,1,1-trichloroethane and toluene on decomposed granite and a clayey soil (Yolo loam) was determined using a newly developed soil micro-column method based on gas chromatography. The micro-column method in general gave lower retardation factors compared to batch experiments, probably due to non-equilibrium sorption processes occurring under rapid gas transport.

1 INTRODUCTION

The soil venting technique, also known as soil vapor extraction, has become the most common in situ method used in removing Volatile Organic Chemicals (VOCs) from contaminated soils. Therefore, it is important to understand the mechanisms of retention and mobility of VOCs in soils during soil venting i.e. under rapid convective-dispersive air flow. Recent studies show that the soil-water content is a governing factor in defining whether clay content or organic carbon content dominates VOC adsorption(Ong and Lion 1991 a,b). In medium wet to water saturated soils, the VOC adsorption will be a function of organic matter content, while direct adsorption onto clay minerals can occur in very dry soils .

Numerical analysis by Culver et al. (1991) and Ong et al. (1992) and experimental evaluation by Steinberg and Kreamer (1993) suggested that vapor adsorption onto soil minerals significantly retards the transport of VOCs under dry conditions. Petersen et al. (1994, 1995) studied the behavior of trichloroethylene and toluene using a batch apparatus and concluded that Henry's law is applicable only for soil-water contents higher than four molecular layers of water coverage around soil particles. At the very low soil-water content range, Ong and Lion (1991a,b) and Petersen et al. (1994) presented data illustrating the non-linear behavior of VOC adsorption. Petersen et al. (1995) also developed a model to determine the shape of the curve relating VOC solid-vapor partition coefficients to soil-water contents lower than four molecular layers of water, valid at low VOC vapor pressures. However, methods for predicting VOC retardation under rapid convective-dispersive gas transport as function of soil-water content have not been developed. The objectives of this study are to determine the effect of soil-water content on VOC retardation under rapid convective-dispersive gas transport and to see the influence of temperature, flow velocity, and soil column dimensions when using our newly-developed convective flow-based micro-column method for measuring VOC retardation (Yamaguchi et al. 1996 a,b).

2 THEORY

The model concept developed by Petersen et al. (1995) was used to describe our measured data for VOC retardation as function of soil-water content. In the model presented below, the point of four molecular layer coverage of water, in agreement with the findings of Ong and Lion (1991 a,b), is the matching point between the linear and non-linear fitting curves.

$$A=\log(k_D'),\ k_D'=\frac{k_D}{k_H}+\frac{w}{k_H\phi\rho},\quad w>w_4 \text{ (linear)} \qquad [1]$$

$$A=(A_0-\beta)e^{-\alpha w}+\beta,\quad w\leq w_4 \text{ (non-linear)} \qquad [2]$$

where Eq. (1) is Henry's law, k_D' is the solid/vapor partition coefficient (cm^3/g), k_D is the aqueous/solid partition coefficient (cm^3/g), k_H is Henry's constant (dimensionless), w is water content (g H_2O/g soil), w_4 is the soil-water content corresponding to four molecular layers of water coverage, ϕ is the aqueous activity coefficient (= 1), ρ is the density of water ($1 g/cm^3$), β is a fitting parameter whose value should minimize the root mean square error (RMSE) of A,

$$RMSE=\sqrt{\left(\frac{1}{n}\right)\sum_{i=1}^{n}\left(A_{measured}-A_{estimated}\right)^2} \qquad [3]$$

where n is number of data points, $A_0 = \log(k_D')$ for w = 0, and α is given as

$$\alpha=-\frac{\ln\left(\frac{A_4-\beta}{A_0-\beta}\right)}{w_4} \qquad [4]$$

Petersen et al. (1994) determined the value of k_H and k_D for toluene to be $k_H = 0.261$ and $k_D = 0.87\ cm^3/g$ in agreement with the value measured by Garbardini and Lion (1991 a, b).

The retardation factor R is given by

$$R=1+\frac{k_D'\rho_b}{\varepsilon} \qquad [5]$$

where ρ_b is the soil bulk density (g/cm^3) and ε is the volumetric soil-air content (total porosity minus volumetric soil-water content).

3 MATERIALS and METHOD

Adsorption of 1,1,1-trichlorethane (TCA) and toluene on Decomposed Granite (0.212-0.425 mm and < 0.425 mm grain size fractions) and a clayey soil (Yolo loam, the <0.425 mm grain size fraction) were investigated using two soil column dimensions of 0.4 cm (diameter) X 50 cm (length), and 1.0 cm X 20 cm. The characteristics of the soils are given in Table 1.

Table 1. Characteristics of the soils used in the analysis.

Soil	Clay/Silt/Sand %	CEC (meq/100 g soil)	Surface Area m^2/g soil	TOC (%)
Yolo loam[(a)]	18/49/33	21.1	80.6	1.05
Decomposed Granite (<0.425 mm)[(b)]	4/5.4/90.6	2.7	11.5	<0.1
Decomposed Granite(0.212-0.425 mm)[(b)]	0.8/1.5/97.7	2.7	10.3	<0.1

(a) Petersen et al. 1994
(b) Yamaguchi et al. 1996b

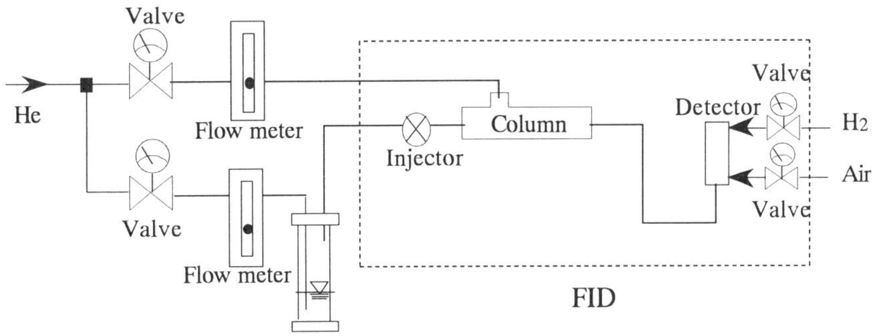

Fig. 1. Illustration of the micro-column apparatus used for the measurements.

VOC measurements were performed on a SHIMADZU GC-8AIF gas chromatograph with flame ionization detector where the soil columns were installed (Fig. 1).

The First Term Method of Yamaguchi et al. (1989) was used to determine dispersion coefficient and flow velocity of methane, as standard gas (CH_4/He=103 ppm), in the dry soils. Helium (99.999%) was used as a carrier gas and the humidity of gas and soil moisture were adjusted during the experiment by a water disperser installed in front of the injector. The soil columns obtained a uniform low moisture content after 16-18 hours. Comparing the retention time for a pulse of methane (non-reacting gas) and the one for a VOC, the VOC retardation factor was determined. Details of the experimental procedure have been described by Yamaguchi et al. (1996 a,b).

4 RESULTS AND DISCUSSION

The influence of temperature increases with decreasing water content. For very low water content the values of R abruptly increase with decreasing temperature probably due to the changes in gas diffusion which controls the movement of VOCs through micropores in the soil (Fig. 2).

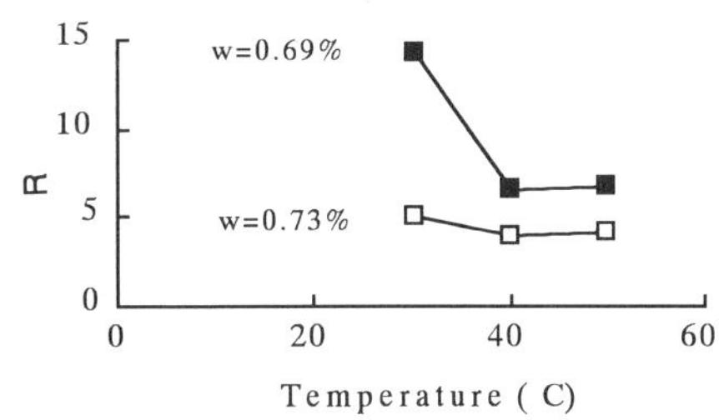

Fig. 2. The influence of temperature on the adsorption (given as retardation factor, R) of TCA on Decomposed Granite (0.212-0.425 mm grain size fraction). Micro-column dimensions 0.4 cm X50 cm.

The retardation of VOC on Decomposed Granite was less affected by the flow velocities in the high flow velocity range and, also, the column dimensions had no effect in the values of retardation factor R (Fig. 3-4). For the soil-water content of 0.73% we observe a slightly decreasing R with increasing velocity (Fig. 3) suggesting the influence of non-equilibrium sorption processes at higher flow velocities.

In our measurements the values of R for Yolo loam under convective flow condition were three times smaller than the values measured using a batch method, and the retardation of VOC on the decomposed granite (DG) was much smaller than on the Yolo loam, mainly

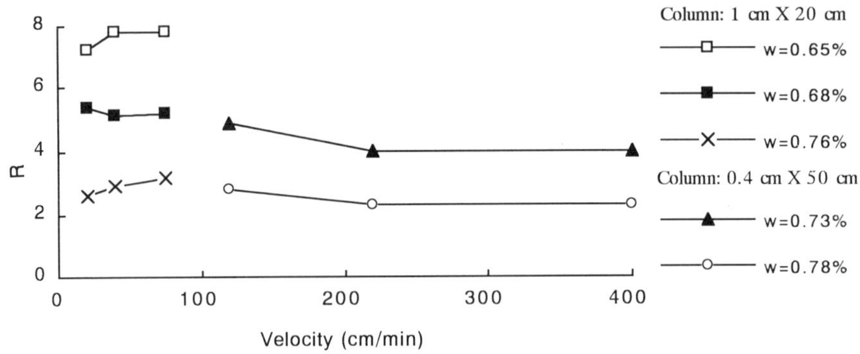

Fig. 3. The influence of flow velocity on adsorption of TCA on Decomposed Granite (0.212-0.425 mm grain size fraction) in two soil columns, and at different soil moisture content (w).

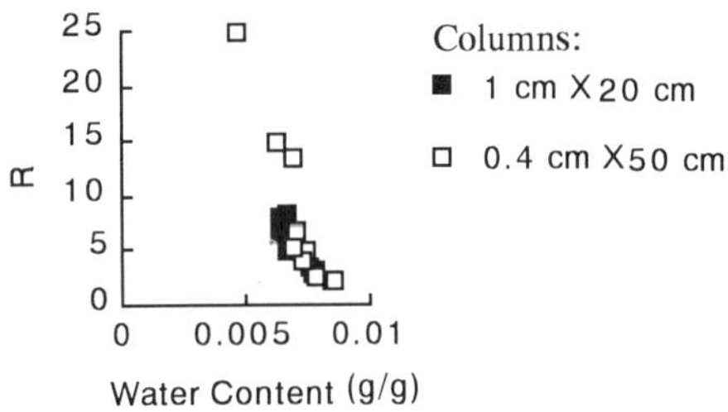

Fig. 4. The influence of column dimensions on the adsorption of TCA on Decomposed Granite (0.212-0.425 mm grain size fraction).

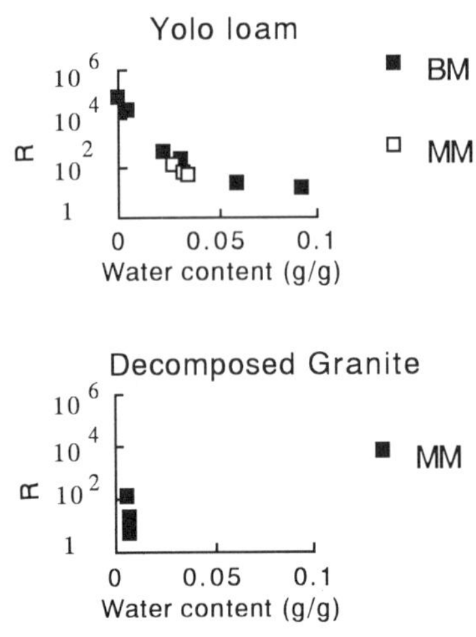

Fig. 5. Toluene retardation factor (R) as function of soil-water content; comparison of batch (BM) and transport-based (MM) measurement methods for a fine-textured soil, Yolo loam, and a coarse-textured soil, Decomposed Granite.

due to the much lower surface area of DG (Fig. 5). We assumed the differences between batch and micro-column measurements to be due to non-equilibrium adsorption processes.

Figure 6 illustrates the two-part model concept (Eqs. 1 and 2) for VOC vapor adsorption as function of soil-water content (modified from Petersen et al. 1995). The linear region of correlation between solid/vapor partition coefficient and water content occurs in the wet soil range where the dissolution of VOCs is described by Henry's law (Eq. 1), and the non-linear region occurs for the very dry water content range where Henry's law is not valid (Eq. 2).

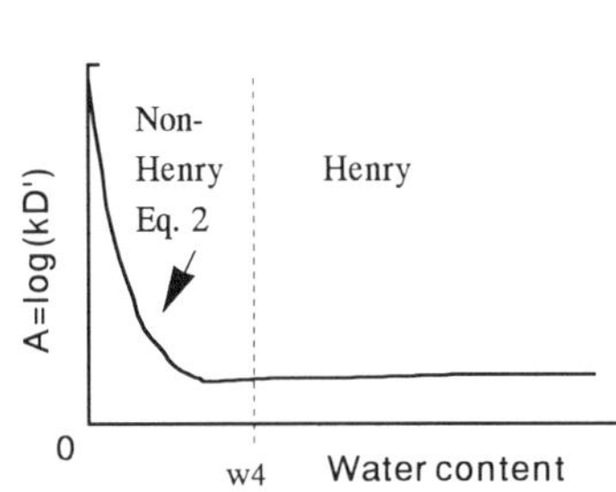

Fig. 6. Illustration of the two-part model concept for VOC vapor adsorption as function of soil-water content (modified from Petersen et al. 1995)

Table 2. Parameter values obtained when fitting Eq. [2] to the measured VOC adsorption data for Yolo loam soil.

Measurement Method:	A_0	w_4	A4	n (a)	β	RMSE(b)	α(c)
Batch (BM)	4.34	0.092	0.549	6	0.38	0.078	34.3
Micro-column (MM)	3.859	0.092	0.549	3	0.52	0.063	43.9

(a) Number of data points
(b) Root Mean Square Error
(c) Fitted value of α assuming that the k_D' value at w = 0 is one-third of the Batch Method value (corresponding to measurements at $0 < w < w_4$), and that the A_4 value (matching point between Henry and Non-Henry region) is the same for the Batch and Micro-column Methods.

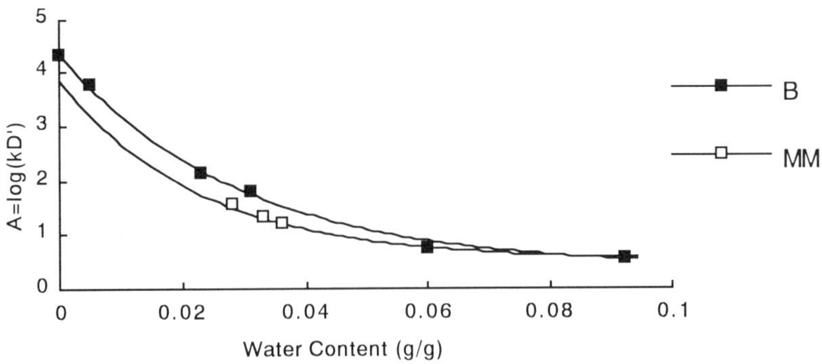

Fig. 7. Model for VOC adsorption in the dry (Non-Henry) water content interval fitted to Toluene solid-vapor partition coefficients (k_D') measured by batch (BM) and transport (MM) methods, respectively.

Figure 7 shows the curve-fitted measurements of VOC adsorption on Yolo loam soil. The parameter values for fitting Eq. 2 to the measured VOC adsorption data are given in Table 2. The model (Eq. 2) accurately described the measured data for both batch and flow experiments in the dry (non-Henry) range of soil-water content. Thus the model of Petersen et al. (1995) seems to be an useful approach indescribing the VOC retardation factor also under rapid convective transport conditions.

5 CONCLUSION

1. The VOC retardation factor for low water content (under 4 molecular layers) was extremely sensitive to variations in soil-water content.
2. The temperature influenced the adsorption of VOCs by controlling their diffusion in soils.
3. The soil column dimensions and flow rates had minor influence on the measured retardation of VOC on Decomposed Granite.
4. The micro-column method for measuring VOC retardation during rapid convective air flow (corresponding to soil-venting systems) gave lower retardation factors compared to batch measurements, probably due to non-equilibrium processes. Thus, the retardation factors measured by the traditionally used batch methods probably overestimates VOC retardation

in soil-venting systems. This should be carefully considered when designing soil-venting systems and/or performing numerical analysis of soil-venting systems.

ACKNOWLEDGMENTS

A travel grant from the Japanese Ministry of Education, Science and Culture (Monbusho International Scientific Research Program: Joint Research, No. 06044158) has greatly enhanced the work and cooperation of the authors and is hereby gratefully acknowledged.

REFERENCES

1. Culver, T.B., C.A. Shoemaker, and L.W. Lion. 1991. Impact of vapor sorption on the subsurface transport of volatile organic compounds: A numerical model and analysis. *Water Resour. Res.* 27:2259-2270.

2. Ong, S.K. and L.W. Lion 1991a. Mechanisms for trichloroethylene vapor sorption onto soil minerals. *J. Environ. Qual.* 20:180-188.

3. Ong, S.K. and L.W. Lion 1991b. Effects of soil properties and moisture on the sorption of trichloroethylene vapor. *Water Res.* 25:29-36

4. Ong, S.K., T.B. Culver, and C. A. Shoemaker. 1992. Effects of soil moisture and physical-chemical properties of organic pollutants on vapor-phase transport in the vadose zone. *J. Contam. Hydrol.* 11:273-290.

5. Petersen, L.W., D.E. Rolston, P. Moldrup, and T. Yamaguchi. 1994. Volatile organic vapor diffusion and adsorption in soils. *J. Environ. Qual.* 23:799-805.

6. Petersen, L.W., P. Moldrup, Y.H. El-Farhan, O.H. Jakobsen, T. Yamaguchi, and D.E. Rolston. 1995. The effect of moisture and soil texture on the adsorption of organic vapors. *J. Environ. Qual.* 24:752-759.

7. Steinberg, S.M. and D.K. Kreamer. 1993. Evaluation of the sorption of volatile organic compounds by unsaturated calcareous soil from the Southern Nevada using inverse gas chromatography. *Environ. Sci. Technol.* 27:883-889.

8. Yamaguchi, T., P. Moldrup, and S. Yokoshi. 1989. Using breakthrough curves for parameter estimation in the convection-dispersion model of solute transport. *Soil Sci. Soc. Am. J.* 53:1635-1641.

9. Yamaguchi, T., H. Inoki, and P. Moldrup. 1996a. GC-Micro-column method for measuring VOC retardation: I. Theory and preliminary tests. *J. of Environ. Systems and Engineering,* JSCE, No. 552/VII-1, 1996-Nov. (in press).

10. Yamaguchi, T., T. Tanaka, R.I. Farcasanu, and P. Moldrup. 1996b. GC-Micro-column method for measuring VOC retardation: II. Retardation of trichloroethane and toluene under various environmental conditions. *J. of Environ. Systems and Engineering,* JSCE, No. 552/VII-1, 1996-Nov. (in press).

Geo-environmental risks and regulations

Environmental Geotechnics, Kamon (ed.) © 1996 Balkema, Rotterdam. ISBN 90 5410 848 7

Stability of soil heaps in cold climate

G.I. Bondarenko
Research Institute of Bases and Underground Structures, Moscow, Russia

S.A. Klevtsov
Rosugol Company, Moscow, Russia

N.V. Sigachov
Joint-stock 'NGS-Orgproekteconomika', Moscow, Russia

ABSTRACT: Geological conditions in the regions with cold climate are especially sensitive to technogenic action of spoil heaps.The problem of rational land–tenure and environmental protection has been considered in the given paper.The results of 20 summer experiments have been given in this paper.Long–term prediction and regulation of spoil heaps behaviour has been proposed,providing equilibrium of ecological system: spoil–heap–natural and technical environment.

ECOLOGICAL AND GEOLOGICAL PLANNING

Mining works and spoil heaps formation are always connected with changes of natural environment.Geological conditions of the regions with cold climate are especially sensitive to the technogenic action of spoil heaps: change of relief, temperature, destruction of vegetable, statical and dynamic loads etc.

However, it is necessary to provide safety for population as well as to minimize negative action of spoil heaps to environment in order to prevent irreversible changes of ecology, destruction of natural and natural–technical systems.The problem of rational nature management should be based on ecological and geological planning of spoil heaps works.Ecological and geological planning is the system of technical and technological solutions for disposition and operation of spoil heaps in geological space, which gives equilbrium to ecological system: spoil–heap–natural and technical conditions.

The choice of disposition and volume of spoil heaps excavation, technology and intensity of excavation works are based on estimation of interaction of spoil heaps and natural technical conditions.The estimation of spoil heaps action to environment is carried out in all stages: design, spoil heaps formation, sliding, finish of mining works and recultivation of land.

During design it is necessary to consider some variants of spoil heaps formation.The choice of optimum variant should be accompanied by long–term prediction of spoil heaps behaviour, by their action to natural conditions and by the planning of protective measures.

Long–term prediction and regulation of behaviour of spoil heaps should be based on the knowledge of he main regularities of deformation of spoil heaps of frozen and thawing soils.

PREDICTION OF STABILITY AND DEFORMATION OF SPOIL HEAPS

Low average annual temperature and precipitations in the regions with cold climate promote deep freezing of spoil heaps.The main peculiarity of these spoil heaps is their frozen state.Deformation of such spoil heaps depends on thermodynamic equilibrium of frozen grounds. Resulting from in–situ, experimental and theoretical investigations the mechanizm and the main reguliarities of deformaton of spoil heaps of frozen and thawing soils have been stated.If the grounds have plastic properties, then long–term deformation as a rheological process (Bondarenko 1994, Vyalov 1978) precedes failure of spoil heaps.

Dependence of stresses and rate of flow may be written by the following equation:

$$\tau = \tau_P+\eta\cdot\gamma = C_P+\sigma\cdot tg\varphi_P+\eta\cdot\gamma \qquad (1)$$

where τ — acting shear stress of weight of ground;

η — coeffcient of plastic viscosity;

γ — relative rate of visco-plastic flow of soil;

C_P,φ_P — design values of cohesion and angle of internal friction.

The deformations are not so large when we have shear stresses in spoil heaps less than design values of strength. These deformations have a damped character and do not lead to failure.

The flow appears when we have stresses, exceeding design shear resistance. This flow, as a rule, leads to failure.

The volumes of spoil heaps are dictated by the condition of their stability or by the permissible rate of displacement.

The condition of stability of spoil heaps in the most general form may be expressed by the following equation:

$$K=\frac{T_h}{T_s}=\frac{tg\ \varphi_p}{tg\ \alpha}+\frac{C_p}{\rho\cdot H\cdot \sin\alpha}>=1 \qquad (2)$$

where T_h — holding forces;

T_s — shearing forces;

ρ — density of ground;

α — angle of slope inclination;

C_p,φ_p — design cohesion and friction;

H — distance to a surface of sliding.

With the increase of spoil heaps volume (thickness of layer) a landsliding pressure appears:

$$\Delta T=T_s-T_h \qquad (3)$$

Deformation of spoil heaps occurs from the displacement of spoil heaps mass along the sliding line and flow.

This pressure leads to flow, if spoil heap is structured by plastic grounds (frozen or clay thawing soils). The rate of flow may be expressed in the most general form by the following equation:

$$V=\frac{\Delta T}{\eta} \qquad (4)$$

At the same time important differences are in the mechanizm of deformation of spoil heaps of frozen and thawing grounds.The process of deformation of spoil heaps of frozen grounds is developed at the negative temperature. The limited-long-term values of soil are taken as design friction and cohesion of frozen soil. Friction, cohesion and viscosity coefficient of frozen soil are determined from long-term tests with account of temperature, granulometric composition, density, ice-content etc.(Bondarenko 1994).

The sliding surface in frozen spoil heaps is often coincided with the izoline of higher temperatures, because cohesion of soil is sharply reduced with a rise of temperature. Therefore engineerng and geological survey should include additionally investigation of geocryological conditions of region and organization of temperature control on spoil heaps. It should be taken into account that pressure of ice and snow reduces long-term stablity of spoil heaps of frozen grounds.In this case calculation of stabilty is carried out from the limited-long-term values of stability of ice or snow (ice or snow with inclusions of spoil heap grounds) and their contact with the underlying rock ground. Creep of heaps begins only, when some volume of the filled frozen grounds (thickness of layer H) was formed. For example, with the inclination of rock base at 48^0, creep of heaps begins with H>10m even at the temperature (-0.5^0C); when we have the temperature -2.5^0C, then even the layer of 35m thickness does not creep (see Fig.1). The increase of temperature to 1^0C can increase the rate of creeping almost in 10 times. Disturbance of stability of spoil heaps of thawing soil occurs as a result of displacement of the over-saturated thawing soil along the underlying frozen layer. The process of deformation of thawing spoil heaps is developed in the same way as the process of solufluction on slopes (Bondarenko1990).

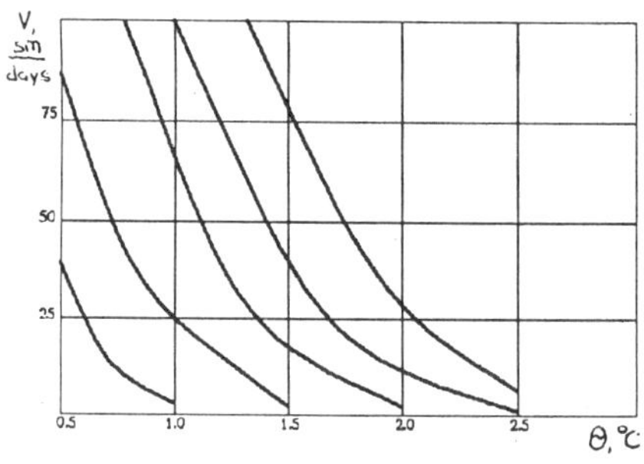

Fig.1. Dependence of rate of creep of frozen grounds heap on the temperature at $\alpha = 48^0$ and with the thickness of layer: 1 – 20m; 2 – 25m; 3 – 30m; 4 – 35m; 5 – 40m.

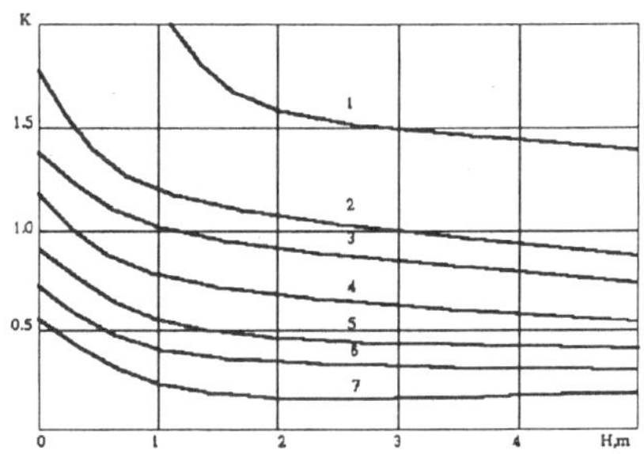

Fig.2. Dependence of coefficient of stability on the depth of thawing of soil at the angles of slope: 1 – 4^0; 2 – 6^0; 3 – 8^0; 4 – 10^0; 5 – 14^0; 6 – 18^0; 7 – 24^0.

It is known, that a thawing soil is creeping on the underlying frozen soil. (Kaplina T.N. 1965; McRoberts E.C., Morgenstern N.R. 1974; Zhikharev L.A. 1967) Design values of friction and cohesion of thawing spoil heaps are determined from the quick shear testing of thawing soil following the method, developed by the author. This method allows determination of shear resistance at the moment of thawing on the boundary with frozen soil, and this makes it possible to take into account humidity, pore pressure, loosening of soil. Coefficient of viscosity of thawing soil is determined from long-term testings. Determination of depth of thawing or depth of disposition of thawing soil layers should precede calculation of stability of spoil heaps and rate of flow of thawing grounds. These depths are determined from thermotechnical calculation or from temperature control. As one can see in Fig.2, with the depth of thawing 4m (sandy loam of fluid consistency) the slopes are stable at the angle of the inclination not more than 4^0. If this layer of soil is moist, then the slope is stable at the angle of inclination more than 20^0. The following stable angles of slope were received from the carried out investigatiions of various clay soils with the depth of thawing 4m depending on the consistency of soil: low plasticity – ($20^0 - 25^0$), soft plasticity – ($16^0 - 19^0$), fluid plasticity – ($10^0 - 15^0$), fluid – ($4^0 - 9^0$). The author made computer programs and nomograms, based on many years investigations of stability and sliding of spoil heaps in the Polar Regions, South Yakutija and the Krasnoyarsky territory, which make it possible to define volume and stable angle of inclination of spoil heaps. Moreover, these nomograms allow to predict behaviour of spoil heaps, accounting changes of properties of spoil heaps grounds (temperature, humidity, granulometric composition etc.) Sometimes displacement of spoil heaps occurs as a result of deformation of frozen and thawing grounds at the same time. To predict such complicated processes it is necessary to make long-term engineering and geological survey.Prediction is formed with account of interaction of frozen and thawing soils in the body of spoil heap.

CONCLUSION

The worked out method of prediction of stability and deformation of spoil heaps in the regions with cold climate makes it possible to provide equilibrium of ecological system: spoil heap – natural-technical environment.So, the problem of rational nature management and environmental protection is solved, and safety of population is provided.

REFERENCES

Bondarenko ,G.I. 1994.Regulation of temperature as a method to control deformation of spoil heaps in Polar

Regions.Proc. IY Int.Symp.on Ground Freezing, Nancy.

Bondarenko G.I.1990. Peculiarities of deformation of thawing soil in spoil heaps. Proc. NIIOSP, No.91, Moscow.

Kaplina T.N. 1965. Cryogenic slope processes. M, Nauka.

McRoberts E.C., Morgenstern N.R. 1974. Thawing Slopes. Canadian Geotech-nical Journal, vol.11, N 4.

Vyalov S.S.1978. Rheological bases of soil mechanics. Vysshaya Shkola, P.H.p.447. Moscow.

Zhikarev L.A. 1967. Reasons and mechanizm of development of solif-luction. M, Nauka.

Environmental Geotechnics, Kamon (ed.) © 1996 Balkema, Rotterdam. ISBN 90 5410 848 7

European trends in remediating contaminated sites

D. D. Genske
Delft University of Technology, Netherlands

ABSTRACT: The historical development of land degradation due to industry is outlined. Surveys on how much land is left derelict and how much it will cost to remediate theses sites are introduced. Tools and techniques to recycle land are discussed and European policies of land remediation are presented.

1 HISTORY

150 years of industrial revolution has changed Europe. Overpopulated industrial belts have developed such as the Ruhr District in Germany, the Belgian Coal Mining District, or the English industry zones north of London.

In the early days of industrialisation no thoughts were spent on the consequences of pollution. Historical documents proof how careless toxic substances were handled. The damage to soil and ground water was considerable increased during the second world war when mines were bombed and plants were destroyed. Air reconnaissance and mapping sorties of the former allies that had been kept confidential until recently allow a historical analysis, and thus the tracing of possible sources of contamination.

Figure 1. Abandoned industrial site.

In the 60 the coal price began to crumble in Europe. It became cheaper to ship coal from open pit mines overseas and Eastern Europe rather than to mine it in the country where it is consumed. Because of that European coal mines have been closing down. Along with that another key industry has been declining: steel mills have gone out of business all over Europe, causing tremendous unemployment figures (in the Ruhr District in the order of 15%) and social unrest.

The vast areas formerly used by coal mines and steel mills, as well as terrain used by secondary industries, are now abandoned (fig. 1). Since they are in most cases severely polluted they remain untouched. Besides contamination the ground is severely disturbed by massive foundations and underground structures. Typical for mining regions is furthermore the occurrence of deep shafts, the stability of which is often not satisfactory, and the variety of problems caused by regional subsidence, including wetland development and structural damage at existing buildings.

Derelict sites in overpopulated urban areas are a *contradictio in adiectu*, especially since high unemployment rate calls for immediate action to re-establish new, innovative, and clean industries. These industries, however, should not be permitted to settle on the remaining virgin land that is allocated for recreation. They shall be established on recycled industrial wasteland.

Now, that vast areas have been recognised as contaminated and spoiled the value of the resource "land" becomes clear. Today, key issues like environmental compatibility, eco-auditing, and sustainable land use dominate the discussion.

2 FACTS AND FIGURES

According to an European survey (Ferber 1995:14) 52.000 hectare are left by the mining industry, 43.000 hectare of which belong to only five EU-Member states, namely Great Britain, France, Spain, Belgium and Germany. In Great Britain an inventory of *all* derelict industrial sites led to a figure of 54.800 hectare in 1991. It is estimated that some 20.000 hectare of industrial sites are abandoned in France, 10.000 hectare in the western German States. In east and west Germany the total number of sites suspected to be contaminated amounts to 140.000, 40 % of which are in the new German States. Beside the derelict sites left by industry other terrains formerly used by military have to be remediated as well. In the country once separated by the Iron Curtain in times of the Cold War some 1 million hectare of land were occupied and are partly still used by military bases. Schmitz (1995) estimates that some 14 % of these sites are considerably contaminated with a complex mix of pollutants ranging from kerosene, lubricants, and paints, to shell fragments, warfare agents, and chemicals. To remediated these sites will cost about 400 billion Deutschmark (US $ 280 billion).

In Germany every day 87 hectare virgin land are consumed by industry (Merkel 1995:15), but only about two hectare are recycled. This development contradicts the idea of a *sustainable* land use; it is the opposite of a "development that meets the needs of the present without compromising the ability of future generations to meet their own needs" (Word Commission on Environment & Development "Our Common Future" United Nations 1987). *Ergo:* land recycling has become a crucial task for our generation and possibly the next generations.

Meanwhile, remediation of degraded land has become a common strategy of the European Union as explained in the "Green Paper on Urban Environment" and the "5th Action Program on Environment Towards Sustainability". Furthermore, a European Environmental Agency has been founded recently. Other initiatives such as an European group dealing with the re-vitalisation of derelict harbour sites have been started. A critical review of some of these activities is given by Williams (1995).

3 TOOLS AND TECHNIQUES

A sustainable land use can only be achieved if land recycling becomes cost effective. Cost effectiveness can only be accomplished if site investigation and remediation techniques are optimised and, furthermore, sophisticated management tools are applied. The basic steps a land recycling project has to run through are depicted in figure 2.

Naturally, the procedure depends much on the question into what the abandoned site shall be converted. Possible options for future utilisation are domestic areas, industry, traffic zones, agriculture, and recreation. The World Expo '98, where some 97 hectare degraded industrial land of Lisbons' eastern outskirts are being converted into exhibition sites, gives a prominent European example of land recycling. The grounds of the Olympic Games 2000 in Sydney, Australia, will also be established on land degraded by industry.

Whatever the future utilisation of a derelict site is, according to the flow diagram of figure 2, three mayor steps can be distinguished: risk assessment, remediation strategy, and realisation of the remediation measures chosen. These steps have already been discussed (Genske & Thein 1994) but shall be repeated briefly in the context of European land recycling strategies. The first step, the risk assessment, involves the documentation of the *status quo*, i.e. the present situation on the site, a historical research on the former utilisation, i.e. the reconstruction of the generations of fabrication plants established on the site, the layout and realisation of a field campaign to investigate the local contamination which is expected to occur in the vicinity of former factories, and finally based on this evaluation the risk assessment report.

The second step, remediation strategy, starts with a feasibility study, the main purpose of which is the discussion of possible remediation plans such as the excavation and dumping of contaminated soil, the confinement of polluted sectors by means of vertical and horizontal barriers, hydraulic measures to clean the contaminated ground water, etc. Criteria to analyse certain remediation strategies are the future use of the site, remediation costs, realisation time, technical feasibility, compatibility with governmental regulations and standards, and public acceptance. The feasibility study will be presented for discussion to all parties involved including the governmental agencies, potential investors, and the public. Of special importance is the involvement of city planners and architects at that early stage of remediation which may take the form of a city planning contest. Based on this it will be decided which remediation strategy will be chosen.

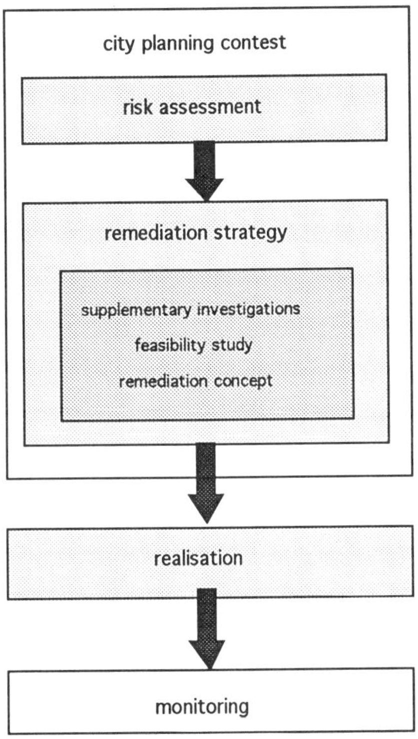

Figure 2. Flow diagram for land remediation projects

The last step is the realisation of the remediation plan, i.e. clearing of the site, removing or confining of contaminated sectors, and re-vitalising the site by building up a new infrastructure. To control the success of clean-up measured the installation of a monitoring system may become necessary, e.g. in the form of observation well to control groundwater quality.

The steps of land recycling in part overlap each other. To optimize the interactions during a land recycling project data bases and information management programs been developed (Dierkes 1995). In practice, however, harmonising all three steps appear to be a difficult task. Maps and profiles are needed to visualise the complex problems of the remediation project and the specific character of a site. It is quite important that this presentation material must be understandable, to the point and, most of all, financiable. Besides that sophisticated management and financing tools are crucial. In the following chapter visualising tools will be discussed and management tools will be touched. Financing tools are left to the financing specialists.

3.1 *Visualising tools*

During the three steps of land recycling specialists from different fields are forced to co-operate in an interdisciplinary way. To clarify crucial aspects of the project thematic maps and sections have to be prepared. It is state-of-the-art to utilise computers for this task. A large number of software packages are on the market ranging from simple CAD programs to more sophisticated applications of Geo-Information Systems (GIS). As described in Genske, Kappernagel, Noll (1994) and Genske, Noll, Trinkaus (1994) there are some tasks in land recycling that impose special demands on the software used:

- In order to track down contamination hot spots and investigate on inhomogenieties in the ground the interpretation of aerial photos as well as historical maps and building permits is necessary. This involves the superposition of photos and historical material with this status quo map of the site, i.e. the map depicting the present situation. Specialised imaging techiques, such as infrared scans to detect differences in soil conditions, vegetation, etc., or stereoscopic analyses of aerial photo pairs may become necessary. In some instances also mapping of special features such as bomb impact craters may be essential since they constitute possible sources of contamination due to the filling material (Dodt et al. 1993). Genske & Heinrich (1996) outlined the capabilities of GIS to cope with these tasks. For 2D-interpretation a GIS-program developed by the International Centre for Aerospace Survey and Earth Science ITC Enschede, Netherlands, is one example of a user friendly software package that satisfies most requirements. Koster & Slop (1994) applied this program to a rehabilitation project in the coal mining district of Kraviná, Czech Republic. Besides the contamination subsidence has been identified as an additional problem. The application of expert systems and knowledge based GIS is the next step in optimising data from land recycling projects (e.g. Hartema et al. 1994).

- The step following the computer aided historical or *multitemporal* analysis is the preparation of an 'intelligent' site investigation program. Based on the desk study carried out in the first step potential locations of contamination and ground disturbance are already known. To interpret the results gained from site investigation geostatistical procedures have been proposed (e.g. Genske et al. 1991). An example is given in figure 3 where a kriging routine was used to locate contaminated sectors. The big advantage of kriging is that besides the map depicting contaminated sectors also a error map is produced that quantifies the prediction error. The superposition of a kriged image with the status quo map enables the site investigators to optimise supplementary field investigation. This consequently results in an increase in cost-effectiveness. It should, however, always be kept in mind that kriging only gives good results for a diffuse type of contamination. Baysian kriging may include discrete boundaries. For discrete search targets such as oil tanks or foundations other statistical methods as outlined in Genske et al. (1991) have to be applied.
- The complexity of remediation projects may require a three dimensional presentation of the subground. Traditionally this is done with profiling. However, profiles can only provide an incomplete idea of the special features relevant for remediation. Full 3D visualisation can be accomplished by a variety of CAD programs that have become available recently. The options for 3D GIS are still limited (Genske & Heinrich 1996), especially if used in combination with knowledge based systems. Most of the programs available are restricted to depicting the geology and perhaps the contaminated sectors as detected during the field investigation.
- A big step towards understanding the complexity of a remediation project is the animation of a 3D-model of the site. With additional voice and sound effects a digital multimedia presentation can be created. This presentation can be stored on video or CD and can be quite useful in explaining complicated aspects to parties with limited expert knowledge like financing partners or the public. In Genske & Thein (1994), Genske, Noll, Olk (1994) an example is presented.

3.2 Management tools

Land recycling is an interdisciplinary task. In order to have the project running smoothly specialists have to co-operate closely to react flexibly on any change in the project such as the sudden discovery of a contamination hot spot in a sector so far considered clean. Any linear command structure combined with hierarchy is doomed to fail, especially under the dictate of the tight time schedule usually compulsory for officially funded projects (Thein & Brüggemann 1995).

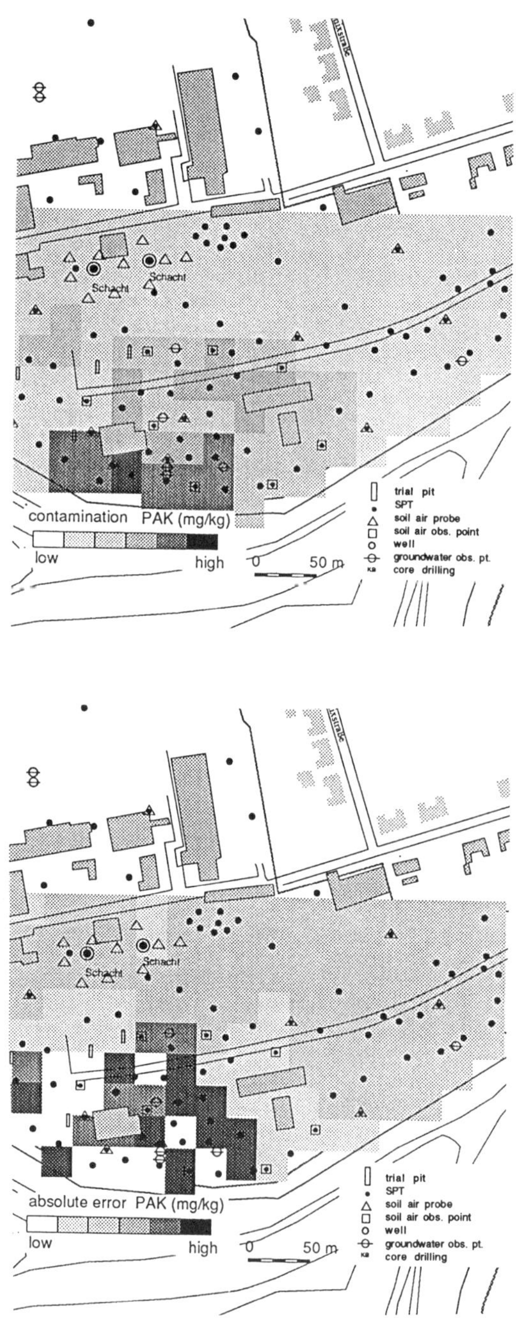

Figure 3. The distribution of polycyclic aromatic hydrocarbons on a derelict industrial site as derived from a kriging analysis. The upper map depicts the contamination grade, the lower map the estmation error (Genske, Noll, Olk 1995).

In Europe an unnoticed but perhaps important trend to decentralise decision making in management processes has taken place. In a new management structure spontaneous formation of self responsible task groups is desired. These task groups interact independently with neighbouring clusters within the company or organisation and the 'outside world'. We are looking at a kill-routine-before-it-kills-you management style that has proven to be successful in enterprises such as the Finish National Cleaning Service SOL (European, 5-11.10.95, pg 32). To my view this management style is urgently needed in tasks as challenging as land recycling. Directed towards the overall remediation goal a step-by-step approach of interacting task groups that follow the 3-step flow scheme as explained earlier appears to be more promising and thus more cost effective as a stiff centralised rule-and-order system. This updates earlier views of the author (Genske & Noll 1994).

One last word on the involvement of the public. According to European Legislation the public generally has to be informed about certain steps within a land recycling project. In Germany, for instance, presentation of the project to the citizen is a compulsory act in the commissioning procedure of a project. In land recycling projects, however, the involvement of citizen may become crucial since contamination and contamination grades have to be discussed openly. The policy to inform citizen about every step in a land recycling project, however, has proven to be far more beneficial than a closed-door-diplomacy. Good experiences has been made in projects of the International Building Fair IBA in the German Ruhr District where derelict contaminated sites are being remediated and re-vitalised. It was observed that the public generally acts co-operative and even contributes valuable ideas to the future utilisation of a site.

4 EUROPEAN TRENDS

Ferber (1995:16) expresses the opinion that three strategies to recycle land can be distinguished in the European Community

- *Remise en état,* i.e. preliminary refurbishment of the site to attract further investment to finance a thorough clean-up. This method is preferred in regions with a low financial profile such as Lothringen in France.
- Revitalising industry, i.e. re-establishment of industry on the abandoned site, thus creating new jobs. The European Community prefers this strategy and subsidises appropriate projects since this will lead to the creation of new long term jobs, thus effectively decreasing unemployment rates. A typical example is the steel mill Riesa in former eastern Germany. This strategy is also in line with the suitable-for-use approach of the German Ministry of Environmental Protection that fears the costs connected with a full restoration to the original condition of a contaminated site.
- Re-dedicating, i.e. using the site for tertiary purposes such as domestic areas, an idea especially suitable for derelict site located in city centres. The costs to satisfy clean-up regulations are, however, quite substantial. Examples of this strategy are the Urban Development Corporation UDC in England, the London Docklands, or certain remediations projects in the German Ruhr District. Another attractive variation of land rededicating is the conception of recreation areas and nature resorts which are urgently needed in the overpopulated industrial regions of Europe. Especially in former subsurface mining districts that suffer from subsidence wetlands and secondary biotops can be integrated into recreation areas (Drecker et al. 1995). Examples of secondary nature resorts can be visited in the German Ruhr District where the *Bundesgartenschau* (a national garden fair) provided the necessary funds to convert abandoned industrial sites. An example from outsite Europe is the 're-greening' of the Sudbury region in Ontario, Canada, where the combination of clear cut, erosion, and air pollution virtually wiped out all vegetation, leaving a 'moonscape' behind with 17.000 hectare of degraded land and 7.000 acid lakes. The Sudbury re-greening program was recognised by the United Nations as an outstanding remediation exercise (Gunn 1995).

Coming back to Europe, an interesting question was discussed by Williams (1995): he asked whether a priority listing as maintained in the United States could be a model for Europe. The United States were perhaps the first country to establish a national fund to remediate contaminated sites. Under the "Comprehensive Environmental Response, Compensation and Liability Act" (CERCLA) of 1980 a federally controlled fund, the so-called "superfund" was introduced. Based on risk assessment criteria specified in CERCLA an national priority listing of contaminated sites to be remediated was set up. According to Williams (1995: 22), however, this national listing in practice has had the effect "of causing the site, plus much of the adjacent area, to be abandoned for any use, and severely depressed property values. Therefore [this] listing is avoided by local government and property developers..."

The approach of the European Union is quite different from the CERCLA legislation. Funds like the European Fund for Regional Development EFRE have been introduced to instigate remediation activities on a regional scale. The funding scheme is structured in a way that the main responsibilities are with the remediation partners involved, i.e. the local community, the investors, and the public. Since only part of the remediation costs are subsidised by the European Union the creation of a variety of management models is stimulated, of which public-private partnership has proven to be a successful model in the Germany.

At this point it shall be mentioned that the evaluation of the abundance of options for subsidies demands expert knowledge. In Germany, according to Schmitz (1995), the sum of European funding possibilities plus national programs amounts to 170 options.

In December 1994 the EU Commission on Environment met with all European Member States including Austria, Sweden, and Finland to face one of the strongest obstacles to a common European policy on land recycling: different legislation, resources, and financing policy hamper a common strategy. Streamlining of legislation is urgently needed to open land recycling to the forthcoming single European Market.

5 CONCLUSION

Three conclusions can be drawn

- Land recycling is an interdisciplinary challenge: specialised tools like Geo-Information Systems are applied to visualise the problem and make it understandable to all parties involved. Management models such as public private partnership make land recycling cost effective. Innovative geotechnology is needed to optimise site investigation, risk assessment, and remedial measures.
- Land recycling is a big market: in former East Germany for instance 1 billion DM (some US $ 700 million) are allocated every year to stimulate land recycling, and that for the next 10 years.
- Land recycling creates jobs: in the German State of North Rhine-Westfalia, for example, the remediation of 525 derelict industrial sites in 217 cities and communities has already been subsidised with about US $ 1.5 billion (over a period from 1980 to 1994). Of the 6.304 hectare 70% were sold mainly to private investors by the end of 1993. It is estimated that some US $ 6 billion have been invested on remediated sites, and - even more important - 87.000 new jobs have been created.

REFERENCES

Dierkes, G. 1995: Industrieflächenrecycling - auch für den Eigentümer sinnvoll. In Genske & Noll (eds) *Brachflächen und Flächenrecycling*: 19-28, Verlag Ernst & Sohn: Berlin.

Dodt, J., D. D. Genske, T. Kappernagel, P. Noll 1993. Digital Evaluation of Contaminated Sites. – *International Conference on Digital Image Processing: Techniques and Application in Civil Engineering*, February 28 - March 5, 1993: Kona, Hawaii .

Drecker P., DD. Genske, K. Heinrich & P. Noll 1995. Subsidence and Wetland Development in the German Ruhr District. *5. International Conference on Land Subsidence, The Hague, Netherlands*, IAHS Press, 413-421. Wallingford, Oxfordshire, England.

Ferber, U. 1995. Flächenrecycling in Europa - Strategien und Empfehlungen, *Brachflächenrecycling / Recycling Derelict Land*, 1/95, 14-18.

Genske, DD. & K. Heinrich 1996. Land Recycling - Tools of Visualization, *Int. Conference on Environmental Geology and Land Use Planning*, Granada, Spain.

Genske, DD., P. Noll & C. Olk 1994. Montanstandorte im Wandel - Konzepte und Beipiele zur Wiedernutzbarmachung. In Genske, Noll & Trinkaus (eds.) *Brachflächenrecycling Taschenbuch*, VGE, Essen.

Genske DD. & J. Thein 1994. Recycling Derelict Land. *1. Int. Congress on Environmental Geotechnics*, 10.-15.07.1994, Edmonton, Canada, 493-498.

Genske, DD. & P. Noll 1994. Reclamation of Abandoned Mining Sites in Germany. *3. Int. Land Reclamation and Mine Drainage Conference*, 25.-29.04.1994, Pittsburg, Pennsylvania, U.S.A.

Genske DD, T. Kappernagel & P. Noll 1994. Computer-Aided Remediation of Contaminated Sites. *7th Congress of the International Association of Engineering Geology*, 4557-4562, 05.-09.09.1994, Lisboa, Portugal.

Genske DD, P. Noll & E. Trinkaus 1994. Land Use, Abuse, and Remediation: Digital Tools of Visualization. *International Conference on Computational Methods in Structural and Geotechnical Engineering*, December 12-14, 1994, Hongkong.

Genske, D., Gillich, M., H. Kories & Ch. Olk 1992. Contaminated Sites - Data Processing, Visualization and Interpretation. *International Conference on Geotechnics & Computers*, 29.09-01.10.1992, Paris.

Genske, DD., Klapperich H., Thein J. & M. Schöpel 1991. Reaktivierung von Industriebrachen. – *Bautechnik*, 12, 416-420, Berlin.

Gunn, JM. 1995. Restoration and Recovery of an Industrial Region. Springer. 358 pg.

Hartema, G. Pulido, JC. & I. Wolff 1994. Erstbewertung von Zechen- und Kokereistandorten mit Hilfe eine Geo-Informationssystems. In Genske & Noll (eds) *Brachflächenrecycling / Recycling Derelict Land*, 1/94, 31-40.

Koster, R. & S. Slob 1995. GIS Application for Mining and Rehabilitation Planning. In Genske & Noll (eds) *Brachflächenrecycling / Recycling Derelict Land*, 2/95, 55-60 and 3/95, 52-56.

Merkel, A. 1995. Ökologische Wirksamkeit bei wirtschaftlicher Rentabilität. Interview with the German Minister of Environmental Affairs. In Genske & Noll (eds) *Brachflächenrecycling / Recycling Derelict Land*, 2/95, 14-15.

Schmitz, C.W. 1995. Möglichkeiten der Finanzierung und Förderung von Entwicklungsmaßnahmen im Flächenrecycling. - Kurzfassungen der Vorträge *Forum Umweltschutz des TÜV Rheinland*, Köln.

Thein, J. & J. Brüggemann 1995. Handlungsstrategien zum Flächenrecycling - der iterative Planungsprozess. In Genske & Noll (eds) *Brachflächen und Flächenrecycling*, 1-17, Verlag Ernst & Sohn, Berlin.

Williams, R.H. 1995. Contaminated Land - A Problem for Europe? In Genske & Noll (eds) *Brachflächenrecycling / Recycling Derelict Land*, 1/95, 19-24.

Environmental Geotechnics, Kamon (ed.) © *1996 Balkema, Rotterdam. ISBN 90 5410 848 7*

Geotechnical recommendations for landfills in Germany

H.L.Jessberger
Ruhr Universität Bochum, Germany
H.Neff
ETN Soil Mechanics Laboratory, Hungen, Germany
E.Gartung
LGA Geotechnical Institute, Nürnberg, Germany

ABSTRACT: Although there is a strong demand to avoid, to reduce and to recycle waste, landfills are and will be indispensable within the waste disposal concept. Regulations by the Federal Government provide the framework for the design, construction and operation of safe landfills in Germany. The regulations are supplemented by Technical Recommendations edited by the German Geotechnical Society. The recommendations cover all geotechnical aspects of landfill design and remedial works. They have been compiled and updated according to the technical development for more than ten years.

1 INTRODUCTION

Like in other parts of the world, in Germany the slogans reduce, reuse, recycle are applied to all residues. During the last few years, legal and educational efforts have lead to a reduction in the generation of waste. Predictions of the amount of waste expected in the future, that were made some years ago for the estimate of the required landfill space, are currently being corrected in all parts of the country, because there is a significant trend towards a reduction of the amount of refuse. In spite of this development, there is still and will always be a demand for the deposition of solid waste in landfills. Landfills are indispensable. In contrast with other structures or production- and processing plants which are used for a limited time, landfills must be erected and operated in such a way that they do not cause any impairment to the well being of the public, for unlimited time. Landfills are structures for eternity. So greatest care has to be taken with respect to the safety of landfill structures and the responsible authorities have issued regulations and recommendations accordingly.

The German Federal Government has issued technical instructions for hazardous waste and for municipal waste landfills and thus established a technical frame, aiming at the same degree of safety in management and deposition of waste in all German States. The titles of the instructions are „Technical Instructions on the Storage, Chemical, Physical and Biological Treatment, Incineration and Landfilling of Waste“ *(TA-A)*, and „Technical Instructions on Recycling, Treatment and other Management of Municipal Waste“ *(TA-SI)*. Some of the German States have supplemented these by their own State instructions, and there are a number of specific regulations on waste management, issued by the pertinent committees.

The geotechnical aspects of solid waste landfills and contaminated land have been compiled by a task group of the German Geotechnical Society, *Deutsche Gesellschaft für Geotechnik*, and edited as technical recommendations under the German title *Empfehlungen des Arbeitskreises „Geotechnik der Deponien und Altlasten“ - GDA*. The topics of the *GDA*-recommendations were discussed by the European Technical Committee ETC 8, ISSMFE, adapted to international conditions and published as „Geotechnics of Landfill Design and Remedial Works, Technical Recommendations - GLR, second edition by Ernst & Sohn, 1993“. New editions of the compiled recommendations are in preparation and will be available in 1997.

2 PRINCIPLES AND INSTRUCTIONS

The landfill structure essentially consists of a large containment with the solid waste body inside. The migration of harmful substances is prevented by several barriers. The geology of the site is an important barrier. The next important barrier is the lining system at the bottom of the landfill. It has to cut off the path of migrating contaminants and must consist of engineered structural components placed under quality control. After closure, the capping system has to be installed. It covers the waste body, prevents ingress of surface water, reduces the emission of gas, odours and dust from the waste, and it facilitates landscaping. Special attention is paid to the properties and the placement of the waste material. The waste body is considered a barrier by itself. The refuse should be in such a condition, that the stability of the landfill is granted, that there is little or no tendency for harmful material to being dissolved and transported with seeping water, and the deformation due to settlements should be predictable and small, so the integrity of the cover would not be impaired in the long term. In summary, the landfill structure forms a multibarrier system (Sief 1986). Each of the barriers has to meet certain technical minimum requirements, independent of the performance of the other barriers.

In Germany, three categories of solid waste landfills are distinguished with respect to the deposited waste material. They differ in some details from each other. The composition and the properties of the waste are the governing criteria for the assignment to the landfill category. Limiting values of harmful substances are listed in the Federal Instructions *(TA-A and TA-SI)*. The delivered waste material is analysed at the entrance of the landfill site to make sure that the acceptance criteria are met.

Refuse with the lowest potential of harmful substances such as mineral waste and construction material or demolition debris are assigned to landfill category I. Category II comprises the majority of solid waste landfills, since typical municipal waste and similar materials with respect to their contents of dangerous substances are assigned to them. Following the philosophy that the waste body itself is an important barrier against the contamination of the environment, very strict criteria have to be met by the solid waste to be assigned to category II landfills. The most stringent criterion is, that in the future, the content of organic carbon of the waste material is not allowed to exceed 5 % by weight. This means, that essentially all domestic waste must be treated in an incinerator. Since there are not yet enough incinerators and treatment plants to meet this requirement at the present time, the regulations allow the deposition of conventional waste in landfills of category II until the year 2002. Waste that contains harmful substances by amounts exceeding the criteria of landfill category II, has to be disposed of in hazardous waste landfills of category III. Those are constructed according to the same basic principles as municipal waste landfills, however, the technical precautions for the safety of the environment are somewhat more involved.

All of the important technical requirements are put down in the Federal Instructions (*TA-A* and *TA-SI*). The philosophy behind the governmental regulations is to apply the state of the art in waste management and deposition of residues in such a way, that the highest possible degree of safety for the environment is reached. At the same time, the regulations should not obstruct further development in the landfilling technique, so the instructions permit alternatives and new methods, provided their functions are equivalent to or better than those of the standard solutions which are specified in greater detail. Since many aspects of the design, construction and operation of solid waste landfills are based upon the geotechnique, the German Geotechnical Society has compiled recommendations since 1985 and continues to update and to amend them. These recommendations address many technical details of design and construction of new landfills as well as of existing older landfills and contaminated land. They constitute the state of the art of environmental geotechnology and have been applied in practice in Germany for more than ten years.

3 GEOTECHNICAL TOPICS

The Technical Recommendations, Geotechnics of Landfills and Remedial Works, are structured under six main headings. Site assessment, principles of design, suitability testing, construction methods, quality assurance and hydraulic measures comprise all of the recommendations issued so far. Under these headlines a wide range of topics are treated. The series of recommendations consists of 53 titles at the present time (November 1995), and it is currently continued. The recommendations are numbered according to the sequence of editing. Brief summaries of the contents of the recommendations are presented under the six headings below.

3.1 *Site assessment*

Site investigation is of great importance for the design of new landfills, for the evaluation of the contamination potential and for the preparation of remedial actions of old landfills as well as of contaminated areas. Since a number of special aspects have to be taken into account with respect to waste material and harmful substances that can be emitted from waste deposits, environmental aspects are focused on, in the description of the methods of investigation of ground and ground water conditions. They comprise desk studies, geological and hydrogeological field investigations, drilling, and geophysical measurements, as well as the determination of the in-situ permeability of the ground. Special attention is paid to sampling methods. In addition to standard procedures of soil and rock sampling, the technique of recovering gas and water samples from the ground, qualified for chemical analyses, is treated.

Since siting of new landfills is becoming increasingly difficult in Germany due to disapproval of the public, fallow land that is available, but that does not meet the geological site requirements at first glance, is examined for its suitability to serve as a landfill site. In some areas of Germany, backfilled open pit mines are taken into consideration for waste landfill sites. The particularities of such situations, the influences of the backfill material and of the backfilling operation are described in one of the recommendations. For the prediction of the long term settlement of a landfill on an open pit mine backfill, large scale loading tests and measurements are needed. The recommendations on this particular subject are based upon extensive studies and experience by Jolas, Pfeiffer and Scheffler (1991).

In many cases it is necessary, to explore the properties of old landfills in order to assess their overall stability, to design dewatering systems or just to find out what the waste body is composed of. Based on the experience of Reuter (1993), recommendations are given for exploratory methods to be applied in waste bodies and for sampling and treatment of waste samples. A methodology is presented for the identification and description as well as for the geotechnical classification of waste material.

3.2 *Principles of design*

In preparing the design of a landfill, a number of points should be given particular consideration. Waste management standards relating to the type and quantity of waste, condition of waste, in certain cases after processing, time schedule of construction and landfilling operations are some of the items mentioned in the recommendations on geotechnical principles in the design of landfills. Especially, it has to be taken into account that the construction of the technical barriers and the landfilling operations are executed at the same time at many landfill sites. So both activities, construction and waste management, have to be co-ordinated. The protection of components already constructed and the requirements of safety for the health of the construction personnel are important boundary conditions to be considered in the design of solid waste landfills. Similar principles have to be followed in the preparation of remedial works on old landfill sites and contaminated areas.

Stability analyses of the landfill body and of the technical barriers are needed for the assessment of safety of a landfill structure. The mechanical properties of the waste material have to be quantified for this purpose, and conventional methods of geotechnical analyses have to be adapted to the particularities of the behaviour of solid waste. Results of scientific research into the properties of wastes carried out during the past 5 years are incorporated into the design recommendations. Apart from structural stability studies, the safety assessment also includes the evaluation of deformations and the associated questions of functional reliability of sealing and dewatering systems. Furthermore, the long term permeation through the basal sealing layer has to be quantified, and finally an overall assessment of the safety of the landfill is made according to the concept developed by Jessberger, Heibrock and Demmert (1993). Since the performance of the landfill structure cannot be predicted with a sufficient degree of confidence on a purely theoretical basis, monitoring is mandatory. Based on extensive practical experience in the supervision of landfills and the accompanying geotechnical interpretation of field measurement results by Neff and Walter (1993), recommendations on settlement and deformation measurements are included in the section on design. This is appropriate, because it is necessary to control the performance of the landfill structure by geotechnical expertise and to take correcting actions, if needed.

The design principles of basal liners and covers are emphasising the mineral components of these systems. Geotextiles are covered to the extent necessary for the selection of the appropriate product, and geomembranes are touched only very briefly. According to the Federal Instructions *TA-A* and *TA-SI,* geomembranes installed in German landfills must be approved by the Federal Institution for Material Research and Testing *(Bundesanstalt für Materialforschung und -prüfung, BAM).* Since the regulations of *BAM* treat all aspects of the application of geomembranes in great detail, there is no need for the recommendations of the German Geotechnical Society to include these aspects. Dewatering systems at the bottom and at the cap are relevant safety components of landfill structures. Rules for their design are presented in the recommendations. The section on the principles of design also contains information on thermal and on microbial treatment of contaminated soil.

3.3 *Suitability testing*

All construction materials and methods have to be tested for their suitability. This applies particularly to soils used in sealing layers. The recommendations list the well known standardised soil mechanics tests for the determination of the characteristic parameters for soil classification, and the tests for the determination of the stress-strain behaviour and strength. The correct assessment of hydraulic conductivity is of great importance for mineral sealing layers in landfills. Since the permeation behaviour of a soil depends on its composition, on the structure and state of compaction, on the state of stress and strain among other factors and on the hydraulic gradient, these boundary conditions have been specified for hydraulic conductivity tests on soils to be used in sealing layers of landfills. Special attention is paid to the mineralogical composition of clays and to changes of the mineral constituents under the influence of chemicals and elevated temperatures that may be encountered at the base of a solid waste landfill. Test liquids, representative of a wide range of leachates, are recommended for the evaluation of the long term performance of compacted clay liners.

The suitability of mineral sealing compounds used for vertical cut off walls to surround landfills or areas with contaminated ground, has to be examined with the same scrutiny as in case of the mineral basal and capping liners. So the recommendations contain specific testing methods for the determination of the composition of sealing compounds, for the evaluation of their characteristics in the fresh and in the hardened state, for checking their permeability, strength and deformation behaviour. It is also important to know the workability and hardening behaviour of the sealing compounds, so appropriate testing methods are recommended accordingly.

Since the international normalisation work does not yet cover all aspects of the application of geotextiles to solid waste landfills, a special section is dedicated to suitability testing for geotextiles and related products. Existing pertinent standards are referred to, and some specialised tests are explained in detail. For example, there is a test for the evaluation of puncture protection. Geomembranes at the bottom of landfills are covered by a coarse grained drainage blanket of 16 to 32 mm diameter grain size. Under the load of the waste body, the coarse grains exert a high local pressure on to the geomembrane and may cause puncturing or excessive local deformation. To prevent failure, the geomembrane is protected by a layer of finer grained soil or by a geosynthetic product. The efficiency of the protecting layer is tested by a loading test that simulates the conditions at the base of a landfill. In order to reach the same degree of safety at all landfills, and to have a common basis for comparison of competing geosynthetic products, it is necessary to have specifications for the suitability tests on puncture protection. The testing method recommended by the German Geotechnical Society has been applied successfully during the last few years.

It was mentioned earlier, that geomembranes for sealing systems of landfills need the approval of the government authorised institution *BAM*. Since the testing methods applied for the approval procedures are specified by *BAM*, the recommendations of the German Geotechnical Society do not contain any specific information on geomembrane testing. However, a testing method for the determination of the frictional resistance at geomembrane surfaces in contact with soil or with other geosynthetic products is included.

The suitability of coarse sand, gravel, crushed rock or other mineral products for drainage blankets in leachate collection or in capping systems has to be tested as well. Such mineral materials are commonly used in road construction, so many of the established testing methods can be employed. However, since the requirements and the milieu in landfills are different than in road construction, additional specific tests have been developed and specified the recommendations for mineral drainage material of waste landfills.

It was mentioned earlier, that the stability and the deformation behaviour of the landfill body are analysed in the design process, and that special methods have been developed for the exploration of waste bodies and for recovering samples of waste material from existing landfills. To provide the designer with waste material parameters, the geotechnical engineer can carry out appropriate laboratory tests on waste samples. Specific recommendations are given accordingly, after the scientific work done in the field of „waste mechanics“ by Jessberger and Kockel (1993).

3.4 *Construction methods*

There are only three recommendations on construction methods up to now. The recommendation on cut-off-wall construction gives a general overview of 9 different systems of construction of vertical barriers. Their principles are listed, the soil conditions where they can be executed, and the construction material involved. More detailed information is presented on diaphragm walls which are quite commonly used in remedial works to surround old waste disposals or areas of contaminated ground. The one-phase-method and the two-phase-method are described. The equipment for various depths of excavation is mentioned, and it is pointed out, that the accuracy of execution of the job has to be monitored by adequate measuring device in order to achieve the required overlap of primary and secondary panels. Other techniques of cut-off-wall construction mentioned are thin walls, intersecting bore pile walls, sheet pile walls, grout curtains, jet grouted curtains and ground freezing. These methods can be used in situations where no perfect barrier is required, or for temporary purposes.

The recommendation on the construction of drainage blankets is of importance, because it accounts for the sensitivity of the coarse grained gravel used for leachate collection systems. Experience shows that coarse gravel tends to being crushed under the load of construction traffic. In order to minimise the amount of finer particles which impair the long term function of the leachate collection system, certain rules must be followed in the placement of mineral drainage material. The recommendations contain a guidance for this work.

The leachate collection systems of older landfills often do not function satisfactorily. If repair work has to be carried out at collection pipes or shafts, it is necessary to dig deep excavations into the waste body all the way down to the basal liner. Such construction activities have to be prepared very carefully. The structural stability of the excavated pit has to be warranted, leachate can be encountered at different elevations and at the bottom, it must be collected safely. Explosive gas is a menace during such work, so gas sensors and ventilation are mandatory. Altogether, it is of utmost importance to take all precautions necessary, to avoid any health risks for the construction personnel. This example shows, that guide lines for the execution of construction work at solid waste landfill sites are needed, because these jobs are much more difficult than common construction work. The committee on the geotechnique of landfill structures of the German Geotechnical Society continues to prepare more recommendations on construction problems.

3.5 *Quality assurance*

To ensure the quality of the overall structure of a landfill, the individual components must meet the quality standards. Quality assurance must relate to both the quality of the materials used and to the quality of the workmanship in accordance with the state of technology. Quality assurance plans have to be prepared as part of the design work. They should specify responsibilities and tasks of the construction supervisors, description of the construction members concerned and processes to be inspected, type and number of quality tests to be undertaken on the construction materials supplied, on the processing and on the completed component.

The recommendations present details of quality assurance and quality control for mineral capping and basal sealing layers, vertical cut-off-walls, geotextiles, drainage blankets and the placement of waste material at zones within the landfill body which are relevant to the overall stability of the landfill structure. The recommendations on quality assurance are in full agreement with the Federal Instructions *TA-A* and *TA-SI*. They present the required details for the compliance with the general quality requirements put down in the governmental regulations.

3.6 *Hydraulic measures*

The principles of theoretical and numerical contaminant migration models are presented. The complexity of the problem is discussed, and it is shown where the application of numerical analyses is useful, and where the limitations of modelling are encountered. It is pointed out, that the determination of reliable transport parameters may be difficult, but that it is important for useful analyses to have sufficient field data. It is concluded, that in planning new landfills and the rehabilitation of abandoned sites, careful use can be made of available numerical models. A simple flow model only offers information on the convective transport of harmful substances. By using advanced mass transport models, influences of diffusion, sorption, advection and degradation can be taken in consideration. Finally, the effectiveness of any rehabilitation measure, and sensible alternatives to it, should be checked by a model calculation. The recommendations contain this general advice and they also list typical magnitudes of transport parameters for preliminary numerical studies.

4 SUMMARY

The German Geotechnical Society has prepared recommendations on the design and construction of solid waste landfills and remedial works. 53 topics have been dealt with up to now, the series of recommendations is continuously updated and amended. During the past decade, extensive research into various geotechnical problems in connection with solid waste landfills, particularly regarding the mechanical properties of waste material and the performance of the technical barriers, has been carried out in Germany. The results of these research efforts have been incorporated in the recommendations. Since the task committee of the German Geotechnical Society is publishing their results annually, the recommendations can be kept up to date, in spite of the rapid technical development in environmental geotechnical engineering. The task committee of the German Geotechnical Society is composed of geotechnical researchers, designers, practising engineers from the construction industry, landfill owners and regulators. The discussion among the experts from these different interest groups is reflected by the recommendations. There is a quick feed back from the practical application of the recommendations to the writing committee. The experience with the recommendations during the past ten years has been very positive, they have contributed substantially to the development of environmental geotechnology in Germany and certainly, their influence in the protection of our environment has been highly beneficial.

REFERENCES

TA Abfall, (1991). *Zweite allgemeine Verwaltungsvorschrift zum Abfallgesetz, Teil 1: Technische Anleitung zur Lagerung, chemisch/physikalischen, biologischen Behandlung, Verbrennung und Ablagerung von besonders überwachungsbedürftigen Abfällen vom 12. März 1991,* GMBL., 42. Jahrg., Nr. 8, S. 139 ff, Carl Heymanns-Verlag, Köln

TA Siedlungsabfall, (1993). *Technische Anleitung zur Verwertung, Behandlung und sonstigen Entsorgung von Siedlungsabfällen,* Bundesanzeiger, 1993

BAM (1992). Bundesanstalt für Materialforschung und -prüfung, Berlin, *Richtlinie für die Zulassung von Kunststoffdichtungsbahnen als Bestandteil einer Kombinationsdichtung für Siedlungs- und Sonderabfalldeponien sowie für Abdichtungen von Altlasten*

GDA (1993). *Empfehlungen des Arbeitskreises „Geotechnik der Deponien und Altlasten"- GDA, 2. Auflage,* Ernst & Sohn, 1993

GLR (1993). *Geotechnics of Landfill Design and Remedial Works, Technical Recommendations - GLR, 2nd Edition,* Ernst & Sohn, 1993

Jolas, P., Pfeiffer, H., Scheffler, H. (1991). *Lastsetzungen von Haldendeponien auf unverdichteten Kippenflächen.* Veröffentlichungen des LGA Grundbauinstituts, Heft 59, S. 71 - 91.

Reuter, R. (1993). *Stabilitätsuntersuchungen bei Altdeponien - eine Grundlage für die Aufstellung von Betriebsplänen und für die Schlußabnahme,* 14. Tagung „Fortschritte der Deponietechnik 1993", Haus der Technik e.V. Essen.

Stief, K. (1986). *Das Multibarrierenkonzept als Grundlage von Planung, Bau, Betrieb und Nachsorge von Deponien.* Müll und Abfall 1/86.

Neff, H., K., Walter, H. (1993). *Handlungs- und Reparaturrichtwerte für die Nachsorge der Siedlungsabfalldeponie Dreieich-Buchschlag der Stadt Frankfurt am Main.* Unveröffentlichtes Gutachten 1991 und unveröffentlichtes EDV-Programm-Paket zur dreidimesionalen Auswertung der Verformungsmessungen.

Jessberger, H., L., Heibrock, G., Demmert, S. (1993). *Entwicklung eines Sicherheitskonzepts für Deponieabdichtungssysteme.* 4. Fachlicher Zwischenbericht BMFT-Verbundvorhaben „Deponiebasisabdichtungssysteme" (1993).

Environmental Geotechnics, Kamon (ed.) © *1996 Balkema, Rotterdam. ISBN 90 5410 848 7*

Evaluation of heavy metals in contaminated soil

Y. Kang & T. Kawachi
Technical Research Institute, Obayashi Corporation, Tokyo, Japan

T. Miura & T. Kosaki
Soil Science Laboratory, Faculty of Agriculture, Kyoto University, Japan

ABSTRACT: In August, 1991, Japan Environmental Agency has set the Environmental Quality Standards for soil pollution (EQSs). The solvent used in the leaching test for EQSs is pure water with a pH of 5.8-6.3. It has no buffer and only reflects the soil pH itself at sampling time. In this study, the suitability of the EQSs leaching test was discussed, and some suggestions are made regarding the EQSs. We examined six contaminated soils by the leaching test, and fractionated heavy metals (Cu, Zn and Cd) into exchangeable, inorganically bound, organically bound, oxide occluded and residual fractions. The results of the leaching test showed that Cd, a substance included in the standards, was undetectable. However, the results of the sequential extraction showed that around 75% of the Cd occurred in the first three fractions which are considered to be in equilibrium with the soil solution. Cd in these fractions may leach out from soil when environmental conditions change. When the pH of the soil was lowered, Cd began to leach out at a pH of around 5. In addition, the value at a pH of 4.2, which could represent the pH of acid rain, was more than 8 times the value of EQSs. It is concluded that the leaching test for EQSs can evaluate the leaching amount of substances from contaminated soil at sampling time, but it cannot estimate the potential dangers of contaminated soil. Hence, use of the EQSs for the purpose of providing clean and safe water may not be altogether adequate. Since soil pollution is an accumulating pollution, evaluation of the potential leachability of the substances is important when considering soil contamination problems. Our suggestion is to introduce total content to the EQSs, and to use it together with the present standards.

1 INTRODUCTION

Heavy metal contamination of soil in Japan can generally be divided into two cases: agricultural land and urban soil. The contamination of agricultural land has been known since the nineteenth century. Copper contamination of soil in the Watarase river region (Gunma, Tochigi and Ibaraki prefectures) and cadmium contamination in the Jintsu river region (Toyama prefecture) are two famous cases. In both of these two cases, soil was polluted by mining activities and rice in these areas contained toxic levels of heavy metals, resulting in deformities among people who consumed it.

Urban soils are polluted with heavy metals principally by the precipitation of those contaminants which are contained in aerosols or suspended particles. Locally, however, the settling of dust from certain sources, the burying of solid wastes, and the infiltration of wastewater also play a part in urban soil pollution (Komai, 1981). In fact, contamination resulting from the latter sources has often been discovered at construction sites in recent years, having been left behind by factories and institutes that have moved to the suburbs. More attention should be paid to urban soils contaminated with the heavy metals, since they directly influence human health.

In 1970, the Japanese Government enacted the "Agricultural Land Soil Pollution Prevention Law" to protect the production of healthy crops. The standard for cadmium was determined based on the effects that the cadmium content of rice had on human health. Cadmium accumulated in rice grown in cadmium-containing soil. Those values for copper and arsenic are set at levels for which the growth of rice plants becomes significantly affected by substances in the soil.

In 1991, Japan's Environmental Agency set the Environmental Quality Standards for soil pollution (EQSs) in order to protect the quality of soil and its ability to purify and provide clean and safe water. Furthermore, the Environmental Agency added fifteen substances to the EQSs and made some of the existing standard values more stringent in 1994. Soil is evaluated by the concentration of leached substances in a test liquid prepared by mixing and shaking soil samples with 10 times as much water by volume as the soil sample weight. These standards apply to almost all sites, including agricultural land. However, the EQSs themselves do not have the power to enforce clean-up measures in accordance with the standards levels, and as of now, there are no laws aimed at minimizing soil pollution except for that which regulating of agricultural land (Gotoh and Udiguchi, 1993; Kawamura and Gotoh, 1995).

The EQSs leaching test is based on that for industrial solid wastes. It has been indicated that the leaching test presents some problems in its use for industrial solid wastes (Hayakawa, 1990). The EQSs suffer from similar problems. The solvent used in the leaching test is pure water with a pH of 5.8-6.3. It has no buffer and only reflects the soil pH itself at the time of sampling. However, soil pH can change under natural conditions. An example of this is the effect of acid rain on soil (Nanbu et al., 1994; Baba et al., 1995) which would lower soils pH and causing the heavy metals to become more soluble. Since the solution in contact with the contaminated soil under natural condition contains organic matter and/or salt, it's buffer must be greater than pure water. Although the EQSs set the reference values for some substances, the standards give no consideration to the total contents and the behaviors of heavy metals in soil. Even the results of the leaching test are lower than the standard values, if the total contents and the behaviors of heavy metals are different, the potential environmental impact would also be different.

To evaluate heavy metals in contaminated soil, contaminated paddy soils were used in this study to investigate the forms and behaviors of copper (Cu), zinc (Zn) and cadmium (Cd), and the leaching patterns of these heavy metals as they relate to changing pH. Furthermore, these soils were evaluated by EQSs, and the suitability of the EQSs leaching test was discussed. Finally some suggestions are made regarding to the EQSs.

MATERIALS AND METHODS

Soil Characteristics

Six samples of contaminated paddy soils were collected from an area impacted by historical mining activities. Some of the characteristics of these soils are given in Table 1. The soil samples were air-dried, passed through a polyethylene sieve, and stored in polyethylene containers.

Experimental Procedures

Total (hydrofluoric acid, perchloric acid and nitric acid decomposed), water-soluble (water/soil = 5; shaking time 2 hours) and 0.1NHCl-soluble (0.1NHCl/soil = 5; shaking time 2 hours) Cu, Zn and Cd were measured with the Inductively Coupled Plasma method (ICP).

The pH Effect on Heavy Metals Leachability. Sample 1 was used for this experiment. The pH of the soil was adjusted approximately to 6.5, 6.0, 5.5, 5.0, 4.5, 4.0, 3.5 and 3.0, and the leaching test for the EQSs was performed. The pH and Cu, Zn and Cd contents of the leachates were measured.

Leaching Test. The leaching test corresponded to the Japan Environmental Agency's extraction procedure. Twenty-five grams of contaminated soils were placed in 500ml shaking bottles, and 250ml of deionized water (pH = 5.8 to 6.3) was added. The bottles were capped with rubber caps and placed on a shaker table at 25 °C for 6 hours. The liquid was decanted and vacuum filtered through 0.45 mm filters. After measuring the pH, the Cu, Zn and Cd contents of the leachates were measured by ICP.

Sequential Extraction. The sequential extraction method proposed by Sadamoto et al. (1994) was chosen for fractionation of Cu, Zn and Cd. Sadamoto et al. fractionated heavy metals into exchangeable (0.05M$CaCl_2$/soil = 1:10, 24 hours), inorganically bound, organically bound, oxides occluded and residual fractions.

Table1 Characterisstics of the soils

Soil	pH	CEC (cmol(+)/kgsoil	Total C (%)	Total N (%)	Sand (%)	Silt (%)	Clay (%)	Free Fe_2O_3 (%)
Sample1	6.4	11.3	1.6	0.12	68.4	17.9	13.7	1.0
Sample2	6.4	10.3	1.5	0.11	68.9	18.5	12.6	1.1
Sample3	7.2	10.8	1.7	0.11	73.4	15.7	10.9	0.7
Sample4	6.7	8.3	1.3	0.09	78.4	12.5	9.1	0.6
Sample5	7.1	10.2	1.6	0.11	73.3	16.1	10.6	0.7
Sample6	7.1	10.1	1.4	0.10	73.9	15.7	10.4	0.7

RESULTS

The total, water-soluble, and 0.1NHCl soluble Cu, Zn and Cd contents of the soils are shown in Table 2, Table 3 and Table 4, respectively. Compared to natural abundance, the total Cu, Zn and Cd contents of the soils used in this study were significantly higher (Table 2).

Table 2 Total Cu, Zn and Cd contents of the soils

Soil	Cu ($\times 10^2 \mu g/g$)	Zn ($\times 10^2 \mu g/g$)	Cd ($\mu g/g$)
Sample 1	2.2	11.4	3.6
Sample 2	2.3	11.5	3.4
Sample 3	1.0	8.9	3.7
Sample 4	0.5	6.7	3.5
Sample 5	1.0	8.4	3.5
Sample 6	0.9	8.4	3.5
Natural abundance	0.30	1.0	0.5

No water-soluble Cd was detected, and both the water-soluble Zn and Cu contents were extremely low (Table 3).

Table 3 Water-soluble Cu, Zn and Cd contents of the soils

Soil	Cu	Zn ($\mu g/gsoil$)	Cd
Sample 1	0.38	0.43	n. d.
Sample 2	0.35	0.35	n. d.
Sample 3	0.22	0.20	n. d.
Sample 4	0.12	0.20	n. d.
Sample 5	0.21	0.17	n. d.
Sample 6	0.21	0.22	n. d.

When these soils were extracted by 0.1NHCl, 63.7% of the total Cd was extracted on the average, indicating that the greater portion of Cd was present in relatively easier dissolving forms.

Table 4 0.1NHCl extracted Cu, Zn and Cd contents of the soils

Soil	Cu ($\times 10 \mu g/g$)	Zn ($\times 10 \mu g/g$)	Cd ($\mu g/g$)
Sample 1	6.5	12.8	2.6
Sample 2	5.9	12.4	2.3
Sample 3	1.9	9.1	2.1
Sample 4	0.8	8.0	2.2
Sample 5	1.6	8.6	2.2
Sample 6	1.4	8.2	1.9

1) The pH Effect on Heavy Metal Leachability

Fig. 1 showed the pH effect on heavy metals leachability. The concentrations of Cu, Zn and Cd were increased with lowering soil pH. The Cd concentration was higher than the standard value (10ng/ml) at about a pH of 5, and more than 8 times that of the standard Cd value leached out from the soil at a pH of 4.2.

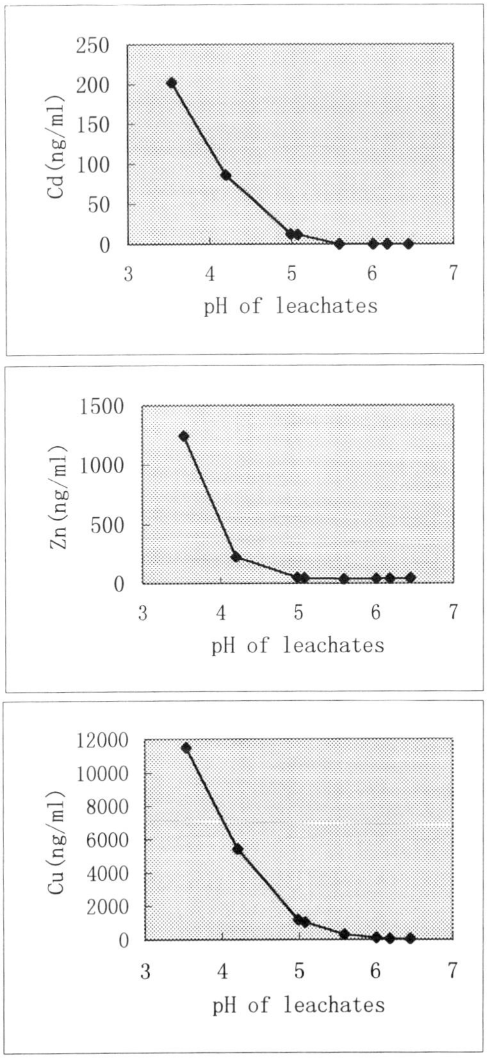

Fig. 1 The pH effect on heavy metals leachability from contaminated soil

2) Forms of Heavy Metals in Contaminated Soils

Table 5 shows the results of sequential extraction of Cu, Zn and Cd. These metals are divided into five fractions: (1) Exchangeable fraction, (2) Inorganically-bound fraction, (3) Organically bound fraction, (4) Free oxides occluded fraction and (5) Residual fraction. The first three are considered to equilibrate with the soil solution, therefore, they may leach out when soil conditions change. The latter two are considered to be insoluble fractions (Sadamoto et al., 1994).

Cu presented in the exchangeable, inorganically-bound and organically-bound fractions occupied about 40% of the total Cu in the soils. In these three fractions, the ratio of organically-bound Cu was the highest, reflecting the property that Cu makes stable coordinate bond with humic acid. The ratio of inorganically-bound Cu was also relatively higher, but that of exchangeable Cu was extremely low. The ratio of residual Cu was about 50% of the total Cu.

Zn is present mainly in oxides occluded and residual fractions. The sum of the ratios of these fractions was around 80% of the total Zn in the soils. Compared to Cu, the percentage of the exchangeable Zn increased somewhat. This is because Zn is a diffusion element, more Zn accumulated in the exchangeable fraction.

In the case of Cd, around 75% of Cd occurred in the exchangeable, inorganically-bound and organically-bound fractions. Compared to Cu and Zn, the ratio of the exchangeable Cd was significantly higher. Sadamoto et al.(1994) reported that this is because the coordinated bond stability of Cd and humic acid is low. They also indicated that since large amounts of Cu derived from contamination was adsorbed by humic acid or free oxides, more Cd occurred in the exchangeable fraction.

Tablc 5 Thc Forms of hcavy metals in contamiated soils

Soils	Exchangeable	Inorganically bound	Organically bound	Oxides occluded	Residuial	Total content
	Cu(μg/gsoil)					
Sample 1	0.2	22	73	14	110	2.2×10^2
Sample 2	0.2	21	72	34	101	2.3×10^2
Sample 3	trace	7	33	9	50	1.0×10^2
Sample 4	trace	3	18	1	27	0.5×10^2
Sample 5	trace	7	32	9	47	1.0×10^2
Sample 6	trace	7	31	9	44	0.9×10^2
	Zn(μg/gsoil)					
Sample 1	4.3	103	126	420	489	11.4×10^2
Sample 2	4.6	96	111	404	536	11.5×10^2
Sample 3	0.5	63	66	356	411	8.9×10^2
Sample 4	1.0	57	62	257	290	6.7×10^2
Sample 5	0.6	59	62	350	372	8.4×10^2
Sample 6	0.5	61	67	351	356	8.4×10^2
	Cd(μg/gsoil)					
Sample 1	0.31	1.9	0.8	0.4	0.2	3.6
Sample 2	0.29	1.6	0.9	0.3	0.3	3.4
Sample 3	0.08	1.5	1.0	0.4	0.7	3.7
Sample 4	0.14	1.5	1.1	0.3	0.5	3.5
Sample 5	0.09	1.4	1.0	0.4	0.6	3.5
Sample 6	0.07	1.5	1.0	0.4	0.5	3.5

3) Results of the Leaching Test

The results of the leaching test are shown in Table 6. The Cu and Zn contents of the leachates were 0.21 — 0.68 μg/ml and 0.41 — 0.78 μg/ml, respectively. However, the concentration of Cd, an EQSs substance, was below the detection level for all soils examined in this study. The pH of the leachates ranged from 6.4-7.2.

Table 6 Cu, Zn and Cd contents in leachates

Soils	Cu	Zn	Cd
		(μg/ml)	
Sample 1	0.78	0.68	n.d.
Sample 2	0.63	0.55	n.d.
Sample 3	0.66	0.36	n.d.
Sample 4	0.74	0.21	n.d.
Sample 5	0.41	0.35	n.d.
Sample 6	0.44	0.34	n.d.

DISCUSSION

When the soils were extracted by the EQSs leaching test, the Cu and Zn concentrations of leachates were very low, and Cd was undetectable. Limited to Cd, the soils used for this study were concluded as harmless, since only Cd is a substance included in the standards. However, as the results of the sequential extraction showed, around 75% of the Cd occurred in the fractions which are considered to be in equilibrium with the soil solution. Cd in these fractions may leach out from soils when environmental conditions change. Although the salt concentration (0.05M $CaCl_2$) of the solvent for the exchangeable fraction was one-tenth that of sea water, a significant amount of Cd was extracted (7-31ng/ml) as the EQSs value for Cd is 10ng/ml. On the other hand, when the pH of the soil (sample 1) was lowered, Cd began to leach out at a pH of around 5. In addition, the value at a pH of 4.2, which could represent the pH of acid rain, was more than 8 times the value of EQSs. The leaching test for EQSs can evaluate the leaching amount of substances from contaminated soil at the time of sampling, but it cannot estimate the potential dangers of contaminated soil. Hence, use of the EQSs for the purpose of providing clean and safe water may not be altogether adequate. Since soil pollution is an accumulating pollution, evaluation of the potential leachability of the substances is important when considering soil contamination problems.

There are two basic considerations for environmental quality standards in the world. One is using total contents to divide the soil contamination into several levels, as is used in Holland (VROM, 1983; Van Den Berg et al., 1993), German (Senatsrensaltung fur Stadtent-wicklung und Umweltschutz, 1990) and Canada (Canadian Council of Ministers of the Environment, 1991); and the other is conducting risk assessments like in the USA (Department of Environmental Protection and Energy, 1992). Japan is a rare case in that it only uses the leaching test. Since the basic consideration of the Japanese environmental quality standards for soil is protecting the water system from pollution, the standards should be evaluated for that purpose. As an important environmental factor, we must consider on-site soil contamination and ultimately work to remediate contaminated soil. As described above, many countries in the world are using the total content of pollutants as their standards. Even in the USA there is a movement to adopt this kind of consideration (U.S.Environmental Protection Agency, 1994). Therefore, our suggestion is to introduce total content to the EQSs, and to use it together with the present standards.

CONCLUSION AND RECOMMENDATION

Since the solvent used in the leaching test for EQSs is pure water with a pH of 5.8-6.3, it only reflects the soil pH itself at sampling time. From the results of this study, it is concluded that the leaching test for EQSs can evaluate the leaching amount of substances from contaminated soil at sampling time, but it cannot estimate the potential dangers of contaminated soil. Hence, use of the EQSs for the purpose of providing clean and safe water may not be altogether adequate. Since soil pollution is an accumulating pollution, evaluation of the potential leachability of the substances is important when considering soil contamination problems. Our recommendation is to introduce total content to the EQSs, and to use it together with the present standards.

REFERENCES

Baba, M., M. Okazaki and T. Hashitani: Effect of acid deposition In forested Catchment In the Western Tokyo, Japan. Water, Air and Soil Pollution, 85, pp.1215-1220, 1995.

Canadian Council of Ministers of the Environment: Interim Canadian Environmental Quality Criteria for Contaminated Sites, Report CCME EPC-CS 34, 1991.

Department of Environmental Protection and Energy:

SITE REMEDIATION PROGRAM, Cleanup Standards for Contaminated Sites, Proposed New Riles, N.J.A.C., 7: 26D, 1992.
Hayakawa, R.: Extraction Procedure for the Evaluation of Industrial Solid Wastes. Water pollution investigation (Japan), Vol.13, No. 3, pp.23-27, 1990.
Gotoh S. and A. Udoguchi: Japan's policies on soil environment protection - History and present status, edited by F. Arendt, C.J. Annokkee, R.Vosman and W.J.van den Brink. Contaminated Soil, pp.3-10, 1993.
Kawamura, K., S. Gotoh: Japan's national soil quality standards as expanded in 1994, edited by W.J. ven den Brink, R. Bosman and F. Arendt, Contaminated Soil, pp. 735-736, 1995.
Komai Y.: Heavy metal pollution in urban soil, edited by Kitagishi K. and I. Yamane. Heavy Metal Pollution in Soils of Japan. pp.193-215.
Nanbu K., T. Kunimatsu and K.Kyuma: Rates of Soil Acidification under Different Patterns of Nitrogen Mineralization. Soil Sci. Plant Nutr. 40(1), pp95-106, 1994.
Sadamoto H., Y.Iimura, T. Honna and S. Yamamoto: Examination of fractionation of heavy metals in soils. Jpn. J. Soil Sci. Plant Nutr. 65, p.645-653,1994.
Senatsrensaltung fur Stadtent-wicklung und Umweltschutz: Bewertungs kriterien fur die beurteilung kontamierter standorte in Berlin (Berlin Listre), 1990.
U.S.Environmental Protection Agency, Region IX: The Proposed 'New' Superfund. U.S. EPA, March 1994.
Van Den Berg, R., Denneman, C.A. and Roels, J. M. : Risk Assessment of Contaminated Soil; Proposals of Adjusted, Tozicologically based Dutch Soil Clean-up Criteria, Contaminated Soil '93, pp.349-364,1993.

Environmental Geotechnics, Kamon (ed.) © *1996 Balkema, Rotterdam. ISBN 90 5410 848 7*

Reliability-based decision analysis for waste containment systems

Arthur S. Kurzydlo
Conestoga-Rovers & Associates, Chicago, Ill., USA
Jamshid Mohammadi
Civil Engineering Department, Illinois Institute of Technology, Chicago, Ill., USA

ABSTRACT: A reliability-based decision analysis for the selection and design of the containment system is shown to be an efficient tool. First, several design scenarios (alternatives) are created based on the preliminary selection of appropriate technologies. Then, reliability analysis for each element of the system is performed. The analysis is based on a defined performance function that incorporates all available data for design and control variables. The uncertainty associated with data quality is also included. Using a decision tree, each alternative is presented through the series of decision-making parameters. System reliability, cost and safety index, (when information on probability distribution is not available) are used as determining factors for the selection of most favorable alternative.

1. INTRODUCTION

The process for selecting a remedy, like waste containment system, generally includes the following steps:
1. Problem definition
2. Objectives formulation
3. Development of alternatives
4. Analysis of alternatives
5. Remedy implementation and monitoring

Problem definition stage identifies the current and future hazards associated with the site. Usually, hazards are identified in terms of sources, pathways and receptors. Every problem definition includes information on release mechanism, environmental transport, potential receptor points, routes of human exposure and routes of environmental exposure.

Objectives are formulated based on the fundamental goal to protect human health and the environment. Since the problems are defined in the context of pathways, the remediation objectives for addressing these problems have to relate specifically to the pathways. The objectives are expressed in measurable terms like contaminant concentration at specific points of exposure. Objectives are established through different methods:
1. Regulatory requirements
2. Site-specific risk assessment
3. Background conditions
4. Technology-based solutions

Applicable or Relevant and Appropriate Requirements (ARARs) from the Superfund program are an example of the regulatory requirement. At certain sites the site-specific risk assessment is used to establish a target level of risk that becomes the remediation objective. The generally recognized levels of acceptable level of lifetime carcinogenic risk (1 x 10E-6) and non-carcinogenic risk are used as target levels. The background condition is used at industrial sites surrounded by heavy industry. There, the risk to receptors may not diminish measurably unless all surrounding properties are cleaned-up to the same target levels. The technology-based standards are applied where the available technology is overly protective.

Once the objectives are established, the remedial alternatives are developed in the following fashion. For each objective a series of general response actions is identified. A general response action specifies only an activity needed for remediation, but not technology. Next, the media are characterized to allow for general sizing of technologies. Information like the volume of soil exceeding the levels identified in a remediation objective or extend of groundwater contamination are collected.

After characterizing the media, potentially applicable response technologies are identified. They are screened using the criteria like, suitability, implementability, performance, environmental and health concerns, institutional concerns and cost. Technologies not eliminated are retained for further consideration. The retained technologies are assembled into a set of alternatives that could be undertaken to remediate the site.

The technologies form a treatment train that together with site controls, such as monitoring, creates an alternative. Once alternatives are assembled the selection process is initiated. However, due to the nature of parameters involved in the decision making process, a probabilistic approach offers a more favorable analysis technique. This paper presents a probabilistic decision analysis (selection) approach for the design of waste containment systems.

2. CONTAINMENT SYSTEM

A typical containment system is a combination of several measures such as vertical barrier wall, interceptor drain, extraction well and impermeable cap. Each element of this system is designed to perform a certain function.

Reliability of the containment system is a function of each component's reliability. Component reliability is evaluated through the performance function. The performance function allows for incorporation of design and control variables into each component expected (probable) level of performance. Several design and control variables for different components are usually evaluated.

2.1 *Design and control variables*

Each element of the containment system is designed, constructed and later evaluated based on several variables. Those variables are related to the material properties, design methods, construction methods, and functional requirements (future use of the site). It is necessary to recognize and analyze those variables, such that each component of the containment system can be modeled. The following is a description of variables for vertical barriers and cover/caps only.

2.1.1 Vertical barriers

Several different types of vertical barriers are currently used in hazardous waste and groundwater containment systems. Among most popular barriers are slurry walls, steel sheet pile walls, vibrating beam walls, grout curtains, deep soil mixing type walls, jet grouted walls, and composite walls.

Design variables for vertical barriers include, (i) common and (ii) specific variables. Common variables are wall permeability, longevity, and deformability. Those variables, in turn are functions of soil characteristics, key-in requirements, associated remedial measures, construction methods, quality control, compatibility with contaminants and/or contaminated groundwater, performance monitoring, laboratory techniques.

Specific variables for slurry walls, for example, include bentonite, attapulgite, concrete, biopolymers or any other backfill material characteristics. For steel sheet piles specific variables include steel corrosion and sealant's (waterstop) compatibility and durability.

2.1.2 Surface water control, capping and liners

Three different technologies, namely surface water control, capping and liners, are generally employed for the containment system. Each of these technologies performs a different role.

Surface water control measures minimize the amount of water flowing onto a site, thus reducing the amount of potential infiltration. Capping minimizes the infiltration of any surface water or direct precipitation. Liners provide ground water protection by inhibiting downward flow of leachate.

Depending on the water balance for the site, a combination of permeable/impermeable cap and impermeable liners are used.

Very often containment systems are designed and constructed for the existing facilities; and as such liners can not be employed. For those facilities a combination of capping, vertical barrier and wells/interceptor drains can provide a means of environmental control.

Many different materials are employed in covers/caps. Depending on the requirements and future function, a more suitable material can be chosen. Currently the most popular cap systems include: compacted soil cover systems, composite covers (soil/geomembranes), asphalt caps, concrete caps. Many other commercial membranes/covers are available for this purpose, and can be used if appropriate.

Certain variables are common for every type of cap. These include; permeability, longevity and deformability. They depend on subgrade or waste characteristics, future function or use, associated remedial measures, construction methods, quality control, laboratory techniques and performance monitoring. In addition, each type of cap has specific variables, such as soil characteristics and compaction requirements for compacted soil.

For composite covers in addition to variables characteristic to the clay covers, geomembrane brings new variables such as, type and characteristics of geomembrane. They are a function of manufacturing method, conformance testing, effects of shipping, handling and storage, placement technique, subgrade condition, temperature effect, wind effect, seaming and joining, destructive testing, nondestructive testing, protection and backfilling.

Asphalt cap will require variables that are dependent upon subgrade stability, subbase quality, binder & surface quality, and maintenance record.

When concrete cap is utilized variables include characteristics like compressive and tensile strength.

2.2 *Analysis of uncertainties*

Evaluation of engineering reliability requires information on uncertainty in terms of standard deviation or coefficient of variation. Uncertainties may be associated with physical phenomena that are inherently random or with predictions and estimations that are made with inadequate information. In the case of the containment system, each element like vertical barrier or cap represents a function of several random variables. Random variables uncertainty is associated with i) inherent variability, ii) prediction error.

In general field or laboratory data is needed to quantify uncertainties. However one may rely on expert judgment in establishing useful data if field or laboratory information is not readily available (Mohammadi, et al., 1991)

3. RELIABILITY EVALUATION

Reliability of vertical barrier, cap and collection/treatment system can be evaluated the same way the reliability of many other engineering systems are evaluated, by formulating the system performance in terms of a capacity and a demand function. Capacity or demand of each measure is a function of several variables as described in the preceding section. The reliability formulation is as follows, a performance function describing the level of performance of a system is represented as (Ang and Tang, 1990):

$$g(x) = g(x_1, x_2, \ldots, x_n) \qquad (1)$$

where $X=(x_1,x_2,\ldots,x_n)$ is a vector of basic state variables, and the function g(x) determines the performance or state of the system.

Therefore, the safe and failure are defined as

$[g(x)>0]$=the "safe state"

and

$[g(x)<0]$=the "failure state"

A design variable can be represented by a probability density function (PDF) which is a probability of a specific value of random variable x in the interval (a,b).

The joint PDF of the design variables x_1, $x_2,\ldots,x_n$, is $f_{(x1,\ldots xn)}(x_1,\ldots x_n)$ and the probability of the safe state is

$$p_s = \int_{g(x)>0} f_x(x)\, dx \qquad (2)$$

Equation 2 is simply the volume integral of $f_x(x)$ over the safe region $g(x)>0$.

Evaluation of such an integral is a difficult task. Therefore alternative and more practical methods are employed.

Calculation of probability of safety requires the knowledge of the distribution. When the distribution is unknown the second-moment formulation can be used.

With the second-moment formulation the reliability may be measured entirely with a function of the first and second moments of the design variables. This function is a safety index, β. It is a convenient approach when no information on the probability distribution is available.

Examples how to evaluate the vertical barriers like the conventional steel sheet pile with sealant or the slurry wall are presented by the authors(Kurzydlo and Mohammadi, 1994, 1995).

Once the reliability of every component of the system is calculated the system reliability is evaluated through the bi-modal bounds method.

The next step is to construct a decision tree for different design scenarios.

4. DECISION TREE

The various components of a decision problem, for example selection of the most suitable containment system, are integrated into a formal layout in the form of a decision tree. The decision tree integrates the relevant components of decision analysis in a systematic manner suitable for an analytical evaluation of the optimal alternative.

Once the decision tree has been established, each alternative's reliability can be calculated. Then, all associated capital costs and operation and maintenance costs are computed.

A major component of the decision tree is a complete layout of various alternatives. Each alternative is expressed via a series of decision-making parameters. Information on each of the parameters (in terms of limits, ranges, potential risk, etc.) is also provided. With such information, one can utilize the approach described in the previous section to readily compute the safety index of the system for a given design alternative. In fact, the reliability analysis is intended as a means to compare the system performance for various potential design alternatives. Along with cost optimization, the safety index (β) can also be used as a determining factor for the selection of the most favorable alternative. In such an analysis, the objective is to identify an alternative with the lowest cost and the highest safety index β.

In this paper the decision tree and cost element evaluation with examples is presented.

The decision tree is build to present design alternatives. Several combinations of different caps and vertical barriers are matched with a specific collection/treatment option.

Then, the cost oriented decision analyses are performed through three different approaches; prior analysis, terminal analysis and preposterior analysis. The prior analysis is based entirely on available information. This information is included in the decision tree, thus the expected monetary value of each alternative may be computed. The terminal analysis employs updated probabilities (probabilities conditional on the experimental outcome) in the computation of the expected monetary value or utility. This analysis relays on new information from experiments or field tests to reduce the levels of uncertainty. But the additional information involves the added cost of time and money. The preposterior analysis allows to evaluate whether the additional information should be acquired. All of the above mentioned analyses were successfully employed in the containment system decision analysis for a site contaminated with DNAPL's. The decision tree included evaluation of several systems made of different vertical barriers, caps and treatment systems. Vertical barriers included in analyses were; slurry wall, steel sheet pile wall with sealant, deep soil mixing type wall and vibrating beam type wall. Two types of caps were evaluated, an asphalt cap and a concrete cap. Treatment systems included conventional pump and treat with activated carbon option and an innovative bioremediation treatment. To present the method a complete analysis is performed for a steel sheet pile wall with asphalt cap and two treatment options.

Figure 1 presents a basic decision tree for the steel sheet pile wall with selected design options for a cap (asphalt cap and concrete cap) and for a treatment system (activated carbon and bioremediation).

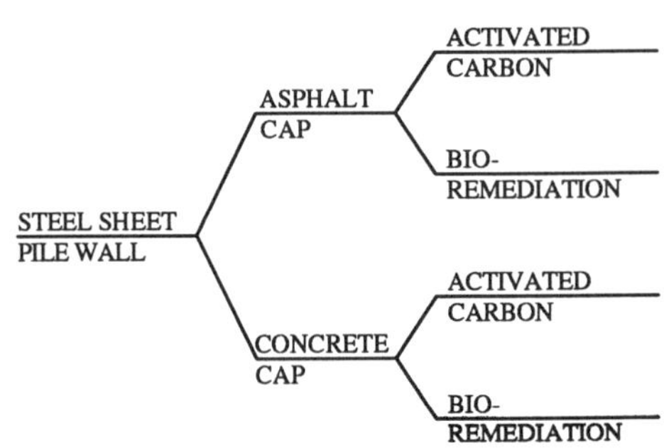

Figure 1. Basic decision tree for the steel sheet pile option

It was assumed that the efficiency of activated carbon (AC) and bioremediation (BIO) could be classified into two levels; high efficiency (EH) and low efficiency (EL). Different stages of a decision tree construction and evaluation are presented here.

First step in the decision tree evaluation was the prior analysis. For that analysis the probabilities for obtaining high (EH) or low (EL) efficiency were estimated to be 70% and 30% for bioremediation and 80% and 20% for activated carbon. It was estimated that the activated carbon option having a high efficiency would have to be operated for approximately 15 years whereas the high efficiency bioremediation with injection would operate for 5 years. The prior analysis decision tree and relative monetary value (RMV) which includes initial cost and present worth of operation and maintenance cost for each of the treatment option are presented on Figure 2.

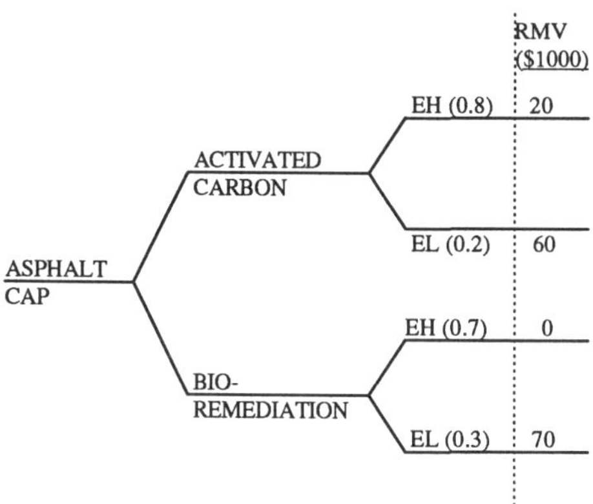

Figure 2. Prior analysis decision tree

The expected monetary value for the two treatment options is:

$$E(AC) = 0.8 \times (20) + 0.2 \times (60) = 28$$
$$E(BIO) = 0.7 \times 0 + 0.3 \times (70) = 21 \qquad (3)$$

Based on the above calculation, the bioremediation option should be the treatment of choice. But, a closer evaluation shows that the activated carbon's relative monetary value is only 33.3% higher. Additionally, the presented earlier assessment of the time required for operation of bioremediation option, based on our experience, could be off by as much as 300%.

The terminal analysis allows for inclusion of additional tests to improve the existing state of information. Between the two treatment options, bioremediation is the one which requires additional study for accurate analysis. The cost for performing the study is approximately \$10,000 and the reliability of the results is as follows: if the efficiency of the proposed bioremediation is actually high (EH), the probability that the biofeasibility will indicate a high efficiency rating, based on expert opinion, is 0.85, and the corresponding probability for low efficiency is 0.15. If the efficiency of the process is actually low (EL), the probabilities that the study will show high or low efficiency ratings are 0.3 and 0.7 respectively. The conditional probabilities are as follows:

$$P(SH/EH) = 0.85, \; P(SL/EH) = 0.15$$
$$P(SH/EL) = 0.3, \quad P(SL/EL) = 0.7 \qquad (4)$$

where SH and SL denote events of high and low efficiency ratings for the study.
The updated probability of EH in the case when study indicates high efficiency rating is given by Bayes' Theorem:

$$P''(EH) = P(EH/SH)$$

$$= \frac{P(SH/EH)P'(EH)}{P(SH/EH)P'(EH) + P(SH/EL)P'(EL)}$$

$$= (0.85 \times 0.7) \;/\; [(0.85 \times 0.7) + (0.3 \times 0.3)]$$
$$= 0.87 \qquad (5)$$

The result shows that the probability of a bioremediation actually being a high efficiency process has increased from 0.7 to 0.87. The corresponding updated probability of EL is 0.13. The expected monetary value of the bioremediation alternative, including the cost for the study, is:

$$E(BIO/SH) = 0.87 \times (10) + 0.13 \times (80)$$
$$= 19.1 \qquad (6)$$

Similarly, if the biofeasibility study indicates low efficiency the updated probability of EH is 0.33 and

EL is 0.66. Thus, the expected monetary value of the bioremediation, including the cost for the study, is:

$$E(BIO/SL) = 0.33 \times (10) + 0.66 \times (80) = 56.1 \quad (7)$$

Since the biofeasibility study adds information to the bioremediation option only, no change in the expected monetary value for the activated carbon option is realized.

The terminal analysis shown above, has improved our understanding of the difference in cost for two treatment options. But, the analysis assumed that the study was performed. Thus, the question was raised, is the study necessary? In other words, what will be the expected monetary value of the bioreemdiation option? To answer this the preposterior analysis was employed. The preposterior analysis allows to calculate the expected monetary value (EMV) for the option with the biofeasibility study. Figure 3 presents the decision tree for preposterior analysis.

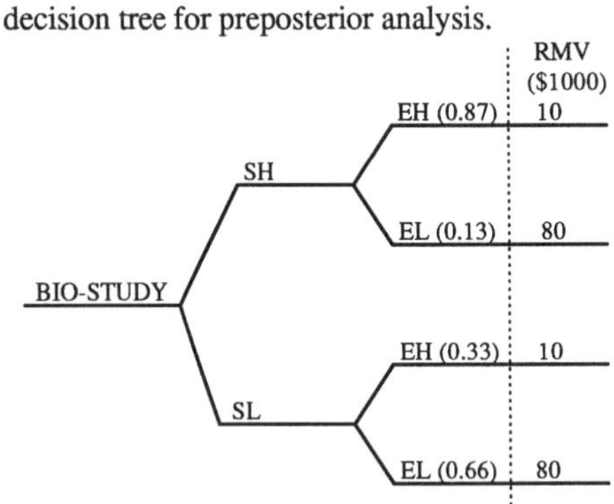

Figure 3. Decision tree for preposterior analysis

First, the probabilities that the study will indicate high efficiency rating (SH) and low efficiency rating (SL) are calculated (Ang and Tang, 1990):

$$P(SH) = P(SH/EH)P'(EH) + P(SH/EL)P'(EL) = 0.685$$
$$P(SL) = P(SL/EH)P'(EH) + P(SL/EL)P'(EL) = 0.315 \quad (8)$$

Using the above calculated values the EMV for the bioremediation with the study is found using the following formula (Ang and Tang, 1990):

$$E(Study) = P(SH) \times E(BIO/SH) + P(SL) \times E(BIO/SL) = 0.685 \times (19.1) + 0.315 \times (56.1) = 30.76 \quad (9)$$

The result indicates that the bioremediation option with the study is equally expensive as the activated carbon option. Therefore, the selection process shall concentrate on reliability of the system and not on the cost. Based on the above, complete system reliability analysis was performed and as a result, the bioremediation option was selected for the proposed containment system.

5. CONCLUSIONS

The realistic approach to the selection and design of the containment systems is presented. The cost and reliability analyses, when incorporated into the decision tree, allows for the selection of the most favorable containment system. The decision analysis should become a very important tool for a design professional, since it is allowing for the selection of the most reliable and cost effective system. It is also important for a nontechnical decision maker, because it can evaluate legal and economic benefits of the system.

REFERENCES

Ang, A.H-S. and Tang, W.H., *Probability Concepts in Engineering Planning and Design, Volume II Decision, Risk and Reliability*. John Wiley and Sons, New York, 1990.

Kurzydlo, A.S. and Mohammadi, J., Reliability Model for the Analysis of Hazardous Waste Containment Systems, *2nd International Symposium on Contamination in Central and Eastern Europe*, Budapest, 1994

Kurzydlo, A.S. and Mohammadi, J., Verical Cutoff Walls -- Reliability Analysis, *5th International Landfill Symposium,* Sardinia, 1995

Mohammadi, J., Longinow, A. and Williams, T., Evaluation of System Reliability Using Expert Opinions, *Structural Safety,* vol. 9, 1991

Environmental Geotechnics, Kamon (ed.) © *1996 Balkema, Rotterdam. ISBN 90 5410 848 7*

Monitoring of land subsidence and estimation of social capital loss

Norihiko Miura
Saga University, Japan

Yoichiro Hachiya
Saga Prefectural Office, Japan

Nobuyuki Kodama, Takashi Tanaka, Nobuo Tajiri & Masatoshi Okada
Fukken Co., Ltd, Hiroshima, Japan

ABSTRACT: Land subsidences in wide area due to the pumping up of groundwater have been occurring in some alluvial plains in Japan, and it has become a social problem. To address the subject of how to regulate the pumping up of groundwater, survey of land subsidence, observation of groundwater, and monitoring of pumping-up water quantities have been conducted to date.

Presented in this paper is a method to grasp in details the situation of the wide-area land subsidence by using aerial photographs. With an analytical plotter system, this method transforms analog data obtained from photographs into wide-area three-dimensional coordinate data. In this study, to grasp the situation of the land subsidence, the distribution of land subsidence in the 21 years in Shiroishi plain, Saga Prefecture, was analyzed based on survey results of 1973 and aerial photographs taken in 1994. Then, the loss in social capital due to the land subsidence was roughly estimated to make clear the importance of regulating the pumping up of groundwater.

1 INTRODUCTION

To date the tendency of land subsidence in a wide area has been grasped based on measured values by direct levelling of survey points provided in the area. Being a method dependent upon point-to-point data, however, it could not grasp the details of land subsidence distribution among the survey points. This time, a digital mapping system was adopted to grasp such details. Using an analytical plotter, this system collects three-dimensional digital data (topographical information in numerical values) from aerial photographs. The situation of land subsidence can be grasped by comparing these data with those of periodical surveys and past surveys.

The precision of photographic surveying by analytical plotter system is closely related to the reduced scale and when it is 1/3,000 - 1/5,000, standard deviation would be in the range from ±5 to ±10 cm (Tanaka et al. 1992, 1994).

In this study, analytical results of the aerial photographs (reduced scale: 1/8,000) taken in February 1994 were compared with the ground height map (1/10,000) of the Saga plain made in May 1973, and the distribution plan of land subsidence in the 21-years and the profiles as of 1973 and as of 1994 were drawn.

The costs of restoring the affected or damaged social capital was defined as the social capital loss, and the magnitude of the effects of the land subsidence upon the society was quantitatively estimated on the basis of the results of the surface-oriented photographic analysis of the subsidence distribution.

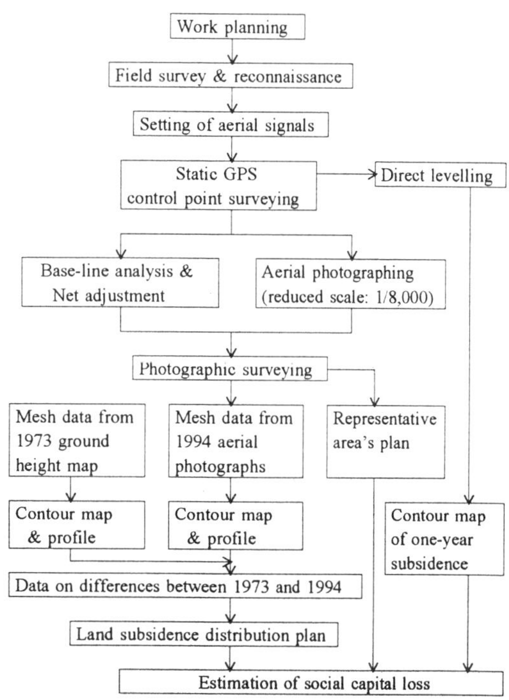

Fig. 1 Procedural flow.

2 DETERMINATION OF WIDE-AREA LAND SUBSIDENCE WITH ANALYTICAL PLOTTER SYSTEM

2.1 *Survey and analysis methods*

Fig. 1 shows the procedural flow from the measurement of wide-area land subsidence with the analytical plotter system through to the estimation of social capital loss.

2.2 *Control surveying and photographing*

Chosen for the survey was the Shiroishi and Ariake towns' area where land subsidence is most remarkable in the Shiroishi plain of Saga Prefecture (Miura et al. 1988, Tohno 1994). Fig. 2 shows the survey area.

The coordinates of aerial signals were determined by using, as postulate data, the national triangulation points and the results of the 1994 direct levelling. Static GPS control point surveying was conducted. Net adjustment was performed with the three-dimensional net adjustment method. The heights of signals were determined by using the geoid relative height map drawn from the results of the direct levelling and GPS control point surveying. The precision of the GPS control point surveying was 1.3 - 6.3 mm in terms of the standard deviations of the basic lines. On the other hand, 21 - 49 mm were the standard deviations of the new points after the three-dimensional net adjustment in case that the coordinates of the four national triangulation points had been fixed.

The aerial photographing with the reduced scale of 1/8,000 was conducted on February 28, 1994. Fig. 2 shows its course and its photographing area was about 1.6 km wide and 9.0 km long.

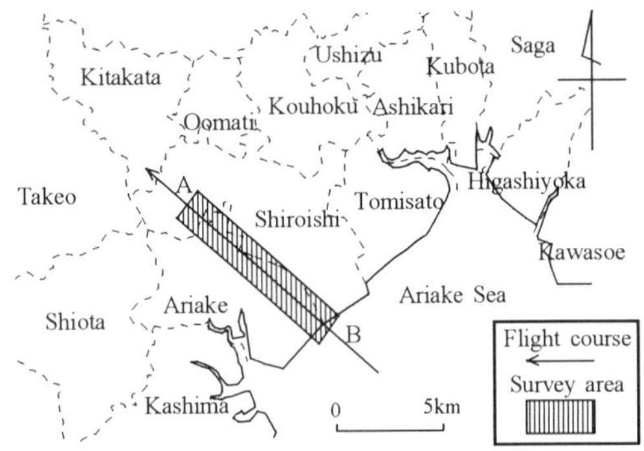

Fig. 2 Survey area and flight course in Shiroishi plain.

2.3 *Acquisition of three-dimensional topographical data and drawing of land-subsidence distribution plan*

By using the analytical plotter system, about 1,000 pieces of three-dimensional data (*X*, *Y* and *Z*) were acquired, at 100 m intervals forming meshes, from the aerial photographs taken in February 1994. The precision (standard deviations) of the photographic surveying was 0.12 m in the east-west direction, 0.11 m in the north-south direction, and 0.12 m in altitude. On the other hand, altitudes were read, at the same points with the above mesh data, from the ground height map of the Saga plain made in May 1973, and three-dimensional mesh data were prepared.

Based on the three-dimensional mesh data of the both years, contour maps (Fig. 3 and Fig. 4), a land-

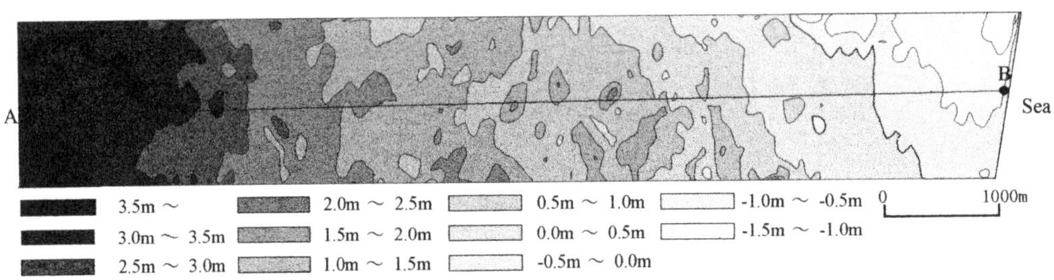

Fig. 3 Contour map (As of May 1973).

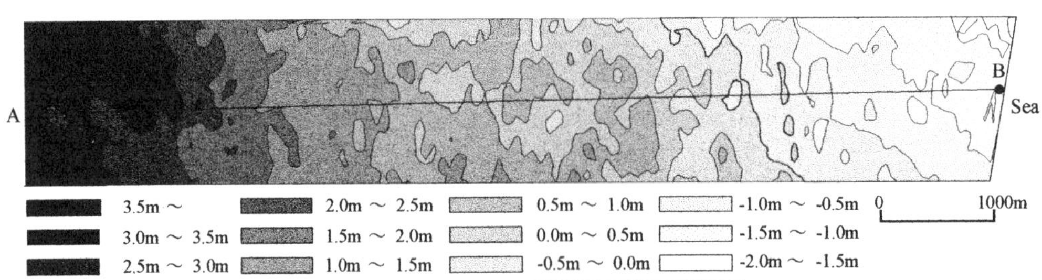

Fig. 4 Contour map (As of February 1994).

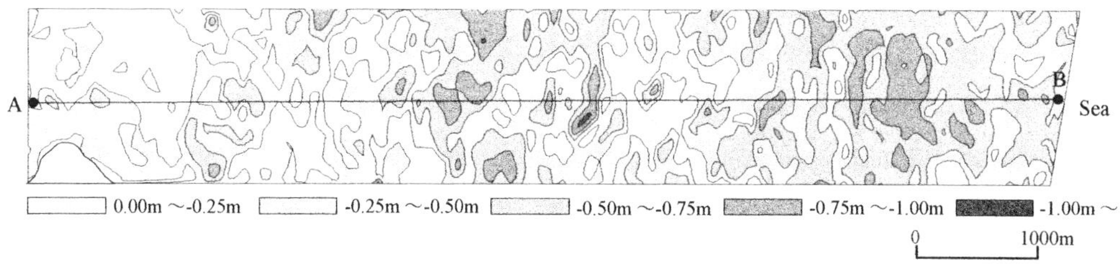

Fig. 5 Land-subsidence distribution plan (May 1973 - February 1994).

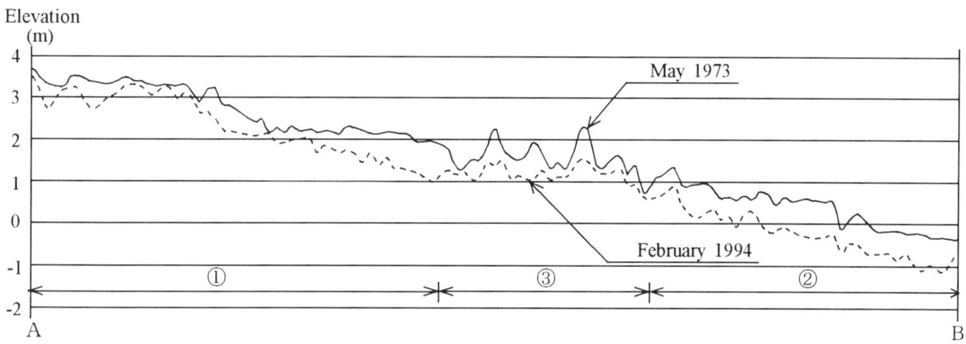

Fig. 6 Cross sectional profile of ground height in Fig.3 and 4.

subsidence distribution plan (Fig. 5), and a cross sectional profile (Fig. 6) were drawn.

The results of the surface-oriented surveying of the area of 13.8 km^2 by the analytical plotter system were as follows (Tanaka et al.1994):

1. The maximum subsidence in this area in the period from 1973 to 1994 reached approximately 1.0 m, and the average subsidence was about 0.5 m.
2. The comparison of the contour maps of 1973 and 1994 (Fig. 3 and Fig. 4) indicates that the line of altitude ±0.0 m moved about 1.0 km toward the land.
3. The profile (Fig. 6) suggests that artificial changes of ground configurations took place in the area ③, but the subsidence in this area is rather small. In the overall view, the relative changes between the ground heights of 1973 and 1994 are small, and the tendency can be observed that the ground is subsiding evenly in the areas ① and ②.

3 ESTIMATION OF SOCIAL CAPITAL LOSS CAUSED BY LAND SUBSIDENCE

3.1 *Classification of social capital loss caused by land subsidence*

Social capital loss caused by land subsidence can be classified as shown in Fig. 7.

As the first trial, a method of estimating the restoration costs in the direct loss category to represent the social capital loss caused by the land subsidence were adopted.

3.2 *Definition and estimation method*

Based on the results of the surface-oriented analysis of the land-subsidence distribution, the magnitude of effects of the land subsidence upon society was

Social capital loss
- · Direct loss
  - · Damage to facilities
  - · Functional disorder of facilities
  - · Costs of restoration works, etc.
- · Indirect loss
  - · Loss of expected benefit due to damage to, and functional disorder of, facilities
  - · Effects of direct loss upon industrial and production activities (Decrease in income and profit, etc.)
  - · Costs of drainage of inner basins, etc.

Fig. 7 Classification of social capital loss caused by land subsidence.

quantitatively estimated, the restoration costs of affected or damaged social capital being defined as the social capital loss.

Adopted in this estimation was a simplified method, where estimated first were the average subsidence and the social capital loss in the representative area shown in the concept chart of Fig. 8, and then the social capital loss in the whole survey area was calculated in the proportion of its average subsidence to that of the representative area.

$$L_1 = L_2 \times (S_1 / S_2) \times (A_1 / A_2) \qquad (1)$$

where the social capital loss in the representative area (L_2) stands for the estimated costs necessary to restore the social capital affected or damaged by the land subsidence in the one-year period from 1994 to 1995. On the other hand, the average subsidence is the weighted average of the subsidence and areas of all the subsidence classes as shown below:

$$S = \sum_i (S_i A_i) / \sum_i A_i \qquad (2)$$

where:

i : Subsidence class
S_i : Median of each subsidence class
A_i : Area of each subsidence class

The social capital loss in the representative area (L_2) was calculated by:

$$L_2 = \sum_j \left(\sum_i S_i a_i / \sum_i a_i \right) C_j \qquad (3)$$

where:

i : Subsidence class
j : Kind of social capital
S_i : Median of each subsidence class
a_i : Area or length of each kind of social capital (roads, waterways, elementary school, residential sites, and cultivated land) in each subsidence class
C_j : Restoration unit cost of each kind of social capital

The survey area comprised the Shiroishi town area and the Ariake town area (Fig. 2 and Fig. 10) of which the aerial photographs were taken and analyzed this time. Chosen as the representative area was a northwestern area around the Suko elementary school (Fig. 9) in the survey area.

The land subsidence in the 21-years was surface-orientedly determined by using the subsidence distribution plan of Fig. 5. To determine the one-year subsidence in the survey area, a subsidence contour map (Fig. 9) was drawn based on the bench-mark data of the 1994 and 1995 direct levelling. On the other hand, a plan of the representative area indicating the boundaries between the kinds of social capital (Fig. 10) was produced by the analytical plotter, and it was overlapped on Fig. 9 to determine the area of each kind of social capital in each subsidence class.

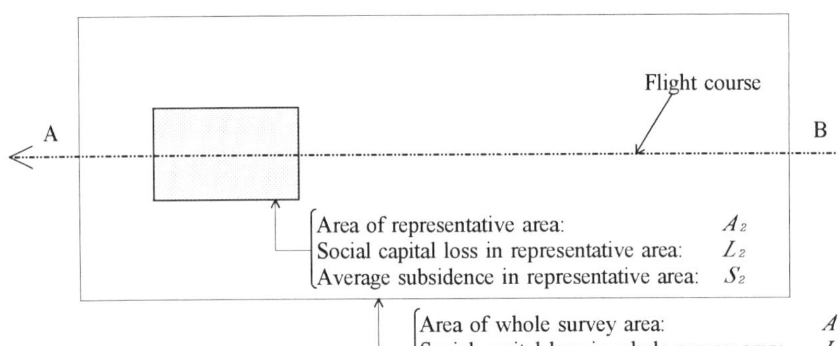

Fig. 8 Concept chart of estimation of social capital loss.

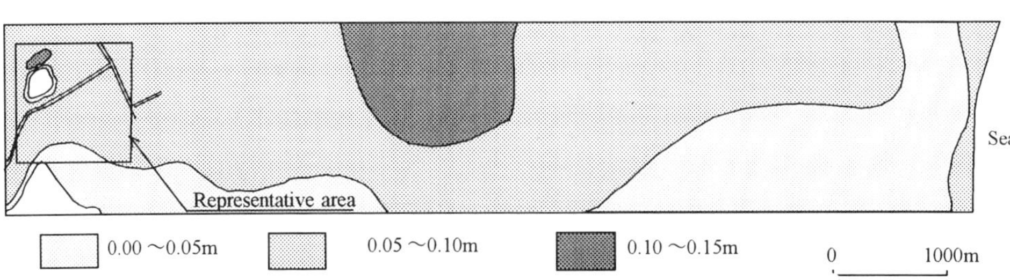

Fig. 9 Subsidence contour map of whole survey area.

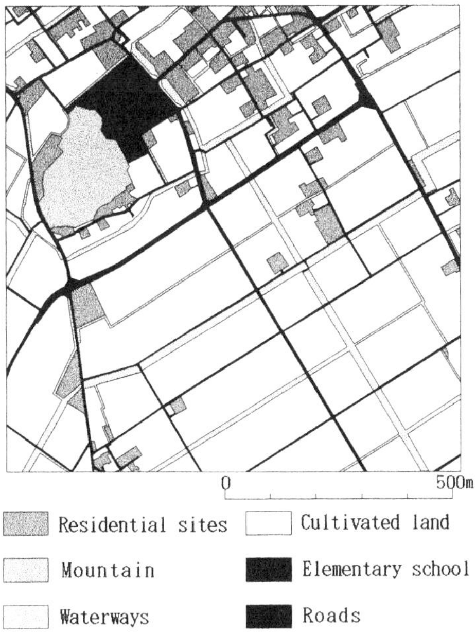

Fig. 10 Plan of representative area.

3.3 *Estimating procedure*

The subsidence in the one-year period from 1994 to 1995 were determined as above-mentioned, and the loss caused by the subsidence was estimated along the following procedure:

1. The average subsidence of the survey area and the representative area were determined by the equation (2).
2. The subsidence contour map of the survey area was overlapped on a topographical map, and the area or length of each kind of social capital within each subsidence class in the representative area was determined.
3. The area or length of each kind of social capital determined in the paragraph 2. was multiplied by the unit-area or unit-length restoration cost to calculate its loss in the representative area.
4. The total social capital loss in the representative area (L_2) was calculated by the equation (3), and then the social capital loss in the survey area (L_1) was calculated by the equation (1).

3.4 *Estimation of social capital loss in representative area(1994-1995)*

Regarding the five kinds of social capital; i.e., roads, waterways, residential sites, cultivated land, and elementary school site, the area or length of each kind in each subsidence class was determined. Then the loss of each kind of social capital was calculated. Table 1 shows losses by the subsidence class and by the kind of social capital.

Obtained from the data of this table were:
Area in the representative area (A_2): 1.0 km^2

Table 1 Social capital loss in representative area.

Social capital	Classification of subsidence Subsidence (m) Median of subsidence (m)	1 0.00-0.05 0.025	2 0.05-0.10 0.075	3 0.10-0.15 0.125
Roads	Area(m^2)	3,512	56,190	1,101
	Work costs *	60	955	19
Waterways **	Length(m)	782	18,446	12
	Area(m^2)	3,082	72,720	52
	Work costs *	422	29,855	32
Elementary school	Area(m^2)	6,200	17,174	0
	Work costs *	884	7,342	0
Residential sites	Area(m^2)	10,200	72,784	2,386
	Work costs *	1,336	28,604	1,563
Cultivated land	Area(m^2)	48,815	663,673	8,897
	Work costs *	7,835	319,559	7,140
Total ***	Area(m^2)	71,809	882,541	12,436
	Work costs *	10,537	386,315	8,754

* unit:¥1,000 ** The work costs of waterways were estimated based on their total length.
*** As the subsidence of the mountains and forests is considered to be 0.00 m in Fig. 10,the areas of classes 1,2 and 3 do not add up to 1.0km^2.

Average subsidence in the rep. area (S_2): 0.07 m
Social capital loss in the rep. area (L_2): ¥406 million

3.5 *Estimation of social capital loss in survey area*

Using the A_2, S_2 and L_2 obtained in section 3.4 and the average subsidence (S_1 = 0.07 m, obtained from Fig. 9) and the area (A_1 = 13.8 km^2) of the survey area, the equation (1) calculated the social capital loss in the survey area in the one-year period from 1994 to 1995 to be approximately ¥5.3 billion.

Then, using average subsidence in the 21-years (S_1 = 0.42 m, obtained from Fig. 5) and the area A_1, the equation (1) calculated the social capital loss in the 21-years period from 1973 to 1994 to be approximately ¥28 billion, or the average annual loss to be about ¥1.4 billion. Incidentally, the unit costs of restoration works used in the above calculation are as of today (1995).

4 CONCLUSION

Land subsidence due to the pumping up of groundwater has been monitored by the analytical plotter system from aerial photographs taken from Shiroishi plain in Saga Prefecture, and the costs of restoring the affected or damaged social capital (social capital loss) were estimated based on land subsidence.

The results of land subsidence monitored by analytical plotter system are as follows:

1. Detailed land subsidence distribution in 13.8 km^2 can be monitored by analytical plotter system compared with direct levelling method. The precision of the system was 0.12 cm in horizontal and 0.12 cm in altitude.
2. The maximum subsidence in the area from 1973 to 1994 reached approximately 1.0 m, and the average was about 0.5 m.
3. The comparison of the contour maps of 1973 and 1994 indicates that the line of altitude ±0.0 m moved about 1.0 km toward the land.
4. In the overall view, the relative changes of ground height between 1973 and 1994 are small, and the tendency can be observed that the ground is subsiding evenly.

The method used this time estimated social capital loss on the based of the restore cost caused by land subsidence. The results of social capital loss estimation are as follows:

1. The average subsidence in one-year period from 1994 to 1995 was 7.0 cm in representative area (1.0 km^2) and 6.6 cm in the whole survey area (13.8 km^2). Social capital loss was estimated at ¥0.4 billion in representative area and ¥5.3 billion in the whole survey area.
2. The average subsidence in survey area in the 21 years from 1973 to 1994 was approximately 42 cm, or about 2 cm a year. The loss in the area was estimated at approximately ¥28 billion in the 21 years, or the average annual loss to be approximately ¥1.4 billion based on today's unit costs.

The former loss might be overestimated since the subsidence in the one-year period from 1994 to 1995 was especially heavy due to abnormal water shortage in hot summer and the change in the land utilization in the 21 years was disregarded.

The fact that these large losses were estimated in the limited area with the simplified method suggests that the social capital loss in whole Saga and Shiroishi plains is sustaining huge losses. From this, the magnitude of the effects of land subsidence upon the society is imaginable. The subjects for the future would be to improve the estimation accuracy by incorporating actual, individual restoration methods into the loss-estimating process, to establish, based on estimation methods of social capital losses, a method to evaluate the curbing effects of regulating the pumping up of groundwater upon social capital losses, and so on.

REFERENCES

Miura, N., Y. Taesiri, A. Sakai and M. Tanaka 1988. Land subsidence and its influence to geotechnical aspects in Saga plain. *Proc. of the Int. Symp. on Shallow Sea and Low Land*, Saga, Japan: 151-158

Tanaka, T., H.T sunezumi, M. Yamamoto and N. Fukuda 1992. Study on the application of aerial photographs to securing topographic changes of land. *Proc. of the ILT Seminar on Problems of Lowland Development, Institute of Lowland Technology*: 287-294. Saga University.

Tanaka, T. and N. Fukuda 1994. Monitoring ground subsidence by aerial photographs. *Lowlands Development and Management*: 461-479. Rotterdam: Balkema.

Tohno, I. 1994. Land subsidence problems in Japan. *Lowlands Development and Management*: 413-437. Rotterdam: Balkema.

Environmental Geotechnics, Kamon (ed.) © 1996 Balkema, Rotterdam. ISBN 90 5410 848 7

The present status of groundwater pollution in Taiwan

Z.C. Moh & K.L. Pan
Moh and Associates, Inc., Taipei, Taiwan

ABSTRACT: Groundwater amounting to 7.1 billion cubic meters is consumed in Taiwan annually. Annual recharge of groundwater, however, is only about 4 billion cubic meters. The indiscriminating and perpetual overpumping practice has resulted in lowering of regional groundwater table, and consequently caused severe land subsidence and seawater intrusion in the coastal areas. The estimated total area affected by ground subsidence is about 1,097 sq. km. Seawater intrusion has increased the chloride content of the groundwater to a degree that it becomes unsuitable for domestic consumption and irrigation. In recent years, the problem is further worsened by the deterioration of groundwater quality due to pollution. There are three major types of pollution: 1) industrial wastes, 2) agricultural wastes, and 3) domestic and municipal wastes. These wastes contribute bacteria, virus, heavy metals, organic and inorganic pollutants to the groundwater. This paper discusses the present status of groundwater pollution in Taiwan which is regarded as in severe conditions.

1. INTRODUCTION

In the last few decades, because of the fast development of industry and increase of population in Taiwan, the amount of waste disposal and wastewater have increased drastically. A lot of the wastes were not or improperly treated and stored. Therefore, effluent from the waste disposal and sewage have contaminated the groundwater and soil resources in many places. This paper reviews the current status of groundwater and soil pollution in Taiwan, the importance of protection of the groundwater resources cannot be overemphasized.

2. PRESENT STATUS OF WATER USAGE IN TAIWAN

The annual precipitation in Taiwan in 1990 was 90.1 billion cu. m. (Fig. 1) (Water Resources Planning Commission, 1992a). Despite the high precipitation rate, the rainfall has unevenly distributed in spatial and time. Twenty-two percent of the annual precipitation, 19.1 billion cu. m. evaporated, 74 percent, 66.7 billion cu. m. was surface runoff, and about 4 percent, 4.3 billion cu. m. recharged to groundwater. For the surface runoff, 3.6 billion cu. m. was captured in reservoirs, 9.1 billion cu. m. was taken from rivers for water supply, and the rest was discharged to the ocean. In 1990, the total water usage in Taiwan was 18.9 billion cu. m.; 12.7 billion cu. m. (67 percent) was from surface water sources and 6.2 billion cu. m. (33 percent) was from groundwater sources. These data indicate that groundwater is an important water resource in Taiwan. In some areas, the amount of pumped groundwater was much more than the amount of recharge. Among the total water usage in Taiwan, 79.5 percent was used for agriculture purpose, 11 and 9.5 percent were used for domestic and industrial purposes, respectively.

3. SOURCES OF POLLUTION

The sources of contamination of groundwater in Taiwan can be classified into four different types. They are industrial wastes, agriculture activities, domestic and municipal wastes, and seawater intrusion .

3.1. *Industrial Wastes*

In 1991, there was about 95,300 industrial factories in Taiwan. Only about 5 percent of them had wastewater or solid waste treatment facility (ROCEPA, 1993). Most of the factories discharged the untreated hazardous waste directly into rivers, municipal sewage systems, and underground injec-

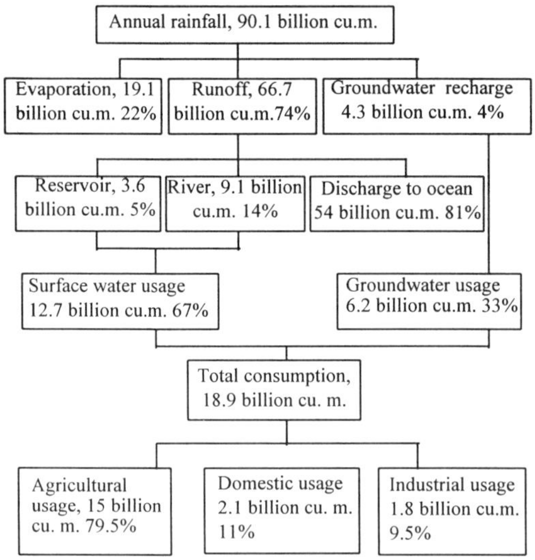

Fig. 1. The usage of water resources in Taiwan in 1990 (modified from Water Resources Planning Commission, 1992a).

tion wells. These discharges have severely contaminated surface and groundwater sources. The industrial wastes production rate was about 22,500 ton/day in Taiwan (ROCEPA, 1994), and the most severe pollution are caused by hazardous wastes. Chemical industry has caused the most pollution for the environment, followed by tannery, metal-plating, apparel, food processing, metal, pulp and paper-mill, and petrochemical industries (ROCEPA, 1991a). Except a small amount of the industrial wastes was incinerated, most of them, include toxic wastes, have been sent to disposal sites and landfills or been disposed in open fields. However, most of the construction of the disposal sites have not met the ROCEPA standards. Leachate from the disposal sites can easily leak into the unconfined aquifer below. The major industrial pollutants include heavy metals, organic and inorganic compounds, and bacteria. Wastewater generally contains organic materials. Thus, biochemical oxygen demand (BOD) is used as an indication of the amount of pollution production. In this paper, pollution production means the amount of discharged BOD after the wastewater is treated.

According to the report of the ROCEPA, in 1992, the estimated total amount of industrial wastewater generated was 2,200 ton BOD/day and the pollution production was about 880 ton BOD/day which was about 41.4 percent of the total pollution production of the year (Shen, 1994). The effective treatment (reduction) rate was about 60 percent for this category.

3.2. *Agriculture Pollution*

The agriculture activities in Taiwan include three major types.

3.2.1. *Crops*

In 1991, the total area of crop field in Taiwan includes 428,000 hectares (ha.) of rice field, 42,000 ha. sugar cane field, and 198,000 ha. for all other types of crops. The use of agriculture chemicals, such as fertilizer, pesticide, and herbicide, have caused soil and groundwater pollution. However, because of the filtering capacity of soil, no serious groundwater contamination caused by this type of agriculture activity has been reported. Nevertheless, the application of chemicals in crop field within the catchment area of reservoirs has caused eutrophication in some reservoirs (Water Resources Planning Commission, 1994).

3.2.2. *Livestocks*

Hog farm is the dominant type of livestock operation in Taiwan. About 10 million hogs were raised in Taiwan in 1992. The waste products from hog farms include high concentrations of organic substances, nitrate (NO_3), phosphate (PO_4), coliform bacteria, and virus. Based on the assumption that each hog produced 50 l/day of wastewater and 100 g BOD/day, the total amount of livestock waste was 1,000 ton BOD/day in 1992. According to the statistic data of ROCEPA, the estimated discharged pollution production was about 500 ton BOD/day which was 23.5 percent of the total pollution production of 1992. The pollution reduction rate after treatment was about 50 percent.

3.2.3. *Aquaculture*

In 1991, there was 53,600 ha. of aquafarm in Taiwan (Water Resources Planning Commission, 1994). Most of the aquafarms are located in the southwestern coastal area. The wastewater generated from the aquafarms, which contains high concentrations of chloride (Cl) and sulfate (SO_4), often infiltrates into the underlying unconfined aquifer and contaminates the groundwater resources. Moreover, aquafarms have seriously overused the groundwater resources for many years which caused depression of groundwater table, ground surface subsidence, and seawater intrusion. Such hazards have caused salinization of soil and groundwater in the southwestern coastal area.

3.3. *Domestic and Municipal Wastes*

Domestic and municipal wastes are the major sources of groundwater pollution in urban areas. The estimated pollution production rate for each person is 40 g BOD/day. With a population of about 20 million, the estimated amount of domestic and municipal waste water was about 800 ton BOD/day in 1992 (Shen , 1994). This kind of wastewater normally contains nitrate, ammonia, coliform bacteria, virus, organic and inorganic compounds. For solid wastes, most of them were transported to landfills or disposed in open fields. However, most of the landfill sites at the present do not meet the ROCEPA standards. Many of them do not have an impermeable layer at the bottom of the landfill, and in many other sites, the impermeable layer(s) was damaged (Lo and Chen, 1989). The characteristics of high water content for the domestic and municipal wastes in Taiwan have made it easier for the leachate to reach the underlying aquifer. Leachate from landfill normally contains heavy metals, coliform bacteria, and organic compounds.

Based on the above data, the total amount of wastewater generated from the above sources in Taiwan was about 3,000 ton BOD/day (Table 1). The amount of discharged pollution products after treatment was 2,120 ton BOD/day in 1992. The total reduction rate was about only 30 percent. As discussed earlier, the reduction rates for the industrial and agriculture waste effluent were 60 and 50 percent, respectively. However, due to lack of sewage treatment system, the reduction rate for municipal wastewater was only about 8 percent. The data indicate that the ROCEPA has more effectively enforced the regulations on industrial and agriculture than on public sewage system. However, there was only 3 percent of the municipal wastes which had been treated. Comparing to the sewage treatment rate of 65 to 97 percent in some western countries, and 15 to 80 percent in some countries of southeastern Asia, the sewage treatment rate in Taiwan is relatively low (Table 2) (Water Resources Planning Commission, 1994). It was estimated that in 1997, 53.7 percent of the total waste production will be from domestic and municipal wastewater (Shen, 1994). Therefore, how to effectively treat the municipal and domestic wastes becomes a critical issue on groundwater pollution control in Taiwan.

3.4. *Seawater Intrusion*

Because of the shortage of surface water resources in Taiwan, the groundwater resources were overused for many years which have caused severe subsidence of ground surfaces. At the present, the total estimated area affected by ground subsidence is about 1,097 sq. km. The maximum spot accumulated subsidence was 2.70 m occurring in Pingtung Hsien, southern Taiwan (Hwang, 1995). In coastal areas, depression of groundwater table has caused seawater intrusion and salinization of soil. Figure 2 indicates that areas of groundwater with specific conductance more than 1,000 mmho/cm are mostly located in where aquafarms are populated, i.e. the coastal area in Yunlin, Chiayi, Tainan, Pingtung, and Ilan Hsiens (Water Resources Planning Commission, 1992b). Groundwater in those areas is no longer suitable for domestic and irrigation purposes.

Salinization of soil and economic reasons have encouraged the farmers to transfer the crop field into aquafarm, particularly in the southwestern coastal area of Taiwan. Such practice has caused a vicious cycle of soil salinization and seawater intrusion. Moreover, the natural resources and regional ecological system have been severely damaged.

4. PRESENT STATUS OF GROUNDWATER AND SOIL POLLUTION

4.1. *Groundwater Quality Standards*

To be able to more effectively use the water resources in Taiwan, the ROCEPA has established a

Table 1. Amount of wastewater generated and discharged in Taiwan in 1992 (modified from ROCEPA, 1993). Unit: BOD ton/day

Amount of pollution / Waste water type	Generated pollution		Pollution production		Reduction rate
Industrial	2200	55%	880	41.5%	60%
Livestock	1000	25%	500	23.5%	50%
Municipal	800	20%	740	35%	8%
Total	3000	100%	2120	100%	30%

Table 2. Comparsion of percentage of dissemination of sewage system in different countries (modified from Water Resrouces Planning Commission, 1994).

Country or area	USA	U.K.	France	Japan	Korea	Singapore	H.K.	Malaysia	Taiwan
Fiscal year	1979	1976	1975	1985	1986	1980	1980	1980	1991
Percent of dissemination	72	97	65	36	25	80	30	15	3

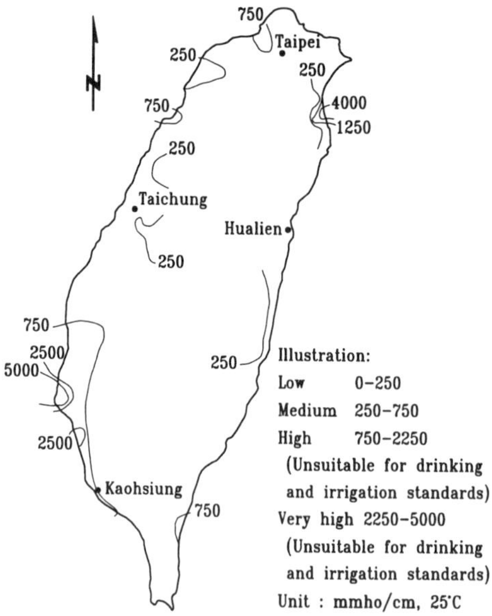

Fig. 2. Contour map of specific conductance of groundwater in Taiwan (July to October, 1989) (Water Resources Planning Commission, 1992b).

Table 3. Proposed classification of groundwater body and associated water quality standards (modified from ROCEPA, 1991b).

Quality \ Classification / Limit	Groundwater A	B	C
TDS mg/l	1000	2000	10000
pH	6.5~8.5	6.5 ~ 9.5	
TOC mg/l	1	3	5
Coliform bacteria MPN/100ml	50	5000	NA
Cyanide mg/l	0.01		NA
Phenols mg/l	0.001		NA
Anionic surfactant mg/l	0.5		NA
NH_3-N mg/l	0.1	0.3	NA
HS mg/l	0.05		NA
P mg/l	0.01		NA
Specific conductivity (μmho/cm, 25° C)	1500	3000	NA
Total hardness	75	150	300
Cl, mg/l	250		1000
Fe, mg/l	0.3	1.0	

regulation to classify the usable water quality and water body. The standards of water quality are defined on the basis of the purpose of water usage, i.e. public, aquaculture, industrial, and irrigation water supply. For example, the first order public water supply has the highest standard which can be used after a basic sterilization treatment. The ROCEPA also divides the surface water bodies into 5 classes. Class A has the highest standard, which is suitable for the first order public water supply. The definition of water quality standards for groundwater is still in progress. A proposed groundwater standard is shown in Table 3 (ROCEPA, 1991b).

4.2. *Current Status of Groundwater Pollution*

In 1991, there were a total of 206,409 water wells in Taiwan, and the Taiwan Provincial Water Conservancy Bureau has 338 long-term observation wells (Water Resources Planning Commission, 1992b). Sheu et al. (1990) analyzed the groundwater quality for drinking purpose from 348 selected wells in northern, central, and southern Taiwan (Table 4). The results indicate that the concentrations of agriculture chemicals were all below the detection limits. Among the items examined, the total microbial count has the highest percentage that did not meet the drinking water standards. These results indicate that most of the selected wells have been polluted. The possible pollution sources are domestic or animal metabolic wastes. For dissolved toxic materials, except Cr, there were a number of samples containing concentrations of Pb, As, CN, Cd, and Hg which exceeded the drinking water standards. The results suggest that for the investigated water wells, there were 53 water wells (15 percent) containing

Table 4. The number of well water samples exceed the drinking water standard in Taiwan in 1989 (modified from Sheu et at., 1990).

Area (well number)	Northern (186)	Central (108)	Southern (54)	Total (348)
Coliform bacteria	0	2	16	18
Total microbial count	144	67	54	265
Pb	5	0	2	7
As	33	0	2	35
Cr	0	0	0	0
CN	1	3	1	5
Cd	5	0	1	6
Hg	1	0	0	1
F	1	2	0	3
NO_3-N	11	6	1	18
Cl	0	7	3	10
TDS	1	6	2	9
Fe	64	26	9	99
Mn	99	55	13	257
Phenols	105	78	38	221
Anionic Surfactant	3	1	0	4
NH_3-N	37	18	5	60
Total hardness	0	7	2	9
pH	64	6	1	71

excessive concentrations of toxic substances and have to be closed immediately. Thirty-eight wells (11 percent) con-tained hazardous substances and the water has to be further treated before use. This investigation indicates that the status of groundwater pollution in Taiwan is in a very serious condition.

4.3. *Present Status of River Pollution*

There are 21 major rivers, 29 minor rivers, and 79 small rivers in Taiwan. Because of the interaction between surface water and groundwater, the status of river pollution can also reflect the groundwater quality around river valleys. The rivers in Taiwan are divided into 4 groups based on the degree of pollution which considers the total dissolved oxygen (TDO), BOD, total suspended solids and nitrate concentration. They are: uncontaminated, slightly contaminated, medium contaminated, and severely contaminated. According to ROCEPA and the Taiwan Provincial EPA, in 1993, 387.7 km of the major and minor rivers were polluted, which was 13.6 percent of the total length of the 50 rivers (ROCEPA, 1994), among them 350.1 km of the rivers (11.9 percent of the total river length) were moderate contaminated.

4.4. *Present Status of Reservoir Pollution*

The Water Resources Planning Commission (1994) has completed a water quality and eutrophication study of 16 major reservoirs in Taiwan. The degree of eutrophication of the surface water body is classified on the basis of Carlson's TSI. The data indicate that the Jihyuehtan Reservoir is the only one of oligotrophication. Feitsui and Wusheh Reservoirs are of mesotrophication, and the other thirteen reservoirs are all in eutrophication status (Fig. 3). These data indicate that most of the reservoirs are polluted.

Water quality of rivers and reservoirs directly affect the quality of public water supply. The results of investigation show that many water sources for public water supply systems are polluted (ROCEPA, 1990). The major pollutants include coliform bacteria, heavy metals, BOD, turbidity, nitrate, and odor. Part of the water treatment plants were forced to close down due to serious water pollution. The pollutant sources are mainly waste waters from industrial, domestic, and landfill leachate.

4.5. *Present Status of Soil Pollution*

Currently, the classification of soil pollution is based on the proposed "Standards and Classification of Heavy Metal Concentrations of Soil in Taiwan" (Table 5) (ROCEPA, 1990). The soil pollution is divided into four degrees based on the concentrations of As, Cd, Cr, Cu, Hg, Ni, Pb, and Zn. Soil of first degree is not contaminated, whereas, soil of fifth degree is severely polluted. During the period of 1982 to 1990, the ROCEPA had conducted a series of studies to investigate the status of soil pollution throughout Taiwan area. The areas of most serious contaminated soil were located in Changhua, Tai-chung, Taoyuan, and Yunlin Hsiens.

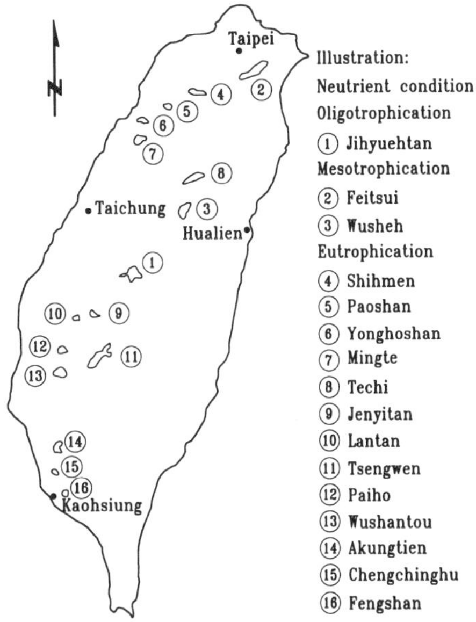

Fig. 3. The neutrient condition of the major reservoirs in Taiwan (Water Resources Planning Commission, 1994).

Table 5. Proposed standards and classification of heavy metal concentrations of soil in Taiwan (modified from ROCEPA, 1993). Unit: mg/l

Classification / Element	Low	Medium	Medium high	High
As	Top Soil <4	4-9	10-60	>60
	Subsoil <4	4-15	15-60	>60
Cd	<0.05	0.05-0.39	0.4-10	>10
Cr	<0.1	0.1-10	11-16	>16
Cu	<11	12-20	21-100	>100
Hg	<0.1	0.1-0.39	0.4-20	>20
Ni	<2	2-10	11-100	>100
Pb	<1	1-15	16-120	>120
Zn	<10	11-25	26-80	>80

It is believed that industrial wastes were the major pollution sources, particularly wastewater from the illegal metal-plating plants. The pollution sources of agriculture land included induced polluted surface or groundwater for irrigation and direct discharge of industrial wastewater. Both practices have caused accumulation of heavy metals in the soil that caused abandonment of farm land. For example, the worst cases occurred in Kuanyin and Luchu area of Taoyuan Hsien. Seventeen and 36 ha. of agriculture land in Kuanyin and Luchu area, respectively, were abandoned due to pollution caused by chemical plants (Water Resources Planning Commission, 1994).

5. PRESENT STATUS OF GROUNDWATER PROTECTION

Protection of groundwater resources has drawn the attention in Taiwan for more than a decade. Studies on groundwater management and pollution have greatly increased in the last 10 years. Recently, many groundwater monitoring well systems, in provincial, hsien, and local scales, were established by the various levels of governments. Groundwater level and quality are monitored and reported on a regular basis. Some individual companies, such as Chinese Petroleum Co. and Taiwan Power Co., also established their own groundwater monitoring systems to monitor the impact on groundwater system caused by their activities.

Regulations have been enforced in the Taipei Basin to prevent further damage of land subsidence caused by overpumping of groundwater since 1970s. Subsidence rate has reduced in the Taipei Basin since the groundwater pumpage has been controlled. Studies also suggest that regulation of groundwater usage must be applied immediately to the coastal area where land subsidence and seawater intrusion have occurred. Recently, some aquifer cleanup actions are in the planning status based on previous monitoring data. In the near future, more actions will be taken place to protect the valuable groundwater resources.

6. CONCLUSIONS

In Taiwan, groundwater has supplied about 33 percent of the total water usage. However, due to overpumping of the water resources and artificial pollution, the amount of available groundwater resources is declining rapidly. The major groundwater pollution sources in Taiwan include effluents from industry, domestic, and landfill. Although agriculture waste has not yet caused serious pollution of groundwater resources, it has significantly affected the surface water quality. Studies indicate that most of the surface water bodies, include rivers and reservoirs, are polluted to a certain degree. Large amount of groundwater and soils are seriously contaminated by industrial wastes and induced seawater intrusion in northern and southwestern Taiwan. The current status of groundwater pollution in Taiwan has indicated the imminent need for strong legal regulations for protection of the valuable water resources as well as implementation of proper remedial measures.

7. ACKNOWLEDGMENT

The writers wish to acknowledge the assistance of Dr. Tyan-ming Chu in preparation of the manuscript.

REFERENCES

Hwang J. S., (1995), Land subsidence and its induced sea water flooding in Taiwan (*in* Chinese): Special Issue of Engineering Environment, No. 6, p.41-61.

Lo, K. M. and Chen, S. C., (1989), Investigation, monitoring, and prevention of groundwater pollution (*in* Chinese): Proceedings of Hydrogeology Conference, Central Geological Survey, MOEA, p. 243-259.

Republic of China Environmental Protection Administration (ROCEPA), (1990), Survey of heavy metals in soil in Taiwan area - final report (*in* Chinese).

ROCEPA, (1991a), Investigation of hazardous wastes at the disposal sites in Taiwan (*in* Chinese).

ROCEPA, (1991b), The feasibility of defining the classification of groundwater body and quality standard in Taiwan area (*in* Chinese).

ROCEPA, (1993), Developments and strategies of waste water management in Taiwan area.

ROCEPA, (1994), 1993 Year book of environmental protection (*in* Chinese).

Shen, S. H., (1994), The prevention of water pollution in Taiwan (*in* Chinese): China Environmental Protection, No. 20, p. 7-13.

Sheu, S. T., Liaw, S. L., Wang, W. C., Lee, K. S., Lee, W. R., Chen, T. Y., and Tsai, S. L., (1990), Investigation of polluted water wells in Taiwan (*in* Chinese): ROCEPA, EPA-77-005-22-119.

Water Resources Planning Commission, (1992a), The water resources in Taiwan (*in* Chinese): MOEA.

Water Resources Planning Commission, (1992b), The groundwater resources in Taiwan (*in* Chinese): MOEA.

Water Resources Planning Commission, (1994), The water resources in Taiwan (*in* Chinese): MOEA.

Environmental Geotechnics, Kamon (ed.) © 1996 Balkema, Rotterdam. ISBN 90 5410 848 7

Radioactive pollution and protection of groundwater

J. Plško, M. Kostolanský & J. Benko
Ekosur, Piešťany, Slovakia

J. Hulla
Slovak Technical University, Bratislava, Slovakia

T. Kovács
NUSI, Bratislava, Slovakia

ABSTRACT: The first nuclear power plant ceased production in 1977. The radionuclides penetrated the groundwater surrounding. Additionally an extensive system for monitoring and protection was constructed. This one enables to observe the groundwater flow and its pollution. The results that are obtained from the field transmission and the numerical modeling are studied. Existing results confirm the fact that the radiological safety is not endangered.

1. INTRODUCTION

We have presented the first information about the influence of the Slovak nuclear power plant (Fig. 1) on the groundwater during the International Conference About the Nuclear Pollution in Prague (Plško, Kostolanský and Polák 1993) and the First International Congress on Environmental Geotechnics in Edmonton (Hulla, Plško and Taha 1994). The information came from the latest facts, that were available at this time and it is possible to characterise them as orientational. At present a more accomplished monitoring and protection system is being constructed.

A more detailed study was made of the hydrogeological conditions and the transmissional characteristics. This provided new information about the resources and contamination conditions of the groundwater pollution. A significant progress was obtained for the numerical modelling for the transmission too.

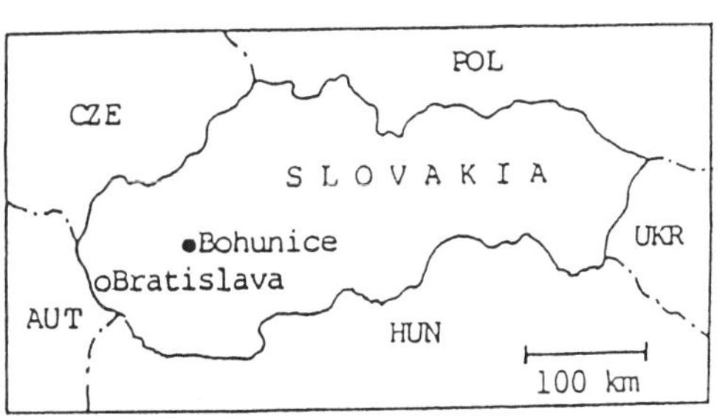

Fig. 1 Slovak nuclear power plants in Jaslovske Bohunice

2. MONITORING AND PROTECTION

At the site are situated three nuclear power plants (NPPs). The monitoring system of the hydrogeological wells is constructed in the surroundings of each NPP, both inside and outside of its territory.

There are, at present, 120 wells at its disposal (some of them are in Fig. 2). It is possible to use almost every well in cases where there is a need for the drawing of the contaminated water. The hydrogeological monitoring is regulary performed as radiological monitoring. The pumping tests and tracer experiments were performed in the many wells.

3. HYDROGEOLOGICAL CONDITIONS

The covering parts of this territory is formed by loess (variable thickness of 2 - 20 m). The underpart of them is formed by a layer of gravel soils with thickness of 15 - 20 m. Deeper layers are formed by clays. The groundwater table is located at the depth of about 20 m, in the territory of the NPPs. It is possible to obtain an idea about the flow direction from the Fig. 2.

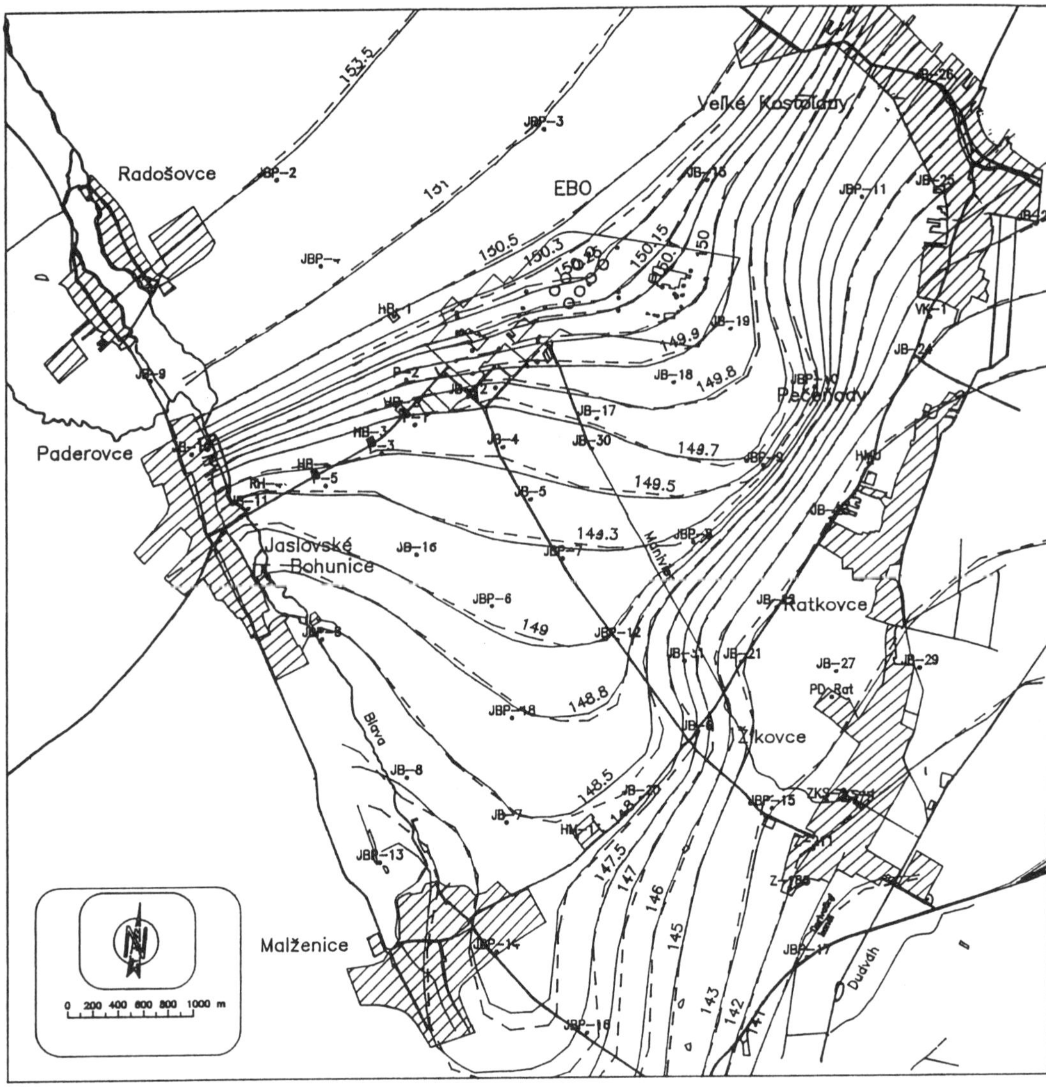

Fig. 2 Isolines of the water levels (---- from the measurements, ——from the model), some observed objects and villages in the nuclear power plants surroundings

These characteristics basically don't change during the year. The isolines of the water levels, that are based on the results from the measurements, are shown by interrupted lines. The modelling isolines obtained by the known program MODFLOW are very near to the isolines arising from the direct measurements of the water levels.

The permeability coefficients from the pumping experiments were for the territory of the NPPs in the range $k = 5.10^{-4}$ up to 2.10^{-2} $m.s^{-1}$, but in surrounding lesser values were obtained (6.10^{-5} up 9.10^{-3} $m.s^{-1}$). The pumping experiments were evaluated by the different methods. Besides them were used tracer and grain methods, too. The correlative dependence is shown between the values which were obtained by the tracer (k_t) and pumping experiments (k_p) in Fig. 3. The least values from the pumping experiments were obtained from

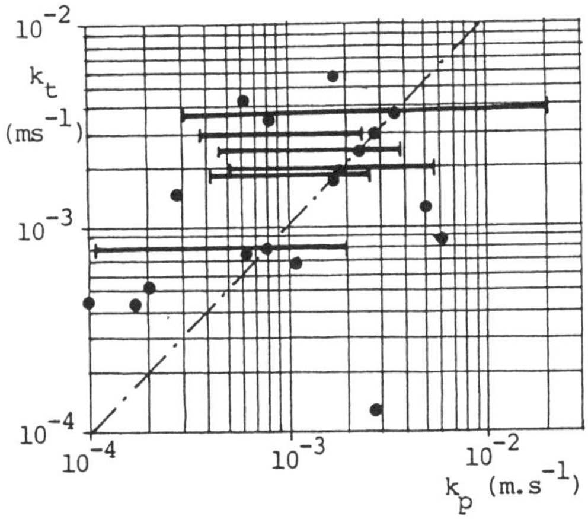

Fig. 3 Correlative dependence of the permeability coefficients determined according to the tracer methods (• k_t) and different evaluated pumping tests (⊢—⊣ k_p)

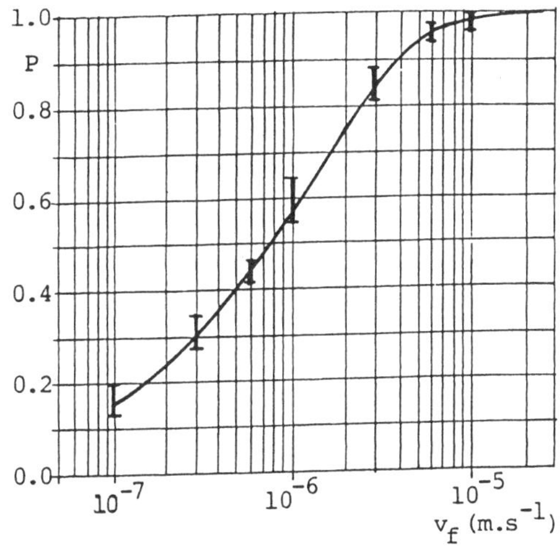

Fig. 4 Empirical values and theoretic distribution function of the filtration velocities (v_f), P - probability

the long-time experiments with the constant stream. The greatest values were obtained from the short-term experiments according to Jacobs.

One-borehole tracer method was used for the determination of the filtration velocity. This method is based on the vertical water flow in the borehole measurements. Vertical flow is determined especially by the connection of the different preassured horizonts. The filtration velocity is calculated according to the following equation

$$v_f = \frac{\Delta q_v}{\alpha\, d\, \Delta h} \qquad (1)$$

Δq_v is either an increase or decrease vertical discharge in the part of the borehole with the high Δh, d is the inside diameter of the perforated tube, α is the drainage influence of the borehole.

The results of the statistical evaluation of the filtration velocities are shown in Fig. 4.

Weibull's theoretical function gives a relatively good realism to the measured values

$$P\{v_f\} = 1 - \exp\left(-\frac{v_f}{\beta}\right)^{\vartheta} \qquad (2)$$

with the coefficient values ϑ = 0,722, β = 1,23 $.10^{-6}$ $m.s^{-1}$.

The contaminated water is penetrated from the resources that are situated near the ground surface, in a vertical direction through the unsaturated zone, that is formed by the loess, inside of the saturated zone, that is formed by the gravel soils with the different content of the sand fractions.

The basic characteristic of the pollutant transmission in the saturated zone is dispersivity. In the NPP territory this was calculated according to the tracer experiment. Its results are given in Fig. 5 together with the data, that was obtained from the literature resources.

Theoretically the longitudinal dispersivity would not be dependent on the transmission way. The results obtained from the field experiments are obvious for the conservative tracers too, these are stable and dont adsorb themselves on the particles of the soils.

The result of this, from the facts that are given above, is that some part of the tracers (respective of the pollutants) should be fixed inside of the dead pores. It is necessary to calculate either with the adsorption (for our conditions for example with the retardation factor R_D = 100) and with constant longitudinal dispersivity (α_L = 0,13 m), or with the variable dispersivity, according the Fig. 5, as well.

We have obtained the acceptable values of the longitudinal dispersivities from the variable

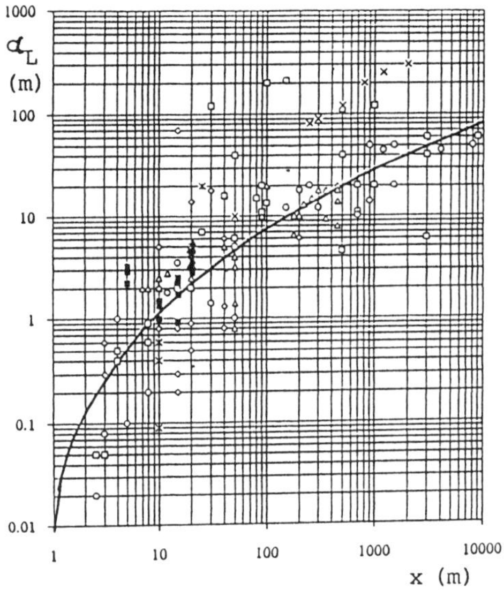

Fig. 5 Dependence of the longitudinal dispersivity (α_L) on the distance (x) from the tracer experiments in Jaslovske Bohunice (■) and from the literature (another marks)

empirical relations, based on the grain size distribution of the cohesionless soils.

4. CONTAMINATION CONDITIONS

On the isolines Fig. 6 are shown the contamination conditions of the groundwater by the tritium (^3H) at the present time. Tritium is the basic characterisation of the radioactive pollution, that is permanently measured. Attention was devoted in the past to other radionuclides, for example ^{60}Co, ^{137}Cs, ^{238}Pu, ^{241}Am, ^{90}Sr, as well. The specific activities were, for these radionuclides, relatively low and outside of the NPP their presence in the groundwater was not detected..

The present condition of the contamination is influenced by the protective drawings. 255 000 m^3 of the water was drawn from the territory of the NPPs during the years 1989 - 1994. This water contained 1,1 mil. MBq of tritium, 86 MBq of ^{60}Co, 0,5 MBq of ^{137}Cs and the other radionuclides, as well.

Existing investigations demonstrate the fact, that the main resource of the radioactive contamination is the first NPP. At this NPP there was a technological accident and ceased production in 1977.

5. TRANSMISSION MODELING

In Fig. 2 and Fig. 7 the last results for the groundwater flow are shown. These are calculated according to mathematical modeling.

In the beginning, the main problems were created by the fact that for the monitoring of the leakages of the pollutants in the groundwater, a monitoring system was not appropriated. There was not any information about the leakage sites, about the qualities of the pollutants and their disposition in the NPPs-territories and surroundings. There was not any appropriate information about the groundwater flow and the contaminants transmission.

The latest facts are obtained and extended gradually and the present conditions are not considered as perfect. It enables us to compare the results from the calculations with the realistic conditions of the pollution and back to derive the situation at the beginning or during the process. It is possible to judge the influence of the water drawing on these ones from the last years. The main meaning resides in the possibilities of modelling various situations, that could arise in the future, together with the efficiency of the protective precautions respectively. The monitoring system is permanently improved in respect of these facts.

The groundwater flow was modelled as a stationary two dimensional system with a free surface. The input data are formed by the following items: discredization of the territory, position of the unpermeable underground, permeability coefficient, boundary conditions and information about the pollution resources. The simulation of the groundwater flow has been done according to the MODFLOW program (Mc Donald and Harbaugh 1988). We have used the optimization program for its verification, with the isolines according to the measurements (Fig. 2). A very good agreement was obtained for the gravel soils.

The program SUTRA (Voss 1984) was used for the modelling of the unstationary disperso - diffusion transmission of tritium in the groundwater. In this program the hydraulical parameters from the MODFLOW program were used.

We have respected tritium half-decay time (12,3 years) and its negative sorption properties without retardation.

In Fig. 7 the results of the modelling analysis are given. These results are relatively in agreement with the actual state of the contam-

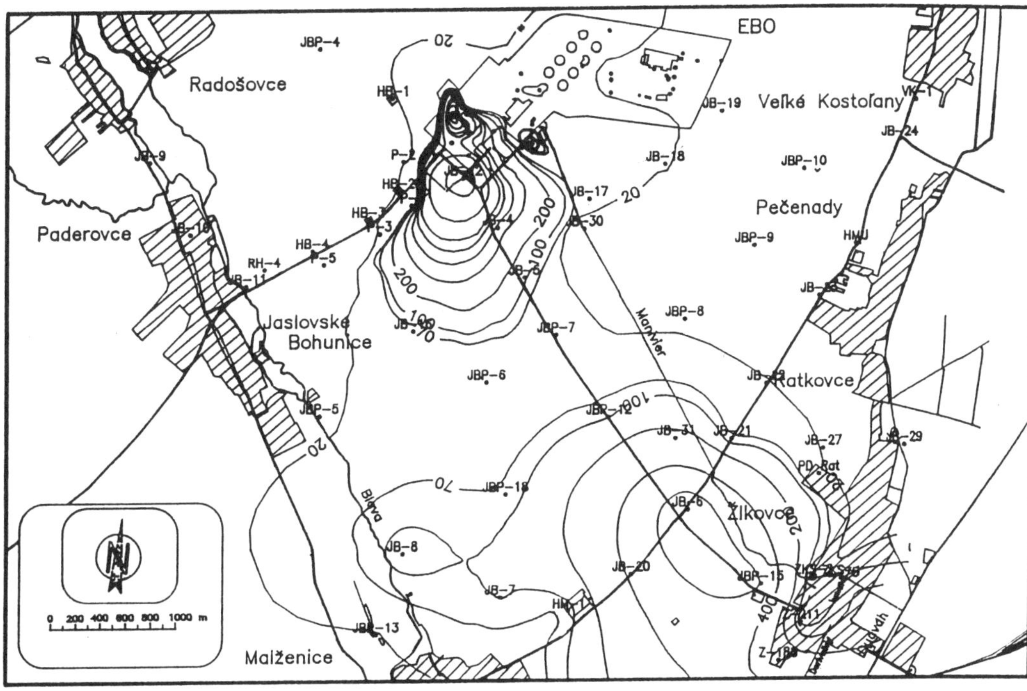

Fig. 6 Isolines for the tritium specific activity (Bq.dm^{-3}) in the groundwater at present

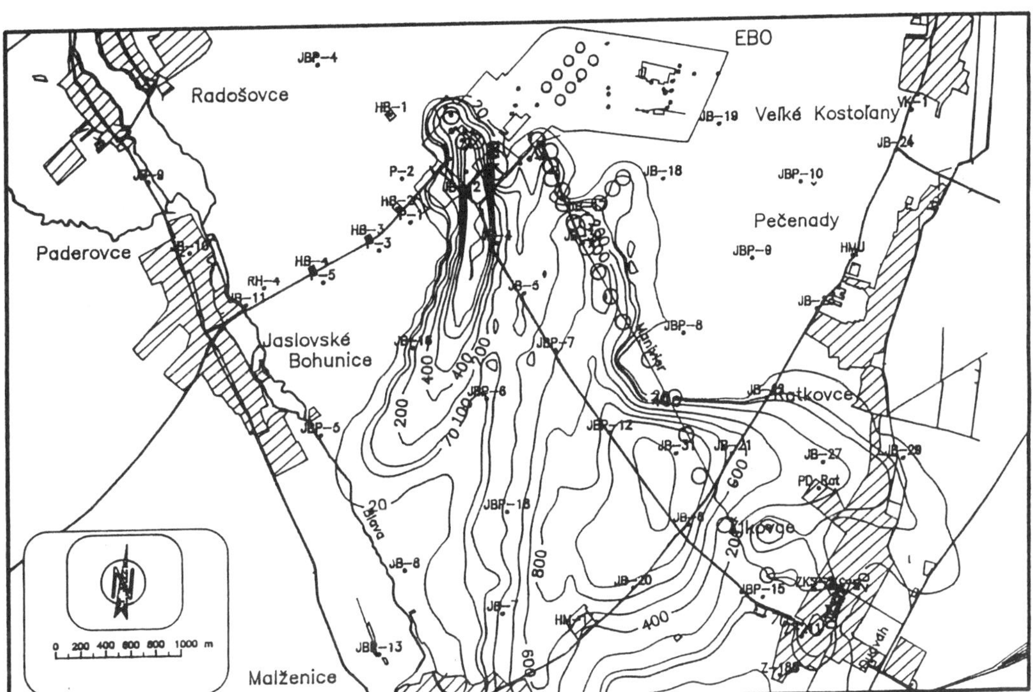

Fig.7 Isolines for the numericaly modelled tritium specific activity (Bq.dm^{-3}) in the groundwater

ination, which is shown in Fig. 6. The results have been obtained for the effective porosity n_{ef} = 0,2, longitudinal dispersivity α_L = 3,1 m, transversal dispersivity α_T = 2,7 m, for the duration of the transfer 17,5 year. The important facts lead to the opinion, from the realised modellings, that the groundwater contamination in the territory of the NPPs was caused not only by the NPP accident in 1977, but other active resources in last ten years, too. The protective pumping of the polluted water in the NPP territory was proved to be very efficient.

6. ASSESSMENT OF THE RADIOLOGICAL RISK

It is possible to observe, according to existing data about the realistic state and model prognosis for the next development of the radiological contamination, that the groundwater contamination caused by tritium has a declining tendency and in the nearest villages will not increase the value 200 $Bq.dm^{-3}$. We introduce for the orientation, that the highest value for the tritium specific activity on the territory of the Slovak Republic was measured in 1963 in the rain, when it reached 270 $Bq.dm^{-3}$ as a result of the nuclear weapons testing.

The low contamination caused by the other radionuclides from the territory of the NPP to the surroundings will not be extended.

The specific activity of the tritium is low in the groundwater in the surrounding villages according to determined and prognostic values as well and from a radiobiological point of view are negligible. The individual doses will not exceed the negligible value 10 μSv/year.

The existing radioactive pollution of the groundwater in the territory of the NPP can not endanger the health of the people neither at present nor in the future.

All limitation parameters, the valid juristic arrangements and international recommendations are substantially higher than their actual values. From these facts it results that the importancy groundwater contamination is low and in the case when this water would be drawn and used, it will not be defective from the radiologicaly point of view.

There are not any regulatory precautions for the people in the surroudings of the NPPs Jaslovske Bohunice. Despite this, other sources of clean drinking water are secured for them.

7. CONCLUSIONS

The extensive hydrogeological investigation on the territory of the NPPs in Jaslovske Bohunice has been carried out by using some special experiments. They determined the real and potential sources for the groundwater contamination as well. The monitoring system enables control of the groundwater quality and to identify the leaking pollutants. The created hydraulic-transport model together with the monitoring system forms an effective tool for the protective system of the NPPs. This system is permanently improved. The remedial provisions are realised in the critical parts and environment. The result for the people, according to the radiological risk prognosis, is that existing radiological groundwater pollution will not endanger health. Special precautions for the surrounding villages are not necessary.

REFERENCES

Hulla,J., Plško,J. & Taha,M. 1994. Influence of nuclear power stations on the groundwater. In. W. D. Carrier III, (ed.): 867 - 872. Richmond: BiTech.

McDonald, M.G. & Harbaugh, A.W. 1988. A modular three dimensional finite diference groundwater flow model. Indiana.

Plško, J., Kostolanský, M. & Polák, R. 1993. Radioactive pollution and protection of groundwaters within the location of NPPs EBO. In. International Conference About the Nuclear Pollution. Prague.

Voss, C. I. 1984. SUTRA. Washington: Scientific Software Group.

Environmental Geotechnics, Kamon (ed.) © 1996 Balkema, Rotterdam. ISBN 90 5410 848 7

Assessment of Kaluga municipal waste-range influence on environment

Eugene N. Samarin & Eugene A. Voznesensky
Department of Engineering Geology, Moscow State University, Russia

Sergey N. Rybakov & Oleg V. Zerkal
Federal Centre for Geoecological Systems, Moscow, Russia

ABSTRACT: The present paper deal with the assessment of negative impact of Zhdamirovsky wast disposal site on the environment. This waste field is the main place of deposition and conservation of solid municipal waste of Kaluga sity. natural conditions of Zhdamirovsky waste polygon location, the geological structure and hydrogeologic conditions of the area were investigated. Simultaneously chemical composetion of waste filtrate probes at some typical sites and that of ground water probes collected from boreholes drilled around of polygon were studied it was shown:

- negative impact of Zhdamirovsky waste disposal site on the environment is, mainly, an inflow of waste filtrate into surface and ground waters. High precipitation results in sufficient increase of soluble contaminants migrating into Gorodionka river;
- characteristic feature of waste field is a presence of narrow paleovalleys filled with well-permeable sediments - ways of ground water contamination;
- data obtained should be taken into consideration for monitoring projects as well as for pollution mitigation methods.

Zhdamirovsky waste field is the main place of deposition and conservation of solid municipal waste of Kaluga city which is a center of the Kaluga region in Russia. Altogether with some big industrial plants, airport, oil reservoir Zhdamirovsky polygon is a part of the Grabtzevsky industrial complex, located in the North-East of Kaluga city (Fig. 1, 2). There is no adequate follow-up of mentioned plants influence on environment during the prolonged industrial development of the Grabtzevsky complex. Nowadays within this area there is constant ingress of contaminating substances into both surface and ground water, including the aquiferous ones which serve for public water supply. Thus, Zhdamirovsky waste field is one of the main sources of contaminant admission to the environment.

1 NATURAL CONDITIONS OF ZHDAMIROVSKY WASTE POLYGON

Considered area is an erosional morainic plain, which had been moderately indented during Holocene period. At present time it has a typical relief for the central region of the European part of Russia as a whole. The polygon, where municipal wastes are deposited, is located in the gully formed by the Vizitchki creek (now it is renamed to Polygonny creek).

In the place of waste deposition and conservation the gully width is up to 80-90 m and its depth equals about 12-13 m. The gully bottom is flat and slightly inclined. The gully slopes are steep with inclination angle being up to 25-30 degrees. In natural conditions both bottom and slopes are forest-covered.

The Polygonny creek flows into Gorodionka river 1.5 km downsteam from wastes depository. Gorodionka river is right tributary of Kaluzhka river which is in turn big left one of Oka river that is one of a main waterstreams in the central European part of Russia.

During the waste field organization draining canals have been constructed to catch the snow and rain waters. Simultaneously the Poligonny creek flowing at the bottom of gully was isolated into a concrete tube 1.5 m in diameter in order to make possible free transition of creek water through waste field. However, filtrate resulting from flooding and waste dehydrating during prolonged

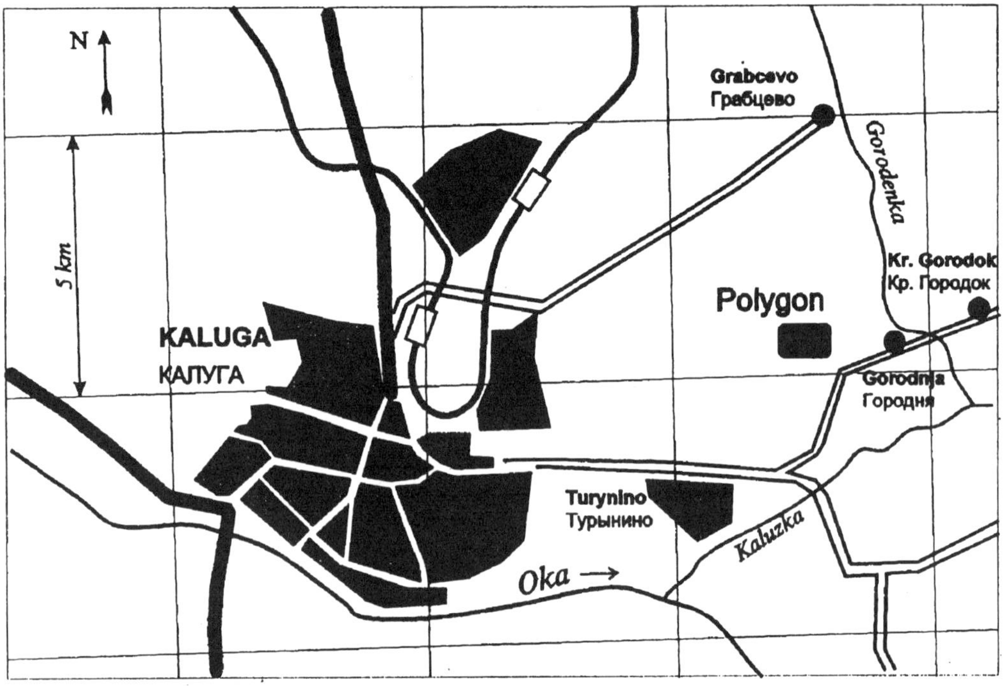

Fig. 1. Geographic location of the Zhdamirovsky waste field polygon.

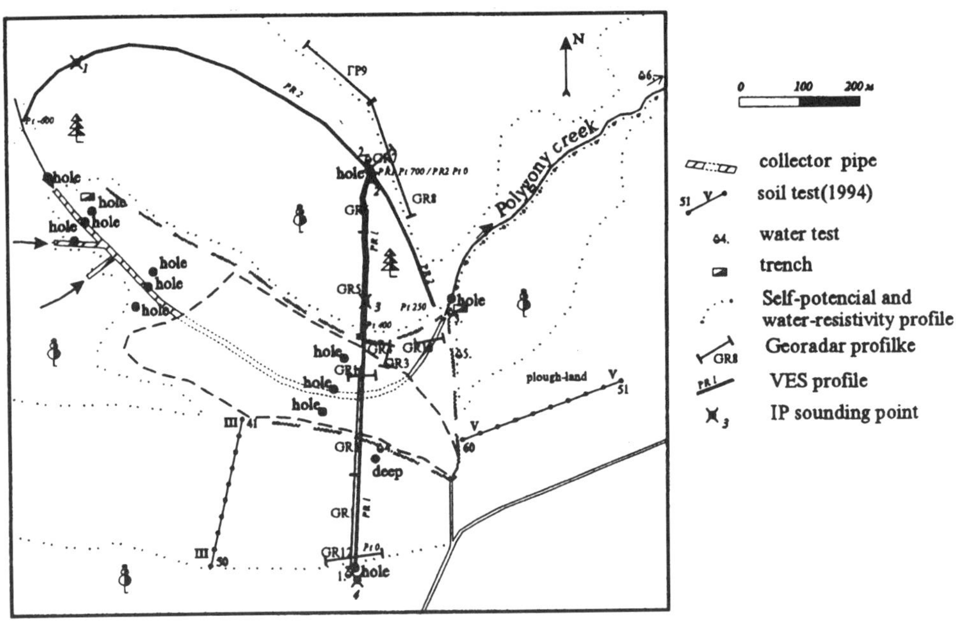

Fig. 2. Scheme of geoenvironmental researches in the Zhdamirovsky waste field polygon.

exploitation period is a main source of various contaminants. This filtrate reaches the Gorodionka river from Polygonny creek. Main pollutants that are contained in it are found in ground waters as well.

2 WASTE FILTRATE AND ITS INFLUENCE ON RIVERS

Waste filtrate chemical composition was studied at some typical sites in summer and winter periods. Filtrate probes were collected from pools on the surface located in waste field and collector outlet. Moreover, collector's probes were sampled both during dry weather and immediatily after rain fallout (Tables 1,2).

The filtrate probes collected from surficial pools during summer period are characterized by highest component content. Cl^-, HCO^{3-}, Na^+ and NH^{4+} predominate in filtrate composition. The content of many heavy metals reaches maximum values in pool water in comparison with other probes. Chemical composition tests carried out with water probes from concrete collector outlet show that this water was contaminated substantially less than pool one. Nevertheless, there are many components in this water with concentration over maximum permissible concentration (MPC). For example, NH^{4+} content is 30 times, Cl^- content is 20 times and total Fe content is 100 times as much as MPC. Content of Li, B, Cr, Cu and W is substantially higher than MPC. Investigation of water probe collected after rain fallout shows that additional rain water decreases total concentration and many polutants. At the same time specific fractions of Mg^{2+}, SO_4^{2-} and NO_2^- as well as of Mn, Sr and Zn remain at the level constant. These components of waste are probably the most soluble ones; therefore, they are primarily taken out by seepage rain water. Comparative chemical composition analysis of probes from the collector outlet in the summer and the winter shows that the total concentration of collector effluent during summer period is five times as much as the one during winter period. This is caused by decrease of infiltration, that is to say the higher is amount of rainfall water, the lower is pollution potential of collector effluent. Thus, filtrate which formes as a result of rainfall water and waste components interaction is a main source of environmental contamination around the Zhdamirovsky waste polygon.

3 THE GEOLOGICAL STRUCTURE AND HYDROGEOLOGIC CONDITIONS

The geologic section of this area consists of carbonate-terrigenous deposits (limestones, dolomites, sands and clays) of Visean and Namurian stages of lower Carboniferous and could be clearly stratified. The former includes Tulean (C_1tl), Alexean (C_1al), Mikhailean (C_1mh) and Venevean (C_1vn) horizons and the latter is divided into Tarusean (C_1tr) and Steshevean (C_1st) ones. In the area of waste deposition Tarusean and Steshevean horizons composing the upper Carboniferous section are represented by yellow-grayish weak and cracked limestone and black clays respectively. The limestone thickness is about 5-14 m and clay one is almost 3-4 m in conformity with drilling data. The Carboniferous deposits are exposed in a river-bed of Polygonny creek.

The Carboniferous limestones and sands are confined to headed aquiferous horizon which is utilized for water supply of Kaluga city. The waters have hydrocarbonate-calcic-magnesium composition with a typical mineralization of 0.27-0.32 g/l. In the loci of waste deposition this aquifier is cut in by drilling at a depth of 76 m. The water head is 24.7 m and water level is 51.3 m.

The Carboniferous clay and limestones are overlayed by Quaternary formations covering the suface of rugged pre-Quaternary rocks. The Quaternary deposits are represented by glacial and fluvioglacial soils with a usual thickness about 20-21 m. Several repeated alterations of relief during periods of glaciations resulted in rather complex geological structure and interlaying of Quaternary formations. However, both palaeorelief and present relief forms are controlled by geological and tectonic conditions of the area. Two systems of depressions are cleary identified. One of them extends in a submeridian direction and another - in a sublatitudinal one. Since the relief has developed during Pleistocene these forms have superimposed each other. The oldest Quaternary deposits include Likonean (Q_1lk), Matveivean (Q_1mk) and Setunean (Q_1st) horizons. They lay on Carboniferous limestones where the pre-Quaternary relief depressions exist. One of these submeridian paleovalleys was discovered by drilling and geophysical data near eastern boundary of Zhdamirovsky waste polygon. The Quaternary deposits had filled this paleovalley in early Pleistocene. They are represented by glacial clay

Table 1. Chemical composition of surface waters.

	Chemical composition, mg/l / mg-eq / %-eq													
	$Na+K^+$	Ca^{2+}	Mg^{2+}	HCO_3^-	SO_4^{2-}	Cl^-	NH_4^+	NO_3^-	NO_2^-	Fe^{2+}	Fe^{3+}	SiO_2	pH	M,mg/l
Collector	96.8	67.5	13.2	597.8	48.6	212.3	150.0	12.2	0.05	0.50	-	9.1	7.7	920.0
N2	4.21	3.37	1.09	9.80	1.01	5.99	8.31	0.20		0.02				
	24.8	19.8	6.4	57.7	5.9	35.2	48.9	1.2		0.1				
Collector	22.3	66.5	30.1	451.4	52.7	103.9	87.5	10.5	1.9	0.5	0.1	6.8	7.66	658.0
N9, (after	0.97	3.32	2.48	7.40	1.1	2.93	4.85	0.17	0.04	0.02				
rain)	8.3	28.5	21.3	63.6	9.46	25.2	41.7	1.5	0.3	0.2				
Collector in	1272.	100.0	30.0	2659.	96.0	887.5	343.0	21.0	-	-	-	18.	7.9	4572
january	0			2								0		
Polygonny	323.1	49.7	14.4	829.6	57.6	339.7	125.0	16.7	1.5	0.80	0.1	9.1	8.15	1509
creek out	14.05	2.48	1.19	13.60	1.2	9.58	6.93	0.27	0.03	0.03				
let N6	56.9	10.1	4.8	55.1	4.9	38.9	28.1	1.1	0.1	0.1				
Gorodionka	32.4	58.7	9.0	207.4	41.1	22.3	0.1	12.2	0.1	0.2	0.1	9.4	7.55	310.0
river, befor	1.41	2.93	0.74	3.40	0.85	0.63		0.2						
creek, N7	27.8	57.7	14.6	66.9	16.7	12.4		3.9						
Gorodionka	12.4	60.7	17.5	280.6	35.4	76.6	50.0	15.5	1.8	0.3	0.1	13.	7.3	470.0
river, afte	0.54	3.03	1.44	4.6	0.74	2.16	2.77	0.25	0.04	0.01		7		
creek, N8	6.9	38.9	18.5	59.0	9.5	27.7	35.6	3.2	0.5	0.1				
filtrate	397.4	105.4	42.2	1018.	97.9	594.5	175.0	14.4	-	0.9	0.2	6.0	7.49	2220
probe, N12	17.28	5.26	3.47	16.7	2.04	16.77	9.7	0.23		0.03				
	48.4	14.7	9.7	46.7	5.7	46.9	27.1	0.6		0.1				

Table 2. Chemical composition (microcomponent) of surface waters, μg/l.

	Pr. N2 creek	Pr.N6 outlet creek	Pr.N9 creek	Pr.N12 filtrate	january filtrate	N9/N2	N6/N2	jan/N2
Li	17	26	16	31	40	0.94	1.53	2.35
B	140	380	96	350	600	0.69	2.71	4.29
Al	9.3	4000	89	160	70	9.57	430.11	7.53
Si	2000	5000	2100	3200	7500	1.05	2.50	3.75
Ti	10	55	7	14	100	0.70	5.50	10.0
Cr	32	57	23	55	220	0.72	1.78	6.88
Mn	33	500	57	680	210	1.73	15.15	6.36
Co	1.9	6.6	1.7	4.8	5.2	0.89	3.47	2.74
Ni	7	18	5.8	30	64	0.83	2.57	9.14
Cu	5.8	14	4.8	14	48	0.83	2.41	8.28
Zn	33	63	9.5	140	52	0.29	1.91	1.58
Sr	190	170	210	510	230	1.11	0.89	1.21
Mo	0.58	2.7	0	7.1	13		4.66	22.41
Sn	12	3.6	4.6	1.7	86	0.38	0.3	7.17
Sb	0.3	0.63	0.35	2.5	1.3	1.17	2.1	4.33
Pb	2.7	12	1.8	6.6	11	0.67	4.44	4.07

mass ($Q_I$lk) with thickness up to 4 m and glacial loams (Q_Ist) with the thickness from 0.5 to 3.3 m. Both masses are divided by a horizon of saturated sand ($Q_I$mt) with of about 5-5.5 m thick. The depth of the upper boundary of the above mentioned glacial formation equals almost 12-13 m and coincides with the depth of modern gully where municipal wastes are accumulated.

The above mentioned sediments are covered by regionally spread Perekhshinean ($Q_I$pr) glacial loams 2-3.5 m thick. These loams are, in turn, overlayed by deposits of Moscowean glaciation period.

Table 3. Chemical composition of ground waters.

	Chemical composition, mg/l / mg-eq / %-eq											
	$Na+K^+$	Ca^{2+}	Mg^{2+}	HCO_3^-	SO_4^{2-}	Cl^-	NH_4^+	NO_3^-	Fe^{2+}	Fe^{3+}	SiO_2	pH	M, mg/l
borehole N10	40.0 1.74 38.9	34.9 1.74 38.9	12.0 0.99 22.1	219.6 3.60 80.5	16.5 0.34 7.6	12.8 0.36 8.1	0.1	10.5 0.17 3.8	0.01	0.3	3.1	7.5	246.0
borehole N10, jan.	28.7	27.0	12.0	190.3	28.8	14.2	-	-	-	-	4.0	7.85	269.7
borehole N9	46.2 2.01 32.0	68.5 3.42 54.5	9.6 0.79 12.6	317.2 5.2 82.8	33.7 0.7 11.2	8.5 0.24 3.8	0.7 0.04 0.6	8.9 0.14 2.2	0.6 0.02 0.3	1.5	10.3	7.5	360.0
borehole N9, jan.	101.2	40.0	6.0	329.4	28.8	28.4	-	-	-	-	5.0	7.35	491.8

The multiphase character of this glaciation resulted in deposition of several different horizons. The lower part of Moscowean section is presented by fluvioglacial sands and sandy loams with the common thickness of about 2.5-4.5 m filling narrow submeridian erosional paleoincisions formed in early - middle Pleistocene on post-Perekhshinean gently sloping glacial flatland. These erosion incisions are discovered in central and eastern parts of polygon and adjacent area. At present time they are buried under the late glacial deposits of Moscovian formation, overlayed in turn by diluvial loams. Total thickness of glacial and diluvial sediments is 3-6 m.

4 INFLUENCE OF BURIED RELIEF ON THE GROUND WATERS CONTAMINATION

In the vicinity of Zhdamirovsky polygon area ground waters in Quaternary deposits are confined in discontinuous sands and sandy silts related to palaeorelief depressions.

The lower Quaternary aquifier is located 14-16 m deep in sands within paleoincision in the eastern part of polygon. Waste filtrate enters ground waters in places where present erosion incision reaches aquiferous sands. This is obvious from the differences in chemical composition of water from 9 boreholes located to the south from waste field and 10 other boreholes located outside of contaminated zone (Table 3, 4).

Upper aquifier in Quaternary soils is located in sands and sandy silts in the lower Moscowean horizon. In the eastern part of waste field ground water table is 6 m deep. The central part - along paleoincision - is characterized by sporadic grond waters in Moscowean deposits. Higher concentration of heavy metals were revealed in permeable zone by chemical investigation. This is an indirect witness of contaminants' migration with ground water flow in Moscowean horizon during the periods of its saturation.

Table 4. Chemical composition (microcomponent) of ground waters, μg/l.

	borehole N 10	borehole N9	b.9/b.10
Li	3.6	4.9	1.36
Si	430	2000	4.65
Ti	2.8	4.2	1.5
Cr	2.6	7.9	3.04
Mn	250	510	2.04
Fe	2300	4900	2.13
Co	0.19	0.72	3.79
Zn	26	5.9	0.23
Sr	40	4	2.10
Pb	0	0.58	

CONCLUSION

Negative impact of Zhdamirovsky waste disposal site on the environment is, mainly, an inflow of waste filtrate into surface and ground waters. High precipitation results in sufficient increase of soluble contaminants migrating into Gorodionka river.

Characteristic feature of waste field is a presence of narrow paleovalleys filled with well-permeable sediments - ways of ground water contamination.

Data obtained should be taken into consideration for monitoring projects as well as for pollution mitigation methods.

Environmental Geotechnics, Kamon (ed.) © 1996 Balkema, Rotterdam. ISBN 90 5410 848 7

Geo-environmental problems related to ground improvement works in Japan

Kohei Sawa
Akashi College of Technology, Japan

Kazuo Aoki
Osaka Institute of Technology, Japan

Takeshi Katsumi
Kyoto University, Uji, Japan

Yasuhisa Enaka
GeoGyro Corporation, Kyoto, Japan

ABSTRACT: This paper reviews the present situation in Japan concerning the effects of ground improvement techniques on geo-environmental problems, and the application of ground improvement techniques for preserving the environment.

1 INTRODUCTION

Many kinds of ground improvement techniques have been developed and applied in order to utilize unsuitable land, such as that composed of soft clay, or ground located in areas adjacent to slopes, because of the dense population and high-level of industrial activity in the narrow strip of flat, habitable land in Japan. Unfortunately, it appears that these ground improvement works have been responsible for creating some of the environmental problems that exist today. However, these works have also contributed to environmental preservation or pollution control.

This paper will assess environmental impact during and after ground improvement execution; contamination of ground and groundwater by cement, lime and grouting chemicals, changes in groundwater flow conditions by ground improvement works and dust raised by soil stabilizers. This paper will also assess some ground improvement techniques for environmental control; countermeasures for contamination and hardening treatment of sediments and turbid water (Table 1).

This paper is the result of research and study carried out by the working group of the Committee on Material Science and Technology for Soil Stabilization (Chair: Prof. K. Nishida, Kansai University) in the Society of Material Science, Japan.

2 CONTAMINATION RESULTING FROM GROUND IMPROVEMENT

2.1 *Cement and lime stabilization*

Soils stabilized by cement or lime are sometimes thought to have a negative environmental impact due to the presence of alkali leachate. Cement and lime stabilization methods are becoming increasingly more important, especially because the recycling of excavated surplus soils by soil stabilization is increasingly required. Assessment and prevention of the alkaline migration is

Table 1 Environmental problem related to ground improvement

Environment	Ground and groundwater	Sea,lake and river	Air
Environmental impact during execution	• Flow-out of grouting chemicals		• Dust due to cement
Environmental impact after execution	• Flow-out of grouting chemicals • Alkaline leachate • Change of groundwater condition	• Alkaline leachate	
Environmental control	• Control of groundwater • Prevention of contamination • Measure of contamination • Utilization of waste materials	• Treatment of sediments • Treatment of turbid water	

therefore becoming more necessary in recent years for various ground conditions and design methods.

The buffer ability of surrounding soil is expected to become the major focus of alkali restraint, as well as carbonation, that is stabilized soil which is continuously carbonated by exposure to air and can be consequently neutralized. Calcium hydroxide ($Ca(OH)_2$) can adsorb the surface of clay particle in exchange for other ions adsorbed, and calcium alkali is therefore confined to a limited area; the surrounding clay soil can be expected to prevent the alkali from migrating and spreading. The degree of alkaline restraint varies with the type of soil, and can be expressed as the alkaline neutralization capacity, as in Table 2 (Miki et al. 1994). This index is useful in evaluating directly the neutralization ability of soil. Evaluation of the combination of the stabilized layer and the natural or man-made cover layer for neutralization, called the filter layer, is also important, and is analyzed based on the conditions described in Fig. 1. Seepage water from the stabilized soil cannot be sufficiently neutralized when the filter layer is high in permeability and low in alkaline neutralization ability (Mishima et al. 1994, Kamon et al. 1995-b). Clayey soil set as the filter layer has an effect on the alkaline migration control due to its high neutralizing ability (Sano et al. 1993). Kamon et al. (1995-b) also showed that the batch method as shown in Table 2 is not sufficient to evaluate the alkaline neutralization ability because some different results are obtained between batch tests and column ones. Besides the neutralization of the filter soil, the stabilized soil layer is required to have low permeability so that the seepage water cannot penetrate into/through stabilized soil.

Some data have been obtained through field performances as well as through laboratory researches (Kitsugi 1989). Where the column of lime group materials penetrated into the underlying soft clay ground, only the soil 10-30 cm from the lime column exhibited high pH, and

Table 2 Alkaline neutralization capacity of soils

The type of soil	Alkaline neutralization capacity (mol/g)
Kanto loam	$1\text{-}3\times10^{-3}$
Kanumatsuchi	2×10^{-3}
Humic soil	1×10^{-3}
Peat	$0.8\text{-}1\times10^{-3}$
Clayey soil	$6\text{-}9\times10^{-4}$
Pit sand	$1\text{-}4\times10^{-4}$
Silty soil	$2\text{-}4\times10^{-4}$
River sand	$0.5\text{-}4\times10^{-4}$
Decomposed granit soil	9×10^{-5}
Shirasu	$2\text{-}4\times10^{-5}$

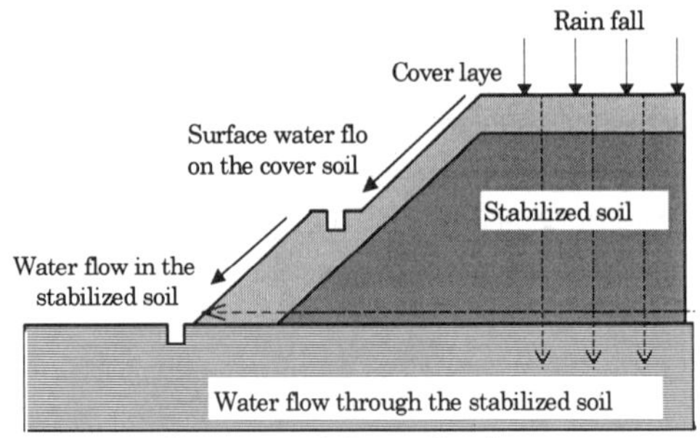

Fig.1 Water flow around the embankment using stabilized soil

the soil more than 30 cm from the column had pH values similar to those of the original ground. On the stabilization of bottom sediments of lake, a sand layer only 3 cm thick on the cement stabilized sediments was shown to be very effective in preventing alkaline migration and because of this, lake water was not only kept neutral but was purified. If the stabilized soil embankment is planned to be vegetated, soil cover and aging will be effective.

2.2 *Chemical grouting*

The chemical grouting method, which has been applied as the emerging countermeasure for temporary work prior to the main work, is expected to become the leading ground improvement technique for permanent substructures. Since the 1950s, much work has been carried out to develop the grouting (injection) method and the materials used. However, in 1974, one serious contamination accident caused by chemical grouting using polyacrilamid triggered the establishment of strict governmental guidelines on chemical grouting, whereby only water glass free of toxins or fluoride is allowed to be used as the grouting chemical and ground-water monitoring is obligatory during execution; the groundwater should be below 8.6 in pH and below 10 ppm in permanganate consumption.

Although only inorganic materials such as cement and water glass have mainly been used as grouting materials since then, there have been some cases in which the grouting has contaminated groundwater. The causes of the contamination can be divided into (1) leaching of the chemical whose reaction has not been completed during grouting, and (2) leaching of the chemical from the hardened materials. Therefore, in order to prevent any possible negative impact on ground water, it is necessary (1) to use safe materials, (2) to minimize the seepage of unreacted materials from the area which is intended to be grouted, and (3) to ensure the grouted materials completely react and harden in the ground.

Some technical methods have been proposed from the aspect of environmental impact. The carbon dioxide injection system, where sodium silicate is grouted into the ground with carbon dioxide, was developed based on the safety of the composition and the alkaline neutralization of carbonic acid gas. The reacted by-products are safe because they consist of only insoluble gel, with the exception of sodium carbonate. Shimada (1992) and Takeuchi et al. (1987) succeeded in developing methods to add carbonic acid gas into the grouting system.

Hydrophobic polyisocyanate grout, which the governmental guidelines do not allow to be used for ordinary grouting work, was developed for special cases such as emergent countermeasure during foundation work (Oshita and Kitano 1992). Positive features of the method are that the unreacted grout is not soluble in water, that compositions of the grout are safe, and that no harmful substances dissolve from the hardened gel, therefore, it has no potential to contaminate groundwater.

3 CHANGE AND CONTROL OF GROUNDWATER CONDITION

The hydraulic aims of ground improvement works have been how to reduce the groundwater table, how to decrease the water pressure on underground structures and how to accurately investigate the reduction in water pressure. Recently, ground subsidence and shortage of groundwater at sites using groundwater have become very serious problems due to dewatering of underground construction works. The obstruction of groundwater flow and salt water intrusion in pure groundwater are also considered to be serious problems.

The ordinary method to prevent dewatering is to recharge the groundwater as shown in Fig.2. Ueda et al. (1987) and Takeshita and Kohno (1994) have studied the optimal control and drainage plan of a group of recharge wells. By optimization method, they proposed the suitable location of discharge wells that reduce the groundwater table and the optimal control of recharge wells that restrict the effects of dewatering and salt water intrusion to the environs.

Drainage bypassing of underground structures is one countermeasure against groundwater flow changes. Underground structures, such as sewers and subways, may obstruct groundwater flow. In order to allow the obstructed ground-water to flow easily around an undergroung structure, a bypassing system consisting of a permeable layer of crushed stone (Fig.3(a)) or a connection pipe (Fig.3(b)) may be used (Nishigaki et al. 1992, Ueda et al. 1987).

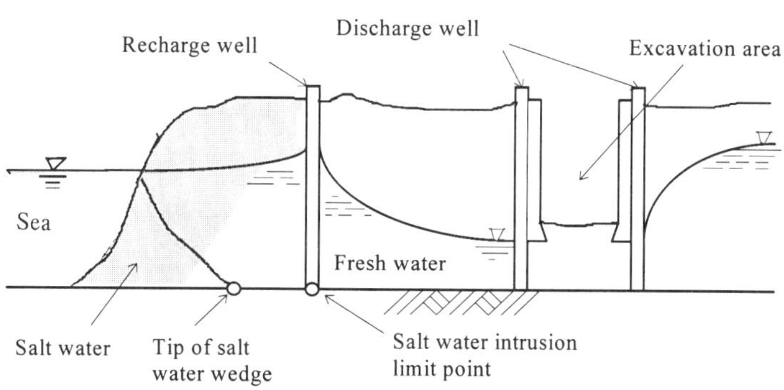

Fig.2 Circumstance of groundwater flow by discharge and recharge wells

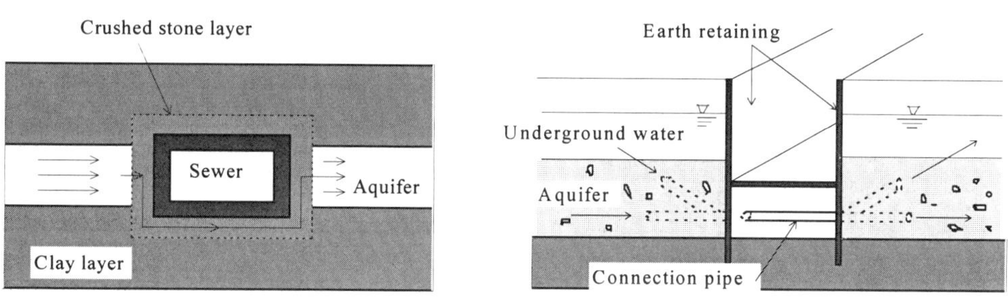

a) Permeable layer (b) Connection pipe

Fig.3 Drainage bypassing underground structure

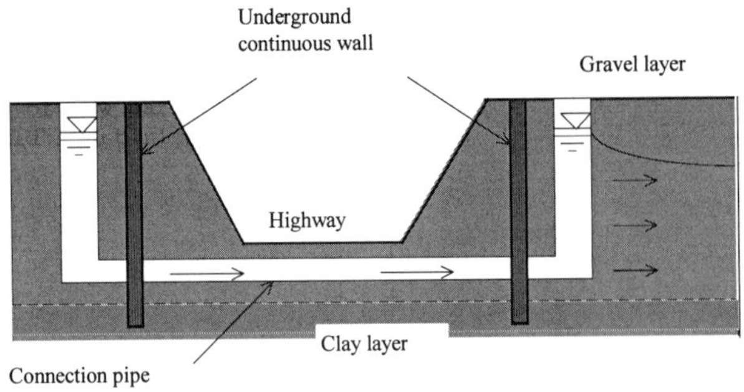

Fig.4 Connection pipe under a highway

Fig.4 shows an example of the restoration method for groundwater in the cut area of a highway construction site that interrupts a gravel aquifer. An underground continuous wall was constructed on the upstream slope shoulder in order to cut off the groundwater flow and to ensure the stability of the slope and highway. The groundwater dammed up by the wall is led downstream through a connection pipe under the highway. Then the groundwater flows from the upstream aquifer and restores the ground-water table in the downstream gravel layer (Uno et al. 1992).

4 DUST FROM SOIL STABILIZERS

As most stabilizers used with shallow soil improvement are powders, much of it is air/windborne and flies everywhere, and is called "dust". This flying dust has become a serious problem at sites near residential areas and for workers health.

One former countermeasure used to control the dust was to moisten the cement or lime with 50-20 % by weight of water. This was difficult to use because it required a lot of stabilizer and could not be stored long term. Another method used was to change the cement or lime into coarse grain, but that was costly and necessi-tated secondary mixing after hydration.

Two new types of anti-dust stabilizer have been developed recently; one is a stabilizer treated with Teflon (poritetra-fluoroethylene), the other is a stabilizer mixed with oil.

The stabilizer treated with Teflon is composed of cement or lime mixed uniformly with 0.03-0.05% of Teflon under temperatures of 50- 320℃ for a few seconds to a few minutes. The fibers produced, which are microns in diameter, catch the grains of stabilizer and prevent the dust from flying (Fig. 5(a)) (Simoda et al. 1992, Nemoto et al. 1991). The stabilizer treated with Teflon has the same effect on improvement as the non-treated stabilizer.

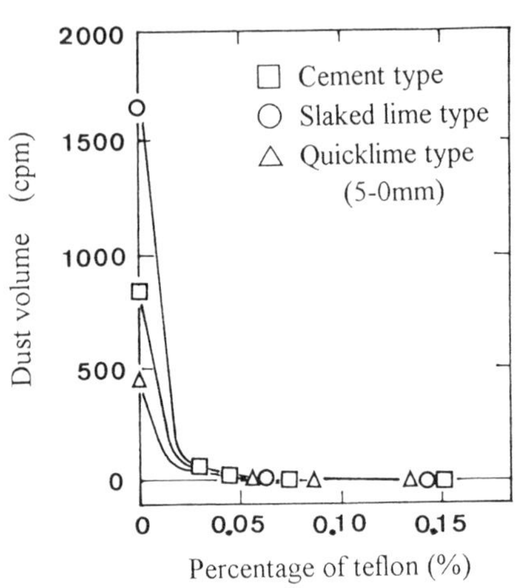

(a) Stabilizer treated with Teflon

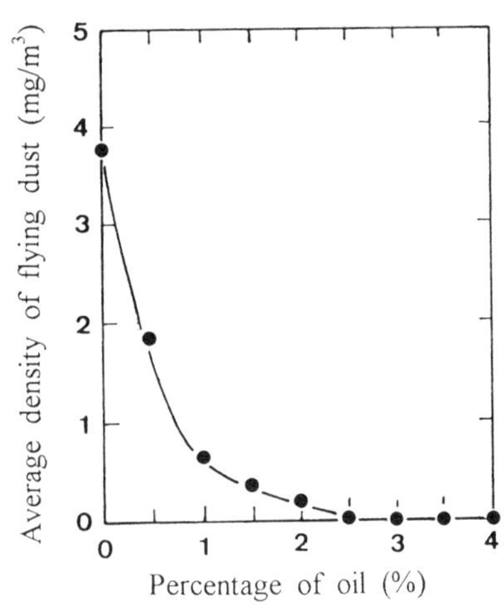

(b) Stabilizer mixed with oil

Fig.5 Amount of flying dust from anti-dust stabilizer

The stabilizer mixed with oil is composed of cement containing small amounts of oil which sticks to the surface of the cement grains (Sawa et al. 1994,1995). The stabilizer containing 1-2 % by weight of oil is able to satisfy the dust level standards of the Japan Society for Occupation Health (Fig.5(b)). The treated soil with the stabilizer containing oil is as strong as or stronger than the stabilizer not containing oil. As the oil dissolved in treated soil is less than 1/30 of the allowable level for the treated waste, it causes less effects on the environment. It is reported that oil is used to fluidized bed combustion coal ash so as to prevent flying dust (Kamon and Katsumi 1995-a).

5 GROUND IMPROVEMENT FOR ENVIRONMENTAL CONTROL

Fig. 6 shows ground improvement techniques applied to the preservation of the environmental system. Countermeasures for contamination, sediments and turbid water will be presented in turn below.

5.1 *Countermeasures for contamination*

As the ground and groundwater are polluted by many kinds of toxic substances, the countermeasures for contaminated ground and groundwater have been developed according to the nature and spreading area of toxic substances.

The main countermeasures for contamination used in Japan are Solidification and Extraction, among others. Solidification by cement or high polymer stabilizer has been applied widely for muddy soil and volcanic ash soil contaminated by heavy metals, and for ground contaminated by waste landfill and ruins of factories. An ordinary method of Extraction is Soil Vapor Extraction, which is an in-situ purification system of ground by vacuum extraction and air injection and is applied to soil polluted by volatile organic compounds (Imamura et al. 1994). New techniques, including Glassing, Bio-remediation and Adsorbed Barriers, are also under development.

5.2 *Countermeasures for sediments*

Soil improvement techniques are also applied to stabilize the bottom sediments of bays and swamps. The sediments are usually composed of high water content, organic matter and noxious substances. Cement group hardening materials are under development and have been applied to sediments. Recently, secondary additives such as Carbonated Aluminate Salt (CAS) materials, gypsum and industrial waste, have been used effectively with many types of muddy soil in order to increase the solidification of materials.

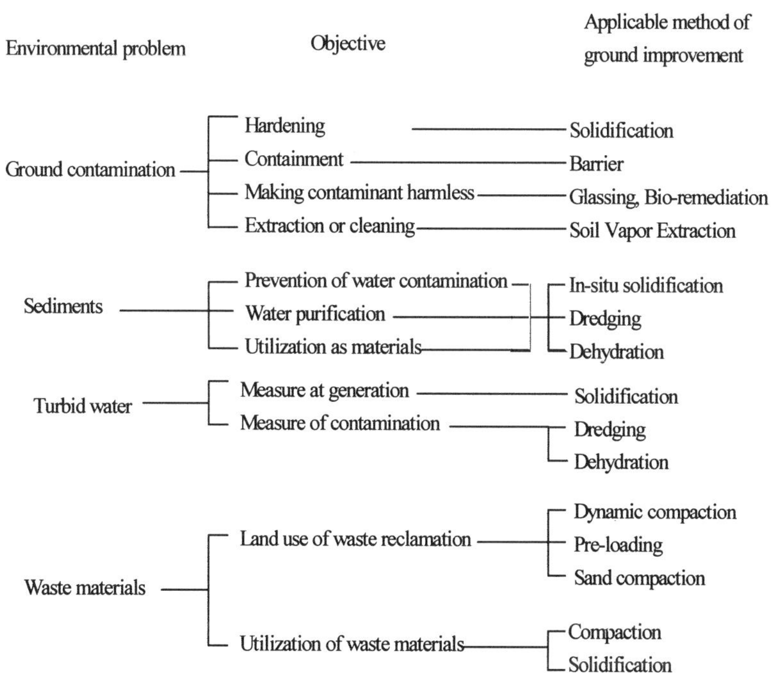

Fig.6 Ground improvement technique applicable to environmental control

5.3 *Countermeasure for turbid water*

Because there is very steep topography and seasonal monsoonal weather conditions in Japan, the surface soil is easily eroded. The eroded soil causes contamination of rivers and the sea. The countermeasures for this contamination have been researched from the view point of slope stability methods. On the main island of Okinawa (southeastern part of Japan), the weathered soil called "Kunigami-maaji" is easily eroded through rainfall and results in turbid reddish brown water that contaminates the ocean and reefs. As a countermeasure to this, the turbid water of Kunigami-maaji is treated with cement group hardening materials, in addition to the other slope stability works such as vegetation (Uehara and Hara 1994).

6 CONCLUSIONS

The geo-environmental problems related to ground improvement techniques in Japan have been assessed. It is necessary to perform and examine more works in this kind of field, in order to decrease the negative effects of ground improvement techniques on the environment, and to increase the positive effects of environmental controls.

ACKNOWLEDGEMENT

The authors would like to acknowledge the helpful discussions with the members of the working group on "Soil Stabilization and Geoenvironment" of the Committee on Material Science and Technology for Soil Stabilization in Society of Material Science, Japan.

REFERENCES

Imamura, S., M. Shimomura & T. Sueoka, 1994. Migration characteristics and remediation efficiency at organic compounds contamination sites of volcanic ash soil layer in Japan, Proc. 1st Inter. Cong. on Environmental Geotechnics: 267-272.

Kamon, M. & T. Katsumi, 1995-a. Utilization of coal fly ash from fluidized bed combustion systems, Geoenvironment 2000, ASCE: 1765-1779.

Kamon, M., T. Katsumi & S. Oyama, 1995-b. Environmental impact and control of alkaline migration by cement stabilized soil, Annuals, DPRI, Kyoto Univ., 38(B-2).

Kitsugi, K., 1989. Environmental aspects accompanied with weak soil ground improvement techniques. Jour. CAJ, 511: 104-115.

Miki, H., N. Mori, & T. Hurusho, 1994. Experimental study on alkaline neutralization ability and alkaline seepage of the soil, Proc. 49th Annual Meeting of JSCE, 3: 1534-1535.

Mishima, N., K. Hoshino, H. Ishihara & K. Terao, 1994. Adherence to pH of soils, Proc. 29th Annual Meeting of JSSMFE; 2327-2328.

Nemoto, N., T. Hayama & T. Shimoda, 1991. Inquiry on the evaluation method of dustless property of soil stabilizer, Proc. 46th Annual Meeting of JSCE: 1052-1053.

Nishigaki, M. 1992, Environmental assessment on groundwater, The Foundation Engineering & Equipment, Vol.20 No.11:25-31

Oshita, T. & M. Kitano, 1992. Influence to underground water with hydrophobic polyisocyanate grout, Proc. 27th Annual Meeting of JSSMFE: 2273-2276.

Sano, M., M. Yamamura, K. Kitsugi & T. Shinkoshi, 1993. Experiment results on the prevention effect within cohesive soil ground on immigration of lime-alkalinity adjacent to lime treated soil mass, Proc. 28th Annual Meeting of JSSMFE; 2585-2588.

Sawa, K., S. Tomohisa, K. Sasabe, A. Koto & K. Suzuki, 1994. On the development of dustless hardening materials containing small amounts of oil, Proc. 1st Symposium on Ground Improvement, JSMS: 65-72.

Sawa, K., S. Tomohisa, A. Koto & K. Suzuki, 1995. On the hardening treatment by dustless hardening material, Proc. 30th Annual Meeting of JGS: 2335-2336.

Shimada, S. 1992. Development of a new injection system for gas-liquid reactions, Soil Improvement, CJMR Vol.9, JSMS, Elsevier Applied Science: 183-196.

Shimoda, T., N. Nemoto & T. Nishimura, 1992. Non-dusty treatment on cement-lime soil stabilization methods, Proc. 19th Japan Road Meeting: 544-545.

Takeshita, Y. & I. Kohno, 1994. Study on optimal pumping design for dewatering in excavation site, Journal of Geotechnical Engineering, JSCE, No.493: 155-163.

Takeuchi K., M. Ishida, S. Kashima, M. Murata & T. Tokoro, 1987. Development of grouting method by carbon dioxide, Proc. 22th Annual Meeting of JSSMFE: 1819-1820.

Ueda, T., K. Fujino, K. Jinno & F. Hirano, 1987. Optimal control of salt water intrusion by group of recharge wells under work of lowering groundwater level, Journal of Hydraulic, Coastal and Environmental Engineering, JSCE, No.381: 45-54.

Uno, T., M. Nishigaki, H. Nagai & M. Yanagida, 1992. Groundwater preservation method by recharge-well system, Proc. 37th JSSMFE Symposium on Soil Engineering and Environmental Problems: 115-120.

Uehara, H. & H. Hara, 1994. Experimental studies on prevention of reddish soils contamination of the sea, Proc. 1st Inter. Cong. on Environmental Geotechnics: 813-818.

Environmental Geotechnics, Kamon (ed.) © 1996 Balkema, Rotterdam. ISBN 90 5410 848 7

Dustless hardening material

Kohei Sawa & Seishi Tomohisa
Akashi College of Technology, Japan

Kazuo Sasabe
Ministry of Construction, Osaka, Japan

Atsushi Koto & Kazuyuki Suzuki
Koto Co. Ltd, Settsu, Osaka, Japan

ABSTRACT: For shallow soil improvement, most of the hardening materials used are powders, and a lot of the powder is air/windborne and flies everywhere, which is called "dust". Recently, the flying dust has become a serious problem at sites near residential areas and for workers' health.

The purpose of this study is to further the development of "dustless hardening materials" that contain small amounts of oil.

As a result, it was found that the new dustless hardening material is very effective with regard to reduction of the amount of dust produced, strength appearance and the amount of oil dissolved from treated soil. Therefore, it can be applied as an effective hardening material, and it poses no hazard to the environment.

1. INTRODUCTION

A number of soil improvement methods have been developed for use at construction sites on soft ground. Shallow soil improvement is one of such methods, and most of the hardening materials used at these sites are powders. As the powders are spread on the ground and mixed with the soil, much of them is air/windborne and flies everywhere, and is refered to as "dust" here. Recently, this flying dust has become a serious problem at sites near residential areas and for workers' health. And as a countermeasure to this problem, hardening material mixed with teflon was developed in order to reduce the amount of dust produced (Nemoto et al. 1991).

The purpose of this study is to further the development of "dustless hardening materials" that contain small amounts of oil (Sawa et al. 1995-a, b). They reduce the tendency of dust to become air/windborne, improve the working environment and keep the area in which they are used clean. The amount of dust produced from the new hardening materials developed, the strength of treated soil and oil dissolution rates from treated soil were investigated by both laboratory work and field tests.

2. EXPERIMENTAL METHODS

The hardening material used in the study was cement for soil improvement (abbreviated as normal agent). And a new hardening material composed of this cement plus a small amount of hydrocarbon compound oil was developed. It was named "the dustless hardening material (abbreviated as dustless agent)". In the case of using both agents, the amount of dust produced, the strength properties of treated soil and oil dissolution rates from treated soil were investigated.

2.1 THE AMOUNT OF DUST

2.1.1 LABORATORY TEST

The amount of dust produced was measured by the device is shown in Fig.1. The experimental procedure was as follows:

(1) Drop the hardening materials (100 grams) from a height of 50 centimeters of a cylindrical device (30cm in diameter and 40cm in height).

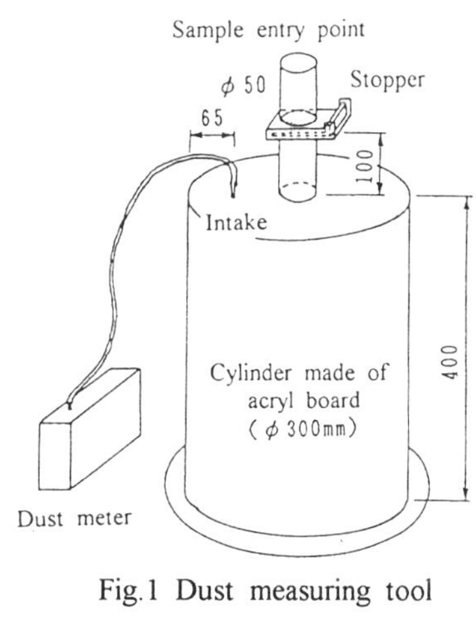

Fig.1 Dust measuring tool

(2) After 1 minute, measure the amount of flying dust using a digital meter for 1 minute at a point 40cm above the bottom.
(3) Convert the dust value of relative density(count per minute) measured by digital meter to weight density(mg/m^3).

2.1.2 FIELD TEST

Field tests were performed using dustless agent and normal one. Fig. 2. shows the experimental field which was a rectangle of 6m width and 14m length, and the location of measurement points of dust volume. It should be noted here that there was a northeasterly breeze of the velocity of 0.7-1.8m/s blowing during the tests.

(1) Measurement of the dust volume was performed at the following 4 construction stages respectively.
(a)Spreading (about 10 minutes): Hardening material was spread on the ground from a container (capacity 1t) using a backhoe.
(b)Traveling (about 1 minute): Backhoe made one pass and return 14m long on the hardening material spread in the field.
(c) Mixing A (about 11 minutes): Stabilizer mixed in the range (2m width, 0.6m depth) of both sides of the field.
(d) Mixing B (about 9 minutes): Mixingfork mixed in the center range (2m width) of the field.

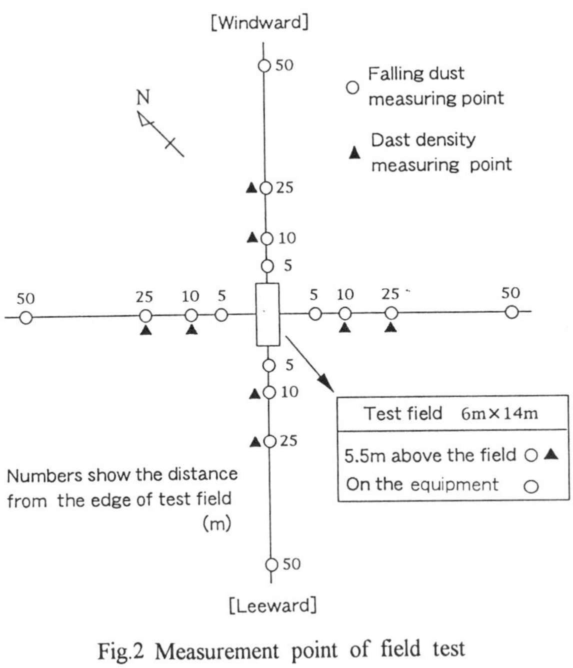

Fig.2 Measurement point of field test

(2) The measurement of dust volume was performed using the following two methods; flying dust in the air (named "flying dust method") and fallen dust in the fixed area (named "falling dust method").
(a) Flying dust method
The amount of flying dust in the air was measured for relative density at points 1.5m above the ground by using a digital dust meter.
(b) Falling dust method
The mass of dust which fell in the collecting box (55×39cm) was measured by weight.

2.2 STRENGTH OF HARDENING TREATED SOIL

The sample, cohesive soil dredged from the Hino River in Shiga Prefecture, was used in this hardening treatment. Its physical properties are shown in Table 1. The treated soils were added to the normal agent or the dustless agent (mixing oil rates are one of the 3%, 7% and 10%) of 3% and 7% by weight of sample respectively. They were made into cylindrical specimens of 5cm in diameter, 10cm in height due to the compaction energy of 5.6cm·kgf/cm^3. After they were aged with constant condition of 20 °C and 90%R.H., an unconfined compression test was performed.

2.3 AMOUNT OF OIL DISSOLVED

The dissolution test, which measures oil dissolved from treated soil, was performed according to the inspection method of oil for industrial wastes in accordance with Environmental Agency Standards. The specimens were cubes which had a side of 1cm in length. The allowable limit of dissolved oil from industrial waste materials has being defined under 100mg/l.

Table 1 Properties of sample

Property	(units)	Value
Density of soil	(g/cm^3)	2.639
Water content	(%)	27.7
Ignition loss	(%)	4.8
Atterberg Lmits	w_L(%)	49.1
	W_P(%)	13.7
	I_P(%)	35.4
Grain size distribution		
Gravel(>5mm)	(%)	1.2
Sand(2~5mm)	(%)	30.7
Silt(75 μm~2mm)	(%)	39.7
Clay(<75 μm)	(%)	28.4
Maximum grain size (mm)		10
Classification of soil		CL

3. RESULTS AND DISCUSSION

3.1 AMOUNT OF DUST

3.1.1 LABORATORY TEST

The relationship between the amount of dust and the rate of oil in the dustless agent is shown in Fig. 3. When the normal agent is mixed with a little oil, the amount of dust produced is reduced. For instance, in the case of normal agent with an oil rate of only about 3%, the amount of dust produced can be reduced to almost zero. In this way, it was found that the dustless agent can be treated to produce a density of flying dust within level 1 (2mg/m^3) and in accordance with the standards set by Japan Society for Occupation Health (Table 2)(SHASEJ 1987).

3.1.2 FIELD TEST

Fig. 4,5 are photographs which show the hardening materials spread by backhoe. It is obvious from these photographs that the amount of dust produced by the dustless hardening agent is very little in comparison with that of the normal agent.

(1) The amount of dust by falling dust method

The amount of dust which fell in the collecting box is shown in Table 3. The volume of dust which is measured at the windward side (northeast) in all stages of construction work was under realizable limit (10mg) regardless of which agent was used. On the other hand, on the leeward side (southwest) and the rectangular sides of the wind (southeast and northwest), when the normal agent was spread, the volume of dust measured was 20mg-550mg at the position of 5m and 20mg at the position of 10m from the edge of the field.

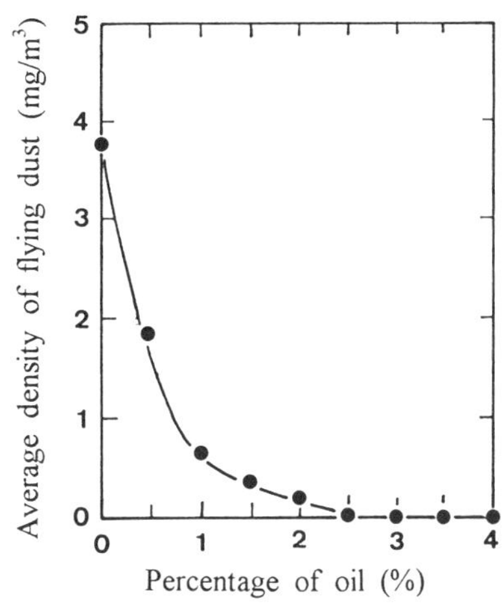

Fig.3 Relationship between the amount of flying dust and percentage of oil

Fig.4 Spreading of normal hardening agent

Fig.5 Spreading of dustless hardening agent

Table 2 Allowable density of flying dust

Level	Dust comporment	Allowable density (mg/m^3)
1	Dust containing 30% over of floating silica, Talk, Alumina, Aluminium, Diatom earth, Sulfide	2
2	Dust containing 30% under of floating silica, Iron, Oxide, Graphite, Carbonblack, Active carbon, Lime	5
3	Others	10

Furthermore, when the dustless hardening agent was spread, the volume of dust measured was only about 20mg, even at the position of 5m from the edge of the field. No dust was detected at positions further away from the field or above the field.

Subsequently, when the nomal agent was mixed, the volume of dust measured was high at 4900mg on the leeward position of 5m from the edge of the field, whereas when the dustless agent was mixed, the amount of dust produced was negligible.

And, the volume of dust found on construction equipment, which has a significant effect on workers health, were 60mg-180mg when the normal agent was spread and mixed. However, in case of the dustless agent, they were found to be under the realizable limit.

(2) Average density of flying dust

Fig.6 shows the results measured of average dust density at respective stages of construction. The volume of dust measured was also very little on the windward side (northeast) in all cases for hardening materials and construction stages.

On the other hand, the dust density of the normal agent on the leeward side (southwest) becomes very high value at even the position of 25m far from the edge of the test field. However, the dust density of the dustless agent had a very low value of 1/8-1/20 in comparison with the normal agent. At the rectangular sides of the wind (southeast and northwest) too, the dust density was extremely small.

The average dust density attributed to using dustless agent in this test was in the range of 0.01-0.14mg/m^3. According to the results of past tests, it is almost similar in value (0.03-0.19mg/m^3) to that measured in supermarkets. Especially, when the machine traveled over the normal agent, the amount of dust at rectangular directions of the wind was small. At the point of the leeward, though, there was a relatively large amount. It is possible that this increase in the volume of dust is caused by concentration to the leeward due to the increase of velocity of the wind during testing, and if so, then, the development of the dustless hardening agent which reduces the volume of dust is very important.

Table 3 Amount of falling dust (mg)

Point of measured	Normal agent		Dustless agent	
	Spreading	Mixing	Spreading	Mixing
North-east (Windward)	——	——	——	——
South-west (Leeward)	30	4900	20	——
South-east	550	——	——	40
North-wes	30	20	20	——
5.5m above the test field	——	——	——	——
On the equipment	180	60	——	——

—— : Under realizable limit

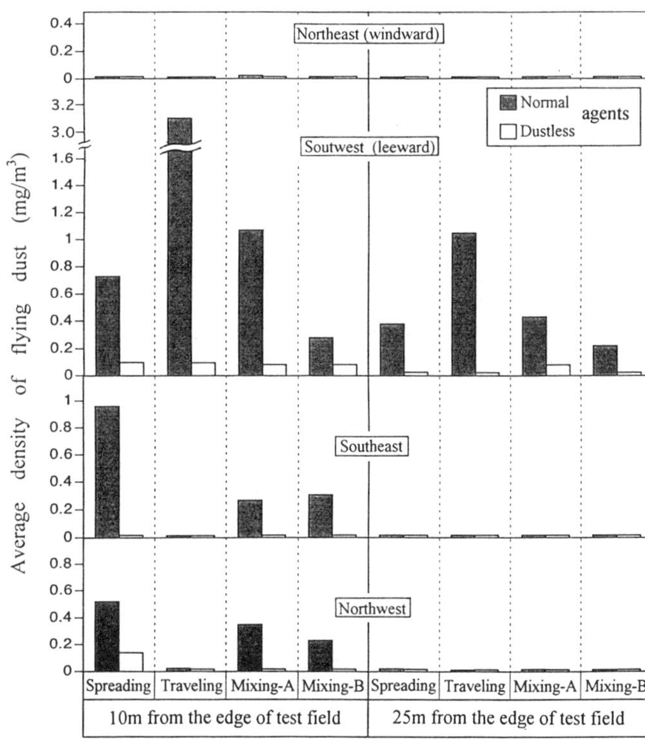

Fig.6 Average density of flying dust under works

(3) Maximum density of flying dust

Fig.7 shows the results measured of maximum dust density which was obtained from continuous measurement at respective stages of construction. The trend of value measured that is based on the direction of the wind was similar to the average dust density in Fig.6. In comparison with allowable limits in Table 2, when the normal agent was used, in all cases except mixing-B, dust density which was measured at points 10m and 25m on the leeward side was larger than the level 1 (2mg/m^3) allowable limit. And, notably, it was even higher than level 3 (10mg/m^3) at the point 10m on the leeward side.

On the other hand, when the dustless agent was used, the largest value of dust density was 1.73 mg/m^3 at mixing-A at the 10m point on the leeward side. And at all stages of construction investigated and at all measurement points, the results measured were within the level 1 (2mg/m^3) allowable limit. Therefore, it is clear that the dustless hardening agent is very effective in reducing the volume of dust.

3.2 STRENGTH PROPERTIES OF TREATED SOIL

Fig.8 shows the relationship between the strength of the treated soil composed of cohesive soil and the dustless agent and the number of days aged. Also, Fig. 9 shows the relationship between strength and the mixing rate of oil in the dustless agent. An increase of strength was obtained during long aging for all kinds of combinations. And when the mixing rate of oil is more rich, the strength obtained increases as to specimens of additive content of 3%, then, that trend is maintained throughout the aging process.

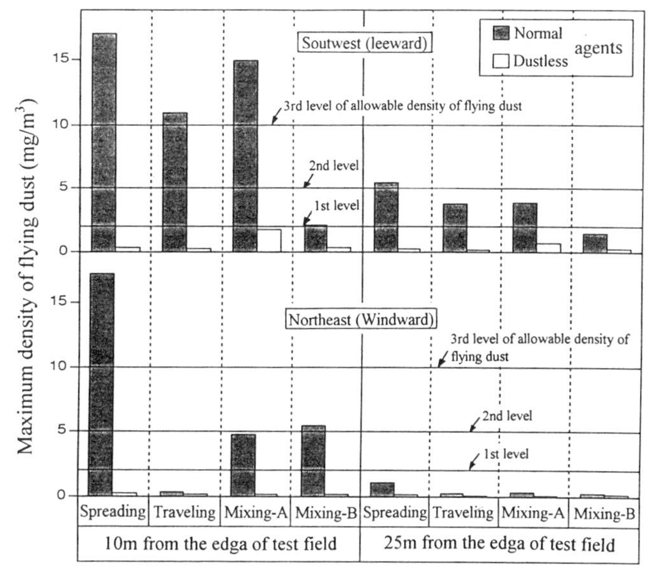

Fig.7 Maximum density of flying dust under works

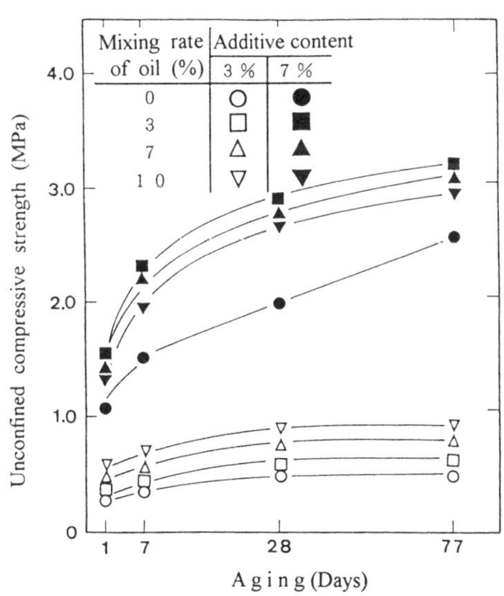

Fig.8 Relationship between the number of aging days and strength

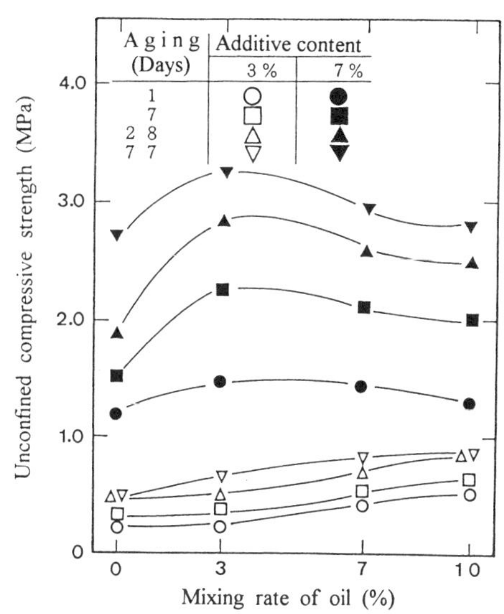

Fig.9 Relationship between mixing rate of oil and strength

On the other hand, for specimens with an additive content of 7%, the increasing strength of the specimens which used dustless agent mixed with oil was remarkable in the early age. As a result, the strength after 28 days of aging was 1.5 times higher than that using the normal agent. Strength development beyond this period of aging, however, decreases with time. When the dustless agent contains 3% oil, it is clear that these specimens obtained the greatest strength.

Fig.10 illustrates the relationship between the additive content of dustless agent and the strength of specimens aged for 7 days. The strength provided for the construction material that can be applied to the lower base course is over 1 MPa. It is possible to save about 1% more of the hardening material that contains the dustless agent owing to that attainment level than that containing the normal agent.

3.3 AMOUNT OF OIL DISSOLVED

After 1 day of aging, the specimens which contained the dustless agent (mixing rate of oil was 5% or 7%) of 3% or 6% by weight of soil respectively, were subjected to a dissolution test.

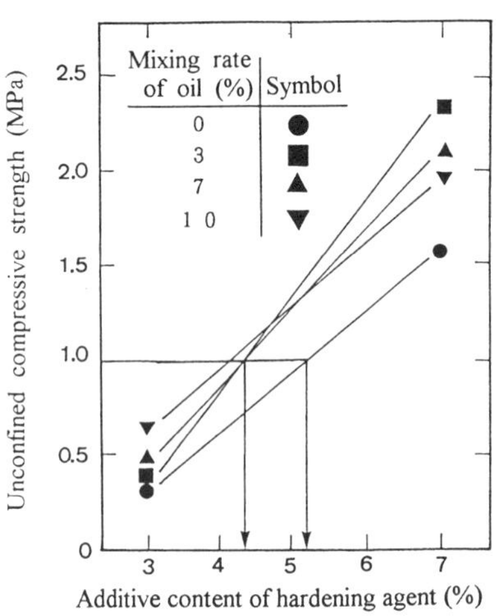

Fig.10 Relationship between additive content of hardening agent and strength

Table 4 Amount of oil dissolved

Mixing rate of oil (%)	Additive content of hardening agent(%)	Oil dissolution (%)
5	3	1.8
	6	3.6
7	3	3.0
	6	5.4

The results of the dissolution test performed on the specimens composed of decomposed granite soil and the dustless agent is shown in Table 4. In combinations consisting of richer oil rate and more hardening materials, the amount of oil dissolved from the treated soil increases. However, it is clear that the amount of oil dissolved is within acceptable limits.

4. CONCLUSIONS

In this study, the effects of a dustless hardening agent, which contains a small amount of oil added to cement for soil improvement, was investigated. The results obtained are summarized as follows :

(1) The amount of dust produced by mixing about 3% wt. of oil into hardening materials is reduced to almost nothing.

(2) In field tests, the amount of dust produced with dustless hardening agents was of very small quantity in comparison with normal hardening agents at all stages of construction work. It was found to be below an acceptable level that provides for a productive working environment.

(3) It is clear that the strength of the treated soil through the application of the dustless hardening agents is the same as that achieved through the application of normal agents. And, notably, when only 3% wt. of oil was mixed with the hardening materials, the strength of specimen was found to be greater than that when only normal hardening agent were used.

(4) The amount of oil dissolved from the treated soil with the dustless agents was minimal even after only 1 day of aging. Thus, it is clear that the effects of the dissolved oil on the treated area are within acceptable limits and pose no hazard to the environment.

REFERENCES

NEMOTO, N., T. HAYAMA and T. SIMODA 1991. Inquiry on the evaluation method of dustless property of soil stabilizer, Proc. the 46th Annual Conf. of JSCE, 3, : 1052-1053.

SAWA, K., S. TOMOHISA, K. SASABE, A. KOTO and K. SUZUKI 1995-a. Development of dustless hardening material containing small amount of oil -on dust-proof properties-, The Society of Materials Science, 44, No.503: 1027-1030.

SAWA, K., S. TOMOHISA, A. KOTO and K. SUZUKI 1995-b. On the hardening treatment by dustless hardening material, Proc., the 30th Japan National Conf. on JGS: 2335-2336.

The Society of Heateing, Air-Conditioning and Sanitary Engieers of Japan 1987. Heateing, Air-Conditioning and Sanitary Engieers Handbook, I, : I-87~I-88.

Environmental Geotechnics, Kamon (ed.) © 1996 Balkema, Rotterdam. ISBN 90 5410 848 7

Risk assessment of Xi'an ground fractures

G.L.Xu
Kyushu University, Fukuoka, Japan

T.Z.Yan & H.M.Tang
China University of Geosciences, Wuhan, People's Republic of China

ABSTRACT: There are 10 ground fractures which have developed in Xi'an, Northwest of China. Because of the high activity rate, Xi'an Ground Fractures (XGF) have become a kind of special geo-environmental hazard, and have brought about innumerable direct losses, indirect losses and non-economic losses. Information value model is adopted to appraise the intensity of the activity of XGF. Risk model, which is defined as the product of the probability of hazard and loss value, was introduced to assess the loss caused by XGF hazard. The expected direct risk losses in the periods of 1994-2000 and 1994-2010 have been predicted in this paper.

1 INTRODUCTION

Xi'an is the heart of Northwest of China. It is also one of the most famous ancient cities in the world. Records of the history of the city spans back to more than 3000 years. There are many famous historic relics, for example, the marvellous figures of warriors and horses in Qing Dynasty.

However, with the development of municipal construction, several geo-environmental hazards or problems, such as crust stability, ground fracturing, landslide, land subsidence, water supply and water containment, have become frequent and intense. Among these hazards or problems, ground fracturing has become the most serious hazard in recent years. After Tangshan calamity earthquake in 1976, innumerous losses, direct losses, indirect losses and non- economic losses, have been caused by XGF.

Much attention has been paid on XGF by geologists, seismologists, hydrogeologists, engineering geologists and even sociologists. Three stages can be divided in the history of research on XGF. Initially, a debate "What the ground fissures(fractures) are" was made around 1976. Then, the distribution of geometry of XGF, such as the strike, length, spacing, has been investigated and confirmed before 1990. After that, the mechanism of XGF has been discussed, and some qualitative and quasi-quantitative models have been developed to predict the activities of XGF(Yan, 1990; Pan et al, 1991; Zhang, 1990). However, the loss caused by XGF has not been touched yet.

The present paper addresses the loss assessment of XGF. First of all, some characteristics of XGF have been summarized. Factors affecting the activity of XGF have been examined. Then, using Information Value Theory, information value was calculated to appraise the intensity of the activity of XGF. Finally, direct loss caused by XGF was estimated using the risk model, and the expected direct loss in the periods 1994-2000 and 1994-2010 have been predicted, respectively.

2 CHARACTERISTICS OF XGF

Based on detail investigation, 10 ground fractures are identified in about 150km² area around Xi'an (Fig.1).

Some characteristics are summarized as follows:

1. Geometric characteristic. All of these fractures strike SWW- NEE, and the general direction is

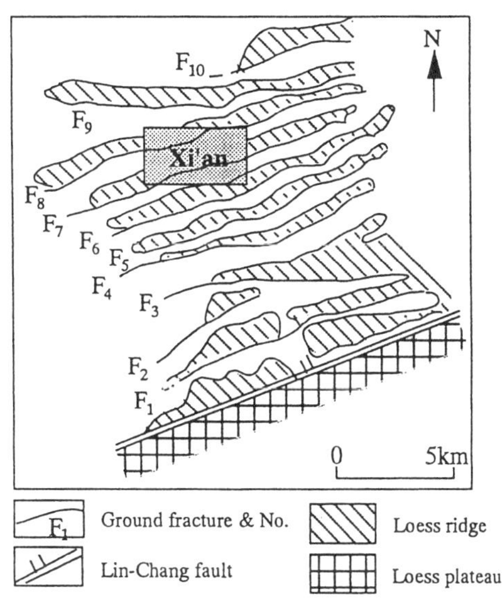

Fig.1 The distribution of Xi'an Ground Fractures

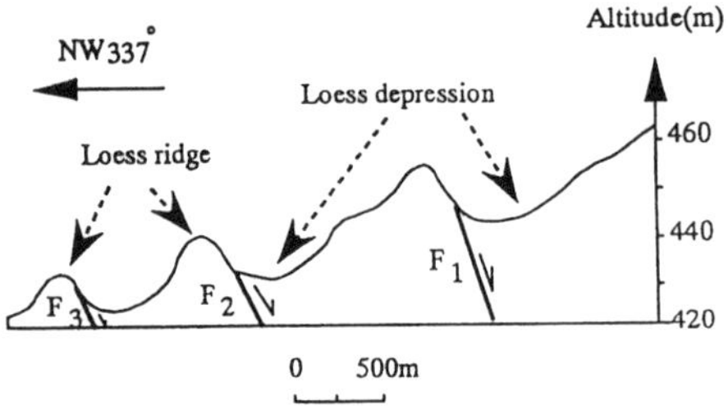

Fig.2 Xi'an ridge and ravine topography

NE70°-NE80°. The length varies from 4km to 14km, and the spacing between two fractures is approximately 1-2km. The dip is almost vertical at near surface, and inclines to south at 75°-60° with depth. The opening , like a wedge , is wider at the surface, and becomes rapidly narrow with depth.

2. Landform characteristics XGF is located at the south side of loess ridge, which is one part of specific ridge and ravine topography(Fig.2). All of the fractures stretch along loess ridges except a small western segment of F_3 and F_5. It has been shown that the cumulative displacement and the rate of activity of XGF are well related to the relative heights of loess ridges(Zhang et al, 1991).

3. Kinematics characteristics With the exception of the western segment of F_2 and F_7, the rate of activity in the eastern segment is generally larger than that in the western segment. Although the activity has a little shear displacement, the movement of XGF is like that of a normal fault, that is, the upper side slips down with respect to the lower side. Fig.3 shows an example of the displacement of F_4 at Xiaozhai. As a matter of fact, the rate of activity varies with fractures, places and time. The vertical displacement, for example, varies between 0 and 49.0mm/a. With the continuous monitoring, the movement of XGF is not a smooth creep, but a series of micro- stick slip.

3 FACTORS AFFECTING XGF

The rate of tectonic activity in Xi'an region is about 3.6mm/y. Obviously, the activity rate of XGF is not commensurate to that of tectonic activity. The activity

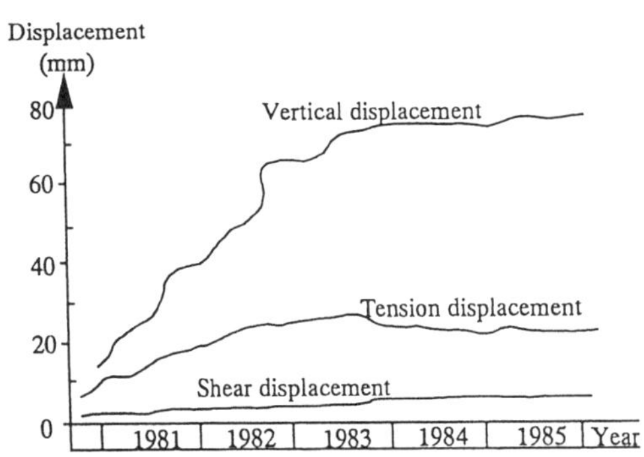

Fig.3 Curve of displacement at Xiaozhao

of XGF is affected not only by internal and external agencies, but also human- being activities.

1. Internal agency From the characteristics of geometry and activities, XGF are intrinsically a kind of tectonic fractures, i.e., a kind of manifestation of internal agency in the surface. Therefore, they are well related to earthquake and the activity of boundary fractures. According to historical records, it is known that the activities of ground fractures coincide with times of earthquakes in Weihe River Basin and in Lin - Chang fault zone. The active period of ground fractures is the active period of earthquakes, the same is observed in the case of the inactive periods (Wang,1985; Pan et al, 1991) (Fig.4). The activity of XGF is so affected by Lin- Chang fault (which is one of the most dominating faults in Xi'an region and constitutes the south boundary of XGF) that XGF are considered as sub-normal faults of Lin-Chang fault in its upper side(Zhang,1990).

The tectonic stress field in Xi'an region is estimated as a potential normal fault stress field, that is, the principal compressional stress orients NE52° with a nearly vertical dip angle, and the principal tensional stress guides NW324° direction and is basically horizontal. That is exactly the stress field of XGF. Therefore, the activity of XGF will be intense or weak depending on whether the stress strengthens or weakens.

2. Meteorological factors Xi'an's climate belongs to temperate zone, and has typical continental climate, i.e., warm and dry in spring, hot and rainy in summer, pleasantly cool and moist in autumn , and cold and seldom rain and snow in winter. Meteorological factors change with seasons. These factors include, precipitation, temperature, evapotranspiration and humidity. According to a number of statistical data, the activity rate or the cumulative vertical displacement of XGF are not correlative to them.

3. Man's activities With the development of science and technology, man's activities become more and more a significant geological agency and result in

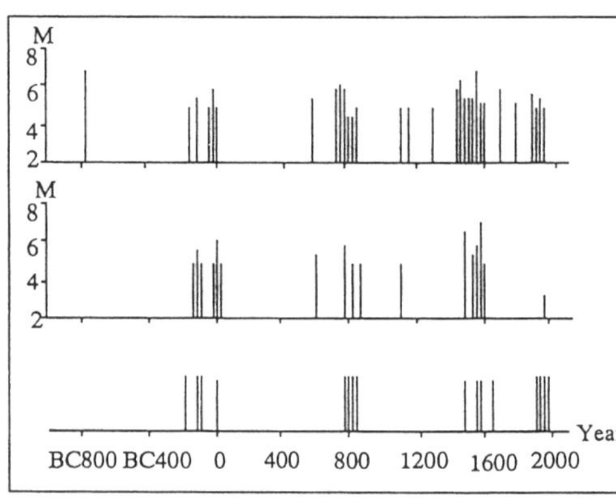

Fig. 4. Historic ground fracturing vs. earthquake
(a) M- t chart in Weihe River Basin
(b) M- t chart in Lin- Chang fault zone
(c) Ground fracturing in Xi'an

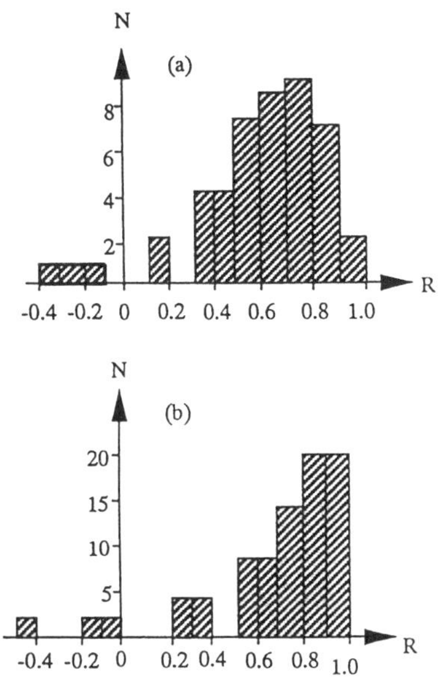

Fig.5 Histogram of coefficient of autocorrelation
(a) Vertical displacement of XGF vs. water pumped
(b) Vertical displacement of XGF vs. water level

very intense XGF activity. They contain mainly, pumping of ground water, irrigation and drainage of surface water, construction and vibration. However, the most devastating activity is pumping of ground water. The maximum head decline of confined water has reached 95m since 1960. Fig.5 plots the histogram of coefficient of autocorrelation between the cumulative vertical displacement of XGF vs. the amount of water pumped and water level.

Therefore, there are multiple factors affecting XGF. The present activity characteristics are the result of interactions between internal agency, external agency and man's activities. Internal agency is the principal factor, the characteristic and magnitude of crustal stress dominate the behavior of activity of XGF. External agency is a subordinate factor. However, man's activities, especially, pumping of ground water, stimulate and amplify the activity of XGF when the internal agency is above a certain level.

4. ACTIVITY INTENSITY OF XGF

Activity intensity of XGF was analyzed in an effort to provide a basis for risk assessment, and to guide the municipal engineering plans and constructions.

Information value, which is calculated by condition probability, is a measure of how close is the relationship between geological factors and object of study . In practice, the frequency of sampling is used instead of the population probability, i.e.,

$$I = \Sigma log \frac{S_j / N_j}{S / N} \quad (1)$$

Table 1 Factor, variable and Information value

Factor	State	Variable	I vaule
Fault	1	X_1	0.880
Height (m)	>10	X_2	0.880
	(6,10]	X_3	0.424
	(4,6]	X_4	0.297
	(2,4]	X_5	0.168
	≤2	X_6	-0.187
Head decline (m)	>90	X_7	0.277
	(70,90]	X_8	0.233
	(50,70]	X_9	0.092
	(30,50]	X_{10}	-0.088
	≤30	X_{11}	-0.682
Land subsidence (mm)	>700	X_{12}	0.301
	(500,700]	X_{13}	0.274
	(300,500]	X_{14}	0.245
	(100,300]	X_{15}	0.079
	≤100	X_{16}	-0.332

where, N is the total number of units; S is the total number of ground fracture units; N_j is the number of units of j th factor; S_j is the number of ground fracture units of j th factor.

Based on the above analysis of factors, prediction factors and variables are chosen as follows:

1. Fault Fault is concealed under earth 1- 7m before ground fracture forms at surface.

2. Tectonic landform Xi'an ridge and ravine topography is essentially related to XGF in geological history. The relative height of loess ridge and loess depression shows the relationship not only with the cumulative displacement of concealed fault, but also with the present activity intensity of ground fracture.

3. Head decline Head decline of confining ground water is calculated with the isopiestic contours of 1968 and 1991. This hydraulic effect influences the activity characteristics of XGF.

4. Land subsidence Land subsidence is also serious in Xi'an. The maximum settlement has surpassed 1.5m, and the maximum rate was recorded at 191mm/a at Xi'an Politics and Law College in 1989. The activity of XGF and land subsidence affect each other. Land subsidence is calculated on the basis of land subsidence iso-contour of 1960-1986.

These factors and variables are listed in Table 1.

In order to predict the activity intensity of XGF, 4345 mesh grids are made based on 1:25000 topographic map. One grid has area 200m × 200m, and the total area of prediction is 173.8km^2.

Thus, information value for each variable can be computed, respectively, from eq.(1) as shown in Table 1. With the principle that information can be additive, the space prediction model of activity intensity of XGF can be written as,

$$I=\Sigma I_j =0.880X_1+0.880X_2+0.424X_3 +0.297X_4+0.168X_5-0.187X_6+0.277X_7 +0.233X_8+0.092X_9-0.088X_{10}-0.682X_{11}$$

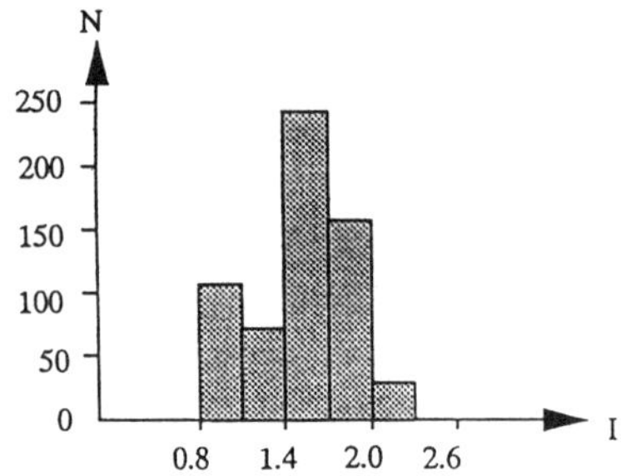

Fig.6 Statistical histogram of Information value

$$+0.301X_{12}+0.274X_{13}+0.245X_{14}$$
$$+0.079X_{15}-0.332X_{16} \quad (2)$$

According to model eq.(2) , information value of each grid can be calculted, and the critical information value is determined as I_c=0.80. Fig.6 gives the statistical histogram of information value which is lager than this critical value.

Then, activity intensity can be classified as ,

I<0.8	Stable
I∈[0.8,1.4)	Low activity
I∈[1.4,1.6)	Moderate activity
I∈[1.6,2.1)	High activity
I≥2.1	Very high activity

Thus, a zoning map of activity intensity(not included) can be compiled. It showed that the precision of the prediction model is as high as 93.3%.

From the prediction model and the zoning map, it can be seen that:

(1) Concealed fault plays a dominate role in XGF, and in turn, it showed that XGF is a kind of tectonic ground fractures.

(2) The more the head losses of confined water, especially greater than 50m, the more effect on XGF. It is known that the change of internal agency is long and slow, whereas the change of head loss is short and fast. Therefore, it is imperative to strictly control the pumping of ground water to slow down the land subsidence and the activity intensity of ground fractures in Xi'an.

(3) The Very high, high, moderate and low activity units in all of ground fractures units are 4.4%, 25.6%, 40.8% and 29.2%, respectively.

(4) The activity intensity in the south is more intense than that in the north, and in the east is more than in the west except for F_7.

5 RISK ASSESSMENT

Risk depends on the intensity and the scope of the spread of the hazard as well as the economic level of a society. Two kinds of losses caused by the activity of XGF, economic and non- economic loss, can be classified like other hazards. Economic loss can be assessed monetarily, and further divided into direct and indirect losses.

1. Direct loss It is caused directly by the activity of XGF, such as the damage of industrial and civil architecture, water and gas supplies, transportation system and communication system.

2. Indirect loss It is the loss caused by the direct loss, such as the stoppage and reduction in production or increases in cost due to stoppage or shortage of water or gas supplies, and public payments for relief and insurance.

3. Non-economic loss It cannot be evaluated in monetary terms. It includes such losses as the loss of life, society functions, the impact on psychology and the deterioration of environment.

Innumerable losses have been caused by XGF. However, since it is difficult to incorporate indirect loss, non-economic loss and direct loss of underground pipes because of the investigation and index, only industrial and civil architecture and road losses in direct loss have been considered in the following risk model.

Risk is defined as the product of probability of event and the value of loss caused (referred from Keller, 1976; Lawrence, 1986). Therefore, the unit of risk is the unit of money, i.e.,

$$R = pC(V) \quad (3)$$

where, R is the risk; p is the probability of event; $C(V)$ is the loss value caused by this event.

Let $p_{1k}(D_j \mid I_i)$ be the mean damage probability of damage grade D_j for i th construction under the intensity I_i of ground fracture at kth site, and $p_{2k}(C \mid D_j)$ be the mean loss probability of damage grade D_j for i th construction under the intensity I_i of ground fracture at kth site. Then , the loss probability of a construction at kth site is the product of these two condition probabilities(Xiao, 1991), i.e.,

$$p = p_{1k}(D_j \mid I_j)\, p_{2k}(C \mid D_j) \quad (4)$$

Thus , the total expected loss risk is

$$R = \Sigma p_k C_k(V)$$
$$= \Sigma p_k A_k C_{ki} \quad (5)$$

where, C_k is the loss value at k th site; A_k is the area or length of construction; C_{ki} is the price for i th construction.

Table 2, 3 and 4 give the damage and loss probabilities and price of constructions, respectively.

Like the information value model, 3136 mesh grids are made based on 1:25000 topographic map. A grid has area 250m×250m, and the total area is 196km^2.

According to the prediction model of activity intensity and risk assessment model, the expected risk loss of each grid can be calculated during 1994- 2000 and 1994- 2010. It should be noted that (1) the loss price is assumed as 10Yuan/m when the industrial and civil architecture is difficult to be defined at the segment of low activity intensity; and (2) the activity of XGF will decrease after 2000(Yan, 1990; Pan et al, 1991) so that the 4 activity intensities are treated as 3 intensities, i.e., high, moderate and low intensities.

The expected risk losses are classified into the following,

R≥100000 Yuan	Grade A
100000>R≥10000 Yuan	Grade B
10000>R≥1000 Yuan	Grade C
R<1000 Yuan	Grade D

Therefore, risk assessment map in 1994- 2000 and 1994- 2010 can be compiled. Fig.7 shows an example of the risk assessment of XGF in 1994-2000. It can be seen that:

(1) risk loss is high where the density of construction is high, and is low at two ends of ground fractures.

(2) the expected risk loss in the middle and west segment of F_2 and F_3, the middle segment of F_4, the middle and east segment of F_5 and F_6 are higher than those in other places.

(3) the expected risk losses with A, B, C and D grades in 1994- 2000 are 16.1%, 23.0%, 16.9% and 44.9%, and in 1994- 2010 are 21.1%, 18.1%, 42.6% and 17.2%, respectively.

(4) the total expected risk loss in 1994- 2000 and 1994- 2010 is 3.73million Yuan and 5.94 million Yuan, respectively, that is, risk loss before and after 2000 is about 0.62 million Yuan and 0.22 million Yuan per year, respectively.

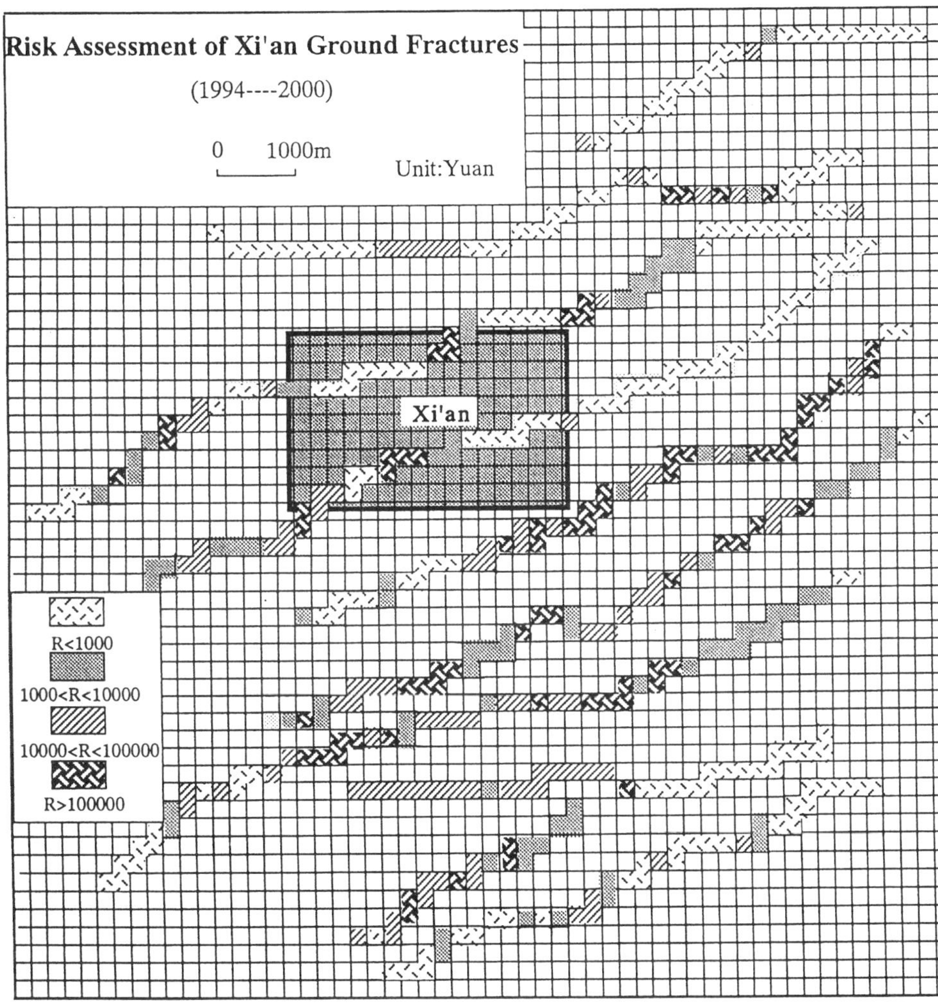

Fig.7 The expected risk loss caused by Xi'an ground fractures between 1994--2000

Table 2 Damage probability with activity intensity

Damage grade	Activity intensity of XGF			
	Low	Moderate	High	Very high
Basiclly well	0.00	0.15	0.35	0.80
Slight damage	0.05	0.30	0.55	0.20
Damage	0.40	0.50	0.10	0.00
Severe damage	0.55	0.05	0.00	0.00

Table3 Loss probability with damage grades

Damage Grade	Basically well	Slight damage	Damage	Severe damage
Loss Percent	0.005	0.05	0.15	0.40

Table 4 Loss prices of constructions Unit: Yuan/ m^2

Construction	Price	Origin
Old house	150	Constrution Bureau
Brick house	200	Constrution Bureau
Multistory build.	270	Constrution Bureau
Factory build.	310	Constrution Bureau
Railway	1000	Communication Office
Road	200	Communication Office
Historic site	10000	Estimation
Farmland	10	Farming & Forestry Bureau

6 CONCLUSION AND REMARKS

1. Xi'an Ground Fractures are the result of interactions between internal and external agency and man's activities. The activity of XGF is essentially the activities of sub- normal faults of Lin- Chang fault, and superimposes the effect of man's activities.

2. Factors affecting XGF are multiple. The characteristic and magnitude of crustal stress dominate the behavior of activity of XGF. However, man's activities, especially the pumping of ground water, stimulate and amplify greatly the activity intensity of XGF and has led to enormous loss in recent years. Internal agency changes are long and slow whereas man's activities effect on XGF is fast and profound. So, it is important to reduce the intensity of man's activities, especially the pumping of ground water.

3. Information value well represents the activity intensity of XGF. The zoning map shows that the activity intensity in the south is more intense than that in the north, and is more intense in the east than in the west except for F_7.

4. The loss caused by XGF is cumulative and increases potentially with the development of society, which is quite different from other hazards, e.g., earthquake. It is estimated that the direct loss from 1994 to 2000 and 2010 is at least 3.7 million and 5.9 million Yuan, respectively. If the indirect loss and non-economic loss are considered, and the promoting of economics and the increasing of man's activities' intensity are also taken into account, the loss caused by XGF hazard is immeasurable. Therefore, it is high time to research and search for measures against XGF hazard.

REFERENCES

Keller, E. A. 1976. Environmental Geology. Charles E. Merill Publishing Co.:51-98

Lawrence Lundgren 1986. Environmental Geology. Prentice- Hall:79-121

Pan, B.T. Xu, G.L. & Zhang,J.M. 1991. Space and time prognosis for Xi'an Ground Fractures. Proc. of Beijing Int. Symp. on Geological Hazards: 20-25.

Wang, J.M. 1985. Periods of ground fracturing and earthquake in Weihe River seismic belt. J. of Seismology: Vol. 7, No.2.

Xiao, G.X. 1991. Quick estimation of loss after earthquake. J. of Disaster: No.4.

Xu, G.L. Zhang, J.M. & Pan, B.T. 1992. Effect of pumping ground water on land subsidence and ground fractures in Xi'an. J. of China Geological Hazards and Control: Vol.3, No.3.

Yan, T. Z. 1988. Recent advances of quantitative prognosis of landslide in China. Proc. of the 5th Symp. on Landslides.

Yan, T.Z. 1990. Prediction of stages of Xi'an land subsidence and ground fractures. Modern Geology: No.5.

Yan, T. Z. 1994. Hydrogeology and Engineering Geology and Environmental Protect. China University of Geosciences Press: 1-22

Zhang, J.M. Pan, B.T. Xu, G.L. & Tan, H.M. 1991. Relationship between Xi'an Ground Fractures and land form. J. of China Geological Hazards and Control: Vol.2,No.2.

Zhang, J.M. 1990. Research on ground fracturing in the region of Xi'an . Northwest University Press.

Environmental Geotechnics, Kamon (ed.) © *1996 Balkema, Rotterdam. ISBN 90 5410 848 7*

Assessment of natural and geotechnics risks in geo-environmental research

Oleg V. Zerkal & Sergey N. Rybakov
Federal Centre for Geoecological Systems, Moscow, Russia

ABSTRACT: Prediction and assessment of natural and geotechnics risks is important for organization of geoecological research. Any risk concept is commonly associated with a hazard one. Hazard as applied by the authors can be subdivided on the basis of a number of indices. Natural and geotechnics risks can be assessed only for existing engineering systems and those under desing. Some approaches as for as assessment of the ecological risk is concerned are set forth in the given paper.

1 IMPORTANCE OF GEOECOLOGICAL RISK ASSESSMENT

When carrying out the geoecological studies a very important problem is the defining of the type of interplay, interrelation and mutual influence of the main three factors - the geological medium, as one of the environment components, the technogenic impact on the medium and the man. Account should be taken of the fact that the geological medium, being in many cases a "carrier" of the main motive forces, determines the origin of unfavourable geoecological situations conditioned by the development of ecologically hazardous geological (endogenic and exogenetic) processes and phenomena and their technogenic analogues. Investigation of the geological processes and phenomena development, their effect on the environment condition, the analysis and prediction of the anthropogenic impact consequences and working out of recommendations for ensuring the safety of the human activities is one of the key trends of the geoecological studies. The organization of such studies requires the solution of a number of problems among which a very important problem is the assessment of natural and technogenic risk. The solution of many problems to a large extent depends on the timely revealing of the hazard and the assessment and forecasting of the geoecological risk level.

Disregarding or inadequate study of the hazard of the development of ecologically dangerous geological processes and inadequate consideration of the geoecological risk when developing the territories offen result in the catastrophic consequences. As an example of erroneous assessment of the geological risk when developing the natural-technogenic system one may consider the situation on the Urtaboz plateau (Tajikistan) in the epicentral zone of the Gissar earthquake in 1989 (Zerkal, 1994). As a result of the undue consideration of the effect of the loess wetting on the slope stability, the irrigation canals were made without any waterproofing. Considerable water losses due to infiltration reaching 2.4-2.6 million m^3 per year have brought about the formation of the aquifer at a depth of 1.5-3 m. In view of the fact that the seismic effect on the stability of the wetted loess masses against sliding was not given due consideration for, the territory was referred to a "no-sliding hazard" one. As a result no engineering protection measures for the given territory have been provided. A consequence of the improper consideration of the geological danger (sliding danger in the given case) was an "unexpected" development of seismogenic slides due to sudden liquefaction during the Gissar earthquake (1989) which was fatal to people and caused drastic damage.

2 GEOLOGICAL (GEOECOLOGICAL) HAZARD

The notion of risk (natural and technogenic) is inseparably linked with the notion of hazard. The geological hazard is considered by many

researchers as a possibility of development of a particular geological process or phenomenon at a give place and for a given time interval and specific characteristics (Dzeker, 1992, Gavrilov, 1995, Sheko, 1994). Such a definition of the "hazard" implies only two categories of assessment when studying the existence of the geological (geoecological) danger:
0 - no hazard,
1 - hazard exists.

In a number of reports the authors suppose that hazard is always linked with a direct damage which impairs the economy development conditions and(or) the people living conditions. However, it is not in every case that development of negative and ecologically hazardous geological processes and phenomena have a direct threat to the man activity. The development of processes and phenomena can engender a remote threat for example, numerous landslide dams in the mountain regions (Schuster, 1994). Such landslide dams which have formed during the last hundred and thousand years when there was no anthropogenic activity, now carry a potential theat. So, we should agree to the fact that the hazard of the development of geological processes is a reality which does not depend on the man-made natural engineering or engineering systems, i.e. it exists and it is not connected with a direct damage.

Taking into account a considerable technogenic effect on the geological medium, the notion "geological hazard" should be broadened. It should also include the danger of the consequences of the technogenic effects on the geological medium (as a part of the environment) which are inauspicious very often and unforeseen in a number of cases. The change for the worse of the natural environment is attended with social and economic damage due to changes in the characteristics of the natural-engineering systems ("structure-natural environment" systems). When the situation takes the worst turn we observe either the degradation of the environment or the failure of the natural-engineering system. Such a broadened interpretation of the notion of hazard is likely to be defined as a "geoecological hazard".

When carrying out the geoecological studies the geological (geoecological) hazard is proposed to be subdivided with regard for a number of its characteristics (Fig. 1) as follows:

Spatial indices (A):

A-1. By manifestation scale - spot (or project), embracing the first km^2, local - (tens, and more seldom hundreds km^2); regional - covering sizable territories comparable with large natural regions or administrative - territorial unities and global danger;

A-2. By manifestation nature - spot, linear and areal;

A-3. By manifestation type - source and overall hazard.

It is evident that the destructive potential increases exponentially as the spatial indices grow.

Time indices (B):

B-1. By the effect time - recurrent, episodical, constant and instantaneous (originating suddenly in a moment);

B-2. By the realization time - remote (potential), real (direct) and arising (catastrophically quick) hazard;

B-3. By the development pattern - rising and decreasing hazard.

The destructive hazard potential rises as the effect period and development rate of the geological processes creating the hazard increase.

The hazard may also be of the primary (C-1) and secondary (C-2) nature; it may be direct (D-1) and indirect (D-2) and it may be manageable (E-1) and unmanageable (E-2).

3 GEOLOGICAL (GEOECOLOGICAL) RISK ASSESSMENT

The possibility of occurrence of the geological (geoecological) hazard characterized by defining the geological (geoecological) risk. In a number of papers the geological risk is thought to be a probability of occurrence and intensification of naturaland technogenic geological processes and phenomena, substituting by the above term the notion of "hazard" (Bakhireva, 1990). The other rather numerous group of researchers links the notion of risk with the "damage" category (Dzeker, 1992, 1994, Gavrilov, 1995, Ragosin, 1995, Sheko, 1994). and considers the geological risk as a probability of losses due to the development of negative ecologically hazard geological processes which is assessed by the product of the hazard quantity and the probability. In the practice of the environment protection activity in the Russian Federation in some cases, for instance, when making the State ecological examination, the notion of risk (ecological risk) is used as an equivalent of the assessment of the technogenic impact on the environment. The plurality of definitions of the notion of the geological (geoecological) risk hampers the elaboration of a common approach for the geoecological studies.

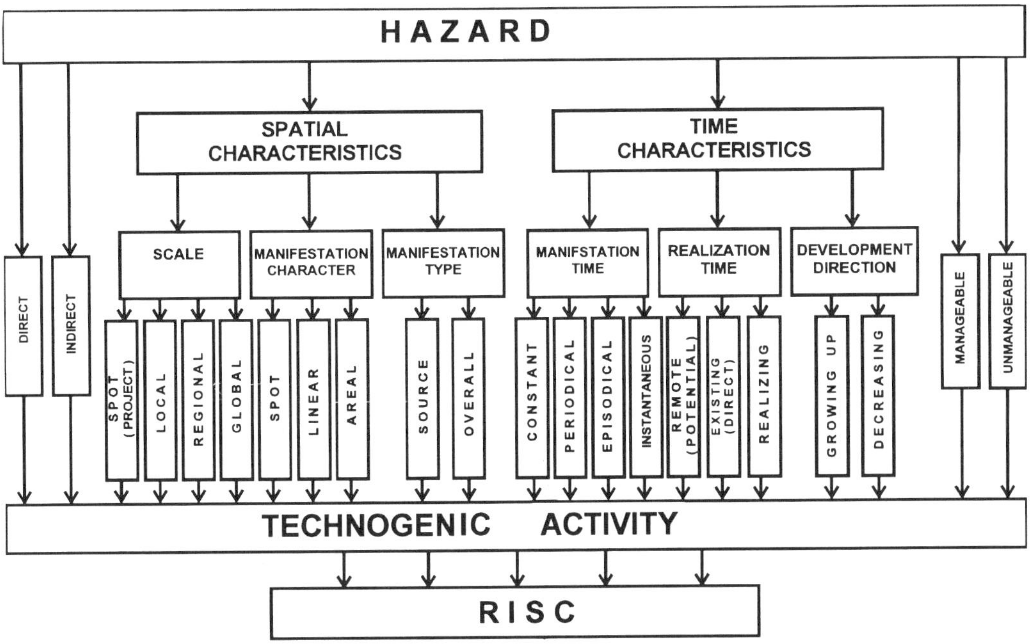

Fig. 1. Types of natural and technogenic hazard.

The concept "risk" may be used only for a project subject to hazard, i.e. when the threat exists. Based on the fact the risk comprises the cognized hazard of the development of one or another geological process or phenomenon, as well as its technogenic analogue when realizing a particular technogenic activity. In other words, when there is no technogenic impact the geoecological risk is also absent, though the natural hazard can exist. Only when a particular effect is realized, the risk for the technogenic activity can appear.

Below are given the basic conditions when the hazard develops into the risk:

- natural conditions which consist in the fallowing: firstly, in gradual accumulation of changes in the geological medium under natural conditions which foster the development of negative ecologically dangerous geological processes; secondly, in the tendency of the natural medium to the development of negative geological processes and phenomena (i.e. regions featuring low natural stability to external effects) and, thirdly, in the potentiality of the flip-flop type processes development;

- natural-technogenic conditions governed either by a drastic gain in the anthropogenic load (impact effect) on the geological medium which brings about the disturbance of the natural balance, or by an accumulation of negative changes in the geological medium as a result of the technogenic activity;

- the technogene (or manufacturing) conditions which promote a greater probability in creating an emergency situation as a result of making use of the obsolete manufacturing processes, complicated technique and manufacturing processes without adequate monitoring.

One should take into acount the risk which accompanies the creation of natural-engineering systems under conditions of development of dangerous natural processes (seismicity, slides and karst phenomenon). In that case the talk is about the tolerable risk which means the inexpediency of making some additional provisions to banish the existing hazard (Fig. 2).

The geological (geoecological) risk represents a realized hazard of development (i.e. probability) of a particular geological process or phenomenon and has the following features:

- variability in time which is connected with the geological medium development and with a change in the natural-technogenic system condition;

- variability in space, governed by the alteration of the geological medium parameters both in plan and section and by changes in the type of economic activity;

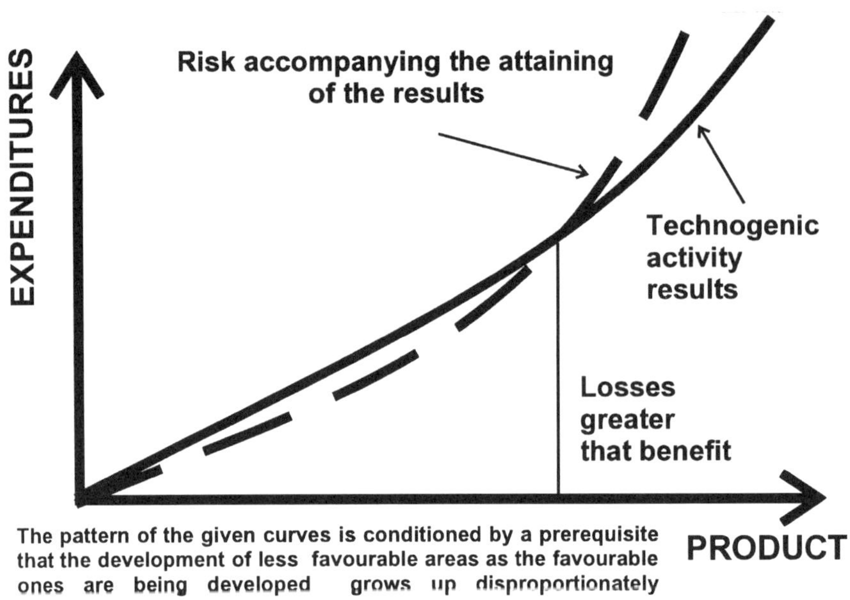

Fig. 2. Expediency of conducting activities to bring down the natural or technogenic risk (according to S.M. Mjagkov with changes).

- acceptability, conditioned by dangerous processes and phenomena development.

The risk assessment should be made with due consideration for the natural hazard indices. The given index is evaluated for specific types of ecologically hazardous processes - risk factors (F) with account for:

- type and kind of a threat produced by them - by the spatial (K_a) and time (K_b) indices, its nature (K_c) and (K_d) and management degree (K_e);
- possible destructive potential (Pt) - social, economic and ecological (worsening of the living conditions and people perish) damage induced as a result of alteration of the characteristics of the natural-technogenic systems and technogenic systems proper;
- probability of realization (P) - probability of the development of the ecologically hazardous geological process obtained as a result of defining the origination factors and prediction of development of the processes and phenomena based on a set of criteria and parameters.

So, for each of the risk factors "i" we'll have:

$$F_i=(K_a+K_b+K_c+K_d+K_e)_i*Pt_i*P_i$$

The physical meaning of the risk quantity is considered as a probability value of realization of the existing hazard in the form of development of ecologically hazardous geological process or phenomenon with damaging of either degree.

The general risk assessment should be evaluated by the summary index of the natural hazard (S) which takes into account the conventional risk value for every given factor.

When carrying out the geoecological studies it is thought expedient to assess the risk proceeding from the data of studying the factor of development of hazardous geological processes and phenomena and their technogenic analogues by the natural conditions worsening level. It is essential to perform at the same time the systematization of the types of the technogenic impact on the geological medium and define a hazard resulting from the opertion of structures and projects depending on the degree of their complexity and responsibility and the probability of technological disasters.

The application of GIS for analyzing the obtained data allow for assessing the geoecological risk when creating natural-technogenic and engineering systems. The assessment of the geological medium parameters which take into account the composition, structure, condition of its individual components (rocks, ground water, geological processes) and define the geoecological state of the environment, is an information basis, the analysis of which with regard for the technogenic load and impact, enables one to create a natural-engineering system model in order to predict its development and to conduct subsequently the assessment of the geological risk. The accuracy of the geoecological risk assessment depends on the scale of a system under consideration, model completeness degree, comlexity of the assessment of natural conditions and the technogenic impact involving the duration and intensity of the natural-technogenic system development.

CONCLUSION

The proposed approach makes it possible to assess the geoecological risk quantity for specific natural-technogenic and engineering systems and take the first step fowards the creation of a complex risk management system to be aimed at retaning the risk within reasonable limits (at the admissible level) at appropriate expenditures or to gain its diminish down to negligible values (safe condition).

REFERENCES

Bakhireva L.V., Osipov V.I., Koff G.L. and Rodina E.E. 1990. Geological and geochemical risk as a criterion of geoecological rating of territories. *Istoria Vzaimodeistvia obshestva i prirodi: facti i conzepzii.* Moscow:129-131 (in russian);

Dzeker E.S. 1992. Geological hazard and risk (methodological studies). *Ing. geologia.* 6:3-10 (in russian);

Dzeker E.S. 1994. Methodological aspects of the problem of geological hazard and risk. *Geoecologia. Ing. geologia. Gidrogeologia. Geokriologia.* 3:3-10 (in russian);

Gavrilov V.V., Zerkal O.V., Utkina I.A. 1995. Natural and technogenic risk assessment for Geoecological survey. *Analysis and Assessment of natural and technogenic risk for the purposes of construction.* Moscow. PNIIIS: 57-62 (in russian);

Mjagkov S.M. 1992. Problems of risk geography. *Vestnik Mosk. Univ. Ser. 5. Geographia.* 4: 3-8 (in russian);

Ragosin A.L. 1995. Current situation and prospects of natural risks analysis and assessment for the purposes of construction. *Analysis and Assessment of natural and technogenic risk for the purposes of construction.* Moscow. PNIIIS: 9-25 (in russian);

Sheko A.I. and Krupoderov B.S. 1994. Hazard and risk assessment of exogenetic geological processes. *Geoecologia. Ing. geologia. Gidrogeologia. Geokriologia.* 3: 11-21 (in russian);

Schuster R.L. 1994. Hazard mitigation for landslide dams. *Proc. 7th Internat. Congr. IAEG.* 3:1441-1450;

Zerkal O.V. 1994. Seismic landslides caused by Gissar earthquake in 1989 (Tajikistan). *Geol. Bull.* 49(2):57-63.

Environmental Geotechnics, Kamon (ed.) © 1996 Balkema, Rotterdam. ISBN 90 5410 848 7

Author index